FINITE ELEMENT PROCEDURES IN ENGINEERING ANALYSIS

PRENTICE-HALL CIVIL ENGINEERING AND ENGINEERING MECHANICS SERIES

FINITE ELEMENT PROCEDURES IN ENGINEERING ANALYSIS

KLAUS-JÜRGEN BATHE

Department of Mechanical Engineering
Massachusetts Institute of Technology

PRENTICE-HALL, INC., *Englewood Cliffs, New Jersey 07632*

Library of Congress Cataloging in Publication Data

Bathe, Klaus-Jürgen.
Finite element procedures in engineering analysis.

Includes bibliographical references and index.
1. Finite element method. 2. Engineering mathematics.
I. Title.
TA347.F5B36 620'.00422 81-12067
ISBN 0-13-317305-4 AACR2

Editorial production supervision and
Interior design by: James M. Chege

Manufacturing buyer: Joyce Levatino
Art production: Bruce Kenselaar

Printed in the United States of America

10 9 8

PRENTICE-HALL INTERNATIONAL, INC., *London*
PRENTICE-HALL OF AUSTRALIA PTY. LIMITED, *Sydney*
PRENTICE-HALL OF CANADA, LTD., *Toronto*
PRENTICE-HALL OF INDIA PRIVATE LIMITED, *New Delhi*
PRENTICE-HALL OF JAPAN, INC., *Tokyo*
PRENTICE-HALL OF SOUTHEAST ASIA PTE. LTD., *Singapore*
WHITEHALL BOOKS LIMITED, *Wellington, New Zealand*

CONTENTS

PREFACE

During the past two decades, the finite element method of analysis has rapidly become a very popular technique for the computer solution of complex problems in engineering. In structures, the method can be understood as an extension of earlier established analysis techniques, in which a structure is represented as an assemblage of discrete truss and beam elements. The same matrix algebra procedures are used, but instead of truss and beam members, finite elements are employed to represent regions of plane stress, plane strain, axisymmetric, three-dimensional, plate, or shell behavior.

In the earliest developments of finite element methods for structural analysis almost all emphasis was directed toward the development of effective finite elements for the solution of specific problems. However, the very broad potential of the method was rapidly recognized and more general techniques for structural analysis were developed while at the same time important applications were found in other fields. At present, the finite element method represents a most general analysis tool and is used in practically all fields of engineering analysis.

The success of the finite element method is based largely on the basic finite element procedures used: the formulation of the problem in variational or weighted residual form, the finite element discretization of this formulation, and the effective solution of the resulting finite element equations. These basic steps are the same whichever problem is considered and provide a general framework and—in conjunction with the use of the digital computer—a quite natural approach to engineering analysis. The final result is a complete numerical process implemented on the digital computer: the formulation of the finite element matrices, the numerical integration to evaluate the matrices, the assemblage of the element matrices into the matrices that correspond to the complete finite element system, and the numerical solution of the system equilibrium equations.

The objective in this book is to present each of the above aspects of finite element analysis and thus to provide a basis for the understanding of the complete solution process when it is applied to problems in solids and structures, heat transfer, seepage, fluid flow and so on.

According to three basic areas in which knowledge is required, the book is divided into three parts. In the first part, important concepts of matrix and linear algebra are presented. Many readers may be familiar with the elementary rules of matrix algebra but specific attention should be given to the concepts of linear algebra, because they provide the foundation for a thorough understanding of the finite element procedures presented later.

The second part of the book comprises the formulation of the finite element method and the numerical procedures used to evaluate the element matrices and the matrices of the complete element assemblage. Starting from elementary concepts the linear and nonlinear analyses of solids and structures, heat transfer, field and fluid mechanics problems are considered. Since the early use of finite element analysis, a large number of finite element formulations have been developed and published. The objective in this book is not to summarize all the finite element formulations and models available, but rather to establish the basic and general principles and describe some finite element procedures that are effective at present.

In the last part of the book, methods for the efficient solution of the finite element equilibrium equations in static and dynamic analyses are presented. This phase of finite element analysis usually comprises most of the computer effort and, consequently, deserves a great deal of attention. It should be noted that because the numerical procedures employed in this phase largely determine the cost of analysis, they also determine whether an analysis can or cannot be performed in practice.

Throughout the presentation, the aim is to establish the finite element procedures using relatively simple and physical concepts, even at the expense of losing some mathematical rigorousness. The principles and procedures are illustrated using about 250 worked-out examples, which are an integral part of the presentation. Some short computer programs are also included to demonstrate the finite element procedures in a compact manner.

My primary aim in writing this book is to provide a tool for teaching finite element procedures to upper-level undergraduate and graduate students in engineering. As such I am using the book at M.I.T. Although the topic is finite element analysis procedures, many of the numerical techniques are quite general and could be employed effectively in almost any method of analysis. In addition, engineers using finite element computer programs should find much valuable information in the text.

A very difficult aspect of writing a book is to give references that appropriately acknowledge the achievements of the various researchers in the field. This is particularly difficult in this book, because the field of finite element analysis has been expanding very rapidly. I would like to extend my sincere apologies to all those researchers whose work I have not referenced to a sufficient degree in this book.

The endeavor to write this book is a result of the excitement and challenge that I have experienced in working in the field of finite element analysis. I am much indebted to Prof. E. L. Wilson of the University of California, Berkeley, —my "finite-element-father"—for his active support of my early research work on finite element methods and his continued interest in my work. Since 1975 the Mechanical Engineering Department of M.I.T. has provided me with a stimulating environment for my teaching, research and writing this book, for which I am very grateful. I also would like to extend very special thanks to Lee W. Ho, who while my teaching assistant at M.I.T. has been very helpful, and to Ioánna Buttlar for her outstanding job in typing the manuscript. Thank you also to James Chege, the editor at Prentice-Hall for his helpful collaboration.

Finally, truly unbounded thanks are due to Zorka, my wife, who with her love and patience supported me in writing this book.

K. J. Bathe

PART I

MATRICES AND LINEAR ALGEBRA

1

ELEMENTARY CONCEPTS OF MATRICES

1.1 INTRODUCTION

The practical use of finite element analysis is based on matrix algebra and the use of the electronic computer, because it is only in matrix form that the complete solution process can be expressed in a compact and elegant manner. The objective in this chapter is to present briefly the fundamentals of matrix algebra that are needed for an understanding of the solution procedures discussed later.[1-3] In the presentation, emphasis is directed to those aspects of matrix algebra that are important in finite element analysis.

From the practical point of view, matrix algebra can be regarded merely as an effective tool with which to manipulate in an elegant manner large amounts of data, and we use this approach in this chapter. Some concepts of linear algebra that are also required for a thorough understanding of the finite element method are presented in Chapter 2.

Consider that the objective of a finite element analysis is to evaluate the displacements at a large number of points of a structure under consideration. Then, once the physical relationships between the required displacements of the structure and the applied loads are known, it is possible, using the concepts of matrix algebra, to express the complete solution process employing a few symbols to identify the various quantities and a few lines to describe the solution process. However, although the solution process is expressed in an elegant manner in terms of matrices and matrix manipulations, it is frequently important to identify the detailed operations that are followed in the matrix solution in order to design more effective solution schemes. Again, this is possible only if the definitions and concepts of matrix algebra are well understood.

1.2 INTRODUCTION TO MATRICES

The effectiveness of using matrices in practical calculations is readily realized by considering the solution of a set of linear simultaneous equations, such as

$$\begin{aligned} 5x_1 - 4x_2 + x_3 \qquad\qquad &= 0 \\ -4x_1 + 6x_2 - 4x_3 + x_4 &= 1 \\ x_1 - 4x_2 + 6x_3 - 4x_4 &= 0 \\ x_2 - 4x_3 + 5x_4 &= 0 \end{aligned} \tag{1.1}$$

where the unknowns are x_1, x_2, x_3, and x_4. Using matrix notation, this set of equations is written as

$$\begin{bmatrix} 5 & -4 & 1 & 0 \\ -4 & 6 & -4 & 1 \\ 1 & -4 & 6 & -4 \\ 0 & 1 & -4 & 5 \end{bmatrix} \begin{bmatrix} x_1 \\ x_2 \\ x_3 \\ x_4 \end{bmatrix} = \begin{bmatrix} 0 \\ 1 \\ 0 \\ 0 \end{bmatrix} \tag{1.2}$$

where it is noted that, rather logically, the coefficients of the unknowns (5, −4, 1, etc.) are grouped together in one array; the left-hand-side unknowns (x_1, x_2, x_3, and x_4) and the right-hand-side known quantities are each grouped together in additional arrays. Although written differently, the relation (1.2) still reads the same way as (1.1). However, using matrix symbols to identify the arrays in (1.2), we can now write the set of simultaneous equations as

$$\mathbf{A}\mathbf{x} = \mathbf{b} \tag{1.3}$$

where $\mathbf{A}$ is the matrix of the coefficients in the set of linear equations, $\mathbf{x}$ is the matrix of unknowns, and $\mathbf{b}$ is the matrix of known quantities; i.e.,

$$\mathbf{A} = \begin{bmatrix} 5 & -4 & 1 & 0 \\ -4 & 6 & -4 & 1 \\ 1 & -4 & 6 & -4 \\ 0 & 1 & -4 & 5 \end{bmatrix}; \qquad \mathbf{x} = \begin{bmatrix} x_1 \\ x_2 \\ x_3 \\ x_4 \end{bmatrix}; \qquad \mathbf{b} = \begin{bmatrix} 0 \\ 1 \\ 0 \\ 0 \end{bmatrix} \tag{1.4}$$

The formal definition of a matrix, as given below, seems now apparent.

Definition: *A* matrix *is an array of ordered numbers. A* general matrix *consists of* mn *numbers arranged in* m *rows and* n *columns, giving the following array:*

$$\mathbf{A} = \begin{bmatrix} a_{11} & a_{12} & \cdots & a_{1n} \\ a_{21} & a_{22} & \cdots & a_{2n} \\ \vdots & \vdots & & \vdots \\ a_{m1} & a_{m2} & \cdots & a_{mn} \end{bmatrix} \tag{1.5}$$

We say that this matrix has order $m \times n$ (m by n). When we have only one row ($m = 1$) or one column ($n = 1$), we call $\mathbf{A}$ also a vector. Matrices are represented in this book using boldface letters, and usually upper case letters when they are not vectors. On the other hand, vectors can be upper case or lower case boldface.

We therefore see that the following are matrices:

$$\begin{bmatrix} 1 \\ 2 \end{bmatrix}; \qquad \begin{bmatrix} 1 & 4 & -5.3 \\ 3 & 2.1 & 6 \end{bmatrix}; \qquad [6.1 \quad 2.2 \quad 3] \tag{1.6}$$

where the first and the last matrices are also column and row vectors, respectively.

A typical element in the ith row and jth column of $\mathbf{A}$ is identified as a_{ij}, e.g., in the first matrix in (1.4), $a_{11} = 5$ and $a_{12} = -4$. Considering the elements a_{ij} in (1.5), we note that the subscript i runs from 1 to m and the subscript j runs from 1 to n. A comma between subscripts will be used when there is any risk of confusion, e.g., $a_{1+r,\, j+s}$, or to denote differentiation (see Chapter 6).

In general, the utility of matrices in practice arises from the fact that we can identify and manipulate an array of many numbers by use of a single symbol.

Relationships among large sets of numbers can be stated in a clear and compact way. Furthermore, if the relationships involve properly defined known and unknown quantities, such as in the solution of a set of simultaneous equations, it is possible to solve for the unknown quantities in a systematic manner. This is where the computer is of so much help, because it can calculate and manipulate the elements of matrices at very high speeds. However, we as the analysts are required to clearly define the variables in a problem, to prescribe the calculation procedures for the evaluation of the elements in the matrices involved, and to design the solution procedures.

1.3 SPECIAL MATRICES

Whenever the elements of a matrix obey a certain law, we can consider the matrix to be of special form. A *real matrix* is a matrix whose elements are all real. A *complex matrix* has elements that may be complex. We shall deal only with real matrices. In addition, the matrix will often be symmetric, which is the property defined next.

> ***Definition:*** *The* transpose *of the* $m \times n$ *matrix* $\mathbf{A}$, *written as* $\mathbf{A}^T$, *is obtained by interchanging the rows and columns in* $\mathbf{A}$. *If* $\mathbf{A} = \mathbf{A}^T$, *it follows that the number of rows and columns in* $\mathbf{A}$ *are equal and that* $a_{ij} = a_{ji}$. *Because* $m = n$, *we say that* $\mathbf{A}$ *is a* square matrix *of order* n, *and because* $a_{ij} = a_{ji}$, *we say that* $\mathbf{A}$ *is a* symmetric matrix. *Note that symmetry implies that* $\mathbf{A}$ *is square but not vice versa; i.e., a square matrix need not be symmetric.*

A symmetric matrix of order 4 was, for example, the coefficient matrix $\mathbf{A}$ in (1.2). We can verify that $\mathbf{A}^T = \mathbf{A}$ by simply checking that $a_{ji} = a_{ij}$ for $i, j = 1, \ldots, 4$.

Another special matrix is the *identity* (*or unit*) *matrix* $\mathbf{I}_n$, which is a square matrix of order n with only zero elements except for its diagonal entries, which are unity. For example, the identity matrix of order 3 is

$$\mathbf{I}_3 = \begin{bmatrix} 1 & 0 & 0 \\ 0 & 1 & 0 \\ 0 & 0 & 1 \end{bmatrix} \tag{1.7}$$

In practical calculations the order of an identity matrix is often implied and the subscript is not written. In analogy with the identity matrix, we also use *identity* (*or unit*) *vectors* of order n, defined as $\mathbf{e}_i$, where the subscript i indicates that the vector is the ith column of an identity matrix.

We shall work abundantly with symmetric banded matrices. Bandedness means that all elements beyond the bandwidth of the matrix are zero. Because $\mathbf{A}$ is symmetric, we can state this condition as

$$a_{ij} = 0 \qquad \text{for } j > i + m_{\mathrm{A}} \tag{1.8}$$

where $2m_{\mathrm{A}} + 1$ *is the bandwidth of* $\mathbf{A}$. As an example, the following matrix is a symmetric banded matrix of order 5. The half-bandwidth m_{A} is 2:

$$\mathbf{A} = \begin{bmatrix} 3 & 2 & 1 & 0 & 0 \\ 2 & 3 & 4 & 1 & 0 \\ 1 & 4 & 5 & 6 & 1 \\ 0 & 1 & 6 & 7 & 4 \\ 0 & 0 & 1 & 4 & 3 \end{bmatrix} \tag{1.9}$$

If the half-bandwidth of a matrix is zero, we have nonzero elements only on the diagonal of the matrix and denote it as a *diagonal matrix*. For example, the identity matrix is a diagonal matrix.

In computer calculations with matrices, we need to use a scheme of storing the elements of the matrices in high-speed storage. An obvious way of storing the elements of a matrix **A** of order $m \times n$ is simply to dimension in the FORTRAN program an array A(M, N), where M $= m$ and N $= n$, and store each matrix element a_{ij} in the storage location A(I,J). However, in many calculations we store in this way unnecessarily many zero elements of **A**, which are never needed in the calculations. Also, if **A** is symmetric, we should probably take advantage of it and store only the upper half of the matrix, including the diagonal elements. In general, only a restricted number of high-speed storage locations will be available, and it is necessary to use an effective storage scheme in order to be able to take into high-speed core the maximum matrix size possible. If the matrix is too large to be contained in high-speed storage, the solution process involves reading and writing on secondary storage, which can add significantly to the solution cost. Fortunately, in finite element analysis, the system matrices are symmetric and banded. Therefore, with an effective storage scheme, rather large-order matrices can be kept in high-speed core.

Let us denote by A(I) the Ith element in the one-dimensional storage array A. A diagonal matrix of order n would simply be stored as shown in Fig. 1.1(a):

$$\mathrm{A(I)} = a_{ii}; \qquad \mathrm{I} = i = 1, \ldots, n \tag{1.10}$$

Consider a banded matrix as shown in Fig. 1.1(b). We will see later that zero elements within the "skyline" of the matrix may be changed to nonzero elements in the solution process; for example, a_{35} may be a zero element but becomes nonzero during the solution process (see Section 8.2.3). Therefore, we allocate storage locations to zero elements within the skyline but do not need to store zero elements that are outside the skyline. The storage scheme that will be used in the finite element solution process is indicated in Fig. 1.1 and is explained further in the Appendix.

1.4 MATRIX EQUALITY, ADDITION, AND MULTIPLICATION BY A SCALAR

We have defined matrices to be ordered arrays of numbers and identified them by single symbols. In order to be able to deal with them in a similar way as we deal with ordinary numbers, it is necessary to define rules corresponding to those which govern equality, addition, subtraction, multiplication, and division of ordinary numbers. We shall simply state the matrix rules and not provide motivation for them. The rationale for these rules will appear later, as it will turn out

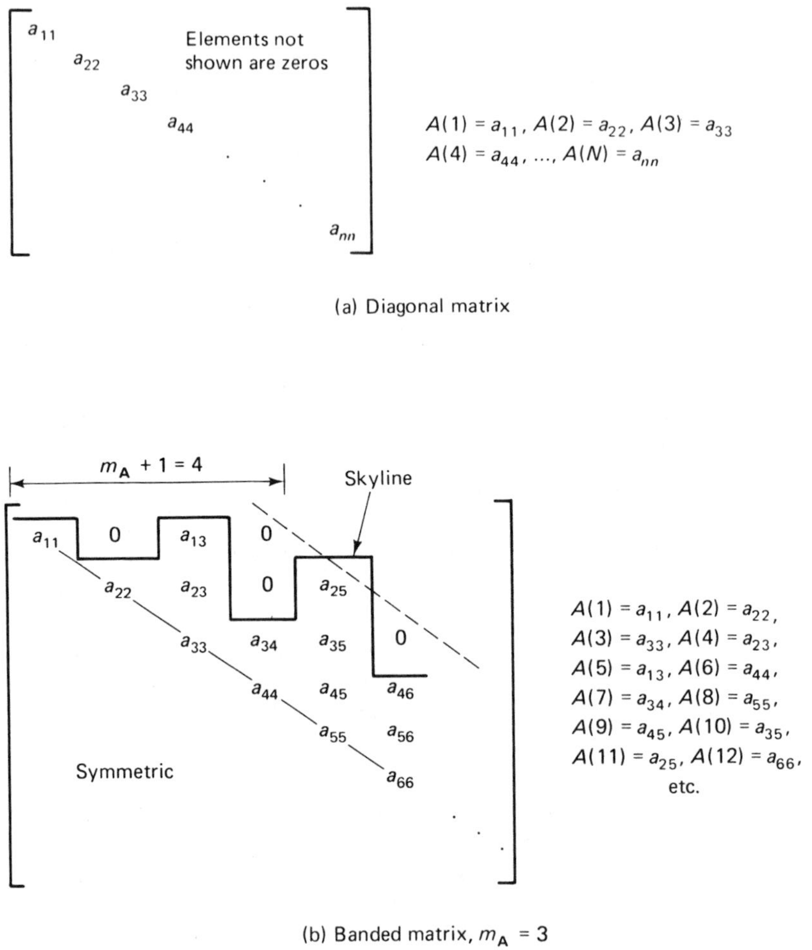

FIGURE 1.1 *Storage of matrix A in a one-dimensional array.*

that these are precisely the rules that are needed to use matrices in the solution of practical problems. For matrix equality, matrix addition, and matrix multiplication by a scalar we provide the following definitions.

Definition: *The matrices* **A** *and* **B** *are* equal *if and only if:*

1. **A** *and* **B** *have the same number of rows and columns.*
2. *All corresponding elements are equal; i.e.,* $a_{ij} = b_{ij}$ *for all* i *and* j.

Definition: *Two matrices* **A** *and* **B** *can be* added *if and only if they have the same number of rows and columns. The addition of the matrices is performed by adding all corresponding elements; i.e., if* a_{ij} *and* b_{ij} *denote a general element of* **A** *and* **B**, *respectively, then* $c_{ij} = a_{ij} + b_{ij}$ *denotes a general element of* **C**, *where* $\mathbf{C} = \mathbf{A} + \mathbf{B}$. *It follows that* **C** *has the same number of rows and columns as* **A** *and* **B**.

EXAMPLE 1.1: Calculate $\mathbf{C} = \mathbf{A} + \mathbf{B}$, where

$$\mathbf{A} = \begin{bmatrix} 2 & 1 & 1 \\ 0.5 & 3 & 0 \end{bmatrix}; \qquad \mathbf{B} = \begin{bmatrix} 3 & 1 & 2 \\ 2 & 4 & 1 \end{bmatrix}$$

Here we have $$\mathbf{C} = \mathbf{A} + \mathbf{B} = \begin{bmatrix} 5 & 2 & 3 \\ 2.5 & 7 & 1 \end{bmatrix}$$

It need be noted that the order in which the matrices are added is not important. The subtraction of matrices is defined in an analogous way.

EXAMPLE 1.2: Calculate $\mathbf{C} = \mathbf{A} - \mathbf{B}$, where $\mathbf{A}$ and $\mathbf{B}$ are given in Example 1.1.

Here we have

$$\mathbf{C} = \mathbf{A} - \mathbf{B} = \begin{bmatrix} -1 & 0 & -1 \\ -1.5 & -1 & -1 \end{bmatrix}$$

From the definition of the subtraction of matrices, it follows that the subtraction of a matrix from itself results into a matrix with zero elements only. Such a matrix is defined to be a *null matrix* $\mathbf{0}$. We turn now to the multiplication of a matrix by a scalar.

Definition: *A matrix is* multiplied *by a scalar by multiplying each matrix element by the scalar; i.e.,* $\mathbf{C} = k\mathbf{A}$ *means that* $c_{ij} = ka_{ij}$.

The following example demonstrates this definition.

EXAMPLE 1.3: Calculate $\mathbf{C} = k\mathbf{A}$, where

$$\mathbf{A} = \begin{bmatrix} 2 & 1 & 1 \\ 0.5 & 3 & 0 \end{bmatrix}; \qquad k = 2$$

We have $$\mathbf{C} = k\mathbf{A} = \begin{bmatrix} 4 & 2 & 2 \\ 1 & 6 & 0 \end{bmatrix}$$

It should be noted that so far all definitions are completely analogous to those which are used in the calculation with ordinary numbers. Furthermore, to add (or subtract) two general matrices of order $n \times m$ requires nm addition (subtraction) operations, and to multiply a general matrix of order $n \times m$ by a scalar requires nm multiplications. Therefore, when the matrices are of special form, such as symmetric and banded, we should take advantage of the situation by only evaluating the elements below the skyline of the matrix $\mathbf{C}$, because all other elements are zero.

1.5 MULTIPLICATION OF MATRICES

We consider two matrices $\mathbf{A}$ and $\mathbf{B}$ and want to find the matrix product $\mathbf{C} = \mathbf{AB}$.

Definition: *Two matrices* $\mathbf{A}$ *and* $\mathbf{B}$ *can be* multiplied *to obtain* $\mathbf{C} = \mathbf{AB}$ *if and only if the number of columns in* $\mathbf{A}$ *is equal to the number of rows in* $\mathbf{B}$. *Assume that* $\mathbf{A}$ *is of order* $p \times m$ *and* $\mathbf{B}$ *is of order* $m \times q$. *Then for each element in* $\mathbf{C}$ *we have*

$$c_{ij} = \sum_{r=1}^{m} a_{ir} b_{rj} \tag{1.11}$$

where $\mathbf{C}$ *is of order* $p \times q$; *i.e., the indices* i *and* j *in (1.11) vary from* 1 *to* p *and* 1 *to* q, *respectively.*

Therefore, to calculate the (i, j)th element in $\mathbf{C}$, we multiply the elements in the ith row of $\mathbf{A}$ by the elements in the jth column of $\mathbf{B}$ and add all individual products. By taking the product of each row in **A** with each column in **B**, it follows that **C** must be of order $p \times q$.

EXAMPLE 1.4: Calculate the matrix product $\mathbf{C} = \mathbf{AB}$, where

$$\mathbf{A} = \begin{bmatrix} 5 & 3 & 1 \\ 4 & 6 & 2 \\ 10 & 3 & 4 \end{bmatrix}; \qquad \mathbf{B} = \begin{bmatrix} 1 & 5 \\ 2 & 4 \\ 3 & 2 \end{bmatrix}$$

We have

$$\begin{aligned} c_{11} &= (5)(1) + (3)(2) + (1)(3) = 14 \\ c_{21} &= (4)(1) + (6)(2) + (2)(3) = 22 \\ c_{31} &= (10)(1) + (3)(2) + (4)(3) = 28 \quad \text{etc.} \end{aligned}$$

Hence we obtain

$$\mathbf{C} = \begin{bmatrix} 14 & 39 \\ 22 & 48 \\ 28 & 70 \end{bmatrix}$$

As can readily be verified, the number of multiplications required in this matrix multiplication is $p \times q \times m$. When we deal with matrices in practice, however, we can often reduce the number of operations by taking advantage of zero elements within the matrices.

EXAMPLE 1.5: Calculate the matrix product $\mathbf{c} = \mathbf{Ab}$, where

$$\mathbf{A} = \begin{bmatrix} 2 & -1 & 0 & 0 \\ & 2 & -1 & 0 \\ & & 2 & -1 \\ \text{symmetric} & & & 1 \end{bmatrix}; \qquad \mathbf{b} = \begin{bmatrix} 4 \\ 1 \\ 2 \\ 3 \end{bmatrix}$$

Here we can take advantage of the fact that the bandwidth of **A** is 3; i.e., $m_{\mathbf{A}} = 1$. Thus, taking into account only the elements within the band of **A**, we have

$$\begin{aligned} c_1 &= (2)(4) + (-1)(1) = 7 \\ c_2 &= (-1)(4) + (2)(1) + (-1)(2) = -4 \\ c_3 &= (-1)(1) + (2)(2) + (-1)(3) = 0 \\ c_4 &= (-1)(2) + (1)(3) = 1 \end{aligned}$$

Hence

$$\mathbf{c} = \begin{bmatrix} 7 \\ -4 \\ 0 \\ 1 \end{bmatrix}$$

As is well known, the multiplication of ordinary numbers is commutative; i.e., $ab = ba$. We need to investigate if the same holds for matrix multiplication. If we consider the matrices

$$\mathbf{A} = \begin{bmatrix} 1 \\ 2 \end{bmatrix}; \qquad \mathbf{B} = [3 \quad 4] \tag{1.12}$$

we have

$$\mathbf{AB} = \begin{bmatrix} 3 & 4 \\ 6 & 8 \end{bmatrix}; \qquad \mathbf{BA} = [11] \tag{1.13}$$

Therefore, the products **AB** and **BA** are not the same, and it follows that matrix multiplication is not commutative. Indeed, depending on the orders of **A** and **B**, the orders of the two product matrices **AB** and **BA** can be different, and the product **AB** may be defined, whereas the product **BA** may not be calculable.

To distinguish the order of multiplication of matrices, we say that in the product **AB**, the matrix **A** premultiplies **B**, or the matrix **B** postmultiplies **A**. Although $\mathbf{AB} \neq \mathbf{BA}$ in general, it may happen that $\mathbf{AB} = \mathbf{BA}$ for special **A** and **B**, in which case we say that **A** and **B** commute.

Although the cummutative law does not hold in matrix multiplication, the distributive law and associative law both are valid. The distributive law states that

$$\mathbf{E} = (\mathbf{A} + \mathbf{B})\mathbf{C} = \mathbf{AC} + \mathbf{BC} \tag{1.14}$$

In other words, we may first add **A** and **B** and then do the multiplication by **C**, or we may first multiply **A** and **B** by **C** and then do the addition. Note that considering the number of operations, the evaluation of **E** by adding **A** and **B** first is much more economical, which is important to remember in the design of an analysis program.

The distributive law is proved using (1.11); namely, using

$$e_{ij} = \sum_{r=1}^{m} (a_{ir} + b_{ir})c_{rj} \tag{1.15}$$

we obtain

$$e_{ij} = \sum_{r=1}^{m} a_{ir}c_{rj} + \sum_{r=1}^{m} b_{ir}c_{rj} \tag{1.16}$$

The associative law states that,

$$\mathbf{G} = (\mathbf{AB})\mathbf{C} = \mathbf{A}(\mathbf{BC}) = \mathbf{ABC} \tag{1.17}$$

in other words, that the order of multiplication is immaterial. The proof is carried out by using the definition of matrix multiplication in (1.11) and calculating in either way a general element of **G**.

Since the associative law holds, in practice, a string of matrix multiplications can be carried out in an arbitrary sequence, and by a clever choice of the sequence, many operations can frequently be saved. The only point that must be remembered when manipulating the matrices is that brackets can be removed or inserted, that powers can be combined, but that the order of multiplication must be preserved.

Consider the following examples to demonstrate the use of the associative and distributive laws in order to simplify a string of matrix multiplications.

EXAMPLE 1.6: Calculate $\mathbf{A}^4$, where

$$\mathbf{A} = \begin{bmatrix} 2 & 1 \\ 1 & 3 \end{bmatrix}$$

One way of evaluating $\mathbf{A}^4$ is to simply calculate

$$\mathbf{A}^2 = \begin{bmatrix} 2 & 1 \\ 1 & 3 \end{bmatrix}\begin{bmatrix} 2 & 1 \\ 1 & 3 \end{bmatrix} = \begin{bmatrix} 5 & 5 \\ 5 & 10 \end{bmatrix}$$

Hence

$$\mathbf{A}^3 = \mathbf{A}^2\mathbf{A} = \begin{bmatrix} 5 & 5 \\ 5 & 10 \end{bmatrix}\begin{bmatrix} 2 & 1 \\ 1 & 3 \end{bmatrix} = \begin{bmatrix} 15 & 20 \\ 20 & 35 \end{bmatrix}$$

and

$$\mathbf{A}^4 = \mathbf{A}^3\mathbf{A} = \begin{bmatrix} 15 & 20 \\ 20 & 35 \end{bmatrix}\begin{bmatrix} 2 & 1 \\ 1 & 3 \end{bmatrix} = \begin{bmatrix} 50 & 75 \\ 75 & 125 \end{bmatrix}$$

Alternatively, we may use

$$\mathbf{A}^4 = \mathbf{A}^2\mathbf{A}^2 = \begin{bmatrix} 5 & 5 \\ 5 & 10 \end{bmatrix}\begin{bmatrix} 5 & 5 \\ 5 & 10 \end{bmatrix} = \begin{bmatrix} 50 & 75 \\ 75 & 125 \end{bmatrix}$$

and save one matrix multiplication.

EXAMPLE 1.7: Evaluate the product $\mathbf{v}^T\mathbf{A}\mathbf{v}$, where

$$\mathbf{A} = \begin{bmatrix} 3 & 2 & 1 \\ 2 & 4 & 2 \\ 1 & 2 & 6 \end{bmatrix}; \qquad \mathbf{v} = \begin{bmatrix} 1 \\ 2 \\ -1 \end{bmatrix}$$

The formal procedure would be to calculate $\mathbf{x} = \mathbf{A}\mathbf{v}$; i.e.,

$$\mathbf{x} = \mathbf{A}\mathbf{v} = \begin{bmatrix} 3 & 2 & 1 \\ 2 & 4 & 2 \\ 1 & 2 & 6 \end{bmatrix}\begin{bmatrix} 1 \\ 2 \\ -1 \end{bmatrix} = \begin{bmatrix} 6 \\ 8 \\ -1 \end{bmatrix}$$

and then calculate $\mathbf{v}^T\mathbf{x}$ to obtain

$$\mathbf{v}^T\mathbf{A}\mathbf{v} = [1 \quad 2 \quad -1]\begin{bmatrix} 6 \\ 8 \\ -1 \end{bmatrix} = 23$$

However, it is more effective to calculate the required product in the following way. First, we write

$$\mathbf{A} = \mathbf{U} + \mathbf{D} + \mathbf{U}^T$$

where $\mathbf{U}$ is a lower triangular matrix and $\mathbf{D}$ is a diagonal matrix,

$$\mathbf{U} = \begin{bmatrix} 0 & 0 & 0 \\ 2 & 0 & 0 \\ 1 & 2 & 0 \end{bmatrix}; \qquad \mathbf{D} = \begin{bmatrix} 3 & 0 & 0 \\ 0 & 4 & 0 \\ 0 & 0 & 6 \end{bmatrix}$$

Hence we have

$$\mathbf{v}^T\mathbf{A}\mathbf{v} = \mathbf{v}^T(\mathbf{U} + \mathbf{D} + \mathbf{U}^T)\mathbf{v}$$
$$\mathbf{v}^T\mathbf{A}\mathbf{v} = \mathbf{v}^T\mathbf{U}\mathbf{v} + \mathbf{v}^T\mathbf{D}\mathbf{v} + \mathbf{v}^T\mathbf{U}^T\mathbf{v}$$

However, $\mathbf{v}^T\mathbf{U}\mathbf{v}$ is a single number and hence $\mathbf{v}^T\mathbf{U}^T\mathbf{v} = \mathbf{v}^T\mathbf{U}\mathbf{v}$, and it follows that

$$\mathbf{v}^T\mathbf{A}\mathbf{v} = 2\mathbf{v}^T\mathbf{U}\mathbf{v} + \mathbf{v}^T\mathbf{D}\mathbf{v} \tag{a}$$

The higher efficiency in the matrix multiplication is obtained by taking advantage of the fact that $\mathbf{U}$ is a lower triangular and $\mathbf{D}$ is a diagonal matrix. Let $\mathbf{x} = \mathbf{U}\mathbf{v}$; then we have

$$x_1 = 0$$
$$x_2 = (2)(1) = 2$$
$$x_3 = (1)(1) + (2)(2) = 5$$

Hence

$$\mathbf{x} = \begin{bmatrix} 0 \\ 2 \\ 5 \end{bmatrix}$$

Next, we obtain

$$\mathbf{v}^T\mathbf{U}\mathbf{v} = \mathbf{v}^T\mathbf{x} = (2)(2) + (-1)(5) = -1$$

Also

$$\mathbf{v}^T\mathbf{D}\mathbf{v} = (1)(1)(3) + (2)(2)(4) + (-1)(-1)(6) = 25$$

Hence using (a) we have $\mathbf{v}^T\mathbf{A}\mathbf{v} = 23$, as before.

Apart from the commutative law, which in general does not hold in matrix multiplications, also the cancellation of matrices in matrix equations cannot be performed, in general, as the cancellation of ordinary numbers. In particular, if $\mathbf{AB} = \mathbf{CB}$, it does not necessarily follow that $\mathbf{A} = \mathbf{C}$. This is easily demonstrated considering a specific case:

$$\begin{bmatrix} 2 & 1 \\ 4 & 0 \end{bmatrix}\begin{bmatrix} 1 \\ 2 \end{bmatrix} = \begin{bmatrix} 4 & 0 \\ 0 & 2 \end{bmatrix}\begin{bmatrix} 1 \\ 2 \end{bmatrix} \tag{1.18}$$

but

$$\begin{bmatrix} 2 & 1 \\ 4 & 0 \end{bmatrix} \neq \begin{bmatrix} 4 & 0 \\ 0 & 2 \end{bmatrix} \tag{1.19}$$

However, it must be noted that $\mathbf{A} = \mathbf{C}$ if the equation $\mathbf{AB} = \mathbf{CB}$ holds for all possible $\mathbf{B}$. Namely, in that case, we simply select $\mathbf{B}$ to be the identity matrix $\mathbf{I}$, and hence $\mathbf{A} = \mathbf{C}$.

It need also be noted that included in the above observation is the fact that if $\mathbf{AB} = \mathbf{0}$, it does not follow that either $\mathbf{A}$ or $\mathbf{B}$ is a null matrix. A specific case demonstrates this observation:

$$\mathbf{A} = \begin{bmatrix} 1 & 0 \\ 2 & 0 \end{bmatrix}; \quad \mathbf{B} = \begin{bmatrix} 0 & 0 \\ 3 & 4 \end{bmatrix}; \quad \mathbf{AB} = \begin{bmatrix} 0 & 0 \\ 0 & 0 \end{bmatrix} \tag{1.20}$$

Some special rules concerning the use of transposed matrices in matrix multiplications need be pointed out. It is noted that the transpose of the product of two matrices $\mathbf{A}$ and $\mathbf{B}$ is equal to the product of the transposed matrices in reverse order; i.e.,

$$(\mathbf{AB})^T = \mathbf{B}^T\mathbf{A}^T \tag{1.21}$$

The proof that (1.21) does hold is obtained using the definition for the evaluation of a matrix product given in (1.11).

Considering the matrix products in (1.21), it need be noted that although $\mathbf{A}$ may be symmetric, $\mathbf{AB}$ is, in general, not symmetric. However, if $\mathbf{A}$ is symmetric, the matrix $\mathbf{B}^T\mathbf{AB}$ is always symmetric. The proof follows using (1.21):

$$(\mathbf{B}^T\mathbf{AB})^T = (\mathbf{AB})^T(\mathbf{B}^T)^T \tag{1.22}$$

$$= \mathbf{B}^T\mathbf{A}^T\mathbf{B} \tag{1.23}$$

But, because $\mathbf{A}^T = \mathbf{A}$, we have

$$(\mathbf{B}^T\mathbf{AB})^T = \mathbf{B}^T\mathbf{AB} \tag{1.24}$$

and hence $\mathbf{B}^T\mathbf{AB}$ is symmetric.

1.6 THE INVERSE MATRIX

We have seen that matrix addition and subtraction are carried out using essentially the same laws as those used in the manipulation of ordinary numbers. However, matrix multiplication is quite different and we have to get used to special rules. With regard to matrix division, it strictly does not exist. Instead, an inverse matrix is defined. We shall define and use the inverse of square matrices only.

Definition: *The* inverse *of a matrix* $\mathbf{A}$ *is denoted by* $\mathbf{A}^{-1}$. *Assume that the inverse exists; then the elements of* $\mathbf{A}^{-1}$ *are such that* $\mathbf{A}^{-1}\mathbf{A} = \mathbf{I}$ *and* $\mathbf{AA}^{-1} = \mathbf{I}$. *A matrix that possesses an inverse is said to be* nonsingular. *A matrix without an inverse is a* singular *matrix.*

As mentioned above, the inverse of a matrix does not need to exist. A trivial example is the null matrix. Assume that the inverse of $\mathbf{A}$ exists. Then we still want to show that either of the conditions $\mathbf{A}^{-1}\mathbf{A} = \mathbf{I}$ or $\mathbf{A}\mathbf{A}^{-1} = \mathbf{I}$ implies the other. Assume that we have evaluated the elements of the matrices $\mathbf{A}_l^{-1}$ and $\mathbf{A}_r^{-1}$ such that $\mathbf{A}_l^{-1}\mathbf{A} = \mathbf{I}$ and $\mathbf{A}\mathbf{A}_r^{-1} = \mathbf{I}$. Then we have

$$\mathbf{A}_l^{-1} = \mathbf{A}_l^{-1}(\mathbf{A}\mathbf{A}_r^{-1}) = (\mathbf{A}_l^{-1}\mathbf{A})\mathbf{A}_r^{-1} = \mathbf{A}_r^{-1} \tag{1.25}$$

and hence $\mathbf{A}_l^{-1} = \mathbf{A}_r^{-1}$.

EXAMPLE 1.8: Evaluate the inverse of the matrix $\mathbf{A}$, where

$$\mathbf{A} = \begin{bmatrix} 2 & -1 \\ -1 & 3 \end{bmatrix}$$

For the inverse of $\mathbf{A}$ we need $\mathbf{A}\mathbf{A}^{-1} = \mathbf{I}$. By trial and error (or otherwise) we find that

$$\mathbf{A}^{-1} = \begin{bmatrix} \frac{3}{5} & \frac{1}{5} \\ \frac{1}{5} & \frac{2}{5} \end{bmatrix}$$

We check that $\mathbf{A}\mathbf{A}^{-1} = \mathbf{I}$ and $\mathbf{A}^{-1}\mathbf{A} = \mathbf{I}$:

$$\mathbf{A}\mathbf{A}^{-1} = \begin{bmatrix} 2 & -1 \\ -1 & 3 \end{bmatrix}\begin{bmatrix} \frac{3}{5} & \frac{1}{5} \\ \frac{1}{5} & \frac{2}{5} \end{bmatrix} = \begin{bmatrix} 1 & 0 \\ 0 & 1 \end{bmatrix}$$

$$\mathbf{A}^{-1}\mathbf{A} = \begin{bmatrix} \frac{3}{5} & \frac{1}{5} \\ \frac{1}{5} & \frac{2}{5} \end{bmatrix}\begin{bmatrix} 2 & -1 \\ -1 & 3 \end{bmatrix} = \begin{bmatrix} 1 & 0 \\ 0 & 1 \end{bmatrix}$$

To calculate the inverse of a product $\mathbf{AB}$, we proceed as follows. Let $\mathbf{G} = (\mathbf{AB})^{-1}$, where $\mathbf{A}$ and $\mathbf{B}$ are both square matrices. Then

$$\mathbf{GAB} = \mathbf{I} \tag{1.26}$$

and postmultiplying by $\mathbf{B}^{-1}$ and $\mathbf{A}^{-1}$, we obtain

$$\mathbf{GA} = \mathbf{B}^{-1} \tag{1.27}$$

$$\mathbf{G} = \mathbf{B}^{-1}\mathbf{A}^{-1} \tag{1.28}$$

Therefore,

$$(\mathbf{AB})^{-1} = \mathbf{B}^{-1}\mathbf{A}^{-1} \tag{1.29}$$

The reader will notice that the same law of matrix reversal was shown to apply when the transpose of a matrix product is calculated.

EXAMPLE 1.9: For the matrices $\mathbf{A}$ and $\mathbf{B}$ given below, check that $(\mathbf{AB})^{-1} = \mathbf{B}^{-1}\mathbf{A}^{-1}$.

$$\mathbf{A} = \begin{bmatrix} 2 & -1 \\ -1 & 3 \end{bmatrix}; \qquad \mathbf{B} = \begin{bmatrix} 3 & 0 \\ 0 & 4 \end{bmatrix}$$

The inverse of $\mathbf{A}$ was used in Example 1.8. The inverse of $\mathbf{B}$ is easy to obtain:

$$\mathbf{B}^{-1} = \begin{bmatrix} \frac{1}{3} & 0 \\ 0 & \frac{1}{4} \end{bmatrix}$$

To check that $(\mathbf{AB})^{-1} = \mathbf{B}^{-1}\mathbf{A}^{-1}$, we need to evaluate $\mathbf{C} = \mathbf{AB}$:

$$\mathbf{C} = \begin{bmatrix} 2 & -1 \\ -1 & 3 \end{bmatrix}\begin{bmatrix} 3 & 0 \\ 0 & 4 \end{bmatrix} = \begin{bmatrix} 6 & -4 \\ -3 & 12 \end{bmatrix}$$

Assume that $\mathbf{C}^{-1} = \mathbf{B}^{-1}\mathbf{A}^{-1}$. Then we would have

$$\mathbf{C}^{-1} = \begin{bmatrix} \frac{1}{3} & 0 \\ 0 & \frac{1}{4} \end{bmatrix}\begin{bmatrix} \frac{3}{5} & \frac{1}{5} \\ \frac{1}{5} & \frac{2}{5} \end{bmatrix} = \begin{bmatrix} \frac{1}{5} & \frac{1}{15} \\ \frac{1}{20} & \frac{1}{10} \end{bmatrix} \tag{a}$$

To check that the matrix given in (a) is indeed the inverse of $\mathbf{C}$, we evaluate $\mathbf{C}^{-1}\mathbf{C}$ and find that

$$\mathbf{C}^{-1}\mathbf{C} = \begin{bmatrix} \frac{1}{5} & \frac{1}{15} \\ \frac{1}{20} & \frac{1}{10} \end{bmatrix}\begin{bmatrix} 6 & -4 \\ -3 & 12 \end{bmatrix} = \mathbf{I}$$

But since $\mathbf{C}^{-1}$ is unique and only the correct $\mathbf{C}^{-1}$ satisfies the relation $\mathbf{C}^{-1}\mathbf{C} = \mathbf{I}$, we indeed have found in (a) the inverse of $\mathbf{C}$, and the relation $(\mathbf{AB})^{-1} = \mathbf{B}^{-1}\mathbf{A}^{-1}$ is satisfied.

In Examples 1.8 and 1.9, the inverse of $\mathbf{A}$ and $\mathbf{B}$ could be found by trial and error. However, to obtain the inverse of a general matrix, we need to have a general algorithm. One way of calculating the inverse of a matrix $\mathbf{A}$ of order n is to solve the n systems of equations

$$\mathbf{AX} = \mathbf{I} \tag{1.30}$$

where $\mathbf{I}$ is the identity matrix of order n and we have $\mathbf{X} = \mathbf{A}^{-1}$. For the solution of each system of equations in (1.30), we can use the algorithms presented in Section 8.2.

The above considerations show that a system of equations could be solved by calculating the inverse of the coefficient matrix; i.e., if we have

$$\mathbf{Ay} = \mathbf{c} \tag{1.31}$$

where $\mathbf{A}$ is of order $n \times n$ and $\mathbf{y}$ and $\mathbf{c}$ are of order $n \times 1$, then

$$\mathbf{y} = \mathbf{A}^{-1}\mathbf{c} \tag{1.32}$$

However, the inversion of $\mathbf{A}$ is very costly and it is much more effective to use Gauss elimination to solve the equations in (1.31) (see Chapter 8). *Indeed, although we may write symbolically that* $\mathbf{y} = \mathbf{A}^{-1}\mathbf{c}$, *to evaluate* $\mathbf{y}$ *we actually use Gauss elimination.*

1.7 PARTITIONING OF MATRICES

To facilitate matrix manipulations and to take advantage of the special form of matrices, it may be useful to partition a matrix into submatrices. A submatrix is a matrix that is obtained from the original matrix by including only the elements of certain rows and columns. The idea is demonstrated using a specific case in which the dashed lines are the lines of partitioning:

$$\mathbf{A} = \left[\begin{array}{c:ccc:cc} a_{11} & a_{12} & a_{13} & a_{14} & a_{15} & a_{16} \\ a_{21} & a_{22} & a_{23} & a_{24} & a_{25} & a_{26} \\ \hdashline a_{31} & a_{32} & a_{33} & a_{34} & a_{35} & a_{36} \end{array}\right] \tag{1.33}$$

It should be noted that each of the partitioning lines must run all across the original matrix. Using the partitioning, matrix $\mathbf{A}$ is written as

$$\mathbf{A} = \begin{bmatrix} \mathbf{A}_{11} & \mathbf{A}_{12} & \mathbf{A}_{13} \\ \mathbf{A}_{21} & \mathbf{A}_{22} & \mathbf{A}_{23} \end{bmatrix} \tag{1.34}$$

where $$\mathbf{A}_{11} = \begin{bmatrix} a_{11} \\ a_{21} \end{bmatrix}; \qquad \mathbf{A}_{12} = \begin{bmatrix} a_{12} & a_{13} & a_{14} \\ a_{22} & a_{23} & a_{24} \end{bmatrix}; \qquad \text{etc.} \tag{1.35}$$

The right-hand side of (1.34) could again be partitioned, such as

$$\mathbf{A} = \left[\begin{array}{c|cc} \mathbf{A}_{11} & \mathbf{A}_{12} & \mathbf{A}_{13} \\ \mathbf{A}_{21} & \mathbf{A}_{22} & \mathbf{A}_{23} \end{array}\right] \tag{1.36}$$

and we may write **A** as

$$\mathbf{A} = [\bar{\mathbf{A}}_1 \quad \bar{\mathbf{A}}_2]; \qquad \bar{\mathbf{A}}_1 = \begin{bmatrix} \mathbf{A}_{11} \\ \mathbf{A}_{21} \end{bmatrix}; \qquad \bar{\mathbf{A}}_2 = \begin{bmatrix} \mathbf{A}_{12} & \mathbf{A}_{13} \\ \mathbf{A}_{22} & \mathbf{A}_{23} \end{bmatrix} \tag{1.37}$$

The partitioning of matrices can be of advantage in saving computer storage; namely, if submatrices repeat, it would only be necessary to store the submatrix once. The same applies in arithmetic. Using submatrices, we may identify a typical operation that is repeated many times. We would then carry out this operation only once and use the result whenever it is needed.

The rules to be used in calculations with partitioned matrices follow from the definition of matrix addition, subtraction, and multiplication. Using partitioned matrices we can add, subtract, or multiply as if the submatrices were ordinary matrix elements, provided the original matrices have been partitioned in such a way that it is permissible to perform the individual submatrix additions, subtractions, or multiplications.

These rules are easily justified and remembered if we keep in mind that *the partitioning of the original matrices is only a device to facilitate the matrix manipulations and does not change any results.*

EXAMPLE 1.10: Evaluate the matrix product $\mathbf{C} = \mathbf{AB}$ of Example 1.4 by using the following partitioning:

$$\mathbf{A} = \left[\begin{array}{cc|c} 5 & 3 & 1 \\ 4 & 6 & 2 \\ \hline 10 & 3 & 4 \end{array}\right]; \qquad \mathbf{B} = \left[\begin{array}{cc} 1 & 5 \\ 2 & 4 \\ \hline 3 & 2 \end{array}\right]$$

Here we have $$\mathbf{A} = \begin{bmatrix} \mathbf{A}_{11} & \mathbf{A}_{12} \\ \mathbf{A}_{21} & \mathbf{A}_{22} \end{bmatrix}; \qquad \mathbf{B} = \begin{bmatrix} \mathbf{B}_1 \\ \mathbf{B}_2 \end{bmatrix}$$

Therefore, $$\mathbf{AB} = \begin{bmatrix} \mathbf{A}_{11}\mathbf{B}_1 + \mathbf{A}_{12}\mathbf{B}_2 \\ \mathbf{A}_{21}\mathbf{B}_1 + \mathbf{A}_{22}\mathbf{B}_2 \end{bmatrix} \tag{a}$$

But $$\mathbf{A}_{11}\mathbf{B}_1 = \begin{bmatrix} 5 & 3 \\ 4 & 6 \end{bmatrix}\begin{bmatrix} 1 & 5 \\ 2 & 4 \end{bmatrix} = \begin{bmatrix} 11 & 37 \\ 16 & 44 \end{bmatrix}$$

$$\mathbf{A}_{12}\mathbf{B}_2 = \begin{bmatrix} 1 \\ 2 \end{bmatrix}[3 \quad 2] = \begin{bmatrix} 3 & 2 \\ 6 & 4 \end{bmatrix}$$

$$\mathbf{A}_{21}\mathbf{B}_1 = [10 \quad 3]\begin{bmatrix} 1 & 5 \\ 2 & 4 \end{bmatrix} = [16 \quad 62]$$

$$\mathbf{A}_{22}\mathbf{B}_2 = [4][3 \quad 2] = [12 \quad 8]$$

Then substituting into (a) we have

$$\mathbf{AB} = \begin{bmatrix} 14 & 39 \\ 22 & 48 \\ 28 & 70 \end{bmatrix}$$

EXAMPLE 1.11: Taking advantage of partitioning, evaluate $\mathbf{c} = \mathbf{Ab}$, where

$$\mathbf{A} = \left[\begin{array}{cc|cc} 4 & 3 & 1 & 2 \\ 3 & 6 & 2 & 1 \\ \hline 1 & 2 & 8 & 6 \\ 2 & 1 & 6 & 12 \end{array}\right]; \qquad \mathbf{b} = \left[\begin{array}{c} 2 \\ 2 \\ \hline 1 \\ 1 \end{array}\right]$$

The only products that we need to evaluate are

$$\mathbf{w}_1 = \begin{bmatrix} 4 & 3 \\ 3 & 6 \end{bmatrix}\begin{bmatrix} 1 \\ 1 \end{bmatrix} = \begin{bmatrix} 7 \\ 9 \end{bmatrix}$$

and

$$\mathbf{w}_2 = \begin{bmatrix} 1 & 2 \\ 2 & 1 \end{bmatrix}\begin{bmatrix} 1 \\ 1 \end{bmatrix} = \begin{bmatrix} 3 \\ 3 \end{bmatrix}$$

We can now construct $\mathbf{c}$:

$$\mathbf{c} = \begin{bmatrix} 2\mathbf{w}_1 + \mathbf{w}_2 \\ 2\mathbf{w}_1 + 2\mathbf{w}_2 \end{bmatrix}$$

or, substituting,

$$\mathbf{c} = \begin{bmatrix} 17 \\ 21 \\ 20 \\ 24 \end{bmatrix}$$

1.8 THE TRACE AND DETERMINANT OF A MATRIX

The trace and determinant of a matrix are only defined if the matrix is square. Both quantities are single numbers, which are evaluated from the elements of the matrix and are therefore functions of the matrix elements.

Definition: *The* trace *of the matrix* $\mathbf{A}$ *is denoted as* $tr(\mathbf{A})$ *and is equal to* $\sum_{i=1}^{n} a_{ii}$, *where* n *is the order of* $\mathbf{A}$.

EXAMPLE 1.12: Calculate the trace of the matrix $\mathbf{A}$ given in Example 1.11.

Here we have

$$\text{tr}(\mathbf{A}) = 4 + 6 + 8 + 12 = 30$$

The determinant of a matrix $\mathbf{A}$ can be defined in terms of the determinants of submatrices of $\mathbf{A}$, and noting that the determinant of a matrix of order 1 is simply the element of the matrix; i.e., if $\mathbf{A} = [a_{11}]$, then $\det \mathbf{A} = a_{11}$.

Definition: *The* determinant *of an* n × n *matrix* $\mathbf{A}$ *is denoted as det* $\mathbf{A}$ *and is defined by the recurrence relation*

$$\det \mathbf{A} = \sum_{j=1}^{n} (-1)^{1+j} a_{1j} \det \mathbf{A}_{1j} \tag{1.38}$$

where $\mathbf{A}_{1j}$ *is the* $(n-1) \times (n-1)$ *matrix obtained by eliminating the 1st row and* j*th column from the matrix* $\mathbf{A}$.

EXAMPLE 1.13: Evaluate the determinant of $\mathbf{A}$, where

$$\mathbf{A} = \begin{bmatrix} a_{11} & a_{12} \\ a_{21} & a_{22} \end{bmatrix}$$

Using the relation in (1.38), we obtain

$$\det \mathbf{A} = (-1)^2 a_{11} \det \mathbf{A}_{11} + (-1)^3 a_{12} \det \mathbf{A}_{12}$$

But
$$\det \mathbf{A}_{11} = a_{22}; \qquad \det \mathbf{A}_{12} = a_{21}$$

Hence
$$\det \mathbf{A} = a_{11}a_{22} - a_{12}a_{21}$$

This relation is the general formula for the determinant of a 2×2 matrix.

It can be shown that to evaluate the determinant of a matrix we may use the recurrence relation given in (1.38) along any row or column, as indicated in Example 1.14.

EXAMPLE 1.14: Evaluate the determinant of the matrix $\mathbf{A}$, where

$$\mathbf{A} = \begin{bmatrix} 2 & 1 & 0 \\ 1 & 3 & 1 \\ 0 & 1 & 2 \end{bmatrix}$$

Using the recurrence relation in (1.38), we obtain

$$\begin{aligned} \det \mathbf{A} = {} & (-1)^2(2) \det \begin{bmatrix} 3 & 1 \\ 1 & 2 \end{bmatrix} \\ & + (-1)^3(1) \det \begin{bmatrix} 1 & 1 \\ 0 & 2 \end{bmatrix} \\ & + (-1)^4(0) \det \begin{bmatrix} 1 & 3 \\ 0 & 1 \end{bmatrix} \end{aligned}$$

We now employ the formula for the determinant of a 2×2 matrix given in Example 1.13, and have

$$\det \mathbf{A} = (2)\{(3)(2) - (1)(1)\} - \{(1)(2) - (0)(1)\} + 0$$

Hence
$$\det \mathbf{A} = 8$$

Let us check that the same result is obtained by using (1.38) along the second row instead of the first row. In this case we have, changing the "1" to "2" in (1.38),

$$\begin{aligned} \det \mathbf{A} = {} & (-1)^3(1) \det \begin{bmatrix} 1 & 0 \\ 1 & 2 \end{bmatrix} \\ & + (-1)^4(3) \det \begin{bmatrix} 2 & 0 \\ 0 & 2 \end{bmatrix} \\ & + (-1)^5(1) \det \begin{bmatrix} 2 & 1 \\ 0 & 1 \end{bmatrix} \end{aligned}$$

Again using the formula given in Example 1.13, we have

$$\det \mathbf{A} = -\{(1)(2) - (0)(1)\} + (3)\{(2)(2) - (0)(0)\} - \{(2)(1) - (1)(0)\}$$

or, as before,

$$\det \mathbf{A} = 8$$

Finally, using (1.38) along the third column, we have

$$\begin{aligned} \det \mathbf{A} = {} & (-1)^4(0) \det \begin{bmatrix} 1 & 3 \\ 0 & 1 \end{bmatrix} \\ & + (-1)^5(1) \det \begin{bmatrix} 2 & 1 \\ 0 & 1 \end{bmatrix} \\ & + (-1)^6(2) \det \begin{bmatrix} 2 & 1 \\ 1 & 3 \end{bmatrix} \end{aligned}$$

and, as before, obtain $\det \mathbf{A} = 8$.

Many theorems are associated with the use of determinants. Typically, the solution of a set of simultaneous equations can be obtained by a series of determinant evaluations.[2] However, from a modern viewpoint, most of the results that are obtained using determinants can be obtained much more effectively using matrix theory. For example, the solution of simultaneous equations using determinants is very inefficient. As we shall see later, a primary value of using determinants lies in the convenient shorthand notation that we can use in the discussion of certain questions, such as the existence of an inverse of a matrix. We shall use determinants in particular in the solution of eigenvalue problems.

Considering the evaluation of the determinant of a matrix, it may be effective to first factorize the matrix into a product of matrices and then use the following result:

$$\det(\mathbf{BC}\ldots\mathbf{F}) = (\det \mathbf{B})(\det \mathbf{C})\ldots(\det \mathbf{F}) \tag{1.39}$$

Relation (1.39) states that the determinant of the product of a number of matrices is equal to the product of the determinants of each matrix. The proof of this result is rather lengthy and clumsy [it is obtained using the determinant definition in (1.38)] and therefore we shall not include it here. We shall use the result in (1.39) abundantly in eigenvalue calculations when the determinant of a matrix, say matrix **A**, is required. The specific decomposition used is $\mathbf{A} = \mathbf{LDL}^T$, where **L** is a lower unit triangular matrix and **D** is a diagonal matrix. In that case,

$$\det \mathbf{A} = \det \mathbf{L} \det \mathbf{D} \det \mathbf{L}^T \tag{1.40}$$

and because $\det \mathbf{L} = 1$, we have

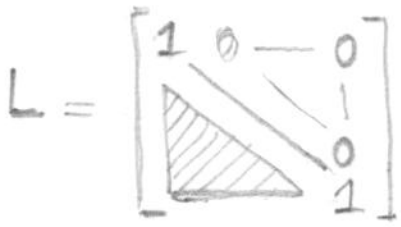

$$\det \mathbf{A} = \prod_{i=1}^{n} d_{ii} \tag{1.41}$$

EXAMPLE 1.15: Using the $\mathbf{LDL}^T$ decomposition, evaluate the determinant of **A**, where **A** is given in Example 1.14.

The procedure to obtain the $\mathbf{LDL}^T$ decomposition of **A** is presented in Section 8.2. Here we simply give **L** and **D**, and it can be verified that $\mathbf{LDL}^T = \mathbf{A}$:

$$\mathbf{L} = \begin{bmatrix} 1 & 0 & 0 \\ \frac{1}{2} & 1 & 0 \\ 0 & \frac{2}{5} & 1 \end{bmatrix}; \qquad \mathbf{D} = \begin{bmatrix} 2 & 0 & 0 \\ 0 & \frac{5}{2} & 0 \\ 0 & 0 & \frac{8}{5} \end{bmatrix}$$

Using (1.41), we obtain

$$\det \mathbf{A} = (2)(\tfrac{5}{2})(\tfrac{8}{5}) = 8$$

This is also the value obtained in Example 1.14.

The determinant and the trace of a matrix are functions of the matrix elements. However, it is important to observe that the off-diagonal elements do not affect the trace of a matrix, whereas the determinant is a function of all the elements in the matrix. Although we can conclude that a large determinant or a large trace means that some matrix elements are large, we cannot conclude that a small determinant or a small trace means that all matrix elements are small.

EXAMPLE 1.16: Calculate the trace and determinant of **A**, where

$$\mathbf{A} = \begin{bmatrix} 1 & 10{,}000 \\ 10^{-4} & 2 \end{bmatrix}$$

Here we have

$$\operatorname{tr}(\mathbf{A}) = 3$$

and $$\det \mathbf{A} = (1)(2) - (10^{-4})(10{,}000)$$

i.e., $$\det \mathbf{A} = 1$$

Hence both the trace and the determinant of **A** are small in relation to the off-diagonal element a_{12}.

REFERENCES

1. C. E. Fröberg, *Introduction to Numerical Analysis*, Addison-Wesley Publishing Company, Inc., Reading, Mass., 1969.
2. B. Noble, *Applied Linear Algebra*, Prentice-Hall, Inc., Englewood Cliffs, N.J., 1969.
3. R. Zurmühl, *Matrizen*, Springer-Verlag, Berlin, 1964.

2

VECTORS, MATRICES, AND TENSORS

2.1 INTRODUCTION

In Chapter 1 we introduced matrices simply as ordered arrays of numbers that are subjected to specific rules of addition, multiplication, and so on. The basic algebra rules will be used extensively in the next chapters, and it is very important to be thoroughly familiar with them. However, we did not discuss so far how to obtain the elements of a matrix from an actual physical problem, and under what conditions, if any, the rules of matrix algebra that have been presented would actually be applicable. In other words, we did not attempt to give a deeper understanding of what a matrix does represent in the calculations. However, such an understanding is very important—indeed essential—if the numerical techniques to be presented later are to be well understood.

The objective in this chapter is to present all the important concepts of linear algebra that need be known for a thorough understanding of the numerical techniques used in finite element analysis.[1-5] However, in addition to developing important knowledge of matrix and linear algebra, it is also hoped that, by means of the concepts presented in this chapter, the somewhat dry aspects of the use of matrices as merely a "bunch of numbers" will be removed. Namely, it is only with an increased understanding of the use of matrices that the solution techniques can be fully appreciated.

2.2 VECTOR SPACES, SUBSPACES, AND THE SPAN OF A MATRIX

In Chapter 1 we defined a vector of order n to be an array of n numbers written in matrix form. We now want to associate a geometrical interpretation to the elements of a vector. Consider as an example a vector of order 3, such as

$$\mathbf{x} = \begin{bmatrix} x_1 \\ x_2 \\ x_3 \end{bmatrix} = \begin{bmatrix} 2 \\ 4 \\ 3 \end{bmatrix} \tag{2.1}$$

We know from elementary geometry that $\mathbf{x}$ represents a geometrical vector in a chosen coordinate system in three-dimensional space. Figure 2.1 shows assumed coordinate axes and the vector corresponding to (2.1) in this system. We should note that the geometric representation of $\mathbf{x}$ in (2.1) depends completely on the coordinate system chosen; in other words, if (2.1) would give the components of a vector in a different coordinate system, then the geometrical representation of $\mathbf{x}$ would be different from the one in Fig. 2.1. Therefore, the coordinates (or components of a vector) alone do not define the actual geometric quantity, but they need to be given together with the specific coordinate system in which they are measured.

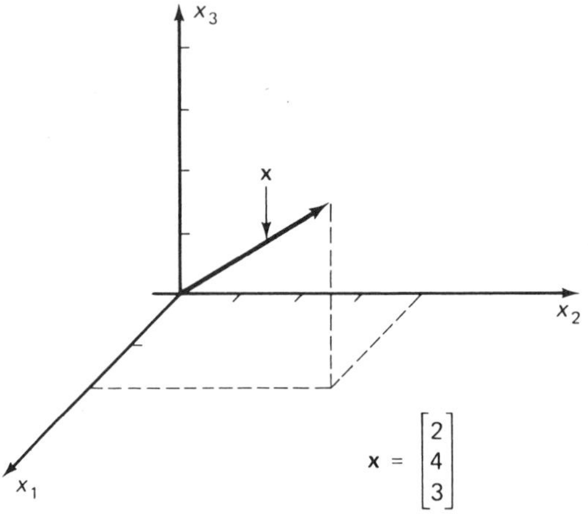

FIGURE 2.1 *Geometric representation of vector* $\mathbf{x}$.

The concepts of three-dimensional geometry generalize to a vector of any finite order n. If $n > 3$, we can no longer obtain a plot of the vector; however, we shall see that mathematically all concepts that pertain to vectors are independent of n. As before, when we considered the specific case $n = 3$, the vector of order n represents a quantity in a specific coordinate system of an n-dimensional space.

Assume that we are dealing with a number of vectors all of order n, which are defined in a fixed coordinate system. Some fundamental concepts that we shall find extremely powerful in the later chapters are summarized in the following definitions and facts.

Definition: *A collection of vectors* $\mathbf{x}_1, \mathbf{x}_2, \ldots, \mathbf{x}_s$ *is said to be* linearly dependent *if there exist numbers* $\alpha_1, \alpha_2, \ldots, \alpha_s$, *which are not all zero, such that*

$$\alpha_1 \mathbf{x}_1 + \alpha_2 \mathbf{x}_2 + \ldots + \alpha_s \mathbf{x}_s = \mathbf{0} \tag{2.2}$$

If the vectors are not linearly dependent, they are called linearly independent *vectors.*

We consider the following examples to clarify the meaning of this definition.

EXAMPLE 2.1: Let $n = 3$ and determine if the vectors $\mathbf{e}_i$, $i = 1, 2, 3$, are linearly dependent or independent.

According to the definition of linear dependency, we need to check if there are constants α_1, α_2, and α_3, not all zero, to satisfy the equation

$$\alpha_1\begin{bmatrix}1\\0\\0\end{bmatrix}+\alpha_2\begin{bmatrix}0\\1\\0\end{bmatrix}+\alpha_3\begin{bmatrix}0\\0\\1\end{bmatrix}=\begin{bmatrix}0\\0\\0\end{bmatrix} \tag{a}$$

But the equations in (a) read

$$\begin{bmatrix}\alpha_1\\\alpha_2\\\alpha_3\end{bmatrix}=\begin{bmatrix}0\\0\\0\end{bmatrix}$$

which is only satisfied if $\alpha_i = 0$, $i = 1, 2, 3$; therefore, the vectors $\mathbf{e}_i$ are linearly independent.

EXAMPLE 2.2: With $n = 4$, investigate where the following vectors are linearly dependent or independent.

$$\mathbf{x}_1=\begin{bmatrix}1\\1\\0\\0.5\end{bmatrix};\quad \mathbf{x}_2=\begin{bmatrix}-1\\0\\1\\0\end{bmatrix};\quad \mathbf{x}_3=\begin{bmatrix}0\\-0.5\\-0.5\\-0.25\end{bmatrix}$$

We need to consider the system of equations

$$\alpha_1\begin{bmatrix}1\\1\\0\\0.5\end{bmatrix}+\alpha_2\begin{bmatrix}-1\\0\\1\\0\end{bmatrix}+\alpha_3\begin{bmatrix}0\\-0.5\\-0.5\\-0.25\end{bmatrix}=\begin{bmatrix}0\\0\\0\\0\end{bmatrix}$$

or, considering each row,

$$\begin{aligned}\alpha_1 - \alpha_2 \qquad\qquad &= 0\\ \alpha_1 \qquad - 0.5\alpha_3 &= 0\\ \alpha_2 - 0.5\alpha_3 &= 0\\ 0.5\alpha_1 \qquad - 0.25\alpha_3 &= 0\end{aligned}$$

where we note that the equations are satisfied for $\alpha_1 = 1$, $\alpha_2 = 1$, and $\alpha_3 = 2$. Therefore, the vectors are linearly dependent.

In the preceding examples, the solution for α_1, α_2, and α_3 could be obtained by inspection. We shall later develop a systematic procedure of checking whether a number of vectors are linearly dependent or independent.

Another way of looking at the problem, which may be more appealing, is to say that the vectors are linearly dependent if any one of them can be expressed in terms of the others. That is, if not all of the α_i in (2.2) are zero, say $\alpha_j \neq 0$, then we can write

$$\mathbf{x}_j = -\sum_{\substack{k=1\\k\neq j}}^{s} \frac{\alpha_k}{\alpha_j}\mathbf{x}_k \tag{2.3}$$

Geometrically, when $n \leq 3$, we could plot the vectors and, if they are linearly dependent, we would be able to plot one vector in terms of multiples of the other vectors. For example, plotting the vectors used in Example 2.1, we immediately observe that none of them can be expressed in terms of multiples of the remaining ones; hence the vectors are linearly independent.

Consider that we are given q vectors of order n, $n \geq q$, which are linearly dependent, but that we only consider any $(q - 1)$ vectors of them. These $(q - 1)$ vectors may still be linearly dependent. However, by continuing to decrease the number of vectors under consideration, we would arrive at p vectors, which are linearly independent, where, in general, $p \leq q$. The other $(q - p)$ vectors can be expressed in terms of the p vectors. We are thus led to the following definition.

Definition: *Assume that we have* p *linearly independent vectors of order* n, *where* $n \geq p$. *These* p *vectors form a* basis *for a* p-dimensional vector space.

We talk about a vector space of dimension p because any vector in the space can be expressed as a linear combination of the p base vectors. We should note that the base vectors for the specific space considered are not unique; linear combinations of them can give another basis for the same space. Specifically, if $p = n$, then a basis for the space considered is $\mathbf{e}_i$, $i = 1, \ldots, n$ (which we call the *natural basis*), from which it also follows that p cannot be larger than n.

Definition: q *vectors, of which* p *vectors are linearly independent, are said to* span *a* p-*dimensional vector space.*

We therefore realize that all the importance lies in the base vectors, since they are the smallest number of vectors that span the space considered. All q vectors can be expressed in terms of the base vectors, however large q may be (and indeed q could be larger than n).

EXAMPLE 2.3: Establish a basis for the space spanned by the three vectors of Example 2.2.

In this case $q = 3$ and $n = 4$. We find by inspection that the two vectors $\mathbf{x}_1$ and $\mathbf{x}_2$ are linearly independent. Hence $\mathbf{x}_1$ and $\mathbf{x}_2$ can be taken as base vectors of the two-dimensional space spanned by $\mathbf{x}_1$, $\mathbf{x}_2$, and $\mathbf{x}_3$. Also, using the result of Example 2.2, we have that $\mathbf{x}_3 = -\frac{1}{2}\mathbf{x}_2 - \frac{1}{2}\mathbf{x}_1$.

Assume that we are given a p-dimensional vector space which we denote as E_p, for which $\mathbf{x}_1, \mathbf{x}_2, \ldots, \mathbf{x}_p$ are chosen base vectors, $p > 1$. Then we may like to consider only all those vectors that can be expressed in terms of $\mathbf{x}_1$ and $\mathbf{x}_2$. But the vectors $\mathbf{x}_1$ and $\mathbf{x}_2$ also form the basis of a vector space that we call E_2. If $p = 2$, we note that E_p and E_2 coincide. We call E_2 a subspace of E_p, the concise meaning of which is defined next.

Definition: *A* subspace *of a vector space is a vector space such that any vector in the subspace is also in the original space. If* $\mathbf{x}_1, \mathbf{x}_2, \ldots, \mathbf{x}_p$ *are the base vectors of the original space, any subset of these vectors forms the* basis *of a subspace; the* dimension *of the subspace is equal to the number of base vectors selected.*

EXAMPLE 2.4: The three vectors $\mathbf{x}_1$, $\mathbf{x}_2$, and $\mathbf{x}_3$ are linearly independent and therefore form the basis of a three-dimensional vector space E_3:

$$\mathbf{x}_1 = \begin{bmatrix} 1 \\ 2 \\ 1 \\ 0 \end{bmatrix}; \qquad \mathbf{x}_2 = \begin{bmatrix} 1 \\ 0 \\ 0 \\ 0 \end{bmatrix}; \qquad \mathbf{x}_3 = \begin{bmatrix} 0 \\ -1 \\ 0 \\ 1 \end{bmatrix} \tag{a}$$

Identify some possible two-dimensional subspaces of E_3.

Using the base vectors in (a), a two-dimensional subspace is formed by any two of the three vectors; e.g., $\mathbf{x}_1$ and $\mathbf{x}_2$ represent a basis for a two-dimensional subspace; $\mathbf{x}_1$ and $\mathbf{x}_3$ are the basis for another two-dimensional subspace; and so on. Indeed, any two linearly independent vectors in E_3 form the basis of a two-dimensional subspace, and it follows that there are an infinite number of two-dimensional subspaces in E_3.

Having considered the concepts of a vector space we way now recognize that the columns of any rectangular marrix $\mathbf{A}$ also span a vector space. We call this space the column space of $\mathbf{A}$. Similarly, the rows of a matrix span a vector space, which we call the row space of $\mathbf{A}$. Conversely, we may assemble any q vectors of order n into a matrix of order $n \times q$. The number of linearly independent vectors used is equal to the dimension of the column space of $\mathbf{A}$. For example, the three vectors in Example 2.4 form the matrix

$$\mathbf{A} = \begin{bmatrix} 1 & 1 & 0 \\ 2 & 0 & -1 \\ 1 & 0 & 0 \\ 0 & 0 & 1 \end{bmatrix} \tag{2.4}$$

Assume that we are given a matrix $\mathbf{A}$ and that we need to calculate the dimension of the column space of $\mathbf{A}$. In other words, we want to evaluate how many columns in $\mathbf{A}$ are linearly independent. The number of linearly independent columns in $\mathbf{A}$ is neither increased nor decreased by taking any linear combinations of them. Therefore, in order to identify the column space of $\mathbf{A}$, we may try to transform the matrix, by linearly combining its columns, to obtain unit vectors $\mathbf{e}_i$. Because unit vectors $\mathbf{e}_i$ with distinct i are linearly independent, the dimension of the column space of $\mathbf{A}$ would be equal to the number of unit vectors that can be obtained. However, it need be noted that we may not be able to actually obtain unit vectors $\mathbf{e}_i$ (see Example 2.5), but the process followed in the transformation of $\mathbf{A}$ will always lead to a form that displays the dimension of the column space.

EXAMPLE 2.5: Calculate the dimension of the column space of the matrix $\mathbf{A}$ formed by the vectors $\mathbf{x}_1$, $\mathbf{x}_2$, and $\mathbf{x}_3$ considered in Example 2.4.

The matrix considered is

$$\mathbf{A} = \begin{bmatrix} 1 & 1 & 0 \\ 2 & 0 & -1 \\ 1 & 0 & 0 \\ 0 & 0 & 1 \end{bmatrix}$$

Writing the second and third columns as the first and second columns, respectively, we obtain

$$\mathbf{A}_1 = \begin{bmatrix} 1 & 0 & 1 \\ 0 & -1 & 2 \\ 0 & 0 & 1 \\ 0 & 1 & 0 \end{bmatrix}$$

Subtracting the first column from the third column, adding twice the second column to the third column, and finally multiplying the second column by (-1), we obtain

$$\mathbf{A}_2 = \begin{bmatrix} 1 & 0 & 0 \\ 0 & 1 & 0 \\ 0 & 0 & 1 \\ 0 & -1 & 2 \end{bmatrix}$$

But we have now reduced the matrix to a form where we can identify that the three columns are linearly independent; i.e., the columns are linear independent because the first three elements in the vectors are the columns of the identity matrix of order 3. However, since we obtained $\mathbf{A}_2$ from $\mathbf{A}$ by interchanging and linearly combining the original columns of $\mathbf{A}$ and thus have not increased in the solution process the space spanned by the columns of the matrix, we have that the dimension of the column space of $\mathbf{A}$ is 3.

In the above presentation we linearly combined the vectors $\mathbf{x}_1, \ldots, \mathbf{x}_q$, which were the columns of $\mathbf{A}$, in order to identify whether they are linearly independent. Alternatively, to find the dimension of the space spanned by a set of vectors $\mathbf{x}_1, \mathbf{x}_2, \ldots, \mathbf{x}_q$, we could use the definition of vector linear independence in (2.2) and consider the set of simultaneous homogeneous equations

$$\alpha_1 \mathbf{x}_1 + \alpha_2 \mathbf{x}_2 + \ldots + \alpha_q \mathbf{x}_q = \mathbf{0} \tag{2.5}$$

which is, in matrix form,

$$\mathbf{A}\boldsymbol{\alpha} = \mathbf{0} \tag{2.6}$$

where $\boldsymbol{\alpha}$ is a vector with elements $\alpha_1, \ldots, \alpha_q$, and the columns of $\mathbf{A}$ are the vectors $\mathbf{x}_1, \mathbf{x}_2, \ldots, \mathbf{x}_q$. The solution for the unknowns $\alpha_1, \ldots, \alpha_q$ is not changed by linearly combining or multiplying any of the rows in the matrix $\mathbf{A}$. Therefore, we may try to reduce $\mathbf{A}$ by multiplying and combining its rows into a matrix, in which columns consist only of unit vectors. This reduced matrix is called the *row-echelon form* of $\mathbf{A}$. The number of unit column vectors in the row-echelon form of $\mathbf{A}$ is equal to the dimension of the column space of $\mathbf{A}$ and, from the preceding discussion, is also equal to the dimension of the row space of $\mathbf{A}$. It follows that the dimension of the column space of $\mathbf{A}$ is equal to the dimension of the row space of $\mathbf{A}$. In other words, the number of linearly independent columns in $\mathbf{A}$ is equal to the number of linearly independent rows in $\mathbf{A}$. This result is summarized in the definition of the rank of a matrix.

Definition: *The* rank *of a matrix* $\mathbf{A}$ *is equal to the dimension of the column space and equal to the dimension of the row space of* $\mathbf{A}$.

EXAMPLE 2.6: Consider the following three vectors:

$$\mathbf{x}_1 = \begin{bmatrix} 1 \\ 2 \\ 1 \\ 3 \\ 4 \\ 3 \end{bmatrix}; \qquad \mathbf{x}_2 = \begin{bmatrix} 3 \\ 1 \\ -2 \\ 4 \\ 2 \\ -1 \end{bmatrix}; \qquad \mathbf{x}_3 = \begin{bmatrix} 2 \\ 3 \\ 1 \\ 5 \\ 6 \\ 4 \end{bmatrix}$$

Use these vectors as the columns of a matrix $\mathbf{A}$ and reduce the matrix to row-echelon form.

We have

$$\mathbf{A} = \begin{bmatrix} 1 & 3 & 2 \\ 2 & 1 & 3 \\ 1 & -2 & 1 \\ 3 & 4 & 5 \\ 4 & 2 & 6 \\ 3 & -1 & 4 \end{bmatrix}$$

Subtracting multiples of the first row from the rows below it in order to obtain the unit vector $\mathbf{e}_1$ in the first column, we obtain

$$\mathbf{A}_1 = \begin{bmatrix} 1 & 3 & 2 \\ 0 & -5 & -1 \\ 0 & -5 & -1 \\ 0 & -5 & -1 \\ 0 & -10 & -2 \\ 0 & -10 & -2 \end{bmatrix}$$

Dividing the second row by (-5) and then subtracting multiples of it from the other rows in order to reduce the second column to the unit vector $\mathbf{e}_2$, we obtain

$$\mathbf{A}_2 = \begin{bmatrix} 1 & 0 & \frac{7}{5} \\ 0 & 1 & \frac{1}{5} \\ 0 & 0 & 0 \\ 0 & 0 & 0 \\ 0 & 0 & 0 \\ 0 & 0 & 0 \end{bmatrix}$$

Hence we can give the following equivalent statements:

1. The solution to $\mathbf{A}\boldsymbol{\alpha} = \mathbf{0}$ is

$$\alpha_1 = -\tfrac{7}{5}\alpha_3$$
$$\alpha_2 = -\tfrac{1}{5}\alpha_3$$

2. The three vectors $\mathbf{x}_1$, $\mathbf{x}_2$, and $\mathbf{x}_3$ are linearly dependent. They form a two-dimensional vector space. The vectors $\mathbf{x}_1$ and $\mathbf{x}_2$ are linearly independent and they form a basis of the two-dimensional space in which $\mathbf{x}_1$, $\mathbf{x}_2$, and $\mathbf{x}_3$ lie.
3. The rank of $\mathbf{A}$ is 2.
4. The dimension of the column space of $\mathbf{A}$ is 2.
5. The dimension of the row space of $\mathbf{A}$ is 2.

2.3 MATRIX REPRESENTATION OF LINEAR TRANSFORMATION

In engineering analysis we generally want to relate certain quantities to some other quantities such as stresses to strains, strains to displacements, forces to displacements, temperatures to heat flows, and so on. Each of these relationships can be regarded as a transformation from one entity to another and can be defined concisely and in a compact manner using matrices.

Let us not specify for the moment the specific transformation that we want to consider and use instead a symbolic notation, thus including in the discussion many possible transformations of interest. Let $\mathcal{x}$ be an element of the entity $\mathfrak{X}$ and $\mathcal{y}$ be an element of the entity $\mathcal{Y}$. Then the transformation of $\mathcal{x}$ into $\mathcal{y}$ is symbolically written as

$$\mathcal{y} = \mathfrak{A}\mathcal{x} \tag{2.7}$$

where $\mathfrak{A}$ is the tranformation operator. If there is a unique element $\mathcal{y}$ for each element $\mathcal{x}$, and vice versa, we say that the operator is nonsingular; otherwise, the operator is singular. The tranformation is linear, if the following two relations hold:

$$\begin{aligned} \mathfrak{A}(c\mathcal{x}) &= c\mathfrak{A}\mathcal{x} \\ \mathfrak{A}(\mathcal{x}_1 + \mathcal{x}_2) &= \mathfrak{A}\mathcal{x}_1 + \mathfrak{A}\mathcal{x}_2 \end{aligned} \tag{2.8}$$

In this section we shall only be concerned with linear transformations.

EXAMPLE 2.7: Consider the operator $\mathfrak{A} = z^2 p'' + 3p'$, where p represents all polynomials in z of degree ≤ 3. Calculate $\mathcal{y} = \mathfrak{A}\mathcal{x}$, where $\mathcal{x} = 3z + 4z^3$. Discuss whether the operator represents a linear transformation and whether it is singular.

Applying the operator $\mathfrak{A}$ to $\mathcal{x}$, we obtain

$$\mathcal{x}' = 3 + 12z^2$$

$$\mathcal{x}'' = 24z$$

and hence

$$\mathcal{y} = 9 + 36z^2 + 24z^3$$

The operator $\mathfrak{A}$ is linear because differentiation is a linear process. Namely, if $p = c\mathcal{x}$, then $p' = (c\mathcal{x})' = c(\mathcal{x})'$; i.e., the constant c can be taken out of the differentiation, and also if $p = \mathcal{x}_1 + \mathcal{x}_2$, then $p' = \mathcal{x}_1' + \mathcal{x}_2'$.

To identify whether $\mathfrak{A}$ is singular, consider a simple element, say $\mathcal{x} = z$. In this case the transformation in (2.7) gives $\mathcal{y} = \mathfrak{A}\mathcal{x} = 3$. Now assume that we are given $\mathcal{y} = 3$. Then we need to ask whether there is a unique element $\mathcal{x}$ corresponding to that $\mathcal{y}$. However, there is no unique element $\mathcal{x}$, because $\mathcal{x} = z$, $\mathcal{x} = z + 3$, $\mathcal{x} = z + 10$, etc., would all give $\mathcal{y} = 3$. Hence the operator $\mathfrak{A}$ is singular.

Although the specific element $\mathcal{y}$ corresponding to an element $\mathcal{x}$ can always be calculated using the explicit definition of $\mathfrak{A}$ as in Example 2.7, a more systematic procedure is needed if the operations are to be performed by the digital computer. This is achieved using the matrix representation of the operator $\mathfrak{A}$.

The first step in the matrix representation of a linear transformation is the selection of two sets of base vectors, one set to represent the elements in $\mathfrak{X}$ and the other set to represent the elements in $\mathcal{Y}$. Let $\mathcal{v}_1, \ldots, \mathcal{v}_n$ be the basis in $\mathfrak{X}$ and $\mathcal{w}_1, \ldots, \mathcal{w}_n$ be the basis in $\mathcal{Y}$. Therefore, each element $\mathcal{x}$ in $\mathfrak{X}$ can be represented as

$$\mathcal{x} = x_1\mathcal{v}_1 + \ldots + x_n\mathcal{v}_n \tag{2.9}$$

where $x_1, \ldots, x_n$ are the coordinates of $\mathcal{x}$, and each element $\mathcal{y}$ in $\mathcal{Y}$ can be represented as

$$\mathcal{y} = y_1\mathcal{w}_1 + \ldots + y_n\mathcal{w}_n \tag{2.10}$$

where $y_1, \ldots, y_n$ are the coordinates of $\mathcal{y}$. The matrix representation of the linear transformation $\mathcal{y} = \mathfrak{A}\mathcal{x}$ is then

$$\mathbf{y} = \mathbf{A}\mathbf{x} \tag{2.11}$$

where $\mathbf{A}$ is a square matrix of order n and $\mathbf{y}$ and $\mathbf{x}$ are vectors that list the coordinates of $\mathcal{y}$ and $\mathcal{x}$, respectively; i.e.,

$$\mathbf{y} = \begin{bmatrix} y_1 \\ \cdot \\ \cdot \\ \cdot \\ y_n \end{bmatrix}; \qquad \mathbf{x} = \begin{bmatrix} x_1 \\ \cdot \\ \cdot \\ \cdot \\ x_n \end{bmatrix} \tag{2.12}$$

The element a_{ij} of the matrix $\mathbf{A}$ is the coordinate y_i obtained in (2.7) when the operator $\mathcal{A}$ is applied to the base vector $\mathcal{v}_j$. Therefore, to calculate the elements of the jth column of $\mathbf{A}$, we transform $\mathcal{x} = \mathcal{v}_j$; i.e., we calculate $\mathcal{y} = \mathcal{A}\mathcal{v}_j$, and represent $\mathcal{y}$ in terms of the base vectors $\mathcal{w}_i$, $i = 1, \ldots, n$. The coefficients of the $\mathcal{w}_i$ are the elements of the jth column in $\mathbf{A}$.

The effectiveness of using a matrix representation of $\mathcal{A}$ lies in that the transformation $\mathcal{y} = \mathcal{A}\mathcal{x}$ can be systematically carried out using (2.11), and, in particular, can readily be programmed for the digital computer.

It should be noted that the rules of matrix algebra defined in the previous sections can be established using matrices as representations of linear transformations.

Consider as an example the multiplication of two linear transformations $\mathcal{A}$ and $\mathcal{B}$, where

$$\mathcal{y} = \mathcal{A}\mathcal{x} \tag{2.13}$$

and

$$\mathcal{z} = \mathcal{B}\mathcal{y} \tag{2.14}$$

It follows that

$$\mathcal{z} = \mathcal{C}\mathcal{x} \tag{2.15}$$

where, symbolically,

$$\mathcal{C} = \mathcal{B}\mathcal{A} \tag{2.16}$$

The matrix representations of (2.13) and (2.14) are, respectively,

$$\mathbf{y} = \mathbf{A}\mathbf{x} \tag{2.17}$$

and

$$\mathbf{z} = \mathbf{B}\mathbf{y} \tag{2.18}$$

From the definition of the elements in $\mathbf{A}$ and $\mathbf{B}$ it follows that the matrix representation of (2.15) is

$$\mathbf{z} = \mathbf{C}\mathbf{x} \tag{2.19}$$

where

$$c_{ij} = \sum_k b_{ik}a_{kj} \tag{2.20}$$

But the relation (2.20) is the matrix multiplication defined earlier in Section 1.5. Similarly, the definitions for matrix addition, subtraction, and the inverse of a matrix given in Chapter 1 would all naturally be arrived at when considering the matrices as representations of linear transformations in chosen bases.

EXAMPLE 2.8: Represent the linear transformation considered in Example 2.7 as a matrix, and perform the transformation $\mathcal{y} = \mathcal{A}\mathcal{x}$, where $\mathcal{y}$ and $\mathcal{x}$ are given in Example 2.7.

Choose the following two bases:

1. $\mathcal{v}_1 = 1$, $\mathcal{v}_2 = z$, $\mathcal{v}_3 = z^2$, $\mathcal{v}_4 = z^3$, and $\mathcal{w}_i = \mathcal{v}_i$ for $i = 1, \ldots, 4$.
2. $\mathcal{v}_1 = 1$, $\mathcal{v}_2 = 1 - z$, $\mathcal{v}_3 = z + 3z^2$, $\mathcal{v}_4 = z^3$, and $\mathcal{w}_i = \mathcal{v}_i$ for $i = 1, \ldots, 4$.

The operator in Example 2.7 was $\mathcal{A} = z^2p'' + 3p'$, where p represents all polynomials in z of degree ≤ 3. To evaluate the elements of the jth column in the matrix representation $\mathbf{A}$ of the operator $\mathcal{A}$, we form $\mathcal{A}\mathcal{v}_j$ and represent the result in terms of the base vectors $\mathcal{w}_j$, $j = 1, \ldots, 4$. Considering the basis in **1**, we have

$$\begin{aligned} \mathcal{A}\mathcal{v}_1 &= 0 \\ \mathcal{A}\mathcal{v}_2 &= 3 = 3\mathcal{w}_1 \\ \mathcal{A}\mathcal{v}_3 &= 6z + 2z^2 = 6\mathcal{w}_2 + 2\mathcal{w}_3 \\ \mathcal{A}\mathcal{v}_4 &= 9z^2 + 6z^3 = 9\mathcal{w}_3 + 6\mathcal{w}_4 \end{aligned}$$

Therefore, the matrix representation of $\mathcal{A}$ in the chosen basis is

$$\mathbf{A} = \begin{bmatrix} 0 & 3 & 0 & 0 \\ 0 & 0 & 6 & 0 \\ 0 & 0 & 2 & 9 \\ 0 & 0 & 0 & 6 \end{bmatrix}$$

To transform the element $\mathcal{x}$, where $\mathcal{x} = 3z + 4z^3$, we find the coordinates of $\mathcal{x}$ in the selected basis:

$$\mathbf{x} = \begin{bmatrix} 0 \\ 3 \\ 0 \\ 4 \end{bmatrix}$$

Then, using $\mathbf{y} = \mathbf{A}\mathbf{x}$, we obtain

$$\mathbf{y} = \begin{bmatrix} 9 \\ 0 \\ 36 \\ 24 \end{bmatrix}$$

which gives $\mathcal{y} = 9 + 36z^2 + 24z^3$, as in Example 2.7.

Using the same procedure to evaluate the matrix representation of $\mathcal{A}$ corresponding to the basis given in **2**, we obtain

$$\bar{\mathbf{A}} = \begin{bmatrix} 0 & -3 & 19 & -3 \\ 0 & 0 & -16 & 3 \\ 0 & 0 & 2 & 3 \\ 0 & 0 & 0 & 6 \end{bmatrix}$$

The coordinate vector of $\mathcal{x} = 3z + 4z^3$ in the new basis is

$$\bar{\mathbf{x}} = \begin{bmatrix} 3 \\ -3 \\ 0 \\ 4 \end{bmatrix}$$

To obtain the coordinate vector of $\mathcal{y}$, we form $\bar{\mathbf{y}} = \mathbf{A}\bar{\mathbf{x}}$ and obtain

$$\bar{\mathbf{y}} = \begin{bmatrix} -3 \\ 12 \\ 12 \\ 24 \end{bmatrix}$$

As a check, we use $\bar{\mathbf{y}}$ to obtain $\mathcal{y}$ and find

$$\mathcal{y} = (-3)(1) + (12)(1 - z) + (12)(z + 3z^2) + (24)(z^3)$$

Hence as before, $\mathcal{y} = 9 + 36z^2 + 24z^3$

2.4 CHANGE OF BASIS

Assume that a transformation has been represented by a matrix **A** in a given basis and that we would like to find the matrix $\bar{\mathbf{A}}$, which represents the same transformation in a different basis. One way of obtaining the elements of the matrix $\bar{\mathbf{A}}$ is to simply use the basic procedure of evaluating the matrix of the operator $\mathcal{A}$ in the new basis as we have done in Example 2.8. However, we can also calculate $\bar{\mathbf{A}}$ from **A** by relating the new basis to the old basis, which has practical advantages. That is, we can change basis without ever going back to the operator $\mathcal{A}$, and evaluate a basis in which the transformation is, from numerical considerations, represented very effectively. Since we use the matrix **A** to relate **x** and **y** by

$$\mathbf{y} = \mathbf{A}\mathbf{x} \tag{2.21}$$

the representation of the linear transformation will be particularly efficient if **A** is diagonal. Fortunately, in finite element analysis, a basis in which the system matrix is diagonal does nearly always exist. However, unfortunately, in most practical calculations we do not know this basis a priori. Therefore, the procedure is to simply represent the transformation first in some convenient basis (i.e., the finite element coordinate basis), and if efficient for the complete solution process, we then transform the matrix into diagonal form. This transformation is carried out in a systematic manner, and as we will show, requires the solution of an eigenproblem. The matrix transformation discussed below is fundamental to the solution of eigenproblems.

Assume that **A** is the $n \times n$ matrix representation of an operator $\mathcal{A}$ in the basis $\mathcal{v}_1, \ldots, \mathcal{v}_n$. Assume that the matrix **A** has been calculated using one basis only, i.e., $\mathcal{w}_i = \mathcal{v}_i$, $i = 1, \ldots, n$, which is the most important case.

The matrix **A** relates **x** to **y** as given in (2.21), where the ith elements of the vectors **x** and **y** (i.e., x_i and y_i) are the multipliers of the base vector $\mathcal{v}_i$. Assume that the new base vectors are $\bar{\mathcal{v}}_i$, $i = 1, \ldots, n$, and that the corresponding coordinates are stored in vectors $\bar{\mathbf{x}}$ and $\bar{\mathbf{y}}$. To obtain the matrix $\bar{\mathbf{A}}$, which relates $\bar{\mathbf{x}}$ to $\bar{\mathbf{y}}$, the first step is to express all new base vectors $\bar{\mathcal{v}}_i$, $i = 1, \ldots, n$, in terms of the old base vectors $\mathcal{v}_i$, $i = 1, \ldots, n$; i.e., we evaluate

$$\bar{\mathcal{v}}_i = \sum_{k=1}^{n} p_{ik}\mathcal{v}_k; \qquad i = 1, \ldots, n \tag{2.22}$$

Using this relation we then have

$$\mathbf{x} = \mathbf{P}\bar{\mathbf{x}}; \qquad \mathbf{y} = \mathbf{P}\bar{\mathbf{y}} \tag{2.23}$$

where **P** is an $n \times n$ matrix with the p_{ik} of (2.22) being element (k, i) in **P**. Substituting the relations in (2.23) into (2.21), we obtain

$$\mathbf{P}\bar{\mathbf{y}} = \mathbf{A}\mathbf{P}\bar{\mathbf{x}} \tag{2.24}$$

Solving for $\bar{\mathbf{y}}$, we obtain

$$\bar{\mathbf{y}} = \mathbf{P}^{-1}\mathbf{A}\mathbf{P}\bar{\mathbf{x}} \tag{2.25}$$

and hence

$$\bar{\mathbf{A}} = \mathbf{P}^{-1}\mathbf{A}\mathbf{P} \tag{2.26}$$

In the above derivation it was assumed that the inverse of **P** exists. This is the case provided that a proper basis $\bar{\mathcal{v}}_i$, $i = 1, \ldots, n$, has been chosen; i.e., all $\bar{\mathcal{v}}_i$ are linearly independent.

It should be noted that the existence of the inverse of $\mathbf{P}$ assures that for any vectors $\mathbf{x}$ and $\mathbf{y}$ there are unique vectors $\bar{\mathbf{x}}$ and $\bar{\mathbf{y}}$ as expressed in (2.23), and vice versa.

In an alternative procedure, to obtain $\bar{\mathbf{A}}$ we may use $\mathbf{x} = \mathbf{P}\bar{\mathbf{x}}$ and $\bar{\mathbf{y}} = \mathbf{Q}\mathbf{y}$, where the elements in $\mathbf{Q}$ are obtained from the relation expressing the base vectors $\mathcal{v}_i$, $i = 1, \ldots, n$, in terms of the base vectors $\bar{\mathcal{v}}_i$, $i = 1, \ldots, n$; i.e., we use

$$\mathcal{v}_i = \sum_{k=1}^{n} q_{ik}\bar{\mathcal{v}}_k \tag{2.27}$$

where q_{ik} is element (k, i) of $\mathbf{Q}$. Then, substituting for $\mathbf{x}$ in (2.21) and evaluating $\bar{\mathbf{y}}$, we obtain

$$\bar{\mathbf{y}} = \mathbf{QAP}\bar{\mathbf{x}} \tag{2.28}$$

and hence

$$\bar{\mathbf{A}} = \mathbf{QAP} \tag{2.29}$$

Because of the uniqueness of the operator $\bar{\mathbf{A}}$, we have

$$\mathbf{Q} = \mathbf{P}^{-1} \tag{2.30}$$

and in (2.27) we essentially evaluated the inverse of $\mathbf{P}$.

EXAMPLE 2.9: Consider the transformation discussed in Example 2.8. Using the basis $\mathcal{v}_1 = 1$, $\mathcal{v}_2 = z$, $\mathcal{v}_3 = z^2$, $\mathcal{v}_4 = z^3$, we obtained

$$\mathbf{A} = \begin{bmatrix} 0 & 3 & 0 & 0 \\ 0 & 0 & 6 & 0 \\ 0 & 0 & 2 & 9 \\ 0 & 0 & 0 & 6 \end{bmatrix}$$

Use a transformation on $\mathbf{A}$ to change basis to $\bar{\mathcal{v}}_1 = 1$, $\bar{\mathcal{v}}_2 = 1 - z$, $\bar{\mathcal{v}}_3 = z + 3z^2$, and $\bar{\mathcal{v}}_4 = z^3$.

Relation (2.22) for each of the base vectors gives

$$\begin{aligned} \bar{\mathcal{v}}_1 &= \mathcal{v}_1 \\ \bar{\mathcal{v}}_2 &= \mathcal{v}_1 - \mathcal{v}_2 \\ \bar{\mathcal{v}}_3 &= \mathcal{v}_2 + 3\mathcal{v}_3 \\ \bar{\mathcal{v}}_4 &= \mathcal{v}_4 \end{aligned}$$

Hence

$$\mathbf{P} = \begin{bmatrix} 1 & 1 & 0 & 0 \\ 0 & -1 & 1 & 0 \\ 0 & 0 & 3 & 0 \\ 0 & 0 & 0 & 1 \end{bmatrix}$$

Similarly, using the relation (2.27), we have

$$\begin{aligned} \mathcal{v}_1 &= \bar{\mathcal{v}}_1 \\ \mathcal{v}_2 &= \bar{\mathcal{v}}_1 - \bar{\mathcal{v}}_2 \\ \mathcal{v}_3 &= -\tfrac{1}{3}\bar{\mathcal{v}}_1 + \tfrac{1}{3}\bar{\mathcal{v}}_2 + \tfrac{1}{3}\bar{\mathcal{v}}_3 \\ \mathcal{v}_4 &= \bar{\mathcal{v}}_4 \end{aligned}$$

and

$$\mathbf{Q} = \begin{bmatrix} 1 & 1 & -\frac{1}{3} & 0 \\ 0 & -1 & \frac{1}{3} & 0 \\ 0 & 0 & \frac{1}{3} & 0 \\ 0 & 0 & 0 & 1 \end{bmatrix}$$

We may check that $\mathbf{QP} = \mathbf{I}$. To obtain $\bar{\mathbf{A}}$ we evaluate (2.29) and obtain

$$\bar{\mathbf{A}} = \begin{bmatrix} 0 & -3 & 19 & -3 \\ 0 & 0 & -16 & 3 \\ 0 & 0 & 2 & 3 \\ 0 & 0 & 0 & 6 \end{bmatrix}$$

This is the same matrix $\bar{\mathbf{A}}$ that we calculated in Example 2.8.

Consider the transformation that corresponds to a change of basis as carried out in Example 2.9; it appears that it requires much numerical effort. First the two matrices $\mathbf{P}$ and $\mathbf{Q}$ (or the inverse of $\mathbf{P}$) need to be established, and then the matrix product in (2.29) must be evaluated. A solution process requiring many transformations could therefore become very expensive. However, in practice special matrices are employed. The advantage of using these matrices lies in the fact that they can be inverted easily (and accurately) and also that the matrix product in (2.26) is calculated with relatively little effort. We will use most frequently the matrices defined in the following section.

2.5 MATRIX REPRESENTATION OF VARIATIONAL FORMULATION

In the previous sections we discussed the matrix representation of a general linear transformation. The resulting matrix $\mathbf{A}$ may be a symmetric, nonsymmetric (see Example 2.8), or even complex matrix. However, in finite element analysis we are concerned primarily with the matrix representation of a variational formulation of the problem under consideration and are thus able to establish symmetric matrices in almost all cases. We present finite element formulations in detail in Chapters 3 to 7, but it is very instructive to briefly focus attention on a formulation here in the light of the procedures given in the previous sections.

As is pointed out later, there are many different finite element formulations. As an example, we consider here the displacement-based formulation, which is most widely used. The basic equation considered is, in this case,

$$\int_{\xi} (\mathcal{L}\, \delta u)^T (\mathcal{S} u)\, d\xi - \int_{\eta} \delta u^T \mathscr{r}\, d\eta = 0 \tag{2.31}$$

where u represents displacement components, $\mathcal{L}$ and $\mathcal{S}$ are operators representing differentiations of u, $\mathscr{r}$ is a forcing function, δ means "variation in," and ξ and η represent appropriate quantities over which is integrated. The matrix representation of (2.31) is

$$\mathbf{A}\mathbf{x} = \mathbf{r} \tag{2.32}$$

where $\mathbf{A}$ is a square symmetric matrix, $\mathbf{x}$ is defined in (2.12), and $\mathbf{r}$ is a forcing vector,

$$\mathbf{r} = \begin{bmatrix} r_1 \\ \cdot \\ \cdot \\ \cdot \\ r_n \end{bmatrix} \tag{2.33}$$

It should be noted that the relation in (2.32) represents a linear transformation in which **x** is related to **r**.

Following the development of Section 2.3, the first step is the selection of base vectors. Here we choose base functions for the displacements u and have

$$u = x_1 u_1 + \dots + x_n u_n \tag{2.34}$$

The element a_{ij} of the matrix **A** is now obtained by substituting into (2.31) for u and δu the jth and ith base functions, i.e., by using in (2.31) $u = u_j$ and $\delta u = \delta u_i$. At the same time, the element r_i is obtained. The only, but important, difference to the procedure used in Section 2.3 is, therefore, that to represent (2.31), we substitute the individual base functions for the displacements and the variations in the displacements, and it is the integral of the resulting product that gives the required element of the matrix **A**. Since the same base functions are employed for u and δu, a symmetric matrix **A** is obtained. The detailed procedures used to evaluate the elements of **A** and **r** are presented in Chapters 3 to 7. But it is instructive to consider here a small example which demonstrates the important points of the process.

EXAMPLE 2.10: The variational formulation used to evaluate the displacements and forces in the bar element shown in Fig. 2.2 is

$$\int_0^L \delta u'(x) EA(x) u'(x)\,dx - \int_0^L A(x)p(x)\delta u(x)\,dx = 0 \tag{a}$$

where $u(x)$ is the longitudinal displacement of the bar, E the Young's modulus of the material, $p(x)$ the applied loading per unit volume at coordinate x, and L and $A(x)$ the length and area of the bar, respectively. Assume as a basis the polynomials x and x^2 and $A = A_0(1 + x/L)$. Evaluate the transformation matrix that relates the coordinates of the selected basis to the applied forces.

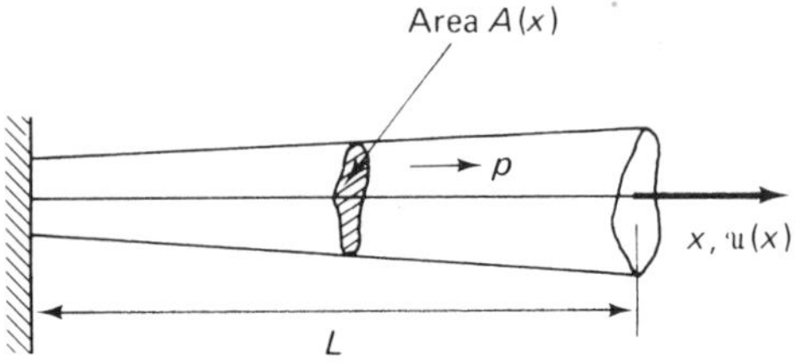

FIGURE 2.2 *Simple bar element.*

To calculate the matrix representation of (a) using the specified basis, we have

$$u = \alpha_1 x + \alpha_2 x^2$$

where α_1 and α_2 are the unknown coordinates. Hence we have

$$u' = \alpha_1 + 2\alpha_2 x$$

and

$$\int_0^L (\delta\alpha_1 + 2x\delta\alpha_2) EA(\alpha_1 + 2x\alpha_2)\,dx - \int_0^L Ap(\delta\alpha_1 x + \delta\alpha_2 x^2)dx = 0 \tag{b}$$

To evaluate element (i, j) of **A**, we consider the first integral and let $\delta\alpha_i = 1$, $\alpha_j = 1$, and all other coordinate variables be zero. Thus we obtain

$$a_{11} = \int_0^L (1)(EA)(1)\,dx$$

$$a_{12} = \int_0^L (1)(EA)(2x)\,dx$$

$$a_{21} = \int_0^L (2x)(EA)(1)\,dx$$

$$a_{22} = \int_0^L (2x)(EA)(2x)\,dx$$

Substituting that $EA = EA_0(1 + x/L)$, we have

$$\mathbf{A} = EA_0\begin{bmatrix} \frac{3}{2}L & \frac{5}{3}L^2 \\ \frac{5}{3}L^2 & \frac{7}{3}L^3 \end{bmatrix}$$

To calculate the ith element of $\mathbf{r}$, we evaluate the second integral in (b) and let $\delta\alpha_i = 1$; thus

$$r_1 = \int_0^L (p)(A)x\,dx$$

$$r_2 = \int_0^L (p)(A)(x^2)\,dx$$

where A is the area of the bar at the coordinate x. Substituting $A = A_0(1 + x/L)$ and using p as a constant, we obtain

$$\mathbf{r} = A_0 p\begin{bmatrix} \frac{5}{6}L^2 \\ \frac{7}{12}L^3 \end{bmatrix}$$

Hence the equations to be solved are

$$\begin{bmatrix} \frac{3}{2}L & \frac{5}{3}L^2 \\ \frac{5}{3}L^2 & \frac{7}{3}L^3 \end{bmatrix}\begin{bmatrix} \alpha_1 \\ \alpha_2 \end{bmatrix} = \begin{bmatrix} \frac{5}{6}L^2 \\ \frac{7}{12}L^3 \end{bmatrix}\frac{p}{E}$$

which gives

$$\begin{bmatrix} \alpha_1 \\ \alpha_2 \end{bmatrix} = \begin{bmatrix} \dfrac{105}{78} \\ -\dfrac{37}{52}\dfrac{1}{L} \end{bmatrix}\frac{pL}{E}$$

Hence

$$u = \frac{105}{78}\frac{pL}{E}x - \frac{37}{52}\frac{p}{E}x^2$$

We now turn to the problem of changing basis when $\mathbf{A}$ has been derived from a variational formulation. Following the development given in Section 2.4, we assume that u is represented in the new basis as

$$u = \bar{x}_1\bar{u}_1 + \ldots + \bar{x}_n\bar{u}_n \tag{2.35}$$

where $\bar{u}_i$ is the ith new base function and $\bar{x}_i$ is the corresponding coordinate. In this case we have

$$\mathbf{x} = \mathbf{P}\bar{\mathbf{x}} \tag{2.36}$$

and element (k, i) of $\mathbf{P}$ is given by p_{ik}, where

$$\bar{u}_i = \sum_{k=1}^{n} p_{ik}u_k \tag{2.37}$$

Noting that we use u^T and u in (2.31), the change of basis now gives

$$\bar{\mathbf{A}}\bar{\mathbf{x}} = \bar{\mathbf{r}} \tag{2.38}$$

where

$$\bar{\mathbf{A}} = \mathbf{P}^T\mathbf{A}\mathbf{P}; \qquad \bar{\mathbf{r}} = \mathbf{P}^T\mathbf{r} \tag{2.39}$$

see (2.26)

To demonstrate the procedure, consider the following example.

EXAMPLE 2.11: Evaluate the matrices **A** and **r** of the problem in Example 2.10 when using as base functions x and $x - x^2$.

1. Calculate **A** and **r** using the basic procedure employed in Example 2.10.
2. Perform the change of basis on the matrices **A** and **r** established in Example 2.10 as given in (2.39).

In the first case we use

$$u = \bar{\alpha}_1 x + \bar{\alpha}_2(x - x^2)$$

and proceed as in Example 2.10 to evaluate $\bar{\mathbf{A}}$ and $\bar{\mathbf{r}}$. Here we have

$$\begin{aligned}
\bar{a}_{11} &= EA_0 \int_0^L (1)\left(1 + \frac{x}{L}\right)(1)\,dx = (\tfrac{3}{2}L)EA_0 \\
\bar{a}_{12} &= EA_0 \int_0^L (1)\left(1 + \frac{x}{L}\right)(1 - 2x)\,dx = (\tfrac{3}{2}L - \tfrac{5}{3}L^2)EA_0 \\
\bar{a}_{21} &= EA_0 \int_0^L (1 - 2x)\left(1 + \frac{x}{L}\right)(1)\,dx = \bar{a}_{12} \\
\bar{a}_{22} &= EA_0 \int_0^L (1 - 2x)\left(1 + \frac{x}{L}\right)(1 - 2x)\,dx = (\tfrac{3}{2}L - \tfrac{10}{3}L^2 + \tfrac{7}{3}L^3)EA_0
\end{aligned} \tag{a}$$

$$\begin{aligned}
\bar{r}_1 &= A_0 p \int_0^L \left(1 + \frac{x}{L}\right)(x)\,dx = A_0 p \tfrac{5}{6}L^2 \\
\bar{r}_2 &= A_0 p \int_0^L \left(1 + \frac{x}{L}\right)(x - x^2)\,dx = A_0 p(\tfrac{5}{6}L^2 - \tfrac{7}{12}L^3)
\end{aligned} \tag{b}$$

Hence the equations to be solved are

$$\begin{bmatrix} \tfrac{3}{2}L & \tfrac{3}{2}L - \tfrac{5}{3}L^2 \\ \tfrac{3}{2}L - \tfrac{5}{3}L^2 & \tfrac{3}{2}L - \tfrac{10}{3}L^2 + \tfrac{7}{3}L^3 \end{bmatrix} \begin{bmatrix} \bar{\alpha}_1 \\ \bar{\alpha}_2 \end{bmatrix} = \begin{bmatrix} \tfrac{5}{6}L^2 \\ \tfrac{5}{6}L^2 - \tfrac{7}{12}L^3 \end{bmatrix} \frac{p}{E}$$

and we obtain

$$\bar{\alpha}_1 = \frac{105}{78}\frac{pL}{E} - \frac{37}{52}\frac{p}{E}$$

$$\bar{\alpha}_2 = \frac{37}{52}\frac{p}{E}$$

Hence
$$u = \left(\frac{105}{78}\frac{pL}{E} - \frac{37}{52}\frac{p}{E}\right)x + \frac{37}{52}\frac{p}{E}(x - x^2)$$

or
$$u = \frac{105}{78}\frac{pL}{E}x - \frac{37}{52}\frac{p}{E}x^2$$

This is the same solution as we obtained in Example 2.10, because the same space is spanned by the base functions used.

To perform the change of basis directly on the matrices **A** and **r** derived in Example 2.10, we need to evaluate first the matrix **P** defined in (2.37). Here we have, with $u_1 = x$, $u_2 = x^2$, $\bar{u}_1 = x$, and $\bar{u}_2 = x - x^2$,

$$\mathbf{P} = \begin{bmatrix} 1 & 1 \\ 0 & -1 \end{bmatrix}$$

Now, using (2.39) with the matrices derived in Example 2.10, we obtain

$$\bar{\mathbf{A}} = EA_0 \begin{bmatrix} 1 & 0 \\ 1 & -1 \end{bmatrix} \begin{bmatrix} \tfrac{3}{2}L & \tfrac{5}{3}L^2 \\ \tfrac{5}{3}L^2 & \tfrac{7}{3}L^3 \end{bmatrix} \begin{bmatrix} 1 & 1 \\ 0 & -1 \end{bmatrix}$$

or
$$\bar{\mathbf{A}} = EA_0\begin{bmatrix} \frac{3}{2}L & \frac{3}{2}L - \frac{5}{3}L^2 \\ \frac{3}{2}L - \frac{5}{3}L^2 & \frac{3}{2}L - \frac{10}{3}L^2 + \frac{7}{3}L^3 \end{bmatrix}$$

and
$$\bar{\mathbf{r}} = A_0 p\begin{bmatrix} 1 & 0 \\ 1 & -1 \end{bmatrix}\begin{bmatrix} \frac{5}{6}L^2 \\ \frac{7}{12}L^3 \end{bmatrix}$$

or
$$\bar{\mathbf{r}} = A_0 p\begin{bmatrix} \frac{5}{6}L^2 \\ \frac{5}{6}L^2 - \frac{7}{12}L^3 \end{bmatrix}$$

Hence we obtained the same matrices $\bar{\mathbf{A}}$ and $\bar{\mathbf{r}}$ as given in (a) and (b).

The automatic change of basis is of particular importance in eigensolution algorithms, as we will indicate in Section 2.7. However, in the solution of standard eigenproblems a change of basis will only be effective if it is of the form $\mathbf{P}^{-1}\mathbf{AP}$ (see Section 2.7), and for this reason orthogonal matrices are used very effectively.

Definition: *A matrix* $\mathbf{P}$ *is an* orthogonal matrix *if* $\mathbf{P}^T\mathbf{P} = \mathbf{PP}^T = \mathbf{I}$. *Hence for an orthogonal matrix, we have* $\mathbf{P}^{-1} = \mathbf{P}^T$.

This definition shows that if an orthogonal matrix is used in the change of basis, we have $\bar{\mathbf{A}} = \mathbf{P}^T\mathbf{AP} = \mathbf{P}^{-1}\mathbf{AP}$. For practical use, there exist some orthogonal matrices that can be constructed and employed easily. Two kinds that we shall employ extensively are introduced below.

An orthogonal matrix very frequently used is the *rotation matrix*

$$\mathbf{P} = \begin{array}{c} \begin{array}{ccccccc} & & & i\text{th} & & j\text{th column} & \end{array} \\ \left[\begin{array}{ccccccccc} 1 & & & & & & & & \\ & \ddots & & & & & & & \\ & & 1 & & & & & & \\ & & & \cos\theta & \cdots & -\sin\theta & & & \\ & & & \vdots & 1 \ddots 1 & \vdots & & & \\ & & & \sin\theta & \cdots & \cos\theta & & & \\ & & & & & & 1 & & \\ & \text{elements not shown} & & & & & & \ddots & \\ & \text{are zeros} & & & & & & & 1 \end{array}\right] \begin{array}{l} \\ \\ \\ i\text{th} \\ \\ j\text{th row} \\ \\ \\ \\ \end{array} \end{array} \tag{2.40}$$

where i and j can be arbitrary but $i \neq j$. The name "rotation matrix" is explained by considering $n = 2$, in which case

$$\mathbf{P} = \begin{bmatrix} \cos\theta & -\sin\theta \\ \sin\theta & \cos\theta \end{bmatrix} \tag{2.41}$$

Referring to Fig. 2.3, we observe that for the base vectors $\mathbf{e}_1$ and $\mathbf{e}_2$, the transformation

$$[\mathbf{v}_1, \mathbf{v}_2] = \mathbf{P}[\mathbf{e}_1, \mathbf{e}_2] \tag{2.42}$$

represents a mere rotation by an angle θ. In the general case, this rotation is carried out in the n-dimensional space.

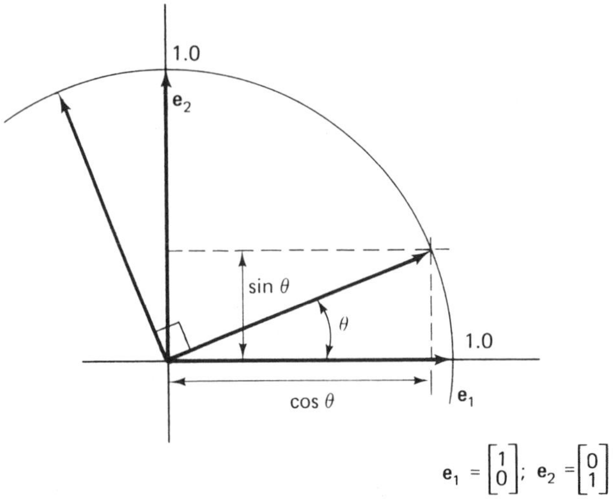

FIGURE 2.3 *Rotation of vectors* $\mathbf{e}_1$ *and* $\mathbf{e}_2$.

EXAMPLE 2.12: Rotate the vector $\mathbf{v}$ through an angle of 45°, where $\mathbf{v} = \begin{bmatrix} 1 \\ 1 \end{bmatrix}$.

The rotation matrix is in this case with $\cos 45° = \sin 45° = \sqrt{2}/2$,

$$\mathbf{P} = \begin{bmatrix} \frac{\sqrt{2}}{2} & -\frac{\sqrt{2}}{2} \\ \frac{\sqrt{2}}{2} & \frac{\sqrt{2}}{2} \end{bmatrix}$$

The rotated vector is given by $\mathbf{Pv}$,

$$\mathbf{Pv} = \begin{bmatrix} \frac{\sqrt{2}}{2} & -\frac{\sqrt{2}}{2} \\ \frac{\sqrt{2}}{2} & \frac{\sqrt{2}}{2} \end{bmatrix} \begin{bmatrix} 1 \\ 1 \end{bmatrix} = \begin{bmatrix} 0 \\ \sqrt{2} \end{bmatrix}$$

It should be noted that the length of the vector $\mathbf{Pv}$ is equal to the length of the vector $\mathbf{v}$. The results obtained can easily be interpreted geometrically, as indicated in Fig. 2.3.

Another orthogonal matrix frequently used is the *reflection matrix*,

$$\mathbf{P} = \mathbf{I} - \alpha \mathbf{v}\mathbf{v}^T; \qquad \alpha = \frac{2}{\mathbf{v}^T\mathbf{v}} \tag{2.43}$$

where $\mathbf{v}$ can be arbitrary. This matrix is called a reflection matrix, because the vector $\mathbf{Pw}$ is the reflection of the vector $\mathbf{w}$ in the plane to which $\mathbf{v}$ is orthogonal. A typical reflection is depicted in Fig. 2.4 for the case $n = 2$ and arbitrary vectors $\mathbf{v}$ and $\mathbf{w}$.

To prove the reflection property when $\mathbf{v}$ has any order n, we use that $\mathbf{w}$ (which is also of order n) has components $\gamma\mathbf{v}$ into the direction $\mathbf{v}$ and $\mathbf{u}$ perpendicular to $\mathbf{v}$. For the component $\gamma\mathbf{v}$, we have

$$\begin{aligned} \mathbf{P}\gamma\mathbf{v} &= \gamma\mathbf{v} - \alpha\mathbf{v}\mathbf{v}^T\gamma\mathbf{v} \\ &= \gamma\left[\mathbf{v} - \frac{2}{\mathbf{v}^T\mathbf{v}}\mathbf{v}(\mathbf{v}^T\mathbf{v})\right] \\ &= -\gamma\mathbf{v} \end{aligned} \tag{2.44}$$

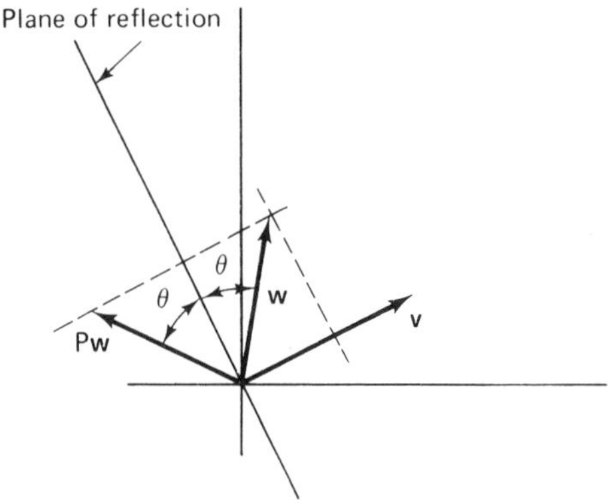

Householder Transformation

FIGURE 2.4 *Reflection of vector* **w**.

Therefore, the component into the direction of **v** has its direction reversed. For the component orthogonal to **v**, we have

$$\begin{aligned} \mathbf{Pu} &= \mathbf{u} - \alpha \mathbf{v}\mathbf{v}^T\mathbf{u} \\ &= \mathbf{u} \end{aligned} \tag{2.45}$$

because $\mathbf{v}^T\mathbf{u} = 0$. Therefore, this component is not transformed. The two cases together prove the general reflection property of the matrix **P**.

EXAMPLE 2.13: Establish the reflection matrix corresponding to the vector $\mathbf{v} = \begin{bmatrix} 1 \\ 1 \end{bmatrix}$ and reflect the vector $\mathbf{w} = \begin{bmatrix} 0 \\ 1 \end{bmatrix}$.

In this case we have

$$\mathbf{P} = \begin{bmatrix} 1 & 0 \\ 0 & 1 \end{bmatrix} - \frac{2}{2}\begin{bmatrix} 1 & 1 \\ 1 & 1 \end{bmatrix}$$

Hence

$$\mathbf{P} = \begin{bmatrix} 0 & -1 \\ -1 & 0 \end{bmatrix}$$

The reflection of the vector **w** is **Pw**, where

$$\mathbf{Pw} = \begin{bmatrix} 0 & -1 \\ -1 & 0 \end{bmatrix}\begin{bmatrix} 0 \\ 1 \end{bmatrix} = \begin{bmatrix} -1 \\ 0 \end{bmatrix}$$

This result can easily be interpreted geometrically, as shown for a general reflection in two-dimensional space in Fig. 2.4. Note that the length of the vector **Pw** is equal to the length of the vector **w**.

2.6 THE DEFINITION OF TENSORS

When considering matrices in engineering analysis, the concept of tensors and their matrix representations can be important. We shall only be concerned with *Cartesian tensors*, which are tensors represented in rectangular Cartesian coordinate frames. Let the coordinates of a point P in one frame be x_i, $i = 1, 2, 3$ and in a second frame be x'_i, $i = 1, 2, 3$. The unprimed and primed coordinates are measured along the base vectors $\mathbf{e}_i$ and $\mathbf{e}'_i$, respectively, as indicated in Fig. 2.5. Let the cosines of the angles between the primed and unprimed base vectors (of the primed and unprimed frames) be p_{ij}, i.e.,

$$p_{ij} = \cos(\mathbf{e}'_i, \mathbf{e}_j) \tag{2.46}$$

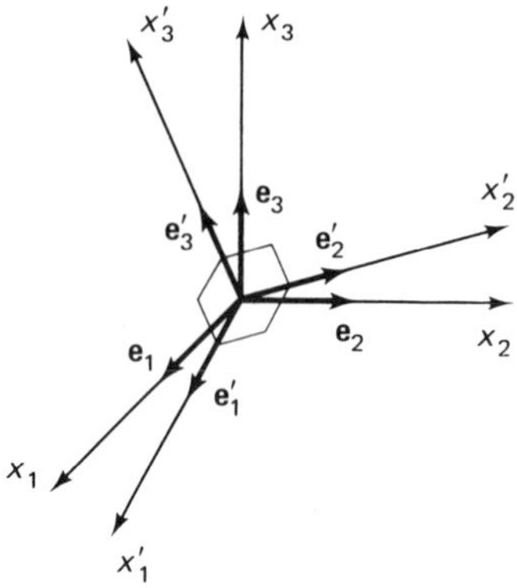

FIGURE 2.5 *Cartesian coordinate systems for definition of tensors.*

so that

$$x_i' = \sum_{j=1}^{3} p_{ij}\, x_j \tag{2.47}$$

or listing the coordinates in the vectors $\mathbf{x}'$ and $\mathbf{x}$,

$$\mathbf{x}' = \mathbf{P}\mathbf{x} \tag{2.48}$$

where $\mathbf{P}$ is an orthogonal matrix, because the base vectors in each coordinate frame are orthogonal to each other. Hence, we also have

$$\mathbf{x} = \mathbf{P}^T\mathbf{x}' \tag{2.49}$$

We may note that (2.49) corresponds to the relation (2.23) used in Section 2.4.

In tensor algebra it is convenient for the purpose of a compact notation to omit in (2.47) the summation sign; i.e., instead of (2.47) we simply write

$$x_i' = p_{ij}\, x_j \tag{2.50}$$

where *the summation on the repeated subscript j over the range of j (here j = 1, 2, 3) is implied.* This convention is referred to as the *summation convention of indicial notation* and is used very effectively to express in a compact manner relations involving tensor quantities (see Chapter 6, where we use this notation extensively). In the following we define what we mean by a tensor.

An entity is called a scalar, vector, or tensor depending on how the components of the entity are defined in the unprimed frame (coordinate system) and how these components transform to the primed frame.

An entity is called a scalar if it has only a single component ϕ in the coordinates x_i and this component does not change under a coordinate transformation,

$$\phi(x_1, x_2, x_3) = \phi'(x_1', x_2', x_3') \tag{2.51}$$

A scalar is also called a tensor of zero order. As an example, the temperature at a point is a scalar.

An entity is called a vector or tensor of first order if it has three components ξ_i in the unprimed frame and three components ξ_i' in the primed frame, and if these components are related by the characteristic law (using the summation convention)

$$\xi_i' = p_{ik}\,\xi_k \tag{2.52}$$

This relation corresponds to a change of basis in the representation of the entity. For example, a force is a first-order tensor as we illustrate in the following example.

EXAMPLE 2.14: The components of a force expressed in the unprimed coordinate system shown in Fig. 2.6 are

$$\mathbf{R} = \begin{bmatrix} 0 \\ 1 \\ \sqrt{3} \end{bmatrix}$$

Evaluate the components of the force in the primed coordinate system of Fig. 2.6.

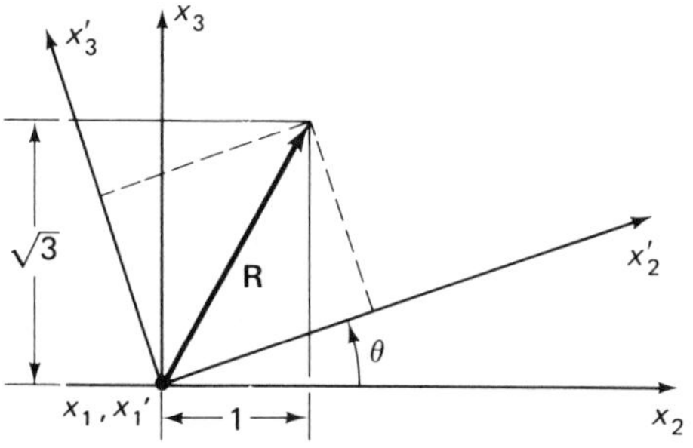

FIGURE 2.6 *Representation of a force in different coordinate systems.*

Here we have using (2.46)

$$\mathbf{P} = \begin{bmatrix} 1 & 0 & 0 \\ 0 & \cos\theta & \sin\theta \\ 0 & -\sin\theta & \cos\theta \end{bmatrix}$$

and then

$$\mathbf{R}' = \mathbf{PR} \tag{a}$$

where $\mathbf{R}'$ gives the components of the force in the primed coordinate system. As a check, if we use $\theta = -30°$ we obtain using (a)

$$\mathbf{R}' = \begin{bmatrix} 0 \\ 0 \\ 2 \end{bmatrix}$$

which is correct because the $\mathbf{e}_3'$-vector is now aligned with the force vector.

Considering vectors in three-dimensional space certain vector algebra is employed effectively.

The scalar (or dot) product of the vectors $\mathbf{u}$ and $\mathbf{v}$, denoted by $\mathbf{u}\cdot\mathbf{v}$ is defined as

$$\mathbf{u}\cdot\mathbf{v} = |\mathbf{u}|\,|\mathbf{v}|\cos\theta \tag{2.53}$$

where $|\mathbf{u}|$ is equal to the length of the vector $\mathbf{u}$, $|\mathbf{u}| = \sqrt{u_i u_i}$. The dot product can be evaluated using the components of the vectors,

$$\mathbf{u}\cdot\mathbf{v} = u_i v_i$$

The vector (or cross) product of the vectors **u** and **v** produces a new vector $\mathbf{w} = \mathbf{u} \times \mathbf{v}$. If **u** and **v** are defined in the Cartesian coordinate frame with unit base vectors $\mathbf{e}_1$, $\mathbf{e}_2$ and $\mathbf{e}_3$, we have

$$\mathbf{w} = \det \begin{bmatrix} \mathbf{e}_1 & \mathbf{e}_2 & \mathbf{e}_3 \\ u_1 & u_2 & u_3 \\ v_1 & v_2 & v_3 \end{bmatrix} \tag{2.54}$$

Figure 2.7 illustrates the vector operations performed in (2.53) and (2.54). It should be noted that the direction of the vector **w** is obtained by the right-hand-rule, i.e., the right-hand thumb points in the direction of **w** when the fingers curl from **u** to **v**.

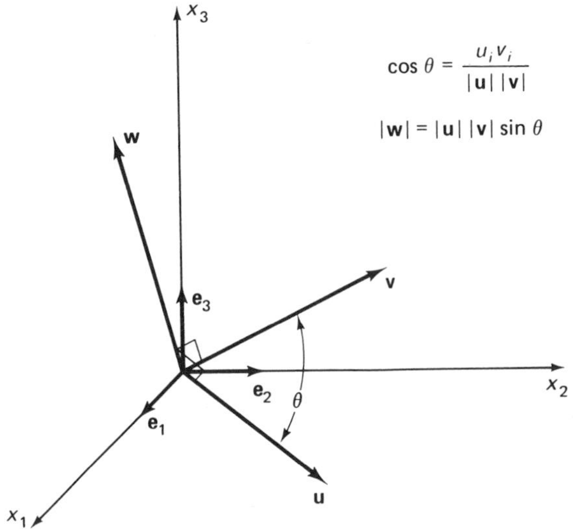

FIGURE 2.7 *Vectors used in products.*

The above vector algebra procedures are frequently employed in finite element analysis to evaluate angles between two given directions and to establish the direction perpendicular to a given plane.

EXAMPLE 2.15: Assume that the vectors **u** and **v** in Fig. 2.7 are

$$\mathbf{u} = \begin{bmatrix} 3 \\ 3 \\ 0 \end{bmatrix} \qquad \mathbf{v} = \begin{bmatrix} 0 \\ 2 \\ 2 \end{bmatrix}$$

Calculate the angle between these vectors, and establish a vector perpendicular to the plane that is defined by these vectors.

Here we have

$$|\mathbf{u}| = 3\sqrt{2}$$
$$|\mathbf{v}| = 2\sqrt{2}$$

Hence

$$\cos\theta = \tfrac{1}{2}$$

and $\theta = 60°$.

A vector perpendicular to the plane defined by **u** and **v** is given by

$$\mathbf{w} = \det \begin{bmatrix} \mathbf{e}_1 & \mathbf{e}_2 & \mathbf{e}_3 \\ 3 & 3 & 0 \\ 0 & 2 & 2 \end{bmatrix}$$

hence

$$\mathbf{w} = \begin{bmatrix} 6 \\ -6 \\ 6 \end{bmatrix}$$

Using $|\mathbf{w}| = \sqrt{w_i w_i}$ we obtain $\quad |\mathbf{w}| = 6\sqrt{3}$

which is also equal to the value obtained using the formula given in Fig. 2.7.

To define a second-order tensor we build on the definition given in (2.52) for a tensor of rank one.

An entity is called a second-order tensor if it has nine components t_{ij}, $i = 1, 2, 3$ and $j = 1, 2, 3$ in the unprimed frame and nine components t'_{ij} in the primed frame and if these components are related by the characteristic law

$$t'_{ij} = p_{ik} p_{jl} t_{kl} \tag{2.55}$$

As in the case of the definition of a first-order tensor, the relation in (2.55) represents a change of basis in the representation of the entity (see Example 2.16).

In the above definitions we assumed that all indices vary from 1 to 3, and these definitions must be modified appropriately if the indices vary from 1 to n, with $n \neq 3$. In engineering analysis we frequently deal only with two-dimensional conditions in which case $n = 2$.

EXAMPLE 2.16: Stress is a tensor of rank two. Assume that the stress at a point measured in an unprimed coordinate frame in a plane stress analysis is

$$\boldsymbol{\tau} = \begin{bmatrix} 1 & -1 \\ -1 & 1 \end{bmatrix}$$

Establish the components of the tensor in the primed coordinate system shown in Fig. 2.8.

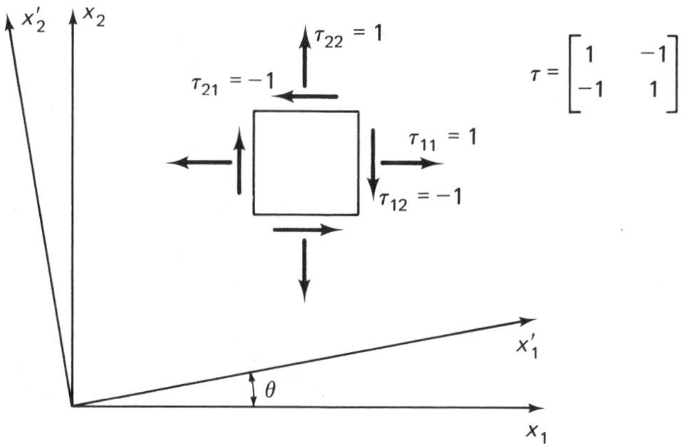

FIGURE 2.8 *Representation of a stress tensor in different coordinate systems.*

Here we use the rotation matrix **P** as in Example 2.14 and the transformation in (2.55) is

$$\boldsymbol{\tau}' = \mathbf{P}\boldsymbol{\tau}\mathbf{P}^T; \qquad \mathbf{P} = \begin{bmatrix} \cos\theta & \sin\theta \\ -\sin\theta & \cos\theta \end{bmatrix}$$

Assume that we are interested in the specific case when $\theta = 45°$. In this case we have

$$\boldsymbol{\tau}' = \tfrac{1}{2}\begin{bmatrix} 1 & 1 \\ -1 & 1 \end{bmatrix}\begin{bmatrix} 1 & -1 \\ -1 & 1 \end{bmatrix}\begin{bmatrix} 1 & -1 \\ 1 & 1 \end{bmatrix} = \begin{bmatrix} 0 & 0 \\ 0 & 2 \end{bmatrix}$$

and we recognize that in this coordinate system the off-diagonal elements of the tensor (shear components) are zero. The primed axes are called the principal coordinate axes and the diagonal elements $\tau'_{11} = 0$ and $\tau'_{22} = 2$ are the principal values of the tensor. We will see in Section 2.7 that the principal tensor values are the eigenvalues of the tensor and that the primed axes define the corresponding eigenvectors.

The above discussion can be directly expanded to also define tensors of higher order than two. In engineering analysis we are, in particular, interested in the constitutive tensors that relate the components of a stress tensor to the components of a strain tensor (see, for example, Sections 4.2.1 and 6.4).

$$\tau_{ij} = C_{ijk\ell}\epsilon_{k\ell} \tag{2.56}$$

The stress and strain tensors are both of rank two, and the constitutive tensor with components $C_{ijk\ell}$ is of rank four, because its components transform in the following way

$$C'_{ijk\ell} = p_{im}p_{jn}p_{kr}p_{\ell s}C_{mnrs} \tag{2.57}$$

The above discussion is only a very brief introduction to the definition and use of tensors. Our objective was merely to introduce the basic concepts of tensors so that later we can work with them (see Chapter 6). The most important aspect of tensors is that the components of a tensor are always represented in a chosen coordinate system, and these components are different when different coordinate systems are employed. It follows from the definition of tensors that if all components of a tensor vanish in one coordinate system, they vanish likewise in any other (admissible) coordinate system. Since the sum and difference of tensors of a given type are tensors of the same type, it also follows that if a tensor equation can be established in one coordinate system, then it must also hold in any other (admissible) coordinate system. This property detaches the fundamental physical relationships between tensors under consideration from the specific reference frame chosen and is the most important ingredient of tensors: *in the analysis of an engineering problem we are concerned with the physics of the problem, and the fundamental physical relationships between the variables involved must be independent of the specific coordinate system chosen*; otherwise, a simple change of the reference system would destroy these relationships and they would have been merely fortuitous. As an example, consider a body subjected to a set of forces. If we can show using one coordinate system that the body is in equilibrium, then we have proven the physical fact that the body is in equilibrium, and this force equilibrium will hold in any other (admissible) coordinate system.

The above discussion also hints at another important consideration in engineering analysis, namely, that for an effective analysis suitable coordinate systems should be chosen, because the effort to express and work with a physical relationship in one coordinate system can be a great deal less than when using another coordinate system. We will see in the discussion of the finite element method (see, for example, Section 4.2) that indeed one important ingredient for the effectiveness of a finite element analysis is the flexibility to choose different coordinate systems for different finite elements (domains) that together idealize the complete structure or continuum.

2.7 THE EIGENPROBLEM $Av = \lambda v$, WITH A BEING A SYMMETRIC MATRIX

In previous sections we discussed how a change of basis can be performed in matrix representations of operators. In finite element analysis we are primarily interested in transformations as applied to symmetric matrices that have been obtained from a variational formulation and we shall assume in the discussion to follow that **A** is symmetric. For example, the matrix **A** may represent the stiffness matrix, mass matrix, or heat capacity matrix of an element assemblage.

There are various important applications (see Examples 2.23, 2.24, and 2.25, and Chapter 9), in which for overall solution effectiveness a change of basis is performed using in the transformation matrix the eigenvectors of the eigenproblem

$$\mathbf{A}\mathbf{v} = \lambda\mathbf{v} \tag{2.58}$$

The problem in (2.58) is a *standard eigenproblem*. If the solution of (2.58) is considered in order to obtain eigenvalues and eigenvectors, the problem $\mathbf{A}\mathbf{v} = \lambda\mathbf{v}$ will be referred to as an *eigenproblem*, whereas if only eigenvalues are to be calculated, $\mathbf{A}\mathbf{v} = \lambda\mathbf{v}$ is called an *eigenvalue problem*. The objective in this section is to discuss the various properties that pertain to the solutions of (2.58).

Let n be the order of the matrix **A**. The first important point is that there exist n nontrivial solutions to (2.58). The word "nontrivial" means here that $\mathbf{v}$ must not be a null vector for which (2.58) is always satisfied. The ith nontrivial solution is given by the *eigenvalue* λ_i and the corresponding *eigenvector* $\mathbf{v}_i$, for which we have

$$\mathbf{A}\mathbf{v}_i = \lambda_i\mathbf{v}_i \tag{2.59}$$

Therefore, each solution consists of an eigenpair, and we write the n solutions as $(\lambda_1, \mathbf{v}_1), (\lambda_2, \mathbf{v}_2), \ldots, (\lambda_n, \mathbf{v}_n)$, where

$$\lambda_1 \leq \lambda_2 \leq \ldots \leq \lambda_n \tag{2.60}$$

We also call all n eigenvalues and eigenvectors the *eigensystem* of **A**.

The proof that there must be n eigenvalues and corresponding eigenvectors can conveniently be obtained by writing (2.58) in the form

$$(\mathbf{A} - \lambda\mathbf{I})\mathbf{v} = \mathbf{0} \tag{2.61}$$

But these equations only have a solution provided that

$$\det(\mathbf{A} - \lambda\mathbf{I}) = 0 \tag{2.62}$$

Unfortunately, the necessity for (2.62) to hold can only be explained once the solution of simultaneous equations has been presented. For this reason we postpone until Section 10.2.2 the discussion that (2.62) is indeed required.

Using (2.62), the eigenvalues of **A** are thus the roots of the polynomial

$$p(\lambda) = \det(\mathbf{A} - \lambda\mathbf{I}) \tag{2.63}$$

This polynomial is called the *characteristic polynomial* of **A**. However, since the order of the polynomial is equal to the order of **A**, we have n eigenvalues, and using (2.61) we obtain n corresponding eigenvectors. It may be noted that the vectors obtained from the solution of (2.61) are only defined within a scalar multiple.

EXAMPLE 2.17: Consider the matrix

$$\mathbf{A} = \begin{bmatrix} -1 & 2 \\ 2 & 2 \end{bmatrix}$$

Show that the matrix has two eigenvalues. Calculate the eigenvalues and eigenvectors.

The characteristic polynomial of **A** is

$$p(\lambda) = \det\begin{bmatrix} -1 - \lambda & 2 \\ 2 & 2 - \lambda \end{bmatrix}$$

Using the procedure given in Section 1.8 to calculate the determinant of a matrix (see Example 1.13), we obtain

$$\begin{aligned} p(\lambda) &= (-1 - \lambda)(2 - \lambda) - (2)(2) \\ &= \lambda^2 - \lambda - 6 \\ &= (\lambda + 2)(\lambda - 3) \end{aligned}$$

The order of the polynomial is 2, and hence there are two eigenvalues. In fact, we have

$$\begin{aligned} \lambda_1 &= -2 \\ \lambda_2 &= 3 \end{aligned}$$

The corresponding eigenvectors are obtained by applying (2.61) at the eigenvalues. Thus we have for λ_1,

$$\begin{bmatrix} -1 - (-2) & 2 \\ 2 & 2 - (-2) \end{bmatrix} \begin{bmatrix} v_1 \\ v_2 \end{bmatrix} = \begin{bmatrix} 0 \\ 0 \end{bmatrix}$$

with the solution $$\mathbf{v}_1 = \begin{bmatrix} 2 \\ -1 \end{bmatrix}$$

Considering λ_2, we have

$$\begin{bmatrix} -1 - 3 & 2 \\ 2 & 2 - 3 \end{bmatrix} \begin{bmatrix} v_1 \\ v_2 \end{bmatrix} = \begin{bmatrix} 0 \\ 0 \end{bmatrix}$$

with the solution $$\mathbf{v}_2 = \begin{bmatrix} \frac{1}{2} \\ 1 \end{bmatrix}$$

An important observation can be made concerning a change of basis on the matrix **A**. Suppose that **A** represents a linear transformation as discussed in Section 2.4. The change of basis is then carried out using

$$\mathbf{v} = \mathbf{P}\bar{\mathbf{v}} \tag{2.64}$$

to substitute for $\mathbf{v}$ in the problem $\mathbf{A}\mathbf{v} = \lambda\mathbf{v}$. We thus obtain

$$\bar{\mathbf{A}}\bar{\mathbf{v}} = \lambda\bar{\mathbf{v}} \tag{2.65}$$

where

$$\bar{\mathbf{A}} = \mathbf{P}^{-1}\mathbf{A}\mathbf{P} \tag{2.66}$$

This transformation is called a *similarity transformation.*

Consider now that $\mathbf{A}$ is the matrix representation of a variational formulation as discussed in Section 2.5. The change of basis using (2.64) now results in

$$\bar{\mathbf{A}}\bar{\mathbf{v}} = \lambda\mathbf{B}\bar{\mathbf{v}} \tag{2.67}$$

where

$$\bar{\mathbf{A}} = \mathbf{P}^T\mathbf{A}\mathbf{P}; \qquad \mathbf{B} = \mathbf{P}^T\mathbf{P} \tag{2.68}$$

The eigenproblem in (2.67) is called a *generalized eigenproblem.* However, since a generalized eigenproblem is more difficult to deal with than a standard problem, the transformation to a generalized problem should be avoided. This is achieved by assuming an orthogonal matrix $\mathbf{P}$, which yields $\mathbf{B} = \mathbf{I}$. If $\mathbf{P}$ is orthogonal, the transformations in (2.65) and (2.67) are referred to as *orthogonal similarity transformations*. We should recall that $\bar{\mathbf{A}} = \mathbf{P}^T\mathbf{A}\mathbf{P}$ is a symmetric matrix.

Considering a change of basis, it should be noted that the problems $\bar{\mathbf{A}}\bar{\mathbf{v}} = \lambda\bar{\mathbf{v}}$ in (2.65) and $\bar{\mathbf{A}}\bar{\mathbf{v}} = \lambda\mathbf{B}\bar{\mathbf{v}}$ in (2.67) have the same eigenvalues as the problem $\mathbf{A}\mathbf{v} = \lambda\mathbf{v}$, whereas the eigenvectors are related as given in (2.64). To show that the eigenvalues are identical, we consider the characteristic polynomials. For the problem in (2.65) we have

$$\bar{p}(\lambda) = \det(\mathbf{P}^{-1}\mathbf{A}\mathbf{P} - \lambda\mathbf{I}) \tag{2.69}$$

which can be written (see Section 1.8)

$$\bar{p}(\lambda) = \det\mathbf{P}^{-1}\det(\mathbf{A} - \lambda\mathbf{I})\det\mathbf{P} \tag{2.70}$$

and hence

$$\bar{p}(\lambda) = \det(\mathbf{A} - \lambda\mathbf{I}) \tag{2.71}$$

Therefore, the characteristic polynomials of the problems $\mathbf{A}\mathbf{v} = \lambda\mathbf{v}$ and $\bar{\mathbf{A}}\bar{\mathbf{v}} = \lambda\bar{\mathbf{v}}$ in (2.65) are the same, from which it follows that the eigenvalues are the same.

Considering the problem in (2.67), we have as its characteristic polynomial,

$$\bar{p}(\lambda) = \det(\mathbf{P}^T\mathbf{A}\mathbf{P} - \lambda\mathbf{P}^T\mathbf{P}) \tag{2.72}$$

which can be written

$$\bar{p}(\lambda) = \det\mathbf{P}^T\det(\mathbf{A} - \lambda\mathbf{I})\det\mathbf{P} \tag{2.73}$$

and therefore

$$\bar{p}(\lambda) = \det\mathbf{P}^T\det\mathbf{P}\;p(\lambda) \tag{2.74}$$

where $p(\lambda)$ is given in (2.63). Hence the characteristic polynomials of the problems $\mathbf{A}\mathbf{v} = \lambda\mathbf{v}$ and $\bar{\mathbf{A}}\bar{\mathbf{v}} = \lambda\mathbf{B}\bar{\mathbf{v}}$ in (2.67) are the same within a multiplier. This means that the eigenvalues of the two problems are identical.

EXAMPLE 2.18: Consider the eigenproblem $\mathbf{A}\mathbf{v} = \lambda\mathbf{v}$, where $\mathbf{A}$ is the matrix calculated in Example 2.10. Use $L = 6$ and $EA_0 = 1$.

1. Establish the characteristic polynomial and calculate the eigenvalues and eigenvectors.
2. Suppose that a change of basis is made using the new basis given in Example 2.11. The eigenproblem is now

$$\mathbf{P}^T\mathbf{A}\mathbf{P}\bar{\mathbf{v}} = \lambda\mathbf{P}^T\mathbf{P}\bar{\mathbf{v}} \tag{a}$$

where $\mathbf{P}$ was defined in Example 2.11. Establish the characteristic polynomial and calculate the eigenvalues and eigenvectors.

3. Assume that a change of basis is performed using an orthogonal matrix $\mathbf{P}$,

$$\mathbf{P} = \begin{bmatrix} \frac{3}{5} & -\frac{4}{5} \\ \frac{4}{5} & \frac{3}{5} \end{bmatrix} \tag{b}$$

Evaluate the characteristic polynomial, the eigenvalues, and eigenvectors of the problem $\mathbf{P}^T\mathbf{A}\mathbf{P}\bar{\mathbf{v}} = \lambda\bar{\mathbf{v}}$.

Using $L = 6$ and $EA_0 = 1$ in the matrix $\mathbf{A}$ of Example 2.10, we have

$$\mathbf{A} = \begin{bmatrix} 9 & 60 \\ 60 & 504 \end{bmatrix}$$

Hence
$$p(\lambda) = \lambda^2 - 513\lambda + 936 \tag{c}$$

and
$$\lambda_1 = 1.83; \qquad \lambda_2 = 511.17$$

$$\mathbf{v}_1 = \begin{bmatrix} 0.991 \\ -0.118 \end{bmatrix}; \qquad \mathbf{v}_2 = \begin{bmatrix} 0.118 \\ 0.991 \end{bmatrix} \tag{d}$$

Next we calculate the matrices in (a) and obtain the eigenproblem

$$\begin{bmatrix} 9 & -51 \\ -51 & 393 \end{bmatrix}\bar{\mathbf{v}} = \lambda\begin{bmatrix} 1 & 1 \\ 1 & 2 \end{bmatrix}\bar{\mathbf{v}} \tag{e}$$

The characteristic polynomial is

$$p(\lambda) = \det\begin{bmatrix} 9 - \lambda & -51 - \lambda \\ -51 - \lambda & 393 - 2\lambda \end{bmatrix}$$

and evaluated we obtain $p(\lambda)$ in (c).

Hence the eigenvalues of (e) are those given in (d). The eigenvectors are obtained using the fact that

$$(\mathbf{P}^T\mathbf{A}\mathbf{P} - \lambda_i\mathbf{P}^T\mathbf{P})\bar{\mathbf{v}}_i = \mathbf{0} \tag{f}$$

Noting that from (f) the eigenvectors can only be calculated within scalar multiples, we can obtain

$$\bar{\mathbf{v}}_1 = \begin{bmatrix} 0.873 \\ 0.118 \end{bmatrix}; \qquad \bar{\mathbf{v}}_2 = \begin{bmatrix} 1.109 \\ -0.991 \end{bmatrix}$$

and we may check that $\mathbf{v}_i = \mathbf{P}\bar{\mathbf{v}}_i$.

Finally, we use the matrix $\mathbf{P}$ defined in (b) to perform a change of basis. In this case we obtain the eigenproblem

$$\begin{bmatrix} 383.4 & 220.8 \\ 220.8 & 129.6 \end{bmatrix}\bar{\mathbf{v}} = \lambda\bar{\mathbf{v}}$$

because $\mathbf{P}^T\mathbf{P} = \mathbf{I}$.

This problem has again the same characteristic polynomial as in (c), and we obtain

$$\bar{\mathbf{v}}_1 = \begin{bmatrix} 0.5 \\ -0.864 \end{bmatrix}; \qquad \bar{\mathbf{v}}_2 = \begin{bmatrix} 0.864 \\ 0.5 \end{bmatrix}$$

where we may again check that $\mathbf{v}_i = \mathbf{P}\bar{\mathbf{v}}_i$.

So far we have shown that there are n eigenvalues and corresponding eigenvectors, but we have not yet discussed the properties of the eigenvalues and vectors.

A first observation is that the eigenvalues are real. Consider the ith eigenpair $(\lambda_i, \mathbf{v}_i)$, for which we have

$$\mathbf{A}\mathbf{v}_i = \lambda_i \mathbf{v}_i \tag{2.75}$$

Assume that $\mathbf{v}_i$ and λ_i are complex, which includes the case of real eigenvalues, and let the elements of $\bar{\mathbf{v}}_i$ and $\bar{\lambda}_i$ be the complex conjugates of the elements of $\mathbf{v}_i$ and λ_i. Then premultiplying (2.75) by $\bar{\mathbf{v}}_i^T$, we obtain

$$\bar{\mathbf{v}}_i^T \mathbf{A}\mathbf{v}_i = \lambda_i \bar{\mathbf{v}}_i^T \mathbf{v}_i \tag{2.76}$$

On the other hand, we also obtain from (2.75),

$$\bar{\mathbf{v}}_i^T \mathbf{A} = \bar{\mathbf{v}}_i^T \bar{\lambda}_i \tag{2.77}$$

and postmultiplying by $\mathbf{v}_i$, we have

$$\bar{\mathbf{v}}_i^T \mathbf{A}\mathbf{v}_i = \bar{\lambda}_i \bar{\mathbf{v}}_i^T \mathbf{v}_i \tag{2.78}$$

But the left-hand sides of (2.76) and (2.78) are the same and thus we have

$$(\lambda_i - \bar{\lambda}_i)\bar{\mathbf{v}}_i^T \mathbf{v}_i = 0 \tag{2.79}$$

Since $\mathbf{v}_i$ is nontrivial, it follows that $\lambda_i = \bar{\lambda}_i$, and hence the eigenvalue must be real. However, then it also follows from (2.61) that the eigenvectors can be made real, because the coefficient matrix $\mathbf{A} - \lambda\mathbf{I}$ is real.

Another important point is that the eigenvectors which correspond to distinct eigenvalues are unique (within scalar multipliers) and *orthogonal*, whereas the eigenvectors corresponding to multiple eigenvalues are not unique, but we can always choose an orthogonal set.

Assume first that the eigenvalues are distinct. In this case we have for two eigenpairs,

$$\mathbf{A}\mathbf{v}_i = \lambda_i \mathbf{v}_i \tag{2.80}$$

and

$$\mathbf{A}\mathbf{v}_j = \lambda_j \mathbf{v}_j \tag{2.81}$$

Premultiplying (2.80) by $\mathbf{v}_j^T$ and (2.81) by $\mathbf{v}_i^T$, we obtain

$$\mathbf{v}_j^T \mathbf{A}\mathbf{v}_i = \lambda_i \mathbf{v}_j^T \mathbf{v}_i \tag{2.82}$$

$$\mathbf{v}_i^T \mathbf{A}\mathbf{v}_j = \lambda_j \mathbf{v}_i^T \mathbf{v}_j \tag{2.83}$$

Taking the transpose in (2.83), we have

$$\mathbf{v}_j^T \mathbf{A}\mathbf{v}_i = \lambda_j \mathbf{v}_j^T \mathbf{v}_i \tag{2.84}$$

and thus from (2.84) and (2.82) we obtain

$$(\lambda_i - \lambda_j)\mathbf{v}_j^T \mathbf{v}_i = 0 \tag{2.85}$$

Since we assumed that $\lambda_i \neq \lambda_j$, it follows that $\mathbf{v}_j^T \mathbf{v}_i = 0$, i.e., that $\mathbf{v}_j$ and $\mathbf{v}_i$ are orthogonal. Furthermore, we can scale the elements of the vector $\mathbf{v}_i$ to obtain that

$$\mathbf{v}_i^T \mathbf{v}_j = \delta_{ij} \tag{2.86}$$

where δ_{ij} = Kronecker delta; i.e., $\delta_{ij} = 1$ when $i = j$, and $\delta_{ij} = 0$ when $i \neq j$. If (2.86) is satisfied, we say that the eigenvectors are *orthonormal*.

It should be noted that the solution of (2.61) yields a vector in which only the relative magnitudes of the elements are defined. If all elements are scaled by the same amount, the new vector would still satisfy (2.61). In effect, the solution of (2.61) yields the direction of the eigenvector, and we use the orthonormality condition in (2.86) to fix the magnitudes of the elements in the vector. Therefore, *when we refer to eigenvectors it is from now on implied that the vectors are orthonormal.*

EXAMPLE 2.19: Check that the vectors calculated in Example 2.17 are orthogonal, and then orthonormalize them.

The orthogonality is checked by forming $\mathbf{v}_1^T\mathbf{v}_2$, which gives

$$\mathbf{v}_1^T\mathbf{v}_2 = (2)(\tfrac{1}{2}) + (-1)(1) = 0$$

Hence the vectors are orthogonal. To orthonormalize the vectors, we need to make the lengths of the vectors equal to 1. Then we have

$$\mathbf{v}_1 = \frac{1}{\sqrt{5}}\begin{bmatrix} 2 \\ -1 \end{bmatrix} \quad \text{or} \quad \mathbf{v}_1 = \frac{1}{\sqrt{5}}\begin{bmatrix} -2 \\ 1 \end{bmatrix}; \qquad \mathbf{v}_2 = \frac{1}{\sqrt{5}}\begin{bmatrix} 1 \\ 2 \end{bmatrix} \quad \text{or} \quad \mathbf{v}_2 = \frac{1}{\sqrt{5}}\begin{bmatrix} -1 \\ -2 \end{bmatrix}$$

We now turn to the case in which multiple eigenvalues are also present. The proof of eigenvector orthonormality given in (2.80) to (2.86) is not possible because for a multiple eigenvalue, λ_i is equal to λ_j in (2.85). Assume that $\lambda_i = \lambda_{i+1} = \ldots = \lambda_{i+m-1}$; i.e., λ_i is an m-times multiple root. Then we can show that it is still always possible to choose m orthonormal eigenvectors that correspond to $\lambda_i, \lambda_{i+1}, \ldots, \lambda_{i+m-1}$. This follows because for a symmetric matrix, we can always establish a complete set of n orthonormal eigenvectors. Corresponding to each distinct eigenvalue we have an *eigenspace* with dimension equal to the multiplicity of the eigenvalue. All eigenspaces are unique and are orthogonal to the eigenspaces that correspond to other distinct eigenvalues. The eigenvectors associated with an eigenvalue provide a basis for the eigenspace, and since the basis is not unique if $m > 1$, the eigenvectors corresponding to a multiple eigenvalue are not unique. The formal proofs of the above statements are an application of the principles discussed earlier and are given in the following examples.

EXAMPLE 2.20: Show that for a symmetric matrix $\mathbf{A}$ of order n, there are always n orthonormal eigenvectors.

Assume that we have calculated an eigenvalue λ_i and corresponding eigenvector $\mathbf{v}_i$. Let us construct an orthonormal matrix $\mathbf{Q}$ whose first column is $\mathbf{v}_i$,

$$\mathbf{Q} = [\mathbf{v}_i \quad \hat{\mathbf{Q}}]; \qquad \mathbf{Q}^T\mathbf{Q} = \mathbf{I}$$

This matrix can always be constructed because the vectors in $\mathbf{Q}$ provide an orthonormal basis for the n-dimensional space in which $\mathbf{A}$ is defined. However, we can now calculate

$$\mathbf{Q}^T\mathbf{A}\mathbf{Q} = \begin{bmatrix} \lambda_i & \mathbf{0} \\ \mathbf{0} & \mathbf{A}_1 \end{bmatrix} \tag{a}$$

where $$\mathbf{A}_1 = \hat{\mathbf{Q}}^T\mathbf{A}\hat{\mathbf{Q}}$$

and $\mathbf{A}_1$ is a full matrix of order $(n-1)$. If $n = 2$, we note that $\mathbf{Q}^T\mathbf{A}\mathbf{Q}$ is diagonal. In that case, if we premultiply (a) by $\mathbf{Q}$ and let $a \equiv \mathbf{A}_1$ we obtain

$$\mathbf{A}\mathbf{Q} = \mathbf{Q}\begin{bmatrix} \lambda_i & 0 \\ 0 & a \end{bmatrix}$$

and hence the vector in $\hat{\mathbf{Q}}$ is the other eigenvector, and a is the other eigenvalue regardless of whether λ_i is a multiple eigenvalue or not.

The complete proof is now obtained by induction. Assume that the statement is true for a matrix of order $(n - 1)$; then we will show that it is also true for a matrix of order n. But since we demonstrated that the statement is true for $n = 2$, it follows that it is true for any n.

The assumption that there are $(n - 1)$ orthonormal eigenvectors for a matrix of order $(n - 1)$ gives

$$\mathbf{Q}_1^T\mathbf{A}_1\mathbf{Q}_1 = \mathbf{\Lambda} \tag{b}$$

where $\mathbf{Q}_1$ is a matrix of the eigenvectors of $\mathbf{A}_1$ and $\mathbf{\Lambda}$ is a diagonal matrix listing the eigenvalues of $\mathbf{A}_1$. However, if we now define

$$\mathbf{S} = \begin{bmatrix} 1 & \mathbf{0} \\ \mathbf{0} & \mathbf{Q}_1 \end{bmatrix}$$

we have

$$\mathbf{S}^T\mathbf{Q}^T\mathbf{A}\mathbf{Q}\mathbf{S} = \begin{bmatrix} \lambda_i & \mathbf{0} \\ \mathbf{0} & \mathbf{\Lambda} \end{bmatrix} \tag{c}$$

Let

$$\mathbf{P} = \mathbf{Q}\mathbf{S}; \qquad \mathbf{P}^T\mathbf{P} = \mathbf{I}$$

Then premultiplying (c) by $\mathbf{P}$ we obtain

$$\mathbf{A}\mathbf{P} = \mathbf{P}\begin{bmatrix} \lambda_i & \mathbf{0} \\ \mathbf{0} & \mathbf{\Lambda} \end{bmatrix}$$

Therefore, under the assumption in (b), the statement is also true for a matrix of order n, which completes the proof.

EXAMPLE 2.21: Show that the eigenvectors corresponding to a multiple eigenvalue of multiplicity m define an m-dimensional space in which each vector is also an eigenvector. This space is called the *eigenspace* corresponding to the eigenvalue considered.

Let λ_i be the eigenvalue of multiplicity m, i.e., we have

$$\lambda_i = \lambda_{i+1} = \ldots = \lambda_{i+m-1}$$

We showed in Example 2.20 that there are m orthonormal eigenvectors $\mathbf{v}_i, \mathbf{v}_{i+1}, \ldots, \mathbf{v}_{i+m-1}$ corresponding to λ_i. These vectors provide the basis of an m-dimensional space. Consider any vector $\mathbf{w}$ in this space, such as

$$\mathbf{w} = \alpha_i\mathbf{v}_i + \alpha_{i+1}\mathbf{v}_{i+1} + \ldots + \alpha_{i+m-1}\mathbf{v}_{i+m-1}$$

where the $\alpha_i, \alpha_{i+1}, \ldots$ are constants. The vector $\mathbf{w}$ is also an eigenvector, because we have

$$\mathbf{A}\mathbf{w} = \alpha_i\mathbf{A}\mathbf{v}_i + \alpha_{i+1}\mathbf{A}\mathbf{v}_{i+1} + \ldots + \alpha_{i+m-1}\mathbf{A}\mathbf{v}_{i+m-1}$$

which gives

$$\mathbf{A}\mathbf{w} = \alpha_i\lambda_i\mathbf{v}_i + \alpha_{i+1}\lambda_i\mathbf{v}_{i+1} + \ldots + \alpha_{i+m-1}\lambda_i\mathbf{v}_{i+m-1} = \lambda_i\mathbf{w}$$

Therefore, any vector $\mathbf{w}$ in the space spanned by the m eigenvectors $\mathbf{v}_i, \mathbf{v}_{i+1}, \ldots, \mathbf{v}_{i+m-1}$ is also an eigenvector. It should be noted that the vector $\mathbf{w}$ will be orthogonal to the eigenvectors that correspond to eigenvalues not equal to λ_i. Hence there is one eigenspace that corresponds to each, distinct or multiple, eigenvalue. The dimension of the eigenspace is equal to the multiplicity of the eigenvalue.

With the main properties of the eigenvalues and eigenvectors of $\mathbf{A}$ having been presented, we can now write the n solutions to $\mathbf{A}\mathbf{v} = \lambda\mathbf{v}$ in various forms. First, we have

$$\mathbf{A}\mathbf{V} = \mathbf{V}\mathbf{\Lambda} \tag{2.87}$$

where $\mathbf{V}$ is a matrix storing the eigenvectors, $\mathbf{V} = [\mathbf{v}_1, \ldots, \mathbf{v}_n]$, and $\mathbf{\Lambda}$ is a diagonal matrix with the corresponding eigenvalues on its diagonal, $\mathbf{\Lambda} = \text{diag}(\lambda_i)$. Using the orthonormality property of the eigenvectors (i.e., $\mathbf{V}^T\mathbf{V} = \mathbf{I}$), we obtain from (2.87) that

$$\mathbf{V}^T\mathbf{A}\mathbf{V} = \mathbf{\Lambda} \tag{2.88}$$

Furthermore, we obtain the *spectral decomposition* of $\mathbf{A}$,

$$\mathbf{A} = \mathbf{V}\mathbf{\Lambda}\mathbf{V}^T \tag{2.89}$$

where it may be convenient to write the spectral decomposition of $\mathbf{A}$ as

$$\mathbf{A} = \sum_{i=1}^{n} \lambda_i \mathbf{v}_i \mathbf{v}_i^T \tag{2.90}$$

It should be noted that each of the above equations represents the solution to the eigenproblem $\mathbf{A}\mathbf{v} = \lambda\mathbf{v}$. Consider the following example.

EXAMPLE 2.22: Establish the relations given in (2.87) to (2.90) for the matrix $\mathbf{A}$ used in Example 2.17.

The eigenvalues and eigenvectors of $\mathbf{A}$ have been calculated in Examples 2.17 and 2.19. Using the information given in those examples, we have, for (2.87),

$$\begin{bmatrix} -1 & 2 \\ 2 & 2 \end{bmatrix}\begin{bmatrix} -\frac{2}{\sqrt{5}} & \frac{1}{\sqrt{5}} \\ \frac{1}{\sqrt{5}} & \frac{2}{\sqrt{5}} \end{bmatrix} = \begin{bmatrix} -\frac{2}{\sqrt{5}} & \frac{1}{\sqrt{5}} \\ \frac{1}{\sqrt{5}} & \frac{2}{\sqrt{5}} \end{bmatrix}\begin{bmatrix} -2 & 0 \\ 0 & 3 \end{bmatrix}$$

for (2.88),

$$\begin{bmatrix} -\frac{2}{\sqrt{5}} & \frac{1}{\sqrt{5}} \\ \frac{1}{\sqrt{5}} & \frac{2}{\sqrt{5}} \end{bmatrix}\begin{bmatrix} -1 & 2 \\ 2 & 2 \end{bmatrix}\begin{bmatrix} -\frac{2}{\sqrt{5}} & \frac{1}{\sqrt{5}} \\ \frac{1}{\sqrt{5}} & \frac{2}{\sqrt{5}} \end{bmatrix} = \begin{bmatrix} -2 & 0 \\ 0 & 3 \end{bmatrix}$$

for (2.89),

$$\begin{bmatrix} -1 & 2 \\ 2 & 2 \end{bmatrix} = \begin{bmatrix} -\frac{2}{\sqrt{5}} & \frac{1}{\sqrt{5}} \\ \frac{1}{\sqrt{5}} & \frac{2}{\sqrt{5}} \end{bmatrix}\begin{bmatrix} -2 & 0 \\ 0 & 3 \end{bmatrix}\begin{bmatrix} -\frac{2}{\sqrt{5}} & \frac{1}{\sqrt{5}} \\ \frac{1}{\sqrt{5}} & \frac{2}{\sqrt{5}} \end{bmatrix}$$

and for (2.90),

$$\mathbf{A} = (-2)\begin{bmatrix} -\frac{2}{\sqrt{5}} \\ \frac{1}{\sqrt{5}} \end{bmatrix}\begin{bmatrix} -\frac{2}{\sqrt{5}} & \frac{1}{\sqrt{5}} \end{bmatrix} + (3)\begin{bmatrix} \frac{1}{\sqrt{5}} \\ \frac{2}{\sqrt{5}} \end{bmatrix}\begin{bmatrix} \frac{1}{\sqrt{5}} & \frac{2}{\sqrt{5}} \end{bmatrix}$$

The relations in (2.88) and (2.89) can be employed effectively in various important applications. The objective in the following examples is to present some solution procedures in which they are used.

EXAMPLE 2.23: Calculate the kth power of a given matrix $\mathbf{A}$; i.e., evaluate $\mathbf{A}^k$. Demonstrate the result using $\mathbf{A}$ of Example 2.17.

One way of evaluating $\mathbf{A}^k$ is to simply calculate $\mathbf{A}^2 = \mathbf{A}\mathbf{A}$, $\mathbf{A}^4 = \mathbf{A}^2\mathbf{A}^2$, etc., as we have done in Example 1.6. However, if k is large, it may be more effective to employ the spectral decomposition of $\mathbf{A}$. Assume that we have calculated the eigenvalues and eigenvectors of $\mathbf{A}$; i.e., we have

$$\mathbf{A} = \mathbf{V}\mathbf{\Lambda}\mathbf{V}^T$$

To calculate $\mathbf{A}^2$, we use $\quad \mathbf{A}^2 = \mathbf{V}\boldsymbol{\Lambda}\mathbf{V}^T\mathbf{V}\boldsymbol{\Lambda}\mathbf{V}^T$

but because $\mathbf{V}^T\mathbf{V} = \mathbf{I}$, we have $\quad \mathbf{A}^2 = \mathbf{V}\boldsymbol{\Lambda}^2\mathbf{V}^T$

Proceeding in the same manner, we thus obtain

$$\mathbf{A}^k = \mathbf{V}\boldsymbol{\Lambda}^k\mathbf{V}^T$$

As an example, let $\mathbf{A}$ be the matrix considered in Example 2.17. Then we have

$$\mathbf{A}^k = \frac{1}{\sqrt{5}}\begin{bmatrix} -2 & 1 \\ 1 & 2 \end{bmatrix}\begin{bmatrix} (-2)^k & 0 \\ 0 & (3)^k \end{bmatrix}\frac{1}{\sqrt{5}}\begin{bmatrix} -2 & 1 \\ 1 & 2 \end{bmatrix}$$

or

$$\mathbf{A}^k = \frac{1}{5}\left[\begin{array}{c:c} (-2)^{k+2} + (3)^k & (-2)^{k+1} + (2)(3)^k \\ \hdashline (-2)^{k+1} + (2)(3)^k & (-2)^k + (4)(3)^k \end{array}\right]$$

It is interesting to note that if the largest absolute value of all the eigenvalues of $\mathbf{A}$ is smaller than 1, we have that $\mathbf{A}^k \longrightarrow \mathbf{0}$ as $k \longrightarrow \infty$. Thus defining

$$\rho(\mathbf{A}) = \max_{\text{all } i} |\lambda_i| \qquad \text{SPECTRAL RADIUS}$$

we have that $\lim_{k\to\infty} \mathbf{A}^k = \mathbf{0}$, provided that $\rho(\mathbf{A}) < 1$. (p62)

EXAMPLE 2.24: Consider the system of differential equations

$$\dot{\mathbf{x}} + \mathbf{A}\mathbf{x} = \mathbf{f}(t) \tag{a}$$

and obtain the solution using the spectral decomposition of $\mathbf{A}$. Demonstrate the result using the matrix $\mathbf{A}$ of Example 2.17 and

$$\mathbf{f}(t) = \begin{bmatrix} e^{-t} \\ 0 \end{bmatrix}; \qquad {}^0\mathbf{x} = \begin{bmatrix} 1 \\ 1 \end{bmatrix}$$

where ${}^0\mathbf{x}$ are the initial conditions.

Substituting $\mathbf{A} = \mathbf{V}\boldsymbol{\Lambda}\mathbf{V}^T$ and premultiplying by $\mathbf{V}^T$, we obtain

$$\mathbf{V}^T\dot{\mathbf{x}} + \boldsymbol{\Lambda}(\mathbf{V}^T\mathbf{x}) = \mathbf{V}^T\mathbf{f}(t)$$

Thus if we define $\mathbf{y} = \mathbf{V}^T\mathbf{x}$, we need to solve the equations

$$\dot{\mathbf{y}} + \boldsymbol{\Lambda}\mathbf{y} = \mathbf{V}^T\mathbf{f}(t)$$

But this is a set of n decoupled differential equations. Consider the rth equation, which is typical:

$$\dot{y}_r + \lambda_r y_r = \mathbf{v}_r^T\mathbf{f}(t)$$

The solution is $\quad y_r(t) = {}^0y_r\, e^{-\lambda_r t} + e^{-\lambda_r t}\int_0^t e^{\lambda_r \tau}\, \mathbf{v}_r^T\mathbf{f}(\tau)\, d\tau$

where ${}^0y_r$ is the value of y_r at time $t = 0$. The complete solution to the system of equations in (a) is

$$\mathbf{x} = \sum_{r=1}^{n} \mathbf{v}_r y_r \tag{b}$$

As an example, we consider the system of differential equations

$$\begin{bmatrix} \dot{x}_1 \\ \dot{x}_2 \end{bmatrix} + \begin{bmatrix} -1 & 2 \\ 2 & 2 \end{bmatrix}\begin{bmatrix} x_1 \\ x_2 \end{bmatrix} = \begin{bmatrix} e^{-t} \\ 0 \end{bmatrix}$$

In this case we have to solve the two decoupled differential equations

$$\dot{y}_1 + (-2)y_1 = 2e^{-t}$$
$$\dot{y}_2 + 3y_2 = e^{-t}$$

with initial conditions

$$ {}^0\mathbf{y} = \mathbf{V}^T {}^0\mathbf{x} = \frac{1}{\sqrt{5}}\begin{bmatrix} 2 & -1 \\ 1 & 2 \end{bmatrix}\begin{bmatrix} 1 \\ 1 \end{bmatrix} = \frac{1}{\sqrt{5}}\begin{bmatrix} 1 \\ 3 \end{bmatrix} $$

We obtain

$$ y_1 = \frac{1}{\sqrt{5}} e^{2t} - \frac{2}{3} e^{-t} $$

$$ y_2 = \frac{3}{\sqrt{5}} e^{-3t} + \frac{1}{2} e^{-t} $$

Thus using (b), we have

$$ \begin{bmatrix} x_1 \\ x_2 \end{bmatrix} = \frac{1}{\sqrt{5}}\left(\begin{bmatrix} 2 \\ -1 \end{bmatrix} y_1 + \begin{bmatrix} 1 \\ 2 \end{bmatrix} y_2\right) $$

$$ = \begin{bmatrix} -\frac{\sqrt{5}}{6} e^{-t} + \frac{3}{5} e^{-3t} + \frac{2}{5} e^{2t} \\ \frac{\sqrt{5}}{3} e^{-t} + \frac{6}{5} e^{-3t} - \frac{1}{5} e^{2t} \end{bmatrix} $$

To conclude the presentation, we may note that by introducing auxiliary variables, higher-order differential equations can be reduced to a system of first-order differential equations. However, the coefficient matrix $\mathbf{A}$ is in that case nonsymmetric.

EXAMPLE 2.25: Using the spectral decomposition of an $n \times n$ symmetric matrix $\mathbf{A}$, evaluate the inverse of the matrix. Demonstrate the result using the matrix $\mathbf{A}$ of Example 2.17.

Assume that we have evaluated the eigenvalues λ_i and corresponding eigenvectors $\mathbf{v}_i$, $i = 1, \ldots, n$, of the matrix $\mathbf{A}$; i.e., we have solved the eigenproblem

$$ \mathbf{A}\mathbf{v} = \lambda\mathbf{v} \qquad \text{(a)} $$

Premultiplying both sides of the relation in (a) by $\lambda^{-1}\mathbf{A}^{-1}$, we obtain the eigenproblem

$$ \mathbf{A}^{-1}\mathbf{v} = \lambda^{-1}\mathbf{v} $$

But this relation shows that the eigenvalues of $\mathbf{A}^{-1}$ are $1/\lambda_i$ and the eigenvectors are $\mathbf{v}_i$, $i = 1, \ldots, n$. Thus using (2.90) for $\mathbf{A}^{-1}$, we have

$$ \mathbf{A}^{-1} = \mathbf{V}\mathbf{\Lambda}^{-1}\mathbf{V}^T $$

or

$$ \mathbf{A}^{-1} = \sum_{i=1}^{n} \left(\frac{1}{\lambda_i}\right)\mathbf{v}_i\mathbf{v}_i^T $$

These equations show that we cannot find the inverse of $\mathbf{A}$ if the matrix has a zero eigenvalue.

As an example we evaluate the inverse of the matrix $\mathbf{A}$ considered in Example 2.17. In this case we have

$$ \mathbf{A}^{-1} = \frac{1}{5}\begin{bmatrix} 2 & 1 \\ -1 & 2 \end{bmatrix}\begin{bmatrix} -\frac{1}{2} & 0 \\ 0 & \frac{1}{3} \end{bmatrix}\begin{bmatrix} 2 & -1 \\ 1 & 2 \end{bmatrix} = \frac{1}{6}\begin{bmatrix} -2 & 2 \\ 2 & 1 \end{bmatrix} $$

Let us now return to the discussion with which we started this section. Comparing (2.88) with (2.67), we find that we perform in (2.88) a change of basis. The new matrix representation of the operator considered is the diagonal matrix $\mathbf{\Lambda}$. Since the vectors in $\mathbf{V}$ are a new basis, they span the n-dimensional space in which $\mathbf{A}$ and $\mathbf{\Lambda}$ are defined, and any vector $\mathbf{w}$ can be expressed as a linear combination of the eigenvectors $\mathbf{v}_i$; i.e., we have

$$\mathbf{w} = \sum_{i=1}^{n} \alpha_i \mathbf{v}_i \tag{2.91}$$

An important observation is that $\mathbf{\Lambda}$ shows directly whether the operator that $\mathbf{\Lambda}$ represents is singular (see Section 2.3), i.e., whether the matrices $\mathbf{A}$ and $\mathbf{\Lambda}$ are singular. Using the definition given in Section 2.3, we find that $\mathbf{\Lambda}$ and hence $\mathbf{A}$ are singular if and only if an eigenvalue is equal to zero, because in that case $\mathbf{\Lambda}^{-1}$ cannot be calculated. In this context it is useful to define some additional terminology. If all eigenvalues are positive, we say that the matrix and the operator that the matrix represents are *positive definite.* If all eigenvalues are greater than or equal to zero, the matrix is *positive semidefinite*; with negative, zero, or positive eigenvalues, the matrix is *indefinite.*

2.8 THE RAYLEIGH QUOTIENT AND THE MINIMAX CHARACTERIZATION OF EIGENVALUES

In the previous section we defined the eigenproblem $\mathbf{A}\mathbf{v} = \lambda\mathbf{v}$ and discussed the basic properties that pertain to the solutions of the problem. The objective in this section is to complement the information given with some very powerful principles.

A number of important principles are derived using the Rayleigh quotient $\rho(\mathbf{v})$, which is defined as

$$\rho(\mathbf{v}) = \frac{\mathbf{v}^T\mathbf{A}\mathbf{v}}{\mathbf{v}^T\mathbf{v}} \tag{2.92}$$

The first observation is that

$$\lambda_1 \leq \rho(\mathbf{v}) \leq \lambda_n \tag{2.93}$$

and it follows that using the definitions given in Section 2.7, we have for any vector $\mathbf{v}$, if $\mathbf{A}$ is positive definite $\rho(\mathbf{v}) > 0$, if $\mathbf{A}$ is positive semidefinite $\rho(\mathbf{v}) \geq 0$, and for $\mathbf{A}$ indefinite $\rho(\mathbf{v})$ can be negative, zero, or positive. For the proof of (2.93) we use that

$$\mathbf{v} = \sum_{i=1}^{n} \alpha_i \mathbf{v}_i \tag{2.94}$$

where $\mathbf{v}_i$ are the eigenvectors of $\mathbf{A}$. Substituting for $\mathbf{v}$ into (2.92) and using that $\mathbf{A}\mathbf{v}_i = \lambda_i \mathbf{v}_i$, $\mathbf{v}_i^T\mathbf{v}_j = \delta_{ij}$, we obtain

$$\rho(\mathbf{v}) = \frac{\lambda_1\alpha_1^2 + \lambda_2\alpha_2^2 + \ldots + \lambda_n\alpha_n^2}{\alpha_1^2 + \ldots + \alpha_n^2} \tag{2.95}$$

Hence, if $\lambda_1 \neq 0$,

$$\rho(\mathbf{v}) = \lambda_1 \frac{\alpha_1^2 + (\lambda_2/\lambda_1)\alpha_2^2 + \ldots + (\lambda_n/\lambda_1)\alpha_n^2}{\alpha_1^2 + \ldots + \alpha_n^2} \tag{2.96}$$

and if $\lambda_n \neq 0$,

$$\rho(\mathbf{v}) = \lambda_n \frac{(\lambda_1/\lambda_n)\alpha_1^2 + (\lambda_2/\lambda_n)\alpha_2^2 + \ldots + \alpha_n^2}{\alpha_1^2 + \ldots + \alpha_n^2} \tag{2.97}$$

But since $\lambda_1 \leq \lambda_2 \leq \ldots \leq \lambda_n$, the relations in (2.95) to (2.97) show that (2.93) holds. Furthermore, it is seen that if $\mathbf{v} = \mathbf{v}_i$, we have $\rho(\mathbf{v}) = \lambda_i$.

Considering the practical use of the Rayleigh quotient, the following property is of particular value. Assume that $\mathbf{v}$ is an approximation to the eigenvector $\mathbf{v}_i$, say with ϵ small we have

$$\mathbf{v} = \mathbf{v}_i + \epsilon \mathbf{x} \tag{2.98}$$

Then the Rayleigh quotient of $\mathbf{v}$ will give an approximation to λ_i of order ϵ^2; i.e.,

$$\rho(\mathbf{v}) = \lambda_i + o(\epsilon^2) \tag{2.99}$$

The notation $o(\epsilon^2)$ means "of order ϵ^2" and indicates that if $\delta = o(\epsilon^2)$, then $|\delta| \leq b\epsilon^2$, where b is a constant.

To prove the above property of the Rayleigh quotient, we substitute for $\mathbf{v}$ from (2.94) into the Rayleigh quotient expression to obtain

$$\rho(\mathbf{v}_i + \epsilon\mathbf{x}) = \frac{(\mathbf{v}_i^T + \epsilon\mathbf{x}^T)\mathbf{A}(\mathbf{v}_i + \epsilon\mathbf{x})}{(\mathbf{v}_i^T + \epsilon\mathbf{x}^T)(\mathbf{v}_i + \epsilon\mathbf{x})} \tag{2.100}$$

or

$$\rho(\mathbf{v}_i + \epsilon\mathbf{x}) = \frac{\mathbf{v}_i^T\mathbf{A}\mathbf{v}_i + 2\epsilon\mathbf{v}_i^T\mathbf{A}\mathbf{x} + \epsilon^2\mathbf{x}^T\mathbf{A}\mathbf{x}}{\mathbf{v}_i^T\mathbf{v}_i + 2\epsilon\mathbf{x}^T\mathbf{v}_i + \epsilon^2\mathbf{x}^T\mathbf{x}} \tag{2.101}$$

However, since $\mathbf{x}$ is an error in $\mathbf{v}_i$, we can write

$$\mathbf{x} = \sum_{\substack{j=1 \\ j\neq i}}^{n} \alpha_j \mathbf{v}_j \tag{2.102}$$

But then using that $\mathbf{v}_i^T\mathbf{v}_j = \delta_{ij}$ and $\mathbf{A}\mathbf{v}_j = \lambda_j\mathbf{v}_j$, we have that $\mathbf{v}_i^T\mathbf{A}\mathbf{x} = 0$ and $\mathbf{x}^T\mathbf{v}_i = 0$, and hence

$$\rho(\mathbf{v}_i + \epsilon\mathbf{x}) = \frac{\lambda_i + \epsilon^2 \sum\limits_{\substack{j=1 \\ j\neq i}}^{n} \alpha_j^2\lambda_j}{1 + \epsilon^2 \sum\limits_{\substack{j=1 \\ j\neq i}}^{n} \alpha_j^2} \tag{2.103}$$

However, using the binomial theorem to expand the denominator in (2.103), we have

$$\rho(\mathbf{v}_i + \epsilon\mathbf{x}) = \left(\lambda_i + \epsilon^2 \sum_{\substack{j=1 \\ j\neq i}}^{n} \alpha_j^2\lambda_j\right)\left[1 - \epsilon^2\left(\sum_{\substack{j=1 \\ j\neq i}}^{n} \alpha_j^2\right) + \epsilon^4\left(\sum_{\substack{j=1 \\ j\neq i}}^{n} \alpha_j^2\right)^2 + \dots\right] \tag{2.104}$$

or

$$\rho(\mathbf{v}_i + \epsilon\mathbf{x}) = \lambda_i + \epsilon^2\left(\sum_{\substack{j=1 \\ j\neq i}}^{n} \alpha_j^2\lambda_j - \lambda_i \sum_{\substack{j=1 \\ j\neq i}}^{n} \alpha_j^2\right) + \text{higher-order terms} \tag{2.105}$$

The relation in (2.99) thus follows. We demonstrate the above results in a small example.

EXAMPLE 2.26: Evaluate the Rayleigh quotients $\rho(\mathbf{v})$ for the matrix $\mathbf{A}$ used in Example 2.17. Using $\mathbf{v}_1$ and $\mathbf{v}_2$ of Example 2.17, consider the following cases:

1. $\mathbf{v} = \mathbf{v}_1 + 2\mathbf{v}_2$; **2.** $\mathbf{v} = \mathbf{v}_1$; **3.** $\mathbf{v} = \mathbf{v}_1 + 0.02\mathbf{v}_2$.

In case **1**, we have

$$\mathbf{v} = \begin{bmatrix} 2 \\ -1 \end{bmatrix} + \begin{bmatrix} 1 \\ 2 \end{bmatrix} = \begin{bmatrix} 3 \\ 1 \end{bmatrix}$$

and thus

$$\rho(\mathbf{v}) = \frac{[3 \quad 1]\begin{bmatrix} -1 & 2 \\ 2 & 2 \end{bmatrix}\begin{bmatrix} 3 \\ 1 \end{bmatrix}}{[3 \quad 1]\begin{bmatrix} 3 \\ 1 \end{bmatrix}} = \frac{1}{2}$$

Recalling that $\lambda_1 = -2$ and $\lambda_2 = 3$, we have, as expected,

$$\lambda_1 \leq \rho(\mathbf{v}) \leq \lambda_2$$

In case **2**, we have

$$\mathbf{v} = \begin{bmatrix} 2 \\ -1 \end{bmatrix}$$

and hence
$$\rho(\mathbf{v}) = \frac{[2 \quad -1]\begin{bmatrix} -1 & 2 \\ 2 & 2 \end{bmatrix}\begin{bmatrix} 2 \\ -1 \end{bmatrix}}{[2 \quad -1]\begin{bmatrix} 2 \\ -1 \end{bmatrix}} = -2$$

and hence, as expected, $\rho(\mathbf{v}) = \lambda_1$.

Finally, in case **3**, we use

$$\mathbf{v} = \begin{bmatrix} 2 \\ -1 \end{bmatrix} + \begin{bmatrix} 0.01 \\ 0.02 \end{bmatrix} = \begin{bmatrix} 2.01 \\ -0.98 \end{bmatrix}$$

and hence
$$\rho(\mathbf{v}) = \frac{[2.01 \quad -0.98]\begin{bmatrix} -1 & 2 \\ 2 & 2 \end{bmatrix}\begin{bmatrix} 2.01 \\ -0.98 \end{bmatrix}}{[2.01 \quad -0.98]\begin{bmatrix} 2.01 \\ -0.98 \end{bmatrix}}$$
$$= -1.99950005$$

Here, we note that $\rho(\mathbf{v}) > \lambda_1$ and that $\rho(\mathbf{v})$ approximates λ_1 more closely than $\mathbf{v}$ approximates $\mathbf{v}_1$.

Having introduced the Rayleigh quotient, we can now proceed to a very important principle, the minimax characterization of eigenvalues. We know from Rayleigh's principle that

$$\rho(\mathbf{v}) \geq \lambda_1 \tag{2.106}$$

where $\mathbf{v}$ is any vector. In other words, if we consider the problem of varying $\mathbf{v}$, we will always have $\rho(\mathbf{v}) \geq \lambda_1$, and the minimum is reached when $\mathbf{v} = \mathbf{v}_1$, in which case $\rho(\mathbf{v}_1) = \lambda_1$. Suppose that we now impose a restriction on $\mathbf{v}$, namely, that $\mathbf{v}$ be orthogonal to a specific vector $\mathbf{w}$, and that we consider the problem of minimizing $\rho(\mathbf{v})$ subject to this restriction. After having calculated the minimum of $\rho(\mathbf{v})$ with the condition $\mathbf{v}^T\mathbf{w} = 0$, we could start varying $\mathbf{w}$ and for each new $\mathbf{w}$ evaluate a new minimum of $\rho(\mathbf{v})$. We would find that the maximum value of all the minimum values evaluated is λ_2. This result can be generalized to the following principle, called the *minimax characterization of eigenvalues*,

$$\lambda_r = \max\left\{\min \frac{\mathbf{v}^T\mathbf{A}\mathbf{v}}{\mathbf{v}^T\mathbf{v}}\right\} \qquad r = 1, \ldots, n \tag{2.107}$$

with $\mathbf{v}$ satisfying $\mathbf{v}^T\mathbf{w}_i = 0$ for $i = 1, \ldots, r-1$. In (2.107) we choose vectors $\mathbf{w}_i$, $i = 1, \ldots, r-1$, and then evaluate the minimum of $\rho(\mathbf{v})$ with $\mathbf{v}$ subject to the condition $\mathbf{v}^T\mathbf{w}_i = 0$, $i = 1, \ldots, r-1$. After having calculated this minimum we vary the vectors $\mathbf{w}_i$ and always evaluate a new minimum. The maximum value that the minima reach is λ_r.

The proof of (2.107) is as follows. Let

$$\mathbf{v} = \sum_{i=1}^{n} \alpha_i \mathbf{v}_i \tag{2.108}$$

and evaluate the right-hand side of (2.107), which we call R,

$$R = \max \left\{ \min \frac{\alpha_1^2 \lambda_1 + \ldots + \alpha_r^2 \lambda_r + \alpha_{r+1}^2 \lambda_{r+1} + \ldots + \alpha_n^2 \lambda_n}{\alpha_1^2 + \ldots + \alpha_r^2 + \alpha_{r+1}^2 + \ldots + \alpha_n^2} \right\} \tag{2.109}$$

The coefficients α_i need to satisfy the conditions

$$\mathbf{w}_j^T \sum_{i=1}^{n} \alpha_i \mathbf{v}_i = 0 \qquad j = 1, \ldots, r-1 \tag{2.110}$$

Rewriting (2.109) we obtain

$$R = \max \left\{ \min \left[\lambda_r - \frac{\begin{array}{c}\alpha_1^2(\lambda_r - \lambda_1) + \ldots + \alpha_{r-1}^2(\lambda_r - \lambda_{r-1}) \\ + \alpha_{r+1}^2(\lambda_r - \lambda_{r+1}) + \ldots + \alpha_n^2(\lambda_r - \lambda_n)\end{array}}{\alpha_1^2 + \ldots + \alpha_r^2 + \alpha_{r+1}^2 + \ldots + \alpha_n^2} \right] \right\} \tag{2.111}$$

But we can now see that for the condition $\alpha_{r+1} = \alpha_{r+2} = \ldots = \alpha_n = 0$, we have

$$R \leq \lambda_r \tag{2.112}$$

and the condition in (2.110) can still be satisfied by a judicious choice for α_r. On the other hand, suppose that we now choose $\mathbf{w}_j = \mathbf{v}_j$ for $j = 1, \ldots, r-1$. This would require that $\alpha_j = 0$ for $j = 1, \ldots, r-1$, and consequently we would have that $R = \lambda_r$, which completes the proof.

A most important property that can be established using the minimax characterization of eigenvalues is the *eigenvalue separation property*. Suppose that in addition to the problem $\mathbf{A}\mathbf{v} = \lambda\mathbf{v}$, we consider the problems

$$\mathbf{A}^{(m)}\mathbf{v}^{(m)} = \lambda^{(m)}\mathbf{v}^{(m)} \tag{2.113}$$

where $\mathbf{A}^{(m)}$ is obtained by omitting the last m rows and columns of $\mathbf{A}$. Hence $\mathbf{A}^{(m)}$ is a square-symmetric matrix of order $(n - m)$. Using also the notation $\mathbf{A}^{(0)} = \mathbf{A}$, $\lambda^{(0)} = \lambda$, $\mathbf{v}^{(0)} = \mathbf{v}$, the eigenvalue separation property states that the eigenvalues of the problem $\mathbf{A}^{(m+1)}\mathbf{v}^{(m+1)} = \lambda^{(m+1)}\mathbf{v}^{(m+1)}$ separate the eigenvalues of the problem in (2.113); i.e., we have

$$\lambda_1^{(m)} \leq \lambda_1^{(m+1)} \leq \lambda_2^{(m)} \leq \lambda_2^{(m+1)} \ldots \leq \lambda_{n-m-1}^{(m)} \leq \lambda_{n-m-1}^{(m+1)} \leq \lambda_{n-m}^{(m)} \qquad \text{for } m = 0, \ldots, n-2 \tag{2.114}$$

For the proof of (2.114) we consider the problems $\mathbf{A}\mathbf{v} = \lambda\mathbf{v}$ and $\mathbf{A}^{(1)}\mathbf{v}^{(1)} = \lambda^{(1)}\mathbf{v}^{(1)}$. If we can show that the eigenvalue separation property holds for these two problems, it will hold also for $m = 1, 2, \ldots, n-2$. Specifically, we therefore want to prove that

$$\lambda_r \leq \lambda_r^{(1)} \leq \lambda_{r+1}; \qquad r = 1, \ldots, n-1 \tag{2.115}$$

Using the minimax characterization, we have that

$$\left.\begin{array}{l} \lambda_{r+1} = \max \left\{ \min \dfrac{\mathbf{v}^T\mathbf{A}\mathbf{v}}{\mathbf{v}^T\mathbf{v}} \right\} \\ \mathbf{v}^T\mathbf{w}_i = 0; \qquad i = 1, \ldots, r; \text{ all } \mathbf{w}_i \text{ arbitrary} \end{array}\right\} \tag{2.116}$$

Similarly, we have that

$$\left.\begin{array}{l} \lambda_r^{(1)} = \max \left\{ \min \dfrac{\mathbf{v}^T\mathbf{A}\mathbf{v}}{\mathbf{v}^T\mathbf{v}} \right\} \\ \mathbf{v}^T\mathbf{w}_i = 0; \qquad i = 1, \ldots, r \\ \mathbf{w}_i \text{ arbitrary for } i = 1, \ldots, r-1 \\ \mathbf{w}_r = \mathbf{e}_n \end{array}\right\} \tag{2.117}$$

where $\mathbf{w}_r$ is constrained to be equal to $\mathbf{e}_n$ to ensure that the last element in $\mathbf{v}$ is zero, because $\mathbf{e}_n$ is the last column of the $n \times n$ identity matrix $\mathbf{I}$. However, since the constraint for λ_{r+1} can be more severe and includes that for $\lambda_r^{(1)}$, we have that

$$\lambda_r^{(1)} \leq \lambda_{r+1} \tag{2.118}$$

To determine λ_r we use

$$\left.\begin{aligned} &\lambda_r = \max\left\{\min \frac{\mathbf{v}^T\mathbf{A}\mathbf{v}}{\mathbf{v}^T\mathbf{v}}\right\} \\ &\mathbf{v}^T\mathbf{w}_i = 0; \qquad i = 1, \ldots, r-1 \\ &\text{all } \mathbf{w}_i \text{ arbitrary} \end{aligned}\right\} \tag{2.119}$$

Comparing the characterizations of $\lambda_r^{(1)}$ and λ_r, i.e., (2.117) with (2.119), we observe that to calculate $\lambda_r^{(1)}$ we have the same constraints as in the calculation of λ_r plus one more (i.e., $\mathbf{v}^T\mathbf{e}_n = 0$), and hence

$$\lambda_r \leq \lambda_r^{(1)} \tag{2.120}$$

But (2.118) and (2.120) together establish the required result given in (2.115).

The eigenvalue separation property now yields the following result. If we write the eigenvalue problems in (2.113) including the problem $\mathbf{A}\mathbf{v} = \lambda\mathbf{v}$ in the form

$$p^{(m)}(\lambda^{(m)}) = \det(\mathbf{A}^{(m)} - \lambda^{(m)}\mathbf{I}); \qquad m = 0, \ldots, n-1 \tag{2.121}$$

where $p^{(0)} = p$, we see that the roots of the polynomials $p(\lambda^{(m+1)})$ separate the roots of the polynomial $p(\lambda^{(m)})$. However, a sequence of polynomials $p_i(x)$, $i = 1, \ldots, q$, form a Sturm sequence if the roots of the polynomial $p_{j+1}(x)$ separate the roots of the polynomial $p_j(x)$. *Hence the eigenvalue separation property states that the characteristic polynomials of the problems* $\mathbf{A}^{(m)}\mathbf{v}^{(m)} = \lambda^{(m)}\mathbf{v}^{(m)}$, $n = 0, 1, \ldots, n-1$, *form a Sturm sequence*. It should be noted that in the presentation we considered all symmetric matrices; i.e., the minimax characterization of eigenvalues and the Sturm sequence property are applicable to positive definite and indefinite matrices. We shall use the Sturm sequence property extensively in later chapters (see Sections 8.2.5, 10.2.2, 11.5, 12.2, and 12.3.4). Consider the following example.

EXAMPLE 2.27: Consider the eigenvalue problem $\mathbf{A}\mathbf{v} = \lambda\mathbf{v}$, where

$$\mathbf{A} = \begin{bmatrix} 5 & -4 & -7 \\ -4 & 2 & -4 \\ -7 & -4 & 5 \end{bmatrix}$$

Evaluate the eigenvalues of $\mathbf{A}$ and of the matrixes $\mathbf{A}^{(m)}$, $m = 1, 2$. Show that the separation property given in (2.114) holds and sketch the characteristic polynomials $p(\lambda)$, $p^{(1)}(\lambda^{(1)})$, and $p^{(2)}(\lambda^{(2)})$.

We have

$$\begin{aligned} p(\lambda) = \det(\mathbf{A} - \lambda\mathbf{I}) &= (5 - \lambda)[(2 - \lambda)(5 - \lambda) - 16] \\ &\quad + 4[-4(5 - \lambda) - 28] - 7[16 + 7(2 - \lambda)] \end{aligned}$$

Hence

$$p(\lambda) = (-6 - \lambda)(6 - \lambda)(12 - \lambda)$$

and the eigenvalues are

$$\lambda_1 = -6, \qquad \lambda_2 = 6, \qquad \lambda_3 = 12$$

Also, $$p^{(1)}(\lambda^{(1)}) = \det(\mathbf{A}^{(1)} - \lambda^{(1)}\mathbf{I}) = (5 - \lambda^{(1)})(2 - \lambda^{(1)}) - 16$$

or $$p^{(1)}(\lambda^{(1)}) = \lambda^{(1)^2} - 7\lambda^{(1)} - 6$$

Hence $$\lambda_1^{(1)} = \tfrac{7}{2} - \tfrac{1}{2}\sqrt{73} = -0.7720$$
$$\lambda_2^{(1)} = \tfrac{7}{2} + \tfrac{1}{2}\sqrt{73} = 7.772$$

Finally, $$p^{(2)}(\lambda^{(2)}) = \det(\mathbf{A}^{(2)} - \lambda^{(2)}\mathbf{I}) = 5 - \lambda^{(2)}$$

Hence $$\lambda_1^{(2)} = 5$$

The separation property holds because

$$\lambda_1 \leq \lambda_1^{(1)} \leq \lambda_2 \leq \lambda_2^{(1)} \leq \lambda_3$$

and $$\lambda_1^{(1)} < \lambda_1^{(2)} < \lambda_2^{(1)}$$

The characteristic polynomials are sketched in Fig. 2.9.

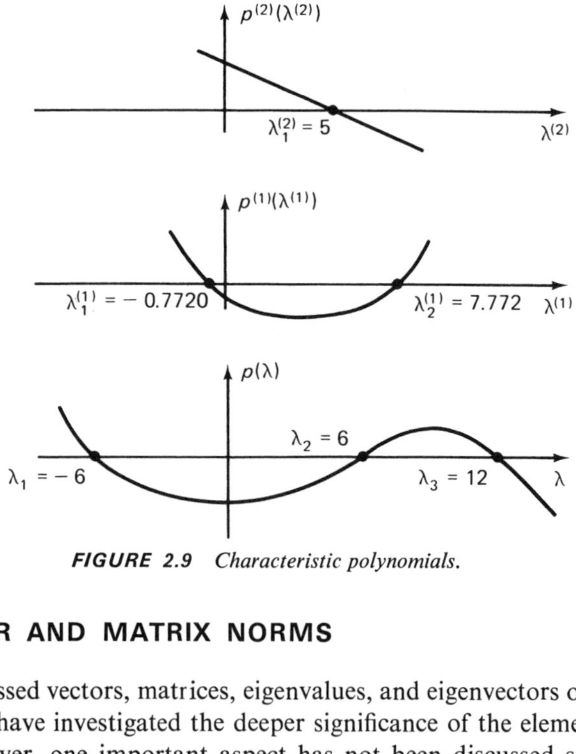

FIGURE 2.9 *Characteristic polynomials.*

2.9 VECTOR AND MATRIX NORMS

We have discussed vectors, matrices, eigenvalues, and eigenvectors of symmetric matrices, and have investigated the deeper significance of the elements in these entities. However, one important aspect has not been discussed so far. If we deal with single numbers, we can identify a number as being large or small. Vectors and matrices are functions of many elements, but we shall also need to measure their "size." Specifically, if single numbers are used in iterative processes, the convergence of a series of numbers, say $x_1, x_2, \ldots, x_k$, to a number x is simply measured by

$$\lim_{k \to \infty} |x_k - x| = 0 \qquad (2.122)$$

or, said in words, convergence is obtained if the residual $y_k = |x_k - x|$

approaches zero as $k \longrightarrow \infty$. Furthermore, if we can find constants $p \geq 1$ and $c > 0$ such that

$$\lim_{k \to \infty} \frac{|x_{k+1} - x|}{|x_k - x|^p} = c \tag{2.123}$$

we say that convergence is "of order p." If $p = 1$, convergence is linear and the rate of convergence is c, in which case c must be smaller than 1.

In iterative solution processes using vectors and matrices we also need a measure of convergence. Realizing that the "size" of a vector or matrix should depend on the magnitude of all elements in the arrays, we arrive at the definition of vector and matrix norms. A norm is a single number which depends on the magnitude of all elements in the vector or matrix.

Definition: *A* norm *of a vector* $\mathbf{v}$ *of order* n *written as* $\|\mathbf{v}\|$ *is a single number. The norm is a function of the elements of* $\mathbf{v}$ *and the following conditions are satisfied:*

1. $\|\mathbf{v}\| \geq 0$ *and* $\|\mathbf{v}\| = 0$ *if and only if* $\mathbf{v} = \mathbf{0}$. (2.124)
2. $\|c\mathbf{v}\| = |c|\,\|\mathbf{v}\|$ *for any scalar* c. (2.125)
3. $\|\mathbf{v} + \mathbf{w}\| \leq \|\mathbf{v}\| + \|\mathbf{w}\|$ *for vectors* $\mathbf{v}$ *and* $\mathbf{w}$. (2.126)

The relation (2.126) is the triangle inequality. The following three vector norms are commonly used, and called the infinity, one and two vector norms:

$$\|\mathbf{v}\|_\infty = \max_i |v_i| \tag{2.127}$$

$$\|\mathbf{v}\|_1 = \sum_{i=1}^{n} |v_i| \tag{2.128}$$

$$\|\mathbf{v}\|_2 = \left(\sum_{i=1}^{n} |v_i|^2\right)^{1/2} \tag{2.129}$$

where (2.129) is known as the *Euclidean vector norm*. Geometrically, this norm is equal to the length of the vector $\mathbf{v}$. All three norms are special cases of the vector norm $\sqrt[p]{\sum_i |v_i|^p}$, where for (2.127), (2.128), and (2.129), $p = \infty$, 1, and 2, respectively. It should be noted that each of the norms in (2.127) to (2.129) satisfies the conditions in (2.124) to (2.126).

We can now measure convergence of a sequence of vectors $\mathbf{x}_1, \mathbf{x}_2, \mathbf{x}_3, \ldots, \mathbf{x}_k$ to a vector $\mathbf{x}$. Namely, for the sequence to converge to $\mathbf{x}$ it is sufficient and necessary that

$$\lim_{k \to \infty} \|\mathbf{x}_k - \mathbf{x}\| = 0 \tag{2.130}$$

for any one of the vector norms. The order of convergence p, and in case $p = 1$, the rate of convergence c, are calculated in an analogous manner as in (2.123) but using norms; i.e., we consider

$$\lim_{k \to \infty} \frac{\|\mathbf{x}_{k+1} - \mathbf{x}\|}{\|\mathbf{x}_k - \mathbf{x}\|^p} = c \tag{2.131}$$

Considering the relationships among the various vector norms, we find that

$$\|\mathbf{v}\|_\infty \leq \|\mathbf{v}\|_1 \leq n\|\mathbf{v}\|_\infty; \qquad \|\mathbf{v}\|_\infty \leq \|\mathbf{v}\|_2 \leq \sqrt{n}\,\|\mathbf{v}\|_\infty \tag{2.132}$$

and can now note that although the actual vector length is calculated using the Euclidean norm, by virtue of relation (2.132), the 1 and ∞ norms also provide some measure of the length of the vector.

EXAMPLE 2.28: Calculate the 1, 2, and ∞ norms of the vector $\mathbf{x}$, and verify the relations in (2.132).

$$\mathbf{x} = \begin{bmatrix} 1 \\ -3 \\ 2 \end{bmatrix}$$

We have

$$\|\mathbf{x}\|_\infty = 3$$
$$\|\mathbf{x}\|_1 = 1 + 3 + 2 = 6$$
$$\|\mathbf{x}\|_2 = \sqrt{1 + 9 + 4} = \sqrt{14}$$

and the relations in (2.132) read

$$3 \le 6 \le (3)(3); \qquad 3 < \sqrt{14} \le (\sqrt{3})(3)$$

In analogy to the definition of a vector norm, we also define a matrix norm.

Definition: *A* norm *of a matrix* $\mathbf{A}$ *of order* $n \times n$, *written as* $\|\mathbf{A}\|$, *is a single number. The norm is a function of the elements of* $\mathbf{A}$, *and the following relations hold:*

1. $\|\mathbf{A}\| \ge 0$ *and* $\|\mathbf{A}\| = 0$ *if and only if* $\mathbf{A} = \mathbf{0}$. (2.133)
2. $\|c\mathbf{A}\| = |c|\,\|\mathbf{A}\|$ *for any scalar* c. (2.134)
3. $\|\mathbf{A} + \mathbf{B}\| \le \|\mathbf{A}\| + \|\mathbf{B}\|$ *for matrices* $\mathbf{A}$ *and* $\mathbf{B}$. (2.135)
4. $\|\mathbf{AB}\| \le \|\mathbf{A}\|\,\|\mathbf{B}\|$ *for matrices* $\mathbf{A}$ *and* $\mathbf{B}$. (2.136)

The relation in (2.135) is the triangle inequality equivalent to (2.126). The additional condition in (2.136), which was not postulated in the definition of a vector norm, must be satisfied, in order to be able to use matrix norms when matrix products occur.

The following are frequently used matrix norms:

$$\|\mathbf{A}\|_\infty = \max_i \sum_{j=1}^{n} |a_{ij}| \tag{2.137}$$

$$\|\mathbf{A}\|_1 = \max_j \sum_{i=1}^{n} |a_{ij}| \tag{2.138}$$

$$\|\mathbf{A}\|_2 = \sqrt{\tilde{\lambda}_n}; \qquad \tilde{\lambda}_n = \text{maximum eigenvalue of } \mathbf{A}^T\mathbf{A} \tag{2.139}$$

where for a symmetric matrix $\mathbf{A}$ we have $\|\mathbf{A}\|_\infty = \|\mathbf{A}\|_1$. The norm $\|\mathbf{A}\|_2$ is called the *spectral norm* of $\mathbf{A}$. Each of the norms given above satisfies the relations in (2.133) to (2.136). The proof that the relation in (2.136) is satisfied for the infinity norm is given in Example 2.30.

EXAMPLE 2.29: Calculate the ∞, 1, and 2 norms of the matrix $\mathbf{A}$, where $\mathbf{A}$ was given in Example 2.27.

The matrix $\mathbf{A}$ considered is

$$\mathbf{A} = \begin{bmatrix} 5 & -4 & -7 \\ -4 & 2 & -4 \\ -7 & -4 & 5 \end{bmatrix}$$

Using the definitions given in (2.137) to (2.139) we have

$$\|\mathbf{A}\|_\infty = 5 + 4 + 7 = 16$$
$$\|\mathbf{A}\|_1 = 5 + 4 + 7 = 16$$

To evaluate $\|\mathbf{A}\|_2$ we first need to calculate $\mathbf{A}^T\mathbf{A}$.

$$\mathbf{A}^T\mathbf{A} = 18\begin{bmatrix} 5 & 0 & -3 \\ 0 & 2 & 0 \\ -3 & 0 & 5 \end{bmatrix}$$

The eigenvalues of $\mathbf{A}^T\mathbf{A}$ are

$$\lambda_1 = 36, \qquad \lambda_2 = 36, \qquad \lambda_3 = 144$$

Hence we have

$$\|\mathbf{A}\|_2 = \sqrt{144} = 12$$

EXAMPLE 2.30: Show that for two matrices **A** and **B**, we have

$$\|\mathbf{AB}\|_\infty \leq \|\mathbf{A}\|_\infty \|\mathbf{B}\|_\infty$$

Using the definition of the ∞ matrix norm in (2.137), we have

$$\|\mathbf{AB}\|_\infty = \max_i \sum_{j=1}^{n} \left| \sum_{k=1}^{n} a_{ik} b_{kj} \right|$$

but then

$$\|\mathbf{AB}\|_\infty \leq \max_i \sum_{j=1}^{n} \sum_{k=1}^{n} |a_{ik}||b_{kj}|$$

$$= \max_i \sum_{k=1}^{n} \left\{ |a_{ik}| \sum_{j=1}^{n} |b_{kj}| \right\}$$

$$\leq \left\{ \max_i \sum_{k=1}^{n} |a_{ik}| \right\} \left\{ \max_k \sum_{j=1}^{n} |b_{kj}| \right\}$$

This proves the desired result.

As in the case of a sequence of vectors, we can now measure the convergence of a sequence of matrices $\mathbf{A}_1, \mathbf{A}_2, \mathbf{A}_3, \ldots, \mathbf{A}_k$ to a matrix **A**. For example, for convergence it is sufficient and necessary that

$$\lim_{k \to \infty} \|\mathbf{A}_k - \mathbf{A}\| = 0 \tag{2.140}$$

for any one of the given matrix norms.

In the definition of a matrix norm we needed relation (2.136) to be able to use norms when we encounter matrix products. Similarly, we also want to use norms when products of matrices with vectors occur. In such a case, in order to obtain useful information by applying norms, we need to employ only specific vector norms with specific matrix norms. Which matrix and vector norms should only be used together is determined by the condition that the following relation hold for any matrix **A** and vector **v**:

$$\|\mathbf{Av}\| \leq \|\mathbf{A}\| \|\mathbf{v}\| \tag{2.141}$$

where $\|\mathbf{Av}\|$ and $\|\mathbf{v}\|$ are evaluated using the vector norm and $\|\mathbf{A}\|$ is evaluated using the matrix norm. We may note the close relationship to the condition (2.136), which was required to hold for a matrix norm. If (2.141) holds for a specific vector and matrix norm, the two norms are said to be compatible, and the matrix norm is said to be subordinate to the vector norm. The 1, 2, and ∞ norms of a matrix, as defined above, are subordinate, respectively, to the 1, 2, and ∞ vector norms given in (2.127) to (2.129). In the following example we give the proof that the ∞ norms are compatible and subordinate. The compatibility of the vector and matrix 1 and 2 norms is proved similarly.

EXAMPLE 2.31: Show that for a matrix $\mathbf{A}$ and vector $\mathbf{v}$, we have

$$\|\mathbf{A}\mathbf{v}\|_\infty \leq \|\mathbf{A}\|_\infty \|\mathbf{v}\|_\infty \tag{a}$$

Using the definitions of the infinity norms, we have

$$\begin{aligned}\|\mathbf{A}\mathbf{v}\|_\infty &= \max_i \left| \sum_{j=1}^{n} a_{ij} v_j \right| \\ &\leq \max_i \sum_{j=1}^{n} |a_{ij}||v_j| \\ &\leq \left\{\max_i \sum_{j=1}^{n} |a_{ij}|\right\}\left\{\max_j |v_j|\right\}\end{aligned}$$

This proves (a).

To show that the equality sign can be reached, we only need to consider the case where $\mathbf{v}$ is a full unit vector and $a_{ij} \geq 0$. In this case, $\|\mathbf{v}\|_\infty = 1$ and $\|\mathbf{A}\mathbf{v}\|_\infty = \|\mathbf{A}\|_\infty$.

In later chapters we shall encounter various applications of norms. One valuable application arises in the calculation of eigenvalues of a matrix. Namely, if we consider the problem $\mathbf{A}\mathbf{v} = \lambda\mathbf{v}$, we obtain, taking norms on both sides,

$$\|\mathbf{A}\mathbf{v}\| = \|\lambda\mathbf{v}\| \tag{2.142}$$

and hence using (2.125) and (2.141), we have

$$\|\mathbf{A}\|\|\mathbf{v}\| \geq |\lambda|\|\mathbf{v}\| \tag{2.143}$$

or

$$|\lambda| \leq \|\mathbf{A}\| \tag{2.144}$$

Therefore, every eigenvalue of $\mathbf{A}$ is in absolute magnitude smaller than or equal to any norm of $\mathbf{A}$. Defining the *spectral radius* $\rho(\mathbf{A})$ as

$$\rho(\mathbf{A}) = \max_i |\lambda_i| \tag{2.145}$$

we have that

$$\rho(\mathbf{A}) \leq \|\mathbf{A}\| \tag{2.146}$$

In practice, the ∞ norm of $\mathbf{A}$ is calculated most conveniently and thus used effectively to estimate an upper bound on the largest absolute value reached by the eigenvalues.

EXAMPLE 2.32: Calculate the spectral radius of the matrix $\mathbf{A}$ considered in Example 2.27. Then show that $\rho(\mathbf{A}) \leq \|\mathbf{A}\|_\infty$.

The spectral radius is equal to $\max |\lambda_i|$. The eigenvalues of $\mathbf{A}$ have been calculated in Example 2.27.

$$\lambda_1 = -6; \qquad \lambda_2 = 6; \qquad \lambda_3 = 12$$

Hence

$$\rho(\mathbf{A}) = 12$$

We calculated in Example 2.29 $\|\mathbf{A}\|_\infty = 16$. Thus the relation $\rho(\mathbf{A}) \leq \|\mathbf{A}\|_\infty$ is satisfied.

REFERENCES

1. K. Hoffman and R. Kunze, *Linear Algebra*, Prentice-Hall, Inc., Englewood Cliffs, N.J., 1961.
2. B. Noble, *Applied Linear Algebra*, Prentice-Hall, Inc., Englewood Cliffs, N.J., 1969.

3. C. E. Fröberg, *Introduction to Numerical Analysis*, Addison-Wesley Publishing Company, Inc., Reading, Mass., 1969.
4. R. Zurmühl, *Matrizen*, Springer-Verlag, Berlin, 1964.
5. J. H. Wilkinson, *The Algebraic Eigenvalue Problem*, Clarendon Press, Oxford, 1965.

PART II

THE FINITE ELEMENT METHOD

3

SOME BASIC CONCEPTS OF ENGINEERING ANALYSIS

3.1 INTRODUCTION

The analysis of an engineering system requires the idealization of the system into a form that can be analyzed, the formulation of the governing equilibrium equations of this idealized system, the solution of the equilibrium equations, and the interpretation of the results. The main objective in this chapter is to discuss classical techniques used for the formulation of the governing equilibrium equations of engineering systems. This discussion should provide a valuable basis for the presentation of finite element procedures in the next chapters. Two categories of systems are considered: discrete and continuous systems.

In the analysis of a discrete system, the actual system response can directly be described by the solution of a finite number of state variables. In this chapter we discuss some general procedures that are employed to obtain the governing equilibrium equations of discrete systems. We consider steady-state, propagation, and eigenvalue problems, and also briefly discuss the nature of the solutions of these problems.

In the analysis of continuous systems, the formulation of the exact equilibrium equations is achieved as in discrete system analysis, but instead of a set of algebraic equations for the unknown state variables, differential equations govern the continuous system response. The exact solution of the differential equations satisfying all boundary conditions is only possible for relatively simple systems, and numerical procedures must in general be employed to predict the system response. These procedures, in essence, reduce the continuous system to a discrete idealization that can be analyzed in the same manner as a discrete physical system. In this chapter we summarize some important classical procedures that are employed to reduce continuous systems to discrete numerical systems. In the next chapters, we then show how these classical procedures provide the basis for modern finite element methods.

In practice, a most important preliminary decision to be made by the analyst is whether an engineering system should be treated as discrete or continuous.

Furthermore, if the system is to be analyzed as a continuous system, the analyst has to decide how the system is best reduced to an appropriate numerical (i.e., discrete) system. It is here that much of the value of finite element procedures can be found; namely, finite element techniques in conjunction with the electronic digital computer have enabled the numerical idealization and solution of continuous systems in a systematic manner, and in effect have made possible the practical extension and application of the classical procedures presented in this chapter to very complex engineering systems.

3.2 ANALYSIS OF DISCRETE SYSTEMS

In this section we deal with *discrete or lumped-parameter* systems, i.e., systems with a finite number of degrees of freedom. The essence of a lumped-parameter system is that the state of the system can be described directly with adequate precision by the magnitudes of a finite number of state variables. The analysis of a discrete system requires the following solution steps:

1. *system idealization:* the actual system is idealized as an assemblage of elements,
2. *element equilibrium:* the equilibrium requirements of each element are established in terms of state variables,
3. *element assemblage:* the element interconnection requirements are invoked to establish a set of simultaneous equations for the unknown state variables, and
4. *solution of response:* the simultaneous equations are solved for the state variables and using the element equilibrium requirements the response of each element can be calculated.

The above steps of solution are followed in the analyses of the different types of problems that we cônsider: steady-state problems, propagation problems, and eigenvalue problems. The objective in this section is to give an introduction on how problems in these particular areas are analyzed and to briefly discuss the nature of the solutions. It should be realized that not all types of problems in engineering are considered; however, a large majority of problems do fall naturally into these problem areas.[1] In the examples of this section we consider structural, electrical, fluid flow and heat transfer problems, and we emphasize that in each of these analyses the same basic steps of solution given above are followed.

3.2.1 Steady-State Problems

The main characteristic of a steady-state problem is that the response of the system does not change with time. Thus, the state variables describing the response of the system under consideration can be obtained from the solution of a set of equations that do not involve time as a variable. In the following examples we illustrate the procedure of analysis in the solution of some problems. Five sample problems are presented:

1. Elastic spring system
2. Heat transfer system
3. Hydraulic network
4. D-C network
5. Nonlinear elastic spring system

The analysis of each problem illustrates the application of the general steps of analysis summarized in Section 3.2. The first four problems involve the analysis of linear systems, whereas the nonlinear elastic spring system responds nonlinearly to the applied loads. All the problems are well defined and a unique solution exists for each system response.

EXAMPLE 3.1: Figure 3.1 shows a system of three rigid carts on a horizontal plane that are interconnected by a system of linear elastic springs. Calculate the displacements of the carts and the forces in the springs for the loading shown.

We perform the analysis by following steps (1) to (4) of Section 3.2. As state variables that characterize the response of the system, we choose the displacements U_1, U_2, and U_3. These displacements are measured from the initial positions of the carts, in which the springs are unstretched. The individual spring elements and their equilibrium requirements are shown in Fig. 3.1(b).

To generate the governing equations for the state variables we invoke the element interconnection requirements, which correspond to the static equilibrium of the three carts:

$$\begin{aligned} F_1^{(1)} + F_1^{(2)} + F_1^{(3)} + F_1^{(4)} &= R_1 \\ F_2^{(2)} + F_2^{(3)} + F_2^{(5)} &= R_2 \\ F_3^{(4)} + F_3^{(5)} &= R_3 \end{aligned} \tag{a}$$

We can now substitute for the element end forces $F_i^{(j)}$; $i = 1, 2, 3$; $j = 1, \ldots, 5$ using the element equilibrium requirements given in Fig. 3.1(b). Here we recognize that corresponding to the displacement components U_1, U_2, and U_3 we can write for element 1,

$$\begin{bmatrix} k_1 & 0 & 0 \\ 0 & 0 & 0 \\ 0 & 0 & 0 \end{bmatrix} \begin{bmatrix} U_1 \\ U_2 \\ U_3 \end{bmatrix} = \begin{bmatrix} F_1^{(1)} \\ 0 \\ 0 \end{bmatrix}$$

or

$$\mathbf{K}^{(1)}\mathbf{U} = \mathbf{F}^{(1)}$$

for element 2,

$$\begin{bmatrix} k_2 & -k_2 & 0 \\ -k_2 & k_2 & 0 \\ 0 & 0 & 0 \end{bmatrix} \begin{bmatrix} U_1 \\ U_2 \\ U_3 \end{bmatrix} = \begin{bmatrix} F_1^{(2)} \\ F_2^{(2)} \\ 0 \end{bmatrix}$$

or $\mathbf{K}^{(2)}\mathbf{U} = \mathbf{F}^{(2)}$ and so on. Hence, the element interconnection requirements in (a) reduce to

$$\mathbf{K}\,\mathbf{U} = \mathbf{R} \tag{b}$$

where

$$\mathbf{U}^T = [U_1 \quad U_2 \quad U_3]$$

$$\mathbf{K} = \begin{bmatrix} (k_1 + k_2 + k_3 + k_4) & -(k_2 + k_3) & -k_4 \\ -(k_2 + k_3) & (k_2 + k_3 + k_5) & -k_5 \\ -k_4 & -k_5 & (k_4 + k_5) \end{bmatrix}$$

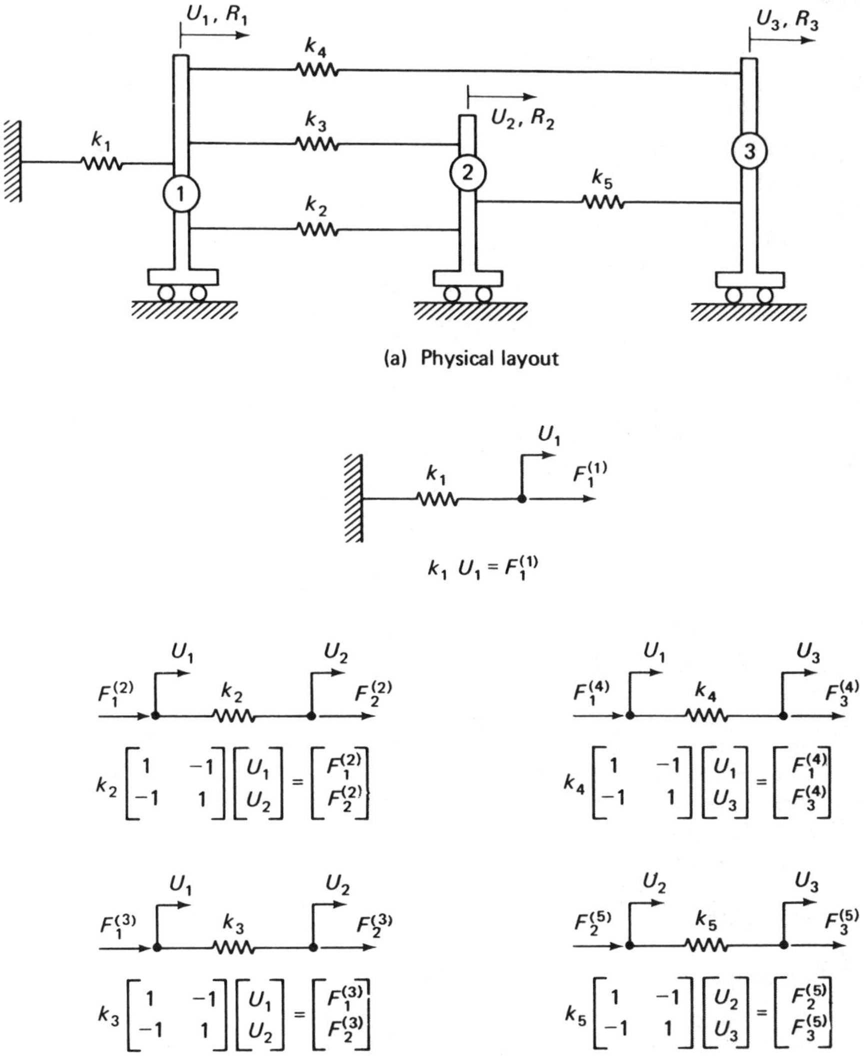

(a) Physical layout

(b) Element equilibrium relations

FIGURE 3.1 System of rigid carts interconnected by linear springs.

and
$$\mathbf{R}^T = [R_1 \quad R_2 \quad R_3]$$

Here it is noted that the coefficient matrix $\mathbf{K}$ can be obtained using

$$\mathbf{K} = \sum_{i=1}^{5} \mathbf{K}^{(i)} \tag{c}$$

where the $\mathbf{K}^{(i)}$ are the element stiffness matrices. The summation process to obtain in (c) the total structure stiffness matrix by direct summation of the element stiffness matrices is referred to as the *direct stiffness method.*

The analysis of the system is completed by solving (b) for the state variables U_1, U_2, and U_3, and then calculating the element forces from the element equilibrium relationships in Fig. 3.1.

EXAMPLE 3.2: A wall is constructed of two homogeneous slabs in contact as shown in Fig. 3.2. In steady-state conditions the temperatures in the wall are characterized by the external surface temperatures θ_1 and θ_3 and the interface temperature θ_2. Establish the equilibrium equations of the problem in terms of these temperatures, when the ambient temperatures θ_0 and θ_4 are known.

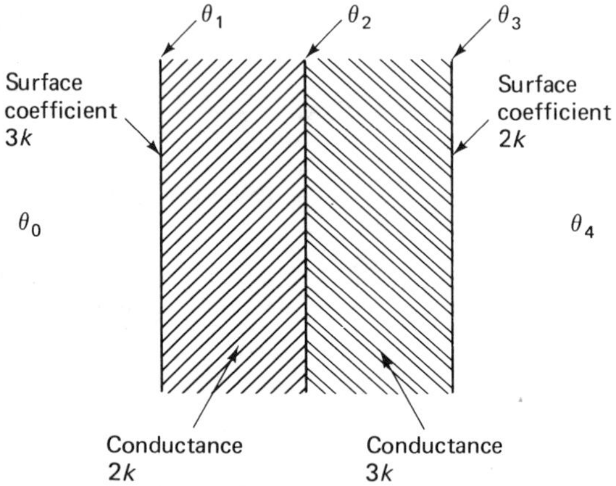

FIGURE 3.2 *Slab subjected to temperature boundary conditions.*

The conductance per unit area for the individual slabs and the surface coefficients are given in Fig. 3.2. The heat conduction law is $q/A = \bar{k}\,\Delta\theta$, where q is the total heat flow, A is the area, $\Delta\theta$ is the temperature drop in the direction of heat flow, and $\bar{k}$ is the conductance or surface coefficient. The state variables are in this analysis θ_1, θ_2, and θ_3. Using the heat conduction law, the element equilibrium equations are:

for the left surface, per unit area,

$$q_1 = 3k(\theta_0 - \theta_1)$$

for the left slab $\quad q_2 = 2k(\theta_1 - \theta_2)$

for the right slab $\quad q_3 = 3k(\theta_2 - \theta_3)$

for the right surface $\quad q_4 = 2k(\theta_3 - \theta_4)$

To obtain the governing equations for the state variables, we invoke the heat flow equilibrium requirement $q_1 = q_2 = q_3 = q_4$. Thus

$$\begin{aligned} 3k(\theta_0 - \theta_1) &= 2k(\theta_1 - \theta_2) \\ 2k(\theta_1 - \theta_2) &= 3k(\theta_2 - \theta_3) \\ 3k(\theta_2 - \theta_3) &= 2k(\theta_3 - \theta_4) \end{aligned}$$

Writing these equations in matrix form we obtain

$$\begin{bmatrix} 5k & -2k & 0 \\ -2k & 5k & -3k \\ 0 & -3k & 5k \end{bmatrix} \begin{bmatrix} \theta_1 \\ \theta_2 \\ \theta_3 \end{bmatrix} = \begin{bmatrix} 3k\theta_0 \\ 0 \\ 2k\theta_4 \end{bmatrix} \tag{a}$$

The above equilibrium equations can be also derived in a systematic manner using a direct stiffness procedure. Using this technique we proceed as in Example 3.1 with the typical element equilibrium relations

$$\bar{k}\begin{bmatrix} 1 & -1 \\ -1 & 1 \end{bmatrix}\begin{bmatrix} \theta_i \\ \theta_j \end{bmatrix} = \begin{bmatrix} q_i \\ q_j \end{bmatrix}$$

where q_i, q_j are the heat flows into the element, and θ_i, θ_j are the element-end temperatures. For the system in Fig. 3.2 we have two conduction elements, (each slab being one element), hence we obtain

$$\begin{bmatrix} 2k & -2k & 0 \\ -2k & 5k & -3k \\ 0 & -3k & 3k \end{bmatrix}\begin{bmatrix} \theta_1 \\ \theta_2 \\ \theta_3 \end{bmatrix} = \begin{bmatrix} 3k\,(\theta_0 - \theta_1) \\ 0 \\ 2k\,(\theta_4 - \theta_3) \end{bmatrix} \tag{b}$$

Since θ_1 and θ_3 are unknown, the equilibrium relations in (b) are rearranged for solution to obtain the relations in (a).

It is interesting to note the analogy between the displacement and force analysis of the spring system in Example 3.1 and the temperature and heat transfer analysis of Example 3.2. The coefficient matrices are very similar in both analyses, and they can both be obtained in a very systematic manner. To emphasize the analogy we give in Fig. 3.3 a spring model that is governed by the coefficient matrix of the heat transfer problem.

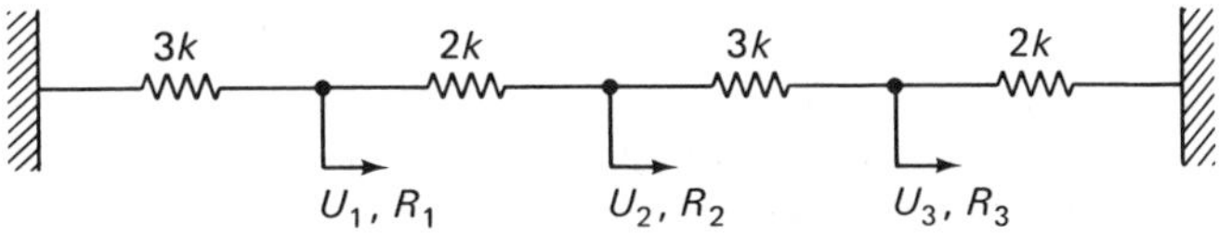

FIGURE 3.3 *Assemblage of springs governed by same coefficient matrix as heat transfer problem in Fig. 3.2.*

We next consider the analyses of a simple flow problem and a simple electrical system, both of which are again analyzed in much the same manner as the spring and heat transfer problems.

EXAMPLE 3.3: Establish the equations that govern the steady-state pressure and flow distributions in the hydraulic network shown in Fig. 3.4. Assume the fluid to be incompressible, and the pressure drop in a branch to be proportional to the flow q through that branch, $\Delta p = Rq$, where R is the branch resistance coefficient.

The elements are in this analysis the individual branches of the pipe network.

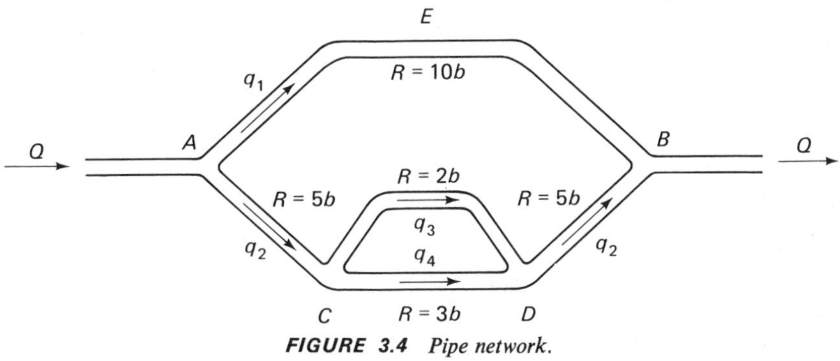

FIGURE 3.4 *Pipe network.*

As unknown state variables that characterize the flow and pressure distribution in the system we select the pressures at A, C and D, which we denote as p_A, p_C, and p_D, and we assume that the pressure at B is zero. Thus, we have for the elements

$$
\begin{aligned}
q_1 &= \frac{p_A}{10b} \qquad\qquad q_3 = \frac{p_C - p_D}{2b} \\
q_2|_{AC} &= \frac{p_A - p_C}{5b} \qquad q_2|_{DB} = \frac{p_D}{5b} \qquad q_4 = \frac{p_C - p_D}{3b}
\end{aligned}
\tag{a}
$$

The element interconnectivity requirements require continuity of flow, hence

$$
\begin{aligned}
Q &= q_1 + q_2 \\
q_2|_{AC} &= q_3 + q_4; \qquad q_2|_{DB} = q_3 + q_4
\end{aligned}
\tag{b}
$$

Substituting from (a) into (b) and writing the resulting equations in matrix form we obtain

$$
\begin{bmatrix} 3 & -2 & 0 \\ -6 & 31 & -25 \\ -1 & 1 & 1 \end{bmatrix}
\begin{bmatrix} p_A \\ p_C \\ p_D \end{bmatrix}
=
\begin{bmatrix} 10bQ \\ 0 \\ 0 \end{bmatrix}
$$

or

$$
\begin{bmatrix} 9 & -6 & 0 \\ -6 & 31 & -25 \\ 0 & -25 & 31 \end{bmatrix}
\begin{bmatrix} p_A \\ p_C \\ p_D \end{bmatrix}
=
\begin{bmatrix} 30bQ \\ 0 \\ 0 \end{bmatrix}
\tag{c}
$$

The analysis of the pipe network is completed by solving from (c) for the pressures p_A, p_C, and p_D and then the element equilibrium relations in (a) can be employed to obtain the flow distributions.

The equilibrium relations in (c) can also be derived—as in the preceding spring and heat transfer examples—using a direct stiffness procedure. Using this technique we proceed as in Example 3.1 with the typical element equilibrium relations

$$
\frac{1}{R}\begin{bmatrix} 1 & -1 \\ -1 & 1 \end{bmatrix}
\begin{bmatrix} p_i \\ p_j \end{bmatrix}
=
\begin{bmatrix} q_i \\ q_j \end{bmatrix}
$$

where q_i, q_j are the fluid flows into the element and p_i, p_j are the element-end pressures.

EXAMPLE 3.4: Consider the D-C network shown in Fig. 3.5. The network with the resistances shown is subjected to the constant voltage inputs E and $2E$ at A and B, respectively. We are to determine the steady-state current distributions in the network.

In this analysis we use as unknown state variables the currents I_1, I_2, and I_3. The system elements are the resistors and the element equilibrium requirements are obtained by applying Ohm's law to the resistors. Namely, for a resistor $\bar{R}$, carrying current I, we have Ohm's law

$$
\Delta E = \bar{R}I
$$

where ΔE is the voltage drop across the resistor.

The element interconnection law to be satisfied is Kirchhoff's voltage law for each closed loop in the network,

$$
\begin{aligned}
2E &= 2RI_1 + 2R(I_1 - I_3) \\
E &= 4R(I_2 - I_3) \\
0 &= 6RI_3 + 4R(I_3 - I_2) + 2R(I_3 - I_1)
\end{aligned}
$$

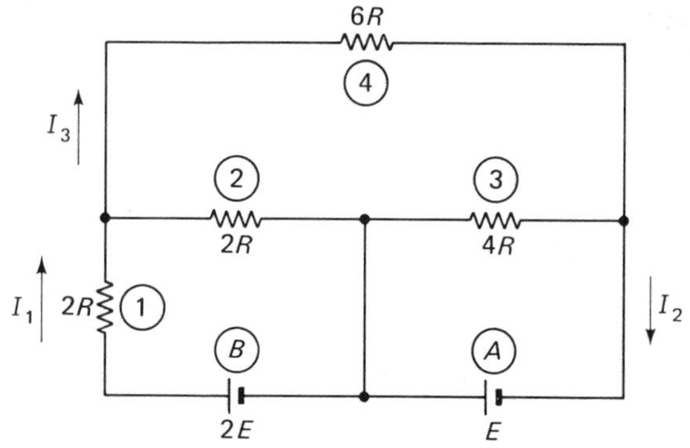

FIGURE 3.5 *D-C network.*

Writing these equations in matrix form, we obtain

$$\begin{bmatrix} 4R & 0 & -2R \\ 0 & 4R & -4R \\ -2R & -4R & 12R \end{bmatrix} \begin{bmatrix} I_1 \\ I_2 \\ I_3 \end{bmatrix} = \begin{bmatrix} 2E \\ E \\ 0 \end{bmatrix} \qquad \text{(a)}$$

The analysis is completed by solving these equations for I_1, I_2 and I_3. Note that the equilibrium equations in (a) could also have been established using a direct stiffness procedure, as in Examples 3.1, 3.2 and 3.3.

We should note once again that the steps of analysis in the above structural, heat transfer, fluid flow and electrical problems are very similar: the basic analogy being possibly best expressed in the use of the direct stiffness procedure for each problem. This indicates already that the same basic numerical procedures will be applicable in the analysis of almost any physical problem (see Chapters 4 and 7).

In each of the above examples we dealt with a linear system; i.e., the coefficient matrix is constant and thus, if the r.h.s. forcing functions are multiplied by a constant α, the system response is also α times as large. We consider in this chapter primarily linear systems, but the same steps for solution summarized above are also applicable in nonlinear analysis, as demonstrated in the following example (see also Chapters 6 and 7).

EXAMPLE 3.5: Consider the spring–cart system of Fig. 3.1, and assume that spring ① has now the nonlinear behavior shown in Fig. 3.6. Discuss how the equilibrium equations given in Example 3.1 have to be modified for this analysis.

As long as $U_1 \leq \Delta y$ the equilibrium equations of Example 3.1 are applicable with $k_1 = k$. However, if the loads are such that $U_1 > \Delta y$, i.e., $F_1^{(1)} > F_y$, we need to use a different value for k_1 and this value depends on the force $F_1^{(1)}$ acting in the element. Denoting the stiffness value by k_s as shown in Fig. 3.6, the response of the

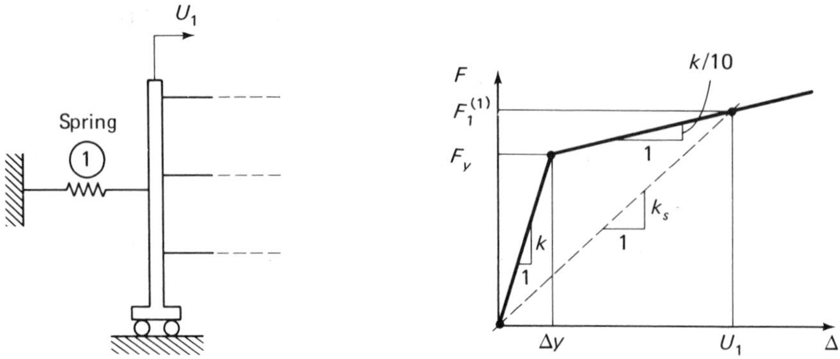

FIGURE 3.6 Spring ① of the cart-spring system of Fig. 3.1 with nonlinear elastic characteristics.

system is described for any load by the equilibrium equations

$$\mathbf{K}_s \mathbf{U} = \mathbf{R} \tag{a}$$

where the coefficient matrix is established exactly as in Example 3.1 but using k_s instead of k_1,

$$\mathbf{K}_s = \begin{bmatrix} (k_s + k_2 + k_3 + k_4) & -(k_2 + k_3) & -k_4 \\ -(k_2 + k_3) & (k_2 + k_3 + k_5) & -k_5 \\ -k_4 & -k_5 & (k_4 + k_5) \end{bmatrix} \tag{b}$$

Although the response of the system can be calculated using the above approach, in which $\mathbf{K}_s$ is referred to as the secant matrix, we will see in Chapter 6 that in general practical analysis we actually use an incremental procedure with a tangent stiffness matrix.

The above analyses demonstrate the general analysis procedure: the selection of unknown state variables that characterize the response of the system under consideration, the identification of elements that together comprise the complete system, the establishment of the element equilibrium requirements, and finally the assemblage of the elements by invoking interelement continuity requirements.

A few observations should be made. Firstly, we need to recognize that there is some choice in the selection of the state variables. For example, in the analysis of the carts of Example 3.1, we could have chosen the unknown forces in the springs as state variables. A second observation is that the equations from which the state variables are calculated can be linear or nonlinear equations and the coefficient matrix can be of general nature. However, it is most desirable to deal with a symmetric positive definite coefficient matrix, because in such cases the solution of the equations is numerically very effective (see Section 8.2).

In general, the physical characteristics of a problem determine whether the numerical solution can actually be cast into a form that leads to a symmetric positive definite coefficient matrix. However, even if possible, a positive definite coefficient matrix is only obtained provided appropriate solution variables are selected, and in a nonlinear analysis an appropriate linearization must be performed in the iterative solution. For this reason, in practice, it is valuable to employ general formulations for whole classes of problems (e.g., structural analysis, heat transfer, and so on—see Sections 4.2 and 7.2) that for any analysis lead to a symmetric and positive definite coefficient matrix.

In the above discussion we employed the direct approach of assembling the system governing equilibrium equations. An important point is that the governing equilibrium equations for the state variables can in some analyses also be obtained using an extremum, or variational formulation. *An extremum problem consists of locating the set (or sets) of values (state variables)* U_i, $i = 1, \ldots n$, *for which a given functional* $\Pi(U_1, \ldots, U_n)$ *is a maximum, minimum or has a saddle point.* The conditions to obtain the equations for the state variables must therefore be

$$\delta\Pi = 0 \quad \text{or} \quad \frac{\partial \Pi}{\partial U_i} = 0 \qquad i = 1, \ldots, n \tag{3.1}$$

where δ stands for "arbitrary variations in the state variables U_i satisfying the boundary conditions," and the second derivatives of Π with respect to the state variables decide whether the solution corresponds to a maximum, minimum, or saddle point. In the analysis of discrete systems we can consider that Π *is defined such that the relations in (3.1) generate the system governing equilibrium equations.** For example, in linear structural analysis, when displacements are used as state variables, Π is the total potential (or total potential energy)

$$\Pi = \mathcal{U} - \mathcal{W} \tag{3.2}$$

where $\mathcal{U}$ is the strain energy of the system and $\mathcal{W}$ is the total potential of the loads. The solution for the state variables corresponds in this case to the minimum of Π.

EXAMPLE 3.6: Consider a simple spring of stiffness k and applied load P to discuss the use of (3.1) and (3.2).

Let u be the displacement of the spring under the load P. We then have

$$\mathcal{U} = \tfrac{1}{2}ku^2; \qquad \mathcal{W} = Pu$$

and

$$\Pi = \tfrac{1}{2}ku^2 - Pu$$

Since u is the only variable, we have

$$\delta\Pi = (ku - P)\delta u$$

which gives the equilibrium equation

$$ku = P \tag{a}$$

Using (a) to evaluate $\mathcal{W}$, we have at equilibrium $\mathcal{W} = ku^2$, i.e., $\mathcal{W} = 2\mathcal{U}$, and $\Pi = -\tfrac{1}{2}ku^2 = -\tfrac{1}{2}Pu$.

EXAMPLE 3.7: Consider the analysis of the system of rigid carts of Example 3.1. Establish Π and invoke the condition in (3.1) to obtain the system of governing equilibrium equations.

Using the notation defined in Example 3.1 we have

$$\mathcal{U} = \tfrac{1}{2}\,\mathbf{U}^T\mathbf{K}\mathbf{U} \tag{a}$$

and

$$\mathcal{W} = \mathbf{U}^T\mathbf{R} \tag{b}$$

* This way we consider a specific variational formulation, as further discussed in Chapters 4 and 7.

where it should be noted that the total strain energy in (a) could also be written as

$$\begin{aligned}\mathcal{U} &= \tfrac{1}{2}\mathbf{U}^T\Big(\sum_{i=1}^{5}\mathbf{K}^{(i)}\Big)\mathbf{U} \\ &= \tfrac{1}{2}\mathbf{U}^T\mathbf{K}^{(1)}\mathbf{U} + \tfrac{1}{2}\mathbf{U}^T\mathbf{K}^{(2)}\mathbf{U} + \ldots . + \tfrac{1}{2}\mathbf{U}^T\mathbf{K}^{(5)}\mathbf{U} \\ &= \mathcal{U}_1 + \mathcal{U}_2 + \ldots + \mathcal{U}_5\end{aligned}$$

where $\mathcal{U}_i$ is the strain energy stored in the ith element.

Using (a) and (b) we now obtain

$$\Pi = \tfrac{1}{2}\mathbf{U}^T\mathbf{K}\mathbf{U} - \mathbf{U}^T\mathbf{R} \tag{c}$$

Applying (3.1) gives

$$\mathbf{K}\mathbf{U} = \mathbf{R}$$

Solving for $\mathbf{U}$ and then substituting into (c) we have that Π corresponding to the displacements at system equilibrium is

$$\Pi = -\tfrac{1}{2}\mathbf{U}^T\mathbf{R}$$

Since the same equilibrium equations are generated using the direct solution approach and the variational approach, we may ask what are the advantages of employing a variational scheme. Assume that for the problem under consideration Π is defined. The equilibrium equations can then be generated by simply adding the contributions from all elements to Π and invoking the stationarity condition in (3.1). In essence, this condition generates automatically the element interconnectivity requirements. Thus, the variational technique can be very effective because the system governing equilibrium equations can be generated "quite mechanically." The advantages of a variational solution are even more pronounced when we consider the numerical analysis of a continuous system (see Section 3.3.2). However, a main disadvantage of a variational solution is that, in general, less physical insight into a problem formulation is obtained than when using the direct approach. Therefore, it may be critical that we interpret physically the system equilibrium equations, once they have been established using a variational approach, in order to identify possible errors in the solution and in order to gain a better understanding of the physical meaning of the equations.

3.2.2 Propagation Problems

The main characteristic of a propagation or dynamic problem is that the response of the system under consideration changes with time. For the analysis of a system, in principle, the same procedures as in the analysis of a steady-state problem are employed, but now the state variables and element equilibrium relations depend on time. The objective of the analysis is to calculate the state variables for all time t.

Before discussing actual propagation problems, consider the case where the time effect on the element equilibrium relations is negligible, but let the load vector be a function of time. In this case the system response is obtained using the equations governing the steady-state response but substituting the time-dependent load or forcing vector for the load vector employed in the steady-state analysis. Since such an analysis is in essence still a steady-state analysis, but with steady-state conditions considered at any time t, the analysis may be referred to as a pseudo steady-state analysis.

In an actual propagation problem, the element equilibrium relations are time-dependent and this accounts for major differences in the response characteristics when compared to steady-state problems. In the following we present two examples that demonstrate the formulation of the governing equilibrium equations of propagation problems. Methods for calculating the solution of these equations are given in Chapter 9.

EXAMPLE 3.8: Consider the system of rigid carts that was analyzed in Example 3.1. Assume that the system loads are time-dependent and establish the equations that govern the dynamic response of the system.

For the analysis we assume that the springs are massless and the carts have masses m_1, m_2, and m_3 (which amounts to lumping the distributed mass of each spring to its two end points). Then, using the information given in Example 3.1 and invoking d'Alembert's principle, the element interconnectivity requirements yield the equations

$$F_1^{(1)} + F_1^{(2)} + F_1^{(3)} + F_1^{(4)} = R_1(t) - m_1\ddot{U}_1$$
$$F_2^{(2)} + F_2^{(3)} + F_2^{(5)} = R_2(t) - m_2\ddot{U}_2$$
$$F_3^{(4)} + F_3^{(5)} = R_3(t) - m_3\ddot{U}_3$$

where
$$\ddot{U}_i = \frac{d^2U_i}{dt^2}, \qquad i = 1, 2, 3$$

Thus we obtain as the system governing equilibrium equations

$$\mathbf{M}\ddot{\mathbf{U}} + \mathbf{K}\mathbf{U} = \mathbf{R}(t) \tag{a}$$

where **K**, **U**, and **R** have been defined in Example 3.1, and **M** is the system mass matrix

$$\mathbf{M} = \begin{bmatrix} m_1 & 0 & 0 \\ 0 & m_2 & 0 \\ 0 & 0 & m_3 \end{bmatrix}$$

The equilibrium equations in (a) represent a system of ordinary differential equations of second order in time. For the solution of these equations it is also necessary that the initial conditions on $\mathbf{U}$ and $\dot{\mathbf{U}}$ be given, i.e., we need to have ${}^0\mathbf{U}$ and ${}^0\dot{\mathbf{U}}$, where

$${}^0\mathbf{U} = \mathbf{U}|_{t=0}; \qquad {}^0\dot{\mathbf{U}} = \dot{\mathbf{U}}|_{t=0}$$

We mentioned above the case of a pseudo steady-state analysis. Considering the response of the carts, such analysis implies that the loads change very slowly and hence mass effects can be neglected. Therefore, to obtain the pseudo steady-state response, the equilibrium equations (a) in Example 3.8 would be solved with $\mathbf{M} = \mathbf{0}$.

EXAMPLE 3.9: Figure 3.7 shows an idealized case of the transient heat flow in an electron tube. A filament is heated to a temperature θ_f by an electric current; heat is convected from the filament to the surrounding gas and is radiated to the wall, which also receives heat by convection from the gas. The wall itself convects heat to the surrounding atmosphere, which is at temperature θ_a. It is required to formulate the system governing heat flow equilibrium equations.

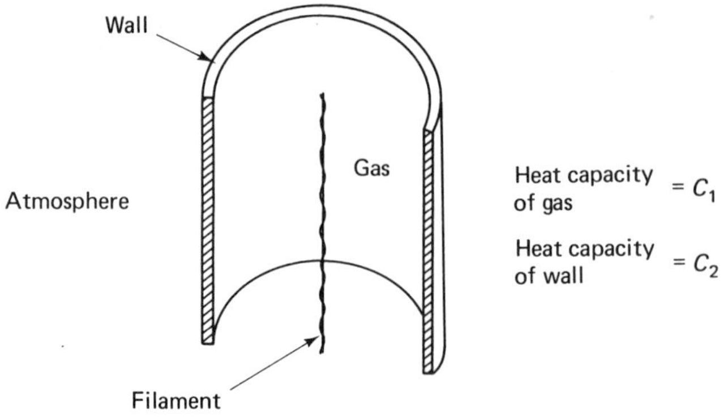

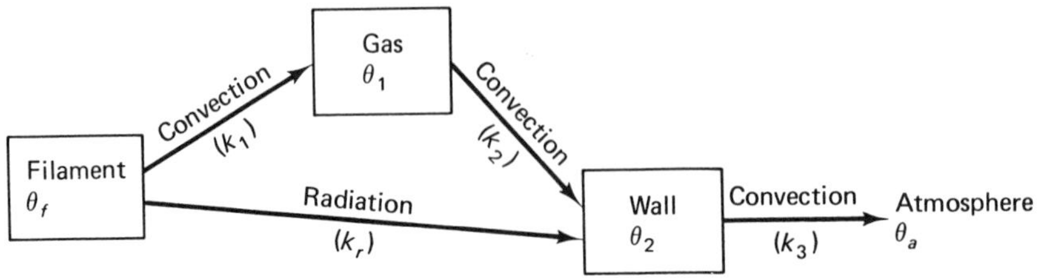

FIGURE 3.7 *Heat transfer idealization of an electron tube.*

In this analysis, we choose as unknown state variables the temperature of the gas, θ_1, and the temperature of the wall, θ_2. The system equilibrium equations are generated by invoking the heat flow equilibrium for the gas and the wall. Using the heat transfer coefficients given in Fig. 3.7, we obtain for the gas

$$C_1 \frac{d\theta_1}{dt} = k_1(\theta_f - \theta_1) - k_2(\theta_1 - \theta_2)$$

and for the wall

$$C_2 \frac{d\theta_2}{dt} = k_r(\theta_f^4 - \theta_2^4) + k_2(\theta_1 - \theta_2) - k_3(\theta_2 - \theta_a)$$

These two equations can be written in matrix form as

$$\mathbf{C}\dot{\boldsymbol{\theta}} + \mathbf{K}\boldsymbol{\theta} = \mathbf{Q} \tag{a}$$

where

$$\mathbf{C} = \begin{bmatrix} C_1 & 0 \\ 0 & C_2 \end{bmatrix}; \qquad \mathbf{K} = \begin{bmatrix} (k_1 + k_2) & -k_2 \\ -k_2 & (k_2 + k_3) \end{bmatrix}$$

$$\boldsymbol{\theta} = \begin{bmatrix} \theta_1 \\ \theta_2 \end{bmatrix}; \qquad \mathbf{Q} = \begin{bmatrix} k_1\theta_f \\ k_r(\theta_f^4 - \theta_2^4) + k_3\theta_a \end{bmatrix}$$

We note that due to the radiation boundary condition the heat flow equilibrium equations are nonlinear in $\boldsymbol{\theta}$. The radiation boundary condition term has here been incorporated in the heat flow load vector $\mathbf{Q}$. The solution of the equations can be carried out as described in Sections 7.2 and 9.6.

Although we considered in the above examples very specific cases, the examples illustrate in a quite general way how propagation problems of discrete systems are formulated for analysis. In essence, the same procedures are employed as in the analysis of steady-state problems, but "time-dependent loads" are generated that are a result of the "resistance to change" of the elements and thus of the complete system. This resistance to change or *inertia* of the system must be considered in a dynamic analysis.

Based on the above arguments and observations, it appears that we can conclude that the analysis of a propagation problem is a very simple extension of the analysis of the corresponding steady-state problem. However, we assumed in the above discussion that the discrete system is given and thus the degrees of freedom or state variables can directly be identified. In practice, the selection of an appropriate discrete system that contains all the important characteristics of the actual physical system is usually not straight-forward, and in general a different discrete model must be chosen for a dynamic response prediction than is chosen for the steady-state analysis. However, the above discussion illustrates that once the discrete model has been chosen for a propagation problem, the formulation of the governing equilibrium equations can proceed in much the same way as in the analysis of a steady-state response, except that inertia loads are generated that act on the system in addition to the externally applied loads (see Section 4.2.1). This observation leads us to anticipate that the procedures for solving the dynamic equilibrium equations of a system are largely based on the techniques that are employed for the solution of steady-state equilibrium equations (see Section 9.2).

3.2.3 Eigenvalue Problems

In our earlier discussion of steady-state and propagation problems we implied the existence of a unique solution for the response of the system. A main characteristic of an eigenvalue problem is that there is no unique solution to the response of the system, and the objective of the analysis is to calculate the various possible solutions. Eigenvalue problems arise in both steady-state and dynamic analyses.

Various different eigenvalue problems can be formulated in engineering analysis. In this book we are primarily concerned with the generalized eigenvalue problem of the form

$$\mathbf{A}\mathbf{v} = \lambda \mathbf{B}\mathbf{v} \tag{3.3}$$

where $\mathbf{A}$ and $\mathbf{B}$ are symmetric matrices, λ is a scalar and $\mathbf{v}$ is a vector. If λ_i and $\mathbf{v}_i$ satisfy (3.3), they are called an eigenvalue and eigenvector, respectively.

In steady-state analysis an eigenvalue problem of the form (3.3) is formulated when it is necessary to investigate the physical stability of the system under consideration. The question that is asked and leads to the eigenvalue problem is as follows: *Assuming that the steady-state solution of the system is known, is there another solution into which the system could bifurcate if it were slightly perturbed from its equilibrium position?* The answer to this question depends on the system under consideration and the loads acting onto the system. We consider a very simple example to demonstrate the basic idea.

EXAMPLE 3.10: Consider the simple cantilever shown in Fig. 3.8. The structure consists of a rotational spring and a rigid lever arm. Predict the response of the structure for the load applications shown in the figure.

We consider first only the steady-state response as discussed in Section 3.2.1. Since the bar is rigid, the cantilever is a single degree of freedom system and we employ as the state variable Δ_v.

In loading condition I, the bar is subjected to a longitudinal tensile force P, and the moment in the spring is zero. Since the bar is rigid, we have

$$\Delta_v = 0 \tag{a}$$

Next consider loading condition II. Assuming small displacements we have in this case

$$\Delta_v = \frac{PL^2}{k} \tag{b}$$

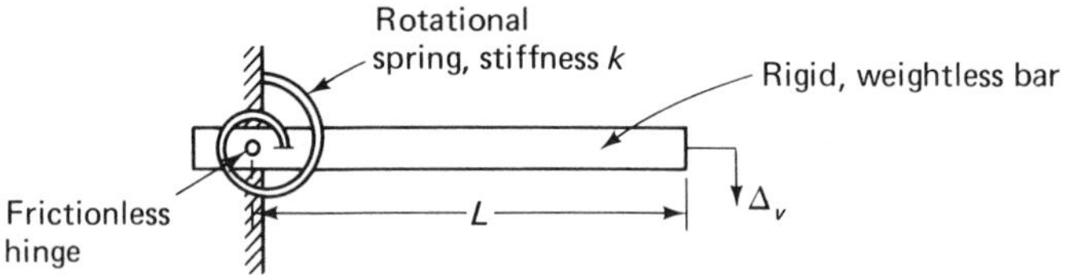

(a) Cantilever beam

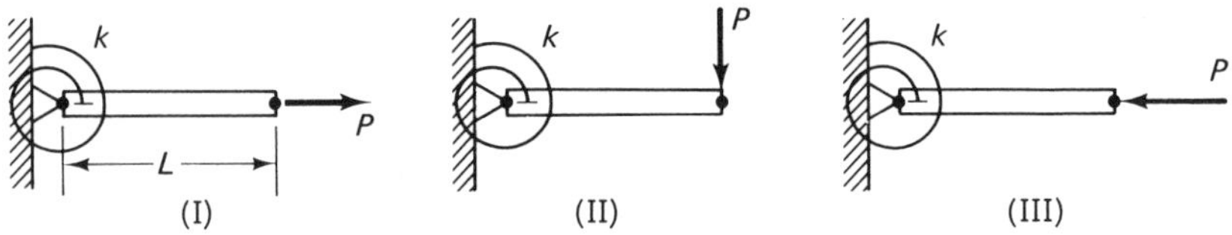

(b) Loading conditions

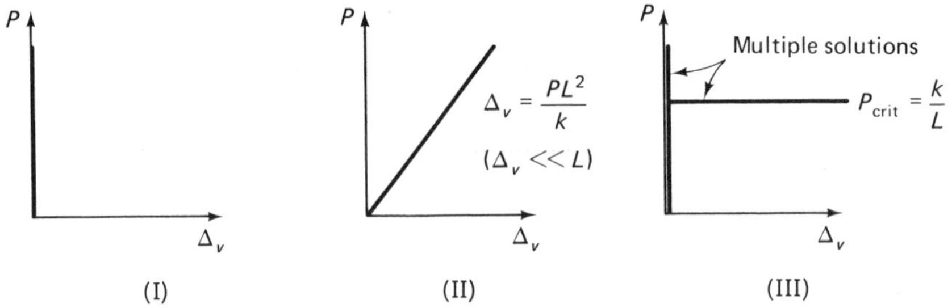

(c) Displacement responses

FIGURE 3.8 *Analysis of a simple cantilever model.*

Finally, for loading condition III we have as in condition I

$$\Delta_v = 0 \tag{c}$$

We now proceed to consider whether the system is stable under the above load applications. To investigate the stability we perturb the structure from the equilibrium positions defined in (a), (b), and (c) and ask whether an additional equilibrium position is possible.

Assume that Δ_v is positive but small in loading conditions I and II. If we write the equilibrium equations taking this displacement into account, we observe that in loading condition I the small nonzero Δ_v cannot be sustained, and in loading condition II the effect of including Δ_v in the analysis is negligible.

Consider next that $\Delta_v > 0$ in loading condition III. In this case, for an equilibrium configuration to be possible with Δ_v nonzero, the following equilibrium equation must be satisfied:

$$P\Delta_v = k\frac{\Delta_v}{L}$$

But this equation is satisfied for any Δ_v, provided $P = k/L$. Hence the critical load P_{crit} at which an equilibrium position in addition to the horizontal one becomes possible is

$$P_{\text{crit}} = \frac{k}{L}$$

In summary, we have

$P < P_{\text{crit}}$	only horizontal position of the bar is possible; equilibrium is stable
$P = P_{\text{crit}}$	horizontal and deflected positions of the bar are possible; horizontal equilibrium position is unstable for $P \geq P_{\text{crit}}$.

To gain an improved understanding of these results we may assume that in addition to the load P shown in Fig. 3.8(b), a small transverse load W is applied as shown in Fig. 3.9(a). If we then perform an analysis of the cantilever model subjected to P and W, the response curves shown schematically in Fig. 3.9(b) are obtained. Thus we observe that the effect of the load W decreases and is constant as P increases in loading conditions I and II, but in loading condition III the transverse displacement Δ_v increases very rapidly as P approaches the critical load, P_{crit}.

The analyses given in Example 3.10 illustrate the main objective of an eigenvalue formulation and solution in instability analysis, namely, to predict whether small disturbances that are imposed on the given equilibrium configuration tend to increase very substantially. The load level at which this situation arises corresponds to the critical loading of the system. In the second solution carried out in Example 3.10 the small disturbance was due to the small load W, which, for example, may simulate the possibility that the horizontal load on the cantilever is not acting perfectly horizontal. In the eigenvalue analysis, we simply assume a deformed configuration and investigate whether there is a load level

(a) Loading conditions (small load W)

$$\Delta_v \text{ due to } W \text{ at } P = 0$$

$$\Delta_v = \frac{WL}{\left(\dfrac{k}{L} + P\right)}$$

(I)

$$\Delta_v \text{ due to } W \text{ at } P = 0$$

$$\Delta_v = \frac{(P + W)L}{\left(\dfrac{k}{L}\right)}$$

(II)

$$\Delta_v \text{ due to } W \text{ at } P = 0$$

$$\Delta_v = \frac{WL}{\left(\dfrac{k}{L} - P\right)} \text{ for } P < P_{crit}$$

(III)

(b) Displacement responses

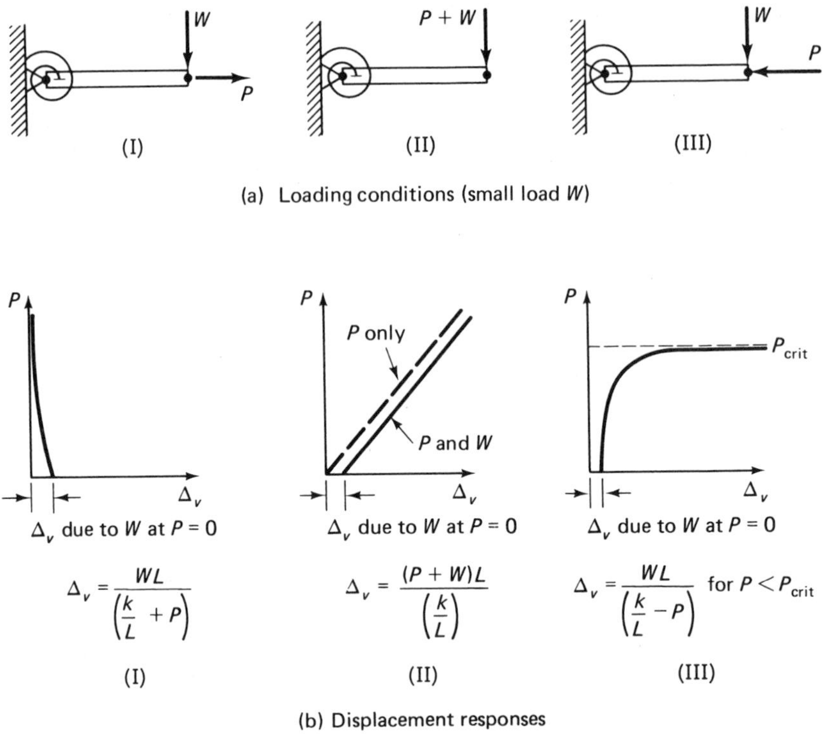

FIGURE 3.9 *Another analysis of the cantilever in Fig. 3.8.*

that indeed admits such a configuration as a possible equilibrium solution. We shall discuss in Section 6.5.2 that the eigenvalue analysis really consists of a linearization of the nonlinear response of the system, and that it depends largely on the system considered whether a reliable critical load is calculated. The eigenvalue solution is particularly applicable in the analysis of "beam-column type situations" of beam, plate, and shell structures.

EXAMPLE 3.11 : Experience shows that in structural analysis the critical load on column-type structures can be assessed appropriately using an eigenvalue problem formulation. Consider the system defined in Fig. 3.10. Establish the eigenvalue problem from which the critical loading on the system can be calculated.

As in the derivation of steady-state equilibrium equations (see Section 3.2.1) we can employ the direct procedure or a variational approach to establish the problem governing equations, and we describe both techniques in this problem solution.

In the direct approach we establish the governing equilibrium equations directly by considering the equilibrium of the structure in its deformed configuration. Referring to Fig. 3.10, the moment equilibrium of bar AB requires that

$$PL \sin (\alpha + \beta) = kU_1 L \cos (\alpha + \beta) + k_r \alpha \tag{a}$$

Similarly, for bars CBA we need

$$PL[\sin (\alpha + \beta) + \sin \beta] = kU_1 L[\cos (\alpha + \beta) + \cos \beta] + kU_2 L \cos \beta \tag{b}$$

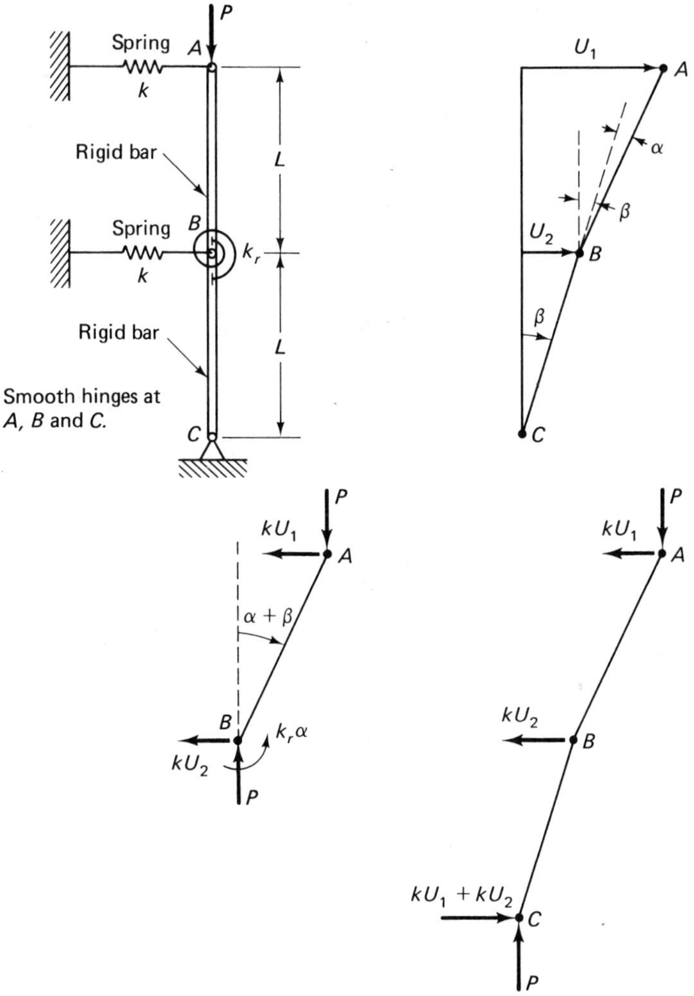

FIGURE 3.10 *Instability analysis of a column.*

As state variables we select U_1 and U_2, which completely describe the structural response. We also assume small displacements, for which

$$L \sin(\alpha + \beta) = U_1 - U_2; \qquad L \sin \beta = U_2$$

$$L \cos(\alpha + \beta) \doteq L; \qquad L \cos \beta \doteq L; \qquad \alpha = \frac{U_1 - 2U_2}{L}$$

Substituting these approximations into (a) and (b) and writing the resulting equations into matrix form, we obtain

$$\begin{bmatrix} kL + \dfrac{k_r}{L} & -2\dfrac{k_r}{L} \\ 2kL & kL \end{bmatrix} \begin{bmatrix} U_1 \\ U_2 \end{bmatrix} = P \begin{bmatrix} 1 & -1 \\ 1 & 0 \end{bmatrix} \begin{bmatrix} U_1 \\ U_2 \end{bmatrix}$$

We can symmetrize the coefficient matrices by multiplying the first row by (-2)

and adding the result to row 2, which gives the eigenvalue problem

$$\begin{bmatrix} kL + \frac{k_r}{L} & -\frac{2k_r}{L} \\ -\frac{2k_r}{L} & kL + \frac{4k_r}{L} \end{bmatrix} \begin{bmatrix} U_1 \\ U_2 \end{bmatrix} = P \begin{bmatrix} 1 & -1 \\ -1 & 2 \end{bmatrix} \begin{bmatrix} U_1 \\ U_2 \end{bmatrix} \qquad \text{(c)}$$

It may be noted that the second equation in (c) can also be obtained by considering the moment equilibrium of bar CB.

Considering next the variational approach, we need to establish the total potential Π of the system in the deformed configuration. Here we have

$$\Pi = \frac{1}{2}kU_1^2 + \frac{1}{2}kU_2^2 + \frac{1}{2}k_r\alpha^2 - PL[1 - \cos(\alpha + \beta) + 1 - \cos\beta] \qquad \text{(d)}$$

As in the direct approach we now assume small displacement conditions. Since we want to derive using (3.1) an eigenvalue problem of form (3.3) in which the coefficient matrices are independent of the state variables, we approximate the trigonometric expressions to second order in the state variables. Thus we use

$$\begin{aligned} \cos(\alpha + \beta) &\doteq 1 - \frac{(\alpha + \beta)^2}{2} \\ \cos\beta &\doteq 1 - \frac{\beta^2}{2} \end{aligned} \qquad \text{(e)}$$

and

$$\alpha + \beta \doteq \frac{U_1 - U_2}{L}, \qquad \alpha \doteq \frac{U_1 - 2U_2}{L}, \qquad \beta \doteq \frac{U_2}{L} \qquad \text{(f)}$$

Substituting from (e) and (f) into (d) we obtain

$$\Pi = \frac{1}{2}kU_1^2 + \frac{1}{2}kU_2^2 + \frac{1}{2}k_r\left(\frac{U_1 - 2U_2}{L}\right)^2 - \frac{P}{2L}(U_1 - U_2)^2 - \frac{P}{2L}U_2^2$$

Applying now the stationarity principle

$$\frac{\partial \Pi}{\partial U_1} = 0 \qquad \frac{\partial \Pi}{\partial U_2} = 0$$

the equations in (c) are obtained.

Considering next dynamic analysis, an eigenvalue problem may need to be formulated in the solution of the dynamic equilibrium equations. In essence, the objective is then to find a mathematical transformation on the state variables that is employed effectively in the solution of the dynamic response (see Section 9.3). Considering the analysis of physical problems it is then most valuable to identify the eigenvalues and vectors with physical quantities (see Section 9.3).

To illustrate how eigenvalue problems are formulated in dynamic analysis, we consider the following examples.

EXAMPLE 3.12: Consider the dynamic analysis of the system of rigid carts discussed in Example 3.8. Assume free vibration conditions and that

$$\mathbf{U} = \boldsymbol{\phi} \sin(\omega t - \psi) \qquad \text{(a)}$$

where $\boldsymbol{\phi}$ is a vector with components independent of time, ω is a circular frequency and ψ is a phase angle. Show that with this assumption an eigenvalue problem of the form given in (3.3) is obtained when searching for a solution of $\boldsymbol{\phi}$ and ω.

The equilibrium equations of the system when considering free vibration conditions are

$$\mathbf{M}\ddot{\mathbf{U}} + \mathbf{K}\mathbf{U} = \mathbf{0} \tag{b}$$

where the matrices **M** and **K** and vector **U** have been defined in Examples 3.1 and 3.8. If **U** given in (a) is to be a solution of the equations in (b), these equations must be satisfied when substituting for **U**,

$$-\omega^2 \mathbf{M}\boldsymbol{\phi} \sin(\omega t - \psi) + \mathbf{K}\boldsymbol{\phi} \sin(\omega t - \psi) = \mathbf{0}$$

Thus, for (a) to be a solution of (b) we obtain the condition,

$$\mathbf{K}\boldsymbol{\phi} = \omega^2 \mathbf{M}\boldsymbol{\phi} \tag{c}$$

which is an eigenvalue problem of form (3.3). We discuss in Section 9.3 the physical characteristics of a solution, ω_i^2 and $\boldsymbol{\phi}_i$, to the problem in (c).

EXAMPLE 3.13: Consider the electric circuit in Fig. 3.11. Determine the eigenvalue problem from which the resonant frequencies and modes can be calculated when $L_1 = L_2 = L$ and $C_1 = C_2 = C$.

Our first objective is to derive the dynamic equilibrium equations of the system. The element equilibrium equation for an inductor is

$$L\frac{dI}{dt} = V \tag{a}$$

where L is the inductance, I is the current through the inductor and V is the voltage drop across the inductor. For a capacitor of capacitance C the equilibrium equation is

$$I = C\frac{dV}{dt} \tag{b}$$

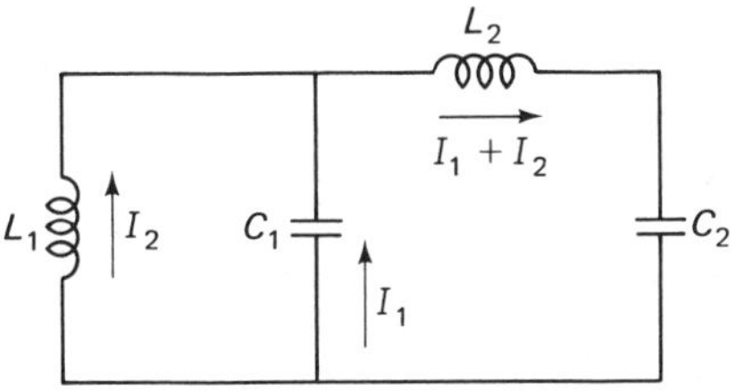

FIGURE 3.11 *Electric circuit.*

As state variables we select the currents I_1 and I_2 shown in Fig. 3.11. The governing equilibrium equations are obtained by invoking the element interconnectivity requirements contained in Kirchhoff's voltage law:

$$\begin{aligned} V_{C_1} + V_{L_2} + V_{C_2} &= 0 \\ V_{L_1} + V_{L_2} + V_{C_2} &= 0 \end{aligned} \tag{c}$$

Differentiating (a) and (c) with respect to time and substituting into (c) with $L_1 = L_2 = L$ and $C_1 = C_2 = C$ we obtain

$$L\begin{bmatrix} 1 & 1 \\ 1 & 2 \end{bmatrix}\begin{bmatrix} \ddot{I}_1 \\ \ddot{I}_2 \end{bmatrix} + \frac{1}{C}\begin{bmatrix} 2 & 1 \\ 1 & 1 \end{bmatrix}\begin{bmatrix} I_1 \\ I_2 \end{bmatrix} = \begin{bmatrix} 0 \\ 0 \end{bmatrix} \tag{d}$$

We note that these equations are quite analogous to the free vibration equilibrium equations of a structural system. Indeed, recognizing the analogy

$$I \longrightarrow \text{displacement}; \qquad \frac{1}{C} \longrightarrow \text{stiffness}; \qquad L \longrightarrow \text{mass}$$

the eigenproblem for the resonant frequencies is established as in Example 3.12 (and an equivalent structural system could be constructed).

3.2.4 On the Nature of Solutions

In the preceding sections we discussed the formulation of steady-state, propagation, and eigenvalue problems, and we gave a number of simple examples. In all cases a system of equations for the unknown state variables was established but not solved. For the solution of the equations we refer to the techniques presented in Chapters 8 to 12. The objective in this section is to discuss briefly the nature of the solutions that are calculated when steady-state, propagation, or eigenvalue problems are considered.

Considering steady-state and propagation problems, it is convenient to distinguish between linear and nonlinear problems. In simple terms, a linear problem is characterized by the fact that the response of the system varies in proportion to the magnitude of the applied loads. All other problems are nonlinear, as discussed in more detail in Section 6.1. To demonstrate in an introductory way some basic response characteristics that are predicted in linear steady-state, propagation, and eigenvalue analyses we consider the following example.

EXAMPLE 3.14: Consider the simple structural system consisting of an assemblage of rigid weightless bars, springs and concentrated masses shown in Fig. 3.12. The elements are connected at A, B, and C using frictionless pins. It is required to analyze the discrete system for the loading indicated, when the initial displacements and velocities are zero.

The response of the system is described by the two state variables U_1 and U_2 shown in Fig. 3.12(c). To decide what kind of analysis is appropriate we need to have sufficient information on the characteristics of the structure and the applied forces F and P. Let us assume that the structural characteristics and the applied forces are such that the displacements of the element assemblage are relatively small,

$$\frac{U_1}{L} < \frac{1}{10} \quad \text{and} \quad \frac{U_2}{L} < \frac{1}{10}$$

We can then assume that

$$\begin{aligned} \cos\alpha &= \cos\beta = \cos(\beta - \alpha) = 1 \\ \sin\alpha &= \alpha; \qquad \sin\beta = \beta \\ \alpha &= \frac{U_1}{L}, \qquad \beta = \frac{U_2 - U_1}{L} \end{aligned} \tag{a}$$

The governing equilibrium equations are derived as in Example 3.11 but including inertia forces (see Example 3.8); thus we obtain

$$\begin{bmatrix} m & 0 \\ 0 & \frac{m}{2} \end{bmatrix} \begin{bmatrix} \ddot{U}_1 \\ \ddot{U}_2 \end{bmatrix} + \begin{bmatrix} \left(5k + \frac{2P}{L}\right) & -\left(2k + \frac{P}{L}\right) \\ -\left(2k + \frac{P}{L}\right) & \left(2k + \frac{P}{L}\right) \end{bmatrix} \begin{bmatrix} U_1 \\ U_2 \end{bmatrix} = \begin{bmatrix} 2F \\ F \end{bmatrix} \tag{b}$$

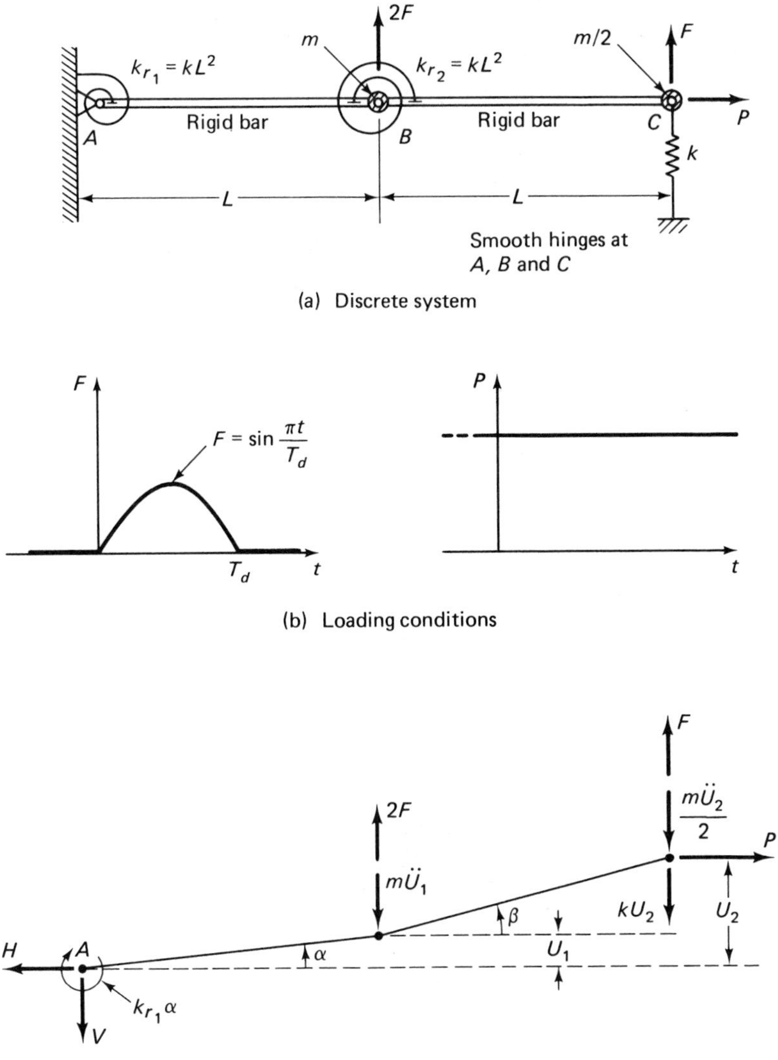

FIGURE 3.12 *A two degree of freedom system.*

The response of the system must depend on the relative values of k, m, and P/L. In order to obtain a measure of whether a static or dynamic analysis must be performed we calculate the natural frequencies of the system. These frequencies are obtained by solving the eigenvalue problem

$$\begin{bmatrix} \left(5k + \frac{2P}{L}\right) & -\left(2k + \frac{P}{L}\right) \\ -\left(2k + \frac{P}{L}\right) & \left(2k + \frac{P}{L}\right) \end{bmatrix} \begin{bmatrix} U_1 \\ U_2 \end{bmatrix} = \omega^2 \begin{bmatrix} m & 0 \\ 0 & \frac{m}{2} \end{bmatrix} \begin{bmatrix} U_1 \\ U_2 \end{bmatrix} \qquad \text{(c)}$$

The solution of (c) gives (see Section 2.7)

$$\omega_1 = \left\{\frac{9k}{2m} + \frac{2P}{mL} - \sqrt{\frac{33k^2}{4m^2} + \frac{8Pk}{m^2L} + \frac{2P^2}{m^2L^2}}\right\}^{1/2}$$

$$\omega_2 = \left\{\frac{9k}{2m} + \frac{2P}{mL} + \sqrt{\frac{33k^2}{4m^2} + \frac{8Pk}{m^2L} + \frac{2P^2}{m^2L^2}}\right\}^{1/2}$$

We note that for constant k/m the natural frequencies (rad/unit time) are a function of P/L and increase with P/L as shown in Fig. 3.13. The ith natural period, T_i, of the system is given by $T_i = 2\pi/\omega_i$, hence

$$T_1 = 2\pi/\omega_1$$
$$T_2 = 2\pi/\omega_2$$

The response of the system depends to a large degree on the duration of load application, when measured on the natural periods of the system. Since P is constant, the duration of load application is measured by T_d. To illustrate the response characteristics of the system, we assume a specific case $k = m = P/L = 1$ and three different values of T_d.

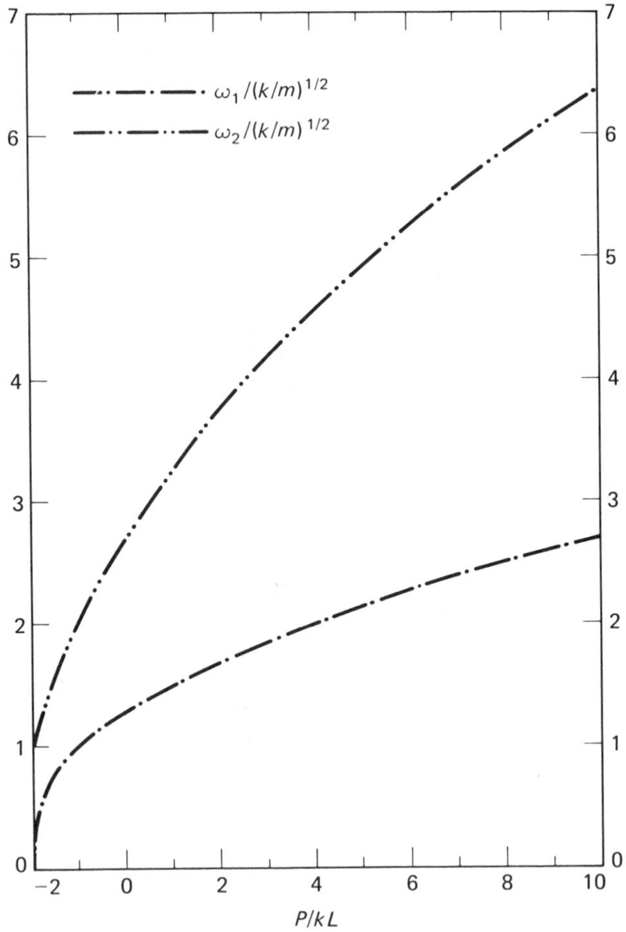

FIGURE 3.13 *Frequencies of structural system of Fig. 3.12.*

Case (i) $T_d = 4T_1$: The response of the system is shown for this case of load application in Fig. 3.14(a). We note that the dynamic response solution is somewhat close to the static response of the system and would be very close if $T_1 \ll T_d$.

Case (ii) $T_d = (T_1 + T_2)/2$: The response of the system is truly dynamic as shown in Fig. 3.14(b). It would be completely inappropriate to neglect the inertia effects.

Case (iii) $T_d = 1/4\ T_2$: In this case the duration of the loading is relatively short compared to the natural periods of the system. The response of the system is truly dynamic and inertia effects must be included in the analysis as shown in Fig. 3.14(c). The response of the system is somewhat close to the response generated assuming impulse conditions and would be very close if $T_2 \gg T_d$.

To identify some conditions for which the structure becomes unstable, we note from (b) that the stiffness of the structure increases with increasing values of P/L (which is the reason why the frequencies increase as P/L increases). Therefore, for the structure to become unstable, we need a negative value of P; i.e., P must be compressive. Let us assume now that P is very slowly decreased (P increases in compression), and that F is very small. In this case a static analysis is appropriate, and we can neglect the force F to obtain from (b) the governing equilibrium equations

$$\begin{bmatrix} 5k & -2k \\ -2k & 2k \end{bmatrix} \begin{bmatrix} U_1 \\ U_2 \end{bmatrix} = \frac{P}{L} \begin{bmatrix} -2 & 1 \\ 1 & -1 \end{bmatrix} \begin{bmatrix} U_1 \\ U_2 \end{bmatrix}$$

The solution of this eigenvalue problem gives two values for P/L. Because of the

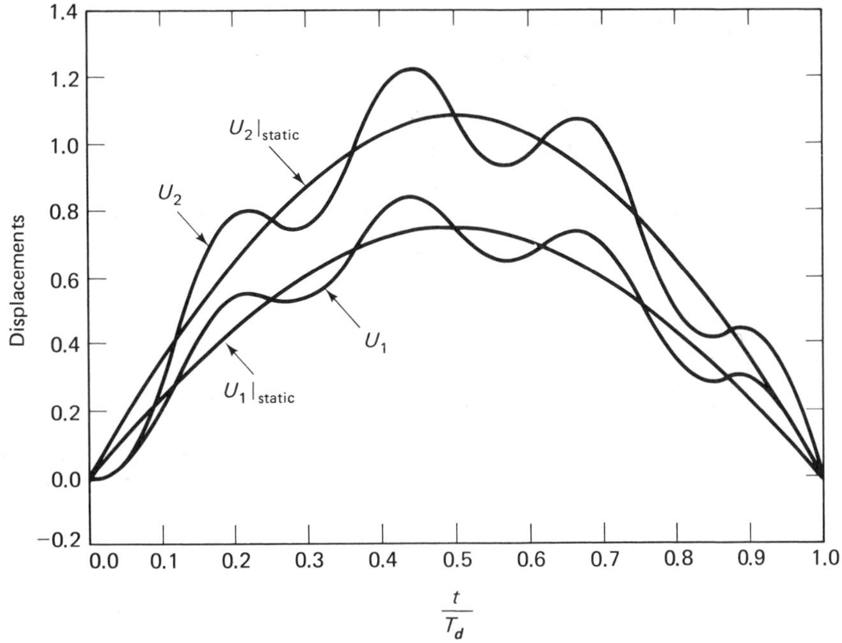

FIGURE 3.14 *(a) Analysis of two degree of freedom system of Fig. 3.12, case i; (b) case ii; (c) case iii (here the actual displacements are obtained by multiplying the given values by $2T_d/\pi$, and the impulse response was calculated using ${}^0U_1 = {}^0U_2 = 0$ and ${}^0\dot{U}_1 = {}^0\dot{U}_2 = 4T_d/\pi$ and setting the external loads to zero).*

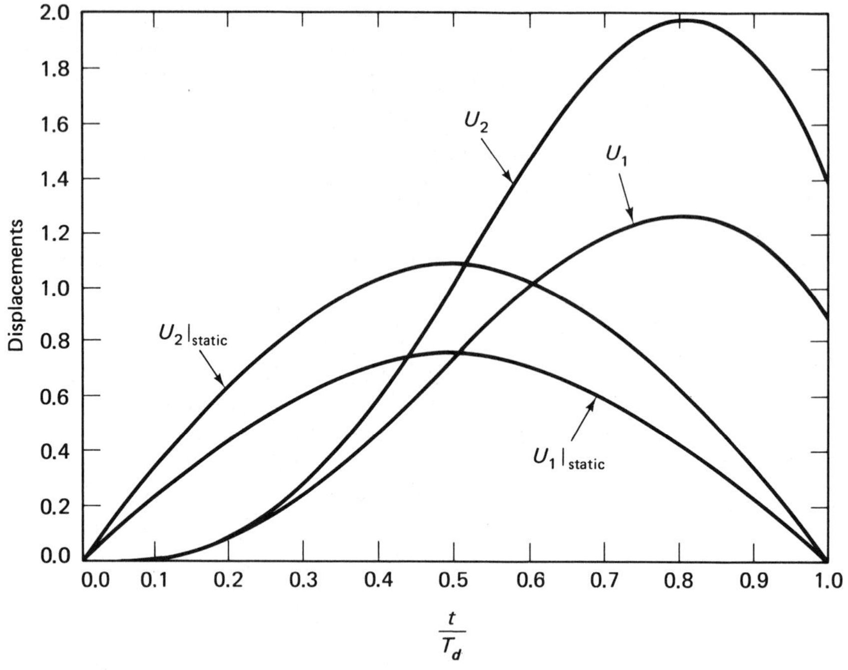

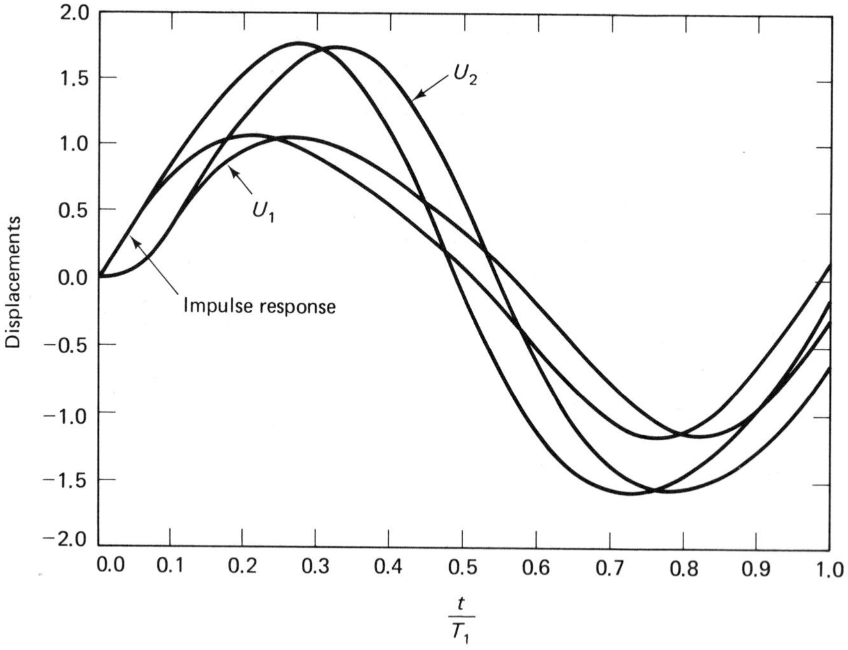

FIGURE 3.14 (*Cont.*)

sign convention for P, the larger eigenvalue gives the critical load

$$P_{\text{crit}} = -2kL$$

It may be noted that this is the load at which the smallest frequency of the system is zero (see Fig. 3.13).

Although we considered in the above example a structural system, most of the solution characteristics presented are also directly observed in the analysis of other types of problems. As shown in an introductory manner in the example, it is important that an analyst be able to decide what kind of analysis is required; whether a steady-state analysis is sufficient or whether a dynamic analysis should be performed, and whether the system may become unstable. We discuss some important factors that influence this decision in Chapters 6 and 9.

In addition to deciding on what kind of analysis should be performed, an appropriate model of the actual physical system must be selected. The characteristics of this model depend on the analysis to be carried out, but considering complex analysis, in essence, *the actual continuous system is reduced to an appropriate discrete system*. In the discussion given so far we assumed that the discrete system was already established and we now turn to the analysis of continuous systems.

3.3 ANALYSIS OF CONTINUOUS SYSTEMS

The basic steps in the analysis of a continuous system are quite similar to those employed in the analysis of a discrete system (see Section 3.2). However, instead of dealing with discrete elements, we focus attention on one (or more) typical differential element(s) with the objective of obtaining differential equations that express the element equilibrium requirements, constitutive relations and element interconnectivity requirements. These differential equations must hold throughout the domain of the system, and they must be supplemented by boundary conditions and, in dynamic analysis, also by initial conditions, before the response of the system can be calculated.

As in the analysis of discrete systems, *two different approaches can be followed to generate the system governing differential equations: the direct method and the variational method.* We discuss both approaches in this section and illustrate the variational procedure in some detail, because this approach can be regarded as the basis of the finite element method.

3.3.1 Differential Formulation

In the differential formulation of a problem we establish the equilibrium and constitutive requirements of (a) typical differential element(s) in terms of state variables. These considerations lead to a system of differential equations in the state variables, and it is possible that all compatibility requirements (i.e., the interconnectivity requirements of the differential elements) are already contained in these differential equations (for example, by the mere fact that the solution is to be continuous). However, in general, the equations must be supplemented by additional differential equations that impose appropriate constraints on the state variables in order that all compatibility requirements be satisfied. Finally, to complete the formulation of the problem, all boundary conditions and in a dynamic analysis the initial conditions are stated.[1-4]

For purposes of mathematical analysis it is expedient to classify problem-governing differential equations. Consider the second-order general partial differential equation in the domain (x, y),

$$A(x, y)\frac{\partial^2 u}{\partial x^2} + 2B(x, y)\frac{\partial^2 u}{\partial x \partial y} + C(x, y)\frac{\partial^2 u}{\partial y^2} = \phi\left((x, y, u, \frac{\partial u}{\partial x}, \frac{\partial u}{\partial y}\right) \tag{3.4}$$

where u is the unknown state variable. Depending on the coefficients in (3.4) the differential equation is elliptic, parabolic or hyperbolic,

$$B^2 - AC \quad \begin{cases} < 0 \text{ elliptic} \\ = 0 \text{ parabolic} \\ > 0 \text{ hyperbolic} \end{cases} \tag{3.5}$$

The classification in (3.5) is established when solving (3.4) using the method of characteristics, because it is then observed that the character of the solutions is distinctly different for the three categories of equations. These differences are also apparent when the differential equations are identified with the different physical problems that they govern. In their simplest form the three types of equations can be identified with the *Laplace equation*, the *heat conduction equation*, and the *wave equation*, respectively. We demonstrate how these equations arise in the solution of physical problems by the following examples.

EXAMPLE 3.15: The idealized dam shown in Fig. 3.15 stands on permeable soil. Establish the differential equation governing the steady-state seepage of water through the soil, and the corresponding boundary conditions.

Considering a typical element of widths dx and dy (and unit thickness) we have that the total flow into the element must be equal to the total flow out of the element. Hence we have

$$(q|_y - q|_{y+dy})\,dx + (q|_x - q|_{x+dx})\,dy = 0$$

or

$$-\frac{\partial q_y}{\partial y}\,dy\,dx - \frac{\partial q_x}{\partial x}\,dx\,dy = 0 \tag{a}$$

Using Darcy's law, the flow is given in terms of the total potential ϕ,

$$q_x = -k\frac{\partial \phi}{\partial x}; \qquad q_y = -k\frac{\partial \phi}{\partial y} \tag{b}$$

where we assumed a uniform permeability k. Substituting from (b) into (a) we obtain the *Laplace equation*

$$k\left(\frac{\partial^2 \phi}{\partial x^2} + \frac{\partial^2 \phi}{\partial y^2}\right) = 0 \tag{c}$$

It may be noted that this same equation is also obtained in heat transfer analysis and in the solution of electrostatic potential and other field problems (see Chapter 7).

The boundary conditions are no-flow boundary conditions in the soil at $x = -\infty$ and $x = +\infty$,

$$\left.\frac{\partial \phi}{\partial x}\right|_{x=-\infty} = 0; \qquad \left.\frac{\partial \phi}{\partial x}\right|_{x=+\infty} = 0 \tag{d}$$

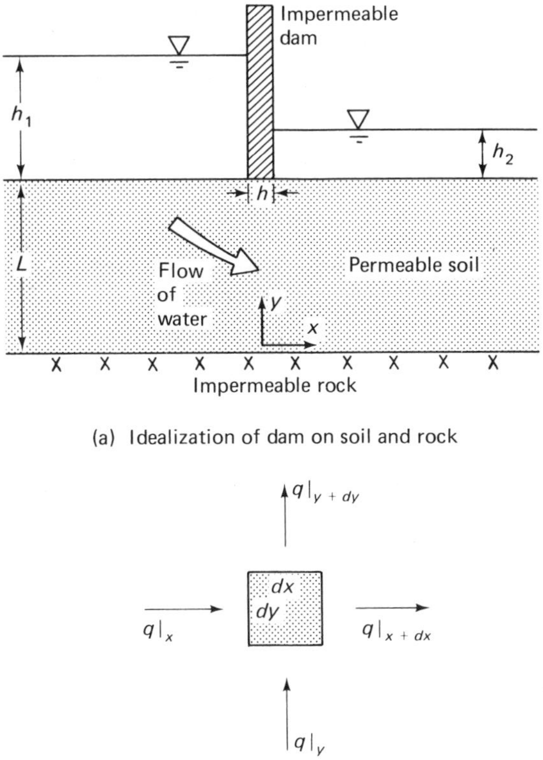

(a) Idealization of dam on soil and rock

(b) Differential element of soil

FIGURE 3.15 *Two-dimensional seepage problem.*

at the rock-soil interface,

$$\left.\frac{\partial \phi}{\partial y}\right|_{y=0} = 0 \tag{e}$$

and at the dam–soil interface

$$\frac{\partial \phi}{\partial y}(x, L) = 0 \qquad \text{for } -\frac{h}{2} \leq x \leq +\frac{h}{2} \tag{f}$$

In addition, the total potential is prescribed at the water–soil interface

$$\phi(x, L)|_{x<-(h/2)} = h_1; \qquad \phi(x, L)|_{x>(h/2)} = h_2 \tag{g}$$

The differential equation in (c) and the boundary conditions in (d) to (g) define the seepage flow steady-state response.

EXAMPLE 3.16: The very long slab shown in Fig. 3.16 is at a constant initial temperature θ_i when the surface at $x = 0$ is suddenly subjected to a constant uniform heat flow input. The surface at $x = L$ of the slab is kept at the temperature θ_i, and the surfaces parallel to the x-z plane are insulated. Assuming one-dimensional heat flow conditions, show that the problem-governing differential equation is the *heat conduction equation*

$$k \frac{\partial^2 \theta}{\partial x^2} = \rho c \frac{\partial \theta}{\partial t}$$

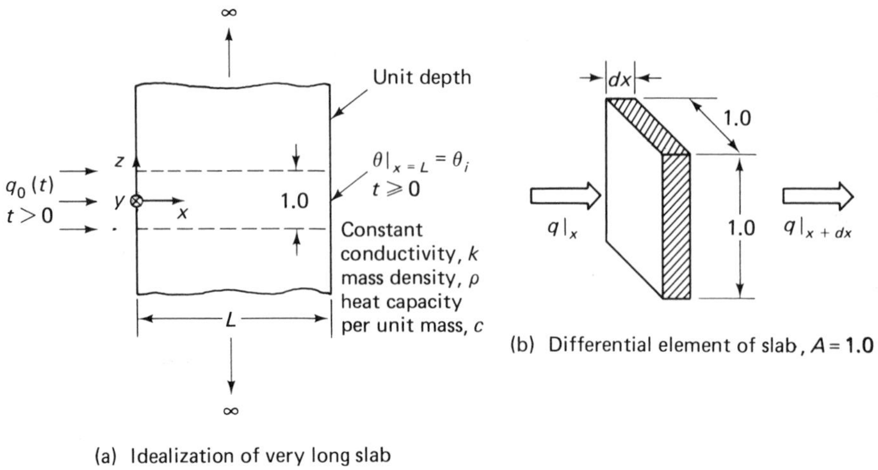

FIGURE 3.16 *One-dimensional heat conduction problem.*

where the parameters are defined in Fig. 3.16, and the temperature θ is the state variable. State also the boundary and initial conditions.

We consider a typical differential element of the slab, see Fig. 3.16(b). The element equilibrium requirement is that the net heat flow input to the element must equal the rate of heat stored in the element, thus

$$qA|_x - \left(qA|_x + A\frac{\partial q}{\partial x}\,dx\right) = \rho A\,dx\,c\,\frac{\partial \theta}{\partial t} \tag{a}$$

The constitutive relation is given by Fourier's law of heat conduction

$$q = -k\,\frac{\partial \theta}{\partial x} \tag{b}$$

Substituting from (b) into (a) we obtain

$$k\,\frac{\partial^2 \theta}{\partial x^2} = \rho c\,\frac{\partial \theta}{\partial t} \tag{c}$$

In this case the element interconnectivity requirements are contained in the assumption that the temperature θ be a continuous function of x and no additional compatibility conditions are applicable.

The boundary conditions are

$$\begin{aligned} \frac{\partial \theta}{\partial x}(0, t) &= -\frac{q_0(t)}{k} \\ \theta\,(L, t) &= \theta_i \end{aligned} \quad ; \qquad t > 0 \tag{d}$$

and the initial condition is $\qquad \theta(x, 0) = \theta_i \qquad$ (e)

The formulation of the problem is now complete, and the solution of (c) subject to the boundary and initial conditions in (d) and (e) yields the temperature response of the slab.

EXAMPLE 3.17: The rod shown in Fig. 3.17 is initially at rest, when a load $R(t)$ is suddenly applied at its free end. Show that the problem-governing differential equation is the *wave equation*

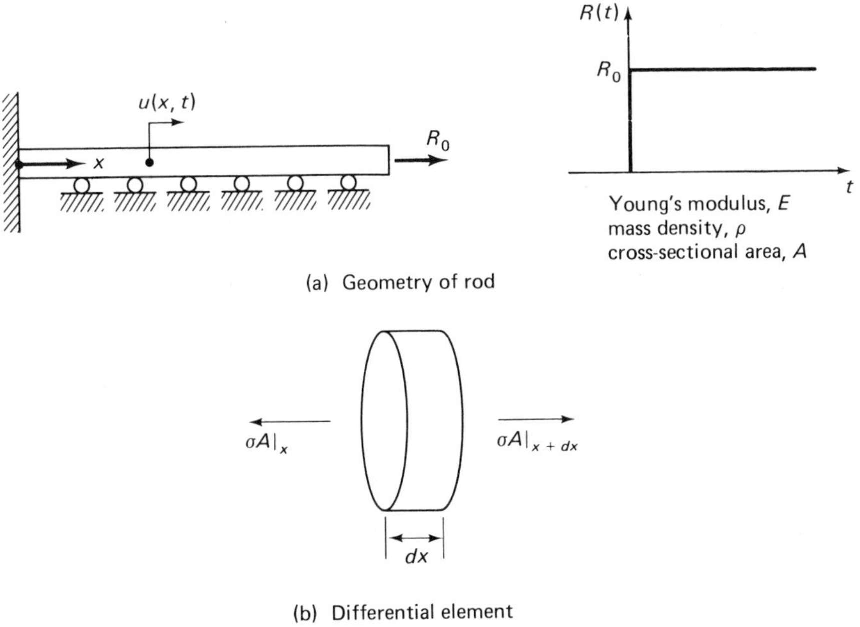

FIGURE 3.17 *Rod subjected to step load.*

$$\frac{\partial^2 u}{\partial x^2} = \frac{1}{c^2}\frac{\partial^2 u}{\partial t^2}, \qquad c = \sqrt{\frac{E}{\rho}}$$

where the variables are defined in Fig. 3.17 and the displacement of the rod, u, is the state variable. Also state the boundary and initial conditions.

The element force equilibrium requirements of a typical differential element are using d'Alembert's principle

$$\sigma A|_x + A\frac{\partial \sigma}{\partial x}\bigg|_x dx - \sigma A|_x = \rho A \frac{\partial^2 u}{\partial t^2}dx \tag{a}$$

The constitutive relation is

$$\sigma = E\frac{\partial u}{\partial x} \tag{b}$$

Combining (a) and (b) we obtain

$$\frac{\partial^2 u}{\partial x^2} = \frac{1}{c^2}\frac{\partial^2 u}{\partial t^2} \tag{c}$$

The element interconnectivity requirements are satisfied because we assume that the displacement u is continuous, and no additional compatibility conditions are applicable.

The boundary conditions are

$$\left.\begin{aligned} u(0, t) &= 0 \\ EA\frac{\partial u}{\partial x}(L, t) &= R_0 \end{aligned}\right\} ; \qquad t > 0 \tag{d}$$

and the initial conditions are

$$\begin{aligned} u(x, 0) &= 0 \\ \frac{\partial u}{\partial t}(x, 0) &= 0 \end{aligned} \tag{e}$$

With the conditions in (d) and (e) the formulation of the problem is complete and (c) can be solved for the displacement response of the rod.

Although we considered in the above examples specific problems that are governed by elliptic, parabolic, and hyperbolic differential equations, the problem formulations illustrate in a quite general way some basic characteristics. In elliptic problems (see Example 3.15) the values of the unknown state variables (or their normal derivatives) are given on the boundary. These problems are for this reason also called *boundary value problems*, where we should note that the solution at a general interior point depends on the data at every point of the boundary. A change in only one boundary value affects the complete solution; for instance, in Example 3.15 the complete solution for ϕ depends on the actual value of h_1. Elliptic differential equations generally govern the steady-state response of systems.

Comparing the governing differential equations given in Examples 3.15 to 3.17 it is noted that in contrast to the elliptic equation, the parabolic and hyperbolic equations (Examples 3.16 and 3.17, respectively) include time as an independent variable and thus define propagation problems. These problems are also called *initial value problems* because the solution depends on the initial conditions. We may note that in analogy to the derivation of the equilibrium equations of discrete systems, the governing differential equations of propagation problems are obtained from the steady-state equations by including the "resistance to change" (inertia) of the differential elements. Conversely, the parabolic and hyperbolic differential equations in Examples 3.16 and 3.17 would become elliptic equations if the time-dependent terms were neglected. This way the initial value problems would be converted to boundary value problems with steady-state solutions.

We stated earlier that the solution of a boundary value problem depends on the data at all points of the boundary. Here lies a significant difference to the analysis of a propagation problem; namely, considering propagation problems the solution at an interior point may depend only on the boundary conditions of part of the boundary and the initial conditions over part of the interior domain.

3.3.2 Variational Formulation

The variational approach of establishing the governing equilibrium equations of a system was already introduced as an alternative to the direct approach when we discussed the analysis of discrete systems (see Section 3.2.1). As described, the essence of the approach is to calculate the total potential Π of the system and to invoke the stationarity of Π, ie. $\delta\Pi = 0$, with respect to the state variables.* We pointed out that the use of the variational technique can be effective in the analysis of discrete systems; however, we shall now observe that the variational approach provides a particularly powerful mechanism for the analysis of continuous systems.[5-8] The main reason for this effectiveness lies in

* In Section 3.2.1 we defined δ to mean "arbitrary variations in the state variables that satisfy the *boundary conditions*." These are to be the *essential boundary conditions* as discussed now.

the way by which some boundary conditions (namely, the natural boundary conditions defined below) can be generated and taken into account when using the variational approach.

To demonstrate the variational formulation in the examples below, we assume that the total potential Π is given, and defer the description of how an appropriate Π can be established until after the presentation of the examples.

The total potential Π is also called the *functional* of the problem. Assume that in the functional the highest derivative of a state variable (with respect to a space coordinate) is of order m; i.e., the operator contains at most mth-order derivatives. Such a problem we call a C^{m-1} *variational problem.* Considering the boundary conditions of the problem, we identify two classes of boundary conditions, called *essential* and *natural boundary conditions.* The essential boundary conditions are also called *geometric boundary conditions*; namely, in structural mechanics, the essential boundary conditions correspond to prescribed displacements and rotations. The order of the derivatives in the essential boundary conditions is, in a C^{m-1} problem, at most $m - 1$ and only these boundary conditions must be satisfied when invoking the stationarity of Π. The second class of boundary conditions, i.e., the natural boundary conditions, are also called the *force boundary conditions*; namely, in structural mechanics, the natural boundary conditions correspond to prescribed boundary forces and moments. The highest derivatives in these boundary conditions are of order m to $2m - 1$. We will see later that this classification of variational problems and associated boundary conditions is also most useful when questions of convergence of numerical solutions are investigated. Consider the following examples.

EXAMPLE 3.18: The functional governing the temperature distribution in the slab considered in Example 3.16 is

$$\Pi = \int_0^L \frac{1}{2}k\left(\frac{\partial \theta}{\partial x}\right)^2 dx - \int_0^L \theta q^B \, dx - \theta_0 q_0 \tag{a}$$

and the essential boundary condition is

$$\theta_L = \theta_i \tag{b}$$

where $\qquad \theta_0 = \theta(0, t) \qquad$ and $\qquad \theta_L = \theta(L, t)$;

q^B is the heat generated per unit volume, and otherwise the same notation as in Example 3.16 is used. Invoke the stationarity condition on Π to derive the governing heat conduction equation and the natural boundary condition.

This is a C^0 variational problem, i.e., the highest derivative in the functional in (a) is of order 1, or $m = 1$. An essential boundary condition, here given in (b), can therefore only correspond to a prescribed temperature, and a natural boundary condition must correspond to a prescribed temperature gradient or boundary heat flow input.

Invoking the stationarity condition $\delta\Pi = 0$ we obtain

$$\int_0^L \left(k\frac{\partial \theta}{\partial x}\right)\left(\delta\frac{\partial \theta}{\partial x}\right) dx - \int_0^L \delta\theta \, q^B \, dx - \delta\theta_0 \, q_0 = 0 \tag{c}$$

where we note that variations and differentiations are performed using the same rules. Noting further that the conductivity k is constant, that

$$\delta \frac{\partial \theta}{\partial x} = \frac{\partial \delta\theta}{\partial x}$$

and using integration by parts we obtain from (c) the following equation:

$$\underbrace{-\int_0^L \left(k\frac{\partial^2\theta}{\partial x^2} + q^B\right)\delta\theta\, dx}_{①} + \underbrace{k\frac{\partial\theta}{\partial x}\bigg|_{x=L}\delta\theta_L}_{②} - \underbrace{\left[k\frac{\partial\theta}{\partial x}\bigg|_{x=0} + q_0\right]\delta\theta_0}_{③} = 0 \qquad \text{(d)}$$

To extract from (d) the governing differential equation and natural boundary condition, we use the argument that the variations on θ are completely arbitrary, except that there can be no variations on the prescribed essential boundary conditions. Hence, because θ_L is prescribed we have $\delta\theta_L = 0$ and term ② in (d) vanishes.

Considering next terms ① and ③, assume that $\delta\theta_0 = 0$ but that $\delta\theta$ is everywhere nonzero (except at $x = 0$, where we have a sudden jump to a zero value). If (d) is to hold for any nonzero $\delta\theta$ we need to have*

$$k\frac{\partial^2\theta}{\partial x^2} + q^B = 0 \qquad \text{(e)}$$

Conversely, assume that $\delta\theta$ is zero everywhere except at $x = 0$, i.e. we have $\delta\theta_0 \neq 0$; then (d) is only valid provided

$$k\frac{\partial\theta}{\partial x}\bigg|_{x=0} + q_0 = 0 \qquad \text{(f)}$$

The expression in (f) represents the natural boundary condition. The governing differential equation of the propagation problem is obtained from (e), specifying here that

$$q^B = -\rho c\frac{\partial\theta}{\partial t} \qquad \text{(g)}$$

Hence (e) reduces to

$$k\frac{\partial^2\theta}{\partial x^2} = \rho c\frac{\partial\theta}{\partial t}$$

We may note that until the heat capacity effect was introduced in the formulation in (g), the equations were derived as if a steady-state problem (or if q^B is time-dependent a pseudo steady-state problem) was being considered. Hence, as noted before, the formulation of the propagation problem can be obtained from the equation governing the steady-state response by simply taking into account the time-dependent "inertia term."

EXAMPLE 3.19: The functional and essential boundary condition governing the wave propagation in the rod considered in Example 3.17 are

$$\Pi = \int_0^L \frac{1}{2}EA\left(\frac{\partial u}{\partial x}\right)^2 dx - \int_0^L u\, f^B\, dx - u_L R \qquad \text{(a)}$$

and

$$u_0 = 0 \qquad \text{(b)}$$

where the same notation as in Example 3.17 is used and $u_0 = u(0, t)$ and $u_L = u(L, t)$, and f^B is the body force per unit length of the rod. Show that by invoking the stationarity condition on Π the governing differential equation of the propagation problem and the corresponding natural boundary condition can be derived.

* We in effect imply here that the limits of integration are not "0 to L" but "0^+ to L^-".

We proceed as in Example 3.18. The stationarity condition $\delta\Pi = 0$ gives

$$\int_0^L \left(EA\frac{\partial u}{\partial x}\right)\left(\delta\frac{\partial u}{\partial x}\right) dx - \int_0^L \delta u\, f^B\, dx - \delta u_L R = 0$$

Writing $\partial\delta u/\partial x$ for $\delta\, \partial u/\partial x$, recalling that EA is constant and using integration by parts yields

$$-\int_0^L \left(EA\frac{\partial^2 u}{\partial x^2} + f^B\right)\delta u\, dx + \left[EA\frac{\partial u}{\partial x}\bigg|_{x=L} - R\right]\delta u_L - EA\frac{\partial u}{\partial x}\bigg|_{x=0}\delta u_0 = 0$$

To obtain now the governing differential equation and natural boundary condition we use, in essence, the same argument as in Example 3.18; i.e., since δu_0 is zero but δu is arbitrary at all other points, we must have

$$EA\frac{\partial^2 u}{\partial x^2} + f^B = 0 \tag{c}$$

and

$$EA\frac{\partial u}{\partial x}\bigg|_{x=L} = R \tag{d}$$

In this problem we have $f^B = -A\rho\, \partial^2 u/\partial t^2$ and hence (c) reduces to the problem governing differential equation,

$$\frac{\partial^2 u}{\partial x^2} = \frac{1}{c^2}\frac{\partial^2 u}{\partial t^2}; \qquad c = \sqrt{\frac{E}{\rho}}$$

The natural boundary condition was stated in (d).

Finally, it may be noted that the problem in (a) and (b) is a C^0 variational problem, i.e. $m = 1$ in this case.

EXAMPLE 3.20: The functional governing static buckling of the column in Fig. 3.18 is

$$\Pi = \frac{1}{2}\int_0^L EI\left(\frac{d^2w}{dx^2}\right)^2 dx - \frac{P}{2}\int_0^L \left(\frac{dw}{dx}\right)^2 dx + \frac{1}{2}kw_L^2 \tag{a}$$

where $w_L = w|_{x=L}$ and the essential boundary conditions are

$$w|_{x=0} = 0, \qquad \frac{dw}{dx}\bigg|_{x=0} = 0 \tag{b}$$

Invoke the stationarity condition $\delta\Pi = 0$ to derive the problem-governing differential equation and the natural boundary conditions.

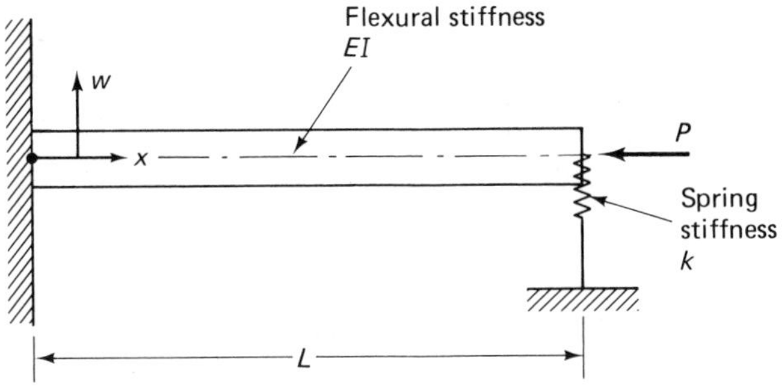

FIGURE 3.18 *Column subjected to a compressive load.*

This problem is a C^1 variational problem, i.e. $m = 2$, because the highest derivative in the functional is of order 2.

The stationarity condition $\delta\Pi = 0$ yields

$$\int_0^L EI\, w''\, \delta w''\, dx - P\int_0^L w'\, \delta w'\, dx + kw_L\, \delta w_L = 0$$

where we use the notation $w' = dw/dx$ and so on. But $\delta w'' = d(\delta w')/dx$, and EI is constant; hence using integration by parts we obtain

$$\int_0^L EI\, w''\, \delta w''\, dx = EI\, w''\, \delta w' |_0^L - EI\int_0^L w'''\, \delta w'\, dx$$

If we continue to use integration by parts on $\int_0^L w'''\, \delta w'\, dx$, and also integrate by parts $\int_0^L w'\, \delta w'\, dx$, we obtain

$$\underbrace{\int_0^L (EI\, w^{\text{iv}} + Pw'')\delta w\, dx}_{①} + \underbrace{[EI\, w''\, \delta w']|_L}_{②} - \underbrace{[EI\, w''\, \delta w']|_0}_{③}$$

$$- \underbrace{[(EI\, w''' + Pw')\delta w]|_L}_{④} + \underbrace{[(EI\, w''' + Pw')\delta w]|_0}_{⑤} + \underbrace{kw_L\, \delta w_L}_{⑥} = 0 \qquad \text{(c)}$$

Since the variations on w and w' must satisfy the essential boundary conditions, we have $\delta w_0 = 0$ and $\delta w_0' = 0$. It follows that terms ③ and ⑤ are zero. The variations on w and w' are arbitrary at all other points, hence to satisfy (c) we conclude using the earlier arguments (see Example 3.18) that the following equations must be satisfied:

term 1: $$EI\, w^{\text{iv}} + Pw'' = 0 \qquad \text{(d)}$$

term 2: $$EI\, w''|_{x=L} = 0 \qquad \text{(e)}$$

terms 4 and 6: $$[EI\, w''' + Pw' - kw]|_{x=L} = 0 \qquad \text{(f)}$$

The problem-governing differential equation is given in (d) and the natural boundary conditions are the relations in (e) and (f). We should note that these boundary conditions correspond to the physical conditions of moment and shear equilibrium at $x = L$.

We have illustrated in the above examples how the problem-governing differential equation and the natural boundary conditions can be derived by invoking the stationarity of the functional of the problem. At this point a number of observations should be made.

First, considering a C^{m-1} variational problem, the order of the highest derivative that is present in the problem-governing differential equation is $2m$. The reason for obtaining a derivative of order $2m$ in the problem-governing differential equation is that integration by parts is employed m times.

A second observation is that the effect of the natural boundary conditions was always included as a potential in the expression for Π. Hence the natural boundary conditions are implicitly contained in Π, whereas the essential boundary conditions have been stated separately.

Our objective in Examples 3.18 to 3.20 was to derive the governing differential equations and natural boundary conditions by invoking the stationarity of a functional, and for this purpose the appropriate functional was given in each case. However, an important question then arises: How can we establish an appropriate functional corresponding to a given problem? The two observations above and the mathematical manipulations in Examples 3.18 to 3.20 suggest that to derive a functional for a given problem we could start with the governing differential equation and natural boundary conditions, assemble them in one appropriate integral equation, and then proceed backwards in the mathematical manipulations. In this derivation it would be necessary to use integration by parts and the final check would be that the stationarity condition on the Π derived does indeed yield the governing differential equations. Although somewhat vague, this procedure can indeed be employed to derive appropriate functionals in many cases. In addition, a number of valuable theorems exist which provide further guidelines in establishing an appropriate functional. In this context, it should also be noted that considering a specific problem, there does not generally exist a unique appropriate functional, but a number of functionals may be applicable. For instance, in the solution of structural mechanics problems, we can employ the principle of minimum potential energy, the principle of minimum complementary energy, the Reissner principle and so on (see Sections 4.2 and 4.3).

Another important observation is that once a functional has been established for a certain class of problems, the functional can be employed to generate the governing equations for *all* problems in that class and therefore provides a general analysis tool. For example, the principle of minimum potential energy is generally applicable to all problems in linear elasticity theory.

Considered simply from a utilitarian point of view, the following observations are made when considering variational formulations.

1. The variational method may provide a relatively easy way to construct the system governing equations. This ease of use of a variational principle depends largely on the fact that in the variational formulation scalar quantities (energies, potentials, and so on) are considered rather than vector quantities (forces, displacements and so on).
2. A variational approach may lead more directly to the system governing equations and boundary conditions. Namely, if a complex system is considered, it is of advantage that some variables that need be included in a direct formulation are not considered in a variational formulation (such as internal forces that do no net work).
3. The variational approach provides some additional insight into a problem, and gives an independent check on the formulation of a problem.
4. For approximate solutions, a larger class of trial functions can be employed in many cases if the analyst operates on the variational formulation rather than on the differential formulation of the problem. The reason is that using a variational formulation, the trial functions need not satisfy all boundary conditions because some boundary conditions are implicitly contained in the functional.

This last consideration has most important consequences and much of the success of the finite element method and the finite difference energy method hinges on the fact that by operating on a variational formulation it is relatively easy to deal with complex boundary conditions. We discuss this point in more detail in the next section and when we discuss the convergence of the finite element method (see Section 4.2.5).

3.3.3 Weighted Residual Methods; Ritz Method

In previous sections we have discussed differential and variational formulations of the governing equilibrium equations of continuous systems. Considering relatively simple systems, these equations can be solved in closed form using techniques of integration, separation of variables, and so on. For more complex systems, approximate procedures of solution must be employed. The objective in this section is to survey some classical techniques in which a family of trial functions is used to obtain an approximate solution. We shall see later that these techniques are very closely related to the finite element method of analysis, and that indeed the finite element method can be regarded as an extension of these classical procedures.

Consider the analysis of a steady-state problem using its differential formulation

$$L_{2m}[\phi] = r \tag{3.6}$$

in which L_{2m} is a linear differential operator, ϕ is the state variable to be calculated, and r is the forcing function. The solution to the problem must also satisfy the boundary conditions

$$B_i[\phi] = q_i\big|_{\text{at boundary } S_i} \qquad i = 1, 2, \ldots \tag{3.7}$$

We shall be concerned, in particular, with symmetric and positive definite operators, which satisfy the symmetry condition

$$\int_D (L_{2m}[u])v\, dD = \int_D (L_{2m}[v])u\, dD \tag{3.8}$$

and the condition of positive definiteness

$$\int_D (L_{2m}[u])u\, dD > 0 \tag{3.9}$$

where D is the domain of the operator, and u and v are any functions that satisfy homogeneous essential and natural boundary conditions. To clarify the meaning of the relations (3.6) to (3.9), we consider the following example.

EXAMPLE 3.21: The steady-state response of the bar shown in Fig. 3.17 is calculated by solving the differential equation

$$-EA\frac{\partial^2 u}{\partial x^2} = 0 \tag{a}$$

subject to the boundary conditions

$$u\big|_{x=0} = 0$$

$$EA\frac{\partial u}{\partial x}\bigg|_{x=L} = R \tag{b}$$

Identify the operators and functions of (3.6) and (3.7) and check whether the operator L_{2m} is symmetric and positive definite.

Comparing (3.6) with (a) we see that in this problem

$$L_{2m} = -EA\frac{\partial^2}{\partial x^2}; \qquad \phi = u; \qquad r = 0$$

Similarly, comparing (3.7) with (b) we obtain

$$B_1 = 1; \qquad q_1 = 0$$

$$B_2 = EA\frac{\partial}{\partial x}; \qquad q_2 = R$$

To identify whether the operator L_{2m} is symmetric and positive definite, we consider the case $R = 0$. This means physically that we are only concerned with the structure itself and not with the loading applied to it. Considering (3.8) we have

$$\begin{aligned}\int_0^L -EA\frac{\partial^2 u}{\partial x^2} v\, dx &= -EA\frac{\partial u}{\partial x} v \Big|_0^L + \int_0^L EA\frac{\partial u}{\partial x}\frac{\partial v}{\partial x}\, dx \\ &= -EA\frac{\partial u}{\partial x} v \Big|_0^L + EAu\frac{\partial v}{\partial x}\Big|_0^L - \int_0^L EA\frac{\partial^2 v}{\partial x^2} u\, dx\end{aligned} \tag{c}$$

Since the boundary conditions are $u = v = 0$ at $x = 0$ and $\partial u/\partial x = \partial v/\partial x = 0$ at $x = L$, we have

$$\int_0^L - EA\frac{\partial^2 u}{\partial x^2} v\, dx = \int_0^L - EA\frac{\partial^2 v}{\partial x^2} u\, dx$$

and the operator is symmetric. We can also directly conclude that the operator is positive definite, because from (c) we obtain

$$\int_0^L - EA\frac{\partial^2 u}{\partial x^2} u\, dx = \int_0^L EA\left(\frac{\partial u}{\partial x}\right)^2 dx$$

In the following we discuss the use of the weighted residual methods and the Ritz method in the solution of linear steady-state problems as considered in (3.6) and (3.7), but the same concepts can also be employed in the analysis of propagation problems and eigenproblems, and in the analysis of nonlinear response (see Examples 3.23 and 3.24).

The basic step in the weighted residual and Ritz analyses is to assume a solution of the form

$$\bar{\phi} = \sum_{i=1}^{n} a_i f_i \tag{3.10}$$

where the f_i are linearly independent trial functions and the a_i are multipliers to be determined in the solution.

Consider first the *weighted residual methods*. These techniques operate directly on (3.6) and (3.7). Using these methods, we choose the functions f_i in (3.10) so as to satisfy all boundary conditions in (3.7) and we then calculate the residual

$$R = r - L_{2m}\left[\sum_{i=1}^{n} a_i f_i\right] \tag{3.11}$$

For the exact solution this residual is, of course, zero. A "good" approximation to the exact solution would imply that R is small at all points of the solution domain. The various weighted residual methods differ in the criteria that they employ to calculate the a_i such that R is small. However, in all techniques we determine the a_i so as to make a weighted average of R vanish.

Galerkin method: In this technique the parameters a_i are determined from the n equations

$$\int_D f_i R\, dD = 0 \qquad i = 1, 2, \ldots, n \tag{3.12}$$

where D is the solution domain.

Least squares method: In this technique the integral of the square of the residual is minimized with respect to the parameters a_i,

$$\frac{\partial}{\partial a_i} \int_D R^2\, dD = 0 \qquad i = 1, 2, \ldots, n$$

Substituting from (3.11) we thus obtain the following n simultaneous equations for the parameters a_i,

$$\int_D R\, L_{2m}[f_i]\, dD = 0 \qquad i = 1, 2, \ldots, n \tag{3.13}$$

Collocation method: In this method the residual R is set equal to zero at n distinct points in the solution domain to obtain n simultaneous equations for the parameters a_i. The location of the n points can be somewhat arbitrary, and a uniform pattern may be appropriate, but usually the analyst should use some judgment to employ appropriate locations.

Subdomain method: The complete domain of solution is subdivided into n subdomains, and the integral of the residual in (3.11) over each subdomain is set equal to zero to generate n equations for the parameters a_i.

An important step in using a weighted residual method is the solution of the simultaneous equations for the parameters a_i. We note that since L_{2m} is a linear operator, in all above procedures a linear set of equations in the parameters a_i is generated. In the Galerkin method, the coefficient matrix is symmetric (and also positive definite) if L_{2m} is a symmetric (and also positive definite) operator. In the least squares method we always generate a symmetric coefficient matrix irrespective of the properties of the operator L_{2m}. However, in the collocation and subdomain methods, nonsymmetric coefficient matrices may be generated. In practical analysis, therefore, the Galerkin and least squares methods are usually preferable.

Using the above weighted residual methods we operate directly on (3.6) and (3.7) to minimize the error between the trial solution in (3.10) and the actual solution to the problem. Considering next the *Ritz analysis method*,[7] the fundamental difference to the weighted residual methods is that in the Ritz method we operate on the functional corresponding to the problem in (3.6) and (3.7). Let Π be the functional of the C^{m-1} variational problem that is equivalent to the differential formulation given in (3.6) and (3.7); in the Ritz method we substitute the trial functions $\bar{\phi}$ given in (3.10) into Π and generate n simultaneous equations for the parameters a_i using the stationarity condition of Π,

$$\frac{\partial \Pi}{\partial a_i} = 0 \qquad i = 1, 2, \ldots, n \tag{3.14}$$

An important consideration is the selection of the trial (or Ritz) functions f_i in (3.10). In the Ritz analysis these functions need only satisfy the essential boundary conditions and not the natural boundary conditions. The reason for this relaxed requirement on the trial functions is that the natural boundary conditions are implicitly contained in the functional Π. Assume that the L_{2m} operator corresponding to the variational problem is symmetric and positive definite. In this case the actual extremum of Π is its minimum, and by invoking (3.14) we minimize (in some sense) the violation of the internal equilibrium requirements and the violation of the natural boundary conditions. Therefore, for convergence in a Ritz analysis, the trial functions need only satisfy the essential boundary conditions, which is a fact that may not be anticipated because we know that the exact solution also satisfies the natural boundary conditions. Actually, assuming a given number of trial functions, it can be expected that in most cases the solution will be more accurate if these functions also satisfy the natural boundary conditions. However, it can be very difficult to find such trial functions and it is generally more effective to use rather a larger number of functions that only satisfy the essential boundary conditions. We demonstrate the use of the Ritz method in the following examples.

EXAMPLE 3.22: Consider the simple bar fixed at one end ($x = 0$) and subjected to a concentrated force at the other end ($x = 180$) as shown in Fig. 3.19. Using the notation given in the figure the total potential of the structure is

$$\Pi = \int_0^{180} \tfrac{1}{2} EA \left(\frac{du}{dx}\right)^2 dx - 100\, u|_{x=180} \tag{a}$$

and the essential boundary condition is $u|_{x=0} = 0$

1. Calculate the exact displacement and stress distributions in the bar.
2. Calculate the displacement and stress distributions using the Ritz method with the following displacement assumptions:

$$u = a_1 x + a_2 x^2 \tag{b}$$

and

$$u = \frac{x u_B}{100} \qquad 0 \le x \le 100$$

$$u = \left(1 - \frac{x - 100}{80}\right) u_B + \left(\frac{x - 100}{80}\right) u_C \qquad 100 \le x \le 180 \tag{c}$$

where u_B and u_C are the displacements at points B and C.

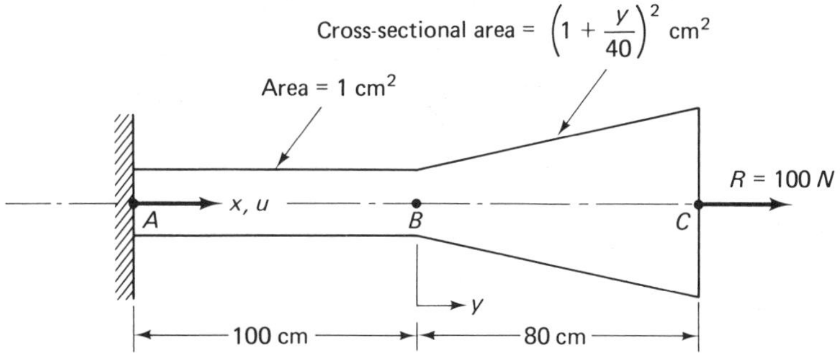

FIGURE 3.19 *Bar subjected to a concentrated end force.*

In order to calculate the exact displacements in the structure, we use the stationarity condition of Π and generate the governing differential equation and the natural boundary condition. We have

$$\delta\Pi = \int_0^{180} \left(EA\frac{du}{dx}\right)\delta\left(\frac{du}{dx}\right)dx - 100\,\delta u|_{x=100} \tag{d}$$

Setting $\delta\Pi = 0$ and using integration by parts we obtain (see Example 3.21)

$$\frac{d}{dx}\left(EA\frac{du}{dx}\right) = 0 \tag{e}$$

$$EA\frac{du}{dx}\bigg|_{x=180} = 100 \tag{f}$$

The solution of (e) subject to the natural boundary condition in (f) and the essential boundary condition $u|_{x=0} = 0$ gives

$$u = \frac{100}{E}x; \qquad 0 \le x \le 100$$

$$u = \frac{10000}{E} + \frac{4000}{E} - \frac{4000}{E\left(1 + \dfrac{x-100}{40}\right)}; \qquad 100 \le x \le 180$$

The exact stresses in the bar are

$$\sigma = 100; \qquad 0 \le x \le 100$$

$$\sigma = \frac{100}{\left(1 + \dfrac{x-100}{40}\right)^2}; \qquad 100 \le x \le 180$$

To perform next the Ritz analyses, we note that the displacement assumptions in (b) and (c) satisfy the essential boundary condition but not the natural boundary condition. Substituting from (b) into (a) we obtain

$$\Pi = \frac{E}{2}\int_0^{100}(a_1 + 2a_2x)^2\,dx + \frac{E}{2}\int_{100}^{180}\left(1 + \frac{x-100}{40}\right)^2(a_1 + 2a_2x)^2\,dx$$
$$- 100u|_{x=180}$$

Invoking that $\delta\Pi = 0$ we obtain the following equations for a_1 and a_2

$$E\begin{bmatrix} 0.4467 & 115.6 \\ 115.6 & 34075.7 \end{bmatrix}\begin{bmatrix} a_1 \\ a_2 \end{bmatrix} = \begin{bmatrix} 18 \\ 3240 \end{bmatrix} \tag{g}$$

and
$$a_1 = \frac{129}{E}; \qquad a_2 = -\frac{0.341}{E}$$

This Ritz analysis yields therefore the approximate solution

$$u = \frac{129}{E}x - \frac{0.341}{E}x^2 \tag{h}$$

$$\sigma = 129 - 0.682x; \qquad 0 \le x \le 180 \tag{i}$$

Using next the Ritz functions in (c) we have

$$\Pi = \frac{E}{2}\int_0^{100}\left(\frac{1}{100}u_B\right)^2 dx + \frac{E}{2}\int_{100}^{180}\left(1 + \frac{x-100}{40}\right)^2\left(-\frac{1}{80}u_B + \frac{1}{80}u_C\right)^2 dx - 100u_C$$

Invoking again $\delta\Pi = 0$ we obtain

$$\frac{E}{240}\begin{bmatrix} 15.4 & -13 \\ -13 & 13 \end{bmatrix}\begin{bmatrix} u_B \\ u_C \end{bmatrix} = \begin{bmatrix} 0 \\ 100 \end{bmatrix}$$

Hence, we now have

$$u_B = \frac{10000}{E}; \qquad u_C = \frac{11846.2}{E}$$

and

$$\sigma = 100; \qquad 0 \leq x \leq 100$$

$$\sigma = \frac{1846.2}{80} = 23.08; \qquad 100 \leq x \leq 180$$

We shall see in Chapter 4 (see Example 4.3) that this Ritz analysis can be regarded to be a finite element analysis.

EXAMPLE 3.23: Consider the slab of Example 3.16. Assume that

$$\theta(t) = \theta_1(t) + \theta_2(t)x + \theta_3(t)x^2 \tag{a}$$

where $\theta_1(t)$, $\theta_2(t)$, and $\theta_3(t)$ are the undetermined parameters. Use the Ritz analysis procedure to generate the governing heat transfer equilibrium equations.

The functional governing the temperature distribution in the slab is (see Example 3.18),

$$\Pi = \int_0^L \frac{1}{2} k \left(\frac{\partial \theta}{\partial x}\right)^2 dx - \int_0^L \theta q^B \, dx - \theta|_{x=0}\, q_0 \tag{b}$$

with the essential boundary condition

$$\theta|_{x=L} = \theta_i$$

Substituting the temperature assumption of (a) into (b) we obtain

$$\Pi = \int_0^L \frac{1}{2} k(\theta_2^2 + 4\theta_2\theta_3 x + 4\theta_3^2 x^2)\, dx - \int_0^L (\theta_1 + \theta_2 x + \theta_3 x^2) q^B \, dx - \theta_1 q_0$$

Invoking the stationarity condition of Π, i.e. $\delta\Pi = 0$, we use

$$\frac{\partial \Pi}{\partial \theta_1} = 0; \qquad \frac{\partial \Pi}{\partial \theta_2} = 0; \qquad \frac{\partial \Pi}{\partial \theta_3} = 0$$

and obtain

$$k \begin{bmatrix} 0 & 0 & 0 \\ 0 & L & L^2 \\ 0 & L^2 & \frac{4}{3}L^3 \end{bmatrix} \begin{bmatrix} \theta_1 \\ \theta_2 \\ \theta_3 \end{bmatrix} = \begin{bmatrix} \int_0^L q^B \, dx + q_0 \\ \int_0^L x q^B \, dx \\ \int_0^L x^2 q^B \, dx \end{bmatrix} \tag{c}$$

In this analysis q_0 varies with time, so that the temperature varies with time, and heat capacity effects can be important. Using

$$q^B = -\rho c \frac{\partial \theta}{\partial t} \tag{d}$$

because no other heat is generated, substituting for θ in (d) from (a), and then substituting into (c) we obtain as the equilibrium equations,

$$k \begin{bmatrix} 0 & 0 & 0 \\ 0 & L & L^2 \\ 0 & L^2 & \frac{4}{3}L^3 \end{bmatrix} \begin{bmatrix} \theta_1 \\ \theta_2 \\ \theta_3 \end{bmatrix} + \rho c \begin{bmatrix} L & \frac{1}{2}L^2 & \frac{1}{3}L^3 \\ \frac{1}{2}L^2 & \frac{1}{3}L^3 & \frac{1}{4}L^4 \\ \frac{1}{3}L^3 & \frac{1}{4}L^4 & \frac{1}{5}L^5 \end{bmatrix} \begin{bmatrix} \dot{\theta}_1 \\ \dot{\theta}_2 \\ \dot{\theta}_3 \end{bmatrix} = \begin{bmatrix} q_0 \\ 0 \\ 0 \end{bmatrix} \tag{e}$$

The final equilibrium equations are now obtained by imposing on the equations in (e) the condition that $\theta|_{x=L} = \theta_i$, i.e.

$$\theta_1(t) + \theta_2(t)L + \theta_3(t)L^2 = \theta_i$$

which can be achieved by expressing θ_1 in (e) in terms of θ_2, θ_3 and θ_i.

EXAMPLE 3.24: Consider the static buckling response of the column in Example 3.20. Assume that

$$w = a_1 x^2 + a_2 x^3 \tag{a}$$

and use the Ritz method to formulate the equations from which we can obtain an approximate buckling load.

The functional governing the problem was given in Example 3.20,

$$\Pi = \tfrac{1}{2}\int_0^L EI\left(\frac{d^2w}{dx^2}\right)^2 dx - \frac{P}{2}\int_0^L \left(\frac{dw}{dx}\right)^2 dx + \frac{1}{2}k(w|_{x=L})^2 \tag{b}$$

We note that the trial function on w in (a) satisfies already the essential boundary conditions (displacement and slope equal to zero at the fixed end). Substituting for w into (b) we obtain

$$\Pi = \tfrac{1}{2}\int_0^L EI(2a_1 + 6a_2 x)^2\, dx - \frac{P}{2}\int_0^L (2a_1 x + 3a_2 x^2)^2\, dx + \tfrac{1}{2}k(a_1 L^2 + a_2 L^3)^2$$

Invoking the stationarity condition $\delta\Pi = 0$, i.e.

$$\frac{\partial \Pi}{\partial a_1} = 0; \qquad \frac{\partial \Pi}{\partial a_2} = 0$$

we obtain

$$\left\{2EI\begin{bmatrix} 2L & 3L^2 \\ 3L^2 & 6L^3 \end{bmatrix} + kL^4\begin{bmatrix} 1 & L \\ L & L^2 \end{bmatrix}\right\}\begin{bmatrix} a_1 \\ a_2 \end{bmatrix} - PL^3\begin{bmatrix} \frac{4}{3} & \frac{3L}{2} \\ \frac{3L}{2} & \frac{9L^2}{5} \end{bmatrix}\begin{bmatrix} a_1 \\ a_2 \end{bmatrix} = \begin{bmatrix} 0 \\ 0 \end{bmatrix}$$

The solution of this eigenproblem gives two values of P for which w in (a) is nonzero. The smaller value of P represents an approximation to the lowest buckling load of the structure.

The weighted residual methods presented above are difficult to apply in practice, because the trial functions must satisfy all—essential and natural—boundary conditions. The functions must fulfill this requirement because we minimize in the use of (3.11) to (3.13) only the error in the satisfaction of the differential equation of equilibrium. Thus there is no mechanism by which an error in the satisfaction of the boundary conditions would be reduced, and only trial functions that satisfy all boundary conditions can be employed. On the other hand, with the Ritz method, which operates on the functional corresponding to the problem considered, the natural boundary conditions are contained in the functional. Hence, in the Ritz solution the error in the satisfaction of both the differential equations of equilibrium and the natural boundary conditions is minimized. As discussed above, the result is that, in the Ritz analysis, we can use functions that only satisfy the essential boundary conditions. This is, for practical analysis, a most significant point and largely accounts for the effectiveness of the displacement-based finite element analysis procedure (see Section 4.2.5).

From the above discussion it appears that it may be feasible to extend the formulation of a weighted residual method by including in the solution for the multipliers a_i a mechanism that would minimize the error in the satisfaction of the natural boundary conditions—that is, if these are not satisfied by the trial functions. Assuming now that the trial functions do not satisfy the natural boundary conditions, an additional error to the value of R given in (3.11) is

$$R_B = \sum_j \left\{q_j - B_j\left[\sum_{i=1}^{n} a_i f_i\right]\right\} \tag{3.15}$$

where we sum in j over all natural boundary conditions. This error must now also be minimized in the weighted residual method used and, in essence, all the procedures employed in the classical techniques summarized above are available. An important and effective technique is the *Galerkin method*, for which, corresponding to (3.12), the governing equations for the function multipliers a_i are now, with S denoting the boundary,

$$\int_D f_i R \, dD + \int_S f_i R_B \, dS = 0 \qquad i = 1, 2, \ldots, n \tag{3.16}$$

The least squares, collocation, and subdomain methods would be extended in an analogous way to include the natural boundary conditions, and it is also possible to use one method on R_B and another method on R.

The Galerkin method in (3.16) is a very general procedure and can be applied in the solution of structural, fluid, heat transfer problems and so on. The method provides also an effective basis for finite element solutions and is in wide use (see Chapter 7). We should note that to apply the Galerkin procedure (or any of the other weighted residual methods) there does not need to exist a functional corresponding to the problem considered.

Although the Galerkin method formulated in (3.16) does now admit the use of trial functions that do not satisfy the natural boundary conditions, when comparing the formulation of the Ritz method and the Galerkin method, we recognize that the Ritz method possesses still some significant advantages over the Galerkin method. First, the functions that we employ in the Galerkin method must be $2m$-times differentiable whereas the functions for the Ritz method need only be m-times differentiable. This may not be a serious disadvantage for the Galerkin method if trial functions are employed that span over the complete domain of the problem. However, considering a finite element solution, in which the complete domain is considered as an assemblage of subdomains it appears that higher continuity requirements on the trial functions between subdomains need be satisfied in the Galerkin method than in the Ritz method. A second disadvantage is that with a symmetric operator L_{2m} the Galerkin method as written in (3.16) leads in general to a nonsymmetric coefficient matrix, whereas the Ritz method always yields a symmetric matrix. Since the solution of a system of equations with a nonsymmetric coefficient matrix is a great deal more expensive, and since we also want to reduce the order to which the trial functions must be differentiable, we endeavor to transform the equations in (3.16) into a more suitable form. This transformation is achieved using integration by parts, which in the case of problems governed by a symmetric positive definite operator L_{2m} then leads to the same equations that are generated in the use of the Ritz method. We demonstrate this fact in the following example.

EXAMPLE 3.25: Consider the simple bar of Example 3.22. Formulate the analysis of the bar with the Galerkin method including the natural boundary condition.

In this case, the relation in (3.16) reduces to

$$\int_0^{180} \bar{u}_i \left\{ E \frac{d}{dx}\left(A \frac{du}{dx} \right) \right\} dx + \left(100 - EA \frac{du}{dx} \right) \bar{u}_i \bigg|_{x=180} = 0 \tag{a}$$

where $\bar{u}_i$ is the ith trial function, for example, using the assumption of Example 3.22,

$$u = a_1 x + a_2 x^2 \tag{b}$$

we have

$$\bar{u}_1 = x; \qquad \bar{u}_2 = x^2 \tag{c}$$

Using integration by parts on (a) we obtain

$$\bar{u}_i EA \frac{du}{dx}\bigg|_0^{180} - \int_0^{180} \left(\frac{d\bar{u}_i}{dx} \right)\left(EA \frac{du}{dx} \right) dx + \left(100 - EA \frac{du}{dx} \right) \bar{u}_i \bigg|_{x=180} = 0$$

Since $\bar{u}_i = 0$ at $x = 0$, we thus obtain

$$\int_0^{180} \left(\frac{d\bar{u}_i}{dx} \right)\left(EA \frac{du}{dx} \right) dx - 100\bar{u}_i \bigg|_{x=180} = 0 \tag{d}$$

However, the use of (d) above is completely equivalent to using the relation (d) in Example 3.22, and with the displacement assumptions given in (b) and (c), the Galerkin solution of the problem yields the relations in (g), (h), and (i) of Example 3.22.

We may also note that the relation in (d) above is in fact the principle of virtual displacements used in Chapter 4 for the formulation of the displacement-based finite element method.

In the above example we considered a symmetric positive definite operator such as may arise in structural and solid mechanics, and in the analysis of heat transfer and field problems. In this case a symmetric coefficient matrix is generated if the integration by parts is performed. Considering an operator that is nonsymmetric, as encountered in the analysis of fluid mechanics problems, integration by parts is in general still useful because the required order on the differentiability of the trial functions is reduced. However, in this case the coefficient matrix that is generated in the Galerkin procedure is nonsymmetric as dictated by the actual physical problem. We discuss the application of the Galerkin method to the analysis of fluid mechanics problems in Section 7.4.

3.4 IMPOSITION OF CONSTRAINTS

The analysis of an engineering problem frequently requires that a specific constraint be imposed on certain solution variables. These constraints may need to be imposed on some continuous solution parameters or on some discrete variables, and may consist of certain continuity requirements, the imposition of specified values for the solution variables, or conditions to be satisfied between certain solution variables. Two widely used procedures are available, namely the Lagrange multiplier method and the penalty method, which we will introduce here briefly. Further applications of these techniques are given in the following chapters (see Sections 4.2.2, 4.3.2, 5.4, and 7.4). Both, the Lagrange multiplier and the penalty methods operate on the variational or weighted residual formulation of the problem under consideration.

For purposes of demonstration, consider the variational formulation of a discrete structural system for a steady-state analysis,

$$\Pi = \tfrac{1}{2}\mathbf{U}^T\mathbf{K}\mathbf{U} - \mathbf{U}^T\mathbf{R} \tag{3.17}$$

with the conditions

$$\frac{\partial \Pi}{\partial U_i} = 0 \text{ (for all } i) \tag{3.18}$$

Assume that we want to impose the displacement at the degree of freedom U_i with

$$U_i = U_i^* \tag{3.19}$$

In the *Lagrange multiplier method* we amend the r.h.s. of (3.17) to obtain

$$\Pi^* = \tfrac{1}{2}\mathbf{U}^T\mathbf{K}\mathbf{U} - \mathbf{U}^T\mathbf{R} + \lambda(U_i - U_i^*) \tag{3.20}$$

where λ is an additional variable, and invoke that $\delta\Pi^* = 0$, which gives

$$\delta\mathbf{U}^T\mathbf{K}\mathbf{U} - \delta\mathbf{U}^T\mathbf{R} + \lambda\delta U_i + \delta\lambda(U_i - U_i^*) = 0 \tag{3.21}$$

Since $\delta\mathbf{U}$ and $\delta\lambda$ are arbitrary, we obtain

$$\begin{bmatrix} \mathbf{K} & \mathbf{e}_i \\ \mathbf{e}_i^T & 0 \end{bmatrix} \begin{bmatrix} \mathbf{U} \\ \lambda \end{bmatrix} = \begin{bmatrix} \mathbf{R} \\ U_i^* \end{bmatrix} \tag{3.22}$$

where $\mathbf{e}_i$ is a vector with all entries equal to zero, except its ith entry equal to one. Hence, the equilibrium equations without a constraint are amended with an additional equation that embodies the constraint condition.

In the *penalty method* we also amend the r.h.s. of (3.17) but without introducing an additional variable. Now we use

$$\Pi^* = \frac{1}{2}\mathbf{U}^T\mathbf{K}\mathbf{U} - \mathbf{U}^T\mathbf{R} + \frac{\alpha}{2}(U_i - U_i^*)^2 \tag{3.23}$$

in which α is a constant of relatively large magnitude, $\alpha \gg \max(k_{ii})$. The condition $\delta\Pi^* = 0$ now yields

$$\delta\mathbf{U}^T\mathbf{K}\mathbf{U} - \delta\mathbf{U}^T\mathbf{R} + \alpha(U_i - U_i^*)\,\delta U_i = 0 \tag{3.24}$$

and

$$[\mathbf{K} + \alpha\mathbf{e}_i\mathbf{e}_i^T]\mathbf{U} = \mathbf{R} + \alpha U_i^*\,\mathbf{e}_i \tag{3.25}$$

Hence, using this technique a large value is added to the ith diagonal element of $\mathbf{K}$ and a corresponding force is added so that the required displacement U_i is approximately equal to U_i^*. This is a general technique that has been used extensively to impose specified displacements. The effectiveness of the method lies in that no additional equation is required, and the bandwidth of the coefficient matrix is preserved (see Section 4.2.2). We demonstrate the Lagrange multiplier method and penalty procedure in the following example.

EXAMPLE 3.26: Use the Lagrange multiplier method and penalty procedure to analyze the simple spring system shown in Fig. 3.20 with the imposed displacement $U_2 = 1/k$.

The governing equilibrium equations without the imposed displacement U_2 are

$$\begin{bmatrix} 2k & -k \\ -k & k \end{bmatrix}\begin{bmatrix} U_1 \\ U_2 \end{bmatrix} = \begin{bmatrix} R_1 \\ R_2 \end{bmatrix} \quad \text{(a)}$$

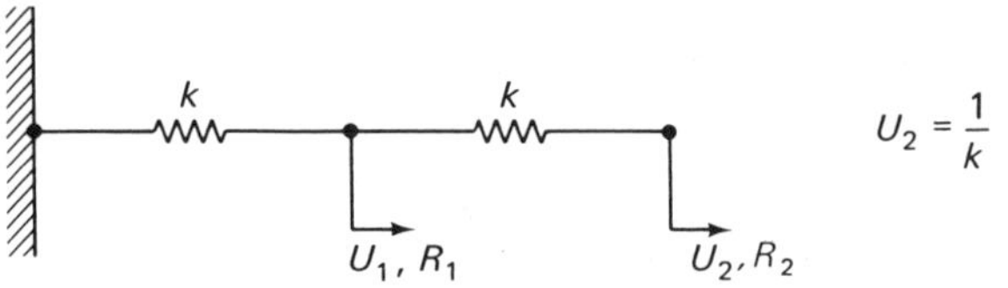

FIGURE 3.20 *A simple spring system.*

The exact solution is obtained by using the relation $U_2 = 1/k$ and solving from the first equation of (a) for U_1,

$$U_1 = \frac{1 + R_1}{2k} \quad \text{(b)}$$

Hence, we also have

$$R_2 = 1 - \frac{1 + R_1}{2}$$

which is the force required at the U_2 degree of freedom to impose $U_2 = 1/k$.

Using the Lagrange multiplier method, the governing equations are

$$\begin{bmatrix} 2k & -k & 0 \\ -k & k & 1 \\ 0 & 1 & 0 \end{bmatrix}\begin{bmatrix} U_1 \\ U_2 \\ \lambda \end{bmatrix} = \begin{bmatrix} R_1 \\ 0 \\ \frac{1}{k} \end{bmatrix} \quad \text{(c)}$$

and we obtain

$$U_1 = \frac{1 + R_1}{2k}; \qquad \lambda = -1 + \frac{1 + R_1}{2}$$

Hence, the solution in (b) is obtained and λ is equal to minus the force that need be applied at the degree of freedom U_2 in order to impose the displacement $U_2 = 1/k$. We may note that with this value of λ the first two equations in (c) reduce to the equations in (a).

Using the penalty method we obtain

$$\begin{bmatrix} 2k & -k \\ -k & (k + \alpha) \end{bmatrix}\begin{bmatrix} U_1 \\ U_2 \end{bmatrix} = \begin{bmatrix} R_1 \\ \frac{\alpha}{k} \end{bmatrix}$$

The solution now depends on α, and we obtain

$$\text{for } \alpha = 10k: \quad U_1 = \frac{11R_1 + 10}{21k} \qquad U_2 = \frac{R_1 + 20}{21k}$$

$$\text{for } \alpha = 100k: \quad U_1 = \frac{101R_1 + 100}{201k} \qquad U_2 = \frac{R_1 + 200}{201k}$$

$$\text{and for } \alpha = 1000k: \quad U_1 = \frac{1001R_1 + 1000}{2001k} \qquad U_2 = \frac{R_1 + 2000}{2001k}$$

In practice, the accuracy obtained using $\alpha = 1000k$ is usually sufficient.

The above example gives only a very elementary demonstration of the use of the Lagrange multiplier method and the penalty procedure. However, some basic observations can already be made that are quite generally applicable. First, we observe that in the Lagrange multiplier method the diagonal elements in the coefficient matrix corresponding to the Lagrange multipliers are zero. Hence, for the solution it is effective to arrange the equations as given in (3.22). Considering the equilibrium equations with the Lagrange multipliers, we also find that these multipliers have the same units as the forcing functions; for example, in (3.22) the Lagrange multipliers are forces.

Using the penalty method, an important consideration is the choice of an appropriate penalty number. In the analysis leading to (3.25) the penalty number α is explicitly specified (such as in Example 3.26), and this is frequently the case (see Section 4.2.2 and 7.4). However, in other analyses, the penalty number is defined by the problem itself using a specific formulation (see Sections 5.4.1 and 5.4.2). The difficulty in the use of a very high penalty number lies in that the coefficient matrix can become ill-conditioned when the off-diagonal elements are multiplied by a large number. If the off-diagonal elements are affected by the penalty number it is necessary to use enough digits in the computer arithmetical operations to ensure an accurate solution of the problem (see Section 8.5).

REFERENCES

1. S. H. Crandall, *Engineering Analysis*, McGraw-Hill Book Company, New York, 1956.
2. R. W. Clough and J. Penzien, *Dynamics of Structures*, McGraw-Hill Book Company, New York, 1975.
3. R. Courant and D. Hilbert, *Methods of Mathematical Physics*, John Wiley & Sons, Inc., New York, 1953.
4. L. Collatz, *The Numerical Treatment of Differential Equations*, Springer-Verlag, Berlin, 1st ed., 1950; 3rd ed., English translation, New York, 1966.
5. R. Courant, "Variational Methods for the Solution of Problems of Equilibrium and Vibrations," *Bulletin of the American Mathematical Society*, Vol. 49, 1943, pp. 1–23.
6. H. L. Langhaar, *Energy Methods in Applied Mechanics*, John Wiley & Sons, Inc., New York, 1962.
7. W. Ritz, "Über eine neue Methode zur Lösung gewisser Variationsprobleme der mathematischen Physik," *Zeitschrift für Angewandte Mathematik und Mechanik*, Vol. 135, Heft 1, 1908, pp. 1–61.
8. J. H. Argyris and S. Kelsey, "Energy Theorems and Structural Analysis." *Aircraft Engineering*, Vols. 26 and 27, 1955.

4

FORMULATION OF THE FINITE ELEMENT METHOD—LINEAR ANALYSIS IN SOLID AND STRUCTURAL MECHANICS

4.1 INTRODUCTION

The development of the finite element method as an analysis tool essentially began with the advent of the electronic digital computer. As discussed in the previous chapter, for the numerical solution of a structural or continuum problem it is basically necessary to establish and solve algebraic equations that govern the response of the system. Using the finite element method on a digital computer, it becomes possible to establish and solve the governing equations for complex problems in a very effective way. It is mainly due to the generality of the structure or continuum that can be analyzed, as well as the relative ease of establishing the governing equations, and for the good numerical properties of the system matrices involved that the finite element method has found wide appeal.

The finite element method was initially developed on a physical basis for the analysis of problems in structural mechanics; however, it was soon recognized that the method can be applied equally well to the solution of many other classes of problems. The objective in this chapter is to discuss finite element analysis procedures for linear problems in solid and structural mechanics. The application of finite element methods to other problem types is discussed in Chapters 6 and 7.

As is often the case with original development, it is rather difficult to quote an exact date on which the finite element method was "invented," but the roots of the method can be traced back to three separate research groups: applied mathematicians, physicists, and engineers.[1-6] Although in principle published already, the finite element method obtained its real impetus from the independent developments carried out by engineers. Important original contributions appeared in papers by Turner et al[5] and Argyris and Kelsey.[6] The name "finite element" was coined in a paper by Clough,[7] in which the technique was presented for plane stress analysis. Since then, a large amount of research has been devoted to the technique, and a very large number of publications on the finite element method is available at present (see for example also ref. [8, 9] for a compilation of references on the finite element method).

Today, the concept of the finite element method is a very broad one. Even when restricting ourselves to the analysis of structural and solid mechanics problems, the method can be used in a variety of different ways. This becomes particularly apparent when variational formulations are discussed. A most important formulation, which is widely used for the solution of practical problems, is the displacement-based finite element method. Practically all major general-purpose analysis programs have been written using this formulation, because of its simplicity, generality, and good numerical properties. In this chapter we shall first concentrate on this formulation, and other formulations, such as the equilibrium, hybrid, and mixed methods will be discussed briefly in Section 4.3. Although most of the discussion concerns the displacement-based finite element method, it should be pointed out that many of the concepts to be presented, and in particular the numerical techniques used for the implementation of the displacement-based finite element method are in many cases directly applicable to other formulations.

4.2 FORMULATION OF THE DISPLACEMENT-BASED FINITE ELEMENT METHOD

The displacement-based finite element method can be regarded as an extension of the displacement method of analysis, which had been used for many years in the analysis of beam and truss structures.[10] The basic steps in the analysis of a beam and truss structure using the displacement method of analysis are:

1. Idealize the total structure as an assemblage of beam and truss elements that are interconnected at the structural joints.
2. Identify the unknown joint displacements that completely define the displacement response of the structural idealization.
3. Establish force balance equations corresponding to the unknown joint displacements and solve these equations.
4. With the beam and truss element end-displacements known, calculate the internal element stress distributions.
5. Interpret the displacements and stresses predicted by the solution of the structural idealization when considering the assumptions used.

In practical analysis and design the most important steps of the complete analysis are the proper idealization of the actual problem, as performed in step (1), and the correct interpretation of the results in step (5). Depending on the complexity of the actual system to be analyzed, considerable knowledge of the characteristics of the system and its mechanical behavior may be required in order to establish an appropriate idealization, as briefly discussed in Sections 4.2.5 and 5.8.3.

The above analysis steps have already been demonstrated to some degree in Chapter 3, but it is instructive to consider another more complex example.

EXAMPLE 4.1: The piping system shown in Fig. 4.1 must be able to carry a large transverse load P when applied accidentally to the flange connecting the small and large diameter pipes. "Analyze this problem."

The study of this problem may require a number of analyses in which the local kinematic behavior of the pipe intersection is properly modeled, the nonlinear material and geometric behaviors are taken into account, the characteristics of the applied load are modeled accurately, and so on. In such a study, it is usually most

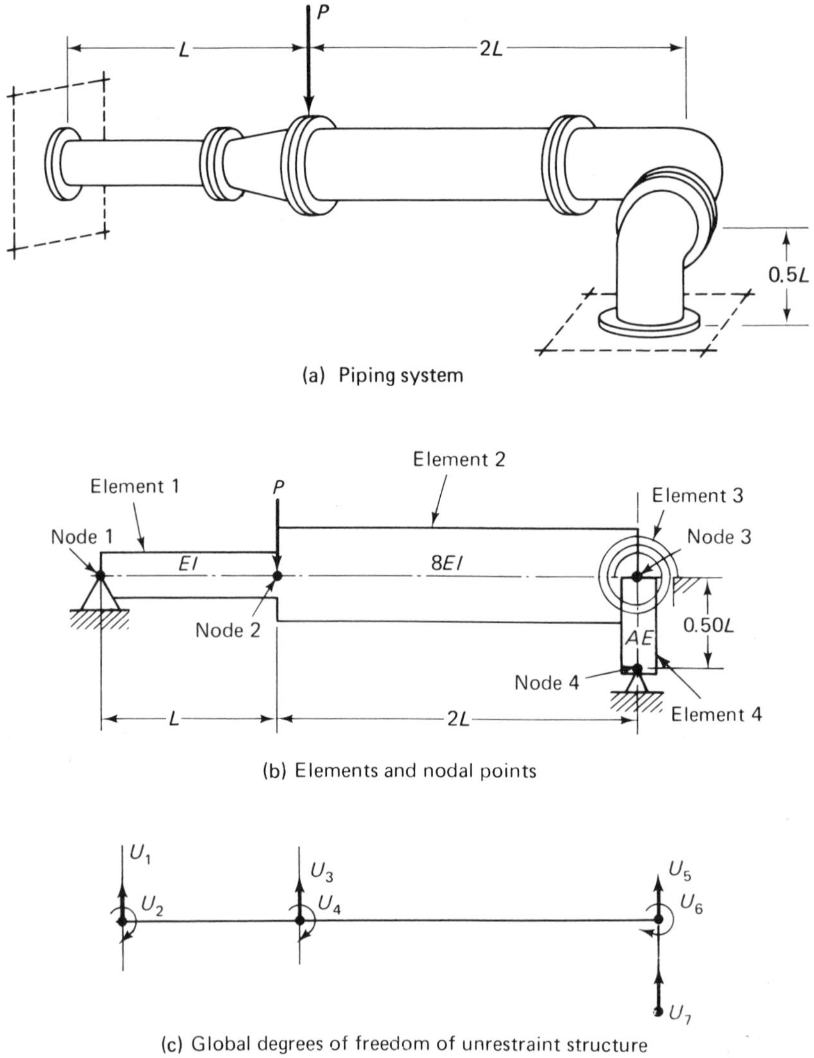

FIGURE 4.1 *Piping system and idealization.*

expedient to start with a simple analysis in which gross assumptions are made and then work towards a more refined model as need arises (see Section 6.5.1).

Assume that in a first analysis we primarily want to calculate the transverse displacement at the flange when the transverse load is applied slowly. In this case it is reasonable to model the structure as an assemblage of beam, truss, and spring elements and perform a static analysis.

The model chosen is shown in Fig. 4.1(b). The structural idealization consists of two beams, one truss, and a spring element. For the analysis of this idealization we first evaluate the element stiffness matrices that correspond to the global structural degrees of freedom shown in Fig. 4.1(c). For the beam, spring, and truss elements, respectively, we have in this case

$$\mathbf{K}_1^e = \frac{EI}{L}\begin{bmatrix} \frac{12}{L^2} & -\frac{6}{L} & -\frac{12}{L^2} & -\frac{6}{L} \\ & 4 & \frac{6}{L} & 2 \\ & & \frac{12}{L^2} & \frac{6}{L} \\ \text{symmetric} & & & 4 \end{bmatrix}; \qquad U_1, U_2, U_3, U_4$$

$$\mathbf{K}_2^e = \frac{EI}{L}\begin{bmatrix} \frac{12}{L^2} & -\frac{12}{L} & -\frac{12}{L^2} & -\frac{12}{L} \\ & 16 & \frac{12}{L} & 8 \\ & & \frac{12}{L^2} & \frac{12}{L} \\ \text{symmetric} & & & 16 \end{bmatrix}; \qquad U_3, U_4, U_5, U_6$$

$$\mathbf{K}_3^e = k_s; \qquad U_6$$

$$\mathbf{K}_4^e = \frac{EA}{L}\begin{bmatrix} 2 & -2 \\ -2 & 2 \end{bmatrix}; \qquad U_5, U_7$$

where the subscript on $\mathbf{K}^e$ indicates the element number, and the global degrees of freedom that correspond to the element stiffnesses are written next to the matrices. It should be noted that in this example the element matrices are independent of direction cosines since the centerlines of the elements are aligned with the global axes. If the local axis of an element would not be in the direction of a global axis, the local element stiffness matrix would have to be transformed to obtain the required global element stiffness matrix (see Example 4.7).

The stiffness matrix of the complete element assemblage is effectively obtained from the stiffness matrices of the individual elements using the *direct stiffness method* (see Examples 3.1 and 4.8). In this procedure the structure stiffness matrix $\mathbf{K}$ is calculated by direct addition of the element stiffness matrices, i.e.,

$$\mathbf{K} = \sum_i \mathbf{K}_i^e$$

where the summation includes all elements. To perform the summation, each element matrix $\mathbf{K}_i^e$ is written as a matrix $\mathbf{K}^{(i)}$ of the same order as the stiffness matrix $\mathbf{K}$, where all entries in $\mathbf{K}^{(i)}$ are zero, except those which correspond to an element degree of freedom. For example, for element 4 we have

$$
\mathbf{K}^{(4)} = \begin{array}{c} \\ 1 \\ 2 \\ 3 \\ 4 \\ 5 \\ 6 \\ 7 \end{array}
\begin{array}{c}
\begin{array}{ccccccc} 1 & 2 & 3 & 4 & 5 & 6 & 7 \leftarrow \text{degree of freedom} \end{array} \\
\begin{bmatrix}
0 & 0 & 0 & 0 & 0 & 0 & 0 \\
0 & 0 & 0 & 0 & 0 & 0 & 0 \\
0 & 0 & 0 & 0 & 0 & 0 & 0 \\
0 & 0 & 0 & 0 & 0 & 0 & 0 \\
0 & 0 & 0 & 0 & \frac{2AE}{L} & 0 & -\frac{2EA}{L} \\
0 & 0 & 0 & 0 & 0 & 0 & 0 \\
0 & 0 & 0 & 0 & -\frac{2AE}{L} & 0 & \frac{2EA}{L}
\end{bmatrix}
\end{array}
$$

Therefore, the stiffness matrix of the structure is

$$
\mathbf{K} = \begin{bmatrix}
\frac{12EI}{L^3} & -\frac{6EI}{L^2} & -\frac{12EI}{L^3} & -\frac{6EI}{L^2} & 0 & 0 & 0 \\
 & \frac{4EI}{L} & \frac{6EI}{L^2} & \frac{2EI}{L} & 0 & 0 & 0 \\
 & & \frac{24EI}{L^3} & -\frac{6EI}{L^2} & -\frac{12EI}{L^3} & -\frac{12EI}{L^2} & 0 \\
 & & & \frac{20EI}{L} & \frac{12EI}{L^2} & \frac{8EI}{L} & 0 \\
 & \text{symmetric} & & & \frac{12EI}{L^3} + \frac{2AE}{L} & \frac{12EI}{L^2} & -\frac{2AE}{L} \\
 & & & & & \frac{16EI}{L} + k_s & 0 \\
 & & & & & & \frac{2AE}{L}
\end{bmatrix}
$$

and the equilibrium equations for the system are

$$\mathbf{KU} = \mathbf{R}$$

where $\mathbf{U}$ is a vector of the system global displacements and $\mathbf{R}$ is a vector of forces acting into the direction of the structure global displacements:

$$\mathbf{U}^T = [U_1, \ldots, U_7]; \qquad \mathbf{R}^T = [R_1, \ldots, R_7]$$

Before solving for the displacements of the structure, we need to impose the boundary conditions that $U_1 = 0$ and $U_7 = 0$. This means that we may consider only five equations in five unknown displacements; i.e.,

$$\bar{\mathbf{K}}\bar{\mathbf{U}} = \bar{\mathbf{R}} \tag{a}$$

where $\bar{\mathbf{K}}$ is obtained by eliminating from $\mathbf{K}$ the first and seventh rows and columns, and $\quad \bar{\mathbf{U}} = [U_2 \quad U_3 \quad U_4 \quad U_5 \quad U_6]; \qquad \bar{\mathbf{R}}^T = [0 \quad -P \quad 0 \quad 0 \quad 0]$

The solution of (a) gives the structure displacements and therefore the element nodal point displacements. The element nodal forces are obtained by multiplying the element stiffness matrices $\mathbf{K}_i^e$ by the element displacements. If the forces at any section of an element are required, we can evaluate them from the element end-forces by use of simple statics.

Considering the analysis results it should be recognized, however, that although the structural idealization in Fig. 4.1(b) was analyzed accurately, *the displacements and stresses are only a prediction of the response of the actual physical structure.* Surely this prediction can only be accurate if the model used was appropriate, and in practice a specific model is in general adequate to predict certain quantities, but inadequate to predict others. For instance, in this analysis the required transverse displacement under the applied load is quite likely predicted accurately using the idealization in Fig. 4.1(b) (provided the load is applied slowly enough, the stresses are small enough not to cause yielding and so on), but the stresses directly under the load are probably predicted very inaccurately. Indeed, a different and more refined finite element model would need to be used in order to calculate accurately the stresses at the pipe intersections.

The above example demonstrates some important ingredients of the displacement method of analysis and the finite element method. As summarized previously, the basic process is that the complete structure is idealized as an assemblage of individual structural elements. The element stiffness matrices corresponding to the global degrees of freedom of the structural idealization are calculated and the total stiffness matrix is formed by the addition of the element stiffness matrices. The solution of the equilibrium equations of the assemblage of elements yields the element displacements, which are then used to calculate the element stresses. Finally, the element displacements and stresses must be interpreted as an estimate of the actual structural behavior, taking into account that a truss and beam idealization was solved.

Considering the analysis of truss and beam assemblages such as in Example 4.1, originally these solutions were not considered to be finite element analyses, because there is one major difference in these solutions when compared to a more general finite element analysis of a two- or three-dimensional problem; namely, in the analysis performed in Example 4.1 the exact element stiffness matrices ("exact" within beam theory) could be calculated. The stiffness properties of a beam element are physically the element end-forces that correspond to unit element end-displacements. These forces can be evaluated by solving the differential equations of equilibrium of the element when it is subjected to the appropriate boundary conditions. Since by virtue of the solution of the differential equations of equilibrium, all three requirements of an exact solution, namely, the stress equilibrium, the compatibility and the constitutive requirements, are fulfilled, the exact element internal displacements and stiffness matrices are calculated. In an alternative approach, these element end-forces could also be evaluated by performing a Ritz analysis based on the stationarity of the total potential of the system. The application of the Ritz method would give the exact element stiffness coefficients if the exact element internal displacements (as calculated in the solution of the differential equations of equilibrium) are used as the Ritz trial functions (see Examples 3.22 and 4.4). However, approximate stiffness coefficients are obtained if other trial functions (which may be more suitable in practice) are employed.

Considering more general two- and three-dimensional finite element analyses, we basically use the Ritz analysis technique with trial functions that approximate the actual displacements, because we do not know the exact displacement functions as in the case of truss and beam elements. The result is that the differential equations of equilibrium are not satisfied in general, but this error is reduced as the finite element idealization of the structure or the continuum is refined. We shall discuss the errors introduced, the convergence requirements and modeling considerations in a finite element analysis in Sections 4.2.5 and 5.8.3 after having presented the general formulation of the displacement-based finite element method. This general formulation is based on the use of the principle of virtual work, which is equivalent to the application of the Ritz method to minimize the total potential of the system, or the use of the Galerkin or least squares techniques (see Section 3.3.3).

4.2.1 General Derivation of Finite Element Equilibrium Equations

Consider the equilibrium of a general three-dimensional body such as in Fig. 4.2. The external forces acting onto the body are surface tractions $\mathbf{f}^S$, body forces $\mathbf{f}^B$, and concentrated forces $\mathbf{F}^i$. These forces include all externally applied forces and reactions and have in general three components corresponding to the three coordinate axes:

$$\mathbf{f}^B = \begin{bmatrix} f_X^B \\ f_Y^B \\ f_Z^B \end{bmatrix}; \qquad \mathbf{f}^S = \begin{bmatrix} f_X^S \\ f_Y^S \\ f_Z^S \end{bmatrix}; \qquad \mathbf{F}^i = \begin{bmatrix} F_X^i \\ F_Y^i \\ F_Z^i \end{bmatrix} \tag{4.1}$$

The displacements of the body from the unloaded configuration are denoted by $\mathbf{U}$, where

$$\mathbf{U}^T = [U \quad V \quad W] \tag{4.2}$$

The strains corresponding to $\mathbf{U}$ are,

$$\boldsymbol{\epsilon}^T = [\epsilon_{XX} \quad \epsilon_{YY} \quad \epsilon_{ZZ} \quad \gamma_{XY} \quad \gamma_{YZ} \quad \gamma_{ZX}] \tag{4.3}$$

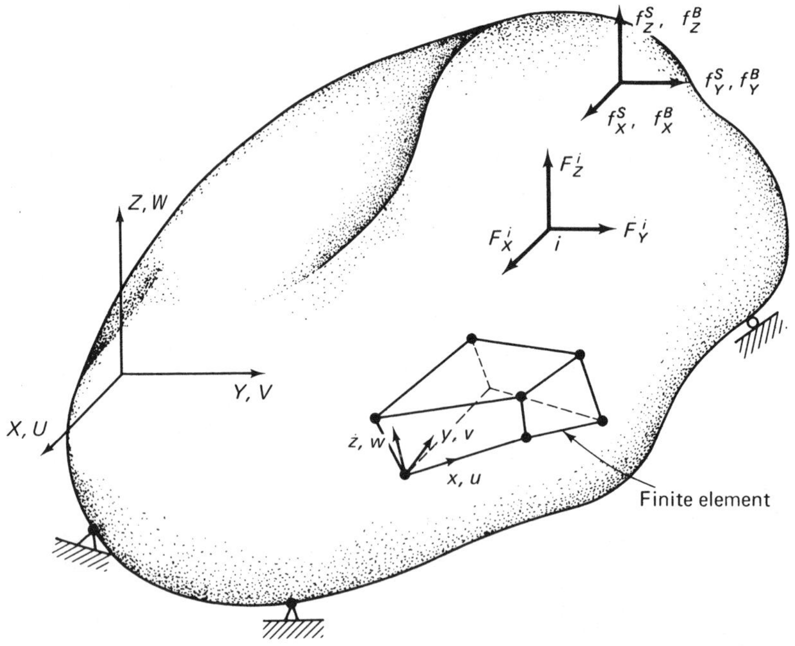

FIGURE 4.2 General three-dimensional body.

and the stresses corresponding to $\boldsymbol{\epsilon}$ are

$$\boldsymbol{\tau}^T = [\tau_{XX} \quad \tau_{YY} \quad \tau_{ZZ} \quad \tau_{XY} \quad \tau_{YZ} \quad \tau_{ZX}] \tag{4.4}$$

Assume that the externally applied forces are given and that we want to solve for the resulting displacements, strains and stresses in (4.2) to (4.4). To calculate the response of the body we could establish the governing differential equations of equilibrium, which would need to be solved subject to appropriate boundary and compatibility conditions (see Section 3.3). In this analysis the differential equations of equilibrium and natural boundary conditions could be established directly by equilibrium considerations on differential elements of the body or by using the stationarity condition of the total potential of the body (see Section 3.3.2).

An equivalent approach to express the equilibrium of the body is to use the *principle of virtual displacements.* This principle states that the equilibrium of the body requires that for any compatible, small virtual displacements (which satisfy the essential boundary conditions, see Section 3.3.2) imposed onto the body, the total internal virtual work is equal to the total external virtual work; i.e., we have

$$\int_V \bar{\boldsymbol{\epsilon}}^T \boldsymbol{\tau}\, dV = \int_V \bar{\mathbf{U}}^T \mathbf{f}^B\, dV + \int_S \bar{\mathbf{U}}^{S^T} \mathbf{f}^S\, dS + \sum_i \bar{\mathbf{U}}^{i^T} \mathbf{F}^i \tag{4.5}$$

The internal virtual work is given on the left side of (4.5), and is equal to the actual stresses $\boldsymbol{\tau}$ going through the virtual strains $\bar{\boldsymbol{\epsilon}}$ (that correspond to the imposed virtual displacements),

$$\bar{\boldsymbol{\epsilon}}^T = [\bar{\epsilon}_{XX} \quad \bar{\epsilon}_{YY} \quad \bar{\epsilon}_{ZZ} \quad \bar{\gamma}_{XY} \quad \bar{\gamma}_{YZ} \quad \bar{\gamma}_{ZX}] \tag{4.6}$$

The external work is given on the right side of (4.5) and is equal to the actual forces $\mathbf{f}^B$, $\mathbf{f}^S$, and $\mathbf{F}^i$ going through the virtual displacements $\bar{\mathbf{U}}$, where

$$\bar{\mathbf{U}}^T = [\bar{U} \quad \bar{V} \quad \bar{W}] \tag{4.7}$$

The superscript S denotes that surface displacements are considered and the superscript i denotes the displacements at the point where the concentrated forces $\mathbf{F}^i$ are applied.

It should be emphasized that the virtual strains used in (4.5) are those corresponding to the imposed body and surface virtual displacements, and that these displacements can be any compatible set of displacements that satisfy the geometric boundary conditions. The equation in (4.5) is an expression of equilibrium, and for different virtual displacements, correspondingly different equations of equilibrium are obtained. However, the equation in (4.5) also contains the compatibility and constitutive requirements if the principle is used in the appropriate manner (such as in the finite element formulation below); namely, the displacements considered should be continuous and compatible and should satisfy the displacement boundary conditions, and the stresses should be evaluated from the strains using the appropriate constitutive relations. Thus, the principle of virtual displacements embodies all requirements that need be fulfilled in the analysis of a problem in solid and structural mechanics. Finally, we may note that although the virtual work equation has been written in (4.5) in the global coordinate system X, Y, Z of the body, it is equally valid in any other system of coordinates.

The principle of virtual displacements can be directly related to the principle that the total potential Π of the system must be stationary (see Section 3.3.2). We study this relationship in the following example.

EXAMPLE 4.2: Show that for a linear elastic continuum the principle of virtual displacements is identical to the principle of stationarity of the total potential.

Assuming a linear elastic continuum, the total potential of the body in Fig. 4.2 is

$$\Pi = \frac{1}{2}\int_V \boldsymbol{\epsilon}^T \mathbf{C} \boldsymbol{\epsilon}\, dV - \int_V \mathbf{U}^T \mathbf{f}^B\, dV - \int_S \mathbf{U}^{S^T} \mathbf{f}^S\, dS - \sum_i \mathbf{U}^{i^T} \mathbf{F}^i \tag{a}$$

where the notation was defined in (4.1) to (4.4),

$$\boldsymbol{\tau} = \mathbf{C}\boldsymbol{\epsilon}$$

and $\mathbf{C}$ is the stress-strain matrix of the material.

Invoking the stationarity of Π, i.e. $\delta\Pi = 0$, and using that $\mathbf{C}$ is symmetric, we obtain

$$\int_V \delta\boldsymbol{\epsilon}^T \mathbf{C} \boldsymbol{\epsilon}\, dV = \int_V \delta\mathbf{U}^T \mathbf{f}^B\, dV + \int_S \delta\mathbf{U}^{S^T} \mathbf{f}^S\, dS + \sum_i \delta\mathbf{U}^{i^T} \mathbf{F}^i \tag{b}$$

However, to evaluate Π in (a) the displacements must satisfy the geometric boundary conditions. Hence in (b) we consider any variations on the displacements that satisfy the geometric boundary conditions, and the corresponding variations in strains. It follows that invoking the stationarity of Π is equivalent to the principle of virtual displacements, and indeed we may write

$$\delta\boldsymbol{\epsilon} \equiv \bar{\boldsymbol{\epsilon}}; \qquad \delta\mathbf{U} \equiv \bar{\mathbf{U}}; \qquad \delta\mathbf{U}^S \equiv \bar{\mathbf{U}}^S; \qquad \delta\mathbf{U}^i \equiv \bar{\mathbf{U}}^i$$

so that (b) reduces to (4.5).

Let us now consider how the principle of virtual displacements is used effectively as a mechanism to generate finite element equations that govern the response of a structure or continuum. We consider first the response of the general three-dimensional body shown in Fig. 4.2, and later specialize this general formulation to specific problems (see Section 4.2.3).

In the finite element analysis we approximate the body in Fig. 4.2 as an assemblage of discrete finite elements with the elements being interconnected at nodal points on the element boundaries. The displacements measured in a local coordinate system x, y, z (to be chosen conveniently) within each element are assumed to be a function of the displacements at the N finite element nodal points. Therefore, for element m we have

$$\mathbf{u}^{(m)}(x, y, z) = \mathbf{H}^{(m)}(x, y, z)\hat{\mathbf{U}} \tag{4.8}$$

where $\mathbf{H}^{(m)}$ is the displacement interpolation matrix, the superscript m denotes element m and $\hat{\mathbf{U}}$ is a vector of the three global displacement components U_i, V_i, and W_i at all nodal points, including those at the supports of the element assemblage; i.e., $\hat{\mathbf{U}}$ is a vector of dimension $3N$,

$$\hat{\mathbf{U}}^T = [U_1 V_1 W_1 \quad U_2 V_2 W_2 \quad \ldots \quad U_N V_N W_N] \tag{4.9a}$$

We may note here that more generally, we write

$$\hat{\mathbf{U}}^T = [U_1\ U_2\ U_3 \ldots U_n] \tag{4.9b}$$

where it is understood that U_i may correspond to a displacement in any direction, which may not even be aligned with a global coordinate axis, and U_i may also signify a rotation when we consider beams, plates, or shells as in Section 4.2.3.

Although all nodal point displacements are listed in $\hat{\mathbf{U}}$, it should be realized that for a given element only the displacements at the nodes of the element affect the displacement and strain distributions within the element. With the assumption on the displacements in (4.8) we can now evaluate the corresponding element strains,

$$\boldsymbol{\epsilon}^{(m)}(x, y, z) = \mathbf{B}^{(m)}(x, y, z)\hat{\mathbf{U}} \tag{4.10}$$

where $\mathbf{B}^{(m)}$ is the strain-displacement matrix; the rows of $\mathbf{B}^{(m)}$ are obtained by appropriately differentiating and combining rows of the matrix $\mathbf{H}^{(m)}$.

The purpose of defining the element displacements and strains in terms of the complete array of finite element assemblage nodal point displacements is probably not obvious now. *However, it will turn out that by using* (4.8) *and* (4.10), *the assemblage process of the element matrices to the structure matrices, referred to as the direct stiffness method* (*see Example* 4.1), *is automatically performed in the finite element application of the principle of virtual displacements.*

The stresses in a finite element are related to the element strains and the element initial stresses, using

$$\boldsymbol{\tau}^{(m)} = \mathbf{C}^{(m)}\boldsymbol{\epsilon}^{(m)} + \boldsymbol{\tau}^{I(m)} \tag{4.11}$$

where $\mathbf{C}^{(m)}$ is the elasticity matrix of element m and $\boldsymbol{\tau}^{I(m)}$ are the element initial stresses. The material law specified in $\mathbf{C}^{(m)}$ for each element can be that of an isotropic or anisotropic material and can vary from element to element.

Using the assumption on the displacements within each finite element, as expressed in (4.8), we can now derive equilibrium equations that correspond to the nodal point displacements of the assemblage of finite elements. First, we rewrite (4.5) as a sum of integrations over the volume and areas of all finite elements; i.e.,

$$\begin{aligned}\sum_m \int_{V^{(m)}} \bar{\boldsymbol{\epsilon}}^{(m)T}\boldsymbol{\tau}^{(m)}\, dV^{(m)} &= \sum_m \int_{V^{(m)}} \bar{\mathbf{U}}^{(m)T}\mathbf{f}^{B(m)}\, dV^{(m)} \\ &\quad + \sum_m \int_{S^{(m)}} \bar{\mathbf{U}}^{S(m)^T}\mathbf{f}^{S(m)}\, dS^{(m)} + \sum_i \bar{\mathbf{U}}^{iT}\mathbf{F}^i\end{aligned} \tag{4.12}$$

where $m = 1, 2, \ldots, k$ and $k =$ number of elements. It is important to note *that the integrations in* (4.12) *are performed over the element volumes and surfaces, and that for convenience we may use different element coordinate systems in the calculations.* Indeed, this is basically the reason why each of the integrals can be evaluated very effectively in general element assemblages. If we substitute into (4.12) for the element displacements, strains, and stresses, using (4.8) to (4.11), we obtain

$$\begin{aligned}\bar{\hat{\mathbf{U}}}^T\left[\sum_m \int_{V^{(m)}} \mathbf{B}^{(m)T}\mathbf{C}^{(m)}\mathbf{B}^{(m)}\, dV^{(m)}\right]\hat{\mathbf{U}} &= \bar{\hat{\mathbf{U}}}^T\Bigg[\left\{\sum_m \int_{V^{(m)}} \mathbf{H}^{(m)T}\mathbf{f}^{B(m)}\, dV^{(m)}\right\} \\ &\quad + \left\{\sum_m \int_{S^{(m)}} \mathbf{H}^{S(m)^T}\mathbf{f}^{S(m)}\, dS^{(m)}\right\} \\ &\quad - \left\{\sum_m \int_{V^{(m)}} \mathbf{B}^{(m)T}\boldsymbol{\tau}^{I(m)}\, dV^{(m)}\right\} + \mathbf{F}\Bigg]\end{aligned} \tag{4.13}$$

where the surface displacement interpolation matrices $\mathbf{H}^{S(m)}$ are obtained from the volume displacement interpolation matrices $\mathbf{H}^{(m)}$ in (4.8) by substituting the element surface coordinates (see Examples 4.6 and 5.12), and $\mathbf{F}$ is a vector of the externally applied forces to the nodes of the element assemblage. We should note that the ith component in $\mathbf{F}$ is the concentrated nodal force, which corresponds to the ith displacement component in $\hat{\mathbf{U}}$. In (4.13) the nodal point displacement vector $\hat{\mathbf{U}}$ of the element assemblage is independent of the element considered and is therefore taken out of the summation signs.

To obtain from (4.13) the equations for the unknown nodal point displacements, we invoke the virtual displacement theorem by imposing unit virtual displacements in turn at all displacement components. In this way we have $\bar{\hat{\mathbf{U}}}^T = \mathbf{I}$ ($\mathbf{I}$ = identity matrix), and denoting from now on the nodal point displacements by $\mathbf{U}$, i.e. letting $\hat{\mathbf{U}} \equiv \mathbf{U}$, the equilibrium equations of the element assemblage corresponding to the nodal point displacements are

$$\mathbf{K}\mathbf{U} = \mathbf{R} \tag{4.14}$$

where

$$\mathbf{R} = \mathbf{R}_B + \mathbf{R}_S - \mathbf{R}_I + \mathbf{R}_C \tag{4.15}$$

The matrix $\mathbf{K}$ is the stiffness matrix of the element assemblage,

$$\mathbf{K} = \sum_m \underbrace{\int_{V^{(m)}} \mathbf{B}^{(m)T}\mathbf{C}^{(m)}\mathbf{B}^{(m)}\, dV^{(m)}}_{= \mathbf{K}^{(m)}} \tag{4.16}$$

The load vector $\mathbf{R}$ includes the effect of the element body forces,

$$\mathbf{R}_B = \sum_m \underbrace{\int_{V^{(m)}} \mathbf{H}^{(m)T}\mathbf{f}^{B(m)}\, dV^{(m)}}_{= \mathbf{R}_B^{(m)}} \tag{4.17}$$

the effect of the element surface forces,

$$\mathbf{R}_S = \sum_m \underbrace{\int_{S^{(m)}} \mathbf{H}^{S(m)T}\mathbf{f}^{S(m)}\, dS^{(m)}}_{= \mathbf{R}_S^{(m)}} \tag{4.18}$$

the effect of the element initial stresses,

$$\mathbf{R}_I = \sum_m \underbrace{\int_{V^{(m)}} \mathbf{B}^{(m)T}\boldsymbol{\tau}^{I(m)}\, dV^{(m)}}_{= \mathbf{R}_I^{(m)}} \tag{4.19}$$

and the concentrated loads,

$$\mathbf{R}_C = \mathbf{F} \tag{4.20}$$

We note that the summation of the element volume integrals in (4.16) expresses the direct addition of the element stiffness matrices $\mathbf{K}^{(m)}$ to obtain the stiffness matrix of the total element assemblage. In the same way, the assemblage body force vector $\mathbf{R}_B$ is calculated by directly adding the element body force vectors $\mathbf{R}_B^{(m)}$; and similarly $\mathbf{R}_S$, $\mathbf{R}_I$, and $\mathbf{R}_C$ are obtained. Therefore, the formulation of the equilibrium equations in (4.14) in the way carried out above includes the assemblage process to obtain the structure matrices from the element matrices, usually referred to as the direct stiffness method. However, in practice, the finite element matrices are calculated in compact form; i.e., they are of the order of the element degrees of freedom and are assembled using the direct stiffness method as described in Examples 4.1 and 4.8 (and the appendix).

Equation (4.14) is a statement of the static equilibrium of the element assemblage. In these equilibrium considerations, the applied forces may vary with time, in which case the displacements vary also with time and (4.14) is a statement of equilibrium for any specific point in time. (In practice, the time-dependent application of loads can thus be used to model multiple-load cases, see Example 4.3.) However, if in actuality the loads are applied rapidly, inertia forces need to be considered; i.e., a truly dynamic problem needs to be solved. Using d'Alembert's principle, we can simply include the element inertia forces as part of the body forces. Assuming that the element accelerations are approximated in the same way as the element displacements in (4.8), the contribution from the total body forces to the load vector **R** is (with the coordinate system X-Y-Z stationary)

$$\mathbf{R}_B = \sum_m \int_{V^{(m)}} \mathbf{H}^{(m)T}[\mathbf{f}^{B(m)} - \rho^{(m)}\mathbf{H}^{(m)}\ddot{\mathbf{U}}]\, dV^{(m)} \tag{4.21}$$

where $\mathbf{f}^{B(m)}$ no longer includes inertia forces, $\ddot{\mathbf{U}}$ lists the nodal point accelerations (i.e., is the second time derivative of **U**), and $\rho^{(m)}$ is the mass density of element m. The equilibrium equations are, in this case,

$$\mathbf{M}\ddot{\mathbf{U}} + \mathbf{K}\mathbf{U} = \mathbf{R} \tag{4.22}$$

where **R** and **U** are time dependent. The matrix **M** is the mass matrix of the structure,

$$\mathbf{M} = \sum_m \underbrace{\int_{V^{(m)}} \rho^{(m)}\mathbf{H}^{(m)T}\mathbf{H}^{(m)}\, dV^{(m)}}_{=\mathbf{M}^{(m)}} \tag{4.23}$$

In actually measured dynamic response of structures it is observed that energy is dissipated during vibration, which, in vibration analysis is usually taken account of by introducing velocity-dependent damping forces. Introducing the damping forces as additional contributions to the body forces, we obtain corresponding to (4.21),

$$\mathbf{R}_B = \sum_m \int_{V^{(m)}} \mathbf{H}^{(m)T}[\mathbf{f}^{B(m)} - \rho^{(m)}\mathbf{H}^{(m)}\ddot{\mathbf{U}} - \kappa^{(m)}\mathbf{H}^{(m)}\dot{\mathbf{U}}]\, dV^{(m)} \tag{4.24}$$

In this case the vectors $\mathbf{f}^{B(m)}$ no longer include inertia and velocity dependent damping forces, $\dot{\mathbf{U}}$ is a vector of the nodal point velocities (i.e., the first time derivative of **U**), and $\kappa^{(m)}$ is the damping property parameter of element m. The equilibrium equations are, in this case,

$$\mathbf{M}\ddot{\mathbf{U}} + \mathbf{C}\dot{\mathbf{U}} + \mathbf{K}\mathbf{U} = \mathbf{R} \tag{4.25}$$

where **C** is the damping matrix of the structure; i.e., formally,

$$\mathbf{C} = \sum_m \underbrace{\int_{V^{(m)}} \kappa^{(m)}\mathbf{H}^{(m)T}\mathbf{H}^{(m)}\, dV^{(m)}}_{=\mathbf{C}^{(m)}} \tag{4.26}$$

In practice it is difficult, if not impossible, to determine for general finite element assemblages the element damping parameters, in particular because the damping properties are frequency dependent. For this reason, the matrix **C** is in general not assembled from element damping matrices, but is constructed using the mass matrix and stiffness matrix of the complete element assemblage together with experimental results on the amount of damping. Some formulations used to construct physically significant damping matrices are described in Section 9.3.3.

To illustrate the above derivation of the finite element equilibrium equations, we consider the following example.

EXAMPLE 4.3 Establish the finite element equilibrium equations of the bar structure shown in Fig. 4.3. Use the two-node bar element idealization given and consider the following two cases:

1. Assume that the loads are applied very slowly when measured on the time constants (natural periods) of the structure.
2. Assume that the loads are applied rapidly. The structure is initially at rest.

In the formulation of the finite element equilibrium equations we use the general equations (4.8) to (4.22), but recognize that the only nonzero stress is the longitudinal stress in the bar. Furthermore, considering the complete bar as an assemblage of 2 two-node bar elements corresponds to assuming a linear displacement variation between the nodal points of each element.

The first step is to construct the matrices $\mathbf{H}^{(m)}$ and $\mathbf{B}^{(m)}$ for $m = 1, 2$. Corresponding to the displacement vector $\mathbf{U}^T = [U_1 U_2 U_3]$ we have

$$\mathbf{H}^{(1)} = \left[\left(1 - \frac{x}{100}\right) \quad \frac{x}{100} \quad 0\right]$$

$$\mathbf{B}^{(1)} = \left[-\frac{1}{100} \quad \frac{1}{100} \quad 0\right]$$

$$\mathbf{H}^{(2)} = \left[0 \quad \left(1 - \frac{x}{80}\right) \quad \frac{x}{80}\right]$$

$$\mathbf{B}^{(2)} = \left[0 \quad -\frac{1}{80} \quad \frac{1}{80}\right]$$

The material property matrices are

$$\mathbf{C}^{(1)} = E; \qquad \mathbf{C}^{(2)} = E$$

where E is Young's modulus of the material. For the volume integrations we need the cross-sectional areas of the elements. We have

$$A^{(1)} = 1 \text{ cm}^2; \qquad A^{(2)} = \left(1 + \frac{x}{40}\right)^2 \text{ cm}^2$$

When the loads are applied very slowly, a static analysis is required, in which the stiffness matrix $\mathbf{K}$ and load vector $\mathbf{R}$ must be calculated. The body forces and loads are given in Fig. 4.3. We therefore have

$$\mathbf{K} = (1)E \int_0^{100} \begin{bmatrix} -\frac{1}{100} \\ \frac{1}{100} \\ 0 \end{bmatrix} \left[-\frac{1}{100} \quad \frac{1}{100} \quad 0\right] dx$$

$$+ E \int_0^{80} \left(1 + \frac{x}{40}\right)^2 \begin{bmatrix} 0 \\ -\frac{1}{80} \\ \frac{1}{80} \end{bmatrix} \left[0 \quad -\frac{1}{80} \quad \frac{1}{80}\right] dx$$

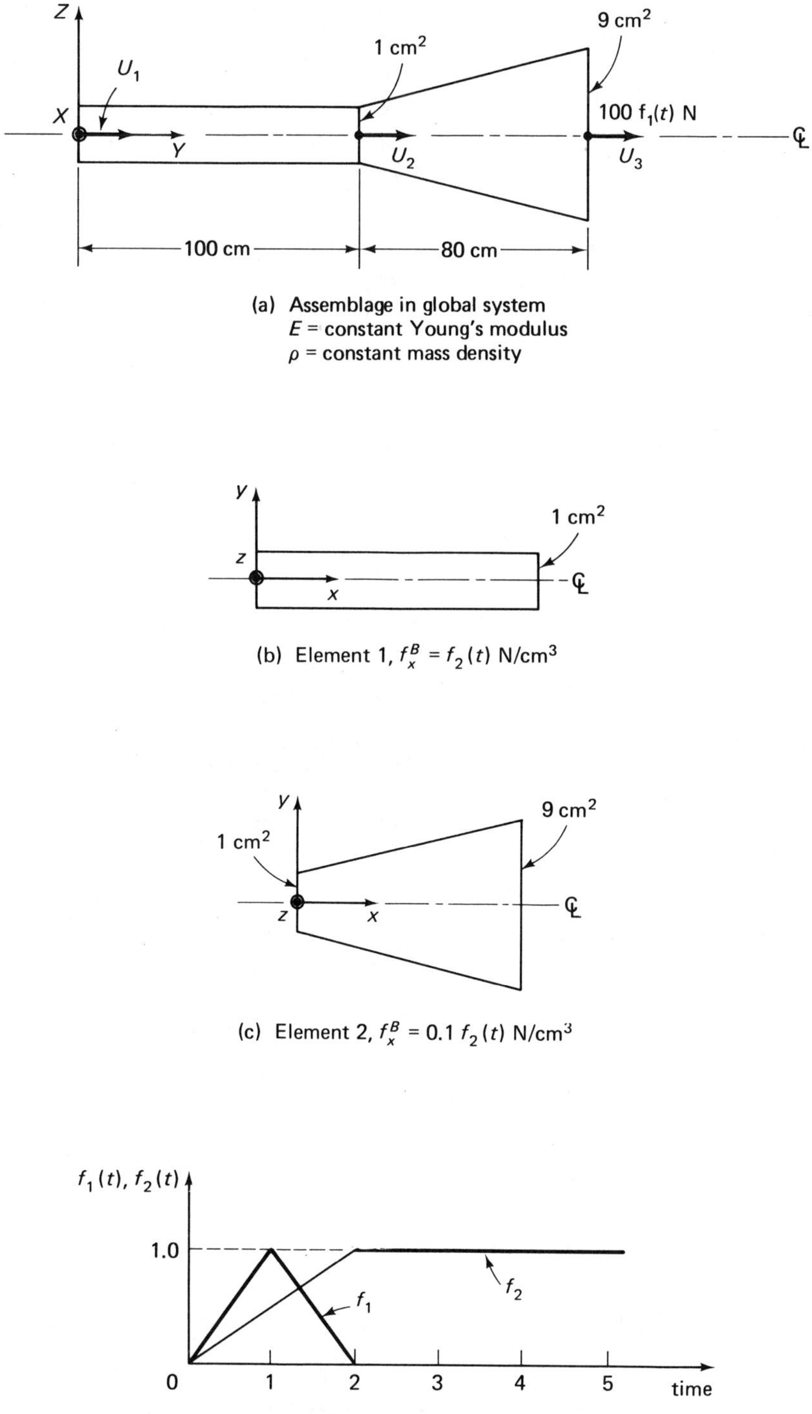

(a) Assemblage in global system
E = constant Young's modulus
ρ = constant mass density

(b) Element 1, $f_x^B = f_2(t)$ N/cm^3

(c) Element 2, $f_x^B = 0.1\, f_2(t)$ N/cm^3

(d) Time variation of loads

FIGURE 4.3 *Two-element bar assemblage.*

or

$$\mathbf{K} = \frac{E}{100}\begin{bmatrix} 1 & -1 & 0 \\ -1 & 1 & 0 \\ 0 & 0 & 0 \end{bmatrix} + \frac{13E}{240}\begin{bmatrix} 0 & 0 & 0 \\ 0 & 1 & -1 \\ 0 & -1 & 1 \end{bmatrix}$$

$$= \frac{E}{240}\begin{bmatrix} 2.4 & -2.4 & 0 \\ -2.4 & 15.4 & -13 \\ 0 & -13 & 13 \end{bmatrix} \tag{a}$$

and also,

$$\mathbf{R}_B = \left\{(1)\int_0^{100}\begin{bmatrix} 1 - \dfrac{x}{100} \\ \dfrac{x}{100} \\ 0 \end{bmatrix}(1)\,dx + \int_0^{80}\left(1 + \frac{x}{40}\right)^2\begin{bmatrix} 0 \\ 1 - \dfrac{x}{80} \\ \dfrac{x}{80} \end{bmatrix}\left(\frac{1}{10}\right)dx\right\}f_2(t)$$

$$= \frac{1}{3}\begin{bmatrix} 150 \\ 186 \\ 68 \end{bmatrix} f_2(t) \tag{b}$$

$$\mathbf{R}_C = \begin{bmatrix} 0 \\ 0 \\ 100 \end{bmatrix} f_1(t) \tag{c}$$

To obtain the solution at a specific time t^*, the vectors $\mathbf{R}_B$ and $\mathbf{R}_C$ must be evaluated corresponding to t^*, and the equation

$$\mathbf{KU}|_{t=t^*} = \mathbf{R}_B|_{t=t^*} + \mathbf{R}_C|_{t=t^*}$$

yields the displacements at t^*. We should note that in this static analysis the displacements at time t^* depend only on the magnitude of the loads at that time and are independent of the loading history.

Considering now the dynamic analysis, we also need to calculate the mass matrix. Using the displacement interpolations and (4.23), we have

$$\mathbf{M} = (1)\rho\int_0^{100}\begin{bmatrix} 1 - \dfrac{x}{100} \\ \dfrac{x}{100} \\ 0 \end{bmatrix}\left[\left(1 - \frac{x}{100}\right) \quad \frac{x}{100} \quad 0\right]dx$$

$$+ \rho\int_0^{80}\left(1 + \frac{x}{40}\right)^2\begin{bmatrix} 0 \\ 1 - \dfrac{x}{80} \\ \dfrac{x}{80} \end{bmatrix}\left[0 \quad \left(1 - \frac{x}{80}\right) \quad \frac{x}{80}\right]dx$$

Hence

$$\mathbf{M} = \frac{\rho}{6}\begin{bmatrix} 200 & 100 & 0 \\ 100 & 584 & 336 \\ 0 & 336 & 1024 \end{bmatrix}$$

Damping was not specified; thus, the equilibrium equations to be solved now are

$$\mathbf{M}\ddot{\mathbf{U}}(t) + \mathbf{KU}(t) = \mathbf{R}_B(t) + \mathbf{R}_C(t)$$

where the stiffness matrix $\mathbf{K}$ and load vectors $\mathbf{R}_B$ and $\mathbf{R}_C$ were already given in (a) to (c). Using now also the initial conditions

$$\mathbf{U}|_{t=0} = \mathbf{0}; \qquad \dot{\mathbf{U}}|_{t=0} = \mathbf{0}$$

these dynamic equilibrium equations need be integrated from time 0 to time t^* in order to obtain the solution at time t^* (see Chapter 9).

Finally, it should be noted that for the solution in the static analysis, a nodal point displacement boundary condition must still be imposed in order to obtain a stable structure, whereas in the dynamic analysis such a displacement boundary condition is not necessary but frequently corresponds to the actual physical situation.

We noted earlier that the analyses of truss and beam assemblages were originally not considered to be finite element analyses, because the "exact" element stiffness matrices can be employed in the analyses. These stiffness matrices are obtained in the application of the principle of virtual displacements, if the assumed displacement interpolations are in fact the "exact" displacements that the element undergoes when subjected to the unit nodal point displacements. Here the word "exact" refers to the fact that imposing these displacements onto the element, all pertinent differential equations of equilibrium and compatibility and the constitutive requirements (and also the boundary conditions) are fully satisfied in static analysis.

Considering the analysis of the truss assemblage in Example 4.3, we obtained the exact stiffness matrix of element 1. However, for element 2 an approximate stiffness matrix was calculated as shown in the next example.

EXAMPLE 4.4: Calculate for element 2 of Example 4.3 the exact element internal displacements that correspond to a unit element end-displacement U_3 and evaluate the corresponding stiffness coefficient. Also show that using the element displacement assumption in Example 4.3 internal element equilibrium is not satisfied.

Consider element 2 with a unit displacement imposed at its right end as shown in Fig. 4.4. The element displacements are calculated by solving the differential equation (see Example 3.22),

$$E\frac{d}{dx}\left(A\frac{du}{dx}\right) = 0 \tag{a}$$

subject to the boundary conditions $u|_{x=0} = 0$ and $u|_{x=80} = 1.0$. Substituting for the area A and integrating the relation in (a), we obtain

$$u = \frac{3}{2}\left(1 - \frac{1}{1 + \dfrac{x}{40}}\right) \tag{b}$$

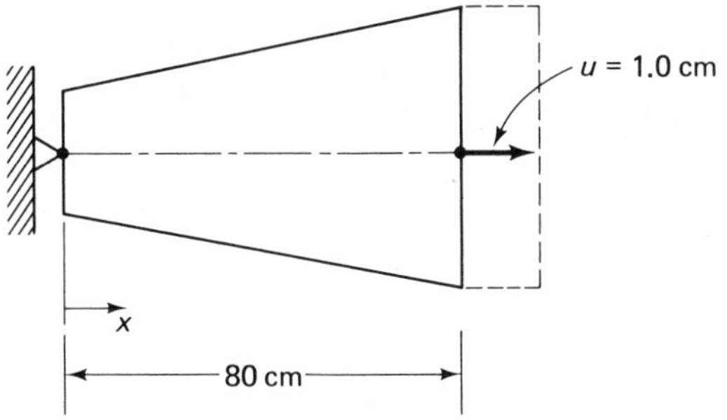

FIGURE 4.4 *Element 2 of bar analyzed in Example 4.3.*

These are the "exact" element internal displacements. The element end-forces required to subject the bar to these displacements are

$$k_{12} = EA \left.\frac{du}{dx}\right|_{x=0}$$
$$k_{22} = EA \left.\frac{du}{dx}\right|_{x=L} \tag{c}$$

Substituting from (b) into (c) we have

$$k_{22} = \frac{3E}{80}; \qquad k_{12} = -\frac{3E}{80}$$

and we note that the stiffness coefficients used in Example 4.3 were larger than those calculated here (see Section 4.2.5 for the explanation).

Finally, using the displacements assumed in Example 4.3 we recognize that the differential equation of equilibrium in (a) is not satisfied,

$$E\frac{d}{dx}\left\{\left(1+\frac{x}{40}\right)^2 \frac{1}{80}\right\} \neq 0$$

Hence, in Example 4.3 the stiffness coefficients have been calculated without satisfying the internal element equilibrium.

Next we consider some illustrative two-dimensional displacement and stress solutions.

EXAMPLE 4.5: Consider the analysis of the cantilever plate shown in Fig. 4.5. To illustrate the analysis technique, use the coarse finite element idealization given in the figure (in a practical analysis more finite elements need to be employed, see Section 5.8.3). Establish the matrices $\mathbf{H}^{(2)}$, $\mathbf{B}^{(2)}$ and $\mathbf{C}^{(2)}$.

The cantilever plate is acting in plane stress conditions. For an isotropic linear elastic material the stress-strain matrix is defined using Young's modulus E and Poisson's ratio ν (see Table 4.3),

$$\mathbf{C}^{(2)} = \frac{E}{1-\nu^2}\begin{bmatrix} 1 & \nu & 0 \\ \nu & 1 & 0 \\ 0 & 0 & \dfrac{1-\nu}{2} \end{bmatrix}$$

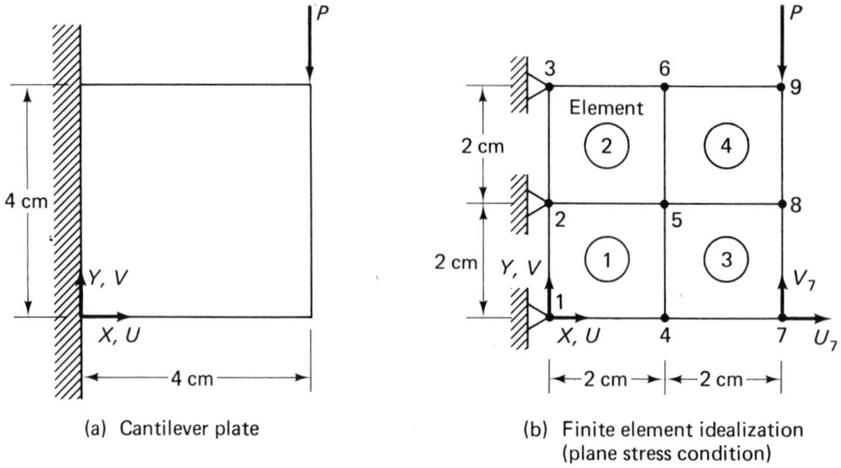

FIGURE 4.5 *Finite element plane stress analysis.*

The displacement transformation matrix $\mathbf{H}^{(2)}$ of element 2 relates the element internal displacements to the nodal point displacements.

$$\begin{bmatrix} u(x, y) \\ v(x, y) \end{bmatrix}^{(2)} = \mathbf{H}^{(2)}\mathbf{U} \tag{a}$$

where $\mathbf{U}$ is a vector listing all nodal point displacements of the structure,

$$\mathbf{U}^T = [U_1 \quad U_2 \quad U_3 \quad U_4 \quad \ldots \quad U_{17} \quad U_{18}] \tag{b}$$

Considering element 2, we recognize that only the displacements at the nodes 6, 3, 2, and 5 affect the displacements in the element. For computational purposes it is convenient to use a convention to number the element nodal points and corresponding element degrees of freedom as shown in Fig. 4.6. In the same figure also the global structure degrees of freedom of the vector $\mathbf{U}$ in (b) are given.

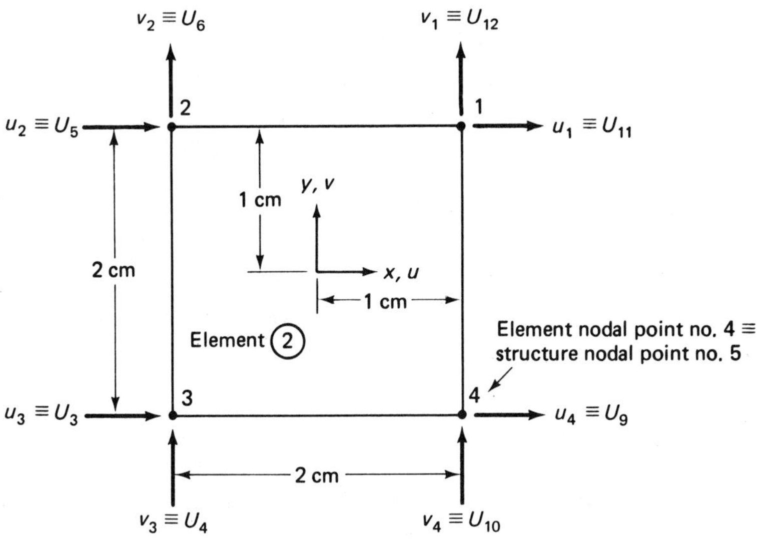

FIGURE 4.6 *Typical two-dimensional four-node element defined in local coordinate system.*

To derive the matrix $\mathbf{H}^{(2)}$ in (a) we recognize that there are four nodal point displacements each to express $u(x, y)$ and $v(x, y)$. Hence we can assume that the local element displacements u and v are given in the following form of polynomials in the local coordinate variables x and y:

$$\begin{aligned} u(x, y) &= \alpha_1 + \alpha_2 x + \alpha_3 y + \alpha_4 xy \\ v(x, y) &= \beta_1 + \beta_2 x + \beta_3 y + \beta_4 xy \end{aligned} \tag{c}$$

The unknown coefficients $\alpha_1, \ldots, \beta_4$, which are also called the generalized coordinates, will be expressed in terms of the unknown element nodal point displacements $u_1, \ldots, u_4$ and $v_1, \ldots, v_4$. Defining

$$\hat{\mathbf{u}}^T = [u_1 \quad u_2 \quad u_3 \quad u_4 \;\vdots\; v_1 \quad v_2 \quad v_3 \quad v_4] \tag{d}$$

we can write (c) in matrix form:

$$\begin{bmatrix} u(x, y) \\ v(x, y) \end{bmatrix} = \boldsymbol{\Phi\alpha} \tag{e}$$

where $$\boldsymbol{\Phi} = \begin{bmatrix} \boldsymbol{\phi} & \mathbf{0} \\ \mathbf{0} & \boldsymbol{\phi} \end{bmatrix}; \qquad \boldsymbol{\phi} = [1 \quad x \quad y \quad xy]$$

and $$\boldsymbol{\alpha}^T = [\alpha_1 \; \alpha_2 \; \alpha_3 \; \alpha_4 \mid \beta_1 \; \beta_2 \; \beta_3 \; \beta_4]$$

Equation (e) must hold for all nodal points of the element; therefore, using (d) we have

$$\hat{\mathbf{u}} = \mathbf{A}\boldsymbol{\alpha} \tag{f}$$

in which $$\mathbf{A} = \begin{bmatrix} \mathbf{A}_1 & \mathbf{0} \\ \mathbf{0} & \mathbf{A}_1 \end{bmatrix}$$

and $$\mathbf{A}_1 = \begin{bmatrix} 1 & 1 & 1 & 1 \\ 1 & -1 & 1 & -1 \\ 1 & -1 & -1 & 1 \\ 1 & 1 & -1 & -1 \end{bmatrix}$$

Solving from (f) for $\boldsymbol{\alpha}$ and substituting into (e) we obtain

$$\mathbf{H} = \boldsymbol{\Phi}\mathbf{A}^{-1} \tag{g}$$

where the fact that no superscript is used on $\mathbf{H}$ indicates that the displacement interpolation matrix is defined corresponding to the element nodal point displacements in (d),

$$\mathbf{H} = \frac{1}{4}\begin{bmatrix} (1+x)(1+y) & (1-x)(1+y) & (1-x)(1-y) & (1+x)(1-y) & 0 & 0 & 0 & 0 \\ 0 & 0 & 0 & 0 & (1+x)(1+y) & (1-x)(1+y) & (1-x)(1-y) & (1+x)(1-y) \end{bmatrix} \tag{h}$$

Let H_{ij} be the (i, j)th element of $\mathbf{H}$, then we have

$$\mathbf{H}^{(2)} = \begin{array}{cccccccccc} & & u_3 & v_3 & u_2 & v_2 & & & u_4 & v_4 \\ U_1 & U_2 & U_3 & U_4 & U_5 & U_6 & U_7 & U_8 & U_9 & U_{10} \\ 0 & 0 & H_{13} & H_{17} & H_{12} & H_{16} & 0 & 0 & H_{14} & H_{18} \\ 0 & 0 & H_{23} & H_{27} & H_{22} & H_{26} & 0 & 0 & H_{24} & H_{28} \end{array}$$

$$\begin{array}{cccccc} u_1 & v_1 \leftarrow \text{element degrees of freedom} & & & & \\ U_{11} & U_{12} & U_{13} & U_{14} & & U_{18} \leftarrow \text{assemblage degrees of freedom} \\ H_{11} & H_{15} & 0 & 0 & \ldots \text{zeros} \ldots & 0 \\ H_{21} & H_{25} & 0 & 0 & \ldots \text{zeros} \ldots & 0 \end{array} \tag{i}$$

The strain-displacement matrix can now directly be obtained from (g). In plane stress conditions the element strains are

$$\boldsymbol{\epsilon}^T = [\epsilon_{xx} \quad \epsilon_{yy} \quad \gamma_{xy}]$$

where $$\epsilon_{xx} = \frac{\partial u}{\partial x}; \qquad \epsilon_{yy} = \frac{\partial v}{\partial y}; \qquad \gamma_{xy} = \frac{\partial u}{\partial y} + \frac{\partial v}{\partial x}$$

Using (g) and recognizing that the elements in $\mathbf{A}^{-1}$ are independent of x and y we obtain

$$\mathbf{B} = \mathbf{E}\mathbf{A}^{-1}$$

where $$\mathbf{E} = \begin{bmatrix} 0 & 1 & 0 & y & 0 & 0 & 0 & 0 \\ 0 & 0 & 0 & 0 & 0 & 0 & 1 & x \\ 0 & 0 & 1 & x & 0 & 1 & 0 & y \end{bmatrix}$$

Hence, the strain-displacement matrix corresponding to the local element degrees of freedom is

$$\mathbf{B} = \frac{1}{4}\begin{bmatrix} (1+y) & -(1+y) & -(1-y) & (1-y) & 0 & 0 & 0 & 0 \\ 0 & 0 & 0 & 0 & (1+x) & (1-x) & -(1-x) & -(1+x) \\ (1+x) & (1-x) & -(1-x) & -(1+x) & (1+y) & -(1+y) & -(1-y) & (1-y) \end{bmatrix} \quad \text{(j)}$$

The matrix **B** could also have been calculated directly by operating on the rows of the matrix **H** in (*h*).

Let B_{ij} be the (i, j)th element of **B**; then we now have

$$\mathbf{B}^{(2)} = \left[\begin{array}{cc|cc|cc|cc|cc|cc|cc|c} 0 & 0 & B_{13} & B_{17} & B_{12} & B_{16} & 0 & 0 & B_{14} & B_{18} & B_{11} & B_{15} & 0 & 0 & 0 \\ 0 & 0 & B_{23} & B_{27} & B_{22} & B_{26} & 0 & 0 & B_{24} & B_{28} & B_{21} & B_{25} & 0 & 0 & \ldots\text{zeroes}\ldots 0 \\ 0 & 0 & B_{33} & B_{37} & B_{32} & B_{36} & 0 & 0 & B_{34} & B_{38} & B_{31} & B_{35} & 0 & 0 & 0 \end{array}\right]$$

where the element degrees of freedom and assemblage degrees of freedom are ordered as in (d) and (b).

EXAMPLE 4.6: A linearly varying surface pressure distribution as shown in Fig. 4.7 is applied to element (m) of an element assemblage. Evaluate the vector $\mathbf{R}_S^{(m)}$ for this element.

The first step in the calculation of $\mathbf{R}_S^{(m)}$ is the evaluation of the matrix $\mathbf{H}^{S(m)}$. This matrix can be established using the same approach as in Example 4.5. For the surface displacements we assume

$$\begin{aligned} u^S &= \alpha_1 + \alpha_2 x + \alpha_3 x^2 \\ v^S &= \beta_1 + \beta_2 x + \beta_3 x^2 \end{aligned} \quad \text{(a)}$$

where (as in Example 4.5) the unknown coefficients $\alpha_1, \ldots, \beta_3$ are evaluated using the nodal point displacements. We thus obtain

$$\begin{bmatrix} u^S(x) \\ v^S(x) \end{bmatrix} = \mathbf{H}^S \,\hat{\mathbf{u}}$$

$$\hat{\mathbf{u}}^T = [u_1 \quad u_2 \quad u_3 \;|\; v_1 \quad v_2 \quad v_3]$$

and

$$\mathbf{H}^S = \begin{bmatrix} \frac{1}{2}x(1+x) & -\frac{1}{2}x(1-x) & (1-x^2) & 0 & 0 & 0 \\ 0 & 0 & 0 & \frac{1}{2}x(1+x) & -\frac{1}{2}x(1-x) & (1-x^2) \end{bmatrix}$$

The vector of surface loads is (with p_1 and p_2 positive)

$$\mathbf{f}^S = \begin{bmatrix} \frac{1}{2}(1+x)p_1^u + \frac{1}{2}(1-x)p_2^u \\ -\frac{1}{2}(1+x)p_1^v - \frac{1}{2}(1-x)p_2^v \end{bmatrix}$$

To obtain $\mathbf{R}_S^{(m)}$ we first evaluate

$$\mathbf{R}_S = 0.5 \int_{-1}^{+1} \mathbf{H}^{S^T}\, \mathbf{f}^S \, dx$$

to obtain

$$\mathbf{R}_S = \frac{1}{3}\begin{bmatrix} p_1^u \\ p_2^u \\ 2(p_1^u + p_2^u) \\ -p_1^v \\ -p_2^v \\ -2(p_1^v + p_2^v) \end{bmatrix}$$

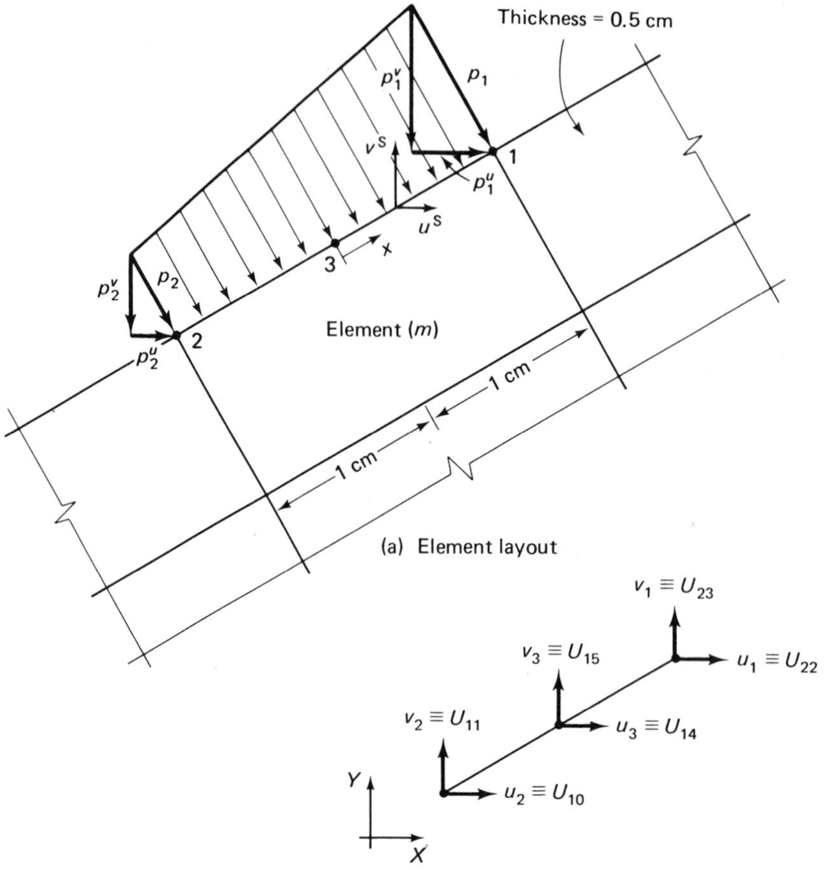

(a) Element layout

(b) Local-global degrees of freedom

FIGURE 4.7 *Pressure loading on element (m).*

Thus, corresponding to the global degrees of freedom given in Fig. 4.7 we have

$$\begin{array}{l} \qquad\qquad\qquad U_{10} \quad U_{11} \quad U_{12} \quad U_{13} \quad U_{14} \qquad\qquad U_{15} \\ \mathbf{R}_S^{(m)T} = \tfrac{1}{3}[0 \ldots 0 \mid p_2^u \quad -p_2^v \mid 0 \quad 0 \mid 2(p_1^u + p_2^u) \quad -2(p_1^v + p_2^v) \mid 0 \ldots \\ \qquad\qquad\qquad\qquad\qquad\qquad U_{22} \quad U_{23} \leftarrow \text{assemblage degrees of freedom} \\ \qquad\qquad\qquad\qquad\qquad \ldots 0 \mid p_1^u \quad -p_2^v \mid 0 \ldots 0] \end{array}$$

The above sample calculations show that in the evaluation of element matrices it is expedient to first establish the matrices corresponding to the local element degrees of freedom. The construction of the finite element matrices, which correspond to the global assemblage degrees of freedom (used in (4.14) to (4.20)), can then directly be achieved by identifying the global degrees of freedom that correspond to the local element degrees of freedom. However, considering the matrices $\mathbf{H}^{(m)}$, $\mathbf{B}^{(m)}$, $\mathbf{K}^{(m)}$ and so on, corresponding to the global assemblage degrees of freedom, only those rows and columns that correspond to element degrees of freedom have nonzero entries, and the main objective in defining these specific matrices was to be able to express the assemblage process of the element matrices in a theoretically elegant manner. In the practical implementation of the finite element method, this elegance is also present, but all element matrices are only calculated corresponding to the element degrees of freedom and are then directly assembled using the correspondence between the local element and global assemblage degrees of freedom. Thus, with (only) the element local nodal point degrees of freedom listed in $\hat{\mathbf{u}}$, we now write (as in Example 4.5)

$$\mathbf{u} = \mathbf{H}\hat{\mathbf{u}} \tag{4.27}$$

where the entries in the vector $\mathbf{u}$ are the element displacements measured in any convenient local coordinate system. We then also have

$$\boldsymbol{\epsilon} = \mathbf{B}\hat{\mathbf{u}} \tag{4.28}$$

Considering the relations in (4.27) and (4.28), the fact that no superscript is used on the interpolation matrices indicates that the matrices are defined with respect to the local element degrees of freedom. Using the relations for the element stiffness matrix, mass matrix and load vector calculations as before, we obtain

$$\mathbf{K} = \int_V \mathbf{B}^T\mathbf{C}\mathbf{B}\, dV \tag{4.29}$$

$$\mathbf{M} = \int_V \rho\mathbf{H}^T\mathbf{H}\, dV \tag{4.30}$$

$$\mathbf{R}_B = \int_V \mathbf{H}^T\mathbf{f}^B\, dV \tag{4.31}$$

$$\mathbf{R}_S = \int_S \mathbf{H}^{S^T}\, \mathbf{f}^S\, dS \tag{4.32}$$

$$\mathbf{R}_I = \int_V \mathbf{B}^T\boldsymbol{\tau}^I\, dV \tag{4.33}$$

where all variables are defined as in (4.14) to (4.19), but corresponding to the local element degrees of freedom. In the derivations and discussions to follow, we shall now refer extensively to the relations in (4.29) to (4.33). Once the matrices given in (4.29) to (4.33) have been calculated, they can be assembled directly using the procedures described in Example 4.8 and Appendix A. In this assemblage process it is assumed that the directions of the element nodal point displacements $\hat{\mathbf{u}}$ in (4.27) are the same as the directions of the global nodal point displacements $\mathbf{U}$. However, in some analyses it is convenient to start the derivation with element nodal point degrees of freedom, $\tilde{\mathbf{u}}$, that are not aligned with the global assemblage degrees of freedom. In this case we have

$$\mathbf{u} = \tilde{\mathbf{H}}\tilde{\mathbf{u}} \tag{4.34}$$

and

$$\tilde{\mathbf{u}} = \mathbf{T}\hat{\mathbf{u}} \tag{4.35}$$

where the matrix $\mathbf{T}$ transforms the degrees of freedom $\hat{\mathbf{u}}$ to the degrees of freedom $\tilde{\mathbf{u}}$, and (4.35) corresponds to a first-order tensor transformation (see Section 2.6); the entries in column j of the matrix $\mathbf{T}$ are the direction cosines of a unit vector corresponding to the jth degree of freedom in $\hat{\mathbf{u}}$, when measured in the directions of the $\tilde{\mathbf{u}}$ degrees of freedom. Substituting (4.35) into (4.34) we obtain

$$\mathbf{H} = \tilde{\mathbf{H}}\mathbf{T} \tag{4.36}$$

Thus, identifying all finite element matrices corresponding to the degrees of freedom $\tilde{\mathbf{u}}$ with a curl placed over them, we obtain from (4.36) and (4.29) to (4.33),

$$\begin{aligned} &\mathbf{K} = \mathbf{T}^T\,\tilde{\mathbf{K}}\mathbf{T}; \qquad \mathbf{M} = \mathbf{T}^T\,\tilde{\mathbf{M}}\mathbf{T} \\ &\mathbf{R}_B = \mathbf{T}^T\,\tilde{\mathbf{R}}_B; \qquad \mathbf{R}_S = \mathbf{T}^T\,\tilde{\mathbf{R}}_S; \qquad \mathbf{R}_I = \mathbf{T}^T\,\tilde{\mathbf{R}}_I \end{aligned} \tag{4.37}$$

It is instructive to note that such transformations are also used when boundary displacements need be imposed that do not correspond to the global assemblage degrees of freedom (see Section 4.2.2). Table 4.1 summarizes some of the notation that we have employed.

TABLE 4.1 *Summary of some notation used.*

(a) $\mathbf{u}^{(m)} = \mathbf{H}^{(m)}\,\hat{\mathbf{U}}$ or $\mathbf{u}^{(m)} = \mathbf{H}^{(m)}\,\mathbf{U}$

where $\mathbf{u}^{(m)}$ = Displacements within element m as a function of the element coordinates.

$\hat{\mathbf{U}}$ = Nodal point displacements of the total element assemblage (from equation (4.14) onwards we simply use $\mathbf{U}$).

(b) $\mathbf{u} = \mathbf{H}\hat{\mathbf{u}}$

where $\mathbf{u} \equiv \mathbf{u}^{(m)}$ and it is implied that a specific element is considered.

$\hat{\mathbf{u}}$ = Nodal point displacements of the element under consideration; the entries of $\hat{\mathbf{u}}$ are those displacements in $\hat{\mathbf{U}}$ that belong to the element.

(c) $\mathbf{u} = \tilde{\mathbf{H}}\tilde{\mathbf{u}}$

where $\tilde{\mathbf{u}}$ = Nodal point displacements of element in a coordinate system other than the global system (in which $\hat{\mathbf{U}}$ is defined).

We demonstrate the above concepts in the following examples.

EXAMPLE 4.7: Establish the matrix $\mathbf{H}$ for the truss element shown in Fig. 4.8. The directions of local and global degrees of freedom are shown in the figure.

Here, we have

$$\begin{bmatrix} u(x) \\ v(x) \end{bmatrix} = \frac{1}{L}\begin{bmatrix} \left(\frac{L}{2} - x\right) & 0 & \left(\frac{L}{2} + x\right) & 0 \\ 0 & \left(\frac{L}{2} - x\right) & 0 & \left(\frac{L}{2} + x\right) \end{bmatrix}\begin{bmatrix} \tilde{u}_1 \\ \tilde{v}_1 \\ \tilde{u}_2 \\ \tilde{v}_2 \end{bmatrix} \tag{a}$$

and

$$\begin{bmatrix} \tilde{u}_1 \\ \tilde{v}_1 \\ \tilde{u}_2 \\ \tilde{v}_2 \end{bmatrix} = \begin{bmatrix} \cos\alpha & \sin\alpha & 0 & 0 \\ -\sin\alpha & \cos\alpha & 0 & 0 \\ 0 & 0 & \cos\alpha & \sin\alpha \\ 0 & 0 & -\sin\alpha & \cos\alpha \end{bmatrix}\begin{bmatrix} u_1 \\ v_1 \\ u_2 \\ v_2 \end{bmatrix}$$

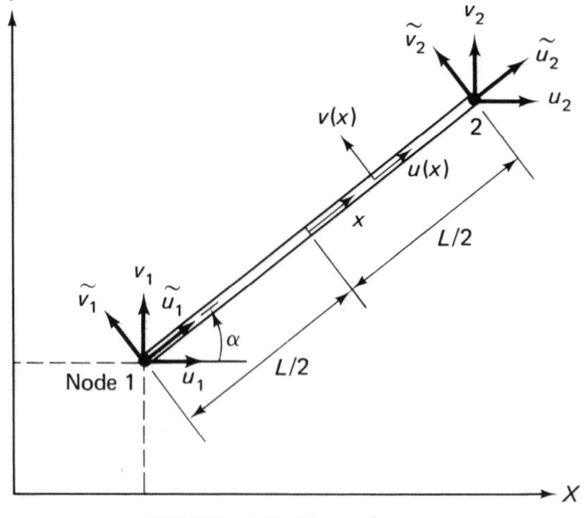

FIGURE 4.8 *Truss element.*

Thus, we have

$$\mathbf{H} = \frac{1}{L}\begin{bmatrix} \left(\frac{L}{2}-x\right) & 0 & \left(\frac{L}{2}+x\right) & 0 \\ 0 & \left(\frac{L}{2}-x\right) & 0 & \left(\frac{L}{2}+x\right) \end{bmatrix}\begin{bmatrix} \cos\alpha & \sin\alpha & 0 & 0 \\ -\sin\alpha & \cos\alpha & 0 & 0 \\ 0 & 0 & \cos\alpha & \sin\alpha \\ 0 & 0 & -\sin\alpha & \cos\alpha \end{bmatrix}$$

It should be noted that for the construction of the strain-displacement matrix **B** (in linear analysis) only the first row of H is required, because only the normal strain $\epsilon_{xx} = \partial u/\partial x$ is considered in the derivation of the stiffness matrix. In practice, it is effective to use only the first row of the matrix $\tilde{\mathbf{H}}$ in (a) and then transform the matrix $\tilde{\mathbf{K}}$ as given in (4.37).

EXAMPLE 4.8: Assume that the element stiffness matrices corresponding to the element displacements shown in Fig. 4.9 have been calculated, and denote the elements as shown Ⓐ, Ⓑ, Ⓒ, and Ⓓ. Assemble these element matrices directly into the global structure stiffness matrix with the displacement boundary conditions shown in Fig. 4.9(a).

In this analysis all element stiffness matrices have already been established corresponding to degrees of freedom that are aligned with the global directions. Therefore, no transformation as given in (4.37) is required and we can directly assemble the complete stiffness matrix. In the assemblage process we recognize that

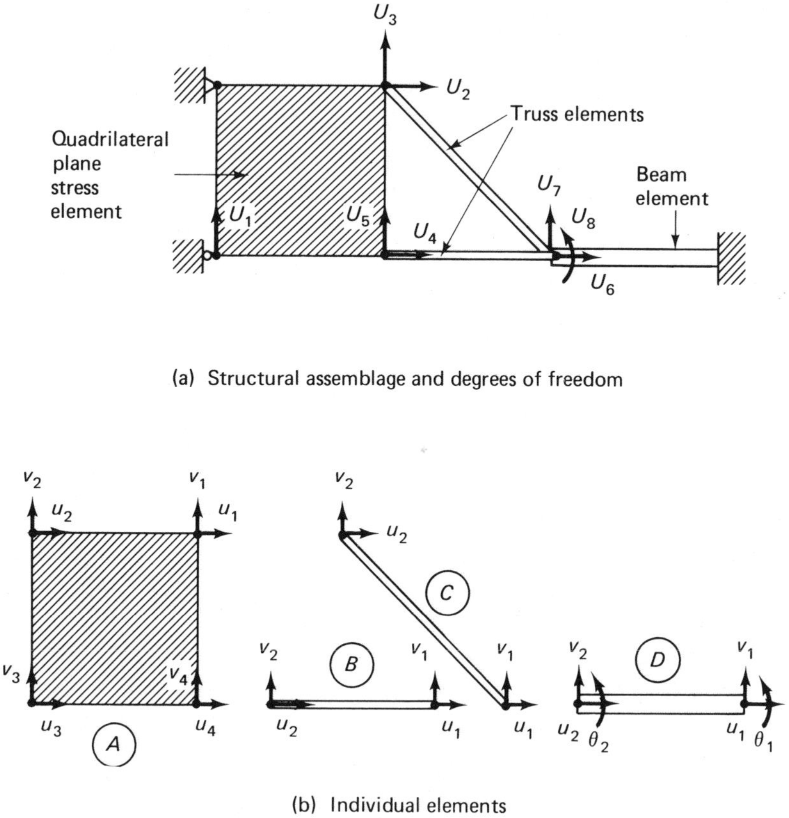

FIGURE 4.9 *A simple element assemblage.*

some of the element degrees of freedom do not appear in the array of assemblage degrees of freedom for which the complete stiffness matrix must be formed, because the displacements are zero at these supports (see the Appendix).

$$
\begin{array}{c}
\begin{array}{cccccccccl}
 & U_2 & U_3 & & & & U_1 & U_4 & U_5 & \text{global displacements} \\
 & u_1 & v_1 & u_2 & v_2 & u_3 & v_3 & u_4 & v_4 & \text{local displacements}
\end{array} \\
\mathbf{K}_A = \left[\begin{array}{cccccc}
a_{11} & a_{12} & \cdots & a_{16} & a_{17} & a_{18} \\
a_{21} & a_{22} & \cdots & a_{26} & a_{27} & a_{28} \\
\vdots & & & & & \vdots \\
 & & \cdots & & & \\
\vdots & & & & & \vdots \\
a_{61} & a_{62} & \cdots & a_{66} & a_{67} & a_{68} \\
a_{71} & a_{72} & \cdots & a_{76} & a_{77} & a_{78} \\
a_{81} & a_{82} & & a_{86} & a_{87} & a_{88}
\end{array}\right]
\begin{array}{ll}
u_1 & U_2 \\
v_1 & U_3 \\
u_2 & \\
v_2 & \\
u_3 & \\
v_3 & U_1 \\
u_4 & U_4 \\
v_4 & U_5
\end{array}
\end{array}
$$

$$
\begin{array}{cccc}
U_6 & U_7 & U_4 & U_5 \\
u_1 & v_1 & u_2 & v_2
\end{array}
\qquad\qquad
\begin{array}{cccc}
U_6 & U_7 & U_2 & U_3 \\
u_1 & v_1 & u_2 & v_2
\end{array}
$$

$$
\mathbf{K}_B = \left[\begin{array}{cccc}
b_{11} & b_{12} & b_{13} & b_{14} \\
b_{21} & b_{22} & b_{23} & b_{24} \\
b_{31} & b_{32} & b_{33} & b_{34} \\
b_{41} & b_{42} & b_{43} & b_{44}
\end{array}\right]
\begin{array}{ll}
u_1 & U_6 \\
v_1 & U_7 \\
u_2 & U_4 \\
v_2 & U_5
\end{array}
\qquad
\mathbf{K}_C = \left[\begin{array}{cccc}
c_{11} & c_{12} & c_{13} & c_{14} \\
c_{21} & c_{22} & c_{23} & c_{24} \\
c_{31} & c_{32} & c_{33} & c_{34} \\
c_{41} & c_{42} & c_{43} & c_{44}
\end{array}\right]
\begin{array}{ll}
u_1 & U_6 \\
v_1 & U_7 \\
u_2 & U_2 \\
v_2 & U_3
\end{array}
$$

$$
\begin{array}{c}
\begin{array}{cccccc}
 & & & U_6 & U_7 & U_8 \\
u_1 & v_1 & \theta_1 & u_2 & v_2 & \theta_2
\end{array} \\
\mathbf{K}_D = \left[\begin{array}{cccccc}
\cdots & \cdot & \cdot & \cdot & \cdot & \cdots \\
\cdots & \cdot & \cdot & \cdot & \cdot & \cdots \\
\cdots & \cdot & \cdot & \cdot & \cdot & \cdots \\
\cdots & \cdot & \cdots & d_{44} & d_{45} & d_{46} \\
\cdots & \cdot & \cdots & d_{54} & d_{55} & d_{56} \\
\cdots & \cdot & \cdots & d_{64} & d_{65} & d_{66}
\end{array}\right]
\begin{array}{ll}
u_1 & \\
v_1 & \\
\theta_1 & \\
u_2 & U_6 \\
v_2 & U_7 \\
\theta_2 & U_8
\end{array}
\end{array}
$$

and the equation $\mathbf{K} = \sum_m \mathbf{K}^{(m)}$ gives

$$
\begin{array}{c}
\begin{array}{cccccccc}
U_1 & U_2 & U_3 & U_4 & U_5 & U_6 & U_7 & U_8
\end{array} \\
\mathbf{K} = \left[\begin{array}{cccccccc}
a_{66} & a_{61} & a_{62} & a_{67} & a_{68} & & \text{zeros} & \\
a_{16} & a_{11}+c_{33} & a_{12}+c_{34} & a_{17} & a_{18} & c_{31} & c_{32} & \\
a_{26} & a_{21}+c_{43} & a_{22}+c_{44} & a_{27} & a_{28} & c_{41} & c_{42} & \\
a_{76} & a_{71} & a_{72} & a_{77}+b_{33} & a_{78}+b_{34} & b_{31} & b_{32} & \\
a_{86} & a_{81} & a_{82} & a_{87}+b_{43} & a_{88}+b_{44} & b_{41} & b_{42} & \\
 & c_{13} & c_{14} & b_{13} & b_{14} & b_{11}+c_{11}+d_{44} & b_{12}+c_{12}+d_{45} & d_{46} \\
 & c_{23} & c_{24} & b_{23} & b_{24} & b_{21}+c_{21}+d_{54} & b_{22}+c_{22}+d_{55} & d_{56} \\
 & & & & & d_{64} & d_{65} & d_{66}
\end{array}\right]
\begin{array}{l}
U_1 \\ U_2 \\ U_3 \\ U_4 \\ U_5 \\ U_6 \\ U_7 \\ U_8
\end{array}
\end{array}
$$

symmetric about diagonal

4.2.2 Imposition of Boundary Conditions

We discussed in Section 3.3.2 that in the analysis of a continuum we have displacement or essential boundary conditions and force or natural boundary conditions. Using the displacement-based finite element method, the force boundary conditions are taken into account in the evaluation of the externally applied nodal point force vector. The vector $\mathbf{R}_C$ assembles the concentrated loads including the reactions, and the vector $\mathbf{R}_S$ contains the effect of the distributed surface loads and distributed reactions.

Assume that the equilibrium equations of a finite element system without the imposition of the displacement boundary conditions as derived in Section 4.2.1 are, neglecting damping,

$$\begin{bmatrix} \mathbf{M}_{aa} & \mathbf{M}_{ab} \\ \mathbf{M}_{ba} & \mathbf{M}_{bb} \end{bmatrix} \begin{bmatrix} \ddot{\mathbf{U}}_a \\ \ddot{\mathbf{U}}_b \end{bmatrix} + \begin{bmatrix} \mathbf{K}_{aa} & \mathbf{K}_{ab} \\ \mathbf{K}_{ba} & \mathbf{K}_{bb} \end{bmatrix} \begin{bmatrix} \mathbf{U}_a \\ \mathbf{U}_b \end{bmatrix} = \begin{bmatrix} \mathbf{R}_a \\ \mathbf{R}_b \end{bmatrix} \tag{4.38}$$

where the $\mathbf{U}_a$ are the unknown displacements and the $\mathbf{U}_b$ are the known, or prescribed, displacements. Solving for $\mathbf{U}_a$ we obtain

$$\mathbf{M}_{aa}\ddot{\mathbf{U}}_a + \mathbf{K}_{aa}\mathbf{U}_a = \mathbf{R}_a - \mathbf{K}_{ab}\mathbf{U}_b - \mathbf{M}_{ab}\ddot{\mathbf{U}}_b \tag{4.39}$$

Hence, in this solution for $\mathbf{U}_a$, only the stiffness and mass matrices of the complete assemblage corresponding to the unknown degrees of freedom $\mathbf{U}_a$ need be assembled (see Example 4.8), but the load vector $\mathbf{R}_a$ must be modified to include the effect of imposed nonzero displacements. Once the displacements $\mathbf{U}_a$ have been evaluated from (4.39), the nodal point forces corresponding to $\mathbf{U}_b$ are obtained using (4.38), i.e., we have

$$\mathbf{R}_b = \mathbf{M}_{ba}\ddot{\mathbf{U}}_a + \mathbf{M}_{bb}\ddot{\mathbf{U}}_b + \mathbf{K}_{ba}\mathbf{U}_a + \mathbf{K}_{bb}\mathbf{U}_b \tag{4.40}$$

The relations in (4.39) and (4.40) represent a formal procedure to evaluate the displacements $\mathbf{U}_a$ and reactions $\mathbf{R}_b$, when the $\mathbf{U}_b$ displacements are known. In using (4.38) we assumed that the displacement components employed in Section 4.2.1 actually contain all prescribed displacements (denoted as $\mathbf{U}_b$ in (4.38)). If this is not the case, we need to identify all prescribed displacements that do not correspond to defined assemblage degrees of freedom and transform the finite element equilibrium equations to correspond to the prescribed displacements. Thus, we write

$$\mathbf{U} = \mathbf{T}\bar{\mathbf{U}} \tag{4.41}$$

where $\bar{\mathbf{U}}$ is the vector of nodal point degrees of freedom in the required directions. The transformation matrix $\mathbf{T}$ is an identity matrix that has been altered by the direction cosines of the components in $\bar{\mathbf{U}}$ measured on the original displacement directions. Using (4.41) in (4.38) we obtain

$$\bar{\mathbf{M}}\ddot{\bar{\mathbf{U}}} + \bar{\mathbf{K}}\bar{\mathbf{U}} = \bar{\mathbf{R}} \tag{4.42}$$

where $$\bar{\mathbf{M}} = \mathbf{T}^T\mathbf{M}\mathbf{T}; \qquad \bar{\mathbf{K}} = \mathbf{T}^T\mathbf{K}\mathbf{T}; \qquad \bar{\mathbf{R}} = \mathbf{T}^T\mathbf{R} \tag{4.43}$$

We should note that the matrix multiplications in (4.43) involve only changes in those columns and rows of $\mathbf{M}$, $\mathbf{K}$, and $\mathbf{R}$ that are actually affected, and that this transformation is equivalent to the calculations performed in (4.37) on a single element matrix. In practice, the transformation is carried out effectively on the element level just prior to adding the element matrices to the matrices of the total assemblage. Figure 4.10 gives the transformation matrices **T** for a typical nodal point in two- and three-dimensional analysis when displacements are constrained in skew directions. The unknown displacements can now be calculated from (4.42) using the procedure of (4.39) and (4.40).

In an alternative approach, the required displacements can also be imposed by adding into the finite element equilibrium equations (4.42) the constraint equations that express the prescribed displacement conditions. Assume that the displacement is to be specified at degree of freedom i, say $\bar{U}_i = b$, then the constraint equation

$$k\bar{U}_i = kb \tag{4.44}$$

is added into the equilibrium equations (4.42), where $k \gg \bar{k}_{ii}$. Therefore, the

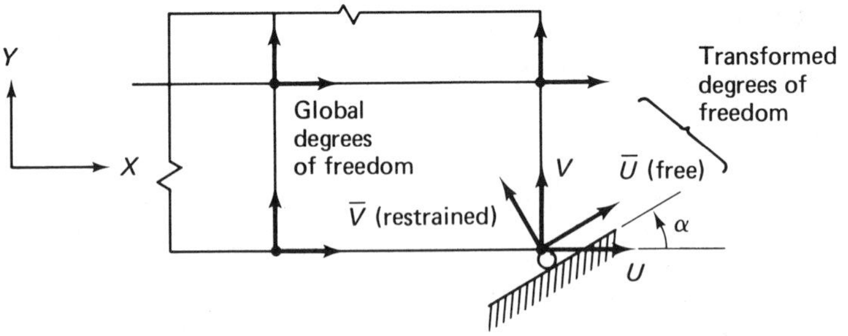

$$\mathbf{T} = \begin{bmatrix} \cos\alpha & -\sin\alpha \\ \sin\alpha & \cos\alpha \end{bmatrix}$$

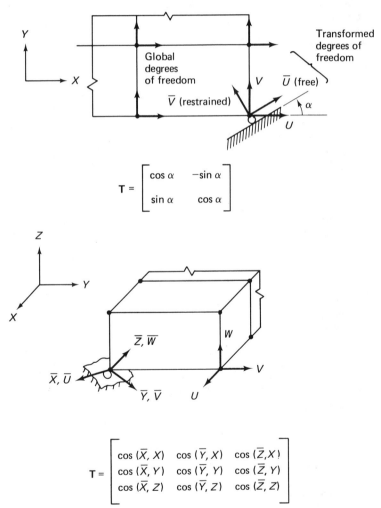

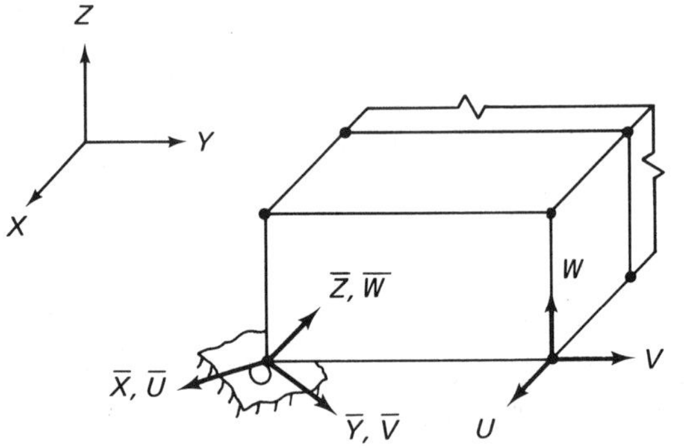

$$\mathbf{T} = \begin{bmatrix} \cos(\bar{X}, X) & \cos(\bar{Y}, X) & \cos(\bar{Z}, X) \\ \cos(\bar{X}, Y) & \cos(\bar{Y}, Y) & \cos(\bar{Z}, Y) \\ \cos(\bar{X}, Z) & \cos(\bar{Y}, Z) & \cos(\bar{Z}, Z) \end{bmatrix}$$

FIGURE 4.10 *Transformation to skew boundary conditions.*

solution of the modified equilibrium equations must now give $\bar{U}_i = b$, and we note that because (4.42) was used, only the diagonal element in the stiffness matrix was affected, resulting in a numerically stable solution (see Section 8.5). Physically, this procedure can be interpreted as adding at the degree of freedom i a spring of large stiffness k, and specifying a load which, because of the relatively flexible element assemblage, produces at this degree of freedom the required displacement b (see Fig. 4.11). Mathematically, the procedure corresponds to an application of the penalty method, discussed in Section 3.4.

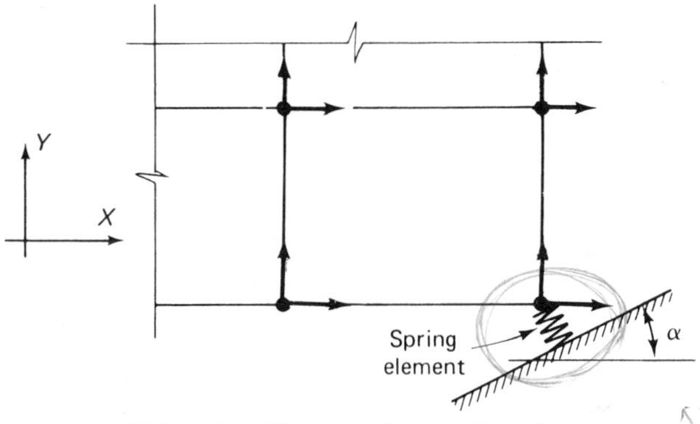

FIGURE 4.11 *Skew boundary condition imposed using spring element.*

In addition to specified nodal point displacement conditions, some nodal point displacements may also be subjected to constraint conditions. Considering (4.22), a typical constraint equation would be

$$U_i = \sum_{j=1}^{r_i} \alpha_{q_j} U_{q_j} \tag{4.45}$$

where the U_i is a dependent nodal point displacement and the U_{q_j} are r_i independent nodal point displacements. Using all constraint equations of the form (4.45) and recognizing that these constraints must hold in the application of the principle of virtual work for the actual nodal point displacements as well as for the virtual displacements, the imposition of the constraints corresponds to a transformation of the form (4.42) and (4.43), in which $\mathbf{T}$ is now a rectangular matrix and $\bar{\mathbf{U}}$ contains all independent degrees of freedom. This transformation corresponds to adding α_{q_j} times the ith columns and rows to the q_jth columns and rows, for $j = 1, \ldots, r_i$ and all i considered. In the actual implementation the transformation is performed effectively on the element level during the assemblage process.

Finally, it should be noted that combinations of the above displacement boundary conditions are possible, where, for example, in (4.45) an independent displacement component may correspond to a skew boundary condition with a specified displacement. We demonstrate the imposition of displacement constraints in the following examples.

EXAMPLE 4.9: Consider the truss assemblage shown in Fig. 4.12. Establish the stiffness matrix of the structure that contains the constraint conditions given.

The independent degrees of freedom are in this analysis U_1, U_2, and U_4. The element stiffness matrices are given in Fig. 4.12, and we recognize that corresponding to (4.45) we have $i = 3$, $\alpha_1 = 2$, and $q_1 = 1$. Establishing the complete

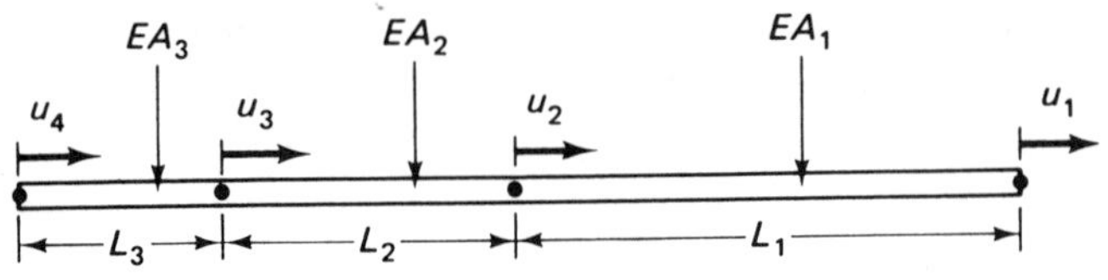

Displacement conditions: $u_3 = 2u_1$
$u_4 = \delta$

$$\mathbf{K}_i = \frac{EA_i}{L_i}\begin{bmatrix} 1 & -1 \\ -1 & 1 \end{bmatrix}$$

FIGURE 4.12 *Truss assemblage.*

stiffness matrix directly during the assemblage process, we have

$$\mathbf{K} = \begin{bmatrix} \frac{EA_1}{L_1} & -\frac{EA_1}{L_1} & 0 \\ -\frac{EA_1}{L_1} & \frac{EA_1}{L_1} & 0 \\ 0 & 0 & 0 \end{bmatrix} + \begin{bmatrix} \frac{4EA_2}{L_2} & -\frac{2EA_2}{L_2} & 0 \\ -\frac{2EA_2}{L_2} & \frac{EA_2}{L_2} & 0 \\ 0 & 0 & 0 \end{bmatrix}$$

$$+ \begin{bmatrix} \frac{4EA_3}{L_3} & 0 & -\frac{2EA_3}{L_3} \\ 0 & 0 & 0 \\ -\frac{2EA_3}{L_3} & 0 & \frac{EA_3}{L_3} \end{bmatrix} + \begin{bmatrix} 0 & 0 & 0 \\ 0 & 0 & 0 \\ 0 & 0 & k \end{bmatrix}$$

where
$$k \gg \frac{EA_3}{L_3}$$

EXAMPLE 4.10: The frame structure shown in Fig. 4.13 is to be analyzed. Use symmetry and constraint conditions to establish a suitable model for analysis.

The complete structure and applied loading display cyclic symmetry, so that only one quarter of the structure need be considered, as shown in Fig. 4.13(b), with the following constraint conditions:

$$u_5 = v_4$$
$$v_5 = -u_4$$
$$\theta_5 = \theta_4$$

This is a simple example to demonstrate how the analysis effort can be reduced considerably through the use of symmetry conditions. In practice, the saving through the use of cyclic symmetry conditions can in some cases be considerable, and indeed may make the analysis only possible.

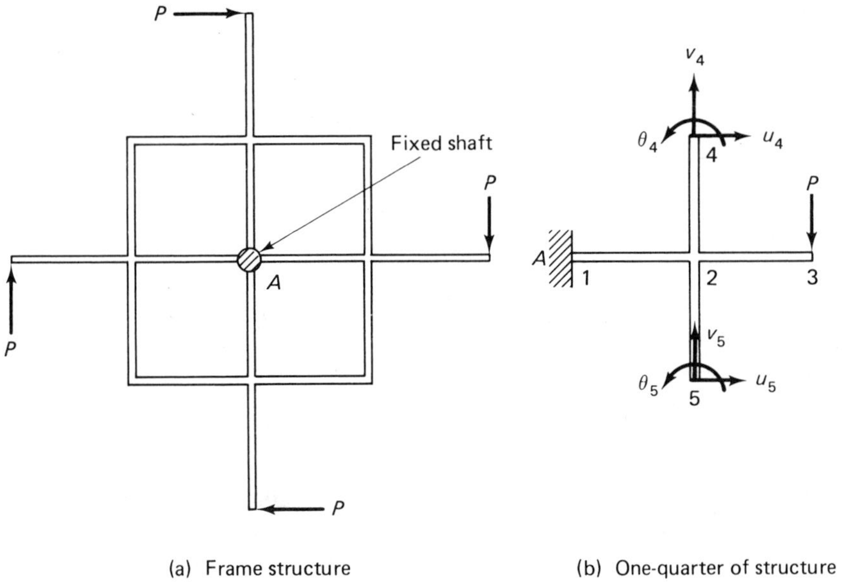

FIGURE 4.13 *Analysis of a cyclicly symmetric structure.*

4.2.3 Generalized Coordinate Models for Specific Problems

In Section 4.2.1 the finite element discretization procedure and derivation of the equilibrium equations was presented in general, i.e., a general three-dimensional body was considered. As shown in Examples 4.2 to 4.7, the general equations derived must be specialized in specific analyses to the specific stress and strain conditions considered. The objective in this section is to discuss and summarize how the finite element matrices that correspond to specific problems can be obtained from the general finite element equations (4.8) to (4.23).

Although in theory any body may be understood to be three-dimensional, for practical analysis it is in many cases imperative to reduce the dimensionality of the problem. The first step in a finite element analysis is therefore to decide what kind of problem is at hand. This decision is based on the assumptions used in the theory of elasticity approach to the solution of specific problems. The classes of problems that are encountered may be summarized as (1) truss, (2) beam, (3) plane stress, (4) plane strain, (5) axisymmetric, (6) plate bending, (7) thin shell, (8) thick shell, and (9) general three-dimensional. For each of these problem cases, the general formulation is applicable; however, only the appropriate displacement, stress and strain variables must be used. These variables are summarized in Tables 4.2 and 4.3 together with the stress-strain matrices to be employed when considering an isotropic material. Figure 4.14 shows various stress and strain conditions considered in the formulation of finite element matrices.[11,12]

TABLE 4.2 Corresponding kinematic and static variables in various problems.

Problem	Displacement Components	Strain Vector $\boldsymbol{\epsilon}^T$	Stress Vector $\boldsymbol{\tau}^T$
Bar	u	$[\epsilon_{xx}]$	$[\tau_{xx}]$
Beam	w	$[\kappa_{xx}]$	$[M_{xx}]$
Plane stress	u, v	$[\epsilon_{xx}\ \epsilon_{yy}\ \gamma_{xy}]$	$[\tau_{xx}\ \tau_{yy}\ \tau_{xy}]$
Plane strain	u, v	$[\epsilon_{xx}\ \epsilon_{yy}\ \gamma_{xy}]$	$[\tau_{xx}\ \tau_{yy}\ \tau_{xy}]$
Axisymmetric	u, v	$[\epsilon_{xx}\ \epsilon_{yy}\ \gamma_{xy}\ \epsilon_{zz}]$	$[\tau_{xx}\ \tau_{yy}\ \tau_{xy}\ \tau_{zz}]$
Three-dimensional	u, v, w	$[\epsilon_{xx}\ \epsilon_{yy}\ \epsilon_{zz}\ \gamma_{xy}\ \gamma_{yz}\ \gamma_{zx}]$	$[\tau_{xx}\ \tau_{yy}\ \tau_{zz}\ \tau_{xy}\ \tau_{yz}\ \tau_{zx}]$
Plate Bending	w	$[\kappa_{xx}\ \kappa_{yy}\ \kappa_{xy}]$	$[M_{xx}\ M_{yy}\ M_{xy}]$

Notation: $\epsilon_{xx} = \dfrac{\partial u}{\partial x},\ \epsilon_{yy} = \dfrac{\partial v}{\partial y},\ \gamma_{xy} = \dfrac{\partial u}{\partial y} + \dfrac{\partial v}{\partial x}, \ldots, \kappa_{xx} = -\dfrac{\partial^2 w}{\partial x^2},\ \kappa_{yy} = -\dfrac{\partial^2 w}{\partial y^2},\ \kappa_{xy} = 2\dfrac{\partial^2 w}{\partial x\, \partial y}$

In Examples 4.2 to 4.7 we developed already some specific finite element matrices. Referring to Example 4.5, in which we considered a plane stress condition, we used for the u and v displacements simple linear polynomial assumptions, where we identified the unknown coefficients in the polynomials as generalized coordinates. The number of unknown coefficients in the polynomials was equal to the number of element nodal point displacements. Expressing the generalized coordinates in terms of the element nodal point displacements, we found that, in general, each polynomial coefficient is not an actual physical displacement but is equal to a linear combination of the element nodal point displacements.

Finite element matrices which are formulated by assuming that the displacements vary in the form of a function whose unknown coefficients are treated as generalized coordinates are referred to as generalized coordinate finite element models. A rather natural class of functions to use for approximating element displacements are polynomials, because polynomials are commonly employed to approximate unknown functions, and the higher the degree of the polynomial, the better is the approximation that we can expect. In addition, polynomials are easy to differentiate; i.e., if the polynomials approximate the displacements of the structure, we can evaluate the strains with relative ease.

Using polynomial displacement assumptions, a very large number of finite elements for practically all problems in structural mechanics have been developed.

The objective in this section is to describe the formulation of a variety of generalized coordinate finite element models that use polynomials to approximate the displacement fields. Other functions would, in principle, be used in the same way and their use can be effective in specific applications (see Example 4.16). In the presentation, emphasis is given to the general formulation rather than to numerically effective finite elements. Therefore, this section serves primarily to enhance our general understanding of the finite element method. The currently most effective finite elements for general application are believed to be the isoparametric and related elements that are described in Chapter 5.

TABLE 4.3 *Generalized stress-strain matrices for isotropic materials and the problems in table 4.2.*

Problem	*Material Matrix* **C**
Bar	E
Beam	EI
Plane stress	$\dfrac{E}{1-\nu^2}\begin{bmatrix} 1 & \nu & 0 \\ \nu & 1 & 0 \\ 0 & 0 & \dfrac{1-\nu}{2} \end{bmatrix}$
Plane strain	$\dfrac{E(1-\nu)}{(1+\nu)(1-2\nu)}\begin{bmatrix} 1 & \dfrac{\nu}{1-\nu} & 0 \\ \dfrac{\nu}{1-\nu} & 1 & 0 \\ 0 & 0 & \dfrac{1-2\nu}{2(1-\nu)} \end{bmatrix}$
Axisymmetric	$\dfrac{E(1-\nu)}{(1+\nu)(1-2\nu)}\begin{bmatrix} 1 & \dfrac{\nu}{1-\nu} & 0 & \dfrac{\nu}{1-\nu} \\ \dfrac{\nu}{1-\nu} & 1 & 0 & \dfrac{\nu}{1-\nu} \\ 0 & 0 & \dfrac{1-2\nu}{2(1-\nu)} & 0 \\ \dfrac{\nu}{1-\nu} & \dfrac{\nu}{1-\nu} & 0 & 1 \end{bmatrix}$
Three-dimensional	$\dfrac{E(1-\nu)}{(1+\nu)(1-2\nu)}\begin{bmatrix} 1 & \dfrac{\nu}{1-\nu} & \dfrac{\nu}{1-\nu} & & & \\ \dfrac{\nu}{1-\nu} & 1 & \dfrac{\nu}{1-\nu} & & & \\ \dfrac{\nu}{1-\nu} & \dfrac{\nu}{1-\nu} & 1 & & & \\ & & & \dfrac{1-2\nu}{2(1-\nu)} & & \\ & & & & \dfrac{1-2\nu}{2(1-\nu)} & \\ & & & & & \dfrac{1-2\nu}{2(1-\nu)} \end{bmatrix}$ (elements not shown are zeros)
Plate bending	$\dfrac{Eh^3}{12(1-\nu^2)}\begin{bmatrix} 1 & \nu & 0 \\ \nu & 1 & 0 \\ 0 & 0 & \dfrac{1-\nu}{2} \end{bmatrix}$

Notation: E = Young's modulus, ν = Poisson's ratio, h = thickness of plate, I = moment of inertia

In the following derivations the displacements of the finite elements are always described in the local coordinate systems shown in Fig. 4.14. Also, since we consider one specific element, we shall leave out the superscript (m) used in Section 4.2.1.

For one-dimensional bar elements (truss elements) we have

$$u(x) = \alpha_1 + \alpha_2 x + \alpha_3 x^2 + \ldots \tag{4.46}$$

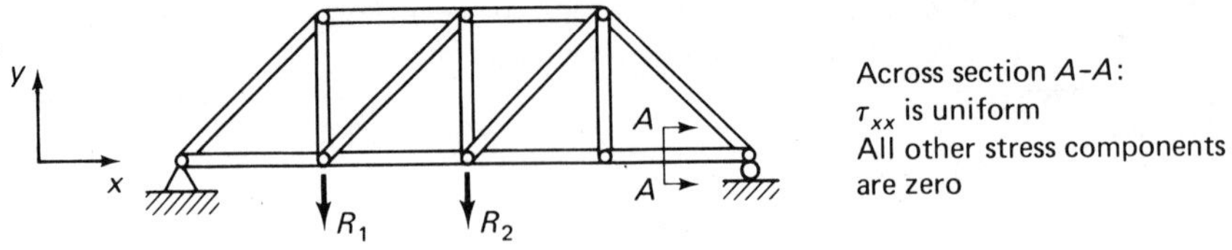

(a) Uniaxial stress condition: frame under concentrated loads

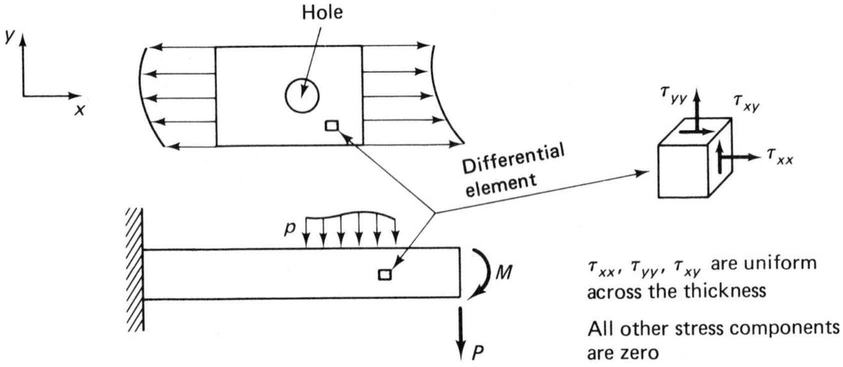

(b) Plane stress conditions: membrane and beam under in-plane actions

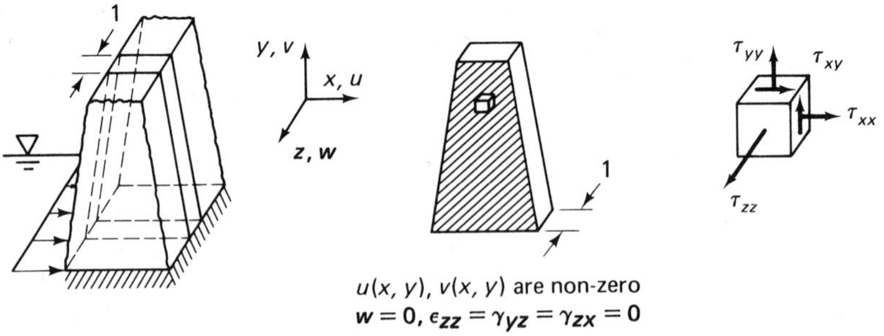

(c) Plane strain condition: long dam subjected to water pressure

FIGURE 4.14 *Various stress and strain conditions with illustrative examples.*

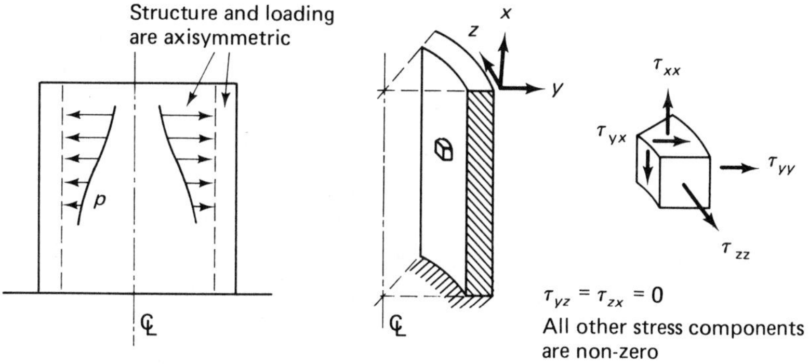

(d) Axisymmetric condition: cylinder under internal pressure

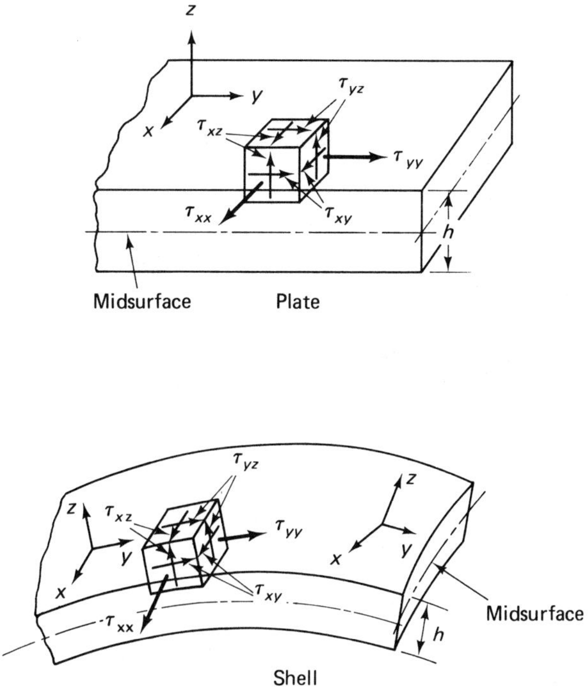

$\tau_{zz} = 0$
All other stress components
are nonzero.

(e) Plate and shell structures

FIGURE 4.14 (*cont.*)

where x varies over the length of the element, u is the local element displacement, and $\alpha_1, \alpha_2, \ldots$ are the generalized coordinates. The displacement expansion in (4.46) would also be used for the transverse and longitudinal displacements of a beam.

For two-dimensional elements (i.e., plane stress, plane strain, and axisymmetric elements), we have for the u and v displacements as a function of the element x and y coordinates,

$$\begin{aligned} u(x, y) &= \alpha_1 + \alpha_2 x + \alpha_3 y + \alpha_4 xy + \alpha_5 x^2 + \ldots \\ v(x, y) &= \beta_1 + \beta_2 x + \beta_3 y + \beta_4 xy + \beta_5 x^2 + \ldots \end{aligned} \tag{4.47}$$

where $\alpha_1, \alpha_2, \ldots$ and $\beta_1, \beta_2, \ldots$ are the generalized coordinates.

In the case of a plate bending element, the transverse deflection w is assumed as a function of the element coordinates x and y; i.e.,

$$w(x, y) = \gamma_1 + \gamma_2 x + \gamma_3 y + \gamma_4 xy + \gamma_5 x^2 + \ldots \tag{4.48}$$

where $\gamma_1, \gamma_2, \ldots$ are the generalized coordinates.

Finally, for elements in which the u, v, and w displacements are measured as a function of the element x, y, and z coordinates, we have, in general,

$$\begin{aligned} u(x, y, z) &= \alpha_1 + \alpha_2 x + \alpha_3 y + \alpha_4 z + \alpha_5 xy + \ldots \\ v(x, y, z) &= \beta_1 + \beta_2 x + \beta_3 y + \beta_4 z + \beta_5 xy + \ldots \\ w(x, y, z) &= \gamma_1 + \gamma_2 x + \gamma_3 y + \gamma_4 z + \gamma_5 xy + \ldots \end{aligned} \tag{4.49}$$

where $\alpha_1, \alpha_2, \ldots, \beta_1, \beta_2, \ldots,$ and $\gamma_1, \gamma_2, \ldots$ are now the generalized coordinates.

As in the discussion of the plane stress element in Example 4.5 the relations (4.46) to (4.49) can be written in matrix form,

$$\mathbf{u} = \boldsymbol{\Phi}\boldsymbol{\alpha} \tag{4.50}$$

where the vector $\mathbf{u}$ corresponds to the displacements used in (4.46) to (4.49), the elements of $\boldsymbol{\Phi}$ are the corresponding polynomial terms, and $\boldsymbol{\alpha}$ is a vector of the generalized coordinates arranged in the appropriate order.

To evaluate the generalized coordinates in terms of the element nodal point displacements, we need to have as many nodal point displacements as assumed generalized coordinates. Then evaluating (4.50) specifically for the nodal point displacements $\hat{\mathbf{u}}$ of the element, we obtain

$$\hat{\mathbf{u}} = \mathbf{A}\boldsymbol{\alpha} \tag{4.51}$$

Assuming that the inverse of $\mathbf{A}$ exists, we have

$$\boldsymbol{\alpha} = \mathbf{A}^{-1}\hat{\mathbf{u}} \tag{4.52}$$

The element strains to be considered depend on the specific problem to be solved. Denoting by $\boldsymbol{\epsilon}$ a generalized strain vector, whose components are given for specific problems in Table 4.2, we have

$$\boldsymbol{\epsilon} = \mathbf{E}\boldsymbol{\alpha} \tag{4.53}$$

where the matrix $\mathbf{E}$ is established using the displacement assumptions in (4.50). A vector of generalized stresses $\boldsymbol{\tau}$ is obtained using the relation

$$\boldsymbol{\tau} = \mathbf{C}\boldsymbol{\epsilon} \tag{4.54}$$

where **C** is a generalized elasticity matrix. The quantities $\boldsymbol{\tau}$ and **C** are defined for some problems in Tables 4.2 and 4.3. We may note that except in bending problems, the generalized $\boldsymbol{\tau}$, $\boldsymbol{\epsilon}$, and **C** matrices are those that are used in the theory of elasticity. The word "generalized" is employed merely to include curvatures and moments as strains and stresses, respectively. The advantage of using curvatures and moments in bending analysis is that in the stiffness evaluation an integration over the thickness of the corresponding element is not required because this stress and strain variation has already been taken into account (see Example 4.11).

Referring to Table 4.3 it should be noted that all stress-strain matrices can be derived from the general three dimensional stress-strain relationship. The plane strain and axisymmetric stress-strain matrices are obtained simply by deleting in the three-dimensional stress-strain matrix the rows and columns that correspond to the zero strain components. The stress-strain matrix for plane stress analysis is then obtained from the axisymmetric stress-strain matrix by using the condition that τ_{zz} is zero (see the program QUADS in Section 5.9). To calculate the generalized stress-strain matrix for plate bending analysis, the stress-strain matrix corresponding to plane stress conditions is used, as shown in the following example.

EXAMPLE 4.11: Derive the stress-strain matrix **C** used for plate bending analysis (see Table 4.3).

The strains at a distance z measured upward from the neutral axis of the plate are

Kirchoff !!

$$\left[-z\frac{\partial^2 w}{\partial x^2} \quad -z\frac{\partial^2 w}{\partial y^2} \quad z\frac{2\partial^2 w}{\partial x \partial y}\right]$$

In plate bending analysis it is assumed that each layer of the plate acts in plane stress condition. Hence integrating the normal stresses in the plate to obtain moments per unit length, the generalized stress-strain matrix is

$$\mathbf{C} = \int_{-h/2}^{+h/2} z^2 \frac{E}{1-\nu^2}\begin{bmatrix} 1 & \nu & 0 \\ \nu & 1 & 0 \\ 0 & 0 & \dfrac{1-\nu}{2} \end{bmatrix}$$

or

$$\mathbf{C} = \frac{Eh^3}{12(1-\nu^2)}\begin{bmatrix} 1 & \nu & 0 \\ \nu & 1 & 0 \\ 0 & 0 & \dfrac{1-\nu}{2} \end{bmatrix}$$

Considering (4.50) to (4.54) we recognize that, in general terms, all relationships for the evaluation of the finite element matrices corresponding to the local finite element nodal point displacements have been defined, and using the notation of Section 4.2.1, we have

$$\begin{aligned} \mathbf{H} &= \mathbf{\Phi} \mathbf{A}^{-1} \\ \mathbf{B} &= \mathbf{E} \mathbf{A}^{-1} \end{aligned} \tag{4.55}$$

Let us now consider briefly various types of finite elements encountered, which are subject to certain static or kinematic assumptions.

Truss and beam elements: Truss and beam elements are very widely used in structural engineering to model, for example, building frames and bridges (see Fig. 4.14(a) for an assemblage of truss elements).

As discussed in Section 4.2.1, the stiffness matrices of these elements can in many cases be calculated by the solution of the differential equations of equilibrium (see Example 4.4), and much literature has been published on such derivations. The results of these derivations have been employed in the displacement method of analysis and the corresponding approximate solution techniques, such as the method of moment distribution. However, it can be effective to evaluate the stiffness matrices using the finite element formulation, i.e., the virtual work principle, in particular, when considering complex beam geometries and geometric nonlinear analysis (see Section 5.4).

Plane stress and plane strain elements: Plane stress elements are employed to model membranes, the in-plane action of beams and plates as shown in Fig. 4.14(b) and so on. In each of these cases a two-dimensional stress situation exists in an x-y plane with the stresses τ_{zz}, τ_{yz} and τ_{zx} equal to zero. Plane strain elements are used to represent a slice (of unit thickness) of a structure in which the strain components ϵ_{zz}, γ_{yz} and γ_{zx} are zero. This situation arises in the analysis of a long dam as illustrated in Fig. 4.14(c).

Axisymmetric elements: Axisymmetric elements are used to model structural components that are rotationally symmetric about an axis. Examples of application are pressure vessels and solid rings. If these structures are also subjected to axisymmetric loads a two-dimensional analysis of a unit radian of the structure yields the complete stress and strain distributions as illustrated in Fig. 4.14(d).

On the other hand, if the axisymmetric structure is loaded nonaxisymmetrically, the choice lies between a fully three-dimensional analysis, in which substructuring (see Section 8.2.4) or cyclic symmetry (see Example 4.10) are used, and a Fourier decomposition of the loads with a Fourier-axisymmetric solution (see Example 4.16).

Plate bending and shell elements: The basic proposition in plate bending and shell analyses is that the structure is thin in one dimension and therefore the following assumptions can be made (see Fig. 4.14(e)):

1. The stress through the thickness (i.e., perpendicular to the mid-surface) of the plate/shell is zero.
2. Material particles that are originally on a straight line perpendicular to the mid-surface of the plate/shell remain on a straight line during deformations. In the Kirchhoff theory, shear deformations are neglected and the straight line remains perpendicular to the mid-surface during deformations. In the Mindlin theory, shear deformations are included and therefore the line originally normal to the mid-surface does in general not remain perpendicular to the mid-surface during the deformations.[12,13]

The first finite elements developed to model thin plates in bending were based on the Kirchhoff plate theory.[14,15] The difficulties in these approaches are that the elements must satisfy the convergence requirements *and* be relatively effective in their applications. Much research effort has been spent on the development of such elements; however, in recent years it has been recognized that more effective elements can frequently be formulated using the Mindlin plate theory (see Section 5.4).

To obtain a shell element a simple approach is to superimpose a plate bending stiffness and a plane-stress membrane stiffness.[16] This way flat shell elements are obtained that can be used to model flat components of shells (e.g. folded plates), and that are also employed to model general curved shells as an assemblage of flat elements. We demonstrate the development of a plate bending element based on the Kirchhoff plate theory and the construction of an associated flat shell element in Examples 4.14 and 4.15.

EXAMPLE 4.12: Discuss the derivation of the displacement and strain-displacement interpolation matrices of the beam shown in Fig. 4.15.

The exact stiffness matrix (within beam theory) of this beam could be evaluated by solving the beam differential equations of equilibrium, which are for the bending behavior

$$\frac{d^2}{d\xi^2}\left(EI\frac{d^2w}{d\xi^2}\right) = 0; \qquad EI = E\frac{bh^3}{12} \tag{a}$$

and for the axial behavior

$$\frac{d}{d\xi}\left(EA\frac{du}{d\xi}\right) = 0; \qquad A = bh \tag{b}$$

where E is Young's modulus. The procedure is to impose a unit end-displacement, with all other end-displacements equal to zero and solve the appropriate differential equation of equilibrium of the beam subject to these boundary conditions. Once the element internal displacements for these boundary conditions have been calculated, appropriate derivatives give the element end-forces which together constitute the column of the stiffness matrix corresponding to the imposed end-

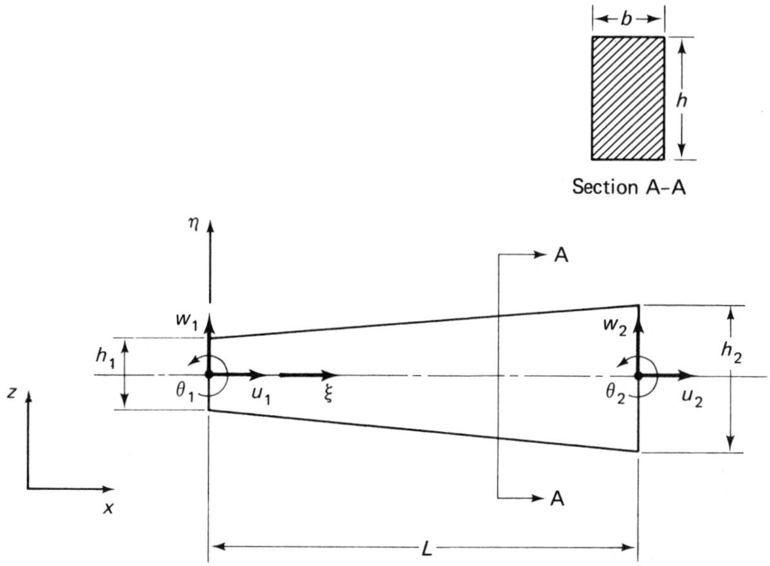

FIGURE 4.15 *Beam element with varying section.*

displacement. It should be noted that this stiffness matrix is only "exact" for static analysis, because in dynamic analysis the stiffness coefficients are frequency-dependent.

Alternatively, the formulation given in (4.16) can be used. The same stiffness matrix as would be evaluated by the above procedure is obtained if the exact element internal displacements (that satisfy (a) and (b)) are employed to construct the strain-displacement matrix. However, in practice it is frequently expedient to use the displacement interpolations that correspond to a uniform cross-section beam, and this yields an approximate stiffness matrix. The approximation is generally adequate when h_2 is not very much larger than h_1 (hence when a sufficiently large number of beam elements is employed to model the complete structure). The errors encountered in the analysis are those discussed in Section 4.2.5, because this formulation corresponds to displacement-based finite element analysis (in which (a) and (b) are only satisfied in the limit as the number of elements used in the idealization increases).

Using the variables defined in Fig. 4.15 and the "exact" displacements (Hermitian functions) corresponding to a prismatic beam we have

$$u = \left(1 - \frac{\xi}{L}\right)u_1 + \frac{6\eta}{L}\left(\frac{\xi}{L} - \frac{\xi^2}{L^2}\right)w_1 - \eta\left(1 - 4\frac{\xi}{L} + 3\frac{\xi^2}{L^2}\right)\theta_1$$
$$+ \frac{\xi}{L}u_2 - \frac{6\eta}{L}\left(\frac{\xi}{L} - \frac{\xi^2}{L^2}\right)w_2 + \eta\left(2\frac{\xi}{L} - 3\frac{\xi^2}{L^2}\right)\theta_2$$

Hence

$$\mathbf{H} = \left[\left(1 - \frac{\xi}{L}\right) \;\vdots\; \frac{6\eta}{L}\left(\frac{\xi}{L} - \frac{\xi^2}{L^2}\right) \;\vdots\; -\eta\left(1 - \frac{4\xi}{L} + 3\frac{\xi^2}{L^2}\right) \;\vdots\; \frac{\xi}{L} \;\vdots\; -\frac{6\eta}{L}\left(\frac{\xi}{L} - \frac{\xi^2}{L^2}\right) \;\vdots\; \eta\left(\frac{2\xi}{L} - 3\frac{\xi^2}{L^2}\right)\right] \qquad \text{(c)}$$

For (c) we ordered the nodal point displacements as follows

$$\hat{\mathbf{u}}^T = [u_1 w_1 \theta_1 \quad u_2 w_2 \theta_2]$$

Considering only normal strains and stresses in the beam, i.e. neglecting shearing deformations, we have as the only strain and stress components

$$\epsilon_{\xi\xi} = \frac{du}{d\xi}; \qquad \tau_{\xi\xi} = E\epsilon_{\xi\xi}$$

and hence

$$\mathbf{B} = \left[-\frac{1}{L} \;\vdots\; \frac{6\eta}{L}\left(\frac{1}{L} - \frac{2\xi}{L^2}\right) \;\vdots\; -\eta\left(\frac{-4}{L} + \frac{6\xi}{L^2}\right) \;\vdots\; \frac{1}{L} \;\vdots\; -\frac{6\eta}{L}\left(\frac{1}{L} - \frac{2\xi}{L^2}\right) \;\vdots\; \eta\left(\frac{2}{L} - \frac{6\xi}{L^2}\right)\right] \qquad \text{(d)}$$

The relations in (c) and (d) can directly be used to evaluate the element matrices defined in (4.29) to (4.33), e.g.,

$$\mathbf{K} = Eb \int_0^L \int_{-h/2}^{h/2} \mathbf{B}^T\mathbf{B}\, d\eta\, d\xi$$

where $$h = h_1 + (h_2 - h_1)\frac{\xi}{L}$$

This formulation can directly be extended to develop the element matrices corresponding to the three-dimensional action of the beam element and to include shear deformations as discussed in ref. [10].

EXAMPLE 4.13: Discuss the derivation of the stiffness, mass, and load matrices of the axisymmetric three-node finite element in Fig. 4.16.

This element was one of the first finite elements developed. For most practical applications, much more effective finite elements are presently available (see Chapter

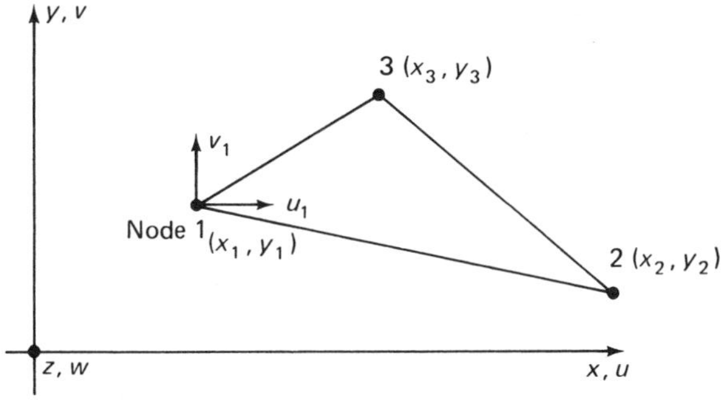

FIGURE 4.16 *Axisymmetric three-node element.*

5), but the element is conveniently used for instructional purposes because the equations to be dealt with are relatively simple.

The displacement assumption used is

$$u(x, y) = \alpha_1 + \alpha_2 x + \alpha_3 y$$
$$v(x, y) = \beta_1 + \beta_2 x + \beta_3 y$$

Therefore, a linear displacement variation is assumed, just as for the derivation of the four-node plane stress element considered in Example 4.5 where the fourth node required that the term xy be included in the displacement assumption. Referring to the derivations carried out in Example 4.5, we can directly establish the following relationships:

$$\begin{bmatrix} u(x, y) \\ v(x, y) \end{bmatrix} = \mathbf{H} \begin{bmatrix} u_1 \\ u_2 \\ u_3 \\ v_1 \\ v_2 \\ v_3 \end{bmatrix}$$

where $$\mathbf{H} = \begin{bmatrix} 1 & x & y & 0 & 0 & 0 \\ 0 & 0 & 0 & 1 & x & y \end{bmatrix} \mathbf{A}^{-1}$$

$$\mathbf{A}^{-1} = \begin{bmatrix} \mathbf{A}_1^{-1} & \mathbf{0} \\ \mathbf{0} & \mathbf{A}_1^{-1} \end{bmatrix}; \qquad \mathbf{A}_1 = \begin{bmatrix} 1 & x_1 & y_1 \\ 1 & x_2 & y_2 \\ 1 & x_3 & y_3 \end{bmatrix}$$

Hence $$\mathbf{A}_1^{-1} = \frac{1}{\det \mathbf{A}_1} \begin{bmatrix} x_2 y_3 - x_3 y_2 & x_3 y_1 - x_1 y_3 & x_1 y_2 - x_2 y_1 \\ y_2 - y_3 & y_3 - y_1 & y_1 - y_2 \\ x_3 - x_2 & x_1 - x_3 & x_2 - x_1 \end{bmatrix}$$

where $$\det \mathbf{A}_1 = x_1(y_2 - y_3) + x_2(y_3 - y_1) + x_3(y_1 - y_2)$$

We may note that det $\mathbf{A}_1$ is only zero if the three element nodal points lie on a straight line. The strains are given in Table 4.2; i.e.,

$$\epsilon_{xx} = \frac{\partial u}{\partial x}; \qquad \epsilon_{yy} = \frac{\partial v}{\partial y}; \qquad \gamma_{xy} = \frac{\partial u}{\partial y} + \frac{\partial v}{\partial x}; \qquad \epsilon_{zz} = \frac{\partial w}{\partial z} = \frac{u}{x}$$

Using the assumed displacement polynomials, we obtain

$$\begin{bmatrix} \epsilon_{xx} \\ \epsilon_{yy} \\ \gamma_{xy} \\ \epsilon_{zz} \end{bmatrix} = \mathbf{B} \begin{bmatrix} u_1 \\ u_2 \\ u_3 \\ v_1 \\ v_2 \\ v_3 \end{bmatrix}; \qquad \mathbf{B} = \begin{bmatrix} 0 & 1 & 0 & 0 & 0 & 0 \\ 0 & 0 & 0 & 0 & 0 & 1 \\ 0 & 0 & 1 & 0 & 1 & 0 \\ \frac{1}{x} & 1 & \frac{y}{x} & 0 & 0 & 0 \end{bmatrix} \mathbf{A}^{-1}$$

Using the relations (4.29) to (4.33) we thus have

$$\mathbf{K} = \mathbf{A}^{-T} \left\{ \int_A \frac{E(1-\nu)}{(1+\nu)(1-2\nu)} \begin{bmatrix} 0 & 0 & 0 & \frac{1}{x} \\ 1 & 0 & 0 & 1 \\ 0 & 0 & 1 & \frac{y}{x} \\ 0 & 0 & 0 & 0 \\ 0 & 0 & 1 & 0 \\ 0 & 1 & 0 & 0 \end{bmatrix} \begin{bmatrix} 1 & \frac{\nu}{1-\nu} & 0 & \frac{\nu}{1-\nu} \\ \frac{\nu}{1-\nu} & 1 & 0 & \frac{\nu}{1-\nu} \\ 0 & 0 & \frac{1-2\nu}{2(1-\nu)} & 0 \\ \frac{\nu}{1-\nu} & \frac{\nu}{1-\nu} & 0 & 1 \end{bmatrix} \begin{bmatrix} 0 & 1 & 0 & 0 & 0 & 0 \\ 0 & 0 & 0 & 0 & 0 & 1 \\ 0 & 0 & 1 & 0 & 1 & 0 \\ \frac{1}{x} & 1 & \frac{y}{x} & 0 & 0 & 0 \end{bmatrix} x\,dx\,dy \right\} \mathbf{A}^{-1} \qquad \text{(a)}$$

where 1 radian of the axisymmetric element is considered in the volume integration. Similarly, we have

$$\mathbf{R}_B = \mathbf{A}^{-T} \int_A \begin{bmatrix} 1 & 0 \\ x & 0 \\ y & 0 \\ 0 & 1 \\ 0 & x \\ 0 & y \end{bmatrix} \begin{bmatrix} f_x^B \\ f_y^B \end{bmatrix} x\,dx\,dy$$

$$\mathbf{R}_I = \mathbf{A}^{-T} \int_A \begin{bmatrix} 0 & 0 & 0 & \frac{1}{x} \\ 1 & 0 & 0 & 1 \\ 0 & 0 & 1 & \frac{y}{x} \\ 0 & 0 & 0 & 0 \\ 0 & 0 & 1 & 0 \\ 0 & 1 & 0 & 0 \end{bmatrix} \begin{bmatrix} \tau_{xx}^I \\ \tau_{yy}^I \\ \tau_{xy}^I \\ \tau_{zz}^I \end{bmatrix} x\,dx\,dy \qquad \text{(b)}$$

$$\mathbf{M} = \rho \mathbf{A}^{-T} \left\{ \int_A \begin{bmatrix} 1 & 0 \\ x & 0 \\ y & 0 \\ 0 & 1 \\ 0 & x \\ 0 & y \end{bmatrix} \begin{bmatrix} 1 & x & y & 0 & 0 & 0 \\ 0 & 0 & 0 & 1 & x & y \end{bmatrix} x\,dx\,dy \right\} \mathbf{A}^{-1}$$

where the mass density ρ is assumed to be constant.

For calculation of the surface load vector $\mathbf{R}_S$, it is expedient in practice to introduce auxiliary coordinate systems that are located along the loaded sides of the element. Assume that the side 2-3 of the element is loaded as shown in Fig. 4.17. The load vector $\mathbf{R}_S$ is then evaluated using as the variable s,

$$\mathbf{R}_S = \int_S \begin{bmatrix} 0 & 0 \\ 1-\dfrac{s}{L} & 0 \\ \dfrac{s}{L} & 0 \\ 0 & 0 \\ 0 & 1-\dfrac{s}{L} \\ 0 & \dfrac{s}{L} \end{bmatrix} \begin{bmatrix} f_x^S \\ f_y^S \end{bmatrix} \left[x_2\left(1-\frac{s}{L}\right) + x_3\frac{s}{L} \right] ds$$

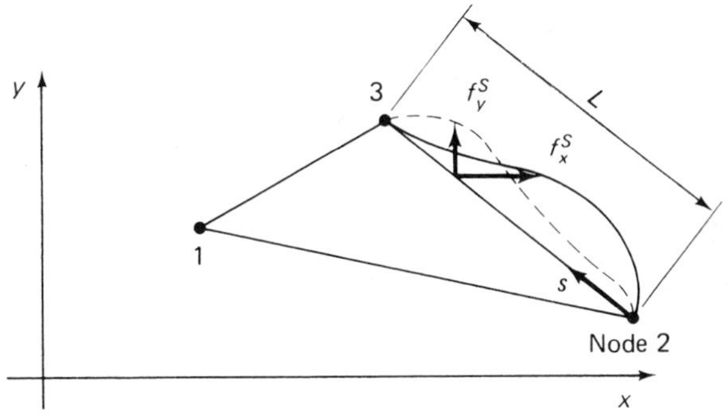

FIGURE 4.17 *Axisymmetric element with surface loading.*

Considering the above finite element matrix evaluations the following observations can be made. First, to evaluate the integrals, it is possible to obtain closed-form solutions; alternatively, numerical integration (discussed in Section 5.7) can be used. Second, we find that the stiffness, mass, and load matrices corresponding to plane stress and plane strain finite elements would simply be obtained by (1) not including the fourth row in the strain-displacement matrix $\mathbf{E}$ used in (a) and (b), (2) employing the appropriate stress-strain matrix $\mathbf{C}$ in (a), and (3) using as the differential volume element $h\,dx\,dy$ instead of $x\,dx\,dy$, where h is the thickness of the element (conveniently taken equal to 1 in plane strain analysis). Therefore, axisymmetric, plane stress, and plane strain analyses can effectively be implemented in a single computer program. Also, the matrix $\mathbf{E}$ shows that constant strain conditions ϵ_{xx}, ϵ_{yy}, and γ_{xy} are assumed in either analysis.

The concept of performing axisymmetric, plane strain, and plane stress analysis in an effective manner in one computer program is, in fact, presented in Section 5.9, where we discuss the efficient implementation of isoparametric finite element analysis.

EXAMPLE 4.14: Derive the matrices $\mathbf{\Phi}(x, y)$, $\mathbf{E}(x, y)$, and $\mathbf{A}$ for the rectangular plate bending element in Fig. 4.18.

This element is one of the first plate bending elements derived, and more effective plate bending elements are already in use (see Chapter 5).

As shown in Fig. 4.18, the plate bending element considered has three degrees of freedom per node. Therefore, it is necessary to have 12 unknown generalized

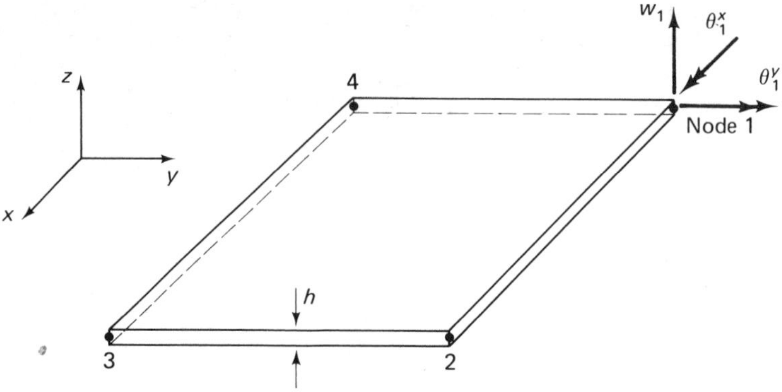

FIGURE 4.18 *Rectangular plate bending element.*

coordinates, $\alpha_1, \ldots, \alpha_{12}$, in the displacement assumption for w. The polynomial used is

$$w = \alpha_1 + \alpha_2 x + \alpha_3 y + \alpha_4 x^2 + \alpha_5 xy + \alpha_6 y^2 + \alpha_7 x^3 + \alpha_8 x^2 y + \alpha_9 xy^2 + \alpha_{10} y^3 + \alpha_{11} x^3 y + \alpha_{12} xy^3$$

Hence

$$\mathbf{\Phi}(x, y) = [1 \quad x \quad y \quad x^2 \quad xy \quad y^2 \quad x^3 \quad x^2y \quad xy^2 \quad y^3 \quad x^3y \quad xy^3] \tag{a}$$

We can now calculate $\partial w/\partial x$ and $\partial w/\partial y$; i.e.,

$$\frac{\partial w}{\partial x} = \alpha_2 + 2\alpha_4 x + \alpha_5 y + 3\alpha_7 x^2 + 2\alpha_8 xy + \alpha_9 y^2 + 3\alpha_{11} x^2 y + \alpha_{12} y^3 \tag{b}$$

and

$$\frac{\partial w}{\partial y} = \alpha_3 + \alpha_5 x + 2\alpha_6 y + \alpha_8 x^2 + 2\alpha_9 xy + 3\alpha_{10} y^2 + \alpha_{11} x^3 + 3\alpha_{12} xy^2 \tag{c}$$

Using the conditions

$$\left.\begin{array}{ll} w_i = (w)_{x_i, y_i}; & \theta_i^x = \left(\dfrac{\partial w}{\partial y}\right)_{x_i, y_i} \\ \theta_i^y = \left(-\dfrac{\partial w}{\partial x}\right)_{x_i, y_i} & \end{array}\right\} i = 1, \ldots, 4$$

we can construct the matrix $\mathbf{A}$; i.e., we obtain

$$\begin{bmatrix} w_1 \\ \cdot \\ \cdot \\ \cdot \\ w_4 \\ \theta_1^x \\ \cdot \\ \cdot \\ \cdot \\ \theta_4^x \\ \theta_1^y \\ \cdot \\ \cdot \\ \cdot \\ \theta_4^y \end{bmatrix} = \mathbf{A} \begin{bmatrix} \alpha_1 \\ \alpha_2 \\ \cdot \\ \cdot \\ \cdot \\ \\ \\ \\ \\ \\ \\ \\ \\ \\ \alpha_{12} \end{bmatrix}$$

where

$$
\mathbf{A} = \begin{bmatrix}
1 & x_1 & y_1 & x_1^2 & x_1y_1 & y_1^2 & x_1^3 & x_1^2y_1 & x_1y_1^2 & y_1^3 & x_1^3y_1 & x_1y_1^3 \\
\cdot & & & & & \cdot & & & & & & \\
\cdot & & & & & \cdot & & & & & & \\
\cdot & & & & & \cdot & & & & & & \\
1 & x_4 & y_4 & x_4^2 & x_4y_4 & y_4^2 & x_4^3 & x_4^2y_4 & x_4y_4^2 & y_4^3 & x_4^3y_4 & x_4y_4^3 \\
0 & 0 & 1 & 0 & x_1 & 2y_1 & 0 & x_1^2 & 2x_1y_1 & 3y_1^2 & x_1^3 & 3x_1y_1^2 \\
\cdot & & & & & \cdot & & & & & & \\
\cdot & & & & & \cdot & & & & & & \\
\cdot & & & & & \cdot & & & & & & \\
0 & 0 & 1 & 0 & x_4 & 2y_4 & 0 & x_4^2 & 2x_4y_4 & 3y_4^2 & x_4^3 & 3x_4y_4^2 \\
0 & -1 & 0 & -2x_1 & -y_1 & 0 & -3x_1^2 & -2x_1y_1 & -y_1^2 & 0 & -3x_1^2y_1 & -y_1^3 \\
\cdot & & & & & \cdot & & & & & & \\
\cdot & & & & & \cdot & & & & & & \\
\cdot & & & & & \cdot & & & & & & \\
0 & -1 & 0 & -2x_4 & -y_4 & 0 & -3x_4^2 & -2x_4y_4 & -y_4^2 & 0 & -3x_4^2y_4 & -y_4^3
\end{bmatrix} \tag{d}
$$

which can be shown to be always nonsingular.

To evaluate the matrix **E**, we recall that in plate bending analysis curvatures and moments are used as generalized strains and stresses (see Tables 4.2 and 4.3) Calculating the required derivatives of (b) and (c), we obtain

$$
\begin{aligned}
-\frac{\partial^2 w}{\partial x^2} &= -2\alpha_4 - 6\alpha_7 x - 2\alpha_8 y - 6\alpha_{11}xy \\
-\frac{\partial^2 w}{\partial y^2} &= -2\alpha_6 - 2\alpha_9 x - 6\alpha_{10} y - 6\alpha_{12}xy \\
2\frac{\partial^2 w}{\partial x \partial y} &= 2\alpha_5 + 4\alpha_8 x + 4\alpha_9 y + 6\alpha_{11}x^2 + 6\alpha_{12}y^2
\end{aligned} \tag{e}
$$

Hence we have

$$
\mathbf{E} = \begin{bmatrix}
0 & 0 & 0 & -2 & 0 & 0 & -6x & -2y & 0 & 0 & -6xy & 0 \\
0 & 0 & 0 & 0 & 0 & -2 & 0 & 0 & -2x & -6y & 0 & -6xy \\
0 & 0 & 0 & 0 & 2 & 0 & 0 & 4x & 4y & 0 & 6x^2 & 6y^2
\end{bmatrix} \tag{f}
$$

With the matrices $\mathbf{\Phi}$, **A**, and **E** given in (a), (d), and (f) and the material matrix **C** in Table 4.3 the element stiffness matrix, mass matrix, and load vectors can now be calculated.

An important consideration in the evaluation of an element stiffness matrix is whether the element is complete and compatible. The element considered in this example is complete as shown in (e); i.e., the element can represent constant curvature states, but the element is not compatible. The compatibility requirements are violated in a number of plate bending elements, meaning that convergence in the analysis is in general not monotonic (see Sections 4.2.5 and 4.3.1).

EXAMPLE 4.15: Discuss the evaluation of the stiffness matrix of a flat rectangular shell element.

A simple rectangular flat shell element can be obtained by superimposing the plate bending behavior considered in Example 4.14 and the plane stress behavior of the element used in Example 4.5. The resulting element is shown in Fig. 4.19. The element can be employed to model assemblages of flat plates (e.g. folded plate structures) and also curved shells. For actual analyses more effective shell elements are available, and we only discuss here the element in Fig. 4.19 in order to demonstrate some basic analysis approaches.

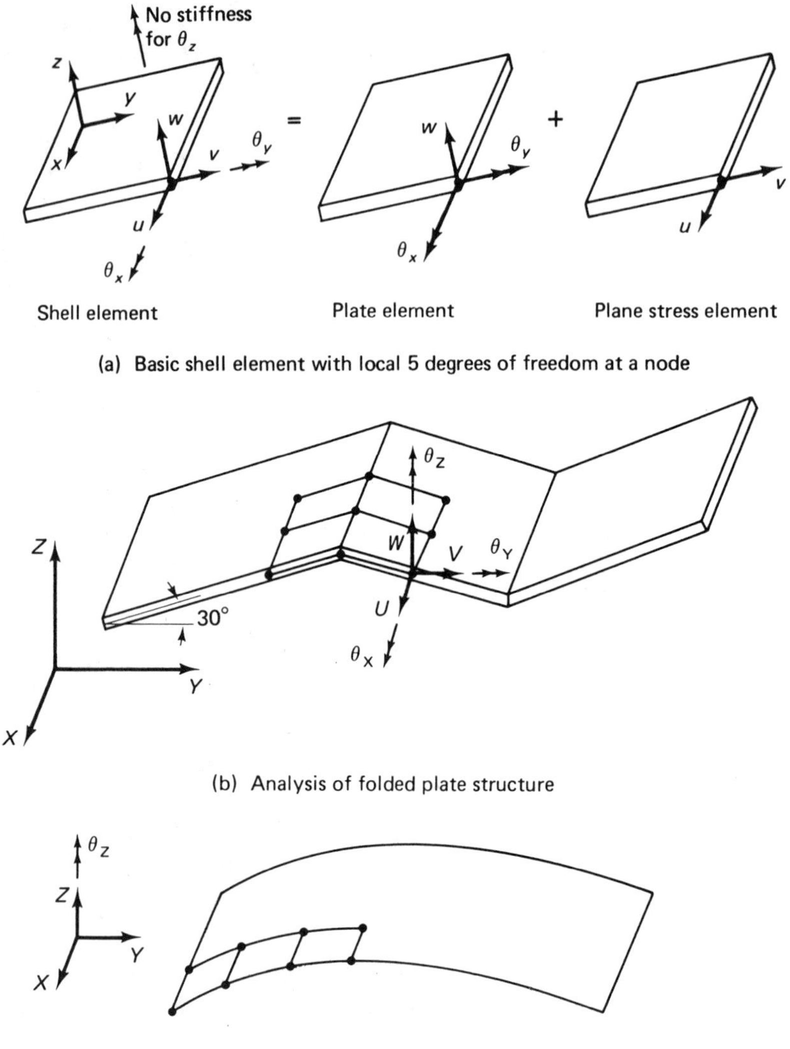

(a) Basic shell element with local 5 degrees of freedom at a node

(b) Analysis of folded plate structure

(c) Analysis of slightly curved shell

FIGURE 4.19 *Use of a flat shell element.*

Let $\tilde{\mathbf{K}}_B$ and $\tilde{\mathbf{K}}_M$ be the stiffness matrices, in the local coordinate system, corresponding to the bending and membrane behavior of the element, respectively. Then the shell element stiffness matrix $\tilde{\mathbf{K}}_S$ is

$$\underset{20\times 20}{\tilde{\mathbf{K}}_S} = \begin{bmatrix} \underset{12\times 12}{\tilde{\mathbf{K}}_B} & \mathbf{0} \\ \mathbf{0} & \underset{8\times 8}{\tilde{\mathbf{K}}_M} \end{bmatrix} \tag{a}$$

The matrices $\tilde{\mathbf{K}}_M$ and $\tilde{\mathbf{K}}_B$ were discussed in Examples 4.5 and 4.14, respectively.

This shell element can now directly be employed in the analysis of a variety of shell structures. Consider the structures in Fig. 4.19 which might be idealized as shown. Since we deal in these analyses with 6 degrees of freedom per node, the element stiffness matrices corresponding to the global degrees of freedom are calculated using the transformation given in (4.37)

$$\underset{24\times 24}{\mathbf{K}_S} = \mathbf{T}^T\tilde{\mathbf{K}}_S^*\mathbf{T} \tag{b}$$

where

$$\underset{24\times 24}{\tilde{\mathbf{K}}_S^*} = \begin{bmatrix} \underset{20\times 20}{\tilde{\mathbf{K}}_S} & \mathbf{0} \\ \mathbf{0} & \underset{4\times 4}{\mathbf{0}} \end{bmatrix} \tag{c}$$

and $\mathbf{T}$ is the transformation matrix between the local and global element degrees of freedom. To define $\tilde{\mathbf{K}}_S^*$ corresponding to 6 degrees of freedom per node, we have amended $\tilde{\mathbf{K}}_S$ on the r.h.s. of (c) to include the stiffness coefficients corresponding to the local rotations θ_z (rotations about the z-axis) at the nodes. These stiffness coefficients have been set in (c) equal to zero. The reason for doing so is that these degrees of freedom have not been included in the formulation of the element; thus the element rotation θ_z at a node is not measured and does not contribute to the strain energy stored in the element.

The solution of a model can be obtained using $\tilde{\mathbf{K}}_S^*$ in (c) as long as the elements surrounding a node are not coplanar. This does not hold for the folded plate model, and considering the analysis of the slightly curved shell in Fig. 4.19(c), the elements may be almost coplanar (depending on the curvature of the shell and the idealization used). In these cases, the global stiffness matrix is singular or ill-conditioned, and difficulties arise in the solution of the global equilibrium equations. To avoid this problem it is possible to add a small stiffness coefficient corresponding to the θ_z rotation, i.e. instead of $\tilde{\mathbf{K}}_S^*$ in (c) we use

$$\tilde{\mathbf{K}}_S^{*\prime} = \begin{bmatrix} \underset{20\times 20}{\tilde{\mathbf{K}}_S} & \mathbf{0} \\ \mathbf{0} & \underset{4\times 4}{k\mathbf{I}} \end{bmatrix} \tag{d}$$

where k is about $\frac{1}{1000}$th of the smallest diagonal element of $\tilde{\mathbf{K}}_S$. The stiffness coefficient k must be large enough to enable the accurate solution of the finite element system equilibrium equations and small enough not to affect the system response significantly. Therefore a large enough number of digits must be used in the floating point arithmetic (see Section 8.5).

A more effective way to circumvent the above problem is to use curved shell elements with 5 degrees of freedom per node where these are defined corresponding to a plane tangent to the mid-surface of the shell. In this case the rotation normal to the shell surface is not a degree of freedom (see Section 5.4.2).

In the above element formulations we used polynomial functions to express the displacements. We should briefly note, however, that for certain applications the use of other functions such as trigonometric expressions can be effective. Trigonometric functions are used in the analysis of axisymmetric structures subjected to nonaxisymmetric loading and in the finite strip method.[17,18] The advantage of the trigonometric functions lies in their orthogonality properties. Namely, if sine and cosine products are integrated over an appropriate interval, the integral can be zero. This then means that there is no coupling in the equilibrium equations between the generalized coordinates that correspond to the sine and cosine functions, and the equilibrium equations can be solved more effectively. In this context it may be noted that the best functions that we could use in the finite element analysis would be given by the eigenvectors of the problem, because they would give a diagonal stiffness matrix. However, these functions are not known, and for general applications, the use of polynomial, trigonometric, or other assumptions for the finite element displacements is most natural. We demonstrate the use of trigonometric functions in the following example.

EXAMPLE 4.16: Figure 4.20 shows an axisymmetric structure subjected to a nonaxisymmetric loading into the radial direction. Discuss the analysis of this structure using the 3-node axisymmetric element of Example 4.13, when the loading is represented as a superposition of Fourier components.

The stress distribution in the structure is three-dimensional and could be calculated using a three-dimensional finite element idealization. However, it is possible to take advantage of the axisymmetric geometry of the structure and, depending on the exact loading applied, reduce the computational effort very significantly.

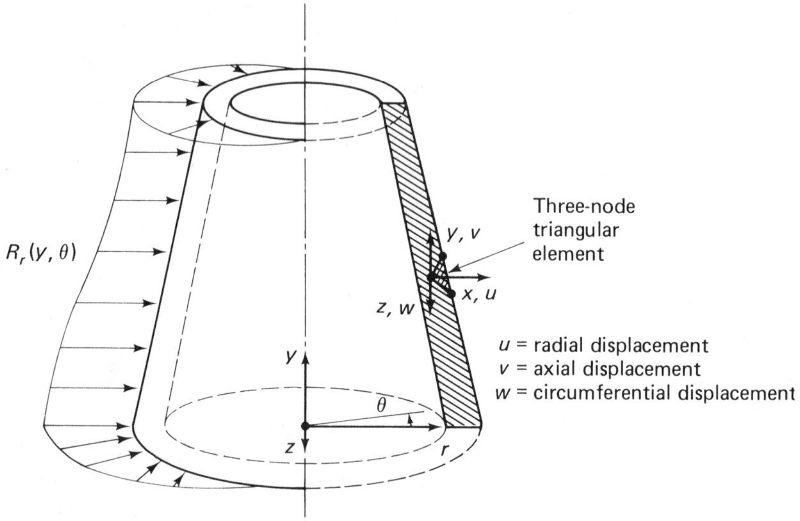

(a) Structure of interest

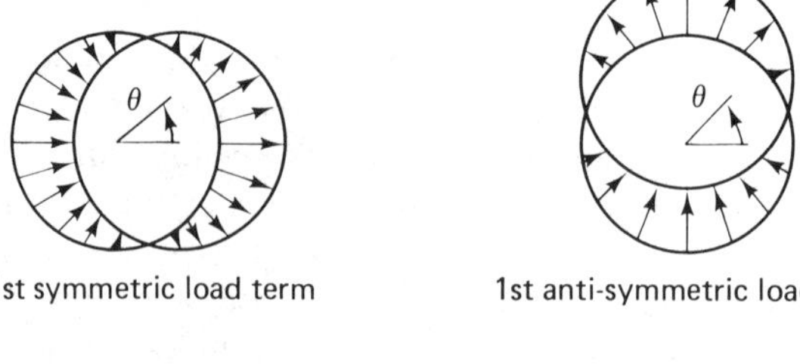

1st symmetric load term 1st anti-symmetric load term

(b) Representation of nonaxisymmetric loading

FIGURE 4.20 *Axisymmetric structure subjected to nonaxisymmetric loading.*

The key point in this analysis is that we expand the externally applied loads $R_r(\theta, y)$ in the Fourier series[17]

$$R_r = \sum_{p=1}^{p_c} R_{pc} \cos p\theta + \sum_{p=1}^{p_s} R_{ps} \sin p\theta \tag{a}$$

where p_c and p_s are the total number of symmetric and antisymmetric load contributions, about $\theta = 0$ respectively. Figure 4.20(b) illustrates the first terms in the expansion of (a).

The complete analysis can now be performed by superimposing the response due to the symmetric and antisymmetric load contributions defined in (a). For example, considering the symmetric response we use for an element

$$\begin{aligned} u(x, y, \theta) &= \sum_{p=1}^{p_c} \cos p\theta\, \mathbf{H}\mathbf{u}^p \\ v(x, y, \theta) &= \sum_{p=1}^{p_c} \cos p\theta\, \mathbf{H}\mathbf{v}^p \\ w(x, y, \theta) &= \sum_{p=1}^{p_c} \sin p\theta\, \mathbf{H}\mathbf{w}^p \end{aligned} \tag{b}$$

where for the triangular elements, referring to Example 4.13,

$$\mathbf{H} = [1 \quad x \quad y]\, \mathbf{A}_1^{-1} \tag{c}$$

and the $\mathbf{u}^p$, $\mathbf{v}^p$ and $\mathbf{w}^p$ are the element unknown generalized nodal point displacements corresponding to mode p.

We should note that we superimpose in (b) the response measured in individual harmonic displacement distributions. Using (b) we can now establish the strain-displacement matrix of the element. Since we are dealing with a three-dimensional stress distribution we use the expression for three-dimensional strain distributions in cylindrical coordinates:

$$\boldsymbol{\epsilon} = \begin{bmatrix} \dfrac{\partial u}{\partial r} \\[2mm] \dfrac{\partial v}{\partial y} \\[2mm] \dfrac{u}{r} + \dfrac{1}{r}\dfrac{\partial w}{\partial \theta} \\[2mm] \dfrac{\partial u}{\partial y} + \dfrac{\partial v}{\partial r} \\[2mm] \dfrac{\partial w}{\partial y} + \dfrac{1}{r}\dfrac{\partial v}{\partial \theta} \\[2mm] \dfrac{\partial w}{\partial r} + \dfrac{1}{r}\dfrac{\partial u}{\partial \theta} - \dfrac{w}{r} \end{bmatrix} \tag{d}$$

where $\quad \boldsymbol{\epsilon}^T = [\epsilon_{rr} \quad \epsilon_{yy} \quad \epsilon_{\theta\theta} \quad \gamma_{ry} \quad \gamma_{y\theta} \quad \gamma_{\theta r}] \qquad$ (e)

Substituting from (b) into (d) we obtain a strain-displacement matrix, $\mathbf{B}_p$, for each value of p and the total strains can be thought of as the superposition of the strain distributions contained in each harmonic.

The unknown nodal point displacements can now be evaluated using the usual procedures. The equilibrium equations corresponding to the generalized nodal point displacements U_i^p, V_i^p, W_i^p, $i = 1, \ldots, N$ (N is equal to the total number of nodes) and $p = 1, \ldots, p_c$, are evaluated as given in (4.14) to (4.18), where we have now

$$\mathbf{U}^T = [\mathbf{U}^{1^T}\, \mathbf{U}^{2^T} \ldots \mathbf{U}^{p_c T}] \tag{f}$$

and $$\mathbf{U}^{pT} = [U_1^p \; V_1^p \; W_1^p \mid U_2^p \ldots W_N^p] \tag{g}$$

In the calculations of $\mathbf{K}$ and $\mathbf{R}_S$ we note that because of the orthogonality properties

$$\begin{aligned} \int_0^{2\pi} \sin n\theta \sin m\theta \, d\theta &= 0 \qquad n \neq m \\ \int_0^{2\pi} \cos n\theta \cos m\theta \, d\theta &= 0 \qquad n \neq m \end{aligned} \tag{h}$$

the stiffness matrices corresponding to the different harmonics are decoupled from each other. Hence, we have the following equilibrium equations for the structure

$$\mathbf{K}^p \mathbf{U}^p = \mathbf{R}_S^p \qquad p = 1, \ldots, p_c \tag{i}$$

where $\mathbf{K}^p$ and $\mathbf{R}_S^p$ are the stiffness matrix and load vector corresponding to the pth harmonic. The solution of the equations in (i) gives the generalized nodal point displacements of each element and (b) then yields all element internal displacements.

In the above displacement solution we considered only the symmetric load contributions. But an analogous analysis can be performed for the antisymmetric load harmonics of (a) by simply replacing in (b) to (i) all sine and cosine terms by cosine and sine terms, respectively. The complete structural response is then obtained by superimposing the displacements corresponding to all harmonics.

Although we have considered in the above discussion only surface loading, the analysis can be extended using the same approach to include body force loading and initial stresses.

Finally, we note that the computational effort required in the analysis is directly proportional to the number of load harmonics used. Hence, the solution procedure is very efficient if the loading can be represented using only a few harmonics (e.g. wind loading), but may be inefficient when many harmonics must be used to represent the loading (e.g. a concentrated force).

4.2.4 Lumping of Structure Properties and Loads

A physical interpretation of the finite element procedure of analysis as presented in the previous section is that the structure properties—stiffness and mass—and the loads, internal and external, are lumped at the discrete nodes of the element assemblage using the virtual work principle. *Because the same interpolation functions are employed in the calculation of the load vectors and the mass matrix as in the evaluation of the stiffness matrix, we say that "consistent" load vectors and a consistent mass matrix are evaluated.* In this case, provided certain conditions are fulfilled (see Section 4.2.5), the finite element solution is a Ritz analysis.

It may now be recognized that instead of performing the integrations leading to the consistent load vector, we may evaluate an approximate load vector by simply adding to the actually applied concentrated nodal forces $\mathbf{R}_C$ additional forces that are in some sense equivalent to the distributed loads on the elements. A somewhat obvious way of constructing approximate load vectors is to calculate the total body and surface forces corresponding to an element and to assign equal parts to the appropriate element nodal degrees of freedom. Consider as an example the rectangular plane stress element in Fig. 4.21 with the variation of the body force shown. The total body force is equal to 2.0, and hence we obtain the lumped body force vector given in the figure.

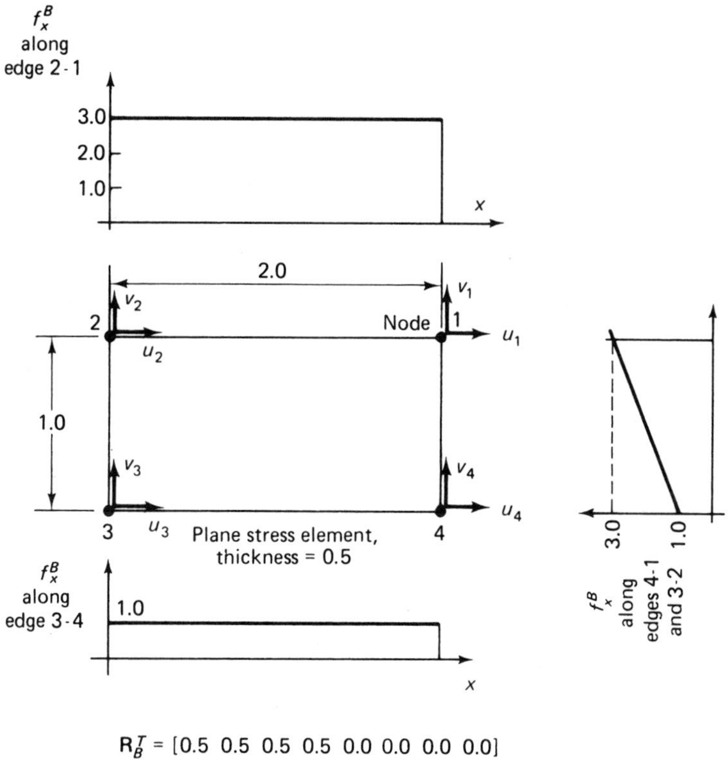

FIGURE 4.21 *Body force distribution and corresponding lumped body force vector* $\mathbf{R}_B$ *of a rectangular element.*

Considering the derivation of an element mass matrix, we recall that the inertia forces have been considered as part of the body forces. Hence we may also establish an approximate mass matrix by lumping equal parts of the total element mass to the nodal points. Realizing that each nodal mass essentially corresponds to the mass of an element-contributing-volume around the node, we note that using this procedure of lumping mass, we assume in essence that the accelerations of the contributing volume to a node are constant and equal to the nodal values.

An important advantage of using a lumped mass matrix is that the matrix is diagonal, and, as will be seen later, the numerical operations for the solution of the dynamic equations of equilibrium are in some cases reduced very significantly.

EXAMPLE 4.17: Evaluate the lumped body force vector and the lumped mass matrix of the truss element assemblage in Fig. 4.3.

The lumped mass matrix is

$$\mathbf{M} = \rho \int_0^{100} (1) \begin{bmatrix} \frac{1}{2} & 0 & 0 \\ 0 & \frac{1}{2} & 0 \\ 0 & 0 & 0 \end{bmatrix} dx + \rho \int_0^{80} \left(1 + \frac{x}{40}\right)^2 \begin{bmatrix} 0 & 0 & 0 \\ 0 & \frac{1}{2} & 0 \\ 0 & 0 & \frac{1}{2} \end{bmatrix} dx$$

or $$\mathbf{M} = \frac{\rho}{3}\begin{bmatrix} 150 & 0 & 0 \\ 0 & 670 & 0 \\ 0 & 0 & 520 \end{bmatrix}$$

Similarly, the lumped body force vector is

$$\mathbf{R}_B = \left(\int_0^{100} (1) \begin{bmatrix} \frac{1}{2} \\ \frac{1}{2} \\ 0 \end{bmatrix} (1)\, dx + \int_0^{80} \left(1 + \frac{x}{40}\right)^2 \begin{bmatrix} 0 \\ \frac{1}{2} \\ \frac{1}{2} \end{bmatrix} \left(\frac{1}{10}\right) dx \right) f_2(t)$$

$$= \frac{1}{3}\begin{bmatrix} 150 \\ 202 \\ 52 \end{bmatrix} f_2(t)$$

It may be noted that, as required, the sums of the elements each in $\mathbf{M}$ and $\mathbf{R}_B$ in both this example and in Example 4.3 are the same.

When using the load lumping procedure it should be recognized that the nodal point loads are, in general, only calculated approximately, and if a coarse finite element mesh is employed the resulting solution may be very inaccurate. Indeed, in some cases when using higher-order finite elements, surprising results are obtained. Figure 4.22 demonstrates such a case (see also Example 5.12).

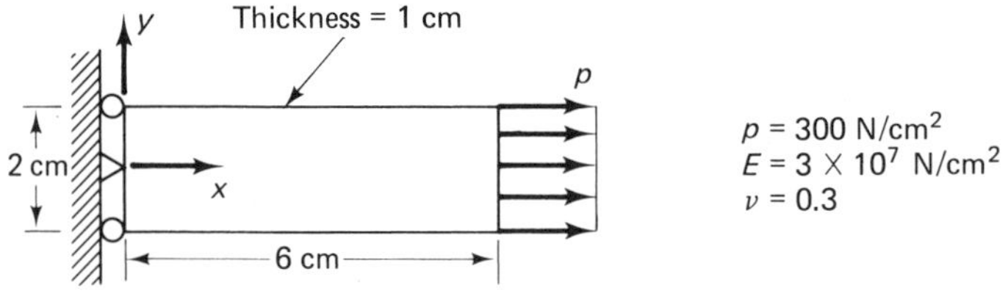

(a) Problem

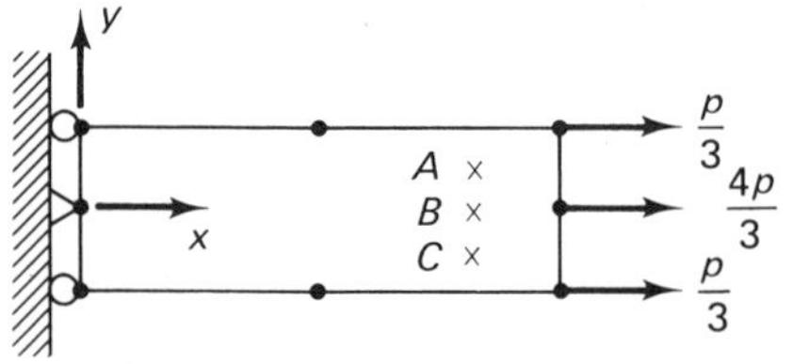

(b) Finite element model with consistent loading

Integration point	τ_{xx}	τ_{yy}	τ_{xy}
A	300.00	0.0	0.0
B	300.00	0.0	0.0
C	300.00	0.0	0.0

(All stresses have units of N/cm^2)

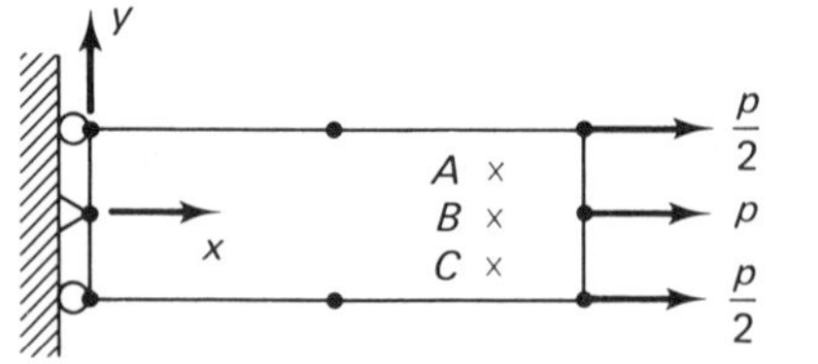

(c) Finite element model with lumped loading

Integration point	τ_{xx}	τ_{yy}	τ_{xy}
A	301.41	−7.85	−24.72
B	295.74	−9.55	0.0
C	301.41	−7.85	24.72

(All stresses have units of N/cm^2)

FIGURE 4.22 *Some sample analysis results with and without consistent loading.*

Considering dynamic analysis, it should be recalled that the inertia effects can be thought of as body forces. Therefore, if a lumped mass matrix is employed, little might be gained by using a consistent load vector; whereas consistent nodal point loads should be used if a consistent mass matrix is employed in the analysis.

4.2.5 Convergence of Analysis Results

Based on the preceding discussion, we can now say that, in general, a finite element analysis requires the idealization of an actual physical problem into a mechanical description and then the finite element solution of that idealization. Figure 4.23 summarizes these concepts. The distinction given in that figure is frequently not recognized in practical analysis, because the differential equations of motion of the mechanical idealization are not dealt with, and indeed the equations may be unknown in the analysis of a complex problem, such as the response prediction of a three-dimensional shell. Instead, in a practical analysis, a finite element idealization of the physical problem is established directly. However, to study the convergence of the finite element solution as the number of elements increases, it is valuable to recognize that a mechanical idealization is actually implied in the finite element representation of the physical problem. Namely, a proper finite element solution should converge (as the number of elements is in-

Actual Physical Problem

Geometric domain
Material
Loading
Boundary conditions

↓

Mechanical Idealization

Kinematics, e.g.,	truss plane stress three-dimensional Kirchhoff plate etc.	Yields: Governing differential equation(s) of motion e.g., $\frac{\partial}{\partial x}\left(EA\frac{\partial u}{\partial x}\right) = -p(x)$
Material, e.g.,	isotropic linear elastic Mooney-Rivlin rubber etc.	
Loading, e.g.,	concentrated centrifugal etc.	
Boundary Conditions, e.g.,	prescribed displacements etc.	

↓

Finite Element Solution

Choice of elements and solution procedures } Yields: Approximate response of mechanical idealization

FIGURE 4.23 Finite element solution process.

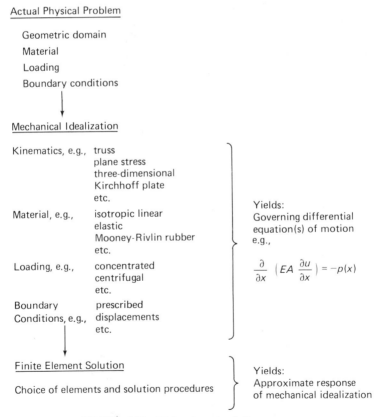

creased) to the analytical (exact) solution of the differential equations that govern the response of the mechanical idealization. Furthermore, this convergence behavior displays all the characteristics of the finite element scheme, because the differential equations of motion of the mechanical idealization express in a very precise and compact manner all basic conditions that the solution variables (stress, displacement, strain, and so on) must satisfy. If the differential equations of motion are not known and/or analytical solutions of the exact response of the mechanical idealization cannot be obtained, the convergence of the finite element solutions can only be measured on the fact that all basic kinematic, static, and constitutive conditions contained in the mechanical idealization must ultimately (at convergence) be satisfied. Therefore, in all discussions of the convergence of finite element solutions we imply that the convergence to the exact response of a mechanical idealization is meant.

Considering the approximate finite element solution to the exact response of the mechanical idealization, we need to recognize that different sources of errors affect the finite element solution results. Table 4.4 summarizes various general sources of errors. Round-off errors are a result of the finite precision arithmetic of the computer used; solution errors in the constitutive modeling are due to the linearization and integration of the constitutive relations; solution errors in the calculation of the dynamic response arise in the numerical integration of the equations of motion or because only a few modes are used in a mode superposition analysis; and solution errors arise when an iterative solution is obtained, because convergence is measured on increments in the solution variables that are small but not zero. In this section, we will only discuss the finite

TABLE 4.4 *Finite element solution errors.*

Error	*Error Occurrence In*	*Section Discussing Error*
Discretization	use of finite element interpolations	4.2.5 4.3.1
Numerical integration in space	evaluation of finite element matrices using numerical integration	5.8.1 6.5.3
Evaluation of constitutive relations	use of nonlinear material models	6.4.2
Solution of dynamic equilibrium equations	direct time integration, mode superposition	9.2 9.4
Solution of finite element equations by iteration	Gauss-Seidel, Newton-Raphson, quasi-Newton methods, eigensolutions	8.4 8.6 9.5 10.4
Round-off	setting-up equations and their solution	8.5

element discretization errors, which are due to the idealization of the structure or continuum by finite elements and the associate interpolation of the solution variables. The other sources of errors are discussed in detail in the sections listed in Table 4.4. Thus, in essence, we consider in this section a model problem in which the other solution errors referred to above do not arise: a linear elastic static problem with the exact calculation of the element matrices and solution of equations, and negligible round-off errors. Depending on the specific displacement-based finite elements used in the analysis of such a problem, we may converge *monotonically* or *nonmonotonically* to the exact solution as the number of finite elements is increased. In the following discussion we consider only the criteria for monotonic convergence of solutions. Finite element analysis conditions that lead to nonmonotonic convergence are discussed in Section 4.3.

To encounter monotonic convergence the elements must be complete and compatible.[19-24] If these conditions are fulfilled, the accuracy of the solution results increases continuously as we continue to refine the finite element mesh. This mesh refinement should be performed by subdividing a previously used element into two or more elements; thus, the old mesh will be "embedded" in the new mesh. This means mathematically that the new space of finite element interpolation functions will contain the previously used space, and as the mesh is refined the dimension of the finite element solution space is continuously increased to contain ultimately the exact solution.

The requirement of completeness of an element means that the displacement functions of the element must be able to represent *the rigid body displacements* and the *constant strain states.*

The rigid body displacements are those displacement modes that the element must be able to undergo as a rigid body without stresses being developed in it. As an example, a two-dimensional plane stress element must be able to translate uniformly in either direction of its plane and to rotate without straining. The reason that the element must be able to undergo these displacements without developing stresses is illustrated in the analysis of the cantilever shown in Fig. 4.24; namely, the element at the tip of the beam must translate and rotate stress-free, because by simple statics the cantilever is not subjected to stresses beyond the point of load application.

The number of rigid body modes that an element must be able to undergo can usually be identified without difficulty by inspection, but it is instructive to note that the number of element rigid body modes is equal to the number of element degrees of freedom minus the number of element straining modes (or natural modes). As an example, a two-noded truss has one straining mode (constant strain state), and thus 1, 3, and 5 rigid body modes in one-, two-and three-dimensional conditions, respectively. For more complex finite elements the individual straining modes and rigid body modes are displayed effectively by representing the stiffness matrix in the basis of eigenvectors. Thus, solving the eigenproblem

$$\mathbf{K}\boldsymbol{\phi} = \lambda\boldsymbol{\phi} \tag{4.56}$$

we have (see Section 2.7)

$$\mathbf{K}\boldsymbol{\Phi} = \boldsymbol{\Phi}\boldsymbol{\Lambda} \tag{4.57}$$

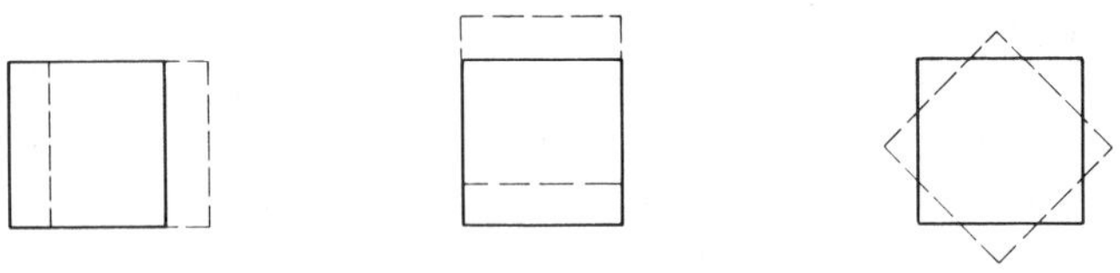

(a) Rigid body modes of a plane stress element

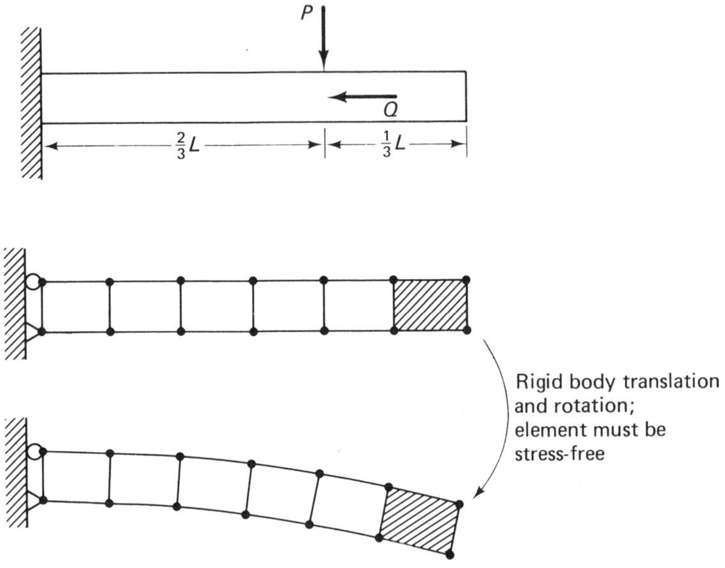

(b) Analysis to illustrate the rigid body mode condition

FIGURE 4.24 *Use of plane stress element in analysis of cantilever.*

where $\mathbf{\Phi}$ is a matrix storing the eigenvectors $\boldsymbol{\phi}_1, \ldots, \boldsymbol{\phi}_n$ and $\mathbf{\Lambda}$ is a diagonal matrix storing the corresponding eigenvalues, $\mathbf{\Lambda} = \text{diag}(\lambda_i)$. Using the eigenvector orthonormality property we thus have

$$\mathbf{\Phi}^T\mathbf{K}\mathbf{\Phi} = \mathbf{\Lambda} \tag{4.58}$$

We may look at $\mathbf{\Lambda}$ as being the stiffness matrix of the element corresponding to the eigenvector displacement modes. The stiffness coefficients $\lambda_1, \ldots, \lambda_n$ directly display how stiff the element is in the corresponding displacement mode. Thus the transformation in (4.58) shows clearly whether the rigid body modes and what additional straining modes are present.* As an example, the eight eigenvectors and corresponding eigenvalues of a four-node element are shown in Fig. 4.25.

The necessity for the constant strain states can physically be understood if we imagine that more and more elements are used in the assemblage to represent the structure. Then in the limit as each element approaches a very small size,

* Note also that, since the finite element analysis overestimates the stiffness, as discussed on p. 174, the "smaller" the eigenvalues the more effective is the element.

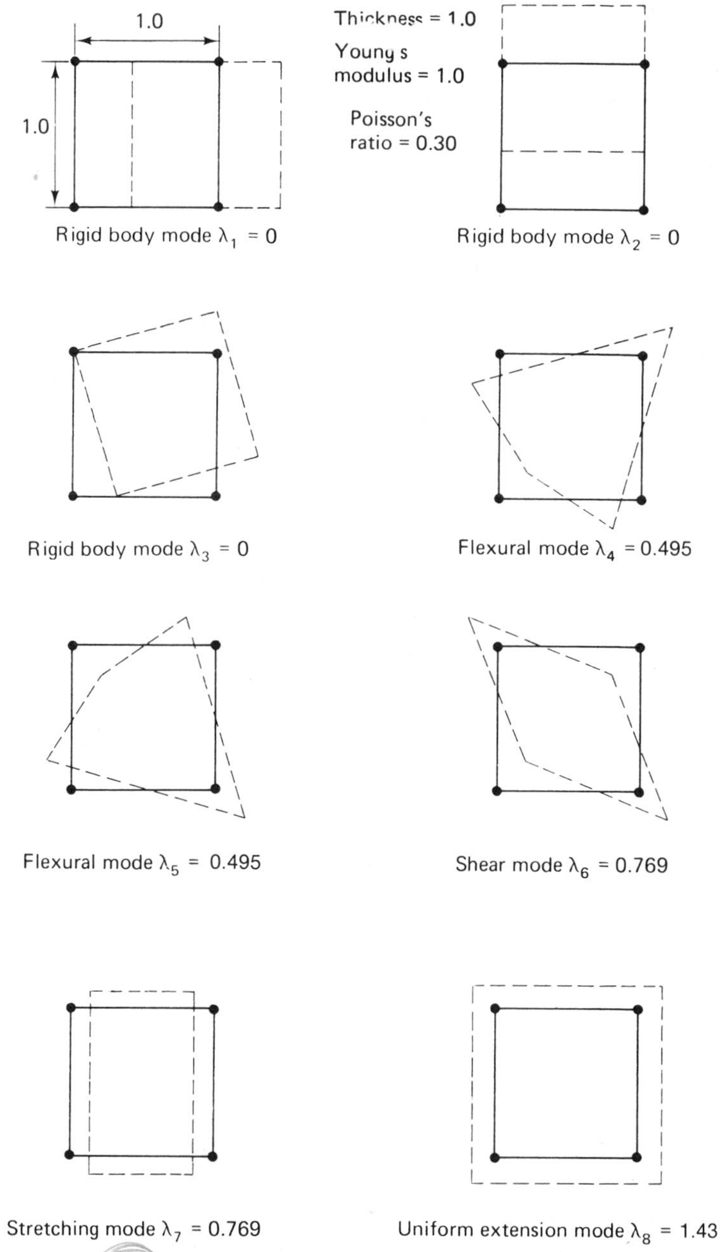

FIGURE 4.25 *Eigenvalues and eigenvectors of four-node plane stress element.*

the strain in each element approaches a constant value, and any complex variation of strain within the structure can be approximated. As an example, the plane stress element used in Fig. 4.24 must be able to represent two constant normal stress conditions and one constant shearing stress condition. Figure 4.25 shows that the element can represent these constant stress conditions and, in addition, contains two flexural straining modes.

The rigid body modes and constant strain states that an element can represent can also directly be identified by studying the element strain-displacement matrix (see Example 4.13).

The requirement of compatibility means that the displacements within the elements and across the element boundaries must be continuous. Physically, compatibility assures that no gaps occur between elements when the assemblage is loaded. When only translational degrees of freedom are defined at the element nodes, only continuity in the displacements u, v, and w, whichever is applicable, must be preserved. However, when rotational degrees of freedom are also defined that are obtained by differentiation of the transverse displacement (such as in the formulation of the plate bending element in Example 4.15), it is also necessary to satisfy element continuity in the corresponding first displacement derivatives. This is a consequence of the kinematic assumption on the displacements over the depth of the plate bending element; namely, the continuity in the displacement w and the derivatives $\partial w/\partial x$ and/or $\partial w/\partial y$ along the respective element edges assures continuity of displacements over the thickness of adjoining elements.

Compatibility is automatically assured between truss and beam elements because they only join at the nodal points, and compatibility is relatively easy to maintain in two-dimensional plane strain, plane stress, and axisymmetric analysis and in three-dimensional analysis, when only u, v, and w degrees of freedom are used as nodal point variables. However, the requirements of compatibility are difficult to satisfy in plate bending analysis, and in particular in thin shell analysis if the rotations are derived from the transverse displacements. For this reason, much emphasis has been directed during recent years towards the development of plate and shell elements, in which the displacements and rotations are independent variables (see Section 5.4). With such elements the compatibility requirements are just as easy to fulfill as in the case of dealing with translational degrees of freedom only.

Whether a specific element is complete and compatible depends on the formulation used, and each formulation need be analyzed individually. Consider the following simple example.

EXAMPLE 4.18: Investigate if the plane stress element used in Example 4.5 is compatible and complete.

We had for the displacements of the element,

$$u(x, y) = \alpha_1 + \alpha_2 x + \alpha_3 y + \alpha_4 xy$$
$$v(x, y) = \beta_1 + \beta_2 x + \beta_3 y + \beta_4 xy$$

Observing that the displacements within an element are continuous, in order to show that the element is compatible we only need to investigate if interelement continuity is also preserved when an element assemblage is loaded. Consider two elements interconnected at two node points (Fig. 4.26) onto which we impose two arbitrary displacements. It follows from the displacement assumptions that the points on the adjoining element edges displace linearly, and therefore continuity between the elements is preserved. Hence the element is compatible.

Considering completeness, the displacement functions show that a rigid body translation in the x direction is achieved if only α_1 is nonzero. Similarly, a rigid

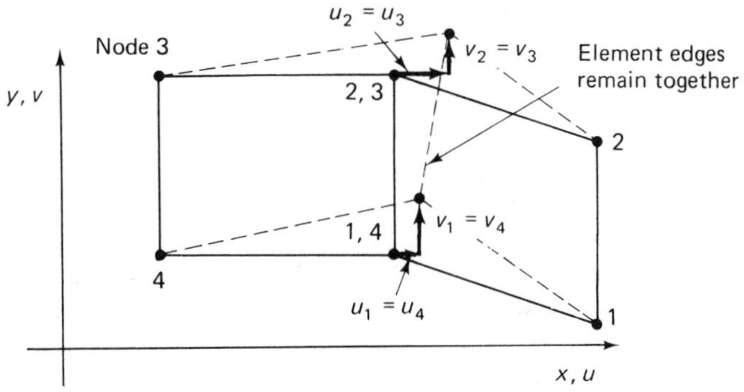

FIGURE 4.26 *Compatibility of plane stress element.*

body displacement in the y direction is imposed by having only β_1 nonzero, and for a rigid body rotation we need that α_3 and β_2 are nonzero only with $\beta_2 = -\alpha_3$. The same conclusion can also be arrived at using the matrix $\mathbf{E}$ that relates the strains to the generalized coordinates (see Example 4.5). This matrix also shows that constant strain states are possible. Therefore, the element is complete.

We observed earlier that the application of the principle of virtual work is identical to using the stationarity condition of the total potential of the system (see Example 4.2). Considering also the discussion of the Ritz method in Section 3.3.3 we can conclude that monotonically convergent displacement-based finite element solutions are really only applications of the Ritz method. In the finite element analysis the Ritz functions are contained in the element displacement interpolation matrices $\mathbf{H}^{(m)}$, $m = 1, 2, \ldots$ and the Ritz parameters are the unknown nodal point displacements stored in $\mathbf{U}$. As we discuss further below, the mathematical conditions on the displacement interpolation functions in the matrices $\mathbf{H}^{(m)}$, in order that the finite element solution be a Ritz analysis, are exactly those that we identified above using physical reasoning. The correspondence between the analysis methods is illustrated in Examples 3.22 and 4.3 and leads to some important practical observations.

Although the classical Ritz analysis procedure and the displacement-based finite element method are theoretically identical, in practice, the finite element method has most important advantages over a conventional Ritz analysis. One disadvantage in the conventional Ritz analysis is that the Ritz functions are defined over the whole region considered. For example, in the analysis of the cantilever in Example 3.24, the Ritz functions spanned from $x = 0$ to $x = L$. Therefore, in the conventional Ritz analysis, the matrix $\mathbf{K}$ is a full matrix, and as pointed out in Section 8.2.3, the numerical operations required for the solution of the resulting algebraic equations are considerable if many functions are used.

A particular difficulty in a conventional Ritz analysis is the selection of appropriate Ritz functions, since the solution is a linear combination of these functions. In order to solve accurately for large displacement or stress gradients, many functions may be needed. However, these functions also unnecessarily span over the regions in which the displacements and stresses vary rather slowly and where not many functions are required.

Another difficulty arises in the conventional Ritz analysis when the total region of interest is made up of subregions with different kinds of strain distributions. As an example, consider a plate that is supported by edge beams and columns. In such a case, the Ritz functions used for one region (e.g., the plate) are not appropriate for the other regions (i.e., the edge beams and columns), and special displacement continuity conditions or boundary relations must be introduced.

The few reasons given already show that the conventional Ritz analysis is, in general, not particularly computer-oriented, except in some cases for the development of special-purpose programs. On the other hand, the finite element method has to a large extent removed the practical difficulties while retaining the advantageous properties of the conventional Ritz method. With regard to the difficulties mentioned above, the selection of Ritz functions is handled by using an adequate element library in the computer program. The use of relatively many functions in regions of high stress and displacement gradients is possible simply by using many elements. The combination of domains with different kinds of strain distributions is possible by using different kinds of elements to idealize the domains.

In general, the classical Ritz analysis method is not an effective tool for the solution of complex problems, but the fact that the displacement-based finite element solutions can be regarded, in theory, as Ritz analyses leads to some important additional insight into the procedure of finite element discretization. In particular, this correspondence has been the basis of much of the mathematical theory of finite element discretization and the basis for the development of an increased understanding and a firm mathematical foundation of the convergence properties of finite element analyses. The important results of these investigations can directly be interpreted and tied together with the convergence conditions on a finite element analysis discussed above using physical reasoning. The relevant thoughts are as follows.

Considering the Ritz method of analysis with the finite element interpolations, we have

$$\Pi = \tfrac{1}{2}\mathbf{U}^T\mathbf{K}\mathbf{U} - \mathbf{U}^T\mathbf{R} \tag{4.59}$$

where Π is the total potential of the system. Invoking the stationarity of Π with respect to the Ritz parameters U_i stored in $\mathbf{U}$, and recognizing that the matrix $\mathbf{K}$ is symmetric, we obtain

$$\mathbf{K}\mathbf{U} = \mathbf{R} \tag{4.60}$$

The solution of (4.60) yields the Ritz parameters and then the displacement solution in the domain considered is

$$\mathbf{u}^{(m)} = \mathbf{H}^{(m)}\mathbf{U}; \qquad m = 1, 2, \ldots \tag{4.61}$$

The relations in (4.59) to (4.61) represent a Ritz analysis provided the functions used satisfy certain conditions. We defined in Section 3.3.2 a C^{m-1} variational problem as one in which the variational indicator of the problem contains derivatives of order m and lower. We then noted that for convergence the Ritz functions must satisfy the essential (or geometric) boundary conditions of the problem involving derivatives up to order $m - 1$, but the functions do not need to satisfy the natural (or force) boundary conditions involving derivatives of order m to $2m - 1$, because these conditions are implicitly contained in the variational indicator Π. Therefore, in order that a finite element solution be a Ritz analysis, the essential boundary conditions must be completely satisfied by the finite element nodal point displacements and the displacement interpolations between the nodal points. However, in the selection of the finite element displacement functions, no special attention need be given to the natural boundary conditions, because these conditions are imposed with the load vector and are satisfied approximately in the Ritz solution. The accuracy with which the natural or force boundary conditions are satisfied depends on the specific Ritz functions employed, but this accuracy can always be increased by using a larger number of functions; i.e. a larger number of finite elements to model the problem.

In the classical Ritz analysis the Ritz functions span over the complete domain considered, whereas in the finite element analysis the individual Ritz functions span only over subdomains (finite elements) of the complete region. Hence, there must be a question as to what conditions must be fulfilled by the finite element interpolations with regard to continuity requirements *between* adjacent subdomains. To answer this question we consider the integrations that need be performed to evaluate the coefficient matrix **K**. We recognize that considering a C^{m-1} problem we need continuity in at least the $(m - 1)$st derivatives of the Ritz trial functions in order that we can perform the integrations across the element boundaries. However, this continuity requirement corresponds entirely to the element compatibility conditions that we discussed earlier; e.g., in the analysis of fully three-dimensional problems only the displacements between elements must be continuous, whereas in the analysis of plate problems formulated using the Kirchhoff plate theory we also need continuity in the first derivatives of the displacement functions.

In summary, therefore, considering a C^{m-1} problem [$C^{m-1} \equiv$ continuity on trial functions and their derivatives up to order $(m - 1)$], in the classical Ritz analysis the trial functions are selected to satisfy exactly all boundary conditions that involve derivatives up to order $(m - 1)$. The same holds in finite element analysis, but in addition, continuity in the trial functions and their derivatives up to order $(m - 1)$ must be satisfied between elements in order for the finite element solution to correspond to a Ritz analysis.

Having determined that the displacement-based finite element method is a modern application of Ritz analysis, it follows that the convergence properties and numerical advantages associated with the Ritz method are also applicable to the finite element method. An important numerical property is that the discretization matrix **K** is positive definite (or can always be rendered positive definite). In Section 2.7 we defined a matrix **A** as positive definite if $\mathbf{u}^T\mathbf{A}\mathbf{u} > 0$ for any vector $\mathbf{u}$ ($\mathbf{u} \neq \mathbf{0}$). In finite element analysis we consider the strain energy of the system, $\mathcal{U} = \frac{1}{2}\mathbf{U}^T\mathbf{K}\mathbf{U}$, but since $\mathcal{U}$ is positive for any vector $\mathbf{U}$ (assuming as before that the structure is properly supported and $\mathbf{U} \neq \mathbf{0}$), we have that **K** is a positive definite matrix.

If the structure analyzed is not supported, there will be a number of linearly independent vectors $\mathbf{U}_1, \mathbf{U}_2, \ldots, \mathbf{U}_q$ for which the expression $\mathbf{U}_i^T\mathbf{K}\mathbf{U}_i$ is equal to zero; i.e., zero strain energy is stored in the system when $\mathbf{U}_i$ is the displacement vector. Each vector $\mathbf{U}_i$ is said to represent a *rigid body mode* of the system;

hence q is the number of rigid body modes present. Since in this case $\mathbf{U}^T\mathbf{K}\mathbf{U} \geq 0$, $\mathbf{K}$ is said to be positive semidefinite.

An important consideration is the study of the convergence properties of the finite element method when the procedure is looked upon as a Ritz analysis. To this end we first need to define "convergence"; namely, we may consider convergence of displacements or stresses, pointwise or in a norm. The following definition is mathematically relatively easy to work with. Let $\mathbf{u}^{\star}$ be the exact solution and $\mathbf{u}^n$ be the solution obtained using n Ritz functions; then the solution converges provided that

$$\int_V \mathcal{L}(\mathbf{u}^{\star} - \mathbf{u}^n)\mathcal{S}(\mathbf{u}^{\star} - \mathbf{u}^n)\, dV \longrightarrow 0 \qquad \text{as} \quad n \longrightarrow \infty \tag{4.62}$$

where $\mathcal{L}$ and $\mathcal{S}$ are the appropriate strain and stress differential operators corresponding to the problem considered. We say that we obtain "convergence in energy," since the energy norm of the residual displacements $\mathbf{u}^{\star} - \mathbf{u}^n$ converges to zero.

In the finite element analysis we have functions that extend piecewise continuously over the domain of the finite elements. Let h be the largest dimension of an element; then the statements $n \longrightarrow \infty$ and $h \longrightarrow 0$ are equivalent. Let $\mathbf{u}^h$ denote the finite element solution obtained when h is the largest dimension of any element. Then the definition of convergence in (4.62) can also be wirtten as

$$\int_V \mathcal{L}(\mathbf{u}^{\star} - \mathbf{u}^h)\mathcal{S}(\mathbf{u}^{\star} - \mathbf{u}^h)\, dV \longrightarrow 0 \qquad \text{as} \quad h \longrightarrow 0 \tag{4.63}$$

We should note that in the limit, convergence in energy is equivalent to convergence in displacements and stresses. However, for any given finite element mesh, the accuracy of the solution will be different when measured by the energy norm, or by the accuracy in the displacements or the stresses obtained.

To obtain further insight into the convergence of the finite element results we may also refer to our discussion in Section 2.5 and identify the Ritz or finite element interpolation functions as basis vectors. Hence, in the finite element or Ritz analysis, we actually find the minimum of the total potential energy within the subspace spanned by the Ritz trial functions. As discussed above, provided that the trial functions satisfy certain continuity conditions, convergence is achieved as $n \longrightarrow \infty$, i.e., as the subspace is enlarged to, finally, an infinite-dimensional space. This convergence appears physically reasonable, since the exact solution should be reached as the space considered increases in dimension. It is also important to note that as the dimension of the subspace considered increases and provided the currently considered subspace contains the previous one, *the minimum of the total potential energy reached will monotonically decrease*. Hence, we talk about monotonic convergence; in practice, this means that starting with a set of trial functions and always adding new functions to those previously used, the solution will continuously be improved. Furthermore, using (4.59) we have $\Pi = -\frac{1}{2}\mathbf{U}^T\mathbf{R}$ and $\mathcal{U} = \frac{1}{2}\mathbf{U}^T\mathbf{R}$ and a decrease in the total potential Π corresponds to an increase in the strain energy $\mathcal{U}$. It follows that in the finite element analysis *the displacements are underestimated and the predicted stiffness is too large.* This means that considering a dynamic analysis the frequencies predicted by the finite element model (using a consistent mass matrix) are upper bounds on the exact frequencies of the problem (see also Section 10.3.2).

In the above discussion we considered the conditions for monotonic convergence of the finite element analysis results, but we did not mention the rate with which convergence occurs. This question has been much researched during recent years.

As must be expected, the rate of convergence depends on the polynomial expansions used in the displacement assumptions. In this context the notions of *complete polynomials* and *spatially isotropic elements*—which have the same displacement expansions in all directions—are useful.

Figure 4.27 shows the polynomial terms that should be included to have complete polynomials in x and y for two-dimensional analysis. It is seen that all possible terms of the form $x^\alpha y^\beta$ are present, where $\alpha + \beta = c$ and c is the degree to which the polynomial is complete. For example, we may note that the element investigated in Example 4.5 uses a polynomial displacement expansion that is complete to degree 1 only. Considering three-dimensional analysis, a figure analogous to Fig. 4.27 can be drawn, in which the variable z would be included.

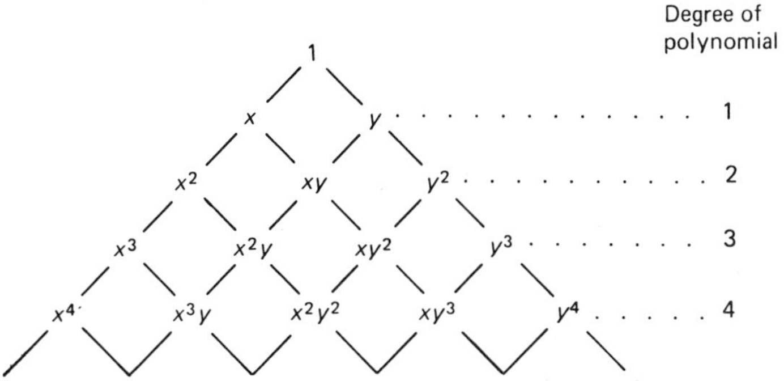

FIGURE 4.27 *Polynomial terms for complete polynomials in two-dimensional analysis.*

As an example of a spatially nonisotropic element, consider a two-dimensional element. With the local element coordinate axes selected, let, just for illustration,

$$\begin{aligned} u(x, y) &= \alpha_1 + \alpha_2 x + \alpha_3 y + \alpha_4 xy \\ v(x, y) &= \beta_1 + \beta_2 x + \beta_3 y + \beta_4 x^2 \end{aligned} \tag{4.64}$$

In this case the element displacements are dependent on the orientation of the coordinate axes, because the polynomial terms associated with α_4 and β_4 are not the same. For example, assume that the x and y axes are interchanged; then clearly a different displacement field is assumed. Therefore, the stiffness matrix of an element will depend on the orientation of the local element coordinate system.

Although it may seem that such directional behavior is always undesirable, actually, in a variety of element formulations, specific local coordinate axes are assumed, and different displacement variations are associated with each axis. The displacement functions are selected based on the kinematic assumptions, which would be used in the continuum mechanics solution. For example, in the formulation of thick shell elements, a lower order displacement variation is used through the thickness of the elements than over the element surfaces.

Similarly, a spar element in bridge analysis may have different displacement expansions in each direction. However, although different displacement functions may be used intentionally for certain directions, for some other directions it may be a requirement that the displacement assumption is independent of the direction considered. Typically, in the analysis of two- and three-dimensional continua with spatially isotropic behavior, geometric invariance is a requirement and results in a restriction on the polynomial terms that may be used in the displacement formulation. In general, a displacement model is spatially isotropic (geometrically invariant) if the same order polynomial terms are used for the displacements u, v, and w if applicable, and in each polynomial term the x, y, and z element coordinates have been appropriately interchanged.

Considering next the problem of estimating the rate of convergence of the displacements, stresses, and the strain energy within a finite element, it is expedient to perform Taylor series analyses. Without going into the mathematical details, we summarize some important results of these analyses.[23] Since a finite element of "dimension h" with a complete displacement expansion of order c can represent displacement variations up to that order exactly, the local error in representing arbitrary displacements with a uniform mesh is estimated to be $o(h^{c+1})$. Also, for a C^{m-1} problem the stresses are calculated by differentiating the displacements m times, and therefore the error in the stresses is $o(h^{c+1-m})$. The "dimension h" used in these convergence rate estimates can approximately be taken to be a characteristic length of the element.

We demonstrate the procedure of a convergence analysis in the following example.

EXAMPLE 4.19: Consider a uniform bar with load $p(x)$ per unit length (Fig. 4.28). Assume that the bar is idealized as an assemblage of bar elements of unequal lengths and that the distributed load application is idealized by the application of lumped loads at the nodes of the element idealization, where each load is equal to the value of the distributed load at the node multiplied by one-half of the lengths of the adjoining elements. This calculation of nodal loads corresponds to a simple lumping of the loads. Discuss the convergence characteristics of the finite element solution.

The stiffness matrix of a bar element of length h is

$$\mathbf{K} = \frac{EA}{h}\begin{bmatrix} 1 & -1 \\ -1 & 1 \end{bmatrix}$$

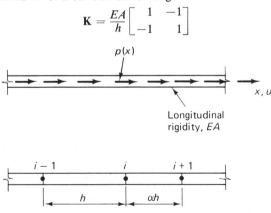

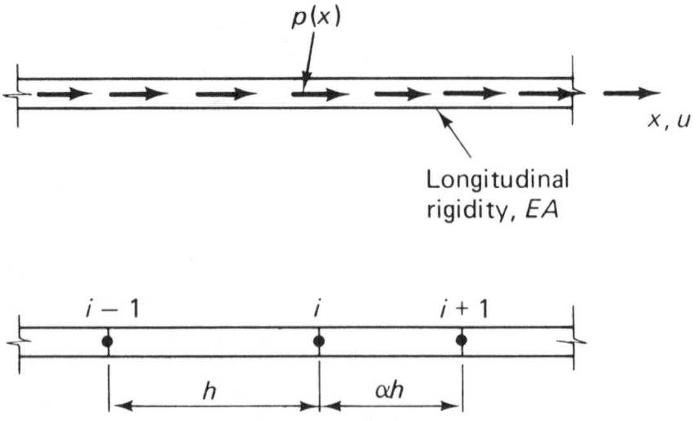

FIGURE 4.28 *Uniform bar under axial loading.*

and, correspondingly, the equilibrium equation at node i is

$$\frac{EA}{h}[-u_{i-1} + u_i] + \frac{EA}{\alpha h}[u_i - u_{i+1}] = \frac{p_i h(1 + \alpha)}{2} \tag{a}$$

The differential equation of equilibrium corresponding to the finite element idealization at node i is obtained by using the Taylor series expansion of the displacements at nodes $i - 1$ and $i + 1$ about node i. We have

$$\begin{aligned} u_{i+1} &= u_i + u_i'(\alpha h) + u_i'' \frac{(\alpha h)^2}{2} + u_i''' \frac{(\alpha h)^3}{6} + \dots \\ u_{i-1} &= u_i - u_i' h + u_i'' \frac{h^2}{2} - u_i''' \frac{h^3}{6} + \dots \end{aligned} \tag{b}$$

Substituting (b) into (a), we obtain

$$u_i'' - \frac{h}{3}(1 - \alpha)u_i''' + \frac{h^2}{12}\frac{1 + \alpha^3}{1 + \alpha} u_i^{\text{iv}} + \dots + \frac{p_i}{EA} = 0$$

However, we recall that the exact differential equation of equilibrium at point i is

$$u_i'' + \frac{p_i}{EA} = 0$$

Therefore, the finite element formulation converges to the exact solution as $h \longrightarrow 0$. Also, for $\alpha \neq 1$, convergence is linear, and for $\alpha = 1$, convergence is quadratic.

4.2.6 Calculation of Stresses

In the previous section we discussed that for monotonic convergence to the exact results ("exact" within the mechanical assumptions made), the elements must be complete and compatible. Using compatible (or conforming) elements means that in the finite element representation of a C^{m-1} variational problem the displacements and their $(m - 1)$st derivatives are continuous across the element boundaries. Hence, for example, in a plane stress analysis the u and v displacements are continuous, and in the analysis of a plate bending problem using the transverse displacement w as the only unknown variable, this displacement w and its derivatives, $\partial w/\partial x$ and $\partial w/\partial y$, are continuous. However, this continuity does not mean that the element stresses are continuous across element boundaries.

The element stresses are calculated using derivatives of the displacements (see (4.10) and (4.11)), and the stresses obtained at an element edge (or face) when calculated in adjacent elements may differ substantially if a coarse finite element idealization is used. The stress differences at the element boundaries decrease as the finite element idealization is refined, and in practice, acceptable results are usually obtained if element boundary stresses are averaged. Frequently, stress distributions are also based on calculating only stresses in the interior of the elements, in particular, because the stresses are predicted to a comparatively high accuracy at specific points (see Section 5.8.2).

For the same mathematical reason that the element stresses are, in general, not continuous across element boundaries, likewise the element stresses at the surface of the structure that is modeled are, in general, not in equilibrium with the externally applied tractions. However, as before, force equilibrium is established in the limit as the number of elements to model the structure increases.

In practice, both checks, the extent to which the stresses are discontinuous across the element boundaries and the extent to which force equilibrium is not satisfied on the surface of the structure, directly indicate whether an appropriate finite element idealization was employed (see Section 5.8.3).

4.3 INCOMPATIBLE, MIXED AND HYBRID FINITE ELEMENT MODELS; FINITE DIFFERENCE METHODS

In previous sections we considered the displacement-based finite element method, and the conditions imposed so far on the assumed displacement (or field) functions were those of completeness and compatibility. If these conditions are satisfied, the calculated solution converges monotonically (i.e., one-sided) to the exact solution. The completeness condition can, in general, be satisfied with relative ease. The compatibility condition can also be satisfied without major difficulties in C^0 problems; for example, in plane stress and plane strain problems or the analysis of three-dimensional solids such as dams. However, in the analysis of bending problems, and, in particular, in shell analyses, continuity of first displacement derivatives along interelement boundaries is difficult to maintain if the solution approach discussed in Section 4.2.3 and in Examples 4.14 and 4.15 is used. (We consider in Section 5.4.2 isoparametric plate and shell elements with which interelement continuity can be maintained as easily as in the analysis of C^0 problems.) Furthermore, considering complex analyses in which completely different finite elements must be used to idealize different regions of the structure, compatibility may be quite impossible to maintain. However, although the compatibility requirements are violated, experience shows that good results are frequently obtained.

The difficulties of developing for certain types of problems compatible displacement-based finite elements that are computationally effective, and the realization that using variational approaches many more finite element procedures can be developed led to a very large research effort. In these activities various classes of new types of elements have been proposed, and the amount of information available on these elements is rather voluminous. We shall not present the various formulations in detail; however, we shall briefly outline some of the major ideas that have been used in the development of these elements. This discussion should strengthen the general understanding of the finite element discretization procedures discussed earlier, and further indicate the generality of the finite element method.

4.3.1 *Incompatible Models*

In practice, a frequently made observation is that satisfactory finite element analysis results have been obtained, although some continuity requirements between elements in the mesh employed were violated. In some instances the nodal point layout was such that interelement continuity was not preserved, and in other cases elements have been used that contain interelement incompatibilities (e.g., see Section 5.6.2). The final result is the same in either case; namely, the displacements or their derivatives between elements are not continuous to the degree necessary that all compatibility conditions discussed in Section 4.2.5 are satisfied.

Considering the implications of these incompatibilities, we recognize that the continuity requirements for a conventional Ritz analysis are not of the same concern in finite element analysis. The functions and derivatives must be, as discussed in Section 4.2.5, continuous within each finite element, which can be satisfied easily, but any singularities in the derivatives across element boundaries do not enter into the calculation of the element stiffness matrices. Therefore, although—within the context of classical Ritz analysis—infinite strain energies would be sampled when integrating across element boundaries, in practice, these contributions are not calculated and thus are simply ignored.

Since in finite element analysis using incompatible (nonconforming) elements the Ritz analysis requirements are not satisfied, the calculated total potential energy is not necessarily an upper bound to the exact total potential energy of the system, and, consequently, monotonic convergence is not assured. However, having relaxed the objective of monotonic convergence in the analysis, we still need to establish conditions that will assure at least a nonmonotonic convergence. Referring to Section 4.2.5, the element completeness condition must always be satisfied, where it may be noted that this condition is not affected by the size of the finite element. On the other hand, the compatibility condition can be relaxed somewhat at the expense of not obtaining a monotonically convergent solution, provided, when relaxing this requirement, the essential ingredients of the completeness condition are not lost. We recall that as the finite element mesh is refined (i.e., the size of the elements gets smaller), each element should approach a constant strain condition. Therefore, the second condition on convergence of an assemblage of incompatible finite elements, where the elements may again be of any size, is that the elements together can represent constant strain conditions. We should note that *this is not a condition on a single individual element but on an assemblage of elements*. That is, although an individual element may be able to represent all constant strain states (i.e., be complete), when the element is used in an assemblage, the incompatibilities between elements may prohibit constant strain states from being represented. We may call this condition the *completeness condition on an element assemblage*.

As a test to investigate whether an assemblage of nonconforming elements is complete, the *patch test* has been proposed.[25,26] In this test a patch of elements is subjected to the boundary nodal point forces that in an exact analysis correspond to constant strain conditions. If the element strains do actually represent the constant strain conditions, the patch test is passed; i.e., the completeness condition is satisfied by the element assemblage. We should note that the success or failure of the patch test may depend strongly on the geometry of the elements used, i.e., on the topology of the element layout. For example, an assemblage of the four-node two-dimensional elements with incompatible modes, discussed in Section 5.6.2 passes the patch test if all elements are rectangular (or parallelograms), but would fail the test otherwise (see Example 4.20).

Although the patch test requires the consideration of an element assemblage, it is desirable to perform the test without actually using the computer.

To this aim we need to subject an individual element of the assemblage to the nodal point displacements or nodal point forces, which by an exact analysis correspond to element constant strain conditions, and check whether the other not imposed quantity also corresponds to constant strain conditions.

EXAMPLE 4.20: Consider the four-node two-dimensional element with incompatible modes in Fig. 4.29 and check whether the patch test is passed. The element under consideration is discussed in Section 5.6.2.

As shown in Fig. 4.29 assume that the displacement field to which the element is subjected in the element assemblage is

$$u^c = ax + b$$

where u^c corresponds to constant strain conditions $\epsilon_{xx}^c = a$, $\epsilon_{yy}^c = 0$, and $\gamma_{xy}^c = 0$. In order to complete the patch test we would afterwards also consider nonzero constant strain conditions for ϵ_{yy}^c and γ_{xy}^c. To investigate whether on imposing u^c onto the finite element, an incompatible mode ϕ_i is activated, we consider the integral $\int_V (\mathcal{L}\phi_i)(\mathcal{S}u^c)\, dV$ with $\mathcal{L}$ and $\mathcal{S}$ being the appropriate strain and stress differential operators (see (4.62)). This is equivalent to evaluating whether constraining forces that correspond to the incompatible mode are required when u_c is imposed onto the element. If no constraining forces corresponding to any one of the possible incompatible modes are present, it follows that the incompatible modes have not been activated and the patch test is passed.

Displacement interpolation functions

$$u = \sum_{i=1}^{4} h_i u_i + \alpha_1 \phi_1 + \alpha_2 \phi_2$$

$$v = \sum_{i=1}^{4} h_i v_i + \alpha_3 \phi_3 + \alpha_4 \phi_4$$

$$\phi_1 = \phi_3 = (1 - x^2); \quad \phi_2 = \phi_4 = (1 - y^2)$$

E = Young's modulus
ν = Poisson's ratio
constant thickness t

$h_1 = \frac{1}{4}(1 + x)(1 + y)$
$h_2 = \frac{1}{4}(1 - x)(1 + y)$
$h_3 = \frac{1}{4}(1 - x)(1 - y)$
$h_4 = \frac{1}{4}(1 + x)(1 - y)$

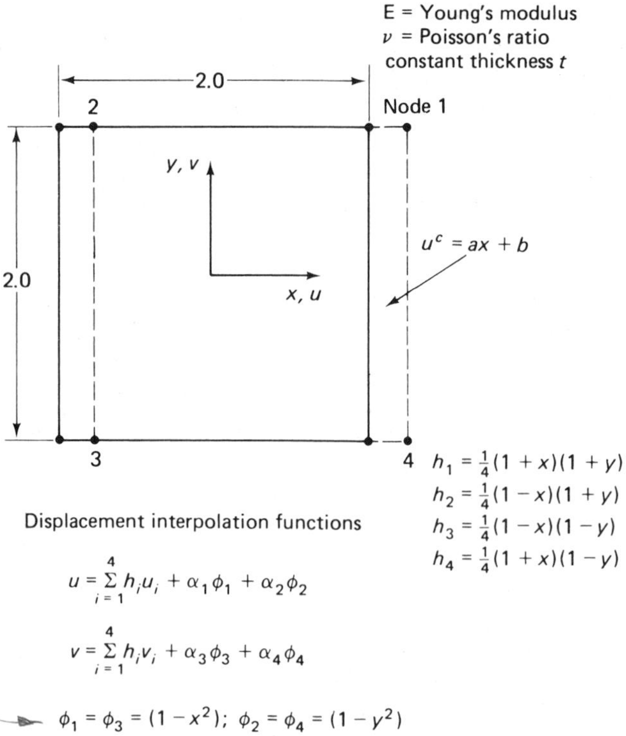

FIGURE 4.29 *Four-node rectangular element with incompatible modes (plane stress conditions).*

For plane stress conditions, we have the material law given in Table 4.3. The evaluation of the integral for the incompatible mode ϕ_1 in Fig. 4.29 gives

$$\frac{Et}{1-\nu^2}\int_A (a)(-2x)dxdy = 0$$

But this is the requirement of the patch test.

Considering ϕ_2, ϕ_3, and ϕ_4 and then displacement fields that correspond to constant strain conditions ϵ_{yy}^c and γ_{xy}^c in a similar manner would complete the patch test. It is found that since the incompatible modes are of the form $(1 - x^2)$ and $(1 - y^2)$, all integrals of the form $\int_V (\mathcal{L}\phi_i)(\mathcal{S}u^c)\, dV$ are zero, and thus the patch test is passed. However, it should be observed that the success of the patch test depends on the geometry of the element under consideration, and indeed, the test is not passed when the element in Fig. 4.29 is not rectangular (or a parallelogram).

The use of the four-node element with incompatible modes corresponds to the use of the eight-node isoparametric element discussed in Section 5.3.1, when the midside nodes of adjacent elements are assigned individual degrees of freedom. Figure 4.30 shows the results obtained in the analysis of a constant

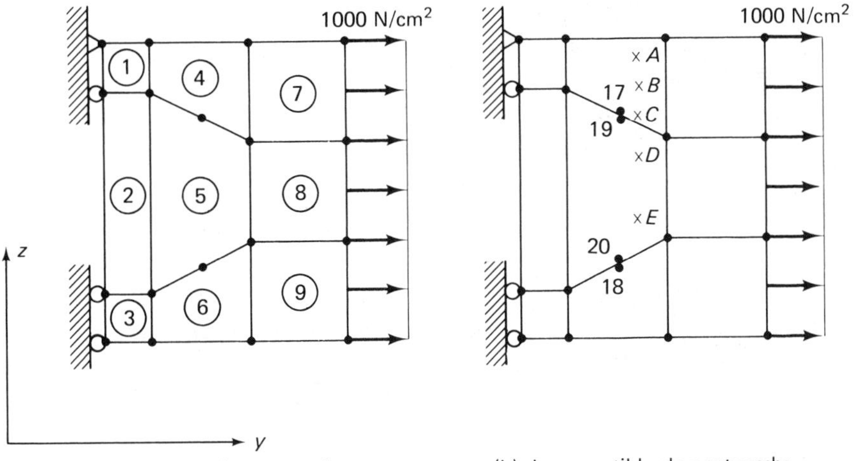

(a) Compatible element mesh; constant stress τ_{yy} = 1000 N/cm^2 in each element

(b) Incompatible element mesh; node 17 belongs to element 4, nodes 19 and 20 belong to element 5, and node 18 belongs to element 6.

τ_{yy} stress predicted by the incompatible element mesh:

Point	τ_{yy} (N/cm^2)
A	1066
B	716
C	359
D	1303
E	1303

FIGURE 4.30 *Effect of displacement incompatibility in stress prediction (3 × 3 Gauss integration).*

one-dimensional stress situation when using a compatible and an incompatible element mesh idealization. We should note the very large errors in the stress predictions using the incompatible element idealization because elements 4, 5, 6 do not satisfy the patch test.

4.3.2 Mixed and Hybrid Models

To formulate the displacement-based finite elements we have used the principle of virtual displacements, which is equivalent to invoking the stationarity of the total potential energy Π (see Example 4.2). The essential theory used can be summarized briefly as follows:

1. We use

$$\Pi = \frac{1}{2}\int_V \boldsymbol{\epsilon}^T \mathbf{C} \boldsymbol{\epsilon}\, dV - \int_V \mathbf{U}^T \mathbf{f}^B\, dV - \int_S \mathbf{U}^{S^T} \mathbf{f}^S\, dS \tag{4.65}$$

2. The equilibrium equations are obtained by invoking the stationarity of Π with respect to the displacements:

$$\int_V \delta\boldsymbol{\epsilon}^T \mathbf{C} \boldsymbol{\epsilon}\, dV = \int_V \delta\mathbf{U}^T \mathbf{f}^B\, dV + \int_S \delta\mathbf{U}^{S^T} \mathbf{f}^S\, dS \tag{4.66}$$

3. The differential equations of equilibrium and the stress (natural) boundary conditions are only satisfied in the limit as the number of elements increases.
4. The strain-displacement compatibility conditions and the displacement (geometric) boundary conditions are satisfied exactly

$$\boldsymbol{\epsilon} = \mathbf{B}\mathbf{U} \tag{4.67}$$
$$\mathbf{U}_a - \mathbf{U}^p = \mathbf{0} \tag{4.68}$$

where the $\mathbf{U}^p$ are the prescribed displacements and the vector $\mathbf{U}_a$ lists the corresponding displacement components of $\mathbf{U}$.

The important point in the formulation (4.65) to (4.68) is that the only solution variables are the displacements which must satisfy the displacement boundary conditions in (4.68) and appropriate interelement conditions. Once we have calculated the displacements, other variables of interest such as strains and stresses can be directly obtained.

In practice, the above displacement-based finite element formulation is used most frequently; however, other techniques have also been employed successfully.[27-35]

A first such technique is based on invoking stationarity of the total complementary energy,

$$\Pi^* = \frac{1}{2}\int_V \boldsymbol{\tau}^T \mathbf{C}^{-1} \boldsymbol{\tau}\, dV - \int_S \mathbf{f}^{S^T} \mathbf{U}^S\, dS \tag{4.69}$$

where—as in (4.65)—the second integral is taken over an appropriate surface S (keeping in mind that some boundary conditions may be imposed later). To use this principle the stress components and corresponding surface tractions are expressed in terms of unknown parameters, and the stress functions must satisfy stress continuity between elements, the differential equations of equilibrium and the stress boundary conditions. Invoking the stationarity of Π^* with respect to the stress parameters, we obtain

$$\int_V \delta\boldsymbol{\tau}^T\mathbf{C}^{-1}\boldsymbol{\tau}\,dV = \int_S \delta\mathbf{f}^{S^T}\mathbf{U}^S\,dS \tag{4.70}$$

and recognize that this is a dual formulation to the displacement-based finite element method; namely, now the stress equilibrium conditions are satisfied exactly, but the strain-displacement compatibility conditions and geometric boundary conditions are only satisfied if (4.70) holds for any arbitrary variations of stresses that satisfy the stress boundary conditions. This means that in a general finite element analysis based on (4.70), the strain and displacement compatibility conditions will only be satisfied in the limit as the number of elements (and hence assumed stress parameters) becomes very large.

The use of (4.70) in finite element analysis appears to be very appealing because we directly solve for the stresses, which are the primary variables of interest. However, in practice, the selection of appropriate stress functions presents some major difficulties and the method is hardly used in general applications.

Some very general finite element formulations are obtained by extending the principles of stationarity of total potential energy or total complementary energy. The objective in these formulations is to relax the conditions to be satisfied by the solution variables, and enlarge the number of solution variables which may include simultaneously displacements, strains and stresses. These formulations are therefore referred to as mixed finite element formulations.

A large number of mixed formulations for the different problems in structural and solid mechanics have been proposed, and our objective here is only to present briefly some of the basic ideas. For this purpose, we consider the formulation of the displacement-based finite element method in (4.65) to (4.68) and recognize that the conditions on the assumed displacement variations in (4.67) and (4.68) can be relaxed provided these conditions are included in Π using Lagrange multipliers (see Section 3.4). Thus, a new and very general functional is

$$\Pi_1 = \Pi + \int_V \boldsymbol{\lambda}_\epsilon^T(\boldsymbol{\epsilon} - \mathbf{B}\mathbf{U})\,dV + \int_{S_u} \boldsymbol{\lambda}_u^T(\mathbf{U}_a - \mathbf{U}^p)\,dS \tag{4.71}$$

where $\boldsymbol{\lambda}_\epsilon$ and $\boldsymbol{\lambda}_u$ are Lagrange multipliers, and S_u is the total surface on which displacements are prescribed. The Lagrange multipliers can be identified as stresses and forces (see Section 3.4), so that (4.71) can be written as

$$\Pi_1 = \Pi - \int_V \boldsymbol{\tau}^T(\boldsymbol{\epsilon} - \mathbf{B}\mathbf{U})\,dV - \int_{S_u} \mathbf{f}^{S^T}(\mathbf{U}_a - \mathbf{U}^p)\,dS \tag{4.72}$$

Invoking the stationarity of Π_1 we now obtain

$$\begin{aligned}\int_V \delta\boldsymbol{\epsilon}^T\mathbf{C}\boldsymbol{\epsilon}\,dV &- \int_V \delta\mathbf{U}^T\mathbf{f}^B\,dV - \int_{S_f} \delta\mathbf{U}^{S^T}\mathbf{f}^S\,dS \\ &- \int_V \delta\boldsymbol{\tau}^T(\boldsymbol{\epsilon} - \mathbf{B}\mathbf{U})\,dV - \int_V \boldsymbol{\tau}^T(\delta\boldsymbol{\epsilon} - \mathbf{B}\delta\mathbf{U})\,dV \\ &- \int_{S_u} \delta\mathbf{f}^{S^T}(\mathbf{U}_a - \mathbf{U}^p)\,dS - \int_{S_u} \mathbf{f}^{S^T}\delta\mathbf{U}_a\,dS = 0\end{aligned} \tag{4.73}$$

where S_f is the surface on which forces are prescribed. This variational formulation may be regarded as a generalization of the principle of virtual displacements, in which the displacement boundary conditions and strain compatibility conditions have been relaxed, and variations are performed on all unknown displacements, strains, stresses, and boundary forces. That this principle is indeed a valid and most general description of the static and kinematic conditions of the body under consideration follows because (4.73) yields—invoking stationarity with respect to all individual unknown variables as in Section 3.3.2—

the stress-strain conditions,

$$\boldsymbol{\tau} = \mathbf{C}\boldsymbol{\epsilon} \tag{4.74}$$

the compatibility conditions,

$$\boldsymbol{\epsilon} = \mathbf{B}\mathbf{U} \tag{4.75}$$

and the equilibrium conditions

$$\begin{aligned} \frac{\partial \tau_{xx}}{\partial x} + \frac{\partial \tau_{yx}}{\partial y} + \frac{\partial \tau_{zx}}{\partial z} + f_x^B = 0 \\ \frac{\partial \tau_{xy}}{\partial x} + \frac{\partial \tau_{yy}}{\partial y} + \frac{\partial \tau_{zy}}{\partial z} + f_y^B = 0 \\ \frac{\partial \tau_{xz}}{\partial x} + \frac{\partial \tau_{yz}}{\partial y} + \frac{\partial \tau_{zz}}{\partial z} + f_z^B = 0 \end{aligned} \tag{4.76}$$

all corresponding to the volume V. For the boundary of the body the prescribed displacement and force conditions on S_u and S_f, respectively, are generated, and the reactions on S_u are equal to the Lagrange multipliers.

The use of (4.73) results in a very general *mixed finite element formulation*, and we may note that by imposing the appropriate constraints on the solution variables this formulation can be employed to deduce directly various different solution approaches, including the displacement-based finite element formulation in (4.65) to (4.68), and the stress-based formulation obtained using Π^* (in 4.69), see, for example, ref. [28]. Although not discussed in detail, we should recognize that the assumed solution variables must satisfy in a finite element analysis appropriate continuity, compatibility and/or equilibrium conditions that depend on the specific formulation employed.

The above brief discussion indicates already that there are various possibilities to formulate finite element models. In each of the above finite element solution approaches the finite element variables would be coupled through adjacent elements, and the algebraic equations governing the response of the complete finite element system contain as unknowns all solution variables that are varied in the functional. Hence, the vector of solution unknowns would contain in the most general case all field variables, i.e. stresses, strains, and displacements.

A computationally more effective approach is frequently given by a hybrid formulation. In these formulations we still use multifield descriptions of the solution, but some field variables are eliminated on the element level prior to the assemblage process. These solution variables will then in general not be inter-element continuous. As in the mixed formulations, a large number of possibilities exist for formulating hybrid solution approaches. A very successful formulation pioneered by Pian[29] and extensively applied to the analysis of plates and shells, fracture and other problems, is the *hybrid stress method.* In this formulation the displacements on the element boundary are assumed so as to provide displacement compatibility between elements, and internal stress variations are assumed that satisfy the differential equations of equilibrium. The stress parameters are eliminated on the element level prior to the assemblage process. Thus, in the solution much of the computational simplicity of the displacement-based finite element formulation is preserved, and an element formulated using the hybrid stress method can directly be incorporated into a computer program that has been written to solve for unknown nodal point displacements only.

The basic assumptions in the hybrid stress method are for the element stresses

$$\boldsymbol{\tau} = \mathbf{P}\boldsymbol{\beta} \tag{4.77}$$

and for the element boundary displacements

$$\mathbf{u}^S = \mathbf{H}^S\,\mathbf{U} \tag{4.78}$$

In (4.77) the matrix $\mathbf{P}$ contains the polynomial terms of the generalized stress parameters stored in the vector $\boldsymbol{\beta}$, and in (4.78) the matrix $\mathbf{H}^S$ is a matrix that interpolates the boundary displacements of the element. The matrix $\mathbf{H}^S$ is constructed as illustrated in Example 4.6 but now all boundaries of an element are considered. Substituting (4.77) and (4.78) into (4.69) we obtain *for an individual element,*

$$\Pi^* = \tfrac{1}{2}\boldsymbol{\beta}^T\left\{\int_V \mathbf{P}^T\mathbf{C}^{-1}\mathbf{P}\,dV\right\}\boldsymbol{\beta} - \boldsymbol{\beta}^T\left\{\int_S \mathbf{P}^{S^T}\mathbf{H}^S\,dS\right\}\mathbf{U} \tag{4.79}$$

where $\mathbf{P}^S$ is interpolating the element surface tractions. The matrix $\mathbf{P}^S$ is constructed from $\mathbf{P}$ by substituting the coordinates corresponding to the surface considered. Invoking the stationarity condition $\delta\Pi^* = 0$ with respect to the stress parameters we obtain

$$\mathbf{E}\boldsymbol{\beta} = \mathbf{G}\mathbf{U} \tag{4.80}$$

where

$$\mathbf{E} = \int_V \mathbf{P}^T\mathbf{C}^{-1}\mathbf{P}\,dV \tag{4.81}$$

$$\mathbf{G} = \int_V \mathbf{P}^{S^T}\mathbf{H}^S\,dS \tag{4.82}$$

Hence

$$\boldsymbol{\beta} = \mathbf{E}^{-1}\mathbf{G}\mathbf{U} \tag{4.83}$$

and substituting into (4.79) we obtain

$$\Pi^* = -\tfrac{1}{2}\mathbf{U}^T\mathbf{G}^T\mathbf{E}^{-1}\mathbf{G}\mathbf{U} \tag{4.84}$$

so that

$$\mathbf{K} = \mathbf{G}^T\mathbf{E}^{-1}\mathbf{G} \tag{4.85}$$

An important step in the actual application of the hybrid stress method is the use of appropriate stress functions. In addition to satisfying the differential equations of equilibrium, the number of functions must be large enough, and the "quality" of the functions must be such that the important stress variations can be represented.[30]

The hybrid stress method is in some regards similar to the use of incompatible displacement modes, which are also statically condensed out prior to the element assemblage process. Indeed, in the formulation of a rectangular four-node plane stress element the same stiffness matrices are calculated using the hybrid stress method or displacement method with incompatible modes. We consider this element in the following example.

EXAMPLE 4.21: For the four-node plane stress element shown in Fig. 4.31 give the matrices $\mathbf{P}$ and $\mathbf{H}^S$ that are used to interpolate the stresses and boundary displacements as defined in (4.77) and (4.78).

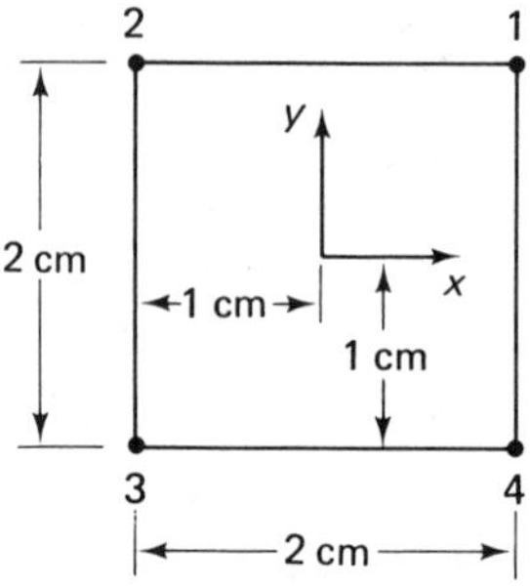

FIGURE 4.31 *Four-node element.*

Considering (4.77) we need in the formulation of hybrid elements at least $(m - r)$ stress functions, where m = number of element displacement degrees of freedom and r = number of physical rigid body modes. Otherwise, the element formulation will contain spurious zero-energy modes. Suitable stress functions for the element in Fig. 4.31 are obtained by assuming bilinear variations in τ_{xx}, τ_{yy}, and τ_{xy}, with the constant stress terms also present,

$$\mathbf{P} = \begin{bmatrix} 1 & 0 & 0 & y & 0 & x & 0 \\ 0 & 1 & 0 & 0 & x & 0 & y \\ 0 & 0 & 1 & 0 & 0 & -y & -x \end{bmatrix}$$

We should note that with this assumption the stresses satisfy the differential equations of equilibrium

$$\frac{\partial \tau_{xx}}{\partial x} + \frac{\partial \tau_{yx}}{\partial y} = 0$$

$$\frac{\partial \tau_{yy}}{\partial y} + \frac{\partial \tau_{xy}}{\partial x} = 0$$

The displacement interpolation matrix $\mathbf{H}^S$ is obtained as in Example 4.6, but now we need to establish one matrix for each element side. Using the displacement vector

$$\mathbf{U}^T = [u_1 u_2 u_3 u_4 \quad v_1 v_2 v_3 v_4]$$

we have

$$\mathbf{H}^S_{1-2} = \frac{1}{2}\begin{bmatrix} (1+x) & (1-x) & 0 & 0 & 0 & 0 & 0 & 0 \\ 0 & 0 & 0 & 0 & (1+x) & (1-x) & 0 & 0 \end{bmatrix}$$

$$\mathbf{H}^S_{2-3} = \frac{1}{2}\begin{bmatrix} 0 & (1+y) & (1-y) & 0 & 0 & 0 & 0 & 0 \\ 0 & 0 & 0 & 0 & 0 & (1+y) & (1-y) & 0 \end{bmatrix}$$

$$\mathbf{H}^S_{3-4} = \frac{1}{2}\begin{bmatrix} 0 & 0 & (1-x) & (1+x) & 0 & 0 & 0 & 0 \\ 0 & 0 & 0 & 0 & 0 & 0 & (1-x) & (1+x) \end{bmatrix}$$

$$\mathbf{H}^S_{4-1} = \frac{1}{2}\begin{bmatrix} (1+y) & 0 & 0 & (1-y) & 0 & 0 & 0 & 0 \\ 0 & 0 & 0 & 0 & (1+y) & 0 & 0 & (1-y) \end{bmatrix}$$

In the above discussion we briefly considered various alternative approaches to the development of finite element models. Since a large number of different finite elements could be considered and used, a natural question is: Which models are most efficient? This question is difficult to answer since many aspects, such as simplicity of formulation, accuracy, numerical stability, convergence, and cost-effectiveness, need to be considered.[36] In particular, we should note that the efficiency of an element may to a large degree be problem-dependent. Furthermore, new ideas concerning the implementation of an element that was believed to be inefficient could, at any time, make the element effective. What may be stated is that at this time the displacement-based isoparametric and related elements described in Chapters 5 and 6 are most effective, *in general*, and are used a great deal in practical analysis, because of their efficient formulation, general applicability, numerical stability, and simple theoretical foundation. However, considering the hybrid and mixed methods there is no doubt that significant potential lies in the further development of these solution approaches and that such developments may ultimately lead to the extensive use of hybrid and mixed finite elements.

4.3.3 Finite Difference Differential and Energy Methods

So far we have considered only actual finite element discretization procedures. However, when studying the finite difference analysis of systems,[37] and comparing the methods used with finite element techniques, we readily recognize the very close relationship between finite difference and finite element procedures. As an example, consider the analysis of the uniform bar in Fig. 4.32 with the governing differential equation

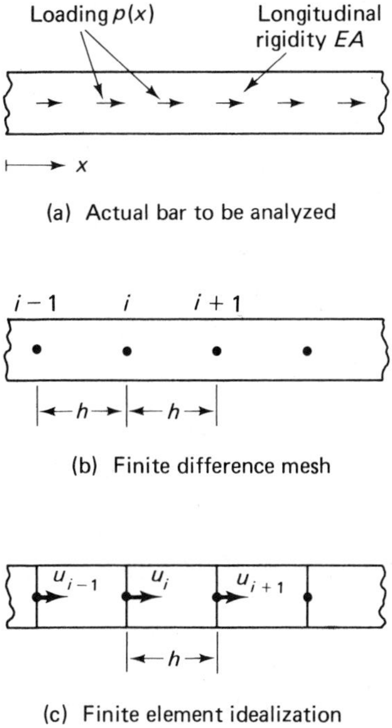

FIGURE 4.32 *Analysis of a bar.*

$$u'' + \frac{p}{EA} = 0 \tag{4.86}$$

Using an equal mesh point spacing and the central difference formula to approximate u'' (see Section 9.2.1) we obtain as the difference equation at point i

$$\frac{EA}{h}(-u_{i+1} + 2u_i - u_{i-1}) = p_i h \tag{4.87}$$

This same equation is also obtained using a finite element idealization of the bar, in which the stiffness matrix of a single bar element is

$$\mathbf{K} = \frac{EA}{h}\begin{bmatrix} 1 & -1 \\ -1 & 1 \end{bmatrix} \tag{4.88}$$

Hence, *finite difference equations can, therefore, also be interpreted as stiffness relations.*[38,39]

A main difficulty in the use of the conventional finite difference method lies in the incorporation of the boundary conditions. Since in the analysis the differential equations of equilibrium of the system are approximated directly by the difference scheme, it is necessary to satisfy in the differencing both the essential and the natural boundary conditions (see Section 3.3.2). This can be difficult to achieve at arbitrary boundaries, since the topology of the finite difference mesh restricts the form of differencing that can be carried out, and it may be difficult to maintain symmetry properties in the coefficient matrix.

The difficulties associated with conventional finite difference analysis have given impetus to the development of a finite difference analysis procedure that is based on the principle of minimum total potential energy, referred to as the *finite difference energy method.*[40,41] In this scheme the displacement derivatives in the total potential energy, Π, of the system are approximated by finite differences, and the minimum condition of Π is used to calculate the unknown displacement parameters. Since the variational formulation of the problem under consideration is employed, only the geometric boundary conditions need be satisfied in the differencing. Furthermore, a symmetric coefficient matrix is always obtained.

As might well be expected, the finite difference energy method is very closely related to the displacement-based finite element method, and in some cases the same algebraic equations are generated. The specific differences between the finite element method and the finite difference energy method lie essentially in the choice of the generalized displacement components but the solution procedures used are identical in both methods (see Example 4.23). Because of the close relationship between the finite element method and the finite difference energy method, it is also possible to identify equivalent physical models that are actually being analyzed when using the finite difference energy procedure.

An advantage of the finite difference energy method lies in the effectiveness with which the coefficient matrix of the algebraic equations can be generated. This is due to the simple scheme of energy integration employed, which has been shown to be equivalent to using incompatible finite elements. However, when comparing the finite difference energy method with the finite element analysis procedure, it must be noted that the finite difference method has been applied to specific problems (e.g., the analysis of shells of revolution) and that finite element programs can also be written for specific classes of problems which are unusually effective. The main advantage of the finite element method is that the procedure can be used effectively in general-purpose analysis programs. To demonstrate the finite difference analysis methods, consider the following examples.

EXAMPLE 4.22: Consider the simply supported beam in Fig. 4.33. Use conventional finite differencing to establish the system equilibrium equations.

The finite difference grid used for the beam analysis is shown in the figure. In the conventional finite difference analysis the differential equation of equilibrium and the geometric and natural boundary conditions are considered; i.e., we approximate by finite differences at each station,

$$EI\frac{d^4w}{dx^4} = q \tag{a}$$

and use the conditions that $w = 0$ and $w'' = 0$ at $x = 0$ and $x = L$.

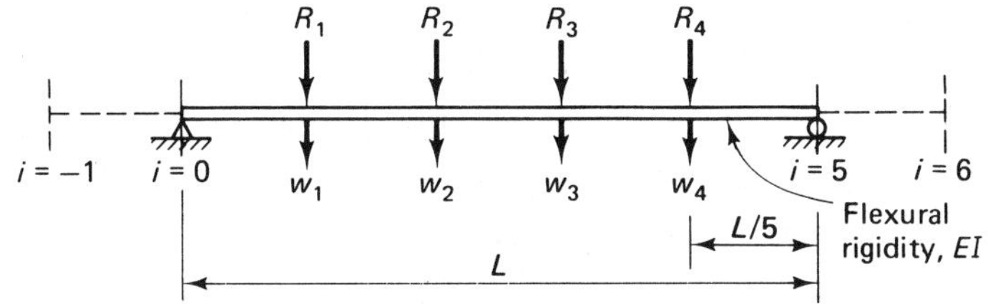

FIGURE 4.33 *Finite difference stations for simply supported beam.*

Using central differencing, (a) is approximated at station i by

$$\frac{EI}{\left(\frac{L}{5}\right)^3}\{w_{i-2} - 4w_{i-1} + 6w_i - 4w_{i+1} + w_{i+2}\} = R_i \tag{b}$$

where $R_i = q_iL/5$ and is the concentrated load applied at station i. The condition that w'' is zero at station i is approximated using

$$w_{i-1} - 2w_i + w_{i+1} = 0 \tag{c}$$

Applying (b) at each finite difference station, $i = 1, 2, 3, 4$, and using condition (c) at the support points, we obtain the system of equations

$$\frac{125EI}{L^3}\begin{bmatrix} 5 & -4 & 1 & 0 \\ -4 & 6 & -4 & 1 \\ 1 & -4 & 6 & -4 \\ 0 & 1 & -4 & 5 \end{bmatrix}\begin{bmatrix} w_1 \\ w_2 \\ w_3 \\ w_4 \end{bmatrix} = \begin{bmatrix} R_1 \\ R_2 \\ R_3 \\ R_4 \end{bmatrix}$$

where the coefficient matrix of the displacement vector can be regarded as a stiffness matrix.

EXAMPLE 4.23: Consider the cantilever beam in Fig. 4.34. Evaluate the tip deflection using the conventional finite difference method and the finite difference energy method.

The finite difference mesh used is shown in the figure. Using the conventional finite difference procedure and central differencing as in Example 4.22, we obtain

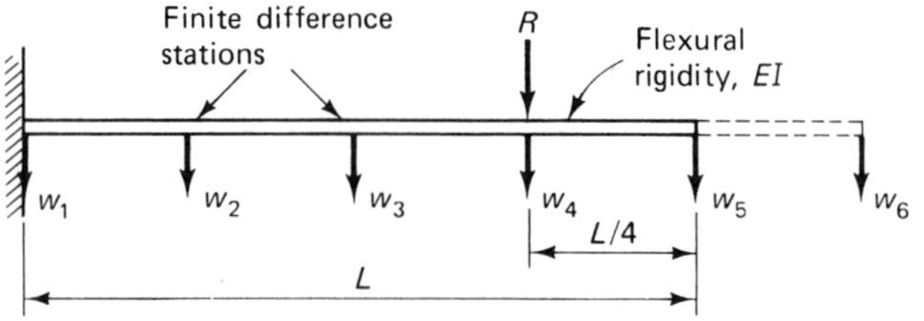

FIGURE 4.34 *Finite difference stations on cantilever beam.*

the equilibrium equations

$$\frac{64EI}{L^3}\begin{bmatrix} 7 & -4 & 1 & 0 \\ -4 & 6 & -4 & 1 \\ 1 & -4 & 5 & -2 \\ 0 & 1 & -2 & 1 \end{bmatrix}\begin{bmatrix} w_2 \\ w_3 \\ w_4 \\ w_5 \end{bmatrix} = \begin{bmatrix} 0 \\ 0 \\ R \\ 0 \end{bmatrix} \tag{a}$$

It may be noted that in addition to the equations employed in Example 4.22, the conditions that $w' = 0$ at the fixed end and $w''' = 0$ at the free end are also used. For w' and w''' equal to zero at station i, we employ, respectively,

$$w_{i+1} - w_{i-1} = 0$$

$$w_{i+2} - 2w_{i+1} + 2w_{i-1} - w_{i-2} = 0$$

Using the finite difference energy method, the total potential energy Π is given as

$$\Pi = \frac{EI}{2}\int_0^L [w''(x)]^2\, dx - Rw(\tfrac{3}{4}L)$$

To evaluate the integral we need to approximate $w''(x)$. Using central differencing we obtain for station i,

$$w_i'' = \frac{1}{\left(\frac{L}{4}\right)^2}(w_{i+1} - 2w_i + w_{i-1}) \tag{b}$$

An approximate solution can now be obtained by evaluating Π at the finite difference stations using (b) and replacing the integral by a summation process; i.e., we use the approximation

$$\Pi = \frac{L}{8}\Pi_1 + \frac{L}{4}(\Pi_2 + \Pi_3 + \Pi_4) + \frac{L}{8}\Pi_5 - Rw_4 \tag{c}$$

where $$\Pi_i = \frac{1}{2}[w_{i-1} \quad w_i \quad w_{i+1}]\begin{bmatrix} 1 \\ -2 \\ 1 \end{bmatrix}\frac{EI}{(L/4)^4}[1 \quad -2 \quad 1]\begin{bmatrix} w_{i-1} \\ w_i \\ w_{i+1} \end{bmatrix}$$

Therefore, we can write, in analogy to the finite element analysis procedure,

$$\Pi_i = \tfrac{1}{2}\mathbf{U}^T\mathbf{B}_i^T\mathbf{C}_i\mathbf{B}_i\mathbf{U}$$

where $\mathbf{B}_i$ is a generalized strain-displacement transformation matrix, $\mathbf{C}_i$ is the stress-strain matrix, and $\mathbf{U}$ is a vector listing all nodal point displacements. Using the direct stiffness method to calculate the total potential energy as given in (c) and employing the condition that the total potential energy is stationary (i.e., $\delta\Pi = 0$), we obtain the equilibrium equations

$$\frac{64EI}{L^3}\begin{bmatrix} 7 & -4 & 1 & & \\ -4 & 6 & -4 & 1 & \\ 1 & -4 & 5.5 & -3 & 0.5 \\ & 1 & -3 & 3 & -1 \\ & & 0.5 & -1 & 0.5 \end{bmatrix}\begin{bmatrix} w_2 \\ w_3 \\ w_4 \\ w_5 \\ w_6 \end{bmatrix} = \begin{bmatrix} 0 \\ 0 \\ R \\ 0 \\ 0 \end{bmatrix} \tag{d}$$

where the condition of zero slope at the fixed end has already been used.

The close similarity between the equilibrium equations in (a) and (d) should be noted. Indeed, if we eliminate w_6 from the equations in (d), we obtain the equations in (a). Hence, using the finite difference energy method and the conventional finite difference method we obtain in this case the same equilibrium equations.

As an example, let $R = 1$, $EI = 10^3$ and $L = 10$. Then we obtain, using the equations in (a) or (d) (without w_6),

$$\mathbf{U} = \begin{bmatrix} 0.023437 \\ 0.078125 \\ 0.14843 \\ 0.21875 \end{bmatrix}$$

The exact answer for the tip deflection is $w_5 = 0.2109375$. Hence the finite difference analysis gives a good approximate solution.

REFERENCES

1. R. Courant, "Variational Methods for the Solution of Problems of Equilibrium and Vibrations," *Bulletin of the American Mathematical Society*, Vol. 49, 1943, pp. 1–23.
2. R. Courant and D. Hilbert, *Methods of Mathematical Physics*, John Wiley & Sons, Inc., New York, N.Y., 1953.
3. J. L. Synge, *The Hypercircle in Mathematical Physics*, Cambridge University Press, London, 1957.
4. S. G. Mikhlin, *Variational Methods in Mathematical Physics*, Pergamon Press, Inc., Elmsford, N.Y., 1964.
5. M. J. Turner, R. W. Clough, H. C. Martin, and L. J. Topp, "Stiffness and Deflection Analysis of Complex Structures," *Journal of Aeronautical Science*, Vol. 23, 1956, pp. 805–823.
6. J. H. Argyris and S. Kelsey, "Energy Theorems and Structural Analysis," *Aircraft Engineering*, Vols. 26 and 27, 1955.
7. R. W. Clough, "The Finite Element in Plane Stress Analysis," *Proceedings, 2nd A.S.C.E. Conference on Electronic Computation*, Pittsburgh, Pa., Sept. 1960.
8. O. C. Zienkiewicz,, *The Finite Element Method*, 3rd ed., McGraw-Hill Book Company, New York, N.Y., 1977.
9. K. H. Huebner, *The Finite Element Method for Engineers*, J. Wiley & Sons, Inc., New York, N.Y., 1975.
10. J. S. Przemieniecki, *Theory of Matrix Structural Analysis*, McGraw-Hill Book Company, New York, N.Y., 1968.
11. S. Timoshenko and J. N. Goodier, *Theory of Elasticity*, McGraw-Hill Book Company, New York, N.Y., 1951.

12. S. Timoshenko and S. Woinowski-Krieger, *Theory of Plates and Shells*, 2nd ed., McGraw-Hill Book Company, New York, N.Y., 1959.
13. R. D. Mindlin, "Influence of Rotary Inertia and Shear of Flexural Motion of Isotropic Elastic Plates," *J. Appl. Mech.*, Vol. 18, pp. 31–38, 1951.
14. R. W. Clough and J. L. Tocher, "Finite Element Stiffness Matrices for the Analysis of Plate Bending," *Proceedings, Conference on Matrix Methods in Structural Mechanics*, Wright-Patterson A.F.B., Ohio, 1965.
15. J. H. Argyris, "Continua and Discontinua," *Proceedings, Conference on Matrix Methods in Structural Mechanics*, Wright-Patterson A.F.B., Ohio, Oct. 1965.
16. R. W. Clough and E. L. Wilson, "Dynamic Finite Element Analysis of Arbitrary Thin Shells," *J. Computers and Structures*, Vol. 1, 1971, pp. 33–56.
17. E. L. Wilson, "Structural Analysis of Axi-symmetric Solids," *A.I.A.A. Journal*, Vol. 3, 1965, pp. 2269–2274.
18. Y. K. Cheung, "Finite Strip Method of Analysis of Elastic Slabs," *Proceedings of the American Society of Civil Engineers*, Vol. 94, EM6, 1968, pp. 1365–1378.
19. R. J. Melosh, "Basis of Derivation of Matrices for the Direct Stiffness Method," *A.I.A.A. Journal*, Vol. 1, 1963, pp. 1631–1637.
20. S. Key, "A Convergence Investigation of the Direct Stiffness Method," Ph.D. thesis, University of Washington, 1966.
21. E. R. de Arantes e Oliveira, "Theoretical Foundations of the Finite Element Method," *International Journal of Solids and Structures*, Vol. 4, 1968, p. 929.
22. C. A. Felippa and R. W. Clough, "The Finite Element Method in Solid Mechanics," *Symposium on Numerical Solutions of Field Problems in Continuum Mechanics*, Durham, N.C., Apr. 1968.
23. G. Strang and G. J. Fix, *An Analysis of the Finite Element Method*, Prentice-Hall, Inc., Englewood Cliffs, N.J., 1973.
24. J. T. Oden and J. N. Reddy, *An Introduction to the Mathematical Theory of Finite Elements*, John Wiley & Sons, Inc., New York, N.Y., 1976.
25. G. P. Bazeley, Y. K. Cheung, B. M. Irons, and O. C. Zienkiewicz, "Triangular Elements in Plate Bending Conforming and Non-Conforming Solutions," *Proceedings, Conference on Matrix Methods in Structural Mechanics*, Wright-Patterson A.F.B., Ohio, 1965.
26. B. M. Irons, O. C. Zienkiewicz, and E. R. de Arantes e Oliveira, "Comments on the Paper: Theoretical Foundations of the Finite Element Method," *International Journal of Solids and Structures*, Vol. 6, 1970, pp. 695–697.
27. E. Reissner, "On a Variational Theorem in Elasticity," *Journal of Mathematics and Physics*, Vol. 29, 1950, pp. 90–95.
28. K. Washizu, *Variational Methods in Elasticity and Plasticity*, Pergamon Press, Inc., Elmsford, N.Y., 1975.
29. T. H. H. Pian, "Derivation of Element Stiffness Matrices by Assumed Stress Distributions," *A.I.A.A. J.*, Vol. 2, 1964,, pp. 1333–1336.
30. T. H. H. Pian and P. Tong, "Basis of Finite Element Methods for Solid Continua," *International Journal for Numerical Methods in Engineering*, Vol. 1, 1969, pp. 3–28.
31. R. L. Taylor, K. S. Pister, and L. R. Herrmann, "On a Variational Theorem for Incompressible and Nearly Incompressible Orthotropic Elasticity," *International Journal of Solids and Structures*, Vol. 4, 1968, pp. 875–883.
32. T. H. H. Pian, "Formulations of Finite Element Methods for Solid Continua," in *Recent Advances in Matrix Methods of Structural Analysis and Design* (R. H. Gallagher, Y. Yamada, and J. T. Oden, eds.), University of Alabama Press, Huntsville, Ala., 1971.

33. W. Wunderlich, "Discretization of Structural Problems by a Generalized Variational Approach," *Proceedings, I.A.S.S. Symposium*, Honolulu, Hawaii, 1971.
34. B. F. de Veubeke and G. Sander, "An Equilibrium Model for Plate Bending," *International Journal of Solids and Structures*, Vol. 4, 1968, pp. 447–468.
35. R. H. Gallagher, *Finite Element Analysis—Fundamentals*, Prentice-Hall, Inc., Englewood Cliffs, N.J., 1975.
36. J. L. Batoz, K. J. Bathe, and L. W. Ho, "A Study of Three-Node Triangular Plate Bending Elements," *Int. J. Num. Meth. in Engineering*, Vol. 15, 1771–1812, 1980.
37. G. E. Forsythe and W. R. Wasow, *Finite Difference Methods for Partial Differential Equations*, John Wiley & Sons, Inc., New York, N.Y., 1960.
38. A. Ghali and K. J. Bathe, "Analysis of Plates Subjected to In-Plane Forces Using Large Finite Elements," *International Association for Bridge and Structural Engineering Bulletin*, Vol. 30-I, 1970, pp. 61–72.
39. A. Ghali and K. J. Bathe, "Analysis of Plates in Bending Using Large Finite Elements," *International Association for Bridge and Structural Engineering Bulletin*, Vol. 30-II, 1970, pp. 29–40.
40. D. Bushnell, "Finite Difference Energy Models Versus Finite Element Models: Two Variational Approaches in One Computer Program," in *Numerical and Computer Methods in Structural Mechanics* (S. J. Fenves, N. Perrone, J. Robinson, and W. C. Schnobrich, etc.) Academic Press, Inc. New York, N.Y., 1973.
41. D. Bushnell and B. O. Almroth, "Finite Difference Energy Method for Nonlinear Shell Analysis," *J. Computers and Structures*, Vol. 1, 1971, p. 361.

5

FORMULATION AND CALCULATION OF ISOPARAMETRIC FINITE ELEMENT MATRICES

5.1 INTRODUCTION

A very important phase of a finite element analysis is the calculation of the finite element matrices. In Chapter 4 we discussed the formulation and calculation of generalized coordinate finite element models. The aim in the presentation of the generalized coordinate finite elements was primarily to enhance the understanding of the finite element method. We already pointed out that in most practical analyses the use of isoparametric finite elements is more effective.[1-14]

Our objective in this chapter is to present the formulation of isoparametric finite elements and describe effective implementations. In the derivation of generalized coordinate finite element models, we used local element coordinate systems x, y, z and assumed the element displacements $u(x, y, z)$, $v(x, y, z)$, and $w(x, y, z)$ in the form of polynomials in x, y, and z with undetermined constant coefficients α_i, β_i, and γ_i, $i = 1, 2, \ldots$, identified as generalized coordinates. It was not possible to associate a priori a physical meaning to the generalized coordinates; however, on evaluation we found that the generalized coordinates are linear combinations of the element nodal point displacements. The principal idea of the isoparametric finite element formulation is to achieve the relationship between the element displacements at any point and the element nodal point displacements directly through the use of *interpolation functions* (also called *shape functions*). This means that the transformation matrix $\mathbf{A}^{-1}$ (see (4.55)) is not evaluated; instead, the element matrices corresponding to the required degrees of freedom are obtained directly.

5.2 ISOPARAMETRIC DERIVATION OF BAR ELEMENT STIFFNESS MATRIX

Consider the example of a bar element to illustrate the procedure of an isoparametric stiffness formulation. In order to simplify the explanation, assume that the bar lies in the global X-coordinate axis, as shown in Fig. 5.1. The first step is to relate the actual global coordinates X to a natural coordinate system with variable r, where $-1 \leq r \leq 1$ (Fig. 5.1). This transformation is given by

$$X = \tfrac{1}{2}(1 - r)X_1 + \tfrac{1}{2}(1 + r)X_2 \tag{5.1}$$

or

$$X = \sum_{i=1}^{2} h_i X_i \tag{5.2}$$

where $h_1 = \frac{1}{2}(1 - r)$ and $h_2 = \frac{1}{2}(1 + r)$ are the interpolation or shape functions. Note that (5.2) establishes a unique relationship between the coordinates X and r on the bar.

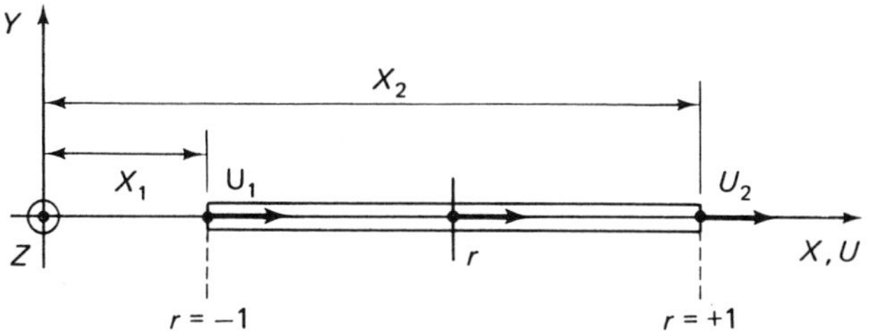

FIGURE 5.1 *Element in global and natural coordinate system.*

The bar global displacements are expressed in the same way as the global coordinates:

$$U = \sum_{i=1}^{2} h_i U_i \tag{5.3}$$

where in this case a linear displacement variation is specified. *The interpolation of the element coordinates and element displacements using the same interpolation functions, which are defined in a natural coordinate system, is the basis of the isoparametric finite element formulation.*

For the calculation of the element stiffness matrix we need to find the element strains $\epsilon = dU/dX$. Here we use

$$\epsilon = \frac{dU}{dr}\frac{dr}{dX} \tag{5.4}$$

where, from (5.3),

$$\frac{dU}{dr} = \frac{U_2 - U_1}{2} \tag{5.5}$$

and using (5.2) we obtain

$$\frac{dX}{dr} = \frac{X_2 - X_1}{2} = \frac{L}{2} \tag{5.6}$$

where L is the length of the bar. Hence, as expected, we have

$$\epsilon = \frac{U_2 - U_1}{L} \tag{5.7}$$

The strain-displacement transformation matrix corresponding to (4.28) is therefore

$$\mathbf{B} = \frac{1}{L}[-1 \quad 1] \tag{5.8}$$

In general, the strain-displacement transformation matrix is a function of the natural coordinates and we therefore evaluate the stiffness matrix volume integral in (4.29) by integrating over the natural coordinates. Following this general procedure, although in this example it is not necessary, we have

$$\mathbf{K} = \frac{AE}{L^2}\int_{-1}^{1}\begin{bmatrix}-1\\ 1\end{bmatrix}[-1 \quad 1]\,J dr \tag{5.9}$$

where the bar area A and modulus of elasticity E have been assumed constant, and J is the Jacobian relating an element length in the global coordinate system to an element length in the natural coordinate system; i.e.,

$$dX = J\,dr \tag{5.10}$$

From (5.6) we have

$$J = \frac{L}{2} \tag{5.11}$$

Then evaluating (5.9) we obtain the well-known matrix

$$\mathbf{K} = \frac{AE}{L}\begin{bmatrix}1 & -1\\ -1 & 1\end{bmatrix} \tag{5.12}$$

As was stated in the introduction, the isoparametric formulation avoids the construction of the transformation matrix $\mathbf{A}^{-1}$. In order to compare the above formulation with the generalized coordinate formulation, we need to solve from (5.1) for r and then substitute for r into (5.3). We obtain

$$r = \frac{X - [(X_1 + X_2)/2]}{L/2} \tag{5.13}$$

and then

$$U = \alpha_0 + \alpha_1 X \tag{5.14}$$

where

$$\left.\begin{aligned}\alpha_0 &= \tfrac{1}{2}(U_1 + U_2) - \frac{X_1 + X_2}{2L}(U_2 - U_1)\\ \alpha_1 &= \frac{1}{L}(U_2 - U_1)\end{aligned}\right\} \tag{5.15}$$

or

$$\boldsymbol{\alpha} = \begin{bmatrix}\dfrac{1}{2} + \dfrac{X_1 + X_2}{2L} & \dfrac{1}{2} - \dfrac{X_1 + X_2}{2L}\\ -\dfrac{1}{L} & \dfrac{1}{L}\end{bmatrix}\mathbf{U} \tag{5.16}$$

where

$$\boldsymbol{\alpha}^T = [\alpha_0 \quad \alpha_1]; \qquad \mathbf{U}^T = [U_1 \quad U_2] \tag{5.17}$$

and the matrix relating in (5.16) $\boldsymbol{\alpha}$ to $\mathbf{U}$ is $\mathbf{A}^{-1}$. It should be noted that in this example the generalized coordinates α_0 and α_1 relate the global element displacements to the global element coordinates.

5.3 FORMULATION OF CONTINUUM ELEMENTS

Considering the calculation of a continuum element, it is in most cases effective to directly calculate the element matrices corresponding to the global degrees of freedom. However, we shall first present the formulation of the matrices that correspond to the element local degrees of freedom, because additional considerations may be necessary when the element matrices that correspond to the global degrees of freedom are calculated directly (see Section 5.3.3). In the following we consider the derivation of the element matrices of straight truss elements, two-dimensional plane stress, plane strain and axisymmetric elements, and three-dimensional elements that all have a variable number of nodes. Typical elements are shown in Fig. 5.2.

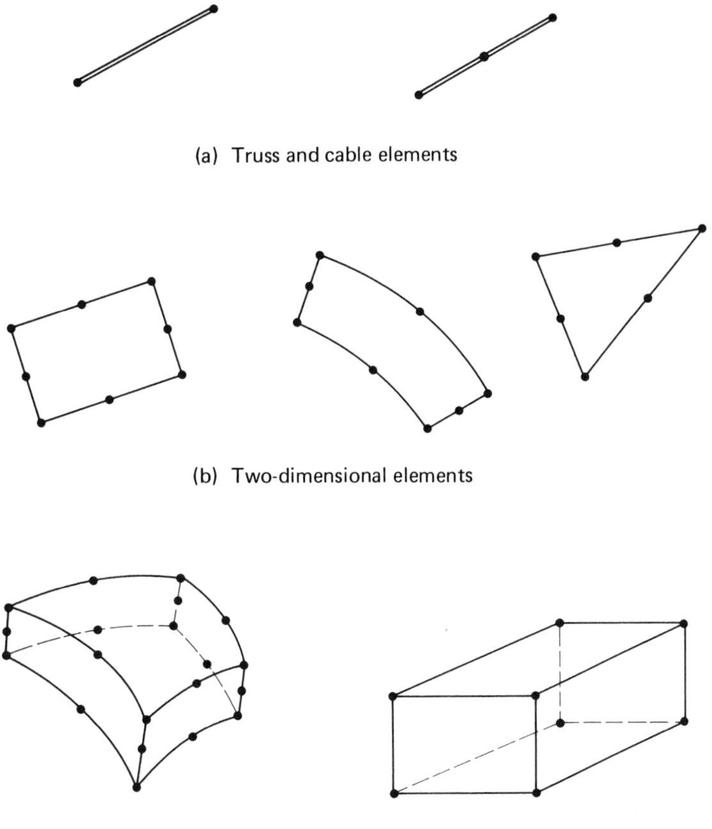

FIGURE 5.2 *Some typical continuum elements.*

5.3.1 Rectangular Elements

The basic procedure in the isoparametric finite element formulation is to express the element coordinates and element displacements in the form of interpolations using the natural coordinate system of the element. This coordinate system is one-, two-, or three-dimensional, depending on the dimensionality of the element. The formulation of the element matrices is the same whether we deal with a one-, two-, or three-dimensional element. For this reason we use in the following general presentation the equations of a three-dimensional element. However, the one- and two-dimensional elements are included by simply using only the relevant coordinate axes and the appropriate interpolation functions.

Considering a general three-dimensional element, the coordinate interpolations are

$$\left.\begin{aligned} x &= \sum_{i=1}^{q} h_i x_i \\ y &= \sum_{i=1}^{q} h_i y_i \\ z &= \sum_{i=1}^{q} h_i z_i \end{aligned}\right\} \tag{5.18}$$

where x, y, and z are the coordinates at any point of the element (here local coordinates), and x_i, y_i, z_i, $i = 1, \ldots, q$, are the coordinates of the q element nodes. The interpolation functions h_i are defined in the natural coordinate system of the element, which has variables r, s, and t that each vary from -1 to $+1$. For one- or two- dimensional elements, only the relevant equations in (5.18) would be employed, and the interpolation functions would depend only on the natural coordinate variables r and r, s, respectively.

The unknown quantities in (5.18) are so far the interpolation functions h_i. The fundamental property of the interpolation function h_i is that its value in the natural coordinate system is unity at node i and is zero at all other nodes. Using these conditions the functions h_i corresponding to a specific nodal point layout could be solved for in a systematic manner. However, it is convenient to construct them by inspection, which is demonstrated in the following simple example.

EXAMPLE 5.1: Construct the interpolation functions corresponding to the three-node truss element in Fig. 5.3.

A first observation is that for the three-node truss element we want interpolation polynomials that involve r^2 as the highest power of r; in other words, the interpolation functions shall be parabolas. The function h_2 can thus be constructed without much effort. Namely, the parabola that satisfies the conditions to be equal to zero at $r = \pm 1$ and equal to 1 at $r = 0$ is given by $(1 - r^2)$. The other two interpolation functions h_1 and h_3 are constructed by superimposing a linear function and a parabola. Consider the interpolation function h_3. Using $\frac{1}{2}(1 + r)$, the conditions that the function shall be zero at $r = -1$ and 1 at $r = +1$ are satisfied. To assure that h_3 is also zero at $r = 0$, we need to use $h_3 = \frac{1}{2}(1 + r) - \frac{1}{2}(1 - r^2)$. The interpolation function h_1 is obtained in a similar manner.

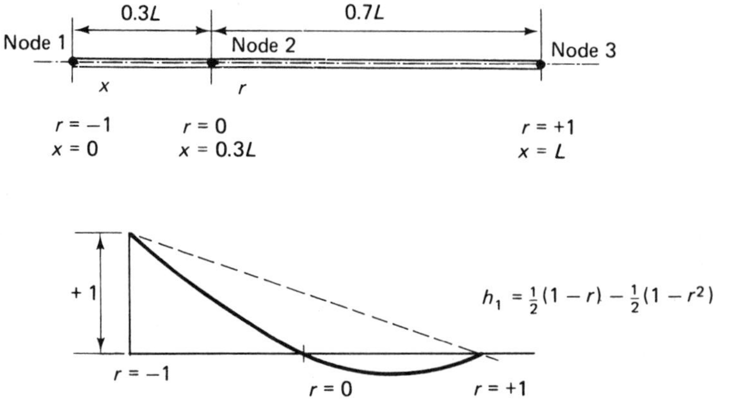

FIGURE 5.3 *One-dimensional interpolation functions of a truss element.*

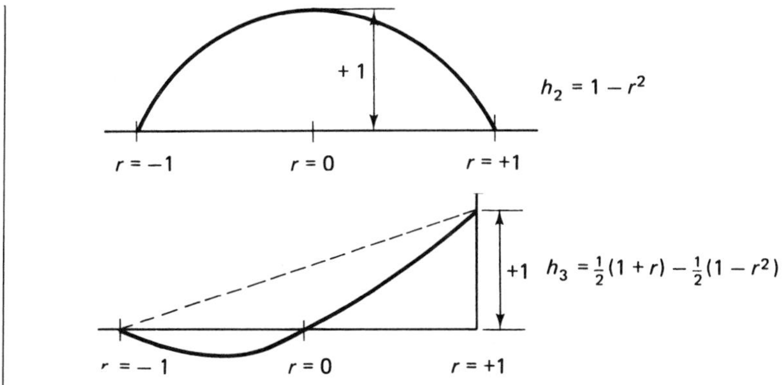

FIGURE 5.3 (*cont.*)

The procedure used in Example 5.1 of constructing the final required interpolation functions suggests an attractive formulation of an element with a variable number of nodes. This formulation is achieved by constructing first the interpolations corresponding to a basic two-node element. The addition of another node then results into an additional interpolation function and a correction to be applied to the already existing interpolation functions. Figure 5.4 gives the interpolation functions of the one-dimensional element considered in Example 5.1, with an additional fourth node possible. As shown the element can have from two to four nodes. We should note that nodes 3 and 4 are now intermediate nodes, because nodes 1 and 2 are used to define the two-node element.

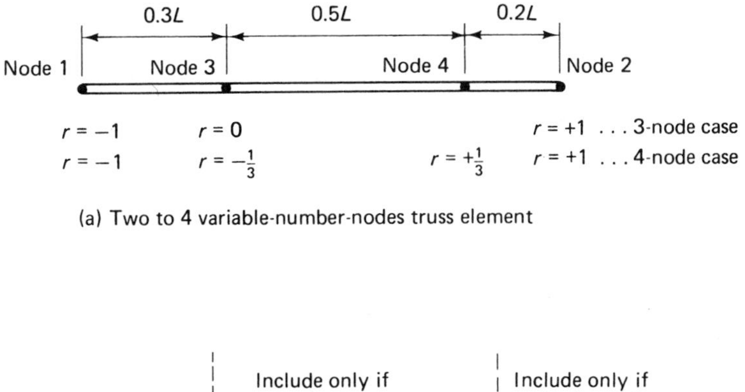

(a) Two to 4 variable-number-nodes truss element

	Include only if node 3 is present	Include only if nodes 3 and 4 are present
$h_1 = \frac{1}{2}(1 - r)$	$-\frac{1}{2}(1 - r^2)$	$+\frac{1}{16}(-9r^3 + r^2 + 9r - 1)$
$h_2 = \frac{1}{2}(1 + r)$	$-\frac{1}{2}(1 - r^2)$	$+\frac{1}{16}(9r^3 + r^2 - 9r - 1)$
$h_3 = (1 - r^2)$		$+\frac{1}{16}(27r^3 + 7r^2 - 27r - 7)$
$h_4 = \frac{1}{16}(-27r^3 - 9r^2 + 27r + 9)$		

(b) Interpolation functions

FIGURE 5.4 *Interpolation functions of two to four variable-number-nodes one-dimensional element.*

This procedure of constructing the element interpolation functions for one-dimensional analysis can be directly generalized for use in two and three dimensions. Figure 5.5 shows the interpolation functions of a four to nine variable-number-nodes two-dimensional element and Fig. 5.6 gives the interpolation functions for three-dimensional eight- to twenty-node elements. The two- and three-dimensional interpolations have been established in a manner analogous to the one-dimensional interpolations, where the basic functions used are, in fact, those already employed in Fig. 5.4. We consider in Figs. 5.5 and 5.6 at most parabolic interpolation, but variable-number-nodes elements with interpolations of higher order could be derived in an analogous way.

The attractiveness of the elements in Figs. 5.4 to 5.6 lies in that the elements can have any number of nodes between the minimum and the maximum, and triangular elements can also be formed (see Section 5.3.2). However, to obtain

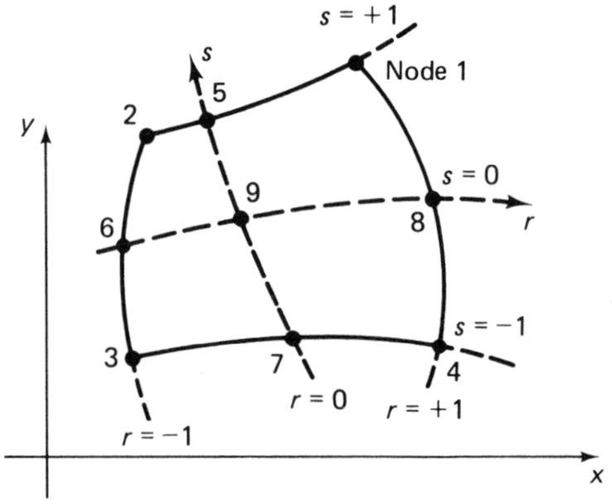

(a) Four to 9 variable-number-nodes two-dimensional element

		Include only if node i is defined				
		$i = 5$	$i = 6$	$i = 7$	$i = 8$	$i = 9$
$h_1 =$	$\frac{1}{4}(1+r)(1+s)$	$-\frac{1}{2}h_5$	....	....	$-\frac{1}{2}h_8$	$-\frac{1}{4}h_9$
$h_2 =$	$\frac{1}{4}(1-r)(1+s)$	$-\frac{1}{2}h_5$	$-\frac{1}{2}h_6$	....	....	$-\frac{1}{4}h_9$
$h_3 =$	$\frac{1}{4}(1-r)(1-s)$	....	$-\frac{1}{2}h_6$	$-\frac{1}{2}h_7$	....	$-\frac{1}{4}h_9$
$h_4 =$	$\frac{1}{4}(1+r)(1-s)$	....	....	$-\frac{1}{2}h_7$	$-\frac{1}{2}h_8$	$-\frac{1}{4}h_9$
$h_5 =$	$\frac{1}{2}(1-r^2)(1+s)$	....	....	....	....	$-\frac{1}{2}h_9$
$h_6 =$	$\frac{1}{2}(1-s^2)(1-r)$	....	....	....	....	$-\frac{1}{2}h_9$
$h_7 =$	$\frac{1}{2}(1-r^2)(1-s)$	....	....	....	....	$-\frac{1}{2}h_9$
$h_8 =$	$\frac{1}{2}(1-s^2)(1+r)$	....	....	....	....	$-\frac{1}{2}h_9$
$h_9 =$	$(1-r^2)(1-s^2)$					

(b) Interpolation functions

FIGURE 5.5 *Interpolation functions of four to nine variable-number-nodes two-dimensional element.*

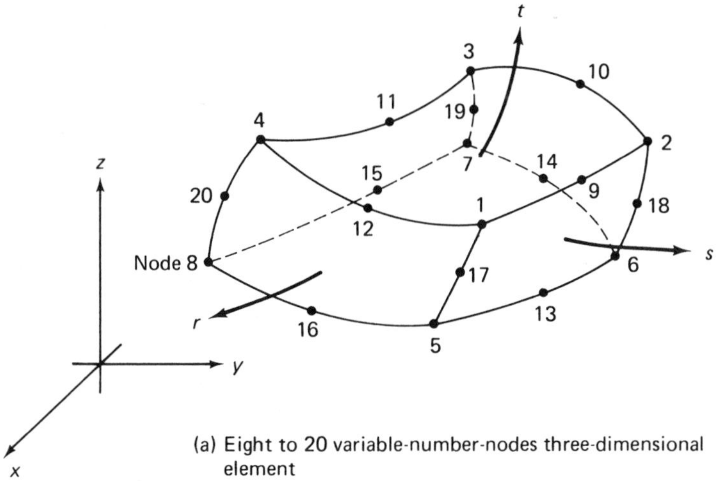

(a) Eight to 20 variable-number-nodes three-dimensional element

$$
\begin{aligned}
h_1 &= g_1 - (g_9 + g_{12} + g_{17})/2 & h_6 &= g_6 - (g_{13} + g_{14} + g_{18})/2 \\
h_2 &= g_2 - (g_9 + g_{10} + g_{18})/2 & h_7 &= g_7 - (g_{14} + g_{15} + g_{19})/2 \\
h_3 &= g_3 - (g_{10} + g_{11} + g_{19})/2 & h_8 &= g_8 - (g_{15} + g_{16} + g_{20})/2 \\
h_4 &= g_4 - (g_{11} + g_{12} + g_{20})/2 & h_j &= g_j \text{ for } j = 9, \ldots, 20 \\
h_5 &= g_5 - (g_{13} + g_{16} + g_{17})/2 & &
\end{aligned}
$$

$g_i = 0$ if node i is not included; otherwise,

$$g_i = G(r, r_i)\, G(s, s_i)\, G(t, t_i)$$

$$
\left.
\begin{aligned}
G(\beta, \beta_i) &= \tfrac{1}{2}(1 + \beta_i \beta) && \text{for } \beta_i = \pm 1 \\
G(\beta, \beta_i) &= (1 - \beta^2) && \text{for } \beta_i = 0
\end{aligned}
\right\} ; \beta = r, s, t
$$

(b) Interpolation functions

FIGURE 5.6 *Interpolation functions of eight to twenty variable-number-nodes three-dimensional element.*

maximum accuracy, the variable-number-nodes elements should be as nearly rectangular (in three-dimensional analysis, rectangular in each local plane) as possible and the noncorner nodes should in general be located at their natural coordinate positions, e.g., for the 9-node two-dimensional element the intermediate side nodes should in general be located at the midpoints between the corner nodes and the ninth node should be at the center of the element (for some exceptions see Section 5.3.2).

Considering the geometry of the two- and three-dimensional elements in Figs. 5.5 and 5.6 we note that by means of the coordinate interpolations in (5.18), the elements can have, without any difficulties, curved boundaries. This is an important advantage over the generalized coordinate finite element formulation. Another important advantage is the ease with which the element displacement functions can be constructed.

In the isoparametric formulation the element displacements are interpolated in the same way as the geometry; i.e., we use

$$u = \sum_{i=1}^{q} h_i u_i; \qquad v = \sum_{i=1}^{q} h_i v_i; \qquad w = \sum_{i=1}^{q} h_i w_i \tag{5.19}$$

where u, v, and w are the local element displacements at any point of the element and u_i, v_i, and w_i, $i = 1, \ldots, q$, are the corresponding element displacements at its nodes. Therefore, it is assumed that to each nodal point coordinate necessary to describe the geometry of the element, there corresponds one nodal point displacement.

To be able to evaluate the stiffness matrix of an element, we need to calculate the strain-displacement transformation matrix. The element strains are obtained in terms of derivatives of element displacements with respect to the local coordinates. Because the element displacements are defined in the natural coordinate system using (5.19), we need to relate the x, y, and z derivatives to the r, s, and t derivatives, where we realize that (5.18) is of the form

$$x = f_1(r, s, t) \qquad y = f_2(r, s, t); \qquad z = f_3(r, s, t); \tag{5.20}$$

where f_i denotes "function of." The inverse relationship is

$$r = f_4(x, y, z); \qquad s = f_5(x, y, z); \qquad t = f_6(x, y, z) \tag{5.21}$$

We require the derivatives $\partial/\partial x$, $\partial/\partial y$, and $\partial/\partial z$ and it seems natural to use the chain rule in the following form:

$$\frac{\partial}{\partial x} = \frac{\partial}{\partial r}\frac{\partial r}{\partial x} + \frac{\partial}{\partial s}\frac{\partial s}{\partial x} + \frac{\partial}{\partial t}\frac{\partial t}{\partial x} \tag{5.22}$$

with similar relationships for $\partial/\partial y$ and $\partial/\partial z$. However, to evaluate $\partial/\partial x$ in (5.22), we need to calculate $\partial r/\partial x$, $\partial s/\partial x$, and $\partial t/\partial x$, which means that the explicit inverse relationships in (5.21) would need to be evaluated. These inverse relationships are in general difficult to establish explicitly, and it is necessary to evaluate the required derivatives in the following way. Using the chain rule, we have

$$\begin{bmatrix} \dfrac{\partial}{\partial r} \\ \dfrac{\partial}{\partial s} \\ \dfrac{\partial}{\partial t} \end{bmatrix} = \begin{bmatrix} \dfrac{\partial x}{\partial r} & \dfrac{\partial y}{\partial r} & \dfrac{\partial z}{\partial r} \\ \dfrac{\partial x}{\partial s} & \dfrac{\partial y}{\partial s} & \dfrac{\partial z}{\partial s} \\ \dfrac{\partial x}{\partial t} & \dfrac{\partial y}{\partial t} & \dfrac{\partial z}{\partial t} \end{bmatrix} \begin{bmatrix} \dfrac{\partial}{\partial x} \\ \dfrac{\partial}{\partial y} \\ \dfrac{\partial}{\partial z} \end{bmatrix} \tag{5.23}$$

or, in matrix notation,

$$\frac{\boldsymbol{\partial}}{\boldsymbol{\partial} \mathbf{r}} = \mathbf{J} \frac{\boldsymbol{\partial}}{\boldsymbol{\partial} \mathbf{x}} \tag{5.24}$$

where $\mathbf{J}$ is the *Jacobian operator* relating the natural coordinate derivatives to the local coordinate derivatives. We should note that the Jacobian operator can easily be found using (5.18). We require $\boldsymbol{\partial}/\boldsymbol{\partial}\mathbf{x}$ and use

$$\frac{\boldsymbol{\partial}}{\boldsymbol{\partial} \mathbf{x}} = \mathbf{J}^{-1} \frac{\boldsymbol{\partial}}{\boldsymbol{\partial} \mathbf{r}} \tag{5.25}$$

which requires that the inverse of $\mathbf{J}$ exists. This inverse exists provided that there is a one-to-one (i.e., unique) correspondence between the natural and the local coordinates of the element, as expressed in (5.20) and (5.21). In most formulations the one-to-one correspondence between the coordinate systems (i.e., to each r, s, and t, there corresponds only one x, y, and z) is obviously given, such as for the elements in Figs. 5.3 to 5.6. However, in cases where the element is much distorted or folds back upon itself, as in Fig. 5.7, the unique relation between the coordinate systems does not exist (see also Section 5.3.2 for singularities in the Jacobian transformation, Example 5.17).

Using (5.19) and (5.25), we evaluate $\partial u/\partial x$, $\partial u/\partial y$, $\partial u/\partial z$, $\partial v/\partial x, \ldots, \partial w/\partial z$ and can therefore construct the strain-displacement transformation matrix $\mathbf{B}$, with

$$\boldsymbol{\epsilon} = \mathbf{B}\hat{\mathbf{u}} \tag{5.26}$$

where $\hat{\mathbf{u}}$ is a vector listing the element nodal point displacements of (5.19), and we note that $\mathbf{J}$ affects the elements in $\mathbf{B}$. The element stiffness matrix corresponding to the local element degrees of freedom is then

$$\mathbf{K} = \int_V \mathbf{B}^T\mathbf{C}\mathbf{B}\, dV \tag{5.27}$$

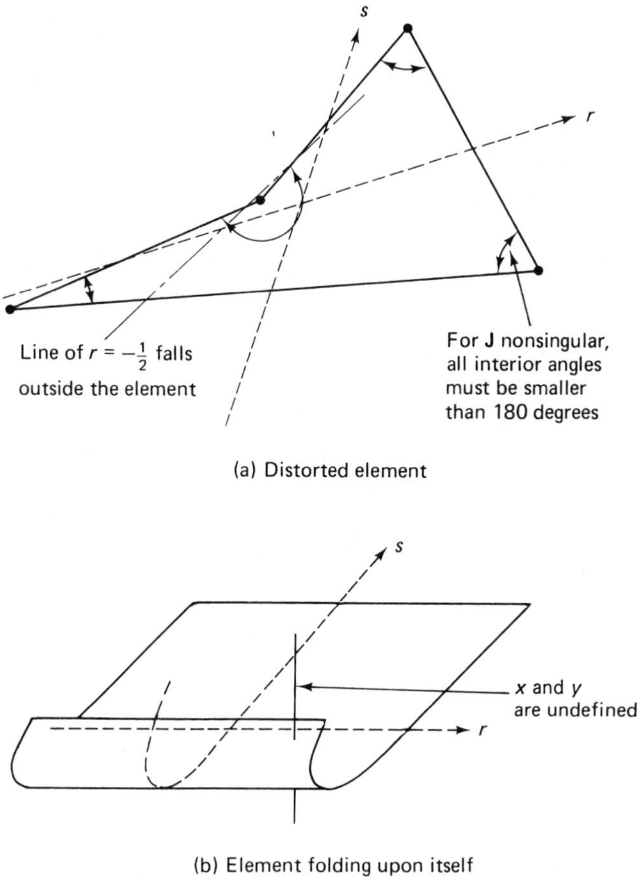

FIGURE 5.7 *Elements with possible singular Jacobian.*

We should note that the elements of $\mathbf{B}$ are functions of the natural coordinates r, s, and t. Therefore, the volume integration extends over the natural coordinate volume, and the volume differential dV need also be written in terms of the natural coordinates. In general, we have

$$dV = \det \mathbf{J} \, dr \, ds \, dt \tag{5.28}$$

where $\det \mathbf{J}$ is the determinant of the Jacobian operator in (5.24).

An explicit evaluation of the volume integral in (5.27) is, in general, not effective; in particular, when higher-order interpolations are used or the element is distorted. Therefore, numerical integration is employed. Indeed, numerical integration must be regarded as an integral part of the isoparametric element matrix evaluations. The details of the numerical integration procedure are described in Section 5.7, but the process can briefly be summarized as follows. First, we write (5.27) in the form

$$\mathbf{K} = \int_V \mathbf{F} \, dr \, ds \, dt \tag{5.29}$$

where $\mathbf{F} = \mathbf{B}^T\mathbf{C}\mathbf{B} \det \mathbf{J}$ and the integration is performed in the natural coordinate system of the element. As stated above, the elements of $\mathbf{F}$ depend on r, s, and t, but the detailed functional relationship is usually not calculated. Using numerical integration, the stiffness matrix is now evaluated as

$$\mathbf{K} = \sum_{i,j,k} \alpha_{ijk} \mathbf{F}_{ijk} \tag{5.30}$$

where $\mathbf{F}_{ijk}$ is the matrix $\mathbf{F}$ evaluated at points r_i, s_j, and the t_k, and the α_{ijk} are given constants which depend on the values of r_i, s_j, and t_k. The sampling points r_i, s_j, and t_k of the function and the corresponding weighting factors α_{ijk} are chosen to obtain maximum accuracy in the integration. Naturally, the integration accuracy can increase as the number of sampling points is increased.

The purpose of the above brief outline of the numerical integration procedure was to complete the description of the general isoparametric formulation. The relative simplicity of the formulation may already be noted. It is the simplicity of the element formulation and the efficiency with which the element matrices can actually be evaluated in the computer that has drawn much attention to the development of the isoparametric and related elements.

The formulation of the element mass matrix and load vectors is now straightforward. Namely, writing the element displacements in the form

$$\mathbf{u}(r, s, t) = \mathbf{H}\hat{\mathbf{u}} \tag{5.31}$$

where $\mathbf{H}$ is a matrix of the interpolation functions, we have, as in (4.30) to (4.33),

$$\mathbf{M} = \int_V \rho \mathbf{H}^T\mathbf{H} \, dV \tag{5.32}$$

$$\mathbf{R}_B = \int_V \mathbf{H}^T\mathbf{f}^B \, dV \tag{5.33}$$

$$\mathbf{R}_S = \int_S \mathbf{H}^{S^T}\mathbf{f}^S \, dS \tag{5.34}$$

$$\mathbf{R}_I = \int_V \mathbf{B}^T\boldsymbol{\tau}^I \, dV \tag{5.35}$$

The above matrices are evaluated using numerical integration as indicated for the stiffness matrix $\mathbf{K}$ in (5.30). In the evaluation we need to use the appropriate function $\mathbf{F}$. To calculate the body force vector $\mathbf{R}_B$ we use $\mathbf{F} = \mathbf{H}^T\mathbf{f}^B \det \mathbf{J}$, for the surface force vector we use $\mathbf{F} = \mathbf{H}^{S^T}\mathbf{f}^S \det \mathbf{J}^S$, for the initial stress load vector we use $\mathbf{F} = \mathbf{B}^T\boldsymbol{\tau}^I \det \mathbf{J}$, and for the mass matrix we have $\mathbf{F} = \rho\mathbf{H}^T\mathbf{H} \det \mathbf{J}$. The weight coefficients α_{ijk} are the same as in the stiffness matrix evaluation if the same order of numerical integration is used. However, in practice, different integration orders may be employed because the required accuracy in the element matrices varies (see Section 5.7).

The above formulation was for one-, two-, or three-dimensional elements. We shall now consider some specific cases and demonstrate the details of the calculation of element matrices.

EXAMPLE 5.2: Derive the displacement interpolation matrix $\mathbf{H}$, strain-displacement interpolation matrix $\mathbf{B}$, and the Jacobian operator $\mathbf{J}$ for the three-node truss element shown in Fig. 5.8.

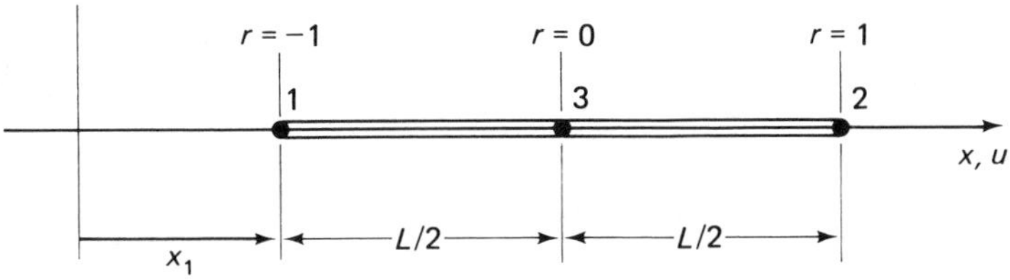

FIGURE 5.8 *Truss element with node 3 at center of element.*

The interpolation functions of the element were already given in Fig. 5.3. Thus we have

$$\mathbf{H} = \left[-\frac{r}{2}(1 - r) \quad \frac{r}{2}(1 + r) \quad (1 - r^2)\right] \tag{a}$$

The strain-displacement matrix $\mathbf{B}$ is obtained by differentiation of $\mathbf{H}$ with respect to r and premultiplying the result by the inverse of the Jacobian operator

$$\mathbf{B} = \mathbf{J}^{-1}[(-\tfrac{1}{2} + r) \quad (\tfrac{1}{2} + r) \quad -2r] \tag{b}$$

To evaluate $\mathbf{J}$ formally we use

$$x = -\frac{r}{2}(1 - r)x_1 + \frac{r}{2}(1 + r)(x_1 + L) + (1 - r^2)\left(x_1 + \frac{L}{2}\right)$$

hence

$$x = x_1 + \frac{L}{2} + \frac{L}{2}r \tag{c}$$

where we may note that because node 3 is at the center of the truss, x is interpolated linearly between nodes 1 and 2. The same result would be obtained using only nodes 1 and 2 for the geometry interpolation. Using now the relation in (c) we have

$$\mathbf{J} = \left[\frac{L}{2}\right] \tag{d}$$

and

$$\mathbf{J}^{-1} = \left[\frac{2}{L}\right], \quad \det \mathbf{J} = \frac{L}{2}.$$

With the relations in (a) to (d), we could now evaluate all finite element matrices and vectors given in (5.27) to (5.35).

EXAMPLE 5.3: Establish the Jacobian operator **J** of the two-dimensional elements shown in Fig. 5.9.

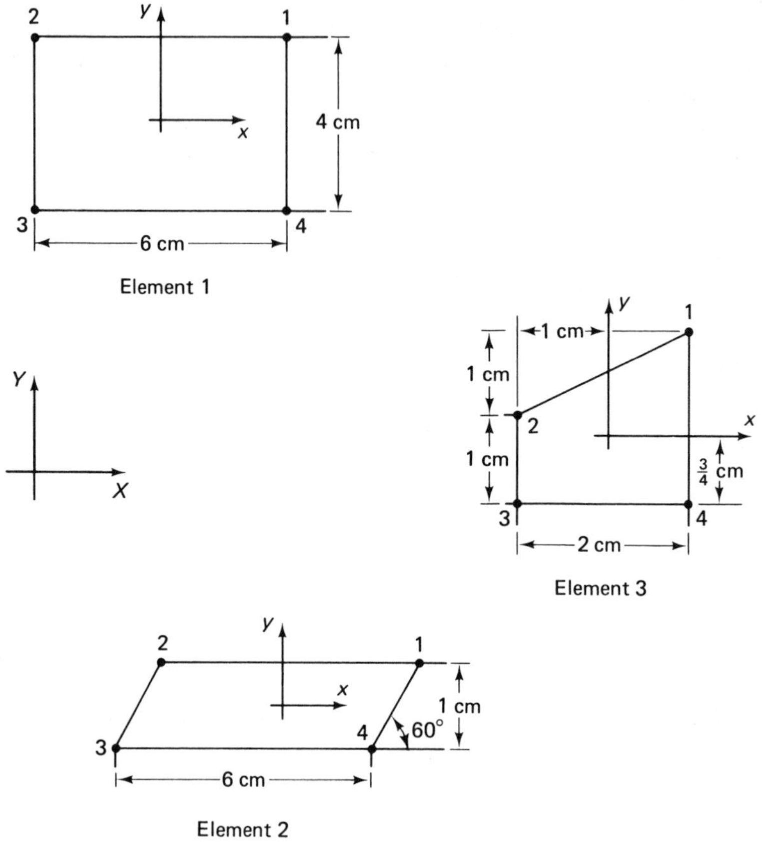

FIGURE 5.9 *Some two-dimensional elements.*

The Jacobian operator is the same for the global X-Y and the local x-y coordinate systems. For convenience we therefore use the local coordinate systems. Substituting into (5.18) and (5.23) using the interpolation functions given in Fig. 5.5, we obtain for element 1:

$$x = 3r; \qquad y = 2s$$

$$\mathbf{J} = \begin{bmatrix} 3 & 0 \\ 0 & 2 \end{bmatrix}$$

Similarly, for element 2, we have

$$\begin{aligned} x = \tfrac{1}{4}\{&(1 + r)(1 + s)(3 + 1/(2\sqrt{3})) + (1 - r)(1 + s)(-(3 - 1/(2\sqrt{3}))) \\ &+ (1 - r)(1 - s)(-(3 + 1/(2\sqrt{3}))) + (1 + r)(1 - s)(3 - 1/(2\sqrt{3}))\} \end{aligned}$$

$$\begin{aligned} y = \tfrac{1}{4}\{&(1 + r)(1 + s)(1/2) + (1 - r)(1 + s)(1/2) + (1 - r)(1 - s)(-1/2) \\ &+ (1 + r)(1 - s)(-1/2)\} \end{aligned}$$

and hence
$$\mathbf{J} = \begin{bmatrix} 3 & 0 \\ \dfrac{1}{2\sqrt{3}} & \dfrac{1}{2} \end{bmatrix}$$

Also, for element 3

$$x = \tfrac{1}{4}\{(1+r)(1+s)(1) + (1-r)(1+s)(-1) + (1-r)(1-s)(-1) + (1+r)(1-s)(+1)\}$$

$$y = \tfrac{1}{4}\{(1+r)(1+s)(5/4) + (1-r)(1+s)(1/4) + (1-r)(1-s)(-3/4) + (1+r)(1-s)(-3/4)\}$$

therefore,

$$\mathbf{J} = \frac{1}{4}\begin{bmatrix} 4 & (1+s) \\ 0 & (3+r) \end{bmatrix}$$

We may recognize that the Jacobian operator of a 2×2 square element is the identity matrix, and that the entries in the operator $\mathbf{J}$ of a general element express the amount of distortion from that 2×2 square element. Since the distortion is constant at any point (r, s) of elements 1 and 2 the operator $\mathbf{J}$ is constant for these elements.

EXAMPLE 5.4: Establish the interpolation functions of the two-dimensional element shown in Fig. 5.10.

The individual functions are obtained by combining the basic linear, parabolic, and cubic interpolations corresponding to the r and s directions. Thus using the functions in Figure 5.4 we obtain

$$\begin{aligned}
h_5 &= \{\tfrac{1}{16}(-27r^3 - 9r^2 + 27r + 9)\}\,\{\tfrac{1}{2}(1+s)\} \\
h_6 &= \{(1-r^2) + \tfrac{1}{16}(27r^3 + 7r^2 - 27r - 7)\}\,\{\tfrac{1}{2}(1+s)\} \\
h_2 &= \{\tfrac{1}{2}(1-r) - \tfrac{1}{2}(1-r^2) + \tfrac{1}{16}(-9r^3 + r^2 + 9r - 1)\}\,\{\tfrac{1}{2}(1+s)\} \\
h_3 &= \tfrac{1}{4}(1-r)(1-s) \\
h_7 &= \tfrac{1}{2}(1-s^2)(1+r) \\
h_4 &= \tfrac{1}{4}(1+r)(1-s) - \frac{h_7}{2} \\
h_1 &= \tfrac{1}{4}(1+r)(1+s) - \tfrac{2}{3}h_5 - \tfrac{1}{3}h_6 - \tfrac{1}{2}h_7
\end{aligned}$$

where h_1 is constructed as indicated in Fig. 5.10.

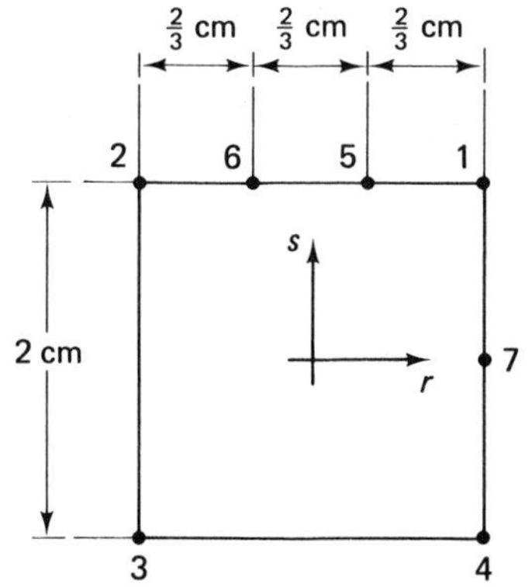

(a) Seven node element

FIGURE 5.10 *A 7-node element.*

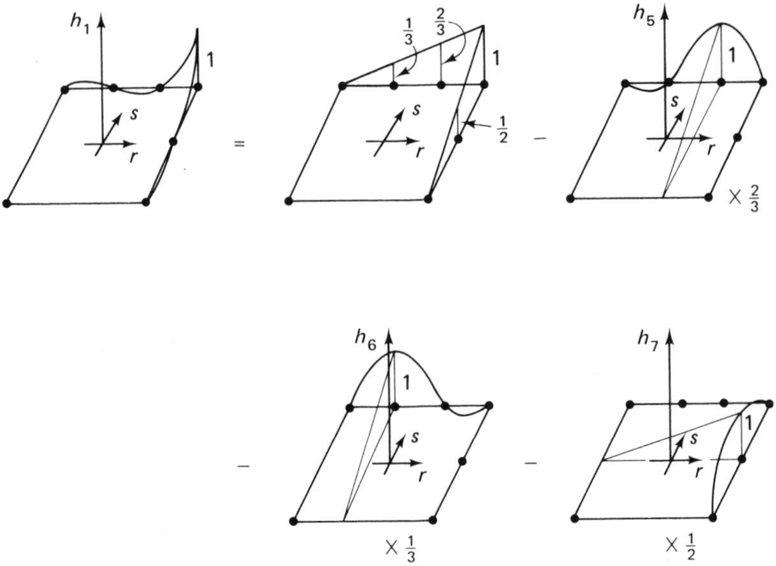

(b) Construction of h_1

FIGURE 5.10 *(cont.)*

EXAMPLE 5.5: Derive the expressions needed for the evaluation of the stiffness matrix of the isoparametric four-node finite element in Fig. 5.11. Assume plane stress or plane strain conditions.

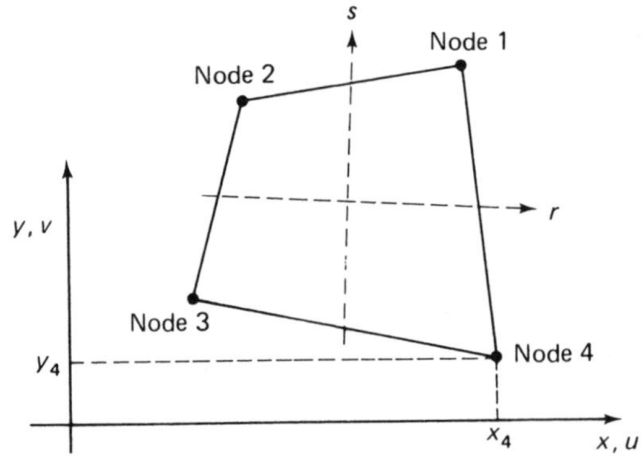

FIGURE 5.11 *Four-node two-dimensional element.*

Using the interpolation functions h_1, h_2, h_3, and h_4 defined in Fig. 5.5, the coordinate interpolation given in (5.18) is, for this element,

$$\begin{aligned} x &= \tfrac{1}{4}(1+r)(1+s)x_1 + \tfrac{1}{4}(1-r)(1+s)x_2 \\ &\quad + \tfrac{1}{4}(1-r)(1-s)x_3 + \tfrac{1}{4}(1+r)(1-s)x_4 \\ y &= \tfrac{1}{4}(1+r)(1+s)y_1 + \tfrac{1}{4}(1-r)(1+s)y_2 \\ &\quad + \tfrac{1}{4}(1-r)(1-s)y_3 + \tfrac{1}{4}(1+r)(1-s)y_4 \end{aligned}$$

The displacement interpolation given in (5.19) is

$$u = \tfrac{1}{4}(1+r)(1+s)u_1 + \tfrac{1}{4}(1-r)(1+s)u_2 + \tfrac{1}{4}(1-r)(1-s)u_3 + \tfrac{1}{4}(1+r)(1-s)u_4$$
$$v = \tfrac{1}{4}(1+r)(1+s)v_1 + \tfrac{1}{4}(1-r)(1+s)v_2 + \tfrac{1}{4}(1-r)(1-s)v_3 + \tfrac{1}{4}(1+r)(1-s)v_4$$

The element strains are given by

$$\boldsymbol{\epsilon}^T = [\epsilon_{xx} \quad \epsilon_{yy} \quad \gamma_{xy}]$$

where $$\epsilon_{xx} = \frac{\partial u}{\partial x}; \qquad \epsilon_{yy} = \frac{\partial v}{\partial y}; \qquad \gamma_{xy} = \frac{\partial u}{\partial y} + \frac{\partial v}{\partial x}$$

To evaluate the displacement derivatives, we need to evaluate (5.23):

$$\begin{bmatrix} \dfrac{\partial}{\partial r} \\ \dfrac{\partial}{\partial s} \end{bmatrix} = \begin{bmatrix} \dfrac{\partial x}{\partial r} & \dfrac{\partial y}{\partial r} \\ \dfrac{\partial x}{\partial s} & \dfrac{\partial y}{\partial s} \end{bmatrix} \begin{bmatrix} \dfrac{\partial}{\partial x} \\ \dfrac{\partial}{\partial y} \end{bmatrix} \quad \text{or} \quad \frac{\boldsymbol{\partial}}{\boldsymbol{\partial} \mathbf{r}} = \mathbf{J} \frac{\boldsymbol{\partial}}{\boldsymbol{\partial} \mathbf{x}}$$

where
$$\frac{\partial x}{\partial r} = \tfrac{1}{4}(1+s)x_1 - \tfrac{1}{4}(1+s)x_2 - \tfrac{1}{4}(1-s)x_3 + \tfrac{1}{4}(1-s)x_4$$
$$\frac{\partial x}{\partial s} = \tfrac{1}{4}(1+r)x_1 + \tfrac{1}{4}(1-r)x_2 - \tfrac{1}{4}(1-r)x_3 - \tfrac{1}{4}(1+r)x_4$$
$$\frac{\partial y}{\partial r} = \tfrac{1}{4}(1+s)y_1 - \tfrac{1}{4}(1+s)y_2 - \tfrac{1}{4}(1-s)y_3 + \tfrac{1}{4}(1-s)y_4$$
$$\frac{\partial y}{\partial s} = \tfrac{1}{4}(1+r)y_1 + \tfrac{1}{4}(1-r)y_2 - \tfrac{1}{4}(1-r)y_3 - \tfrac{1}{4}(1+r)y_4$$

Therefore, for any value r and s, $-1 \leq r \leq +1$ and $-1 \leq s \leq +1$, we can form the Jacobian operator $\mathbf{J}$ by using the above expressions for $\partial x/\partial r$, $\partial x/\partial s$, and $\partial y/\partial r$, $\partial y/\partial s$. Assume that we evaluate $\mathbf{J}$ at $r = r_i$ and $s = s_j$ and denote the operator by $\mathbf{J}_{ij}$ and its determinant by det $\mathbf{J}_{ij}$. Then we have

$$\begin{bmatrix} \dfrac{\partial}{\partial x} \\ \dfrac{\partial}{\partial y} \end{bmatrix}_{\substack{\text{at } r=r_i \\ s=s_j}} = \mathbf{J}_{ij}^{-1} \begin{bmatrix} \dfrac{\partial}{\partial r} \\ \dfrac{\partial}{\partial s} \end{bmatrix}_{\substack{\text{at } r=r_i \\ s=s_j}}$$

To evaluate the element strains we use

$$\frac{\partial u}{\partial r} = \tfrac{1}{4}(1+s)u_1 - \tfrac{1}{4}(1+s)u_2 - \tfrac{1}{4}(1-s)u_3 + \tfrac{1}{4}(1-s)u_4$$
$$\frac{\partial u}{\partial s} = \tfrac{1}{4}(1+r)u_1 + \tfrac{1}{4}(1-r)u_2 - \tfrac{1}{4}(1-r)u_3 - \tfrac{1}{4}(1+r)u_4$$
$$\frac{\partial v}{\partial r} = \tfrac{1}{4}(1+s)v_1 - \tfrac{1}{4}(1+s)v_2 - \tfrac{1}{4}(1-s)v_3 + \tfrac{1}{4}(1-s)v_4$$
$$\frac{\partial v}{\partial s} = \tfrac{1}{4}(1+r)v_1 + \tfrac{1}{4}(1-r)v_2 - \tfrac{1}{4}(1-r)v_3 - \tfrac{1}{4}(1+r)v_4$$

Therefore,

$$\begin{bmatrix} \dfrac{\partial u}{\partial x} \\ \dfrac{\partial u}{\partial y} \end{bmatrix}_{\substack{\text{at } r=r_i \\ s=s_j}} = \tfrac{1}{4}\mathbf{J}_{ij}^{-1} \begin{bmatrix} 1+s_j & 0 & -(1+s_j) & 0 & -(1-s_j) & 0 & 1-s_j & 0 \\ 1+r_i & 0 & 1-r_i & 0 & -(1-r_i) & 0 & -(1+r_i) & 0 \end{bmatrix} \hat{\mathbf{u}} \quad \text{(a)}$$

and

$$\begin{bmatrix} \dfrac{\partial v}{\partial x} \\ \\ \dfrac{\partial v}{\partial y} \end{bmatrix}_{\substack{\text{at } r=r_i \\ s=s_j}} = \tfrac{1}{4}\mathbf{J}_{ij}^{-1}\begin{bmatrix} 0 & 1+s_j & 0 & -(1+s_j) & 0 & -(1-s_j) & 0 & 1-s_j \\ 0 & 1+r_i & 0 & 1-r_i & 0 & -(1-r_i) & 0 & -(1+r_i) \end{bmatrix}\hat{\mathbf{u}} \quad \text{(b)}$$

where $$\hat{\mathbf{u}}^T = [u_1 \quad v_1 \quad u_2 \quad v_2 \quad u_3 \quad v_3 \quad u_4 \quad v_4]$$

Evaluating the relations in (a) and (b), we can establish the strain-displacement transformation matrix at the point (r_i, s_j); i.e., we obtain

$$\boldsymbol{\epsilon}_{ij} = \mathbf{B}_{ij}\hat{\mathbf{u}}$$

where the subscripts i and j indicate that the strain-displacement transformation is evaluated at the points r_i and s_j. For example, if $x = r$, $y = s$, i.e., the stiffness matrix of a square element is required that has side lengths equal to 2, the Jacobian operator is the identity matrix, and hence

$$\mathbf{B}_{ij} = \tfrac{1}{4}\begin{bmatrix} 1+s_j & 0 & -(1+s_j) & 0 & -(1-s_j) & 0 & 1-s_j & 0 \\ 0 & 1+r_i & 0 & 1-r_i & 0 & -(1-r_i) & 0 & -(1+r_i) \\ 1+r_i & 1+s_j & 1-r_i & -(1+s_j) & -(1-r_i) & -(1-s_j) & -(1+r_i) & 1-s_j \end{bmatrix}$$

The matrix $\mathbf{F}_{ij}$ in (5.30) is now simply

$$\mathbf{F}_{ij} = \mathbf{B}_{ij}^T\mathbf{C}\mathbf{B}_{ij} \det \mathbf{J}_{ij}$$

where the material property matrix $\mathbf{C}$ is given in Table 4.3. In the case of plane stress or plane strain conditions, we integrate in the r, s plane and assume that the function $\mathbf{F}$ is constant through the thickness of the element. The stiffness matrix of the element is therefore

$$\mathbf{K} = \sum_{i,j} t_{ij}\, \alpha_{ij}\, \mathbf{F}_{ij}$$

where t_{ij} is the thickness of the element at the sampling point r_i, s_j ($t_{ij} = 1.0$ in plane strain analysis). With the matrices $\mathbf{F}_{ij}$ given as above and the weighting factors α_{ij} being determined, the required stiffness matrix can readily be evaluated.

For the actual implementation it should be noted that in the evaluation of $\mathbf{J}_{ij}$ and of the matrices defining the displacement derivatives in (a) and (b), only the eight possible derivatives of the interpolation functions $h_1, \ldots, h_4$ are required. Therefore, it is expedient to calculate these derivatives corresponding to the point (r_i, s_j) once at the start of the evaluation of $\mathbf{B}_{ij}$ and use them whenever they are required.

It should also be realized that considering the specific point (r_i, s_j), the relations in (a) and (b) may be written, respectively, as

$$\left.\begin{aligned} \frac{\partial u}{\partial x} &= \sum_{i=1}^{4} \frac{\partial h_i}{\partial x} u_i \\ \frac{\partial u}{\partial y} &= \sum_{i=1}^{4} \frac{\partial h_i}{\partial y} u_i \end{aligned}\right\} \quad \text{(c)}$$

and

$$\left.\begin{aligned} \frac{\partial v}{\partial x} &= \sum_{i=1}^{4} \frac{\partial h_i}{\partial x} v_i \\ \frac{\partial v}{\partial y} &= \sum_{i=1}^{4} \frac{\partial h_i}{\partial y} v_i \end{aligned}\right\} \quad \text{(d)}$$

Hence we have

$$\mathbf{B} = \begin{bmatrix} \frac{\partial h_1}{\partial x} & 0 & \frac{\partial h_2}{\partial x} & 0 & \frac{\partial h_3}{\partial x} & 0 & \frac{\partial h_4}{\partial x} & 0 \\ 0 & \frac{\partial h_1}{\partial y} & 0 & \frac{\partial h_2}{\partial y} & 0 & \frac{\partial h_3}{\partial y} & 0 & \frac{\partial h_4}{\partial y} \\ \frac{\partial h_1}{\partial y} & \frac{\partial h_1}{\partial x} & \frac{\partial h_2}{\partial y} & \frac{\partial h_2}{\partial x} & \frac{\partial h_3}{\partial y} & \frac{\partial h_3}{\partial x} & \frac{\partial h_4}{\partial y} & \frac{\partial h_4}{\partial x} \end{bmatrix} \qquad \text{(e)}$$

where it is implied that in (c) and (d), the derivatives are evaluated at point (r_i, s_j), and therefore in (e), we have, in fact, the matrix $\mathbf{B}_{ij}$.

EXAMPLE 5.6: Derive the expressions needed for the evaluation of the mass matrix of the element considered in Example 5.5.

The mass matrix of the element is given by

$$\mathbf{M} = \sum_{i,j} \alpha_{ij}\ t_{ij}\ \mathbf{F}_{ij}$$

where

$$\mathbf{F}_{ij} = \rho_{ij}\mathbf{H}_{ij}^T\mathbf{H}_{ij} \det \mathbf{J}_{ij}$$

and $\mathbf{H}_{ij}$ is the displacement interpolation matrix. The displacement interpolation functions for u and v of the four-node element have been given in Example 5.5, and we have

$$\mathbf{H}_{ij} = \tfrac{1}{4}\begin{bmatrix} (1+r_i)(1+s_j) & 0 & (1-r_i)(1+s_j) & 0 & (1-r_i)(1-s_j) & 0 & (1+r_i)(1-s_j) & 0 \\ 0 & (1+r_i)(1+s_j) & 0 & (1-r_i)(1+s_j) & 0 & (1-r_i)(1-s_j) & 0 & (1+r_i)(1-s_j) \end{bmatrix}$$

The determinant of the Jacobian matrix, $\det \mathbf{J}_{ij}$, was given in Example 5.5, and ρ_{ij} is the mass density at the sampling point r_i, s_j. Therefore, all required variables for the evaluation of the mass matrix have been defined.

EXAMPLE 5.7: Derive the expressions needed for the evaluation of the body force vector $\mathbf{R}_B$ and the initial stress vector $\mathbf{R}_I$ of the element considered in Example 5.5.

These vectors are obtained using the matrices $\mathbf{H}_{ij}$, $\mathbf{B}_{ij}$, and $\mathbf{J}_{ij}$ defined in Examples 5.5 and 5.6; i.e., we have

$$\mathbf{R}_B = \sum_{i,j} \alpha_{ij}\ t_{ij}\ \mathbf{H}_{ij}^T\ \mathbf{f}_{ij}^B \det \mathbf{J}_{ij}$$

$$\mathbf{R}_I = \sum_{i,j} \alpha_{ij}\ t_{ij}\ \mathbf{B}_{ij}^T\ \boldsymbol{\tau}_{ij}^I \det \mathbf{J}_{ij}$$

where $\mathbf{f}_{ij}^B$ and $\boldsymbol{\tau}_{ij}^I$ are the body force vector and initial stress vector evaluated at the integration sampling points.

EXAMPLE 5.8: Derive the expressions needed in the calculation of the surface force vector $\mathbf{R}_S$, when the element edge 1-2 of the four-node isoparametric element considered in Example 5.5 is loaded as shown in Fig. 5.12.

The first step is to establish the displacement interpolations. Since $s = +1$ at the edge 1-2, we have, using the interpolation functions given in Example 5.5,

$$u^S = \tfrac{1}{2}(1+r)u_1 + \tfrac{1}{2}(1-r)u_2$$
$$v^S = \tfrac{1}{2}(1+r)v_1 + \tfrac{1}{2}(1-r)v_2$$

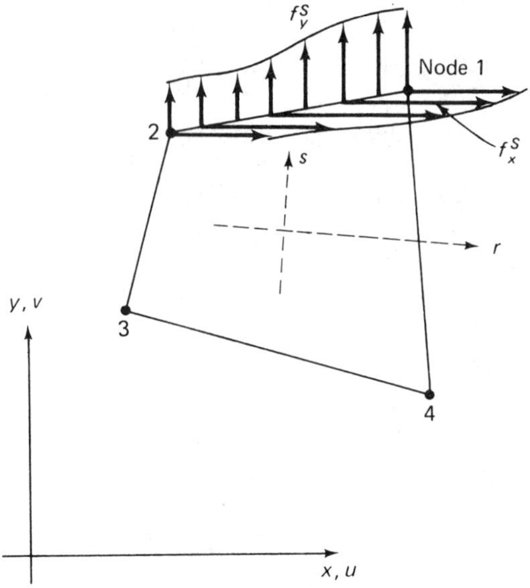

FIGURE 5.12 Pressure distribution along edge 1-2 of a four-node element.

Hence to evaluate $\mathbf{R}_S$ in (5.34) we can use

$$\mathbf{H}^S = \begin{bmatrix} \frac{1}{2}(1+r) & 0 & \frac{1}{2}(1-r) & 0 & 0 & 0 & 0 & 0 \\ 0 & \frac{1}{2}(1+r) & 0 & \frac{1}{2}(1-r) & 0 & 0 & 0 & 0 \end{bmatrix}$$

and $$\mathbf{f}^S = \begin{bmatrix} f_x^S \\ f_y^S \end{bmatrix}$$

where f_x^S and f_y^S are the x and y components of the applied surface force. These components may have been given as a function of r.

For the evaluation of the integral in (5.34), we also need the differential surface area dS expressed in the r, s natural coordinate system. If t_r is the thickness, $dS = t_r\, dl$, where dl is a differential length,

$$dl = \det \mathbf{J}^S\, dr; \qquad \det \mathbf{J}^S = \left\{ \left(\frac{\partial x}{\partial r} \right)^2 + \left(\frac{\partial y}{\partial r} \right)^2 \right\}^{1/2}$$

But the derivatives $\partial x/\partial r$ and $\partial y/\partial r$ have been given in Example 5.5. Using $s = +1$, we have, in this case,

$$\frac{\partial x}{\partial r} = \frac{x_1 - x_2}{2}; \qquad \frac{\partial y}{\partial r} = \frac{y_1 - y_2}{2}$$

Although the vector $\mathbf{R}_S$ could in this case be evaluated in a closed form solution (provided that the functions used in $\mathbf{f}^S$ are simple), in order to keep generality in the program that calculates $\mathbf{R}_S$, it is expedient to use numerical integration. This way, variable-number-nodes elements can be implemented in an elegant manner in one program. Thus, using the notation defined in this section, we have

$$\mathbf{R}_S = \sum_i \alpha_i t_i \mathbf{F}_i$$

$$\mathbf{F}_i = \mathbf{H}_i^{S^T} \mathbf{f}_i^S \det \mathbf{J}_i^S$$

It is noted that in this case only one-dimensional numerical integration is required, because s is not a variable.

EXAMPLE 5.9: Explain how the expressions given in Examples 5.5 to 5.7 need be modified when the element considered is an axisymmetric element.

In this case two modifications are necessary. Firstly, we consider one radian of the structure. Hence, the thickness to be employed in all integrations is that corresponding to one radian, which means that at an integration point the thickness is equal to the radius at that point:

$$t_{ij} = \sum_{k=1}^{4} h_k \Big|_{r_i s_j} x_k \tag{a}$$

Secondly, it is recognized that also circumferential strains and stresses are developed (see Table 4.2). Hence, the strain-displacement matrix must be augmented by one row for the hoop strain u/R, i.e. we have

$$\mathbf{B} = \begin{bmatrix} \cdots & & & & & & & \cdots \\ \frac{h_1}{t} & 0 & \frac{h_2}{t} & 0 & \frac{h_3}{t} & 0 & \frac{h_4}{t} & 0 \end{bmatrix} \tag{b}$$

where the first three rows have already been defined in Example 5.5 and t is equal to the radius. To obtain the strain-displacement matrix at integration point (i, j) we use (a) to evaluate t and substitute into (b).

EXAMPLE 5.10: Calculate the nodal point forces of the four-node axisymmetric finite element shown in Fig. 5.13 when the element is subjected to centrifugal loading.

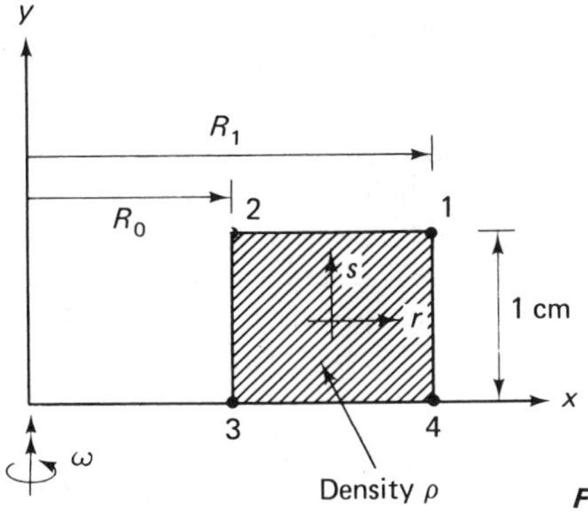

FIGURE 5.13 Four-node axisymmetric element rotating at angular velocity ω (rad/sec)

Here we want to evaluate

$$\mathbf{R}_B = \int_V \mathbf{H}^T \mathbf{f}^B \, dV$$

where

$$f_x^B = \rho\omega^2 R, \qquad f_y^B = 0 \qquad y = \tfrac{1}{2}(1+s)$$

$$R = \tfrac{1}{2}(1-r)R_0 + \tfrac{1}{2}(1+r)R_1 = x$$

(1)

$$\mathbf{H} = \begin{bmatrix} h_1 & 0 & h_2 & 0 & h_3 & 0 & h_4 & 0 \\ 0 & h_1 & 0 & h_2 & 0 & h_3 & 0 & h_4 \end{bmatrix}; \qquad \mathbf{J} = \begin{bmatrix} \frac{R_1 - R_0}{2} & 0 \\ 0 & \frac{1}{2} \end{bmatrix}$$

and the h_i are defined in Fig. 5.5. Also, considering one radian

$$dV = \det \mathbf{J} \, dr \, ds \, R = \left(\frac{R_1 - R_0}{4}\right) dr \, ds \left[\frac{R_1 + R_0}{2} + \frac{R_1 - R_0}{2} r\right]$$

(2) ditte $t = R.\alpha$ $(\alpha = 1)$

Hence

$$\mathbf{R}_B = \frac{\rho\omega^2(R_1 - R_0)}{64}\int_{r=-1}^{+1}\int_{s=-1}^{+1}\begin{bmatrix}(1+r)(1+s) & 0\\ 0 & (1+r)(1+s)\\ (1-r)(1+s) & 0\\ 0 & (1-r)(1+s)\\ (1-r)(1-s) & 0\\ 0 & (1-r)(1-s)\\ (1+r)(1-s) & 0\\ 0 & (1+r)(1-s)\end{bmatrix}$$

$$[(R_1 + R_0) + (R_1 - R_0)r]^2\begin{bmatrix}1\\0\end{bmatrix} dr\, ds$$

If we let $A = R_1 + R_0$ and $B = R_1 - R_0$, we have

$$\mathbf{R}_B = \frac{\rho\omega^2 B}{64}\begin{bmatrix}\frac{2}{3}(6A^2 + 4AB + 2B^2)\\ 0\\ \frac{2}{3}(6A^2 - 4AB + 2B^2)\\ 0\\ \frac{2}{3}(6A^2 - 4AB + 2B^2)\\ 0\\ \frac{2}{3}(6A^2 + 4AB + 2B^2)\\ 0\end{bmatrix}$$

EXAMPLE 5.11: The four-node plane stress element shown in Fig. 5.14 is subjected to the given temperature distribution. If the temperature corresponding to the stress-free state is θ_0, evaluate the nodal point forces to which the element must be subjected in order that there are no nodal point displacements.

In this case we have for the total stresses, due to mechanical strains $\boldsymbol{\epsilon}$ and thermal strains $\boldsymbol{\epsilon}^{\text{th}}$,

$$\boldsymbol{\tau} = \mathbf{C}(\boldsymbol{\epsilon} - \boldsymbol{\epsilon}^{\text{th}}) \tag{a}$$

where $\epsilon_{xx}^{\text{th}} = \alpha(\theta - \theta_0)$, $\epsilon_{yy}^{\text{th}} = \alpha(\theta - \theta_0)$, $\gamma_{xy}^{\text{th}} = 0$. If the nodal point displacements are zero, we have $\boldsymbol{\epsilon} = \mathbf{0}$, and the stresses due to the thermal strains can be thought of as initial stresses. Thus, the nodal point forces are

$$\mathbf{R}_I = \int_V \mathbf{B}^T \boldsymbol{\tau}^I \, dV$$

$$\boldsymbol{\tau}^I = -\frac{E\alpha}{1-\nu^2}\begin{bmatrix}1 & \nu & 0\\ \nu & 1 & 0\\ 0 & 0 & \frac{1-\nu}{2}\end{bmatrix}\begin{bmatrix}1\\1\\0\end{bmatrix}\left\{\left(\sum_{i=1}^{4} h_i\theta_i\right) - \theta_0\right\}$$

and the h_i are the interpolation functions defined in Fig. 5.5. Also,

$$\mathbf{J} = \begin{bmatrix}2 & 0\\ 0 & 1.5\end{bmatrix}; \qquad \mathbf{J}^{-1} = \begin{bmatrix}\frac{1}{2} & 0\\ 0 & \frac{2}{3}\end{bmatrix}; \qquad \det \mathbf{J} = 3$$

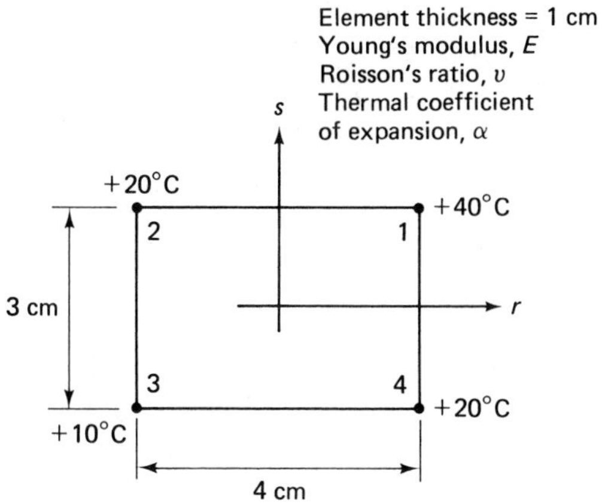

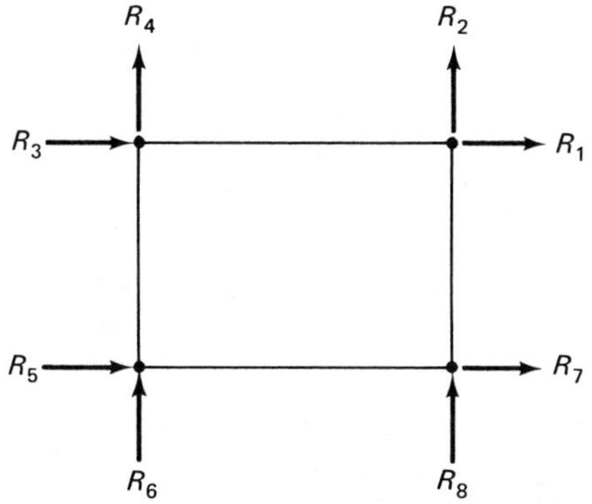

FIGURE 5.14 *Nodal point forces due to initial temperature distribution.*

$$
\mathbf{B} = \begin{bmatrix} \frac{(1+s)}{8} & 0 & -\frac{(1+s)}{8} & 0 & -\frac{(1-s)}{8} & 0 & \frac{(1-s)}{8} & 0 \\ 0 & \frac{(1+r)}{6} & 0 & \frac{(1-r)}{6} & 0 & -\frac{(1-r)}{6} & 0 & -\frac{(1+r)}{6} \\ \frac{(1+r)}{6} & \frac{(1+s)}{8} & \frac{(1-r)}{6} & -\frac{(1+s)}{8} & -\frac{(1-r)}{6} & -\frac{(1-s)}{8} & -\frac{(1+r)}{6} & \frac{(1-s)}{8} \end{bmatrix}
$$

Hence

$$\mathbf{R}_I = \int_{-1}^{+1}\int_{-1}^{+1} - \begin{bmatrix} \frac{(1+s)}{8} & 0 & \frac{(1+r)}{6} \\ 0 & \frac{(1+r)}{6} & \frac{(1+s)}{8} \\ -\frac{(1+s)}{8} & 0 & \frac{(1-r)}{6} \\ 0 & \frac{(1-r)}{6} & -\frac{(1+s)}{8} \\ -\frac{(1-s)}{8} & 0 & -\frac{(1-r)}{6} \\ 0 & -\frac{(1-r)}{6} & -\frac{(1-s)}{8} \\ \frac{(1-s)}{8} & 0 & -\frac{(1+r)}{6} \\ 0 & -\frac{(1+r)}{6} & \frac{(1-s)}{8} \end{bmatrix} \begin{bmatrix} (1+\nu) \\ (1+\nu) \\ 0 \end{bmatrix} \frac{E\alpha}{1-\nu^2}$$

$$[2.5(s+3)(r+3) - \theta_0]3\, dr\, ds$$

$$\mathbf{R}_I = -\frac{E\alpha}{(1-\nu)} \begin{bmatrix} 37.5 - 1.5\theta_0 \\ 50 - 2\theta_0 \\ -37.5 + 1.5\theta_0 \\ 40 - 2\theta_0 \\ -30 + 1.5\theta_0 \\ -40 + 2\theta_0 \\ +30 - 1.5\theta_0 \\ -50 + 2\theta_0 \end{bmatrix}$$

It should be noted that the calculation of the initial stress force vector as performed above is one typical step in a thermal stress analysis. In a complete thermal stress analysis the temperatures are calculated as described in Section 7.2, the element load vectors due to the thermal effects are evaluated as illustrated in this example, and the solution of the equilibrium equations of the complete element assemblage then yields the nodal point displacements. The element mechanical strains $\boldsymbol{\epsilon}$ are evaluated from the nodal point displacements and then using (a) the final element stresses are calculated.

EXAMPLE 5.12: Consider the elements in Fig. 5.15. Evaluate the consistent nodal point forces corresponding to the surface loading (assuming that the nodal point forces are positive when acting into the direction of the pressure).

Here we want to evaluate

$$\mathbf{R}_S = \int_S \mathbf{H}^{S^T}\mathbf{f}^S\, dS$$

Consider first the two-dimensional interpolations. Since $s = +1$ at the edge 1-2, we have, using the interpolation functions for the eight-node element (see Fig. 5.5),

$$\begin{aligned} h_5 &= \tfrac{1}{2}(1-r^2)(1+s)|_{s=+1} = (1-r^2) \\ h_1 &= \tfrac{1}{4}(1+r)(1+s)(r+s-1)|_{s=+1} = \tfrac{1}{2}r(1+r) \\ h_2 &= \tfrac{1}{4}(1-r)(1+s)(s-r-1)|_{s=+1} = -\tfrac{1}{2}r(1-r) \end{aligned}$$

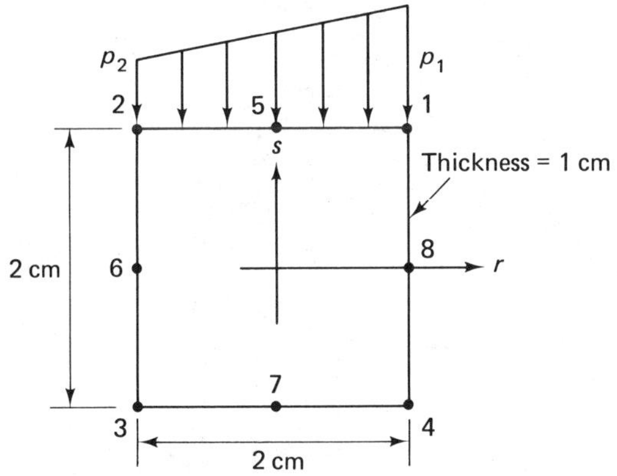

(a) Two-dimensional element subjected to linearly varying pressure along one side

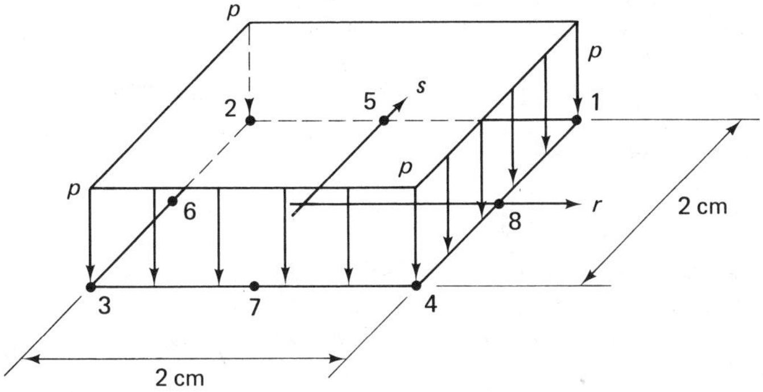

(b) Flat surface of three-dimensional element subjected to constant pressure p

FIGURE 5.15 *Two- and three-dimensional elements subjected to pressure loading.*

which are equal to the interpolation functions of the 3-node bar in Fig. 5.3. Hence

$$\begin{bmatrix} u^S \\ v^S \end{bmatrix} = \begin{bmatrix} \frac{1}{2}r(1+r) & 0 & -\frac{1}{2}r(1-r) & 0 & (1-r^2) & 0 \\ 0 & \frac{1}{2}r(1+r) & 0 & -\frac{1}{2}r(1-r) & 0 & (1-r^2) \end{bmatrix} \begin{bmatrix} u_1 \\ v_1 \\ u_2 \\ v_2 \\ u_5 \\ v_5 \end{bmatrix}$$

Also $$\mathbf{f}^S = \begin{bmatrix} f_r^S \\ f_s^S \end{bmatrix} = \begin{bmatrix} 0 \\ \frac{1}{2}(1+r)p_1 + \frac{1}{2}(1-r)p_2 \end{bmatrix}; \qquad \det \mathbf{J}^S = 1$$

Hence

$$\mathbf{R}_S = \int_{-1}^{+1} \frac{t}{2} \begin{bmatrix} r(1+r) & \\ 0 & r(1+r) \\ -r(1-r) & 0 \\ 0 & -r(1-r) \\ 2(1-r^2) & 0 \\ 0 & 2(1-r^2) \end{bmatrix} \frac{1}{2} \begin{bmatrix} 0 \\ (1+r)p_1 + (1-r)p_2 \end{bmatrix} dr$$

$$\mathbf{R}_S = \frac{1}{3} \begin{bmatrix} 0 \\ p_1 \\ 0 \\ p_2 \\ 0 \\ 2(p_1 + p_2) \end{bmatrix} \tag{a}$$

For the three-dimensional element we proceed similarly. Since the surface is flat and the loading is normal to it, only the nodal point forces normal to the surface are nonzero (see also (a) above). Also, by symmetry, we know that the forces at nodes 1, 2, 3, 4 and 5, 6, 7, 8 are equal, respectively. Using the interpolation functions of Fig. 5.5, we have for the force at node 1:

$$R_1 = p \int_{-1}^{+1} \int_{-1}^{+1} \tfrac{1}{4}(1+r)(1+s)(r+s-1)\, dr\, ds = -\tfrac{1}{3}p$$

and for the force at node 5,

$$R_5 = p \int_{-1}^{+1} \int_{-1}^{+1} \tfrac{1}{2}(1-r^2)(1+s)\, dr\, ds = \tfrac{4}{3}p$$

The total pressure loading on the surface is $4p$, which, as a check, is equal to the sum of all the nodal point forces. However, it should be noted that the consistent nodal point forces at the corners of the element act into the opposite direction of the pressure!

EXAMPLE 5.13: Calculate the deflection u_A of the structural model shown in Fig. 5.16.

Because of the symmetry and boundary conditions we only need to evaluate the stiffness coefficient corresponding to u_A. Here we have for the four-node element

$$\mathbf{J} = \begin{bmatrix} 4 & 0 \\ 0 & 3 \end{bmatrix}; \qquad \mathbf{B} = \frac{1}{48} \begin{bmatrix} & 3(1-s) & \\ \cdots & 0 & \cdots \\ & -4(1+r) & \end{bmatrix}$$

$$k_{77} = \int_{-1}^{+1} \int_{-1}^{+1} \left(\frac{1}{48}\right)^2 \frac{E}{1-\nu^2} [3(1-s) \mid 0 \mid -4(1+r)] \begin{bmatrix} 3(1-s) \\ 3\nu(1-s) \\ -2(1-\nu)(1+r) \end{bmatrix} (12)(0.1)\, dr\, ds$$

or $$k_{77} = 1{,}336{,}996.34 \text{ N/cm}$$

Also, the stiffness of the truss is AE/L, or

$$k = \frac{(1)(30{,}000)}{8} = 3{,}750{,}000 \text{ N/cm}$$

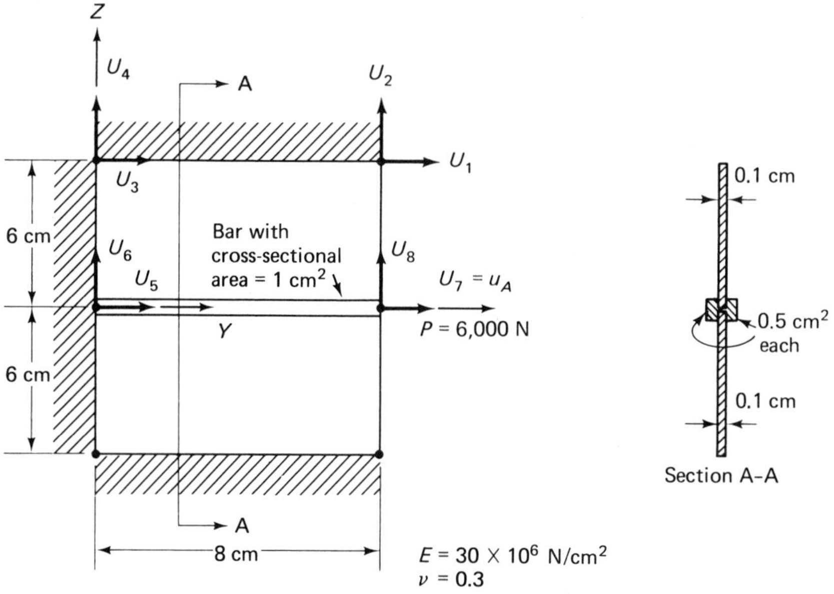

FIGURE 5.16 *A simple structural model.*

Hence $$k_{\text{total}} = 6.424 \times 10^6 \text{ N/cm}$$

and $$u_A = 9.34 \times 10^{-4} \text{ cm}$$

EXAMPLE 5.14: Consider the five-node element in Fig. 5.17. Evaluate the consistent nodal point forces corresponding to the stresses given.

Using the interpolation functions of Fig. 5.5, we can evaluate the strain-displacement matrix of the element:

$$\mathbf{B} = \frac{1}{8}\left[\begin{array}{cccccccccc} (1+s) & 0 & -s(1+s) & 0 & s(1-s) & 0 & (1-s) & 0 & -2(1-s^2) & 0 \\ 0 & 2(1+r) & 0 & 2(1-r)(1+2s) & 0 & -2(1-r)(1-2s) & 0 & -2(1+r) & 0 & -8(1-r)s \\ 2(1+r) & (1+s) & 2(1-r)(1+2s) & -s(1+s) & -2(1-r)(1-2s) & s(1-s) & -2(1+r) & (1-s) & -8(1-r)s & -2(1-s^2) \end{array}\right]$$

where we used that $$\mathbf{J} = \begin{bmatrix} 2 & 0 \\ 0 & 1 \end{bmatrix}$$

The required nodal point forces can now be evaluated using (5.35); hence

$$\mathbf{R}_I = \int_{-1}^{+1}\int_{-1}^{+1} \mathbf{B}^T \begin{bmatrix} 0 \\ 10 \\ 20 \end{bmatrix} (2)\, dr\, ds$$

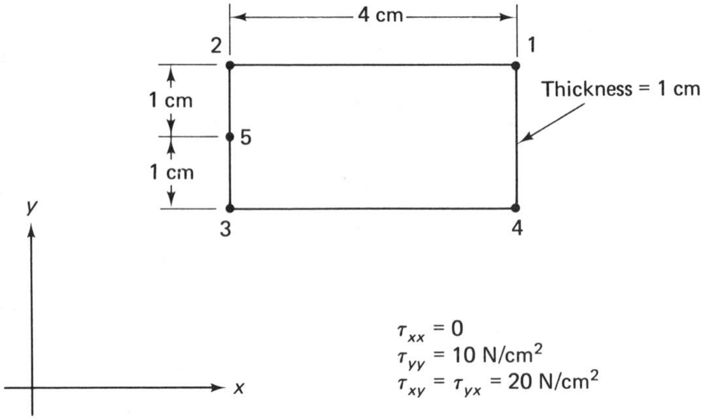

FIGURE 5.17 *Five-node element with stresses given.*

which gives

$$\mathbf{R}_I^T = [40 \quad 40 \quad 40 \quad \tfrac{40}{3} \quad -40 \quad -\tfrac{80}{3} \quad -40 \quad 0 \quad 0 \quad -\tfrac{80}{3}]$$

It should be noted that the forces in this vector are also equal to the nodal point consistent forces that correspond to the (constant) surface stresses, which are in equilibrium with the internal stresses given in Fig. 5.17.

5.3.2 Triangular Elements

In the previous section we discussed rectangular isoparametric elements which can be used to model very general geometries. However, in some cases the use of triangular or wedge elements may be attractive.

Since the elements discussed in Section 5.3.1 can be distorted, as shown for example in Fig. 5.2, a natural way of generating triangular elements appears to be to simply distort the basic rectangular element into the required triangular form, see Fig. 5.18. This is achieved in practice by assigning the same global node to two corner nodes of the element. We demonstrate this procedure in the following example.

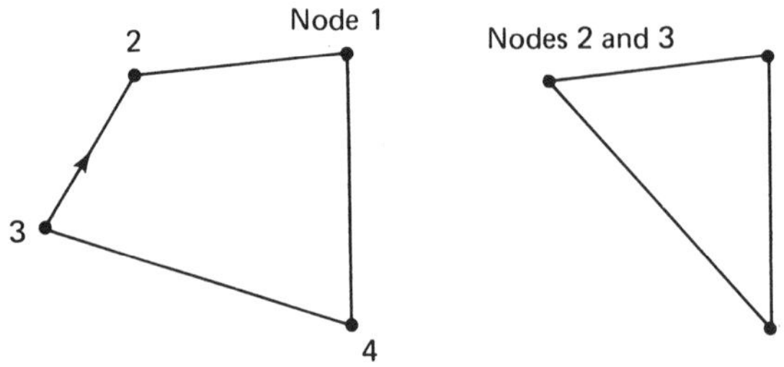

(a) Degeneration of four-node to three-node two-dimensional element

FIGURE 5.18 *Degenerate forms of four- and eight-node elements of Figs. 5.5 and 5.6.*

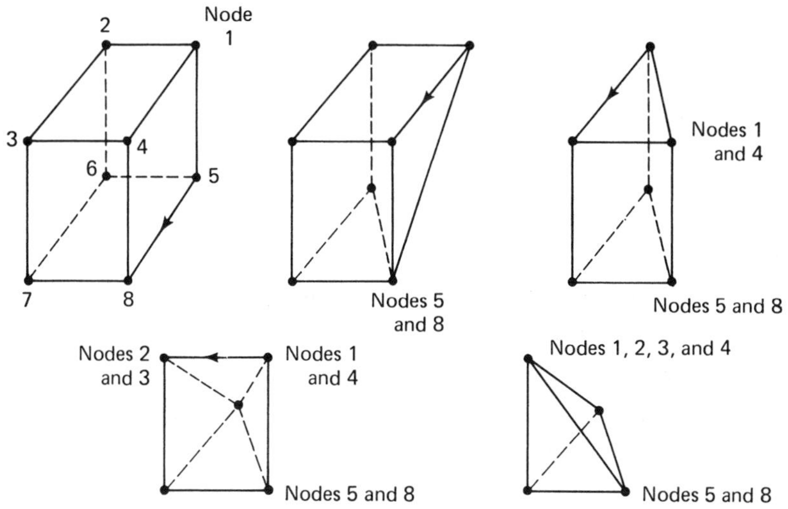

(b) Degenerate forms of eight-node three-dimensional element

EXAMPLE 5.15: Show that by collapsing the side 1-2 of the four-node quadrilateral element in Fig. 5.19 a constant strain triangle is obtained.

Using the interpolation functions of Fig. 5.5, we have

$$
\begin{aligned}
x &= \tfrac{1}{4}(1 + r)(1 + s)x_1 + \tfrac{1}{4}(1 - r)(1 + s)x_2 + \tfrac{1}{4}(1 - r)(1 - s)x_3 \\
&\quad + \tfrac{1}{4}(1 + r)(1 - s)x_4 \\
y &= \tfrac{1}{4}(1 + r)(1 + s)y_1 + \tfrac{1}{4}(1 - r)(1 + s)y_2 + \tfrac{1}{4}(1 - r)(1 - s)y_3 \\
&\quad + \tfrac{1}{4}(1 + r)(1 - s)y_4
\end{aligned}
$$

Thus, using the conditions $x_1 = x_2$ and $y_1 = y_2$ we obtain

$$
\begin{aligned}
x &= \tfrac{1}{2}(1 + s)x_2 + \tfrac{1}{4}(1 - r)(1 - s)x_3 + \tfrac{1}{4}(1 + r)(1 - s)x_4 \\
y &= \tfrac{1}{2}(1 + s)y_2 + \tfrac{1}{4}(1 - r)(1 - s)y_3 + \tfrac{1}{4}(1 + r)(1 - s)y_4
\end{aligned}
$$

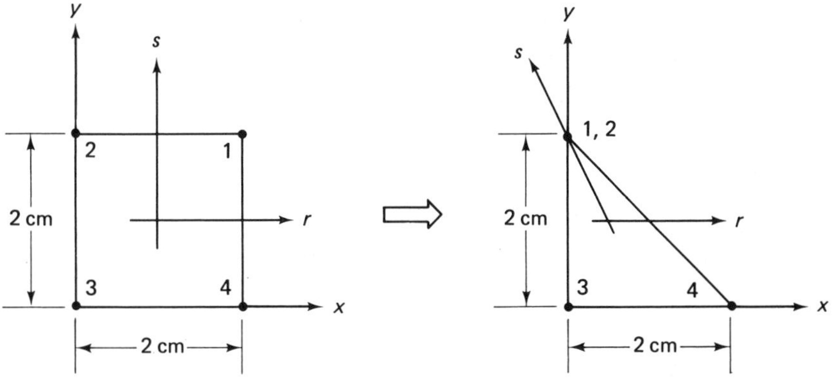

FIGURE 5.19 *Collapsing a four-node element to a triangular element.*

and hence with the nodal coordinates given in Fig. 5.19

$$x = \tfrac{1}{2}(1 + r)(1 - s)$$
$$y = (1 + s)$$

It follows that

$$\frac{\partial x}{\partial r} = \frac{1}{2}(1 - s) \qquad \frac{\partial y}{\partial r} = 0 \quad ; \qquad \mathbf{J} = \frac{1}{2}\begin{bmatrix} (1 - s) & 0 \\ -(1 + r) & 2 \end{bmatrix}; \qquad \mathbf{J}^{-1} = \begin{bmatrix} \frac{2}{1 - s} & 0 \\ \frac{1 + r}{1 - s} & 1 \end{bmatrix}$$
$$\frac{\partial x}{\partial s} = -\frac{1}{2}(1 + r) \qquad \frac{\partial y}{\partial s} = 1$$

Using the isoparametric assumption, we also have

$$u = \tfrac{1}{2}(1 + s)u_2 + \tfrac{1}{4}(1 - r)(1 - s)u_3 + \tfrac{1}{4}(1 + r)(1 - s)u_4$$
$$v = \tfrac{1}{2}(1 + s)v_2 + \tfrac{1}{4}(1 - r)(1 - s)v_3 + \tfrac{1}{4}(1 + r)(1 - s)v_4$$
$$\partial u/\partial r = -\tfrac{1}{4}(1 - s)u_3 + \tfrac{1}{4}(1 - s)u_4; \qquad \partial v/\partial r = -\tfrac{1}{4}(1 - s)v_3 + \tfrac{1}{4}(1 - s)v_4$$
$$\partial u/\partial s = \tfrac{1}{2}u_2 - \tfrac{1}{4}(1 - r)u_3 - \tfrac{1}{4}(1 + r)u_4; \qquad \partial v/\partial s = \tfrac{1}{2}v_2 - \tfrac{1}{4}(1 - r)v_3 - \tfrac{1}{4}(1 + r)v_4$$

$$\begin{bmatrix} \frac{\partial}{\partial x} \\ \frac{\partial}{\partial y} \end{bmatrix} = \mathbf{J}^{-1} \begin{bmatrix} \frac{\partial}{\partial r} \\ \frac{\partial}{\partial s} \end{bmatrix}$$

Hence

$$\begin{bmatrix} \frac{\partial u}{\partial x} \\ \frac{\partial u}{\partial y} \end{bmatrix} = \begin{bmatrix} \frac{2}{1 - s} & 0 \\ \frac{1 + r}{1 - s} & 1 \end{bmatrix} \begin{bmatrix} 0 & 0 & -\frac{1}{4}(1 - s) & 0 & \frac{1}{4}(1 - s) & 0 \\ \frac{1}{2} & 0 & -\frac{1}{4}(1 - r) & 0 & -\frac{1}{4}(1 + r) & 0 \end{bmatrix} \begin{bmatrix} u_2 \\ v_2 \\ u_3 \\ v_3 \\ u_4 \\ v_4 \end{bmatrix}$$

and

$$\begin{bmatrix} \frac{\partial u}{\partial x} \\ \frac{\partial u}{\partial y} \end{bmatrix} = \begin{bmatrix} 0 & 0 & -\frac{1}{2} & 0 & \frac{1}{2} & 0 \\ \frac{1}{2} & 0 & -\frac{1}{2} & 0 & 0 & 0 \end{bmatrix} \begin{bmatrix} u_2 \\ \cdot \\ \cdot \\ \cdot \\ u_4 \\ v_4 \end{bmatrix}$$

Similarly

$$\begin{bmatrix} \frac{\partial v}{\partial x} \\ \frac{\partial v}{\partial y} \end{bmatrix} = \begin{bmatrix} 0 & 0 & 0 & -\frac{1}{2} & 0 & \frac{1}{2} \\ 0 & \frac{1}{2} & 0 & -\frac{1}{2} & 0 & 0 \end{bmatrix} \begin{bmatrix} u_2 \\ \cdot \\ \cdot \\ \cdot \\ u_4 \\ v_4 \end{bmatrix}$$

So we obtain

$$\boldsymbol{\epsilon} = \begin{bmatrix} 0 & 0 & -\frac{1}{2} & 0 & \frac{1}{2} & 0 \\ 0 & \frac{1}{2} & 0 & -\frac{1}{2} & 0 & 0 \\ \frac{1}{2} & 0 & -\frac{1}{2} & -\frac{1}{2} & 0 & \frac{1}{2} \end{bmatrix} \begin{bmatrix} u_2 \\ v_2 \\ u_3 \\ v_3 \\ u_4 \\ v_4 \end{bmatrix}$$

For any values of u_2, v_2, u_3, v_3, and u_4, v_4 the strain vector is constant and independent of r, s. Thus, the triangular element is a constant strain triangle.

In the above example we considered only one specific case. However, using the same approach it can be shown that collapsing any one side of a four-node element will always result in a constant strain triangle.

Considering the process of collapsing an element side, it is interesting to note that in the formulation used in Example 5.15 the matrix $\mathbf{J}$ is singular at $s = +1$, but this singularity disappears when the strain-displacement matrix is calculated. A practical consequence is that if in a computer program the general formulation of the four-node element is employed to generate a constant strain triangle (as done in Example 5.15), the stresses should not be calculated at the two local nodes that have been assigned the same global node. (Since the stresses are constant throughout the element, they are conveniently evaluated at the center of the element, i.e. at $r = 0$, $s = 0$.)

The same procedure can also be employed in three-dimensional analysis in order to obtain, from the basic eight-node element, wedge or tetrahedral elements. The procedure is illustrated in Fig. 5.18 and in the following example.

EXAMPLE 5.16: Show that the three-dimensional tetrahedral element generated in Fig. 5.20 from the eight-node three-dimensional brick element is a constant strain element.

Here we proceed as in Example 5.15. Thus, using the interpolation functions of the brick element (see Fig. 5.6) and substituting the nodal point coordinates of the tetrahedron, we obtain

$$x = \tfrac{1}{4}(1 + r)(1 - s)(1 - t)$$
$$y = \tfrac{1}{2}(1 + s)(1 - t)$$
$$z = 1 + t$$

Hence

$$\mathbf{J} = \begin{bmatrix} \frac{1}{4}(1 - s)(1 - t) & 0 & 0 \\ -\frac{1}{4}(1 + r)(1 - t) & \frac{1}{2}(1 - t) & 0 \\ -\frac{1}{4}(1 + r)(1 - s) & -\frac{1}{2}(1 + s) & 1 \end{bmatrix};$$

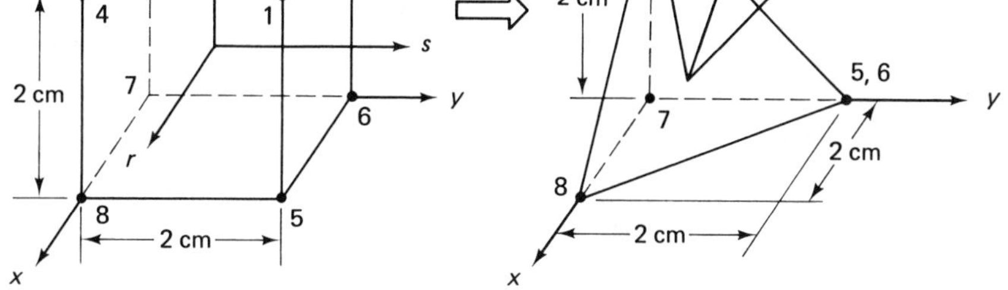

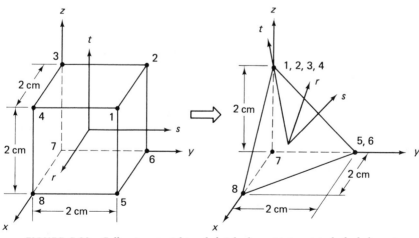

FIGURE 5.20 *Collapsing an eight-node brick element into a tetrahedral element.*

$$
\mathbf{J}^{-1} = \begin{bmatrix} \dfrac{4}{(1-s)(1-t)} & 0 & 0 \\[2ex] \dfrac{2(1+r)}{(1-s)(1-t)} & \dfrac{2}{(1-t)} & 0 \\[2ex] \dfrac{2(1+r)}{(1-s)(1-t)} & \dfrac{(1+s)}{(1-t)} & 1 \end{bmatrix} \tag{a}
$$

Using the same interpolation functions for u, and the conditions that $u_1 = u_2 = u_3 = u_4$ and $u_5 = u_6$, we obtain

$$
u = h_4^* u_4 + h_5^* u_5 + h_7^* u_7 + h_8^* u_8
$$

with

$$
h_4^* = \tfrac{1}{2}(1+t); \qquad h_5^* = \tfrac{1}{4}(1+s)(1-t);
$$
$$
h_7^* = \tfrac{1}{8}(1-r)(1-s)(1-t); \qquad h_8^* = \tfrac{1}{8}(1+r)(1-s)(1-t)
$$

Similarly, we also have

$$
v = h_4^* v_4 + h_5^* v_5 + h_7^* v_7 + h_8^* v_8
$$
$$
w = h_4^* w_4 + h_5^* w_5 + h_7^* w_7 + h_8^* w_8
$$

Evaluating now the derivatives of the displacements u, v, and w with respect to r, s, and t, and using $\mathbf{J}^{-1}$ of (a), we obtain

$$
\begin{bmatrix} \dfrac{\partial u}{\partial x} \\[1.5ex] \dfrac{\partial v}{\partial y} \\[1.5ex] \dfrac{\partial w}{\partial z} \\[1.5ex] \dfrac{\partial u}{\partial y} + \dfrac{\partial v}{\partial x} \\[1.5ex] \dfrac{\partial v}{\partial z} + \dfrac{\partial w}{\partial y} \\[1.5ex] \dfrac{\partial u}{\partial z} + \dfrac{\partial w}{\partial x} \end{bmatrix} = \left[\begin{array}{ccc:ccc:ccc:ccc}
0 & 0 & 0 & 0 & 0 & 0 & -\frac{1}{2} & 0 & 0 & \frac{1}{2} & 0 & 0 \\
0 & 0 & 0 & 0 & \frac{1}{2} & 0 & 0 & -\frac{1}{2} & 0 & 0 & 0 & 0 \\
0 & 0 & \frac{1}{2} & 0 & 0 & 0 & 0 & 0 & -\frac{1}{2} & 0 & 0 & 0 \\
0 & 0 & 0 & \frac{1}{2} & 0 & 0 & -\frac{1}{2} & -\frac{1}{2} & 0 & 0 & \frac{1}{2} & 0 \\
0 & \frac{1}{2} & 0 & 0 & 0 & \frac{1}{2} & 0 & -\frac{1}{2} & -\frac{1}{2} & 0 & 0 & 0 \\
\frac{1}{2} & 0 & 0 & 0 & 0 & 0 & -\frac{1}{2} & 0 & -\frac{1}{2} & 0 & 0 & \frac{1}{2}
\end{array}\right] \begin{bmatrix} u_4 \\ v_4 \\ w_4 \\ \hdashline u_5 \\ v_5 \\ w_5 \\ \hdashline u_7 \\ v_7 \\ w_7 \\ \hdashline u_8 \\ v_8 \\ w_8 \end{bmatrix}
$$

Hence, the strains are constant for any nodal point displacements, which means that the element can only represent constant strain conditions.

The process of collapsing an element side, or in three-dimensional analysis a number of element sides, may directly yield a desired element, but when higher-order two- or three-dimensional elements are employed some special considerations may be necessary regarding the interpolation functions used. Specifically, when the lower-order elements displayed in Fig. 5.18 are employed, spatially isotropic triangular and wedge elements are automatically generated, but this is not necessarily the case when using higher-order elements, i.e. rectangular and brick elements with more than four and eight nodes in two- and three-dimensional analyses, respectively. In other words, for two-dimensional analysis, although the higher-order element is spatially isotropic (geometrically invariant) in the 2×2 square natural coordinate system, the generated triangular element is in general not spatially isotropic.

As an example, we consider the six-node triangular two-dimensional element obtained by collapsing one side of an eight- node element as shown in Fig. 5.21. If the triangular element has sides of equal length, we may want the element to

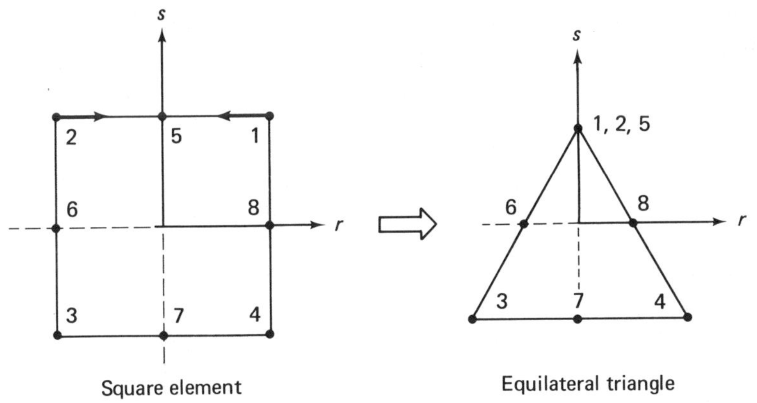

FIGURE 5.21 *Collapsing an eight-node element into a triangle.*

be spatially isotropic (see Section 4.2.5), i.e., we wish the internal element displacements u and v to vary in the same manner for each corner nodal displacement and each midside nodal displacement, respectively. However, the interpolation functions that are generated for the triangle when the side 1-2-5 of the square is simply collapsed do not fulfill the requirement that we should be able to change the numbering of the vertices without a change in the displacement assumptions. In order to fulfill this requirement corrections need be applied to the interpolation functions of the nodes 3, 4, and 7 to obtain the final interpolations, h_i^*, of the triangular element:

$$\begin{aligned}
h_1^* &= \tfrac{1}{2}(1+s) - \tfrac{1}{2}(1-s^2) \\
h_3^* &= \tfrac{1}{4}(1-r)(1-s) - \tfrac{1}{4}(1-s^2)(1-r) - \tfrac{1}{4}(1-r^2)(1-s) + \Delta h \\
h_4^* &= \tfrac{1}{4}(1+r)(1-s) - \tfrac{1}{4}(1-r^2)(1-s) - \tfrac{1}{4}(1-s^2)(1+r) + \Delta h \\
h_6^* &= \tfrac{1}{2}(1-s^2)(1-r) \\
h_7^* &= \tfrac{1}{2}(1-r^2)(1-s) - 2\Delta h \\
h_8^* &= \tfrac{1}{2}(1-s^2)(1+r)
\end{aligned} \tag{5.36}$$

where we added the appropriate interpolations given in Fig. 5.5 and

$$\Delta h = \frac{(1-r^2)(1-s^2)}{8} \tag{5.37}$$

Thus, to generate higher-order triangular elements by collapsing sides of square elements it may be necessary to apply a correction to the interpolation functions used.[15]

In the above considerations, we assumed that a spatially isotropic element is desirable, because the element is to be employed in a finite element assemblage that is used to predict a somewhat homogeneous stress field. However, in some cases, very specific stress variations are to be predicted and in such analyses a spatially non-isotropic element may be more effective. One area of analysis in which specific spatially non-isotropic elements are employed is the field of fracture mechanics. Here it is known that specific stress singularities exist at crack tips, and for the calculation of stress concentration factors or limit loads the use of finite elements that contain the required stress singularities can be effective. Various elements of this sort have been designed, but very simple and attractive elements can be obtained by distorting the higher-order isoparametric elements.[16-19] Figure 5.22 shows two-dimensional isoparametric elements that have been employed with much success in linear and nonlinear fracture mechanics, because they contain the $1/\sqrt{R}$ and $1/R$ stress singularities, respectively. We should note that these elements have the interpolation functions given in (5.36) but with $\Delta h = 0$. The same node shifting and side collapsing procedures can also be employed with higher-order three-dimensional elements in order to generate the required three-dimensional stress singularities.[17,19] We demonstrate the procedure of node shifting to generate a stress singularity in the following example.

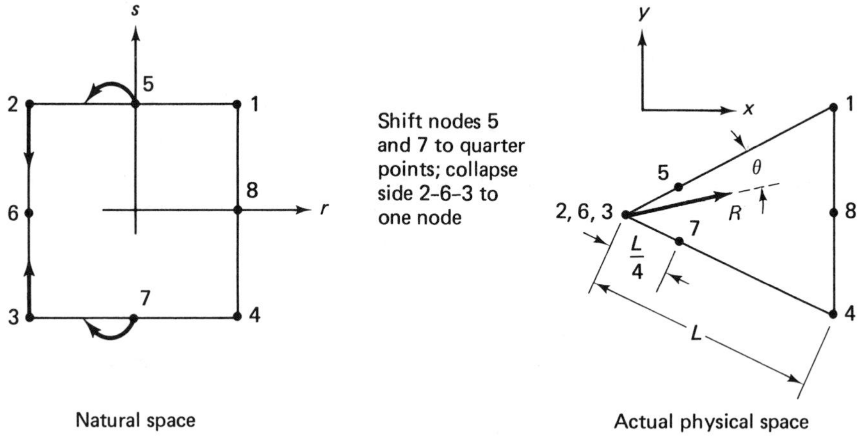

(a) Quarter-point triangular element with $\dfrac{1}{\sqrt{R}}$ stress singularity at node (2-6-3)

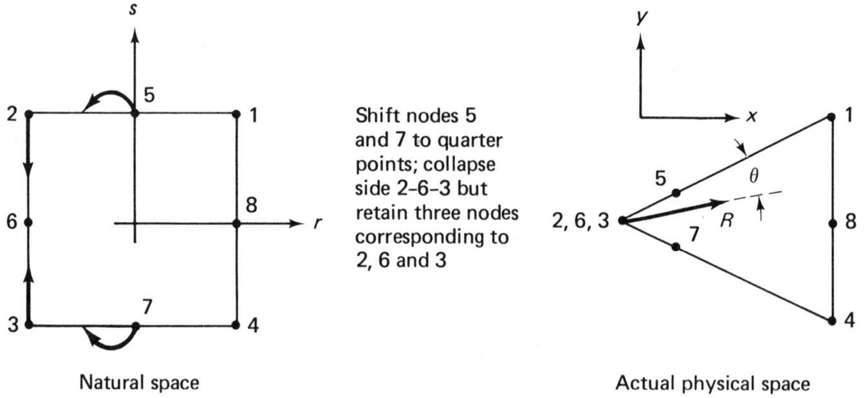

(b) Quarter-point triangular element with $\dfrac{1}{R}$ stress singularity at nodes 2, 6, and 3.

FIGURE 5.22 *Two-dimensional distorted (quarter point) isoparametric elements useful in fracture mechanics. Stress singularities are within the element for any angle θ. (Note that in (a) the one node (2-6-3) has two degrees of freedom, and that in (b) nodes 2, 3 and 6* each *have two degrees of freedom. In case (b) the shifting of nodes 5 and 7 to the quarter points is frequently not performed in practice.)*

EXAMPLE 5.17: Consider the three-node truss element in Fig. 5.23. Show that when node 3 is specified to be at the quarter point, the stress has a singularity of $1/\sqrt{x}$ at node 1.

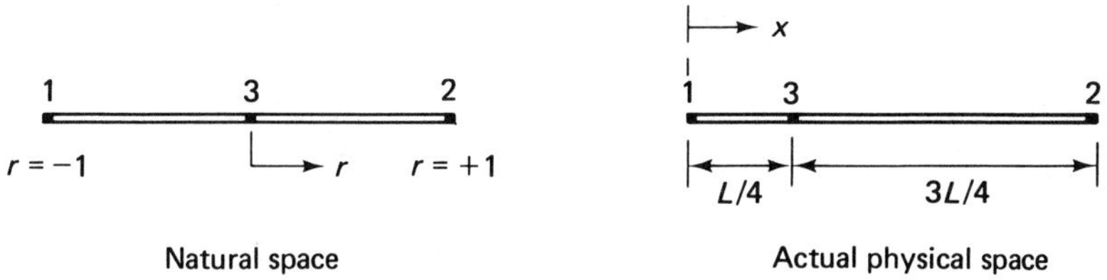

FIGURE 5.23 *Quarter-point one-dimensional element.*

We considered a three-node truss already in Example 5.2. Proceeding as before we now have

$$x = \frac{r}{2}(1 + r)L + (1 - r^2)\frac{L}{4}$$

or

$$x = \frac{L}{4}(1 + r)^2 \tag{a}$$

Hence

$$\mathbf{J} = \left[\frac{L}{2} + \frac{r}{2}L\right]$$

and the strain-displacement matrix is (using (b) of Example 5.2)

$$\mathbf{B} = \left(\frac{1}{L/2 + rL/2}\right)\left[\left(-\frac{1}{2} + r\right) \quad \left(\frac{1}{2} + r\right) \quad -2r\right] \tag{b}$$

To show the $1/\sqrt{x}$ singularity we need to express r in terms of x. Using (a) we have

$$r = 2\sqrt{\frac{x}{L}} - 1$$

Substituting this value for r into (b) we obtain

$$\mathbf{B} = \left[\left(\frac{2}{L} - \frac{3}{2\sqrt{L}}\frac{1}{\sqrt{x}}\right) \quad \left(\frac{2}{L} - \frac{1}{2\sqrt{L}}\frac{1}{\sqrt{x}}\right) \quad \left(\frac{2}{\sqrt{L}}\frac{1}{\sqrt{x}} - \frac{4}{L}\right)\right]$$

Hence at $x = 0$ the quarter-point element in Fig. 5.23 has a stress singularity of order $1/\sqrt{x}$.

Although the procedure of distorting a rectangular isoparametric element to generate a triangular element can be effective in some cases as discussed above, triangular elements (and in particular spatially isotropic elements) can be constructed directly by using area coordinates. Considering the triangle in Fig. 5.24, the position of a typical interior point P with coordinates x and y is defined by the area coordinates

$$L_1 = \frac{A_1}{A}; \qquad L_2 = \frac{A_2}{A}; \qquad L_3 = \frac{A_3}{A} \tag{5.38}$$

where the areas A_i, $i = 1, 2, 3$ are defined in the figure and A is the total area of the triangle. Thus, we also have

$$L_1 + L_2 + L_3 = 1 \tag{5.39}$$

Since element strains are obtained by taking derivatives with respect to the Cartesian coordinates, we need a relation that gives the triangular coordinates in terms of the coordinates x and y. Here, we have

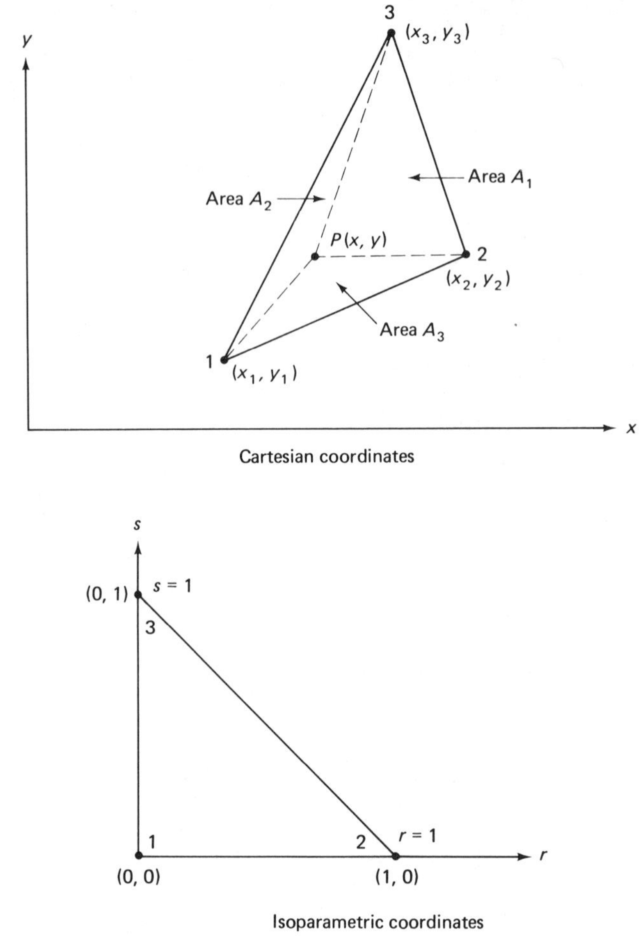

FIGURE 5.24 *Description of three-node triangle.*

$$x = L_1 x_1 + L_2 x_2 + L_3 x_3 \tag{5.40}$$

$$y = L_1 y_1 + L_2 y_2 + L_3 y_3 \tag{5.41}$$

because these relations hold at points 1, 2 and 3 and x and y vary linearly in-between. Using (5.39) to (5.41) we have

$$\begin{bmatrix} 1 \\ x \\ y \end{bmatrix} = \begin{bmatrix} 1 & 1 & 1 \\ x_1 & x_2 & x_3 \\ y_1 & y_2 & y_3 \end{bmatrix} \begin{bmatrix} L_1 \\ L_2 \\ L_3 \end{bmatrix} \tag{5.42}$$

which gives $$L_i = \frac{1}{2A}(a_i + b_i x + c_i y); \qquad i = 1, 2, 3$$

where $$2A = x_1 y_2 + x_2 y_3 + x_3 y_1 - y_1 x_2 - y_2 x_3 - y_3 x_1$$

$$\begin{aligned} a_1 &= x_2 y_3 - x_3 y_2 & a_2 &= x_3 y_1 - x_1 y_3 & a_3 &= x_1 y_2 - x_2 y_1 \\ b_1 &= y_2 - y_3 & b_2 &= y_3 - y_1 & b_3 &= y_1 - y_2 \\ c_1 &= x_3 - x_2 & c_2 &= x_1 - x_3 & c_3 &= x_2 - x_1 \end{aligned} \tag{5.43}$$

As must have been expected, these L_i are equal to the interpolation functions of a constant strain triangle. Thus, in summary we have for the three-node triangular element in Fig. 5.24,

$$u = \sum_{i=1}^{3} h_i u_i \qquad x \equiv \sum_{i=1}^{3} h_i x_i$$
$$v = \sum_{i=1}^{3} h_i v_i \qquad y \equiv \sum_{i=1}^{3} h_i y_i \tag{5.44}$$

where $h_i = L_i$, $i = 1, 2, 3$, and the h_i are functions of the coordinates x and y.

Using the relations in (5.44), the various finite element matrices of (5.27) to (5.35) can directly be evaluated. However, just as in the formulation of the rectangular elements (see Section 5.3.1), in practice, it is frequently expedient to use a natural coordinate space in order to describe the element coordinates and displacements. Using the natural coordinate system shown in Fig. 5.24, we have

$$h_1 = 1 - r - s; \qquad h_2 = r; \qquad h_3 = s \tag{5.45}$$

and the evaluation of the element matrices involves now a Jacobian transformation. Furthermore, all integrations are carried out over the natural coordinates; i.e., the r-integrations go from 0 to 1 and the s-integrations go from 0 to $(1 - r)$.

EXAMPLE 5.18: Using the isoparametric natural coordinate system of Fig. 5.24, establish the displacement and strain-displacement interpolation matrices of a three-node triangular element with

$$x_1 = 0 \qquad x_2 = 4 \qquad x_3 = 1$$
$$y_1 = 0 \qquad y_2 = 0 \qquad y_3 = 3$$

In this case we have using (5.44)

$$x = 4r + s$$
$$y = 3s$$

Hence using (5.23)
$$\mathbf{J} = \begin{bmatrix} 4 & 0 \\ 1 & 3 \end{bmatrix};$$

and
$$\frac{\partial}{\partial \mathbf{x}} = \frac{1}{12}\begin{bmatrix} 3 & 0 \\ -1 & 4 \end{bmatrix}\frac{\partial}{\partial \mathbf{r}}$$

It follows that
$$\mathbf{H} = \left[\begin{array}{cc|cc|cc} (1-r-s) & 0 & r & 0 & s & 0 \\ 0 & (1-r-s) & 0 & r & 0 & s \end{array}\right]$$

and
$$\mathbf{B} = \frac{1}{12}\left[\begin{array}{cc|cc|cc} -3 & 0 & 3 & 0 & 0 & 0 \\ 0 & -3 & 0 & -1 & 0 & 4 \\ -3 & -3 & -1 & 3 & 4 & 0 \end{array}\right]$$

By analogy to the formulation of higher-order rectangular elements, we can also directly formulate higher-order triangular elements. Namely, using the natural coordinate system of Fig. 5.24 which reduces to

$$L_1 = 1 - r - s \qquad L_2 = r \qquad L_3 = s \tag{5.46}$$

where the L_i are the area coordinates of the "unit triangle," the interpolation functions of a 3 to 6 variable-number-nodes element are given in Fig. 5.25. These functions are constructed in the usual way; namely, h_i must be unity at node i and zero at all other nodes (see Example 5.1). The interpolation functions of still higher-order triangular elements are obtained in a similar manner.*

* It is interesting to note that the functions of the 6-node triangle in Fig. 5.25 are exactly those given in (5.36), provided the variables r and s in Fig. 5.25 are replaced by $\frac{1}{2}(1 + r)$ and $\frac{1}{4}(1 - r)(1 + s)$, respectively, in order to account for the different natural coordinate systems. Hence, the correction Δh in (5.36) can be evaluated from the functions in Fig. 5.25.

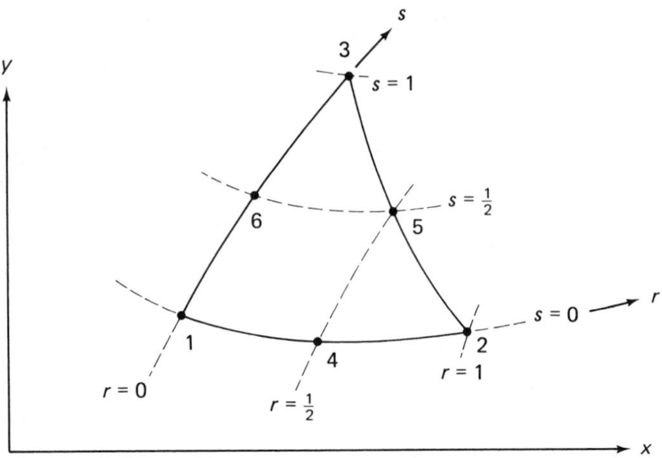

(a) Coordinate system and nodal points

Include only if node i is defined

		$i = 4$	$i = 5$	$i = 6$
$h_1 =$	$1-r-s$	$-\frac{1}{2}h_4$	....	$-\frac{1}{2}h_6$
$h_2 =$	r	$-\frac{1}{2}h_4$	$-\frac{1}{2}h_5$	
$h_3 =$	s	....	$-\frac{1}{2}h_5$	$-\frac{1}{2}h_6$
$h_4 =$	$4r(1-r-s)$			
$h_5 =$	$4rs$			
$h_6 =$	$4s(1-r-s)$			

(b) Interpolation functions

FIGURE 5.25 *Interpolation functions of three to six variable-number-nodes two-dimensional triangle.*

Using the above approach we can now also directly construct the interpolation functions of three-dimensional tetrahedral elements. First, we note that in analogy to (5.46) we now employ *volume coordinates*

$$\begin{aligned} L_1 &= 1 - r - s - t \qquad & L_2 &= r \\ L_3 &= s & L_4 &= t \end{aligned} \tag{5.47}$$

where we may check that $L_1 + L_2 + L_3 + L_4 = 1$. The L_i in (5.47) are the interpolation functions of the four-node element in Fig. 5.26 in its natural space. The interpolation functions of a 4 to 10 three-dimensional variable-number-nodes element are given in Fig. 5.27.

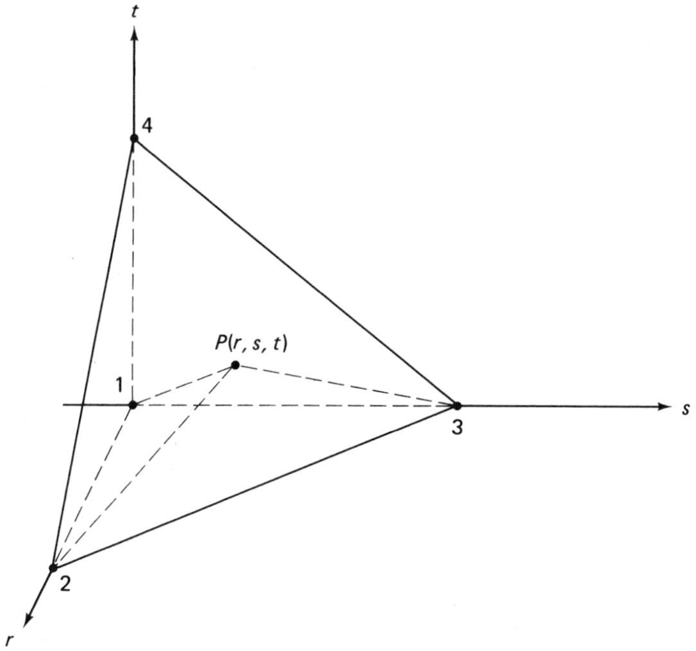

FIGURE 5.26 *Natural coordinate system of tetrahedral element.*

To evaluate the element matrices, it is necessary to include the Jacobian transformation as given in (5.24) and to perform the r-integrations from 0 to 1, the s-integrations from 0 to $(1 - r)$, and the t-integrations from 0 to $(1 - r - s)$. As for the rectangular elements, these integrations are carried out effectively in general analysis using numerical integration, but the integration rules employed are different from those used for rectangular elements (see Section 5.7.4).

5.3.3 Element Matrices in Global Coordinate System

So far we considered the calculation of isoparametric element matrices that correspond to local element degrees of freedom. In the evaluation we used local element coordinates x, y, and z, whichever are applicable, and local element degrees of freedom u_i, v_i, and w_i. However, we may note that for the two-dimensional element considered in Examples 5.5 to 5.7 the element matrices could have been evaluated using the global coordinate variables X and Y, and the global nodal point displacements U_i and V_i. Indeed, in the calculations presented, the x and y local coordinates and u and v local displacement components needed simply to be replaced by the X and Y global coordinates and U and V global displacement components, respectively. In such case the matrices would have corresponded directly to the global displacement components.

In general, the calculation of the element matrices should be carried out in the global coordinate system, using global displacement components if the number of natural coordinate variables is equal to the number of global variables.

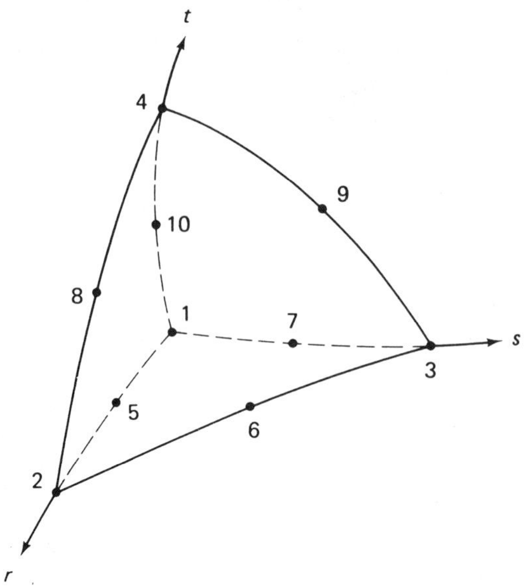

(a) Coordinate system and nodal points

		Include only if node i is defined					
		$i = 5$	$i = 6$	$i = 7$	$i = 8$	$i = 9$	$i = 10$
$h_1 =$	$1-r-s-t$	$-\frac{1}{2}h_5$	....	$-\frac{1}{2}h_7$	....	....	$-\frac{1}{2}h_{10}$
$h_2 =$	r	$-\frac{1}{2}h_5$	$-\frac{1}{2}h_6$	....	$-\frac{1}{2}h_8$		
$h_3 =$	s	....	$-\frac{1}{2}h_6$	$-\frac{1}{2}h_7$	....	$-\frac{1}{2}h_9$	
$h_4 =$	t	....	....	....	$-\frac{1}{2}h_8$	$-\frac{1}{2}h_9$	$-\frac{1}{2}h_{10}$
$h_5 =$	$4r(1-r-s-t)$						
$h_6 =$	$4rs$						
$h_7 =$	$4s(1-r-s-t)$						
$h_8 =$	$4rt$						
$h_9 =$	$4st$						
$h_{10} =$	$4t(1-r-s-t)$						

(b) Interpolation functions

FIGURE 5.27 *Interpolation functions of four to ten variable-number-nodes three-dimensional tetrahedral element.*

Typical examples are two-dimensional elements that are defined in a global plane, and the three-dimensional element in Fig. 5.6. In these cases the Jacobian operator in (5.24) is a square matrix, which can be inverted as required in (5.25), and the element matrices correspond directly to the global displacement components.

In those cases where the order of the global coordinate system is higher than the order of the natural coordinate system, it is usually most expedient to calculate first the element matrices in the local coordinate system and corresponding to local displacement components. Afterwards, the matrices must be transformed in the usual manner to the global displacement system. Examples are the truss element or the plane stress element when they are oriented arbitrarily in three-dimensional space. However, alternatively, we may include the transformation to the global displacement components directly in the formulation. This is accomplished by introducing a transformation that expresses in the displacement interpolation the local nodal point displacements in terms of the global components.

EXAMPLE 5.19: Evaluate the element stiffness matrix of the truss element in Fig. 5.28 using directly global nodal point displacements.

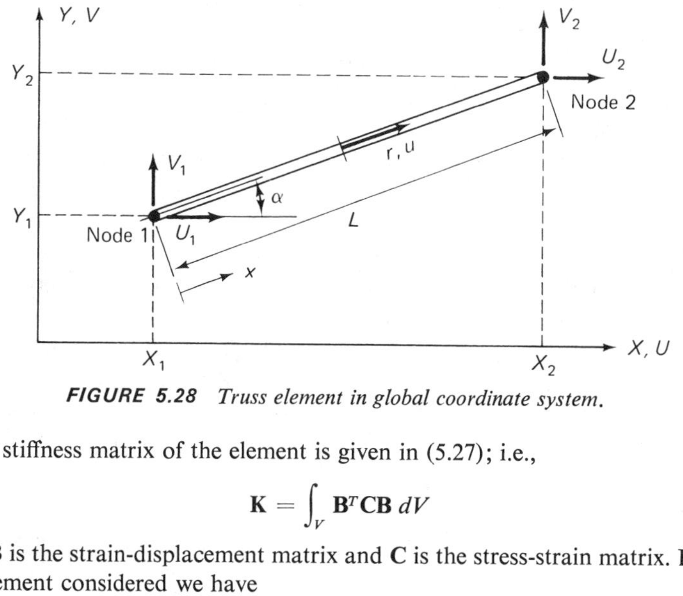

FIGURE 5.28 *Truss element in global coordinate system.*

The stiffness matrix of the element is given in (5.27); i.e.,

$$\mathbf{K} = \int_V \mathbf{B}^T\mathbf{C}\mathbf{B}\, dV$$

where $\mathbf{B}$ is the strain-displacement matrix and $\mathbf{C}$ is the stress-strain matrix. For the truss element considered we have

$$u = [\cos\alpha \ \sin\alpha]\begin{bmatrix} \frac{1}{2}(1-r)U_1 + \frac{1}{2}(r+1)U_2 \\ \frac{1}{2}(1-r)V_1 + \frac{1}{2}(r+1)V_2 \end{bmatrix}$$

Then using that the strain $\epsilon = \partial u/\partial x$, which expressed in the natural coordinate system is $\epsilon = (2/L)\, \partial u/\partial r$ (see Section 5.2), we can write the strain-displacement transformation corresponding to the displacement vector $\mathbf{U}^T = [U_1 \ \ V_1 \ \ U_2 \ \ V_2]$ as

$$\mathbf{B} = \frac{1}{L}[\cos\alpha \quad \sin\alpha \quad \cos\alpha \quad \sin\alpha]\begin{bmatrix} -1 & & & \\ & -1 & \text{zeros} & \\ & \text{zeros} & 1 & \\ & & & 1 \end{bmatrix}$$

Also, as given in Section 5.2, we have

$$dV = \frac{AL}{2}\, dr \qquad \text{and} \qquad \mathbf{C} = E$$

Substituting the relations for **B**, **C**, and dV and evaluating the integral, we obtain

$$\mathbf{K} = \frac{AE}{L}\begin{bmatrix} \cos^2\alpha & \cos\alpha\sin\alpha & -\cos^2\alpha & -\cos\alpha\sin\alpha \\ \sin\alpha\cos\alpha & \sin^2\alpha & -\sin\alpha\cos\alpha & -\sin^2\alpha \\ -\cos^2\alpha & -\cos\alpha\sin\alpha & \cos^2\alpha & \cos\alpha\sin\alpha \\ -\sin\alpha\cos\alpha & -\sin^2\alpha & \sin\alpha\cos\alpha & \sin^2\alpha \end{bmatrix}$$

5.4 FORMULATION OF STRUCTURAL ELEMENTS

The concepts of geometry and displacement interpolations that have been employed in the formulation of two- and three-dimensional continuum elements can also be employed in the evaluation of beam, plate, and shell structural element matrices. However, whereas in the formulation of the continuum elements the displacements u, v, w (whichever applicable) are interpolated in terms of nodal point displacements of the same kind, in the formulation of structural elements, the displacements u, v, and w are interpolated in terms of midsurface displacements and rotations. We will show that this procedure corresponds in essence to a "continuum isoparametric element formulation with displacement constraints." (In addition, there is the major assumption that the stress normal to the mid-surface is zero.) This can be interpreted as using a higher degree of interpolation on the geometry than on the displacements and for this reason the structural elements are also referred to as *superparametric elements* (see Section 5.6), or as *degenerate isoparametric elements.*

Considering the formulation of structural elements, we have already discussed briefly in Section 4.2.3 how beam, plate, and shell elements can be formulated using the Kirchhoff plate theory, in which shear deformations are neglected. In these formulations it is difficult to satisfy interelement continuity on displacements and edge rotations because the plate (or shell) rotations are calculated from the transverse displacements. Furthermore, using an assemblage of flat elements to represent a shell structure, a relatively large number of elements may be required in order to represent the shell geometry to sufficient accuracy.

Our objective in this section is to discuss an alternative approach to formulating beam, plate, and shell elements. The basis of this method is a theory that includes the effects of shear deformations. With this theory, the displacements and rotations of the mid-surface normals are independent and the interelement continuity conditions on these quantities can be satisfied directly as in the analysis of continua. In addition, if the concepts of isoparametric interpolation are employed, the geometry of curved shell surfaces is interpolated and can be represented to a high degree of accuracy. In the following sections we discuss first the formulation of beam elements, where we can demonstrate in detail the basic principles used, and we then present the formulation of general plate and shell elements.

5.4.1 Beam Elements

The basic assumption in beam-bending analysis excluding shear deformations is that a normal to the mid-surface (neutral axis) of the beam remains straight during deformation and that its angular rotation is equal to the slope of the beam mid-surface. This kinematic assumption, illustrated in Fig. 5.29(a), leads to the well-known beam-bending governing differential equation in which the transverse displacement w is the only variable (see Example 3.20). Therefore, using beam elements formulated with this theory, displacement continuity between elements requires that w and dw/dx be continuous.

Considering now beam-bending analysis with the effect of shear deformations, we retain the assumption that a plane section originally normal to

Boundary conditions between beam elements

$$w\bigg|_{x^{-0}} = w\bigg|_{x^{+0}} \quad ; \quad \frac{dw}{dx}\bigg|_{x^{-0}} = \frac{dw}{dx}\bigg|_{x^{+0}}$$

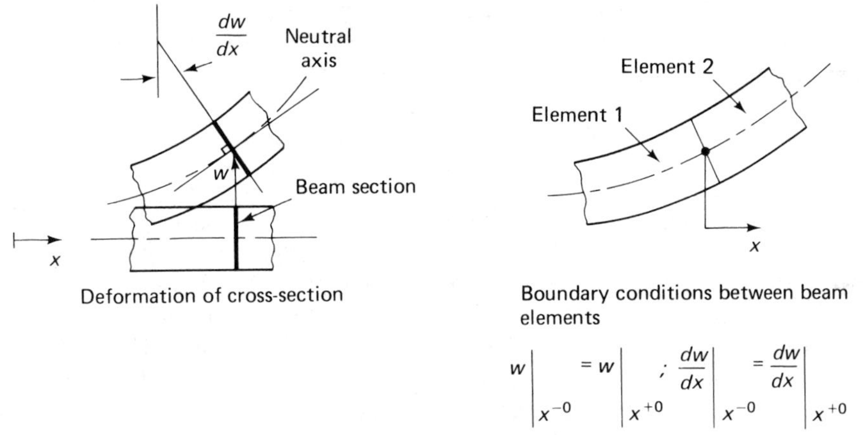

(a) Beam deformations excluding shear effect

Boundary conditions between beam elements

$$w\bigg|_{x^{-0}} = w\bigg|_{x^{+0}}$$

$$\beta\bigg|_{x^{-0}} = \beta\bigg|_{x^{+0}}$$

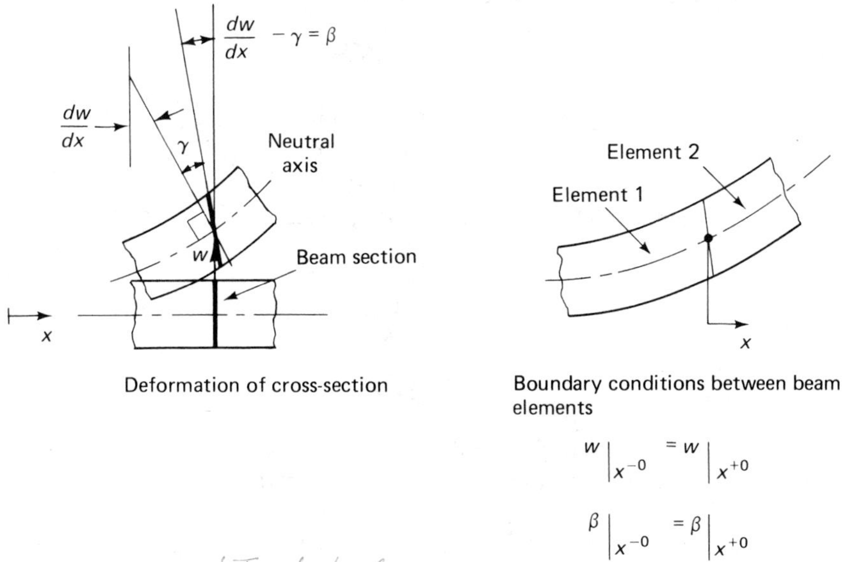

{ Timoshenko Beam

(b) Beam deformations including shear effect

FIGURE 5.29 *Beam deformation assumptions.*

the neutral axis remains plane, but because of shear deformations this section does not remain normal to the neutral axis. As illustrated in Fig. 5.29(b) the total rotation of the plane originally normal to the neutral axis of the beam is due to the rotation of the tangent to the neutral axis and the shear deformation,[20]

$$\beta = \frac{dw}{dx} - \gamma \tag{5.48}$$

where γ is a constant shearing strain across the section. Since the actual shearing stress and strain vary over the section (e.g. a parabolic distribution of shearing stress pertains to a rectangular section), the shearing strain γ in (5.48) is an equivalent constant strain on a corresponding shear area A_s,

$$\tau = \frac{V}{A_s}; \qquad \gamma = \frac{\tau}{G}; \qquad k = \frac{A_s}{A} \tag{5.49}$$

where V is the shear force at the section considered. The shear correction factor k can be evaluated using the condition that the constant shear stress in (5.49) when acting on A_s must yield the same shear strain energy as the actual shearing stress acting on the actual cross-sectional area A of the beam.

The finite element formulation of a beam element with the assumption in (5.48) can be obtained using the basic virtual work expressions in (4.1) to (4.20), or the principle of stationarity of the total potential energy (see Example 4.2). In the following we consider first, for illustrative purposes, the formulation of the beam element matrices corresponding to the simple beam element in Fig. 5.30 using the total potential energy approach, and we discuss afterwards the formulation of more general three-dimensional beam elements using the principle of virtual displacements.

Figure 5.30 shows the two-dimensional rectangular cross-section beam considered. The total potential energy of the beam is

$$\Pi = \frac{EI}{2}\int_0^L \left(\frac{d\beta}{dx}\right)^2 dx + \frac{GAk}{2}\int_0^L \left(\frac{dw}{dx} - \beta\right)^2 dx - \int_0^L pw\, dx - \int_0^L m\beta\, dx \tag{5.50}$$

where p and m are the transverse and moment loading per unit length. The first two integrals on the r.h.s. of (5.50) represent the strain energy corresponding to bending and shearing deformations, and the last two integrals represent the potential of the external loads. The stationarity condition $\delta\Pi = 0$ using (5.50) yields

$$\begin{aligned} EI\int_0^L \left(\frac{d\beta}{dx}\right)\delta\left(\frac{d\beta}{dx}\right) dx + GAk\int_0^L \left(\frac{dw}{dx} - \beta\right)\delta\left(\frac{dw}{dx} - \beta\right) dx \\ - \int_0^L p\,\delta w\, dx - \int_0^L m\,\delta\beta\, dx = 0 \end{aligned} \tag{5.51}$$

This relation is in fact the principle of virtual displacements corresponding to the beam element deformations. Using now the interpolations

$$w = \sum_{i=1}^{q} h_i w_i; \qquad \beta = \sum_{i=1}^{q} h_i \theta_i \tag{5.52}$$

where q is equal to the number of nodes used and the h_i are the one-dimensional interpolation functions listed in Fig. 5.4, we can directly employ the concepts of

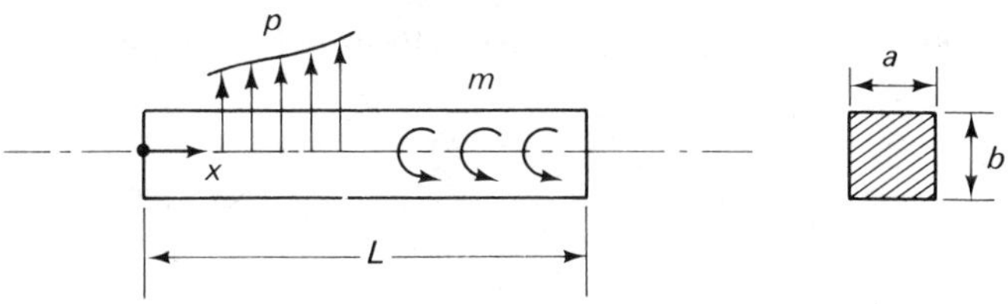

(a) Beam with applied loading
E = Young's modulus, G = shear modulus
$k = \frac{5}{6}$, $A = ab$, $I = \dfrac{ab^3}{12}$

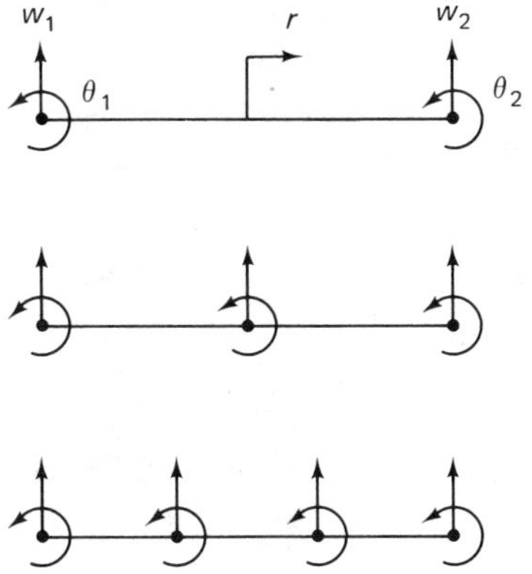

(b) Two, three and four-node models; $\theta_i = \beta_i$, $i = 1, \ldots, q$
(Interpolation functions are given in Fig. 5.4)

FIGURE 5.30 *Formulation of two-dimensional beam element.*

the isoparametric formulations discussed in Section 5.3 to establish all relevant element matrices. Let

$$
\begin{aligned}
w &= \mathbf{H}_w \hat{\mathbf{u}}; \qquad & \beta &= \mathbf{H}_\beta \hat{\mathbf{u}} \\
\frac{\partial w}{\partial x} &= \mathbf{B}_w \hat{\mathbf{u}}; \qquad & \frac{\partial \beta}{\partial x} &= \mathbf{B}_\beta \hat{\mathbf{u}}
\end{aligned}
\tag{5.53}
$$

where

$$
\begin{aligned}
\hat{\mathbf{u}}^T &= [w_1 \ldots w_q \quad \theta_1 \ldots \theta_q] \\
\mathbf{H}_w &= [h_1 \ldots h_q \quad 0 \ldots 0] \\
\mathbf{H}_\beta &= [0 \ldots 0 \quad h_1 \ldots h_q]
\end{aligned}
\tag{5.54}
$$

and

$$
\begin{aligned}
\mathbf{B}_w &= J^{-1}\left[\frac{\partial h_1}{\partial r} \ldots \frac{\partial h_q}{\partial r} \quad 0 \cdots 0\right] \\
\mathbf{B}_\beta &= J^{-1}\left[0 \cdots 0 \quad \frac{\partial h_1}{\partial r} \ldots \frac{\partial h_q}{\partial r}\right]
\end{aligned}
\tag{5.55}
$$

where $J = \partial x / \partial r$, then we have for a single element

$$
\mathbf{K} = EI \int_{-1}^{1} \mathbf{B}_\beta^T \mathbf{B}_\beta \det J \, dr + GAk \int_{-1}^{1} (\mathbf{B}_w - \mathbf{H}_\beta)^T (\mathbf{B}_w - \mathbf{H}_\beta) \det J \, dr \tag{5.56}
$$

and

$$
\mathbf{R} = \int_{-1}^{1} \mathbf{H}_w^T p \det J \, dr + \int_{-1}^{1} \mathbf{H}_\beta^T m \det J \, dr \tag{5.57}
$$

Also, in dynamic analysis the mass matrix can be calculated using the d'Alembert principle (see (4.21)); hence

$$\mathbf{M} = \int_{-1}^{1} \begin{bmatrix} \mathbf{H}_w \\ \mathbf{H}_\beta \end{bmatrix}^T \begin{bmatrix} \rho ab & 0 \\ 0 & \dfrac{\rho ab^3}{12} \end{bmatrix} \begin{bmatrix} \mathbf{H}_w \\ \mathbf{H}_\beta \end{bmatrix} \det J \, dr \tag{5.58}$$

In the above evaluations we are using the natural coordinate system of the beam because this is effective in the formulation of more general beam, plate, and shell elements. However, when considering a straight beam of constant cross-section, the integrals can also be evaluated efficiently without using the natural coordinate system as demonstrated in the following example.

EXAMPLE 5.20: Evaluate the detailed expressions for the calculation of the stiffness matrix and the load vector of the three-node beam element shown in Fig. 5.31.

The interpolation functions to be used are listed in Fig. 5.4. These functions are given in terms of r and yield

$$x = \sum_{i=1}^{3} h_i x_i$$

Using $x_1 = 0$, $x_2 = L$, $x_3 = L/2$, we obtain

$$x = \frac{L}{2}(1 + r)$$

Hence, the interpolation functions in terms of x are

$$h_1 = \frac{2x^2}{L^2} - \frac{3x}{L} + 1$$

$$h_2 = \frac{2x^2}{L^2} - \frac{x}{L}$$

$$h_3 = \frac{4x}{L} - \frac{4x^2}{L^2}$$

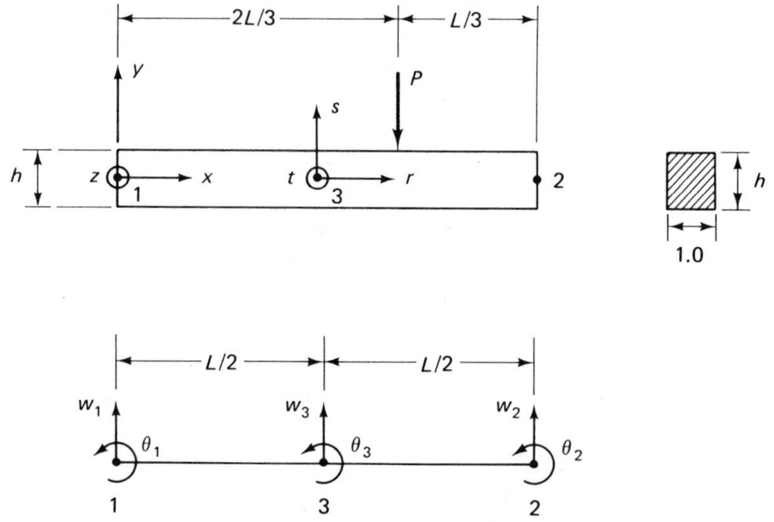

FIGURE 5.31 *Three-node beam element considered in Examples 5.20 and 5.22.*

Using the notation $(\ )' \equiv \dfrac{\partial}{\partial x}$ it follows that

$$h'_1 = \frac{4x}{L^2} - \frac{3}{L}$$

$$h'_2 = \frac{4x}{L^2} - \frac{1}{L}$$

$$h'_3 = \frac{4}{L} - \frac{8x}{L^2}$$

Hence with the degrees of freedom ordered as in (5.54) we have

$$\mathbf{K} = \frac{Eh^3}{12}\int_0^L \begin{bmatrix} 0 \\ 0 \\ 0 \\ h'_1 \\ h'_2 \\ h'_3 \end{bmatrix} [0 \quad 0 \quad 0 \quad h'_1 \quad h'_2 \quad h'_3]\, dx$$

$$+ \frac{5Gh}{6}\int_0^L \begin{bmatrix} h'_1 \\ h'_2 \\ h'_3 \\ -h_1 \\ -h_2 \\ -h_3 \end{bmatrix} [h'_1 \quad h'_2 \quad h'_3 \quad -h_1 \quad -h_2 \quad -h_3]\, dx$$

and
$$\mathbf{R}^T = -P[-\tfrac{1}{9} \quad \tfrac{2}{9} \quad \tfrac{8}{9} \quad 0 \quad 0 \quad 0]$$

The element of Fig. 5.30 can be employed effectively in the analysis of moderately thick and thin beam structures. However, in the analysis of thin beams, the three- or four-node element must be employed, because the two-node element does not represent the bending and shearing deformations to sufficient accuracy.

In order to obtain some insight into the behavior of the beam elements when the beam becomes thin, consider the total potential $\tilde{\Pi}$,

$$\tilde{\Pi} = \int_0^L \left(\frac{d\beta}{dx}\right)^2 dx + \frac{GAk}{EI}\int_0^L \left(\frac{dw}{dx} - \beta\right)^2 dx \tag{5.59}$$

which is obtained by neglecting the load contributions in (5.50) and dividing by $\frac{1}{2}EI$. The relation in (5.59) shows the relative importance of the bending and shearing contributions to the stiffness matrix of an element. The bending and shearing contributions are represented by the first and second terms on the r.h.s. of (5.59), respectively, where we note that the factor GAk/EI in the shearing term can be very large when a thin element is considered. This factor can be interpreted as a penalty number (see Section 3.4) as the element becomes very thin, i.e., we can write

$$\tilde{\Pi} = \int_0^L \left(\frac{d\beta}{dx}\right)^2 dx + \alpha\int_0^L \left(\frac{dw}{dx} - \beta\right)^2 dx; \qquad \alpha = \frac{GAk}{EI} \tag{5.60}$$

where $\alpha \longrightarrow \infty$ as $h \longrightarrow 0$. However, this means that as the beam becomes thin the constraint of zero shear deformations, i.e., $dw/dx = \beta$ with $\gamma = 0$, will be approached.

The above argument holds for the actual continuous model which is governed by the stationarity condition on Π and the appropriate boundary conditions. Considering now the finite element representation, it is important that the finite element displacement assumptions on β and w admit that for large values of α the shearing deformations can be small throughout the domain of the element. If by virtue of the assumptions used on w and β the shearing deformations cannot be small everywhere, then the element stiffness will grossly overestimate the actual structural stiffness because the penalty number is a multiplier to the erroneous shearing deformations. This is precisely what happens when the two-node beam element of Fig. 5.30 is used, which therefore should not be employed in the analysis of thin beam structures. This conclusion is also applicable to the use of the low-order plate and shell elements discussed in Section 5.4.2. The phenomenon that the thin elements are very stiff has been referred to as "element locking."[21,22]

Considering the above structural elements, we therefore conclude that, unless the low-order element behavior is "somehow" improved, only high-order elements should be used in the analysis of thin structures. Since the use of low-order elements can be attractive from a cost viewpoint, a considerable amount of research has been spent on finding remedies to improve the behavior of the low-order elements for thin structures. Two procedures have found wide appeal: *the use of selective or reduced integration*[21-23] and *the use of a discrete Kirchhoff theory.*[24,25]

In the above discussion of the beam elements, we assumed that all integrations are performed exactly. The reduced and selective integration procedures are based on the premise that by not integrating the shear strain energy exactly, the effect of "element locking" can be avoided. The difficulty in this procedure is to assure that the element does not contain any spurious zero-energy modes due to the special integration scheme used (which would render the element unreliable) and yet the element must possess good accuracy characteristics. We discuss the approach of reduced and selective integration further in Sections 5.8.1 and 6.5.3.

The procedure used in the discrete Kirchhoff theory to avoid element "locking" is to assume that the shearing strains are so small that the shear strain energy need not be included in the total potential energy functional. However, since the bending strains are given only in terms of the rotations of the cross-sections and we are interpolating these and the transverse displacements, additional equations are needed to relate the nodal point displacements and rotations. These equations are established by invoking the condition that the shearing strains are zero at specific points in the domain of the element. Thus, the basis of the discrete Kirchhoff element formulations is that the Kirchhoff hypothesis is satisfied at discrete points of the element (which, if a large enough number of points is selected, approximates the condition that the hypothesis is satisfied throughout the entire element). The discrete Kirchhoff analysis procedure is most effective in the formulation of low-order elements and we demonstrate the technique in the following simple analysis.

EXAMPLE 5.21: Consider the cantilever beam shown in Fig. 5.32. Idealize the cantilever using one two-node element and predict the tip displacement. Also, predict the tip deflection when using a discrete Kirchhoff formulation.

The stiffness matrix of a two-node beam element can directly be inferred from the results of Example 5.20 and the interpolation functions listed in Fig. 5.4:

$$\mathbf{K} = \frac{Eh^3}{12}\int_0^L \begin{bmatrix} 0 \\ 0 \\ -\dfrac{1}{L} \\ \dfrac{1}{L} \end{bmatrix} \begin{bmatrix} 0 & 0 & -\dfrac{1}{L} & \dfrac{1}{L} \end{bmatrix} dx$$

$$+ \frac{5Gh}{6}\int_0^L \begin{bmatrix} -\dfrac{1}{L} \\ \dfrac{1}{L} \\ -\left(1 - \dfrac{x}{L}\right) \\ -\dfrac{x}{L} \end{bmatrix} \begin{bmatrix} -\dfrac{1}{L} & \dfrac{1}{L} & -\left(1 - \dfrac{x}{L}\right) & -\dfrac{x}{L} \end{bmatrix} dx$$

With $G = \frac{1}{2}E$ we have

$$\mathbf{K} = \frac{E}{12}\begin{bmatrix} \dfrac{5h}{L} & -\dfrac{5h}{L} & \dfrac{5h}{2} & \dfrac{5h}{2} \\ & \dfrac{5h}{L} & -\dfrac{5h}{2} & -\dfrac{5h}{2} \\ & \text{sym.} & \left(\dfrac{h^3}{L} + \dfrac{5hL}{3}\right) & \left(\dfrac{5hL}{6} - \dfrac{h^3}{L}\right) \\ & & & \left(\dfrac{h^3}{L} + \dfrac{5hL}{3}\right) \end{bmatrix}$$

Since $w_1 = \theta_1 = 0$, the governing equilibrium equations for this problem are

$$\frac{E}{12}\begin{bmatrix} \dfrac{5h}{L} & -\dfrac{5h}{2} \\ -\dfrac{5h}{2} & \left(\dfrac{h^3}{L} + \dfrac{5hL}{3}\right) \end{bmatrix} \begin{bmatrix} w_2 \\ \theta_2 \end{bmatrix} = \begin{bmatrix} -P \\ 0 \end{bmatrix}$$

which gives

$$\delta = \frac{12PL}{5hE}\left(1 + \frac{5hL}{4\left(\dfrac{h^3}{L} + \dfrac{5hL}{12}\right)}\right)$$

The tip displacement predicted using elementary beam theory is $\delta = 4PL^3/Eh^3$; hence the solution obtained with the two-node element is very inaccurate. (It is interesting to note here that the response using one four-node element with $h/L = \frac{1}{100}$ is $\delta = 4.0002PL^3/Eh^3$.)

However, the solution using the two-node element idealization can be improved significantly with the discrete Kirchhoff theory in the formulation. Firstly, the shear strain energy contribution is neglected in the calculation of the stiffness matrix. Hence, using (5.50) we now have

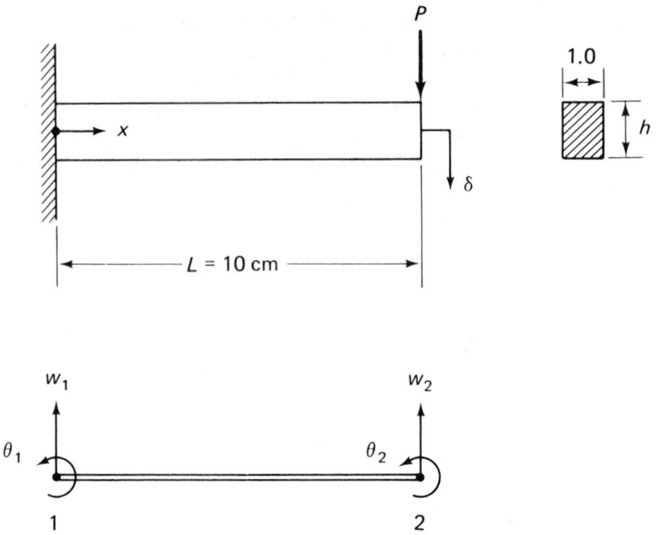

FIGURE 5.32 *Analysis of cantilever beam using one two-node element.*

$$\mathcal{U} = \frac{1}{2}[\theta_1 \quad \theta_2] \frac{Eh^3}{12L} \begin{bmatrix} 1 & -1 \\ -1 & 1 \end{bmatrix} \begin{bmatrix} \theta_1 \\ \theta_2 \end{bmatrix}$$

In the cantilever problem considered $\theta_1 = 0$. The second step in using the discrete Kirchhoff formulation is to invoke the constraint that $\gamma = 0$ in order to relate θ_2 to w_2. Using $\gamma = 0$, i.e. $dw/dx = \beta$, at the midpoint of the element gives

$$\theta_2 = \frac{2w_2}{L}$$

and

$$\mathcal{U} = \frac{1}{2}[w_2] \frac{Eh^3}{12L} \left[\frac{4}{L^2}\right] [w_2]$$

Hence, we now obtain

$$\delta = \frac{3PL^3}{Eh^3}$$

which is significantly closer to the elementary beam theory solution.

The above discussion and example solution provides only a very brief introduction to the basic concepts employed in discrete Kirchhoff formulations. A particularly difficult step in these formulations is the choice of the appropriate constraints between the nodal point variables. However, the discrete Kirchhoff theory has been used with much success in the formulation of various beam, plate and shell elements, and in particular for the derivation of very effective low-order elements.[25]

The formulation of the two-dimensional beam element of Fig. 5.30 given in (5.50) to (5.58) is a special case of the formulation of more general three-dimensional beam elements. In order to illustrate this point and show how the isoparametric interpolation concepts can be applied to formulate more general beam elements, we consider the three-dimensional beam of rectangular cross-section in Fig. 5.33. In the derivation we assume that an accurate representation of the torsional rigidity is not required. Namely, the torsional interpolation functions that

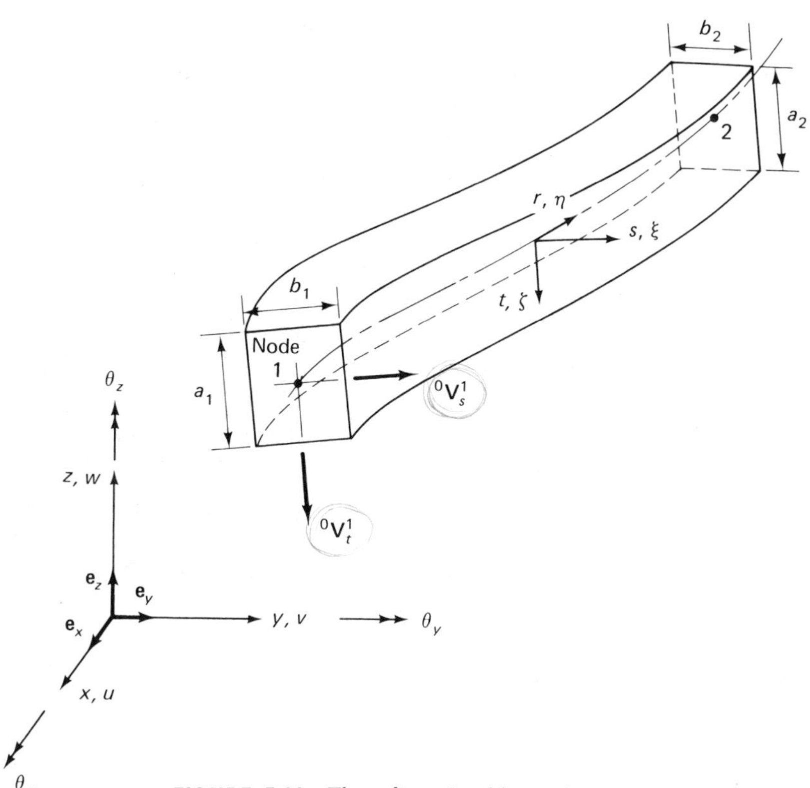

FIGURE 5.33 *Three-dimensional beam element.*

we will use in the formulation correspond to the exact torsional displacements of a circular section and should be amended for an accurate torsional description of rectangular cross-sections.* In practice, complex sections such as L and I cross-sections and pipe elbows need be considered. These elements are also more difficult to deal with because again deformation patterns in addition to those used in the formulation given below can be important in the description of the behavior of these beams. For example, in the formulation of a curved pipe element the ovalization of the cross-section must in general be taken into account.[26] However, although we consider in the following only a special three-dimensional beam element the discussion is important because the procedures used in this formulation represent the basic and general steps that are followed also in the formulation of beams of other cross-sectional shapes.

The basic kinematic assumption used in the formulation of the three-dimensional beam element is the same as that employed in the formulation of the two-dimensional element of Fig. 5.30; namely, that plane sections originally normal to the center line axis remain plane and undistorted under deformation, but not necessarily normal to this axis. Based on this assumption, we can formulate the beam element matrices directly using the isoparametric element procedures discussed in Section 5.3.

* The required amendment is that warping displacements must be interpolated.

Using the natural coordinates r, s, and t, the Cartesian coordinates of a point in the element with q nodal points are, before and after deformations:

$$\begin{aligned}
{}^{\ell}x(r, s, t) &= \sum_{k=1}^{q} h_k\,{}^{\ell}x_k + \frac{t}{2}\sum_{k=1}^{q} a_k\,h_k\,{}^{\ell}V_{tx}^k + \frac{s}{2}\sum_{k=1}^{q} b_k\,h_k\,{}^{\ell}V_{sx}^k \\
{}^{\ell}y(r, s, t) &= \sum_{k=1}^{q} h_k\,{}^{\ell}y_k + \frac{t}{2}\sum_{k=1}^{q} a_k\,h_k\,{}^{\ell}V_{ty}^k + \frac{s}{2}\sum_{k=1}^{q} b_k\,h_k\,{}^{\ell}V_{sy}^k \\
{}^{\ell}z(r, s, t) &= \sum_{k=1}^{q} h_k\,{}^{\ell}z_k + \frac{t}{2}\sum_{k=1}^{q} a_k\,h_k\,{}^{\ell}V_{tz}^k + \frac{s}{2}\sum_{k=1}^{q} b_k\,h_k\,{}^{\ell}V_{sz}^k
\end{aligned} \tag{5.61}$$

where the $h_k(r)$ are the interpolation functions summarized in Fig. 5.4 and

${}^{\ell}x,\ {}^{\ell}y,\ {}^{\ell}z$ = Cartesian coordinates of any point in the element
${}^{\ell}x_k,\ {}^{\ell}y_k,\ {}^{\ell}z_k$ = Cartesian coordinates of nodal point k
$a_k,\ b_k$ = Cross-sectional dimensions of the beam at nodal point k
${}^{\ell}V_{tx}^k,\ {}^{\ell}V_{ty}^k,\ {}^{\ell}V_{tz}^k$ = Components of unit vector ${}^{\ell}\mathbf{V}_t^k$ in direction t at nodal point k*
${}^{\ell}V_{sx}^k,\ {}^{\ell}V_{sy}^k,\ {}^{\ell}V_{sz}^k$ = Components of unit vector ${}^{\ell}\mathbf{V}_s^k$ in direction s at nodal point k*

and the left superscript ℓ denotes the configuration of the element; i.e., $\ell = 0$ denotes the original configuration, whereas $\ell = 1$ corresponds to the configuration in the deformed position. Thus, to obtain the displacement components at any point of the element we have

$$\begin{aligned}
u(r, s, t) &= {}^{1}x - {}^{0}x \\
v(r, s, t) &= {}^{1}y - {}^{0}y \\
w(r, s, t) &= {}^{1}z - {}^{0}z
\end{aligned} \tag{5.62}$$

and substituting from (5.61) we obtain

$$\begin{aligned}
u(r, s, t) &= \sum_{k=1}^{q} h_k\,u_k + \frac{t}{2}\sum_{k=1}^{q} a_k\,h_k\,V_{tx}^k + \frac{s}{2}\sum_{k=1}^{q} b_k\,h_k\,V_{sx}^k \\
v(r, s, t) &= \sum_{k=1}^{q} h_k\,v_k + \frac{t}{2}\sum_{k=1}^{q} a_k\,h_k\,V_{ty}^k + \frac{s}{2}\sum_{k=1}^{q} b_k\,h_k\,V_{sy}^k \\
w(r, s, t) &= \sum_{k=1}^{q} h_k w_k + \frac{t}{2}\sum_{k=1}^{q} a_k\,h_k\,V_{tz}^k + \frac{s}{2}\sum_{k=1}^{q} b_k\,h_k\,V_{sz}^k
\end{aligned} \tag{5.63}$$

where $$\mathbf{V}_t^k = {}^{1}\mathbf{V}_t^k - {}^{0}\mathbf{V}_t^k; \qquad \mathbf{V}_s^k = {}^{1}\mathbf{V}_s^k - {}^{0}\mathbf{V}_s^k \tag{5.64}$$

Finally, we express the vectors $\mathbf{V}_t^k$ and $\mathbf{V}_s^k$ in terms of rotations about the Cartesian axes x, y, z:

$$\begin{aligned}
\mathbf{V}_t^k &= \boldsymbol{\theta}_k \times {}^{0}\mathbf{V}_t^k \\
\mathbf{V}_s^k &= \boldsymbol{\theta}_k \times {}^{0}\mathbf{V}_s^k
\end{aligned} \tag{5.65}$$

where $\boldsymbol{\theta}_k$ is a vector listing the nodal point rotations at nodal point k (see Fig. 5.33):

$$\boldsymbol{\theta}_k = \begin{bmatrix} \theta_x^k \\ \theta_y^k \\ \theta_z^k \end{bmatrix} \tag{5.66}$$

Using (5.61) to (5.66) we have all the basic equations necessary to establish the displacement and strain interpolation matrices that are employed in the evaluation of the beam element matrices as expressed in (5.27) to (5.35). The

* Note that these vectors are not necessarily exactly normal to the neutral axis of the beam.

terms in the displacement interpolation matrix **H** are obtained by substituting (5.65) into (5.63). To evaluate the strain-displacement matrix, we recognize that for the beam the only strain components of interest are the longitudinal strain $\epsilon_{\eta\eta}$ and transverse shear strains $\gamma_{\eta\xi}$ and $\gamma_{\eta\zeta}$, where η, ξ, and ζ are convected (body-attached) coordinate axes (see Fig. 5.33). Thus, we seek a relation of the form

$$\begin{bmatrix} \epsilon_{\eta\eta} \\ \gamma_{\eta\xi} \\ \gamma_{\eta\zeta} \end{bmatrix} = \sum_{k=1}^{q} \mathbf{B}_k \, \hat{\mathbf{u}}_k \tag{5.67}$$

where

$$\hat{\mathbf{u}}_k^T = [u_k \; v_k \; w_k \quad \theta_x^k \, \theta_y^k \, \theta_z^k] \tag{5.68}$$

and the matrices $\mathbf{B}_k$, $k = 1, \ldots, q$ constitute together the matrix **B**,

$$\mathbf{B} = [\mathbf{B}_1, \quad \ldots, \quad \mathbf{B}_q] \tag{5.69}$$

Following the usual procedure of isoparametric finite element formulation, we have using (5.63)

$$\begin{bmatrix} \dfrac{\partial u}{\partial r} \\[2ex] \dfrac{\partial u}{\partial s} \\[2ex] \dfrac{\partial u}{\partial t} \end{bmatrix} = \sum_{k=1}^{q} \begin{bmatrix} \dfrac{\partial h_k}{\partial r}[1 & (g)_{1i}^k & (g)_{2i}^k & (g)_{3i}^k] \\[2ex] h_k\,[0 & (\hat{g})_{1i}^k & (\hat{g})_{2i}^k & (\hat{g})_{3i}^k] \\[2ex] h_k\,[0 & (\bar{g})_{1i}^k & (\bar{g})_{2i}^k & (\bar{g})_{3i}^k] \end{bmatrix} \begin{bmatrix} u_k \\ \theta_x^k \\ \theta_y^k \\ \theta_z^k \end{bmatrix} \tag{5.70}$$

and the derivatives of v and w are obtained by simply substituting v and w for u. In (5.70) we have $i = 1$ for u, $i = 2$ for v and $i = 3$ for w, and we employ the notation:

$$(\hat{\mathbf{g}})^k = \frac{b_k}{2} \begin{bmatrix} 0 & -{}^0V_{sz}^k & {}^0V_{sy}^k \\ {}^0V_{sz}^k & 0 & -{}^0V_{sx}^k \\ -{}^0V_{sy}^k & {}^0V_{sx}^k & 0 \end{bmatrix} \tag{5.71}$$

$$(\bar{\mathbf{g}})^k = \frac{a_k}{2} \begin{bmatrix} 0 & -{}^0V_{tz}^k & {}^0V_{ty}^k \\ {}^0V_{tz}^k & 0 & -{}^0V_{tx}^k \\ -{}^0V_{ty}^k & {}^0V_{tx}^k & 0 \end{bmatrix} \tag{5.72}$$

$$(g)_{ij}^k = s(\hat{g})_{ij}^k + t(\bar{g})_{ij}^k \tag{5.73}$$

To obtain the displacement derivatives corresponding to the coordinate axes x, y, and z, we employ the Jacobian transformation

$$\frac{\partial}{\partial \mathbf{x}} = \mathbf{J}^{-1} \frac{\partial}{\partial \mathbf{r}} \tag{5.74}$$

where the Jacobian matrix, **J**, contains the derivatives of the coordinates x, y, z with respect to the natural coordinates r, s, and t (see (5.23)). Substituting from (5.70) into (5.74) we obtain

$$\begin{bmatrix} \dfrac{\partial u}{\partial x} \\[2ex] \dfrac{\partial u}{\partial y} \\[2ex] \dfrac{\partial u}{\partial z} \end{bmatrix} = \sum_{k=1}^{q} \begin{bmatrix} J_{11}^{-1} \dfrac{\partial h_k}{\partial r} & (G1)_{i1}^k & (G2)_{i1}^k & (G3)_{i1}^k \\[2ex] J_{21}^{-1} \dfrac{\partial h_k}{\partial r} & (G1)_{i2}^k & (G2)_{i2}^k & (G3)_{i2}^k \\[2ex] J_{31}^{-1} \dfrac{\partial h_k}{\partial r} & (G1)_{i3}^k & (G2)_{i3}^k & (G3)_{i3}^k \end{bmatrix} \begin{bmatrix} u_k \\ \theta_x^k \\ \theta_y^k \\ \theta_z^k \end{bmatrix} \tag{5.75}$$

and, again, the derivatives oı v and w are obtained by simply substituting v and w for u. In (5.75) we employ the notation

$$(Gm)_{in}^{k} = (J_{n1}^{-1}\,(g)_{mi}^{k})\frac{\partial h_k}{\partial r} + (J_{n2}^{-1}\,(\hat{g})_{mi}^{k} + J_{n3}^{-1}\,(\bar{g})_{mi}^{k})h_k \tag{5.76}$$

Using the displacement derivatives in (5.75) we can now calculate the elements of $\mathbf{B}_k$ by establishing the strain components corresponding to the x, y, z axes and transforming these components to the local strains $\epsilon_{\eta\eta}$, $\gamma_{\eta\xi}$ and $\gamma_{\eta\zeta}$.

The corresponding stress-strain law to be employed in the formulation is then (using k as the shear correction factor),

$$\begin{bmatrix} \tau_{\eta\eta} \\ \tau_{\eta\xi} \\ \tau_{\eta\zeta} \end{bmatrix} = \begin{bmatrix} E & 0 & 0 \\ 0 & Gk & 0 \\ 0 & 0 & Gk \end{bmatrix} \begin{bmatrix} \epsilon_{\eta\eta} \\ \gamma_{\eta\xi} \\ \gamma_{\eta\zeta} \end{bmatrix} \tag{5.77}$$

Considering the formulation in (5.61) to (5.77), it should be recognized that the element can be arbitrarily curved and the cross-sectional dimensions can change along its axis. The width, height, and location of the element neutral axis are interpolated along the element, where it should be noted that if the element is straight and its cross-sectional area is constant, considerable simplifications can result in the calculation of the element matrices (see Example 5.22). Finally, it should also be noted that in addition to representing a general approach to the linear analysis of beam structures, the above formulation is particularly useful as the basis of nonlinear large displacement analysis of beam structures. As discussed in Section 6.3.4, in those analyses initially straight beam elements become curved and distorted, and these deformations can be modeled accurately using the formulation presented above.

EXAMPLE 5.22: Show that the application of the general formulation in (5.61) to (5.77) to the beam element in Fig. 5.31 reduces to the use of (5.51).

For the application of the general relations in (5.61) to (5.77) we refer to Figs. 5.31 and 5.33 and thus have

$$ {}^0\mathbf{V}_s = \begin{bmatrix} 0 \\ 1 \\ 0 \end{bmatrix}; \qquad {}^0\mathbf{V}_t = \begin{bmatrix} 0 \\ 0 \\ 1 \end{bmatrix}; \qquad a_k = 1, \quad b_k = h, \quad k = 1, 2, 3$$

Hence, the relations in (5.61) reduce to

$$\begin{aligned} {}^0x &= \sum_{k=1}^{3} h_k\, {}^0x_k \\ {}^0y &= \frac{s}{2}h \\ {}^0z &= \frac{t}{2} \end{aligned}$$

We next evaluate (5.65) to obtain (see Section 2.6)

$$\mathbf{V}_t^k = \det \begin{bmatrix} \mathbf{e}_x & \mathbf{e}_y & \mathbf{e}_z \\ \theta_x^k & \theta_y^k & \theta_z^k \\ 0 & 0 & 1 \end{bmatrix}$$

or

$$\mathbf{V}_t^k = \theta_y^k\, \mathbf{e}_x - \theta_x^k\, \mathbf{e}_y \tag{a}$$

and
$$\mathbf{V}_s^k = \det \begin{bmatrix} \mathbf{e}_x & \mathbf{e}_y & \mathbf{e}_z \\ \theta_x^k & \theta_y^k & \theta_z^k \\ 0 & 1 & 0 \end{bmatrix}$$

or
$$\mathbf{V}_s^k = -\theta_z^k \, \mathbf{e}_x + \theta_x^k \, \mathbf{e}_z \tag{b}$$

The relations in (a) and (b) correspond to the three-dimensional action of the beam. We allow only rotations about the z-axis, in which case

$$\mathbf{V}_t^k = \mathbf{0}; \qquad \mathbf{V}_s^k = -\theta_z^k \, \mathbf{e}_x$$

Furthermore, we assume that the nodal points can only displace into the y-direction. Hence, (5.63) yields the displacement assumptions,

$$u(r, s) = -\frac{sh}{2} \sum_{k=1}^{3} h_k \, \theta_z^k \tag{c}$$

$$v(r) = \sum_{k=1}^{3} h_k \, v_k \tag{d}$$

where we note that u is only a function of r, s and v is only a function of r. These relations are identical to the displacement assumptions used before, but with the more conventional beam displacement notation we identified the transverse displacement and section rotation at a nodal point with w_k and θ_k instead of v_k and θ_z^k.

Using now (5.70) we obtain

$$\begin{bmatrix} \dfrac{\partial u}{\partial r} \\ \dfrac{\partial u}{\partial s} \end{bmatrix} = \sum_{k=1}^{3} \begin{bmatrix} -\dfrac{sh}{2} \dfrac{\partial h_k}{\partial r} \\ -\dfrac{h}{2} h_k \end{bmatrix} \theta_z^k$$

$$\begin{bmatrix} \dfrac{\partial v}{\partial r} \\ \dfrac{\partial v}{\partial s} \end{bmatrix} = \sum_{k=1}^{3} \begin{bmatrix} \dfrac{\partial h_k}{\partial r} \\ 0 \end{bmatrix} v_k$$

These relations could also directly be obtained by differentiating the displacements in (c) and (d). Since

$$\mathbf{J} = \begin{bmatrix} \dfrac{L}{2} & 0 \\ 0 & \dfrac{h}{2} \end{bmatrix}; \qquad \mathbf{J}^{-1} = \begin{bmatrix} \dfrac{2}{L} & 0 \\ 0 & \dfrac{2}{h} \end{bmatrix}$$

we obtain
$$\begin{bmatrix} \dfrac{\partial u}{\partial x} \\ \dfrac{\partial u}{\partial y} \end{bmatrix} = \sum_{k=1}^{3} \begin{bmatrix} -\dfrac{h}{2} \dfrac{2}{L} s \dfrac{\partial h_k}{\partial r} \\ -h_k \end{bmatrix} \theta_z^k \tag{e}$$

and
$$\begin{bmatrix} \dfrac{\partial v}{\partial x} \\ \dfrac{\partial v}{\partial y} \end{bmatrix} = \sum_{k=1}^{3} \begin{bmatrix} \dfrac{2}{L} \dfrac{\partial h_k}{\partial r} \\ 0 \end{bmatrix} v_k \tag{f}$$

To analyze the response of the beam in Fig. 5.31 we now use the principle of virtual work (see (4.12)) with the appropriate strain measures:

$$\int_{-1}^{+1}\int_{-1}^{+1} [\bar{\epsilon}_{xx} \quad \bar{\gamma}_{xy}] \begin{bmatrix} E & 0 \\ 0 & Gk \end{bmatrix} \begin{bmatrix} \epsilon_{xx} \\ \gamma_{xy} \end{bmatrix} \det \mathbf{J}\, ds\, dr = P\bar{v}\big|_{r=\frac{1}{3}} \tag{g}$$

where

$$\epsilon_{xx} = \frac{\partial u}{\partial x}; \qquad \bar{\epsilon}_{xx} = \frac{\partial \bar{u}}{\partial x}$$

$$\gamma_{xy} = \frac{\partial u}{\partial y} + \frac{\partial v}{\partial x}; \qquad \bar{\gamma}_{xy} = \frac{\partial \bar{u}}{\partial y} + \frac{\partial \bar{v}}{\partial x}$$

Considering now the relations in (e), (f), (g), and (5.51) we recognize that (g) corresponds to (5.51) if we use $\beta \equiv \theta_z$, and $w \equiv v$.

In our discussion so far we have considered continuum elements and beam elements separately. However, the very close relationship between these elements should be recognized; the only differences are the kinematic assumption that plane sections initially normal to the neutral axis remain plane, and the stress assumption that stresses normal to the neutral axis are zero. In the beam formulation presented, the kinematic assumption was directly incorporated in the basic geometry and displacement interpolations and the stress assumption was used in the stress-strain law. Since these two assumptions are the only two basic differences between the beam and continuum elements, it appears that the structural element matrices can also be derived from the continuum element matrices by degeneration. We demonstrate this feature in the following example.

EXAMPLE 5.23: Assume that the strain-displacement matrix of a four-node plane stress element has been derived. Show how the strain-displacement matrix of a two-node beam element can be established.

Figure 5.34 shows the plane stress element with its degrees of freedom and the beam element for which we want to establish the strain-displacement matrix. Consider node 2 of the beam element and nodes 2 and 3 of the plane stress element. The entries in the strain-displacement matrix of the plane stress element are

$$\begin{matrix} & u_2^* & v_2^* & u_3^* & v_3^* & \\ & \downarrow & \downarrow & \downarrow & \downarrow & \end{matrix}$$

$$\mathbf{B}^* = \left[\begin{array}{c|cc|cc|c} \cdots & -\frac{1}{2L}(1+s) & 0 & -\frac{1}{2L}(1-s) & 0 & \cdots \\ & 0 & \frac{1}{2t}(1-r) & 0 & -\frac{1}{2t}(1-r) & \\ & \frac{1}{2t}(1-r) & -\frac{1}{2L}(1+s) & -\frac{1}{2t}(1-r) & -\frac{1}{2L}(1-s) & \end{array}\right] \tag{a}$$

Using now the beam deformation assumptions we have the following kinematic constraints

$$\begin{aligned} u_2^* &= u_2 - \frac{t}{2}\theta_2 \\ u_3^* &= u_2 + \frac{t}{2}\theta_2 \\ v_2^* &= v_2; \qquad v_3^* = v_2 \end{aligned} \tag{b}$$

These constraints are now substituted to obtain from the elements of $\mathbf{B}^*$ in (a) the elements of the strain-displacement matrix of the beam. Using the rows of $\mathbf{B}^*$, we

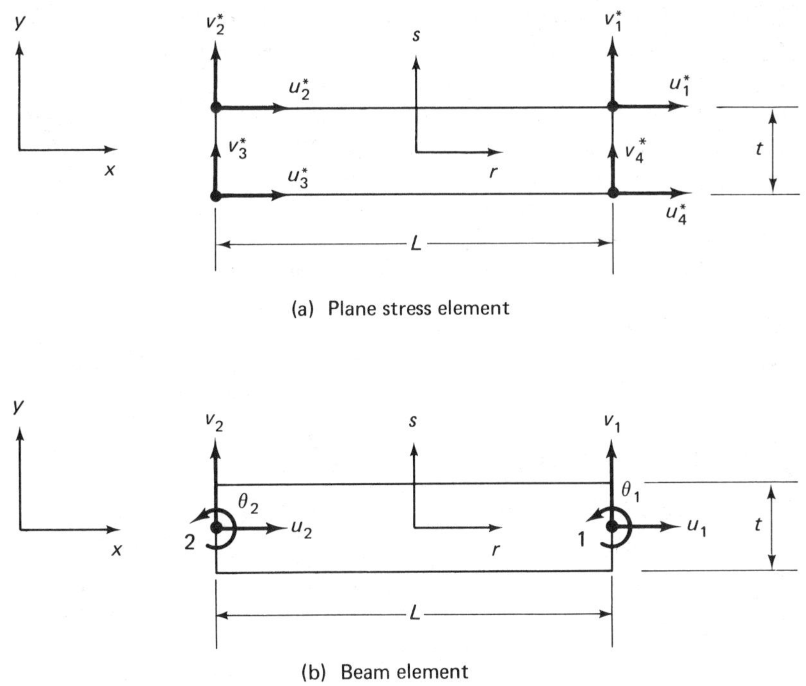

FIGURE 5.34 *Derivation of beam element from plane stress element.*

have with (b),

$$-\frac{1}{2L}(1+s)u_2^* - \frac{1}{2L}(1-s)u_3^* = -\frac{1}{2L}(1+s)\left(u_2 - \frac{t}{2}\theta_2\right) - \frac{1}{2L}(1-s)\left(u_2 + \frac{t}{2}\theta_2\right) \qquad \text{(c)}$$

$$\frac{1}{2t}(1-r)v_2^* - \frac{1}{2t}(1-r)v_3^* = \frac{1}{2t}(1-r)v_2 - \frac{1}{2t}(1-r)v_2 \qquad \text{(d)}$$

$$\frac{1}{2t}(1-r)u_2^* - \frac{1}{2L}(1+s)v_2^* - \frac{1}{2t}(1-r)u_3^* - \frac{1}{2L}(1-s)v_3^*$$
$$= \frac{1}{2t}(1-r)\left(u_2 - \frac{t}{2}\theta_2\right) - \frac{1}{2L}(1+s)v_2 - \frac{1}{2t}(1-r)\left(u_2 + \frac{t}{2}\theta_2\right) - \frac{1}{2L}(1-s)v_2 \qquad \text{(e)}$$

The relations on the r.h.s. of (c), (d), and (e) comprise the entries of the beam strain-displacement matrix

$$\mathbf{B} = \left[\begin{array}{c|ccc} & \overset{u_2}{\downarrow} & \overset{v_2}{\downarrow} & \overset{\theta_2}{\downarrow} \\ & -\frac{1}{L} & 0 & \frac{t}{2L}s \\ \cdots & 0 & 0 & 0 \\ & 0 & -\frac{1}{L} & -\frac{1}{2}(1-r) \end{array}\right]$$

However, the first and third row entries are those that are also obtained using the beam formulation of (5.61) to (5.76). We should note that the zeros in the second row of **B** only express the fact that the strain ϵ_{yy} is not included in the formulation. This strain is actually equal to $-\nu\,\epsilon_{xx}$, because the stress τ_{yy} is zero.

The formulation of a structural element using the approach discussed in the above example is computationally inefficient, and is certainly not recommended for general analysis. However, it is instructive to study this approach and recognize that the structural element matrices can in principle be obtained from continuum element matrices by imposing the appropriate static and kinematic assumptions. Moreover, this formulation directly suggests the construction of transition elements that can be used in an effective manner to couple structural and continuum elements without the use of constraint equations (see Fig. 5.35(a)). To demonstrate the formulation of transition elements, we consider in the following example a simple transition beam element.

EXAMPLE 5.24: Establish the displacement and strain-displacement interpolation matrices of the transition element shown in Fig. 5.35.

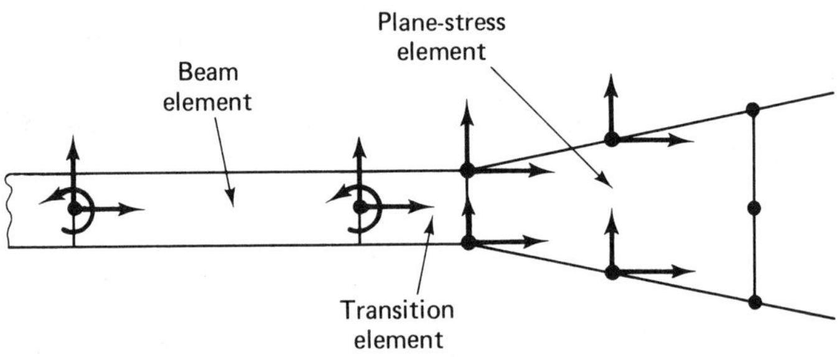

(a) Beam transition element connecting beam and plane stress elements

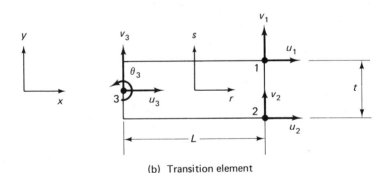

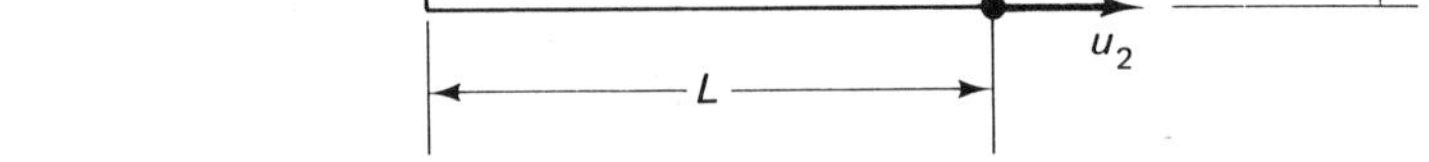

(b) Transition element

FIGURE 5.35 *Two-dimensional beam transition element.*

We define the nodal point displacement vector of the element as

$$\hat{\mathbf{u}}^T = [u_1 \quad v_1 \quad u_2 \quad v_2 \quad u_3 \quad v_3 \quad \theta_3] \tag{a}$$

Since at $r = +1$ we have plane stress element degrees of freedom, the interpolation functions corresponding to nodes 1 and 2 are (see Fig. 5.5)

$$h_1 = \tfrac{1}{4}(1 + r)(1 + s); \qquad h_2 = \tfrac{1}{4}(1 + r)(1 - s)$$

Node 3 is a beam node and the interpolation function is (see Fig. 5.4)

$$h_3 = \tfrac{1}{2}(1 - r)$$

The displacements of the element are thus

$$u(r, s) = h_1 u_1 + h_2 u_2 + h_3 u_3 - \frac{t}{2} s\, h_3 \theta_3$$

$$v(r, s) = h_1 v_1 + h_2 v_2 + h_3 v_3$$

Hence, corresponding to the displacement vector in (a) we have

$$\mathbf{H} = \begin{bmatrix} h_1 & 0 & h_2 & 0 & h_3 & 0 & -\frac{t}{2} s h_3 \\ 0 & h_1 & 0 & h_2 & 0 & h_3 & 0 \end{bmatrix}$$

The coordinate interpolation is the same as that of the four-node plane stress element:

$$x(r, s) = \frac{1}{2}(1 + r)L$$

$$y(r, s) = \frac{s}{2} t$$

Hence $$\mathbf{J} = \begin{bmatrix} \frac{L}{2} & 0 \\ 0 & \frac{t}{2} \end{bmatrix}; \qquad \mathbf{J}^{-1} = \begin{bmatrix} \frac{2}{L} & 0 \\ 0 & \frac{2}{t} \end{bmatrix}$$

Using (5.25) we thus obtain

$$\mathbf{B} = \begin{bmatrix} \frac{1}{2L}(1 + s) & 0 & \frac{1}{2L}(1 - s) & 0 & -\frac{1}{L} & 0 & \frac{t}{2L} s \\ 0 & \frac{1}{2t}(1 + r) & 0 & -\frac{1}{2t}(1 + r) & 0 & 0 & 0 \\ \frac{1}{2t}(1 + r) & \frac{1}{2L}(1 + s) & -\frac{1}{2t}(1 + r) & \frac{1}{2L}(1 - s) & 0 & -\frac{1}{L} & -\frac{1}{2}(1 - r) \end{bmatrix}$$

We may finally note that the last three columns of the **B**-matrix could also have been derived as described in Example 5.23.

The above presented beam elements represent an alternative to the classical Hermitian beam elements (see Example 4.12), and it must be questioned under what circumstances these beam elements are effective, since for a cubic displacement description twice as many degrees of freedom are required. There is no doubt that in the linear analysis of straight thin beams, the Hermitian elements are much more effective. However, the use of the above elements can be efficient when shear deformations are important, in the analysis of curved structures, in the analysis of stiffened shells (because the beam elements representing the stiffeners are compatible with the shell elements discussed in the next section) and in geometric nonlinear analysis, which is discussed in Section 6.3.4.

5.4.2 Plate and Shell Elements

The same procedures that have been employed in the previous section to formulate beam elements can also directly be used to establish effective plate and shell elements.[21-25,27,28] In the following description we proceed as before; namely, we first discuss the formulation of plate elements, which may be regarded to be simple flat shell elements, and then we proceed to summarize the formulation of general shell elements.

The general shell element formulation presented later contains as a special case the calculation of plate elements that are based on the theory of plates with transverse shear deformations included. This theory uses the assumption that "particles of the plate originally on a line that is normal to the undeformed middle surface remain on a straight line during deformation, but this line is not necessarily normal to the deformed middle surface." With this assumption, the displacement components of a point of coordinates x, y, z are in the small displacement bending theory:

$$u = z\beta_x(x, y); \qquad v = -z\beta_y(x, y); \qquad \text{and} \quad w = w(x, y) \tag{5.78}$$

where w is the transverse displacement, β_x and β_y are the rotations of the normal to the undeformed middle surface in the x-z and y-z planes, respectively, see Fig. 5.36. It is instructive to note that in the Kirchhoff plate theory excluding shear deformations, $\beta_x = -w_{,x}$ and $\beta_y = w_{,y}$.

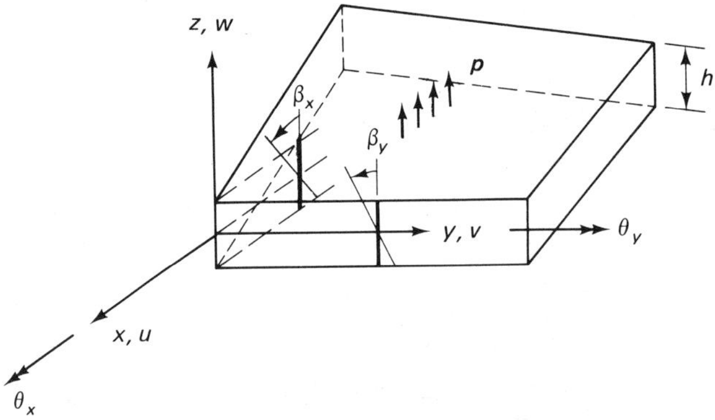

FIGURE 5.36 Deformation assumptions in analysis of plate including shear deformations.

Considering the plate in Fig. 5.36 the bending strains ϵ_{xx}, ϵ_{yy}, γ_{xy} vary linearly through the plate thickness and are given by the curvatures of the plate using (5.78),

$$\begin{bmatrix} \epsilon_{xx} \\ \epsilon_{yy} \\ \gamma_{xy} \end{bmatrix} = z \begin{bmatrix} \dfrac{\partial \beta_x}{\partial x} \\ -\dfrac{\partial \beta_y}{\partial y} \\ \dfrac{\partial \beta_x}{\partial y} - \dfrac{\partial \beta_y}{\partial x} \end{bmatrix} \tag{5.79}$$

whereas the transverse shear strains are assumed to be constant through the thickness

$$\begin{bmatrix} \gamma_{yz} \\ \gamma_{zx} \end{bmatrix} = \begin{bmatrix} \dfrac{\partial w}{\partial y} - \beta_y \\ \dfrac{\partial w}{\partial x} + \beta_x \end{bmatrix} \tag{5.80}$$

The state of stress in the plate corresponds to plane stress conditions, i.e.,

$\tau_{zz} = 0$. Considering an isotropic material we can thus write

$$\begin{bmatrix} \tau_{xx} \\ \tau_{yy} \\ \tau_{xy} \end{bmatrix} = z\frac{E}{1-\nu^2}\begin{bmatrix} 1 & \nu & 0 \\ \nu & 1 & 0 \\ 0 & 0 & \frac{1-\nu}{2} \end{bmatrix}\begin{bmatrix} \frac{\partial \beta_x}{\partial x} \\ -\frac{\partial \beta_y}{\partial y} \\ \frac{\partial \beta_x}{\partial y} - \frac{\partial \beta_y}{\partial x} \end{bmatrix} \tag{5.81}$$

$$\begin{bmatrix} \tau_{yz} \\ \tau_{zx} \end{bmatrix} = \frac{E}{2(1+\nu)}\begin{bmatrix} \frac{\partial w}{\partial y} - \beta_y \\ \frac{\partial w}{\partial x} + \beta_x \end{bmatrix} \tag{5.82}$$

To establish the element equilibrium equations we now proceed as in the formulation of the special two-dimensional beam element of rectangular cross-section (see (5.50) to (5.58)). Considering the plate element the expression for the total potential Π is, with p equal to the transverse loading per unit area,

$$\begin{aligned} \Pi = \frac{1}{2}\int_A \int_{-h/2}^{h/2} [\epsilon_{xx} \quad \epsilon_{yy} \quad \gamma_{xy}]\begin{bmatrix} \tau_{xx} \\ \tau_{yy} \\ \tau_{xy} \end{bmatrix} dz\, dA \\ + \frac{k}{2}\int_A \int_{-h/2}^{h/2} [\gamma_{yz} \quad \gamma_{zx}]\begin{bmatrix} \tau_{yz} \\ \tau_{zx} \end{bmatrix} dz\, dA - \int_A wp\, dA \end{aligned} \tag{5.83}$$

where k is again a constant to account for the actual nonuniformity of the shearing stresses. Substituting from (5.79) to (5.82) into (5.83) we thus obtain

$$\Pi = \tfrac{1}{2}\int_A \boldsymbol{\kappa}^T \mathbf{C}_b \boldsymbol{\kappa}\, dA + \tfrac{1}{2}\int_A \boldsymbol{\gamma}^T \mathbf{C}_s \boldsymbol{\gamma}\, dA - \int_A wp\, dA \tag{5.84}$$

where

$$\boldsymbol{\kappa} = \begin{bmatrix} \frac{\partial \beta_x}{\partial x} \\ -\frac{\partial \beta_y}{\partial y} \\ \frac{\partial \beta_x}{\partial y} - \frac{\partial \beta_y}{\partial x} \end{bmatrix}; \qquad \boldsymbol{\gamma} = \begin{bmatrix} \frac{\partial w}{\partial y} - \beta_y \\ \frac{\partial w}{\partial x} + \beta_x \end{bmatrix} \qquad (5.85)\ (5.86)$$

$$\mathbf{C}_b = \frac{Eh^3}{12(1-\nu^2)}\begin{bmatrix} 1 & \nu & 0 \\ \nu & 1 & 0 \\ 0 & 0 & \frac{1-\nu}{2} \end{bmatrix}; \qquad \mathbf{C}_s = \frac{Ehk}{2(1+\nu)}\begin{bmatrix} 1 & 0 \\ 0 & 1 \end{bmatrix} \tag{5.87}$$

Equilibrium requires that Π is stationary, i.e., $\delta\Pi = 0$ where it must be recognized that w, β_x, and β_y are independent variables. Hence, in the finite element analysis of an assemblage of elements, we only need to enforce interelement continuity on w, β_x and β_y, which can readily be achieved in the same way as in the isoparametric finite element analysis of solids.

Using that the total potential must be stationary, we obtain

$$\int_A \delta\boldsymbol{\kappa}^T \mathbf{C}_b \boldsymbol{\kappa}\, dA + \int_A \delta\boldsymbol{\gamma}^T \mathbf{C}_s \boldsymbol{\gamma}\, dA - \int_A \delta w\; p\, dA = 0 \tag{5.88}$$

which may be regarded as the principle of virtual displacements for the plate element.

For the finite element analysis we use

$$w = \sum_{i=1}^{q} h_i w_i; \qquad \beta_x = \sum_{i=1}^{q} h_i \theta_y^i$$
$$\beta_y = \sum_{i=1}^{q} h_i \theta_x^i \tag{5.89}$$

where the h_i are the interpolation functions and q is the number of nodes of the element. With these interpolations we can now proceed in the usual way, and all concepts pertaining to the isoparametric finite elements discussed earlier are directly applicable. For example, some interpolation functions applicable to the formulation of plate elements are listed in Fig. 5.5, and triangular elements can be established as discussed in Section 5.3.2. Since the interpolation functions are given in terms of the isoparametric coordinates (r, s), we can also directly calculate the matrices of plate elements that are curved in their plane (to model, for example, a circular plate).

We demonstrate the formulation of a simple 4-node element in the following example.

EXAMPLE 5.25: Establish the expressions used in the evaluation of the stiffness matrix of the four-node plate element shown in Fig. 5.37.

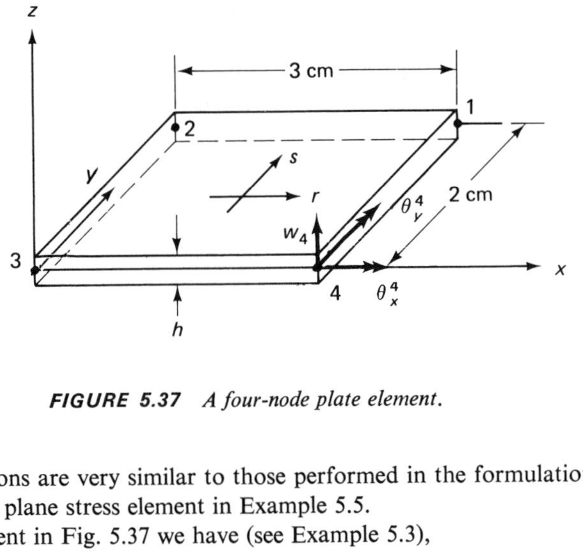

FIGURE 5.37 *A four-node plate element.*

The calculations are very similar to those performed in the formulation of the two-dimensional plane stress element in Example 5.5.

For the element in Fig. 5.37 we have (see Example 5.3),

$$\mathbf{J} = \begin{bmatrix} \frac{3}{2} & 0 \\ 0 & 1 \end{bmatrix}$$

and then, using the interpolation functions defined in Fig. 5.5:

$$\begin{bmatrix} \dfrac{\partial w}{\partial x} \\ \dfrac{\partial w}{\partial y} \end{bmatrix} = \frac{1}{4} \begin{bmatrix} \dfrac{2}{3} & 0 \\ 0 & 1 \end{bmatrix} \begin{bmatrix} (1+s) & -(1+s) & -(1-s) & (1-s) \\ (1+r) & (1-r) & -(1-r) & -(1+r) \end{bmatrix} \begin{bmatrix} w_1 \\ w_2 \\ w_3 \\ w_4 \end{bmatrix}$$

with analogous expressions for the derivatives of β_x and β_y. Thus, if we use the following notation

$$\boldsymbol{\kappa}(r, s) = \mathbf{B}_\kappa \hat{\mathbf{u}}$$
$$\boldsymbol{\gamma}(r, s) = \mathbf{B}_\gamma \hat{\mathbf{u}}$$
$$w(r, s) = \mathbf{H}_w \hat{\mathbf{u}}$$

where $\hat{\mathbf{u}}^T = [w_1 \quad \theta_x^1 \quad \theta_y^1 \quad w_2 \ldots \theta_y^4]$

We have
$$\mathbf{B}_\kappa = \begin{bmatrix} 0 & 0 & \frac{1}{6}(1+s) & \vdots & & \frac{1}{6}(1-s) \\ 0 & -\frac{1}{4}(1+r) & 0 & \vdots & \cdots & 0 \\ 0 & -\frac{1}{6}(1+s) & \frac{1}{4}(1+r) & \vdots & & -\frac{1}{4}(1+r) \end{bmatrix}$$

$$\mathbf{B}_\gamma = \begin{bmatrix} \frac{1}{4}(1+r) & -\frac{1}{4}(1+r)(1+s) & 0 & \vdots & \cdots & 0 \\ \frac{1}{6}(1+s) & 0 & \frac{1}{4}(1+r)(1+s) & \vdots & & \frac{1}{4}(1+r)(1-s) \end{bmatrix}$$

$$\mathbf{H}_w = \tfrac{1}{4}[(1+r)(1+s) \mid \ldots \mid (1+r)(1-s) \mid 0 \mid 0]$$

The element stiffness matrix is then

$$\mathbf{K} = \tfrac{3}{2}\int_{-1}^{+1}\int_{-1}^{+1} (\mathbf{B}_\kappa^T \mathbf{C}_b \mathbf{B}_\kappa + \mathbf{B}_\gamma^T \mathbf{C}_s \mathbf{B}_\gamma)\, dr\, ds \tag{a}$$

and the consistent load vector is

$$\mathbf{R}_S = \tfrac{3}{2}\int_{-1}^{+1}\int_{-1}^{+1} \mathbf{H}_w^T p \, dr\, ds \tag{b}$$

where the integrals in (a) and (b) are evaluated using numerical integration (see Section 5.7).

The above isoparametric elements are effective if used as high-order elements and specifically the 9-node and 16-node quadrilateral Lagrangian elements can be employed efficiently. As for the beam element considered above, the low-order elements are much too stiff in modeling thin plates. However, the behavior of these elements can again be improved by neglecting the shear strain energy in the formulation and imposing the Kirchhoff assumption that the shear deformations are zero at discrete points of the element.

As in the discussion of the beam element in Section 5.4.1, where we regarded the formulation of the matrices corresponding to a two-dimensional beam of rectangular cross-section as a special case of a general three-dimensional formulation, we now can regard the formulation of the plate elements above as a special case of a formulation of a general three-dimensional shell element. In this general shell element formulation, we employ the concepts discussed already in Section 5.4.1, and we follow the same formulative steps.

As in the formulation of the continuum and beam elements, we can formulate the shell element for a variable number of nodes. Figure 5.38 shows a nine-node shell element. Using the natural coordinates r, s, and t, the Cartesian coordinates of a point in the element with q nodal points are, before and after deformations:

$$\begin{aligned} {}^{\ell}x(r, s, t) &= \sum_{k=1}^{q} h_k\, {}^{\ell}x_k + \frac{t}{2}\sum_{k=1}^{q} a_k h_k\, {}^{\ell}V_{nx}^k \\ {}^{\ell}y(r, s, t) &= \sum_{k=1}^{q} h_k\, {}^{\ell}y_k + \frac{t}{2}\sum_{k=1}^{q} a_k h_k\, {}^{\ell}V_{ny}^k \\ {}^{\ell}z(r, s, t) &= \sum_{k=1}^{q} h_k\, {}^{\ell}z_k + \frac{t}{2}\sum_{k=1}^{q} a_k h_k\, {}^{\ell}V_{nz}^k \end{aligned} \tag{5.90}$$

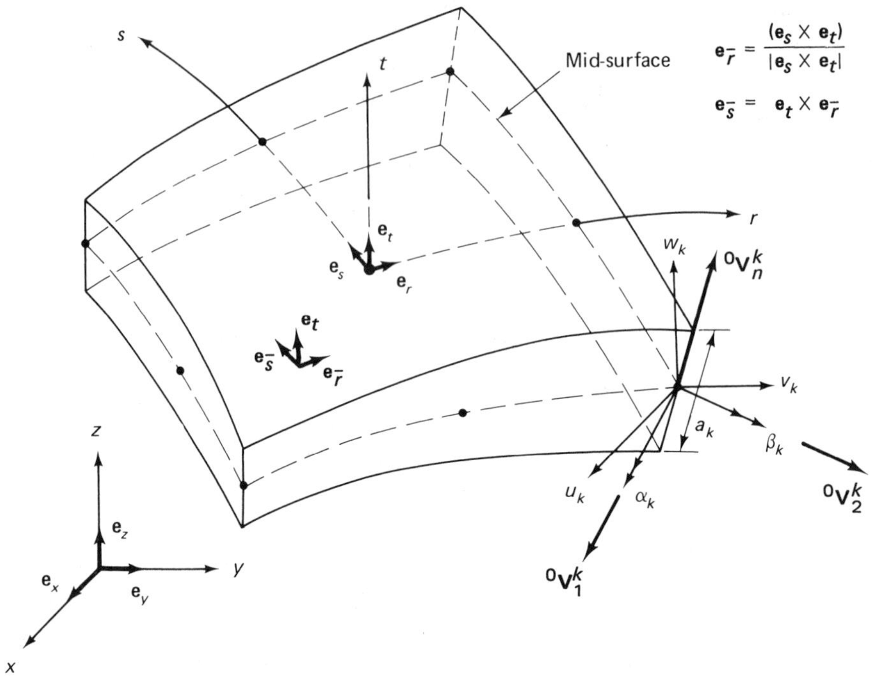

FIGURE 5.38 *Nine-node shell element. Note that the unit vectors* $\mathbf{e}_r$, $\mathbf{e}_s$, $\mathbf{e}_t$ *are in general not orthogonal, whereas* $\mathbf{e}_{\bar{r}}$, $\mathbf{e}_{\bar{s}}$, $\mathbf{e}_t$ *are orthogonal.*

where the $h_k(r, s)$ are the interpolation functions summarized in Fig. 5.5 and

$^\ell x, {}^\ell y, {}^\ell z$ = Cartesian coordinates of any point in the element

$^\ell x_k, {}^\ell y_k, {}^\ell z_k$ = Cartesian coordinates of nodal point k

a_k = Thickness of shell in t direction at nodal point k

$^\ell V_{nx}^k, {}^\ell V_{ny}^k, {}^\ell V_{nz}^k$ = Components of unit vector $^\ell \mathbf{V}_n^k$ "normal" to the shell mid-surface in direction t at nodal point k*

and the left superscript ℓ denotes, as in the general beam formulation, the configuration of the element; i.e. $\ell = 0$ and 1 denote the original and final configurations of the shell element. Hence using (5.90), the displacement components are

$$\begin{aligned} u(r, s, t) &= \sum_{k=1}^{q} h_k\, u_k + \frac{t}{2} \sum_{k=1}^{q} a_k\, h_k\, V_{nx}^k \\ v(r, s, t) &= \sum_{k=1}^{q} h_k\, v_k + \frac{t}{2} \sum_{k=1}^{q} a_k\, h_k\, V_{ny}^k \\ w(r, s, t) &= \sum_{k=1}^{q} h_k\, w_k + \frac{t}{2} \sum_{k=1}^{q} a_k\, h_k\, V_{nz}^k \end{aligned} \tag{5.91}$$

where $\mathbf{V}_n^k$ stores the increments in the direction cosines of ${}^0\mathbf{V}_n^k$.

$$\mathbf{V}_n^k = {}^1\mathbf{V}_n^k - {}^0\mathbf{V}_n^k \tag{5.92}$$

The components of $\mathbf{V}_n^k$ can be expressed in terms of rotations at the nodal point k; however, there is no unique way of proceeding. An efficient way is to

* We call $^\ell\mathbf{V}_n^k$ the *normal* vector although it may not be exactly normal to the mid-surface of the shell in the original configuration, and in the final configuration (for example, because of shear deformations).

define two unit vectors ${}^0\mathbf{V}_1^k$ and ${}^0\mathbf{V}_2^k$ that are orthogonal to ${}^0\mathbf{V}_n^k$:

$$
{}^0\mathbf{V}_1^k = \frac{(\mathbf{e}_y \times {}^0\mathbf{V}_n^k)}{|\mathbf{e}_y \times {}^0\mathbf{V}_n^k|} \tag{5.93a}
$$

where $\mathbf{e}_y$ is a unit vector into the direction of the y-axis. (For the special case ${}^0\mathbf{V}_n^k$ parallel to $\mathbf{e}_y$, we may simply use ${}^0\mathbf{V}_1^k$ equal to $\mathbf{e}_z$.) We can now obtain ${}^0\mathbf{V}_2^k$,

$$
{}^0\mathbf{V}_2^k = {}^0\mathbf{V}_n^k \times {}^0\mathbf{V}_1^k \tag{5.93b}
$$

Let α_k and β_k be the rotations of the normal vector ${}^0\mathbf{V}_n^k$ about the vectors ${}^0\mathbf{V}_1^k$ and ${}^0\mathbf{V}_2^k$. We then have, because α_k and β_k are small angles,

$$
\mathbf{V}_n^k = -{}^0\mathbf{V}_2^k\,\alpha_k + {}^0\mathbf{V}_1^k\beta_k \tag{5.94}
$$

This relationship can readily be proven when ${}^0\mathbf{V}_1 = \mathbf{e}_x$, ${}^0\mathbf{V}_2 = \mathbf{e}_y$ and ${}^0\mathbf{V}_n = \mathbf{e}_z$, but since these vectors are tensors, the relationship must also hold in general (see Section 2.6). Substituting from (5.94) into (5.91) we thus obtain

$$
\begin{aligned}
u(r, s, t) &= \sum_{k=1}^{q} h_k\, u_k + \frac{t}{2} \sum_{k=1}^{q} a_k\, h_k(-{}^0V_{2x}^k\,\alpha_k + {}^0V_{1x}^k\,\beta_k)\\
v(r, s, t) &= \sum_{k=1}^{q} h_k\, v_k + \frac{t}{2} \sum_{k=1}^{q} a_k\, h_k(-{}^0V_{2y}^k\,\alpha_k + {}^0V_{1y}^k\,\beta_k)\\
w(r, s, t) &= \sum_{k=1}^{q} h_k\, w_k + \frac{t}{2} \sum_{k=1}^{q} a_k\, h_k(-{}^0V_{2z}^k\,\alpha_k + {}^0V_{1z}^k\,\beta_k)
\end{aligned} \tag{5.95}
$$

With the element displacements and coordinates defined in (5.95) and (5.90) we can now proceed as usual with the evaluation of the element matrices. The entries in the displacement interpolation matrix $\mathbf{H}$ of the shell element are given in (5.95), and the entries in the strain-displacement interpolation matrix can be calculated using the procedures described already in the formulation of the beam element (see Section 5.4.1).

To evaluate the strain-displacement matrix, we obtain from (5.95)

$$
\begin{bmatrix} \dfrac{\partial u}{\partial r} \\[2ex] \dfrac{\partial u}{\partial s} \\[2ex] \dfrac{\partial u}{\partial t} \end{bmatrix} = \sum_{k=1}^{q} \begin{bmatrix} \dfrac{\partial h_k}{\partial r}[1 \quad t g_{1x}^k \quad t g_{2x}^k] \\[2ex] \dfrac{\partial h_k}{\partial s}[1 \quad t g_{1x}^k \quad t g_{2x}^k] \\[2ex] h_k\,[0 \quad g_{1x}^k \quad g_{2x}^k] \end{bmatrix} \begin{bmatrix} u_k \\[2ex] \alpha_k \\[2ex] \beta_k \end{bmatrix} \tag{5.96}
$$

and the derivatives of v and w are given by simply substituting for u and x the variables v, y and w, z, respectively. In (5.96) we use the notation

$$
\mathbf{g}_1^k = -\tfrac{1}{2}\, a_k\, {}^0\mathbf{V}_2^k; \qquad \mathbf{g}_2^k = \tfrac{1}{2}\, a_k\, {}^0\mathbf{V}_1^k \tag{5.97}
$$

To obtain the displacement derivatives corresponding to the Cartesian coordinates x, y, z, we use the standard transformation

$$
\frac{\boldsymbol{\partial}}{\boldsymbol{\partial} \mathbf{x}} = \mathbf{J}^{-1} \frac{\boldsymbol{\partial}}{\boldsymbol{\partial} \mathbf{r}} \tag{5.98}
$$

where the Jacobian matrix $\mathbf{J}$ contains the derivatives of the coordinates x, y, z with respect to the natural coordinates r, s, and t. Substituting from (5.96) into (5.98), we obtain

$$\begin{bmatrix} \frac{\partial u}{\partial x} \\ \frac{\partial u}{\partial y} \\ \frac{\partial u}{\partial z} \end{bmatrix} = \sum_{k=1}^{q} \begin{bmatrix} \frac{\partial h_k}{\partial x} & g_{1x}^k G_x^k & g_{2x}^k G_x^k \\ \frac{\partial h_k}{\partial y} & g_{1x}^k G_y^k & g_{2x}^k G_y^k \\ \frac{\partial h_k}{\partial z} & g_{1x}^k G_z^k & g_{2x}^k G_z^k \end{bmatrix} \begin{bmatrix} u_k \\ \alpha_k \\ \beta_k \end{bmatrix} \tag{5.99}$$

and the derivatives of v and w are obtained in an analogous manner. In (5.99) we have

$$\frac{\partial h_k}{\partial x} = J_{11}^{-1} \frac{\partial h_k}{\partial r} + J_{12}^{-1} \frac{\partial h_k}{\partial s}$$

$$G_x^k = t\left(J_{11}^{-1} \frac{\partial h_k}{\partial r} + J_{12}^{-1} \frac{\partial h_k}{\partial s}\right) + J_{13}^{-1} h_k$$

where J_{ij}^{-1} is element (i, j) of $\mathbf{J}^{-1}$, and so on.

With the displacement derivatives defined in (5.99) we can now directly assemble the strain-displacement matrix **B** of a shell element. Assuming that the rows in this matrix correspond to all six global Cartesian strain components, $\epsilon_{xx}, \epsilon_{yy}, \ldots, \gamma_{zx}$, the entries in **B** are constructed in the usual way (see Section 5.3), but then the stress-strain law must contain the shell assumption that the stress normal to the shell surface is zero. Thus, if $\boldsymbol{\tau}$ and $\boldsymbol{\epsilon}$ store the Cartesian stress and strain components, we use

$$\boldsymbol{\tau} = \mathbf{C}_{sh}\boldsymbol{\epsilon} \tag{5.100}$$

where

$$\boldsymbol{\tau}^T = [\tau_{xx} \quad \tau_{yy} \quad \tau_{zz} \quad \tau_{xy} \quad \tau_{yz} \quad \tau_{zx}];$$
$$\boldsymbol{\epsilon}^T = [\epsilon_{xx} \quad \epsilon_{yy} \quad \epsilon_{zz} \quad \gamma_{xy} \quad \gamma_{yz} \quad \gamma_{zx}]$$

$$\mathbf{C}_{sh} = \mathbf{Q}_{sh}^T\left(\frac{E}{1-\nu^2}\begin{bmatrix} 1 & \nu & 0 & 0 & 0 & 0 \\ & 1 & 0 & 0 & 0 & 0 \\ & & 0 & 0 & 0 & 0 \\ & & & \frac{1-\nu}{2} & 0 & 0 \\ & \text{symmetric} & & & k\,\frac{1-\nu}{2} & 0 \\ & & & & & k\,\frac{1-\nu}{2} \end{bmatrix}\right)\mathbf{Q}_{sh} \tag{5.101}$$

and $\mathbf{Q}_{sh}$ represents a matrix that transforms the stress-strain law from an $\bar{r}$, $\bar{s}$, t Cartesian shell-aligned coordinate system to the global Cartesian coordinate system. The elements of the matrix $\mathbf{Q}_{sh}$ are obtained from the direction cosines of the $\bar{r}$, $\bar{s}$, t coordinate axes measured in the x, y, z coordinate directions,

$$\mathbf{Q}_{sh} = \begin{bmatrix} l_1^2 & m_1^2 & n_1^2 & l_1 m_1 & m_1 n_1 & n_1 l_1 \\ l_2^2 & m_2^2 & n_2^2 & l_2 m_2 & m_2 n_2 & n_2 l_2 \\ l_3^2 & m_3^2 & n_3^2 & l_3 m_3 & m_3 n_3 & n_3 l_3 \\ 2l_1 l_2 & 2m_1 m_2 & 2n_1 n_2 & l_1 m_2 + l_2 m_1 & m_1 n_2 + m_2 n_1 & n_1 l_2 + n_2 l_1 \\ 2l_2 l_3 & 2m_2 m_3 & 2n_2 n_3 & l_2 m_3 + l_3 m_2 & m_2 n_3 + m_3 n_2 & n_2 l_3 + n_3 l_2 \\ 2l_3 l_1 & 2m_3 m_1 & 2n_3 n_1 & l_3 m_1 + l_1 m_3 & m_3 n_1 + m_1 n_3 & n_3 l_1 + n_1 l_3 \end{bmatrix} \tag{5.102}$$

$$\text{where}\quad \begin{array}{lll} l_1 = \cos(\mathbf{e}_x, \mathbf{e}_{\bar{r}}) & m_1 = \cos(\mathbf{e}_y, \mathbf{e}_{\bar{r}}) & n_1 = \cos(\mathbf{e}_z, \mathbf{e}_{\bar{r}}) \\ l_2 = \cos(\mathbf{e}_x, \mathbf{e}_{\bar{s}}) & m_2 = \cos(\mathbf{e}_y, \mathbf{e}_{\bar{s}}) & n_2 = \cos(\mathbf{e}_z, \mathbf{e}_{\bar{s}}) \\ l_3 = \cos(\mathbf{e}_x, \mathbf{e}_t) & m_3 = \cos(\mathbf{e}_y, \mathbf{e}_t) & n_3 = \cos(\mathbf{e}_z, \mathbf{e}_t) \end{array}$$

and the relation in (5.101) corresponds to a fourth-order tensor transformation as described in Section 2.6.

It follows that in the analysis of a general shell the matrix $\mathbf{Q}_{sh}$ may have to be evaluated anew at each integration point that is employed in the numerical integration of the stiffness matrix (see Section 5.7). However, when special shells are considered and, in particular, when a plate is analyzed, the transformation matrix and the stress-strain matrix $\mathbf{C}_{sh}$ need only be evaluated at specific points and can then be employed repetitively. For example, in the analysis of an assemblage of flat plates, the stress-strain matrix $\mathbf{C}_{sh}$ need only be calculated once for each flat structural part.

In the above formulation we assumed that the strain-displacement matrix is formulated corresponding to the Cartesian strain components, which can be directly established using the derivatives in (5.99). Alternatively, we could calculate the strain components corresponding to coordinate axes aligned with the shell element midsurface and establish a strain-displacement matrix for these strain components, as we did in the formulation of the general beam element in Section 5.4.1. The relative computational efficiency of these two approaches depends on whether it is more effective to transform the strain components (which always differ at the integration points) or to transform the stress-strain law.

It is instructive to compare the above shell element formulation with a formulation in which flat elements with a superimposed plate bending and membrane stress behavior are employed (see Section 4.2.3). To identify the differences, assume that the general shell element is used as a flat element in the modeling of a shell; then the stiffness matrix of this element could also be obtained by superimposing the plate bending stiffness matrix derived in (5.78) to (5.88) (see Example 5.23) and the plane stress stiffness matrix discussed in Section 5.3. Thus, in this case, the general shell element reduces to a plate bending element plus a plane stress element, but a computational difference lies in the fact that these element matrices are calculated by integrating numerically only in the *r-s* element midplanes, whereas in the shell element stiffness calculation numerical integration is also performed in the *t*-direction (unless the general formulation is modified for this special case).

The major advantage of the general shell element formulation can now be recognized; namely, using this formulation the geometry of any shell surface can be represented accurately and all displacement compatibility conditions between elements are satisfied directly and in an effective manner. This generality of the analysis capabilities is increased still further if the element is also implemented as a transition element, which can be formulated analogously to the beam transition element discussed in Section 5.4.1. As illustrated in Fig. 5.39, with a shell transition element it is possible to model shell intersections and shell-to-solid transitions using compatible element idealizations without the use of special constraint equations. The above features of generality and accuracy in the representation of a shell geometry can be particularly important in

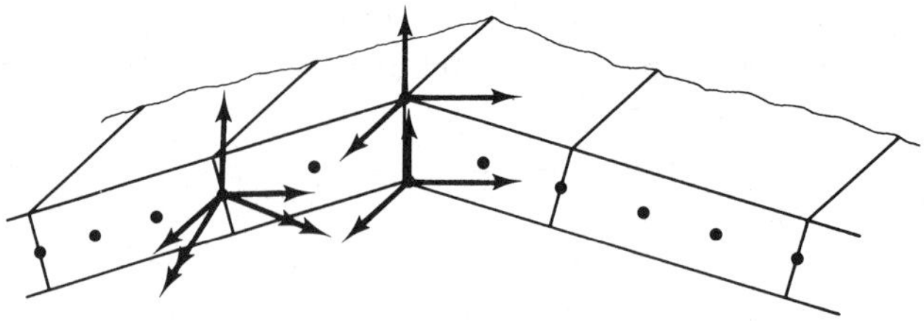

(a) Shell intersections

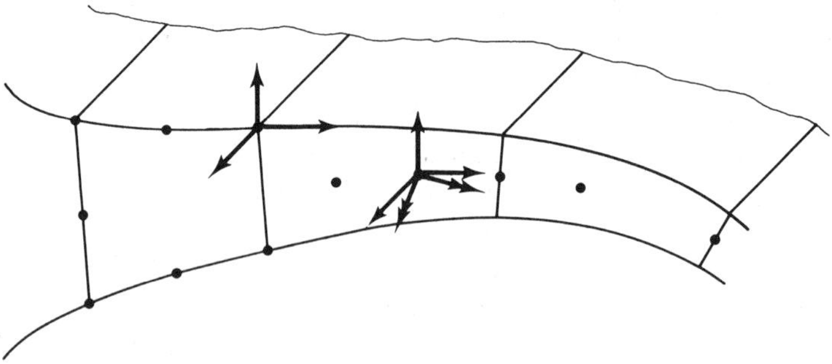

(b) Solid to shell intersection

FIGURE 5.39 *Use of shell transition elements.*

the geometric nonlinear analysis of shell structures, in which any change in the shell geometry must be accounted for accurately. We discuss the formulation of the general shell element for nonlinear analysis in Section 6.3.5.

5.5 CONVERGENCE CONSIDERATIONS

In Section 4.2.5 we discussed the requirements for monotonic convergence of a finite element analysis. The two requirements are that the elements must be compatible and complete. We now want to investigate whether the continuum and structural isoparametric element formulations satisfy these convergence criteria.

5.5.1 Continuum Elements

To investigate the compatibility of an element assemblage, we need to consider each edge, or rather face, between adjacent elements. For compatibility it is necessary that the coordinates and the displacements of the elements at the common face be the same. This is the case if the elements have the same nodes on the common face and the coordinates and displacements along the common face are in each element defined by the same interpolation functions. Examples of adjacent elements that preserve compatibility, and that do not, are shown in Fig. 5.40.

Completeness requires that the rigid body displacements and constant strain states be possible. One way to investigate whether these criteria are satisfied for an isoparametric element is to follow the considerations given in Section 4.2.5. However, we now want to get more insight into the specific conditions that per-

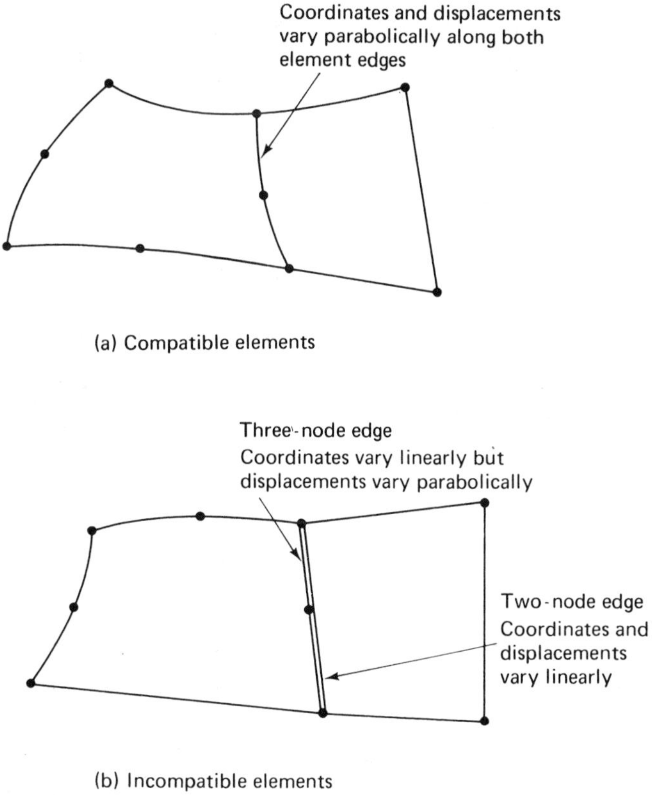
Coordinates and displacements vary parabolically along both element edges
(a) Compatible elements
Three-node edge
Coordinates vary linearly but displacements vary parabolically
Two-node edge
Coordinates and displacements vary linearly
(b) Incompatible elements

FIGURE 5.40 *Compatible and incompatible two-dimensional elements.*

tain to the isoparametric formulation of a continuum element. For this purpose we consider in the following a three-dimensional continuum element because the one- and two-dimensional elements can be regarded as special cases of these three-dimensional considerations. For the rigid body and constant strain states to be possible, the following displacements defined in the local element coordinate system must be contained in the isoparametric formulation

$$\left.\begin{aligned} u &= a_1 + b_1 x + c_1 y + d_1 z \\ v &= a_2 + b_2 x + c_2 y + d_2 z \\ w &= a_3 + b_3 x + c_3 y + d_3 z \end{aligned}\right\} \tag{5.103}$$

where the a_j, b_j, c_j and d_j, $j = 1, 2, 3$, are constants. The nodal point displacements corresponding to this displacement field are

$$\left.\begin{aligned} u_i &= a_1 + b_1 x_i + c_1 y_i + d_1 z_i \\ v_i &= a_2 + b_2 x_i + c_2 y_i + d_2 z_i \\ w_i &= a_3 + b_3 x_i + c_3 y_i + d_3 z_i \end{aligned}\right\} \tag{5.104}$$

where $i = 1, \ldots, q$ and $q =$ number of nodes. To show that the displacements in (5.103) are possible when the isoparametric formulation is employed, assume that the nodal point displacements of the element are given by (5.104). We should find that, with these nodal point displacements, the displacements in the isoparametric formulation are actually those given in (5.103). In the isoparametric formulation we have the displacement interpolation

$$u = \sum_{i=1}^{q} h_i u_i; \qquad v = \sum_{i=1}^{q} h_i v_i; \qquad w = \sum_{i=1}^{q} h_i w_i$$

which using (5.104) reduces to

$$\left.\begin{aligned} u &= a_1 \sum_{i=1}^{q} h_i + b_1 \sum_{i=1}^{q} h_i x_i + c_1 \sum_{i=1}^{q} h_i y_i + d_1 \sum_{i=1}^{q} h_i z_i \\ v &= a_2 \sum_{i=1}^{q} h_i + b_2 \sum_{i=1}^{q} h_i x_i + c_2 \sum_{i=1}^{q} h_i y_i + d_2 \sum_{i=1}^{q} h_i z_i \\ w &= a_3 \sum_{i=1}^{q} h_i + b_3 \sum_{i=1}^{q} h_i x_i + c_3 \sum_{i=1}^{q} h_i y_i + d_3 \sum_{i=1}^{q} h_i z_i \end{aligned}\right\} \tag{5.105}$$

Since in the isoparametric formulation the coordinates are interpolated in the same way as the displacements, we can use (5.18) to obtain from (5.105)

$$\begin{aligned} u &= a_1 \sum_{i=1}^{q} h_i + b_1 x + c_1 y + d_1 z \\ v &= a_2 \sum_{i=1}^{q} h_i + b_2 x + c_2 y + d_2 z \\ w &= a_3 \sum_{i=1}^{q} h_i + b_3 x + c_3 y + d_3 z \end{aligned} \tag{5.106}$$

The displacements defined in (5.106), however, are the same as those given in (5.104), provided that for any point in the element

$$\sum_{i=1}^{q} h_i = 1 \tag{5.107}$$

The relation in (5.107) is the condition on the interpolation functions for the completeness requirements to be satisfied. We may note that (5.107) is certainly satisfied at the nodes of an element, because the interpolation function h_i has been constructed to be unity at node i with all other interpolation functions $h_j, j \neq i$, being zero at that node; but in order that an isoparametric element be properly constructed, the condition must be satisfied for all points in the element.

In the above discussion, we have considered only a three-dimensional continuum element, but the conclusions are also directly applicable to the other isoparametric continuum element formulations. For the one- or two-dimensional continuum elements we simply include only the appropriate displacement and coordinate interpolations in the relations (5.103) to (5.107). We demonstrate the covergence considerations in the following example.

EXAMPLE 5.26: Investigate whether the requirements for monotonic convergence are satisfied for the variable-number-nodes elements in Figs. 5.5 and 5.6.

Compatibility is maintained between element edges in two-dimensional analysis and element faces in three-dimensional analysis, provided that the same number of nodes is used on connecting element edges and faces. A typical compatible element layout is shown in Fig. 5.40(a).

The second requirement for monotonic convergence is the completeness condition. Considering first the basic four-node two-dimensional element, we recognize that the arguments leading to the condition in (5.107) are directly applicable, provided only the x and y coordinates and u and v displacements are considered.

Evaluating $\sum_{i=1}^{4} h_i$ for the element we find

$$\tfrac{1}{4}(1 + r)(1 + s) + \tfrac{1}{4}(1 - r)(1 + s) + \tfrac{1}{4}(1 - r)(1 - s) + \tfrac{1}{4}(1 + r)(1 - s) = 1$$

Hence the basic four-node element is complete. We now study the interpolation functions given in Fig. 5.5 for the variable-number-nodes element and find that the total contribution that is added to the basic four-node interpolation functions is always zero for whichever additional node is included. Hence any one of the possible elements defined by the variable number of nodes in Fig. 5.5 is complete. The proof for the three-dimensional elements in Fig. 5.6 is carried out in an analogous manner.

It follows, therefore, that the variable-number-nodes continuum elements satisfy the requirements for monotonic convergence.

5.5.2 Structural Elements

The compatibility requirements between structural elements are satisfied in the same way as when using continuum elements (see Fig. 5.41). Furthermore, the use of transition elements as described in Section 5.4.1 enables also a

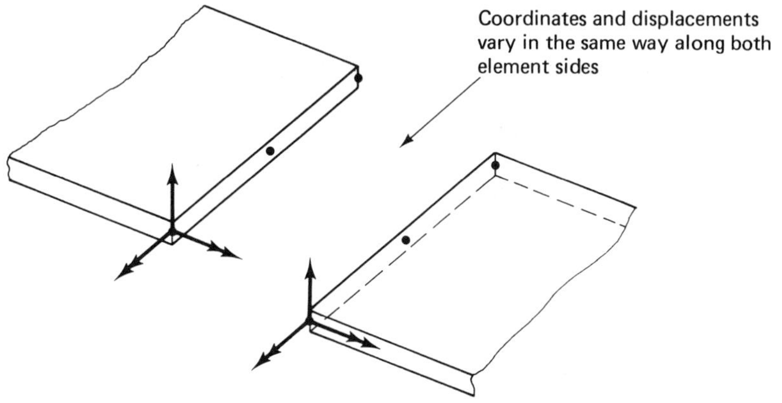

(a) Compatible element layout

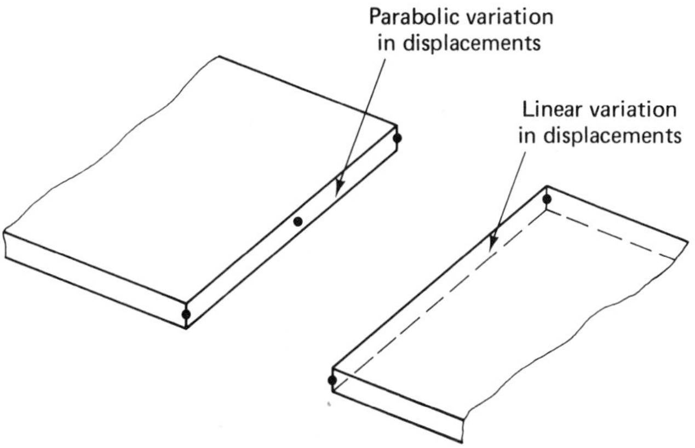

(b) Incompatible element layout

FIGURE 5.41 *Compatible and incompatible plate element layouts.*

compatible transition between structural and continuum elements. However, some additional considerations are necessary when studying the completeness of a structural or transition element.

We discussed in Section 5.4.1 (see Example 5.23) that the structural elements could be derived from the continuum elements, although such a derivation would in practice be computationally ineffective. The fact that in theory the structural elements can be considered as "continuum elements subject to certain constraints" renders the completeness considerations for the continuum elements somewhat applicable to the structural elements—but only within the limits of the constraints imposed. Therefore, the effect of these constraints must be investigated and each structural element must be subjected to a more detailed convergence analysis. We give an analysis of this type in the following example.

EXAMPLE 5.27: Examine whether the convergence requirements are satisfied for the three-node beam element shown in Fig. 5.42(a) and discussed in Example 5.20.

The element is compatible because w and β continuity is ensured between elements.

The condition given in (5.107) requires that $\sum_{i=1}^{3} h_i = 1$, which is satisfied here.

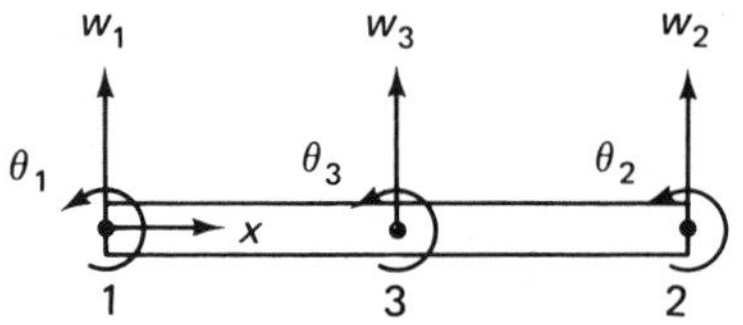

(a) Degrees of freedom of beam element

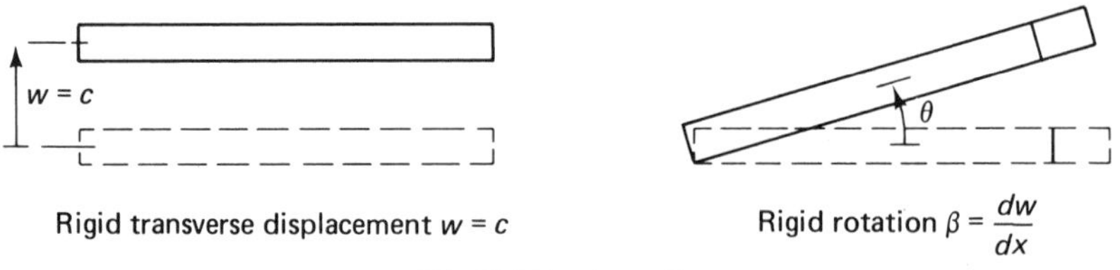

(b) Rigid body motions

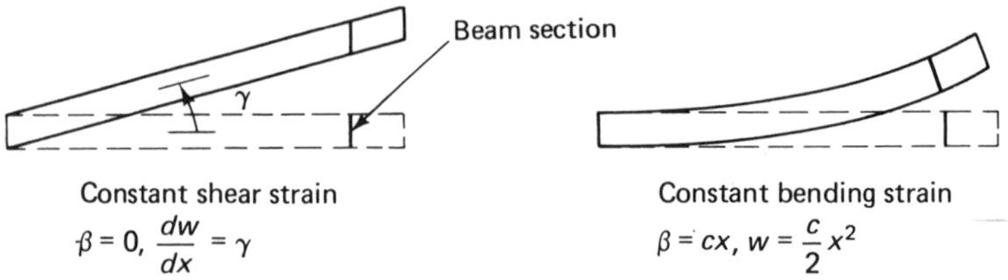

(c) Constant strain states

FIGURE 5.42 *Convergence requirements of three-node beam.*

However, since this is a structural element, we need to investigate in detail how the conditions of completeness are contained in the formulation.

The element must be able to undergo two rigid body modes (see Fig. 5.42(b)); a rigid transverse displacement and a rigid rotation. The rigid transverse displacement is obtained using $w_i = c$, $i = 1, 2, 3$, where c is a constant. The rigid rotation is obtained using $w_1 = 0$, $w_2 = \theta L$, $w_3 = \theta L/2$ and $\beta_i = \theta$, $i = 1, 2, 3$, where θ is a constant. Hence the rigid body mode criterion is satisfied.

The constant strain states of interest are a constant shear strain and a constant bending strain (see Fig. 5.42(c)). For a constant shear strain we want $\beta = 0$ and dw/dx equal to a constant. This state is obtained when for the element $\beta_i = 0$, $i = 1, 2, 3$ and $w_1 = 0$, $w_2 = \gamma L$, $w_3 = \gamma L/2$. For a constant bending strain we need $d\beta/dx$ equal to a constant and $\beta = dw/dx$. This state is reached when $\beta = cx$ and $w = cx^2/2$, which corresponds to the element nodal point variables $\beta_1 = 0$, $\beta_2 = cL$, $\beta_3 = cL/2$ and $w_1 = 0$, $w_2 = cL^2/2$, $w_3 = cL^2/8$. Hence, the two constant strain states of interest are contained in the element formulation and the element is therefore complete.

Using the information of the above example, it is interesting to note that the two-node beam element (i.e. the element in Fig. 5.42 without the center node) cannot represent the constant bending strain condition without shearing strains, because β and w vary at most linearly over the element. This observation shows once more why the two-node beam element cannot be used in the analysis of thin structures. (However, the constant bending strain condition would be represented in a discrete Kirchhoff theory element, because the shear deformations are not included in the formulation (see Example 5.21).)

5.6 ASSOCIATED ELEMENT FAMILIES

In the isoparametric finite element derivations of continuum elements, we assumed that the element coordinates are interpolated in exactly the same way as the element displacements. We may, however, find it expedient in the solution process to interpolate the coordinates different from the displacements, and certain displacement components different from the others. Thus, there appear to be various possibilities of constructing new elements, and in fact the structural elements presented in Section 5.4 represent already a group of such new elements. In the following sections some considerations pertaining to such elements are discussed.

5.6.1 Use of Different Interpolations

In the discussion of the structural elements in Section 5.4, we showed that these elements can be thought of as isoparametric continuum elements with displacement constraints (and the transverse normal stresses were also set to zero). Thus, in this case the element geometry is, in essence, interpolated to a higher degree than the displacements. Elements with a higher degree of interpolation on the coordinates than on the displacements are also called, in general, *superparametric elements*. Considering the convergence properties of such elements, each element formulation must be studied individually in order to identify whether the formulation satisfies the specific compatibility and completeness requirements of the problem considered.

Another class of elements is obtained when the coordinates are interpolated to a lower degree than the displacements. Typically, for the straight-sided eight-node quadrilateral in Fig. 5.43 the displacements u and v are interpolated parabolically, whereas the coordinates x and y of the element need only be interpolated linearly because the element edges are straight. Elements with a lower degree of interpolation on the coordinates than on the displacements are called *subparametric elements.* These continuum elements satisfy the monotonic convergence requirements, because, firstly, compatibility is satisfied for the same reasoning that is used in the isoparametric formulation. Secondly, a sub-parametric element is complete because the associated isoparametric element, which is based only on the coordinate interpolations used, is complete. This follows because the addition of higher-order displacement interpolations does not prevent the element from still undergoing the rigid body modes and constant strain states.

In the above formulations we assumed that each node with displacement degrees of freedom is assigned one degree of freedom corresponding to each unknown displacement. However, in some analyses it is advantageous to interpolate some unknown state variables to a higher degree than the others. In this case, a node may be assigned degrees of freedom corresponding to all state variables or a smaller number. This procedure is employed effectively, for example, in the finite element analysis of fluid flow (see Section 7.4).

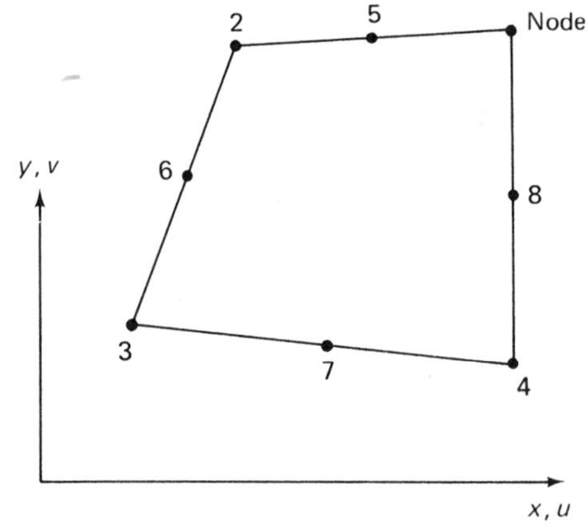

$$x = \sum_{i=1}^{4} h_i x_i; \quad y = \sum_{i=1}^{4} h_i y_i$$

$$u = \sum_{i=1}^{8} h_i u_i; \quad v = \sum_{i=1}^{8} h_i v_i$$

FIGURE 5.43 *Two-dimensional subparametric element.*

5.6.2 Addition of Incompatible Modes

An important procedure to improve the behavior of isoparametric elements is the introduction of incompatible modes.[29,30] Consider the simplest case, the addition of incompatible modes to a four-node two-dimensional continuum element. In this case the coordinate and displacement interpolations used are, respectively,

$$x = \sum_{i=1}^{4} h_i x_i; \qquad y = \sum_{i=1}^{4} h_i y_i \tag{5.108}$$

and

$$\left.\begin{aligned} u &= \sum_{i=1}^{4} h_i u_i + \alpha_1(1 - r^2) + \alpha_2(1 - s^2) \\ v &= \sum_{i=1}^{4} h_i v_i + \beta_1(1 - r^2) + \beta_2(1 - s^2) \end{aligned}\right\} \tag{5.109}$$

where the $h_1 \ldots, h_4$ are as defined in Fig. 5.5 or Example 5.5. The $\alpha_1, \alpha_2, \beta_1$, and β_2 are additional degrees of freedom, with a higher-order displacement interpolation than is present in the h_i. If the element considered is rectangular, the additional degrees of freedom allow the element to represent a constant bending moment, and the element will perform much better in many stress analyses.

Considering the actual evaluation of the element stiffness matrix, we note that using the interpolation in (5.109) a matrix of order 12 is obtained, and that the degrees of freedom $\alpha_1, \alpha_2, \beta_1$, and β_2 are not associated with an element nodal point. To reduce the matrix to order 8, the additional degrees of freedom are eliminated using static condensation (see Section 8.2.4), which is equivalent to minimizing the element total potential energy with respect to the variables $\alpha_1, \alpha_2, \beta_1$, and β_2. We shall discuss in Section 5.8.1 that the same objective is pursued by using reduced order or selective numerical integration.

Since the additional degrees of freedom are not associated with nodal point degrees of freedom, in the analysis using the element, displacement incompatibilities arise between elements. Therefore, the element does not satisfy the requirements for monotonic convergence discussed in Section 4.2.5, and an important question is whether the element does converge in a nonmonotonic manner. Referring to Section 4.3.1 where we discussed the patch test, the results are that *the four-node incompatible element converges provided that the element is a rectangle* (*or parallelogram*) (see Example 4.20).

Although the use of the variable-number-nodes elements given in Figs. 5.5 and 5.6 is considered to be most effective, it should be noted that in three-dimensional analysis, the use of incompatible modes can in some cases decrease the analysis cost considerably. An element that has been used with much success is the eight-node brick element with incompatible modes analogous to those used in (5.109); i.e., the displacement interpolations employed are

$$\left.\begin{aligned} u &= \sum_{i=1}^{8} h_i u_i + \alpha_1(1 - r^2) + \alpha_2(1 - s^2) + \alpha_3(1 - t^2) \\ v &= \sum_{i=1}^{8} h_i v_i + \beta_1(1 - r^2) + \beta_2(1 - s^2) + \beta_3(1 - t^2) \\ w &= \sum_{i=1}^{8} h_i w_i + \gamma_1(1 - r^2) + \gamma_2(1 - s^2) + \gamma_3(1 - t^2) \end{aligned}\right\} \tag{5.110}$$

where the $\alpha_1, \alpha_2, \ldots, \gamma_3$ are displacement components that are not associated with nodal points and are eliminated using static condensation. In analogy to the two-dimensional element discussed above, the three-dimensional brick should be rectangular in all three planes. Observing this rule, the element has been employed successfully in a variety of three-dimensional and shell analyses. Finally, it should be noted that the concept of introducing incompatible modes to increase the effectiveness of an element is quite general, and can also be employed in the formulation of higher-order elements.

5.7 NUMERICAL INTEGRATION

An important aspect of isoparametric and related finite element analysis is the required numerical integration. The required matrix integrals in the finite element calculations have been written as

$$\int \mathbf{F}(r)\,dr, \qquad \int \mathbf{F}(r, s)\,dr\,ds, \qquad \int \mathbf{F}(r, s, t)\,dr\,ds\,dt \tag{5.111}$$

in the one-, two-, and three-dimensional cases, respectively. It was stated that these integrals are in practice evaluated numerically using

$$\left.\begin{aligned} \int \mathbf{F}(r)\,dr &= \sum_i \alpha_i \mathbf{F}(r_i) + \mathbf{R}_n \\ \int \mathbf{F}(r, s)\,dr\,ds &= \sum_{i,j} \alpha_{ij} \mathbf{F}(r_i, s_j) + \mathbf{R}_n \\ \int \mathbf{F}(r, s, t)\,dr\,ds\,dt &= \sum_{i,j,k} \alpha_{ijk} \mathbf{F}(r_i, s_j, t_k) + \mathbf{R}_n \end{aligned}\right\} \tag{5.112}$$

where the summations extend over all i, j, and k specified, the α_i, α_{ij}, and α_{ijk} are weighting factors, and $\mathbf{F}(r_i)$, $\mathbf{F}(r_i, s_j)$, and $\mathbf{F}(r_i, s_j, t_k)$ are the matrices $\mathbf{F}(r)$, $\mathbf{F}(r, s)$, and $\mathbf{F}(r, s, t)$ evaluated at the points specified in the arguments. The matrices $\mathbf{R}_n$ are error matrices, which, in practice, are usually not evaluated. Therefore, we use

$$\left.\begin{aligned} \int \mathbf{F}(r)\,dr &= \sum_i \alpha_i \mathbf{F}(r_i) \\ \int \mathbf{F}(r, s)\,dr\,ds &= \sum_{i,j} \alpha_{ij} \mathbf{F}(r_i, s_j) \\ \int \mathbf{F}(r, s, t)\,dr\,ds\,dt &= \sum_{i,j,k} \alpha_{ijk} \mathbf{F}(r_i, s_j, t_k) \end{aligned}\right\} \tag{5.113}$$

The purpose in this section and Section 5.8.1 is to present the theory and practical implications of numerical integrations. An important point is the integration accuracy that is needed, i.e., the number of integration points required in the element formation.

As presented above, in finite element analysis we integrate matrices, which means that each element of the matrix considered is integrated individually. Hence, for the derivation of the numerical integration formulas we can consider a typical element of a matrix which we denote as F.

Consider the one-dimensional case first, i.e., the integration of $\int_a^b F(r)\,dr$. In an isoparametric element calculation we would actually have $a = -1$ and $b = +1$.

The numerical integration of $\int_a^b F(r)\,dr$ is essentially based on passing a polynomial $\psi(r)$ through given values of $F(r)$ and then using $\int_a^b \psi(r)\,dr$ as an approximation to $\int_a^b F(r)\,dr$. The number of evaluations of $F(r)$ and the positions of the sampling points in the interval from a to b determines how well $\psi(r)$ approximates $F(r)$ and hence the error of the numerical integration.

5.7.1 Interpolation Using a Polynomial

Assume that $F(r)$ has been evaluated at the $n + 1$ distinct points $r_0, r_1, \ldots, r_n$ to obtain $F_0, F_1, \ldots, F_n$, respectively, and that a polynomial $\psi(r)$ is to be passed through these data. Then there is a unique polynomial $\psi(r)$ given as

$$\psi(r) = a_0 + a_1 r + a_2 r^2 + \ldots + a_n r^n \tag{5.114}$$

Using the condition that $\psi(r) = F(r)$ at the $n + 1$ interpolating points, we have

$$\mathbf{F} = \mathbf{Va} \tag{5.115}$$

where

$$\mathbf{F} = \begin{bmatrix} F_0 \\ F_1 \\ \cdot \\ \cdot \\ \cdot \\ F_n \end{bmatrix}; \qquad \mathbf{a} = \begin{bmatrix} a_0 \\ a_1 \\ \cdot \\ \cdot \\ \cdot \\ a_n \end{bmatrix} \tag{5.116}$$

and $\mathbf{V}$ is the *Vandermonde matrix*,

$$\mathbf{V} = \begin{bmatrix} 1 & r_0 & r_0^2 & \cdots & r_0^n \\ 1 & r_1 & r_1^2 & \cdots & r_1^n \\ \cdot & \cdot & \cdot & & \cdot \\ \cdot & \cdot & \cdot & & \cdot \\ \cdot & \cdot & \cdot & & \cdot \\ 1 & r_n & r_n^2 & \cdots & r_n^n \end{bmatrix} \tag{5.117}$$

Since $\det \mathbf{V} \neq 0$, provided that the r_i are distinct points, we have a unique solution for $\mathbf{a}$.

However, a more convenient way to obtain $\psi(r)$ is to use *Lagrangian interpolation*. First, we recall that the $(n + 1)$ functions $1, r, r^2, \ldots, r^n$ form an $(n + 1)$-dimensional vector space, say V_n, in which $\psi(r)$ is an element (see Section 2.2). Since the coordinates $a_0, a_1, a_2, \ldots, a_n$ of $\psi(r)$ are relatively difficult to evaluate, we seek to use a different basis for the space V_n, in which the coordinates of $\psi(r)$ are more easily evaluated. This basis is provided by the fundamental polynomials of Lagrangian interpolation, given as

$$l_j(r) = \frac{(r - r_0)(r - r_1)\ldots(r - r_{j-1})(r - r_{j+1})\ldots(r - r_n)}{(r_j - r_0)(r_j - r_1)\ldots(r_j - r_{j-1})(r_j - r_{j+1})\ldots(r_j - r_n)} \tag{5.118}$$

where

$$l_j(r_i) = \delta_{ij} \tag{5.119}$$

where δ_{ij} is the Kronecker delta; i.e., $\delta_{ij} = 1$ for $i = j$, and $\delta_{ij} = 0$ for $i \neq j$. Using the property in (5.119), the coordinates of the base vectors are simply the values of $F(r)$, and the polynomial $\psi(r)$ is

$$\psi(r) = F_0 l_0(r) + F_1 l_1(r) + \ldots + F_n l_n(r) \tag{5.120}$$

EXAMPLE 5.28: Establish the interpolating polynomial $\psi(r)$ for the function $F(r) = 2^r - r$ when the data at the points $r = 0$, 1, and 3 are used. In this case $r_0 = 0, r_1 = 1, r_2 = 3$, and $F_0 = 1, F_1 = 1, F_2 = 5$.

In the first approach we use the relation in (5.115) to calculate the unknown coefficients a_0, a_1, and a_2 of the polynomial $\psi(r) = a_0 + a_1 r + a_2 r^2$. In this case we have

$$\begin{bmatrix} 1 & 0 & 0 \\ 1 & 1 & 1 \\ 1 & 3 & 9 \end{bmatrix} \begin{pmatrix} a_0 \\ a_1 \\ a_2 \end{pmatrix} = \begin{pmatrix} 1 \\ 1 \\ 5 \end{pmatrix}$$

The solution gives $a_0 = 1$, $a_1 = -\frac{2}{3}$, $a_2 = \frac{2}{3}$, and therefore $\psi(r) = 1 - \frac{2}{3}r + \frac{2}{3}r^2$.

If Lagrangian interpolation is employed, we use the relation in (5.120) which in this case gives

$$\psi(r) = (1)\frac{(r-1)(r-3)}{(-1)(-3)} + (1)\frac{(r)(r-3)}{(1)(-2)} + (5)\frac{(r)(r-1)}{(3)(2)}$$

or, as before,

$$\psi(r) = 1 - \tfrac{2}{3}r + \tfrac{2}{3}r^2$$

5.7.2 *The Newton–Cotes Formulas (one-dimensional integration)*

Having established an interpolating polynomial $\psi(r)$, we can now obtain an approximation to the integral $\int_a^b F(r)\,dr$. In Newton–Cotes integration, it is assumed that the sampling points of F are spaced at equal distances, and we define

$$r_0 = a, \qquad r_n = b, \qquad h = \frac{b-a}{n} \tag{5.121}$$

Using Lagrangian interpolation to obtain $\psi(r)$ as an approximation to $F(r)$, we have

$$\int_n^b F(r)\,dr = \sum_{i=0}^{n} \left\{\int_a^b l_i(r)\,dr\right\} F_i + R_n \tag{5.122}$$

or, evaluated,

$$\int_a^b F(r)\,dr = (b-a)\sum_{i=0}^{n} C_i^n F_i + R_n \tag{5.123}$$

where R_n is the remainder and the C_i^n are the *Newton–Cotes constants* for numerical integration with n sampling points.

The Newton–Cotes constants and corresponding remainder terms have been published[31] and are summarized in Table 5.1 for $n = 1$ to 6. The case $n = 1$ and $n = 2$ are the well-known *trapezoidal rule* and *Simpson formula.* We note that the formulas for $n = 3$ and $n = 5$ have the same order of accuracy as the formulas for $n = 2$ and $n = 4$, respectively. For this reason, the even formulas with $n = 2$ and 4 are used in practice.

TABLE 5.1 *Newton–Cotes numbers and error estimates.*

Number of Intervals n	C_0^n	C_1^n	C_2^n	C_3^n	C_4^n	C_5^n	C_6^n	Upper Bound on Error R_n as a Function of the Derivative of F
1	$\frac{1}{2}$	$\frac{1}{2}$						$10^{-1}(b-a)^3F^{II}(r)$
2	$\frac{1}{6}$	$\frac{4}{6}$	$\frac{1}{6}$					$10^{-3}(b-a)^5F^{IV}(r)$
3	$\frac{1}{8}$	$\frac{3}{8}$	$\frac{3}{8}$	$\frac{1}{8}$				$10^{-3}(b-a)^5F^{IV}(r)$
4	$\frac{7}{90}$	$\frac{32}{90}$	$\frac{12}{90}$	$\frac{32}{90}$	$\frac{7}{90}$			$10^{-6}(b-a)^7F^{VI}(r)$
5	$\frac{19}{288}$	$\frac{75}{288}$	$\frac{50}{288}$	$\frac{50}{288}$	$\frac{75}{288}$	$\frac{19}{288}$		$10^{-6}(b-a)^7F^{VI}(r)$
6	$\frac{41}{840}$	$\frac{216}{840}$	$\frac{27}{840}$	$\frac{272}{840}$	$\frac{27}{840}$	$\frac{216}{840}$	$\frac{41}{840}$	$10^{-9}(b-a)^9F^{VIII}(r)$

EXAMPLE 5.29: Evaluate the Newton–Cotes constants when the interpolating polynomial is of order 2; i.e., $\psi(r)$ is a parabola.

In this case we have

$$\int_a^b F(r)\,dr \doteq$$

$$\int_a^b \left[F_0\frac{(r-r_1)(r-r_2)}{(r_0-r_1)(r_0-r_2)} + F_1\frac{(r-r_0)(r-r_2)}{(r_1-r_0)(r_1-r_2)} + F_2\frac{(r-r_0)(r-r_1)}{(r_2-r_0)(r_2-r_1)}\right]dr$$

Using that $r_0 = a$, $r_1 = a + h$, $r_2 = a + 2h$, where $h = (b - a)/2$, the evaluation of the integral gives

$$\int_a^b F(r)\,dr = \frac{b-a}{6}(F_0 + 4F_1 + F_2)$$

Hence the Newton–Cotes constants are as given in Table 5.1 for the case $n = 2$.

EXAMPLE 5.30: Use Simpson's rule to integrate $\int_0^3 (2^r - r)\,dr$.

In this case $n = 2$ and $h = \frac{3}{2}$. Therefore, $r_0 = 0$, $r_1 = \frac{3}{2}$, $r_2 = 3$, and $F_0 = 1$, $F_1 = 1.328427$, $F_2 = 5$, and we obtain

$$\int_0^3 (2^r - r)\,dr = \frac{3}{6}[(1)(1) + (4)(1.328427) + (1)(5)]$$

or

$$\int_0^3 (2^r - r)\,dr \doteq 5.656854$$

The exact result is

$$\int_0^3 (2^r - r)\,dr = 5.598868$$

Hence the error is

$$R = 0.057986$$

However, using the upper bound value on the error we would have

$$R < \frac{(3-0)^5}{1000}(\ln 2)^4(2^r) = 0.448743$$

To obtain greater accuracy in the integration using the Newton–Cotes formulas we need to employ a smaller interval h; i.e., include more evaluations of the function to be integrated. Then we have the choice between two different strategies: we may use a higher-order Newton–Cotes formula, or, alternatively, employ the lower-order formula in a repeated manner, in which case the integration procedure is referred to as a *composite formula*. Consider the following example.

EXAMPLE 5.31: Increase the accuracy of the integration in Example 5.30 by using half the interval spacing.

In this case we have $h = \frac{3}{4}$ and the required function values are $F_0 = 1$, $F_1 = 0.931792$, $F_2 = 1.328427$, $F_3 = 2.506828$, and $F_4 = 5$. The choice lies now between using the higher-order Newton–Cotes formula with $n = 4$ or applying the Simpson's rule twice, i.e., to the first two intervals and then to the second two intervals. Using the Newton–Cotes formula with $n = 4$, we obtain

$$\int_0^3 (2^r - r)\,dr \doteq \tfrac{3}{90}(7F_0 + 32F_1 + 12F_2 + 32F_3 + 7F_4)$$

Hence
$$\int_0^3 (2^r - r)\,dr \doteq 5.599232$$

On the other hand, using Simpson's rule twice, we consider

$$\int_0^3 (2^r - r)\,dr \doteq \int_0^{3/2} (2^r - r)\,dr + \int_{3/2}^3 (2^r - r)\,dr$$

The integration is performed using

$$\int_0^{3/2} (2^r - r)\,dr \doteq \frac{\frac{3}{2} - 0}{6}(F_0 + 4F_1 + F_2)$$

where F_0, F_1, and F_2 are the function values at $r = 0$, $r = \frac{3}{4}$, and $r = \frac{3}{2}$, respectively; i.e.,

$$F_0 = 1; \qquad F_1 = 0.931792; \qquad F_2 = 1.328427$$

Hence we use
$$\int_0^{3/2} (2^r - r)\,dr \doteq 1.513899 \tag{a}$$

Next we need to evaluate

$$\int_{3/2}^3 (2^r - r)dr \doteq \frac{3 - \frac{3}{2}}{6}(F_0 + 4F_1 + F_2)$$

where F_0, F_1, and F_2 are the function values at $r = \frac{3}{2}$, $r = \frac{9}{4}$, and $r = 3$, respectively,

$$F_0 = 1.328427; \qquad F_1 = 2.506828; \qquad F_2 = 5$$

Hence we have
$$\int_{3/2}^3 (2^r - r)\,dr \doteq 4.088935 \tag{b}$$

Adding the results of (a) and (b), we obtain

$$\int_0^3 (2^r - r)\,dr \doteq 5.602834$$

The use of a composite formula has a number of advantages over the application of high-order Newton–Cotes formulas. A composite formula, such as the repetitive use of the Simpson rule, is easy to employ. Convergence is assured as the interval of sampling decreases, and, in practice, a sampling interval could be used that varies from one application of the basic formula to the next. This is particularly advantageous when there are discontinuities in the function to be integrated. For these reasons, in practice, composite formulas are commonly used.

EXAMPLE 5.32: Use a composite formula that employs Simpson's rule to evaluate the integral $\int_{-1}^{+13} F(r)\,dr$ of the function $F(r)$ in Fig. 5.44.

This function is best integrated by considering three intervals of integration, as follows:

$$\int_{-1}^{13} F\,dr = \int_{-1}^{2} (r^3 + 3)\,dr + \int_{2}^{9} [10 + (r - 1)^{1/3}]\,dr + \int_{9}^{13} [\tfrac{1}{128}(13 - r)^5 + 4]\,dr$$

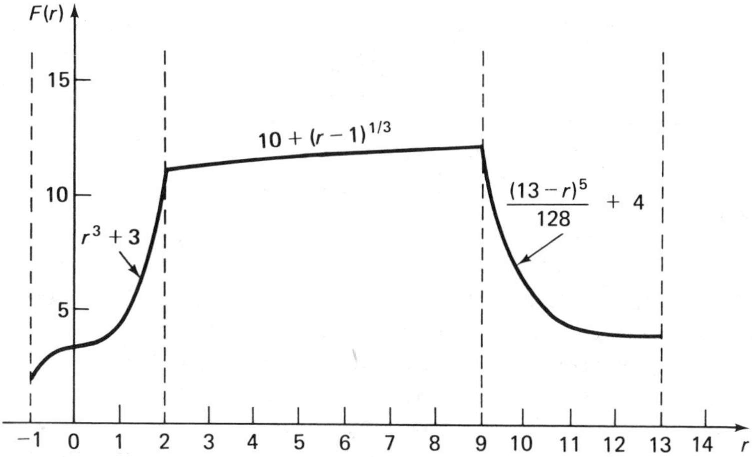

FIGURE 5.44 *Function F(r) integrated in Example 5.32.*

We evaluate each of the three integrals using Simpson's rule and have

$$\int_{-1}^{2} (r^3 + 3)dr \doteq \frac{2 - (-1)}{6}[(1)(2) + (4)(3.125) + (1)(11)]$$

or

$$\int_{-1}^{2} (r^3 + 3)dr \doteq 12.75$$

$$\int_{2}^{9} [10 + (r - 1)^{1/3}]\,dr \doteq \frac{9 - 2}{6}[(1)(11) + (4)(11.650964) + (1)(12)]$$

or

$$\int_{2}^{9} [10 + (r - 1)^{1/3}]\,dr \doteq 81.204498$$

$$\int_{9}^{13} [\tfrac{1}{128}(13 - r)^5 + 4]\,dr \doteq \frac{13 - 9}{6}[(1)(12) + (4)(4.25) + (1)(4)]$$

or $\int_9^{13} [\frac{1}{128}(13 - r)^5 + 4]\, dr \doteq 22$

Hence $$\int_{-1}^{13} F\, dr \doteq 12.75 + 81.204498 + 22$$

or $$\int_{-1}^{13} F\, dr \doteq 115.954498$$

5.7.3 The Gauss Formulas (one-dimensional integration)

The basic integration schemes that we considered so far used equally spaced sampling points, although the basic methods could be employed to construct procedures that allow varying the interval of sampling; i.e., the composite formulas have been introduced. The methods considered so far are effective when measurements of an unknown function to be integrated have been taken at equal intervals. However, in the integration of finite element matrices, a subroutine is called to evaluate the unknown function F at given points, and these points may be anywhere on the element. No additional difficulties arise if the sampling points are not equally spaced. Therefore, it seems natural to try to improve the accuracy that can be obtained for a given number of function evaluations by also optimizing the positions of the sampling points. A very important numerical integration procedure in which both the positions of the sampling points and the weights have been optimized is *Gauss quadrature*. The basic assumption in Gauss numerical integration is that

$$\int_a^b F(r)\, dr = \alpha_1 F(r_1) + \alpha_2 F(r_2) + \ldots + \alpha_n F(r_n) + R_n \tag{5.124}$$

where both the weights $\alpha_1, \ldots, \alpha_n$ and the sampling points $r_1, \ldots, r_n$ are variables. It should be recalled that in the derivation of the Newton–Cotes formulas, only the weights were unknown, and they have been determined from the integration of a polynomial $\psi(r)$, which passed through equally spaced sampling points of the function $F(r)$. We now calculate also the positions of the sampling points and, therefore, have $2n$ unknowns to determine a higher-order integration scheme.

In analogy to the derivation of the Newton–Cotes formulas, we use an interpolating polynomial $\psi(r)$ of the form given in (5.120):

$$\psi(r) = \sum_{j=1}^{n} F_j l_j(r) \tag{5.125}$$

where n samplings points are now considered, $r_1, \ldots, r_n$, which are still unknown. For the determination of the values $r_1, \ldots, r_n$, we define a function $P(r)$,

$$P(r) = (r - r_1)(r - r_2) \ldots (r - r_n) \tag{5.126}$$

which is a polynomial of order n. We note that $P(r) = 0$ at the sampling points $r_1, \ldots, r_n$. Therefore, we can write

$$F(r) = \psi(r) + P(r)(\beta_0 + \beta_1 r + \beta_2 r^2 + \ldots) \tag{5.127}$$

Integrating $F(r)$ we obtain

$$\int_a^b F(r)\,dr = \sum_{j=1}^{n} F_j\left[\int_a^b l_j(r)\,dr\right] + \sum_{j=0}^{\infty} \beta_j\left[\int_a^b r^j P(r)\,dr\right] \tag{5.128}$$

where it need be noted that in the first integral on the right of (5.128), functions of order $(n-1)$ and lower are integrated, and in the second integral the functions that are integrated are of order n and higher. The unknown values r_j, $j = 1, 2, \ldots, n$, can now be determined using the conditions

$$\int_a^b P(r) r^k\,dr = 0 \qquad k = 0, 1, 2, \ldots, n-1 \tag{5.129}$$

Then, since the polynomial $\psi(r)$ passes through n sampling points of $F(r)$ and $P(r)$ vanishes at these points, the conditions in (5.129) mean that the required integral $\int_a^b F(r)\,dr$ is approximated by integrating a polynomial of order $2n-1$ instead of $F(r)$.

In summary, using the Newton–Cotes formulas we use $(n+1)$ equally spaced sampling points and integrate exactly a polynomial of order at most n. On the other hand, in Gauss quadrature we require n unequally spaced sampling points and integrate exactly a polynomial of order at most $(2n-1)$. Polynomials of orders less than n and $(2n-1)$, respectively, for the two cases would also be integrated exactly.

To determine the sampling points and the integration weights, we realize that they depend on the interval a to b. However, to make the calculations general, we can consider a natural interval from -1 to $+1$ and deduce the sampling points and weights for any interval. Namely, if r_i is a sampling point and α_i is the weight for the interval -1 to $+1$, the corresponding sampling point and weight in the integration from a to b are

$$\frac{a+b}{2} + \frac{b-a}{2} r_i \quad \text{and} \quad \frac{b-a}{2}\alpha_i$$

respectively.

For the above reasons, consider an interval from -1 to $+1$. The sampling points are determined from (5.129) with $a = -1$ and $b = +1$. To calculate the integration weights we substitute for $F(r)$ in (5.124) the interpolating polynomial $\psi(r)$ from (5.125) and perform the integration. It should be noted that because the sampling points have been determined, the polynomial $\psi(r)$ is known, and hence

$$\alpha_j = \int_{-1}^{+1} l_j(r)\,dr; \qquad j = 1, 2, \ldots, n \tag{5.130}$$

The sampling points and weights for the interval -1 to $+1$ have been published and are reproduced in Table 5.2 for values $n = 1$ to 6.[32]

The coefficients in Table 5.2 can be calculated directly using (5.129) and (5.130) (see Example 5.33). However, for larger n the solution becomes cumbersome and it is expedient to use Legendre polynomials to solve for the coefficients, which are thus referred to as Gauss–Legendre coefficients.

TABLE 5.2 *Sampling points and weights in Gauss–Legendre numerical integration (interval −1 to +1).*

n	r_i	α_i
1	0. (15 zeros)	2. (15 zeros)
2	±0.57735 02691 89626	1.00000 00000 00000
3	±0.77459 66692 41483	0.55555 55555 55556
	0.00000 00000 00000	0.88888 88888 88889
4	±0.86113 63115 94053	0.34785 48451 37454
	±0.33998 10435 84856	0.65214 51548 62546
5	±0.90617 98459 38664	0.23692 68850 56189
	±0.53846 93101 05683	0.47862 86704 99366
	0.00000 00000 00000	0.56888 88888 88889
6	±0.93246 95142 03152	0.17132 44923 79170
	±0.66120 93864 66265	0.36076 15730 48139
	±0.23861 91860 83197	0.46791 39345 72691

EXAMPLE 5.33: Derive the sampling points and weights for two-point Gauss quadrature.

In this case $P(r) = (r - r_1)(r - r_2)$ and (5.129) gives the two equations

$$\int_{-1}^{+1} (r - r_1)(r - r_2)\, dr = 0$$

$$\int_{-1}^{+1} (r - r_1)(r - r_2) r dr = 0$$

Solving, we obtain $$r_1 r_2 = -\tfrac{1}{3}$$

and $$r_1 + r_2 = 0$$

Hence $$r_1 = -\frac{1}{\sqrt{3}}; \qquad r_2 = +\frac{1}{\sqrt{3}}$$

The corresponding weights are obtained using (5.130), which in this case gives

$$\alpha_1 = \int_{-1}^{+1} \frac{r - r_2}{r_1 - r_2}\, dr$$

$$\alpha_2 = \int_{-1}^{+1} \frac{r - r_1}{r_2 - r_1}\, dr$$

Since $r_2 = -r_1$, we obtain $\alpha_1 = \alpha_2 = 1.0$.

EXAMPLE 5.34: Use two-point Gauss quadrature to evaluate the integral $\int_0^3 (2^r - r)\, dr$ considered in Examples 5.30 and 5.31.

Using two-point Gauss quadrature, we obtain, from (5.124),

$$\int_0^3 (2^r - r)\, dr \doteq \alpha_1 F(r_1) + \alpha_2 F(r_2) \tag{a}$$

where α_1, α_2 and r_1, r_2 are weights and sampling points, respectively. Since the interval is from 0 to 3, we need to determine the values α_1, α_2, r_1, and r_2 from the values given in Table 5.2,

$$\alpha_1 = \tfrac{3}{2}(1); \qquad \alpha_2 = \tfrac{3}{2}(1)$$

$$r_1 = \tfrac{3}{2}\left(1 - \frac{1}{\sqrt{3}}\right); \qquad r_2 = \tfrac{3}{2}\left(1 + \frac{1}{\sqrt{3}}\right)$$

where $1/\sqrt{3} = 0.5773502692$. Thus

$$F(r_1) = 0.91785978 \qquad F(r_2) = 2.78916389$$

and (a) gives

$$\int_0^3 (2^r - r)\, dr \doteq 5.56053551$$

The Gauss–Legendre integration procedure is commonly used in isoparametric finite element analysis. However, it should be noted that other integration schemes, in which both the weights and sampling positions were varied to obtain maximum accuracy, have been derived also.[31-33] In addition, for the specific objectives in finite element analysis, the basic Gauss quadrature has been applied in modified form as described in Section 5.8.1.

5.7.4 Integrations in Two and Three Dimensions

So far we have considered the integration of a one-dimensional function $F(r)$. However, two- and three-dimensional integrals need be evaluated in two- and three-dimensional finite element analyses. Considering the evaluation of rectangular elements, we can apply the above one-dimensional integration formulas successively in each direction.* As in the analytical evaluation of multi-dimensional integrals, in this procedure, successively, the innermost integral is evaluated by keeping the variables corresponding to the other integrals constant. Therefore, we have for a two-dimensional integral

$$\int_{-1}^{+1}\int_{-1}^{+1} F(r, s)\, dr\, ds = \sum_i \alpha_i \int_{-1}^{+1} F(r_i, s)\, ds \tag{5.131}$$

or

$$\int_{-1}^{+1}\int_{-1}^{+1} F(r, s)\, dr\, ds = \sum_{i,j} \alpha_i \alpha_j F(r_i, s_j) \tag{5.132}$$

and corresponding to (5.113), $\alpha_{ij} = \alpha_i \alpha_j$, where α_i and α_j are the integration weights for one-dimensional integration. Similarly, for a three-dimensional integral,

$$\int_{-1}^{+1}\int_{-1}^{+1}\int_{-1}^{+1} F(r, s, t)\, dr\, ds\, dt = \sum_{i,j,k} \alpha_i \alpha_j \alpha_k F(r_i, s_j, t_k) \tag{5.133}$$

and $\alpha_{ijk} = \alpha_i \alpha_j \alpha_k$. We should note that *it is not necessary in the numerical integration to use the same quadrature rule in the two or three dimensions; i.e., we can employ different numerical integration schemes in the r, s, and t directions.*

EXAMPLE 5.35: Given that the (i, j)th element of a stiffness matrix $\mathbf{K}$ is $\int_{-1}^{+1}\int_{-1}^{+1} r^2 s^2\, dr\, ds$. Evaluate the integral $\int_{-1}^{+1}\int_{-1}^{+1} r^2 s^2\, dr\, ds$ using (1) Simpson's rule in both r and s, (2) Gauss quadrature in both r and s, and (3) Gauss quadrature in r and Simpson's rule in s.

* This results in much generality of the integration, but for special cases somewhat less costly procedures can be designed.[34]

1. Using Simpson's rule, we have

$$\int_{-1}^{+1}\int_{-1}^{+1} r^2 s^2\, dr\, ds = \int_{-1}^{+1} \tfrac{1}{3}[(1)(1) + (4)(0) + (1)(1)]s^2\, ds$$

$$= \int_{-1}^{+1} \tfrac{2}{3}s^2\, ds = \tfrac{1}{3}[(1)(\tfrac{2}{3}) + (4)(0) + (1)(\tfrac{2}{3})] = \tfrac{4}{9}$$

2. Using two-point Gauss quadrature, we have

$$\int_{-1}^{+1}\int_{-1}^{+1} r^2 s^2\, dr\, ds = \int_{-1}^{+1} \left[(1)\left(\frac{1}{\sqrt{3}}\right)^2 + (1)\left(\frac{1}{\sqrt{3}}\right)^2\right] s^2\, ds$$

$$= \int_{-1}^{+1} \tfrac{2}{3}s^2\, ds = \tfrac{2}{3}\left[(1)\left(\frac{1}{\sqrt{3}}\right)^2 + (1)\left(\frac{1}{\sqrt{3}}\right)^2\right] = \tfrac{4}{9}$$

3. Finally, using Gauss quadrature in r and Simpson's rule in s, we have

$$\int_{-1}^{+1}\int_{-1}^{+1} \left[(1)\left(\frac{1}{\sqrt{3}}\right)^2 + (1)\left(\frac{1}{\sqrt{3}}\right)^2\right] s^2\, ds$$

$$= \int_{-1}^{+1} \tfrac{2}{3}s^2\, ds = \tfrac{1}{3}[(1)(\tfrac{2}{3}) + (4)(0) + (1)(\tfrac{2}{3})] = \tfrac{4}{9}$$

We should note that the above numerical integrations are exact because both integration schemes, i.e., Simpson's rule and two-point Gauss quadrature, integrate a parabola exactly.

The above procedure is directly applicable to the evaluation of matrices of rectangular elements in which all integration limits are -1 to $+1$. Hence, in the evaluation of a two-dimensional finite element, the integrations can be carried out for each entry of the stiffness and mass matrices and load vectors as illustrated in Example 5.35. Based on the information given in Table 5.2 some common Gauss quadrature rules for two-dimensional analysis are summarized in Table 5.3.

Considering next the evaluation of triangular and tetrahedral element matrices, however, the procedure given in Example 5.35 is not applicable directly, because now the integration limits involve the variables themselves. A great deal of research has been spent on the development of suitable integration formulas for triangular domains, and here too formulas of the Newton–Cotes type[35] and of the Gauss quadrature type are available.[36,37] As in the solution of rectangular domains, the Gauss quadrature rules are in general more efficient because they yield a higher integration accuracy for the same number of evaluations. Table 5.4 lists the integration stations and integration weights of the Gauss integration formulas published by Cowper.[37]

5.8 PRACTICAL CONSIDERATIONS IN ISOPARAMETRIC ELEMENT CALCULATIONS

In previous sections we have presented the basic relations used to derive isoparametric element matrices. However, the proper application of these relations requires that certain choices be made, such as which element to select, what kind and order of numerical integration to use, and so on. Our objective in this section is to discuss some of these practical and important considerations.

<table>
<tr><th>Integration order</th><th>Degree of precision</th><th>Location of integration points</th></tr>
<tr><td>2 × 2</td><td>3</td><td>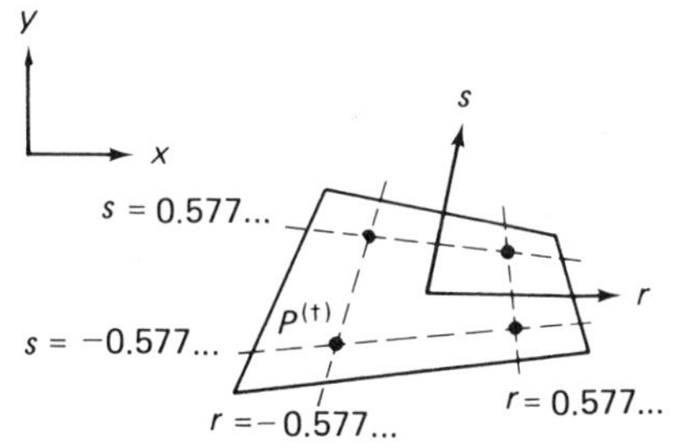
</td></tr>
<tr><td>3 × 3</td><td>5</td><td>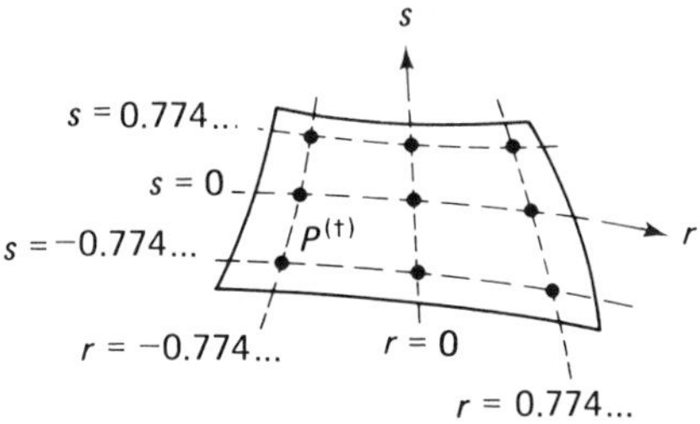
</td></tr>
<tr><td>4 × 4</td><td>7</td><td>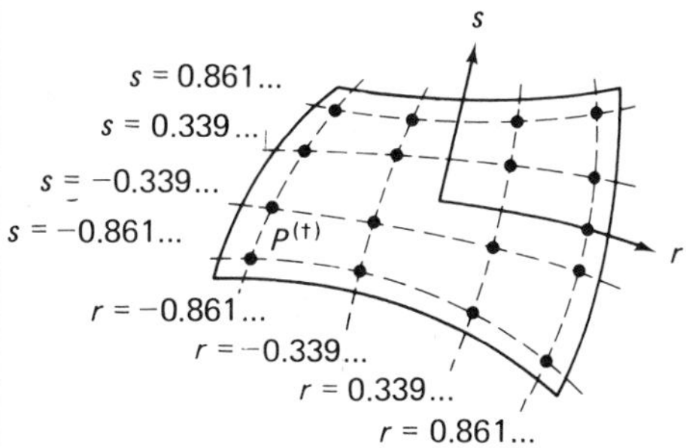
</td></tr>
</table>

(†)The location of any integration point in the x-y coordinate system is given by : $x_p = \Sigma_i h_i(r_p, s_p)x_i$ and $y_p = \Sigma_i h_i(r_p, s_p)y_i$

TABLE 5.3 *Gauss numerical integrations over rectangular domains. The integration weights are given in Table 5.2 using (5.132).*

5.8.1 Use of Numerical Integration

In the practical use of the numerical integration procedures presented in the previous section, basically two questions arise; namely, what kind of integration scheme to use, and what order to select. We pointed out that using the Newton–Cotes formulas, $(n + 1)$ function evaluations are required to integrate without error a polynomial of order n. On the other hand, if Gauss quadrature is used, a polynomial of order $(2n - 1)$ is integrated exactly with only n function evaluations. In finite element analysis a large number of function evaluations directly

Integration order	Degree of precision	Integration points	r-coordinates	s-coordinates	Weights
3-point	2	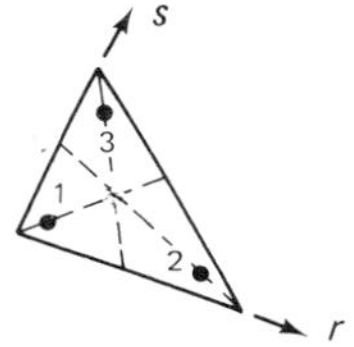	$r_1 = 0.16666\ 66666\ 667$ $r_2 = 0.66666\ 66666\ 667$ $r_3 = r_1$	$s_1 = r_1$ $s_2 = r_1$ $s_3 = r_2$	$w_1 = 0.33333\ 33333\ 333$ $w_2 = w_1$ $w_3 = w_1$
7-point	5	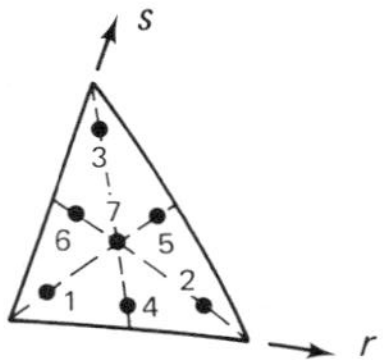	$r_1 = 0.10128\ 65073\ 235$ $r_2 = 0.79742\ 69853\ 531$ $r_3 = r_1$ $r_4 = 0.47014\ 20641\ 051$ $r_5 = r_4$ $r_6 = 0.05971\ 58717\ 898$ $r_7 = 0.33333\ 33333\ 333$	$s_1 = r_1$ $s_2 = r_1$ $s_3 = r_2$ $s_4 = r_6$ $s_5 = r_4$ $s_6 = r_4$ $s_7 = r_7$	$w_1 = 0.12593\ 91805\ 448$ $w_2 = w_1$ $w_3 = w_1$ $w_4 = 0.13239\ 41527\ 885$ $w_5 = w_4$ $w_6 = w_4$ $w_7 = 0.225$
13-point	7	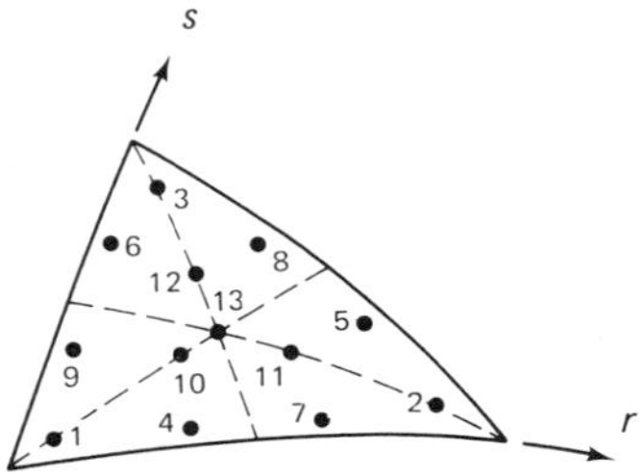	$r_1 = 0.06513\ 01029\ 022$ $r_2 = 0.86973\ 97941\ 956$ $r_3 = r_1$ $r_4 = 0.31286\ 54960\ 049$ $r_5 = 0.63844\ 41885\ 698$ $r_6 = 0.04869\ 03154\ 253$ $r_7 = r_5$ $r_8 = r_4$ $r_9 = r_6$ $r_{10} = 0.26034\ 59660\ 790$ $r_{11} = 0.47930\ 80678\ 419$ $r_{12} = r_{10}$ $r_{13} = 0.33333\ 33333\ 333$	$s_1 = r_1$ $s_2 = r_1$ $s_3 = r_2$ $s_4 = r_6$ $s_5 = r_4$ $s_6 = r_5$ $s_7 = r_6$ $s_8 = r_5$ $s_9 = r_4$ $s_{10} = r_{10}$ $s_{11} = r_{10}$ $s_{12} = r_{11}$ $s_{13} = r_{13}$	$w_1 = 0.05334\ 72356\ 088$ $w_2 = w_1$ $w_3 = w_1$ $w_4 = 0.07711\ 37608\ 903$ $w_5 = w_4$ $w_6 = w_4$ $w_7 = w_4$ $w_8 = w_4$ $w_9 = w_4$ $w_{10} = 0.17561\ 52574\ 332$ $w_{11} = w_{10}$ $w_{12} = w_{10}$ $w_{13} = -0.14957\ 00444\ 677$

TABLE 5.4 *Gauss numerical integrations over triangular domains; here* $\iint F\,dr\,ds = \frac{1}{2}\sum w_i F(r_i, s_i)$.

increases the cost of analysis, and the use of Gauss quadrature is attractive. However, the Newton–Cotes formulas may be efficient in nonlinear analysis for the reasons discussed in Section 6.5.3.

Having selected a numerical integration scheme, the order of numerical integration to be used in the evaluation of the various finite element integrals needs to be determined. The choice of the order of numerical integration is important in practice, because, firstly, the cost of analysis increases when a higher order integration is employed, and, secondly, using a different integration order, the results can be affected by a very large amount. These considerations are particularly important in three-dimensional analysis.

The matrices to be evaluated by numerical integration are the stiffness matrix $\mathbf{K}$, the mass matrix $\mathbf{M}$, the body force vector $\mathbf{R}_B$, the initial stress vector $\mathbf{R}_I$, and the surface load vector $\mathbf{R}_S$. In general, the appropriate integration order depends on the matrix that is evaluated and the specific finite element considered. To demonstrate the important aspects, consider the Gauss numerical integration order required to evaluate the matrices of the variable-number-nodes elements discussed in Sections 5.3 and 5.4.

A first observation in the selection of the order of numerical integration is that, in theory, using a high-enough order, all matrices are evaluated exactly. On the other hand, using too low an order of integration, the matrices may be evaluated very inaccurately and, in fact, the problem solution may not be possible. For example, considering an element stiffness matrix, if the order of numerical integration is too low, the matrix can have a larger number of zero eigenvalues than the number of physical rigid body modes. Hence, for a successful solution of the equilibrium equations of the element assemblage, it would be necessary that the deformation modes corresponding to all zero eigenvalues of the element be properly restrained in the assemblage of finite elements, because otherwise the structure stiffness matrix would be singular. A simple example would be the evaluation of the stiffness matrix of a three-node truss element. If one-point Gauss numerical integration is used, the row and column corresponding to the degree of freedom at the midnode of the element would be null vectors, which may result in a structure stiffness matrix that is singular. Therefore, the integration order need in general be higher than a certain limit.

The integration order required to evaluate a specific element matrix without error can be determined by studying the order of the function to be integrated. In the case of the stiffness matrix, we need to evaluate

$$\mathbf{K} = \int_V \mathbf{B}^T\mathbf{C}\mathbf{B} \det \mathbf{J}\, dV \tag{5.134}$$

where $\mathbf{C}$ is a constant material property matrix; $\mathbf{B}$ the strain-displacement matrix in the natural coordinate system r, s, t; $\det \mathbf{J}$ the determinant of the Jacobian transforming local (or global) to natural coordinates (see Section 5.3); and the integration is performed over the element volume in the natural coordinate system. The matrix function $\mathbf{F}$ to be integrated is, therefore,

$$\mathbf{F} = \mathbf{B}^T\mathbf{C}\mathbf{B} \det \mathbf{J} \tag{5.135}$$

The matrices $\mathbf{J}$ and $\mathbf{B}$ have been defined in Sections 5.3 and 5.4.

A case for which the order of the variables in **F** can be evaluated with relative ease arises when the four-node two-dimensional element studied in Example 5.5 is used as a rectangular or parallelogram element. It is instructive to consider this case in detail because the procedure of evaluating the required integration order is displayed clearly.

EXAMPLE 5.36: Evaluate the required Gauss numerical integration order for the calculation of the stiffness matrix of a four-node rectangular element.

The integration order to be used depends on the order of the variables r and s in **F** defined in 5.135. For a rectangular element with sides $2a$ and $2b$, we can write

$$x = ar$$
$$y = bs$$

and consequently the Jacobian matrix **J** is

$$\mathbf{J} = \begin{bmatrix} a & 0 \\ 0 & b \end{bmatrix}$$

Since the elements of **J** are constant, referring to the information given in Example 5.5, the elements of the strain-displacement matrix **B** are therefore functions of r or s only. But the determinant of **J** is also constant; hence

$$\mathbf{F} = f(r^2, rs, s^2)$$

where f denotes "function of."

Using two-point Gauss numerical integration in the r and s directions, all functions in r and s involving at most cubic terms are integrated without error; i.e., for integration order n, the order of r and s integrated exactly is $(2n - 1)$. Hence two-point Gauss integration is adequate.

In an analogous manner, the required integration order to evaluate exactly the stiffness matrices, mass matrices, and element load vectors of other elements can be assessed. In this context it should be noted that the Jacobian matrix is not constant for nonparallelogram element shapes, which may mean that quite high an integration order would be required to evaluate the element matrices exactly.

We discussed in Sections 3.3.3 and 4.2.5 that provided the convergence criteria are satisfied, the displacement formulation of finite element analysis yields a lower bound on the "exact" strain energy of the system considered; i.e., physically, a displacement formulation results in overestimating the system stiffness. Therefore, by not evaluating the element stiffness matrices exactly in the numerical integration, in fact, better results can be obtained provided that the error in the numerical integration compensates appropriately for the overestimation of structural stiffness due to the finite element discretization.[23,38,39] In other words, a *reduction of the numerical integration order* from the order that is required to evaluate the element stiffness matrix exactly should lead in many cases to improved results. This is indeed the case, and a great deal of research effort has been spent to evaluate the optimum integration order and scheme for isoparametric finite element analysis. In addition to merely using a *reduced integration* order, it may well be beneficial to use *selective integration*, i.e. to integrate the different strain terms with different orders of integration.[21,40] In certain analyses drastic improvements have been obtained; however, it is instructive to summarize the difficulties encountered in the search for the optimum integration scheme, and make some comments on the practical use of reduced and selective integration.

A first observation is that the bounding property of the displacement analysis is lost. This is in practical analysis usually not a serious shortcoming. However, an important observation is that if the order of integration is too low, the solution obtained may be meaningless. In particular, as pointed out already, if the integration order for the evaluation of an element stiffness matrix is too low to include all displacement modes, the rank of the element matrix will be smaller than if evaluated exactly. This causes solution difficulties if the element is not provided with sufficient stiffness restraint in the assemblage of elements; i.e., the stiffness matrix of the total element assemblage could be singular and is more frequently ill-conditioned. Using certain elements that contain spurious zero energy modes some "rules" have been proposed that should be followed so that no solution instabilities develop.[21] However, in a practical analysis with complex boundary conditions and the use of different kinds of elements to model a structure, these rules may only be of limited value in preventing an ill-conditioned set of equilibrium equations.

It has also been argued that the lowest order of numerical integration required for convergence is that order which would evaluate the volume of the element under consideration without error. However, this rule has to be used with great care. For example, in the formulation of a three-node truss element, using one-point Gauss integration we would evaluate the volume exactly, but, as pointed out earlier, the row and column corresponding to the degree of freedom at the midnode of the element are null vectors!

In summary, considering the use of reduced and selective integrations, it is important that with any special integration scheme the following criteria are satisfied:

1. the element does not contain any spurious zero energy modes (i.e. the rank of the element stiffness matrix is not smaller than evaluated exactly); and
2. the element contains the required constant strain states.

The condition in (1) assures that the governing finite element equations can be solved and that no spurious mechanisms develop in the solution. If condition (2) is also fulfilled we know that the completeness condition is satisfied and therefore convergence as discussed in Sections 4.2.5 and 4.3 can be expected. Hence the above two criteria should, in general, be satisfied for a reliable finite element solution.

If an element based on reduced or selective integration is employed that violates one of the criteria given above, it is necessary to use the element with much care. This is particularly the case in an actual practical analysis where the *reliability of the analysis results is of utmost concern.* The confident use of a reduced or selective integration scheme that does not satisfy the above criteria must largely depend on experience that has been accumulated when analyzing similar structures with high-order numerical integration, or the knowledge of another reliable numerical solution, or the availability of experimental results.

Here it should be noted that in practical analysis reduced integration is in essence always used implicitly whenever the distortion of a finite element is such that with the assigned order of integration the exact stiffness matrix is not evaluated.

In addition to the above considerations, we also need to realize that the order of convergence and the spatial isotropy or directionality of an element stiffness matrix can directly be affected by the integration scheme used. Hence, if by virtue of a specific element selection a certain directionality is purposely introduced in the finite element idealization (see Section 5.8.3), we need to preserve this directionality in the numerically evaluated element stiffness matrices—and vice versa.

In the above discussion we focussed attention on the evaluation of the element stiffness matrices. Considering the element force vectors, it is usually good practice to employ the same integration scheme and the same order of integration as for the stiffness matrices. However, in the evaluation of an element mass matrix, it should be recognized that for a lumped mass matrix only the volume of the element need be evaluated correctly, whereas for the exact evaluation of a consistent mass matrix a higher-order integration may be necessary than in the calculation of the stiffness matrix.

EXAMPLE 5.37: Evaluate the stiffness and mass matrices and the body force vector of element 2 of Example 4.3 using Gauss numerical integration.

The expressions to be integrated have been derived in Example 4.3,

$$\mathbf{K} = E\int_0^{80}\left(1+\frac{x}{40}\right)^2\begin{bmatrix}-\dfrac{1}{80}\\[2mm] \dfrac{1}{80}\end{bmatrix}\begin{bmatrix}-\dfrac{1}{80} & \dfrac{1}{80}\end{bmatrix}dx \tag{a}$$

$$\mathbf{M} = \rho\int_0^{80}\left(1+\frac{x}{40}\right)^2\begin{bmatrix}1-\dfrac{x}{80}\\[2mm] \dfrac{x}{80}\end{bmatrix}\begin{bmatrix}\left(1-\dfrac{x}{80}\right) & \dfrac{x}{80}\end{bmatrix}dx \tag{b}$$

$$\mathbf{R}_B = \frac{1}{10}\int_0^{80}\left(1+\frac{x}{40}\right)^2\begin{bmatrix}1-\dfrac{x}{80}\\[2mm] \dfrac{x}{80}\end{bmatrix}dx \tag{c}$$

The expressions in (a) and (c) are integrated exactly with two-point integration, whereas the evaluation of the integral in (b) requires three-point integration. A higher-order integration is required in the evaluation of the mass matrix, because this matrix is obtained from the displacement interpolation functions, whereas the stiffness matrix is calculated using derivatives of the displacement functions.

Using 1, 2, and 3 point Gauss integration to evaluate (a), (b), and (c) we obtain:

1-point integration

$$\mathbf{K} = \frac{12E}{240}\begin{bmatrix}1 & -1\\ -1 & 1\end{bmatrix};\qquad \mathbf{M} = \frac{\rho}{6}\begin{bmatrix}480 & 480\\ 480 & 480\end{bmatrix};\qquad \mathbf{R}_B = \frac{1}{6}\begin{bmatrix}96\\ 96\end{bmatrix}$$

2-point integration

$$\mathbf{K} = \frac{13E}{240}\begin{bmatrix} 1 & -1 \\ -1 & 1 \end{bmatrix}; \qquad \mathbf{M} = \frac{\rho}{6}\begin{bmatrix} 373.3 & 346.7 \\ 346.7 & 1013.3 \end{bmatrix}; \qquad \mathbf{R}_B = \frac{1}{6}\begin{bmatrix} 72 \\ 136 \end{bmatrix}$$

3-point integration

$$\mathbf{K} = \frac{13E}{240}\begin{bmatrix} 1 & -1 \\ -1 & 1 \end{bmatrix}; \qquad \mathbf{M} = \frac{\rho}{6}\begin{bmatrix} 384 & 336 \\ 336 & 1024 \end{bmatrix}; \qquad \mathbf{R}_B = \frac{1}{6}\begin{bmatrix} 72 \\ 136 \end{bmatrix}$$

It is interesting to note that with too low an order of integration the total mass of the element and the total load to which the element is subjected are not taken fully into account.

As discussed above, a suitable integration order for the evaluation of element matrices can be established by studying the order of the polynomials that need be integrated and referring to various considerations concerning a stable and convergent solution. Table 5.5 summarizes the results of such an

	Element	Reliable integration order	Reduced integration used in practice (with spurious zero energy mode(s))
4-node	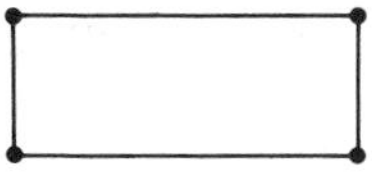	2 X 2	—
4-node distorted	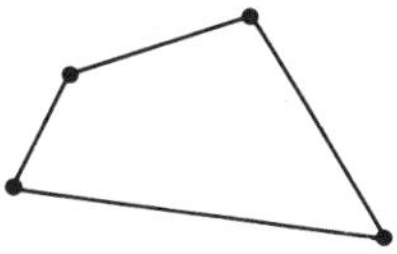	2 X 2	—
8-node	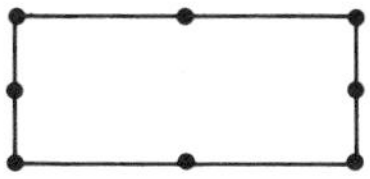	3 X 3	2 X 2
8-node distorted	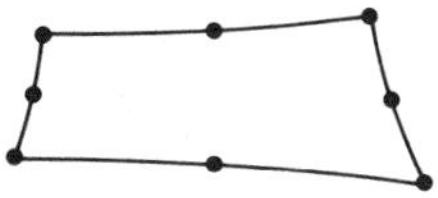	3 X 3	2 X 2
9-node	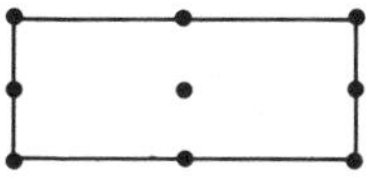	3 X 3	2 X 2
9-node distorted	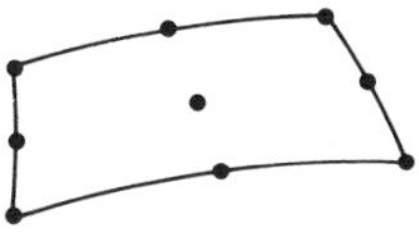	3 X 3	2 X 2

TABLE 5.5 *Gauss numerical integration orders in evaluation of stiffness matrices of two-dimensional rectangular elements (use of Table 5.2).*

16-node	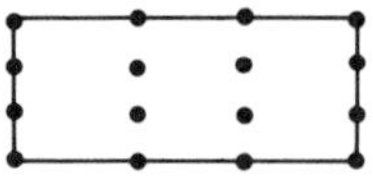	4 × 4	3 × 3
16-node distorted	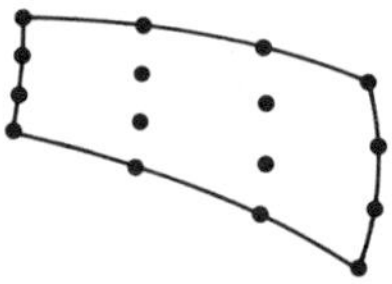	4 × 4	3 × 3

TABLE 5.5 (*cont.*)

analysis for the appropriate integration orders in the evaluation of the stiffness matrices of two-dimensional elements. This table is presented here in order to give specific guidelines for the choice of the Gauss numerical integration orders in plane stress, plane strain, and axisymmetric analyses. However, the information given in the table is also valuable in deducing appropriate orders of integration for the calculation of the stiffness matrices of other elements.

EXAMPLE 5.38: Discuss the required integration order for the evaluation of the shell and three-dimensional elements shown in Fig. 5.45.

Consider first the shell element. The integration in the r-s plane corresponds to the evaluation of the nine-node element in Table 5.5. Also, the displacements vary

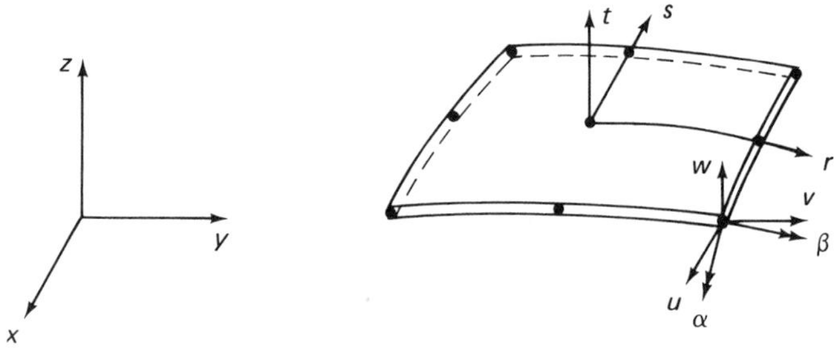

(a) Nine-node shell element

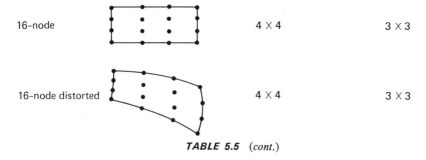

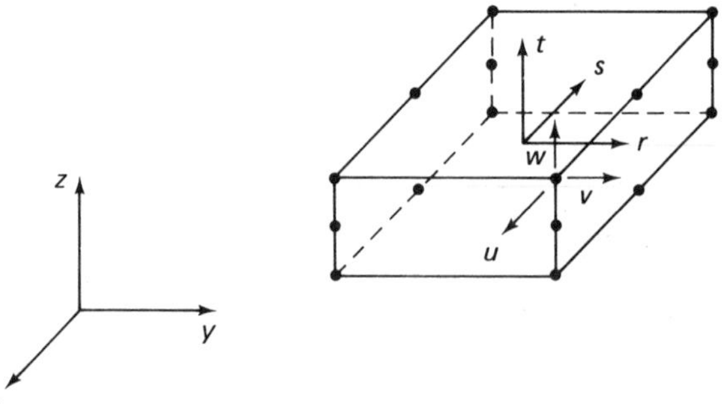

(b) Three-dimensional solid element

FIGURE 5.45 *Elements considered in Example 5.38.*

linearly with t. Hence, if the element were flat and rectangular we would evaluate the exact stiffness matrix with $3 \times 3 \times 2$ Gauss integration corresponding to the r, s, t axes. In general, this integration order will also be effective when the element is used in distorted form to model a curved shell.

The required integration order for the evaluation of the stiffness matrix of the three-dimensional solid element can also be deduced from the information given in Table 5.5. The displacements vary linearly in the r-direction; hence two-point integration is sufficient. In the t-s planes, i.e. at r equal to a constant, the element displacements correspond to those of the eight-node element in Table 5.5. Hence, $2 \times 3 \times 3$ Gauss integration is required to evaluate the element stiffness matrix exactly.

In Table 5.5 we list a reduced integration order that is used in practice and that may for a given finite element mesh result in significantly improved displacement and stress predictions. However, this integration order yields element stiffness matrices that display one or more spurious zero energy modes, which may also be the cause of an unstable (or very inaccurate) solution. As an example, consider the analysis of the simple element assemblage in Fig. 5.46.

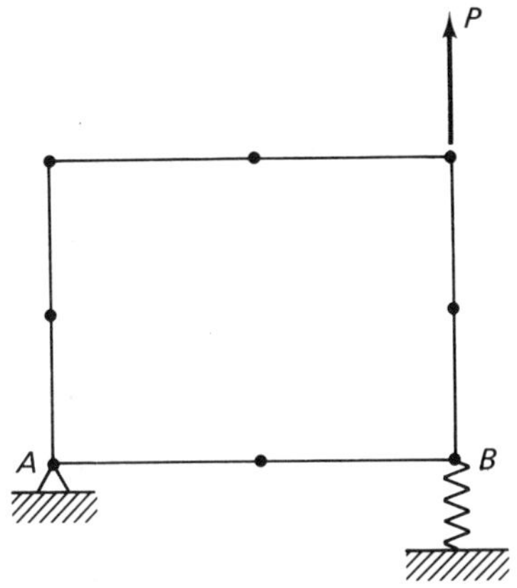

FIGURE 5.46 *Eight-node plane stress element supported at B by a spring.*

Using 2×2 Gauss integration for the eight-node element, the solution is unstable (the calculated nodal point displacements are very large and have no resemblance with the actual solution), whereas using 3×3 Gauss integration (which yields the exact stiffness matrix of the eight-node element), the exact solution—a rigid body rotation about point A—is calculated.

Finally, considering the information given in Table 5.5, we should note that using the "reliable" integration order we evaluate the stiffness matrices of rectangular elements exactly, whereas the matrices of the distorted elements are, in general, only approximated, but the given integration orders should be sufficient.*

In the above discussion we referred specifically to the evaluation of the element matrices of rectangular elements. But the basic observations summarized above are also applicable to the numerical evaluation of the matrices of trian-

* Note here that the element distortions may lower significantly the order of strain interpolations (see Section 5.3), and a lower integration order than for the undistorted element may actually be appropriate (also to prevent locking of a plate or shell element, see Section 5.4.1).

gular elements for which, however, a different integration scheme is employed (see Table 5.4).

5.8.2 Calculation of Stresses

Assuming that the nodal point displacements of a finite element idealization have been calculated, the stresses within an element are evaluated using the relation

$$\boldsymbol{\tau} = \mathbf{CBU} + \boldsymbol{\tau}^I \tag{5.136}$$

which was already used in (4.11). *The relation in* (5.136) *gives the stresses at any point of the element*, but in practice, the element stresses are only calculated and printed at some specific points. These may be the center of the element, the nodal point locations or the numerical integration points used in the evaluation of the element stiffness matrix. Considering the evaluation of the element stresses, we want to summarize some important observations, some of which we stated already in Section 4.2.6.

Assume that an element idealization is employed which satisfies all convergence criteria. This does not mean that the predicted stresses are in general continuous across the element boundaries. Hence, considering a compatible element mesh, we would find that the displacements are continuous from element to element, but *the stress components are not continuous unless a stress field is analyzed which is contained in the element formulation*. A simple analysis that demonstrates this observation is described in Fig. 5.47. When the two eight-node element model of a cantilever is subjected to a bending moment the exact stresses (within beam theory) are predicted, because the eight-node element can represent a constant bending moment exactly (see Example 5.39). Hence, there is no stress discontinuity between the elements. However, the model cannot represent exactly the linearly varying bending moment to be predicted in the analysis of Fig. 5.47(b), which therefore results in stress discontinuities between the elements. The theoretical reason for the stress discontinuities is that only continuity in the displacements, and not their derivatives, is imposed in the analysis (see Sections 4.2.5 and 5.5). For the same reason, imposed stress boundary conditions are only approximately satisfied at a specific boundary point (but they are satisfied in an integral sense).

Based on the convergence considerations discussed earlier (see Section 4.2.5), we note once more that the stress discontinuities between elements and the violation of the local stress boundary conditions will decrease as the element mesh is refined. Therefore, the magnitude of the stress discontinuities between elements can be employed in practice as a measure to indicate whether the finite element idealization has to be refined.

Another observation in the stress calculations is that the stresses at some points in an element can be significantly more accurate when compared with the exact solution than at other points. In particular, it has been observed that the stresses may be considerably more accurate at the Gauss integration points than at the nodal points of an element.[41]

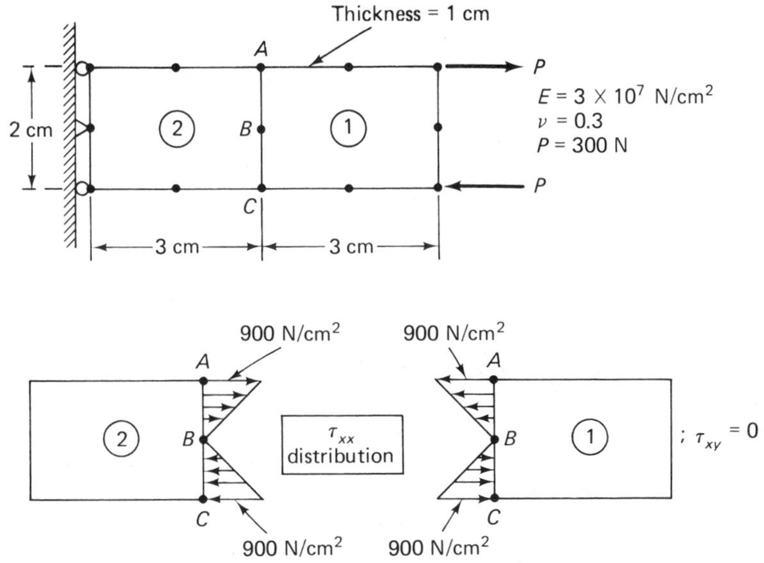

(a) Cantilever subjected to bending moment and finite element solution

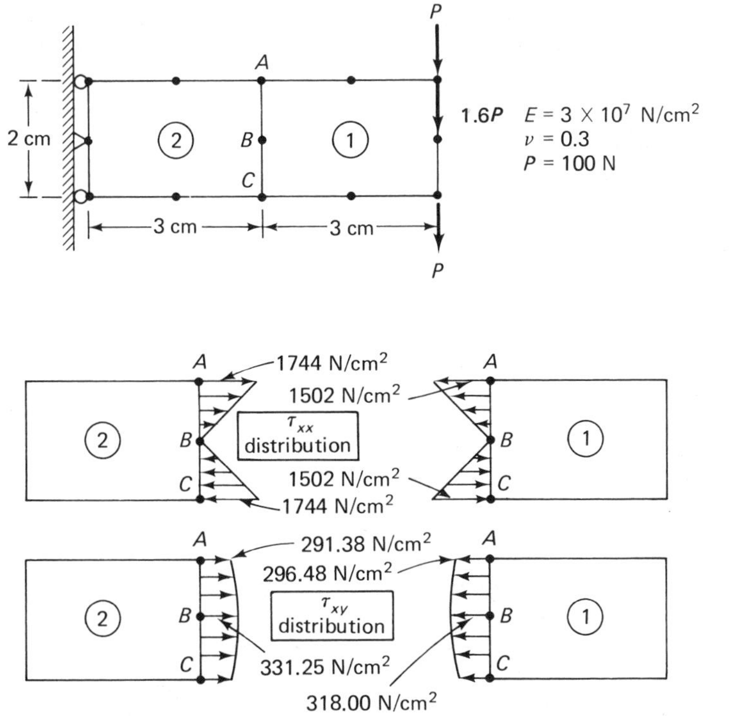

(b) Cantilever subjected to tip-shear force and finite element solution

FIGURE 5.47 *Predicted longitudinal stress distribution in analysis of cantilever.*

The objective in practice is usually to obtain the "best" stress predictions possible once the nodal point displacements have been evaluated. For this purpose, if the difference between the element boundary stresses is not too large, it may be appropriate to simply average them. In an alternative approach, the stresses are only calculated within the elements and then a least squares fit or other interpolation procedure is employed to predict the stresses at the element boundaries or other desired points.[42]

5.8.3 Some Modeling Considerations

The establishment of an appropriate finite element model of an actual practical problem depends to a large degree on the following factors: the understanding of the physical problem including a qualitative knowledge of the structural response to be predicted, a thorough knowledge of the basic principles of mechanics, and a good understanding of the finite element procedures available for analysis. Hence, in essence, all the material presented in the previous and the following sections represents important basic knowledge in order that an appropriate finite element model for the analysis of a specific problem can be established. It is our objective in this section to point out a few additional modeling considerations that frequently arise in practice.

A first consideration is the choice of finite elements to be used. Considering the various types of problems that require analysis, it is quite impossible to say that a specific element will always be most effective. However, if we restrict our choice to the isoparametric elements discussed in this chapter, some recommendations can be given.

Table 5.6 summarizes the elements that are usually effective in an actual analysis. We may note that—except for the two-node truss element which is effective because of its simplicity and versatility—the other recommended ele-

Type of Problem	Element	
Truss or cable	2-node	
Two-dimensional Plane stress Plane strain Axisymmetric	8-node or 9-node	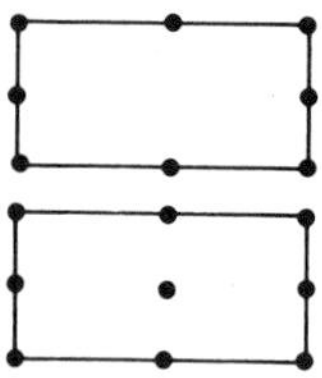
Three-dimensional	20-node	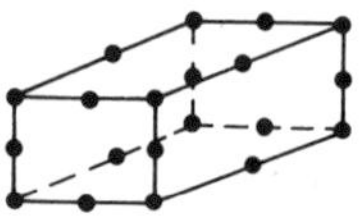

TABLE 5.6 *Isoparametric elements usually effective in analysis.*

Three-dimensional beam	3-node or 4-node	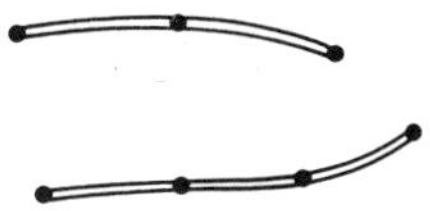
Plate	9-node	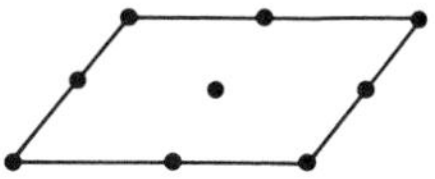
Shell	9-node or 16-node	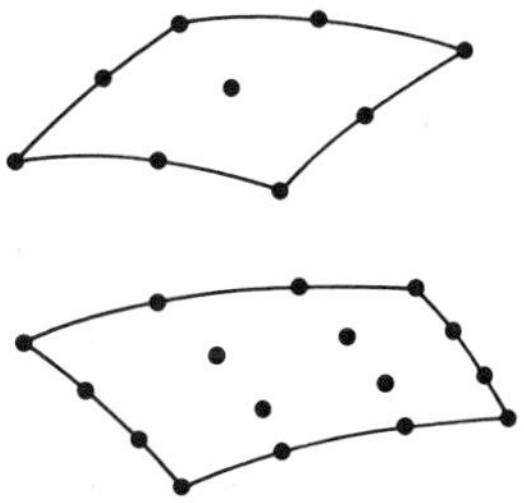

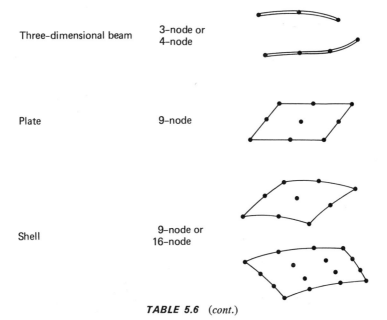

TABLE 5.6 *(cont.)*

ments are higher-order elements, and in particular, the use of parabolic elements is recommended. The reason is that with these elements the cost of using an element and its predictive capability are well-balanced. The predictive capabilities of the parabolic elements can model exactly a constant bending moment, which is important whenever a bending-type stress situation is to be analyzed. We demonstrate this feature in the following example.

EXAMPLE 5.39: Consider the eight-node element in Fig. 5.48 when the element is subjected to the nodal point displacements,

$$u_1 = -\delta, \quad u_2 = +\delta, \quad u_3 = -\delta, \quad u_4 = +\delta$$
$$v_5 = -\frac{\delta}{2}, \quad v_7 = -\frac{\delta}{2} \qquad \text{(a)}$$

and all other nodal point displacements are equal to zero. Evaluate the element strains.

To establish the element strains we evaluate first the element displacements corresponding to the nodal point displacements given in (a). Using

$$u = \sum_{i=1}^{8} h_i u_i; \qquad v = \sum_{i=1}^{8} h_i v_i$$

where the interpolation functions h_i are given in Fig. 5.5, we obtain

$$u = -rs\delta \qquad v = -\tfrac{1}{2}(1 - r^2)\delta \qquad \text{(b)}$$

The element strains are then, since $x \equiv r$, and $y \equiv s$:

$$\epsilon_{xx} = -s\delta; \qquad \epsilon_{yy} = 0; \qquad \gamma_{xy} = 0$$

We note that these strains are those of a beam in pure bending (with $\nu = 0$).

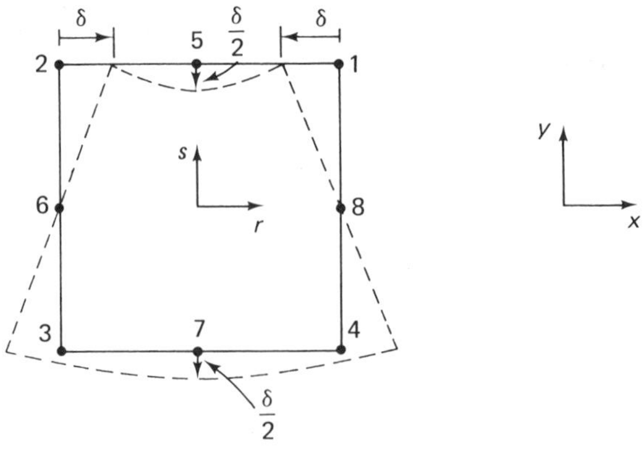

FIGURE 5.48 *2 by 2 eight-node element subjected to bending deformations.*

Table 5.6 also shows that for the plate and shell analyses the nine-node and sixteen-node Lagrangian elements are recommended. It is important to use these elements instead of the eight- and twelve-node elements which do not have the internal nodes and "lock" when the element thickness to length ratio becomes small (see Section 5.4).

We show in Table 5.6 only rectangular elements, but in essence the above recommendations are also applicable to the use of triangular elements; i.e., the parabolic and cubic elements are usually most effective.

Once a certain element has been chosen, we must decide on the total element layout to be used in the analysis. In this process the element sizes and the number of elements must be selected, and we must decide on the use of mesh grading. The decisions to be made are very problem-dependent but again some general guidelines can be given.

We discussed in Section 5.8.2 the calculation of element stresses and pointed out that, considering an area of a structure in which the exact solution would predict stress continuity, the stresses of a finite element analysis would not (necessarily) be continuous between elements unless a very fine finite element idealization is used. Namely, the stress discontinuities between elements decrease as the finite element idealization is refined and convergence to the "exact" solution is reached. These observations led to the conclusion that the amount of stress discontinuities between elements are a measure of the "appropriateness" of the finite element idealization used. Hence, the overall objective in the design of a finite element mesh is that in an area where high solution accuracy is required, the stress discontinuities between elements should be small, whereas larger stress discontinuities can be tolerated away from the areas of interest. The actual amount of stress discontinuities that can be tolerated depends on the accuracy required in the analysis.

The above objective may require some mesh grading; i.e., a finer finite element mesh in certain areas than in others, and this may require the use of transition zones. Figure 5.49 shows some typical transitions with compatible element layouts. Figure 5.49(a) illustrates with a simple example how the use of the variable-number-nodes elements can be effective in constructing transition regions. As another means, constraint equations are frequently useful to preserve compatibility between elements in a transition region. A simple application of constraint equations is shown in Fig. 5.49(b).

Another consideration in the design of a finite element layout is that, except when explicit advantage of element distortions can be taken (see Section 5.3.2), the performance of the isoparametric elements is generally best when they are used without distortions; for example, the rectangular elements should be

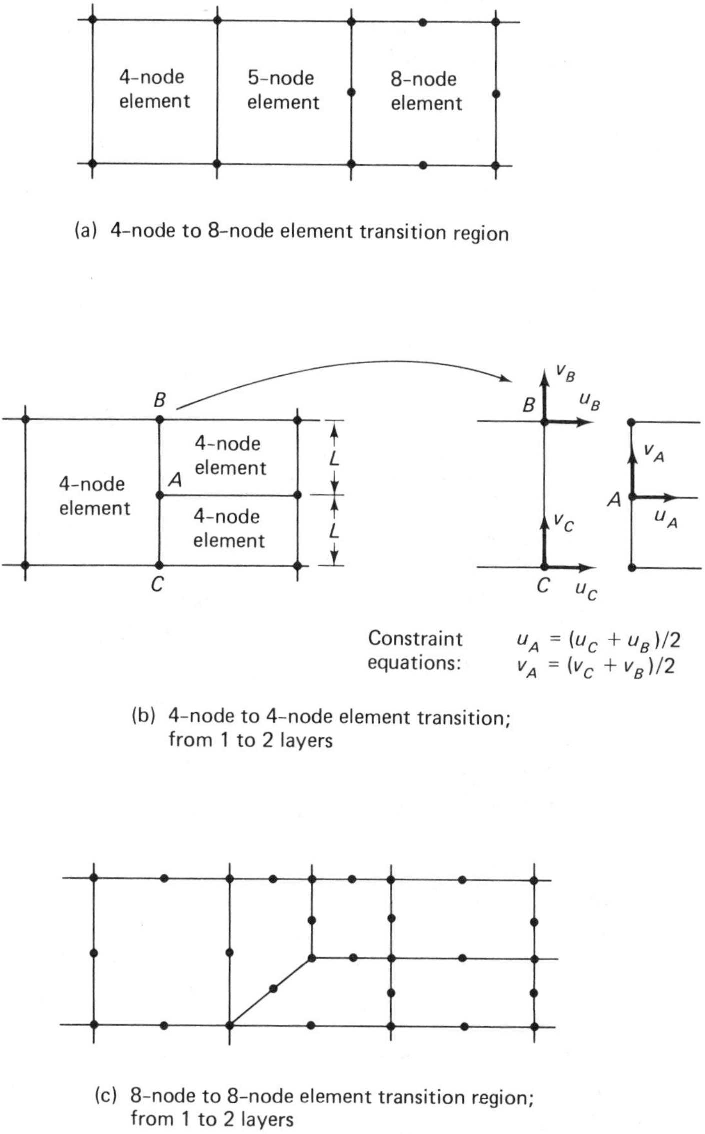

FIGURE 5.49 *Some transitions with compatible element layouts.*

"truly rectangular" with element interior angles of 90 degrees, straight element sides and the noncorner nodes located at the physical positions that correspond to their r, s, and t values.[43] In this case the Jacobian operator between the natural and physical coordinate systems is a diagonal matrix with constant entries and hence does not affect the order of strain interpolations.

The effect of distorting elements on the accuracy of the analysis results depends to a large degree on the problem considered and the elements used. Although not desirable, in practice distortions of elements cannot be avoided, such as in the modeling of boundaries and transition regions. If the analysis results must be very accurate in these areas, a comparatively larger number of elements must be employed to model these regions in order to compensate for the loss in predictive capabilities of the distorted elements.

Since element distortions often cannot be avoided in practice—and similarly for incompatibilities between elements—a frequent question is: What is the effect of the element distortions and incompatibilities in one area of the mesh on the response that is predicted in another area of this mesh? The precise effect is again dependent on the specific problem considered, and the elements and mesh layout used. However, a major guideline in answering this question is provided by the use of the *St. Venant principle*.[44] Considering a finite element mesh, based on this principle we can expect that the effect of element incompatibilities and distortions will in general be small at a "reasonable" distance from these elements. The actual influence can, however, only be measured using the guidelines given above or by comparison with a more accurate solution.

Finally, we note that in order to assess the actual accuracy of a finite element analysis, without the knowledge of the exact solution to the problem it may be necessary to rerun the analysis with a finer finite element mesh until the changes in the analysis results are sufficiently small. At this point it is assumed that the finite element solution has converged to the "exact" solution of the mechanical model which, it is assumed, represents accurately enough the actual physical problem (see Section 4.2.5). The selection of optimal meshes in these successive finite element solutions is currently subject to much research effort.[45]

5.9 COMPUTER PROGRAM IMPLEMENTATION OF ISOPARAMETRIC FINITE ELEMENTS

In Section 5.3 we discussed the isoparametric finite element formulation and showed the specific expressions needed in the calculation of four-node plane stress (or plane strain) elements (see Example 5.5). An important advantage of isoparametric element calculations is the similarity between the calculations of different elements. For example, the calculation of three-dimensional elements is a relatively simple extension from the calculation of two-dimensional elements. Also, in one subroutine, elements with a variety of nodal point configurations can be calculated if an algorithm for selecting the appropriate interpolation functions is used (see Section 5.3).

The purpose of this section is to give an actual computer program for the calculation of the stiffness matrix of four-node isoparametric elements. In essence, SUBROUTINE QUADS given below is the computer program implementation of the procedures presented in Example 5.5. In addition to the plane stress and plane strain analysis capability, axisymmetric conditions can also be analyzed. It is believed that by showing the actual computer implementation of the element, the relative ease of implementing isoparametric elements is demonstrated best. The input and output variables and the flow of the program are described by means of comment cards.

```
      SUBROUTINE QUADS(NEL,ITYPE,NINT,THIC,YM,PR,XX,S,IOUT)            QUAD0001
C                                                                      QUAD0002
C . . . . . . . . . . . . . . . . . . . . . . . . . . . . . . . .     QUAD0003
C .                                                              .     QUAD0004
C .   P R O G R A M                                              .     QUAD0005
C .      TO CALCULATE ISOPARAMETRIC QUADRILATERAL ELEMENT STIFFNESS .  QUAD0006
C .      MATRIX FOR AXISYMMETRIC, PLANE STRESS AND PLANE STRAIN  .     QUAD0007
C .      CONDITIONS                                              .     QUAD0008
C .                                                              .     QUAD0009
C . - - INPUT VARIABLES - -                                      .     QUAD0010
C .      NEL         = NUMBER OF ELEMENT                         .     QUAD0011
C .      ITYPE       = ELEMENT TYPE                              .     QUAD0012
C .                      EQ.0 = AXISYMMETRIC                     .     QUAD0013
C .                      EQ.1 = PLANE STRAIN                     .     QUAD0014
C .                      EQ.2 = PLANE STRESS                     .     QUAD0015
C .      NINT        = GAUSS NUMERICAL INTEGRATION ORDER         .     QUAD0016
C .      THIC        = THICKNESS OF ELEMENT                      .     QUAD0017
C .      YM          = YOUNG*S MODULUS                           .     QUAD0018
C .      PR          = POISSON*S RATIO                           .     QUAD0019
C .      XX (2,4)    = ELEMENT NODE COORDINATES                  .     QUAD0020
C .      S (8,8)     = STIFFNESS MATRIX ON SOLUTION EXIT         .     QUAD0021
C .      IOUT        = OUTPUT PRINTING FILE                      .     QUAD0022
C .                                                              .     QUAD0023
C . - - OUTPUT - -                                               .     QUAD0024
C .      S (8,8)     = STIFFNESS MATRIX                          .     QUAD0025
C .                                                              .     QUAD0026
C . . . . . . . . . . . . . . . . . . . . . . . . . . . . . . . .     QUAD0027
      IMPLICIT REAL*8 (A-H,O-Z)                                        QUAD0028
      SQRT(X)=DSQRT(X)                                                 QUAD0029
      ABS (X)=DABS (X)                                                 QUAD0030
C . . . . . . . . . . . . . . . . . . . . . . . . . . . . . . . .     QUAD0031
C .      THIS PROGRAM IS USED IN SINGLE PRECISION ARITHMETIC ON CDC .  QUAD0032
C .      EQUIPMENT AND DOUBLE PRECISION ARITHMETIC ON IBM OR UNIVAC .  QUAD0033
C .      MACHINES.FOR SINGLE OR DOUBLE PRECISION ARITHMETIC ACTIVATE. QUAD0034
C .      DEACTIVATE OR ADJUST ABOVE AND DATA XG/WGT CARDS.       .     QUAD0035
C . . . . . . . . . . . . . . . . . . . . . . . . . . . . . . . .     QUAD0036
C                                                                      QUAD0037
      DIMENSION D(4,4),B(4,8),XX(2,4),S(8,8),XG(4,4),WGT(4,4),DB(4)    QUAD0038
C                                                                      QUAD0039
C     MATRIX XG STORES GAUSS - LEGENDRE SAMPLING POINTS                QUAD0040
C                                                                      QUAD0041
      DATA XG/  0.D0,    0.D0,    0.D0,    0.D0,    -.5773502691896D0, QUAD0042
     1 .5773502691896D0,    0.D0,    0.D0,  -.7745966692415D0,  0.D0,  QUAD0043
     2 .7745966692415D0,   0.D0,   -.8611363115941D0,                 QUAD0044
     3 -.3399810435849D0,   .3399810435849D0,   .8611363115941D0 /    QUAD0045
C                                                                      QUAD0046
C     MATRIX WGT STORES GAUSS - LEGENDRE WEIGHTING FACTORS             QUAD0047
C                                                                      QUAD0048
      DATA WGT /  2.D0,    0.D0,    0.D0,    0.D0,    1.D0,    1.D0,    QUAD0049
     1 0.D0,    0.D0,  .5555555555556D0,   .8888888888889D0,           QUAD0050
     2 .5555555555556D0,   0.D0,   .3478548451375D0,   .6521451548625D0, QUAD0051
     3 .6521451548625D0,   .3478548451375D0 /                         QUAD0052
C                                                                      QUAD0053
C     O B T A I N   S T R E S S - S T R A I N   L A W                  QUAD0054
C                                                                      QUAD0055
      F=YM/(1. + PR)                                                   QUAD0056
      G=F*PR/(1. - 2.*PR)                                              QUAD0057
      H=F + G                                                          QUAD0058
C                                                                      QUAD0059
C     PLANE STRAIN ANALYSIS                                            QUAD0060
C                                                                      QUAD0061
      D (1,1) =H                                                       QUAD0062
      D (1,2) =G                                                       QUAD0063
      D (1,3) =0.                                                      QUAD0064
      D (2,1) =G                                                       QUAD0065
      D (2,2) =H                                                       QUAD0066
      D (2,3) =0.                                                      QUAD0067
      D (3,1) =0.                                                      QUAD0068
      D (3,2) =0.                                                      QUAD0069
      D (3,3) =F/2.                                                    QUAD0070
      IF (ITYPE.EQ.1) THIC=1.                                          QUAD0071
      IF (ITYPE.EQ.1) GO TO 20                                         QUAD0072
C                                                                      QUAD0073
C     AXISYMMETRIC ANALYSIS                                            QUAD0074
C                                                                      QUAD0075
      D (1,4) =G                                                       QUAD0076
      D (2,4) =G                                                       QUAD0077
      D (3,4) =0.                                                      QUAD0078
      D (4,1) =G                                                       QUAD0079
```

```
      D(4,2)=G                                                        QUAD0080
      D(4,3)=0.                                                       QUAD0081
      D(4,4)=H                                                        QUAD0082
      IF(ITYPE.EQ.0)GO TO 20                                          QUAD0083
C                                                                     QUAD0084
C     FOR PLANE STRESS ANALYSIS CONDENSE STRESS-STRAIN MATRIX         QUAD0085
C                                                                     QUAD0086
      DO 10 I=1,3                                                     QUAD0087
      A=D(I,4)/D(4,4)                                                 QUAD0088
      DO 10 J=I,3                                                     QUAD0089
      D(I,J)=D(I,J) - D(4,J)*A                                        QUAD0090
   10 D(J,I)=D(I,J)                                                   QUAD0091
C                                                                     QUAD0092
C     C A L C U L A T E   E L E M E N T   S T I F F N E S S           QUAD0093
C                                                                     QUAD0094
   20 DO 30 I=1,8                                                     QUAD0095
      DO 30 J=1,8                                                     QUAD0096
   30 S(I,J)=0.                                                       QUAD0097
      IST=3                                                           QUAD0098
      IF (ITYPE.EQ.0) IST=4                                           QUAD0099
      DO 80 LX=1,NINT                                                 QUAD0100
      RI=XG(LX,NINT)                                                  QUAD0101
      DO 80 LY=1,NINT                                                 QUAD0102
      SI=XG(LY,NINT)                                                  QUAD0103
C                                                                     QUAD0104
C     EVALUATE DERIVATIVE OPERATOR B AND THE JACOBIAN DETERMINANT DET QUAD0105
C                                                                     QUAD0106
      CALL STDM(XX,B,DET,RI,SI,XBAR,NEL,ITYPE,IOUT)                   QUAD0107
C                                                                     QUAD0108
C     ADD CONTRIBUTION TO ELEMENT STIFFNESS                          QUAD0109
C                                                                     QUAD0110
      IF (ITYPE.GT.0) XBAR=THIC                                       QUAD0111
      WT=WGT(LX,NINT)*WGT(LY,NINT)*XBAR*DET                           QUAD0112
      DO 70 J=1,8                                                     QUAD0113
      DO 40 K=1,IST                                                   QUAD0114
      DB(K)=0.0                                                       QUAD0115
      DO 40 L=1,IST                                                   QUAD0116
   40 DB(K)=DB(K)+D(K,L)*B(L,J)                                       QUAD0117
      DO 60 I=J,8                                                     QUAD0118
      STIFF=0.0                                                       QUAD0119
      DO 50 L=1,IST                                                   QUAD0120
   50 STIFF=STIFF+B(L,I)*DB(L)                                        QUAD0121
   60 S(I,J)=S(I,J)+STIFF*WT                                          QUAD0122
   70 CONTINUE                                                        QUAD0123
   80 CONTINUE                                                        QUAD0124
C                                                                     QUAD0125
      DO 90 J=1,8                                                     QUAD0126
      DO 90 I=J,8                                                     QUAD0127
   90 S(J,I)=S(I,J)                                                   QUAD0128
C                                                                     QUAD0129
      RETURN                                                          QUAD0130
C                                                                     QUAD0131
      END                                                             QUAD0132


      SUBROUTINE STDM(XX,B,DET,R,S,XBAR,NEL,ITYPE,IOUT)              STDM0001
C                                                                     STDM0002
C . . . . . . . . . . . . . . . . . . . . . . . . . . . . . . . . .   STDM0003
C .                                                               .   STDM0004
C .   P R O G R A M                                               .   STDM0005
C .      TO EVALUATE THE STRAIN-DISPLACEMENT TRANSFORMATION MATRIX B. STDM0006
C .      AT POINT (R,S) FOR A QUADRILATERAL ELEMENT               .   STDM0007
C .                                                               .   STDM0008
C . . . . . . . . . . . . . . . . . . . . . . . . . . . . . . . . .   STDM0009
      IMPLICIT REAL*8(A-H,O-Z)                                        STDM0010
      DIMENSION XX(2,4),B(4,8),H(4),P(2,4),XJ(2,2),XJI(2,2)           STDM0011
C                                                                     STDM0012
      RP = 1.0 + R                                                    STDM0013
      SP = 1.0 + S                                                    STDM0014
      RM = 1.0 - R                                                    STDM0015
      SM = 1.0 - S                                                    STDM0016
C                                                                     STDM0017
C     INTERPOLATION FUNCTIONS                                         STDM0018
C                                                                     STDM0019
      H(1) = 0.25* RP* SP                                             STDM0020
      H(2) = 0.25* RM* SP                                             STDM0021
      H(3) = 0.25* RM* SM                                             STDM0022
      H(4) = 0.25* RP* SM                                             STDM0023
C                                                                     STDM0024
C     NATURAL COORDINATE DERIVATIVES OF THE INTERPOLATION FUNCTIONS   STDM0025
C                                                                     STDM0026
C        1. WITH RESPECT TO R                                        STDM0027
C                                                                     STDM0028
      P(1,1) = 0.25* SP                                               STDM0029
      P(1,2) = - P(1,1)                                               STDM0030
      P(1,3) = - 0.25* SM                                             STDM0031
      P(1,4) = - P(1,3)                                               STDM0032
C                                                                     STDM0033
C        2. WITH RESPECT TO S                                        STDM0034
C                                                                     STDM0035
      P(2,1) = 0.25* RP                                               STDM0036
      P(2,2) = 0.25* RM                                               STDM0037
      P(2,3) = - P(2,2)                                               STDM0038
      P(2,4) = - P(2,1)                                               STDM0039
C                                                                     STDM0040
```

```
C      EVALUATE THE JACOBIAN MATRIX AT POINT (R,S)                      STDM0041
C                                                                       STDM0042
   10 DO 30 I=1,2                                                       STDM0043
      DO 30 J=1,2                                                       STDM0044
      DUM = 0.0                                                         STDM0045
      DO 20 K=1,4                                                       STDM0046
   20 DUM=DUM+P(I,K)*XX(J,K)                                            STDM0047
   30 XJ(I,J)=DUM                                                       STDM0048
C                                                                       STDM0049
C      COMPUTE THE DETERMINANT OF THE JACOBIAN MATRIX AT POINT (R,S)    STDM0050
C                                                                       STDM0051
      DET = XJ(1,1)* XJ(2,2) - XJ(2,1)* XJ(1,2)                         STDM0052
      IF (DET.GT.0.00000001) GO TO 40                                   STDM0053
      WRITE (IOUT,2000) NEL                                             STDM0054
      STOP                                                              STDM0055
C                                                                       STDM0056
C      COMPUTE INVERSE OF THE JACOBIAN MATRIX                           STDM0057
C                                                                       STDM0058
   40 DUM=1./DET                                                        STDM0059
      XJI(1,1) = XJ(2,2)* DUM                                           STDM0060
      XJI(1,2) =-XJ(1,2)* DUM                                           STDM0061
      XJI(2,1) =-XJ(2,1)* DUM                                           STDM0062
      XJI(2,2) = XJ(1,1)* DUM                                           STDM0063
C                                                                       STDM0064
C      EVALUATE GLOBAL DERIVATIVE OPERATOR B                            STDM0065
C                                                                       STDM0066
      K2=0                                                              STDM0067
      DO 60 K=1,4                                                       STDM0068
      K2=K2 + 2                                                         STDM0069
      B(1,K2-1) = 0.                                                    STDM0070
      B(1,K2 ) = 0.                                                     STDM0071
      B(2,K2-1) = 0.                                                    STDM0072
      B(2,K2 ) = 0.                                                     STDM0073
      DO 50 I=1,2                                                       STDM0074
      B(1,K2-1) = B(1,K2-1) + XJI(1,I) * P(I,K)                         STDM0075
   50 B(2,K2 ) = B(2,K2 ) + XJI(2,I) * P(I,K)                           STDM0076
      B(3,K2 ) = B(1,K2-1)                                              STDM0077
   60 B(3,K2-1) = B(2,K2 )                                              STDM0078
C                                                                       STDM0079
C      IN CASE OF PLANE STRAIN CR PLANE STRESS ANALYSIS DO NOT INCLUDE  STDM0080
C      THE NORMAL STRAIN COMPONENT                                      STDM0081
C                                                                       STDM0082
      IF (ITYPE.GT.0) RETURN                                           STDM0083
C                                                                       STDM0084
C      COMPUTE THE RADIUS AT POINT (R,S)                                STDM0085
C                                                                       STDM0086
      XBAR=0.0                                                          STDM0087
      DO 70 K=1,4                                                       STDM0088
   70 XBAR=XBAR + H(K)*XX(1,K)                                          STDM0089
C                                                                       STDM0090
C      EVALUATE THE HOOP STRAIN-DISPLACEMENT RELATION                   STDM0091
C                                                                       STDM0092
      IF(XBAR.GT.0.00000001)GO TO 90                                    STDM0093
C                                                                       STDM0094
C      FOR THE CASE OF ZERO RADIUS EQUATE RADIAL TO HOOP STRAIN         STDM0095
C                                                                       STDM0096
      DO 80 K=1,8                                                       STDM0097
   80 B(4,K)=B(1,K)                                                     STDM0098
      RETURN                                                            STDM0099
C                                                                       STDM0100
C      NON-ZERO RADIUS                                                  STDM0101
C                                                                       STDM0102
   90 DUM=1./XBAR                                                       STDM0103
      K2=0                                                              STDM0104
      DO 100 K=1,4                                                      STDM0105
      K2=K2 + 2                                                         STDM0106
      B(4,K2 ) = 0.                                                     STDM0107
  100 B(4,K2-1)=H(K)*DUM                                                STDM0108
      RETURN                                                            STDM0109
C                                                                       STDM0110
 2000 FORMAT (10H0*** ERROR,                                            STDM0111
     1        52H ZERO OR NEGATIVE JACOBIAN DETERMINANT FOR ELEMENT (,I4, STDM0112
     2        1H) )                                                     STDM0113
C                                                                       STDM0114
      END                                                              STDM0115
```

REFERENCES

1. I. C. Taig, "Structural Analysis by the Matrix Displacement Method," English Electric Aviation Report S017, 1961.

2. B. M. Irons, "Engineering Application of Numerical Integration in Stiffness Method," *A.I.A.A. Journal*, Vol. 4, 1966, pp. 2035–2037.

3. B. M. Irons, "Numerical Integration Applied to Finite Element Methods," Conference on the Use of Digital Computers in Structural Engineering, University of Newcastle, England, 1966.

4. B. M. Irons and O. C. Zienkiewicz, "The Isoparametric Finite Element System—

a New Concept in Finite Element Analysis," *Proceedings, Conference on Recent Advances in Stress Analysis*, Royal Aeronautical Society, London, 1968.

5. J. G. Ergatoudis, "Iso-parametric Finite Elements in Two- and Three-Dimensional Analysis," Ph.D. dissertation, University of Wales, Swansea, 1968.

6. J. G. Ergatoudis, B. M. Irons, and O. C. Zienkiewicz, "Curved, Iso-parametric, Quadrilateral Elements for Finite Element Analysis," *International Journal of Solids and Structures*, Vol. 4, 1968, pp. 31–42.

7. J. G. Ergatoudis, B. M. Irons, and O. C. Zienkiewicz, "Three-Dimensional Analysis of Arch Dams and Their Foundations," *Symposium on Arch Dams*, Institute of Civil Engineering, London, Mar. 1968.

8. R. W. Clough, "Comparison of Three Dimensional Elements," *Symposium on Applied Finite Element Methods in Civil Engineering*, Vanderbilt University, Nashville, 1969, pp. 1–26.

9. O. C. Zienkiewicz, B. M. Irons, J. G. Ergatoudis, S. Ahmad, and F. C. Scott, "Isoparametric and Associated Element Families for Two- and Three-Dimensional Analysis," *Finite Element Methods in Stress Analysis* (I. Holand and K. Bell, eds.), Tapir, Trondheim, 1969, Chap. 13.

10. S. Ahmad, "Curved Finite Elements in the Analysis of Solid, Shell and Plate Structures," Ph.D. dissertation, University of Wales, Swansea, 1969.

11. S. Ahmad, B. M. Irons, and O. C. Zienkiewicz, "Analysis of Thick and Thin Shell Structures by Curved Elements," *International Journal for Numerical Methods in Engineering*, Vol. 2, 1970, pp. 419–451.

12. S. Ahmad, B. M. Irons, and O. C. Zienkiewicz, "Curved Thick Shell and Membrane Elements with Particular Reference to Axi-Symmetric Problems," *Proceedings, 2nd Conference on Matrix Methods in Structural Mechanics*, Wright-Patterson A.F.B., Ohio, 1968.

13. W. Pilkey, K. Saczalski, and H. Schaeffer (eds.), *Structural Mechanics Computer Programs*, University Press of Virginia, Charlottesville, Va., 1974.

14. K. J. Bathe, "ADINA—A Finite Element Program for Automatic Dynamic Incremental Nonlinear Analysis," Report AVL 82448–1, M.I.T., Sept. 1975 (rev. Dec. 1978).

15. R. E. Newton, "Degeneration of Brick-Type Isoparametric Elements," *Int. J. Num. Meth. Eng.*, Vol. 7, 1973, pp. 579–581.

16. R. D. Henshell and K. G. Shaw, "Crack Tip Finite Elements are Unnecessary," *Int. J. Num. Meth. Eng.*, Vol. 9, 1975, pp. 495–507.

17. R. S. Barsoum, "On the Use of Isoparametric Finite Elements in Linear Fracture Mechanics," *Int. J. Num. Meth. in Eng.*, Vol. 10, 1976, pp. 25–37.

18. R. S. Barsoum, "Triangular Quarter-Point Elements on Elastic and Perfectly-Plastic Crack Tip Elements," *Int. J. Num. Meth. Eng.*, Vol. 11, 1977, pp. 85–98.

19. R. S. Barsoum, "A Degenerate Solid Element for Linear Fracture Analysis of Plate Bending and General Shells," *Int. J. Num. Meth. in Eng.*, Vol. 10, pp. 551–564, 1976.

20. R. D. Mindlin, "Influence of Rotary Inertia and Shear on Flexural Motion of Isotropic Elastic Plates," *J. Appl. Mech.*, Vol. 18, pp. 31–38, 1951.

21. T. J. R. Hughes, R. L. Taylor, and W. Kanoknukulchai, "A Simple and Efficient Finite Element for Plate Bending," *Int. J. Num. Meth. in Eng.*, Vol. 11, 1977, pp. 1529–1543.

22. O. C. Zienkiewicz, J. Bauer, K. Morgan, and E. Onate, "A Simple and Efficient Element for Axisymmetric Shells," *Int. J. Num. Meth. in Eng.*, Vol. 11, 1977, pp. 1545–1558.

23. E. D. Pugh, E. Hinton, and O. C. Zienkiewicz, "A Study of Quadrilateral Plate Bending Elements with Reduced Integration," *Int. J. Num. Meth. in Eng.*, Vol. 12, 1978, pp. 1059–1078.

24. J. A. Stricklin, W. Haisler, P. Tisdale, and R. Gunderson, "A Rapidly Converging Triangular Plate Element," *A.I.A.A. J.*, Vol. 7, 1969, pp. 180–181.

25. J. L. Batoz, K. J. Bathe, and L. W. Ho, "A Study of Three-node Triangular Plate Bending Elements," *Int. J. Num. Methods in Eng.*, Vol. 15, pp. 1771–1812, 1980.

26. K. J. Bathe and C. Almeida, "A Simple and Effective Pipe Elbow Element—Linear Analysis," *J. Applied Mech.*, Vol. 47, 1980, pp. 93–100.

27. E. Ramm, "A Plate/Shell Element for Large Deflection and Rotations," in *Formulations and Computational Algorithms in Finite Element Analysis*, (K. J. Bathe, J. T. Oden, and W. Wunderlich, eds.), M.I.T. Press, 1977.

28. K. J. Bathe and S. Bolourchi, "A Geometric and Material Nonlinear Plate and Shell Element," *J. Computers and Structures*, Vol. 11, pp. 23–48, 1980.

29. E. L. Wilson, R. L. Taylor, W. Doherty, and J. Ghaboussi, "Incompatible Displacement Models," in *Numerical and Computer Methods in Structural Mechanics* (S. J. Fenves, N. Perrone, J. Robinson, and W. C. Schnobrich, eds.), Academic Press, Inc., New York, 1973.

30. H. H. Dovey, "Extension of Three-Dimensional Analysis to Shell Structures Using the Finite Element Idealization," Report UC SESM 74–2, Department of Civil Engineering, University of California, Berkeley, January 1974.

31. C. E. Fröberg, *Introduction to Numerical Analysis*, Addison-Wesley Publishing Company, Reading, Mass., 1969.

32. A. N. Loxan, N. Davids, and A. Levenson, "Table of the Zeros of the Legendre Polynomials of Order 1–16 and the Weight Coefficients for Gauss' Mechanical Quadrature Formula," *Bulletin of the American Mathematical Society*, Vol. 48, 1942, pp. 739–743.

33. A. H. Stroud and D. Secrest, *Gaussian Quadrature Formulas*, Prentice-Hall, Inc., Englewood Cliffs, N.J., 1966.

34. B. M. Irons, "Quadrature Rules for Brick-Based Finite Elements," *International Journal for Numerical Methods in Engineering*, Vol. 3, 1971, pp. 293–294.

35. P. Silvester, "Newton–Cotes Quadrature Formulae for N-dimensional Simplexes," *Proc. 2nd Can. Congr. Appl. Mech.*, Waterloo, Canada, 1969.

36. P. C. Hammer, O. P. Marlowe, and A. H. Stroud, "Numerical Integration over Simplexes and Cones," *Math. Tabl. Natn. Res. Coun. Wash.*, Vol. 10, 1956, pp. 130–137.

37. G. R. Cowper, "Gaussian Quadrature Formulas for Triangles," *Int. J. Num. Meth. in Eng.*, Vol. 7, 1973, pp. 405–408.

38. W. P. Doherty, E. L. Wilson, and R. L. Taylor, "Stress Analysis of Axisymmetric Solids Utilizing Higher Order Quadrilateral Finite Elements," Report UC SESM 69–3, Structural Eng. Lab., Univ. of California, Berkeley, 1969.

39. O. C. Zienkiewicz, R. L. Taylor, and J. M. Too, "Reduced Integration Techniques in General Analysis of Plates and Shells," *International Journal for Numerical Methods in Engineering*, Vol. 3, 1971, pp. 275–290.

40. S. F. Pawsey and R. W. Clough, "Improved Numerical Integration of Thick Shell Finite Elements," *International Journal for Numerical Methods in Engineering*, Vol. 3, 1971, pp. 575–586.

41. J. Barlow, "Optimal Stress Locations in Finite Element Models", *Int. J. Num. Meth. Eng.*, Vol. 10, 1976, pp. 243–251.

42. E. Hinton and J. S. Campbell, "Local and Global Smoothing of Discontinuous

Finite Element Functions Using Least Squares Method," *Int. J. Num. Meth. in Eng.*, Vol. 8, 1979, pp. 461–480.

43. G. Strang and G. J. Fix, *An Analysis of the Finite Element Method*, Prentice-Hall, Inc., Englewood Cliffs, N.J., 1973.

44. S. Timoshenko and J. N. Goodier, *Theory of Elasticity*, 3rd ed., McGraw-Hill. Book Company, New York, N.Y., 1970.

45. I. Babuška, "Finite Element Workshop 1980," Technical Note BN-940, Laboratory for Numerical Analysis, University of Maryland, May 1980.

6

FINITE ELEMENT NONLINEAR ANALYSIS IN SOLID AND STRUCTURAL MECHANICS

6.1 INTRODUCTION TO NONLINEAR ANALYSIS

In the finite element formulation given in Section 4.2, we assumed that the displacements of the finite element assemblage are infinitesimally small and that the material is linearly elastic. In addition, we also assumed that the nature of the boundary conditions remains unchanged during the application of the loads on the finite element assemblage. With these assumptions, the finite element equilibrium equations derived were for static analysis

$$\mathbf{KU} = \mathbf{R} \tag{6.1}$$

These equations correspond to a *linear* analysis of a structural problem because the displacement response $\mathbf{U}$ is a linear function of the applied load vector $\mathbf{R}$; i.e., if the loads are $\alpha\mathbf{R}$ instead of $\mathbf{R}$, where α is a constant, the corresponding displacements are $\alpha\mathbf{U}$. With this not the case we perform a *nonlinear* analysis.

The linearity of a response prediction rests on the assumptions stated above and it is instructive to identify in detail where these assumptions have entered the equilibrium equations in (6.1). The fact that the displacements must be small has entered into the evaluation of the matrix $\mathbf{K}$ and load vector $\mathbf{R}$, because all integrations have been performed over the original volume of the finite elements, and the strain-displacement matrix $\mathbf{B}$ of each element was assumed to be constant and independent of the element displacements. The assumption of a linear elastic material is implied in the use of a constant stress–strain matrix $\mathbf{C}$, and, finally, the assumption that the boundary conditions remain unchanged is reflected in the use of constant constraint relations (see (4.39) to (4.45)) for the complete response. If during loading a displacement boundary condition should change, e.g., a degree of freedom which was free becomes restrained at a certain load level, the response is only linear prior to the change in boundary condition. This situation arises, for example, in the analysis of a contact problem (see Example 6.2).

The above discussion of the basic assumptions used in a linear analysis defines what we mean by a nonlinear analysis and also suggests how to categorize different nonlinear analyses. Table 6.1 gives a classification that is used very conveniently in practical nonlinear analysis because this classification considers separately material nonlinear effects and kinematic nonlinear effects. If an actual problem is to be analyzed, for formulative and computational reasons, the problem is effectively described and analyzed as a problem in one of the categories given in Table 6.1. The formulations listed in the table are those that we shall discuss in detail, but we note that other formulations may be applicable to the solution of a problem as well.[1-9]

TABLE 6.1 *Classification of nonlinear analyses.*

Type of Analysis	*Description*	*Typical Formulation Used*	*Stress and Strain Measures*
Materially-nonlinear-only	Infinitesimal displacements and strains, stress–strain relation is nonlinear	Materially nonlinear only (M.N.O.)	Engineering stress and strain
Large displacements, large rotations, but small strains	Displacements and rotations of fibers are large, but fiber extensions and angle changes between fibers are small; the stress–strain relation may be linear or nonlinear	Total Lagrangian (T.L.) Updated Lagrangian (U.L.)	Second Piola–Kirchhoff stress, Green–Lagrange strain Cauchy stress, Almansi strain
Large displacements, large rotations, and large strains	Fiber extensions and angle changes between fibers are large, fiber displacements and rotations may also be large; the stress–strain relation may be linear or nonlinear	Updated Lagrangian Jaumann (U.L.J.) Total Lagrangian (T.L.)	Jaumann stress rate, velocity strain Second Piola–Kirchhoff stress, Green–Lagrange strain

Figure 6.1 gives an illustration of the types of problems that are encountered, as listed in Table 6.1. We should note that in a materially-nonlinear-only analysis, the nonlinear effect lies in the nonlinear stress-strain relation. The displacements and strains are infinitesimally small; therefore, the usual engineering stress and strain measures can be employed in the response description. Considering the large displacement but small strain conditions, we note that in essence the material is subjected to infinitesimally small strains measured in a body-attached coordinate frame x'-y', while this frame undergoes large rigid body displacements and rotations. The stress-strain relation of the material can be linear or nonlinear.

As shown in Fig. 6.1 and Table 6.1, the most general analysis case is the one in which the material is subjected to large displacements and large strains. In this case the stress-strain relation is also usually nonlinear.

$$\sigma = P/A$$
$$\epsilon = \sigma/E$$
$$\Delta = \epsilon L$$

$\epsilon < 0.04$

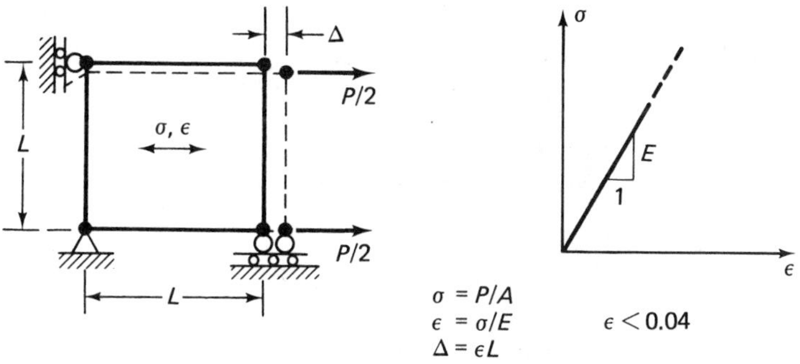

(a) Linear elastic (infinitesimal displacements)

$$\sigma = P/A$$
$$\epsilon = \frac{\sigma_Y}{E} + \frac{\sigma - \sigma_Y}{E_T}$$
$$\epsilon < 0.04$$

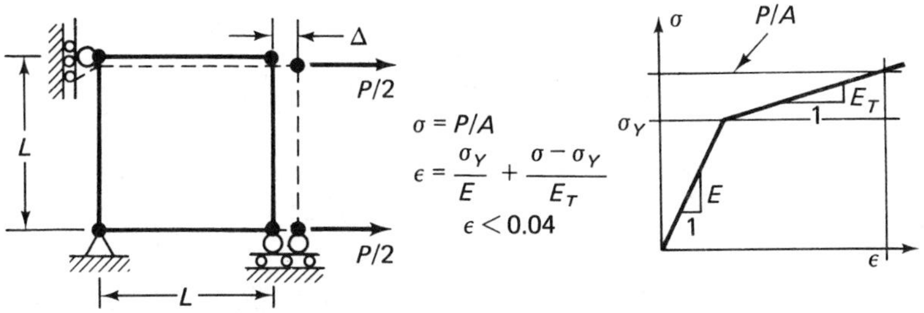

(b) Materially-nonlinear-only (infinitesimal displacements, but nonlinear stress-strain relation)

$\epsilon' < 0.04$
$\Delta' = \epsilon' L$

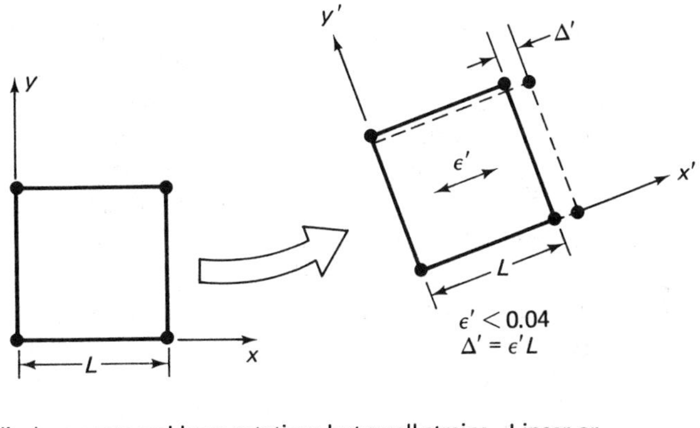

(c) Large displacements and large rotations but small strains. Linear or nonlinear material behavior

FIGURE 6.1 *Classification of analyses.*

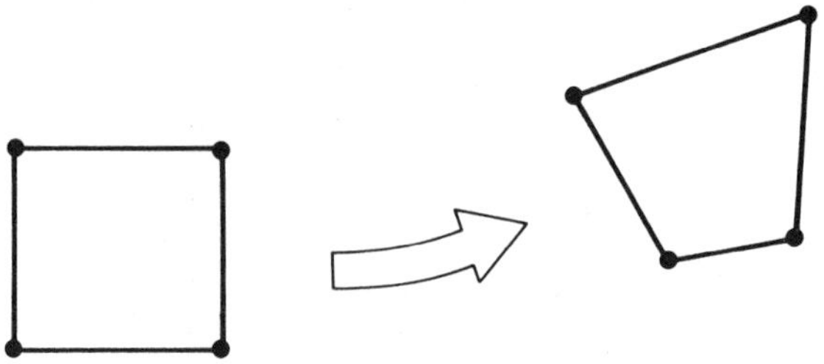

(d) Large displacements, large rotations and large strains. Linear or nonlinear material behavior

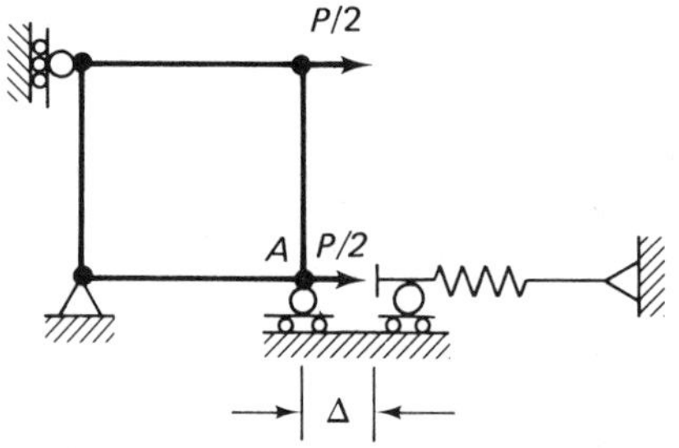

(e) Change in boundary condition at displacement Δ

FIGURE 6.1 *(cont.)*

In addition to the analysis categories listed in Table 6.1, Fig. 6.1 illustrates another type of nonlinear analysis, namely, the analysis of problems in which the boundary conditions change during the motion of the body under consideration. This situation arises in particular in the analysis of contact problems, of which a simple example is given in Fig. 6.1(e). In general, this change in boundary condition may be encountered in any one of the analyses summarized in Table 6.1.

In actual analysis, it is necessary to decide whether a problem falls into one or the other category of analysis, and this dictates which formulation will be used to describe the actual physical situation. Conversely, we may say that by the use of a specific formulation, a model of the actual physical situation is assumed, and the choice of formulation is part of the complete modeling process. Surely, the use of the most general large strain formulation "will always be correct"; however, the use of a more restrictive formulation may be computationally more effective and may also provide more insight into the response prediction.

Before we discuss the general formulations of nonlinear analyses, it is instructive to consider first two simple examples that demonstrate some of the features listed in Table 6.1.

EXAMPLE 6.1: A bar rigidly supported at both ends is subjected to an axial load as shown in Fig. 6.2(a). The stress–strain relation and the load versus time curve relation are given in Figs. 6.2(b) and (c), respectively. Assuming that the displacements and strains are small, and that the load is applied slowly, calculate the displacement at the point of load application.

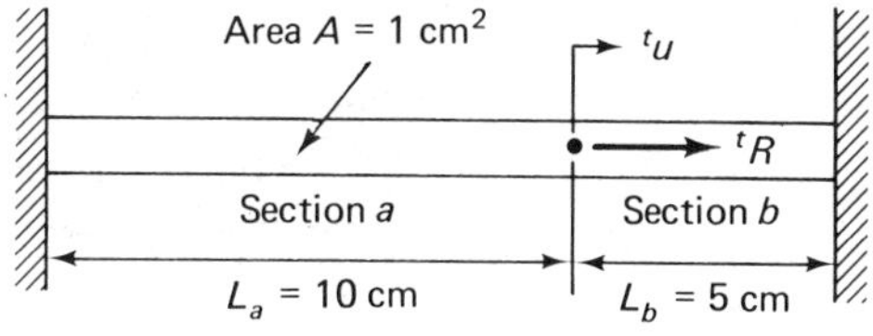

(a) Simple bar structure

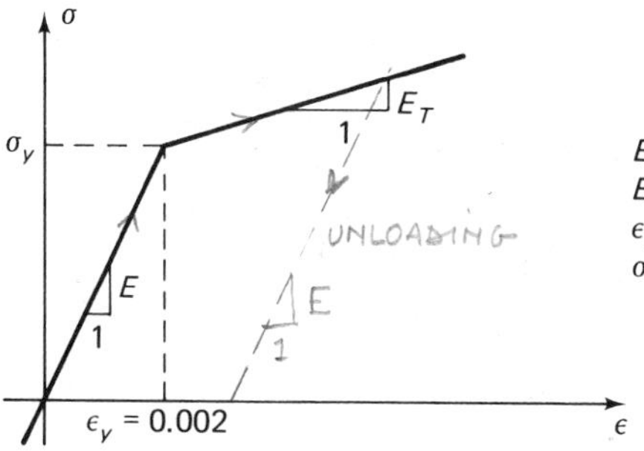

$E = 10^7\ \text{N/cm}^2$
$E_T = 10^5\ \text{N/cm}^2$
ϵ_y = Yield strain
σ_y = Yield stress

(b) Stress-strain relation (in tension and compression)

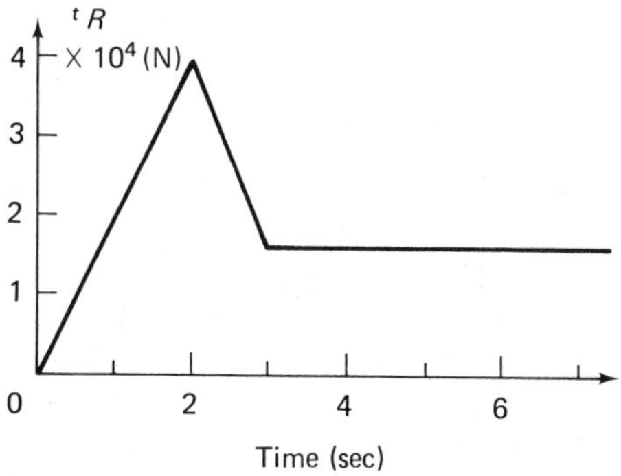

(c) Load variation

FIGURE 6.2 *Analysis of simple bar structure.*

Since the load is applied slowly and the displacements and strains are small we calculate the response of the bar using a static analysis with material nonlinearities only. Then we have for sections a and b, the strain relations

$$^t\epsilon_a = \frac{^tu}{L_a}; \qquad {}^t\epsilon_b = -\frac{^tu}{L_b} \tag{a}$$

the equilibrium relations, $\quad {}^tR + {}^t\sigma_b A = {}^t\sigma_a A \qquad$ (b)

and the constitutive relations, under loading conditions,

$$\begin{aligned} {}^t\epsilon &= {}^t\sigma/E && \text{in the elastic region} \\ {}^t\epsilon &= \epsilon_y + ({}^t\sigma - \sigma_y)/E_T && \text{in the plastic region} \end{aligned} \tag{c}$$

and in unloading, $\quad \Delta\epsilon = \dfrac{\Delta\sigma}{E}$

In these relations the superscript "t" denotes "at time t."

(i) *Both sections a and b are elastic*

During the initial phase of load application both sections a and b are elastic. Then we have using (a) to (c)

$$^tR = EA\,{}^tu\left(\frac{1}{L_a} + \frac{1}{L_b}\right)$$

and substituting the values given in Fig. 6.2 we obtain

$$^tu = \frac{^tR}{3 \times 10^6}$$

with $\quad {}^t\sigma_a = \dfrac{^tR}{3A}; \qquad {}^t\sigma_b = -\dfrac{2}{3}\dfrac{^tR}{A} \qquad$ (d)

(ii) *Section a is elastic while section b is plastic*

Section b will become plastic at time t^* when, using (d),

$$^{t^*}R = \tfrac{3}{2}\sigma_y A$$

Afterwards we therefore have

$$\begin{aligned} {}^t\sigma_a &= E\frac{^tu}{L_a} \\ {}^t\sigma_b &= -E_T\left(\frac{^tu}{L_b} - \epsilon_y\right) - \sigma_y \end{aligned} \tag{e}$$

Using (e) we therefore have for $t \geq t^*$

$$^tR = \frac{EA\,{}^tu}{L_a} + \frac{E_TA\,{}^tu}{L_b} - E_T\,\epsilon_y A + \sigma_y A$$

and thus

$$\begin{aligned} {}^tu &= \frac{^tR/A + E_T\epsilon_y - \sigma_y}{\dfrac{E}{L_a} + \dfrac{E_T}{L_b}} \\ &= \left(\frac{^tR}{1.02 \times 10^6} - 1.9412 \times 10^{-2}\right) \end{aligned}$$

We may note that section a would become plastic when ${}^t\sigma_a = \sigma_y$ or ${}^tR = 4.02 \times 10^4$N. Since the load does not reach this value (see Fig. 6.2(c)), section a remains elastic throughout the response history.

(iii) *In unloading both sections act elastically*

we have
$$\Delta u = \frac{\Delta R}{EA\left(\frac{1}{L_a} + \frac{1}{L_b}\right)}$$

The calculated response is depicted in Fig. 6.2(d).

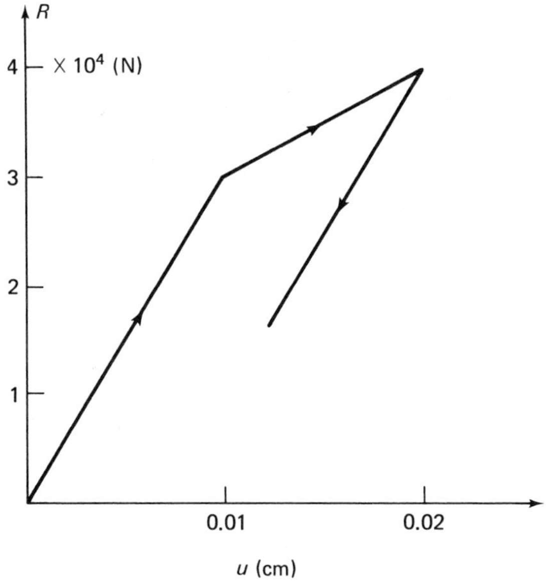

(d) Calculated response

FIGURE 6.2 *(cont.)*

EXAMPLE 6.2: A pretensioned cable is subjected to a transverse load midway between the supports as shown in Fig. 6.3(a). A spring is placed below the load at a distance w_{gap}. Assume that the displacements are small so that the force in the cable remains constant, and that the load is applied slowly. Calculate the displacement under the load as a function of the load intensity.

As in Example 6.1 we neglect inertia forces and assume small displacements. As long as the displacement ${}^tw$ under the load is smaller than w_{gap}, vertical equilibrium requires for small ${}^tw$,

$$ {}^tR = 2H\frac{{}^tw}{L} \tag{a}$$

Once the displacement is larger than w_{gap}, the following equilibrium equation holds

$$ {}^tR = 2H\frac{{}^tw}{L} + k({}^tw - w_{gap}) \tag{b}$$

Figure 6.3(c) shows graphically the force displacement relations given in (a) and (b).

We should note that in this analysis we neglected the elasticity of the cable; therefore the response is calculated using only the equilibrium equations in (a) and (b), and the only nonlinearity is due to the contact condition established when ${}^tw \geq w_{gap}$.

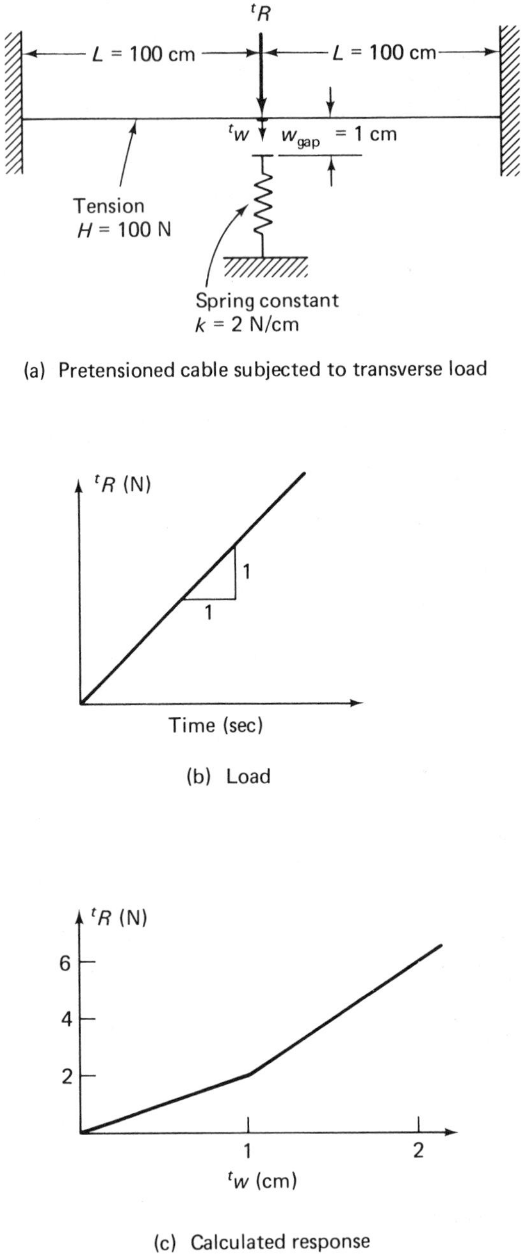

FIGURE 6.3 *Analysis of pretensioned cable with a spring support.*

Although the above examples represent the solution of two very simple problems, the analysis procedures used do display some important general features. The basic problem in a general nonlinear analysis is to find the state of equilibrium of a body corresponding to the applied loads. Assuming that the externally applied loads are described as a function of time, as in Examples 6.1

and 6.2, the equilibrium conditions of a system of finite elements representing the body under consideration can be expressed as

$$ {}^{t}\mathbf{R} - {}^{t}\mathbf{F} = \mathbf{0} \tag{6.2} $$

where the vector ${}^{t}\mathbf{R}$ lists the externally applied nodal point forces in the configuration at time t, and the vector ${}^{t}\mathbf{F}$ lists the nodal point forces that correspond to the element stresses in this configuration. Hence, using the notation of Chapter 4, relations (4.14) to (4.20), we have

$$ {}^{t}\mathbf{R} = {}^{t}\mathbf{R}_{B} + {}^{t}\mathbf{R}_{S} + {}^{t}\mathbf{R}_{C} \tag{6.3} $$

and we have $\mathbf{R}_{I} = {}^{t}\mathbf{F}$,

$$ {}^{t}\mathbf{F} = \sum_{m} \int_{{}^{t}V^{(m)}} {}^{t}\mathbf{B}^{(m)T}\, {}^{t}\boldsymbol{\tau}^{(m)}\, {}^{t}dV^{(m)} \tag{6.4} $$

where in a general large deformation analysis the stresses as well as the volume of the body at time t are unknown.

The relation in (6.2) must express the equilibrium of the system in the current deformed geometry taking due account of all nonlinearities. Also, in a dynamic analysis, the vector ${}^{t}\mathbf{R}$ would include the inertia and damping forces, as discussed in Section 4.2.1.

Considering the solution of the nonlinear response, we recognize that the equilibrium relation in (6.2) must be satisfied throughout the complete history of load application; i.e. the time variable t may take on any value from zero to the maximum time of interest (see Examples 6.1 and 6.2). In a static analysis without time effects other than the definition of the load level (e.g., without creep effects, see Section 6.4.2), time is only a convenient variable which denotes different intensities of load applications and correspondingly different configurations. However, in a dynamic analysis and in static analysis with material time effects, the time variable is an actual variable to be properly included in the modeling of the actual physical situation. Based on these considerations, we may have already realized that the use of the time variable to describe the load application and history of solution represents a very general approach and corresponds to our earlier assertion that a "dynamic analysis is basically a static analysis including inertia effects."

As for the analysis results to be calculated, in many analyses only the stresses and displacements reached at specific load levels or at specific times are required. In some nonlinear static analyses the equilibrium configurations corresponding to those load levels can be calculated without also solving for other equilibrium configurations. However, when the analysis includes path-dependent nonlinear geometric or material conditions, or time-dependent phenomena, the equilibrium relations in (6.2) need be solved for the complete time range of interest. This response calculation is effectively carried out using a step-by-step incremental analysis, which reduces to a one-step analysis if in a static time-independent solution the total load is applied all together and only the configuration corresponding to that load is calculated. However, we shall see that for computational reasons, in practice, even the analysis of such a case frequently requires an incremental solution with a number of load steps to finally reach the total applied load.

The basic approach in an incremental step-by-step solution is to assume that the solution for the discrete time t is known, and that the solution for the discrete time $t + \Delta t$ is required, where Δt is a suitably chosen time increment. Hence, considering (6.2) at time $t + \Delta t$ we have

$$^{t+\Delta t}\mathbf{R} - {}^{t+\Delta t}\mathbf{F} = \mathbf{0} \tag{6.5}$$

where the left superscript denotes "at time $t + \Delta t$." Since the solution is known at time t, we can write

$$^{t+\Delta t}\mathbf{F} = {}^{t}\mathbf{F} + \mathbf{F} \tag{6.6}$$

where $\mathbf{F}$ is the increment in nodal point forces corresponding to the increment in element displacements and stresses from time t to time $t + \Delta t$. This vector can be approximated using a tangent stiffness matrix, ${}^{t}\mathbf{K}$, which corresponds to the geometric and material conditions at time t,

$$\mathbf{F} \doteq {}^{t}\mathbf{K}\mathbf{U} \tag{6.7}$$

where $\mathbf{U}$ is a vector of incremental nodal point displacements. Substituting (6.7) and (6.6) into (6.5) we obtain

$${}^{t}\mathbf{K}\mathbf{U} = {}^{t+\Delta t}\mathbf{R} - {}^{t}\mathbf{F} \tag{6.8}$$

and solving for $\mathbf{U}$, we can calculate an approximation to the displacements at time $t + \Delta t$,

$$^{t+\Delta t}\mathbf{U} \doteq {}^{t}\mathbf{U} + \mathbf{U} \tag{6.9}$$

The exact displacements at time $t + \Delta t$ are those that correspond to the applied loads ${}^{t+\Delta t}\mathbf{R}$. We calculate in (6.9) only an approximation to these displacements because (6.7) was used.

Having evaluated an approximation to the displacements corresponding to time $t + \Delta t$, we can now solve for an approximation to the stresses and corresponding nodal point forces at time $t + \Delta t$, and could then proceed to the next time increment calculations. However, because of the assumption in (6.7) such a solution may be subject to very significant errors and, depending on the time or load step sizes used, may indeed be unstable. In practice, it is therefore frequently necessary to iterate until the solution of (6.5) is obtained to sufficient accuracy.

A widely used iteration procedure is the *modified Newton iteration*. This iterative scheme can be derived from the Newton–Raphson method for the solution of a system of nonlinear equations, as will be discussed in detail in Section 8.6, but can also be presented as an extension of the simple incremental analysis given above. In the following discussions of finite element formulations for nonlinear analysis, we shall always formulate the equations for a solution using the modified Newton method, because the method is quite effective in many analyses, and contains the basic solution steps used in practically all incremental solution strategies. Various other procedures are discussed in Section 8.6 along with some practical guidelines on the use of the individual methods.

The equations used in the modified Newton iteration are, for $i = 1, 2, 3, \ldots,$

$${}^{t}\mathbf{K}\,\Delta\mathbf{U}^{(i)} = {}^{t+\Delta t}\mathbf{R} - {}^{t+\Delta t}\mathbf{F}^{(i-1)} \tag{6.10}$$

$$^{t+\Delta t}\mathbf{U}^{(i)} = {}^{t+\Delta t}\mathbf{U}^{(i-1)} + \Delta\mathbf{U}^{(i)} \tag{6.11}$$

with the initial conditions

$$^{t+\Delta t}\mathbf{U}^{(0)} = {}^{t}\mathbf{U}; \qquad {}^{t+\Delta t}\mathbf{F}^{(0)} = {}^{t}\mathbf{F} \tag{6.12}$$

These equations are derived in a formal mathematical manner in Section 8.6. However, it is instructive here to identify the physical process on which, in essence, the equilibrium iteration is based. In the first iteration, the relations in (6.10) and (6.11) reduce to the equations (6.8) and (6.9) discussed earlier. Then, in subsequent iterations, the latest estimates for the nodal point displacements are used to evaluate the corresponding element stresses and nodal point forces ${}^{t+\Delta t}\mathbf{F}^{(i-1)}$. Furthermore, the out-of-balance load vector ${}^{t+\Delta t}\mathbf{R} - {}^{t+\Delta t}\mathbf{F}^{(i-1)}$ corresponds to a load vector that is not yet balanced by element stresses, and hence an increment in the nodal point displacements is required. This updating of the nodal point displacements in the iteration is continued until the out-of-balance loads and incremental displacements are small.

In the actual practical use of the iteration procedure given in (6.10) to (6.12), the convergence properties of the iteration are most important and appropriate convergence criteria need be employed (see Section 8.6). We now demonstrate the use of the above equations in some illustrative examples.

EXAMPLE 6.3: Idealize the simple arch structure shown in Fig. 6.4(a) as an assemblage of two bar elements. Assume that the force in one bar element is given by ${}^tF_{\text{bar}} = k\delta$, where k is a constant and δ is the elongation of the bar at time t. (The assumption that k is constant is likely to be valid only for small deformations

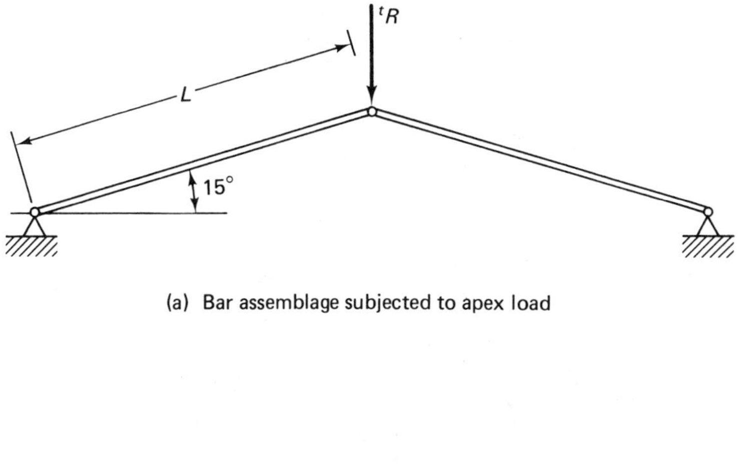

(a) Bar assemblage subjected to apex load

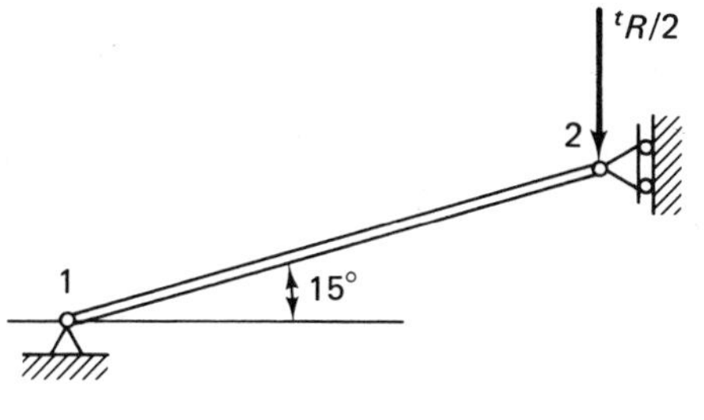

(b) Simple model using one bar (truss) element, nodes 1 and 2

FIGURE 6.4 *A simple arch structure.*

in the bar, but we use this assumption in order to simplify the analysis.) Establish the equilibrium relation (6.5) for this problem.

This is a large displacement problem, and the response is calculated by focussing attention on the equilibrium of the bar assemblage in the configuration corresponding to a typical time t. Using symmetry as shown in Fig. 6.4(b) and (c), we have

$$(L - {}^t e) \cos {}^t\beta = L \cos 15^\circ$$
$$(L - {}^t e) \sin {}^t\beta = L \sin 15^\circ - {}^t\Delta$$

hence

$${}^t e = L - \sqrt{L^2 - 2L\,{}^t\Delta \sin 15^\circ + {}^t\Delta^2}$$
$$\sin {}^t\beta = \frac{L \sin 15^\circ - {}^t\Delta}{L - {}^t e}$$

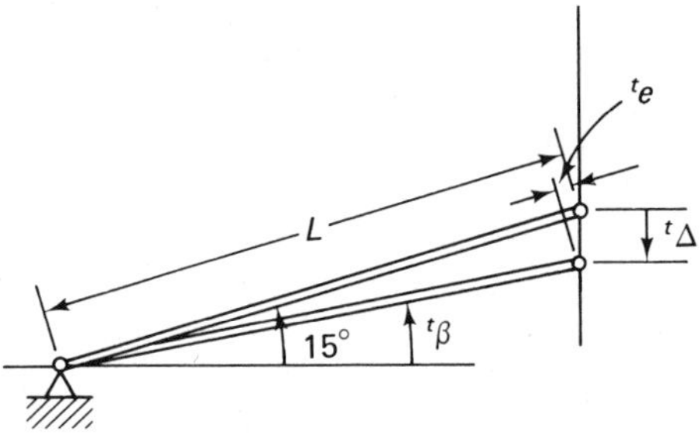

(c) Geometric variables in typical configuration

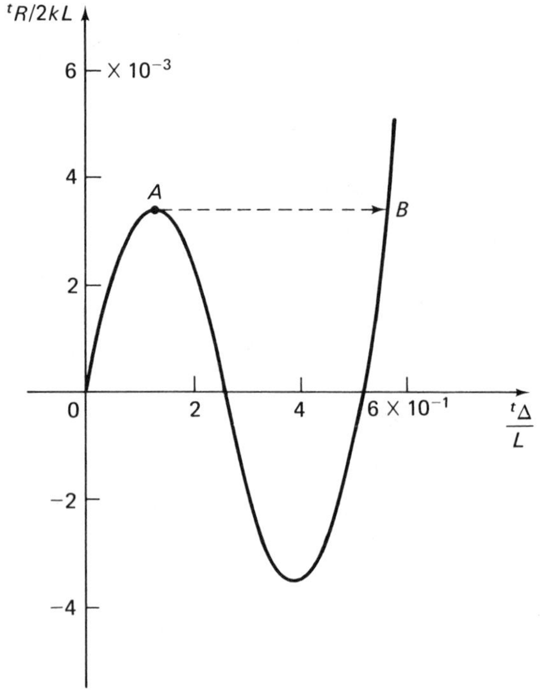

(d) Load-displacement relation

FIGURE 6.4 *(cont.)*

Equilibrium at time t requires that

$$2\,{}^tF_{\text{bar}} \sin {}^t\beta = {}^tR$$

hence the relation in (6.5) is

$$\frac{{}^tR}{2kL} = \left\{ -1 + \frac{1}{\sqrt{1 - 2\dfrac{{}^t\Delta}{L}\sin 15^\circ + \left(\dfrac{{}^t\Delta}{L}\right)^2}} \right\} \left(\sin 15^\circ - \frac{{}^t\Delta}{L} \right) \tag{a}$$

Figure 6.4(d) shows the force displacement relationship established in (a). It should be noted that, for a given load level, between points A and B we have two possible displacement configurations. If the structure is loaded with ${}^tR$ monotonically increasing, the displacement path with snap-through from A to B in Fig. 6.4(d) is likely to be followed in an actual physical situation.

EXAMPLE 6.4: Calculate the response of the bar assemblage considered in Examplè 6.1 using the relations in (6.10) to (6.12). Use two equal load steps to reach the maximum load application.

The iterative equations are in this analysis

$$\begin{aligned} ({}^tK_a + {}^tK_b)\Delta u^{(i)} &= {}^{t+\Delta t}R - {}^{t+\Delta t}F_a^{(i-1)} - {}^{t+\Delta t}F_b^{(i-1)} \\ {}^{t+\Delta t}u^{(i)} &= {}^{t+\Delta t}u^{(i-1)} + \Delta u^{(i)} \end{aligned} \tag{a}$$

with

$$\begin{gathered} {}^{t+\Delta t}u^{(0)} = {}^tu \\ {}^{t+\Delta t}F_a^{(0)} = {}^tF_a; \qquad {}^{t+\Delta t}F_b^{(0)} = {}^tF_b \\ {}^tK_a = {}^tCA/L_a; \qquad {}^tK_b = {}^tCA/L_b \end{gathered} \tag{b}$$

where

$${}^tC \begin{cases} = E & \text{if section is elastic} \\ = E_T & \text{if section is plastic} \end{cases}$$

for an elastic section

$${}^{t+\Delta t}F^{(i-1)} = EA\,{}^{t+\Delta t}\epsilon^{(i-1)} \tag{c}$$

for a plastic section

$${}^{t+\Delta t}F^{(i-1)} = A\{E_T({}^{t+\Delta t}\epsilon^{(i-1)} - \epsilon_y) + \sigma_y\} \tag{d}$$

and the strains in the sections are

$$\begin{aligned} {}^{t+\Delta t}\epsilon_a^{(i-1)} &= {}^{t+\Delta t}u^{(i-1)}/L_a \\ {}^{t+\Delta t}\epsilon_b^{(i-1)} &= {}^{t+\Delta t}u^{(i-1)}/L_b \end{aligned} \tag{e}$$

In the first load step, we have $t = 0$ and $\Delta t = 1$. Thus, the application of the relations in (a) to (e) gives:

$t = 1$

$$({}^0K_a + {}^0K_b)\Delta u^{(1)} = {}^1R - {}^1F_a^{(0)} - {}^1F_b^{(0)}$$

$$\Delta u^{(1)} = \frac{2 \times 10^4}{10^7(\frac{1}{10} + \frac{1}{5})} = 6.6667 \times 10^{-3} \text{ cm}$$

$(i = 1)$

$$\begin{aligned} {}^1u^{(1)} &= {}^1u^{(0)} + \Delta u^{(1)} = 6.6667 \times 10^{-3} \text{ cm} \\ {}^1\epsilon_a^{(1)} &= {}^1u^{(1)}/L_a = 6.6667 \times 10^{-4} < \epsilon_y \longrightarrow \text{section } a \text{ is elastic} \\ {}^1\epsilon_b^{(1)} &= {}^1u^{(1)}/L_b = 1.3333 \times 10^{-3} < \epsilon_y \longrightarrow \text{section } b \text{ is elastic} \\ {}^1F_a^{(1)} &= 6.6667 \times 10^3 \text{ N} \\ {}^1F_b^{(1)} &= 1.3333 \times 10^4 \text{ N} \end{aligned}$$

$$\begin{aligned} ({}^0K_a + {}^0K_b)\Delta u^{(2)} &= {}^1R - {}^1F_a^{(1)} - {}^1F_b^{(1)} \\ &= 0 \end{aligned}$$

$\therefore$ Convergence is achieved in 1 iteration

$${}^1u = 6.6667 \times 10^{-3} \text{ cm}$$

$t = 2$

$$^1K_a = \frac{EA}{L_a}; \qquad ^1K_b = \frac{EA}{L_b}$$

$$^2F_a^{(0)} = {}^1F_a; \qquad ^2F_b^{(0)} = {}^1F_b$$

$$(^1K_a + {}^1K_b)\Delta u^{(1)} = {}^2R - {}^2F_a^{(0)} - {}^2F_b^{(0)}$$

$$\Delta u^{(1)} = \frac{(4 \times 10^4) - (6.6667 \times 10^3) - (1.3333 \times 10^4)}{10^7(\frac{1}{10} + \frac{1}{5})}$$

$$= 6.6667 \times 10^{-3} \text{ cm}$$

$$(i = 1) \qquad {}^2u^{(1)} = {}^2u^{(0)} + \Delta u^{(1)} = 1.3333 \times 10^{-2} \text{ cm}$$

$$^2\epsilon_a^{(1)} = 1.3333 \times 10^{-3} < \epsilon_y \longrightarrow \text{section } a \text{ is elastic}$$

$$^2\epsilon_b^{(1)} = 2.6667 \times 10^{-3} > \epsilon_y \longrightarrow \text{section } b \text{ is plastic}$$

$$^2F_a^{(1)} = 1.3333 \times 10^4 \text{ N}$$

$$^2F_b^{(1)} = \{E_T(^2\epsilon_b^{(1)} - \epsilon_y) + \sigma_y\}A = 2.0067 \times 10^4 \text{ N}$$

$$(^1K_a + {}^1K_b)\Delta u^{(2)} = {}^2R - {}^2F_a^{(1)} - {}^2F_b^{(1)}$$

$$\Delta u^{(2)} = 2.2 \times 10^{-3} \text{ cm}$$

$$(i = 2) \qquad {}^2u^{(2)} = {}^2u^{(1)} + \Delta u^{(2)} = 1.5533 \times 10^{-2} \text{ cm}$$

$$^2\epsilon_a^{(2)} = 1.5533 \times 10^{-3} < \epsilon_y$$

$$^2\epsilon_b^{(2)} = 3.1066 \times 10^{-3} > \epsilon_y$$

$$\therefore {}^2F_a^{(2)} = 1.5533 \times 10^4 \text{ N}$$

$$^2F_b^{(2)} = 2.0111 \times 10^4 \text{ N}$$

$$(^1K_a + {}^1K_b)\Delta u^{(3)} = {}^2R - {}^2F_a^{(2)} - {}^2F_b^{(2)}$$

$$\Delta u^{(3)} = 1.4521 \times 10^{-3} \text{ cm}$$

The procedure is repeated and the results of successive iterations are tabulated below.

i	$\Delta u^{(i)}$ (*cm*)	$^2u^{(i)}$ (*cm*)
3	1.4521×10^{-3}	1.6985×10^{-2}
4	9.5832×10^{-4}	1.7944×10^{-2}
5	6.3249×10^{-4}	1.8576×10^{-2}
6	4.1744×10^{-4}	1.8994×10^{-2}
7	2.7551×10^{-4}	1.9269×10^{-2}

After 7 iterations, we have

$$^2u \doteq {}^2u^{(7)} = 1.9269 \times 10^{-2} \text{ cm}$$

6.2 FORMULATION OF THE CONTINUUM MECHANICS INCREMENTAL EQUATIONS OF MOTION

The objective in the introductory discussion of nonlinear analysis in Section 6.1 was to describe various nonlinearities and the form of the basic finite element equations that are used to analyze the nonlinear response of a structural system. To show the procedure of analysis, we simply stated the finite element equations, discussed their solution and gave a physical argument why the nonlinear response is appropriately predicted using these equations. We demonstrated the applicability of the approach in the solution of two very simple problems, merely to give some insight into the steps of analysis used. In each of these analyses the applicable finite element matrices and vectors have been developed using physical arguments.

The physical approach of analysis used in Examples 6.1 and 6.2 is very instructive and yields insight into the analysis; however, when considering a more complex solution, a consistent continuum-mechanics-based approach should be employed to develop the governing finite element equations. The objective in this section is to present the governing continuum mechanics equations for a displacement-based finite element solution. As in Section 4.2.1, we use the principle of virtual work but now include the possibility that the body considered undergoes large displacements and rotations, and large strains and that the stress–strain relationship is nonlinear. The governing continuum mechanics equations to be presented can therefore be regarded as an extension of the basic equation given in (4.5). In the linear analysis of a general body, the equation in (4.5) was used as the basis for the development of the governing linear finite element equations (given in (4.14) to (4.20)). Considering the nonlinear analysis of a general body, after having developed suitable continuum mechanics equations, we will proceed in a completely analogous manner to establish the nonlinear finite element equations that govern the nonlinear response of the body (see Section 6.3).

6.2.1 The Basic Problem

In Section 6.1 we underlined the fact that in a nonlinear analysis the equilibrium of the body considered must be established in the current configuration. We also pointed out that in general it is necessary to employ an incremental formulation, and that a time variable is used to conveniently describe the loading and the motion of the body.

In the development to follow, we consider the motion of a general body in a stationary Cartesian coordinate system, as shown in Fig. 6.5, and assume that the body can experience large displacements, large strains and a nonlinear constitutive response. The aim is to evaluate the equilibrium positions of the complete body at the discrete time points $0, \Delta t, 2\,\Delta t, 3\,\Delta t, \ldots$, where Δt is an increment in time. To develop the solution strategy, assume that the solutions for the static and kinematic variables for all time steps from time 0 to time t, inclusive, have been obtained. Then the solution process for the next required equilibrium position corresponding to time $t + \Delta t$ is typical and would be applied repetitively until the complete solution path has been solved for. Hence, in the analysis we follow all particles of the body in their motion, from the original to the final configuration of the body, which means that we adopt *a Lagrangian* (*or material*) *formulation* of the problem.[1-4] This approach stands in contrast to an *Eulerian formulation* usually used in the analysis of fluid mechanics problems, in which attention is focused on the motion of the material through a stationary control volume. Considering the analysis of solids and structures, a Lagrangian formulation usually represents a more natural and effective analysis approach than an Eulerian formulation. For example, using an Eulerian formulation of a structural problem with large displacements, new control volumes have to be created (because the boundaries of the solid change continuously) and the nonlinearities in the convective acceleration terms are difficult to deal with (see Section 7.4).

In our Lagrangian incremental analysis approach we express the equilibrium of the body at time $t + \Delta t$ using the principle of virtual displacements. Using tensor notation (see Section 2.6) this principle requires that

$$\int_{^{t+\Delta t}V} {}^{t+\Delta t}\tau_{ij}\, \delta_{t+\Delta t}e_{ij}\, {}^{t+\Delta t}dV = {}^{t+\Delta t}\mathcal{R} \tag{6.13}$$

where the ${}^{t+\Delta t}\tau_{ij}$ are the Cartesian components of the *Cauchy stress tensor*, the ${}_{t+\Delta t}e_{ij}$ are the Cartesian components of an *infinitesimal strain tensor*, and the δ means "variation in," i.e.,

$$\delta_{t+\Delta t}e_{ij} = \delta\frac{1}{2}\left(\frac{\partial u_i}{\partial^{t+\Delta t}x_j} + \frac{\partial u_j}{\partial^{t+\Delta t}x_i}\right) \equiv \frac{1}{2}\left(\frac{\partial \delta u_i}{\partial^{t+\Delta t}x_j} + \frac{\partial \delta u_j}{\partial^{t+\Delta t}x_i}\right)$$

It should be noted that the Cauchy stresses are "body internal forces per unit area" in the configuration at time $t + \Delta t$, and the infinitesimal strain components are also referred to this as yet unknown configuration. Since the variation in the strain components is equivalent to the use of virtual strains (see Section 4.2.1), we recognize that the l.h.s. of (6.13) is the virtual work performed when the body is subjected to a virtual displacement at time $t + \Delta t$. The corresponding external virtual work is ${}^{t+\Delta t}\mathcal{R}$, with

$${}^{t+\Delta t}\mathcal{R} = \int_{^{t+\Delta t}V} {}^{t+\Delta t}f_i^B\, \delta u_i\, {}^{t+\Delta t}dV + \int_{^{t+\Delta t}S} {}^{t+\Delta t}f_i^S\, \delta u_i^S\, {}^{t+\Delta t}dS \tag{6.14}$$

where the ${}^{t+\Delta t}f_i^B$ and ${}^{t+\Delta t}f_i^S$ are the components of the externally applied body and surface force vectors, respectively, and δu_i is the ith component of the virtual displacement vector.

FIGURE 6.5 *Motion of body in stationary Cartesian coordinate system.*

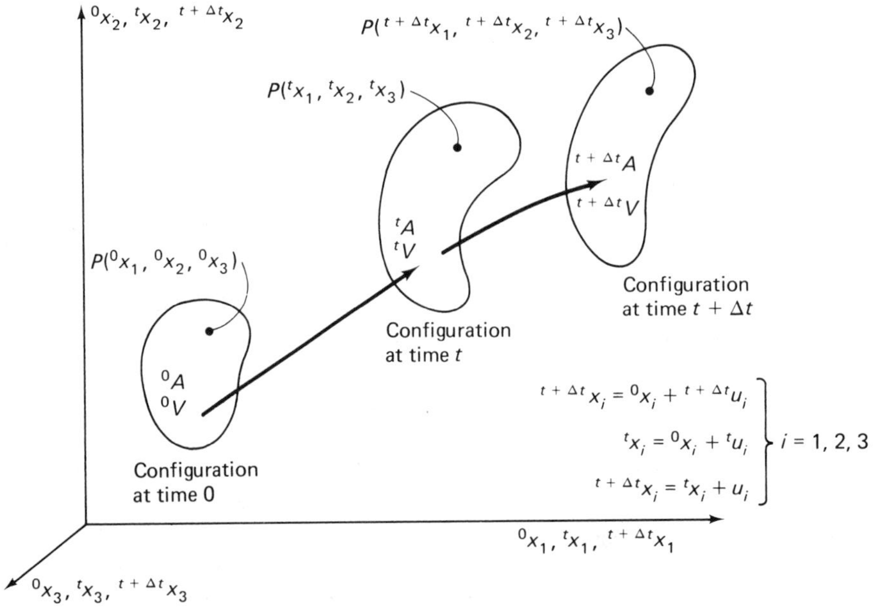

It should be noted that *the virtual work principle stated in (6.13) is simply an application of the equation in (4.5) (used in linear analysis) to the body considered in the configuration at time $t + \Delta t$*. Therefore, all previous discussions and results pertaining to the use of the virtual work principle in linear analysis are now directly applicable, with the current configuration at time $t + \Delta t$ being considered.*

A fundamental difficulty in the general application of (6.13) is that the configuration of the body at time $t + \Delta t$ is unknown. This is an important difference compared with linear analysis in which it is assumed that the displacements are infinitesimally small so that the configuration of the body does not change. The continuous change in the configuration of the body entails some important consequences for the development of an incremental analysis procedure. For example, an important consideration must be that the Cauchy stresses at time $t + \Delta t$ cannot be obtained by simply adding to the Cauchy stresses at time t a stress increment which is due only to the straining of the material. Namely, the calculation of the Cauchy stresses at time $t + \Delta t$ must also take into account the rigid body rotation of the material, because the components of the Cauchy stress tensor also change when the material is only subjected to a rigid body rotation.

The fact that the configuration of the body changes continuously in a large deformation analysis is dealt with in an elegant manner by using appropriate stress and strain measures and constitutive relations, as discussed in detail in the next sections.

Considering the discussions to follow, we recognize that a difficult point in the presentation of continuum mechanics relations for general large deformation analysis is the use of an effective notation, because there are many different quantities that need be dealt with. The symbols used should display all necessary information but should do so in a compact manner in order that the equations can be read with relative ease. For an effective use of a notation, an understanding of the convention employed is most helpful and for this purpose we summarize here briefly some basic facts and conventions used in our notation.

In our analysis we consider the motion of the body in a fixed (stationary) Cartesian coordinate system as displayed in Fig. 6.5. All kinematic and static variables are measured in this coordinate system, and throughout our description we use tensor notation.

The coordinates of a generic point P in the body at time 0 are ${}^0x_1$, ${}^0x_2$, ${}^0x_3$; at time t they are ${}^tx_1$, ${}^tx_2$, ${}^tx_3$; and at time $t + \Delta t$ they are ${}^{t+\Delta t}x_1$, ${}^{t+\Delta t}x_2$, ${}^{t+\Delta t}x_3$, where the left superscripts refer to the configuration of the body and the subscripts to the coordinate axes.

The notation for the displacements of the body is similar to the notation for the coordinates; namely, at time t the displacements are ${}^tu_i$, $i = 1, 2, 3$ and at time $t + \Delta t$ the displacements are ${}^{t+\Delta t}u_i$, $i = 1, 2, 3$; therefore we have

$$\left.\begin{aligned} {}^tx_i &= {}^0x_i + {}^tu_i \\ {}^{t+\Delta t}x_i &= {}^0x_i + {}^{t+\Delta t}u_i \end{aligned}\right\} \quad i = 1, 2, 3$$

* We may imagine that, considering the moving body, a picture is taken at time $t + \Delta t$ and then the principle of virtual displacements is applied to the state of the body in that picture.

The increments in the displacements from time t to time $t + \Delta t$ are denoted as

$$u_i = {}^{t+\Delta t}u_i - {}^{t}u_i; \qquad i = 1, 2, 3$$

During motion of the body, its volume, surface area, mass density, stresses, and strains are changing continuously. We denote the specific mass, area, and volume of the body at times 0, t, and $t + \Delta t$ as ${}^0\rho, {}^t\rho, {}^{t+\Delta t}\rho$; ${}^0A, {}^tA, {}^{t+\Delta t}A$; and ${}^0V, {}^tV, {}^{t+\Delta t}V$, respectively.

Since the configuration of the body at time $t + \Delta t$ is not known, we will refer applied forces, stresses, and strains to a known equilibrium configuration. In analogy to the notation used for coordinates and displacements, a left superscript indicates in which configuration the quantity (body force, surface traction, stress, etc.) occurs; in addition, a left subscript indicates the configuration with respect to which the quantity is measured. For example, the surface and body force components at time $t + \Delta t$, but measured in configuration 0, are ${}^{t+\Delta t}_{0}f_i^S, {}^{t+\Delta t}_{0}f_i^B$, $i = 1, 2, 3$. Here we have the exception that if the quantity under consideration occurs in the same configuration in which it is also measured, the left subscript may not be used; e.g., for the Cauchy stresses we have

$${}^{t+\Delta t}\tau_{ij} \equiv {}^{t+\Delta t}_{t+\Delta t}\tau_{ij}$$

In the formulation of the governing equilibrium equations we also need to consider derivatives of displacements and coordinates. In our notation a comma denotes differentiation with respect to the coordinate following, and the left subscript denoting time indicates the configuration in which this coordinate is measured; thus we have, for example,

$${}^{t+\Delta t}_{0}u_{i,j} = \frac{\partial\, {}^{t+\Delta t}u_i}{\partial\, {}^0x_j}$$

and

$${}^{0}_{t+\Delta t}x_{m,n} = \frac{\partial\, {}^0x_m}{\partial\, {}^{t+\Delta t}x_n}$$

Using these conventions we shall define new symbols when they are first encountered.

6.2.2 Stress and Strain Tensors

We mentioned briefly in the previous section that in a large deformation analysis special attention must be given to the fact that the configuration of the body is changing continuously. This change in configuration can be dealt with in an elegant manner by defining auxiliary stress and strain measures. The objective in their definition is to express the internal virtual work in (6.13) in terms of an integral over a volume that is known and to be able to incrementally decompose the stresses and strains in an effective manner. There are various different stress and strain tensors that, in principle, could be used. However, if the objective is to obtain an effective overall finite element solution procedure, only a few stress and strain measures need be considered. In the following we first define the specific stress and strain tensors that are deemed to be most effective in an incremental formulation of the virtual work principle, and then we discuss their properties, characteristics, and the reasons for choosing them among the large number of possible stress and strain measures.

A stress measure that we will use abundantly is the *2nd Piola-Kirchhoff stress tensor*. At time t the 2nd Piola-Kirchhoff stress referred to the configuration at time 0 is defined as

$$ {}_0^tS_{ij} = \frac{{}^0\rho}{{}^t\rho}\, {}_t^0x_{i,m}\, {}^t\tau_{mn}\, {}_t^0x_{j,n} \tag{6.15}$$

where ${}_t^0x_{i,m} = \partial\,{}^0x_i/\partial\,{}^tx_m$, and ${}^0\rho/{}^t\rho$ represents the ratio of the mass densities at time 0 and time t. Alternatively, we may also write

$$ {}^t\tau_{mn} = \frac{{}^t\rho}{{}^0\rho}\, {}_0^tx_{m,i}\, {}_0^tS_{ij}\, {}_0^tx_{n,j} \tag{6.16}$$

where ${}_0^tx_{m,i}$ is now element (m, i) of the *deformation gradient* ${}_0^t\mathbf{X}$:

$$ {}_0^t\mathbf{X} = ({}_0\boldsymbol{\nabla}\, {}^t\mathbf{x}^T)^T \tag{6.17}$$

$$ {}_0\boldsymbol{\nabla} = \begin{bmatrix} \dfrac{\partial}{\partial\,{}^0x_1} \\ \dfrac{\partial}{\partial\,{}^0x_2} \\ \dfrac{\partial}{\partial\,{}^0x_3} \end{bmatrix}; \qquad {}^t\mathbf{x}^T = [{}^tx_1 \quad {}^tx_2 \quad {}^tx_3] \tag{6.18}$$

The mass density ratio in (6.15) can be evaluated since the mass of the particles considered is conserved:

$$ \int_{{}^tV} {}^t\rho\, d^tx_1\, d^tx_2\, d^tx_3 = \int_{{}^0V} {}^0\rho\, d^0x_1\, d^0x_2\, d^0x_3 \tag{6.19}$$

But $\qquad d^tx_1\, d^tx_2\, d^tx_3 = (\det {}_0^t\mathbf{X})\, d^0x_1\, d^0x_2\, d^0x_3$

and since the relation in (6.19) must hold for any arbitrary number of particles, we have

$$ {}^0\rho = {}^t\rho \det {}_0^t\mathbf{X} \tag{6.20}$$

Hence, once the deformation gradient is known, the change in mass density can readily be calculated, and, provided the Cauchy stresses are known, the corresponding 2nd Piola–Kirchhoff stresses can be evaluated using a purely kinematic transformation. Since the Cauchy stress tensor is symmetric, the relations in (6.15) and (6.16) show that the 2nd Piola–Kirchhoff stress tensor is also symmetric.

EXAMPLE 6.5: Consider the element in Fig. 6.6. Evaluate the deformation gradient and the mass density corresponding to the configuration at time t.

The displacement interpolation functions for this element were given in Fig. 5.5. Since the ${}^0x_1$, ${}^0x_2$ axes correspond to the r, s axes, respectively, we have

$$ h_1 = \tfrac{1}{4}(1 + {}^0x_1)(1 + {}^0x_2) \qquad h_2 = \tfrac{1}{4}(1 - {}^0x_1)(1 + {}^0x_2)$$
$$ h_3 = \tfrac{1}{4}(1 - {}^0x_1)(1 - {}^0x_2) \qquad h_4 = \tfrac{1}{4}(1 + {}^0x_1)(1 - {}^0x_2)$$

and

$$ \frac{\partial h_1}{\partial\,{}^0x_1} = \frac{1}{4}(1 + {}^0x_2) \qquad \frac{\partial h_2}{\partial\,{}^0x_1} = -\frac{1}{4}(1 + {}^0x_2)$$
$$ \frac{\partial h_3}{\partial\,{}^0x_1} = -\frac{1}{4}(1 - {}^0x_2) \qquad \frac{\partial h_4}{\partial\,{}^0x_1} = \frac{1}{4}(1 - {}^0x_2)$$
$$ \frac{\partial h_1}{\partial\,{}^0x_2} = \frac{1}{4}(1 + {}^0x_1) \qquad \frac{\partial h_2}{\partial\,{}^0x_2} = \frac{1}{4}(1 - {}^0x_1)$$
$$ \frac{\partial h_3}{\partial\,{}^0x_2} = -\frac{1}{4}(1 - {}^0x_1) \qquad \frac{\partial h_4}{\partial\,{}^0x_2} = -\frac{1}{4}(1 + {}^0x_1)$$

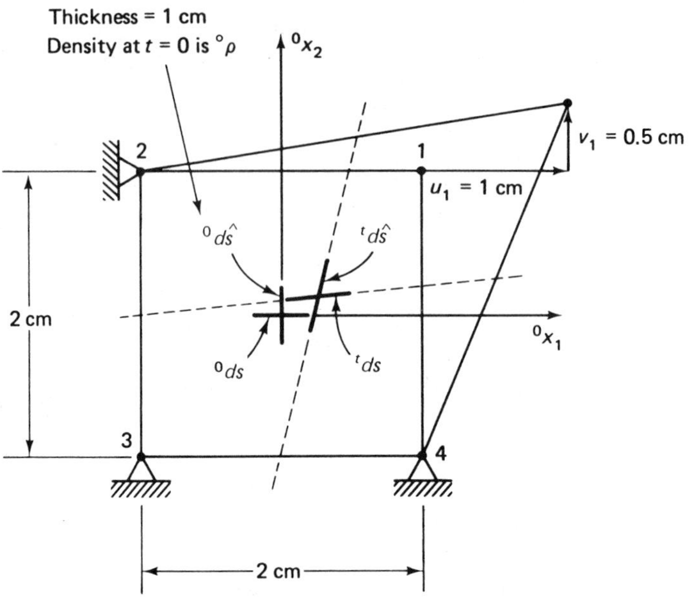

FIGURE 6.6 Four-node element subjected to large deformations.

Now we use that
$$^tx_i = \sum_{k=1}^{4} h_k\, {}^tx_i^k$$

and hence
$$\frac{\partial\, {}^tx_i}{\partial\, {}^0x_j} = \sum_{k=1}^{4} \left(\frac{\partial h_k}{\partial\, {}^0x_j}\right) {}^tx_i^k$$

The nodal point coordinates at time t are

$$\begin{array}{llll} {}^tx_1^1 = 2 & {}^tx_2^1 = 1.5 & {}^tx_1^2 = -1 & {}^tx_2^2 = 1 \\ {}^tx_1^3 = -1 & {}^tx_2^3 = -1 & {}^tx_1^4 = 1 & {}^tx_2^4 = -1 \end{array}$$

Hence

$$\frac{\partial\, {}^tx_1}{\partial\, {}^0x_1} = \frac{1}{4}[(1 + {}^0x_2)(2) - (1 + {}^0x_2)(-1) - (1 - {}^0x_2)(-1) + (1 - {}^0x_2)(1)]$$
$$= \frac{1}{4}(5 + {}^0x_2)$$

and
$$\frac{\partial\, {}^tx_1}{\partial\, {}^0x_2} = \frac{1}{4}(1 + {}^0x_1); \qquad \frac{\partial\, {}^tx_2}{\partial\, {}^0x_1} = \frac{1}{8}(1 + {}^0x_2)$$
$$\frac{\partial\, {}^tx_2}{\partial\, {}^0x_2} = \frac{1}{8}(9 + {}^0x_1)$$

so that the deformation gradient is,

$${}_0^t\mathbf{X} = \frac{1}{4}\begin{bmatrix} (5 + {}^0x_2) & (1 + {}^0x_1) \\ \frac{1}{2}(1 + {}^0x_2) & \frac{1}{2}(9 + {}^0x_1) \end{bmatrix}$$

and using (6.20) the mass density in the deformed configuration is

$${}^t\rho = \frac{32\, {}^0\rho}{(5 + {}^0x_2)(9 + {}^0x_1) - (1 + {}^0x_1)(1 + {}^0x_2)}$$

A most important property of the 2nd Piola–Kirchhoff stress tensor is that the components of the tensor are invariant under a rigid body rotation of the material; i.e. they do not change if the material undergoes a rigid body motion. We demonstrate this property in the following simple example.

EXAMPLE 6.6: Figure 6.7 shows a four-node element in the configuration at time 0. The element is subjected to a stress (initial stress) of ${}^{0}\tau_{11}$. Assume that the element is rotated in time 0 to time Δt as a rigid body through a large angle θ and that the stress in a body-attached coordinate system does not change. Hence the magnitude of ${}^{\Delta t}\bar{\tau}_{11}$ shown in Fig. 6.7 is equal to ${}^{0}\tau_{11}$. Show that the components of the 2nd Piola–Kirchhoff stress tensor did not change as a result of the rigid body rotation.

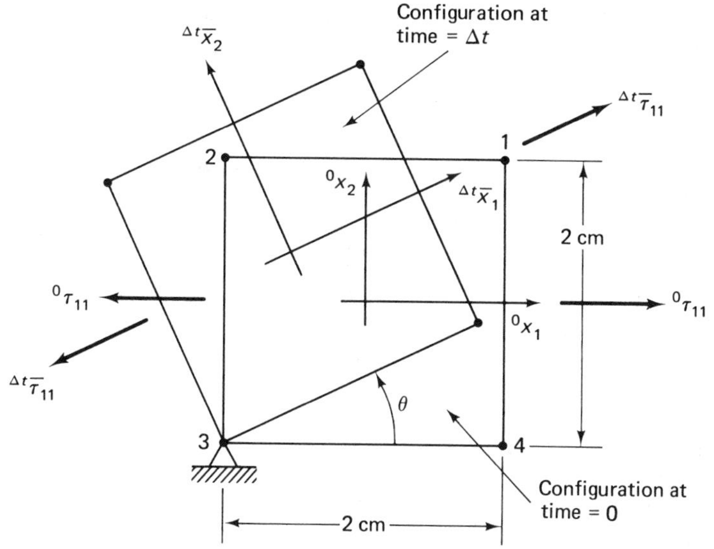

FIGURE 6.7 *Four-node element with initial stress subjected to large rotation.*

The 2nd Piola–Kirchhoff stress tensor at time 0 is equal to the Cauchy stress tensor, because the element deformations are zero,

$$ {}_{0}^{0}\mathbf{S} = \begin{bmatrix} {}^{0}\tau_{11} & 0 \\ 0 & 0 \end{bmatrix} \tag{a} $$

The components of the Cauchy stress tensor at time Δt expressed in the coordinate axes ${}^{0}x_1$, ${}^{0}x_2$ are,

$$ {}^{\Delta t}\boldsymbol{\tau} = \begin{bmatrix} \cos\theta & -\sin\theta \\ \sin\theta & \cos\theta \end{bmatrix} \begin{bmatrix} {}^{\Delta t}\bar{\tau}_{11} & 0 \\ 0 & 0 \end{bmatrix} \begin{bmatrix} \cos\theta & \sin\theta \\ -\sin\theta & \cos\theta \end{bmatrix} \tag{b} $$

This transformation corresponds to a second-order tensor transformation of the components ${}^{\Delta t}\bar{\tau}_{ij}$ from the body-attached coordinate frame ${}^{\Delta t}\bar{x}_1$, ${}^{\Delta t}\bar{x}_2$ to the stationary coordinate frame ${}^{0}x_1$, ${}^{0}x_2$ (see Section 2.6).

The relation between the Cauchy stresses and the 2nd Piola–Kirchhoff stresses at time Δt is according to (6.16)

$$ {}^{\Delta t}\boldsymbol{\tau} = \frac{{}^{\Delta t}\rho}{{}^{0}\rho}\, {}_{0}^{\Delta t}\mathbf{X}\, {}_{0}^{\Delta t}\mathbf{S}\, {}_{0}^{\Delta t}\mathbf{X}^{T} \tag{c} $$

where in this case ${}^{\Delta t}\rho/{}^{0}\rho = 1$. The deformation gradient can be evaluated as in

Example 6.5, where we note that the nodal point coordinates at time Δt are

$$\begin{aligned} {}^{\Delta t}x_1^1 &= 2\cos\theta - 1 - 2\sin\theta & {}^{\Delta t}x_2^1 &= 2\sin\theta - 1 + 2\cos\theta \\ {}^{\Delta t}x_1^2 &= -1 - 2\sin\theta & {}^{\Delta t}x_2^2 &= 2\cos\theta - 1 \\ {}^{\Delta t}x_1^3 &= -1 & {}^{\Delta t}x_2^3 &= -1 \\ {}^{\Delta t}x_1^4 &= 2\cos\theta - 1 & {}^{\Delta t}x_2^4 &= 2\sin\theta - 1 \end{aligned}$$

Hence, using the derivatives of the interpolation functions given in Example 6.5, we have

$${}^{\Delta t}_{0}\mathbf{X} = \frac{1}{4}\left[\begin{array}{c|c} (1 + {}^0x_2)(2\cos\theta - 1 - 2\sin\theta) & (1 + {}^0x_1)(2\cos\theta - 1 - 2\sin\theta) \\ -(1 + {}^0x_2)(-1 - 2\sin\theta) & +(1 - {}^0x_1)(-1 - 2\sin\theta) \\ -(1 - {}^0x_2)(-1) & -(1 - {}^0x_1)(-1) \\ +(1 - {}^0x_2)(2\cos\theta - 1) & -(1 + {}^0x_1)(2\cos\theta - 1) \\ \hline (1 + {}^0x_2)(2\sin\theta - 1 + 2\cos\theta) & (1 + {}^0x_1)(2\sin\theta - 1 + 2\cos\theta) \\ -(1 + {}^0x_2)(2\cos\theta - 1) & +(1 - {}^0x_1)(2\cos\theta - 1) \\ -(1 - {}^0x_2)(-1) & -(1 - {}^0x_1)(-1) \\ +(1 - {}^0x_2)(2\sin\theta - 1) & -(1 + {}^0x_1)(2\sin\theta - 1) \end{array}\right]$$

or

$${}^{\Delta t}_{0}\mathbf{X} = \begin{bmatrix} \cos\theta & -\sin\theta \\ \sin\theta & \cos\theta \end{bmatrix} \tag{d}$$

Substituting now from (b) and (d) into (c) we obtain

$${}^{\Delta t}_{0}\mathbf{S} = \begin{bmatrix} {}^{\Delta t}\bar{\tau}_{11} & 0 \\ 0 & 0 \end{bmatrix} \tag{e}$$

But, since ${}^{\Delta t}\bar{\tau}_{11}$ is equal to ${}^0\tau_{11}$, the relations in (a) and (e) show that the components of the 2nd Piola–Kirchhoff stress tensor did not change during the rigid body rotation. The reason for there being no change in the 2nd Piola–Kirchhoff stress tensor is that the deformation gradient corresponds in this case to the rotation matrix that is used in the transformation of (b) (see Example 6.12).

There has been much discussion about the physical nature of the 2nd Piola–Kirchhoff stress tensor. However, although it is possible to relate the transformation on the Cauchy stress tensor in (6.15) to some geometry arguments as discussed in the example below, it should be recognized that the 2nd Piola–Kirchhoff stresses have little physical meaning and, in practice, Cauchy stresses must be calculated.

EXAMPLE 6.7: Figure 6.8 shows a generic body in the configurations at times 0 and t. Let ${}^t d\mathbf{T}$ be the actual force on a surface area ${}^t dS$ in the configuration at time t, and let us define a (fictitious) force

$${}^0 d\mathbf{T} = {}^0_t\mathbf{X}\, {}^t d\mathbf{T}; \qquad {}^0_t\mathbf{X} = \left[\frac{\partial\, {}^0x_i}{\partial\, {}^tx_j}\right] \tag{a}$$

which acts on the surface area ${}^0 dS$, where ${}^0 dS$ has become ${}^t dS$ and ${}^0_t\mathbf{X}$ is the inverse of the deformation gradient, ${}^0_t\mathbf{X} = {}^t_0\mathbf{X}^{-1}$. Show that the 2nd Piola-Kirchhoff stresses measured in the original configuration are the stress components corresponding to ${}^0 d\mathbf{T}$.

Let the unit normals to the surface areas ${}^0 dS$ and ${}^t dS$ be ${}^0\mathbf{n}$ and ${}^t\mathbf{n}$, respectively. Force equilibrium (of the wedge ABC in Fig. 6.8) in the configuration at time t requires that

$${}^t d\mathbf{T} = {}^t\boldsymbol{\tau}^T\, {}^t\mathbf{n}\, {}^t dS \tag{b}$$

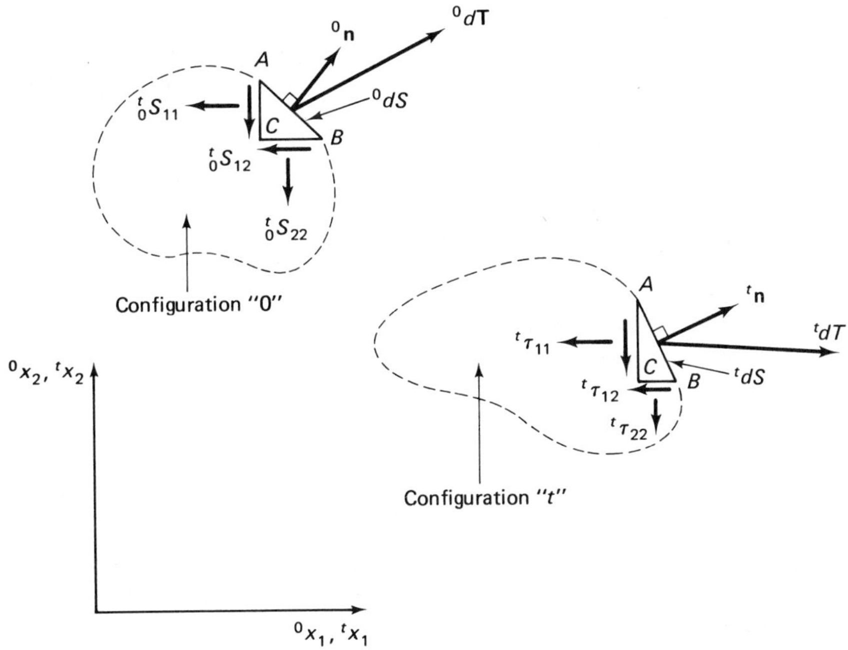

FIGURE 6.8 *Second Piola-Kirchhoff and Cauchy stresses in two-dimensional action.*

and similarly corresponding to the configuration at time 0 we use

$$ {}^0d\mathbf{T} = {}_0^t\mathbf{S}^T \, {}^0\mathbf{n} \, {}^0dS \tag{c} $$

The relations in (b) and (c) are referred to as *Cauchy's formula*. However, it can be shown that the following kinematic relationship exists

$$ {}^t\mathbf{n} \, {}^tdS = \frac{{}^0\rho}{{}^t\rho} \, {}_t^0\mathbf{X}^T \, {}^0\mathbf{n} \, {}^0dS \tag{d} $$

This relation is referred to as *Nanson's formula*.[1] Now using (a) to (d) we obtain

$$ {}_0^t\mathbf{S}^T \, {}^0\mathbf{n} \, {}^0dS = {}_t^0\mathbf{X} \, {}^t\boldsymbol{\tau}^T \frac{{}^0\rho}{{}^t\rho} \, {}_t^0\mathbf{X}^T \, {}^0\mathbf{n} \, {}^0dS $$

or

$$ \left({}_0^t\mathbf{S}^T - \frac{{}^0\rho}{{}^t\rho} \, {}_t^0\mathbf{X} \, {}^t\boldsymbol{\tau}^T \, {}_t^0\mathbf{X}^T\right) {}^0\mathbf{n} \, {}^0dS = \mathbf{0} $$

However, this relationship must hold for any surface area and also any "interior surface area" that could be created by a cut in the body. Hence, the normal ${}^0\mathbf{n}$ is arbitrary and can be chosen to be in succession equal to the unit coordinate vectors. It follows that

$$ {}_0^t\mathbf{S} = \frac{{}^0\rho}{{}^t\rho} \, {}_t^0\mathbf{X} \, {}^t\boldsymbol{\tau} \, {}_t^0\mathbf{X}^T $$

where we used the property that the matrices ${}^t\boldsymbol{\tau}$ and ${}_0^t\mathbf{S}$ are symmetric.

Finally, we may interpret the force defined in (a); namely, we should note that the force ${}^0d\mathbf{T}$ which is balanced by the 2nd Piola–Kirchhoff stresses is in the same way related to the actual force ${}^td\mathbf{T}$ as an original fibre in ${}^0dS$ is related to its final position,

$$ d\,{}^0\mathbf{x} = {}_t^0\mathbf{X} \, d\,{}^t\mathbf{x} $$

We may therefore say that in using (a) to obtain ${}^0d\mathbf{T}$ the force ${}^td\mathbf{T}$ is "stretched and rotated" in the same way as $d\,{}^t\mathbf{x}$ is stretched and rotated to obtain $d\,{}^0\mathbf{x}$.

In (6.15) we defined the 2nd Piola–Kirchhoff stress tensor at time t referred to the original configuration of the body; however, we could also have used another reference configuration. In particular, in our incremental formulations we shall also use the 2nd Piola–Kirchhoff stresses at time $t + \Delta t$ referred to the configuration at time t.

Another stress measure that is effectively used in some formulations is the *Jaumann stress rate tensor*, defined as

$$ {}^t\overset{\nabla}{\tau}_{ij} = {}^t\dot{\tau}_{ij} - {}^t\tau_{ip}\,{}^t\Omega_{pj} - {}^t\tau_{jp}\,{}^t\Omega_{pi} \tag{6.21}$$

where ${}^t\dot{\tau}_{ij}$ is a Cartesian component of the time derivative of the Cauchy stress tensor evaluated at time t, and ${}^t\Omega_{ij}$ is a Cartesian component of the *spin tensor*,

$$ {}^t\Omega_{ij} = \frac{1}{2}\left(\frac{\partial\,{}^t\dot{u}_j}{\partial\,{}^tx_i} - \frac{\partial\,{}^t\dot{u}_i}{\partial\,{}^tx_j}\right) \tag{6.22}$$

Physically, the spin tensor represents the angular velocity of the material.

EXAMPLE 6.8: Consider the *velocity gradient* $\partial\,{}^t\dot{u}_i/\partial\,{}^tx_j$ at a point in a two-dimensional analysis. Show how the spin tensor given in (6.22) can be related to this tensor.

In two-dimensional analysis we can decompose the velocity gradient into a symmetric and an antisymmetric tensor:

$$\begin{bmatrix} \dfrac{\partial\,{}^t\dot{u}_1}{\partial\,{}^tx_1} & \dfrac{\partial\,{}^t\dot{u}_1}{\partial\,{}^tx_2} \\[2ex] \dfrac{\partial\,{}^t\dot{u}_2}{\partial\,{}^tx_1} & \dfrac{\partial\,{}^t\dot{u}_2}{\partial\,{}^tx_2} \end{bmatrix} = \begin{bmatrix} \dfrac{\partial\,{}^t\dot{u}_1}{\partial\,{}^tx_1} & \dfrac{1}{2}\left(\dfrac{\partial\,{}^t\dot{u}_1}{\partial\,{}^tx_2} + \dfrac{\partial\,{}^t\dot{u}_2}{\partial\,{}^tx_1}\right) \\[2ex] \dfrac{1}{2}\left(\dfrac{\partial\,{}^t\dot{u}_2}{\partial\,{}^tx_1} + \dfrac{\partial\,{}^t\dot{u}_1}{\partial\,{}^tx_2}\right) & \dfrac{\partial\,{}^t\dot{u}_2}{\partial\,{}^tx_2} \end{bmatrix} - \begin{bmatrix} 0 & \dfrac{1}{2}\left(\dfrac{\partial\,{}^t\dot{u}_2}{\partial\,{}^tx_1} - \dfrac{\partial\,{}^t\dot{u}_1}{\partial\,{}^tx_2}\right) \\[2ex] \dfrac{1}{2}\left(\dfrac{\partial\,{}^t\dot{u}_1}{\partial\,{}^tx_2} - \dfrac{\partial\,{}^t\dot{u}_2}{\partial\,{}^tx_1}\right) & 0 \end{bmatrix} \tag{a}$$

where the first term on the r.h.s. of (a) represents the *velocity strain tensor* (see (6.24)) and the second term is the spin tensor. To obtain a physical understanding of these tensors, consider the motion during an infinitesimally small time increment from time t, so that

$$\int \frac{\partial\,{}^t\dot{u}_i}{\partial\,{}^tx_j}\,dt = \frac{\partial u_i}{\partial\,{}^tx_j}$$

Let us assume first that the motion corresponds to a rigid body rotation as illustrated in Fig. 6.9(a). In this case we have

$$\alpha = \frac{1}{2}\left(\frac{\partial u_2}{\partial\,{}^tx_1} - \frac{\partial u_1}{\partial\,{}^tx_2}\right)$$

$${}^t\Omega_{12} = \dot{\alpha}$$

and the velocity strain tensor is zero. Hence this motion is completely described by the spin tensor, where ${}^t\Omega_{12}$ gives the angular velocity about the ${}^tx_3$-axis.

On the other hand, if we have the motion illustrated in Fig. 6.9(b), the spin tensor is zero because there is no rigid body rotation, and the elements in the velocity strain tensor give the rate of instantaneous straining of the material.

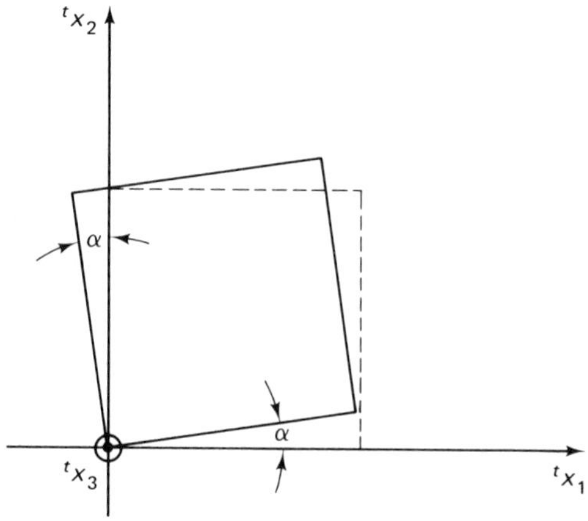

(a) Rigid body motion ${}^t\Omega_{12}$ is equal to the time rate of change of α

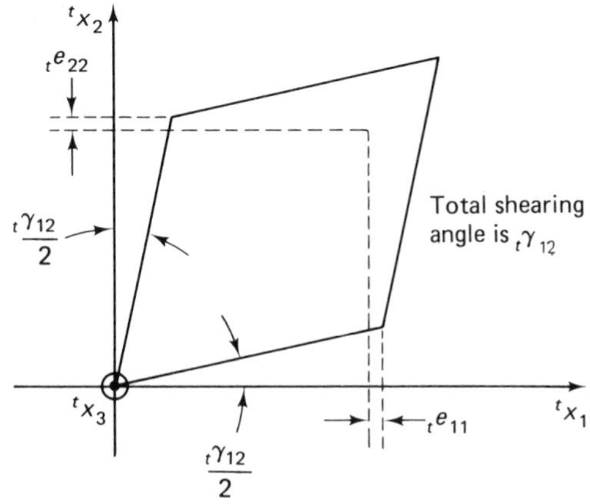

(b) Deformation without rigid body motion; shearing strain is ${}_t\gamma_{12}$, and normal strains are ${}_te_{11}$, ${}_te_{22}$

FIGURE 6.9 *Illustration of spin tensor.*

Considering the Jaumann stress rate tensor in (6.21), we recognize that this tensor is also symmetric, and that if ${}^t\overset{\nabla}{\tau}_{ij}$ is zero, the relation in (6.21) properly evaluates the change in Cauchy stress due to a rigid body rotation of the material. We demonstrate this important feature in the following example.

EXAMPLE 6.9: The two-dimensional element shown in Fig. 6.10 is at time 0 subjected to a stress 10 N/cm² and an angular velocity ω. Assume that the element rotates as a rigid body through an angle of 90 degrees with the constant velocity ω. Use the relation in (6.21) to evaluate the stresses in the element at the end of that rotation.

The stress tensor at time 0 is

$$ {}^0\boldsymbol{\tau} = 10\begin{bmatrix} 1 & 0 \\ 0 & 0 \end{bmatrix} \tag{a} $$

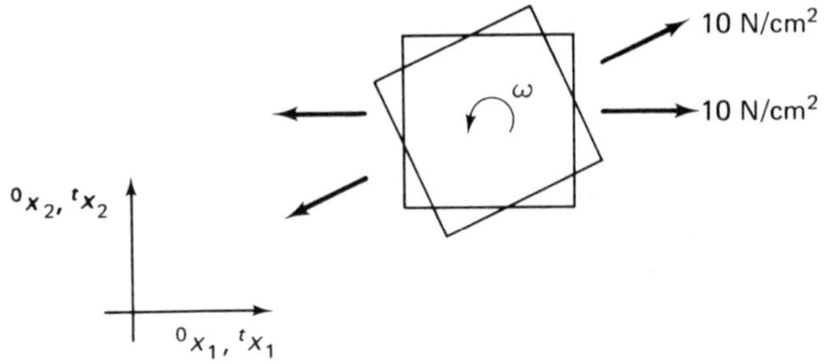

FIGURE 6.10 *Element with stress equal to 10 N/cm² and constant angular velocity ω rad/sec.*

Since the element rotates through 90 degrees as a rigid body we know that the exact stress tensor (corresponding to the stationary coordinate axes ${}^t x_1$, ${}^t x_2$) at the end of that rotation is

$$ {}^t\boldsymbol{\tau}\Big|_{\substack{\text{after}\\ 90^\circ\text{ rotation}}} = 10\begin{bmatrix} 0 & 0 \\ 0 & 1 \end{bmatrix} \tag{b}$$

Hence, the integration of (6.21) with the initial conditions given in (a) should yield (b).

Let us use the Euler forward integration method (see Section 9.6), in which case we assume that

$$ {}^t\dot{\boldsymbol{\tau}} = \frac{{}^{t+\Delta t}\boldsymbol{\tau} - {}^t\boldsymbol{\tau}}{\Delta t} $$

Hence (6.21) can be written as, with ${}^t\overset{\nabla}{\tau}_{ij} = 0$,

$$ {}^{t+\Delta t}\boldsymbol{\tau} = {}^t\boldsymbol{\tau} + {}^t\boldsymbol{\tau}^T\begin{bmatrix} 0 & \alpha \\ -\alpha & 0 \end{bmatrix} + \begin{bmatrix} 0 & -\alpha \\ \alpha & 0 \end{bmatrix}{}^t\boldsymbol{\tau} \tag{c}$$

where
$$\alpha = \frac{\pi}{2n}$$

if we want to use n steps to rotate the element through 90°. Operating on (c) with the initial conditions in (a) we obtain the following results:

$n = 10$

$$ {}^{10}\boldsymbol{\tau} = 10\begin{bmatrix} -0.304 & 0.06 \\ 0.06 & 1.304 \end{bmatrix} $$

$n = 50$

$$ {}^{50}\boldsymbol{\tau} = 10\begin{bmatrix} -0.052 & 0.002 \\ 0.002 & 1.052 \end{bmatrix} $$

$n = 150$

$$ {}^{150}\boldsymbol{\tau} = 10\begin{bmatrix} -0.017 & 0.0002 \\ 0.0002 & 1.017 \end{bmatrix} $$

$n = 500$

$$ {}^{500}\boldsymbol{\tau} = 10\begin{bmatrix} -0.005 & 0.00002 \\ 0.00002 & 1.005 \end{bmatrix} $$

Hence, we note that using the Euler forward integration method a very large number of steps are required to obtain an accurate approximation to the exact solution given in (b); and in practice the solution should be established for a given total angular rotation by simply transforming the stress as was done to obtain the result in (b).

It is instructive at this point to compare the 2nd Piola–Kirchhoff stress and Jaumann stress rate tensors. Both stress measures are functions only of the deformations of the material and are not affected by rigid body motions: they are therefore, respectively, an *objective stress tensor* and an *objective stress rate tensor*. In practice, an important difference between the two stress measures is that the 2nd Piola–Kirchhoff stress is a total stress which in an elastic or hyperelastic constitutive behavior can be calculated from the total current strains (see Section 6.4.1), whereas the Jaumann stress rate is related to the rate of straining, which means that an integration process is always required to evaluate the current Cauchy stresses. This specific difference already indicates under what conditions it is more effective to employ one stress measure rather than the other. Namely, if the constitutive relations are not path-dependent and thus do not require an integration process, the use of the 2nd Piola–Kirchhoff stress tensor is usually more effective, whereas for analysis of path-dependent materials the use of the Jaumann stress rate tensor should be considered (see Section 6.4).

Having presented and briefly discussed the stress tensors that we will be using in our incremental formulations, we now turn to the discussion of appropriate strain measures. The strain tensor used with the 2nd Piola–Kirchhoff stress tensor is the *Green–Lagrange strain tensor* defined as

$$ {}_0^t\epsilon_{ij} = \tfrac{1}{2}({}_0^tu_{i,j} + {}_0^tu_{j,i} + {}_0^tu_{k,i}\,{}_0^tu_{k,j}) \tag{6.23}$$

in which ${}_0^tu_{i,j} = \partial^t u_i/\partial^0 x_j$. The strain measure used with the Jaumann stress rate tensor is the *velocity strain tensor*, or rate of deformation tensor (see Example 6.8),

$$ {}^t\dot{e}_{ij} = \frac{1}{2}\left(\frac{\partial^t\dot{u}_i}{\partial^t x_j} + \frac{\partial^t\dot{u}_j}{\partial^t x_i}\right) \tag{6.24}$$

These strain measures satisfy the same conditions that the corresponding stress measures satisfy, namely, they are symmetric tensors and they are objective tensors, which means that rigid body motions of the material do not change their components (see Examples 6.8 and 6.13). In addition, they fulfill one further requirement, which is a necessary condition for using them in our incremental formulations: *these strain tensors are energetically conjugate to the corresponding stress tensors discussed earlier*; i.e., let

$$\int {}^t\dot{e}_{ij}\,dt = {}_te_{ij}$$

then $\quad {}^t\tau_{ij}\,\delta_t e_{ij} =$ Virtual work at time t per unit current volume

$\quad {}_0^tS_{ij}\,\delta\,{}_0^t\epsilon_{ij} =$ Virtual work at time t per unit original volume

Hence, the total internal virtual work can be calculated using as stress measures either the Cauchy or the 2nd Piola–Kirchhoff stress tensors provided the energy conjugate strain tensors are employed and the integrations are performed over the current and original volumes, respectively. We give the proof that the 2nd Piola–Kirchhoff stress and Green–Lagrange strain tensors are energy conjugate in (6.36) to (6.37) after the development of some fundamental kinematic relationships pertaining to the Green–Lagrange strain tensor.

The components of the Green–Lagrange strain tensor were given in (6.23). However, if we focus attention on a line element of length ${}^0ds$ in its original configuration which has taken on the length ${}^tds$ in the current configuration, we can also define the Green–Lagrange strain tensor with components ${}_0^t\epsilon_{ij}$ in the following way:

$$ {}_0^t\epsilon_{ij}\, d^0x_i\, d^0x_j = \tfrac{1}{2}\{({}^tds)^2 - ({}^0ds)^2\}^* \tag{6.25} $$

in which

$$ ({}^tds)^2 = d^tx_i\, d^tx_i \tag{6.26} $$

$$ ({}^0ds)^2 = d^0x_i\, d^0x_i \tag{6.27} $$

Since

$$ d^tx_i = \frac{\partial^t x_i}{\partial^0 x_j}\, d^0x_j \tag{6.28} $$

we can write

$$ ({}^tds)^2 = d^0\mathbf{x}^T\, {}_0^t\mathbf{X}^T\, {}_0^t\mathbf{X}\, d^0\mathbf{x} \tag{6.29} $$

where ${}_0^t\mathbf{X}$ is the deformation gradient defined in (6.17). Also, in (6.29) we have

$$ d^0\mathbf{x} = \begin{bmatrix} d^0x_1 \\ d^0x_2 \\ d^0x_3 \end{bmatrix} \tag{6.30} $$

With (6.30) we can now write the l.h.s. of (6.25) as follows:

$$ {}_0^t\epsilon_{ij}\, d^0x_i\, d^0x_j = d^0\mathbf{x}^T\, {}_0^t\boldsymbol{\epsilon}\, d^0\mathbf{x} \tag{6.31} $$

in which ${}_0^t\epsilon_{ij}$ is element (i, j) of the matrix ${}_0^t\boldsymbol{\epsilon}$,

$$ {}_0^t\boldsymbol{\epsilon} = \tfrac{1}{2}({}_0^t\mathbf{C} - \mathbf{I}) \tag{6.32} $$

and ${}_0^t\mathbf{C}$ is the *Cauchy–Green deformation tensor*† defined as

$$ {}_0^t\mathbf{C} = {}_0^t\mathbf{X}^T\, {}_0^t\mathbf{X} \tag{6.33} $$

The relation in (6.32) could be regarded as another definition of the Green-Lagrange strain tensor, this time using the Cauchy–Green deformation tensor ${}_0^t\mathbf{C}$ as the fundamental quantity. To obtain from (6.32) the relation in (6.23) we substitute for ${}_0^t\mathbf{C}$ and ${}_0^t\mathbf{X}$ using (6.33) and (6.17),

$$ \begin{aligned} {}_0^t\boldsymbol{\epsilon} &= \tfrac{1}{2}({}_0^t\mathbf{X}^T\, {}_0^t\mathbf{X} - \mathbf{I}) \\ &= \tfrac{1}{2}\{[{}_0\boldsymbol{\nabla}({}^0\mathbf{x} + {}^t\mathbf{u})^T][{}_0\boldsymbol{\nabla}({}^0\mathbf{x} + {}^t\mathbf{u})^T]^T - \mathbf{I}\} \\ &= \tfrac{1}{2}\{[\mathbf{I} + {}_0\boldsymbol{\nabla}\, {}^t\mathbf{u}^T][\mathbf{I} + {}_0\boldsymbol{\nabla}\, {}^t\mathbf{u}^T]^T - \mathbf{I}\} \\ &= \tfrac{1}{2}\{{}_0\boldsymbol{\nabla}\, {}^t\mathbf{u}^T + ({}_0\boldsymbol{\nabla}\, {}^t\mathbf{u}^T)^T + ({}_0\boldsymbol{\nabla}\, {}^t\mathbf{u}^T)({}_0\boldsymbol{\nabla}\, {}^t\mathbf{u}^T)^T\} \end{aligned} \tag{6.34} $$

Thus the relations in (6.23), (6.25), and (6.32) are all equivalent, and can all be employed to evaluate the components of the Green–Lagrange strain tensor.

EXAMPLE 6.10: Consider the four-node element shown in Fig. 6.6. Evaluate the components of the Green-Lagrange strain tensor when the element is subjected to the given displacements.

Since we evaluated already the deformation gradient in Example 6.5, it is now convenient to use (6.32) and (6.33) to calculate the Green–Lagrange strains. Thus, we obtain

$$ {}_0^t\boldsymbol{\epsilon} = \frac{1}{2}\left[\frac{1}{16}\left[\begin{array}{c:c} (5+{}^0x_2)^2+\frac{1}{4}(1+{}^0x_2)^2 & (5+{}^0x_2)(1+{}^0x_1)+\frac{1}{4}(1+{}^0x_2)(9+{}^0x_1) \\ \hdashline (1+{}^0x_1)(5+{}^0x_2)+\frac{1}{4}(9+{}^0x_1)(1+{}^0x_2) & (1+{}^0x_1)^2+\frac{1}{4}(9+{}^0x_1)^2 \end{array}\right] - \begin{bmatrix} 1 & 0 \\ 0 & 1 \end{bmatrix}\right] $$

* Note that, in accordance, with the notation $\partial^t x_i/\partial^0 x_j$ we use here the notation $d^0x_i \equiv d({}^0x_i)$; hence the left time superscript is on the "x_i". However, ${}^0ds$ is stretched to ${}^tds$; hence the left time superscript is on the "ds".

† ${}_0^t\mathbf{C}$ is more precisely called the *right* Cauchy-Green deformation tensor, since there is also a left Cauchy–Green deformation tensor[1] which, however, we do not use in this book.

The Cauchy–Green deformation tensor ${}_0^t\mathbf{C}$ and the deformation gradient ${}_0^t\mathbf{X}$ have some important properties that we study in the following two examples.

EXAMPLE 6.11: The *stretch* ${}^t\lambda$ of a line element of a general body in motion is defined as ${}^t\lambda = {}^t ds/{}^0 ds$ where ${}^0 ds$ and ${}^t ds$ are the original and current lengths of the element as shown in Fig. 6.11. Prove that

$$ {}^t\lambda = ({}^0\mathbf{n}^T \, {}_0^t\mathbf{C} \, {}^0\mathbf{n})^{1/2} \tag{a} $$

where ${}^0\mathbf{n}$ is a vector of the direction cosines of the line element at time 0. Also

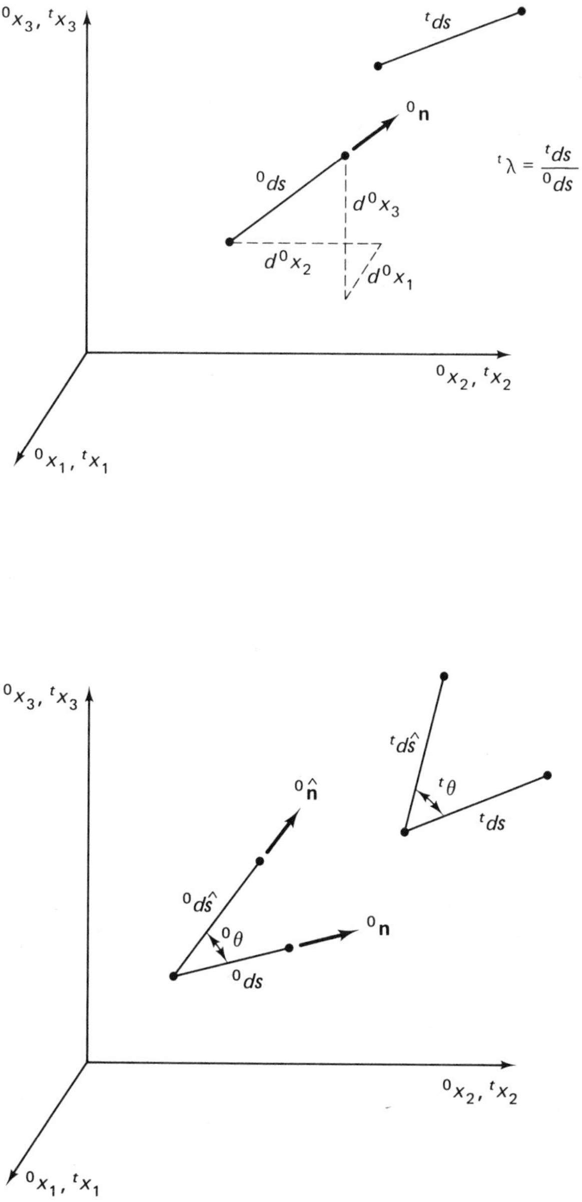

FIGURE 6.11 *Stretch and rotation of line elements.*

prove that considering two line elements emanating from the same material point, the angle ${}^t\theta$ between the line elements at time t is given by

$$\cos {}^t\theta = \frac{{}^0\mathbf{n}^T \, {}^t_0\mathbf{C} \, {}^0\hat{\mathbf{n}}}{{}^t\lambda \, {}^t\hat{\lambda}} \tag{b}$$

where the hat denotes the second line element (see Fig. 6.11).

As an example, apply the formulas in (a) and (b) to evaluate the stretches of the specific line elements ${}^0ds$ and ${}^td\hat{s}$ shown in Fig. 6.6 and evaluate also the angular distortion between them.

To prove (a), we recognize that

$$({}^tds)^2 = d{}^t\mathbf{x}^T \, d{}^t\mathbf{x}; \qquad d{}^t\mathbf{x} = {}^t_0\mathbf{X} \, d{}^0\mathbf{x}$$

so that using (6.33)

$$({}^tds)^2 = d{}^0\mathbf{x}^T \, {}^t_0\mathbf{C} \, d{}^0\mathbf{x}$$

Hence

$${}^t\lambda^2 = \frac{d{}^0\mathbf{x}^T}{{}^0ds} \, {}^t_0\mathbf{C} \, \frac{d{}^0\mathbf{x}}{{}^0ds}$$

and since

$${}^0\mathbf{n} = \frac{d{}^0\mathbf{x}}{{}^0ds}$$

we have

$${}^t\lambda = ({}^0\mathbf{n}^T \, {}^t_0\mathbf{C} \, {}^0\mathbf{n})^{1/2}$$

To prove (b) we use that (see Section 2.6)

$$d{}^t\mathbf{x}^T \, d{}^t\hat{\mathbf{x}} = ({}^tds)({}^td\hat{s}) \cos {}^t\theta$$

Hence

$$\cos {}^t\theta = \frac{d{}^0\mathbf{x}^T \, {}^t_0\mathbf{X}^T \, {}^t_0\hat{\mathbf{X}} \, d{}^0\hat{\mathbf{x}}}{({}^tds)({}^td\hat{s})} \tag{c}$$

Since ${}^t_0\mathbf{X} \equiv {}^t_0\hat{\mathbf{X}}$ (it is the deformation gradient at the location of the differential line elements), we obtain from (c)

$$\cos {}^t\theta = \frac{{}^0\mathbf{n}^T \, {}^t_0\mathbf{C} \, {}^0\hat{\mathbf{n}}}{{}^t\lambda \, {}^t\hat{\lambda}}$$

It should be noted that the relations in (a) and (b) show that when ${}^t_0\mathbf{C} = \mathbf{I}$ the stretches of the line elements are equal to one and the angle between line elements has not changed during the motion. Hence, when the Cauchy–Green deformation tensor is equal to the identity matrix, the motion can have been at most a rigid body motion.

If we apply (a) and (b) to the line elements depicted in Fig. 6.6, we obtain at ${}^0x_1 = 0$, ${}^0x_2 = 0$ (see Example 6.5)

$${}^t_0\mathbf{C} = \frac{1}{16}\begin{bmatrix} 25.25 & 7.25 \\ 7.25 & 21.25 \end{bmatrix}$$

$${}^0\mathbf{n} = \begin{bmatrix} 1 \\ 0 \end{bmatrix}; \qquad {}^0\hat{\mathbf{n}} = \begin{bmatrix} 0 \\ 1 \end{bmatrix}$$

Hence, using (a)

$${}^t\lambda = 1.256; \qquad {}^t\hat{\lambda} = 1.152$$

and using (b)

$$\cos {}^t\theta = 0.313, \qquad {}^t\theta = 71.75^\circ$$

Therefore, the angular distortion between the line elements ${}^0ds$ and ${}^0d\hat{s}$ due to the motion from time 0 to time t is 18.25 degrees.

EXAMPLE 6.12: Show that the deformation gradient ${}^t_0\mathbf{X}$ can always be decomposed as follows:

$${}^t_0\mathbf{X} = {}^t_0\mathbf{R} \, {}^t_0\mathbf{U} \tag{a}$$

where ${}^t_0\mathbf{R}$ is an orthogonal (rotation) matrix and ${}^t_0\mathbf{U}$ is a stretch (symmetric) matrix.

To prove the relationship in (a), we consider the Cauchy–Green deformation tensor ${}_0^t\mathbf{C}$ and represent this tensor in its principal coordinate axes. For this purpose we solve the eigenproblem

$$ {}_0^t\mathbf{C}\mathbf{p} = \lambda \mathbf{p} \tag{b} $$

The complete solution of (b) can be written as (see Section 2.7)

$$ {}_0^t\mathbf{C}\mathbf{P} = \mathbf{P}\ {}_0^t\mathbf{C}' $$

where the columns of $\mathbf{P}$ are the eigenvectors of ${}_0^t\mathbf{C}$ and ${}_0^t\mathbf{C}'$ is a diagonal matrix storing the corresponding eigenvalues. We also have

$$ \mathbf{P}^T\ {}_0^t\mathbf{C}\mathbf{P} = {}_0^t\mathbf{C}' \tag{c} $$

and ${}_0^t\mathbf{C}'$ is the representation of the Cauchy–Green deformation tensor in its principal coordinate axes. The representation of the deformation gradient in this coordinate system, ${}_0^t\mathbf{X}'$, is similarly obtained:

$$ {}_0^t\mathbf{X}' = \mathbf{P}^T\ {}_0^t\mathbf{X}\mathbf{P} \tag{d} $$

where we note that (c) and (d) are really tensor transformations from the original to a new coordinate system (see Section 2.6).

Using the above relations and that ${}_0^t\mathbf{C} = {}_0^t\mathbf{X}^T\ {}_0^t\mathbf{X}$ we have

$$ {}_0^t\mathbf{C}' = {}_0^t\mathbf{X}'^T\ {}_0^t\mathbf{X}' $$

and we note that the matrix

$$ {}_0^t\mathbf{R}' = {}_0^t\mathbf{X}'\ ({}_0^t\mathbf{C}')^{-1/2} $$

is an orthogonal matrix, i.e., ${}_0^t\mathbf{R}'^T\ {}_0^t\mathbf{R}' = \mathbf{I}$

Hence, we can write

$$ {}_0^t\mathbf{X}' = {}_0^t\mathbf{R}'\ {}_0^t\mathbf{U}' \tag{e} $$

where

$$ {}_0^t\mathbf{U}' = ({}_0^t\mathbf{C}')^{1/2} $$

and to evaluate ${}_0^t\mathbf{U}'$ we use the positive values of the square roots of the diagonal elements of ${}_0^t\mathbf{C}'$.

The relation in (e) is the decomposition of the deformation gradient ${}_0^t\mathbf{X}'$ into the product of the orthogonal matrix ${}_0^t\mathbf{R}'$ and the stretch matrix ${}_0^t\mathbf{U}'$. This decomposition has been accomplished in the principal axes of ${}_0^t\mathbf{C}$, but is also valid in any other (admissible) coordinate system because the deformation gradient is a tensor (see Section 2.6). Indeed, we can now directly obtain ${}_0^t\mathbf{R}$ and ${}_0^t\mathbf{U}$ corresponding to the decomposition in (a); i.e.,

$$ {}_0^t\mathbf{R} = \mathbf{P}\ {}_0^t\mathbf{R}'\ \mathbf{P}^T $$
$$ {}_0^t\mathbf{U} = \mathbf{P}\ {}_0^t\mathbf{U}'\ \mathbf{P}^T $$

where we used the inverse of the transformation employed in (d).

Two important consequences of the properties (discussed in Examples 6.11 and 6.12) of the tensors ${}_0^t\mathbf{C}$ and ${}_0^t\mathbf{X}$ are that the 2nd Piola–Kirchhoff stress components and the Green–Lagrange strain components are invariant under rigid body motions. The 2nd Piola–Kirchhoff stress components and Green–Lagrange strain components do not change when the body is subjected to a rigid body motion, because in this case the deformation gradient reduces to a rotation matrix, and the Cauchy–Green deformation tensor is the identity matrix. To demonstrate the invariance of the 2nd Piola–Kirchhoff stress components under a rigid body rotation, we considered a specific case in Example 6.6. The fact that the Green-Lagrange strain components are zero for and do not change with a rigid body motion follows directly from (6.32) and (a) of Example 6.12, and we now demonstrate this property.

EXAMPLE 6.13: A four-node element is stretched until time t, and then undergoes without distortion a large rigid body rotation from time t to time $t + \Delta t$ as depicted in Fig. 6.12. Show explicitly that for the element the components of the Green–Lagrange strain tensor at time t and time $t + \Delta t$ are exactly equal.

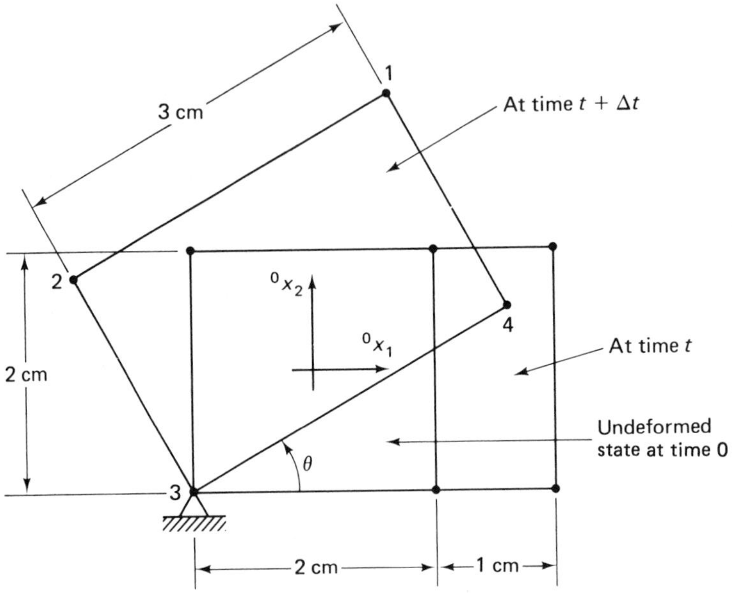

FIGURE 6.12 *Element subjected to large rigid body rotation after initial stretch.*

The Green–Lagrange strain components at time t can be evaluated by inspection using (6.25):

$$ {}_0^t\epsilon_{22} = 0; \qquad {}_0^t\epsilon_{12} = {}_0^t\epsilon_{21} = 0 $$

and

$$ {}_0^t\epsilon_{11} = \frac{1}{2}\,\frac{{}^t ds^2 - {}^0 ds^2}{{}^0 ds^2} $$

$$ = \frac{1}{2}\,\frac{(3)^2 - (2)^2}{(2)^2} $$

$$ = \frac{5}{8} $$

Hence

$$ {}_0^t\boldsymbol{\epsilon} = \begin{bmatrix} \frac{5}{8} & 0 \\ 0 & 0 \end{bmatrix} $$

Alternatively, we can use (6.32), where we first evaluate the deformation gradient as in Example 6.5:

$$ {}_0^t\mathbf{X} = \begin{bmatrix} \frac{3}{2} & 0 \\ 0 & 1 \end{bmatrix} $$

Hence

$$ {}_0^t\mathbf{C} = \begin{bmatrix} \frac{9}{4} & 0 \\ 0 & 1 \end{bmatrix} $$

and as before

$$ {}_0^t\boldsymbol{\epsilon} = \begin{bmatrix} \frac{5}{8} & 0 \\ 0 & 0 \end{bmatrix} \qquad \text{(a)} $$

After the rigid body rotation the nodal point coordinates are

Node	${}^{t+\Delta t}x_1$	${}^{t+\Delta t}x_2$
1	$3\cos\theta - 1 - 2\sin\theta$	$3\sin\theta - 1 + 2\cos\theta$
2	$-1 - 2\sin\theta$	$2\cos\theta - 1$
3	-1	-1
4	$3\cos\theta - 1$	$3\sin\theta - 1$

Thus, using again the procedure of Example 6.5 to evaluate the deformation gradient we obtain

$$
{}^{t+\Delta t}_{0}\mathbf{X} = \frac{1}{4}\left[\begin{array}{l|l}
(1+{}^0x_2)(3\cos\theta-1-2\sin\theta) & (1+{}^0x_1)(3\cos\theta-1-2\sin\theta) \\
-(1+{}^0x_2)(-1-2\sin\theta) & +(1-{}^0x_1)(-1-2\sin\theta) \\
-(1-{}^0x_2)(-1) & -(1-{}^0x_1)(-1) \\
+(1-{}^0x_2)(3\cos\theta-1) & -(1+{}^0x_1)(3\cos\theta-1) \\
\hline
(1+{}^0x_2)(3\sin\theta-1+2\cos\theta) & (1+{}^0x_1)(3\sin\theta-1+2\cos\theta) \\
-(1+{}^0x_2)(2\cos\theta-1) & +(1-{}^0x_1)(2\cos\theta-1) \\
-(1-{}^0x_2)(-1) & -(1-{}^0x_1)(-1) \\
+(1-{}^0x_2)(3\sin\theta-1) & -(1+{}^0x_1)(3\sin\theta-1)
\end{array}\right] \tag{b}
$$

or

$$
{}^{t+\Delta t}_{0}\mathbf{X} = \begin{bmatrix} \frac{3}{2}\cos\theta & -\sin\theta \\ \frac{3}{2}\sin\theta & \cos\theta \end{bmatrix} \tag{c}
$$

In reference to the discussion given in Example 6.12 we may note that this deformation gradient can be written as

$$
{}^{t+\Delta t}_{0}\mathbf{X} = {}^{t+\Delta t}_{0}\mathbf{R}\;{}^{t+\Delta t}_{0}\mathbf{U} \tag{d}
$$

where

$$
{}^{t+\Delta t}_{0}\mathbf{R} = \begin{bmatrix} \cos\theta & -\sin\theta \\ \sin\theta & \cos\theta \end{bmatrix}; \qquad {}^{t+\Delta t}_{0}\mathbf{U} = \begin{bmatrix} \frac{3}{2} & 0 \\ 0 & 1 \end{bmatrix}
$$

This decomposition certainly corresponds to the actual physical situation, in which we measured a stretch in the ${}^0x_1$ direction and then a rotation. Therefore, we could have established ${}^{t+\Delta t}_{0}\mathbf{X}$ using (d) instead of performing all the calculations leading to (b) and thus (c)!

Using (d) and (6.33) we obtain

$$
{}^{t+\Delta t}_{0}\mathbf{C} = \begin{bmatrix} \frac{9}{4} & 0 \\ 0 & 1 \end{bmatrix}
$$

and thus using (6.32) we have

$$
{}^{t+\Delta t}_{0}\boldsymbol{\epsilon} = \begin{bmatrix} \frac{5}{8} & 0 \\ 0 & 0 \end{bmatrix} \tag{e}
$$

Hence ${}^{t}_{0}\boldsymbol{\epsilon}$ in (a) is equal to ${}^{t+\Delta t}_{0}\boldsymbol{\epsilon}$ in (e), which shows that the Green–Lagrange strain components did not change as a result of the rigid body rotation.

Using the kinematic relationships discussed above, we can now also prove the following important relationship between the Green–Lagrange strain tensor and the deformation tensor

$$
\delta\, {}^{t}_{0}\epsilon_{ij} = \frac{\partial\, {}^t x_m}{\partial\, {}^0 x_i}\frac{\partial\, {}^t x_n}{\partial\, {}^0 x_j}\,\delta\, {}_{t}e_{mn} \tag{6.35}
$$

where δ stands for "variation in" or "infinitesimally small virtual increment in." We give this proof in the following example.

EXAMPLE 6.14: Prove the following relationship:

$$\delta\,{}_0^t\epsilon_{ij} = \frac{\partial\,{}^tx_m}{\partial\,{}^0x_i}\frac{\partial\,{}^tx_n}{\partial\,{}^0x_j}\,\delta_t e_{mn} \tag{a}$$

where
$$\delta\,{}_0^t\epsilon_{ij} = \frac{1}{2}\left(\frac{\partial\,\delta u_i}{\partial\,{}^0x_j} + \frac{\partial\,\delta u_j}{\partial\,{}^0x_i} + \frac{\partial\,\delta u_k}{\partial\,{}^0x_i}\frac{\partial\,{}^tu_k}{\partial\,{}^0x_j} + \frac{\partial\,{}^tu_k}{\partial\,{}^0x_i}\frac{\partial\,\delta u_k}{\partial\,{}^0x_j}\right),$$

$$\delta_t e_{mn} = \frac{1}{2}\left(\frac{\partial\,\delta u_m}{\partial\,{}^tx_n} + \frac{\partial\,\delta u_n}{\partial\,{}^tx_m}\right)$$

and δu_i is a variation (or virtual increment) in the displacement ${}^tu_i$.

To prove the relationship in (a) we use (6.25) to (6.27); hence we have

$$2\,{}_0^t\epsilon_{ij}\,d\,{}^0x_i\,d\,{}^0x_j = d\,{}^tx_i\,d\,{}^tx_i - d\,{}^0x_i\,d\,{}^0x_i$$

Since $\delta d\,{}^0x_i = 0$ we obtain

$$2\,\delta\,{}_0^t\epsilon_{ij}\,d\,{}^0x_i\,d\,{}^0x_j = 2(\delta d\,{}^tx_i)d\,{}^tx_i$$

But
$$\delta d\,{}^tx_i = d\,\delta\,{}^tx_i = d\,\delta u_i = \frac{\partial\,\delta u_i}{\partial\,{}^tx_\ell}\,d\,{}^tx_\ell$$

so that
$$2\delta\,{}_0^t\epsilon_{ij}\,d\,{}^0x_i\,d\,{}^0x_j = 2\frac{\partial\,\delta u_i}{\partial\,{}^tx_\ell}\,d\,{}^tx_\ell\,d\,{}^tx_i$$

or
$$2\delta\,{}_0^t\epsilon_{ij}\,d\,{}^0x_i\,d\,{}^0x_j = 2(\delta_t e_{mn})\,d\,{}^tx_m\,d\,{}^tx_n$$

Hence
$$2\delta\,{}_0^t\epsilon_{ij}\,d\,{}^0x_i\,d\,{}^0x_j = 2(\delta_t e_{mn})\frac{\partial\,{}^tx_m}{\partial\,{}^0x_i}\frac{\partial\,{}^tx_n}{\partial\,{}^0x_j}\,d\,{}^0x_i\,d\,{}^0x_j$$

Since this relationship must hold for any arbitrary $d\,{}^0x_i$ and $d\,{}^0x_j$, we have proven the relation in (a).

We can now show that the 2nd Piola–Kirchhoff stress tensor is *energetically conjugate* to the Green–Lagrange strain tensor. Considering the integral $\int_{^0V} {}_0^tS_{ij}\,\delta\,{}_0^t\epsilon_{ij}\,{}^0dV$ and substituting from (6.15) and (6.35) we obtain

$$\begin{aligned}\int_{^0V} {}_0^tS_{ij}\,\delta\,{}_0^t\epsilon_{ij}\,{}^0dV &= \int_{^0V}\frac{{}^0\rho}{{}^t\rho}\,{}_t^0x_{i,m}\,{}_t^0x_{j,n}\,{}^t\tau_{mn}\,{}_0^tx_{k,i}\,{}_0^tx_{\ell,j}\,\delta_t e_{k\ell}\,{}^0dV \\ &= \int_{^0V}\frac{{}^0\rho}{{}^t\rho}\,{}^t\tau_{mn}\,\delta_t e_{mn}\,{}^0dV\end{aligned} \tag{6.36}$$

because
$${}_0^tx_{k,\ell}\,{}_t^0x_{\ell,m} = \delta_{km} \quad \text{(Kronecker delta)}$$

and since ${}^0\rho\,{}^0dV = {}^t\rho\,{}^tdV$ we have

$$\int_{^0V} {}_0^tS_{ij}\,\delta\,{}_0^t\epsilon_{ij}\,{}^0dV = \int_{^tV} {}^t\tau_{mn}\,\delta_t e_{mn}\,{}^tdV \tag{6.37}$$

The virtual work expression in (6.36) using the 2nd Piola–Kirchhoff stress and Green–Lagrange strain tensors is the basis of the incremental Lagrangian formulations used in the nonlinear analysis of solids and structures. In this section we have defined the stress and strain components that are employed; the use of appropriate constitutive relations is discussed in Section 6.4.

6.2.3 Total and Updated Lagrangian Formulations, Materially-Nonlinear-Only Analysis

We discussed in Sections 6.1 and 6.2.1 the basic difficulties and the solution approach when a general nonlinear problem is analyzed, and we concluded that, for an effective incremental analysis, appropriate stress and strain measures need be employed. This led in Section 6.2.2 to the presentation of some stress and strain tensors that are very effectively employed in practice, and then to the principle of virtual displacements expressed in terms of 2nd Piola–Kirchhoff stresses and Green–Lagrange strains. We now use this fundamental result in the development of two general incremental formulations of nonlinear problems.

The basic equation that we want to solve is relation (6.13), which expresses the equilibrium and compatibility requirements of the general body considered in the configuration corresponding to time $t + \Delta t$. (The constitutive equations also enter (6.13), namely in the calculation of the stresses). Since in general the body can undergo large displacements and large strains, and the constitutive relations are nonlinear, the relation in (6.13) cannot be solved directly; however, an approximate solution can be obtained by referring all variables to a previously calculated known equilibrium configuration, and linearizing the resulting equation. This solution can then be improved by iteration.

To develop the governing equation for the approximate solution obtained by linearization we recall that the solutions for times $0, \Delta t, 2\Delta t, \ldots, t$ have already been calculated and that we can employ (6.37) to refer the stresses and strains to one of these known equilibrium configurations. Hence, in principle, any one of the already calculated equilibrium configurations could be used. In practice, however, the choice lies essentially between two formulations which have been termed total Lagrangian (T.L.) and updated Lagrangian (U.L.) formulations.[9–11] The T.L. formulation has also been referred to as the Lagrangian formulation. In this solution scheme all static and kinematic variables are referred to the initial configuration at time 0. The U.L. formulation is based on the same procedures that are used in the T.L. formulation, but in the solution all static and kinematic variables are referred to the configuration at time t. *Both the T.L. and U.L. formulations include all kinematic nonlinear effects due to large displacements, large rotations and large strains, but whether the large strain behavior is modeled appropriately depends on the constitutive relations specified (see Section 6.4).* The only advantage of using one formulation rather than the other lies in its greater numerical efficiency.

Using (6.37), in the T.L. formulation we consider the basic equation

$$\int_{{}^0V} {}^{t+\Delta t}_{\ \ 0}S_{ij}\, \delta\, {}^{t+\Delta t}_{\ \ 0}\epsilon_{ij}\, {}^0dV = {}^{t+\Delta t}\mathcal{R} \qquad (6.38)$$

whereas in the U.L. formulation we consider {choice of reference}

$$\int_{{}^tV} {}^{t+\Delta t}_{\ \ t}S_{ij}\, \delta\, {}^{t+\Delta t}_{\ \ t}\epsilon_{ij}\, {}^t dV = {}^{t+\Delta t}\mathcal{R} \qquad (6.39)$$

in which ${}^{t+\Delta t}\mathcal{R}$ is the external virtual work given in (6.14). As pointed out above, approximate solutions to (6.38) and (6.39) can be obtained by linearizing the relations. Tables 6.2 and 6.3 summarize the relations used to arrive at the linearized equations of motion in the T.L. and U.L. formulations. These

TABLE 6.2 *Total Lagrangian formulation.*

1. *Equation of Motion*

$$\int_{^0V} {}^{t+\Delta t}_{0}S_{ij}\, \delta\, {}^{t+\Delta t}_{0}\epsilon_{ij}\, {}^0dV = {}^{t+\Delta t}\mathcal{R}$$

where

$${}^{t+\Delta t}_{0}S_{ij} = \frac{{}^0\rho}{{}^{t+\Delta t}\rho}\, {}^{0}_{t+\Delta t}x_{i,m}\, {}^{t+\Delta t}\tau_{mn}\, {}^{0}_{t+\Delta t}x_{j,n}; \qquad \delta\, {}^{t+\Delta t}_{0}\epsilon_{ij} = \delta \tfrac{1}{2}({}^{t+\Delta t}_{0}u_{i,j} + {}^{t+\Delta t}_{0}u_{j,i} + {}^{t+\Delta t}_{0}u_{k,i}\, {}^{t+\Delta t}_{0}u_{k,j})$$

2. *Incremental Decompositions*

(a) Stresses

$${}^{t+\Delta t}_{0}S_{ij} = {}^{t}_{0}S_{ij} + {}_{0}S_{ij}$$

(b) Strains

$${}^{t+\Delta t}_{0}\epsilon_{ij} = {}^{t}_{0}\epsilon_{ij} + {}_{0}\epsilon_{ij}; \qquad {}_{0}\epsilon_{ij} = {}_{0}e_{ij} + {}_{0}\eta_{ij}$$

$${}_{0}e_{ij} = \tfrac{1}{2}({}_{0}u_{i,j} + {}_{0}u_{j,i} + {}^{t}_{0}u_{k,i}\, {}_{0}u_{k,j} + {}_{0}u_{k,i}\, {}^{t}_{0}u_{k,j}); \qquad {}_{0}\eta_{ij} = \tfrac{1}{2}\, {}_{0}u_{k,i}\, {}_{0}u_{k,j}$$

3. *Equation of Motion with Incremental Decompositions*

Noting that $\delta\, {}^{t+\Delta t}_{0}\epsilon_{ij} = \delta\, {}_{0}\epsilon_{ij}$ the equation of motion is:

$$\int_{^0V} {}_{0}S_{ij}\, \delta\, {}_{0}\epsilon_{ij}\, {}^0dV + \int_{^0V} {}^{t}_{0}S_{ij}\, \delta\, {}_{0}\eta_{ij}\, {}^0dV = {}^{t+\Delta t}\mathcal{R} - \int_{^0V} {}^{t}_{0}S_{ij}\, \delta\, {}_{0}e_{ij}\, {}^0dV$$

4. *Linearization of Equation of Motion*

Using the approximations ${}_{0}S_{ij} = {}_{0}C_{ijrs}\, {}_{0}e_{rs}$, $\delta\, {}_{0}\epsilon_{ij} = \delta\, {}_{0}e_{ij}$
we obtain as approximate equation of motion:

$$\int_{^0V} {}_{0}C_{ijrs}\, {}_{0}e_{rs}\, \delta\, {}_{0}e_{ij}\, {}^0dV + \int_{^0V} {}^{t}_{0}S_{ij}\, \delta\, {}_{0}\eta_{ij}\, {}^0dV = {}^{t+\Delta t}\mathcal{R} - \int_{^0V} {}^{t}_{0}S_{ij}\, \delta\, {}_{0}e_{ij}\, {}^0dV$$

TABLE 6.3 *Updated Lagrangian formulation.*

1. *Equation of Motion*

$$\int_{{}^tV} {}^{t+\Delta t}_{\ \ \ t}S_{ij}\, \delta\, {}^{t+\Delta t}_{\ \ \ t}\epsilon_{ij}\, {}^t dV = {}^{t+\Delta t}\mathcal{R}$$

where

$${}^{t+\Delta t}_{\ \ \ t}S_{ij} = \frac{{}^t\rho}{{}^{t+\Delta t}\rho}\, {}^{\ \ \ \ \ \ t}_{t+\Delta t}x_{i,m}\, {}^{t+\Delta t}\tau_{mn}\, {}^{\ \ \ \ \ \ t}_{t+\Delta t}x_{j,n}; \qquad \delta\, {}^{t+\Delta t}_{\ \ \ t}\epsilon_{ij} = \delta \tfrac{1}{2}({}_tu_{i,j} + {}_tu_{j,i} + {}_tu_{k,i}\, {}_tu_{k,j})$$

2. *Incremental Decompositions*

(a) Stresses

$${}^{t+\Delta t}_{\ \ \ t}S_{ij} = {}^t\tau_{ij} + {}_tS_{ij} \qquad (\text{note that } {}^t_tS_{ij} \equiv {}^t\tau_{ij})$$

(b) Strains

$${}^{t+\Delta t}_{\ \ \ t}\epsilon_{ij} = {}_t\epsilon_{ij}; \qquad {}_t\epsilon_{ij} = {}_te_{ij} + {}_t\eta_{ij}$$

$${}_te_{ij} = \tfrac{1}{2}({}_tu_{i,j} + {}_tu_{j,i}); \qquad {}_t\eta_{ij} = \tfrac{1}{2}\, {}_tu_{k,i}\, {}_tu_{k,j}$$

3. *Equation of Motion with Incremental Decompositions*

The equation of motion is:

$$\int_{{}^tV} {}_tS_{ij}\, \delta_t\epsilon_{ij}\, {}^tdV + \int_{{}^tV} {}^t\tau_{ij}\, \delta_t\eta_{ij}\, {}^tdV = {}^{t+\Delta t}\mathcal{R} - \int_{{}^tV} {}^t\tau_{ij}\, \delta_te_{ij}\, {}^tdV$$

4. *Linearization of Equation of Motion*

Using the approximations ${}_tS_{ij} = {}_tC_{ijrs}\, {}_te_{rs}$, $\delta_t\epsilon_{ij} = \delta_te_{ij}$
we obtain as approximate equation of motion:

$$\int_{{}^tV} {}_tC_{ijrs}\, {}_te_{rs}\, \delta_te_{ij}\, {}^tdV + \int_{{}^tV} {}^t\tau_{ij}\, \delta_t\eta_{ij}\, {}^tdV = {}^{t+\Delta t}\mathcal{R} - \int_{{}^tV} {}^t\tau_{ij}\, \delta_te_{ij}\, {}^tdV$$

equilibrium equations are in the T.L. formulation

$$\int_{^0V} {}_0C_{ijrs}\, {}_0e_{rs}\, \delta_0 e_{ij}\, {}^0dV + \int_{^0V} {}_0^tS_{ij}\, \delta_0\eta_{ij}\, {}^0dV = {}^{t+\Delta t}\mathcal{R} - \int_{^0V} {}_0^tS_{ij}\, \delta_0 e_{ij}\, {}^0dV \tag{6.40}$$

and in the U.L. formulation

$$\int_{^tV} {}_tC_{ijrs}\, {}_te_{rs}\, \delta_t e_{ij}\, {}^tdV + \int_{^tV} {}^t\tau_{ij}\, \delta_t\eta_{ij}\, {}^tdV = {}^{t+\Delta t}\mathcal{R} - \int_{^tV} {}^t\tau_{ij}\, \delta_t e_{ij}\, {}^tdV \tag{6.41}$$

where ${}_0C_{ijrs}$ and ${}_tC_{ijrs}$ are the incremental material property tensors at time t referred to the configurations at times 0 and t, respectively. The derivation of ${}_0C_{ijrs}$ and ${}_tC_{ijrs}$ for various materials is discussed in Section 6.4. We may also note that in (6.40) and (6.41) ${}_0^tS_{ij}$ and ${}^t\tau_{ij}$ are the known 2nd Piola–Kirchhoff and Cauchy stresses at time t; and ${}_0e_{ij}$, ${}_0\eta_{ij}$ and ${}_te_{ij}$, ${}_t\eta_{ij}$ are the linear and non-linear incremental strains which are referred to the configurations at times 0 and t, respectively.

Comparing the U.L. and T.L. formulations in Tables 6.2 and 6.3 we observe once more that they are quite analogous and that, in fact, the only theoretical difference between the two formulations lies in the choice of different reference configurations for the kinematic and static variables. Indeed, if in the numerical solution the appropriate constitutive tensors are employed, identical results are obtained (see Section 6.4).

It should also be noted that the incremental decompositions of the stresses and strains used in the tables are only possible because in each solution approach all stresses and strains are referred to the same configuration.

The choice of using either the U.L. or the T.L. formulation in a finite element solution depends, in practice, on their relative numerical effectiveness, which in turn depends on the finite element and the constitutive law used (see Section 6.4). However, one general observation can be made considering Tables 6.2 and 6.3; namely, that the incremental linear strains ${}_0e_{ij}$ in the T.L. formulation contain an initial displacement effect that leads to a more complex strain-displacement matrix than in the U.L. formulation.

The linearized equations in (6.40) and (6.41) contain on the r.h.s. the external virtual work ${}^{t+\Delta t}\mathcal{R}$ defined in (6.14). This expression also depends in general on the surface area and the volume of the body under consideration. However, for simplicity of discussion we assume for the moment that the loading is deformation-independent, a very important form of such loading being concentrated forces whose directions and intensitites are independent of the structural response. Later we shall discuss how to include in the analysis deformation-dependent loading (see (6.49) to (6.50)).

The relations in (6.40) and (6.41) can be employed to calculate an increment in the displacements, which then is used to evaluate approximations to the displacements, strains, and stresses corresponding to time $t + \Delta t$. The displacement approximations corresponding to $t + \Delta t$ are simply obtained by adding the calculated increments to the displacements of time t, and the strain approximations are evaluated from the displacements using the available kinematic relations (e.g. relation (6.23) in the T.L. formulation). However, the evaluation of the stresses corresponding to time $t + \Delta t$ depends on the specific constitutive relations used and is discussed in detail in Section 6.4.

Assuming that the approximate displacements, strains and thus stresses have been obtained, we can now check into how much difference there is between the internal virtual work when evaluated with the calculated static and kinematic variables for time $t + \Delta t$ and the external virtual work. Denoting the approximate values with a superscript (1) in anticipation that an iteration will in general be necessary, the error due to linearization is, in the T.L. formulation,

$$\text{Error} = {}^{t+\Delta t}\mathcal{R} - \int_{{}^0V} {}^{t+\Delta t}_{\ \ 0}S_{ij}^{(1)}\ \delta^{t+\Delta t}_{\ \ 0}\epsilon_{ij}^{(1)}\ {}^0dV \tag{6.42}$$

and in the U.L. formulation,

$$\text{Error} = {}^{t+\Delta t}\mathcal{R} - \int_{{}^{t+\Delta t}V^{(1)}} {}^{t+\Delta t}\tau_{ij}^{(1)}\ \delta_{t+\Delta t}e_{ij}^{(1)}\ {}^{t+\Delta t}dV \tag{6.43}$$

We should note that the r.h.s. of (6.42) and (6.43) are equivalent to the r.h.s. of (6.40) and (6.41), respectively, but in each case the current configurations with the corresponding stress and strain variables are employed. The correspondence in the U.L. formulation can be seen directly, but considering the T.L. formulation it must be recognized that $\delta_0 e_{ij}$ is equivalent to $\delta^{t+\Delta t}_{\ \ 0}\epsilon_{ij}$ when the same current displacements are used.

The above considerations show that the r.h.s. in (6.40) and (6.41) represent an "out-of-balance virtual work" prior to the calculation of the increments in the displacements, whereas the r.h.s. of (6.42) and (6.43) represent the "out-of-balance virtual work" after the solution, as the result of the linearizations performed. In order to further reduce the "out-of-balance virtual work" we need to perform an iteration in which the above solution step is repeated until the difference between the external virtual work and the internal virtual work is negligible within a certain convergence measure. Using the T.L. formulation the equation solved repetitively, for $k = 1, 2, 3, \ldots,$ is

$$\int_{{}^0V} {}_0C_{ijrs}\,\Delta_0 e_{rs}^{(k)}\ \delta_0 e_{ij}\ {}^0dV + \int_{{}^0V} {}^t_0S_{ij}\ \delta\Delta_0\eta_{ij}^{(k)}\ {}^0dV = {}^{t+\Delta t}\mathcal{R} - \int_{{}^0V} {}^{t+\Delta t}_{\ \ 0}S_{ij}^{(k-1)}\ \delta^{t+\Delta t}_{\ \ 0}\epsilon_{ij}^{(k-1)}\ {}^0dV \tag{6.44}$$

and using the U.L. formulation the equation considered is

$$\int_{{}^tV} {}_tC_{ijrs}\,\Delta_t e_{rs}^{(k)}\ \delta_t e_{ij}\ {}^tdV + \int_{{}^tV} {}^t\tau_{ij}\ \delta\Delta_t\eta_{ij}^{(k)}\ {}^tdV = {}^{t+\Delta t}\mathcal{R} - \int_{{}^{t+\Delta t}V^{(k-1)}} {}^{t+\Delta t}\tau_{ij}^{(k-1)}\ \delta_{t+\Delta t}e_{ij}^{(k-1)}\ {}^{t+\Delta t}dV \tag{6.45}$$

where the case $k = 1$ corresponds to the relations in (6.40) and (6.41), and the displacements are updated as follows

$${}^{t+\Delta t}u_i^{(k)} = {}^{t+\Delta t}u_i^{(k-1)} + \Delta u_i^{(k)}; \qquad {}^{t+\Delta t}u^{(0)} = {}^tu \tag{6.46}$$

The relations in (6.44) to (6.46) correspond to a modified Newton iteration already introduced in Section 6.1. The iteration is performed without updating the components of the constitutive and stress tensors on the l.h.s., which, as we shall see in the next section, corresponds to the use of a constant tangent stiffness matrix during the iteration.

In an overview of this section, we may note once more a very important point. Our objective is to solve the equilibrium relation in (6.13), which can be regarded as an extension of the virtual work principle used in linear analysis. We saw that for a general incremental analysis, certain stress and strain measures can be employed effectively, and this led to a transformation of (6.13) into an updated and a total Lagrangian form. The linearization of these equations then resulted in the relations (6.44) and (6.45). It is most important to recognize that the solution of either (6.44) or (6.45) corresponds entirely to the solution of the relation in (6.13). Namely, provided that the appropriate constitutive relations are employed, identical numerical results are obtained using either (6.44) or (6.45) for solution, and, as mentioned earlier, whether to use the T.L. or the U.L. formulation depends in practice only on the relative numerical effectiveness of the two solution approaches.

So far we have assumed that the loading is deformation-independent and can be specified prior to the incremental analysis. Thus, we assumed that the expression in (6.14) can be evaluated using

$$ {}^{t+\Delta t}\mathcal{R} = \int_{^0V} {}^{t+\Delta t}_{\ \ 0}f_i^B \, \delta u_i \, {}^0dV + \int_{^0S} {}^{t+\Delta t}_{\ \ 0}f_i^S \, \delta u_i^S \, {}^0dS \tag{6.47} $$

which is only possible for certain types of loading, such as concentrated loading that does not change direction as a function of the deformations. Using the displacement-based isoparametric elements, another important loading condition that can be modeled with (6.47) is the inertia force loading to be included in dynamic analysis. In this case we have

$$ \int_{^{t+\Delta t}V} {}^{t+\Delta t}\rho \; {}^{t+\Delta t}\ddot{u}_i \, \delta u_i \; {}^{t+\Delta t}dV = \int_{^0V} {}^0\rho \; {}^{t+\Delta t}\ddot{u}_i \, \delta u_i \; {}^0dV \tag{6.48} $$

and hence the mass matrix can be evaluated using the initial configuration of the body. The practical consequence is that in a dynamic analysis the mass matrices of isoparametric elements can be calculated prior to the step-by-step solution.

Assume now that the external virtual work is deformation-dependent and cannot be evaluated using (6.47). If in this case the load (or time) step is small enough, the external virtual work can frequently be approximated to sufficient accuracy using the intensity of loading corresponding to time $t + \Delta t$, but integrating over the volume and area last calculated in the iteration

$$ \int_{^{t+\Delta t}V} {}^{t+\Delta t}f_i^B \, \delta u_i \; {}^{t+\Delta t}dV \doteq \int_{^{t+\Delta t}V^{(k-1)}} {}^{t+\Delta t}f_i^B \, \delta u_i \; {}^{t+\Delta t}dV \tag{6.49} $$

and

$$ \int_{^{t+\Delta t}S} {}^{t+\Delta t}f_i^S \, \delta u_i^S \; {}^{t+\Delta t}dS \doteq \int_{^{t+\Delta t}S^{(k-1)}} {}^{t+\Delta t}f_i^S \, \delta u_i^S \; {}^{t+\Delta t}dS \tag{6.50} $$

In order to obtain a more accurate approximation, the unknown incremental displacements in (6.49) and (6.50) could be linearized, but this yields a non-symmetric contribution to the stiffness matrix, and is hardly effective computationally if the load steps are small and equilibrium iterations are employed.[9]

The total and updated Lagrangian formulations are incremental continuum mechanics equations that include all nonlinear effects due to large displacements, large strains, and material nonlinearities; however, in practice, it is often sufficient to account for nonlinear material effects only. In this case, the nonlinear strain components and any updating of surface areas and volumes are neglected in the formulations. Therefore, (6.44) and (6.45) reduce to the same equation of motion, namely,

$$\int_V C_{ijrs}\,\Delta e_{rs}^{(k)}\,\delta e_{ij}\,dV = {}^{t+\Delta t}\mathcal{R} - \int_V {}^{t+\Delta t}\sigma_{ij}^{(k-1)}\,\delta e_{ij}\,dV \tag{6.51}$$

where ${}^{t+\Delta t}\sigma_{ij}^{(k-1)}$ is the actual physical stress at time $t + \Delta t$ and iteration $(k - 1)$. In this analysis we assume that the volume of the body does not change and therefore ${}^{t+\Delta t}_{\ 0}S_{ij} \equiv {}^{t+\Delta t}\tau_{ij} \equiv {}^{t+\Delta t}\sigma_{ij}$, and there can be no deformation-dependent loading. Since no kinematic nonlinearities are considered in (6.51), it also follows that if the material is linear elastic, the relation in (6.51) is identical to the principle of virtual work discussed in Section 4.2.1 and would lead to a linear finite element solution.

In the above formulations we assumed that the proposed iteration does converge, so that the incremental analysis can actually be carried out. We discuss this question in detail in Section 8.6. Furthermore, we assumed in the formulation that a static analysis is performed or a dynamic analysis is sought with an implicit time integration scheme (see Section 9.2). If a dynamic analysis is to be performed using an explicit time integration method the governing continuum mechanics equations are, using the T.L. formulation,

$$\int_{{}^0V} {}^{t}_{0}S_{ij}\,\delta\,{}^{t}_{0}\epsilon_{ij}\,{}^0dV = {}^{t}\mathcal{R} \tag{6.52}$$

implicit → equilibrium at t + Δt

using the U.L. formulation,

$$\int_{{}^tV} {}^{t}\tau_{ij}\,\delta_t e_{ij}\,{}^t dV = {}^{t}\mathcal{R} \tag{6.53}$$

explicit → equilibrium at t

and using the materially-nonlinear-only analysis,

$$\int_V {}^{t}\sigma_{ij}\,\delta e_{ij}\,dV = {}^{t}\mathcal{R} \tag{6.54}$$

where the stress and strain tensors are as defined previously, and equilibrium is considered at time t. In these analyses the external virtual work must include the inertia forces corresponding to time t, and the incremental solution corresponds to a marching-forward algorithm without equilibrium iterations. The details of the actual step-by-step solution are discussed in Section 9.2.1.

6.3 ISOPARAMETRIC FINITE ELEMENT DISCRETIZATION

In the previous section we presented the general incremental continuum mechanics equations that form the basis of general nonlinear displacement-based finite element analysis. We use these equations now to develop the governing finite element equations.

The basic steps in the derivation of the governing finite element equations are the same as those used in linear analysis: the selection of the interpolation functions and the interpolation of the element coordinates and displacements with these functions in the governing continuum mechanics equations. By invoking the principle of virtual displacements for each of the nodal point displacements in turn, the governing finite element equations are obtained. As in linear analysis, we only need to consider a single element of a specific type in this derivation, because the governing equilibrium equations of an assemblage of elements can be constructed using the direct stiffness procedure.

Considering the element coordinate and displacement interpolations, we should recognize that it is important to employ the same interpolations for the coordinates and displacements at any and all times during the motion of the element. Since the new element coordinates are obtained by adding the element displacements to the original coordinates, it follows that the use of the same interpolations for the displacements and coordinates represents a consistent solution approach, and means that all discussions on convergence requirements in Sections 4.2.5 and 5.5 are directly applicable to the incremental analysis. In particular, it is then assured that an assemblage of elements which are displacement-compatible across element boundaries in the original configuration will preserve this compatibility in all subsequent configurations.

In Section 6.2.3 we derived the governing continuum mechanics equations for use in a modified Newton iteration. Substituting now the element coordinate and displacement interpolations into these equations as we did in linear analysis, we obtain—for a single element or for an assemblage of elements—

in materially-nonlinear-only analysis:

static analysis:

$$ {}^{t}\mathbf{K}\,\Delta\mathbf{U}^{(i)} = {}^{t+\Delta t}\mathbf{R} - {}^{t+\Delta t}\mathbf{F}^{(i-1)} \tag{6.55}$$

dynamic analysis, implicit time integration:

$$ \mathbf{M}\,{}^{t+\Delta t}\ddot{\mathbf{U}}^{(i)} + {}^{t}\mathbf{K}\,\Delta\mathbf{U}^{(i)} = {}^{t+\Delta t}\mathbf{R} - {}^{t+\Delta t}\mathbf{F}^{(i-1)} \tag{6.56}$$

dynamic analysis, explicit time integration:

$$ \mathbf{M}\,{}^{t}\ddot{\mathbf{U}} = {}^{t}\mathbf{R} - {}^{t}\mathbf{F} \tag{6.57}$$

using the T.L. formulation:

static analysis:

$$ ({}_{0}^{t}\mathbf{K}_{L} + {}_{0}^{t}\mathbf{K}_{NL})\,\Delta\mathbf{U}^{(i)} = {}^{t+\Delta t}\mathbf{R} - {}_{0}^{t+\Delta t}\mathbf{F}^{(i-1)} \tag{6.58}$$

dynamic analysis, implicit time integration:

$$ \mathbf{M}\,{}^{t+\Delta t}\ddot{\mathbf{U}}^{(i)} + ({}_{0}^{t}\mathbf{K}_{L} + {}_{0}^{t}\mathbf{K}_{NL})\,\Delta\mathbf{U}^{(i)} = {}^{t+\Delta t}\mathbf{R} - {}_{0}^{t+\Delta t}\mathbf{F}^{(i-1)} \tag{6.59}$$

dynamic analysis, explicit time integration:

$$ \mathbf{M}\,{}^{t}\ddot{\mathbf{U}} = {}^{t}\mathbf{R} - {}_{0}^{t}\mathbf{F} \tag{6.60}$$

and using the U.L. formulation:

static analysis:

$$ ({}_{t}^{t}\mathbf{K}_{L} + {}_{t}^{t}\mathbf{K}_{NL})\,\Delta\mathbf{U}^{(i)} = {}^{t+\Delta t}\mathbf{R} - {}_{t+\Delta t}^{t+\Delta t}\mathbf{F}^{(i-1)} \tag{6.61}$$

dynamic analysis, implicit time integration:

$$ \mathbf{M}\,{}^{t+\Delta t}\ddot{\mathbf{U}}^{(i)} + ({}_{t}^{t}\mathbf{K}_{L} + {}_{t}^{t}\mathbf{K}_{NL})\,\Delta\mathbf{U}^{(i)} = {}^{t+\Delta t}\mathbf{R} - {}_{t+\Delta t}^{t+\Delta t}\mathbf{F}^{(i-1)} \tag{6.62}$$

dynamic analysis, explicit time integration:

$$ \mathbf{M}\,{}^{t}\ddot{\mathbf{U}} = {}^{t}\mathbf{R} - {}_{t}^{t}\mathbf{F} \tag{6.63}$$

where $\mathbf{M}$ = Time-independent mass matrix

$^{t}\mathbf{K}$ = Linear strain incremental stiffness matrix, not including the initial displacement effect

${}_0^t\mathbf{K}_L, {}_t^t\mathbf{K}_L$ = Linear strain incremental stiffness matrices

${}_0^t\mathbf{K}_{NL}, {}_t^t\mathbf{K}_{NL}$ = Nonlinear strain (geometric or initial stress) incremental stiffness matrices

$^{t+\Delta t}\mathbf{R}$ = Vector of externally applied nodal point loads at time $t+\Delta t$; this vector is also used at time t in explicit time integration

$^{t}\mathbf{F}, {}_0^t\mathbf{F}, {}_t^t\mathbf{F}$ = Vectors of nodal point forces equivalent to the element stresses at time t; these vectors are also employed corresponding to time $t+\Delta t$ and iteration $(i-1)$

$\Delta\mathbf{U}^{(i)}$ = Vector of increments in the nodal point displacements in iteration i, $^{t+\Delta t}\mathbf{U}^{(i)} = {}^{t+\Delta t}\mathbf{U}^{(i-1)} + \Delta\mathbf{U}^{(i)}$

$^{t}\ddot{\mathbf{U}}$ = Vector of nodal point accelerations at time t; this vector is also employed corresponding to time $t+\Delta t$ and iteration i.

In the above finite element discretization we have assumed that damping effects are negligible or can be modeled in the nonlinear constitutive relationships (for example, by use of a strain-rate dependent material law). We also assumed that the externally applied loads are deformation-independent, and thus the load vector corresponding to all load (or time) steps can be calculated prior to the incremental analysis. If the loads include deformation-dependent components, it is necessary to update and iterate on the load vector as discussed in Section 6.2.3.

The above finite element matrices are evaluated as in linear analysis. Table 6.4 summarizes—for a single element—the basic integrals considered and the corresponding matrix evaluations. It may be noted that in the table only the nodal point force vectors corresponding to the internal element stresses at time t are considered, since the vectors corresponding to time $t+\Delta t$ and iteration i are calculated in an analogous manner. The following notation is used for the calculation of the element matrices:

$\mathbf{H}^S, \mathbf{H}$ = Surface- and volume displacement interpolation matrices

${}^{t+\Delta t}_{\ 0}\mathbf{f}^S\ {}^{t+\Delta t}_{\ 0}\mathbf{f}^B$ = Vectors of surface and body forces defined per unit area and per unit volume of the element at time 0

$\mathbf{B}_L, {}_0^t\mathbf{B}_L, {}_t^t\mathbf{B}_L$ = Linear strain-displacement transformation matrices; $\mathbf{B}_L$ is equal to ${}_0^t\mathbf{B}_L$ when the initial displacement effect is neglected

${}_0^t\mathbf{B}_{NL}, {}_t^t\mathbf{B}_{NL}$ = Nonlinear strain-displacement transformation matrices

$\mathbf{C}$ = Stress-strain material property matrix (incremental or total)

${}_0\mathbf{C}, {}_t\mathbf{C}$ = Incremental stress-strain material property matrices

$^{t}\boldsymbol{\tau}, {}^{t}\hat{\boldsymbol{\tau}}$ = Matrix and vector of Cauchy stresses

${}_0^t\mathbf{S}, {}_0^t\hat{\mathbf{S}}$ = Matrix and vector of 2nd Piola–Kirchhoff stresses

$^{t}\hat{\boldsymbol{\Sigma}}$ = Vector of stresses in materially-nonlinear-only analysis

The above matrices used in Table 6.4 depend on the specific element considered. The displacement interpolation matrices are simply assembled as in linear analysis from the displacement interpolation functions. In the following sections we discuss the calculation of the strain-displacement transformation matrices and give the stress matrices and vectors pertaining to the elements that we considered earlier for linear analysis in Chapter 5. The discussion is abbreviated, because the basic numerical procedures employed in the calculation of

TABLE 6.4 *Finite element matrices.*

Analysis Type	*Integral*	*Matrix Evaluation*
In all analyses	$\int_{^0V} {}^0\rho\, {}^{t+\Delta t}\ddot{u}_i\, \delta u_i\, {}^0dV$	$\mathbf{M}\, {}^{t+\Delta t}\ddot{\hat{\mathbf{u}}} = {}^0\rho\left(\int_{^0V} \mathbf{H}^T \mathbf{H}\, {}^0dV\right) {}^{t+\Delta t}\ddot{\hat{\mathbf{u}}}$
	${}^{t+\Delta t}\mathcal{R} = \int_{^0S} {}^{t+\Delta t}_{\ 0}f_i^S\, \delta u_i^S\, {}^0dS + \int_{^0V} {}^{t+\Delta t}_{\ 0}f_i^B\, \delta u_i\, {}^0dV$	${}^{t+\Delta t}\mathbf{R} = \int_{^0S} \mathbf{H}^{S^T}\, {}^{t+\Delta t}_{\ 0}\mathbf{f}^S\, {}^0dS + \int_{^0V} \mathbf{H}^T\, {}^{t+\Delta t}_{\ 0}\mathbf{f}^B\, {}^0dV$
Materially-nonlinear-only	$\int_V C_{ijrs}\, e_{rs}\, \delta e_{ij}\, dV$	${}^t\mathbf{K}\hat{\mathbf{u}} = \left(\int_V \mathbf{B}_L^T \mathbf{C}\, \mathbf{B}_L\, dV\right)\hat{\mathbf{u}}$
	$\int_V {}^t\sigma_{ij}\, \delta e_{ij}\, dV$	${}^t\mathbf{F} = \int_V \mathbf{B}_L^T\, {}^t\hat{\mathbf{\Sigma}}\, dV$
Total Lagrangian formulation	$\int_{^0V} {}_0C_{ijrs}\, {}_0e_{rs}\, \delta_0 e_{ij}\, {}^0dV$	${}^t_0\mathbf{K}_L\, \hat{\mathbf{u}} = \left(\int_{^0V} {}^t_0\mathbf{B}_L^T\, {}_0\mathbf{C}\, {}^t_0\mathbf{B}_L\, {}^0dV\right)\hat{\mathbf{u}}$
	$\int_{^0V} {}^t_0S_{ij}\, \delta_0\eta_{ij}\, {}^0dV$	${}^t_0\mathbf{K}_{NL}\, \hat{\mathbf{u}} = \left(\int_{^0V} {}^t_0\mathbf{B}_{NL}^T\, {}^t_0\mathbf{S}\, {}^t_0\mathbf{B}_{NL}\, {}^0dV\right)\hat{\mathbf{u}}$
	$\int_{^0V} {}^t_0S_{ij}\, \delta_0 e_{ij}\, {}^0dV$	${}^t_0\mathbf{F} = \int_{^0V} {}^t_0\mathbf{B}_L^T\, {}^t_0\hat{\mathbf{S}}\, {}^0dV$
Updated Lagrangian formulation	$\int_{^tV} {}_tC_{ijrs}\, {}_te_{rs}\, \delta_t e_{ij}\, {}^tdV$	${}^t_t\mathbf{K}_L\, \hat{\mathbf{u}} = \left(\int_{^tV} {}^t_t\mathbf{B}_L^T\, {}_t\mathbf{C}\, {}^t_t\mathbf{B}_L\, {}^tdV\right)\hat{\mathbf{u}}$
	$\int_{^tV} {}^t\tau_{ij}\, \delta_t\eta_{ij}\, {}^tdV$	${}^t_t\mathbf{K}_{NL}\, \hat{\mathbf{u}} = \left(\int_{^tV} {}^t_t\mathbf{B}_{NL}^T\, {}^t\boldsymbol{\tau}\, {}^t_t\mathbf{B}_{NL}\, {}^tdV\right)\hat{\mathbf{u}}$
	$\int_{^tV} {}^t\tau_{ij}\, \delta_t e_{ij}\, {}^tdV$	${}^t_t\mathbf{F} = \int_{^tV} {}^t_t\mathbf{B}_L^T\, {}^t\hat{\boldsymbol{\tau}}\, {}^tdV$

the nonlinear finite element matrices are those that we have covered already. For example, we consider again variable-number-nodes elements whose interpolation functions were previously given. As before, the displacement interpolations and strain-displacement matrices are expressed in terms of the isoparametric coordinates. Thus, the integrations indicated in Table 6.4 are performed as explained in Section 5.7.

In the following discussion we consider only the U.L. and T.L. formulations, because the matrices of the materially-nonlinear-only analysis can directly be obtained from these formulations, and we are only concerned with the required kinematic expressions. The evaluation of the material property matrices of the elements depends on the material model used. These considerations are discussed in Section 6.4.

6.3.1 Truss and Cable Elements

As discussed previously in Section 4.2.3 a truss element is a structural member capable of transmitting stresses only in the direction normal to the cross-section. It is assumed that this normal stress is constant over the cross-sectional area, and that during deformation the area itself remains constant. This assumption means that only small strain conditions can be analyzed.

In the following we consider a truss element that has an arbitrary orientation in space and is described by two to four nodes, as shown in Fig. 6.13. The global coordinates of the nodal points of the element are at time 0, ${}^0x_1^k$, ${}^0x_2^k$, ${}^0x_3^k$ and at time t, ${}^tx_1^k$, ${}^tx_2^k$, ${}^tx_3^k$, where $k = 1, \ldots, N$, with N equal to the number of nodes ($2 \leq N \leq 4$). These nodal point coordinates are assumed to determine the spatial configuration of the truss at time 0 and time t using

$$ {}^0x_1(r) = \sum_{k=1}^{N} h_k\, {}^0x_1^k \qquad {}^0x_2(r) = \sum_{k=1}^{N} h_k\, {}^0x_2^k \qquad {}^0x_3(r) = \sum_{k=1}^{N} h_k\, {}^0x_3^k \tag{6.64} $$

and

$$ {}^tx_1(r) = \sum_{k=1}^{N} h_k\, {}^tx_1^k \qquad {}^tx_2(r) = \sum_{k=1}^{N} h_k\, {}^tx_2^k \qquad {}^tx_3(r) = \sum_{k=1}^{N} h_k\, {}^tx_3^k \tag{6.65} $$

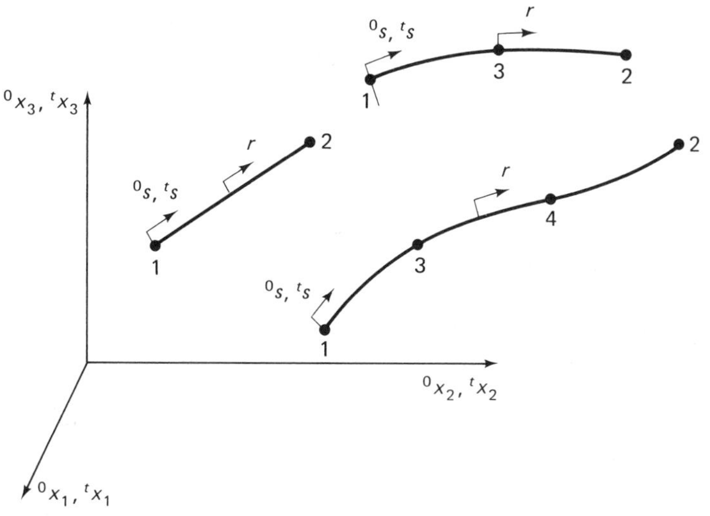

FIGURE 6.13 *Two- to four-node truss element.*

where the interpolation functions $h_k(r)$ have been defined in Fig. 5.4. Using (6.64) and (6.65) it follows that

$$^t u_i(r) = \sum_{k=1}^{N} h_k \, {}^t u_i^k \tag{6.66}$$

and

$$u_i(r) = \sum_{k=1}^{N} h_k \, u_i^k, \; i = 1, 2, 3 \tag{6.67}$$

Since for the truss element the only stress is the normal stress on its cross-sectional area, we only consider the corresponding longitudinal strain. Denoting the local element longitudinal strain by a curl, we have in the U.L. formulation,

$$_t\tilde{e}_{11} = \frac{d\,{}^t x_i}{{}^t ds} \frac{du_i}{{}^t ds} \qquad _t\tilde{\eta}_{11} = \frac{1}{2} \frac{du_k}{{}^t ds} \frac{du_k}{{}^t ds} \tag{6.68}$$

where ${}^t s(r)$ is the arc length at time t at the material point ${}^t x_1(r)$, ${}^t x_2(r)$, ${}^t x_3(r)$ given by

$$^t s(r) = \sum_{k=1}^{N} h_k \, {}^t s^k \tag{6.69}$$

To develop the strain-displacement matrices we define

$$^t\mathbf{x}^T = [{}^t x_1^1 \quad {}^t x_2^1 \quad {}^t x_3^1 \ldots {}^t x_1^N \quad {}^t x_2^N \quad {}^t x_3^N] \tag{6.70}$$

$$\hat{\mathbf{u}}^T = [u_1^1 \quad u_2^1 \quad u_3^1 \ldots u_1^N \quad u_2^N \quad u_3^N] \tag{6.71}$$

$$\mathbf{H} = [h_1 \mathbf{I}_3 \mid \ldots h_N \mathbf{I}_3]; \qquad \mathbf{I}_3 = \begin{bmatrix} 1 & 0 & 0 \\ 0 & 1 & 0 \\ 0 & 0 & 1 \end{bmatrix} \tag{6.72}$$

$$\mathbf{H}_{,r} = \left[\frac{dh_1}{dr} \mathbf{I}_3 \,\middle|\, \ldots \frac{dh_N}{dr} \mathbf{I}_3 \right] \tag{6.73}$$

and

$$^t J^{-1} = \frac{dr}{{}^t ds}$$

Hence

$$_t\tilde{e}_{11} = \frac{1}{{}^t J^2} \, {}^t\mathbf{x}^T \, \mathbf{H}_{,r}^T \, \mathbf{H}_{,r} \, \hat{\mathbf{u}}$$

so that

$$^t_t\mathbf{B}_L = \frac{1}{{}^t J^2} \, {}^t\mathbf{x}^T \, \mathbf{H}_{,r}^T \, \mathbf{H}_{,r} \tag{6.74}$$

Also,

$$2\,_t\tilde{\eta}_{11} = \left(\frac{1}{{}^t J} \mathbf{H}_{,r} \hat{\mathbf{u}} \right)^T \left(\frac{1}{{}^t J} \mathbf{H}_{,r} \hat{\mathbf{u}} \right)$$

so that

$$^t_t\mathbf{B}_{NL} = \frac{1}{{}^t J} \mathbf{H}_{,r} \tag{6.75}$$

The stress matrix and vector are simply the Cauchy stress ${}^t\tilde{\tau}_{11}$.

We should note that we obtain in this formulation the element matrices directly corresponding to the global displacement components (measured in the stationary coordinate system ${}^0x_1$, ${}^0x_2$, ${}^0x_3$). This is due to the fact that we are using in (6.70) and (6.71) the coordinates and displacements of the element in this coordinate system. Alternatively, and in particular for a straight two-node element, a local coordinate system aligned with the element could be employed to evaluate the element matrices, after which a transformation to the global coordinate system is necessary (see Section 4.2.1 and Example 6.15).

The strain-displacement matrices in (6.74) and (6.75) have been developed using the U.L. formulation. However, since we assume small strain-conditions during which the cross-sectional area, length, and hence the volume of the truss element remain constant during deformation, it follows that the strain-displacement matrices of the T.L. formulations are identically equal. This must be the case because the element stiffness matrices and force vectors (given in Table 6.4) of the two formulations must be equal. We demonstrate the evaluation of the element matrices and a simple application in the following examples.

EXAMPLE 6.15: Consider the two-node truss element shown in Fig. 6.14. Discuss the evaluation of the stiffness matrix and force vector of the element in the configuration at time t. Assume small strain but large rotation conditions.

Perhaps the most straightforward approach is to introduce a local body-attached coordinate system ${}^t\tilde{x}_1 - {}^t\tilde{x}_2$, evaluate the strain-displacement matrices in this coordinate system, and then transform the resulting matrices to obtain the stiffness matrix and force vector corresponding to the stationary system ${}^0x_1$, ${}^0x_2$.

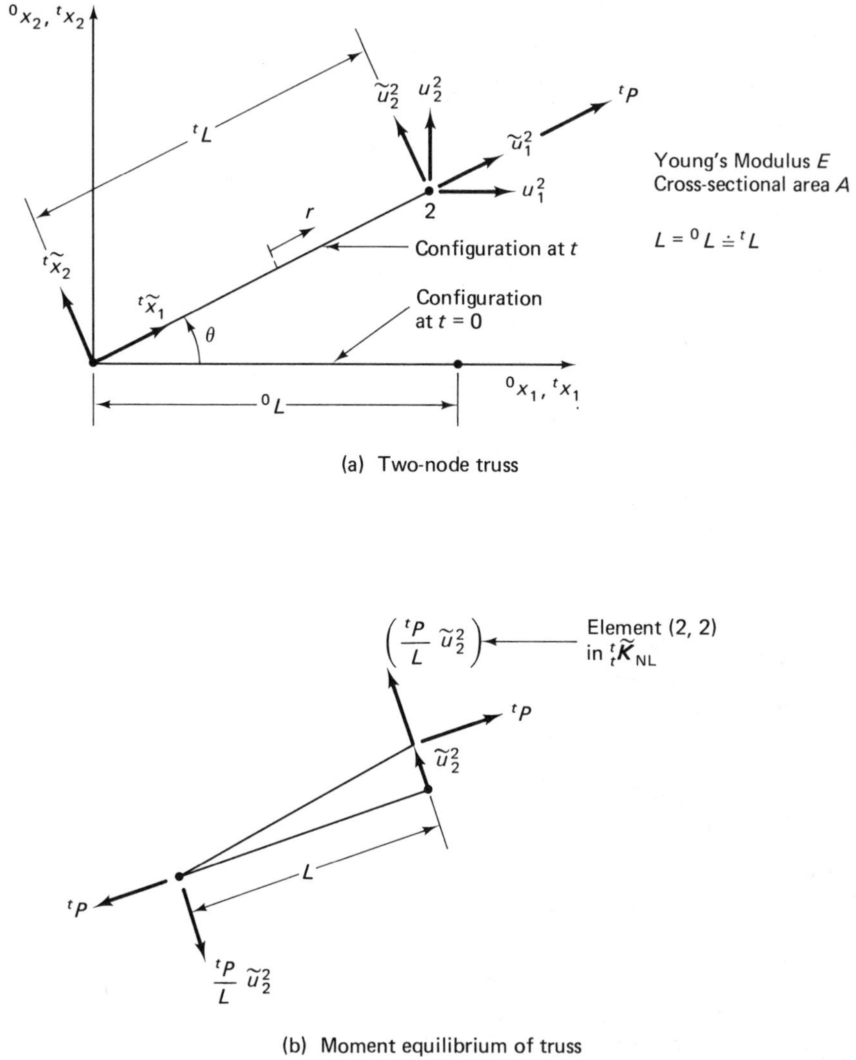

FIGURE 6.14 *Formulation of two-node truss element.*

In this case we apply (6.70) to (6.75) using the nodal point coordinates in the system ${}^t\tilde{x}_1$, ${}^t\tilde{x}_2$ and we obtain

$$
{}^t\tilde{\mathbf{x}}^T = [0 \quad 0 \quad L \quad 0] \tag{a}
$$

$$
\tilde{\mathbf{H}} = \frac{1}{2}\begin{bmatrix} 1-r & 0 & 1+r & 0 \\ 0 & 1-r & 0 & 1+r \end{bmatrix}
$$

$$
\tilde{\mathbf{H}}_{,r} = \frac{1}{2}\begin{bmatrix} -1 & 0 & 1 & 0 \\ 0 & -1 & 0 & 1 \end{bmatrix}
$$

$$
{}^tJ = \frac{L}{2}
$$

Hence
$$
{}^t_t\tilde{\mathbf{B}}_L = \frac{1}{L}[-1 \quad 0 \quad 1 \quad 0]
$$

and
$$
{}^t_t\tilde{\mathbf{B}}_{NL} = \frac{1}{L}\begin{bmatrix} -1 & 0 & 1 & 0 \\ 0 & -1 & 0 & 1 \end{bmatrix}
$$

But,
$$
{}^t_t\tilde{\mathbf{K}}_L = EA\int_{-1}^{+1} {}^t_t\tilde{\mathbf{B}}_L^T \, {}^t_t\tilde{\mathbf{B}}_L \, {}^tJ \, dr
$$

$$
{}^t_t\tilde{\mathbf{K}}_{NL} = {}^tP\int_{-1}^{+1} {}^t_t\tilde{\mathbf{B}}_{NL}^T \, {}^t_t\tilde{\mathbf{B}}_{NL} \, {}^tJ \, dr
$$

$$
{}^t_t\tilde{\mathbf{F}} = \int_{-1}^{+1} {}^t_t\tilde{\mathbf{B}}_L^T \, {}^tP \, {}^tJ \, dr
$$

so that
$$
{}^t_t\tilde{\mathbf{K}}_L = \frac{EA}{L}\begin{bmatrix} 1 & 0 & -1 & 0 \\ 0 & 0 & 0 & 0 \\ -1 & 0 & 1 & 0 \\ 0 & 0 & 0 & 0 \end{bmatrix} \tag{b}
$$

$$
{}^t_t\tilde{\mathbf{K}}_{NL} = \frac{{}^tP}{L}\begin{bmatrix} 1 & 0 & -1 & 0 \\ 0 & 1 & 0 & -1 \\ -1 & 0 & 1 & 0 \\ 0 & -1 & 0 & 1 \end{bmatrix} \tag{c}
$$

$$
{}^t_t\tilde{\mathbf{F}} = {}^tP\begin{bmatrix} -1 \\ 0 \\ 1 \\ 0 \end{bmatrix} \tag{d}
$$

We may note that the linear strain stiffness matrix in (b) is simply the stiffness matrix of a linear element in the coordinate system ${}^t\tilde{x}_1$, ${}^t\tilde{x}_2$. This matrix does not define a stiffness corresponding to the transverse incremental displacements $\tilde{u}_2^1$, $\tilde{u}_2^2$. On the other hand, such stiffness is provided by the nonlinear strain stiffness matrix ${}^t_t\tilde{\mathbf{K}}_{NL}$.

Considering the elements in ${}^t_t\tilde{\mathbf{K}}_{NL}$ we note that usually $|{}^tP| \ll EA$, so that a reasonable approximation in the equilibrium equations (6.61) is

$$
{}^t_t\tilde{\mathbf{K}}_{NL} \doteq \frac{{}^tP}{L}\begin{bmatrix} 0 & 0 & 0 & 0 \\ 0 & 1 & 0 & -1 \\ 0 & 0 & 0 & 0 \\ 0 & -1 & 0 & 1 \end{bmatrix}
$$

The entries in this matrix can be established by simple equilibrium considerations, as demonstrated in Fig. 6.14(b). Similarly, the force vector ${}^t_t\tilde{\mathbf{F}}$ in (d) can be directly obtained from basic equilibrium considerations.

To obtain from the expressions in (b), (c), and (d) the matrices corresponding to the stationary axes we use the transformations (see Section 4.2, relations (4.37))

$$ {}^t_t\mathbf{K}_L = \mathbf{T}^T\, {}^t_t\tilde{\mathbf{K}}_L\, \mathbf{T}; \qquad {}^t_t\mathbf{K}_{NL} = \mathbf{T}^T\, {}^t_t\tilde{\mathbf{K}}_{NL}\, \mathbf{T}; \qquad {}^t_t\mathbf{F} = \mathbf{T}^T\, {}^t_t\tilde{\mathbf{F}} \tag{e} $$

where

$$ \mathbf{T} = \begin{bmatrix} \cos\theta & \sin\theta & 0 & 0 \\ -\sin\theta & \cos\theta & 0 & 0 \\ 0 & 0 & \cos\theta & \sin\theta \\ 0 & 0 & -\sin\theta & \cos\theta \end{bmatrix} $$

It is interesting to note that because of the nature of the nonlinear strain contribution,

$$ {}^t_t\tilde{\mathbf{K}}_{NL} = {}^t_t\mathbf{K}_{NL} $$

which means that for the truss element the nonlinear strain stiffness matrix in the stationary coordinate system ${}^0x_1$, ${}^0x_2$ is independent of the orientation of the element.

A second approach to evaluate the required matrices is to use (6.70) to (6.75) with the element coordinates in the stationary coordinate system. Thus, instead of (a) we now use

$$ {}^t\mathbf{x}^T = L[0 \quad 0 \quad \cos\theta \quad \sin\theta]; \qquad \mathbf{H} = \tilde{\mathbf{H}} $$

and ${}^tJ$ is unchanged. Therefore we obtain using (6.74) and (6.75)

$$ {}^t_t\mathbf{B}_L = \frac{1}{L}[-\cos\theta \quad -\sin\theta \quad \cos\theta \quad \sin\theta] \tag{f} $$

$$ {}^t_t\mathbf{B}_{NL} = \frac{1}{L}\begin{bmatrix} -1 & 0 & 1 & 0 \\ 0 & -1 & 0 & 1 \end{bmatrix} $$

Hence, as expected, the evaluation of ${}^t_t\mathbf{K}_L$, ${}^t_t\mathbf{K}_{NL}$ and ${}^t_t\mathbf{F}$ gives the same result as obtained in the above approach, and we note that the ${}^t_t\mathbf{B}_L$ matrix in (f) already incorporates the effect of the transformation matrix $\mathbf{T}$ used in (e).

Finally, it is instructive to study the use of the total Lagrangian formulation. In this case we need to express the strains ${}_0e_{11}$ and ${}_0\eta_{11}$ given in Table 6.2 in terms of the element displacement interpolation functions. Since the truss undergoes displacements only in the ${}^0x_1 - {}^0x_2$ plane we have

$$ {}_0e_{11} = \frac{\partial u_1}{\partial\, {}^0x_1} + \frac{\partial\, {}^tu_1}{\partial\, {}^0x_1}\frac{\partial u_1}{\partial\, {}^0x_1} + \frac{\partial\, {}^tu_2}{\partial\, {}^0x_1}\frac{\partial u_2}{\partial\, {}^0x_1} $$

and

$$ {}_0\eta_{11} = \frac{1}{2}\left\{\left(\frac{\partial u_1}{\partial\, {}^0x_1}\right)^2 + \left(\frac{\partial u_2}{\partial\, {}^0x_1}\right)^2\right\} \tag{g} $$

But using ${}^tu_i = \sum_{k=1}^{2} h_k\, {}^tu_i^k$ we obtain

$$ \frac{\partial\, {}^tu_1}{\partial\, {}^0x_1} = \cos\theta - 1; \qquad \frac{\partial\, {}^tu_2}{\partial\, {}^0x_1} = \sin\theta $$

Since we assume that the length and area of the element do not change we also have ${}^0J = \frac{L}{2}$, and hence

$$ {}_0e_{11} = \frac{1}{L}\left\{[-1 \quad 0 \quad 1 \quad 0] - (1 - \cos\theta)[-1 \quad 0 \quad 1 \quad 0] + \sin\theta\,[0 \quad -1 \quad 0 \quad 1]\right\}\begin{bmatrix} u_1^1 \\ u_2^1 \\ u_1^2 \\ u_2^2 \end{bmatrix} $$

We may note that in this case (see Table 6.2)

$$ {}_0^t\mathbf{B}_{L0} = \frac{1}{L}[-1 \quad 0 \quad 1 \quad 0] $$

and the strain-displacement matrix due to the initial displacement effect is

$$ {}_0^t\mathbf{B}_{L1} = \frac{1}{L}[(1 - \cos\theta) \quad -\sin\theta \quad -(1 - \cos\theta) \quad \sin\theta] \tag{h} $$

The complete strain displacement matrix is

$$ {}_0^t\mathbf{B}_L = \frac{1}{L}[-\cos\theta \quad -\sin\theta \quad \cos\theta \quad \sin\theta] $$

which is the same as in (f). Considering the above derivation we note that the initial displacement effect expressed in (h) accounts for the transformation matrix used in (e).

The nonlinear strain-displacement matrix is obtained from (g) in the same way as (6.75) was derived:

$$ {}_0^t\mathbf{B}_{NL} = \frac{1}{{}_0J}\mathbf{H}_{,r}; \qquad \mathbf{H} = \tilde{\mathbf{H}} $$

It follows that the stiffness matrix and force vector calculated in the T.L. formulation are exactly the same as have been obtained in the U.L. formulation.

EXAMPLE 6.16: Establish the equilibrium equations used in the nonlinear analysis of the simple arch structure considered in Example 6.3 when the modified Newton–Raphson iteration is used for solution (see (6.10)).

As in Example 6.3 we idealize the structure using one truss element, see Fig. 6.4(b). Since the displacements at node 1 are zero we only need to consider the displacements at node 2. Using the derivations given in Example 6.15, we thus have

$$ {}_t^t\tilde{\mathbf{K}}_L = \frac{EA}{L}\begin{bmatrix} 1 & 0 \\ 0 & 0 \end{bmatrix} $$

$$ {}_t^t\tilde{\mathbf{K}}_{NL} = \frac{{}^tP}{L}\begin{bmatrix} 1 & 0 \\ 0 & 1 \end{bmatrix} $$

$$ {}_t^t\tilde{\mathbf{F}} = {}^tP\begin{bmatrix} 1 \\ 0 \end{bmatrix} $$

and then using (e) of Example 6.15,

$$ {}_t^t\mathbf{K}_L = \frac{EA}{L}\begin{bmatrix} (\cos{}^t\beta)^2 & \sin{}^t\beta\cos{}^t\beta \\ \sin{}^t\beta\cos{}^t\beta & (\sin{}^t\beta)^2 \end{bmatrix} \tag{a} $$

$$ {}_t^t\mathbf{K}_{NL} = \frac{{}^tP}{L}\begin{bmatrix} 1 & 0 \\ 0 & 1 \end{bmatrix} \tag{b} $$

$$ {}_t^t\mathbf{F} = {}^tP\begin{bmatrix} \cos{}^t\beta \\ \sin{}^t\beta \end{bmatrix} \tag{c} $$

We may point out once more that we assumed in Example 6.3 and in the above derivations that $k = EA/L$ is constant throughout the response.

The matrices in (a) to (c) above correspond to the global displacements ${}^tu_1$ and ${}^tu_2$ at node 2. However ${}^tu_1$ is zero, hence the governing equilibrium equation corresponding to (6.10) is

$$ \left[\frac{EA}{L}(\sin{}^t\beta)^2 + \frac{{}^tP}{L}\right]\Delta u_2^{(i)} = -\frac{{}^{t+\Delta t}R}{2} - {}^{t+\Delta t}P^{(i-1)}\sin\left({}^{t+\Delta t}\beta^{(i-1)}\right) $$

where ${}^{t+\Delta t}R/2$ is positive as shown in Fig. 6.4(b), and ${}^{t+\Delta t}P^{(i-1)}$ is the force in the bar (tensile force being positive) corresponding to the displacements at time $t+\Delta t$ and iteration $(i - 1)$.

6.3.2 Two-Dimensional Axisymmetric, Plane Strain and Plane Stress Elements

For the derivation of the required matrices and vectors, we consider a typical two-dimensional element in its configuration at time 0 and at time t, as illustrated for a nine-node element in Fig. 6.15. The global coordinates of the nodal points of the element are at time 0, ${}^0x_1^k$, ${}^0x_2^k$, and at time t, ${}^tx_1^k$, ${}^tx_2^k$, where $k = 1, 2, \ldots, N$, and N denotes the total number of element nodes, as discussed in Section 5.3.

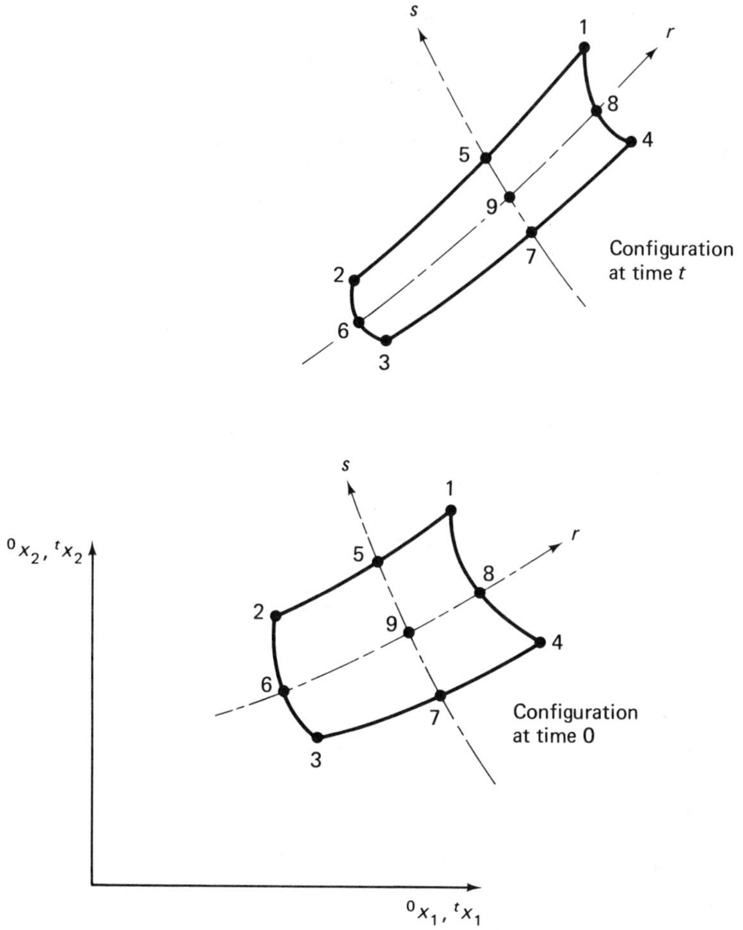

FIGURE 6.15 *Two-dimensional element shown in the global ${}^tx_1-{}^tx_2$ plane.*

Using the interpolation concepts discussed in Section 5.3, we have at time 0,

$$ {}^0x_1 = \sum_{k=1}^{N} h_k\, {}^0x_1^k; \qquad {}^0x_2 = \sum_{k=1}^{N} h_k\, {}^0x_2^k \tag{6.76}$$

and at time t $\qquad {}^t x_1 = \sum_{k=1}^{N} h_k \, {}^t x_1^k; \qquad {}^t x_2 = \sum_{k=1}^{N} h_k \, {}^t x_2^k \qquad (6.77)$

where the h_k are the interpolation functions presented in Fig. 5.5.

Since we use the isoparametric finite element discretization, the element displacements are interpolated in the same way as the geometry, i.e.,

$$ {}^t u_1 = \sum_{k=1}^{N} h_k \, {}^t u_1^k; \qquad {}^t u_2 = \sum_{k=1}^{N} h_k \, {}^t u_2^k \tag{6.78}$$

$$ u_1 = \sum_{k=1}^{N} h_k \, u_1^k; \qquad u_2 = \sum_{k=1}^{N} h_k \, u_2^k \tag{6.79}$$

The evaluation of strains requires the following derivatives:

$$\frac{\partial\, {}^t u_i}{\partial\, {}^0 x_j} = \sum_{k=1}^{N} \left(\frac{\partial h_k}{\partial\, {}^0 x_j}\right) {}^t u_i^k \tag{6.80}$$

$$\frac{\partial u_i}{\partial\, {}^0 x_j} = \sum_{k=1}^{N} \left(\frac{\partial h_k}{\partial\, {}^0 x_j}\right) u_i^k \qquad \begin{matrix} i = 1, 2 \\ j = 1, 2 \end{matrix} \tag{6.81}$$

$$\frac{\partial u_i}{\partial\, {}^t x_j} = \sum_{k=1}^{N} \left(\frac{\partial h_k}{\partial\, {}^t x_j}\right) u_i^k \tag{6.82}$$

These derivatives are calculated in the same way as in linear analysis, i.e., using a Jacobian transformation. As an example, consider briefly the evaluation of the derivatives in (6.82). The other derivatives are obtained in an analogous manner.

The chain rule relating ${}^t x_1$, ${}^t x_2$, to r, s derivatives is written as

$$\begin{bmatrix} \dfrac{\partial}{\partial r} \\[2ex] \dfrac{\partial}{\partial s} \end{bmatrix} = {}^t\mathbf{J} \begin{bmatrix} \dfrac{\partial}{\partial\, {}^t x_1} \\[2ex] \dfrac{\partial}{\partial\, {}^t x_2} \end{bmatrix}$$

in which

$${}^t\mathbf{J} = \begin{bmatrix} \dfrac{\partial\, {}^t x_1}{\partial r} & \dfrac{\partial\, {}^t x_2}{\partial r} \\[2ex] \dfrac{\partial\, {}^t x_1}{\partial s} & \dfrac{\partial\, {}^t x_2}{\partial s} \end{bmatrix}$$

Inverting the Jacobian operator $\mathbf{J}$, we obtain

$$\begin{bmatrix} \dfrac{\partial}{\partial\, {}^t x_1} \\[2ex] \dfrac{\partial}{\partial\, {}^t x_2} \end{bmatrix} = \frac{1}{\det {}^t\mathbf{J}} \begin{bmatrix} \dfrac{\partial\, {}^t x_2}{\partial s} & -\dfrac{\partial\, {}^t x_2}{\partial r} \\[2ex] -\dfrac{\partial\, {}^t x_1}{\partial s} & \dfrac{\partial\, {}^t x_1}{\partial r} \end{bmatrix} \begin{bmatrix} \dfrac{\partial}{\partial r} \\[2ex] \dfrac{\partial}{\partial s} \end{bmatrix}$$

where the Jacobian determinant is

$$\det {}^t\mathbf{J} = \frac{\partial\, {}^t x_1}{\partial r}\frac{\partial\, {}^t x_2}{\partial s} - \frac{\partial\, {}^t x_1}{\partial s}\frac{\partial\, {}^t x_2}{\partial r}$$

and the derivatives of the coordinates with respect to r and s are obtained as usual using (6.77), e.g.

$$\frac{\partial\, {}^t x_1}{\partial r} = \sum_{k=1}^{N} \frac{\partial h_k}{\partial r} \, {}^t x_1^k$$

With all required derivatives being defined it is now possible to establish the strain-displacement transformation matrices for the elements. Table 6.5 gives the required matrices for the U.L. and T.L. formulations. In the numerical

TABLE 6.5 *Matrices used in the two-dimensional element formulations.*

A. *Total Lagrangian Formulation*

1. *Incremental Strains*

$$
\begin{aligned}
{}_0\epsilon_{11} &= {}_0u_{1,1} + {}^t_0u_{1,1}\,{}_0u_{1,1} + {}^t_0u_{2,1}\,{}_0u_{2,1} + \tfrac{1}{2}(({}_0u_{1,1})^2 + ({}_0u_{2,1})^2)\\
{}_0\epsilon_{22} &= {}_0u_{2,2} + {}^t_0u_{1,2}\,{}_0u_{1,2} + {}^t_0u_{2,2}\,{}_0u_{2,2} + \tfrac{1}{2}(({}_0u_{1,2})^2 + ({}_0u_{2,2})^2)\\
{}_0\epsilon_{12} &= \tfrac{1}{2}({}_0u_{1,2} + {}_0u_{2,1}) + \tfrac{1}{2}({}^t_0u_{1,1}\,{}_0u_{1,2} + {}^t_0u_{2,1}\,{}_0u_{2,2} + {}^t_0u_{1,2}\,{}_0u_{1,1} + {}^t_0u_{2,2}\,{}_0u_{2,1}) + \tfrac{1}{2}({}_0u_{1,1}\,{}_0u_{1,2} + {}_0u_{2,1}\,u_{2,2})\\
{}_0\epsilon_{33} &= \frac{u_1}{{}^0x_1} + \frac{{}^tu_1u_1}{({}^0x_1)^2} + \frac{1}{2}\left(\frac{u_1}{{}^0x_1}\right)^2 \qquad \text{(for axisymmetric analysis)}
\end{aligned}
$$

where $\displaystyle {}_0u_{i,j} = \frac{\partial u_i}{\partial\, {}^0x_j}; \qquad {}^t_0u_{i,j} = \frac{\partial\, {}^tu_i}{\partial\, {}^0x_j}$

2. *Linear Strain-Displacement Transformation Matrix*

Using ${}_0\mathbf{e} = {}^t_0\mathbf{B}_L\,\hat{\mathbf{u}}$

where ${}_0\mathbf{e}^T = [{}_0e_{11} \quad {}_0e_{22} \quad 2{}_0e_{12} \quad {}_0e_{33}]; \qquad \hat{\mathbf{u}}^T = [u_1^1 \quad u_2^1 \quad u_1^2 \quad u_2^2 \quad \cdots \quad u_1^N \quad u_2^N]$

and ${}^t_0\mathbf{B}_L = {}^t_0\mathbf{B}_{L0} + {}^t_0\mathbf{B}_{L1}$

$$
{}^t_0\mathbf{B}_{L0} = \begin{bmatrix}
{}_0h_{1,1} & 0 & {}_0h_{2,1} & 0 & {}_0h_{3,1} & 0 & \cdots & {}_0h_{N,1} & 0\\
0 & {}_0h_{1,2} & 0 & {}_0h_{2,2} & 0 & {}_0h_{3,2} & \cdots & 0 & {}_0h_{N,2}\\
{}_0h_{1,2} & {}_0h_{1,1} & {}_0h_{2,2} & {}_0h_{2,1} & {}_0h_{3,2} & {}_0h_{3,1} & \cdots & {}_0h_{N,2} & {}_0h_{N,1}\\
\dfrac{h_1}{{}^0\bar{x}_1} & 0 & \dfrac{h_2}{{}^0\bar{x}_1} & 0 & \dfrac{h_3}{{}^0\bar{x}_1} & 0 & \cdots & \dfrac{h_N}{{}^0\bar{x}_1} & 0
\end{bmatrix}
$$

where $\displaystyle {}_0h_{k,j} = \frac{\partial h_k}{\partial\, {}^0x_j}; \qquad u_j^k = {}^{t+\Delta t}u_j^k - {}^tu_j^k; \qquad {}^0\bar{x}_1 = \sum_{k=1}^{N} h_k\, {}^0x_1^k;$

N = number of nodes

TABLE 6.5 *Matrices used in the two-dimensional element formulations. (continued)*

and

$$
{}_0^t\mathbf{B}_{L1} = \begin{bmatrix}
\ell_{11}\, {}_0h_{1,1} & \ell_{21}\, {}_0h_{1,1} & \ell_{11}\, {}_0h_{2,1} & \ell_{21}\, {}_0h_{2,1} & \cdots & \ell_{11}\, {}_0h_{N,1} & \ell_{21}\, {}_0h_{N,1} \\
\ell_{12}\, {}_0h_{1,2} & \ell_{22}\, {}_0h_{1,2} & \ell_{12}\, {}_0h_{2,2} & \ell_{22}\, {}_0h_{2,2} & \cdots & \ell_{12}\, {}_0h_{N,2} & \ell_{22}\, {}_0h_{N,2} \\
(\ell_{11}\, {}_0h_{1,2} + \ell_{12}\, {}_0h_{1,1}) & (\ell_{21}\, {}_0h_{1,2} + \ell_{22}\, {}_0h_{1,1}) & (\ell_{11}\, {}_0h_{2,2} + \ell_{12}\, h_{2,1}) & (\ell_{21}\, {}_0h_{2,2} + \ell_{22}\, {}_0h_{2,1}) & \cdots & (\ell_{11}\, {}_0h_{N,2} + \ell_{12}\, {}_0h_{N,1}) & (\ell_{21}\, {}_0h_{N,2} + \ell_{22}\, {}_0h_{N,1}) \\
\ell_{33}\dfrac{h_1}{{}^0\bar{x}_1} & 0 & \ell_{33}\dfrac{h_2}{{}^0\bar{x}_1} & 0 & & \ell_{33}\dfrac{h_N}{{}^0\bar{x}_1} & 0
\end{bmatrix}
$$

where $\ell_{11} = \sum_{k=1}^{N} {}_0h_{k,1}\, {}^tu_1^k; \quad \ell_{22} = \sum_{k=1}^{N} {}_0h_{k,2}\, {}^tu_2^k; \quad \ell_{21} = \sum_{k=1}^{N} {}_0h_{k,1}\, {}^tu_2^k; \quad \ell_{12} = \sum_{k=1}^{N} {}_0h_{k,2}\, {}^tu_1^k; \quad \ell_{33} = \dfrac{\sum_{k=1}^{N} h_k\, {}^tu_1^k}{{}^0\bar{x}_1}$

3. *Nonlinear Strain-Displacement Transformation Matrix*

$$
{}_0^t\mathbf{B}_{NL} = \begin{bmatrix}
{}_0h_{1,1} & 0 & {}_0h_{2,1} & 0 & {}_0h_{3,1} & 0 & \cdots & {}_0h_{N,1} & 0 \\
{}_0h_{1,2} & 0 & {}_0h_{2,2} & 0 & {}_0h_{3,2} & 0 & \cdots & {}_0h_{N,2} & 0 \\
0 & {}_0h_{1,1} & 0 & {}_0h_{2,1} & 0 & {}_0h_{3,1} & \cdots & 0 & {}_0h_{N,1} \\
0 & {}_0h_{1,2} & 0 & {}_0h_{2,2} & 0 & {}_0h_{3,2} & \cdots & 0 & {}_0h_{N,2} \\
\dfrac{h_1}{{}^0\bar{x}_1} & 0 & \dfrac{h_2}{{}^0\bar{x}_1} & 0 & \dfrac{h_3}{{}^0\bar{x}_1} & 0 & \cdots & \dfrac{h_N}{{}^0\bar{x}_1} & 0
\end{bmatrix}
$$

4. *2nd Piola–Kirchhoff Stress Matrix and Vector*

$$
{}_0^t\mathbf{S} = \begin{bmatrix}
{}_0^tS_{11} & {}_0^tS_{12} & 0 & 0 & 0 \\
{}_0^tS_{21} & {}_0^tS_{22} & 0 & 0 & 0 \\
0 & 0 & {}_0^tS_{11} & {}_0^tS_{12} & 0 \\
0 & 0 & {}_0^tS_{21} & {}_0^tS_{22} & 0 \\
0 & 0 & 0 & 0 & {}_0^tS_{33}
\end{bmatrix}; \qquad
{}_0^t\hat{\mathbf{S}} = \begin{bmatrix}
{}_0^tS_{11} \\ {}_0^tS_{22} \\ {}_0^tS_{12} \\ {}_0^tS_{33}
\end{bmatrix}
$$

TABLE 6.5 *Matrices used in the two-dimensional element formulations. (continued)*

B. *Updated Lagrangian Formulation*

1. *Incremental Strains*

$$
\begin{aligned}
{}_t\epsilon_{11} &= {}_tu_{1,1} + \tfrac{1}{2}(({}_tu_{1,1})^2 + ({}_tu_{2,1})^2)\\
{}_t\epsilon_{22} &= {}_tu_{2,2} + \tfrac{1}{2}(({}_tu_{1,2})^2 + ({}_tu_{2,2})^2)\\
{}_t\epsilon_{12} &= \tfrac{1}{2}({}_tu_{1,2} + {}_tu_{2,1}) + \tfrac{1}{2}({}_tu_{1,1}\,{}_tu_{1,2} + {}_tu_{2,1}\,{}_tu_{2,2})\\
{}_t\epsilon_{33} &= \frac{u_1}{{}^tx_1} + \frac{1}{2}\left(\frac{u_1}{{}^tx_1}\right)^2 \qquad \text{(for axisymmetric analysis)}
\end{aligned}
$$

where $\quad {}_tu_{i,j} = \dfrac{\partial u_i}{\partial\, {}^tx_j}$

2. *Linear Strain-Displacement Transformation Matrix*

Using $\quad {}_t\mathbf{e} = {}^t_t\mathbf{B}_L\,\hat{\mathbf{u}}$

where $\quad {}_t\mathbf{e}^T = [{}_te_{11} \quad {}_te_{22} \quad 2{}_te_{12} \quad {}_te_{33}]; \qquad \hat{\mathbf{u}}^T = [u_1^1 \quad u_2^1 \quad u_1^2 \quad u_2^2 \quad \cdots \quad u_1^N \quad u_2^N]$

$$
{}^t_t\mathbf{B}_L = \begin{bmatrix}
{}_th_{1,1} & 0 & {}_th_{2,1} & 0 & {}_th_{3,1} & 0 & \cdots & {}_th_{N,1} & 0\\
0 & {}_th_{1,2} & 0 & {}_th_{2,2} & 0 & {}_th_{3,2} & \cdots & 0 & {}_th_{N,2}\\
{}_th_{1,2} & {}_th_{1,1} & {}_th_{2,2} & {}_th_{2,1} & {}_th_{3,2} & {}_th_{3,1} & \cdots & {}_th_{N,2} & {}_th_{N,1}\\
\dfrac{h_1}{{}^t\bar{x}_1} & 0 & \dfrac{h_2}{{}^t\bar{x}_1} & 0 & \dfrac{h_3}{{}^t\bar{x}_1} & 0 & \cdots & \dfrac{h_N}{{}^t\bar{x}_1} & 0
\end{bmatrix}
$$

where $\quad {}_th_{k,j} = \dfrac{\partial h_k}{\partial\, {}^tx_j}; \qquad u_j^k = {}^{t+\Delta t}u_j^k - {}^tu_j^k; \qquad {}^t\bar{x}_1 = \sum_{k=1}^{N} h_k\, {}^tx_1^k; \qquad N = \text{Number of nodes}$

3. *Nonlinear Strain-Displacement Transformation Matrix*

$$
{}^t_t\mathbf{B}_{NL} = \begin{bmatrix}
{}_th_{1,1} & 0 & {}_th_{2,1} & 0 & {}_th_{3,1} & 0 & \cdots & {}_th_{N,1} & 0\\
{}_th_{1,2} & 0 & {}_th_{2,2} & 0 & {}_th_{3,2} & 0 & \cdots & {}_th_{N,2} & 0\\
0 & {}_th_{1,1} & 0 & {}_th_{2,1} & 0 & {}_th_{3,1} & \cdots & 0 & {}_th_{N,1}\\
0 & {}_th_{1,2} & 0 & {}_th_{2,2} & 0 & {}_th_{3,2} & \cdots & 0 & {}_th_{N,2}\\
\dfrac{h_1}{{}^t\bar{x}_1} & 0 & \dfrac{h_2}{{}^t\bar{x}_1} & 0 & \dfrac{h_3}{{}^t\bar{x}_1} & 0 & \cdots & \dfrac{h_N}{{}^t\bar{x}_1} & 0
\end{bmatrix}
$$

TABLE 6.5 *Matrices used in the two-dimensional element formulations. (continued)*

4. *Cauchy Stress Matrix and Stress Vector*

$$
{}^{t}\boldsymbol{\tau} = \begin{bmatrix} {}^{t}\tau_{11} & {}^{t}\tau_{12} & 0 & 0 & 0 \\ {}^{t}\tau_{21} & {}^{t}\tau_{22} & 0 & 0 & 0 \\ 0 & 0 & {}^{t}\tau_{11} & {}^{t}\tau_{12} & 0 \\ 0 & 0 & {}^{t}\tau_{21} & {}^{t}\tau_{22} & 0 \\ 0 & 0 & 0 & 0 & {}^{t}\tau_{33} \end{bmatrix}; \qquad {}^{t}\hat{\boldsymbol{\tau}} = \begin{bmatrix} {}^{t}\tau_{11} \\ {}^{t}\tau_{22} \\ {}^{t}\tau_{12} \\ {}^{t}\tau_{33} \end{bmatrix}
$$

integration these matrices are evaluated at the Gauss integration points (see Section 5.7.3)

As we pointed out earlier, the choice between the T.L. and U.L. formulations essentially depends on their relative numerical effectiveness. Table 6.5 shows that all matrices of the two formulations have corresponding patterns of zero elements, except that ${}_0^t\mathbf{B}_L$ is a full matrix whereas ${}_t^t\mathbf{B}_L$ is sparse. The strain-displacement transformation matrix ${}_0^t\mathbf{B}_L$ is full because of the initial displacement effect in the linear strain terms (see Tables 6.2 and 6.3). Therefore, the calculation of the matrix product ${}_t^t\mathbf{B}_L^T\,{}_t\mathbf{C}\,{}_t^t\mathbf{B}_L$ in the U.L. formulation requires less time than the matrix product ${}_0^t\mathbf{B}_L^T\,{}_0\mathbf{C}\,{}_0^t\mathbf{B}_L$ in the T.L. formulation.

The second numerical difference between the two formulations is that in the T.L. formulation all derivatives of interpolation functions are with respect to the initial coordinates, whereas in the U.L. formulation all derivatives are with respect to the coordinates at time t. Therefore, in the T.L. formulation the derivatives could be calculated only once in the first load step, and stored on back-up storage for use in all subsequent load steps. However, in practice, the use of tape or disc to store and retrieve the required derivatives is generally more costly than simply to recalculate them in each time step. In addition, the required back-up storage is a problem-size governing factor, because saturation of back-up storage may be reached. It is primarily for this reason that in a computer implementation the derivatives of the interpolation functions are in general best recalculated in each time step.

EXAMPLE 6.17: Establish the matrices ${}_0^t\mathbf{B}_{L0}$, ${}_0^t\mathbf{B}_{L1}$ and ${}_0^t\mathbf{B}_{NL}$ corresponding to the T.L. formulation for the two-dimensional plane strain element shown in Fig. 6.16.

In this case we can directly use the information given in Table 6.5 with

$$\begin{aligned} {}^tu_1^1 &= 1 & {}^tu_2^1 &= 0.5 \\ {}^tu_1^2 &= 0 & {}^tu_2^2 &= 0.5 \\ {}^tu_1^3 &= 0 & {}^tu_2^3 &= 0 \\ {}^tu_1^4 &= 1 & {}^tu_2^4 &= 0 \end{aligned} \qquad {}^0\mathbf{J} = \begin{bmatrix} \frac{3}{2} & 0 \\ 0 & \frac{3}{2} \end{bmatrix}$$

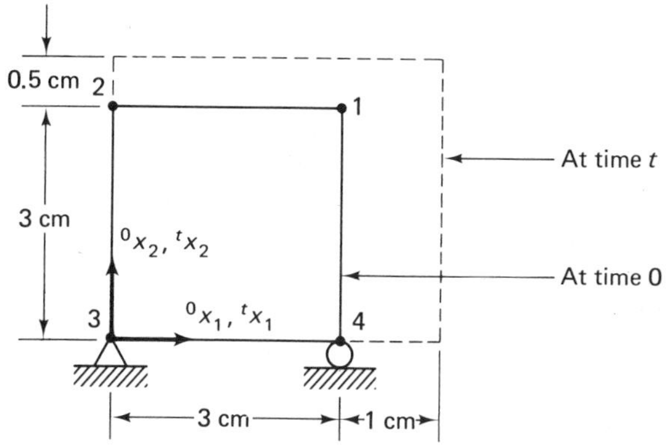

FIGURE 6.16 *Four-node plane strain element in large displacement/large strain conditions.*

The interpolation functions of the four-node element are given in Fig. 5.5 (and the required derivatives have been given in Example 5.5) so that we obtain

$$
{}_0^t\mathbf{B}_{L0} = \frac{1}{6}\begin{bmatrix} (1+s) & 0 & -(1+s) & 0 & -(1-s) & 0 & (1-s) & 0 \\ 0 & (1+r) & 0 & (1-r) & 0 & -(1-r) & 0 & -(1+r) \\ (1+r) & (1+s) & (1-r) & -(1+s) & -(1-r) & -(1-s) & -(1+r) & (1-s) \end{bmatrix}
$$

To evaluate ${}_0^t\mathbf{B}_{L1}$ we also need the l_{ij} values, where

$$
\ell_{11} = \sum_{k=1}^{4} {}_0h_{k,1}\, {}^tu_1^k = \tfrac{2}{3}\{h_{1,r}\, {}^tu_1^1 + h_{4,r}\, {}^tu_1^4\} = \tfrac{1}{3}
$$

$$
\ell_{12} = \sum_{k=1}^{4} {}_0h_{k,2}\, {}^tu_1^k = \tfrac{2}{3}\{h_{1,s}\, {}^tu_1^1 + h_{4,s}\, {}^tu_1^4\} = 0
$$

$$
\ell_{21} = \sum_{k=1}^{4} {}_0h_{k,1}\, {}^tu_2^k = \tfrac{2}{3}\{h_{1,r}\, {}^tu_2^1 + h_{2,r}\, {}^tu_2^2\} = 0
$$

$$
\ell_{22} = \sum_{k=1}^{4} {}_0h_{k,2}\, {}^tu_2^k = \tfrac{2}{3}\{h_{1,s}\, {}^tu_2^1 + h_{2,s}\, {}^tu_2^2\} = \tfrac{1}{6}
$$

Hence, we have

$$
{}_0^t\mathbf{B}_{L1} = \frac{1}{36}\begin{bmatrix} 2(1+s) & 0 & -2(1+s) & 0 & -2(1-s) & 0 & 2(1-s) & 0 \\ 0 & (1+r) & 0 & (1-r) & 0 & -(1-r) & 0 & -(1+r) \\ 2(1+r) & (1+s) & 2(1-r) & -(1+s) & -2(1-r) & -(1-s) & -2(1+r) & (1-s) \end{bmatrix}
$$

The nonlinear strain-displacement matrix can also directly be constructed using the derivatives of the interpolation functions and the Jacobian matrix:

$$
{}_0^t\mathbf{B}_{NL} = \frac{1}{6}\begin{bmatrix} (1+s) & 0 & -(1+s) & 0 & -(1-s) & 0 & (1-s) & 0 \\ (1+r) & 0 & (1-r) & 0 & -(1-r) & 0 & -(1+r) & 0 \\ 0 & (1+s) & 0 & -(1+s) & 0 & -(1-s) & 0 & (1-s) \\ 0 & (1+r) & 0 & (1-r) & 0 & -(1-r) & 0 & -(1+r) \end{bmatrix}
$$

6.3.3 Three-Dimensional Solid Elements

The evaluation of the matrices required in three-dimensional isoparametric finite element analysis is accomplished using in essence the same procedures as in two-dimensional analysis. Thus, referring to Section 6.3.2 we simply note that for a typical element we now use the coordinate and displacement interpolations,

$$
{}^0x_i = \sum_{k=1}^{N} h_k\, {}^0x_i^k; \qquad {}^tx_i = \sum_{k=1}^{N} h_k\, {}^tx_i^k \qquad i = 1, 2, 3 \tag{6.83}
$$

$$
{}^tu_i = \sum_{k=1}^{N} h_k\, {}^tu_i^k; \qquad u_i = \sum_{k=1}^{N} h_k\, u_i^k \qquad i = 1, 2, 3 \tag{6.84}
$$

where the element interpolation functions h_k have been given in Fig. 5.6. Using (6.83) and (6.84) in the same way as in two-dimensional analysis we can develop the relevant element matrices used in the T.L. and U.L. formulations for three-dimensional analysis. Table 6.6 summarizes these matrices. It is instructive to note the similarity between these matrices and those given in Table 6.5 for two-dimensional analysis.

6.3.4 Beam Elements

A large number of beam elements have been proposed for general nonlinear analysis. However, beam elements based on an isoparametric formulation can be particularly attractive because of the consistent formulation that is used, their generality in use, and in many cases, their computational efficiency.

TABLE 6.6 *Matrices used in the three-dimensional element formulations.*

A. *Total Lagrangian Formulation*

1. *Incremental Strains*

$$ {}_0\epsilon_{ij} = \tfrac{1}{2}({}_0u_{i,j} + {}_0u_{j,i}) + \tfrac{1}{2}({}_0^t u_{k,i}\, {}_0u_{k,j} + {}_0u_{k,i}\, {}_0^t u_{k,j}) + \tfrac{1}{2}({}_0u_{k,i}\, {}_0u_{k,j}) \qquad i = 1, 2, 3;\; j = 1, 2, 3;\; k = 1, 2, 3 $$

where $ {}_0u_{i,j} = \dfrac{\partial u_i}{\partial^0 x_j} $

2. *Linear Strain-Displacement Transformation Matrix*

Using $ {}_0\mathbf{e} = {}_0^t\mathbf{B}_L\, \hat{\mathbf{u}} $

where $ {}_0\mathbf{e}^T = [{}_0e_{11} \quad {}_0e_{22} \quad {}_0e_{33} \quad 2{}_0e_{12} \quad 2{}_0e_{23} \quad 2{}_0e_{13}]; $

$ \hat{\mathbf{u}}^T = [u_1^1\, u_2^1\, u_3^1 \quad u_1^2\, u_2^2\, u_3^2 \quad \cdots \quad u_1^N\, u_2^N\, u_3^N] $

$ {}_0^t\mathbf{B}_L = {}_0^t\mathbf{B}_{L0} + {}_0^t\mathbf{B}_{L1} $

$$ {}_0^t\mathbf{B}_{L0} = \begin{bmatrix} {}_0h_{1,1} & 0 & 0 & {}_0h_{2,1} & \cdots & 0 \\ 0 & {}_0h_{1,2} & 0 & 0 & \cdots & 0 \\ 0 & 0 & {}_0h_{1,3} & 0 & \cdots & {}_0h_{N,3} \\ {}_0h_{1,2} & {}_0h_{1,1} & 0 & {}_0h_{2,2} & \cdots & 0 \\ 0 & {}_0h_{1,3} & {}_0h_{1,2} & 0 & \cdots & {}_0h_{N,2} \\ {}_0h_{1,3} & 0 & {}_0h_{1,1} & {}_0h_{2,3} & \cdots & {}_0h_{N,1} \end{bmatrix} $$

where $ {}_0h_{k,j} = \dfrac{\partial h_k}{\partial^0 x_j}; \qquad u_j^k = {}^{t+\Delta t}u_j^k - {}^t u_j^k $

$$ {}_0^t\mathbf{B}_{L1} = \begin{bmatrix} \ell_{11}\, {}_0h_{1,1} & \ell_{21}\, {}_0h_{1,1} & \ell_{31}\, {}_0h_{1,1} & \ell_{11}\, {}_0h_{2,1} & \cdots & \ell_{31}\, {}_0h_{N,1} \\ \ell_{12}\, {}_0h_{1,2} & \ell_{22}\, {}_0h_{1,2} & \ell_{32}\, {}_0h_{1,2} & \ell_{12}\, {}_0h_{2,2} & \cdots & \ell_{32}\, {}_0h_{N,2} \\ \ell_{13}\, {}_0h_{1,3} & \ell_{23}\, {}_0h_{1,3} & \ell_{33}\, {}_0h_{1,3} & \ell_{13}\, {}_0h_{2,3} & \cdots & \ell_{33}\, {}_0h_{N,3} \\ (\ell_{11}\, {}_0h_{1,2} + \ell_{12}\, {}_0h_{1,1}) & (\ell_{21}\, {}_0h_{1,2} + \ell_{22}\, {}_0h_{1,1}) & (\ell_{31}\, {}_0h_{1,2} + \ell_{32}\, {}_0h_{1,1}) & (\ell_{11}\, {}_0h_{2,2} + \ell_{12}\, {}_0h_{2,1}) & \cdots & (\ell_{31}\, {}_0h_{N,2} + \ell_{32}\, {}_0h_{N,1}) \\ (\ell_{12}\, {}_0h_{1,3} + \ell_{13}\, {}_0h_{1,2}) & (\ell_{22}\, {}_0h_{1,3} + \ell_{23}\, {}_0h_{1,2}) & (\ell_{32}\, {}_0h_{1,3} + \ell_{33}\, {}_0h_{1,2}) & (\ell_{12}\, {}_0h_{2,3} + \ell_{1,3}\, {}_0h_{22}) & \cdots & (\ell_{32}\, {}_0h_{N,3} + \ell_{33}\, {}_0h_{N,2}) \\ (\ell_{11}\, {}_0h_{1,3} + \ell_{13}\, {}_0h_{1,1}) & (\ell_{21}\, {}_0h_{1,3} + \ell_{23}\, {}_0h_{1,1}) & (\ell_{31}\, {}_0h_{1,3} + \ell_{33}\, {}_0h_{1,1}) & (\ell_{11}\, {}_0h_{2,3} + \ell_{13}\, {}_0h_{2,1}) & \cdots & (\ell_{31}\, {}_0h_{N,3} + \ell_{33}\, {}_0h_{N,1}) \end{bmatrix} $$

where $ \ell_{ij} = \sum_{k=1}^{N} {}_0h_{k,j}\, {}^t u_i^k $

TABLE 6.6 *Matrices used in the three-dimensional element formulations. (continued)*

3. *Nonlinear Strain-Displacement Transformation Matrix*

$$ {}_0^t\mathbf{B}_{NL} = \begin{bmatrix} {}_0^t\tilde{\mathbf{B}}_{NL} & \tilde{\mathbf{0}} & \tilde{\mathbf{0}} \\ \tilde{\mathbf{0}} & {}_0^t\tilde{\mathbf{B}}_{NL} & \tilde{\mathbf{0}} \\ \tilde{\mathbf{0}} & \tilde{\mathbf{0}} & {}_0^t\tilde{\mathbf{B}}_{NL} \end{bmatrix}; \qquad \tilde{\mathbf{0}} = \begin{bmatrix} 0 \\ 0 \\ 0 \end{bmatrix} $$

where

$$ {}_0^t\tilde{\mathbf{B}}_{NL} = \begin{bmatrix} {}_0h_{1,1} & 0 & 0 & {}_0h_{2,1} & \cdots & {}_0h_{N,1} \\ {}_0h_{1,2} & 0 & 0 & {}_0h_{2,2} & \cdots & {}_0h_{N,2} \\ {}_0h_{1,3} & 0 & 0 & {}_0h_{2,3} & \cdots & {}_0h_{N,3} \end{bmatrix} $$

4. *2nd Piola–Kirchhoff Stress Matrix and Vector*

$$ {}_0^t\mathbf{S} = \begin{bmatrix} {}_0^t\tilde{\mathbf{S}} & \bar{\mathbf{0}} & \bar{\mathbf{0}} \\ \bar{\mathbf{0}} & {}_0^t\tilde{\mathbf{S}} & \bar{\mathbf{0}} \\ \bar{\mathbf{0}} & \bar{\mathbf{0}} & {}_0^t\tilde{\mathbf{S}} \end{bmatrix}; \qquad \bar{\mathbf{0}} = \begin{bmatrix} 0 & 0 & 0 \\ 0 & 0 & 0 \\ 0 & 0 & 0 \end{bmatrix} $$

$$ {}_0^t\hat{\mathbf{S}}^T = [{}_0^tS_{11} \quad {}_0^tS_{22} \quad {}_0^tS_{33} \quad {}_0^tS_{12} \quad {}_0^tS_{23} \quad {}_0^tS_{13}] $$

where

$$ {}_0^t\tilde{\mathbf{S}} = \begin{bmatrix} {}_0^tS_{11} & {}_0^tS_{12} & {}_0^tS_{13} \\ {}_0^tS_{21} & {}_0^tS_{22} & {}_0^tS_{23} \\ {}_0^tS_{31} & {}_0^tS_{32} & {}_0^tS_{33} \end{bmatrix} $$

B. *Updated Lagrangian Formulation*

1. *Incremental Strains*

$$ {}_t\epsilon_{11} = {}_tu_{1,1} + \tfrac{1}{2}(({}_tu_{1,1})^2 + ({}_tu_{2,1})^2 + ({}_tu_{3,1})^2) $$
$$ {}_t\epsilon_{22} = {}_tu_{2,2} + \tfrac{1}{2}(({}_tu_{1,2})^2 + ({}_tu_{2,2})^2 + ({}_tu_{3,2})^2) $$
$$ {}_t\epsilon_{33} = {}_tu_{3,3} + \tfrac{1}{2}(({}_tu_{1,3})^2 + ({}_tu_{2,3})^2 + ({}_tu_{3,3})^2) $$
$$ {}_t\epsilon_{12} = \tfrac{1}{2}({}_tu_{1,2} + {}_tu_{2,1}) + \tfrac{1}{2}({}_tu_{1,1}\,{}_tu_{1,2} + {}_tu_{2,1}\,{}_tu_{2,2} + {}_tu_{3,1}\,{}_tu_{3,2}) $$
$$ {}_t\epsilon_{23} = \tfrac{1}{2}({}_tu_{2,3} + {}_tu_{3,2}) + \tfrac{1}{2}({}_tu_{1,2}\,{}_tu_{1,3} + {}_tu_{2,2}\,{}_tu_{2,3} + {}_tu_{3,2}\,{}_tu_{3,3}) $$
$$ {}_t\epsilon_{13} = \tfrac{1}{2}({}_tu_{1,3} + {}_tu_{3,1}) + \tfrac{1}{2}({}_tu_{1,1}\,{}_tu_{1,3} + {}_tu_{2,1}\,{}_tu_{2,3} + {}_tu_{3,1}\,{}_tu_{3,3}) $$

where $\quad {}_tu_{i,j} = \dfrac{\partial u_i}{\partial\,{}^tx_j}$

TABLE 6.6 *Matrices used in the three-dimensional element formulations. (continued)*

2. *Linear Strain-Displacement Transformation Matrix*

Using $\quad {}_t\mathbf{e} = {}^t_t\mathbf{B}_L\,\hat{\mathbf{u}}$

where $\quad {}_t\mathbf{e}^T = [{}_te_{11} \quad {}_te_{22} \quad {}_te_{33} \quad 2{}_te_{12} \quad 2{}_te_{23} \quad 2{}_te_{13}]$

$\hat{\mathbf{u}}^T = [u_1^1 \quad u_2^1 \quad u_3^1 \quad u_1^2 \quad u_2^2 \quad u_3^2 \quad \cdots \quad u_1^N \quad u_2^N \quad u_3^N]$

$$
{}^t_t\mathbf{B}_L = \begin{bmatrix} {}_th_{1,1} & 0 & 0 & {}_th_{2,1} & \cdots & 0 \\ 0 & {}_th_{1,2} & 0 & 0 & \cdots & 0 \\ 0 & 0 & {}_th_{1,3} & 0 & \cdots & {}_th_{N,3} \\ {}_th_{1,2} & {}_th_{1,1} & 0 & {}_th_{2,2} & \cdots & 0 \\ 0 & {}_th_{1,3} & {}_th_{1,2} & 0 & \cdots & {}_th_{N,2} \\ {}_th_{1,3} & 0 & {}_th_{1,1} & {}_th_{2,3} & \cdots & {}_th_{N,1} \end{bmatrix}
$$

where $\quad {}_th_{k,j} = \dfrac{\partial h_k}{\partial\, {}^tx_j}; \qquad u_j^k = {}^{t+\Delta t}u_j^k - {}^tu_j^k; \qquad N = \text{Number of nodes}$

3. *Nonlinear Strain-Displacement Transformation Matrix*

$$
{}^t_t\mathbf{B}_{NL} = \begin{bmatrix} {}^t_t\tilde{\mathbf{B}}_{NL} & \tilde{\mathbf{0}} & \tilde{\mathbf{0}} \\ \tilde{\mathbf{0}} & {}^t_t\tilde{\mathbf{B}}_{NL} & \tilde{\mathbf{0}} \\ \tilde{\mathbf{0}} & \tilde{\mathbf{0}} & {}^t_t\tilde{\mathbf{B}}_{NL} \end{bmatrix} \qquad \tilde{\mathbf{0}} = \begin{bmatrix} 0 \\ 0 \\ 0 \end{bmatrix}
$$

where

$$
{}^t_t\tilde{\mathbf{B}}_{NL} = \begin{bmatrix} {}_th_{1,1} & 0 & 0 & {}_th_{2,1} & \cdots & {}_th_{N,1} \\ {}_th_{1,2} & 0 & 0 & {}_th_{2,2} & \cdots & {}_th_{N,2} \\ {}_th_{1,3} & 0 & 0 & {}_th_{2,3} & \cdots & {}_th_{N,3} \end{bmatrix}
$$

4. *Cauchy Stress Matrix and Stress Vector*

$$
{}^t\boldsymbol{\tau} = \begin{bmatrix} {}^t\bar{\boldsymbol{\tau}} & \bar{\mathbf{0}} & \bar{\mathbf{0}} \\ \bar{\mathbf{0}} & {}^t\bar{\boldsymbol{\tau}} & \bar{\mathbf{0}} \\ \bar{\mathbf{0}} & \bar{\mathbf{0}} & {}^t\bar{\boldsymbol{\tau}} \end{bmatrix} \qquad {}^t\hat{\boldsymbol{\tau}} = \begin{bmatrix} {}^t\tau_{11} \\ {}^t\tau_{22} \\ {}^t\tau_{33} \\ {}^t\tau_{12} \\ {}^t\tau_{23} \\ {}^t\tau_{13} \end{bmatrix} \qquad \bar{\mathbf{0}} = \begin{bmatrix} 0 & 0 & 0 \\ 0 & 0 & 0 \\ 0 & 0 & 0 \end{bmatrix}
$$

where

$$
{}^t\bar{\boldsymbol{\tau}} = \begin{bmatrix} {}^t\tau_{11} & {}^t\tau_{12} & {}^t\tau_{13} \\ {}^t\tau_{21} & {}^t\tau_{22} & {}^t\tau_{23} \\ {}^t\tau_{31} & {}^t\tau_{32} & {}^t\tau_{33} \end{bmatrix}
$$

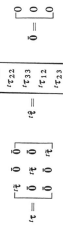

We consider here the calculation of the element matrices partaining to the large displacement/rotation behavior of the beam with rectangular cross-sectional area that we discussed already in Section 5.4.1 (see Fig. 5.33). Figure 6.17 shows a typical element in the original configuration and the position at time t. To describe the element behavior we use the same assumptions that we employed in the linear analysis, namely that plane sections normal to the neutral axis remain plane and that only the longitudinal stress and two shear stresses are nonzero, but the displacements and rotations of the element can now be arbitrarily large. The element strains are still assumed to be small, which means that the cross-sectional area does not change. This is an appropriate assumption for most geometrically nonlinear analyses of beam-type structures.

Using the general continuum mechanics equations for nonlinear analysis

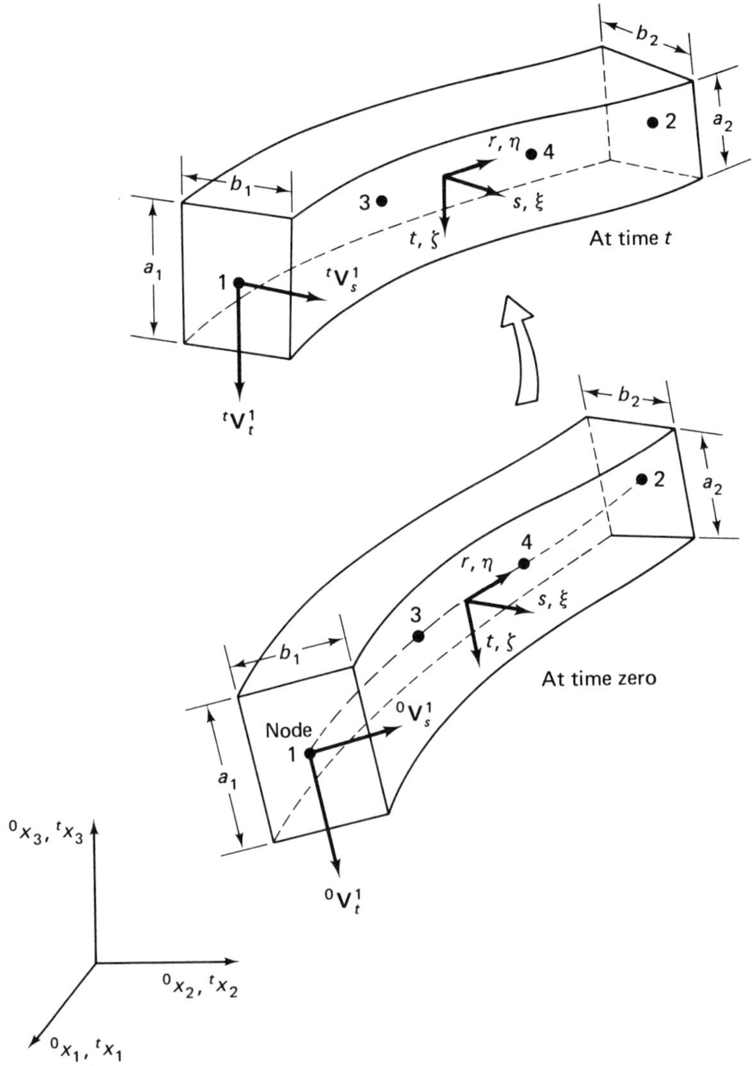

FIGURE 6.17 *Beam element undergoing large displacements and rotations.*

presented in Section 6.2.3, the calculation of the beam element matrices for nonlinear analysis represents a direct extension of the formulation given in Section 5.4.1. The calculations are performed as in the evaluation of the matrices of the elements with displacement degrees of freedom only (see Sections 5.3.1 and 6.3.3).

Using the same notation as in Section 5.4.1, the geometry of the bending element at time t is given by

$$ {}^t x_i = \sum_{k=1}^{N} h_k \, {}^t x_i^k + \frac{t}{2} \sum_{k=1}^{N} a_k h_k \, {}^t V_{ti}^k + \frac{s}{2} \sum_{k=1}^{N} b_k h_k \, {}^t V_{si}^k \qquad i = 1, 2, 3 \qquad (6.85) $$

where the coordinates of a typical point in the beam are ${}^t x_1, {}^t x_2, {}^t x_3$. If we apply (6.85) at times 0, t, and $t + \Delta t$, the displacement components are

$$ {}^t u_i = {}^t x_i - {}^0 x_i $$

and

$$ u_i = {}^{t+\Delta t} x_i - {}^t x_i \qquad (6.86) $$

Substituting (6.85) into (6.86) we obtain expressions for the displacement components in terms of the nodal point displacements and changes in the direction cosines of the nodal point vectors; i.e.,

$$ {}^t u_i = \sum_{k=1}^{N} h_k \, {}^t u_i^k + \frac{t}{2} \sum_{k=1}^{N} a_k h_k ({}^t V_{ti}^k - {}^0 V_{ti}^k) + \frac{s}{2} \sum_{k=1}^{N} b_k h_k ({}^t V_{si}^k - {}^0 V_{si}^k) \qquad (6.87) $$

and

$$ u_i = \sum_{k=1}^{N} h_k u_i^k + \frac{t}{2} \sum_{k=1}^{N} a_k h_k V_{ti}^k + \frac{s}{2} \sum_{k=1}^{N} b_k h_k V_{si}^k \qquad (6.88) $$

where

$$ \begin{aligned} V_{ti}^k &= {}^{t+\Delta t} V_{ti}^k - {}^t V_{ti}^k \\ V_{si}^k &= {}^{t+\Delta t} V_{si}^k - {}^t V_{si}^k \end{aligned} \qquad (6.89) $$

The relation in (6.87) is directly employed to evaluate the total displacements, but in order to use (6.88) the changes in the direction cosines V_{ti}^k and V_{si}^k are first expressed in terms of rotations about the global stationary Cartesian axes,

$$ \begin{aligned} \mathbf{V}_t^k &= \boldsymbol{\theta}_k \times {}^t \mathbf{V}_t^k \\ \mathbf{V}_s^k &= \boldsymbol{\theta}_k \times {}^t \mathbf{V}_s^k \end{aligned} \qquad (6.90) $$

These relations hold provided that the angles are small.† However, having calculated the angles in the finite element solution we can evaluate the direction cosines of the new nodal point vectors more accurately using

$$ \begin{aligned} {}^{t+\Delta t} \mathbf{V}_t^k &= {}^t \mathbf{V}_t^k + \int_{\boldsymbol{\theta}_k} d\boldsymbol{\theta}_k \times {}^\tau \mathbf{V}_t^k \\ {}^{t+\Delta t} \mathbf{V}_s^k &= {}^t \mathbf{V}_s^k + \int_{\boldsymbol{\theta}_k} d\boldsymbol{\theta}_k \times {}^\tau \mathbf{V}_s^k \end{aligned} \qquad (6.91) $$

The integrations in (6.91) can be carried out using, for example, Euler forward integration, where it is noted that using only one integration step corresponds to the use of (6.90).*

The relations given in (6.85) to (6.91) are the basic interpolations and expressions that are used to establish the strain-displacement interpolation matrices for geometric nonlinear analysis. Consider first the U.L. formulation. The displacement derivatives corresponding to the r, s, and t axes are

* Note that using (6.91) the unit lengths of the vectors ${}^\tau \mathbf{V}_s^k$, ${}^\tau \mathbf{V}_t^k$ must be preserved.

† The use of this linear relationship means that the exact linear strain-displacement matrix is obtained but the nonlinear strain-displacement matrix (see Table 6.7) does not include all second-order incremental rotation effects.

$$
\begin{bmatrix} u_{i,r} \\ u_{i,s} \\ u_{i,t} \end{bmatrix} = \sum_{k=1}^{N} \begin{bmatrix} h_{k,r} & [1 & {}^t(g)_{1i}^k & {}^t(g)_{2i}^k & {}^t(g)_{3i}^k] \\ h_k & [0 & {}^t(\hat{g})_{1i}^k & {}^t(\hat{g})_{2i}^k & {}^t(\hat{g})_{3i}^k] \\ h_k & [0 & {}^t(\bar{g})_{1i}^k & {}^t(\bar{g})_{2i}^k & {}^t(\bar{g})_{3i}^k] \end{bmatrix} \begin{bmatrix} u_i^k \\ \theta_1^k \\ \theta_2^k \\ \theta_3^k \end{bmatrix} \tag{6.92}
$$

where we use the notation

$$
{}^t(\hat{\mathbf{g}})^k = \frac{1}{2} b_k \begin{bmatrix} 0 & -{}^tV_{s3}^k & {}^tV_{s2}^k \\ {}^tV_{s3}^k & 0 & -{}^tV_{s1}^k \\ -{}^tV_{s2}^k & {}^tV_{s1}^k & 0 \end{bmatrix}
$$

$$
{}^t(\bar{\mathbf{g}})^k = \frac{1}{2} a_k \begin{bmatrix} 0 & -{}^tV_{t3}^k & {}^tV_{t2}^k \\ {}^tV_{t3}^k & 0 & -{}^tV_{t1}^k \\ -{}^tV_{t2}^k & {}^tV_{t1}^k & 0 \end{bmatrix}
$$

and

$$
{}^t(g)_{ij}^k = s\,{}^t(\hat{g})_{ij}^k + t\,{}^t(\bar{g})_{ij}^k
$$

The displacement derivatives corresponding to the axes ${}^tx_i$, $i = 1, 2, 3$ are now obtained using the Jacobian transformation

$$
\frac{\boldsymbol{\partial}}{\boldsymbol{\partial}\,{}^t\mathbf{x}} = {}^t\mathbf{J}^{-1} \frac{\boldsymbol{\partial}}{\boldsymbol{\partial}\mathbf{r}} \tag{6.93}
$$

where the Jacobian matrix, ${}^t\mathbf{J}$, contains the derivatives of the current coordinates ${}^tx_i$, $i = 1, 2, 3$ with respect to the natural coordinates r, s, and t. Substituting from (6.92) into (6.93) we obtain

$$
\begin{bmatrix} \dfrac{\partial u_i}{\partial\,{}^tx_1} \\[2ex] \dfrac{\partial u_i}{\partial\,{}^tx_2} \\[2ex] \dfrac{\partial u_i}{\partial\,{}^tx_3} \end{bmatrix} = \sum_{k=1}^{N} \begin{bmatrix} {}_th_{k,1} & {}_t(G1)_{i1}^k & {}_t(G2)_{i1}^k & {}_t(G3)_{i1}^k \\[2ex] {}_th_{k,2} & {}_t(G1)_{i2}^k & {}_t(G2)_{i2}^k & {}_t(G3)_{i2}^k \\[2ex] {}_th_{k,3} & {}_t(G1)_{i3}^k & {}_t(G2)_{i3}^k & {}_t(G3)_{i3}^k \end{bmatrix} \begin{bmatrix} u_i^k \\ \theta_1^k \\ \theta_2^k \\ \theta_3^k \end{bmatrix} \tag{6.94}
$$

where ${}_th_{k,i} = {}^tJ_{i1}^{-1}\, h_{k,r}$; ${}_t(Gm)_{in}^k = ({}^tJ_{n1}^{-1}\, {}^t(g)_{mi}^k)h_{k,r} + ({}^tJ_{n2}^{-1}(\hat{g})_{mi}^k + {}^tJ_{n3}^{-1}(\bar{g})_{mi}^k)h_k$

The expressions in (6.92) to (6.94) are completely analogous to the relations (5.70) to (5.75) used in linear analysis, but in the U.L. formulation all differentiations are performed with respect to the current configuration at time t.

The linear and nonlinear strain displacement matrices of the beam element corresponding to the U.L. formulation can now be evaluated using the approach employed in linear analysis. Namely, using (6.94) the strain components are calculated corresponding to the global axes and are then transformed to obtain the strain components corresponding to the local beam axes. Since the element stiffness matrix is evaluated using numerical integration, the transformation from global to local strain components must be performed during the numerical integration at each integration point. Table 6.7 summarizes the calculations and gives the stress matrix and vector that correspond to the strain-displacement matrices.

Considering next the T.L. formulation, we recognize that, first, derivatives analogous to those used in the U.L. formulation are required, but the derivatives are taken with respect to the coordinates at time 0. In addition, however, in order to include the initial displacement effect, the derivatives of the displace-

TABLE 6.7 *Matrices used in the beam element formulations.*

A. *Total Lagrangian Formulation*

1. *Linear Strain-Displacement Transformation Matrix*

$$
\begin{aligned}
{}_0\bar{\mathbf{e}} &= {}^t_0\bar{\mathbf{B}}_L\,\hat{\mathbf{u}}\\
{}^t_0\bar{\mathbf{B}}_L &= {}^t\mathbf{T}\,{}^t_0\mathbf{B}_L; \qquad {}^t_0\mathbf{B}_L = {}^t_0\mathbf{B}_{L0} + {}^t_0\mathbf{B}_{L1}\\
{}_0\bar{\mathbf{e}}^T &= [{}_0e_{\eta\eta} \quad 2{}_0e_{\eta\xi} \quad 2{}_0e_{\eta\zeta}]\\
\hat{\mathbf{u}}^T &= [u_1^1 \quad u_2^1 \quad u_3^1 \quad \theta_1^1 \quad \theta_2^1 \quad \theta_3^1; \quad u_1^2 \quad u_2^2 \ldots; \quad u_1^N \quad u_2^N \quad u_3^N \quad \theta_1^N \quad \theta_2^N \quad \theta_3^N]
\end{aligned}
$$

$$
{}^t_0\mathbf{B}_{L0} = \left[\begin{array}{c|cccccc|c}
\cdots & {}_0h_{k,1} & 0 & 0 & {}_0(G1)_{11}^k & {}_0(G2)_{11}^k & {}_0(G3)_{11}^k & \cdots\\
& 0 & {}_0h_{k,2} & 0 & {}_0(G1)_{22}^k & {}_0(G2)_{22}^k & {}_0(G3)_{22}^k & \\
& 0 & 0 & {}_0h_{k,3} & {}_0(G1)_{33}^k & {}_0(G2)_{33}^k & {}_0(G3)_{33}^k & \\
& {}_0h_{k,2} & {}_0h_{k,1} & 0 & {}_0(G1)_{12}^k + {}_0(G1)_{21}^k & {}_0(G2)_{12}^k + {}_0(G2)_{21}^k & {}_0(G3)_{12}^k + {}_0(G3)_{21}^k & \\
& 0 & {}_0h_{k,3} & {}_0h_{k,2} & {}_0(G1)_{23}^k + {}_0(G1)_{32}^k & {}_0(G2)_{23}^k + {}_0(G2)_{32}^k & {}_0(G3)_{23}^k + {}_0(G3)_{32}^k & \\
& {}_0h_{k,3} & 0 & {}_0h_{k,1} & {}_0(G1)_{13}^k + {}_0(G1)_{31}^k & {}_0(G2)_{13}^k + {}_0(G2)_{31}^k & {}_0(G3)_{13}^k + {}_0(G3)_{31}^k &
\end{array}\right]
$$

for nodal point k

$$
{}_0h_{k,i} = {}^0J_{i1}^{-1}\frac{\partial h_k}{\partial r}; \qquad {}_0(Gm)_{in}^k = ({}^0J_{n1}^{-1}\,(g)_{mi}^k)\frac{\partial h_k}{\partial r} + ({}^0J_{n2}^{-1}\,(\hat{g})_{mi}^k + {}^0J_{n3}^{-1}\,(\bar{g})_{mi}^k)h_k
$$

$$
{}^t_0\mathbf{B}_{L1} = \left[\begin{array}{c|cccccc|c}
\cdots & \ell_{11}\,{}_0h_{k,1} & \ell_{21}\,{}_0h_{k,1} & \ell_{31}\,{}_0h_{k,1} & (\Phi1)_{11}^k & (\Phi2)_{11}^k & (\Phi3)_{11}^k & \cdots\\
& \ell_{12}\,{}_0h_{k,2} & \ell_{22}\,{}_0h_{k,2} & \ell_{32}\,{}_0h_{k,2} & (\Phi1)_{22}^k & (\Phi2)_{22}^k & (\Phi3)_{22}^k & \\
& \ell_{13}\,{}_0h_{k,3} & \ell_{23}\,{}_0h_{k,3} & \ell_{33}\,{}_0h_{k,3} & (\Phi1)_{33}^k & (\Phi2)_{33}^k & (\Phi3)_{33}^k & \\
& (\ell_{11}\,{}_0h_{k,2} + \ell_{12}\,{}_0h_{k,1}) & (\ell_{21}\,{}_0h_{k,2} + \ell_{22}\,{}_0h_{k,1}) & (\ell_{31}\,{}_0h_{k,2} + \ell_{32}\,{}_0h_{k,1}) & (\Phi1)_{12}^k + (\Phi1)_{21}^k & (\Phi2)_{12}^k + (\Phi2)_{21}^k & (\Phi3)_{12}^k + (\Phi3)_{21}^k & \\
& (\ell_{12}\,{}_0h_{k,3} + \ell_{13}\,{}_0h_{k,2}) & (\ell_{22}\,{}_0h_{k,3} + \ell_{23}\,{}_0h_{k,2}) & (\ell_{32}\,{}_0h_{k,3} + \ell_{33}\,{}_0h_{k,2}) & (\Phi1)_{23}^k + (\Phi1)_{32}^k & (\Phi2)_{23}^k + (\Phi2)_{32}^k & (\Phi3)_{23}^k + (\Phi3)_{32}^k & \\
& (\ell_{11}\,{}_0h_{k,3} + \ell_{13}\,{}_0h_{k,1}) & (\ell_{21}\,{}_0h_{k,3} + \ell_{23}\,{}_0h_{k,1}) & (\ell_{31}\,{}_0h_{k,3} + \ell_{33}\,{}_0h_{k,1}) & (\Phi1)_{13}^k + (\Phi1)_{31}^k & (\Phi2)_{13}^k + (\Phi2)_{31}^k & (\Phi3)_{13}^k + (\Phi3)_{31}^k &
\end{array}\right]
$$

for nodal point k

$$
\ell_{ij} = \frac{\partial^t u_i}{\partial^0 x_j}; \qquad (\Phi m)_{jn}^k = \sum_{i=1}^{3} \ell_{ij}\,{}_0(Gm)_{in}^k
$$

TABLE 6.7 *Matrices used in the beam element formulations. (continued)*

$$
{}^t\mathbf{T} = \begin{bmatrix} {}^tV_{r1}^2 & {}^tV_{r2}^2 & {}^tV_{r3}^2 & {}^tV_{r1}\,{}^tV_{r2} & {}^tV_{r2}\,{}^tV_{r3} & {}^tV_{r1}\,{}^tV_{r3} \\ 2\,{}^tV_{r1}\,{}^tV_{s1} & 2\,{}^tV_{r2}\,{}^tV_{s2} & 2\,{}^tV_{r3}\,{}^tV_{s3} & {}^tV_{r2}\,{}^tV_{s1} + {}^tV_{r1}\,{}^tV_{s2} & {}^tV_{r3}\,{}^tV_{s2} + {}^tV_{r2}\,{}^tV_{s3} & {}^tV_{r3}\,{}^tV_{s1} + {}^tV_{r1}\,{}^tV_{s3} \\ 2\,{}^tV_{r1}\,{}^tV_{t1} & 2\,{}^tV_{r2}\,V_{t2} & 2\,{}^tV_{r3}\,{}^tV_{t3} & {}^tV_{r2}\,{}^tV_{t1} + {}^tV_{r1}\,{}^tV_{t2} & {}^tV_{r3}\,{}^tV_{t2} + {}^tV_{r2}\,{}^tV_{t3} & {}^tV_{r3}\,{}^tV_{t1} + {}^tV_{r1}\,{}^tV_{t3} \end{bmatrix}
$$

$$
{}^t\mathbf{V}_r = {}^t\mathbf{V}_s \times {}^t\mathbf{V}_t
$$

2. *Nonlinear Strain-Displacement Transformation Matrix*

$$
{}^t_0\bar{\mathbf{B}}_{NL} = {}^t\tilde{\mathbf{T}}^T\,{}^t_0\mathbf{B}_{NL}; \qquad {}^t\tilde{\mathbf{T}} = \begin{bmatrix} \mathbf{T}^* & & \\ & \mathbf{T}^* & \\ & & \mathbf{T}^* \end{bmatrix} \quad \text{where } \mathbf{T}^* = \begin{bmatrix} {}^tV_{r1} & {}^tV_{s1} & {}^tV_{t1} \\ {}^tV_{r2} & {}^tV_{s2} & {}^tV_{t2} \\ {}^tV_{r3} & {}^tV_{s3} & {}^tV_{t3} \end{bmatrix}; \qquad {}^t\mathbf{V}_r = {}^t\mathbf{V}_s \times {}^t\mathbf{V}_t
$$

$$
{}^t_0\mathbf{B}_{NL} = \left[\cdots \;\middle|\; \begin{matrix} {}_0h_{k,1} & 0 & 0 & {}_0(G1)^k_{11} & {}_0(G2)^k_{11} & {}_0(G3)^k_{11} \\ {}_0h_{k,2} & 0 & 0 & {}_0(G1)^k_{12} & {}_0(G2)^k_{12} & {}_0(G3)^k_{12} \\ {}_0h_{k,3} & 0 & 0 & {}_0(G1)^k_{13} & {}_0(G2)^k_{13} & {}_0(G3)^k_{13} \\ 0 & {}_0h_{k,1} & 0 & {}_0(G1)^k_{21} & {}_0(G2)^k_{21} & {}_0(G3)^k_{21} \\ 0 & {}_0h_{k,2} & 0 & {}_0(G1)^k_{22} & {}_0(G2)^k_{22} & {}_0(G3)^k_{22} \\ 0 & {}_0h_{k,3} & 0 & {}_0(G1)^k_{23} & {}_0(G2)^k_{23} & {}_0(G3)^k_{23} \\ 0 & 0 & {}_0h_{k,1} & {}_0(G1)^k_{31} & {}_0(G2)^k_{31} & {}_0(G3)^k_{31} \\ 0 & 0 & {}_0h_{k,2} & {}_0(G1)^k_{32} & {}_0(G2)^k_{32} & {}_0(G3)^k_{32} \\ 0 & 0 & {}_0h_{k,3} & {}_0(G1)^k_{33} & {}_0(G2)^k_{33} & {}_0(G3)^k_{33} \end{matrix} \;\middle|\; \cdots \right]
$$

for nodal point k

3. *2nd Piola–Kirchhoff Stress Matrix*

$$
{}^t_0\bar{\mathbf{S}} = \begin{bmatrix} {}^t_0\tilde{\mathbf{S}} & & \\ & {}^t_0\tilde{\mathbf{S}} & \\ & & {}^t_0\tilde{\mathbf{S}} \end{bmatrix} \qquad \text{where } {}^t_0\tilde{\mathbf{S}} = \begin{bmatrix} {}^t_0S_{\eta\eta} & & \text{sym.} \\ {}^t_0S_{\eta\xi} & 0 & \\ {}^t_0S_{\eta\zeta} & 0 & 0 \end{bmatrix}
$$

(Note that instead of ${}^t\mathbf{T}$ and ${}^t\tilde{\mathbf{T}}$ we could use ${}^0\mathbf{T}$ and ${}^0\tilde{\mathbf{T}}$)

TABLE 6.7 *Matrices used in the beam element formulations. (continued)*

4. *Stress Vector*

$$
{}_0^t\hat{\bar{\mathbf{S}}} = \begin{bmatrix} {}_0^tS_{\eta\eta} \\ {}_0^tS_{\eta\xi} \\ {}_0^tS_{\eta\zeta} \end{bmatrix}
$$

B. *Updated Lagrangian Formulation*

1. *Linear Strain-Displacement Transformation Matrix*

$$
{}_t\bar{\mathbf{e}} = {}_t^t\bar{\mathbf{B}}_L\,\hat{\mathbf{u}}
$$

$$
{}_t^t\bar{\mathbf{B}}_L = {}^t\mathbf{T}\,{}_t^t\mathbf{B}_L
$$

$$
{}_t\bar{\mathbf{e}}^T = [{}_te_{\eta\eta} \quad 2{}_te_{\eta\xi} \quad 2{}_te_{\eta\zeta}]
$$

$$
\hat{\mathbf{u}}^T = [u_1^1 \quad u_2^1 \quad u_3^1 \quad \theta_1^1 \quad \theta_2^1 \quad \theta_3^1; \quad u_1^2 \quad u_2^2 \quad \ldots; \quad u_1^N \quad u_2^N \quad u_3^N \quad \theta_1^N \quad \theta_2^N \quad \theta_3^N]
$$

$$
{}_t^t\mathbf{B}_L = \left[\begin{array}{c|cccccc|c}
 & {}_th_{k,1} & 0 & 0 & {}_t(G1)_{11}^k & {}_t(G2)_{11}^k & {}_t(G3)_{11}^k & \\
 & 0 & {}_th_{k,2} & 0 & {}_t(G1)_{22}^k & {}_t(G2)_{22}^k & {}_t(G3)_{22}^k & \\
\cdots & 0 & 0 & {}_th_{k,3} & {}_t(G1)_{33}^k & {}_t(G2)_{33}^k & {}_t(G3)_{33}^k & \cdots \\
 & {}_th_{k,2} & {}_th_{k,1} & 0 & {}_t(G1)_{12}^k + {}_t(G1)_{21}^k & {}_t(G2)_{12}^k + {}_t(G2)_{21}^k & {}_t(G3)_{12}^k + {}_t(G3)_{21}^k & \\
 & 0 & {}_th_{k,3} & {}_th_{k,2} & {}_t(G1)_{23}^k + {}_t(G1)_{32}^k & {}_t(G2)_{23}^k + {}_t(G2)_{32}^k & {}_t(G3)_{23}^k + {}_t(G3)_{32}^k & \\
 & {}_th_{k,3} & 0 & {}_th_{k,1} & {}_t(G1)_{13}^k + {}_t(G1)_{31}^k & {}_t(G2)_{13}^k + {}_t(G2)_{31}^k & {}_t(G3)_{13}^k + {}_t(G3)_{31}^k &
\end{array}\right]
$$

for nodal point k

$$
{}_th_{k,i} = \frac{\partial h_k}{\partial {}^tx_i} = {}^tJ_{i1}^{-1}\frac{\partial h_k}{\partial r} \qquad {}_t(Gm)_{in}^k = ({}^tJ_{n1}^{-1}\,(g)_{mi}^k)\frac{\partial h_k}{\partial r} + ({}^tJ_{n2}^{-1}\,(\hat{g})_{mi}^k + {}^tJ_{n3}^{-1}\,(\bar{g})_{mi}^k)h_k
$$

$$
{}^t\mathbf{T} = \begin{bmatrix}
{}^tV_{r1}^2 & {}^tV_{r2}^2 & {}^tV_{r3}^2 & {}^tV_{r1}\,{}^tV_{r2} & {}^tV_{r2}\,{}^tV_{r3} & {}^tV_{r1}\,{}^tV_{r3} \\
2{}^tV_{r1}\,{}^tV_{s1} & 2{}^tV_{r2}\,{}^tV_{s2} & 2{}^tV_{r3}\,{}^tV_{s3} & {}^tV_{r2}\,{}^tV_{s1} + {}^tV_{r1}\,{}^tV_{s2} & {}^tV_{r3}\,{}^tV_{s2} + {}^tV_{r2}\,{}^tV_{s3} & {}^tV_{r3}\,{}^tV_{s1} + {}^tV_{r1}\,{}^tV_{s3} \\
2{}^tV_{r1}\,{}^tV_{t1} & 2{}^tV_{r2}\,{}^tV_{t2} & 2{}^tV_{r3}\,{}^tV_{t3} & {}^tV_{r2}\,{}^tV_{t1} + {}^tV_{r1}\,{}^tV_{t2} & {}^tV_{r3}\,{}^tV_{t2} + {}^tV_{r2}\,{}^tV_{t3} & {}^tV_{r3}\,{}^tV_{t1} + {}^tV_{r1}\,{}^tV_{t3}
\end{bmatrix}
$$

$$
{}^t\mathbf{V}_r = {}^t\mathbf{V}_s \times {}^t\mathbf{V}_t
$$

TABLE 6.7 *Matrices used in the beam element formulations. (continued)*

2. *Nonlinear Strain-Displacement Transformation Matrix*

$$
{}_t^t\bar{\mathbf{B}}_{NL} = {}^t\tilde{\mathbf{T}}^T\,{}_t^t\mathbf{B}_{NL}; \qquad {}^t\tilde{\mathbf{T}} = \begin{bmatrix} \mathbf{T}^* & & \\ & \mathbf{T}^* & \\ & & \mathbf{T}^* \end{bmatrix} \qquad \text{where } \mathbf{T}^* = \begin{bmatrix} {}^tV_{r1} & {}^tV_{s1} & {}^tV_{t1} \\ {}^tV_{r2} & {}^tV_{s2} & {}^tV_{t2} \\ {}^tV_{r3} & {}^tV_{s3} & {}^tV_{t3} \end{bmatrix}; \qquad {}^t\mathbf{V}_r = {}^t\mathbf{V}_s \times {}^t\mathbf{V}_t
$$

$$
{}_t^t\mathbf{B}_{NL} = \left[\begin{array}{c|ccccccc|c}
 & {}_th_{k,1} & 0 & 0 & {}_t(G1)^k_{11} & {}_t(G2)^k_{11} & {}_t(G3)^k_{11} & & \\
 & {}_th_{k,2} & 0 & 0 & {}_t(G1)^k_{12} & {}_t(G2)^k_{12} & {}_t(G3)^k_{12} & & \\
 & {}_th_{k,3} & 0 & 0 & {}_t(G1)^k_{13} & {}_t(G2)^k_{13} & {}_t(G3)^k_{13} & & \\
 & 0 & {}_th_{k,1} & 0 & {}_t(G1)^k_{21} & {}_t(G2)^k_{21} & {}_t(G3)^k_{21} & & \\
\cdots & 0 & {}_th_{k,2} & 0 & {}_t(G1)^k_{22} & {}_t(G2)^k_{22} & {}_t(G3)^k_{22} & & \cdots \\
 & 0 & {}_th_{k,3} & 0 & {}_t(G1)^k_{23} & {}_t(G2)^k_{23} & {}_t(G3)^k_{23} & & \\
 & 0 & 0 & {}_th_{k,1} & {}_t(G1)^k_{31} & {}_t(G2)^k_{31} & {}_t(G3)^k_{31} & & \\
 & 0 & 0 & {}_th_{k,2} & {}_t(G1)^k_{32} & {}_t(G2)^k_{32} & {}_t(G3)^k_{32} & & \\
 & 0 & 0 & {}_th_{k,3} & {}_t(G1)^k_{33} & {}_t(G2)^k_{33} & {}_t(G3)^k_{33} & &
\end{array}\right]
$$

for nodal point k

3. *Cauchy Stress Matrix and Stress Vector*

$$
{}^t\bar{\boldsymbol{\tau}} = \begin{bmatrix} {}^t\tilde{\boldsymbol{\tau}} & & \\ & {}^t\tilde{\boldsymbol{\tau}} & \\ & & {}^t\tilde{\boldsymbol{\tau}} \end{bmatrix} \qquad \text{where } {}^t\tilde{\boldsymbol{\tau}} = \begin{bmatrix} {}^t\tau_{\eta\eta} & & \text{sym.} \\ {}^t\tau_{\eta\xi} & 0 & \\ {}^t\tau_{\eta\zeta} & 0 & 0 \end{bmatrix}
$$

4. *Stress Vector*

$$
{}^t\hat{\bar{\boldsymbol{\tau}}}^T = [{}^t\tau_{\eta\eta} \quad {}^t\tau_{\eta\xi} \quad {}^t\tau_{\eta\zeta}]
$$

ments at time t with respect to the original coordinates are needed. These derivatives are evaluated using (6.87) to obtain

$$\begin{aligned}\frac{\partial^t u_i}{\partial^0 x_j} &= \sum_{k=1}^{N} {}^0J_{j1}^{-1}\, h_{k,r}\Big[{}^t u_i^k + \frac{s}{2} b_k({}^tV_{si}^k - {}^0V_{si}^k) + \frac{t}{2} a_k({}^tV_{ti}^k - {}^0V_{ti}^k)\Big] \\ &+ \tfrac{1}{2}\sum_{k=1}^{N} {}^0J_{j2}^{-1}\, b_k\, h_k({}^tV_{si}^k - {}^0V_{si}^k) \\ &+ \tfrac{1}{2}\sum_{k=1}^{N} {}^0J_{j3}^{-1}\, a_k\, h_k({}^tV_{ti}^k - {}^0V_{ti}^k)\end{aligned} \tag{6.95}$$

Using (6.94) with the derivatives taken with respect to ${}^0x_1$, ${}^0x_2$, and ${}^0x_3$ and (6.95), the linear and nonlinear strain-displacement matrices of the beam element can now be evaluated as in the U.L. formulation. Table 6.7 summarizes the calculations that are performed.

It should be noted that this beam element formulation admits very large displacements and rotations, and has an important advantage when compared with the formulation of a straight beam element based on Hermitian displacement interpolations: all individual displacement components are expressed using the same functions because the displacement expressions are derived from the geometry interpolations. Thus, there is no directionality in the displacement interpolations. Also, the change in the geometry of the beam structure with increasing deformations is modeled more accurately than using straight Hermitian-based beam elements. However, the beam element of Table 6.7 must in general be employed with three or four nodes (see Section 5.4.1), which makes the calculation of the element matrices more expensive. Therefore, although the element presents much generality for modeling beam-type structures, in practice, the use of a Hermitian-based beam element can be computationally more effective in the analysis of some specific problems. We demonstrate the formulation of a simple two-node element in the following example.

EXAMPLE 6.18: Consider the two-node beam element shown in Fig. 6.18. Evaluate the coordinate and displacement interpolations and derivatives that are required for the calculation of the strain-displacement matrices of the U.L. and T.L. formulations.

This two-node beam element is not effective in practice because the element "locks" when it is thin. Hence, three- or four-node elements should actually be used, but we consider the element here for instructive purposes. The formulations of the higher-order elements would be achieved in an analogous way, but using more nodes and the corresponding interpolation functions.

Using the variables in Fig. 6.18 we have corresponding to (6.85),

$$\begin{aligned}{}^tx_1 &= \left(\frac{1-r}{2}\right){}^tx_1^1 + \left(\frac{1+r}{2}\right){}^tx_1^2 - \frac{sh}{2}\left(\frac{1-r}{2}\right)\sin{}^t\theta_1 - \frac{sh}{2}\left(\frac{1+r}{2}\right)\sin{}^t\theta_2 \\ {}^tx_2 &= \left(\frac{1-r}{2}\right){}^tx_2^1 + \left(\frac{1+r}{2}\right){}^tx_2^2 + \frac{sh}{2}\left(\frac{1-r}{2}\right)\cos{}^t\theta_1 + \frac{sh}{2}\left(\frac{1+r}{2}\right)\cos{}^t\theta_2 \\ {}^0x_1 &= \left(\frac{1+r}{2}\right){}^0L \\ {}^0x_2 &= \frac{sh}{2}\end{aligned}$$

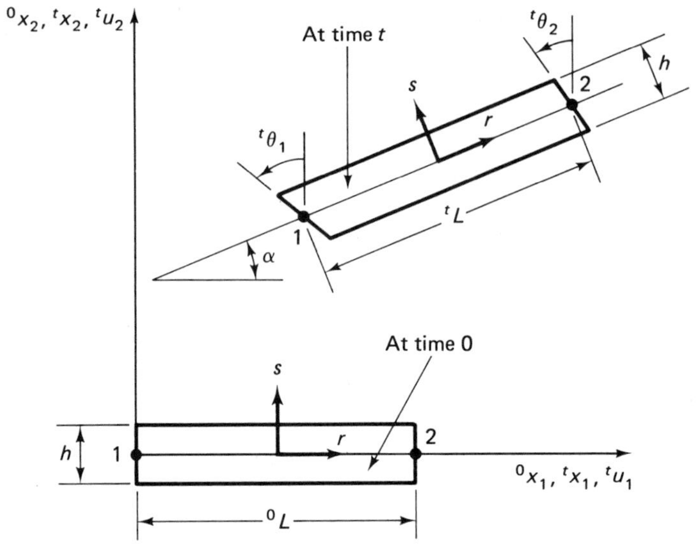

FIGURE 6.18 *Two-node beam element in large displacements and rotations.*

Hence, the displacement components at any point are at time t,

$$
\begin{aligned}
{}^tu_1 &= \left(\frac{{}^tx_1^1 + {}^tx_1^2 - {}^0L}{2}\right) + \left(\frac{{}^tx_1^2 - {}^tx_1^1 - {}^0L}{2}\right)r \\
&\qquad - \frac{sh}{2}\left[\left(\frac{1-r}{2}\right)\sin {}^t\theta_1 + \left(\frac{1+r}{2}\right)\sin {}^t\theta_2\right] \\
{}^tu_2 &= \left(\frac{{}^tx_2^1 + {}^tx_2^2}{2}\right) + \left(\frac{{}^tx_2^2 - {}^tx_2^1}{2}\right)r \\
&\qquad + \frac{sh}{2}\left[\left(\frac{1-r}{2}\right)\cos {}^t\theta_1 + \left(\frac{1+r}{2}\right)\cos {}^t\theta_2 - 1\right]
\end{aligned}
$$

Considering the U.L. formulation first, we have

$$
\begin{bmatrix} u_{i,r} \\ \\ u_{i,s} \end{bmatrix} = \left[\begin{array}{c|c} \left(-\frac{1}{2}\right)[1 \quad {}^t(g)_i^1] & \left(\frac{1}{2}\right)[1 \quad {}^t(g)_i^2] \\ \left(\frac{1-r}{2}\right)[0 \quad {}^t(\hat{g})_i^1] & \left(\frac{1+r}{2}\right)[0 \quad {}^t(\hat{g})_i^2] \end{array}\right] \begin{bmatrix} u_i^1 \\ \theta_1 \\ \hline u_i^2 \\ \theta_2 \end{bmatrix} \tag{a}
$$

where

$$
\begin{aligned}
{}^t(g)_i^k &= s(\hat{g})_i^k \\
{}^t(\hat{g})_1^k &= -\frac{h}{2}\cos {}^t\theta_k \\
{}^t(\hat{g})_2^k &= -\frac{h}{2}\sin {}^t\theta_k
\end{aligned}
$$

The required derivatives for the Jacobian are:

$$
\begin{aligned}
\frac{\partial {}^tx_1}{\partial r} &= \frac{L\cos\alpha}{2} - \frac{sh}{4}[\sin {}^t\theta_2 - \sin {}^t\theta_1] \\
\frac{\partial {}^tx_1}{\partial s} &= \left(-\frac{h}{2}\right)\left[\left(\frac{1-r}{2}\right)\sin {}^t\theta_1 + \left(\frac{1+r}{2}\right)\sin {}^t\theta_2\right] \\
\frac{\partial {}^tx_2}{\partial r} &= \frac{L\sin\alpha}{2} + \frac{sh}{4}[\cos {}^t\theta_2 - \cos {}^t\theta_1] \\
\frac{\partial {}^tx_2}{\partial s} &= \left(\frac{h}{2}\right)\left[\left(\frac{1-r}{2}\right)\cos {}^t\theta_1 + \left(\frac{1+r}{2}\right)\cos {}^t\theta_2\right]
\end{aligned}
$$

where we assumed ${}^tL \doteq {}^0L = L$.

The above expressions are all those required for the construction of the linear and nonlinear strain displacement transformation matrices of the U.L. formulation.

Next, we consider the T.L. formulation. Here we use (a) and

$$ {}^{0}\mathbf{J} = \begin{bmatrix} {}^{0}L/2 & 0 \\ 0 & h/2 \end{bmatrix} \tag{b} $$

Also, the initial displacement effect is taken into account using the derivatives

$$ \begin{aligned} {}^{t}_{0}u_{1,1} &= (\cos\alpha - 1) - \frac{sh}{2L}(\sin{}^{t}\theta_2 - \sin{}^{t}\theta_1) \\ {}^{t}_{0}u_{1,2} &= -\left(\frac{1-r}{2}\right)\sin{}^{t}\theta_1 - \left(\frac{1+r}{2}\right)\sin{}^{t}\theta_2 \\ {}^{t}_{0}u_{2,1} &= \sin\alpha + \frac{sh}{2L}[\cos{}^{t}\theta_2 - \cos{}^{t}\theta_1] \\ {}^{t}_{0}u_{2,2} &= \left(\frac{1-r}{2}\right)\cos{}^{t}\theta_1 + \left(\frac{1+r}{2}\right)\cos{}^{t}\theta_2 - 1 \end{aligned} \tag{c} $$

where we again assumed ${}^{t}L \doteq {}^{0}L = L$.

The relations in (a), (b), and (c) are used to construct the strain-displacement matrices.

6.3.5 *Plate and Shell Elements*

A great many types of plate and shell elements have been proposed for the nonlinear analysis of plates, specific shells, and general shell structures. However, as with the beam element discussed in the previous section, the isoparametric formulation of plate and shell elements for nonlinear analysis is very attractive, because the formulation is both consistent and general, and the elements can be employed in an effective manner for the analysis of a variety of plates and shells.[12–14] Namely, as in the linear analysis, no specific shell theory is employed in the formulation so that the shell elements are applicable, in principle, to the analysis of any plate and shell structure.

Considering a plate undergoing large deflections, we recognize that as soon as the plate has deflected significantly, the action of the structure is really that of a shell; i.e. the structure is now curved and both membrane and bending stresses are significant. Therefore, in the discussion below we only consider a general shell element, where we imply that if the element is initially flat it represents a plate.

In the following we consider the nonlinear formulation of the variable-number-nodes isoparametric shell element that we discussed for linear analysis in Section 5.4.2. Figure 6.19 shows a typical nine-node element in its original position and its configuration at time t. The element behavior is based on the same assumptions that are employed in linear analysis, namely that lines which are originally normal to the midsurface of the shell remain straight during the element deformations and that no transverse normal stress is developed. However, the nonlinear formulation does admit arbitrarily large displacements and rotations of the shell element.

The U.L. and T.L. formulations of the shell element are based on the general continuum mechanics equations presented in Section 6.2.3 and are a direct extension of the formulation for linear analysis. Also, the calculations of the

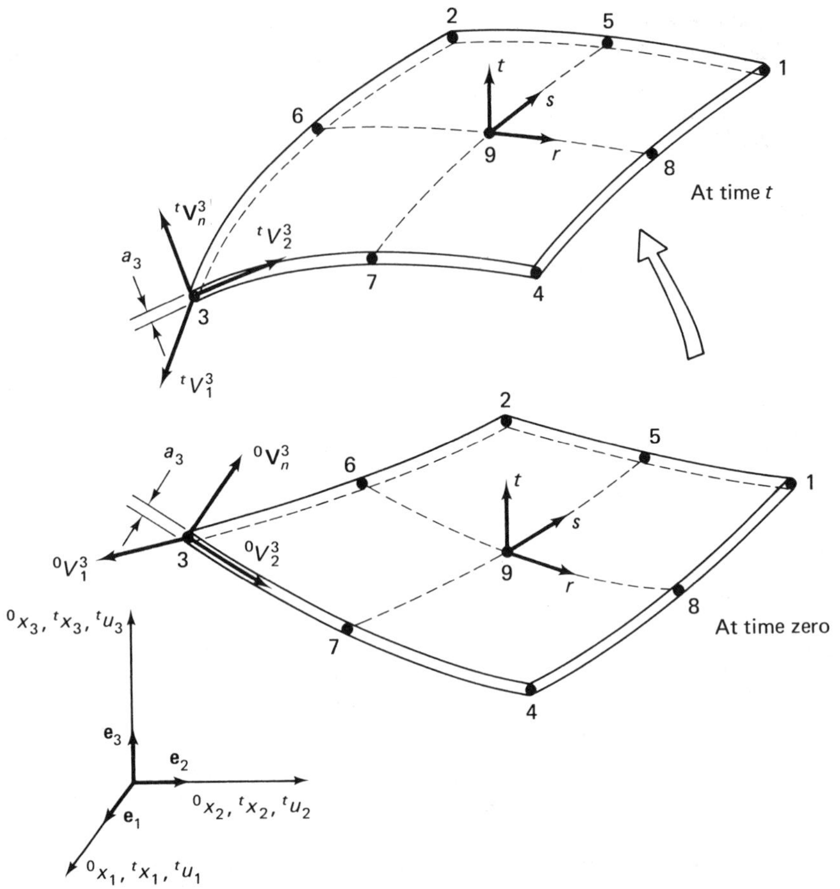

FIGURE 6.19 *Shell element undergoing large displacements and rotations.*

element matrices follow closely those used for the other elements (see Sections 6.3.1 to 6.3.4).

Using the same notation as in Section 5.4.2, the coordinates of a generic point in the shell element now undergoing very large displacements and rotations are,

$$ {}^tx_i = \sum_{k=1}^{N} h_k\, {}^tx_i^k + \frac{t}{2}\sum_{k=1}^{N} a_k\, h_k\, {}^tV_{ni}^k \tag{6.96}$$

Using (6.96) at time 0, t, and $t + \Delta t$, we thus have

$$ {}^tu_i = {}^tx_i - {}^0x_i \tag{6.97}$$

and

$$ u_i = {}^{t+\Delta t}x_i - {}^tx_i \tag{6.98}$$

Substituting from (6.96) into (6.97) and (6.98), we obtain:

$$ {}^tu_i = \sum_{k=1}^{N} h_k\, {}^tu_i^k + \frac{t}{2}\sum_{k=1}^{N} a_k\, h_k\,({}^tV_{ni}^k - {}^0V_{ni}^k) \tag{6.99}$$

and

$$ u_i = \sum_{k=1}^{N} h_k\, u_i^k + \frac{t}{2}\sum_{k=1}^{N} a_k\, h_k\, V_{ni}^k \tag{6.100}$$

where

$$ V_{ni}^k = {}^{t+\Delta t}V_{ni}^k - {}^tV_{ni}^k \tag{6.101}$$

The relation in (6.99) is directly employed to evaluate the total element displacements, but to apply (6.100) we first express the changes in the direction cosines of the shell normals, given by V_{ni}^k, in terms of the rotations about two vectors that are orthogonal to ${}^t\mathbf{V}_n^k$. These two vectors ${}^t\mathbf{V}_1^k$ and ${}^t\mathbf{V}_2^k$ are defined as in linear analysis:

$$ {}^t\mathbf{V}_1^k = (\mathbf{e}_2 \times {}^t\mathbf{V}_n^k)/|\mathbf{e}_2 \times {}^t\mathbf{V}_n^k| \tag{6.102} $$

where we set ${}^t\mathbf{V}_1^k$ equal to $\mathbf{e}_3$ if ${}^t\mathbf{V}_n^k$ is parallel to $\mathbf{e}_2$, and

$$ {}^t\mathbf{V}_2^k = {}^t\mathbf{V}_n^k \times {}^t\mathbf{V}_1^k \tag{6.103} $$

Let α_k and β_k be the rotations of the normal vector about the vectors ${}^t\mathbf{V}_1^k$ and ${}^t\mathbf{V}_2^k$ from the configuration at time t to the configuration at time $t + \Delta t$. Then we have approximately (for small incremental angles α_k and β_k)

$$ \mathbf{V}_n^k = -{}^t\mathbf{V}_2^k\,\alpha_k + {}^t\mathbf{V}_1^k\,\beta_k\,\dagger \tag{6.104} $$

Substituting from (6.104) into (6.100) we thus obtain the incremental internal element displacements in terms of the nodal point incremental displacements and rotations,

$$ u_i = \sum_{k=1}^{N} h_k\, u_i^k + \frac{t}{2} \sum_{k=1}^{N} a_k\, h_k[-{}^tV_{2i}^k\,\alpha_k + {}^tV_{1i}^k\,\beta_k] \tag{6.105} $$

The finite element solution will yield the nodal point variables u_i^k, α_k, and β_k, which can then be employed to evaluate accurately ${}^{t+\Delta t}\mathbf{V}_n^k$,

$$ {}^{t+\Delta t}\mathbf{V}_n^k = {}^t\mathbf{V}_n^k + \int_{\alpha_k,\beta_k} -{}^\tau\mathbf{V}_2^k\,d\alpha_k + {}^\tau\mathbf{V}_1^k\,d\beta_k \tag{6.106} $$

This integration may be carried out accurately using, for example, the Euler forward method with a sufficiently large number of integration intervals, where it may be noted that the use of only one interval of integration corresponds to the assumption in (6.104).*

The relations in (6.96), (6.99), and (6.105) are the basic equations used in the evaluation of the strain-displacement matrices of the shell element. Consider first the U.L. formulation. In this case we use (6.105) to obtain:

$$ \begin{bmatrix} u_{i,r} \\ u_{i,s} \\ u_{i,t} \end{bmatrix} = \sum_{k=1}^{N} \begin{bmatrix} h_{k,r}[1 & t\,{}^tg_{1i}^k & t\,{}^tg_{2i}^k] \\ h_{k,s}[1 & t\,{}^tg_{1i}^k & t\,{}^tg_{2i}^k] \\ h_k\;[0 & {}^tg_{1i}^k & {}^tg_{2i}^k] \end{bmatrix} \begin{bmatrix} u_i^k \\ \alpha_k \\ \beta_k \end{bmatrix} \tag{6.107} $$

where we use the notation:

$$ {}^tg_{1i}^k = -\tfrac{1}{2}a_k\,{}^tV_{2i}^k $$
$$ {}^tg_{2i}^k = \tfrac{1}{2}a_k\,{}^tV_{1i}^k $$

To obtain the displacement derivatives corresponding to the axes ${}^tx_i$, $i = 1, 2, 3$, we now employ the Jacobian transformation,

$$ \frac{\partial}{\partial\,{}^t\mathbf{x}} = \mathbf{J}^{-1}\frac{\partial}{\partial \mathbf{r}} \tag{6.108} $$

where the Jacobian matrix, ${}^t\mathbf{J}$, contains the derivatives of the current coordinates ${}^tx_i$, $i = 1, 2, 3$, of (6.96) with respect to the natural coordinates r, s, and t.

* Here we must take care that *within* the solution step (or rather from the configuration at which the stiffness matrix was defined) the convention for the definition of ${}^\tau\mathbf{V}_1^k$ and ${}^\tau\mathbf{V}_2^k$ does not suddenly change because ${}^\tau\mathbf{V}_n^k$ has "moved passed" $\mathbf{e}_2$. Also, using (6.106) the unit length of the normal vector must be preserved.

† The use of this linear relationship means that the exact linear strain-displacement matrix is obtained but the nonlinear strain-displacement matrix (see Table 6.8) does not include all second-order incremental rotation effects.

Substituting from (6.107) into (6.108) we obtain

$$\begin{bmatrix} \dfrac{\partial u_i}{\partial^t x_1} \\ \dfrac{\partial u_i}{\partial^t x_2} \\ \dfrac{\partial u_i}{\partial^t x_3} \end{bmatrix} = \sum_{k=1}^{N} \begin{bmatrix} {}_t h_{k,1} & {}^t g_{1i}^k \, {}_t G_1^k & {}^t g_{2i}^k \, {}_t G_1^k \\ {}_t h_{k,2} & {}^t g_{1i}^k \, {}_t G_2^k & {}^t g_{2i}^k \, {}_t G_2^k \\ {}_t h_{k,3} & {}^t g_{1i}^k \, {}_t G_3^k & {}^t g_{2i}^k \, {}_t G_3^k \end{bmatrix} \begin{bmatrix} u_i^k \\ \alpha_k \\ \beta_k \end{bmatrix} \tag{6.109}$$

where

$$ {}_t h_{k,i} = {}^t J_{i1}^{-1} \, h_{k,r} + {}^t J_{i2}^{-1} \, h_{k,s}$$

$$ {}_t G_i^k = t({}^t J_{i1}^{-1} h_{k,r} + {}^t J_{i2}^{-1} \, h_{k,s}) + {}^t J_{i3}^{-1} \, h_k$$

and ${}^t J_{ij}^{-1}$ is the element (i, j) of the matrix ${}^t\mathbf{J}^{-1}$ in (6.108).

With the displacement derivatives defined in (6.109) we can now directly assemble the strain-displacement matrices ${}_t^t\mathbf{B}_L$ and ${}_t^t\mathbf{B}_{NL}$. Table 6.8 gives these matrices and also defines the corresponding stress matrix ${}^t\boldsymbol{\tau}$ and stress vector ${}^t\hat{\boldsymbol{\tau}}$.

Considering next the T.L. formulation, the derivatives of the incremental displacements are also required, but now with respect to the original coordinates. Thus, we use the relation in (6.109), but with the coordinates ${}^0x_i$ instead of ${}^tx_i$. In addition, the derivatives of the total displacements with respect to the original coordinates are needed in order to include the initial displacement effect. These derivatives are calculated using (6.99) to obtain

$$\frac{\partial^t u_i}{\partial^0 x_j} = \sum_{k=1}^{N} {}_0 h_{k,j} \, {}^t u_i^k + \sum_{k=1}^{N} \frac{a_k}{2} (t \, {}_0 h_{k,j} + {}^0 J_{j3}^{-1} h_k)({}^t V_{ni}^k - {}^0 V_{ni}^k) \tag{6.110}$$

where the derivatives ${}_0 h_{k,j}$ can be evaluated using (6.108) corresponding to time 0.

The strain-displacement trānsformation matrices ${}_0^t\mathbf{B}_L$ and ${}_0^t\mathbf{B}_{NL}$ can now be obtained using (6.109) corresponding to the coordinates ${}^0x_i$, $i = 1, 2, 3$, and the relation in (6.110). Table 6.8 summarizes the evaluation of these matrices and defines also the corresponding stress matrix and stress vector.

It may be noted once more that the shell element discussed above is a very general element since no specific shell theory has been employed in its formulation. In fact, the use of the general incremental virtual work equation with only the two basic shell assumptions—that lines originally normal to the shell mid-surface remain straight and that the trānsverse normal stress remains zero—is equivalent to using a general nonlinear shell theory. This generality in the formulation is fully preserved by using the isoparametric interpolation of the shell geometry and shell displacements. An important feature in this interpolation is the use of the normal to the shell mid-surface, which makes it possible for the element to undergo arbitrary large displacements and rotations. This generality in the formulation is further extended if the element is also implemented as a transition element, as discussed in Sections 5.4.1 and 5.4.2. Using the shell and transition elements to model shell intersections and shell-to-solid regions, fully compatible element idealizations of many practical shell structures can be constructed and these element idealizations remain compatible throughout the motion of the structure. Furthermore, the shell element can be employed in a compatible manner with the beam element of Section 6.3.4, since both elements

TABLE 6.8 *Matrices used in the shell element formulations.*

A. *Total Lagrangian Formulation*

1. *Linear Strain-Displacement Transformation Matrix*

$$
{}_0\mathbf{e} = {}_0^t\mathbf{B}_L\hat{\mathbf{u}}; \qquad {}_0^t\mathbf{B}_L = {}_0^t\mathbf{B}_{L0} + {}_0^t\mathbf{B}_{L1}
$$

$$
{}_0\mathbf{e}^T = [{}_0e_{11} \quad {}_0e_{22} \quad {}_0e_{33} \quad 2{}_0e_{12} \quad 2{}_0e_{23} \quad 2{}_0e_{13}]
$$

$$
\hat{\mathbf{u}}^T = [u_1^1 \quad u_2^1 \quad u_3^1 \quad \alpha_1 \quad \beta_1; \quad u_1^2 \quad u_2^2 \quad \ldots.; \quad u_1^N \quad u_2^N \quad u_3^N \quad \alpha_N \quad \beta_N]
$$

$$
{}_0^t\mathbf{B}_{L0} = \left[\cdots \begin{array}{|ccccc|} {}_0h_{k,1} & 0 & 0 & {}^tg_{11}^k\,{}_0G_1^k & {}^tg_{21}^k\,{}_0G_1^k \\ 0 & {}_0h_{k,2} & 0 & {}^tg_{12}^k\,{}_0G_2^k & {}^tg_{22}^k\,{}_0G_2^k \\ 0 & 0 & {}_0h_{k,3} & {}^tg_{13}^k\,{}_0G_3^k & {}^tg_{23}^k\,{}_0G_3^k \\ {}_0h_{k,2} & {}_0h_{k,1} & 0 & ({}^tg_{11}^k\,{}_0G_2^k + {}^tg_{12}^k\,{}_0G_1^k) & ({}^tg_{21}^k\,{}_0G_2^k + {}^tg_{22}^k\,{}_0G_1^k) \\ 0 & {}_0h_{k,3} & {}_0h_{k,2} & ({}^tg_{12}^k\,{}_0G_3^k + {}^tg_{13}^k\,{}_0G_2^k) & ({}^tg_{22}^k\,{}_0G_3^k + {}^tg_{23}^k\,{}_0G_2^k) \\ {}_0h_{k,3} & 0 & {}_0h_{k,1} & ({}^tg_{11}^k\,{}_0G_3^k + {}^tg_{13}^k\,{}_0G_1^k) & ({}^tg_{21}^k\,{}_0G_3^k + {}^tg_{23}^k\,{}_0G_1^k) \end{array} \cdots \right]
$$

for nodal point k

$$
{}_0^t\mathbf{B}_{L1} = \left[\cdots \begin{array}{|ccccc|} \ell_{11}\,{}_0h_{k,1} & \ell_{21}\,{}_0h_{k,1} & \ell_{31}\,{}_0h_{k,1} & \phi_{11}^k\,{}_0G_1^k & \phi_{21}^k\,{}_0G_1^k \\ \ell_{12}\,{}_0h_{k,2} & \ell_{22}\,{}_0h_{k,2} & \ell_{32}\,{}_0h_{k,2} & \phi_{12}^k\,{}_0G_2^k & \phi_{22}^k\,{}_0G_2^k \\ \ell_{13}\,{}_0h_{k,3} & \ell_{23}\,{}_0h_{k,3} & \ell_{33}\,{}_0h_{k,3} & \phi_{13}^k\,{}_0G_3^k & \phi_{23}^k\,{}_0G_3^k \\ (\ell_{11}\,{}_0h_{k,2} + \ell_{12}\,{}_0h_{k,1}) & (\ell_{21}\,{}_0h_{k,2} + \ell_{22}\,{}_0h_{k,1}) & (\ell_{31}\,{}_0h_{k,2} + \ell_{32}\,{}_0h_{k,1}) & (\phi_{12}^k\,{}_0G_1^k + \phi_{11}^k\,{}_0G_2^k) & (\phi_{22}^k\,{}_0G_1^k + \phi_{21}^k\,{}_0G_2^k) \\ (\ell_{12}\,{}_0h_{k,3} + \ell_{13}\,{}_0h_{k,2}) & (\ell_{22}\,{}_0h_{k,3} + \ell_{23}\,{}_0h_{k,2}) & (\ell_{32}\,{}_0h_{k,3} + \ell_{33}\,{}_0h_{k,2}) & (\phi_{13}^k\,{}_0G_2^k + \phi_{12}^k\,{}_0G_3^k) & (\phi_{23}^k\,{}_0G_2^k + \phi_{22}^k\,{}_0G_3^k) \\ (\ell_{11}\,{}_0h_{k,3} + \ell_{13}\,{}_0h_{k,1}) & (\ell_{21}\,{}_0h_{k,3} + \ell_{23}\,{}_0h_{k,1}) & (\ell_{31}\,{}_0h_{k,3} + \ell_{33}\,{}_0h_{k,1}) & (\phi_{13}^k\,{}_0G_1^k + \phi_{11}^k\,{}_0G_3^k) & (\phi_{23}^k\,{}_0G_1^k + \phi_{21}^k\,{}_0G_3^k) \end{array} \cdots \right]
$$

for nodal point k

where

$$
\ell_{ij} = \frac{\partial^t u_i}{\partial^0 x_j}; \qquad \phi_{ij}^k = \sum_{m=1}^{3} {}^tg_{im}^k\,\ell_{mj}
$$

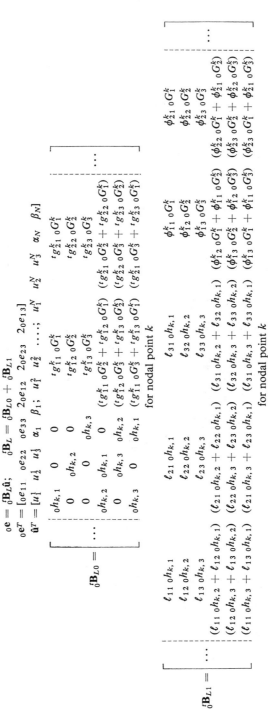

TABLE 6.8 *Matrices used in the shell element formulations. (continued)*

2. *Nonlinear Strain-Displacement Transformation Matrix*

$$
{}_0^t\mathbf{B}_{NL} = \left[\begin{array}{c|ccccc|c}
 & {}_0h_{k,1} & 0 & 0 & {}^tg_{11}^k\,{}_0G_1^k & {}^tg_{21}^k\,{}_0G_1^k & \\
 & 0 & {}_0h_{k,1} & 0 & {}^tg_{12}^k\,{}_0G_1^k & {}^tg_{22}^k\,{}_0G_1^k & \\
 & 0 & 0 & {}_0h_{k,1} & {}^tg_{13}^k\,{}_0G_1^k & {}^tg_{23}^k\,{}_0G_1^k & \\
 & {}_0h_{k,2} & 0 & 0 & {}^tg_{11}^k\,{}_0G_2^k & {}^tg_{21}^k\,{}_0G_2^k & \\
\cdots & 0 & {}_0h_{k,2} & 0 & {}^tg_{12}^k\,{}_0G_2^k & {}^tg_{22}^k\,{}_0G_2^k & \cdots \\
 & 0 & 0 & {}_0h_{k,2} & {}^tg_{13}^k\,{}_0G_2^k & {}^tg_{23}^k\,{}_0G_2^k & \\
 & {}_0h_{k,3} & 0 & 0 & {}^tg_{11}^k\,{}_0G_3^k & {}^tg_{21}^k\,{}_0G_3^k & \\
 & 0 & {}_0h_{k,3} & 0 & {}^tg_{12}^k\,{}_0G_3^k & {}^tg_{22}^k\,{}_0G_3^k & \\
 & 0 & 0 & {}_0h_{k,3} & {}^tg_{13}^k\,{}_0G_3^k & {}^tg_{23}^k\,{}_0G_3^k &
\end{array} \right]
$$

for nodal point k

3. *Stress Matrix*

$$
{}_0^t\mathbf{S} = \begin{bmatrix} {}_0^tS_{11}\mathbf{I}_3 & & \text{symm.} \\ {}_0^tS_{12}\mathbf{I}_3 & {}_0^tS_{22}\mathbf{I}_3 & \\ {}_0^tS_{13}\mathbf{I}_3 & {}_0^tS_{23}\mathbf{I}_3 & {}_0^tS_{33}\mathbf{I}_3 \end{bmatrix}
$$

where

$$
\mathbf{I}_3 = \begin{bmatrix} 1 & 0 & 0 \\ 0 & 1 & 0 \\ 0 & 0 & 1 \end{bmatrix}
$$

4. *Stress Vector*

$$
{}_0^t\hat{\mathbf{S}}^T = [{}_0^tS_{11} \quad {}_0^tS_{22} \quad {}_0^tS_{33} \quad {}_0^tS_{12} \quad {}_0^tS_{23} \quad {}_0^tS_{13}]
$$

B. *Updated Lagrangian Formulation*

1. *Linear Strain-Displacement Transformation Matrix*

$$
{}_t\mathbf{e} = {}^t_t\mathbf{B}_L \hat{\mathbf{u}}
$$

$$
{}_t\mathbf{e}^T = [{}_te_{11} \quad {}_te_{22} \quad {}_te_{33} \quad 2{}_te_{12} \quad 2{}_te_{23} \quad 2{}_te_{13}]
$$

$$
\hat{\mathbf{u}}^T = [u_1^1 \quad u_2^1 \quad u_3^1 \quad \alpha_1 \quad \beta_1; \quad u_1^2 \quad u_2^2 \quad \ldots.; \quad u_1^N \quad u_2^N \quad u_3^N \quad \alpha_N \quad \beta_N]
$$

$$
{}^t_t\mathbf{B}_L = \left[\cdots \left| \begin{array}{ccccc}
{}_th_{k,1} & 0 & 0 & {}^tg_{11}^k\,{}_tG_1^k & {}^tg_{21}^k\,{}_tG_1^k \\
0 & {}_th_{k,2} & 0 & {}_tg_{12}^k\,{}_tG_2^k & {}^tg_{22}^k\,{}_tG_2^k \\
0 & 0 & {}_th_{k,3} & {}^tg_{13}^k\,{}_tG_3^k & {}^tg_{23}^k\,{}_tG_3^k \\
{}_th_{k,2} & {}_th_{k,1} & 0 & ({}^tg_{11}^k\,{}_tG_2^k + {}^tg_{12}^k\,{}_tG_1^k) & ({}^tg_{21}^k\,{}_tG_2^k + {}^tg_{22}^k\,{}_tG_1^k) \\
0 & {}_th_{k,3} & {}_th_{k,2} & ({}^tg_{12}^k\,{}_tG_3^k + {}^tg_{13}^k\,{}_tG_2^k) & ({}^tg_{22}^k\,{}_tG_3^k + {}^tg_{23}^k\,{}_tG_2^k) \\
{}_th_{k,3} & 0 & {}_th_{k,1} & ({}^tg_{11}^k\,{}_tG_3^k + {}^tg_{13}^k\,{}_tG_1^k) & ({}^tg_{21}^k\,{}_tG_3^k + {}^tg_{23}^k\,{}_tG_1^k)
\end{array} \right| \cdots \right]
$$

for nodal point k

2. *Nonlinear Strain-Displacement Transformation Matrix*

$$
{}^t_t\mathbf{B}_{NL} = \left[\cdots \left| \begin{array}{ccccc}
{}_th_{k,1} & 0 & 0 & {}^tg_{11}^k\,{}_tG_1^k & {}^tg_{21}^k\,{}_tG_1^k \\
0 & {}_th_{k,1} & 0 & {}^tg_{12}^k\,{}_tG_1^k & {}^tg_{22}^k\,{}_tG_1^k \\
0 & 0 & {}_th_{k,1} & {}^tg_{13}^k\,{}_tG_1^k & {}^tg_{23}^k\,{}_tG_1^k \\
{}_th_{k,2} & 0 & 0 & {}^tg_{11}^k\,{}_tG_2^k & {}^tg_{21}^k\,{}_tG_2^k \\
0 & {}_th_{k,2} & 0 & {}^tg_{12}^k\,{}_tG_2^k & {}^tg_{22}^k\,{}_tG_2^k \\
0 & 0 & {}_th_{k,2} & {}^tg_{13}^k\,{}_tG_2^k & {}^tg_{23}^k\,{}_tG_2^k \\
{}_th_{k,3} & 0 & 0 & {}^tg_{11}^k\,{}_tG_3^k & {}^tg_{21}^k\,{}_tG_3^k \\
0 & {}_th_{k,3} & 0 & {}^tg_{12}^k\,{}_tG_3^k & {}^tg_{22}^k\,{}_tG_3^k \\
0 & 0 & {}_th_{k,3} & {}^tg_{13}^k\,{}_tG_3^k & {}^tg_{23}^k\,{}_tG_3^k
\end{array} \right| \cdots \right]
$$

for nodal point k

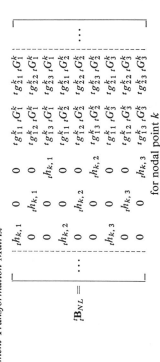

TABLE 6.8 *Matrices used in the shell element formulations. (continued)*

3. *Stress Matrix*

$$ {}^{t}\boldsymbol{\tau} = \begin{bmatrix} {}^{t}\tau_{11}\mathbf{I}_3 & & \text{sym.} \\ {}^{t}\tau_{12}\mathbf{I}_3 & {}^{t}\tau_{22}\mathbf{I}_3 & \\ {}^{t}\tau_{13}\mathbf{I}_3 & {}^{t}\tau_{23}\mathbf{I}_3 & {}^{t}\tau_{33}\mathbf{I}_3 \end{bmatrix} $$

where

$$ \mathbf{I}_3 = \begin{bmatrix} 1 & 0 & 0 \\ 0 & 1 & 0 \\ 0 & 0 & 1 \end{bmatrix} $$

4. *Stress Vector*

$$ {}^{t}\hat{\boldsymbol{\tau}}^{T} = [{}^{t}\tau_{11} \quad {}^{t}\tau_{22} \quad {}^{t}\tau_{33} \quad {}^{t}\tau_{12} \quad {}^{t}\tau_{23} \quad {}^{t}\tau_{13}] $$

were formulated using the same basic equations. Therefore, these two elements can be employed together to model stiffened plate and shell structures.

In the above discussion we pointed out the generality of the above shell element; however, it should also be mentioned that the use of this general element can be relatively expensive compared to the use of other elements, in particular when these have been developed specifically for the analysis of certain shell problems. This difference in cost-effectiveness may at present be significant in the analysis of specific problems, but we can expect it to become less pronounced as the formulation of the general shell element is further refined through ongoing research efforts.

6.4 USE OF CONSTITUTIVE RELATIONS

In Section 6.3 we discussed the evaluation of the displacement and strain-displacement relations for various elements. We pointed out that the evaluation of these kinematic relations is quite straightforward and that they yield an accurate description of the geometric nonlinear behavior of an element because the basis of these derivations is the general virtual work principle. As a consequence, the kinematic descriptions of the two- and three-dimensional elements admit the accurate representation of large displacements, large rotations, and large strains; i.e., no kinematic assumptions whatsoever have been made, whereas the truss, beam, and shell element matrices have been formulated to allow arbitrary large displacements and rotations but only small strains.*

The kinematic descriptions in the element formulations are therefore very general. However, it must be noted that in order for a formulation of an element to be applicable to a specific response prediction, *it is also necessary to use appropriate constitutive descriptions.* Clearly, the finite element equilibrium equations contain the displacement and strain-displacement matrices plus the constitutive matrix of the material (see Table 6.4). Therefore, *in order for a formulation to be applicable to a certain response prediction, it is imperative that both the kinematic and the constitutive descriptions be appropriate.* For example, assume that the T.L. formulation is employed to describe the kinematic behavior of a two-dimensional element and a material law is used which is only formulated for small strain conditions. In this case the analysis can only model small strain conditions, although the T.L. kinematic formulation does admit large strains.

The objective in this section is to present and discuss some fundamental observations pertaining to the use of material laws in nonlinear finite element analysis. A great deal of different material laws are employed in practice, and we shall not attempt to survey and summarize these models. Instead, our only objectives are to discuss the stress and strain tensors that are used effectively with certain classes of material models and to present some important general observations pertaining to the material model implementations and their use.

The three classes of models that we consider in the following sections are those with which we are widely concerned in practice, namely, elastic, elastic-

* For example, for the structural elements to be applicable to large strain computations, in the U.L. formulations the cross-sectional areas (truss, beam and shell thicknesses) would have to be updated.

plastic and creep material models. Some basic properties of these material descriptions are given in Table 6.9, which provides a very brief overview of the major classes of material behavior.

TABLE 6.9 *Overview of some material descriptions.*

Material Model	*Characteristics*	*Examples*
Elastic linear or nonlinear elastic	Stress is a function of strain only; same stress path on loading as on unloading. $${}^{t}\sigma_{ij} = {}^{t}C_{ijrs}\, {}^{t}e_{rs}$$ linear elastic: ${}^{t}C_{ijrs}$ is constant; nonlinear elastic: ${}^{t}C_{ijrs}$ varies as a function of strain	Almost all materials provided the stresses are small enough: steels, cast iron, glass, rock, wood, and so on before yielding or fracture
Hyperelastic	Stress is calculated from a strain energy functional W, $${}_{0}^{t}S_{ij} = \frac{\partial W}{\partial\, {}_{0}^{t}\epsilon_{ij}}$$	Rubbers, e.g., Mooney-Rivlin model[4]
Hypoelastic	Stress increments are calculated from strain increments $$d\sigma_{ij} = C_{ijrs}\, de_{rs}$$ The material moduli C_{ijrs} are defined as functions of stress, strain, fracture criteria, loading and unloading parameters, maximum strains reached, and so on.	Concrete model, curve description model in ADINA[11]
Elastic-plastic	Linear elastic behavior until yield, use of yield condition, flow rule, and hardening rule to calculate stress and plastic strain increments; plastic strain increments are instantaneous.	Metals, soils, rocks, when subjected to high stresses
Creep	Time effect of increasing strains under constant load, or decreasing stresses under constant deformations; creep strain increments are non-instantaneous.	Metals at high temperatures
Viscoplasticity	Combined plastic and creep effects	Polymers, metals

6.4.1 Elastic Material Behavior

A simple and widely used elastic material description for large deformation analysis is obtained by generalizing the linear elastic relations summarized in Chapter 4 (see Table 4.3) to the T.L. formulation:

$${}_{0}^{t}S_{ij} = {}_{0}^{t}C_{ijrs}\, {}_{0}^{t}\epsilon_{rs} \tag{6.111}$$

where the ${}_{0}^{t}S_{ij}$ and ${}_{0}^{t}\epsilon_{rs}$ are components of the 2nd Piola–Kirchhoff stress and Green–Lagrange strain tensors, and the ${}_{0}^{t}C_{ijrs}$ are components of the constant elasticity tensor. Considering three-dimensional stress conditions we have

$$ {}_0^tC_{ijrs} = \lambda\, \delta_{ij}\, \delta_{rs} + \mu(\delta_{ir}\, \delta_{js} + \delta_{is}\, \delta_{jr}) \tag{6.112}$$

where λ and μ are the Lamé constants and δ_{ij} is the Kronecker delta,

$$\lambda = \frac{Ev}{(1+v)(1-2v)}; \qquad \mu = \frac{E}{2(1+v)}$$

$$\delta_{ij} = \begin{cases} 0; & i \neq j \\ 1; & i = j \end{cases}$$

The components of the elasticity tensor given in (6.112) are identical to the values given in Table 4.3.

Considering this material description we can make a number of important observations. We recognize that in infinitesimal displacement analysis, the relation in (6.111) reduces to the description used in linear elastic analysis, because under these conditions the stress and strain variables reduce to the engineering stress and strain measures. However, an important observation is that *in large displacement and large rotation but small strain analysis*, the relation in (6.111) provides a natural material description, because the 2nd Piola–Kirchhoff stress and Green–Lagrange strain tensors are invariant under rigid body rotations (see Section 6.2.2 and Examples 6.6 and 6.13). Thus, only the actual straining of the material will yield an increase in the components of the stress tensor, and as long as this material straining (accompained by large rotations and displacements) is small, the use of the relation in (6.111) is completely equivalent to using Hooke's law in infinitesimal displacement conditions.

The fundamental observation that "the stress state is invariant under rigid body motions if the 2nd Piola–Kirchhoff stress and Green–Lagrange strain tensors are used in (6.111)" is important not only for elastic analysis. Indeed, this observation implies that any material description which has been developed for infinitesimal displacement analysis using engineering stress and strain measures can directly be employed in large displacement and large rotation but small strain analysis, provided 2nd Piola–Kirchhoff stresses and Green–Lagrange strains are used. Figure 6.20 illustrates this fundamental fact. A practical consequence is, for example, that elastic-plastic and creep material models (see Section 6.4.2) can directly be employed for large displacement, large rotation, and small strain analysis by simply substituting 2nd Piola–Kirchhoff stresses and Green–Lagrange strains for the engineering stress and strain measures.

The use of the constant constitutive coefficients given in (6.111) is only appropriate for real materials provided that the material strains are relatively small. At larger strain levels the generalized Hooke's law does not represent an appropriate material description because the stress-strain relation is nonlinear. In nonlinear elasticity we still use the relation in (6.111), but now the components of the constitutive tensor are given in terms of certain material constants and they are a function of the material strains. We consider as an example an isotropic material, in which case the stress-strain relation can frequently be defined using a strain energy function W, given in terms of the Green–Lagrange strains. Such a material is referred to as a hyperelastic material and we have[1]

$$ {}_0^tS_{ij} = \frac{\partial W}{\partial\, {}_0^t\epsilon_{ij}} \tag{6.113}$$

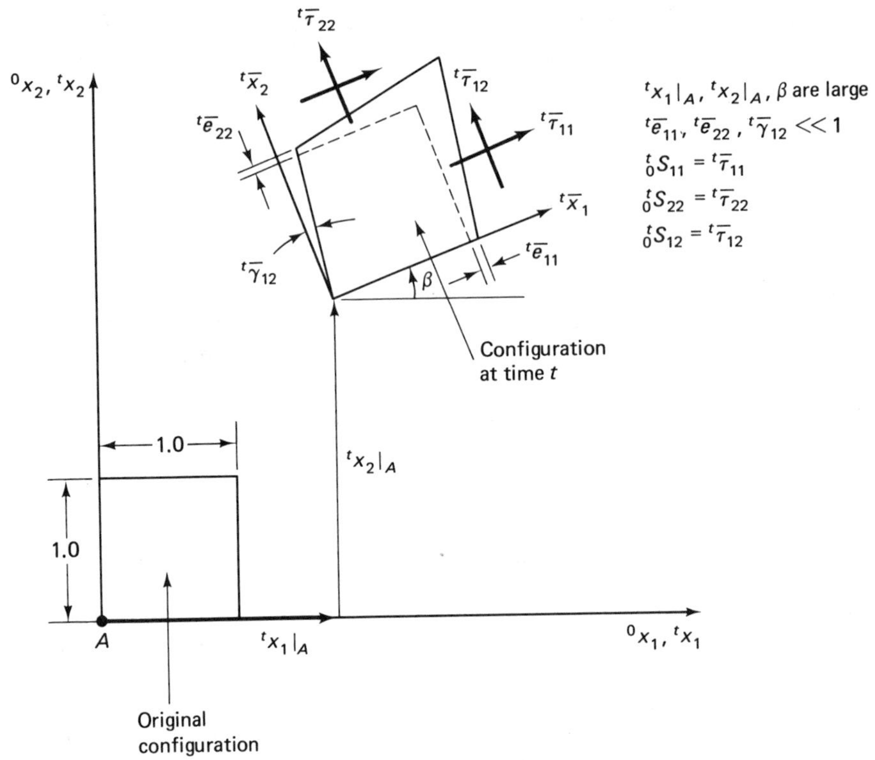

FIGURE 6.20 *Large displacement/large rotation but small strain conditions.*

On studying the possible structure of the strain energy function (or elastic potential function) W and invoking the requirement of coordinate frame indifference we can conclude that W can only be a function of the strain tensor invariants (since these are independent of the coordinate system chosen). Therefore, we have $W(I_1, I_2, I_3)$ where

$$\begin{aligned} I_1 &= {}_0^t\epsilon_{ii} \\ I_2 &= \tfrac{1}{2}\,{}_0^t\epsilon_{ij}\,{}_0^t\epsilon_{ij} - \tfrac{1}{2}I_1^2 \\ I_3 &= \det {}_0^t\boldsymbol{\epsilon} \end{aligned} \tag{6.114}$$

Of particular interest are rubber-like materials that can be assumed to be incompressible, in which case we have

$$\frac{{}^t\rho}{{}^0\rho} = 1 \tag{6.115}$$

or using (6.20) and (6.33)

$$\det {}_0^t\mathbf{X} = 1; \qquad \det {}_0^t\mathbf{C} = 1 \tag{6.116}$$

Once an appropriate function W has been determined by experiments we can use (6.113) to evaluate the stress-strain relation, which for an incompressible material must also contain the condition in (6.115).

In the above discussion we considered the stress-strain description given in (6.111) which implicitly assumes that a T.L. formulation is used to analyze the physical problem. Let us now assume that we want to employ a U.L. formulation in the analysis but that we are given the constitutive relationship in (6.111). In this case we can write, substituting (6.111) into (6.37),

$$\int_{^0V} {}_0^tC_{ijrs}\, {}_0^t\epsilon_{ij}\, \delta\, {}_0^t\epsilon_{rs}\, {}^0dV = {}^t\mathcal{R} \tag{6.117}$$

Thus, if we define a new constitutive tensor

$${}_t^tC_{mnpq} = \frac{{}^t\rho}{{}^0\rho}\, {}_0^tx_{m,i}\, {}_0^tx_{n,j}\, {}_0^tC_{ijrs}\, {}_0^tx_{p,r}\, {}_0^tx_{q,s} \tag{6.118}$$

meaning that

$${}_0^tC_{ijrs} = \frac{{}^0\rho}{{}^t\rho}\, {}_t^0x_{i,m}\, {}_t^0x_{j,n}\, {}_t^tC_{mnpq}\, {}_t^0x_{r,p}\, {}_t^0x_{s,q} \tag{6.119}$$

and if we use (6.35), we recognize that (6.117) can be written as

$$\int_{^tV} {}_t^tC_{mnpq}\, {}_t^t\epsilon_{mn}\, \delta_t e_{pq}\, {}^tdV = {}^t\mathcal{R} \tag{6.120}$$

where the ${}_t^t\epsilon_{mn}$ are the components of the *Almansi strain tensor*,

$${}_t^t\epsilon_{mn} = {}_t^0x_{i,m}\, {}_t^0x_{j,n}\, {}_0^t\epsilon_{ij} \tag{6.121}$$

As with the Green–Lagrange strain tensor, the Almansi strain tensor can also be defined in a number of different but completely equivalent ways.* Namely, we also have corresponding to (6.23),

$${}_t^t\epsilon_{ij} = \tfrac{1}{2}({}_t^tu_{i,j} + {}_t^tu_{j,i} - {}_t^tu_{k,i}\, {}_t^tu_{k,j}) \tag{6.122}$$

and corresponding to (6.25)

$${}_t^t\epsilon_{ij}\, d^tx_i\, d^tx_j = \tfrac{1}{2}\{({}^tds)^2 - ({}^0ds)^2\} \tag{6.123}$$

EXAMPLE 6.19: Prove that the definitions of the Almansi strain tensor given in (6.121), (6.122), and (6.123) are all equivalent.

The relation in (6.121) can be written in matrix form as

$${}_t^t\boldsymbol{\epsilon} = {}_t^0\mathbf{X}^T\, {}_0^t\boldsymbol{\epsilon}\, {}_t^0\mathbf{X} \tag{a}$$

But using (6.32) to substitute for ${}_0^t\boldsymbol{\epsilon}$ in (a) and recognizing that

$${}_t^0\mathbf{X}\ \ {}_0^t\mathbf{X} = \mathbf{I}$$

we obtain

$${}_t^t\boldsymbol{\epsilon} = \tfrac{1}{2}(\mathbf{I} - {}_t^0\mathbf{X}^T\, {}_t^0\mathbf{X}) \tag{b}$$

Substituting now for ${}_t^0\mathbf{X}$ using $\quad {}_t^0\mathbf{X} = [{}_t\boldsymbol{\nabla}\, {}^0\mathbf{x}^T]^T$

where, in accordance with (6.18),

$${}_t\boldsymbol{\nabla} = \begin{bmatrix} \dfrac{\partial}{\partial^t x_1} \\ \dfrac{\partial}{\partial^t x_2} \\ \dfrac{\partial}{\partial^t x_3} \end{bmatrix}; \qquad {}^0\mathbf{x}^T = [{}^0x_1 \quad {}^0x_2 \quad {}^0x_3]$$

and following the procedure in (6.34) we obtain

$${}_t^t\boldsymbol{\epsilon} = \tfrac{1}{2}\{\mathbf{I} - [{}_t\boldsymbol{\nabla}({}^t\mathbf{x}^T - {}^t\mathbf{u}^T)][{}_t\boldsymbol{\nabla}({}^t\mathbf{x}^T - {}^t\mathbf{u}^T)]^T\}$$

Since

$${}_t\boldsymbol{\nabla}\, {}^t\mathbf{x}^T = \mathbf{I}$$

* However, in contrast to the Green–Lagrange strain tensor, the components of the Almansi strain tensor are not invariant under a rigid body rotation of the material.

we thus obtain $\quad {}_t^t\boldsymbol{\epsilon} = \frac{1}{2}\{\mathbf{I} - [\mathbf{I} - {}_t\nabla\, {}^t\mathbf{u}^T][\mathbf{I} - {}_t\nabla\, {}^t\mathbf{u}^T]^T\}$

or
$$ {}_t^t\boldsymbol{\epsilon} = \frac{1}{2}\{{}_t\nabla\, {}^t\mathbf{u}^T + ({}_t\nabla\, {}^t\mathbf{u}^T)^T - ({}_t\nabla\, {}^t\mathbf{u}^T)({}_t\nabla\, {}^t\mathbf{u}^T)^T\} \tag{c}$$

and the components of ${}_t^t\boldsymbol{\epsilon}$ in (c) are the relations in (6.122).

To show that (6.123) also holds, we use the relation in (b) to obtain

$$ d^t\mathbf{x}^T\, {}_t^t\boldsymbol{\epsilon}\, d^t\mathbf{x} = \frac{1}{2}(d^t\mathbf{x}^T\, d^t\mathbf{x} - d^0\mathbf{x}^T\, d^0\mathbf{x}) \tag{d}$$

because $\quad d^0\mathbf{x} = {}_t^0\mathbf{X}\, d^t\mathbf{x}$

But (d) can also be written as

$$ d^t\mathbf{x}^T\, {}_t^t\boldsymbol{\epsilon}\, d^t\mathbf{x} = \frac{1}{2}\{({}^t ds)^2 - ({}^0 ds)^2\} \tag{e}$$

because $\quad d^t\mathbf{x}^T\, d^t\mathbf{x} = ({}^t ds)^2; \qquad d^0\mathbf{x}^T\, d^0\mathbf{x} = ({}^0 ds)^2$

and (e) is equivalent to (6.123).

The relations in (6.117) to (6.120) show that a U.L. formulation can be employed when the constitutive tensor ${}_0^tC_{ijrs}$ is given, provided the Cauchy stress components are calculated using

$$ {}^t\tau_{mn} = {}_t^tC_{mnpq}\, {}_t^t\epsilon_{pq} \tag{6.124}$$

and the components of the constitutive tensor ${}_t^tC_{mnpq}$ are evaluated using (6.118). Considering a finite element analysis, if these conditions are fulfilled, identical numerical results are predicted for any level of stress and strain using either the T.L. or U.L. formulation. Furthermore, we can directly conclude that the reverse is also true; i.e., if the constitutive components ${}_t^tC_{mnpq}$ are given, a T.L. formulation may be employed and will yield the same results as the U.L. formulation provided the components ${}_0^tC_{ijrs}$ are evaluated using (6.119).

In practice it is most natural to employ that formulation for which the constitutive relations are given explicitly; i.e. if the components ${}_0^tC_{ijrs}$ are given, the T.L. formulation is used for the finite element formulation, and if the components ${}_t^tC_{mnpq}$ are available, the U.L. formulation is employed. This is usually effective because otherwise the transformations in (6.118) and (6.119) are required and these must be performed at each element integration point during the numerical integration of the element stiffness matrices and force vectors.

The distinction between the constitutive components corresponding to the U.L. and T.L. formulations is usually very important, because completely erroneous results can be generated if a kinematic formulation is employed with an inappropriate constitutive relation. However, in one practical area of analysis a special situation arises, namely, when the generalized Hooke's law is employed for an isotropic material in a small strain analysis. In this case, the relations in (6.112) can be used or, alternatively, we surely could also employ

$$ {}^t\tau_{ij} = {}_t^tC_{ijrs}\, {}_t^t\epsilon_{rs} \tag{6.125}$$

where
$$ {}_t^tC_{ijrs} = \lambda\, \delta_{ij}\, \delta_{rs} + \mu(\delta_{ir}\, \delta_{js} + \delta_{is}\, \delta_{jr}) \tag{6.126}$$

$$ \lambda = \frac{E\nu}{(1+\nu)(1-2\nu)} \qquad \mu = \frac{E}{2(1+\nu)}$$

and δ_{ij} is the Kronecker delta.

It should be noted that in (6.112) and (6.126) the same Young's modulus and Poisson's ratio are used (which have been evaluated experimentally under small strain conditions), and that there now appears to be a question as to whether the generalized Hooke's law applies to the T.L. or the U.L. formulation. Figure 6.21 illustrates the different results that are generated in the U.L. and T.L. formulations when the *same* values for Young's modulus and Poisson's ratio are used. However, we already pointed out that in practice the generalized Hooke's law is only applicable to small strain (but large displacement and large rotation) analysis, and in this case the use of either (6.112) or (6.126) with the same material constants in the T.L. and U.L. formulations, respectively, yields practically the same analysis results. The reason is that considering large displacements and rotations but small strains, the transformations on the constitutive tensors given in (6.118) and (6.119) reduce to mere rotations (see Example 6.12). Therefore, since the material is assumed to be isotropic, the transformations do not change the components of the constitutive tensors and the use of either (6.111) and (6.112) or (6.125) and (6.126) to characterize the material response is quite equivalent.

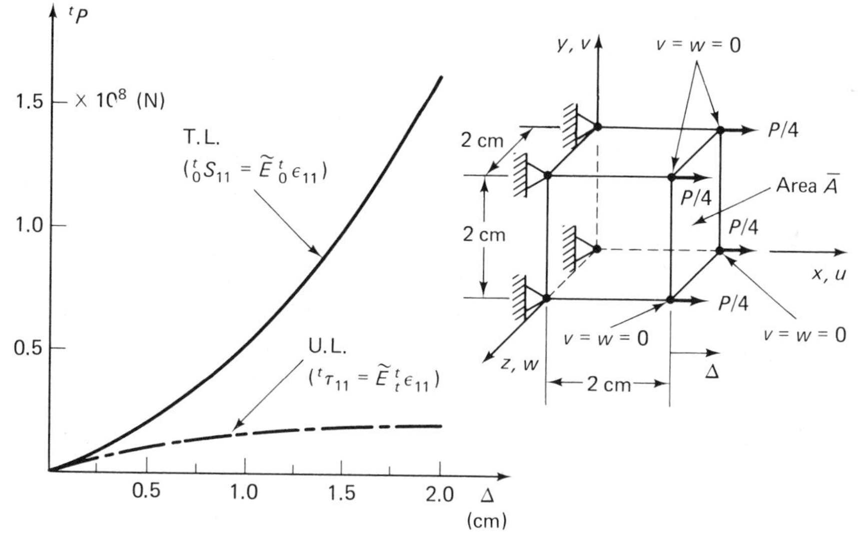

(a) $P - \Delta$ response of 8-node element under uniform loading ($E = 10^7$ N/cm^2, $\nu = 0.30$)

Stress-strain relation:	$\widetilde{E} = \dfrac{E(1-\nu)}{(1+\nu)(1-2\nu)}$
Deformation gradient:	${}_0^tX_{11} = {}^tL/{}^0L$
Green-Lagrange strain:	${}_0^t\epsilon_{11} = \frac{1}{2}\left\{({}^tL/{}^0L)^2 - 1\right\}$
Almansi strain:	${}_t^t\epsilon_{11} = \frac{1}{2}\left\{1 - ({}^0L/{}^tL)^2\right\}$
Conservation of mass:	${}^0\rho/{}^t\rho = {}^tL/{}^0L$
Cauchy stress:	${}^t\tau_{11} = {}^tP/\overline{A}$
2nd Piola-Kirchhoff stress:	${}_0^tS_{11} = ({}^tP/\overline{A})({}^0L/{}^tL)$

(b) Basic relations

FIGURE 6.21 *One-dimensional response analysis.*

Finally, we should note that we did not discuss the use of the Jaumann stress rate tensor in elastic material descriptions. The reason is that the use of a total stress–total strain description as discussed above is more natural and computationally more effective in elastic analysis. An inefficiency arises in the description of an elastic constitutive response using stress and strain rates because an integration is required during the incremental load (or time) steps, as discussed in more detail in Section 6.4.2 (see also Example 6.9).

6.4.2 Inelastic Material Behavior with Special Reference to Elasto-Plasticity and Creep

A fundamental observation comparing elastic and inelastic analysis is that in elastic solutions the total stress can be evaluated from the total strain alone (as given in (6.111) and (6.124)), whereas in an inelastic response calculation the total stress at time t also depends on the stress and strain history. Typical inelastic phenomena are elasto-plasticity, creep and viscoplasticity,[1,2] and a very large number of material models have been developed in order to characterize such a material response. Our objective is again not to summarize or survey the models available, but rather to present some of the basic finite element procedures that are employed in inelastic response calculations. We are mainly concerned with some formulations and numerical procedures that are employed efficiently, and the general approach followed in inelastic finite element analysis. Since a very important field is elastic-plastic analysis, we primarily consider a classical elasto-plasticity material description in our discussion of the general finite element procedures used. Most of these general techniques could also be employed in the finite element implementation of other inelastic material models, although for specific models some additional considerations may be necessary.

Considering the incremental analysis of inelastic response, there are basically three kinematic formulations that are employed effectively:

Materially-nonlinear-only analysis. This formulation assumes infinitesimally small displacements and rotations, and considers only material nonlinearities. As long as the material is elastic, the solution using this formulation is identical to the linear elastic solution discussed in Section 4.2.

Total Lagrangian formulation. As discussed in Section 6.2.3, the kinematic assumptions permit large displacements, large rotations and large strains. However, considering elastic-plastic analysis this formulation is most effectively employed only for large displacement, large rotation but small strain analysis. Assuming small strains the material model implementation used for materially-nonlinear-only analysis can directly be employed in the T.L. formulation for large displacement and large rotation analysis by simply substituting in the material characterization the 2nd Piola–Kirchhoff stresses and Green–Lagrange strains for the small displacement engineering stress and strain measures. The use of the T.L. formulation in inelastic analysis is therefore a direct extension of an elastic analysis using the T.L. formulation in small strain but large displacement and large rotation conditions. Considering the computer implementation, it is important to note that in this case the same subroutine which is employed in materially-nonlinear-only analysis is also used in the T.L. formulation. No modification to the subroutine is required, except that 2nd Piola–Kirchhoff stresses and Green–Lagrange strains, instead of the engineering stress and strain measures, are passed to the program.

Updated Lagrangian Jaumann stress rate (U.L.J.) formulation. The U.L.J. formulation is applicable to general elastic-plastic analysis with large displacements and rotations and large strains. The formulation is effective in large strain analysis, because the stress and strain measures used are those that describe the material response in a natural way. The basic equations employed in the U.L.J. formulation are the incremental virtual work equation of the U.L. formulation (to calculate the incremental displacements, see (6.41)) and the Jaumann stress rate equation (to calculate the Cauchy stresses, see (6.21)). Therefore, in the formulation we always operate on the Cauchy stresses (the true physical stresses), which can directly be employed to describe the material behavior. The only difference from materially-nonlinear-only analysis is that the configuration and volume of the body under consideration are not constant. This necessitates the use of a true stress-logarithmic strain diagram for the definition of the uniaxial material response.

As discussed above, the large displacement and large rotation (T.L. formulation) and large strain (U.L.J. formulation) analysis of elastic-plastic response is a relatively simple extension of a materially-nonlinear-only analysis. In the following we therefore consider only small displacement conditions, but Table 6.10 summarizes, based on the above discussion, which equations would be employed in the T.L. and U.L.J. formulations.

Consider now the construction of the stress-strain matrix in elasto-plasticity. Using the usual approach of flow theory to describe the elastic-plastic material

TABLE 6.10 *Elasto-plasticity formulations.*

$$\int_V C^{EP}_{ijrs}\, e_{ij}\, \delta e_{rs}\, dV = {}^{t+\Delta t}\mathcal{R} - \int_V {}^t\sigma_{ij}\, \delta e_{ij}\, dV$$

$${}^tF({}^t\sigma_{ij}, {}^t\kappa) = 0; \qquad de^P_{ij} = {}^t\lambda \frac{\partial\, {}^tF}{\partial\, {}^t\sigma_{ij}}; \qquad d\sigma_{ij} = C^E_{ijrs}(de_{rs} - de^P_{rs})$$

$${}^{t+dt}\sigma_{ij} = {}^t\sigma_{ij} + d\sigma_{ij}$$

(a) materially-nonlinear-only formulation (infinitesimally small displacements)

$$\int_{{}^0V} {}_0C^{EP}_{ijrs}\, {}_0e_{ij}\, \delta_0 e_{rs}\, {}^0dV + \int_{{}^0V} {}^t_0S_{ij}\, \delta_0\eta_{ij}\, {}^0dV = {}^{t+\Delta t}\mathcal{R} - \int_{{}^0V} {}^t_0S_{ij}\, \delta_0 e_{ij}\, {}^0dV$$

$${}^tF({}^t_0S_{ij}, {}^t\kappa) = 0; \qquad d_0\epsilon^P_{ij} = {}^t\lambda \frac{\partial\, {}^tF}{\partial\, {}^t_0S_{ij}}; \qquad d_0S_{ij} = {}_0C^E_{ijrs}(d_0\epsilon_{rs} - d_0\epsilon^P_{rs})$$

$${}^{t+dt}_{\ \ \ 0}S_{ij} = {}^t_0S_{ij} + d_0S_{ij}$$

(b) Total Lagrangian formulation (large displacements and large rotations, but small strains, ${}^t_0\bar{\epsilon}^P = \sum \sqrt{\tfrac{2}{3}\, d_0\epsilon^P_{ij}\, d_0\epsilon^P_{ij}} < 2\%$)

$$\int_{{}^tV} {}_tC^{EP}_{ijrs}\, {}_te_{ij}\, \delta_t e_{rs}\, {}^tdV + \int_{{}^tV} {}^t\tau_{ij}\, \delta_t\eta_{ij}\, {}^tdV = {}^{t+\Delta t}\mathcal{R} - \int_{{}^tV} {}^t\tau_{ij}\, \delta_t e_{ij}\, {}^tdV$$

$${}^tF({}^t\tau_{ij}, {}^t\kappa) = 0; \qquad d_te^P_{ij} = {}^t\lambda \frac{\partial\, {}^tF}{\partial\, {}^t\tau_{ij}}; \qquad {}^t\overset{\nabla}{\tau}_{ij}\, dt = C^E_{ijrs}(d_te_{rs} - d_te^P_{rs})$$

$${}^{t+dt}\tau_{ij} = {}^t\tau_{ij} + {}^t\overset{\nabla}{\tau}_{ij}\, dt + {}^t\tau_{ip}\, {}^t\Omega_{pj}\, dt + {}^t\tau_{jp}\, {}^t\Omega_{pi}\, dt$$

(c) Updated Lagrangian Jaumann stress rate (U.L.J.) formulation (large displacements, large rotations, large strains) behavior, three properties in addition to the elastic stress-strain relations characterize the material behavior:

1. *a yield condition*, which specifies the state of multiaxial stress corresponding to the start of plastic flow;
2. *a flow rule*, which relates the plastic strain increments to the current stresses and the stress increments subsequent to yielding; and
3. *a hardening rule*, which specifies how the yield condition is modified during plastic flow.

Considering isothermal conditions and isotropic hardening, the (initial and subsequent) yield condition can be written at time t as

$$^tF(^t\sigma_{ij}, {}^t\kappa) = 0 \tag{6.127}$$

where $^t\kappa$ is a state variable which depends on the plastic strains $^te^P_{ij}$. If in addition we restrict the analysis to an associated flow rule, the function ^tF in (6.127) is used to calculate the plastic strain increments

$$de^P_{ij} = {}^t\lambda \frac{\partial {}^tF}{\partial {}^t\sigma_{ij}} \tag{6.128}$$

where $^t\lambda$ is a scalar still to be determined. The relation in (6.128) is referred to as the normality rule. Since during plastic deformations $^tF = 0$, we also have

$$\frac{\partial {}^tF}{\partial {}^t\sigma_{ij}} d\sigma_{ij} + \frac{\partial {}^tF}{\partial {}^te^P_{ij}} de^P_{ij} = 0$$

where $d\sigma_{ij}$ and de^P_{ij} are differential increments in stresses and plastic strains.

For convenience of presentation let

$$^tq_{ij} = \frac{\partial {}^tF}{\partial {}^t\sigma_{ij}}; \qquad {}^tp_{ij} = -\frac{\partial {}^tF}{\partial {}^te^P_{ij}} \tag{6.129}$$

and to use matrix notation we also define

$$^t\mathbf{q}^T = [{}^tq_{11} \quad {}^tq_{22} \quad {}^tq_{33} \quad 2{}^tq_{12} \quad 2{}^tq_{23} \quad 2{}^tq_{31}]$$
$$^t\mathbf{p}^T = [{}^tp_{11} \quad {}^tp_{22} \quad {}^tp_{33} \quad {}^tp_{12} \quad {}^tp_{23} \quad {}^tp_{31}]$$

The stress increments are evaluated using

$$d\boldsymbol{\sigma} = \mathbf{C}^E(d\mathbf{e} - d\mathbf{e}^P) \tag{6.130}$$

where $\mathbf{C}^E$ is the elastic stress-strain matrix defined in Table 4.3. Using (6.128) to (6.130), we can now evaluate the scalar $^t\lambda$,

$$^t\lambda = \frac{{}^t\mathbf{q}^T\, \mathbf{C}^E\, d\mathbf{e}}{{}^t\mathbf{p}^T\, {}^t\mathbf{q} + {}^t\mathbf{q}^T\, \mathbf{C}^E\, {}^t\mathbf{q}} \tag{6.131}$$

Then substituting from (6.128) and (6.131) into (6.130), we obtain

$$d\boldsymbol{\sigma} = \mathbf{C}^{EP}\, d\mathbf{e} \tag{6.132}$$

where the matrix $\mathbf{C}^{EP'}$ represents the instantaneous elastic-plastic stress-strain matrix:

$$\mathbf{C}^{EP} = \mathbf{C}^E - \frac{\mathbf{C}^E\, {}^t\mathbf{q}\, (\mathbf{C}^E\, {}^t\mathbf{q})^T}{{}^t\mathbf{p}^T\, {}^t\mathbf{q} + {}^t\mathbf{q}^T\, \mathbf{C}^E\, {}^t\mathbf{q}} \tag{6.133}$$

The above stress-strain law depends on the yield function ^tF used, and depending on the material characteristics to be modeled, different yield conditions are employed in practice. We consider here the von Mises yield condition with isotropic hardening which is used, for example, to model steels.[1,2]

The von Mises yield surface is a cylinder in the principal stress space as shown in Fig. 6.22, and is given by

$$^tF = \tfrac{1}{2}{}^ts_{ij}\,{}^ts_{ij} - {}^t\kappa \tag{6.134}$$

in which the ${}^ts_{ij}$ are the deviatoric stresses

$$^ts_{ij} = {}^t\sigma_{ij} - \frac{{}^t\sigma_{mm}}{3}\,\delta_{ij}; \qquad {}^t\sigma_{mm} = \sum_m {}^t\sigma_{mm} \tag{6.135}$$

and

$$^t\kappa = \tfrac{1}{3}{}^t\sigma_y^2 \tag{6.136}$$

with ${}^t\sigma_y$ being the yield stress at time t. This yield stress is a function of the plastic work per unit volume ${}^tW_P$:

$$^tW_P = \int_0^{{}^te_{ij}^P} {}^t\sigma_{ij}\, de_{ij}^P \tag{6.137}$$

Evaluating ${}^tq_{ij}$ and ${}^tp_{ij}$ we obtain

$$^tq_{ij} = {}^ts_{ij}; \qquad {}^tp_{ij} = {}^tH\,{}^t\sigma_{ij} \tag{6.138}$$

where

$$^tH = \frac{2}{3}{}^t\sigma_y \frac{d\sigma_y}{dW_P} \tag{6.139}$$

The variable ${}^tH$ is zero in perfect plasticity. For work-hardening materials usually data from a simple tension test are known and then

$$^tH = \frac{2}{3}\left(\frac{E\,E_T}{E - E_T}\right)$$

where E_T is the strain-hardening modulus illustrated in Fig. 6.22.

The above relations are substituted into (6.133) to obtain the stress-strain matrix summarized in Table 6.11. In this table the stress-strain matrix for

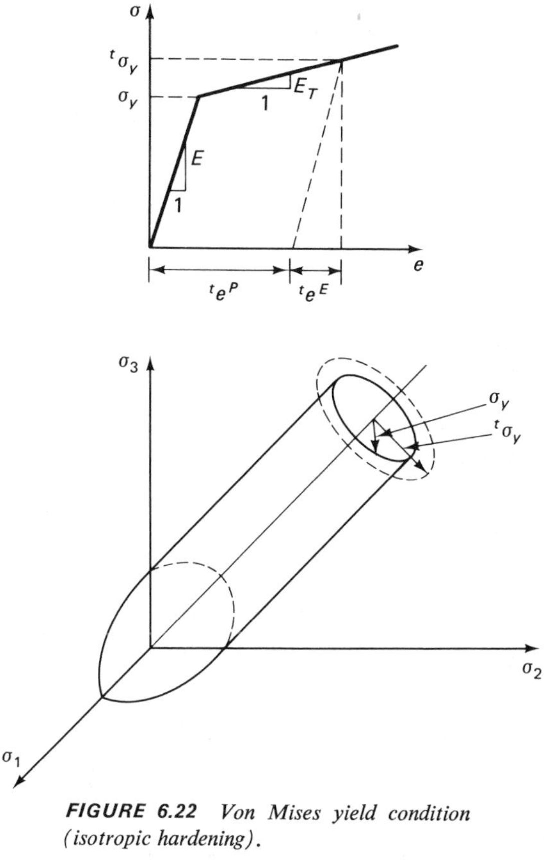

FIGURE 6.22 *Von Mises yield condition (isotropic hardening).*

TABLE 6.11 *Stress-strain matrix in three-dimensional elastic-plastic analysis, von Mises yield condition, material isotropic hardening.*

$$
\mathbf{C}^{EP} = \frac{E}{1+\nu}\begin{bmatrix}
\dfrac{1-\nu}{1-2\nu} - \beta\, {}^t s_{11}^2 & \dfrac{\nu}{1-2\nu} - \beta\, {}^t s_{11}\, {}^t s_{22} & \dfrac{\nu}{1-2\nu} - \beta\, {}^t s_{11}\, {}^t s_{33} & -\beta\, {}^t s_{11}\, {}^t s_{12} & -\beta\, {}^t s_{11}\, {}^t s_{23} & -\beta\, {}^t s_{11}\, {}^t s_{13} \\
 & \dfrac{1-\nu}{1-2\nu} - \beta\, {}^t s_{22}^2 & \dfrac{\nu}{1-2\nu} - \beta\, {}^t s_{22}\, {}^t s_{33} & -\beta\, {}^t s_{22}\, {}^t s_{12} & -\beta\, {}^t s_{22}\, {}^t s_{23} & -\beta\, {}^t s_{22}\, {}^t s_{13} \\
 & & \dfrac{1-\nu}{1-2\nu} - \beta\, {}^t s_{33}^2 & -\beta\, {}^t s_{33}\, {}^t s_{12} & -\beta\, {}^t s_{33}\, {}^t s_{23} & -\beta\, {}^t s_{33}\, {}^t s_{13} \\
 & & & \tfrac{1}{2} - \beta\, {}^t s_{12}^2 & -\beta\, {}^t s_{12}\, {}^t s_{23} & -\beta\, {}^t s_{12}\, {}^t s_{13} \\
 & \text{symmetric} & & & \tfrac{1}{2} - \beta\, {}^t s_{23}^2 & -\beta\, {}^t s_{23}\, {}^t s_{13} \\
 & & & & & \tfrac{1}{2} - \beta\, {}^t s_{13}^2
\end{bmatrix}
$$

where

$$
\beta = \frac{3}{2}\frac{1}{{}^t\sigma_y^2}\left(\frac{1}{1 + \dfrac{2}{3}\dfrac{E\,E_T}{E - E_T}\dfrac{1+\nu}{E}}\right)
$$

three-dimensional stress conditions is given. In two-dimensional analysis the stress-strain matrix for axisymmetric and plane strain conditions would simply be obtained by only evaluating the applicable rows and columns of the matrix in Table 6.11, and the stress-strain matrix of plane stress analysis would be evaluated using static condensation on the stress-strain matrix of axisymmetric stress conditions (see Section 8.2.4 and Program QUADS in Section 5.9).

The above formulation of the elastic-plastic stress-strain matrix is a very specific formulation that is subject to a considerable number of assumptions; namely, flow theory with an associated flow rule has been used, isothermal conditions have been assumed, the von Mises yield condition was employed, and so on. The implications of these assumptions are very important when the model is employed in an actual practical analysis. However, our objective in the above discussion was merely to demonstrate the main features of how a typical plasticity model can be implemented in a finite element solution. Various other plasticity models are discussed, for example, in references[15–17]. We demonstrate the use of another important yield surface, namely the Drucker–Prager yield condition[18] for the modeling of geological materials, in the example below.

EXAMPLE 6.20: Using the Drucker–Prager yield criterion, the yield function is defined as (see Fig. 6.23)

$$^tF = 3\alpha\, {}^t\sigma_m + {}^t\bar{\sigma} - k$$

where $\quad {}^t\bar{\sigma}^2 = \frac{1}{2}\, {}^ts_{ij}\, {}^ts_{ij}; \qquad {}^t\sigma_m = {}^t\sigma_{ii/3}$

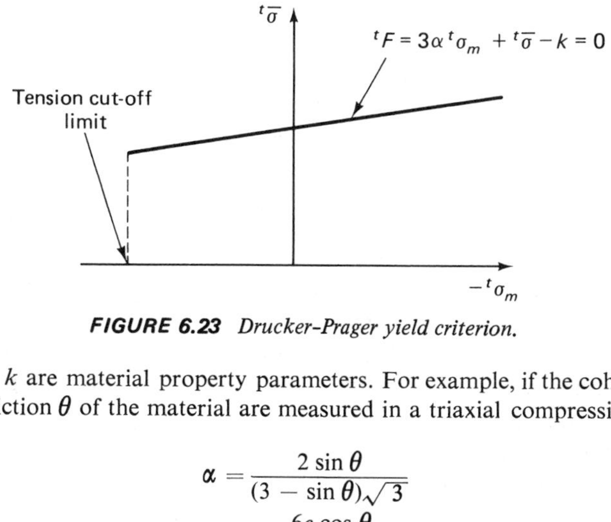

FIGURE 6.23 *Drucker–Prager yield criterion.*

and α and k are material property parameters. For example, if the cohesion c and angle of friction θ of the material are measured in a triaxial compression test, we have

$$\alpha = \frac{2 \sin \theta}{(3 - \sin \theta)\sqrt{3}}$$

$$k = \frac{6c \cos \theta}{(3 - \sin \theta)\sqrt{3}}$$

Consider the case of perfect plasticity—i.e., c and θ are constant—and derive the plasticity stress-strain relation for plane strain analysis.

Using (6.129) we have in this case

$$^tp_{ij} = 0 \tag{a}$$

and

$$^tq_{ij} = \alpha\, \delta_{ij} + \frac{1}{2\,{}^t\bar{\sigma}}\, {}^ts_{ij} \tag{b}$$

In the analysis of geological materials using the Drucker–Prager model, we usually employ the elastic shear and bulk moduli:

$$G = \frac{E}{2(1+\nu)}, \qquad K = \frac{1}{3}\frac{E}{(1-2\nu)} \tag{c}$$

Substituting from (a), (b), and (c) into (6.133) we thus obtain

$$\mathbf{C}^{EP} = \begin{bmatrix} K + \frac{4}{3}G - (\beta_1\, {}^t s_{11} + \beta_2)^2 & & \text{symmetric} \\ K - \frac{2}{3}G - (\beta_1\, {}^t s_{11} + \beta_2)(\beta_1\, {}^t s_{22} + \beta_2) & K + \frac{4}{3}G - (\beta_1\, {}^t s_{22} + \beta_2)^2 & \\ -(\beta_1\, {}^t s_{11} + \beta_2)(\beta_1\, {}^t s_{12}) & -(\beta_1\, {}^t s_{22} + \beta_2)(\beta_1\, {}^t s_{12}) & G - (\beta_1\, {}^t s_{12})^2 \end{bmatrix}$$

$$\beta_1 = \frac{G/{}^t\bar{\sigma}}{(G + 9K\alpha^2)^{1/2}}; \qquad \beta_2 = \frac{3K\alpha}{(G + 9K\alpha^2)^{1/2}}$$

The finite element formulation of the von Mises plasticity model discussed above demonstrates the main features of an inelastic analysis. A major characteristic is that the stress–strain matrix is dependent on the current conditions of the stresses, strains, and other state variables plus their history. It follows that to obtain an accurate response prediction it is necessary to employ the relation (6.132) in an integration process. Specifically, once the approximations to the displacement increments have been evaluated using (6.8) and (6.9), the stresses corresponding to time $t + \Delta t$ are calculated using

$${}^{t+\Delta t}\boldsymbol{\sigma} = {}^t\boldsymbol{\sigma} + \int_{{}^t\mathbf{e}}^{{}^{t+\Delta t}\mathbf{e}} \mathbf{C}^{EP}\, d\mathbf{e} \tag{6.140}$$

and equivalent evaluations are employed in the T.L. and U.L.J. formulations (see Table 6.10). The solution for the stresses in (6.140) is still approximate because the displacement and hence strain increments are only approximate. As discussed earlier, to solve for an accurate response prediction, equilibrium iterations are in general required (see (6.51)), in which case the relation in (6.140) is generalized to the equation,

$${}^{t+\Delta t}\boldsymbol{\sigma}^{(i-1)} = {}^t\boldsymbol{\sigma} + \int_{{}^t\mathbf{e}}^{{}^{t+\Delta t}\mathbf{e}^{(i-1)}} \mathbf{C}^{EP}\, d\mathbf{e} \tag{6.141}$$

and similarly the corresponding relations given in Table 6.10 for the T.L. and U.L.J. formulations are generalized.

The general observation is, therefore, that in inelastic analysis an integration of the stresses need be carried out during each equilibrium iteration. This integration can, in principle, be performed using a large number of integration procedures; however, in practice an efficient method should be employed. If an explicit integration technique is used, such as the Euler forward method, the solution is simply obtained by forward integrating over a sufficient number of subincrements in order to obtain the required accuracy in the integration process. Table 6.12 summarizes the use of the Euler forward integration for the solution of (6.140) (see step (h)). On the other hand, if an implicit integration method is employed, such as the Euler backward method, iteration for the stresses corresponding to the end of each subincrement is, in general, necessary (see Section 9.2 for the use of implicit integration methods). In this case, therefore, we are in general concerned with two iteration cycles: iterations on all

TABLE 6.12 Solution algorithm for elastic-plastic stress calculation.

Given: STRAIN = Total strains at time $t + \Delta t$
SIG = Total stresses at time t
EPS = Total strains at time t

The procedure below is used to calculate the total stresses TAU at time $t+\Delta t$. (We should note that the elastic-plastic incremental stress-strain law can be calculated, if the stresses are known.)

(a) Calculate the strain increment DELEPS:

DELEPS = STRAIN − EPS

(b) Calculate the stress increment DELSIG, assuming elastic behavior:

DELSIG = C^E * DELEPS

(c) Calculate TAU:

TAU = SIG + DELSIG

(d) With TAU as the state of stress, determine the value of the yield function F.

(e) If F(TAU) ≤ 0, elastic behavior assumption holds (loading elastically, neutral loading, or unloading). Hence STRESS = TAU, and *we return*
If F(TAU) > 0, *we continue*

(f) If the previous state of stress was plastic (as indicated by a flag) set RATIO = 0 and go to step (g). Otherwise, there is a transition from elastic to plastic, and RATIO, which is the portion of incremental strain taken elastically has to be determined. The variable RATIO is determined from the equation

F{SIG + RATIO * DELSIG} = 0

since at the stress SIG + RATIO * DELSIG the yield function F becomes equal to zero and yielding is initiated.

(g) Redefine TAU as the stress at start of yield

TAU = SIG + RATIO * DELSIG

and calculate the elastic-plastic strain increment

DEPS = (1 − RATIO) * DELEPS

(h) To obtain the final stresses, which include the effect of the complete strain increment DELEPS we need to add to TAU the stresses corresponding to the elastic-plastic strain increment DEPS. Since the material law is dependent on the current stresses, DEPS is divided into subincrements DDEPS and TAU is updated for each interval by the increments in stresses corresponding to the increments in elastic-plastic strains in that interval. In the calculations, the stress-strain matrix corresponding to the latest available stress conditions is used; i.e. we calculate

TAU ← TAU + C^{EP} * DDEPS

for all elastic-plastic strain subincrements DDEPS.

---

finite element equations in order to evaluate corrections to the vector of nodal point displacements (see (6.51)), and iterations during the integration of (6.141) in order to evaluate accurate stresses, ${}^{t+\Delta t}\boldsymbol{\sigma}^{(i-1)}$.

The use of implicit integration in the calculation of the stresses can be particularly important in the analysis of creep problems.[19,20] In this case, we calculate the stresses using instead of (6.140) and (6.141) the following relations

$$ {}^{t+\Delta t}\boldsymbol{\sigma} = {}^{t}\boldsymbol{\sigma} + \int_{{}^{t}\mathbf{e}}^{{}^{t+\Delta t}\mathbf{e}} \mathbf{C}^E \, d(\mathbf{e} - \mathbf{e}^C) $$

or rather

$$ {}^{t+\Delta t}\boldsymbol{\sigma}^{(i-1)} = {}^{t}\boldsymbol{\sigma} + \int_{{}^{t}\mathbf{e}}^{{}^{t+\Delta t}\mathbf{e}^{(i-1)}} \mathbf{C}^E \, d(\mathbf{e} - \mathbf{e}^C) \tag{6.142} $$

where the creep strains $d\mathbf{e}^c$ are evaluated from a generalization of uniaxial creep laws. A widely used technique is the strain-hardening method,[16] but other methods can also be effective.

In order to illustrate the integration of (6.142) using the strain-hardening method, first consider the conditions at time t. We calculate the effective stress

$$ {}^t\bar{\sigma} = \sqrt{\tfrac{3}{2}\,{}^ts_{ij}\,{}^ts_{ij}} \tag{6.143} $$

and the effective creep strain

$$ {}^t\bar{e}^C = \sqrt{\tfrac{2}{3}\,{}^te_{ij}^C\,{}^te_{ij}^C} \tag{6.144} $$

where we note that in uniaxial stress conditions ${}^t\bar{\sigma} = {}^t\sigma_{11}$ and ${}^t\bar{e}^C = {}^te_{11}^C$, because the material is assumed to be incompressible in creep. Assume now that the uniaxial creep law to be used is

$$ {}^te^C = a_0\,{}^t\sigma^{a_1}\,t^{a_2} \tag{6.145} $$

where a_0, a_1, and a_2 are given constants, and the variable t represents time. Other creep laws could be employed in a similar manner. The generalization of (6.145) to multiaxial conditions is achieved by substituting the effective stress and effective creep strain into (6.145) to obtain an equation for the effective time $\bar{t}$,

$$ {}^t\bar{e}^C = a_0({}^t\bar{\sigma})^{a_1}\,(\bar{t})^{a_2} \tag{6.146} $$

In practice, the calculation of $\bar{t}$ is frequently performed using Newton–Raphson iteration (see Section 8.6). Once the effective time $\bar{t}$ is known, we evaluate the effective creep strain rate

$$ {}^t\dot{\bar{e}}^C = a_0a_2({}^t\bar{\sigma})^{a_1}\,(\bar{t})^{a_2-1} \tag{6.147} $$

and calculate a constant ${}^t\gamma$ given by

$$ {}^t\gamma = \frac{3}{2}\frac{{}^t\dot{\bar{e}}^C}{{}^t\bar{\sigma}} \tag{6.148} $$

A differential increment in the creep strains at time t can now be evaluated using ${}^t\gamma$,

$$ d\mathbf{e}^C = ({}^t\gamma\mathbf{D}\,{}^t\boldsymbol{\sigma})\,dt \tag{6.149} $$

where $\mathbf{D}$ is a matrix that operates on ${}^t\boldsymbol{\sigma}$ to obtain deviatoric stresses; i.e., in three-dimensional analysis we have

$$ \mathbf{D} = \begin{bmatrix} \frac{2}{3} & -\frac{1}{3} & -\frac{1}{3} & & & \\ & \frac{2}{3} & -\frac{1}{3} & & & \\ & & \frac{2}{3} & & & \\ & & & 1 & & \\ & \text{sym.} & & & 1 & \\ & & & & & 1 \end{bmatrix} \tag{6.150} $$

The relation in (6.149) could now be employed using the Euler forward method (possibly with subincrements as demonstrated in Table 6.12 for elastic-plastic analysis) to evaluate the integral in (6.142). However, an analysis of this solution algorithm, as well as practical experience, show that the solution procedure is numerically unstable unless a small enough time increment Δt is employed. A "rule of thumb" for selecting the time step Δt is that the maximum creep strain increment must be smaller than $(1/a_1)$th, conservatively one-tenth, of the current elastic strain. Since the Euler forward integration may require in practice a very small time step, it is frequently more effective to use an implicit technique that is unconditionally stable. Such a method provides a stable integration for any time step size Δt, which therefore need be selected based only on considerations of solution accuracy and convergence in the iterations (see Section 9.4).

The solution difficulties in analyses involving creep arise because of the relatively large exponent a_1 on the stresses. Therefore, it is reasonable to integrate the stresses implicitly and evaluate all other solution variables correspondingly. In practice, the α-method also employed in heat transfer analysis (see Section 9.6) has proven to be simple and effective.[20] In this method we use

$$^{t+\alpha\Delta t}\boldsymbol{\sigma} = (1 - \alpha)\,{}^{t}\boldsymbol{\sigma} + \alpha\,{}^{t+\Delta t}\boldsymbol{\sigma} \tag{6.151}$$

with $0 \leq \alpha \leq 1$. The method reduces to the Euler forward method when $\alpha = 0$ and is an implicit technique when $\alpha > 0$, because the unknown stresses $^{t+\Delta t}\boldsymbol{\sigma}$ enter into the solution. As in heat transfer analysis the method is unconditionally stable when $\alpha \geq \frac{1}{2}$. The procedure is with $\alpha = 1$ the Euler backward method.

As pointed out earlier, when an implicit method is employed to solve (6.142), iteration must in general be used. A simple scheme would be to employ successive substitutions for the solution of $^{t+\Delta t}\boldsymbol{\sigma}^{(i-1)}$ (where i denotes the iteration counter for the solution of the total finite element equilibrium equations and is now constant). In this case we operate on

$$^{t+\Delta t}\boldsymbol{\sigma}^{(i-1)}_{(k)} = {}^{t}\boldsymbol{\sigma} + \mathbf{C}^{E}\left\{\mathbf{e}^{(i-1)} - \Delta t\,{}^{t+\alpha\Delta t}\gamma^{(i-1)}_{(k-1)}\,\mathbf{D}\,{}^{t+\alpha\Delta t}\boldsymbol{\sigma}^{(i-1)}_{(k-1)}\right\} \tag{6.152}$$
$$k = 1, 2, \ldots$$

in which $^{t+\alpha\Delta t}\boldsymbol{\sigma}^{(i-1)}_{(0)} = {}^{t}\boldsymbol{\sigma}$ and $^{t+\alpha\Delta t}\gamma^{(i-1)}_{(k-1)}$ is evaluated as summarized in (6.143) to (6.148) but corresponding to the stresses $^{t+\alpha\Delta t}\boldsymbol{\sigma}^{(i-1)}_{(k-1)}$. Convergence in this iteration can be measured using

$$\frac{\|{}^{t+\Delta t}\boldsymbol{\sigma}^{(i-1)}_{(k)} - {}^{t+\Delta t}\boldsymbol{\sigma}^{(i-1)}_{(k-1)}\|_2}{\|{}^{t+\Delta t}\boldsymbol{\sigma}^{(i-1)}_{(k)}\|_2} \leq \text{ctol} \tag{6.153}$$

where it is assumed that $\|{}^{t+\Delta t}\boldsymbol{\sigma}^{(i-1)}_{(k)}\|_2$ is nonzero and ctol is a convergence tolerance (see Section 2.9 for the definition of the Euclidean norm $\|\mathbf{a}\|_2$ of a vector $\mathbf{a}$).

Unfortunately, in practice the simple scheme of successive substitution is a very slowly converging process and diverges unless the time step Δt is relatively small. Therefore, the iteration process must be stabilized and accelerated and this can be achieved using a Newton–Raphson iteration. In addition, it can also be effective to use subincrements in the integration of (6.142), in which case the relation in (6.152) is applied to each time step subincrement. The use of time step subincrements results in more accuracy in the stress prediction and less convergence difficulties. The details of such a solution procedure are given in ref. [20].

In the above discussion we considered separately effects of plasticity and creep. However, in some analyses it is necessary to deal with combined thermal, creep, and plastic effects. Based on a number of assumptions, the stresses are in this case evaluated using

$$ {}^{t+\Delta t}\boldsymbol{\sigma} = {}^{t}\boldsymbol{\sigma} + \int_{{}^{t}\mathbf{e}}^{{}^{t+\Delta t}\mathbf{e}} {}^{\tau}\mathbf{C}^{E}\, d(\mathbf{e} - \mathbf{e}^{P} - \mathbf{e}^{C} - \mathbf{e}^{TH}) \tag{6.154} $$

in which ${}^{\tau}\mathbf{C}^{E}$ is now a function of the temperature ${}^{\tau}\theta$, and $\mathbf{e}^{P}$, $\mathbf{e}^{C}$ and $\mathbf{e}^{TH}$ are the plastic, creep, and thermal strain increments. Using the α-method, the integration in (6.154) is carried out as follows:

$$ {}^{t+\Delta t}\boldsymbol{\sigma} = {}^{t+\Delta t}\mathbf{C}^{E}\,[(\mathbf{e} - \mathbf{e}^{P} - \mathbf{e}^{C} - \mathbf{e}^{TH}) + ({}^{t}\mathbf{e} - {}^{t}\mathbf{e}^{P} - {}^{t}\mathbf{e}^{C} - {}^{t}\mathbf{e}^{TH})] \tag{6.155} $$

in which $\mathbf{e}$ is the total incremental strain, $\mathbf{e} = {}^{t+\Delta t}\mathbf{e} - {}^{t}\mathbf{e}$, and

$$ \mathbf{e}^{P} = \Delta t({}^{t+\alpha\Delta t}\bar{\lambda})\mathbf{D}({}^{t+\alpha\Delta t}\boldsymbol{\sigma}) \tag{6.156} $$

$$ \mathbf{e}^{C} = \Delta t({}^{t+\alpha\Delta t}\gamma)\mathbf{D}({}^{t+\alpha\Delta t}\boldsymbol{\sigma}) \tag{6.157} $$

and

$$ e_{ij}^{TH} = ({}^{t+\Delta t}\alpha\, {}^{t+\Delta t}\theta - {}^{t}\alpha\, {}^{t}\theta)\, \delta_{ij} \tag{6.158} $$

where ${}^{t}\alpha$ is the coefficient of thermal expansion at temperature ${}^{t}\theta$.* The details for the evaluation of the variables ${}^{t+\alpha\Delta t}\bar{\lambda}$ and ${}^{t+\alpha\Delta t}\gamma$ used in (6.156) and (6.157) are given in ref. [20]. Using (6.155) the solution for ${}^{t+\Delta t}\boldsymbol{\sigma}$ in (6.154) is obtained in the same way as ${}^{t+\Delta t}\boldsymbol{\sigma}$ in (6.142) was evaluated (see the above discussion).

In the above presentation of inelastic analysis procedures we considered the effects of elasto-plasticity and creep with emphasis on the general analysis approach followed. Considering these models we did not discuss various computational details, although they are important ingredients of an overall effective nonlinear analysis scheme. As an example, special attention must be given in elastic-perfectly-plastic analysis that the stress solution remains on the yield surface, a problem on which much research has been expended to develop satisfactory procedures.[20-23] A discussion of such computational details would require a rather lengthy exposition and goes beyond our objectives in this section.

6.5 SOME PRACTICAL CONSIDERATIONS

We already pointed out in Section 5.8 that the establishment of an appropriate finite element model for the analysis of an actual engineering problem is to a large degree based on a sufficient understanding of the problem under consideration and a thorough knowledge of the finite element procedures available for analysis. This observation is even more applicable to nonlinear analysis than to linear analysis, because quite a few more solution variables have to be judiciously selected, such as the nonlinear kinematic formulations, material models, and incremental solution strategies. The fact that all these variables can only be chosen once the complete analysis problem is understood sufficiently well implies that, depending on the experience of the analyst, and unless a routine nonlinear problem is considered, a number of preliminary analyses may be required before a final and appropriate finite element model is established.

The objective in this section is to briefly discuss some important practical aspects pertaining to the construction and solution of an appropriate nonlinear finite element model.

* Note that α without a superscript denotes here the integration parameter defined in (6.151).

6.5.1 The General Approach to Nonlinear Analysis

In an actual engineering analysis, it is good practice that a nonlinear analysis of a problem is always preceded by a linear analysis, so that the nonlinear analysis is considered an extension of the complete analysis process beyond the assumptions of linear analysis. Based on the linear response solution, the analyst is able to predict which nonlinearities will be significant and how to account for these nonlinearities most appropriately. Namely, the linear analysis results indicate the regions where geometric nonlinearities may be significant and where the material exceeds its elastic limit. Also, in the linear analysis the finite element idealization chosen should be tested for its appropriateness using the modeling considerations discussed in Section 5.8. Since a nonlinear analysis involves a great deal more expense than a linear analysis, it is important that prior to the nonlinear response solution a cost-effective finite element idealization be established.

Based on the above considerations, we recognize that all modeling considerations pertaining to the linear analysis of a problem are also relevant to its nonlinear analysis. Hence, the discussion of Section 5.8 provides a number of considerations for the selection of an appropriate finite element mesh for nonlinear analysis. However, there are some important additional considerations.

When performing a nonlinear analysis, there is frequently an unfortunate trend to select immediately a large number of elements and the most general nonlinear formulations available for modeling the problem. The engineering time used to prepare the model is large, the computer time that is needed for the analysis of the model is also very significant and usually a voluminous amount of information is generated that cannot be fully absorbed and interpreted. If there are significant modeling or program input errors, it may also happen that the analyst "gives up in despair" during the course of the analysis, because a relatively large amount of money has already been spent on the analysis, no significant results are as yet available, and the analyst is unable to realistically estimate how much further expense there would be until significant results could be produced.

The important point is that the whole approach to nonlinear analysis briefly described above cannot be recommended. Instead, a finite element model should first be established with as few elements as possible and yet containing all the important characteristics of the problem. After a variety of linear analyses have been performed that provide insight into the problem under consideration and the behavior of the finite element model, allowance for some nonlinearities—and not necessarily immediately for all nonlinearities that can be anticipated—should be made by choosing appropriate nonlinear formulations and material models for certain elements. Here it should be noted that by employing the formulations discussed in Sections 5.3, 5.4, and 6.3, finite elements formulated using the linear analysis assumptions, the materially-nonlinear-only formulation and the T.L. and U.L. formulations can all be used together in one finite element idealization. If this finite element mesh is a compatible mesh in linear analysis, the elements will also remain displacement-compatible in nonlinear analysis. The subdivision of the complete finite element idealization into elements governed by different nonlinear formulations merely means that in the analysis different nonlinearities are being accounted for in different parts of the structure. Once the effect of these nonlinearities has been studied, the analyst may decide to allow for additional nonlinearities in the analysis. This is achieved in the next analyses by assigning new kinematic nonlinear formulations and material models to some of the elements. An effective procedure for introducing different kinds of nonlinearities into the analysis is the use of linear and nonlinear element groups[11].

The complete process of analysis can be likened to a series of laboratory experiments in which different assumptions are made in each experiment—in the finite element analysis these experiments are performed on the computer with a finite element program.

The advantages of starting with a linear analysis after which judiciously selected nonlinear analyses are performed are that, firstly, the effect of each nonlinearity introduced can more easily be explained, secondly, confidence in the analysis results can be established and, thirdly, useful information is accumulated throughout the period of analysis.

In addition to the general recommendations for an appropriate approach to nonlinear analysis given above, some details pertaining to nonlinear analysis are important and are briefly discussed in the following sections.

6.5.2 Collapse and Buckling Analyses

The objective of a nonlinear analysis is in many cases to estimate the maximum load that a structure can support prior to structural instability or collapse. In the analysis the load distribution on the structure is known but the load magnitude that the structure can sustain is unknown.[24–28]

One way to calculate the maximum load which the structure can carry safely is to simply perform an incremental analysis using the nonlinear formulations. In a collapse analysis the equations of equilibrium are for each load (or time) step using, for example, the T.L. formulation,

$$({}_0^t\mathbf{K}_L + {}_0^t\mathbf{K}_{NL})\,\Delta\mathbf{U}^{(i)} = ({}^{t+\Delta t}\beta)\,{}^{\Delta t}\mathbf{R} - {}^{t+\Delta t}_{\;\;0}\mathbf{F}^{(i-1)} \qquad i = 1, 2, \ldots \tag{6.159}$$

where ${}^{\Delta t}\mathbf{R}$ is the load vector for the first load step and ${}^{t+\Delta t}\beta$ is the variable which scales ${}^{\Delta t}\mathbf{R}$ to obtain the loads corresponding to time $t + \Delta t$. The collapse of the structure is reached when for a small load increment the displacements become relatively large, which means that the overall stiffness of the structure becomes small. Numerically, this means that some pivot elements in the triangular factorization of the tangent stiffness matrix become very small, until finally a zero pivot element is encountered, i.e. the tangent stiffness matrix is singular (its determinant is zero, see Section 8.2.5). If equilibrium iterations are performed in the incremental solution, we may also observe convergence difficulties as the collapse load is approached (see Section 8.6).

Considering next the buckling of structures, we introduced in Section 3.2.3 the essence of a linearized buckling analysis, in which we solve the problem

$$\det(\mathbf{K} + \lambda\mathbf{K}_G) = 0 \tag{6.160}$$

In (6.160) $\mathbf{K}$ and $\mathbf{K}_G$ are, respectively, the linear and nonlinear strain stiffness matrices that usually correspond to the initial configuration of the structure.

Hence using the notation in this chapter, we have $\mathbf{K} = {}^0_0\mathbf{K}_L$ and $\mathbf{K}_G = {}^{\Delta t}_{0}\mathbf{K}_{NL}$, and we assume that at buckling,

$$ {}^0_0\mathbf{K}_L + \lambda\, {}^{\Delta t}_{0}\mathbf{K}_{NL} \doteq {}^t_0\mathbf{K}_L + {}^t_0\mathbf{K}_{NL} \tag{6.161} $$

where the r.h.s. matrix is the coefficient matrix of (6.159). Using (6.160) we thus obtain as the buckling load

$$ \mathbf{R}_{\text{buckling}} = \lambda_1\, {}^{\Delta t}\mathbf{R} \tag{6.162} $$

where λ_1 is the smallest eigenvalue of the problem considered in (6.160).

The assumptions used in this linearized buckling analysis are displayed in (6.161); namely, we assume that the linear strain stiffness matrix does not change appreciably prior to buckling and the nonlinear strain stiffness matrix is simply a multiple of its initial value. The linearized buckling analysis is therefore appropriately used to predict the load level at which a structure becomes unstable if the pre-buckling displacements and their effects are negligible.

Figure 6.24 illustrates schematically the response of some structures that collapse or buckle. Note that the buckling response of the column is calculated using an incremental analysis on a structural model with a small imperfection which can either be in the geometry (by perturbing the element nodal points slightly from the values of a "perfect" structure) or in the load application. In Fig. 6.24(b) the measure of the imperfection is the value β, where $\beta \geq 0$. For larger values of β, the incremental analysis amounts to a collapse analysis of the structure. In all cases the analysis procedures discussed in this chapter are directly applicable, but in the calculation of a post-buckling response the stiffness matrix may not be positive definite (an increase in the displacements may correspond to a decrease in the loads) and an iteration simultaneously with the magnitudes of the displacements and the loads may be necessary. Note that the incremental analyses can include elastic-plastic and creep effects, and eigensolutions of the form given in (6.160) where now the matrices $\mathbf{K}$ and $\mathbf{K}_G$ may correspond to the matrices ${}^{t-\Delta t}_{0}\mathbf{K}$ and $\Delta\mathbf{K} = {}^t_0\mathbf{K} - {}^{t-\Delta t}_{0}\mathbf{K}$ at any load level considered.

Although a large number of procedures to predict the buckling of a complex structure are already available, a complete buckling analysis can represent a major task. In a very general analysis the incremental solution is carried out including the geometric and material nonlinearities, and at every load (or time) step it is checked whether a collapse or bifurcation into a displacement pattern with associated large deformations is possible. Once buckling has been initiated it may further be necessary to calculate the postbuckling response. An even more difficult task may be the dynamic buckling analysis of a complex shell structure and this response solution may well be beyond the state of the current analysis and computing capabilities available.[28]

6.5.3 Element Distortions

We mentioned in Section 5.8.3 that finite elements are in general most effective in the prediction of displacements and stresses when they are undistorted. The same applies in nonlinear analysis; however, there is the additional consideration that elements which are initially rectangular or triangular in general become distorted in geometrically nonlinear analysis.

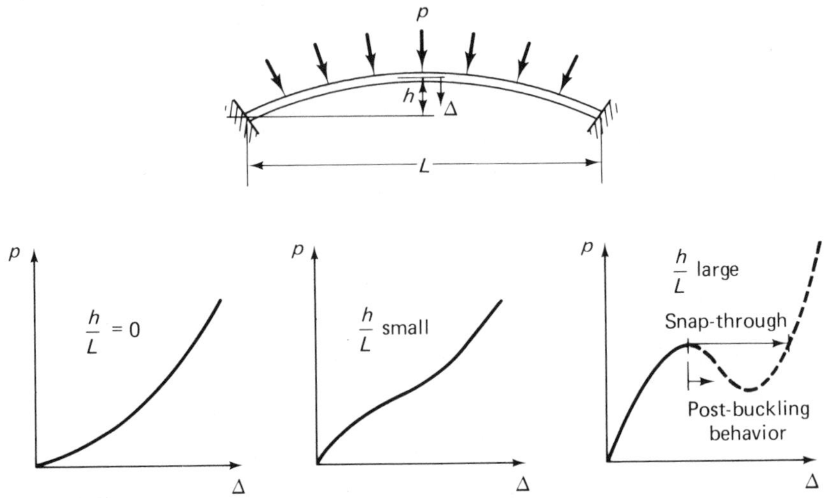

(a) Response of a thin plate/shell

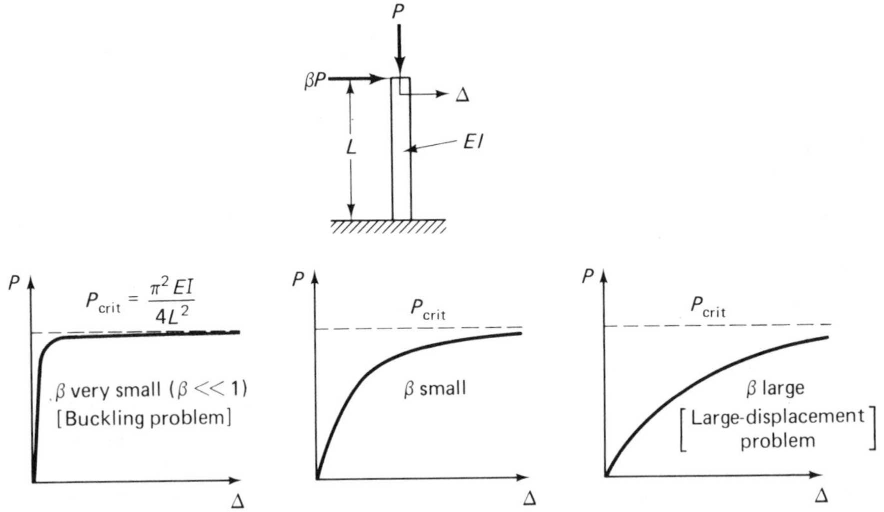

(b) Response of a column

FIGURE 6.24 *Instability and collapse analyses.*

Using the T.L. and U.L. formulations, the principle of virtual displacements is applied to each individual element corresponding to the current configuration instead of the initial configuration used in linear analysis. Thus, element distortions must be expected to affect the accuracy of the nonlinear response prediction in a manner similar to that in linear analysis, but now the considerations concerning the element distortions in linear analysis summarized in Section 5.8.3 are applicable throughout the response history of the mesh. In an analysis it is therefore necessary to monitor the changing shape of each element, and if element distortions become significant, a different and finer mesh may be required for the geometrically nonlinear analysis than for the linear analysis. Alternatively, mesh rezoning could be used at certain critical times in order to keep more closely to a uniform element layout with rectangular or triangular elements. However, such rezoning is not a straightforward task because it necessitates the calculation of the displacements of the nodal points at their new locations and the element stresses, strains, and other state variables at the new Gauss integration points, which must all be achieved without significant loss of accuracy in the response prediction.

We should also point out that the above considerations are equally applicable to the T.L. and the U.L. formulations, because, except for certain constitutive assumptions both formulations are completely equivalent (see Section 6.4.1). In the T.L. formulation the integrations are performed over the initial undistorted element shape, but the distortions enter into the use of the 2nd Piola–Kirchhoff stresses and the initial displacement matrix ${}_0^t\mathbf{B}_{L1}$.

6.5.4 Order of Numerical Integration

To select the appropriate numerical integration scheme and order of integration in nonlinear analysis, some specific considerations beyond those discussed already in Section 5.8.1 are important.

Based on the preceding discussion on the distortions of the elements that can occur in geometric nonlinear analysis and the information given in Section 5.8.1 on the required integration order for uniformly shaped and distorted elements, we can directly conclude that in some—but not frequent—cases of geometric nonlinear analysis a higher integration order should be employed than is used in linear analysis.

A higher integration order than used in linear analysis is also frequently required in the analysis of materially nonlinear response simply in order for the analysis to capture the onset and spread of the materially nonlinear conditions accurately enough. Specifically, since the material nonlinearities are only measured at the integration points of the elements, the use of a relatively low integration order may mean that the spread of the materially nonlinear conditions through the element is not represented accurately. This consideration is particularly important in the materially nonlinear analysis of beam, plate, and shell structures and also leads to conclude that the Newton–Cotes methods may be effective (say for integration in certain directions) because then integration points for stiffness and stress evaluations are on the boundaries of the elements (e.g. on the top and bottom surfaces of a shell).

Figure 6.25 gives the results of using different Gauss integrations in the analysis of an eight-node plane stress element. The element is subjected to an increasing bending moment and the numerically predicted response is compared with the response calculated using beam theory. This analysis illustrates that to predict the materially nonlinear response accurately a higher integration order in the thickness direction of the beam is required than in linear analysis. Another example that demonstrates the effect of using different integration orders in materially nonlinear analysis is given below.

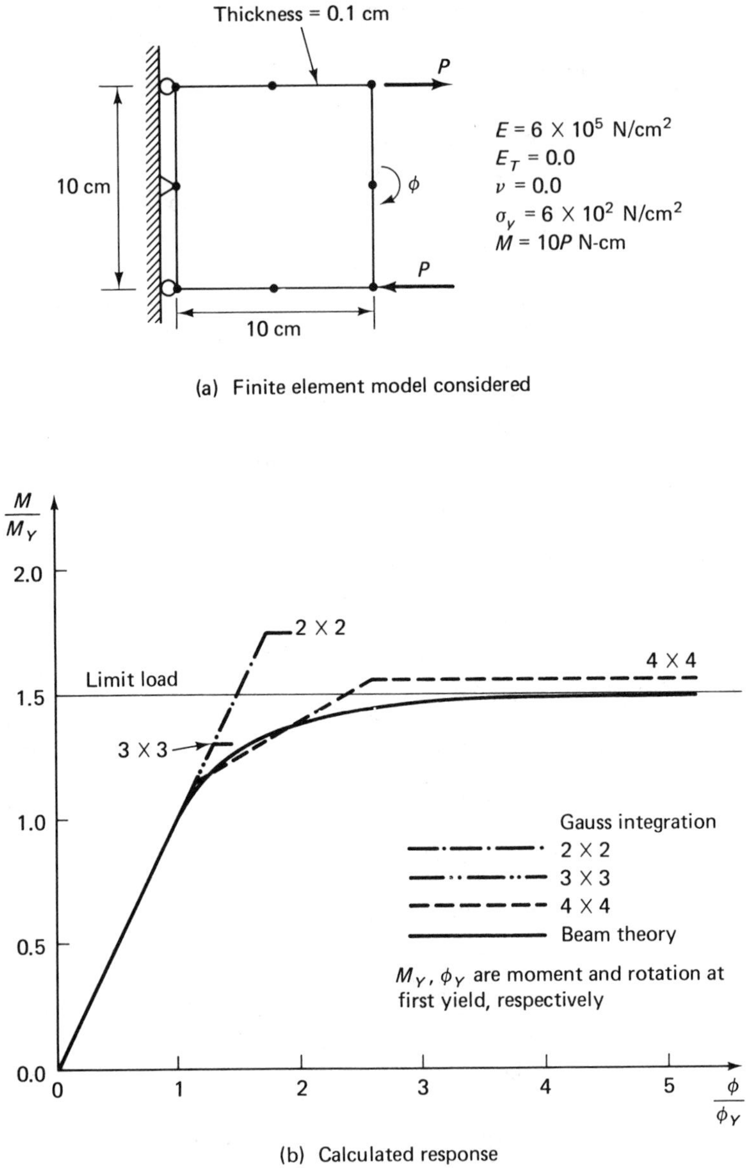

FIGURE 6.25 *Effect of integration order in elastic-plastic analysis of beam section.*

EXAMPLE 6.21: Consider element 2 of Example 4.3, and assume that in an elastic-plastic analysis the stresses at time t in the element are such that the tangent moduli of the material are equal to $E/100$ for $0 \leq x \leq 40$ and equal to E for $40 < x \leq 80$ as illustrated in Fig. 6.26. Evaluate the tangent stiffness matrix ${}^t\mathbf{K}$ using one-, two-, three-, and four-point Gauss integration, and compare these results with the exact stiffness matrix. Consider only material nonlinearities.

For the evaluation of the matrix ${}^t\mathbf{K}$ we use the information given in Example 4.3 and in Table 5.6. Thus, we obtain the following results:

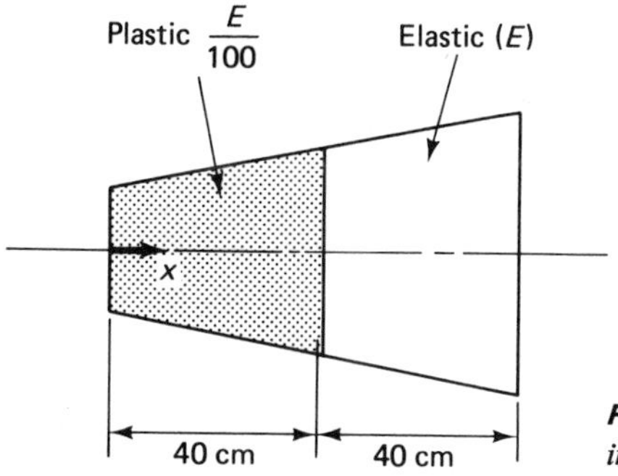

FIGURE 6.26 *Element 2 of Example 4.3 in elastic-plastic conditions.*

one-point integration

$$ {}^t\mathbf{K} = 2 \times 40\begin{bmatrix} -\frac{1}{80} \\ \frac{1}{80} \end{bmatrix} \frac{E}{100}\begin{bmatrix} -\frac{1}{80} & \frac{1}{80} \end{bmatrix}(1 + 1)^2 = 0.0005E\begin{bmatrix} 1 & -1 \\ -1 & 1 \end{bmatrix} $$

two-point integration

$$ \begin{aligned} {}^t\mathbf{K} &= 1 \times 40\begin{bmatrix} -\frac{1}{80} \\ \frac{1}{80} \end{bmatrix} \frac{E}{100}\begin{bmatrix} -\frac{1}{80} & \frac{1}{80} \end{bmatrix}\left(1 + 1 - \frac{1}{\sqrt{3}}\right)^2 \\ &\quad + 1 \times 40\begin{bmatrix} -\frac{1}{80} \\ \frac{1}{80} \end{bmatrix} E\begin{bmatrix} -\frac{1}{80} & \frac{1}{80} \end{bmatrix}\left(1 + 1 + \frac{1}{\sqrt{3}}\right)^2 \\ &= 0.04164E\begin{bmatrix} 1 & -1 \\ -1 & 1 \end{bmatrix} \end{aligned} $$

three-point integration

$$ \begin{aligned} {}^t\mathbf{K} &= \frac{5}{9}40\begin{bmatrix} -\frac{1}{80} \\ \frac{1}{80} \end{bmatrix} \frac{E}{100}\begin{bmatrix} -\frac{1}{80} & \frac{1}{80} \end{bmatrix}(1 + 1 - \sqrt{3/5})^2 \\ &\quad + \frac{8}{9}40\begin{bmatrix} -\frac{1}{80} \\ \frac{1}{80} \end{bmatrix} \frac{E}{100}\begin{bmatrix} -\frac{1}{80} & \frac{1}{80} \end{bmatrix}(1 + 1)^2 \\ &\quad + \frac{5}{9}40\begin{bmatrix} -\frac{1}{80} \\ \frac{1}{80} \end{bmatrix} E\begin{bmatrix} -\frac{1}{80} & \frac{1}{80} \end{bmatrix}(1 + 1 + \sqrt{3/5})^2 \\ {}^t\mathbf{K} &= 0.02700E\begin{bmatrix} 1 & -1 \\ -1 & 1 \end{bmatrix} \end{aligned} $$

four-point integration

	r_i	α_i
$n = 4$	$\pm 0.8611\ldots$	$0.3478\ldots$
	$\pm 0.3399\ldots$	$0.6521\ldots$

$$ \begin{aligned} {}^t\mathbf{K} &= 0.3478\ldots(40)\begin{bmatrix} -\frac{1}{80} \\ \frac{1}{80} \end{bmatrix} \frac{E}{100}\begin{bmatrix} -\frac{1}{80} & \frac{1}{80} \end{bmatrix}(1 + 1 - 0.8611\ldots)^2 \\ &\quad + \ldots \\ {}^t\mathbf{K} &= 0.04026E\begin{bmatrix} 1 & -1 \\ -1 & 1 \end{bmatrix} \end{aligned} $$

The exact stiffness matrix is

$$
{}^{t}\mathbf{K} = \begin{bmatrix} -\frac{1}{80} \\ \frac{1}{80} \end{bmatrix} \frac{E}{100} [-\tfrac{1}{80} \quad \tfrac{1}{80}] \left\{ \int_0^{40} \left(1 + \frac{y}{40}\right)^2 dy + \int_{40}^{80} 100 \left(1 + \frac{y}{40}\right)^2 dy \right\}
$$

$$
= \begin{bmatrix} -\frac{1}{80} \\ \frac{1}{80} \end{bmatrix} E [-\tfrac{1}{80} \quad \tfrac{1}{80}] \left\{ \frac{40}{300} \left(1 + \frac{y}{40}\right)^3 \Bigg|_0^{40} + \frac{40}{3} \left(1 + \frac{y}{40}\right)^3 \Bigg|_{40}^{80} \right\}
$$

$$
{}^{t}\mathbf{K} = 0.03973E \begin{bmatrix} 1 & -1 \\ -1 & 1 \end{bmatrix}
$$

It is interesting to note that in this case the two-point integration yields more accurate results than the three-point integration and a good approximation to the exact stiffness matrix is obtained using four-point integration.

In the above discussion we pointed out that for an accurate response prediction in nonlinear analysis frequently a higher-order integration is required than in linear analysis, because the elements may distort and the nonlinear material conditions should be sampled to sufficient accuracy. This observation is directly in contrast to the advocation of reduced integration as discussed in Section 5.8.1. However, we also pointed out in Section 5.8.1 that the use of reduced integration is at present still largely based on computational experience. The ultimate check on whether the use of reduced integration is appropriate in an analysis is frequently still a re-analysis of the same problem without reduced integration and possibly using a finer finite element mesh. In many linear analyses this check need not be performed, because enough experience has been accumulated about the structural response to be expected, and therefore it is effective to use reduced integration. However, considering nonlinear analysis, this computational experience is largely unavailable and is much more difficult to acquire, so that in a practical analysis the use of high-order integration is also for this reason usually preferable. High-order numerical integration reduces the number of possible uncertainties that can enter into the interpretation of the results of a complex analysis, and should therefore be used until sufficient experience with the response of the structure has been obtained. At that time the use of reduced integration together with less elements to idealize the structure may be considered.

Instead of using uniform reduced integration, a selective integration scheme may be considered. Such a scheme can be effective, although, as already pointed out in Section 5.8.1, the scheme should not under any deformation conditions introduce spurious zero energy modes into an element. A specific selective integration scheme that has been found very useful in some elastic perfectly-plastic analyses is the integration of the bulk energy terms with one order lower than the deviatoric energy terms.[29-31] This scheme was derived for incompressible analysis using the penalty method (see also Section 7.4) and is useful in plastic analysis because in perfect plasticity (a stress-strain law without strain-hardening), using the von Mises yield condition and flow rule, the material behavior, incompressible in the plastic strains, is difficult to analyze.

Much research attention has lately been devoted to the development of selective integration schemes with emphasis on the analysis of incompressible media and plates and shells,[30-34] and the close relationship of such schemes with mixed methods has been established.[30,31,34] The effective and reliable use of selective integration methods is particularly difficult in nonlinear analysis (see Section 5.8.1) but these methods do provide a significant potential for improvements in the currently available analysis procedures.

REFERENCES

1. L. E. Malvern, *Introduction to the Mechanics of a Continuous Medium*, Prentice-Hall, Inc., Englewood Cliffs, N.J., 1969.
2. Y. C. Fung, *Foundations of Solid Mechanics*, Prentice-Hall, Inc., Englewood Cliffs, N.J., 1965.
3. W. Prager, *Introduction to the Mechanics of Continua*, Dover, Publications Inc., New York, N.Y., 1973.
4. J. T. Oden, *Finite Elements of Nonlinear Continua*, McGraw-Hill Book Company, New York, 1972.
5. H. D. Hibbitt, P. V. Marcal, and J. R. Rice, "Finite Element Formulation for Problems of Large Strain and Large Displacements," *Int. J. Solids Struct.*, Vol. 6, pp. 1069–1086, 1970.
6. S. Yaghmai and E. P. Popov, "Incremental Analysis of Large Deflections of Shells of Revolution," *Int. J. Solids and Struct.*, Vol. 7, pp. 1375–1393, 1971.
7. C. A. Felippa and P. Sharifi, "Computer Implementation of Nonlinear Finite Element Analysis," *Proceedings, Symposium ASME*, Detroit, November, 1973.
8. S. W. Key, "A Finite Element Procedure for the Large Deformation Dynamic Response of Axisymmetric Solids," *J. Comp. Meth. in Appl. Mech. and Eng.*, Vol. 4, 1974, pp. 195–218.
9. K. J. Bathe, E. Ramm, and E. L. Wilson, "Finite Element Formulations for Large Deformation Dynamic Analysis," *Int. J. Numerical Methods in Engineering*, Vol. 9, pp. 353–386, 1975.
10. K. J. Bathe, "An Assessment of Current Finite Element Analysis of Nonlinear Problems in Solid Mechanics," in *Numerical Solution of Partial Differential Equations-III*, (B. Hubbard, ed.), Academic Press, Inc., New York, N.Y., 1976.
11. K. J. Bathe, "ADINA—A Finite Element Program for Automatic Dynamic Incremental Nonlinear Analysis," Report 82448-1, Acoustics and Vibration Lab., Mechanical Engineering Dept., M.I.T., September 1975 (rev. Dec. 1978).
12. E. Ramm, "A Plate/Shell Element for Large Deflections and Rotations," in *Formulations and Computational Algorithms in Finite Element Analysis*, (K. J. Bathe, J. T. Oden, and W. Wunderlich, eds.), M.I.T. Press, Cambridge, Mass., 1977.
13. B. Kråkeland, "Large Displacement Analysis of Shells Considering Elastic-Plastic and Elasto-Viscoplastic Materials." Report No. 776, The Norwegian Institute of Technology, The Univ. of Trondheim, Norway, Dec. 1977.
14. K. J. Bathe and S. Bolourchi, "A Geometric and Material Nonlinear Plate and Shell Element," *J. Computers and Structures*, Vol. 11, 1980, pp. 23–48.
15. G. C. Nayak and O. C. Zienkiewicz, "Elasto-Plastic Stress Analysis. A Generalization for Various Constitutive Relations Including Strain Softening," *Int. J. Num. Meth. in Eng.*, Vol. 3, pp. 113–135, 1972.
16. C. E. Pugh, J. M. Corum, K. C. Liu, and W. L. Greenstreet, "Currently Recommended Constitutive Equations for Inelastic Design Analysis of FFTF Components," Report No. TM-3602, Oak Ridge National Laboratory, Oak Ridge, Tennessee, 1972.
17. C. S. Desai and J. T. Christian, (eds.), *Numerical Methods in Geotechnical Engineering*, McGraw Hill Book Company, New York, N.Y., 1977.

18. D. C. Drucker and W. Prager, "Soil Mechanics and Plastic Analysis or Limit Design," *Quart. Appl. Math.*, Vol. 10, No. 2, pp. 157–165, 1952.
19. C. A. Anderson, "An Investigation of the Steady Creep of a Spherical Cavity in a Half Space," *Journal of Applied Mechanics*, Vol. 98, No. 2, June 1976, pp. 254–258.
20. M. D. Snyder and K. J. Bathe, "A Solution Procedure for Thermo-Elastic-Plastic and Creep Problems," *J. Nuclear Engineering and Design*, Vol. 64, 1981, pp. 49–80.
21. R. D. Krieg and D. B. Krieg, "Accuracies of Numerical Solution Methods for the Elastic-Perfectly Plastic Model," *Journal of Pressure Vessel Technology*, Vol. 99, No. 4, Nov. 1977, pp. 510–515.
22. D. Bushnell, "A Subincremental Strategy for Solving Problems Involving Large Deflections, Plasticity, and Creep," in *Constitutive Equations in Viscoplasticity: Computational and Engineering Aspects*, (J. A. Stricklin and K. J. Saczalski, eds.), American Society of Mechanical Eng., AMD Vol. 20, 1976.
23. M. D. Snyder and K. J. Bathe, "Finite Element Analysis of Thermo-Elastic-Plastic and Creep Response" Report 82448–10, Acoustics and Vibration Laboratory, Dept. of Mechanical Engineering, M.I.T., Dec. 1980.
24. R. Brendel and E. Ramm, "Linear and Nonlinear Stability Analysis of Cylindrical Shells," *Proceedings, Conference on Eng. Appl. of the Finite Element Method*, Computas, Høvik, Norway, May 1979.
25. P. G. Bergan and T. Søreide, "Solution of Large Displacement and Instability Problems using the Current Stiffness Parameters," *Proc. Int. Conf. on Finite Elements in Nonlinear Mechanics*, Geilo, Norway, 1977, Tapir, 1978.
26. P. Sharifi and E. P. Popov, "Nonlinear Buckling Analysis of Sandwich Arches," *Proc. ASCE, J. Eng. Mech. Div.*, 1970, pp. 1397–1412.
27. K. J. Bathe, "Static and Dynamic Geometric and Material Nonlinear Analysis using ADINA," Report AVL 82448–2, Mechanical Eng. Dept., M.I.T., May 1976, rev. May 1977.
28. T. Ishizaki and K. J. Bathe, "On Finite Element Large Displacement and Elastic-Plastic Dynamic Analysis of Shell Structures," *J. Computers and Structures*, Vol. 12, pp. 309–318, 1980.
29. S. Key, "HONDO—A Finite Element Computer Program for the Large Deformation Dynamic Response of Axisymmetric Solids," Report SLA-74-0039, Sandia Laboratories, Albuquerque, New Mexico, Jan. 1974.
30. M. Bercovier, "Perturbation of Mixed Variational Problems. Application to Mixed Finite Element Methods", *R.A.I.R.O. (Numerical Analysis)*, Vol. 12, pp. 211–236, 1978.
31. D. S. Malkus and T. J. R. Hughes, "Mixed Finite Element Methods—Reduced and Selective Integration Techniques: A Unification of Concepts," *Computer Methods in Applied Mechanics and Engineering*, Vol. 15, No. 1, 1978, pp. 63–81.
32. J. T. Oden, "A Theory of Penalty Methods for Finite Element Approximation of Highly Nonlinear Problems in Continuum Mechanics", *J. Comp. Struct.* Vol. 8, pp. 445–449, 1978.
33. M. Bercovier, Y. Hasbani, Y. Gilon, and K. J. Bathe, "On a Finite Element Procedure for Nonlinear Incompressible Elasticity", *Proceedings, Symposium on Hybrid and Mixed Finite Elements*, Georgia Institute of Technology, April, 1981.
34. A. K. Noor and J. M. Peters, "Mixed Models and Reduced/Selective Integration Displacement Models for Nonlinear Analysis of Curved Beams," *International Journal for Numerical Methods in Engineering*, Vol. 17, 1981, pp. 615–631.

7

FINITE ELEMENT ANALYSIS OF HEAT TRANSFER, FIELD PROBLEMS, AND FLUID FLOW

7.1 INTRODUCTION

In the preceding chapters we considered the finite element formulation and solution of problems in stress analysis of solids and structural systems. However, finite element analysis procedures have also gained increasing importance in the solution of nonstructural problems, in particular, in heat transfer, field problems, and fluid flow. Indeed the power and wide applicability of finite element techniques can only be recognized fully when considering also the analysis of other than structural problems.

The possiblities for application of finite element methods to nonstructural problems are very large and are currently under active research and development. Our objective in the following sections is to discuss only briefly the application of the finite element method to the solution of various heat transfer, fluid flow, and field problems, in which the methods can already be employed effectively. In addition to presenting some practical solution procedures, emphasis is directed to the general techniques that are employed and to demonstrate the commonality among the various problem formulations. In this way we hope that the applicability of finite element procedures to the solution of problems not discussed here becomes apparent, and the inherent potential in the procedures for analysis of even more complex problems (e.g. coupled fluid flow-structural problems, general non-Newtonian flow conditions) is realized.

7.2 HEAT TRANSFER ANALYSIS

In the study of finite element analysis of heat transfer problems it is instructive to first recall the differential and variational equations that govern the heat transfer conditions to be analyzed. These equations provide the basis for the finite element solution of a heat transfer problem.

7.2.1 Governing Classical Heat Transfer Equations

Consider a three-dimensional body in heat transfer conditions as shown in Fig. 7.1.[1-5] In the analysis of the heat transfer conditions, we assume that the material obeys Fourier's law of heat conduction;

$$q_x = -k_x \frac{\partial \theta}{\partial x}; \qquad q_y = -k_y \frac{\partial \theta}{\partial y}; \qquad q_z = -k_z \frac{\partial \theta}{\partial z}$$

where q_x, q_y, and q_z are the heat flows conducted per unit area, θ is the temperature of the body, and k_x, k_y, k_z are the thermal conductivities corresponding to the principal axes x, y, and z. Considering the heat flow equilibrium in the interior of the body we thus obtain

$$\frac{\partial}{\partial x}\left(k_x \frac{\partial \theta}{\partial x}\right) + \frac{\partial}{\partial y}\left(k_y \frac{\partial \theta}{\partial y}\right) + \frac{\partial}{\partial z}\left(k_z \frac{\partial \theta}{\partial z}\right) = -q^B \tag{7.1}$$

where q^B is the rate of heat generated per unit volume. On the surfaces of the body the following conditions must be satisfied:

$$\theta \big|_{S_1} = \theta_e \tag{7.2}$$

$$k_n \frac{\partial \theta}{\partial n}\bigg|_{S_2} = q^S \tag{7.3}$$

where θ_e is the environmental temperature, k_n is the body thermal conductivity, n denotes the direction of the normal (pointing outward) to the surface, and q^S is the heat flow input on the surface of the body.

A number of important assumptions apply to the use of (7.1) to (7.3). A primary assumption is that the material particles of the body are at rest, and thus we consider the heat conduction conditions in solids and structures. If the heat transfer in a moving fluid is to be analyzed, it is necessary to include in (7.1) a term allowing for the convective heat transfer through the medium. This term leads to a nonsymmetric coefficient matrix (as in the Eulerian formulation of fluid flow, see Section 7.4), and in the numerical solution special attention

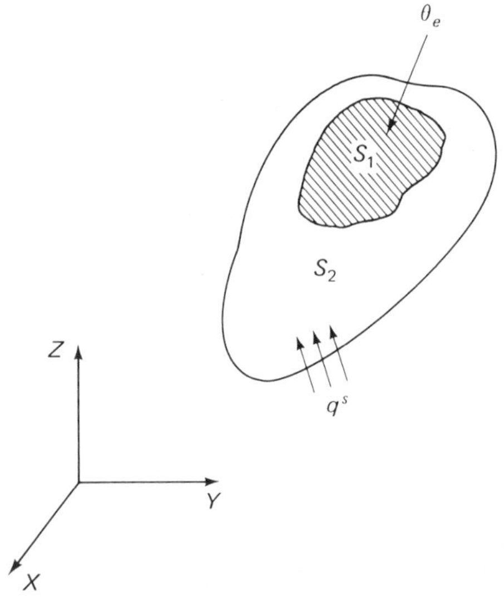

FIGURE 7.1 *Body subjected to heat transfer.*

must be given to the convective heat transfer in order to avoid numerical instabilities. Another assumption is that the heat transfer conditions can be analyzed decoupled from the stress conditions. This assumption is valid in many structural analyses, but may not be appropriate, for example, in the analysis of metal forming processes. A further assumption is that there are no phase changes and latent heat effects. However, we will consider in the following formulation that the material parameters are temperature-dependent.

A variety of boundary conditions are encountered in heat transfer analysis:

Temperature Conditions

The temperature may be prescribed at specific points and surfaces of the body, denoted by S_1 in (7.2).

Heat Flow Conditions

The heat flow input may be prescribed at specific points and surfaces of the body. These heat flow boundary conditions are specified in (7.3).

Convection Boundary Conditions

Included in (7.3) are convection boundary conditions, where

$$q^S = h(\theta_e - \theta^S) \tag{7.4}$$

and h is the convection coefficient, which may be temperature-dependent.

Radiation Boundary Conditions

Radiation boundary conditions are also specified in (7.3) with

$$q^S = \kappa(\theta_r - \theta^S) \tag{7.5}$$

where θ_r is the temperature of the external radiative source and κ is a coefficient:

$$\kappa = h_r(\theta_r^2 + (\theta^S)^2)(\theta_r + \theta^S) \tag{7.6}$$

The variable h_r is determined from the Stefan–Boltzmann constant, the emissivity of the radiant and absorbing materials and the geometric view factors.[5]

In addition to these boundary conditions also the temperature initial conditions must in a transient analysis be specified.

For the development of a finite element solution scheme either a Galerkin formulation operating on the differential equation of equilibrium or a variational formulation of the heat transfer problem can be employed (see Section 3.3). In the variational formulation a functional Π is defined such that when invoking the stationarity of Π, the governing differential equations (7.1) to (7.3) are obtained. For the three-dimensional body in Figure 7.1 the functional governing heat conduction is

$$\begin{aligned}\Pi = &\int_V \tfrac{1}{2}\left\{k_x\left(\frac{\partial\theta}{\partial x}\right)^2 + k_y\left(\frac{\partial\theta}{\partial y}\right)^2 + k_z\left(\frac{\partial\theta}{\partial z}\right)^2\right\} dV \\ &- \int_V \theta q^B\, dV - \int_{S_2} \theta^S q^S\, dS - \sum_i \theta^i Q^i\end{aligned} \tag{7.7}$$

where the same notation as in (7.1) is employed and the Q^i are concentrated heat flow inputs. Using the condition of stationarity of Π we obtain (because θ

is the only variable)

$$\int_V \delta\boldsymbol{\theta}'^T \mathbf{k}\boldsymbol{\theta}' \, dV = \int_V \delta\theta \, q^B \, dV + \int_{S_2} \delta\theta^S q^S \, dS + \sum_i \delta\theta^i Q^i \tag{7.8}$$

where

$$\boldsymbol{\theta}'^T = \begin{bmatrix} \frac{\partial\theta}{\partial x} & \frac{\partial\theta}{\partial y} & \frac{\partial\theta}{\partial z} \end{bmatrix} \tag{7.9}$$

$$\mathbf{k} = \begin{bmatrix} k_x & 0 & 0 \\ 0 & k_y & 0 \\ 0 & 0 & k_z \end{bmatrix} \tag{7.10}$$

and "δ" denotes "variation in." In order to derive the governing differential equations given in (7.1) to (7.3) it is necessary to use integration by parts on the relations in (7.8) (see Section 3.3.2).

It is instructive to compare the governing heat flow equation in (7.8) with the virtual work equation employed in stress analysis. Specifically, since the variation in θ can be understood as a virtual quantity, i.e. $\delta\theta \equiv \bar{\theta}$, it follows that the equation governing rate of heat transfer in (7.8) is analogous to the principle of virtual displacements for stress analysis. In terms of the equilibrium considerations used in stress analysis, the relation in (7.8) is a statement of equilibrium of heat flow; i.e., (7.8) states that the rate of heat transfer by conduction shall be equal to the rate of heat generation. It is this physical interpretation that adds insight into the heat transfer formulation and leads to a comparison with the formulation used in stress analysis. Because of the analogy of the governing equations, the concepts discussed in the finite element solution of stress analysis are directly applicable to the finite element solution of (7.8). In particular, the procedures used to construct the finite element governing equilibrium equations in stress analysis and the considerations of convergence and accuracy can be used directly (see Chapters 4 and 5). A main difference is that only one unknown variable, the temperature θ, must be calculated.

The heat conduction problem considered above is equivalent to a static stress analysis problem when time effects on the temperature distribution are not included; i.e., steady-state conditions are assumed. However, with significant heat flow input changes specified, it is important to include a term that takes account of the rate at which heat is stored within the material. This rate of heat absorption is given by

$$q^c = c\,\dot{\theta} \tag{7.11}$$

where c is the material heat capacity per unit volume and q^c can be understood to be part of the rate of heat generated q^B in (7.8) (in the same way as the inertia forces were taken to be part of the body forces in stress analysis).

7.2.2 Step-by-Step Incremental Equations

The relation in (7.8) expresses the heat flow equilibrium at all times of interest. For a general solution scheme of both linear and nonlinear, steady-state and transient problems we aim to develop incremental equilibrium equations. As in an incremental finite element stress analysis (see Section 6.1), assume that the conditions at time t have been calculated, and that the temperatures are to be determined for time $t + \Delta t$, where Δt is the time increment. (In steady-state analysis, Δt is used to determine the externally applied heat flow increment, but is otherwise a dummy variable). Considering steady-state conditions or transient analysis with the Euler backward implicit time integration method (see Section 9.6 also for other time integration schemes), the heat flow equilibrium is considered at time $t + \Delta t$ in order to solve for the temperatures at time $t + \Delta t$:

$$\int_V \bar{\boldsymbol{\theta}}'^T \, {}^{t+\Delta t}\mathbf{k} \, {}^{t+\Delta t}\boldsymbol{\theta}' \, dV$$

$$= {}^{t+\Delta t}\mathcal{Q} + \int_{S_c} \bar{\theta}^S \, {}^{t+\Delta t}h({}^{t+\Delta t}\theta_e - {}^{t+\Delta t}\theta^S) \, dS + \int_{S_r} \bar{\theta}^S \, {}^{t+\Delta t}\kappa({}^{t+\Delta t}\theta_r - {}^{t+\Delta t}\theta^S) \, dS \quad (7.12)$$

where the superscript $t + \Delta t$ denotes "at time $t + \Delta t$," S_c and S_r are the surface areas with convection and radiation boundary conditions respectively, and ${}^{t+\Delta t}\mathcal{Q}$ corresponds to further external heat flow input to the system at time $t + \Delta t$. The quantity ${}^{t+\Delta t}\mathcal{Q}$ includes the effects of surface heat flow inputs, ${}^{t+\Delta t}q^S$, that are not included in the convection and radiation boundary conditions and the internal heat generation, ${}^{t+\Delta t}q^B$,

$${}^{t+\Delta t}\mathcal{Q} = \int_V \bar{\theta} \, {}^{t+\Delta t}q^B \, dV + \int_{S_2} \bar{\theta}^S \, {}^{t+\Delta t}q^S \, dS \quad (7.13)$$

where in transient analysis

$${}^{t+\Delta t}q^B \Leftarrow {}^{t+\Delta t}q^B - {}^{t+\Delta t}c \, {}^{t+\Delta t}\dot{\theta} \quad (7.14)$$

and ${}^{t+\Delta t}q^B$ on the r.h.s. of (7.14) is the rate of heat generation excluding the heat capacity effect.

As pointed out above, in steady-state analysis and implicit time integration of transient response, the heat flow equilibrium relation in (7.12) is employed. However, using the Euler forward explicit time integration method to predict transient response, the heat flow equilibrium is considered at time t in order to solve for the temperatures at time $t + \Delta t$ (see Section 9.6). Hence, in explicit time integration of transient response, the heat flow equilibrium equation employed is the relation in (7.12) applied at time t (i.e., the superscript "$t + \Delta t$" is replaced by the superscript "t").

Considering the general heat flow equilibrium relation in (7.12), it should be noted that in linear analysis ${}^{t+\Delta t}\mathbf{k}$ and ${}^{t+\Delta t}h$ are constant and radiation boundary conditions are not included. Hence the relation in (7.12) can be rearranged to obtain

$$\int_V \bar{\boldsymbol{\theta}}'^T \, \mathbf{k} \, {}^{t+\Delta t}\boldsymbol{\theta}' \, dV + \int_{S_c} \bar{\theta}^S \, h \, {}^{t+\Delta t}\theta^S \, dS = {}^{t+\Delta t}\mathcal{Q} + \int_{S_c} \bar{\theta}^S \, h \, {}^{t+\Delta t}\theta_e \, dS \quad (7.15)$$

and it is possible to solve directly for the unknown temperatures ${}^{t+\Delta t}\theta$.

In general nonlinear heat transfer analysis the relation in (7.12) is a nonlinear equation in the unknown temperatures at time $t + \Delta t$. An approximate solution for these temperatures can be obtained by linearizing (7.12) as summarized in Table 7.1. As in stress analysis (see Section 6.1), this linearization can be understood to be the first step of a Newton–Raphson iteration for heat flow equilibrium in which

$${}^{t+\Delta t}\theta^{(i)} = {}^{t+\Delta t}\theta^{(i-1)} + \Delta\theta^{(i)} \quad (7.16)$$

with the initial condition ${}^{t+\Delta t}\theta^{(0)} = {}^{t}\theta$, and $\Delta\theta^{(1)}$ is the temperature increment,

TABLE 7.1 *Linearization of nonlinear heat flow equilibrium equation.*

1. *Equilibrium equation at time* $t + \Delta t$

$$\int_V \bar{\boldsymbol{\theta}}'^T\, {}^{t+\Delta t}\mathbf{k}\, {}^{t+\Delta t}\boldsymbol{\theta}'\, dV = {}^{t+\Delta t}Q + \int_{S_c} \bar{\theta}^S\, {}^{t+\Delta t}h({}^{t+\Delta t}\theta_e - {}^{t+\Delta t}\theta^S)\, dS + \int_{S_r} \bar{\theta}^S\, {}^{t+\Delta t}\kappa({}^{t+\Delta t}\theta_r - {}^{t+\Delta t}\theta^S)\, dS$$

2. *Linearization of equation*

We use: ${}^{t+\Delta t}\theta = {}^t\theta + \theta$; ${}^{t+\Delta t}\boldsymbol{\theta}' = {}^t\boldsymbol{\theta}' + \boldsymbol{\theta}'$; ${}^{t+\Delta t}\mathbf{k} = {}^t\mathbf{k} + \mathbf{k}$; ${}^{t+\Delta t}h = {}^th + h$; ${}^{t+\Delta t}\kappa = {}^t\kappa + \kappa$

and assume: ${}^{t+\Delta t}\mathbf{k}\, {}^{t+\Delta t}\boldsymbol{\theta}' = {}^t\mathbf{k}\, {}^t\boldsymbol{\theta}' + {}^t\mathbf{k}\, \boldsymbol{\theta}'$; ${}^{t+\Delta t}h({}^{t+\Delta t}\theta_e - {}^{t+\Delta t}\theta^S) = {}^th({}^{t+\Delta t}\theta_e - {}^t\theta^S - \theta^S)$

${}^{t+\Delta t}\kappa({}^{t+\Delta t}\theta_r - {}^{t+\Delta t}\theta^S) = {}^t\kappa({}^{t+\Delta t}\theta_r - {}^t\theta^S - \theta^S)$

Substituting into the equation of heat flow equilibrium we obtain

$$\int_V \bar{\boldsymbol{\theta}}'^T\, {}^t\mathbf{k}\, \boldsymbol{\theta}'\, dV + \int_{S_c} \bar{\theta}^S\, {}^th\, \theta^S\, dS + \int_{S_r} \bar{\theta}^S\, {}^t\kappa\, \theta^S\, dS = {}^{t+\Delta t}Q + \int_{S_c} \bar{\theta}^S\, {}^th({}^{t+\Delta t}\theta_e - {}^t\theta^S)\, dS$$
$$+ \int_{S_r} \bar{\theta}^S\, {}^t\kappa({}^{t+\Delta t}\theta_r - {}^t\theta^S)\, dS - \int_V \bar{\boldsymbol{\theta}}'^T\, {}^t\mathbf{k}\, {}^t\boldsymbol{\theta}'\, dV$$

θ, used in Table 7.1. In a full Newton–Raphson iteration the accurate solution of (7.12) would be obtained by using (7.16) and updating in each iteration the coefficient matrix (see Section 8.6 for the analogous calculations in stress analysis). However, in practice, the modified Newton iteration is frequently effective, in which the coefficient matrix is not updated and we solve for $i = 1, 2, \ldots$

$$\int_V \bar{\boldsymbol{\theta}}'^T\, {}^t\mathbf{k}\, \Delta\boldsymbol{\theta}'^{(i)}\, dV + \int_{S_c} \bar{\theta}^S\, {}^th\, \Delta\theta^{S(i)}\, dS + \int_{S_r} \bar{\theta}^S\, {}^t\kappa\, \Delta\theta^{S(i)}\, dS$$
$$= {}^{t+\Delta t}Q + \int_{S_c} \bar{\theta}^S\, {}^{t+\Delta t}h^{(i-1)}({}^{t+\Delta t}\theta_e - {}^{t+\Delta t}\theta^{S(i-1)})\, dS \tag{7.17}$$
$$+ \int_{S_r} \bar{\theta}^S\, {}^{t+\Delta t}\kappa^{(i-1)}({}^{t+\Delta t}\theta_r - {}^{t+\Delta t}\theta^{S(i-1)})\, dS - \int_V \bar{\boldsymbol{\theta}}'^T\, {}^{t+\Delta t}\mathbf{k}^{(i-1)}\, {}^{t+\Delta t}\boldsymbol{\theta}'^{(i-1)}\, dV$$

where ${}^{t+\Delta t}h^{(i-1)}$, ${}^{t+\Delta t}\kappa^{(i-1)}$, and ${}^{t+\Delta t}\mathbf{k}^{(i-1)}$ are the convection and radiation coefficients and the conductivity constitutive matrix that correspond to the temperature ${}^{t+\Delta t}\theta^{(i-1)}$.

7.2.3 Finite Element Discretization of Heat Transfer Equations

The finite element solution of the heat transfer governing equations is obtained using procedures analogous to those employed in stress analysis. We consider first the analysis of steady-state conditions. Assume that the complete body under consideration has been idealized as an assemblage of finite elements; then, in analogy to stress analysis we have at time $t + \Delta t$ for element m

$$\begin{aligned} {}^{t+\Delta t}\theta^{(m)} &= \mathbf{H}^{(m)}\, {}^{t+\Delta t}\boldsymbol{\theta} \\ {}^{t+\Delta t}\theta^{S(m)} &= \mathbf{H}^{S(m)}\, {}^{t+\Delta t}\boldsymbol{\theta} \\ {}^{t+\Delta t}\boldsymbol{\theta}'^{(m)} &= \mathbf{B}^{(m)}\, {}^{t+\Delta t}\boldsymbol{\theta} \end{aligned} \tag{7.18}$$

where the superscript (m) denotes element m and ${}^{t+\Delta t}\boldsymbol{\theta}$ is a vector of all nodal point temperatures at time $t + \Delta t$,

$${}^{t+\Delta t}\boldsymbol{\theta}^T = [{}^{t+\Delta t}\theta_1\ {}^{t+\Delta t}\theta_2\ \ldots\ {}^{t+\Delta t}\theta_n] \tag{7.19}$$

The matrices $\mathbf{H}^{(m)}$ and $\mathbf{B}^{(m)}$ are the element temperature and temperature-gradient interpolation matrices, respectively, and the matrix $\mathbf{H}^{S(m)}$ is the surface temperature interpolation matrix. We evaluate in (7.18) the element temperatures and temperature gradients at time $t + \Delta t$, but the same interpolation matrices are also employed to calculate the element temperature conditions at any other time, and hence for incremental temperatures and incremental temperature gradients.

Linear Steady-State Conditions

Using the relations in (7.18) and substituting into (7.15) we obtain the finite element governing equations in linear heat transfer analysis:

$$(\mathbf{K}^k + \mathbf{K}^c)\,{}^{t+\Delta t}\boldsymbol{\theta} = {}^{t+\Delta t}\mathbf{Q} + {}^{t+\Delta t}\mathbf{Q}^e \tag{7.20}$$

where $\mathbf{K}^k$ is the conductivity matrix,

$$\mathbf{K}^k = \sum_m \int_{V^{(m)}} \mathbf{B}^{(m)T}\,\mathbf{k}^{(m)}\,\mathbf{B}^{(m)}\,dV^{(m)} \tag{7.21}$$

and $\mathbf{K}^c$ is the convection matrix,

$$\mathbf{K}^c = \sum_m \int_{S_c^{(m)}} h^{(m)}\,\mathbf{H}^{S(m)T}\,\mathbf{H}^{S(m)}\,dS^{(m)} \tag{7.22}$$

The nodal point heat flow input vector ${}^{t+\Delta t}\mathbf{Q}$ is given by

$${}^{t+\Delta t}\mathbf{Q} = {}^{t+\Delta t}\mathbf{Q}_B + {}^{t+\Delta t}\mathbf{Q}_S + {}^{t+\Delta t}\mathbf{Q}_C \tag{7.23}$$

where

$${}^{t+\Delta t}\mathbf{Q}_B = \sum_m \int_{V^{(m)}} \mathbf{H}^{(m)T}\,{}^{t+\Delta t}q^{B(m)}\,dV^{(m)} \tag{7.24}$$

$${}^{t+\Delta t}\mathbf{Q}_S = \sum_m \int_{S_2^{(m)}} \mathbf{H}^{S(m)T}\,{}^{t+\Delta t}q^{S(m)}\,dS^{(m)} \tag{7.25}$$

and ${}^{t+\Delta t}\mathbf{Q}_C$ is a vector of concentrated nodal point heat flow input. The nodal point heat flow contribution ${}^{t+\Delta t}\mathbf{Q}^e$ is due to the convection boundary conditions. Using the element surface temperature interpolations to define the environmental temperature ${}^{t+\Delta t}\theta_e$ on the element surfaces in terms of the given nodal point environmental temperatures, ${}^{t+\Delta t}\boldsymbol{\theta}_e$, we have

$${}^{t+\Delta t}\mathbf{Q}^e = \sum_m \int_{S_c^{(m)}} h^{(m)}\mathbf{H}^{S(m)T}\,\mathbf{H}^{S(m)}\,{}^{t+\Delta t}\boldsymbol{\theta}_e\,dS^{(m)} \tag{7.26}$$

The above formulation is effectively used with the variable-number-nodes isoparametric finite elements discussed in Chapter 5. We demonstrate the calculation of the element matrices in the following example.

EXAMPLE 7.1: Consider the four-node isoparametric element in Fig. 7.2. Discuss the calculation of the conductivity matrix $\mathbf{K}^k$, convection matrix $\mathbf{K}^c$ and heat flow input vectors ${}^{t+\Delta t}\mathbf{Q}_B$ and ${}^{t+\Delta t}\mathbf{Q}^e$.

For the evaluation of the above matrices we need the matrices $\mathbf{H}$, $\mathbf{B}$, $\mathbf{H}^S$ and $\mathbf{k}$. The temperature interpolation matrix $\mathbf{H}$ is composed of the interpolation functions defined in Fig. 5.5,

$$\mathbf{H} = \tfrac{1}{4}[(1+r)(1+s)\quad (1-r)(1+s)\quad (1-r)(1-s)\quad (1+r)(1-s)]$$

We obtain $\mathbf{H}^S$ by evaluating $\mathbf{H}$ at $r = 1$, so that

$$\mathbf{H}^S = \tfrac{1}{2}[(1+s)\quad 0\quad 0\quad (1-s)]$$

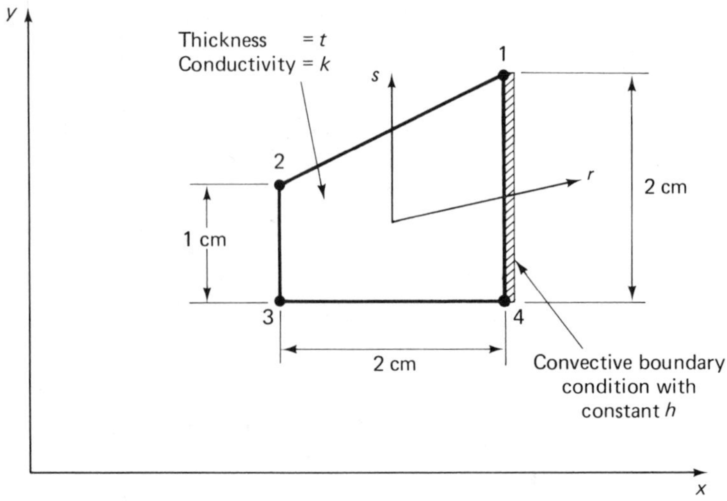

FIGURE 7.2 4-node element in heat transfer conditions.

To evaluate **B** we first evaluate the Jacobian operator **J** (see Example 5.3):

$$\mathbf{J} = \begin{bmatrix} 1 & \dfrac{1+s}{4} \\ 0 & \dfrac{3+r}{4} \end{bmatrix}$$

Hence

$$\mathbf{B} = \frac{1}{4}\begin{bmatrix} 1 & -\left(\dfrac{1+s}{3+r}\right) \\ 0 & \dfrac{4}{(3+r)} \end{bmatrix}\begin{bmatrix} (1+s) & -(1+s) & -(1-s) & (1-s) \\ (1+r) & (1-r) & -(1-r) & -(1+r) \end{bmatrix}$$

$$= \frac{1}{4(3+r)}\begin{bmatrix} 2(1+s) & -4(1+s) & 2(2s-r-1) & 2(2+r-s) \\ 4(1+r) & 4(1-r) & -4(1-r) & -4(1+r) \end{bmatrix}$$

Finally, we have

$$\mathbf{k} = \begin{bmatrix} k & 0 \\ 0 & k \end{bmatrix}$$

The element matrices can now be evaluated using numerical integration as in the analysis of solids and structures (see Chapter 5).

Nonlinear Steady-State Conditions

For general nonlinear analysis the temperature and temperature gradient interpolations of (7.18) are substituted into the heat flow equilibrium relation (7.17) to obtain

$$({}^t\mathbf{K}^k + {}^t\mathbf{K}^c + {}^t\mathbf{K}^r)\,\Delta\boldsymbol{\theta}^{(i)} = {}^{t+\Delta t}\mathbf{Q} + {}^{t+\Delta t}\mathbf{Q}^{c(i-1)} + {}^{t+\Delta t}\mathbf{Q}^{r(i-1)} - {}^{t+\Delta t}\mathbf{Q}^{k(i-1)} \tag{7.27}$$

where the nodal point temperatures at the end of iteration i are

$${}^{t+\Delta t}\boldsymbol{\theta}^{(i)} = {}^{t+\Delta t}\boldsymbol{\theta}^{(i-1)} + \Delta\boldsymbol{\theta}^{(i)} \tag{7.28}$$

The matrices and vectors used in (7.27) are directly obtained from the individual terms used in (7.17) and are defined in Table 7.2. The nodal point heat flow input vector ${}^{t+\Delta t}\mathbf{Q}$ was already defined in (7.23).

TABLE 7.2 *Finite element matrices in nonlinear heat transfer analysis.*

Integral	Finite Element Evaluation
$\int_V \bar{\boldsymbol{\theta}}'^T \, {}^t\mathbf{k} \, \Delta\boldsymbol{\theta}'^{(i)} \, dV$	${}^t\mathbf{K}^k \, \Delta\boldsymbol{\theta}^{(i)} = \left(\sum_m \int_{V^{(m)}} \mathbf{B}^{(m)T} \, {}^t\mathbf{k}^{(m)} \, \mathbf{B}^{(m)} \, dV^{(m)} \right) \Delta\boldsymbol{\theta}^{(i)}$
$\int_{S_c} \bar{\theta}^S \, {}^t h \, \Delta\theta^{S(i)} \, dS$	${}^t\mathbf{K}^c \, \Delta\boldsymbol{\theta}^{(i)} = \left(\sum_m \int_{S_c^{(m)}} {}^t h^{(m)} \, \mathbf{H}^{S(m)T} \, \mathbf{H}^{S(m)} \, dS^{(m)} \right) \Delta\boldsymbol{\theta}^{(i)}$
$\int_{S_r} \bar{\theta}^S \, {}^t\kappa \, \Delta\theta^{S(i)} \, dS$	${}^t\mathbf{K}^r \, \Delta\boldsymbol{\theta}^{(i)} = \left(\sum_m \int_{S_r^{(m)}} {}^t\kappa^{(m)} \, \mathbf{H}^{S(m)T} \, \mathbf{H}^{S(m)} \, dS^{(m)} \right) \Delta\boldsymbol{\theta}^{(i)}$
$\int_{S_c} \bar{\theta}^S \, {}^{t+\Delta t}h^{(i-1)} ({}^{t+\Delta t}\theta_e - {}^{t+\Delta t}\theta^{S(i-1)}) \, dS$	${}^{t+\Delta t}\mathbf{Q}^{c(i-1)} = \sum_m \int_{S_c^{(m)}} {}^{t+\Delta t}h^{(m)(i-1)} \, \mathbf{H}^{S(m)T} \, [\mathbf{H}^{S(m)}({}^{t+\Delta t}\boldsymbol{\theta}_e - {}^{t+\Delta t}\boldsymbol{\theta}^{(i-1)})] \, dS^{(m)}$
$\int_{S_r} \bar{\theta}^S \, {}^{t+\Delta t}\kappa^{(i-1)} ({}^{t+\Delta t}\theta_r - {}^{t+\Delta t}\theta^{S(i-1)}) \, dS$	${}^{t+\Delta t}\mathbf{Q}^{r(i-1)} = \sum_m \int_{S_r^{(m)}} {}^{t+\Delta t}\kappa^{(m)(i-1)} \, \mathbf{H}^{S(m)T} \, [\mathbf{H}^{S(m)}({}^{t+\Delta t}\boldsymbol{\theta}_r - {}^{t+\Delta t}\boldsymbol{\theta}^{(i-1)})] \, dS^{(m)}$
$\int_V \bar{\boldsymbol{\theta}}'^T \, {}^{t+\Delta t}\mathbf{k}^{(i-1)} \, {}^{t+\Delta t}\boldsymbol{\theta}'^{(i-1)} \, dV$	${}^{t+\Delta t}\mathbf{Q}^{k(i-1)} = \sum_m \int_{V^{(m)}} \mathbf{B}^{(m)T} [{}^{t+\Delta t}\mathbf{k}^{(m)(i-1)} \, \mathbf{B}^{(m)} \, {}^{t+\Delta t}\boldsymbol{\theta}^{(i-1)}] dV^{(m)}$

Specified Temperatures

In addition to convection and radiation boundary conditions, nodal point temperature conditions may also be specified. These boundary conditions can be incorporated in the same way as known nodal point displacements are prescribed in stress analysis.

A common procedure is to substitute the known nodal point temperatures in the heat flow equilibrium equations (7.20) and (7.27), and delete the corresponding equations from those to be solved (see Section 4.2.2). However, an effective way to impose nodal point temperatures can be the procedure that is employed to impose convection boundary conditions. Namely, by assigning a very large value of convection coefficient h, where h is much larger than the conductivity of the material, the surface nodal point temperature will be equal to the prescribed environmental nodal point temperature.

EXAMPLE 7.2: Establish the governing finite element equations for the analysis of the infinite parallel-sided slab shown in Fig. 7.3. Consider steady-state conditions and use only one parabolic one-dimensional element to model the slab. (In practice, depending on the temperature gradient to be predicted by the analysis, many more elements may be needed.)

The governing equations for this problem are obtained from (7.27):

$$({}^t\mathbf{K}^k + {}^t\mathbf{K}^c)\, \Delta\boldsymbol{\theta}^{(i)} = {}^{t+\Delta t}\mathbf{Q} + {}^{t+\Delta t}\mathbf{Q}^{c(i-1)} - {}^{t+\Delta t}\mathbf{Q}^{k(i-1)} \tag{a}$$

where

$${}^t\mathbf{K}^k = \int_V \mathbf{B}^T\, {}^tk\mathbf{B}\, dV$$

$${}^t\mathbf{K}^c = \int_{S_c} {}^th\, \mathbf{H}^{S^T}\, \mathbf{H}^S\, dS$$

$${}^{t+\Delta t}\mathbf{Q}^{c(i-1)} = \int_{S_c} {}^{t+\Delta t}h^{(i-1)}\mathbf{H}^{S^T}[\mathbf{H}^S({}^{t+\Delta t}\boldsymbol{\theta}_e - {}^{t+\Delta t}\boldsymbol{\theta}^{S(i-1)})]\, dS$$

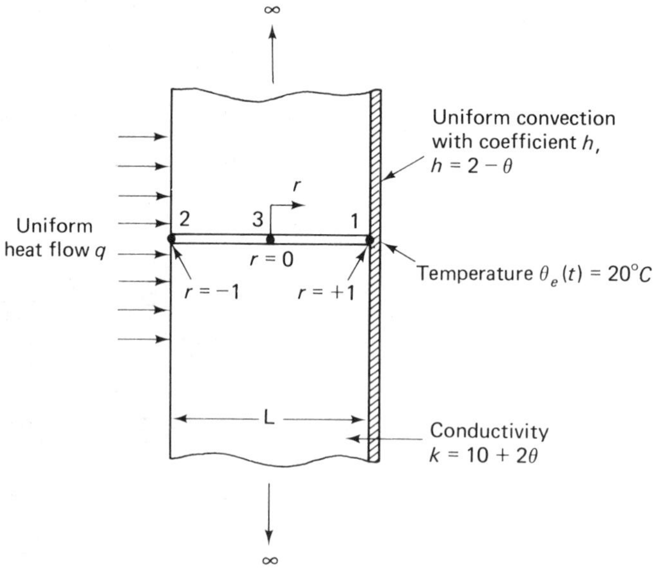

FIGURE 7.3 *Analysis of an infinite slab.*

$$^{t+\Delta t}\mathbf{Q}^{k(i-1)} = \int_V \mathbf{B}^T[^{t+\Delta t}k^{(i-1)}\mathbf{B}\ ^{t+\Delta t}\boldsymbol{\theta}^{(i-1)}]\, dV$$

$$^{t+\Delta t}\mathbf{Q}^T = [0 \quad q \quad 0]$$

and

$$\Delta\boldsymbol{\theta}^{(i)T} = [\Delta\theta_1^{(i)} \quad \Delta\theta_2^{(i)} \quad \Delta\theta_3^{(i)}]$$

$$^{t+\Delta t}\boldsymbol{\theta}^{(i-1)T} = [^{t+\Delta t}\theta_1^{(i-1)} \quad ^{t+\Delta t}\theta_2^{(i-1)} \quad ^{t+\Delta t}\theta_3^{(i-1)}]$$

$$^{t+\Delta t}\boldsymbol{\theta}_e^T = [20 \quad 0 \quad 0]$$

$$^{t+\Delta t}\boldsymbol{\theta}^{S(i-1)T} = [^{t+\Delta t}\theta_1^{(i-1)} \quad 0 \quad 0]$$

For the one-dimensional parabolic element we use the interpolation functions h_1, h_2 and h_3 of Fig. 5.4 to construct $\mathbf{H}$,

$$\mathbf{H} = [\tfrac{1}{2}r(1+r) \quad -\tfrac{1}{2}r(1-r) \quad (1-r^2)]$$

and $\mathbf{H}^S$ corresponding to node 1 is equal to $\mathbf{H}$ evaluated at $r = +1$,

$$\mathbf{H}^S = [1 \quad 0 \quad 0]$$

We also have $J = L/2$, hence

$$\mathbf{B} = \frac{2}{L}\left[\frac{1}{2}(1+2r) \quad -\frac{1}{2}(1-2r) \quad -2r\right]$$

Also, the conductivity of the material is given by

$$^tk = 10 + 2\sum_{i=1}^{3} h_i\, ^t\theta_i$$

and for the convection coefficient we have

$$^th|_{r=+1} = 2 - {}^t\theta_1$$

With the quantities above defined we can now evaluate all matrices in (a) and perform the temperature analysis. Note that we are using $S_c = 1$ and $V = (1) \times (L)$.

Transient Analysis

As mentioned earlier, in transient heat transfer analysis the heat capacity effects can be included in the analysis as part of the rate of heat generated. If the Euler backward implicit time integration is employed the heat flow equilibrium equations used are obtained directly from the equations governing steady-state conditions. Namely, using for element m,

$$\dot{\boldsymbol{\theta}}^{(m)}(x, y, z) = \mathbf{H}^{(m)}(x, y, z)\dot{\boldsymbol{\theta}}(t) \tag{7.29}$$

and now using (7.14) we have in (7.27)

$$^{t+\Delta t}\mathbf{Q}_B = \sum_m \int_{V^{(m)}} \mathbf{H}^{(m)T}(^{t+\Delta t}q^{B(m)} - {}^{t+\Delta t}c^{(m)}\,\mathbf{H}^{(m)}\, ^{t+\Delta t}\dot{\boldsymbol{\theta}})\, dV^{(m)} \tag{7.30}$$

where $^{t+\Delta t}q^{B(m)}$ no longer includes the rate at which heat is stored within the material. Hence, the finite element heat flow equilibrium equations considered in transient conditions are, in linear analysis,

$$\mathbf{C}\, ^{t+\Delta t}\dot{\boldsymbol{\theta}} + (\mathbf{K}^k + \mathbf{K}^c)\, ^{t+\Delta t}\boldsymbol{\theta} = {}^{t+\Delta t}\mathbf{Q} + {}^{t+\Delta t}\mathbf{Q}^e \tag{7.31}$$

and in nonlinear analysis (without linearizing the heat capacity effect, see Section 9.6)

$$\begin{aligned} &^{t+\Delta t}\mathbf{C}^{(i)}\, ^{t+\Delta t}\dot{\boldsymbol{\theta}}^{(i)} + (^t\mathbf{K}^k + {}^t\mathbf{K}^c + {}^t\mathbf{K}^r)\, \Delta\boldsymbol{\theta}^{(i)} \\ &\quad = {}^{t+\Delta t}\mathbf{Q} + {}^{t+\Delta t}\mathbf{Q}^{c(i-1)} + {}^{t+\Delta t}\mathbf{Q}^{r(i-1)} - {}^{t+\Delta t}\mathbf{Q}^{k(i-1)} \end{aligned} \tag{7.32}$$

where $\mathbf{C}$, ${}^{t+\Delta t}\mathbf{C}^{(i)}$ are the heat capacity matrices,

$$\mathbf{C} = \sum_m \int_{V^{(m)}} \mathbf{H}^{(m)T} c^{(m)} \mathbf{H}^{(m)} \, dV^{(m)};$$
$${}^{t+\Delta t}\mathbf{C}^{(i)} = \sum_m \int_{V^{(m)}} \mathbf{H}^{(m)T} \, {}^{t+\Delta t}c^{(m)\,(i)} \, \mathbf{H}^{(m)} \, dV^{(m)} \tag{7.33}$$

In analogy to dynamic analysis, the matrices defined in (7.33) are consistent heat capacity matrices, because the same interpolations are employed for the temperatures as for the time derivatives of temperatures. Following the concepts of displacement analysis it is also possible to use a lumped heat capacity matrix and lumped heat flow input vector, which are evaluated by simply lumping heat capacities and heat flow inputs, using appropriate contributory areas, to the element nodes (see Section 4.2.4).

Using explicit time integration the solution for the unknown temperatures at time $t + \Delta t$ is obtained by considering heat flow equilibrium at time t. Using the relation in (7.12) at time t and substituting the finite element interpolations for temperatures, temperature gradients, and time derivatives of temperatures, we obtain in linear and nonlinear analysis, respectively,

$$\mathbf{C}\,{}^t\dot{\boldsymbol{\theta}} = {}^t\mathbf{Q} + {}^t\mathbf{Q}^c - {}^t\mathbf{Q}^k$$
$${}^t\mathbf{C}\,{}^t\dot{\boldsymbol{\theta}} = {}^t\mathbf{Q} + {}^t\mathbf{Q}^c + {}^t\mathbf{Q}^r - {}^t\mathbf{Q}^k \tag{7.34}$$

where the nodal point heat flow input vectors on the right of (7.34) are defined in Table 7.2 (but the superscript $(i - 1)$ is not used and "$t + \Delta t$" is replaced by "t"). The solution using explicit time integration is only effective when a lumped heat capacity matrix is employed, as discussed further in Section 9.6.

7.3 ANALYSIS OF FIELD PROBLEMS

The heat transfer governing equations discretized in Section 7.2 using finite element procedures are directly applicable to a number of field problems. This analogy is summarized in Table 7.3.[7-10] Hence, it follows that, in practice, if a finite element computer program is available for heat transfer analysis, this same program can also be employed directly for a variety of other analyses by simply operating on the appropriate field variables.[11] We consider in the following a few field problems in some more detail.

7.3.1 Seepage

The equations discussed in Section 7.2.1 are directly applicable in seepage analysis provided confined flow conditions are considered.[8] In this case the boundary surfaces and boundary conditions are all known. If unconfined flow conditions are considered we need to solve also for the position of the free surface.[12]

The basic seepage law used in the analysis is Darcy's law, which gives the flow through the porous medium in terms of the gradient of the total potential ϕ,

$$q_x = -k_x \frac{\partial \phi}{\partial x}, \qquad q_y = -k_y \frac{\partial \phi}{\partial y}, \qquad q_z = -k_z \frac{\partial \phi}{\partial z}$$

TABLE 7.3 Analogies in analysis of field problems.

Problem	*Variable θ*	*Constants* k_x, k_y, k_z	*Input* q^B	*Input* q^S
Heat transfer	Temperature	Thermal conductivity	Internal heat generation	Prescribed heat flow
Seepage	Total head	Permeability	Internal flow generation	Prescribed flow condition
Torsion	Stress function	(Shear modulus)$^{-1}$	(2) × (Angle of twist)	—
Inviscid, incompressible irrotational flow	Potential function	1	Source or sink	Prescribed velocity
Electric conduction	Voltage	Electric conductivity	Internal current source	Prescribed current
Electrostatic field analysis	Field potential	Permittivity	Charge density	Prescribed field

Continuity of flow conditions then results in the equation

$$\frac{\partial}{\partial x}\left(k_x \frac{\partial \phi}{\partial x}\right) + \frac{\partial}{\partial y}\left(k_y \frac{\partial \phi}{\partial y}\right) + \frac{\partial}{\partial z}\left(k_z \frac{\partial \phi}{\partial z}\right) = -q^B \tag{7.35}$$

where k_x, k_y, and k_z are the permeabilities of the medium and q^B is the flow generated per unit volume. The boundary conditions are those of a prescribed total potential ϕ on the surface S_1,

$$\phi|_{S_1} = \phi_e \tag{7.36}$$

and of a prescribed flow condition along the surface S_2,

$$k_n \frac{\partial \phi}{\partial n}\bigg|_{S_2} = q^S \tag{7.37}$$

In (7.35) to (7.37) we are employing the same notation as in (7.1) to (7.3), and on comparing these sets of equations we find a complete analogy between the heat transfer conditions considered in Section 7.2 and the seepage conditions considered here. Figure 7.4 illustrates a finite element analysis of a seepage problem.

7.3.2 Incompressible Inviscid Flow

Consider a fluid in irrotational two-dimensional flow conditions.[7,13] In this case the spin tensor vanishes, so that (see 6.22)

$$\frac{\partial \dot{u}}{\partial y} - \frac{\partial \dot{v}}{\partial x} = 0 \tag{7.38}$$

where $\dot{u}$ and $\dot{v}$ are the fluid velocities into the x and y directions, respectively.

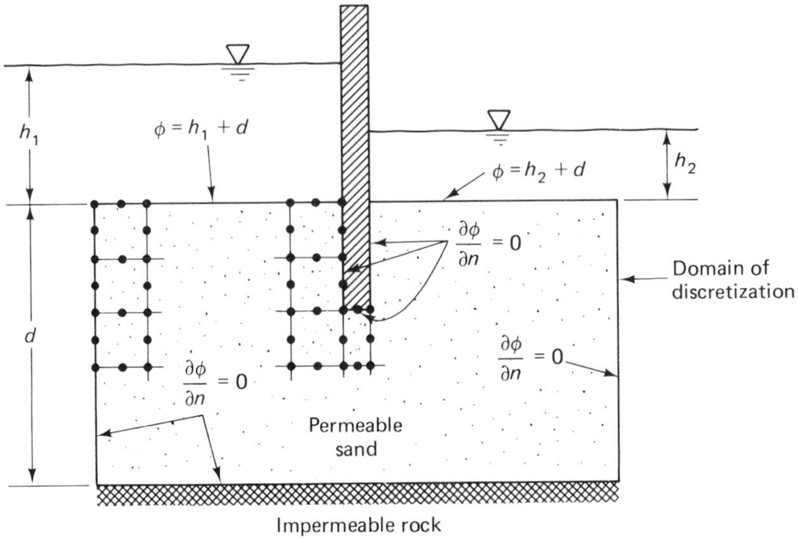

FIGURE 7.4 *Analysis of seepage conditions under a dam.*

The continuity equation also to be satisfied (see Section 7.4) reduces to

$$\frac{\partial \dot{u}}{\partial x} + \frac{\partial \dot{v}}{\partial y} = 0 \tag{7.39}$$

To solve (7.38) and (7.39) we define a potential function $\phi(x, y)$ such that

$$\dot{u} = \frac{\partial \phi}{\partial x}; \qquad \dot{v} = \frac{\partial \phi}{\partial y} \tag{7.40}$$

The relation in (7.38) is then identically satisfied and (7.39) reduces to

$$\frac{\partial^2 \phi}{\partial x^2} + \frac{\partial^2 \phi}{\partial y^2} = 0 \tag{7.41}$$

Using (7.40) we impose all boundary normal velocities $\dot{v}_n$ with

$$\frac{\partial \phi}{\partial n} = \dot{v}_n \tag{7.42}$$

In addition, we need to prescribe an arbitrary value of ϕ at an arbitrary point, because the solution of (7.41) and (7.42) can only be determined once one value of ϕ is fixed.

The solution of (7.41) with the boundary conditions is analogous to the solution of a heat transfer problem. Once the potential function ϕ has been evaluated we can use Bernoulli's equation to calculate the pressure distributions in the fluid.[6] Figure 7.5 illustrates a finite element analysis of the flow around an object in a channel.

7.3.3 Torsion

With the introduction of a stress function, ϕ, the elastic torsional behavior of a shaft is governed by the equation[10]

$$\frac{\partial^2 \phi}{\partial x^2} + \frac{\partial^2 \phi}{\partial y^2} + 2G\theta = 0 \tag{7.43}$$

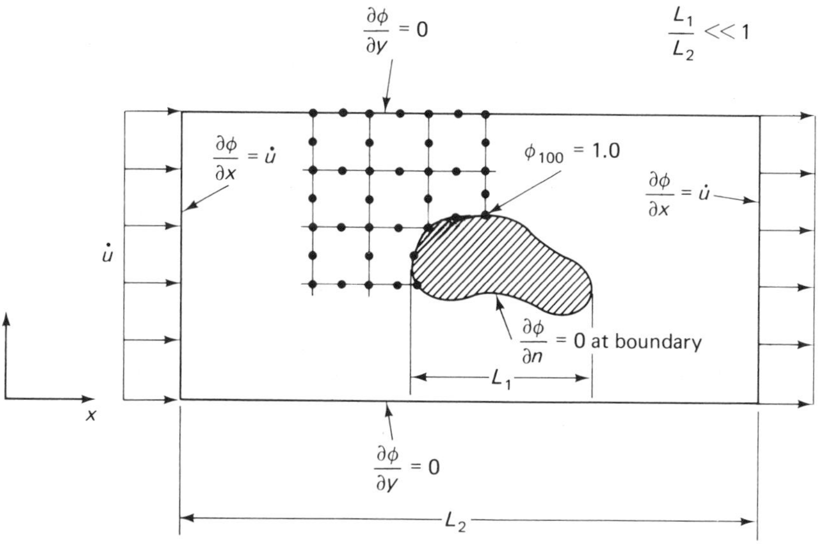

FIGURE 7.5 *Analysis of flow in a channel with an island.*

where θ is the angle of twist per unit length and G is the shear modulus of the shaft material. The shear stress components at any point can be calculated using

$$\tau_{zx} = \frac{\partial \phi}{\partial y}; \qquad \tau_{zy} = -\frac{\partial \phi}{\partial x} \tag{7.44}$$

and the applied torque is given by

$$T = 2 \int_A \phi \, dA \tag{7.45}$$

where A is the cross-sectional area of the shaft. The boundary condition on ϕ is that ϕ is zero on the boundary of the shaft. Hence, the heat transfer equations in (7.1) and (7.2) also govern the torsional behavior of a shaft provided the appropriate field variables are used.

EXAMPLE 7.3: Evaluate the torsional rigidity of a square shaft using the two different finite element meshes shown in Fig. 7.6.

Using the analogy between this torsional problem and a heat transfer analysis for which the governing finite element equations have been stated in Section 7.2.3, we now want to solve

$$\left[\sum_m \int_{V^{(m)}} \mathbf{B}^{(m)T} \mathbf{k}^{*(m)} \mathbf{B}^{(m)} \, dV^{(m)}\right] \boldsymbol{\phi} = \sum_m \int_{V^{(m)}} \mathbf{H}^{(m)T} \theta \, dV^{(m)} \tag{a}$$

where

$$\phi^{(m)} = \mathbf{H}^{(m)} \boldsymbol{\phi}$$
$$\boldsymbol{\phi}'^{(m)} = \mathbf{B}^{(m)} \boldsymbol{\phi}$$
$$\boldsymbol{\phi}^T = [\phi_1 \quad \phi_2 \quad \dots \quad \phi_n]$$
$$\mathbf{k}^{*(m)} = \begin{bmatrix} \frac{1}{2G} & 0 \\ 0 & \frac{1}{2G} \end{bmatrix}$$

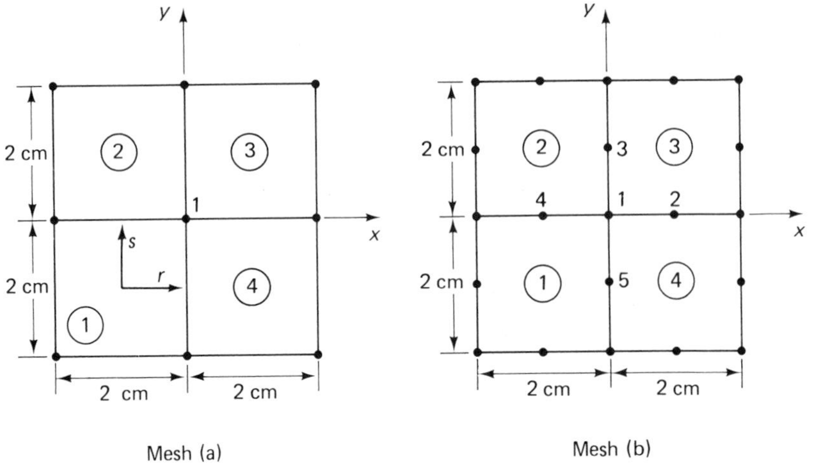

***FIGURE 7.6** Finite element meshes used in calculation of torsional rigidity of a square shaft.*

$$\boldsymbol{\phi}'^{(m)} = \begin{bmatrix} -\tau_{zy} \\ \tau_{zx} \end{bmatrix} = \mathbf{B}^{(m)} \boldsymbol{\phi}$$

and

$$T = \sum_m \int_{A^{(m)}} 2\phi^{(m)} \, dA^{(m)} \tag{b}$$

In each of the analysis cases we need to consider only one element because of symmetry conditions. Using mesh (a) we have for element 1,

$$\phi = \tfrac{1}{4}(1 + r)(1 + s)\phi_1$$

and

$$\begin{bmatrix} \dfrac{\partial \phi}{\partial r} \\ \dfrac{\partial \phi}{\partial s} \end{bmatrix} = \frac{1}{4} \begin{bmatrix} (1 + s) \\ (1 + r) \end{bmatrix} \phi_1$$

Hence the equations in (a) reduce to (considering a unit length of shaft)

$$\left(4 \int_{-1}^{+1} \int_{-1}^{+1} \frac{1}{4}[(1 + s) \quad (1 + r)] \frac{1}{2G} \frac{1}{4} \begin{bmatrix} (1 + s) \\ (1 + r) \end{bmatrix} \det \mathbf{J} \, dr \, ds \right) \phi_1$$

$$= 4\theta \int_{-1}^{+1} \int_{-1}^{+1} \frac{1}{4}(1 + r)(1 + s) \det \mathbf{J} \, dr \, ds; \qquad \det \mathbf{J} = 1$$

or

$$\frac{1}{3G}\phi_1 = \theta$$

Hence

$$\phi_1 = 3G\theta$$

Using (b) we thus obtain

$$T = 4 \int_{-1}^{+1} \int_{-1}^{+1} \tfrac{1}{2}(1 + r)(1 + s)(3G\theta) \, dr \, ds = 24G\theta$$

so that

$$\frac{T}{\theta} = 24G$$

Considering next mesh (b) we recognize that the ϕ values on the boundary are zero, and that for element 1 we have $\phi_4 = \phi_5$. Hence, we have only two unknowns

ϕ_1 and ϕ_4 to be calculated. The interpolation functions for the eight-node element are given in Fig. 5.5. Proceeding in the same way as in the analysis with mesh (a) we obtain

$$\phi_1 = 2.157G\theta$$
$$\phi_4 = 1.921G\theta$$
$$T = 35.2G\theta$$

so that

$$\frac{T}{\theta} = 35.2G$$

The exact solution of (7.43) for T/θ is $36.1G$. Hence, the analysis with mesh (a) gives an error of 33.5% whereas the analysis with mesh (b) yields a result of only 2.5% error.*

7.4 ANALYSIS OF VISCOUS INCOMPRESSIBLE FLUID FLOW

The most common approach followed in the analysis of fluid mechanics problems is to develop the governing differential equations for the specific flow, geometry and boundary conditions under consideration and then solve these equations using finite difference procedures.[13,14] Since the problems encountered in fluid mechanics are of a very complex and varied nature it is—at least at this time—almost imperative to concentrate on the special flow conditions to be analyzed and abandon a very general approach of analysis. However, it is desirable to use within a specific category of fluid flow an analysis approach as general as possible, and it is for this reason that finite element analysis procedures can be valuable. An important area in which finite element analysis procedures are applied successfully to an increasing extent is the analysis of viscous incompressible fluid flow.[15-20] The finite element formulation of these problems follows the approach used in the analysis of solids, but some specific difficulties arise that are special to fluid flow.

In order to identify the similarities and differences in the basic finite element formulations of problems in solid (see Chapter 6) and fluid mechanics, consider the summary of the governing continuum mechanics equations given in Table 7.4. In this table and in the discussion to follow we use the indicial notation employed in Chapter 6. Considering the kinematics of a viscous fluid flow, the fluid particles can undergo very large motions and, for a general description, it is effective to use an Eulerian formulation. The essence of this formulation is that we focus attention on a stationary control volume and that we use this volume to measure the equilibrium and mass continuity of the fluid particles. This means that in the Eulerian formulation a separate equation is written to express the mass conservation relation—a condition which is embodied in the determinant of the deformation gradient when using a Lagrangian formulation. It further means that the inertia forces involve convective terms which in the

* Note that this finite element analysis underestimates the stress function values ϕ (for an imposed value of twist θ) and thus yields a lower bound on the torsional rigidity, whereas a displacement and stress analysis using the procedures of Chapter 5 would yield an upper bound on T/θ (provided the monotonic convergence requirements of Section 4.2.5 are fulfilled).

TABLE 7.4 *Basic continuum mechanics equations used in Lagrangian and Eulerian formulations.*

Lagrangian Formulation	*Eulerian Formulation*
Geometric Representation 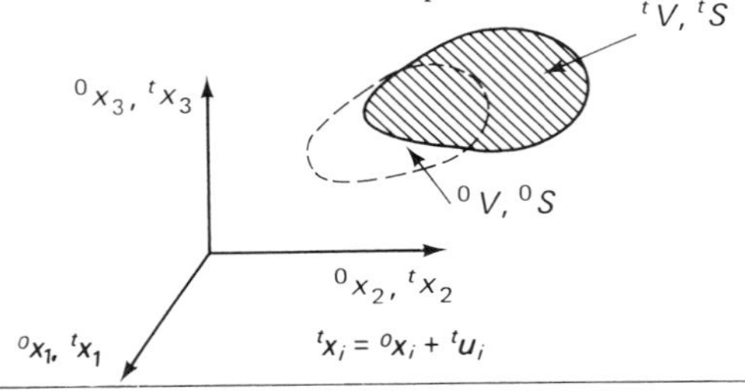 ${}^tx_i = {}^0x_i + {}^tu_i$	Geometric Representation ${}^0x_i = {}^tx_i \equiv x_i$ 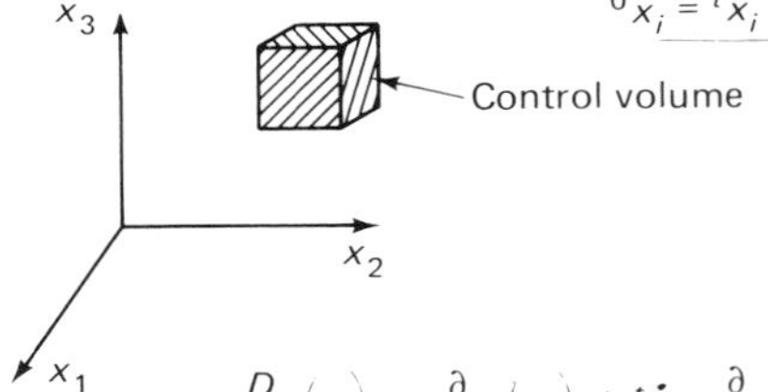 $\frac{D}{Dt}(\;) = \frac{\partial}{\partial t}(\;) + {}^t\dot{u}_j \frac{\partial}{\partial x_j}(\;)$
Conservation of Mass $m = \int_{^0V} {}^0\rho\, {}^0dV = \int_{^tV} {}^t\rho\, {}^tdV \Longrightarrow \frac{{}^0\rho}{{}^t\rho} = \det({}^t_0\mathbf{X})$	Conservation of Mass $m = \int_{^0V} {}^0\rho\, {}^0dV = \int_{^tV} {}^t\rho\, {}^tdV \Longrightarrow \frac{D}{Dt}({}^t\rho) + {}^t\rho \frac{\partial {}^t\dot{u}_i}{\partial x_i} = 0$ Incompressible flow: ${}^t\dot{u}_{i,i} = 0$
Equation of Motion $\frac{\partial}{\partial {}^tx_j}({}^t\tau_{ji}) + {}^t\tilde{f}_i^B = 0; \qquad {}^t\tilde{f}_i^B = {}^tf_i^B - {}^t\rho\, {}^t\ddot{u}_i$	Equation of Motion $\frac{\partial}{\partial x_j}({}^t\tau_{ji}) + {}^t\tilde{f}_i^B = 0; \qquad {}^t\tilde{f}_i^B = {}^tf_i^B - {}^t\rho \frac{D}{Dt}({}^t\dot{u}_i)$
Principle of Virtual Displacements $\int_{^tV} {}^t\tau_{ij}\, \delta_t e_{ij}\, {}^tdV = \int_{^tV} {}^t\tilde{f}_i^B\, \delta u_i\, {}^tdV + \int_{^tS} {}^tf_i^S\, \delta u_i^S\, {}^tdS$ Total Lagrangian formulation: $\int_{^0V} {}^t_0S_{ij}\, \delta {}^t_0\epsilon_{ij}\, {}^0dV = {}^t\mathcal{R}$	Principle of Virtual Velocities $\int_V {}^t\tau_{ij}\, \delta\dot{e}_{ij}\, dV = \int_V {}^t\tilde{f}_i^B\, \delta\dot{u}_i\, dV + \int_S {}^tf_i^S\, \delta\dot{u}_i^S\, dS$ ${}^t\tau_{ij} = -{}^tp\, \delta_{ij} + 2\mu\, {}^t\dot{e}_{ij}; \qquad {}^t\dot{e}_{ij} = \frac{1}{2}({}^t\dot{u}_{i,j} + {}^t\dot{u}_{j,i})$

numerical solution result in a nonsymmetric coefficient matrix that depends on the velocities to be calculated. Finally, we note another difference from a finite element formulation of a solid mechanics problem: the stress components are given in terms of a pressure (which is present without fluid motion) plus the constitutive relation multiplied by the fluid velocity strains. Since the pressure appears on the same side as the velocity strains special considerations are necessary in the formulation.

The advantage of an Eulerian formulation lies in the use of simple stress and strain rate measures; namely, those that we are using in infinitesimal displacement analysis except that velocities must be calculated instead of displacements.

Two different approaches of analysis can be employed effectively as described briefly below.

7.4.1 Velocity-Pressure Formulation

Using the information given in Table 7.4, the governing equilibrium equation of the fluid considered is

$$\int_V {}^t\tau_{ij}\, \delta \dot{e}_{ij}\, dV = \int_V {}^t\tilde{f}_i^B\, \delta \dot{u}_i\, dV + \int_S {}^t f_i^S\, \delta \dot{u}_i^S\, dS \tag{7.46}$$

in which

$${}^t\tilde{f}_i^B = {}^t f_i^B - \rho \frac{D}{Dt}\, {}^t\dot{u}_i \tag{7.47}$$

and

$${}^t\tau_{ij} = -{}^t p\, \delta_{ij} + 2\mu\, {}^t\dot{e}_{ij}; \qquad {}^t\dot{e}_{ij} = \tfrac{1}{2}({}^t\dot{u}_{i,j} + {}^t\dot{u}_{j,i}) \tag{7.48}$$

where ${}^t p$ is the pressure and μ is the fluid viscosity. The continuity equation expressed in integral form is

$$\int_V \delta {}^t p\, {}^t\dot{u}_{i,i}\, dV = 0 \tag{7.49}$$

This equation is generated using the Galerkin method and recognizing that the pressure in the fluid is the appropriate weighting function on the volumetric velocity strains. (Note that this way the dimensions of (7.46) and (7.49) are those of power.) The principle of virtual velocities in (7.46) is quite analogous to the principle of virtual displacements used in stress analysis (see Table 7.4) and we can use the usual procedures to develop the finite element equations governing the fluid response. Since the unknowns are the pressure and the velocities, we use the following interpolations for a typical finite element:

$${}^t p = \sum_{k=1}^{q_p} h_k^*\, {}^t p_k; \qquad x_i = \sum_{k=1}^{q_g} \tilde{h}_k\, x_i^k; \qquad {}^t\dot{u}_i = \sum_{k=1}^{q} h_k\, {}^t\dot{u}_i^k \tag{7.50}$$

where q_p equals the number of nodes used in the pressure interpolation and q is the number of nodes used to interpolate the velocities, $q > q_p$. Considering the geometry interpolation we may have q_g equal to q_p or q, hence $\tilde{h}_k = h_k^*$ or $\tilde{h}_k = h_k$.

We use a higher-order interpolation on the velocities than on the pressure so as to have the same order of approximation on the spatial derivatives of the velocities as we have on the pressure. We proceed this way because multiples of the velocity strains are added to the pressure to obtain the stresses (see (7.48)).

Substituting now from (7.50) into (7.46) to (7.49), we obtain for a system of finite elements the equations expressing equilibrium,

$$\mathbf{M}\,{}^t\ddot{\mathbf{U}} + (\mathbf{K} + {}^t\mathbf{K}_c){}^t\dot{\mathbf{U}} + \mathbf{K}_{Vp}\,{}^t\mathbf{P} = {}^t\mathbf{R}_B + {}^t\mathbf{R}_S \tag{7.51}$$

and continuity,

$$\mathbf{K}_{Vp}^T\,{}^t\dot{\mathbf{U}} = \mathbf{0} \tag{7.52}$$

where for a single element with nodal point velocities ${}^t\dot{\hat{\mathbf{u}}}$,

$$\mathbf{M} = \rho \int_V \mathbf{H}^T \mathbf{H}\, dV \tag{7.53}$$

$$\mathbf{K} = \int_V \mathbf{B}^T \mathbf{C}' \mathbf{B}\, dV \tag{7.54}$$

$${}^t\mathbf{K}_c = \rho \int_V \mathbf{H}^T(\mathbf{B}_c\,{}^t\dot{\hat{\mathbf{u}}})\mathbf{H}\, dV \tag{7.55}$$

$$\mathbf{K}_{Vp} = -\int_V \mathbf{B}_V^T \mathbf{H}_p\, dV \tag{7.56}$$

$${}^t\mathbf{R}_B = \int_V \mathbf{H}^T\,{}^t\mathbf{f}^B\, dV \tag{7.57}$$

$${}^t\mathbf{R}_S = \int_S \mathbf{H}^{S^T}\,{}^t\mathbf{f}^S\, dS \tag{7.58}$$

The matrices **M**, **H**, and **B** are identical to those that we defined in the analysis of solid mechanics problems (see Section 5.3). However, the coefficient matrices $\mathbf{K}$, ${}^t\mathbf{K}_c$, and $\mathbf{K}_{Vp}$ are specific to the fluid mechanics formulation, and we have in three-dimensional analysis:

$$\mathbf{C}' = \mu \begin{bmatrix} 2 & & & & & \\ & 2 & & \text{zeros} & & \\ & & 2 & & & \\ & & & 1 & & \\ & \text{zeros} & & & 1 & \\ & & & & & 1 \end{bmatrix} \tag{7.59}$$

$$\mathbf{B}_c\,{}^t\dot{\hat{\mathbf{u}}} = \begin{bmatrix} \dfrac{\partial\,{}^t\dot{u}_1}{\partial x_1} & \dfrac{\partial\,{}^t\dot{u}_1}{\partial x_2} & \dfrac{\partial\,{}^t\dot{u}_1}{\partial x_3} \\ \dfrac{\partial\,{}^t\dot{u}_2}{\partial x_1} & \dfrac{\partial\,{}^t\dot{u}_2}{\partial x_2} & \dfrac{\partial\,{}^t\dot{u}_2}{\partial x_3} \\ \dfrac{\partial\,{}^t\dot{u}_3}{\partial x_1} & \dfrac{\partial\,{}^t\dot{u}_3}{\partial x_2} & \dfrac{\partial\,{}^t\dot{u}_3}{\partial x_3} \end{bmatrix} \tag{7.60}$$

$$\mathbf{B}_V = \mathbf{m}\mathbf{B}; \qquad \mathbf{m} = [1 \quad 1 \quad 1 \quad 0 \quad 0 \quad 0] \tag{7.61}$$

and $\mathbf{H}_p$ is the pressure interpolation matrix.

We should note that the equations in (7.51) and (7.52) are always nonlinear if the matrix ${}^t\mathbf{K}_c$ is included in the formulation. Hence, a linear problem is only solved in the case of creeping flow.

The complete set of equations in (7.51) and (7.52) can be written as

$$\begin{bmatrix} \mathbf{M} & \mathbf{0} \\ \mathbf{0} & \mathbf{0} \end{bmatrix} \begin{bmatrix} {}^t\ddot{\mathbf{U}} \\ {}^t\mathbf{P} \end{bmatrix} + \begin{bmatrix} (\mathbf{K} + {}^t\mathbf{K}_c) & \mathbf{K}_{Vp} \\ \mathbf{K}_{Vp}^T & \mathbf{0} \end{bmatrix} \begin{bmatrix} {}^t\dot{\mathbf{U}} \\ {}^t\mathbf{P} \end{bmatrix} = \begin{bmatrix} {}^t\mathbf{R}_B + {}^t\mathbf{R}_S \\ \mathbf{0} \end{bmatrix} \tag{7.62}$$

and we see that the coefficient matrix is nonsymmetric and contains zeros on the diagonal. However, provided appropriate interpolations are employed for the velocities and pressures and the appropriate boundary and initial conditions are specified, no difficulties are encountered in the solution of (7.62). Two effective elements for two-dimensional analysis are shown in Fig. 7.7.

Finally, we should note that the formulation given above can be interpreted as a formulation in which a Lagrange multiplier (the pressure in the fluid) is used to enforce the incompressibility constraint. The form of the equations in (7.62) is for this reason similar to those discussed in Section 3.4.

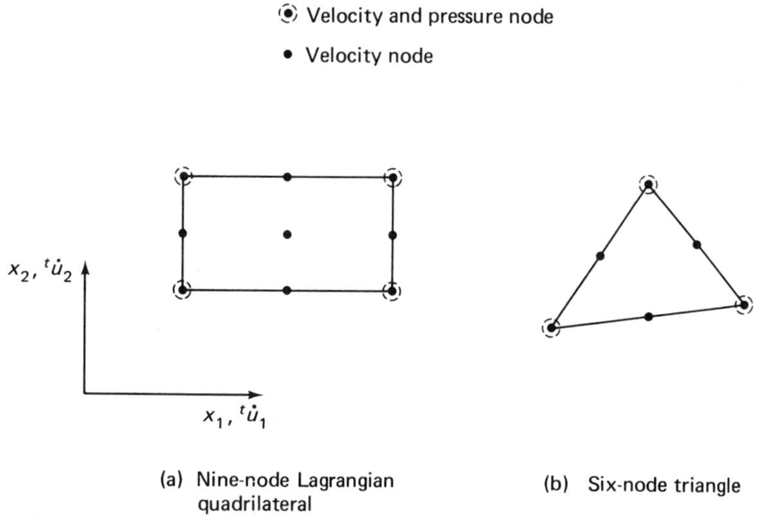

FIGURE 7.7 *Elements used in analysis of incompressible viscous flow; interpolation functions are given in Figs. 5.5 and 5.25.*

7.4.2 Penalty Method Formulation

The basic step in the penalty formulation of viscous incompressible flow is the elimination of the pressure term in the equilibrium equations using[17,18]

$$ {}^tp = -\lambda\, {}^t\dot{u}_{i,i} \tag{7.63}$$

where λ is the penalty parameter and can be interpreted as a bulk-stiffness. It can be formally shown that if we choose λ to be a relatively large number, the continuity equation ${}^t\dot{u}_{i,i} = 0$ will be approximately satisfied in the finite element solution.[21] Substituting from (7.63) into (7.46) we obtain

$$\int_V (\lambda\, {}^t\dot{u}_{i,i}\, \delta_{ij} + 2\mu\, {}^t\dot{e}_{ij})\, \delta\dot{e}_{ij}\, dV = \int_V \delta\dot{u}_i\, {}^tf_i^B\, dV + \int_S \delta\dot{u}_i^S\, {}^tf_i^S\, dS \tag{7.64}$$

with the velocities as the only unknowns. Hence the governing equations for a system of finite elements are now

$$\mathbf{M}\, {}^t\ddot{\mathbf{U}} + (\mathbf{K} + {}^t\mathbf{K}_c + \mathbf{K}_p)\, {}^t\dot{\mathbf{U}} = {}^t\mathbf{R}_B + {}^t\mathbf{R}_S \tag{7.65}$$

where the matrices have already been defined in Section 7.4.1, and for a single element we may consider the use of

$$\mathbf{K}_p = \lambda \int_V \mathbf{B}_V^T\, \mathbf{B}_V\, dV \tag{7.66}$$

The coefficient matrix in (7.65) is still nonsymmetric due to the matrix ${}^t\mathbf{K}_c$ but the diagonal elements are nonzero. Also, the number of unknowns have been reduced because the pressure does not appear as a variable in (7.65).

The approach of analysis used here is that the continuity condition is imposed on the equilibrium equations as a constraint using the penalty method (see Section 3.4). Comparing this formulation with the velocity-pressure formulation presented in Section 7.4.1, and recalling that we interpolated the pressure to a lower degree than the velocities, it appears that we should now evaluate the matrix $\mathbf{K}_p$ in (7.66) with a lower integration order than the matrix $\mathbf{K}$ in (7.65). Indeed, it has been shown mathematically that the use of a lower integration order is necessary for a convergent solution.[18,21] For example, using the nine-node (straight-sided) Lagrangian element in Fig. 7.7 the matrix $\mathbf{K}_p$ should be evaluated using 2×2 Gauss numerical integration which corresponds to a linear interpolation of the pressure (that is discontinuous between elements).

The above procedure is quite appealing but the difficulty lies in defining an appropriate matrix $\mathbf{K}_p$ for a given element. Much research has been performed to study the detailed relations between velocity-pressure and penalty method formulations of specific elements, and to establish effective matrices $\mathbf{K}_p$ for higher-order distorted elements in order to render the penalty analysis approach more generally applicable.[22,23]

Finally, we may note that in practice the penalty parameter should be chosen to be about two to four orders of magnitude larger than the other stiffness terms. With this choice the numerical solution is affected little by the exact value of the penalty parameter, but a large enough number of digits must be used in the floating point arithmetic to solve (7.65) (see Section 8.5).

REFERENCES

1. O. C. Zienkiewicz and Y. K. Cheung, "Finite Elements in the Solution of Field Problems," *The Engineer*, Sept. 24, 1964.
2. E. L. Wilson and R. E. Nickell, "Application of the Finite Element Method to Heat Conduction Analysis," *J. Nuclear Engineering and Design*, Vol. 4, pp. 276–286, October 1966.
3. E. L. Wilson, K. J. Bathe, and F. E. Peterson, "Finite Element Analysis of Linear and Nonlinear Heat Transfer," *J. Nuclear Engineering and Design*, Vol. 29, pp. 110–124, 1974.
4. K. J. Bathe and M. R. Khoshgoftaar, "Finite Element Formulation and Solution of Nonlinear Heat Transfer," *J. Nuclear Engineering and Design*, Vol. 51, 1979, pp. 389–401.
5. H. S. Carslaw and J. C. Jaeger, *Conduction of Heat in Solids*, 2nd ed., Oxford University Press, 1969.
6. W. M. Rohsenow and W. Y. Choi, *Heat, Mass and Momentum Transfer*, Prentice-Hall, Inc., Englewood Cliffs, N.J., 1961.
7. Y. C. Fung, *A First Course in Continuum Mechanics*, Prentice-Hall, Inc., Englewood Cliffs, N.J., 1967.
8. A Verruijt, *Theory of Groundwater Flow*, Gordon and Breach Science Publishers, New York, N.Y., 1970.
9. S. Ramo, J. R. Whinnery, and T. Van Duzer, *Fields and Waves in Communication Electronics*, John Wiley & Sons, New York, N.Y., 1965.

10. Y. C. Fung, *Foundations of Solid Mechanics*, Prentice-Hall, Inc., Englewood Cliffs, N.J., 1965.
11. K. J. Bathe, "ADINAT—A Finite Element Program for Automatic Dynamic Incremental Nonlinear Analysis of Temperatures," Report 82448–1, Acoustics and Vibration Lab., Mechanical Engineering Dept., M.I.T., Sept. 1975, rev. Dec. 1978.
12. K. J. Bathe and M. R. Khoshgoftaar, "Finite Element Free Surface Seepage Analysis without Mesh Iteration," *Int. J. Anal. and Num. Meth. in Geomechanics*, Vol. 3, pp. 13–22, 1979.
13. C. Y. Chow, *An Introduction to Computational Fluid Mechanics*, John Wiley & Sons, Inc., New York, N.Y., 1979.
14. P. J. Roache, *Computational Fluid Dynamics*, Hermosan Publishers, Albuquerque, N.M., 1972.
15. J. T. Oden, O. C. Zienkiewicz, R. H. Gallagher, and C. Taylor (eds.), *Finite Elements in Fluids*, Vol. I and II, John Wiley and Sons, Inc., New York, N.Y., 1975.
16. D. K. Gartling and E. B. Becker, "Finite Element Analysis of Viscous Incompressible Fluid Flow," *J. Comp. Meth. Appl. Mech. and Eng.*, Vol. 8, pp. 51–60, Vol. 8, pp. 127–138, 1976.
17. T. J. R. Hughes, W. K. Liu, and A. Brooks, "Review of Finite Element Analysis of Incompressible Viscous Flows by the Penalty Function Formulation," *J. Computational Physics*, Vol. 30, 1979, pp. 1–60.
18. M. Bercovier and M. Engelman, "A Finite Element for the Numerical Solution of Viscous Incompressible Flow," *J. Computational Physics*, Vol. 30, 1979, pp. 181–201.
19. T. J. R. Hughes (ed.), "Finite Element Methods for Convection Dominated Flows," *American Society of Mechanical Engineers*, AMD, Vol. 34, 1979.
20. K. J. Bathe and V. Sonnad, "On Finite Element Analysis of Compressible and Incompressible Temperature-Dependent Fluid Flow," *Proceedings, AIAA 5th Computational Fluid Dynamics Conference*, Palo Alto, Calif., June 1981.
21. M. Bercovier, "Perturbation of Mixed Variational Problems. Application to Mixed Finite Element Methods," *R.A.I.R.O.* (*Numerical Analysis*), Vol. 12, pp. 211–236 (1978).
22. M. Bercovier, Y. Hasbani, Y. Gilon, and K. J. Bathe, "On a Finite Element Procedure for Nonlinear Incompressible Elasticity," *Proceedings, Symposium on Hybrid and Mixed Finite Elements*, Georgia Institute of Technology, April, 1981.
23. M. Engelman, R. L. Sani, P. M. Gresho, and M. Bercovier, "Consistent versus Reduced Integration Penalty Methods for Incompressible Media Using Several Old and New Elements," *Int. J. Num. Meth. in Fluids*, in press.

PART III

SOLUTION OF FINITE ELEMENT EQUILIBRIUM EQUATIONS

8

SOLUTION OF EQUILIBRIUM EQUATIONS IN STATIC ANALYSIS

8.1 INTRODUCTION

So far we have considered the derivation and calculation of the equilibrium equations of a finite element system. This included the selection and calculation of efficient elements, and the efficient assemblage of the element matrices to the global finite element system matrices. However, the overall effectiveness of an analysis depends to a large degree on the numerical procedures used for the solution of the system equilibrium equations. As discussed in Section 4.2.5, the accuracy of the analysis can, in general, be improved, if a more refined finite element mesh is used. Therefore, in practice, an analyst tends to employ larger and larger finite element systems to approximate the actual structure. However, this means that the cost of an analysis and, in fact, its practical feasibility depend to a considerable degree on the algorithms available for the solution of the resulting systems of equations. Because of the requirement that large systems be solved, much research effort has gone into optimizing the equation solution algorithms. During the early use of the finite element method, equations of the order 100 were in many cases considered of large order. Currently, equations of the order 10,000 are solved without much difficulty.

Depending on the kind and number of elements used in the assemblage and on the topology of the finite element mesh, in a linear static analysis the time for solution of the equilibrium equations can be a large percentage of the total solution time, whereas in dynamic analysis or in nonlinear analysis, this percentage may be still higher. Therefore, if inappropriate techniques for the solution of the equilibrium equations are used, the total cost of analysis is affected a great deal, and indeed the cost may be many times, say 100 times, larger than is necessary.

In addition to considering the actual computer effort that is spent on the solution of the equilibrium equations, it is important to realize that an analysis may, in fact, not be possible if inappropriate numerical procedures are used. This may be the case because the analysis is simply too costly using the slow solution methods. But, more seriously, the analysis may not be possible because the solution procedures are unstable. We will observe that the stability of the solution procedures is particularly important in dynamic analysis.

In this chapter we are concerned with the solution of the simultaneous equations that arise in the static analysis of structures and solids, and we discuss first at length (see Sections 8.2 to 8.5) the solution of the equations that arise in linear analysis,

$$\mathbf{K}\mathbf{U} = \mathbf{R} \tag{8.1}$$

where $\mathbf{K}$ is the stiffness matrix, $\mathbf{U}$ is the displacement vector, and $\mathbf{R}$ is the load vector of the finite element system. Since $\mathbf{R}$ and $\mathbf{U}$ may be functions of time t, we may also consider (8.1) as the dynamic equilibrium equations of a finite element system in which inertia and velocity-dependent damping forces have been neglected. It should be realized that since velocities and accelerations do not enter (8.1), we can evaluate the displacements at any time t independent of the displacement history, which is not the case in dynamic analysis (see Chapter 9). However, the relation between the solution of (8.1) and an actual dynamic analysis suggests that the algorithms used for the evaluation of $\mathbf{U}$ in (8.1) may also be employed as part of the solution algorithms used in dynamic analysis. This is indeed the case; *we will see in the following chapters that the procedures discussed here will be the basis of the algorithms employed for eigen-solutions and direct step-by-step integrations*. Furthermore, as we noted already in Chapter 6 and is further discussed in Section 8.6, the solution of (8.1) also represents a very important basic step of solution in a nonlinear analysis. Therefore, a detailed study of the procedures used to solve (8.1) is very important.

Although we consider in this chapter explicitly the solution of the equilibrium equations that arise in the analysis of solids and structures, the techniques are quite general and are entirely and directly applicable to all those analyses that lead to symmetric (positive definite) coefficient matrices (see Chapters 3 and 7). The only sets of equations presented earlier whose solution we do not really consider are those arising in the analysis of incompressible, viscous fluid flow (see Section 7.4)—because in such analysis a nonsymmetric coefficient matrix is obtained—but even in that case most basic concepts and procedures given below are still applicable.

Essentially, there are two different classes of methods for the solution of the equations in (8.1): direct solution techniques and iterative solution methods. In a direct solution the equations in (8.1) are solved using a number of steps and operations that are predetermined in an exact manner, whereas iteration is used when an iterative solution method is employed. Either solution scheme will be seen to have certain advantages; however, in almost all applications the direct methods are currently most effective. For this reason, the larger part of our discussion is devoted to effective direct solution procedures.

8.2 DIRECT SOLUTIONS USING ALGORITHMS BASED ON GAUSS ELIMINATION

The most effective direct solution techniques currently used are basically applications of Gauss elimination, which Gauss proposed over a century ago.[1-11] However, although the basic Gauss solution scheme can be applied to almost any set of simultaneous linear equations, the effectiveness in finite element analysis depends on the specific properties of the finite element stiffness matrix: symmetry, positive definiteness, and bandedness.

In the following we consider first the Gauss elimination procedure as it is used in the solution of positive definite, symmetric, and banded systems. We briefly consider the solution of symmetric, indefinite systems in Section 8.2.5.

8.2.1 Introduction to Gauss Elimination

We propose to introduce the Gauss solution procedure by studying the solution of the equations $\mathbf{KU} = \mathbf{R}$ derived in Example 4.22 with the parameters $L = 5$, $EI = 1$, i.e.,

$$\begin{bmatrix} 5 & -4 & 1 & 0 \\ -4 & 6 & -4 & 1 \\ 1 & -4 & 6 & -4 \\ 0 & 1 & -4 & 5 \end{bmatrix} \begin{bmatrix} U_1 \\ U_2 \\ U_3 \\ U_4 \end{bmatrix} = \begin{bmatrix} 0 \\ 1 \\ 0 \\ 0 \end{bmatrix} \tag{8.2}$$

In this case the stiffness matrix $\mathbf{K}$ corresponds to a simply supported beam with four translational degrees of freedom, as shown in Fig. 8.1. (We should recall that the equilibrium equations have been derived by finite differences; but, in this case, they have the same properties as in finite element analysis.)

To solve the equations in (8.2) using Gauss elimination, we proceed in the following systematic steps:

Step 1: Subtract a multiple of equation 1 from equations 2 and 3 to obtain zero elements in the first column of $\mathbf{K}$. This means that $(-\frac{4}{5})$ times the first row is subtracted from the second row, and $(\frac{1}{5})$ times the first row is subtracted from the third row. The resulting equations are

$$\begin{bmatrix} 5 & -4 & 1 & 0 \\ 0 & \frac{14}{5} & -\frac{16}{5} & 1 \\ 0 & -\frac{16}{5} & \frac{29}{5} & -4 \\ 0 & 1 & -4 & 5 \end{bmatrix} \begin{bmatrix} U_1 \\ U_2 \\ U_3 \\ U_4 \end{bmatrix} = \begin{bmatrix} 0 \\ 1 \\ 0 \\ 0 \end{bmatrix} \tag{8.3}$$

Step 2: Considering next the equations in (8.3), subtract $(-\frac{16}{14})$ times the second equation from the third equation and $(\frac{5}{14})$ times the second equation from the fourth equation. The resulting equations are

$$\begin{bmatrix} 5 & -4 & 1 & 0 \\ 0 & \frac{14}{5} & -\frac{16}{5} & 1 \\ 0 & 0 & \frac{15}{7} & -\frac{20}{7} \\ 0 & 0 & -\frac{20}{7} & \frac{65}{14} \end{bmatrix} \begin{bmatrix} U_1 \\ U_2 \\ U_3 \\ U_4 \end{bmatrix} = \begin{bmatrix} 0 \\ 1 \\ \frac{8}{7} \\ -\frac{5}{14} \end{bmatrix} \tag{8.4}$$

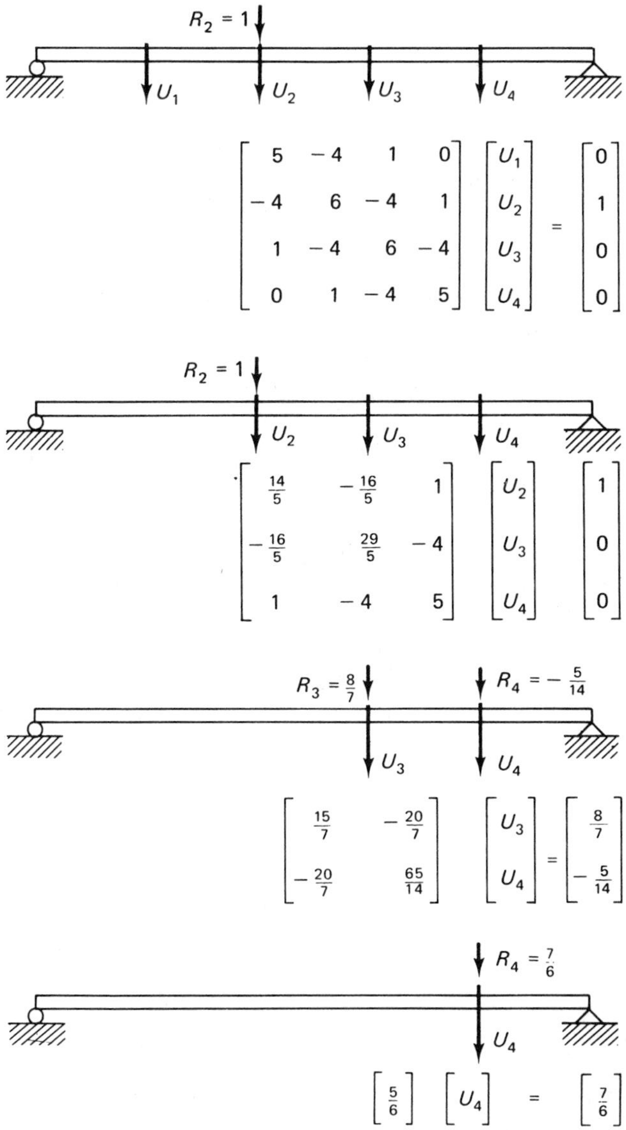

FIGURE 8.1 Physical systems considered in the Gauss elimination solution of the simply supported beam.

Step 3: Subtract $(-\frac{20}{15})$ times the third equation from the fourth equation in (8.4). This gives

$$\begin{bmatrix} 5 & -4 & 1 & 0 \\ 0 & \frac{14}{5} & -\frac{16}{5} & 1 \\ 0 & 0 & \frac{15}{7} & -\frac{20}{7} \\ 0 & 0 & 0 & \frac{5}{6} \end{bmatrix} \begin{bmatrix} U_1 \\ U_2 \\ U_3 \\ U_4 \end{bmatrix} = \begin{bmatrix} 0 \\ 1 \\ \frac{8}{7} \\ \frac{7}{6} \end{bmatrix} \tag{8.5}$$

Using (8.5) we can now simply solve for the unknowns U_4, U_3, U_2, and U_1:

$$
\begin{aligned}
U_4 &= \frac{\frac{7}{6}}{\frac{5}{6}} = \tfrac{7}{5}; \qquad U_3 = \frac{\frac{8}{7} - (-\frac{20}{7})U_4}{\frac{15}{7}} = \tfrac{12}{5} \\
U_2 &= \frac{1 - (-\frac{16}{5})U_3 - (1)U_4}{\frac{14}{5}} = \tfrac{13}{5} \\
U_1 &= \frac{0 - (-4)\frac{13}{5} - (1)\frac{12}{5} - (0)\frac{7}{5}}{5} = \tfrac{8}{5}
\end{aligned}
\tag{8.6}
$$

The procedure in the solution is therefore to subtract in step number i in succession multiples of equation i from equations $i + 1, i + 2, \ldots, n$, where $i = 1, 2, \ldots, n - 1$. In this way the coefficient matrix $\mathbf{K}$ of the equations is reduced to upper triangular form, i.e., a form in which all elements below the diagonal elements are zero. Starting with the last equation, it is then possible to solve for all unknowns in the order $U_n, U_{n-1}, \ldots, U_1$.

It is important to note that at the end of step i the lower right submatrix of order $n - i$ [indicated by dotted lines in (8.3), (8.4), and (8.5)] is symmetric. Therefore, the elements above and including the diagonal can give all elements of the coefficient matrix at all times of the solution. We will see in Section (8.2.3) that in the computer implementation we only work with the upper triangular part of the matrix.

Another important observation is that the above solution assumes in step i a nonzero ith diagonal element in the current coefficient matrix. Namely, it is the nonzero value of the ith diagonal element in the coefficient matrix that makes it possible to reduce the elements below it to zero. Also, in the back-substitution for the solution of the displacements we again divide by the diagonal elements of the coefficient matrix. Fortunately, in the analysis of displacement-based finite element systems, all diagonal elements of the coefficient matrix are positive at all times of the solution, which is another property that makes the application of the Gauss elimination procedure very effective. [This property is not necessarily preserved when the stiffness matrices are derived using hybrid or mixed formulations, or by finite differences (see Section 4.3).] We will prove in Section 8.2.5 that the diagonal elements must remain larger than zero, but this property is also observed by identifying the physical process that is carried out in the Gauss elimination.

In order to identify the physical process corresponding to the mathematical operations in Gauss elimination, we note first that the operations on the coefficient matrix $\mathbf{K}$ are independent of the elements in the load vector $\mathbf{R}$. Therefore, we consider only the operations on the coefficient matrix $\mathbf{K}$. For ease of explanation, let us again use the above example. In this case, $\mathbf{K}$ in (8.2) is the stiffness matrix of the beam element that corresponds to the four degrees of freedom in Fig. 8.1. Since no loads are applied, we now have

$$
\begin{bmatrix} 5 & -4 & 1 & 0 \\ -4 & 6 & -4 & 1 \\ 1 & -4 & 6 & -4 \\ 0 & 1 & -4 & 5 \end{bmatrix}
\begin{bmatrix} U_1 \\ U_2 \\ U_3 \\ U_4 \end{bmatrix}
=
\begin{bmatrix} 0 \\ 0 \\ 0 \\ 0 \end{bmatrix}
\tag{8.7}
$$

Using the condition given by the first equation, i.e.,

$$5U_1 - 4U_2 + U_3 = 0$$

we can write

$$U_1 = \tfrac{4}{5}U_2 - \tfrac{1}{5}U_3 \tag{8.8}$$

and eliminate U_1 from the three equations remaining in (8.7). We thus obtain

$$\begin{aligned} -4(\tfrac{4}{5}U_2 - \tfrac{1}{5}U_3) + 6U_2 - 4U_3 + U_4 &= 0 \\ (\tfrac{4}{5}U_2 - \tfrac{1}{5}U_3) - 4U_2 + 6U_3 - 4U_4 &= 0 \\ U_2 - 4U_3 + 5U_4 &= 0 \end{aligned}$$

or, in matrix form,

$$\begin{bmatrix} \frac{14}{5} & -\frac{16}{5} & 1 \\ -\frac{16}{5} & \frac{29}{5} & -4 \\ 1 & -4 & 5 \end{bmatrix} \begin{bmatrix} U_2 \\ U_3 \\ U_4 \end{bmatrix} = \begin{bmatrix} 0 \\ 0 \\ 0 \end{bmatrix} \tag{8.9}$$

Comparing (8.9) with (8.3), we observe that the coefficient matrix in (8.9) is actually the lower right 3×3 submatrix of the coefficient matrix in (8.3). However, we obtained the coefficient matrix of (8.9) by using (8.7) and the condition in (8.8), which expresses that no force is applied at the degree of freedom 1 of the beam. It follows that the coefficient matrix in (8.9) is, in fact, the stiffness matrix of the beam that corresponds to the degrees of freedom 2, 3, and 4 when no force is applied at the degree of freedom 1, i.e., when the degree of freedom 1 has been released. By the same reasoning, we have obtained in (8.4) the stiffness matrix of the beam when the first two degrees of freedom have been released; and in (8.5), the element (4, 4) of the coefficient matrix represents the stiffness matrix of the beam corresponding to degree of freedom 4 when the degrees of freedom 1, 2, and 3 have all been released.

So far we have assumed that no loads are applied to the structure, since the operations on the stiffness matrix are independent of the operations on the load vector. In case the load vector is not a null vector, the elimination of degrees of freedom proceeds as above, but in this case the equation used to eliminate a displacement variable from the remaining equations involves a load term. In the elimination the effect of this load is carried over to the other degrees of freedom. We may identify that the elimination including the load terms is precisely the Gauss elimination procedure. Therefore, in summary, the physical significance of the Gauss elimination procedure is that n stiffness matrices of order $n, n-1, \ldots, 1$, corresponding to the $(n-i)$ last degrees of freedom, $i = 0, 1, 2, \ldots, n-1$, respectively, are established. In addition, the appropriate load vectors corresponding to the n stiffness matrices are calculated. The unknown displacements are then obtained by considering in succession the systems with only one, two, . . . , degrees of freedom (these correspond to the last, two last, . . . , of the original number of degrees of freedom). Figure 8.1 shows the physical systems considered in the Gauss elimination solution of the simply supported beam.

We can now explain why, because of physical reasons, all diagonal elements in the Gauss elimination procedure must remain positive. This follows because the ith diagonal element is the stiffness at degree of freedom i when the first $(i-1)$ degrees of freedom of the structure have been released, and this stiffness should be positive. If a zero (or negative) diagonal element occurs in the Gauss elimination,

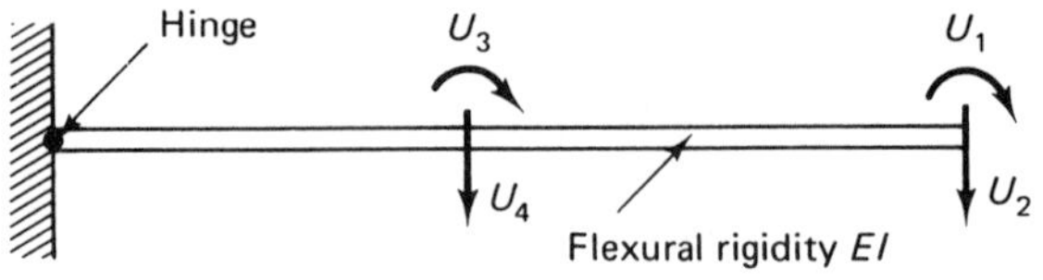

FIGURE 8.2 *Example of an unstable structure.*

the structure is not stable. An example of such a case is shown in Fig. 8.2, where after the release of degrees of freedom U_1, U_2, and U_3, the last diagonal element is zero.

So far we implied that the elimination proceeds in succession from the first to the $(n-1)$st degree of freedom. However, we may in the same way perform the elimination backward (i.e., from the last to the second degree of freedom) or we may choose any desirable order.

EXAMPLE 8.1: Obtain the solution to the equilibrium equations of the beam in Fig. 8.1 by eliminating the displacement variables in the order U_3, U_2, U_4.

We may either write down the individual equations during the elimination process, or directly perform Gauss elimination in the prescribed order. In the latter case we obtain, eliminating U_3,

$$\begin{bmatrix} \frac{29}{6} & -\frac{10}{3} & 0 & \frac{2}{3} \\ -\frac{10}{3} & \frac{10}{3} & 0 & -\frac{5}{3} \\ 1 & -4 & 6 & -4 \\ \frac{2}{3} & -\frac{5}{3} & 0 & \frac{7}{3} \end{bmatrix} \begin{bmatrix} U_1 \\ U_2 \\ U_3 \\ U_4 \end{bmatrix} = \begin{bmatrix} 0 \\ 1 \\ 0 \\ 0 \end{bmatrix}$$

Next, we eliminate U_2 and obtain

$$\begin{bmatrix} \frac{3}{2} & 0 & 0 & -1 \\ -\frac{10}{3} & \frac{10}{3} & 0 & -\frac{5}{3} \\ 1 & -4 & 6 & -4 \\ -1 & 0 & 0 & \frac{3}{2} \end{bmatrix} \begin{bmatrix} U_1 \\ U_2 \\ U_3 \\ U_4 \end{bmatrix} = \begin{bmatrix} 1 \\ 1 \\ 0 \\ \frac{1}{2} \end{bmatrix}$$

and finally we eliminate U_4,

$$\begin{bmatrix} \frac{5}{6} & 0 & 0 & 0 \\ -\frac{10}{3} & \frac{10}{3} & 0 & -\frac{5}{3} \\ 1 & -4 & 6 & -4 \\ -1 & 0 & 0 & \frac{3}{2} \end{bmatrix} \begin{bmatrix} U_1 \\ U_2 \\ U_3 \\ U_4 \end{bmatrix} = \begin{bmatrix} \frac{4}{3} \\ 1 \\ 0 \\ \frac{1}{2} \end{bmatrix}$$

The solution for the displacements is now obtained as follows:

$$U_1 = \tfrac{8}{5}$$

$$U_4 = \frac{\frac{1}{2} - (-1)\frac{8}{5}}{\frac{3}{2}} = \tfrac{7}{5}$$

$$U_2 = \frac{1 - (-\frac{10}{3})\frac{8}{5} - (-\frac{5}{3})\frac{7}{5}}{\frac{10}{3}} = \tfrac{13}{5}$$

$$U_3 = \frac{0 - (1)\frac{8}{5} - (-4)\frac{7}{5} - (-4)\frac{13}{5}}{6} = \tfrac{12}{5}$$

which is the solution already obtained earlier.

8.2.2 *The Gauss Elimination Solution*

We have seen in the preceding section that the basic procedure of the Gauss elimination solution is to reduce the coefficient matrix of the equations to an upper triangular matrix from which the unknown displacements **U** can be calculated by a back-substitution. We now want to formalize the solution procedure using appropriate matrix operations. An additional important purpose of the discussion is to introduce a notation, which can be used throughout the following presentations. The actual computer implementation is given in the next section.

Considering the operations performed in the Gauss elimination solution presented in the preceding section, the reduction of the stiffness matrix **K** to upper triangular form can be written

$$\mathbf{L}_{n-1}^{-1} \dots \mathbf{L}_2^{-1}\mathbf{L}_1^{-1}\mathbf{K} = \mathbf{S} \tag{8.10}$$

where **S** is the final upper triangular matrix, and

$$\mathbf{L}_i^{-1} = \begin{bmatrix} 1 & & & & & \text{elements} & & \\ & \cdot & & & & \text{not shown} & & \\ & & \cdot & & & \text{are zeros*} & & \\ & & & \cdot & & & & \\ & & & & 1 & & & \\ & & & & -l_{i+1,i} & \cdot & & \\ & & & & -l_{i+2,i} & & & \\ & & & & \cdot & & \cdot & \\ & & & & \cdot & & & \\ & & & & \cdot & & & \cdot \\ & & & & -l_{ni} & & & & 1 \end{bmatrix}; \qquad l_{i+j,i} = \frac{k_{i+j,i}^{(i)}}{k_{ii}^{(i)}} \tag{8.11}$$

The elements $l_{i+j,i}$ are the Gauss multiplying factors, and the right superscript (i) indicates that an element of the matrix $\mathbf{L}_{i-1}^{-1} \dots \mathbf{L}_2^{-1}\mathbf{L}_1^{-1}\mathbf{K}$ is used.

We now note that $\mathbf{L}_i$ is obtained by simply reversing the signs of the off-diagonal elements in $\mathbf{L}_i^{-1}$. Therefore, we obtain

$$\mathbf{K} = \mathbf{L}_1\mathbf{L}_2 \dots \mathbf{L}_{n-1}\mathbf{S} \tag{8.12}$$

where

$$\mathbf{L}_i = \begin{bmatrix} 1 & & & & & & & & \\ & \cdot & & & & & & & \\ & & \cdot & & & & & & \\ & & & \cdot & & & & & \\ & & & & 1 & & & & \\ & & & & l_{i+1,i} & \cdot & & & \\ & & & & l_{i+2,i} & & \cdot & & \\ & & & & \cdot & & & \cdot & \\ & & & & \cdot & & & & \cdot \\ & & & & \cdot & & & & & \cdot \\ & & & & l_{ni} & & & & & & 1 \end{bmatrix} \tag{8.13}$$

Hence we can write

$$\mathbf{K} = \mathbf{LS} \tag{8.14}$$

* Throughout this book, elements not shown in matrices are zero.

where $\mathbf{L} = \mathbf{L}_1\mathbf{L}_2 \ldots \mathbf{L}_{n-1}$, or

$$\mathbf{L} = \begin{bmatrix} 1 & & & & & & & \\ l_{21} & 1 & & & & & & \\ l_{31} & l_{32} & 1 & & & & & \\ l_{41} & l_{42} & & 1 & & & & \\ \cdot & \cdot & & & \cdot & & & \\ \cdot & \cdot & & & & \cdot & & \\ \cdot & \cdot & & & & & \cdot & \\ & & & & & & 1 & \\ l_{n1} & \cdot & & \cdot & & \cdot & l_{n,n-1} & 1 \end{bmatrix} \tag{8.15}$$

Since $\mathbf{S}$ is an upper triangular matrix and the diagonal elements are the pivots in the Gauss elimination, we can write $\mathbf{S} = \mathbf{D}\tilde{\mathbf{S}}$ where $\mathbf{D}$ is a diagonal matrix storing the diagonal elements of $\mathbf{S}$; i.e., $d_{ii} = s_{ii}$. Substituting for $\mathbf{S}$ into (8.14) and noting that $\mathbf{K}$ is symmetric and the decomposition is unique we obtain $\tilde{\mathbf{S}} = \mathbf{L}^T$, and hence

$$\mathbf{K} = \mathbf{L}\mathbf{D}\mathbf{L}^T \tag{8.16}$$

It is this $\mathbf{LDL}^T$ decomposition of $\mathbf{K}$ that can be used effectively to obtain the solution of the equations in (8.1) in the following two steps:

$$\mathbf{L}\mathbf{V} = \mathbf{R} \tag{8.17}$$

$$\mathbf{D}\mathbf{L}^T\mathbf{U} = \mathbf{V} \tag{8.18}$$

where in (8.17) the load vector $\mathbf{R}$ is reduced to obtain $\mathbf{V}$,

$$\mathbf{V} = \mathbf{L}_{n-1}^{-1} \ldots \mathbf{L}_2^{-1}\mathbf{L}_1^{-1}\mathbf{R} \tag{8.19}$$

and in (8.18) the solution $\mathbf{U}$ is obtained by a back-substitution,

$$\mathbf{L}^T\mathbf{U} = \mathbf{D}^{-1}\mathbf{V} \tag{8.20}$$

In the implementation the vector $\mathbf{V}$ is frequently calculated at the same time as the matrices $\mathbf{L}_i^{-1}$ are established. This was done in the example solution of the simply supported beam in Section 8.2.1.

It should be noted that in practice the matrix multiplications to obtain $\mathbf{L}$ in (8.15) and $\mathbf{V}$ in (8.19) are not formally carried out, but that $\mathbf{L}$ and $\mathbf{V}$ are established by directly modifying $\mathbf{K}$ and $\mathbf{R}$. This is discussed further in the next section, in which the computer implementation of the solution procedure is presented. However, before proceeding, consider the example of Section 8.2.1 for the derivation of the matrices defined above.

EXAMPLE 8.2: Establish the matrices $\mathbf{L}_i^{-1}$, $i = 1, 2, 3$, $\mathbf{S}$, $\mathbf{L}$, and $\mathbf{D}$ and the vector $\mathbf{V}$ corresponding to the stiffness matrix and the load vector of the simply supported beam treated in Section 8.2.1.

Using the information given in Section 8.2.1, we can directly write down the required matrices:

$$\mathbf{L}_1^{-1} = \begin{bmatrix} 1 & & & \\ \frac{4}{5} & 1 & & \\ -\frac{1}{5} & 0 & 1 & \\ 0 & 0 & 0 & 1 \end{bmatrix}; \qquad \mathbf{L}_2^{-1} = \begin{bmatrix} 1 & & & \\ 0 & 1 & & \\ 0 & \frac{8}{7} & 1 & \\ 0 & -\frac{5}{14} & 0 & 1 \end{bmatrix}$$

$$\mathbf{L}_3^{-1} = \begin{bmatrix} 1 & & & \\ 0 & 1 & & \\ 0 & 0 & 1 & \\ 0 & 0 & \frac{4}{3} & 1 \end{bmatrix}; \qquad \mathbf{S} = \begin{bmatrix} 5 & -4 & 1 & 0 \\ & \frac{14}{5} & -\frac{16}{5} & 1 \\ & & \frac{15}{7} & -\frac{20}{7} \\ & & & \frac{5}{6} \end{bmatrix}$$

where the matrix $\mathbf{L}_i^{-1}$ stores in the ith column the multipliers that were used in the elimination of the ith equation, and the matrix $\mathbf{S}$ is the upper triangular matrix obtained in (8.5). The matrix $\mathbf{D}$ is a diagonal matrix with the pivot elements on its diagonal. In this case

$$\mathbf{D} = \begin{bmatrix} 5 & & & \\ & \frac{14}{5} & & \\ & & \frac{15}{7} & \\ & & & \frac{5}{6} \end{bmatrix}$$

To obtain $\mathbf{L}$ we use (8.15); hence

$$\mathbf{L} = \begin{bmatrix} 1 & & & \\ -\frac{4}{5} & 1 & & \\ \frac{1}{5} & -\frac{8}{7} & 1 & \\ 0 & \frac{5}{14} & -\frac{4}{3} & 1 \end{bmatrix}$$

and we can check that $\mathbf{S} = \mathbf{D}\mathbf{L}^T$.

The vector $\mathbf{V}$ was obtained in (8.5):

$$\mathbf{V} = \begin{bmatrix} 0 \\ 1 \\ \frac{8}{7} \\ \frac{7}{6} \end{bmatrix}$$

8.2.3 Computer Implementation of Gauss Elimination —The Active Column Solution

A very important aspect in the computer implementation of the Gauss solution procedure is that a minimum solution time should be used. In addition, the high-speed storage requirements should be as small as possible to avoid the use of back-up storage. However, for large systems it will nevertheless be necessary to use back-up storage,[12] and for this reason it should also be possible to modify the solution algorithm for effective out-of-core solution.

As briefly pointed out in Section 1.3 and further discussed in the Appendix, an advantage of finite element analysis is that the stiffness matrix of the element assemblage is not only symmetric and positive definite but is also banded; i.e., $k_{ij} = 0$ for $j > i + m_K$, where m_K is the half-bandwidth of the system. The fact that in finite element analysis all nonzero elements are clustered around the diagonal of the system matrices greatly reduces the total number of operations and the high-speed storage required in the equation solution. However, this property depends on the nodal point numbering of the finite element mesh, and the analyst must take care to obtain an effective nodal point numbering (see the Appendix).

Assume that for a given finite element assemblage a specific nodal point numbering has been determined and the corresponding column heights and the stiffness matrix **K** have been calculated (as discussed in Section A.2.3). The $\mathbf{LDL}^T$ decomposition of **K** can be obtained effectively by considering each column in turn; i.e., although the Gauss elimination is carried out by rows, the final elements of **D** and **L** are calculated by columns. Using $d_{11} = k_{11}$, the algorithm for the calculation of the elements l_{ij} and d_{jj} in the jth column is, for $j = 2, \ldots, n$,

$$\left.\begin{aligned} g_{m_j,j} &= k_{m_j,j} \\ g_{ij} &= k_{ij} - \sum_{r=m_m}^{i-1} l_{ri} g_{rj} \qquad i = m_j + 1, \ldots, j-1 \end{aligned}\right\} \tag{8.21}$$

where m_j is the row number of the first nonzero element in column j and $m_m = \max\{m_i, m_j\}$ (see Fig. A.3). The variables m_i, $i = 1, \ldots, n$ define the skyline of the matrix; also, the values $(i - m_i)$ are the column heights and the maximum column height is the half-bandwidth $m_{\mathbf{K}}$. The elements g_{ij} in (8.21) are only defined as intermediate quantities, and the calculation is completed using

$$l_{ij} = \frac{g_{ij}}{d_{ii}} \qquad i = m_j, \ldots, j-1 \tag{8.22}$$

$$d_{jj} = k_{jj} - \sum_{r=m_j}^{j-1} l_{rj} g_{rj} \tag{8.23}$$

It should be noted that the summations in (8.21) and (8.23) do not involve multiplications with zero elements outside the skyline of the matrix, and that the l_{ij} are elements of the matrix $\mathbf{L}^T$ rather than of **L**. We refer to the solution algorithm given in (8.21) to (8.23) (actually with (8.24) and (8.25)) as the *active column solution or the skyline* (*or column*) *reduction method.*

Considering the storage arrangements in the reduction, the element l_{ij} when calculated for use in (8.23) immediately replaces g_{ij} and d_{jj} replaces k_{jj}. Therefore, at the end of the reduction we have elements d_{jj} in the storage locations previously used by k_{jj}, and l_{rj} is stored in the locations of $k_{rj}, j > r$.

In order to get familiar with the solution algorithm, we consider the following examples.

EXAMPLE 8.3: Use the solution algorithm given in (8.21) to (8.23) to calculate the triangular factors **D** and $\mathbf{L}^T$ of the stiffness matrix of the beam considered in Example 8.2.

The initial elements considered are, when written in their respective matrix locations,

$$\begin{bmatrix} 5 & -4 & 1 & \\ & 6 & -4 & 1 \\ & & 6 & -4 \\ & & & 5 \end{bmatrix}$$

with $m_1 = 1$, $m_2 = 1$, $m_3 = 1$, and $m_4 = 2$. Using (8.21) to (8.23) we obtain, for $j = 2$,

$$d_{11} = k_{11} = 5$$
$$g_{12} = k_{12} = -4$$

$$l_{12} = \frac{g_{12}}{d_{11}} = -\tfrac{4}{5}$$

$$d_{22} = k_{22} - l_{12}g_{12} = 6 - (-4)(-\tfrac{4}{5}) = \tfrac{14}{5}$$

and thus the resulting matrix elements are now, using a dotted line to separate the reduced from the unreduced columns,

$$\begin{bmatrix} 5 & -\frac{4}{5} & \vdots & 1 & \\ & \frac{14}{5} & \vdots & -4 & 1 \\ & & \vdots & 6 & -4 \\ & & \vdots & & 5 \end{bmatrix}$$

Next, we obtain, for $j = 3$,

$$g_{13} = k_{13} = 1$$

$$g_{23} = k_{23} - l_{12}g_{13} = -4 - (-\tfrac{4}{5})(1) = -\tfrac{16}{5}$$

$$l_{13} = \frac{g_{13}}{d_{11}} = \tfrac{1}{5}$$

$$l_{23} = \frac{g_{23}}{d_{22}} = \frac{-\frac{16}{5}}{\frac{14}{5}} = -\tfrac{8}{7}$$

$$d_{33} = k_{33} - l_{13}g_{13} - l_{23}g_{23} = 6 - (\tfrac{1}{5})(1) - (-\tfrac{8}{7})(-\tfrac{16}{5}) = \tfrac{15}{7}$$

and the resulting matrix elements are

$$\begin{bmatrix} 5 & -\frac{4}{5} & \frac{1}{5} & \vdots & \\ & \frac{14}{5} & -\frac{8}{7} & \vdots & 1 \\ & & \frac{15}{7} & \vdots & -4 \\ & & & \vdots & 5 \end{bmatrix}$$

Finally, we have, for $j = 4$,

$$g_{24} = k_{24} = 1$$

$$g_{34} = k_{34} - l_{23}g_{24} = -4 - (-\tfrac{8}{7})(1) = -\tfrac{20}{7}$$

$$l_{24} = \frac{g_{24}}{d_{22}} = \frac{1}{\frac{14}{5}} = \frac{5}{14}$$

$$l_{34} = \frac{g_{34}}{d_{33}} = \frac{-\frac{20}{7}}{\frac{15}{7}} = -\tfrac{4}{3}$$

$$d_{44} = k_{44} - l_{24}g_{24} - l_{34}g_{34} = 5 - (\tfrac{5}{14})(1) - (-\tfrac{4}{3})(-\tfrac{20}{7}) = \tfrac{5}{6}$$

and the final elements stored are

$$\begin{bmatrix} 5 & -\frac{4}{5} & \frac{1}{5} & \\ & \frac{14}{5} & -\frac{8}{7} & \frac{5}{14} \\ & & \frac{15}{7} & -\frac{4}{3} \\ & & & \frac{5}{6} \end{bmatrix}$$

We should note that the elements of **D** are stored on the diagonal and the elements l_{ij} have replaced the elements $k_{ij}, j > i$.

Although the solution procedure could already be well demonstrated in Example 8.3, the importance of using the column reduction method could not be shown, since the skyline coincides with the band. The effectiveness of the skyline reduction scheme is more apparent in the factorization of the following matrix.

EXAMPLE 8.4: Use the solution algorithm given in (8.21) to (8.23) to evaluate the triangular factors **D** and $\mathbf{L}^T$ of the stiffness matrix **K**, where

$$\mathbf{K} = \begin{bmatrix} 2 & -2 & & & -1 \\ & 3 & -2 & & 0 \\ & & 5 & -3 & 0 \\ & \text{symmetric} & & 10 & 4 \\ & & & & 10 \end{bmatrix}$$

For this matrix we have $m_1 = 1$, $m_2 = 1$, $m_3 = 2$, $m_4 = 3$, and $m_5 = 1$.

The algorithm gives, in this case with $d_{11} = 2$, for $j = 2$,

$$g_{12} = k_{12} = -2$$
$$l_{12} = \frac{g_{12}}{d_{11}} = \frac{-2}{2} = -1$$
$$d_{22} = k_{22} - l_{12}g_{12} = 3 - (-1)(-2) = 1$$

and thus the resulting matrix elements are

$$\left[\begin{array}{cc:ccc} 2 & -1 & & & -1 \\ & 1 & -2 & & 0 \\ & & 5 & -3 & 0 \\ & & & 10 & 4 \\ & & & & 10 \end{array}\right]$$

For $j = 3$,

$$g_{23} = k_{23} = -2$$
$$l_{23} = \frac{g_{23}}{d_{22}} = \frac{-2}{1} = -2$$
$$d_{33} = k_{33} - l_{23}g_{23} = 5 - (-2)(-2) = 1$$

and the coefficient array is now

$$\left[\begin{array}{ccc:cc} 2 & -1 & & & -1 \\ & 1 & -2 & & 0 \\ & & 1 & -3 & 0 \\ & & & 10 & 4 \\ & & & & 10 \end{array}\right]$$

For $j = 4$,

$$g_{34} = k_{34} = -3$$
$$l_{34} = \frac{g_{34}}{d_{33}} = \frac{-3}{1} = -3$$
$$d_{44} = k_{44} - l_{34}g_{34} = 10 - (-3)(-3) = 1$$

and the resulting matrix elements are

$$\left[\begin{array}{cccc:c} 2 & -1 & & & -1 \\ & 1 & -2 & & 0 \\ & & 1 & -3 & 0 \\ & & & 1 & 4 \\ & & & & 10 \end{array}\right]$$

Finally we have, for $j = 5$,

$$g_{15} = k_{15} = -1$$
$$g_{25} = k_{25} - l_{12}g_{15} = 0 - (-1)(-1) = -1$$

$$
\begin{aligned}
g_{35} &= k_{35} - l_{23}g_{25} = 0 - (-2)(-1) = -2 \\
g_{45} &= k_{45} - l_{34}g_{35} = +4 - (-3)(-2) = -2 \\
l_{15} &= \frac{g_{15}}{d_{11}} = \frac{-1}{2} = -\frac{1}{2} \\
l_{25} &= \frac{g_{25}}{d_{22}} = \frac{-1}{1} = -1 \\
l_{35} &= \frac{g_{35}}{d_{33}} = \frac{-2}{1} = -2 \\
l_{45} &= \frac{g_{45}}{d_{44}} = \frac{-2}{1} = -2 \\
d_{55} &= k_{55} - l_{15}g_{15} - l_{25}g_{25} - l_{35}g_{35} - l_{45}g_{45} \\
&= 10 - (-\tfrac{1}{2})(-1) - (-1)(-1) - (-2)(-2) - (-2)(-2) = \tfrac{1}{2}
\end{aligned}
$$

and the final matrix elements are

$$
\begin{bmatrix}
2 & -1 & & & -\frac{1}{2} \\
 & 1 & -2 & & -1 \\
 & & 1 & -3 & -2 \\
 & & & 1 & -2 \\
 & & & & \frac{1}{2}
\end{bmatrix}
$$

As in Example 8.3, we have the elements of **D** and $\mathbf{L}^T$ replacing the elements k_{ii} and $k_{ij}, j > i$, of the original matrix **K**, respectively.

In the preceding discussion we considered only the decomposition of the stiffness matrix **K**, which constitutes the main part of the equation solution. Once the **L**, **D** factors of **K** have been obtained, the solution for **U** is calculated using (8.19) and (8.20), where it may be noted that the reduction of **R** in (8.19) can be performed at the same time as the stiffness matrix **K** is decomposed or may be carried out separately afterwards. The equation to be used is similar to (8.23); i.e., we have $V_1 = R_1$ and calculate for $i = 2, \ldots, n$,

$$
V_i = R_i - \sum_{r=m_i}^{i-1} l_{ri} V_r \tag{8.24}
$$

where R_i and V_i are the ith elements of **R** and **V**. Considering the storage arrangements, the element V_i replaces R_i.

The back-substitution in (8.20) is performed by evaluating successively U_n, $U_{n-1}, \ldots, U_1$. This is achieved by first calculating $\bar{\mathbf{V}}$, where $\bar{\mathbf{V}} = \mathbf{D}^{-1}\mathbf{V}$. Then using that $\bar{\mathbf{V}}^{(n)} = \bar{\mathbf{V}}$, we have $U_n = \bar{V}_n^{(n)}$, and we calculate for $i = n, \ldots, 2$,

$$
\left.
\begin{aligned}
\bar{V}_r^{(i-1)} &= \bar{V}_r^{(i)} - l_{ri}U_i; \qquad r = m_i, \ldots, i-1 \\
U_{i-1} &= \bar{V}_{i-1}^{(i-1)}
\end{aligned}
\right\} \tag{8.25}
$$

where the superscript $(i - 1)$ indicates that the element is calculated in the evaluation of U_{i-1}. It should be noted that $\bar{V}_k^{(j)}$ for all j is stored in the storage location of V_k, i.e., the original storage location of R_k.

EXAMPLE 8.5: Use the algorithm given in (8.24) and (8.25) to calculate the solution to the problem **KU** = **R**, when **K** is the stiffness matrix considered in

Example 8.4 and

$$\mathbf{R} = \begin{bmatrix} 0 \\ 1 \\ 0 \\ 0 \\ 0 \end{bmatrix}$$

In the solution we employ the **D**, $\mathbf{L}^T$ factors of **K** calculated in Example 8.4. Using (8.24) for the forward reduction we obtain

$$\begin{aligned} V_1 &= R_1 = 0 \\ V_2 &= R_2 - l_{12}V_1 = 1 - 0 = 1 \\ V_3 &= R_3 - l_{23}V_2 = 0 - (-2)(1) = 2 \\ V_4 &= R_4 - l_{34}V_3 = 0 - (-3)(2) = 6 \\ V_5 &= R_5 - l_{15}V_1 - l_{25}V_2 - l_{35}V_3 - l_{45}V_4 \\ &= 0 - 0 - (-1)(1) - (-2)(2) - (-2)(6) = 17 \end{aligned}$$

Immediately after calculation of V_i, the element replaces R_i. Thus we now have in the vector that initially stored the loads,

$$\mathbf{V} = \begin{bmatrix} 0 \\ 1 \\ 2 \\ 6 \\ 17 \end{bmatrix}$$

The first step in the back-substitution is to evaluate $\bar{\mathbf{V}}$, where $\bar{\mathbf{V}} = \mathbf{D}^{-1}\mathbf{V}$. Here we obtain

$$\bar{\mathbf{V}} = \begin{bmatrix} 0 \\ 1 \\ 2 \\ 6 \\ 34 \end{bmatrix} \tag{a}$$

and thus $\qquad U_5 = \bar{V}_5 = 34$

Now we use (8.25) with $\bar{\mathbf{V}}^{(5)} = \bar{\mathbf{V}}$ of (a). Hence we obtain, for $i = 5$,

$$\begin{aligned} \bar{V}_1^{(4)} &= \bar{V}_1^{(5)} - l_{15}U_5 = 0 - (-\tfrac{1}{2})(34) = 17 \\ \bar{V}_2^{(4)} &= \bar{V}_2^{(5)} - l_{25}U_5 = 1 - (-1)(34) = 35 \\ \bar{V}_3^{(4)} &= \bar{V}_3^{(5)} - l_{35}U_5 = 2 - (-2)(34) = 70 \\ \bar{V}_4^{(4)} &= \bar{V}_4^{(5)} - l_{45}U_5 = 6 - (-2)(34) = 74 \end{aligned}$$

and $\qquad U_4 = \bar{V}_4^{(4)} = 74$

For $i = 4$, $\qquad \bar{V}_3^{(3)} = \bar{V}_3^{(4)} - l_{34}U_4 = 70 - (-3)(74) = 292$

and $\qquad U_3 = \bar{V}_3^{(3)} = 292$

For $i = 3$, $\qquad \bar{V}_2^{(2)} = \bar{V}_2^{(3)} - l_{23}U_3 = 35 - (-2)(292) = 619$

and $\qquad U_2 = \bar{V}_2^{(2)} = 619$

For $i = 2$, $\qquad \bar{V}_1^{(1)} = \bar{V}_1^{(2)} - l_{12}U_2 = 17 - (-1)(619) = 636$

and $\qquad U_1 = \bar{V}_1^{(1)} = 636$

The elements stored in the vector that initially stored the loads are after step $i = 5, 4, 3, 2$, respectively:

$$\begin{bmatrix} 17 \\ 35 \\ 70 \\ 74 \\ 34 \end{bmatrix}; \quad \begin{bmatrix} 17 \\ 35 \\ 292 \\ 74 \\ 34 \end{bmatrix}; \quad \begin{bmatrix} 17 \\ 619 \\ 292 \\ 74 \\ 34 \end{bmatrix}; \quad \begin{bmatrix} 636 \\ 619 \\ 292 \\ 74 \\ 34 \end{bmatrix}$$

where the last vector gives the solution **U**.

Considering the effectiveness of the active column solution algorithm, it should be noted that for a specific matrix **K** the algorithm gives an efficient solution, because no operations are performed on zero elements outside the skyline, which also implies that only the elements below the skyline need be stored. However, the total number of operations performed is not an absolute minimum, because, in addition all those multiplications could be skipped in (8.21) to (8.25) for which l_{ri} or g_{rj} is zero. This skipping, though, would require time-consuming, additional testing and logic in the equation solution. In this context, it should also be realized that we consider a specific matrix **K** and that by a different nodal point numbering a different stiffness matrix is obtained, which might yield a more effective solution.

In practice, it is important to be able to estimate the number of operations that are carried out in the solution, because this enables the analyst to estimate the computational cost for a specific analysis. Define one operation to consist of one multiplication (or division), which is nearly always followed by an addition. For an operation count estimate we consider a system with constant column heights; i.e., a half-bandwidth m_K such that $m_K = i - m_i$ for all i, $i > m_K$. In this case the number of operations required for the $\mathbf{LDL}^T$ decomposition of **K** are approximately $n\{m_K + (m_K - 1) + \ldots + 1\} \doteq \frac{1}{2} n m_K^2$, and for the reduction and back-substitution of a load vector, an additional number of approximately $2nm_K$ operations are needed. In practice, systems with constant column heights are encountered rather seldom. However, the above values can still be employed to estimate the approximate number of operations required, if a "mean" column height is used. Also, an additional purpose of performing operation counts is to compare the effectiveness of different solution schemes, and here considering systems with constant column heights does yield some insight.

The solution algorithm in (8.21) to (8.25) has been presented in two-dimensional matrix notation; e.g., element (r, j) of **K** has been denoted by k_{rj}. Also, to demonstrate the working of the algorithm, in Examples 8.3 and 8.4 the elements considered in the reduction have been displayed in their corresponding matrix locations. However, in actual computer solution, the active columns of the matrix **K** are stored effectively in a one-dimensional array. Assume that the storage scheme discussed in the Appendix is used; i.e., the pertinent elements of **K** are stored in the one-dimensional array A of length NWK and the addresses of the diagonal elements of **K** are stored in MAXA. An effective subroutine that uses the algorithm presented above (i.e., the relations in (8.21) to (8.25)) but operates on the stiffness matrix using this storage scheme is given next.

Subroutine COLSOL: Program COLSOL is an active column solver to obtain the **LDL**^{T} factorization of a stiffness matrix or reduce and back-substitute the load vector. The complete process gives the solution of the finite element equilibrium equations. The argument variables and use of the subroutine are defined by means of comment cards in the program.

```
      SUBROUTINE COLSOL(A,V,MAXA,NN,NWK,NNM,KKK,IOUT)               CLSL0001
C . . . . . . . . . . . . . . . . . . . . . . . . . . . . . . . .  CLSL0002
C .                                                              .  CLSL0003
C .                                                              .  CLSL0004
C .    P R O G R A M                                             .  CLSL0005
C .       TO SOLVE FINITE ELEMENT STATIC EQUILIBRIUM EQUATIONS IN.  CLSL0005
C .       CORE, USING COMPACTED STORAGE AND COLUMN REDUCTION SCHEME. CLSL0006
C .                                                              .  CLSL0007
C .    - - INPUT VARIABLES - -                                   .  CLSL0008
C .       A(NWK)    = STIFFNESS MATRIX STORED IN COMPACTED FORM  .  CLSL0009
C .       V(NN)     = RIGHT-HAND-SIDE LOAD VECTOR                .  CLSL0010
C .       MAXA(NNM) = VECTOR CONTAINING ADDRESSES OF DIAGONAL    .  CLSL0011
C .                   ELEMENTS OF STIFFNESS MATRIX IN A          .  CLSL0012
C .       NN        = NUMBER OF EQUATIONS                        .  CLSL0013
C .       NWK       = NUMBER OF ELEMENTS BELOW SKYLINE OF MATRIX .  CLSL0014
C .       NNM       = NN + 1                                     .  CLSL0015
C .       KKK       = INPUT FLAG                                 .  CLSL0016
C .           EQ. 1 TRIANGULARIZATION OF STIFFNESS MATRIX        .  CLSL0017
C .           EQ. 2 REDUCTION AND BACK-SUBSTITUTION OF LOAD VECTOR. CLSL0018
C .       IOUT      = NUMBER OF OUTPUT DEVICE                    .  CLSL0019
C .                                                              .  CLSL0020
C .    - - OUTPUT - -                                            .  CLSL0021
C .       A(NWK)    = D AND L - FACTORS OF STIFFNESS MATRIX      .  CLSL0022
C .       V(NN)     = DISPLACEMENT VECTOR                        .  CLSL0023
C .                                                              .  CLSL0024
C . . . . . . . . . . . . . . . . . . . . . . . . . . . . . . . .  CLSL0025
      IMPLICIT REAL*8 (A-H,O-Z)                                     CLSL0026
C                                                                   CLSL0027
C .       THIS PROGRAM IS USED IN SINGLE PRECISION ARITHMETIC ON .  CLSL0028
C .       CDC EQUIPMENT AND DOUBLE PRECISION ARITHMETIC ON IBM  .  CLSL0029
C .       OR UNIVAC MACHINES .ACTIVATE,DEACTIVATE OR ADJUST ABOVE.  CLSL0030
C .       CARD FOR SINGLE OR DOUBLE PRECISION ARITHMETIC        .  CLSL0031
C . . . . . . . . . . . . . . . . . . . . . . . . . . . . . . . .  CLSL0032
      DIMENSION A(NWK),V(1),MAXA(1)                                 CLSL0033
C                                                                   CLSL0034
C     PERFORM L*D*L(T) FACTORIZATION OF STIFFNESS MATRIX            CLSL0035
C                                                                   CLSL0036
      IF (KKK-2) 40,150,150                                         CLSL0037
   40 DO 140 N=1,NN                                                 CLSL0038
      KN=MAXA(N)                                                    CLSL0039
      KL=KN+1                                                       CLSL0040
      KU=MAXA(N+1) - 1                                              CLSL0041
      KH=KU - KL                                                    CLSL0042
      IF (KH) 110,90,50                                             CLSL0043
   50 K=N-KH                                                        CLSL0044
      IC=0                                                          CLSL0045
      KLT=KU                                                        CLSL0046
      DO 80 J=1,KH                                                  CLSL0047
      IC=IC + 1                                                     CLSL0048
      KLT=KLT - 1                                                   CLSL0049
      KI=MAXA(K)                                                    CLSL0050
      ND=MAXA(K+1) - KI - 1                                         CLSL0051
      IF (ND) 80,80,60                                              CLSL0052
   60 KK=MIN0(IC,ND)                                                CLSL0053
      C=0.                                                          CLSL0054
      DO 70 L=1,KK                                                  CLSL0055
   70 C=C+A(KI+L)*A(KLT+L)                                          CLSL0056
      A(KLT)=A(KLT) - C                                             CLSL0057
   80 K=K+1                                                         CLSL0058
   90 K=N                                                           CLSL0059
      B=0.                                                          CLSL0060
      DO 100 KK=KL,KU                                               CLSL0061
      K=K - 1                                                       CLSL0062
      KI=MAXA(K)                                                    CLSL0063
      C=A(KK)/A(KI)                                                 CLSL0064
      B=B + C*A(KK)                                                 CLSL0065
  100 A(KK)=C                                                       CLSL0066
      A(KN)=A(KN) - B                                               CLSL0067
  110 IF (A(KN)) 120,120,140                                        CLSL0068
  120 WRITE(IOUT,2000) N,A(KN)                                      CLSL0069
      STOP                                                          CLSL0070
  140 CONTINUE                                                      CLSL0071
      RETURN                                                        CLSL0072
C                                                                   CLSL0073
C     REDUCE RIGHT-HAND-SIDE LOAD VECTOR                            CLSL0074
C                                                                   CLSL0075
  150 DO 180 N=1,NN                                                 CLSL0076
      KL=MAXA(N) + 1                                                CLSL0077
      KU=MAXA(N+1) - 1                                              CLSL0078
      IF (KU-KL) 180,160,160                                        CLSL0079
  160 K=N                                                           CLSL0080
      C=0.                                                          CLSL0081
      DO 170 KK=KL,KU                                               CLSL0082
      K=K - 1                                                       CLSL0083
```

```
  170 C=C+A(KK)*V(K)
      V(N)=V(N) - C
  180 CONTINUE
C
C     BACK-SUBSTITUTE
C
      DO 200 N=1,NN
      K=MAXA(N)
  200 V(N)=V(N)/A(K)


      IF (NN.EQ.1) RETURN
      N=NN
      DO 230 L=2,NN
      KL=MAXA(N) + 1
      KU=MAXA(N+1) - 1
      IF (KU-KL) 230,210,210
  210 K=N
      DO 220 KK=KL,KU
      K=K - 1
  220 V(K)=V(K)-A(KK)*V(N)
  230 N=N-1
      RETURN
 2000 FORMAT(//48H STOP - STIFFNESS MATRIX NOT POSITIVE DEFINITE    ,//
     1            32H NONPOSITIVE PIVOT FOR EQUATION  ,I4,//
     2            10H PIVOT =   ,E20.12 )
      END
```

8.2.4 *Cholesky Factorization, Static Condensation, Substructures, and Frontal Solution*

In addition to the $\mathbf{LDL}^T$ decomposition described in the preceding sections, various other schemes are used that are closely related. All methods are applications of the basic Gauss elimination procedure.

In the *Cholesky factorization* the stiffness matrix is decomposed as follows:[7,8]

$$\mathbf{K} = \tilde{\mathbf{L}}\tilde{\mathbf{L}}^T \tag{8.26}$$

where

$$\tilde{\mathbf{L}} = \mathbf{L}\mathbf{D}^{1/2} \tag{8.27}$$

Therefore, the Cholesky factors could be calculated from the $\mathbf{D}$ and $\mathbf{L}$ factors, but, more generally, the elements of $\tilde{\mathbf{L}}$ are calculated directly. An operation count shows that slightly more operations are required in the equation solution if the Cholesky factorization is used rather than the $\mathbf{LDL}^T$ decomposition. In addition, the Cholesky factorization is only suitable for the solution of positive definite systems, for which all diagonal elements d_{ii} are positive, because otherwise complex arithmetic would be required. On the other hand, the $\mathbf{LDL}^T$ decomposition can also be used effectively on indefinite systems (see Section 8.2.5).

Considering a main use of the Cholesky factorization, it should be noted that the decompostition is used effectively in the transformation of a generalized eigenproblem to the standard form (see Section 10.2.5).

EXAMPLE 8.6: Calculate the Cholesky factor $\tilde{\mathbf{L}}$ of the stiffness matrix $\mathbf{K}$ of the simply supported beam treated in Section 8.2.1 and in Examples 8.1, 8.2, and 8.3.

The $\mathbf{L}$ and $\mathbf{D}$ factors of the beam stiffness matrix have been given in Example 8.2. Rounding to three significant decimals, we have

$$\mathbf{L} = \begin{bmatrix} 1.000 & & & 0 \\ -0.800 & 1.000 & & \\ 0.200 & -1.143 & 1.000 & \\ 0.000 & 0.357 & -1.333 & 1.000 \end{bmatrix}; \qquad \mathbf{D} = \begin{bmatrix} 5.000 & & & \\ & 2.800 & & \\ & & 2.143 & \\ & & & 0.833 \end{bmatrix}$$

Hence

$$\tilde{\mathbf{L}} = \begin{bmatrix} 1.000 & & & \\ -0.800 & 1.000 & & \\ 0.200 & -1.143 & 1.000 & \\ 0.000 & 0.357 & -1.333 & 1.000 \end{bmatrix} \begin{bmatrix} 2.236 & & & \\ & 1.673 & & \\ & & 1.464 & \\ & & & 0.913 \end{bmatrix}$$

$$\text{or} \quad \tilde{\mathbf{L}} = \begin{bmatrix} 2.236 & & & \\ -1.789 & 1.673 & & \\ 0.447 & -1.912 & 1.464 & \\ 0 & 0.597 & -1.952 & 0.913 \end{bmatrix}$$

An algorithm that in some cases can effectively be used in the solution of the equilibrium equations is *static condensation.*[13-15] The name "static condensation" was coined in dynamic analysis, for which the solution technique is demonstrated in Section 10.3.1. *Static condensation is employed to reduce the number of element degrees of freedom and thus, in effect, to perform part of the solution of the total finite element system equilibrium equations prior to assembling the structure matrices* **K** and **R**. Consider the three-node truss element in Fig. 8.3. Since the degree of freedom at the midnode does not correspond to a degree of freedom of any other element, we can eliminate it to obtain the element stiffness matrix that corresponds to the degrees of freedom 1 and 3 only. The elimination of the degree of freedom 2 is carried out using, in essence, Gauss elimination, as presented in Section 8.2.1 (see Example 8.1).

In order to establish the equations used in static condensation, we assume that the stiffness matrix and corresponding displacement and force vectors of the element under consideration are partitioned into the form

$$\begin{bmatrix} \mathbf{K}_{aa} & \mathbf{K}_{ac} \\ \mathbf{K}_{ca} & \mathbf{K}_{cc} \end{bmatrix} \begin{bmatrix} \mathbf{U}_a \\ \mathbf{U}_c \end{bmatrix} = \begin{bmatrix} \mathbf{R}_a \\ \mathbf{R}_c \end{bmatrix} \tag{8.28}$$

where $\mathbf{U}_a$ and $\mathbf{U}_c$ are the vectors of displacements to be retained and condensed out, respectively. The matrices $\mathbf{K}_{aa}$, $\mathbf{K}_{ac}$, and $\mathbf{K}_{cc}$, and vectors $\mathbf{R}_a$ and $\mathbf{R}_c$ correspond to the displacement vectors $\mathbf{U}_a$ and $\mathbf{U}_c$.

Using the second matrix equation in (8.28), we obtain

$$\mathbf{U}_c = \mathbf{K}_{cc}^{-1}(\mathbf{R}_c - \mathbf{K}_{ca}\mathbf{U}_a) \tag{8.29}$$

The relation in (8.29) is used to substitute for $\mathbf{U}_c$ into the first matrix equation in (8.28) to obtain the condensed equations

$$(\mathbf{K}_{aa} - \mathbf{K}_{ac}\mathbf{K}_{cc}^{-1}\mathbf{K}_{ca})\mathbf{U}_a = \mathbf{R}_a - \mathbf{K}_{ac}\mathbf{K}_{cc}^{-1}\mathbf{R}_c \tag{8.30}$$

Comparing (8.30) with the Gauss solution scheme introduced in Section 8.2.1, it is seen that static condensation is, in fact, Gauss elimination on the degrees of freedom $\mathbf{U}_c$ (see Example 8.7). In practice, therefore, static condensation is carried out effectively by using Gauss elimination sequentially on each degree of freedom to be condensed out, instead of following through the formal matrix procedure given in (8.28) to (8.30), where it is valuable to keep the physical meaning of Gauss elimination in mind (see Section 8.2.1). Since the system stiffness matrix is obtained by direct addition of the element stiffness matrices, we realize that when condensing out internal element degrees of freedom, in fact, part of the total Gauss solution is already carried out on the element level.

The advantage of using static condensation on the element level is that the order of the system matrices is reduced, which may mean that use of back-up storage is prevented. In addition, if subsequent elements are identical, the stiffness matrix of only the first element need be derived, and performing static condensation on the element internal degrees of freedom also reduces the computer effort required. It should be noted, though, that if static condensation is actually carried out for each element (and no advantage is taken of possible identical finite elements), the total effort involved in the static condensation on all element stiffness matrices plus the Gauss elimination solution of the resulting assembled equilibrium equations is, in fact, the same as using Gauss elimination on the system equations established from the uncondensed element stiffness matrices.

EXAMPLE 8.7: The stiffness matrix of the truss element in Fig. 8.3 is given below. Use static condensation as given in (8.28) to (8.30) to condense out the internal element degree of freedom. Then use Gauss elimination directly on the internal degree of freedom.

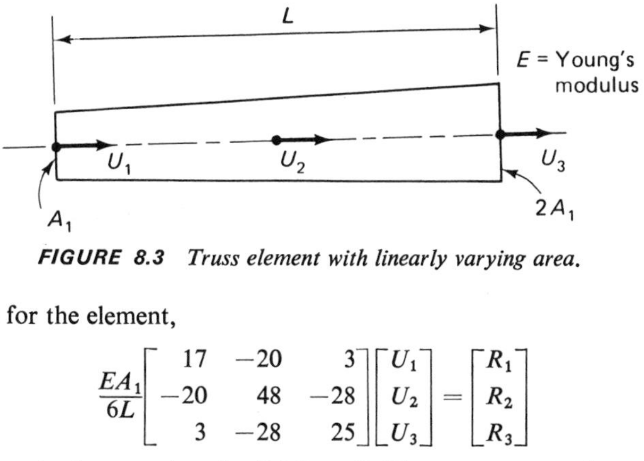

FIGURE 8.3 *Truss element with linearly varying area.*

We have for the element,

$$\frac{EA_1}{6L}\begin{bmatrix} 17 & -20 & 3 \\ -20 & 48 & -28 \\ 3 & -28 & 25 \end{bmatrix}\begin{bmatrix} U_1 \\ U_2 \\ U_3 \end{bmatrix} = \begin{bmatrix} R_1 \\ R_2 \\ R_3 \end{bmatrix} \tag{a}$$

In order to apply the equations in (8.28) to (8.30), we rearrange the equations in (a) to obtain

$$\frac{EA_1}{6L}\begin{bmatrix} 17 & 3 & -20 \\ 3 & 25 & -28 \\ -20 & -28 & 48 \end{bmatrix}\begin{bmatrix} U_1 \\ U_3 \\ U_2 \end{bmatrix} = \begin{bmatrix} R_1 \\ R_3 \\ R_2 \end{bmatrix}$$

The relation in (8.30) now gives

$$\frac{EA_1}{6L}\left\{\begin{bmatrix} 17 & 3 \\ 3 & 25 \end{bmatrix} - \begin{bmatrix} -20 \\ -28 \end{bmatrix}[\tfrac{1}{48}][-20 \quad -28]\right\}\begin{bmatrix} U_1 \\ U_3 \end{bmatrix} = \begin{bmatrix} R_1 + \frac{20}{48}R_2 \\ R_3 + \frac{28}{48}R_2 \end{bmatrix}$$

or

$$\frac{13}{9}\frac{EA_1}{L}\begin{bmatrix} 1 & -1 \\ -1 & 1 \end{bmatrix}\begin{bmatrix} U_1 \\ U_3 \end{bmatrix} = \begin{bmatrix} R_1 + \frac{5}{12}R_2 \\ R_3 + \frac{7}{12}R_2 \end{bmatrix}$$

Also, (8.29) yields

$$U_2 = \frac{1}{24}\left(\frac{3L}{EA_1}R_2 + 10U_1 + 14U_3\right) \tag{b}$$

Using Gauss elimination directly on (a) for U_2, we obtain

$$\frac{EA_1}{6L}\begin{bmatrix} 17-\frac{(20)(20)}{48} & 0 & 3-\frac{(20)(28)}{48} \\ -20 & 48 & -28 \\ 3-\frac{(20)(28)}{48} & 0 & 25-\frac{(28)(28)}{48} \end{bmatrix}\begin{bmatrix} U_1 \\ U_2 \\ U_3 \end{bmatrix} = \begin{bmatrix} R_1 + \frac{20}{48}R_2 \\ R_2 \\ R_3 + \frac{28}{48}R_2 \end{bmatrix} \tag{c}$$

But separating the equations for U_1 and U_3 from the equation for U_2, we can rewrite the relation in (c) as

$$\frac{13}{9}\frac{EA_1}{L}\begin{bmatrix} 1 & -1 \\ -1 & 1 \end{bmatrix}\begin{bmatrix} U_1 \\ U_3 \end{bmatrix} = \begin{bmatrix} R_1 + \frac{5}{12}R_2 \\ R_3 + \frac{7}{12}R_2 \end{bmatrix}$$

and
$$U_2 = \frac{1}{24}\left[\frac{3L}{EA_1}R_2 + 10U_1 + 14U_3\right]$$

which are the relations obtained using the formal static condensation procedure.

EXAMPLE 8.8: Use the stiffness matrix for the three degree of freedom truss element in Example 8.7 to establish the equilibrium equations of the structure shown in Fig. 8.4. Use Gauss elimination directly on degrees of freedom U_2 and U_4. Show that the resulting equilibrium equations are identical to those obtained when the two degree of freedom truss element stiffness matrix derived in Example 8.7 (the internal degree of freedom has been condensed out) is used to assemble the stiffness matrix corresponding to U_1, U_3, and U_5.

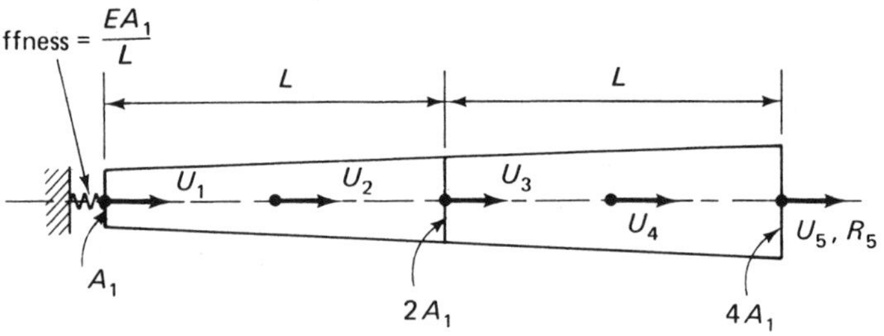

FIGURE 8.4 *Structure composed of two truss elements of Fig. 8.3 and a spring support.*

The stiffness matrix of the three element assemblage in Fig. 8.3 is obtained using the direct stiffness method; i.e., we calculate

$$\mathbf{K} = \sum_{m=1}^{3} \mathbf{K}^{(m)} \tag{a}$$

where
$$\mathbf{K}^{(1)} = \frac{EA_1}{6L}\begin{bmatrix} 6 & 0 & 0 & 0 & 0 \\ 0 & 0 & 0 & 0 & 0 \\ 0 & 0 & 0 & 0 & 0 \\ 0 & 0 & 0 & 0 & 0 \\ 0 & 0 & 0 & 0 & 0 \end{bmatrix}$$

$$\mathbf{K}^{(2)} = \frac{EA_1}{6L}\begin{bmatrix} 17 & -20 & 3 & 0 & 0 \\ -20 & 48 & -28 & 0 & 0 \\ 3 & -28 & 25 & 0 & 0 \\ 0 & 0 & 0 & 0 & 0 \\ 0 & 0 & 0 & 0 & 0 \end{bmatrix}$$

$$\mathbf{K}^{(3)} = \frac{EA_1}{6L}\begin{bmatrix} 0 & 0 & 0 & 0 & 0 \\ 0 & 0 & 0 & 0 & 0 \\ 0 & 0 & 34 & -40 & 6 \\ 0 & 0 & -40 & 96 & -56 \\ 0 & 0 & 6 & -56 & 50 \end{bmatrix}$$

Hence the equilibrium equations of the structure are

$$\frac{EA_1}{6L}\begin{bmatrix} 23 & -20 & 3 & 0 & 0 \\ -20 & 48 & -28 & 0 & 0 \\ 3 & -28 & 59 & -40 & 6 \\ 0 & 0 & -40 & 96 & -56 \\ 0 & 0 & 6 & -56 & 50 \end{bmatrix}\begin{bmatrix} U_1 \\ U_2 \\ U_3 \\ U_4 \\ U_5 \end{bmatrix} = \begin{bmatrix} 0 \\ 0 \\ 0 \\ 0 \\ R_5 \end{bmatrix}$$

Using Gauss elimination on degrees of freedom U_2 and U_4, we obtain

$$\frac{EA_1}{6L}\begin{bmatrix} 23 - \frac{(20)(20)}{48} & 0 & 3 - \frac{(20)(28)}{48} & 0 & 0 \\ -20 & 48 & -28 & 0 & 0 \\ 3 - \frac{(20)(28)}{48} & 0 & 59 - \frac{(28)(28)}{48} - \frac{(40)(40)}{96} & 0 & 6 - \frac{(40)(56)}{96} \\ 0 & 0 & -40 & 96 & -56 \\ 0 & 0 & 6 - \frac{(40)(56)}{96} & 0 & 50 - \frac{(56)(56)}{96} \end{bmatrix}\begin{bmatrix} U_1 \\ U_2 \\ U_3 \\ U_4 \\ U_5 \end{bmatrix} = \begin{bmatrix} 0 \\ 0 \\ 0 \\ 0 \\ R_5 \end{bmatrix} \quad \text{(b)}$$

Now, extracting the equilibrium equations corresponding to degrees of freedom 1, 3, and 5 and degrees of freedom 2 and 4 separately, we have

$$\frac{13}{9}\frac{EA_1}{L}\begin{bmatrix} \frac{22}{13} & -1 & 0 \\ -1 & 3 & -2 \\ 0 & -2 & 2 \end{bmatrix}\begin{bmatrix} U_1 \\ U_3 \\ U_5 \end{bmatrix} = \begin{bmatrix} 0 \\ 0 \\ R_5 \end{bmatrix} \quad \text{(c)}$$

and

$$\begin{aligned} U_2 &= \tfrac{1}{12}[5U_1 + 7U_3] \\ U_4 &= \tfrac{1}{12}[5U_3 + 7U_5] \end{aligned} \quad \text{(d)}$$

However, using the two degree of freedom truss element stiffness matrix derived in Example 8.7 to directly assemble the structure stiffness matrix corresponding to degrees of freedom 1, 3, and 5, we use as element stiffness matrices in (a),

$$\mathbf{K}^{(1)} = \frac{13}{9}\frac{EA_1}{L}\begin{bmatrix} \frac{9}{13} & 0 & 0 \\ 0 & 0 & 0 \\ 0 & 0 & 0 \end{bmatrix}$$

$$\mathbf{K}^{(2)} = \frac{13}{9}\frac{EA_1}{L}\begin{bmatrix} 1 & -1 & 0 \\ -1 & 1 & 0 \\ 0 & 0 & 0 \end{bmatrix} \quad \text{(e)}$$

$$\mathbf{K}^{(3)} = \frac{13}{9}\frac{EA_1}{L}\begin{bmatrix} 0 & 0 & 0 \\ 0 & 2 & -2 \\ 0 & -2 & 2 \end{bmatrix} \quad \text{(f)}$$

and obtain the stiffness matrix in (c). Also, the relation (b) of Example 8.7 corresponds to relations (d) in this example. It should be noted that the total effort to solve the equilibrium equations using the condensed truss element stiffness matrix is less than when the original three degree of freedom element stiffness matrix is used, because in the first case the internal degree of freedom was statically condensed out only once, whereas in (b) the element internal degree of freedom is, in fact, statically condensed out twice. The direct solution using the condensed element stiffness matrices in (e) and (f) is, however, only possible because these stiffness matrices are multiples of each other.

As indicated in Example 8.8, it can be particularly effective to employ static condensation when the same element is used many times. An application of this concept is employed in *substructure analysis*, in which the total structure is considered to be an assemblage of substructures.[16-22] Each substructure, in turn, is idealized as an assemblage of finite elements, and all internal degrees of freedom are statically condensed out. The total structure stiffness is formed by assembling the condensed substructure stiffness matrices. Therefore, *in effect, a substructure is used in the same way as an individual finite element with internal degrees of freedom that are statically condensed out prior to the element assemblage process.* If many substructures are identical, it is effective to establish a library of substructures, from which a condensed total structure stiffness matrix is formed.

It should be noted that the unreduced complete structure stiffness matrix is never calculated in substructure analysis, and *input data are required only for each substructure in the library plus information on the assemblage of the substructures* to make up the complete structure. Typical applications of finite element analysis using substructuring are found in the analysis of buildings and ship hulls, where the substructure technique has allowed economical analysis of very large finite element systems.

As a simple example of substructuring, we refer to Example 8.8, in which each substructure is simply composed of one element, and the uncondensed and condensed stiffness matrix of a typical substructure was given in Example 8.7.

The effectiveness of analysis using the basic substructure concept described above can in many cases still be improved upon by defining different levels of substructures; i.e., since each substructure can be looked upon as a "super-finite element," it is possible to define second, third, etc., levels of substructuring. In a similar procedure, two substructures are always combined to define the next-higher-level substructure until the final substructure is, in fact, the actual structure under consideration. The procedure may be employed in one-, two-, or three-dimensional analysis and, as pointed out earlier, is indeed only an effective application of Gauss elimination, in which advantage is taken of the repetition of submatrices, which are the stiffness matrices that correspond to the identical substructures. The possibility of using the solution procedure effectively therefore depends on whether the structure is made up of repetitive substructures, and this is the reason the procedure can be very effective in special-purpose programs.

EXAMPLE 8.9: Use substructuring to evaluate the stiffness matrix and the load vector corresponding to the end nodal point degrees of freedom U_1 and U_9 of the bar in Fig. 8.5.

The basic element of which the bar is composed is the three degree of freedom truss element considered in Example 8.7. The equilibrium equations of the element

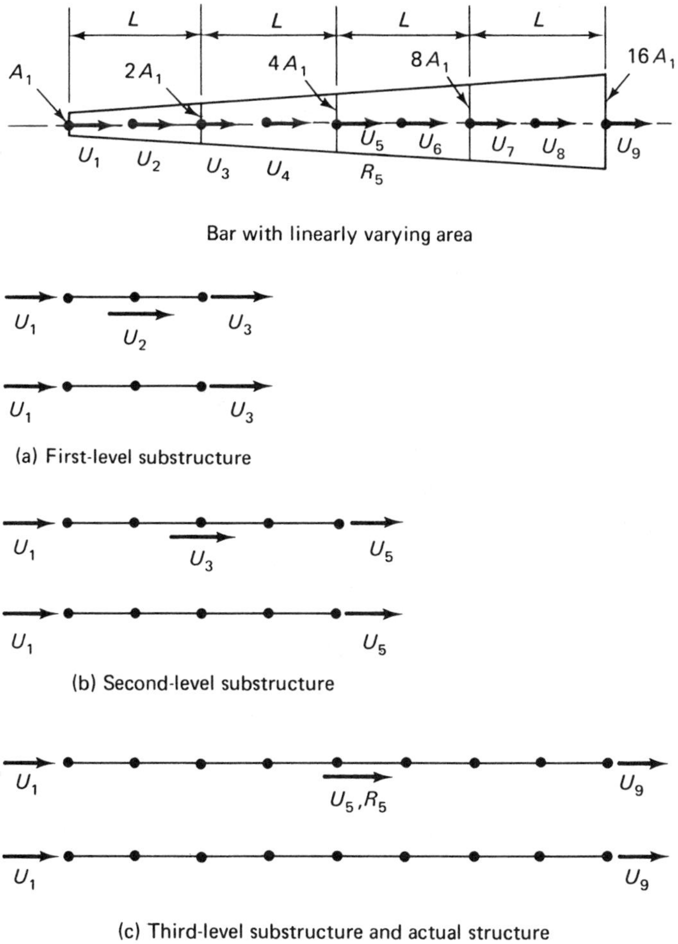

FIGURE 8.5 *Analysis of bar using substructuring.*

corresponding to the two degrees of freedom U_1 and U_3 as shown in Fig. 8.5 are

$$\frac{13}{9}\frac{A_1E}{L}\begin{bmatrix} 1 & -1 \\ -1 & 1 \end{bmatrix}\begin{bmatrix} U_1 \\ U_3 \end{bmatrix} = \begin{bmatrix} R_1 + \frac{5}{12}R_2 \\ R_3 + \frac{7}{12}R_2 \end{bmatrix} \tag{a}$$

Since the internal degree of freedom U_2 has been statically condensed out to obtain the equilibrium relations in (a), we may regard the two degree of freedom element as a first-level substructure. We should recall that once U_1 and U_3 have been calculated, we can evaluate U_2 using the relation (b) in Example 8.7:

$$U_2 = \frac{1}{24}\left[\frac{3L}{EA_1}R_2 + 10U_1 + 14U_3\right] \tag{b}$$

It is now effective to evaluate a second-level substructure corresponding to degrees of freedom U_1 and U_5 of the bar. For this purpose we use the stiffness matrix and load vector in (a) to evaluate the equilibrium relations corresponding to U_1, U_3, and U_5:

$$\frac{13}{9}\frac{A_1E}{L}\begin{bmatrix} 1 & -1 & 0 \\ -1 & 3 & -2 \\ 0 & -2 & 2 \end{bmatrix}\begin{bmatrix} U_1 \\ U_3 \\ U_5 \end{bmatrix} = \begin{bmatrix} R_1 + \frac{5}{12}R_2 \\ R_3 + \frac{7}{12}R_2 + \frac{5}{12}R_4 \\ R_5 + \frac{7}{12}R_4 \end{bmatrix} \tag{c}$$

The relation to calculate U_4 is similar to the one in (b):

$$U_4 = \frac{1}{24}\left[\frac{3L}{2EA_1} R_4 + 10U_3 + 14U_5\right]$$

Using Gauss elimination on the equations in (c) to condense out U_3, we obtain

$$\frac{13}{9}\frac{A_1 E}{L}\begin{bmatrix} \frac{2}{3} & 0 & -\frac{2}{3} \\ -1 & 3 & -2 \\ -\frac{3}{2} & 0 & \frac{2}{3} \end{bmatrix}\begin{bmatrix} U_1 \\ U_3 \\ U_5 \end{bmatrix} = \begin{bmatrix} R_1 + \frac{22}{36}R_2 + \frac{1}{3}R_3 + \frac{5}{36}R_4 \\ R_3 + \frac{7}{12}R_2 + \frac{5}{12}R_4 \\ \frac{14}{36}R_2 + \frac{2}{3}R_3 + \frac{31}{36}R_4 + R_5 \end{bmatrix}$$

or

$$\left(\frac{2}{3}\right)\left(\frac{13}{9}\right)\frac{A_1 E}{L}\begin{bmatrix} 1 & -1 \\ -1 & 1 \end{bmatrix}\begin{bmatrix} U_1 \\ U_5 \end{bmatrix} = \begin{bmatrix} R_1 + \frac{22}{36}R_2 + \frac{1}{3}R_3 + \frac{5}{36}R_4 \\ \frac{14}{36}R_2 + \frac{2}{3}R_3 + \frac{31}{36}R_4 + R_5 \end{bmatrix} \tag{d}$$

and

$$U_3 = \frac{1}{3}\left[\frac{9}{13}\frac{L}{A_1 E}\left(R_3 + \frac{7}{12}R_2 + \frac{5}{12}R_4\right) + U_1 + 2U_5\right]$$

We should note that the stiffness matrix of the second-level substructure in (d) is simply $\frac{2}{3}$ times the stiffness matrix of the first-level substructure in (a). Therefore, we could continue to build up even higher-level substructures in an analogous manner; i.e., the stiffness matrix of the nth-level substructure would simply be a factor times the stiffness matrix given in (a).

In most cases loads are only applied at the boundary degrees of freedom between substructures, such as in this example. Using the stiffness matrix of the second-level substructure to assemble the stiffness matrix of the complete bar, and assembling the actual load vector for this example, we obtain

$$\frac{2}{3}\left(\frac{13}{9}\right)\frac{A_1 E}{L}\begin{bmatrix} 1 & -1 & 0 \\ -1 & 5 & -4 \\ 0 & -4 & 4 \end{bmatrix}\begin{bmatrix} U_1 \\ U_5 \\ U_9 \end{bmatrix} = \begin{bmatrix} 0 \\ R_5 \\ 0 \end{bmatrix}$$

Eliminating U_5 we have

$$\left(\frac{4}{5}\right)\left(\frac{2}{3}\right)\left(\frac{13}{9}\right)\frac{A_1 E}{L}\begin{bmatrix} 1 & -1 \\ -1 & 1 \end{bmatrix}\begin{bmatrix} U_1 \\ U_9 \end{bmatrix} = \begin{bmatrix} \frac{1}{5}R_5 \\ \frac{4}{5}R_5 \end{bmatrix}$$

where the stiffness matrix is simply the third-level substructure stiffness matrix corresponding to the algorithm given above. We also have

$$U_5 = \frac{1}{5}\left[\frac{27}{26}\frac{L}{A_1 E} R_5 + U_1 + 4U_9\right]$$

To solve for specific displacements, it is necessary to impose boundary conditions on the bar, hence obtain U_1 and U_9, and then obtain the internal bar displacements using previously derived relations. It should be noted that corresponding relations must also be employed to evaluate U_6 to U_8.

So far, we have not mentioned how to proceed in the solution if the total system matrix cannot be contained in high-speed storage.[12,23-34] If substructuring is used, it is effective to keep the size of each uncondensed substructure stiffness matrix small enough so that the static condensation of the internal degrees of freedom can be carried out in high-speed core. Therefore, back-up storage would mainly be required to store the required information for the calculation of the displacements of the substructure internal nodes as expressed in (8.29).

However, it may be necessary to use multilevel substructuring (i.e., to define substructures of substructures) in order that the final equations to be solved can be taken into high-speed storage.

In general, it is important to use back-up storage effectively, because a great deal of reading and writing can be very expensive and indeed may limit the system size that can be solved, because not enough back-up storage may be available. In out-of-core solutions the particular scheme used for solving the system equilibrium equations is largely coupled with the specific procedure employed to assemble the element stiffness matrices to the global structure stiffness matrix. In many programs the structure stiffness matrix is assembled prior to performing the Gauss solution. In the program ADINA the equations are assembled as described in the Appendix, but in blocks that can be taken into high-speed core. The block sizes (number of columns per block) are automatically established in the program and depend on the high-speed storage available.[26] The solution of the system equations is then obtained in an effective manner (using subroutine COLSOL of Section 8.2.3 modified for out-of-core solution) by first reducing the blocks of the stiffness matrix and load vectors consecutively, and then performing the back-substitution. Similar procedures are presently used in many different structural analysis programs.[23-31]

Instead of first assembling the complete structure stiffness matrix, we may assemble and reduce the equations at the same time. In this case back-up storage for the total unreduced stiffness matrix is not required. A specific solution scheme called the *frontal solution method* has been used effectively.[32-34] In the solution procedure only those equations that are actually required for the elimination of a specific degree of freedom are assembled, the degree of freedom considered is statically condensed out, and so on.

As an example, consider the analysis of the plane stress finite element idealization of the sheet in Fig. 8.6. There are two equations associated with each node of the finite element mesh, namely, the equations corresponding to the U and V displacements, respectively. In the frontal solution scheme the equations are statically condensed out in the order of the elements; i.e., the first equations considered could be those corresponding to nodes 1, 2, ... To be able to eliminate the degrees of freedom of node 1 it is only necessary to assemble the final equations that correspond to that node. This means that only the stiffness matrix of element 1 need be calculated after which the degrees of freedom corresponding to node 1 are statically condensed out. Next (for the elimination of the equations corresponding to node 2) the final equations corresponding to the degrees of freedom at node 2 are required, meaning that the stiffness matrix of element 2 must be calculated and added to the previously reduced matrix. Now the degrees of freedom corresponding to node 2 are statically condensed out; and so on.

It may now be realized that *the complete procedure, in effect, consists of statically condensing out one degree of freedom after the other, and always assembling only those equations (or, rather, element stiffness matrices) that are actually required during the specific condensation to be performed.* The finite elements that must be considered for the static condensation of the equations corresponding to one specific node define the wave front at that time, as shown in Fig. 8.6.

In principle, the frontal solution is Gauss elimination and the important aspect is the specific computer implementation. Since the equations are as-

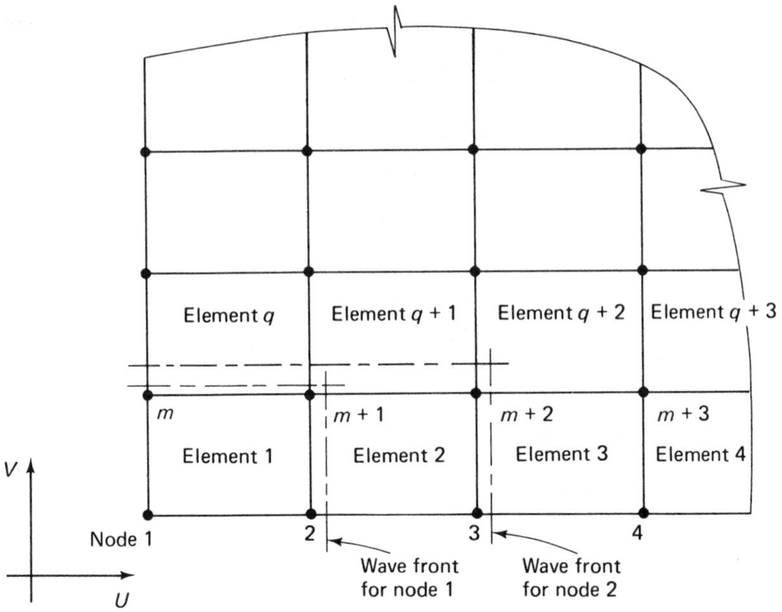

FIGURE 8.6 *Frontal solution of plane stress finite element idealization.*

sembled in the order of the elements, the length of the wave front and therefore the half-bandwidth dealt with are determined by the element numbering. Therefore, an effective ordering of the elements is necessary, and we note that *if the element numbering in the frontal solution corresponds to the nodal point numbering in the active column solution* (*see Section 8.2.3*), *the same number of basic* (*i.e. excluding indexing*) *numerical operations is performed in both solutions.* An advantage of the wave front solution is that elements can be added with relative ease, because no nodal point renumbering is necessary to preserve a small bandwidth. But a disadvantage is that if the wave front is large, the total high-speed storage required may well exceed the storage that is available, in which case additional out-of-core operations are required, that suddenly decrease the effectiveness of the method by a great amount. Also, the active column solution is implemented in a compact stand-alone solver that is independent of the finite elements processed, whereas a frontal solver is intimately coupled to the finite elements and may require more indexing in the solution.

8.2.5 Solution of Equations with Symmetric Nonpositive Definite Coefficient Matrices

In the Gauss elimination discussed so far, we implicitly assumed that the stiffness matrix **K** is positive definite, namely that the structure considered is properly restrained and stable. As we discussed in Sections 2.7 and 2.8, positive definiteness of the stiffness matrix means that for any displacement vector **U**, we have

$$\mathbf{U}^T\mathbf{K}\mathbf{U} > 0 \tag{8.31}$$

Since $\frac{1}{2}\mathbf{U}^T\mathbf{K}\mathbf{U}$ is the strain energy stored in the system for the displacement vector $\mathbf{U}$ (8.31) expresses that for any displacement vector $\mathbf{U}$, the strain energy of a system with a positive definite stiffness matrix must be positive.

It should be noted that the stiffness matrix of a finite element is not positive definite unless the element has been properly restrained; i.e., the rigid body motions have been suppressed. Instead, the stiffness matrix of an unrestrained finite element is positive semidefinite,

$$\mathbf{U}^T\mathbf{K}\mathbf{U} \geq 0 \tag{8.32}$$

where $\mathbf{U}^T\mathbf{K}\mathbf{U} = 0$ when $\mathbf{U}$ corresponds to a rigid body mode. Considering the finite element assemblage process, it should be realized that positive semidefinite element matrices are added to obtain the positive semidefinite stiffness matrix corresponding to the complete structure. The stiffness matrix of the structure is then rendered positive definite by eliminating the rows and columns that correspond to appropriate restraint degrees of freedom, i.e., by eliminating the possibility for the structure to undergo rigid body motions.

It is instructive to consider in more detail the meaning of positive definiteness of the structure stiffness matrix. In Section 2.7, we discussed the representation of a matrix by its eigenvalues and eigenvectors. Following the development given in Section 2.7, the eigenproblem for the stiffness matrix $\mathbf{K}$ can be written

$$\mathbf{K}\boldsymbol{\phi} = \lambda\boldsymbol{\phi} \tag{8.33}$$

The solution to (8.33) are the eigenpairs $(\lambda_i, \boldsymbol{\phi}_i)$, $i = 1, \ldots, n$, and the complete solution can be written

$$\mathbf{K}\boldsymbol{\Phi} = \boldsymbol{\Phi}\boldsymbol{\Lambda}$$

where $\boldsymbol{\Phi}$ is a matrix of the orthonormalized eigenvectors, $\boldsymbol{\Phi} = [\boldsymbol{\phi}_1, \ldots, \boldsymbol{\phi}_n]$, and $\boldsymbol{\Lambda}$ is a diagonal matrix of the corresponding eigenvalues, $\boldsymbol{\Lambda} = \text{diag}(\lambda_i)$. Since $\boldsymbol{\Phi}^T\boldsymbol{\Phi} = \boldsymbol{\Phi}\boldsymbol{\Phi}^T = \mathbf{I}$, we also have

$$\boldsymbol{\Phi}^T\mathbf{K}\boldsymbol{\Phi} = \boldsymbol{\Lambda} \tag{8.34}$$

and

$$\mathbf{K} = \boldsymbol{\Phi}\boldsymbol{\Lambda}\boldsymbol{\Phi}^T \tag{8.35}$$

Referring to Section 2.8, we recall that λ_i of $\mathbf{K}$ represents the minimum that can be reached by the Rayleigh quotient when an orthonormality constraint is satisfied on the eigenvectors $\boldsymbol{\phi}_1, \ldots, \boldsymbol{\phi}_{i-1}$:

$$\left.\begin{aligned} \lambda_i &= \min\left\{\frac{\boldsymbol{\phi}^T\mathbf{K}\boldsymbol{\phi}}{\boldsymbol{\phi}^T\boldsymbol{\phi}}\right\} \\ \text{with} \qquad \boldsymbol{\phi}^T\boldsymbol{\phi}_r = 0; &\qquad \text{for } r = 1, 2, \ldots, i-1 \end{aligned}\right\} \tag{8.36}$$

Therefore, $\frac{1}{2}\lambda_1$ *is the minimum strain energy that can be stored in the element assemblage and the corresponding displacement vector is* $\boldsymbol{\phi}_1$. For a positive definite system stiffness matrix, we therefore have $\lambda_1 > 0$. On the other hand, for the stiffness matrix of an unrestrained system, we have $\lambda_1 = \lambda_2 = \ldots = \lambda_m = 0$, where m is the number of rigid body modes present, $m < n$. As the system is restrained, the number of eigenvalues of $\mathbf{K}$ is decreased by one for each degree of freedom that is eliminated, and a zero eigenvalue is lost if the restraint results in the elimination of a rigid body mode.

EXAMPLE 8.10: Identify whether the deletion of the four degrees of freedom of the plane stress element in Fig. 8.7 results in the elimination of the rigid body modes.

The plane stress element has three rigid body modes: (1) uniform horizontal translation, (2) uniform vertical translation, and (3) in-plane rotation. Consider the sequential deletion of the degrees of freedom, as shown in Fig. 8.7. The deletion of U_4 results into eliminating the horizontal translation rigid body mode. Similarly, the deletion of V_4 results into the deletion of the vertical translation rigid body mode. However, deleting V_1 in addition does not result into the elimination of the last rigid body mode; i.e., the in-plane rotation rigid body mode is only eliminated with the additional deletion of U_2. Therefore, the deletion of U_4 and V_4, and V_1 and U_2 eliminates all rigid body modes of the element, although we should note that, in fact, by the deletion of U_4, V_4, and U_2 alone we would achieve the same result.

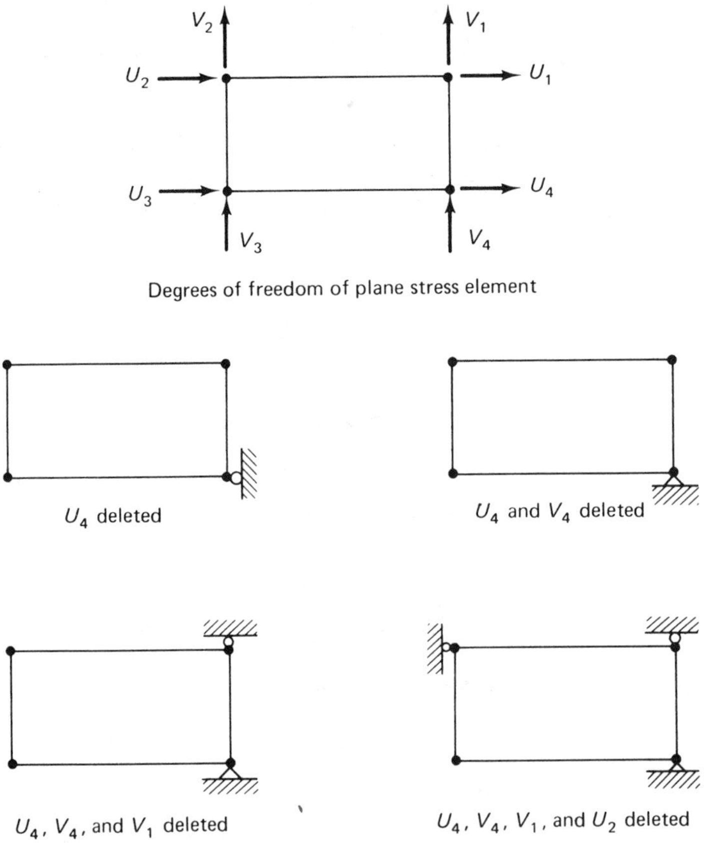

FIGURE 8.7 *Deletion of degrees of freedom of plane stress element.*

The transformation performed in (8.34) has important meaning. Considering the relation and referring to Section 2.5, we realize that in (8.34) a change of basis is performed. The new basis vectors are the finite element interpolations corresponding to the eigenvectors of $\mathbf{K}$, and in this basis the operator is represented by a diagonal matrix with the eigenvalues of $\mathbf{K}$ on its diagonal. We may therefore look at $\mathbf{\Lambda}$ as being the stiffness matrix of the system when the finite element displacement functions used in the virtual work theorem in (4.5) are those corresponding to nodal point displacements $\boldsymbol{\phi}_i$, $i = 1, \ldots, n$, instead of unit nodal point displacements U_i, $i = 1, \ldots, n$ (see Section 4.2.1). The relation in (8.34) is therefore a statement of virtual work resulting into a diagonal stiffness matrix. If the system considered is properly restrained, all stiffness coefficients in $\boldsymbol{\Lambda}$ are positive; i.e., the stiffness matrix $\boldsymbol{\Lambda}$ (and hence $\mathbf{K}$) is positive definite, whereas for an unrestrained system some diagonal elements in $\boldsymbol{\Lambda}$ are zero.

Before studying the solution of nonpositive definite systems of equations, another most important observation should be discussed. In Section 2.8 we introduced the Sturm sequence property of the leading principal minors of a matrix. We should note here the physical meaning of the Sturm sequence. *Let $\mathbf{K}^{(r)}$ be the matrix of order $n - r$ obtained by deleting from $\mathbf{K}$ the last r rows and columns, and consider the eigenproblem*

$$\mathbf{K}^{(r)}\boldsymbol{\phi}^{(r)} = \lambda^{(r)}\boldsymbol{\phi}^{(r)} \tag{8.37}$$

where $\boldsymbol{\phi}^{(r)}$ is a vector of order $n - r$. We say that (8.37) is the eigenproblem of the rth associated constraint problem of the problem $\mathbf{K}\boldsymbol{\phi} = \lambda\boldsymbol{\phi}$. Then we have shown in Section 2.8 that the eigenvalues of the $(r + 1)$st constraint problem separate those of the rth constraint problem,

$$\lambda_1^{(r)} \leq \lambda_1^{(r+1)} \leq \lambda_2^{(r)} \leq \lambda_2^{(r+1)} \ldots \leq \lambda_{n-r-1}^{(r)} \leq \lambda_{n-r-1}^{(r+1)} \leq \lambda_{n-r}^{(r)} \tag{8.38}$$

As an example, the eigenproblems of the simply supported beam discussed in Section 8.2.1 and of its associated constraint problems can be considered. Figure 8.8 shows the eigenvalues calculated and, in particular, displays their separation property. We should note that as we proceed from the $(r + 1)$st constraint problem to the rth constraint problem by including the $(n - r)$th degree of freedom, the new system has an eigenvalue smaller than (or equal to) the smallest eigenvalue of the $(r + 1)$st constraint problem, and also an eigenvalue larger than (or equal to) the largest eigenvalue of the $(r + 1)$st constraint problem.

Using the separation property of the eigenvalues, and realizing that any rows and columns may be interchanged at convenience to become the last rows and columns in the matrix $\mathbf{K}$, it follows that if the stiffness matrix corresponding to the n degrees of freedom is positive definite (i.e., $\lambda_1 > 0$), then any stiffness matrix obtained by deleting any rows and corresponding columns is also positive definite. Furthermore, the smallest eigenvalue of the new matrix can only have increased, and the largest eigenvalue can only have decreased. This conclusion applies also if the matrix $\mathbf{K}$ is positive semidefinite and would apply if it were indefinite since we showed the eigenvalue separation theorem to be applicable to all symmetric matrices.

We shall encounter the use of the Sturm sequence property of the leading principal minors more extensively in the design of eigenvalue solution algorithms (see Chapter 12). However, in the following we use the property to yield more insight into the solution of a set of simultaneous equations with a symmetric positive definite, positive semidefinite, and indefinite coefficient matrix. A symmetric indefinite coefficient matrix will be encountered in the solution of eigenproblems.

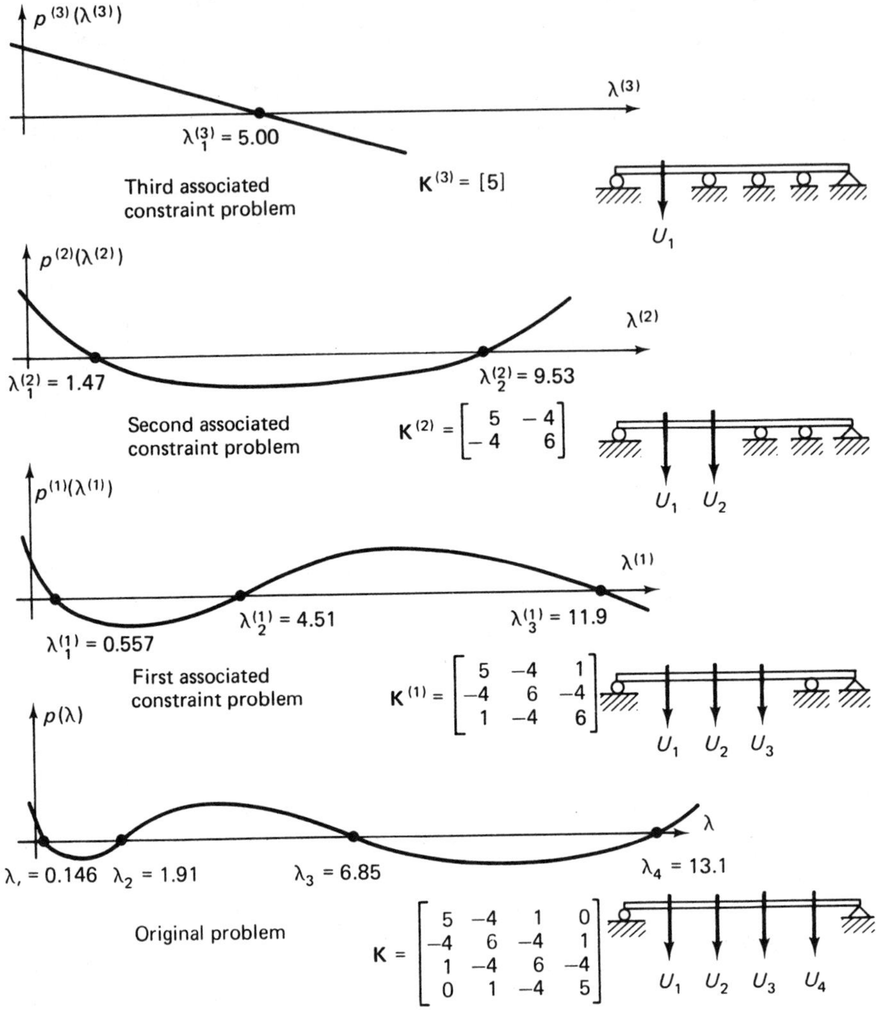

FIGURE 8.8 *Eigenvalue solutions of simply supported beam and of associated constraint problems.*

Using physical reasoning, we showed in Sections 8.2.1 and 8.2.2 that if $\mathbf{K}$ is the stiffness matrix of a properly restrained structure, we can factorize $\mathbf{K}$ into the form

$$\mathbf{K} = \mathbf{LDL}^T \tag{8.39}$$

where $\mathbf{L}$ is a lower unit triangular matrix and $\mathbf{D}$ is a diagonal matrix, with $d_{ii} > 0$. It follows (see Section 1.8) that

$$\begin{aligned} \det \mathbf{K} &= \det \mathbf{L} \det \mathbf{D} \det \mathbf{L}^T \\ &= \prod_{i=1}^{n} d_{ii} > 0 \end{aligned} \tag{8.40}$$

This result can also be obtained by considering the characteristic polynomial of $\mathbf{K}$, defined as

$$p(\lambda) = \det (\mathbf{K} - \lambda \mathbf{I}) \tag{8.41}$$

Since λ_1 is the smallest root of $p(\lambda)$ and $\lambda_1 > 0$ for $\mathbf{K}$ being positive definite, it follows that $\det \mathbf{K} > 0$. However, it does not yet follow that $d_{ii} > 0$ for all i.

In order to formally prove that $d_{ii} > 0$, $i = 1, \ldots, n$, when $\mathbf{K}$ is positive definite, and to identify what happens during the factorization of $\mathbf{K}$, it is expedient to compare the triangular factors of $\mathbf{K}$ with those of $\mathbf{K}^{(i)}$, where $\mathbf{K}^{(i)}$ is the stiffness matrix of the ith associated constraint problem. Assuming that the factors $\mathbf{L}$ and $\mathbf{D}$ of $\mathbf{K}$ have been calculated, we have, for the associated constraint problems,

$$\mathbf{K}^{(i)} = \mathbf{L}^{(i)}\mathbf{D}^{(i)}\mathbf{L}^{(i)T} \qquad i = 1, \ldots, n-1 \tag{8.42}$$

where $\mathbf{L}^{(i)}$ and $\mathbf{D}^{(i)}$ are analogously the factors of $\mathbf{K}^{(i)}$. Since $\mathbf{L}$ is a lower-unit triangular matrix and $\mathbf{D}$ is a diagonal matrix, the factors $\mathbf{L}^{(i)}$ and $\mathbf{D}^{(i)}$ are obtained from $\mathbf{L}$ and $\mathbf{D}$, respectively, by eliminating the last i rows and columns. Therefore, $\mathbf{L}^{(i)}$ and $\mathbf{D}^{(i)}$ are the leading principal submatrices of $\mathbf{L}$ and $\mathbf{D}$, respectively, and they are actually evaluated in the factorization of $\mathbf{K}$. However, because $\lambda_1^{(i)} > 0$, it now follows that we can use the argument in (8.39) and (8.40) starting with $i = n - 1$ to show that $d_{ii} > 0$ for all i. Therefore, the factorization of $\mathbf{K}$ into $\mathbf{LDL}^T$ is indeed possible if $\mathbf{K}$ is positive definite. We demonstrate the analysis in the following example.

EXAMPLE 8.11: Consider the simply supported beam in Fig. 8.8 and the associated constraint problems. The same beam was used in Section 8.2.1. Establish the $\mathbf{L}^{(i)}$ and $\mathbf{D}^{(i)}$ factors of the matrices $\mathbf{K}^{(i)}$, $i = 1, 2, 3$, and show that d_{ii} must be greater than zero because $\lambda_1 > 0$.

The required triangular factorizations are

$$[5] = [1][5][1] \tag{a}$$

$$\begin{bmatrix} 5 & -4 \\ -4 & 6 \end{bmatrix} = \begin{bmatrix} 1 & 0 \\ -\frac{4}{5} & 1 \end{bmatrix} \begin{bmatrix} 5 & 0 \\ 0 & \frac{14}{5} \end{bmatrix} \begin{bmatrix} 1 & -\frac{4}{5} \\ 0 & 1 \end{bmatrix} \tag{b}$$

$$\begin{bmatrix} 5 & -4 & 1 \\ -4 & 6 & -4 \\ 1 & -4 & 6 \end{bmatrix} = \begin{bmatrix} 1 & 0 & 0 \\ -\frac{4}{5} & 1 & 0 \\ \frac{1}{5} & -\frac{8}{7} & 1 \end{bmatrix} \begin{bmatrix} 5 & 0 & 0 \\ 0 & \frac{14}{5} & 0 \\ 0 & 0 & \frac{15}{7} \end{bmatrix} \begin{bmatrix} 1 & -\frac{4}{5} & \frac{1}{5} \\ 0 & 1 & -\frac{8}{7} \\ 0 & 0 & 1 \end{bmatrix} \tag{c}$$

where the matrices $\mathbf{L}^{(i)}$ and $\mathbf{D}^{(i)}$ are obtained from the $\mathbf{L}$ and $\mathbf{D}$ factors given in Example 8.2 by eliminating the last i rows and columns. (As a check, we may want to calculate the product of the matrices on the right sides of the relations in (a), (b), and (c) to obtain the matrices on the left sides.)

Considering the elements d_{ii}, we have that $\lambda_1^{(3)} \geq \lambda_1^{(2)} \geq \lambda_1^{(1)} \geq \lambda_1 > 0$. But using the relation (a), we have $\lambda_1^{(3)} = d_{11}$; hence $d_{11} > 0$. Next we consider $\mathbf{K}^{(2)}$. Since $\lambda_1^{(2)} > 0$, we have, using (7.39) and (7.40), that $d_{11}d_{22} > 0$, which means that $d_{22} > 0$. Similarly, considering $\mathbf{K}^{(1)}$, we have $\lambda_1^{(1)} > 0$, and hence $d_{11}d_{22}d_{33} > 0$, from which it follows that $d_{33} > 0$. Finally, considering $\mathbf{K}$, we have $\lambda_1 > 0$, and hence $d_{11}d_{22}d_{33}d_{44} > 0$. Therefore, $d_{44} > 0$ also.

Assume next that the matrix $\mathbf{K}$ is the stiffness matrix of a finite element assemblage that is unrestrained. In this case, $\mathbf{K}$ is positive semidefinite, $\lambda = 0.0$ is a root, and $\det \mathbf{K}$ is zero, which, using (8.40), means that d_{ii} for some i must be zero. Therefore, the factorization of $\mathbf{K}$ as shown in the preceding sections is, in general, not possible because a pivot element will be zero. It is again instructive to consider the associated constraint problems. When $\mathbf{K}$ is positive semidefinite, we note that the characteristic polynomial corresponding to $\mathbf{K}^{(i)}$ will have a zero eigenvalue, and this zero eigenvalue will be retained in all matrices $\mathbf{K}^{(i-1)}, \ldots, \mathbf{K}$. This follows because, firstly, the Sturm sequence property assures that the smallest eigenvalue of $\mathbf{K}^{(i-1)}$ is smaller or equal to the smallest eigenvalue of $\mathbf{K}^{(i)}$, and secondly, $\mathbf{K}$ has no negative eigenvalues. For the ith associated constraint problem, we will therefore have

$$\det (\mathbf{L}^{(i)}\mathbf{D}^{(i)}\mathbf{L}^{(i)T}) = 0 \tag{8.43}$$

from which it follows that an element of $\mathbf{D}^{(i)}$ is zero. However, assuming that the zero root only occurs in the ith associated constraint problem [i.e., det $(\mathbf{L}^{(r)}\mathbf{D}^{(r)}\mathbf{L}^{(r)T}) > 0$ for $r > i$], it follows that $d_{n-i,n-i}$ is zero. In summary, therefore, if $\mathbf{K}$ is positive semidefinite, the factorization of $\mathbf{K}$ into $\mathbf{LDL}^T$ (i.e., the Gauss elimination process) will break down at the time a zero diagonal element d_{kk} is encountered, which means that the $(n - k)$th associated constraint problem with a zero eigenvalue prevents the continuation of the factorization process.

In the case of a positive semidefinite matrix, a zero diagonal element must be encountered at some stage of the factorization. However, considering the decomposition of an indefinite matrix (i.e., some of the eigenvalues of the matrix are negative and some are positive), a zero diagonal element is encountered only if one of the associated constraint problems has a zero eigenvalue. Namely, as in the case of a positive semidefinite matrix, d_{kk} is zero if the $(n - k)$th associated constraint problem has a zero eigenvalue. However, if none of the associated constraint problems has a zero eigenvalue, all elements d_{ii} are nonzero and, in exact arithmetic, no difficulties are encountered in the factorization. We shall discuss the decomposition of indefinite coefficient matrices further in the solution of eigenproblems (see Sections 11.4.2 and 12.2). Figure 8.9 shows typical cases that use the simply supported beam of Fig. 8.8 on spring supports for which, in the decompositions, we would and would not encounter a zero diagonal element (see Example 8.13).

Assume that a zero diagonal element d_{ii} is encountered in the Gauss elimination. To be able to proceed with the solution, it is necessary to interchange the ith row with another row, say the jth row, where $j > i$. The new diagonal element should not be zero, and to increase solution accuracy, it should be large (see Section 8.5). This row interchange corresponds to a rearranging of the equations, where it should be noted that the row interchange results in the coefficient matrix no longer being symmetric. On the other hand, symmetry would be preserved if we were to interchange not only the ith and jth rows but also the corresponding columns to obtain a nonzero diagonal element in row i, which is not always possible (see Example 10.4). In effect, the interchange of columns and rows corresponds to a rearranging of the associated constraint problems in such a way that these have nonzero eigenvalues.

The remedy of row interchanges assumes that it can be arranged that the new diagonal element is nonzero. In fact, this will always be the case, unless the matrix has a zero eigenvalue of multiplicity m and $i = n - m + 1$. In this case the matrix is singular and d_{ii} is zero, but all other elements of the last m rows of the upper triangular factor of the coefficient matrix are also

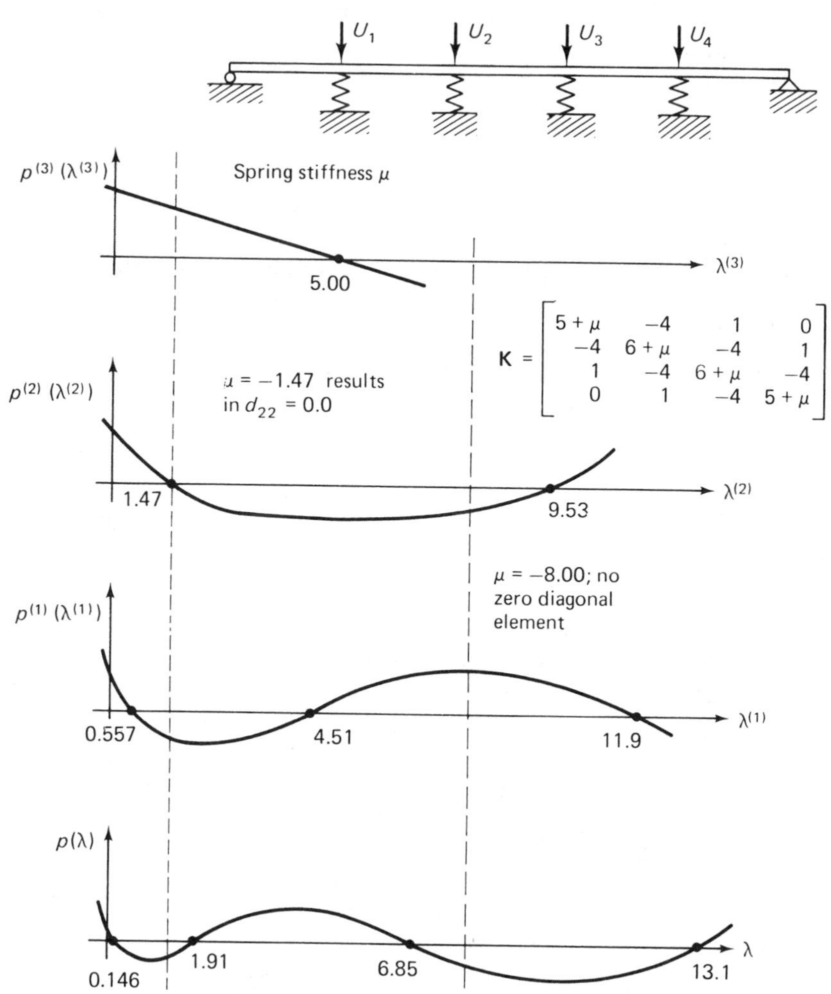

FIGURE 8.9 *Simply supported beam on spring supports; negative spring stiffness can result in solution difficulties.*

zero, and the factorization of the matrix has already been completed. In other words, since the number of rows that are made up of zero elements only is equal to the multiplicity m of the zero eigenvalue, the last $m - 1$ matrices $\mathbf{L}_{n-1}^{-1}$, $\mathbf{L}_{n-2}^{-1}, \ldots, \mathbf{L}_{n-m+1}^{-1}$ in (8.10) cannot and need not be calculated. We shall therefore be able to solve for m linearly independent solutions by assuming appropriate values for the m last entries of the solution vector.

Of particular interest is the case $\mathbf{K}$ representing the stiffness matrix of an unrestrained finite element assemblage. The interchange of rows and corresponding columns during the Gauss elimination always results, in this case, in associated constraint problems with positive definite stiffness matrices, until the elements in the last rows of the reduced stiffness matrix (i.e., $\mathbf{S}$ in (8.10)) all become zero. The number of rows that are made up of zero elements is equal to the number of rigid body modes in the system.

EXAMPLE 8.12: Consider the beam element in Fig. 8.10(a). The stiffness matrix of the element is

$$\mathbf{K} = \begin{bmatrix} 12 & -6 & -12 & -6 \\ -6 & 4 & 6 & 2 \\ -12 & 6 & 12 & 6 \\ -6 & 2 & 6 & 4 \end{bmatrix}$$

Show that Gauss elimination results into the third and fourth row to consist of zero elements only, and evaluate formally the rigid body mode displacements.

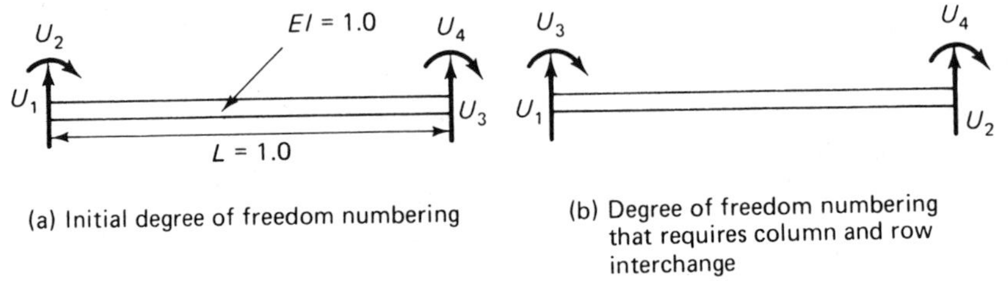

U_2
$EI = 1.0$
U_4
U_1
U_3
$L = 1.0$
(a) Initial degree of freedom numbering
U_3
U_4
U_1
U_2
(b) Degree of freedom numbering that requires column and row interchange

FIGURE 8.10 Beam element with two rigid body modes.

Using the procedure in (8.10), we expect to arrive at a matrix **S** with its last two rows consisting of zero elements only, because the beam element has two rigid body modes, corresponding to vertical translation and rotation. We have

$$\mathbf{L}_1^{-1} = \begin{bmatrix} 1 & & & \\ \frac{1}{2} & 1 & 0 & \\ 1 & 0 & 1 & \\ \frac{1}{2} & 0 & 0 & 1 \end{bmatrix}$$

Hence

$$\mathbf{L}_1^{-1}\mathbf{K} = \begin{bmatrix} 12 & -6 & -12 & -6 \\ 0 & 1 & 0 & -1 \\ 0 & 0 & 0 & 0 \\ 0 & -1 & 0 & 1 \end{bmatrix}$$

Then

$$\mathbf{L}_2^{-1} = \begin{bmatrix} 1 & & & \\ 0 & 1 & & \\ 0 & 0 & 1 & \\ 0 & 1 & 0 & 1 \end{bmatrix}$$

and

$$\mathbf{S} = \mathbf{L}_2^{-1}\mathbf{L}_1^{-1}\mathbf{K} = \begin{bmatrix} 12 & -6 & -12 & -6 \\ 0 & 1 & 0 & -1 \\ 0 & 0 & 0 & 0 \\ 0 & 0 & 0 & 0 \end{bmatrix} \qquad \text{(a)}$$

Therefore, as expected, the last two rows in **S** consist of zero elements, and $\mathbf{L}_3^{-1}$ cannot and need not be calculated. We should also note that if the numbering of the degrees of freedom of the beam element were initially as in Fig. 8.10(b), we would need to interchange rows and columns 2 with rows and columns 3 in order to be able to continue with the triangularization. This, however, is equivalent to using the degree of freedom numbering that we were concerned with in the first place, i.e., the numbering in Fig. 8.10(a).

Using the matrix **S** in (a), we can now formally evaluate the rigid body mode displacements of the beam, i.e., solve the equations $\mathbf{KU} = \mathbf{0}$ to obtain two linearly independent solutions for **U**.

First assume that $U_4 = 1$ and $U_3 = 0$; then using **S**, we obtain

$$12U_1 - 6U_2 = 6$$
$$U_2 = 1$$

Hence $\quad U_1 = 1, \quad U_2 = 1, \quad U_3 = 0, \quad U_4 = 1$

Then assume that $U_4 = 0$, $U_3 = 1$, to obtain

$$12U_1 - 6U_2 = 12$$
$$U_2 = 0$$

Hence $\quad U_1 = 1, \quad U_2 = 0, \quad U_3 = 1, \quad U_4 = 0$

It should be noted that the rigid body displacement vectors are not unique, but instead the two-dimensional subspace spanned by any two linearly independent rigid body mode vectors is unique. Therefore, any rigid body displacement vector can be written as

$$\begin{bmatrix} U_1 \\ U_2 \\ U_3 \\ U_4 \end{bmatrix} = \alpha_1 \begin{bmatrix} 1 \\ 1 \\ 0 \\ 1 \end{bmatrix} + \alpha_2 \begin{bmatrix} 1 \\ 0 \\ 1 \\ 0 \end{bmatrix}$$

where α_1 and α_2 are constants.

In the above discussion we investigated the possibility of obtaining the solution to a system of simultaneous equations by referring to the eigenvalues of the coefficient matrix and of the matrices derived from it by deleting the last rows and columns. However, we should also realize that the Gauss elimination determines implicitly the rank of the coefficient matrix; i.e., the dimension of the row and column space of the matrix is obtained (see Section 2.2). Namely, assuming that interchanges are performed in the Gauss elimination on a general coefficient matrix, the final number of nonzero rows in the reduced matrix is equal to the rank of the matrix. This follows because at each stage of the Gauss elimination a multiple of one row is subtracted from another row. Therefore, if the last i rows have only zero elements, it follows that only $n - i$ rows were linearly independent. Considering an element stiffness matrix, we therefore conclude that a stiffness matrix of order n, in which i rigid body modes have not been eliminated, has rank $n - i$. We should note that the determination of the rank of a matrix by Gauss elimination is applicable to indefinite matrices but that interchanges must be carried out.

EXAMPLE 8.13: Use the Gauss elimination procedure to factorize the matrix $\mathbf{K} - \lambda_1^{(2)}\mathbf{I}$ into an upper and lower triangular matrix, where **K** is the stiffness matrix considered in Example 8.2 and $\lambda_1^{(2)}$ is the smallest eigenvalue of the eigenproblem corresponding to the second associated constraint problem. Thus determine the rank of $\mathbf{K} - \lambda_1^{(2)}\mathbf{I}$.

We have

$$\mathbf{K} = \begin{bmatrix} 5 & -4 & 1 & 0 \\ -4 & 6 & -4 & 1 \\ 1 & -4 & 6 & -4 \\ 0 & 1 & -4 & 5 \end{bmatrix}; \qquad \mathbf{K}^{(2)} = \begin{bmatrix} 5 & -4 \\ -4 & 6 \end{bmatrix}; \qquad \lambda_1^{(2)} = 5.5 - \tfrac{1}{2}\sqrt{65}$$

Hence

$$\mathbf{K} - \lambda_1^{(2)}\mathbf{I} = \begin{bmatrix} 3.53 & -4 & 1 & 0 \\ -4 & 4.53 & -4 & 1 \\ 1 & -4 & 4.53 & -4 \\ 0 & 1 & -4 & 3.53 \end{bmatrix}$$

To set to zero all subdiagonal elements in the first column, we evaluate $\mathbf{L}_1^{-1}(\mathbf{K} - \lambda_1^{(2)}\mathbf{I})$, where

$$\mathbf{L}_1^{-1} = \begin{bmatrix} 1 & & & \\ 1.13 & 1 & & \\ -0.283 & & 1 & \\ 0 & & & 1 \end{bmatrix}$$

and hence

$$\mathbf{L}_1^{-1}(\mathbf{K} - \lambda_1^{(2)}\mathbf{I}) = \begin{bmatrix} 3.53 & -4 & 1 & 0 \\ 0 & 0 & -2.87 & 1 \\ 0 & -2.87 & 4.25 & -4 \\ 0 & 1 & -4 & 3.53 \end{bmatrix} \tag{a}$$

To proceed with the Gauss factorization we need to interchange the second row with either the third or fourth row. Switching rows 2 and 3, we have

$$\bar{\mathbf{L}}_1^{-1}(\bar{\mathbf{K}} - \lambda_1^{(2)}\bar{\mathbf{I}}) = \begin{bmatrix} 3.53 & -4 & 1 & 0 \\ 0 & -2.87 & 4.25 & -4 \\ 0 & 0 & -2.87 & 1 \\ 0 & 1 & -4 & 3.53 \end{bmatrix}$$

where

$$\bar{\mathbf{K}} = \begin{bmatrix} 5 & -4 & 1 & 0 \\ 1 & -4 & 6 & -4 \\ -4 & 6 & -4 & 1 \\ 0 & 1 & -4 & 5 \end{bmatrix}; \qquad \bar{\mathbf{L}}_1^{-1} = \begin{bmatrix} 1 & & & \\ -0.283 & 1 & & \\ 1.13 & & 1 & \\ 0 & & & 1 \end{bmatrix};$$

$$\bar{\mathbf{I}} = \begin{bmatrix} 1 & 0 & 0 & 0 \\ 0 & 0 & 1 & 0 \\ 0 & 1 & 0 & 0 \\ 0 & 0 & 0 & 1 \end{bmatrix}$$

We can now complete the factorization, using

$$\mathbf{L}_2^{-1} = \begin{bmatrix} 1 & & & \\ & 1 & & \\ & 0 & 1 & \\ & 0.349 & & 1 \end{bmatrix}$$

$$\text{We obtain} \quad \mathbf{L}_2^{-1}\bar{\mathbf{L}}_1^{-1}(\bar{\mathbf{K}} - \lambda_1^{(2)}\bar{\mathbf{I}}) = \begin{bmatrix} 3.53 & -4 & 1 & 0 \\ 0 & -2.87 & 4.25 & -4 \\ 0 & 0 & -2.87 & 1 \\ 0 & 0 & -2.52 & 2.14 \end{bmatrix}$$

$$\text{and} \quad \mathbf{L}_3^{-1} = \begin{bmatrix} 1 & & & \\ & 1 & & \\ & & 1 & \\ & & -0.878 & 1 \end{bmatrix}$$

Hence, defining $\mathbf{S} = \mathbf{L}_3^{-1}\mathbf{L}_2^{-1}\bar{\mathbf{L}}_1^{-1}(\bar{\mathbf{K}} - \lambda_1^{(2)}\bar{\mathbf{I}})$, we have

$$\mathbf{S} = \begin{bmatrix} 3.53 & -4 & 1 & 0 \\ & -2.87 & 4.25 & -4 \\ & & -2.87 & 1 \\ & & & 1.26 \end{bmatrix}$$

Furthermore, we have $(\bar{\mathbf{K}} - \lambda_1^{(2)}\bar{\mathbf{I}}) = \mathbf{LS}$, where

$$\mathbf{L} = \begin{bmatrix} 1 & & & \\ 0.283 & 1 & & \\ -1.13 & 0 & 1 & \\ 0 & -0.349 & 0.878 & 1 \end{bmatrix}$$

We should note that the matrix $\bar{\mathbf{K}} - \lambda_1^{(2)}\bar{\mathbf{I}}$ is nonsymmetric, and therefore we cannot write $\mathbf{S} = \mathbf{DL}^T$. However, if we would have interchanged not only rows but also the corresponding columns, the coefficient matrix would have remained symmetric. To determine the rank of $\mathbf{K} - \lambda_1^{(2)}\mathbf{I}$, we note that no zero rows have been obtained in $\mathbf{S}$; hence the rank of $\mathbf{K} - \lambda_1^{(2)}\mathbf{I}$ is four.

Now, interchanging the second and third rows and columns of the matrix in (a), we obtain

$$\bar{\mathbf{L}}_1^{-1}(\hat{\mathbf{K}} - \lambda_1^{(2)}\mathbf{I}) = \begin{bmatrix} 3.53 & 1 & -4 & 0 \\ 0 & 4.25 & -2.87 & -4 \\ 0 & -2.87 & 0 & 1 \\ 0 & -4 & 1 & 3.53 \end{bmatrix}$$

$$\text{where} \quad \hat{\mathbf{K}} = \begin{bmatrix} 5 & 1 & -4 & 0 \\ 1 & 6 & -4 & -4 \\ -4 & -4 & 6 & 1 \\ 0 & --4 & 1 & 5 \end{bmatrix}; \qquad \bar{\mathbf{L}}_1^{-1} = \begin{bmatrix} 1 & & & \\ -0.283 & 1 & & \\ 1.13 & & 1 & \\ 0 & & & 1 \end{bmatrix}$$

$$\text{In the next step we use} \quad \mathbf{L}_2^{-1} = \begin{bmatrix} 1 & & & \\ & 1 & & \\ & 0.675 & 1 & \\ & 0.942 & & 1 \end{bmatrix}$$

$$\text{and obtain} \quad \mathbf{L}_2^{-1}\bar{\mathbf{L}}_1^{-1}(\hat{\mathbf{K}} - \lambda_1^{(2)}\mathbf{I}) = \begin{bmatrix} 3.53 & 1 & -4 & 0 \\ 0 & 4.25 & -2.87 & -4 \\ 0 & 0 & -1.94 & -1.70 \\ 0 & 0 & -1.70 & -0.24 \end{bmatrix}$$

The factorization is completed using

$$\mathbf{L}_3^{-1} = \begin{bmatrix} 1 & & & \\ & 1 & & \\ & & 1 & \\ & & -0.878 & 1 \end{bmatrix}$$

$$\mathbf{L}_3^{-1}\mathbf{L}_2^{-1}\bar{\mathbf{L}}_1^{-1}(\hat{\mathbf{K}} - \lambda_1^{(2)}\mathbf{I}) = \begin{bmatrix} 3.53 & 1 & -4 & 0 \\ & 4.25 & -2.87 & -4 \\ & & -1.94 & -1.70 \\ & & & 1.26 \end{bmatrix}$$

Hence we have $(\hat{\mathbf{K}} - \lambda_1^{(2)}\mathbf{I}) = \mathbf{LS}$, where

$$\mathbf{L} = \begin{bmatrix} 1 & & & \\ 0.283 & 1 & & \\ -1.13 & -0.675 & 1 & \\ 0 & -0.942 & 0.878 & 1 \end{bmatrix}; \qquad \mathbf{S} = \begin{bmatrix} 3.53 & 1 & -4 & 0 \\ & 4.25 & -2.87 & -4 \\ & & -1.94 & -1.70 \\ & & & 1.26 \end{bmatrix}$$

and, as may be checked, $\mathbf{S} = \mathbf{DL}^T$.

So far, we have implicitly assumed in the discussion that all calculations are performed in exact arithmetic, although in some example solutions, the calculations have been performed to an accuracy of only a few digits. In general, using the digital computer in practical calculations, diagonal elements exactly equal to zero are hardly ever generated in the Gauss triangular factorization of a stiffness matrix. Instead, owing to the finite precision arithmetic of the computer, diagonal elements which are small in relation to the other diagonal elements and which cause large multipliers would nearly always be calculated. Therefore, a breakdown of Gauss elimination as considered above will hardly ever occur. However, if small diagonal elements are used, the errors in the factorization can be large, and this is really what we are interested in. The estimation of errors in the solution of simultaneous equations is discussed further in Section 8.5.

8.3 DIRECT SOLUTIONS USING ORTHOGONAL MATRICES

In Section 8.2.2 we have been writing Gauss elimination as premultiplication of $\mathbf{K}$ by matrices $\mathbf{L}_i^{-1}$, $i = 1, 2, \ldots, n-1$, where n is the order of $\mathbf{K}$ and $\mathbf{L}_i^{-1}$ was defined in (8.11). In fact, each matrix $\mathbf{L}_i^{-1}$ is a product of elementary matrices

$$\mathbf{L}_i^{-1} = \mathbf{L}_{n,i}^{-1}\,\mathbf{L}_{n-1,i}^{-1} \ldots \mathbf{L}_{i+1,i}^{-1} \tag{8.44}$$

so that (8.10) becomes

$$\mathbf{L}_{n,n-1}^{-1} \ldots (\mathbf{L}_{n,2}^{-1}\,\mathbf{L}_{n-1,2}^{-1} \ldots \mathbf{L}_{3,2}^{-1})(\mathbf{L}_{n,1}^{-1}\,\mathbf{L}_{n-1,1}^{-1} \ldots \mathbf{L}_{2,1}^{-1})\mathbf{K} = \mathbf{S} \tag{8.45}$$

where

$$\mathbf{L}_{k,i}^{-1} = \begin{bmatrix} 1 & & & & & \\ & \ddots & & & & \\ & & 1 & & & \\ & & 0 & \ddots & & \\ & & \vdots & & & \\ & & 0 & & & \\ & & -l_{ki} & & \ddots & \\ & & 0 & & & \\ & & \vdots & & \ddots & \\ & & 0 & & & 1 \end{bmatrix}; \quad k = i+1, \ldots, n \tag{8.46}$$

and the Gauss multiplying factor $l_{i+j,i}$ was defined in (8.11). This definition showed that the reduction cannot be continued if $k_{ii}^{(i)}$ is zero, and numerical difficulties should be expected if $k_{ii}^{(i)}$ is small, or rather $l_{i+j,i}$ is large.

In practical finite element analysis using a direct solution scheme, the Gauss solution algorithm is nearly always used in one form or the other, because the procedure is cost-effective and numerically sufficiently stable. However, it should be realized that a direct solution can also be achieved using other techniques. The objective in this section is to briefly summarize two additional direct solution methods.

The basic function of each elementary matrix $\mathbf{L}_{j,i}^{-1}$ is to reduce an element of the stiffness matrix (considered in step i) to zero. Specifically, the elementary matrix $\mathbf{L}_{j,i}^{-1}$ reduces the element $k_{ji}^{(i)}$ to zero. As might be expected, there is no reason why instead of elementary matrices we could not use orthogonal matrices to reduce the element $k_{ji}^{(i)}$ to zero.[7] In Section 2.5 we defined rotation and reflection matrices as two different orthogonal matrices, and both can be used to reduce $\mathbf{K}$ into upper triangular form.

8.3.1 The Givens Factorization

In the Givens factorization, rotation matrices are used to reduce $\mathbf{K}$ into upper triangular form. In effect, each elementary matrix $\mathbf{L}_{j,i}^{-1}$ in (8.45) is replaced by a rotation matrix, so that we have

$$\mathbf{P}_G^T \mathbf{K} = \mathbf{S}_G \tag{8.47}$$

where
$$\mathbf{P}_G^T = \mathbf{P}_{n,n-1}^T \ldots (\mathbf{P}_{n,2}^T \mathbf{P}_{n-1,2}^T \ldots \mathbf{P}_{3,2}^T)(\mathbf{P}_{n,1}^T \mathbf{P}_{n-1,1}^T \ldots \mathbf{P}_{2,1}^T) \tag{8.48}$$

and $\mathbf{S}_G$ is an upper triangular matrix; the subscript G denotes that $\mathbf{S}_G$ is not equal to $\mathbf{S}$ in (8.10). The solution of the system of equations $\mathbf{KU} = \mathbf{R}$ in (8.1) is obtained using the decomposition of $\mathbf{K}$ established in (8.47),

$$\mathbf{K} = \mathbf{P}_G \mathbf{S}_G \tag{8.49}$$

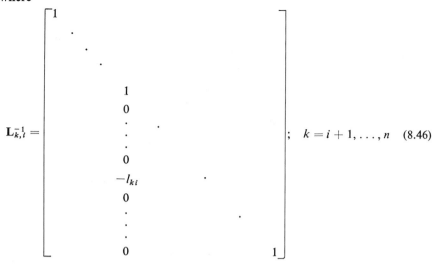

Hence, analogously to the Gauss elimination solution, we need to reduce the load vector $\mathbf{R}$,

$$
\mathbf{P}_G \mathbf{V} = \mathbf{R} \tag{8.50}
$$

from which

$$
\mathbf{V} = \mathbf{P}_G^T \mathbf{R} \tag{8.51}
$$

and obtain the solution $\mathbf{U}$ by a back-substitution,

$$
\mathbf{S}_G \mathbf{U} = \mathbf{V} \tag{8.52}
$$

An important aspect is the evaluation of the rotation matrices $\mathbf{P}_{j,i}^T$, which must reduce to zero the elements in positions $(j, i), j > i$, of the coefficient matrix. A typical rotation matrix is given by

$$
\mathbf{P}_{j,i}^T = \left[\begin{array}{ccccccccccccccc}
1 &&&&&&&&&&&&&& \\
& \cdot &&&&&&&&&&&&& \\
&& \cdot &&&&&&&&&&&& \\
&&& 1 &&&&&&&&&&& \\
&&&& \overset{i\text{th}}{\cos\theta} &&&&&& \overset{j\text{th column}}{\sin\theta} &&&& \\
&&&&& 1 &&&&&&&&& \\
&&&&&& \cdot &&&&&&&& \\
&&&&&&& \cdot &&&&&&& \\
&&&&&&&& \cdot &&&&&& \\
&&&&&&&&& 1 &&&&& \\
&&&& -\sin\theta &&&&&& \cos\theta &&&& \\
&&&&&&&&&&& 1 &&& \\
&&&&&&&&&&&& \cdot && \\
&&&&&&&&&&&&& \cdot & \\
&&&&&&&&&&&&&& 1
\end{array}\right]
\begin{array}{l}
\\ \\ \\ \\
\text{— } i\text{th} \\
\\ \\ \\ \\ \\
\text{— } j\text{th row} \\
\\ \\ \\ \\
\end{array}
\tag{8.53}
$$

where θ is chosen in such way as to reduce to zero element (j, i).

Before deriving a typical matrix $P_{j,i}^T$, we should note the order in which the elements in the stiffness matrix are zeroed; namely, (8.48) shows that we use

$$
\begin{aligned}
& i = 1, \qquad j = 2, 3, \ldots, n \\
\text{and then} \qquad & i = 2, \qquad j = 3, 4, \ldots, n \\
& \cdot \\
& \cdot \\
& \cdot \\
& i = n - 1, \qquad j = n
\end{aligned}
$$

Consider now the construction of the matrix $\mathbf{P}_{2,1}^T$, which is typical. The purpose of evaluating the matrix product $\tilde{\mathbf{K}} = \mathbf{P}_{2,1}^T \mathbf{K}$ is to reduce element (2, 1) in the original stiffness matrix to zero; i.e., we have

$$
\tilde{\mathbf{K}} = \begin{bmatrix}
\cos\theta & \sin\theta &&&&& \\
-\sin\theta & \cos\theta &&&&& \\
&& 1 &&&& \\
&&& \cdot &&& \\
&&&& \cdot && \\
&&&&& \cdot & \\
&&&&&& 1
\end{bmatrix}
\begin{bmatrix}
k_{11} & k_{12} & k_{13} & \cdots \\
k_{21} & k_{22} & k_{23} & \\
k_{31} & k_{32} & k_{33} & \\
\cdot &&& \\
\cdot &&& \\
\cdot &&&
\end{bmatrix}
\tag{8.54}
$$

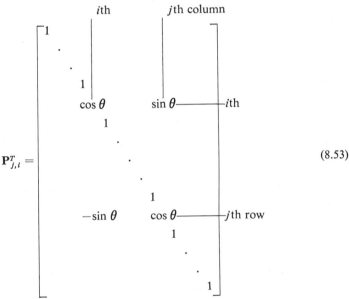

or

$$\tilde{\mathbf{K}} = \begin{bmatrix} \tilde{k}_{11} & \tilde{k}_{12} & \tilde{k}_{13} & \cdots \\ 0 & \tilde{k}_{22} & \tilde{k}_{23} & \\ \tilde{k}_{31} & \tilde{k}_{32} & \tilde{k}_{33} & \\ \cdot & & & \\ \cdot & & & \\ \cdot & & & \end{bmatrix}$$

The angle θ is chosen from the equation

$$-k_{11} \sin \theta + k_{21} \cos \theta = 0 \tag{8.55}$$

Hence

$$\cos \theta = \frac{k_{11}}{(k_{11}^2 + k_{21}^2)^{1/2}}; \qquad \sin \theta = \frac{k_{21}}{(k_{11}^2 + k_{21}^2)^{1/2}} \tag{8.56}$$

and if $k_{11}^2 + k_{21}^2 = 0$, no transformation is required; i.e., we use

$$\cos \theta = 1$$
$$\sin \theta = 0$$

It should be noted that it is not required that $k_{11} \neq 0.0$, as in the Gauss elimination procedure.

The premultiplication of the stiffness matrix $\mathbf{K}$ by the matrix $\mathbf{P}_{2,1}^T$ corresponds to linearly combining the first two rows of $\mathbf{K}$; i.e., for the elements of the resulting matrix $\tilde{\mathbf{K}}$, we have

$$\left.\begin{aligned} \tilde{k}_{1j} &= k_{1j} \cos \theta + k_{2j} \sin \theta \\ \tilde{k}_{2j} &= -k_{1j} \sin \theta + k_{2j} \cos \theta \end{aligned}\right\} \; j = 1, \ldots, n \tag{8.57}$$
$$\tilde{k}_{ij} = k_{ij} \qquad i = 3, \ldots, n; \qquad j = 1, \ldots, n$$

which completes the operations with $\mathbf{P}_{2,1}^T$.

The elimination of the other subdiagonal elements, i.e., the evaluation of the other rotation matrices and the required matrix products, is carried out in an analogous way. We demonstrate the procedure in the following example.

EXAMPLE 8.14: Use the Givens factorization to obtain the solution to the system of equations considered in Section 8.2.1, which rewritten are

$$\begin{bmatrix} 5 & -4 & 1 & 0 \\ -4 & 6 & -4 & 1 \\ 1 & -4 & 6 & -4 \\ 0 & 1 & -4 & 5 \end{bmatrix} \begin{bmatrix} U_1 \\ U_2 \\ U_3 \\ U_4 \end{bmatrix} = \begin{bmatrix} 0 \\ 1 \\ 0 \\ 0 \end{bmatrix}$$

Using the Givens method, we have for $i = 1, j = 2$,

$$\mathbf{P}_{2,1}^T = \begin{bmatrix} 0.780869 & -0.624696 & & \\ 0.624695 & 0.780869 & & \\ & & 1 & \\ & & & 1 \end{bmatrix}$$

Hence

$$\mathbf{P}_{2,1}^T \mathbf{K} = \begin{bmatrix} 6.403124 & -6.871646 & 3.279649 & -0.624695 \\ 0 & 2.186433 & -2.498780 & 0.780869 \\ 1 & -4 & 6 & -4 \\ 0 & 1 & -4 & 5 \end{bmatrix};$$

$$\mathbf{P}_{2,1}^T\mathbf{R} = \begin{bmatrix} -0.624695 \\ 0.780869 \\ 0 \\ 0 \end{bmatrix}$$

For the next step $i = 1, j = 3$, we use

$$\mathbf{P}_{3,1}^T = \begin{bmatrix} 0.988024 & 0 & 0.154303 & 0 \\ 0 & 1 & 0 & 0 \\ -0.154303 & 0 & 0.988024 & 0 \\ 0 & 0 & 0 & 1 \end{bmatrix}$$

Hence

$$\mathbf{P}_{3,1}^T\mathbf{P}_{2,1}^T\mathbf{K} = \begin{bmatrix} 6.480741 & -7.406561 & 4.166190 & -1.234427 \\ 0 & 2.186433 & -2.498780 & 0.780869 \\ 0 & -2.891776 & 5.422080 & -3.855702 \\ 0 & 1 & -4 & 5 \end{bmatrix}$$

$$\mathbf{P}_{3,1}^T\mathbf{P}_{2,1}^T\mathbf{R} = \begin{bmatrix} -0.617213 \\ 0.780869 \\ 0.096393 \\ 0 \end{bmatrix}$$

Writing down the results for the next steps:

$$i = 2, j = 3; \qquad \cos\theta = 0.603103, \sin\theta = -0.797664$$

$$\mathbf{P}_{3,2}^T\mathbf{P}_{3,1}^T\mathbf{P}_{2,1}^T\mathbf{K} = \begin{bmatrix} 6.480741 & -7.406561 & 4.166190 & -1.234427 \\ 0 & 3.625308 & -5.832017 & 3.547497 \\ 0 & 0 & 1.276885 & -1.702513 \\ 0 & 1 & -4 & 5 \end{bmatrix}$$

$$\mathbf{P}_{3,2}^T\mathbf{P}_{3,1}^T\mathbf{P}_{2,1}^T\mathbf{R} = \begin{bmatrix} -0.617213 \\ 0.394055 \\ 0.681005 \\ 0 \end{bmatrix}$$

$$i = 2, j = 4; \qquad \cos\theta = 0.963998, \sin\theta = 0.265908$$

$$\mathbf{P}_{4,2}^T\mathbf{P}_{3,2}^T\mathbf{P}_{3,1}^T\mathbf{P}_{2,1}^T\mathbf{K} = \begin{bmatrix} 6.480741 & -7.406561 & 4.166190 & -1.234427 \\ 0 & 3.760699 & -6.685687 & 4.748357 \\ 0 & 0 & 1.276885 & -1.702513 \\ 0 & 0 & -2.305214 & 3.876950 \end{bmatrix}$$

$$\mathbf{P}_{4,2}^T\mathbf{P}_{3,2}^T\mathbf{P}_{3,1}^T\mathbf{P}_{2,1}^T\mathbf{R} = \begin{bmatrix} -0.617213 \\ 0.379869 \\ 0.681005 \\ -0.104782 \end{bmatrix}$$

In the last step we have

$$i = 3, j = 4; \qquad \cos\theta = 0.484544, \sin\theta = -0.874767$$

and obtain

$$\mathbf{P}_G^T\mathbf{K} = \begin{bmatrix} 6.480741 & -7.406561 & 4.166190 & -1.234427 \\ 0 & 3.760699 & -6.685687 & 4.748357 \\ 0 & 0 & 2.635231 & -4.216370 \\ 0 & 0 & 0 & 0.389249 \end{bmatrix} \tag{a}$$

$$\mathbf{P}_G^T\mathbf{R} = \begin{bmatrix} -0.617213 \\ 0.379869 \\ 0.421637 \\ 0.544949 \end{bmatrix} \tag{b}$$

The relations in (a) and (b) give $\mathbf{S}_G$ and $\mathbf{V}$, respectively. Using (8.52) for the back-substitution, we obtain

$$\mathbf{U} = \begin{bmatrix} 1.6 \\ 2.6 \\ 2.4 \\ 1.4 \end{bmatrix}$$

which is the solution already obtained in Section 8.2.1.

It should be noted that $\mathbf{S}_G = \mathbf{P}_G^T\mathbf{K}$ given in (a) is different from $\mathbf{S}$ obtained in Example 8.2, and that symmetry in the coefficient matrix is lost during the Givens reduction process.

8.3.2 The Householder Factorization

In the Householder reduction of $\mathbf{K}$ to upper triangular form, reflection matrices are used; i.e., we evaluate

$$\mathbf{P}_H^T\,\mathbf{K} = \mathbf{S}_H \tag{8.58}$$

where

$$\mathbf{P}_H^T = \mathbf{P}_{n-1}^T \ldots \mathbf{P}_2^T\,\mathbf{P}_1^T \tag{8.59}$$

and $\mathbf{S}_H$ is an upper triangular matrix. We shall prove in Example 8.15 that the upper triangular matrices $\mathbf{S}_G$ and $\mathbf{S}_H$ and orthogonal matrices $\mathbf{P}_G$ and $\mathbf{P}_H$ used in (8.47) and (8.58), respectively, are "essentially" the same. The matrices $\mathbf{P}_i^T$, $i = 1, \ldots, n-1$, are reflection matrices constructed in such a way that $\mathbf{P}_i^T$ reduces to zero the subdiagonal elements in column i of the current stiffness matrix. More specifically,

$$\mathbf{P}_i^T = \left[\begin{array}{c|c} \mathbf{I}_{i-1} & \mathbf{0} \\ \hline \mathbf{0} & \tilde{\mathbf{P}}_i \end{array}\right] \tag{8.60}$$

where $\mathbf{I}_{i-1}$ is the identity matrix of order $i-1$. Also,

$$\tilde{\mathbf{P}}_i = \mathbf{I} - \theta\mathbf{w}_i\mathbf{w}_i^T; \qquad \theta = \frac{2}{\mathbf{w}_i^T\mathbf{w}_i}$$

where $\tilde{\mathbf{P}}_i$ is a matrix and $\mathbf{w}_i$ is a vector of order $n - i + 1$. Since $\tilde{\mathbf{P}}_i$ is symmetric, we have that $\mathbf{P}_i$ is symmetric and $\mathbf{P}_i^T = \mathbf{P}_i$.

Once the factorization of $\mathbf{K}$ has been obtained, where $\mathbf{K} = \mathbf{P}_H\mathbf{S}_H$, the solution of the equations $\mathbf{KU} = \mathbf{R}$ is carried out as expressed in (8.50) to (8.52) (and replacing $\mathbf{S}_G$ by $\mathbf{S}_H$ and $\mathbf{P}_G$ by $\mathbf{P}_H$).

To demonstrate the reduction procedure of $\mathbf{K}$, consider the first step, i.e., the construction of $\mathbf{P}_1$, which is typical. Let

$$\tilde{\mathbf{K}} = \mathbf{P}_1\mathbf{K} \tag{8.61}$$

and partition $\mathbf{K}$ as follows, $$\mathbf{K} = [\mathbf{k}_1 \mid \mathbf{K}_1] \tag{8.62}$$

where $\mathbf{k}_1$ is the first column of $\mathbf{K}$. We also have

$$\mathbf{P}_1 = \mathbf{I} - \theta \mathbf{w}_1 \mathbf{w}_1^T \tag{8.63}$$

Since the requirement is that the subdiagonal elements in column 1 of $\tilde{\mathbf{K}}$ are zero, we have the condition that $\mathbf{P}_1\mathbf{k}_1$ is a vector with only the first element nonzero. This condition is fulfilled by choosing the vector $\mathbf{w}_1$ in the following way,

$$(\mathbf{I} - \theta \mathbf{w}_1 \mathbf{w}_1^T)\mathbf{k}_1 = \pm \|\mathbf{k}_1\|_2 \mathbf{e}_1 \tag{8.64}$$

where $\mathbf{e}_1$ is the n-dimensional unit vector. The relation (8.64) corresponds to a reflection of the vector $\mathbf{k}_1$ in the n-dimensional space in such a way that only the component corresponding to direction 1 is nonzero and is equal to the length of $\mathbf{k}_1$ (see Example 2.13).

To solve for $\mathbf{w}_1$, we obtain, from (8.64),

$$\mathbf{k}_1 - \bar{\theta}\mathbf{w}_1 = \pm \|\mathbf{k}_1\|_2 \mathbf{e}_1 \tag{8.65}$$

where $\bar{\theta}$ is a constant, $\bar{\theta} = \theta \mathbf{w}_1^T \mathbf{k}_1$. Since $\mathbf{w}_1$ determines only the direction of the reflection (see Fig. 2.4), the vector may have arbitrary length and, for convenience, we set $\bar{\theta} = 1.0$. We thus obtain

$$\mathbf{w}_1 = \mathbf{k}_1 + \text{sign}\,(k_{11})\,\|\mathbf{k}_1\|_2 \mathbf{e}_1 \tag{8.66}$$

where the $+$ or $-$ sign in (8.65) is chosen with the objective that the absolute value of the first element in $\mathbf{w}_1$ is by $\|\mathbf{k}_1\|_2$ larger than the absolute value of k_{11}.

Considering now the calculation of $\tilde{\mathbf{K}}$ in (8.61), we should note that $\mathbf{P}_1$ is actually never evaluated explicitly, but we calculate

$$\mathbf{v}_1^T = \mathbf{w}_1^T \mathbf{K} \tag{8.67}$$

and then $$\tilde{\mathbf{K}} = \mathbf{K} - \theta(\mathbf{w}_1 \mathbf{v}_1^T) \tag{8.68}$$

The other reflection matrices, $\mathbf{P}_2, \ldots, \mathbf{P}_{n-1}$, and required matrix products are obtained in a manner analogous to the evaluation of $\mathbf{P}_1$ and $\tilde{\mathbf{K}}$. We demonstrate the factorization in Example 8.16.

EXAMPLE 8.15: Show that the upper triangular matrices $\mathbf{S}_G$ and $\mathbf{S}_H$ and orthogonal matrices $\mathbf{P}_G$ and $\mathbf{P}_H$ are "essentially" the same. The terminology "essentially" means that the rows in $\mathbf{S}_G$ and $\mathbf{S}_H$ and columns in $\mathbf{P}_G$ and $\mathbf{P}_H$ are the same apart from a multiplier of -1.

The relations in (8.49) and (8.58) give

$$\mathbf{K} = \mathbf{P}_G \mathbf{S}_G$$

and $$\mathbf{K} = \mathbf{P}_H \mathbf{S}_H$$

Hence $$\mathbf{P}_G \mathbf{S}_G = \mathbf{P}_H \mathbf{S}_H$$

or $$\mathbf{P}_H^T \mathbf{P}_G = \mathbf{S}_H \mathbf{S}_G^{-1} = \mathbf{U} \tag{a}$$

where $\mathbf{U}$ must be an upper triangular matrix because $\mathbf{S}_H$ and $\mathbf{S}_G^{-1}$ are upper triangular matrices. We assumed in (a) that $\mathbf{S}_G$ is nonsingular, which is the case if $\mathbf{K}$ is positive definite. Using (a) we obtain

$$\mathbf{U}^T \mathbf{U} = \mathbf{I} \tag{b}$$

because $\mathbf{P}_H$ and $\mathbf{P}_G$ are orthogonal matrices; i.e.,

$$(\mathbf{P}_H^T\mathbf{P}_G)^T(\mathbf{P}_H^T\mathbf{P}_G) = \mathbf{I}$$

However, if we now equate elements of the matrices on either side of the relation in (b), we find that $\mathbf{U}$ must be a diagonal matrix with its diagonal elements $+1$ or -1. Using the relation in (a), we therefore conclude that $\mathbf{P}_H$ and $\mathbf{P}_G$ are equal except for a multiplier of -1 applied to their columns, and $\mathbf{S}_H$ and $\mathbf{S}_G$ are equal except for a multiplier of -1 applied to their rows.

EXAMPLE 8.16: Use the Householder factorization to solve the equations considered in Example 8.14.

In the first step we have

$$\mathbf{w}_1 = \begin{bmatrix} 11.480741 \\ -4.000000 \\ 1.000000 \\ 0 \end{bmatrix}$$

and thus

$$\mathbf{P}_1^T\mathbf{K} = \begin{bmatrix} -6.480741 & 7.406561 & -4.166190 & 1.234427 \\ 0 & 2.025845 & -2.200050 & 0.569914 \\ 0 & -3.006461 & 5.550012 & -3.892478 \\ 0 & 1.000000 & -4.000000 & 5.000000 \end{bmatrix}$$

$$\mathbf{P}_1^T\mathbf{R} = \begin{bmatrix} 0.617213 \\ 0.784957 \\ 0.053761 \\ 0 \end{bmatrix}$$

In the second step we obtain

$$\mathbf{w}_2 = \begin{bmatrix} 5.786544 \\ -3.006461 \\ 1.000000 \end{bmatrix}$$

$$\mathbf{P}_2^T\mathbf{P}_1^T\mathbf{K} = \begin{bmatrix} -6.480741 & 7.406561 & -4.166190 & 1.234427 \\ 0 & -3.760699 & 6.685687 & -4.748357 \\ 0 & 0 & 0.933332 & -1.129313 \\ 0 & 0 & -2.464414 & 4.080924 \end{bmatrix}$$

$$\mathbf{P}_2^T\mathbf{P}_1^T\mathbf{R} = \begin{bmatrix} 0.617213 \\ -0.379869 \\ 0.658958 \\ -0.201299 \end{bmatrix}$$

And finally we have

$$\mathbf{w}_3 = \begin{bmatrix} 3.568564 \\ -2.464414 \end{bmatrix}$$

$$\mathbf{P}_3^T\mathbf{P}_2^T\mathbf{P}_1^T\mathbf{K} = \begin{bmatrix} -6.480741 & 7.406561 & -4.166190 & 1.234427 \\ 0 & -3.760699 & 6.685687 & -4.748357 \\ 0 & 0 & -2.635231 & 4.216370 \\ 0 & 0 & 0 & 0.389249 \end{bmatrix}$$

$$\mathbf{P}_3^T\mathbf{P}_2^T\mathbf{P}_1^T\mathbf{R} = \begin{bmatrix} 0.617213 \\ -0.379869 \\ -0.421637 \\ 0.544949 \end{bmatrix}$$

Now back-substituting, we obtain

$$\mathbf{U} = \begin{bmatrix} 1.6 \\ 2.6 \\ 2.4 \\ 1.4 \end{bmatrix}$$

As mentioned previously, the Givens and Householder factorizations are not frequently used in practice, primarily because the solution procedures are uneconomical when compared to the Gauss reduction scheme. During the factorization of a stiffness matrix, the symmetry and bandwidth properties are lost. An advantage of both the Givens and Householder factorizations is that they are numerically very stable and might be used effectively in the solution of ill-conditioned sets of equations. However, the main purpose of presenting the Givens and Householder factorization procedures was to introduce the use of rotation and reflection matrices, because we shall see in Chapter 11 that rotation and reflection matrices are used effectively in the solution of eigenproblems.

8.4 THE GAUSS–SEIDEL ITERATIVE SOLUTION

At this time practically all finite element large-scale computer programs use some form of Gauss elimination to solve the equilibrium equations $\mathbf{KU} = \mathbf{R}$. However, it is interesting to note that during the initial developments of the finite element method, iterative solution algorithms have been employed extensively, and much research has been spent on improving various iterative solution schemes.[4-6,35-44] A basic disadvantage of an iterative solution is that the time of solution can only be estimated very approximately, because the number of iterations required for convergence depends on the condition number of the matrix $\mathbf{K}$ and whether effective acceleration factors are used. A direct solution using one of the techniques given in Section 8.2 is nearly always most effective. However, at least for instructive purposes, it is important to be familiar also with an iterative solution method that has been used extensively, namely, the Gauss-Seidel method. The objective in this section is to present briefly the basic formulas of the Gauss–Seidel iterative solution together with a physical interpretation of the procedure.

In the iterative solution of $\mathbf{KU} = \mathbf{R}$, it is necessary to use an initial estimate for the displacements $\mathbf{U}$, say $\mathbf{U}^{(1)}$, which, if no better value is known, may be a null vector. In the Gauss-Seidel iteration we then evaluate for $s = 1, 2, \ldots$:

$$U_i^{(s+1)} = k_{ii}^{-1}\left\{R_i - \sum_{j=1}^{i-1} k_{ij}U_j^{(s+1)} - \sum_{j=i+1}^{n} k_{ij}U_j^{(s)}\right\} \tag{8.69}$$

where $U_i^{(s)}$ and R_i are the ith component of $\mathbf{U}$ and $\mathbf{R}$, and s indicates the cycle of iteration. Alternatively, we may write in matrix form,

$$\mathbf{U}^{(s+1)} = \mathbf{K}_D^{-1}\{\mathbf{R} - \mathbf{K}_L\mathbf{U}^{(s+1)} - \mathbf{K}_L^T\mathbf{U}^{(s)}\} \tag{8.70}$$

where $\mathbf{K}_D$ is a diagonal matrix, $\mathbf{K}_D = \text{diag}\,(k_{ii})$, and $\mathbf{K}_L$ is a lower triangular matrix with the elements k_{ij} such that:

$$\mathbf{K} = \mathbf{K}_L + \mathbf{K}_D + \mathbf{K}_L^T \tag{8.71}$$

The iteration is continued until the change in the current estimate of the displacement vector is small enough, i.e., until

$$\frac{\|\mathbf{U}^{(s+1)} - \mathbf{U}^{(s)}\|_2}{\|\mathbf{U}^{(s+1)}\|_2} < \epsilon \tag{8.72}$$

where ϵ is the convergence tolerance. The number of iterations depends on the "quality" of the starting vector $\mathbf{U}^{(1)}$ and on the conditioning of the matrix $\mathbf{K}$. But it is important to note that the iteration will always converge, provided that $\mathbf{K}$ is positive definite. Furthermore, the rate of convergence can be increased using overrelaxation, in which case the iteration is as follows,

$$\mathbf{U}^{(s+1)} = \mathbf{U}^{(s)} + \beta\mathbf{K}_D^{-1}\{\mathbf{R} - \mathbf{K}_L\mathbf{U}^{(s+1)} - \mathbf{K}_D\mathbf{U}^{(s)} - \mathbf{K}_L^T\mathbf{U}^{(s)}\} \tag{8.73}$$

where β is the overrelaxation factor. The optimum value of β depends on the matrix $\mathbf{K}$ but is usually between 1.3 and 1.9.

EXAMPLE 8.17: Use the Gauss-Seidel iteration to solve the system of equations considered in Section 8.2.1.

The equations to be solved are

$$\begin{bmatrix} 5 & -4 & 1 & 0 \\ -4 & 6 & -4 & 1 \\ 1 & -4 & 6 & -4 \\ 0 & 1 & -4 & 5 \end{bmatrix} \begin{bmatrix} U_1 \\ U_2 \\ U_3 \\ U_4 \end{bmatrix} = \begin{bmatrix} 0 \\ 1 \\ 0 \\ 0 \end{bmatrix}$$

In the solution we use (8.73), which in this case becomes

$$\begin{bmatrix} U_1 \\ U_2 \\ U_3 \\ U_4 \end{bmatrix}^{(s+1)} = \begin{bmatrix} U_1 \\ U_2 \\ U_3 \\ U_4 \end{bmatrix}^{(s)} + \beta \begin{bmatrix} \frac{1}{5} & & & \\ & \frac{1}{6} & & \\ & & \frac{1}{6} & \\ & & & \frac{1}{5} \end{bmatrix}$$

$$\times \left\{ \begin{bmatrix} 0 \\ 1 \\ 0 \\ 0 \end{bmatrix} - \begin{bmatrix} 0 & & & \\ -4 & 0 & & \\ 1 & -4 & 0 & \\ 0 & 1 & -4 & 0 \end{bmatrix} \begin{bmatrix} U_1 \\ U_2 \\ U_3 \\ U_4 \end{bmatrix}^{(s+1)} - \begin{bmatrix} 5 & -4 & 1 & 0 \\ & 6 & -4 & 1 \\ & & 6 & -4 \\ & & & 5 \end{bmatrix} \begin{bmatrix} U_1 \\ U_2 \\ U_3 \\ U_4 \end{bmatrix}^{(s)} \right\}$$

We use as an initial guess $\quad \mathbf{U}^{(1)} = \begin{bmatrix} 0 \\ 0 \\ 0 \\ 0 \end{bmatrix}$

Consider first the solution without overrelaxation, i.e., $\beta = 1$. We obtain

$$\begin{bmatrix} U_1 \\ U_2 \\ U_3 \\ U_4 \end{bmatrix}^{(2)} = \begin{bmatrix} 0 \\ 0.167 \\ 0.111 \\ 0.0556 \end{bmatrix}; \qquad \begin{bmatrix} U_1 \\ U_2 \\ U_3 \\ U_4 \end{bmatrix}^{(3)} = \begin{bmatrix} 0.111 \\ 0.305 \\ 0.222 \\ 0.116 \end{bmatrix}$$

Using the convergence limit in (8.72) with $\epsilon = 0.001$, we have convergence after 104 iterations and

$$\begin{bmatrix} U_1 \\ U_2 \\ U_3 \\ U_4 \end{bmatrix}^{(104)} = \begin{bmatrix} 1.59 \\ 2.59 \\ 2.39 \\ 1.39 \end{bmatrix}$$

with the exact solution being

$$\begin{bmatrix} U_1 \\ U_2 \\ U_3 \\ U_4 \end{bmatrix} = \begin{bmatrix} 1.60 \\ 2.60 \\ 2.40 \\ 1.40 \end{bmatrix}$$

We now vary β and recalculate the solution. The following table gives the number of iterations required for convergence with $\epsilon = 0.001$ as a function of β:

β	1.0	1.1	1.2	1.3	1.4	1.5	1.6	1.7	1.8	1.9
Number of iterations	104	88	74	61	49	37	23	30	43	82

Hence, for this example, we find that the minimum number of iterations is required at $\beta = 1.6$.

It is instructive to identify the physical process that is followed in the solution procedure. For this purpose we note that on the right-hand side of (8.73) we evaluate an out-of-balance force corresponding to degree of freedom i,

$$Q_i^{(s)} = R_i - \sum_{j=1}^{i-1} k_{ij} U_j^{(s+1)} - \sum_{j=i}^{n} k_{ij} U_j^{(s)} \tag{8.74}$$

and then calculate an improved value for the corresponding displacement component, $U_i^{(s+1)}$, using

$$U_i^{(s+1)} = U_i^{(s)} + \beta k_{ii}^{-1} Q_i^{(s)} \tag{8.75}$$

where $i = 1, \ldots, n$. Assuming that $\beta = 1$, the correction to $U_i^{(s)}$ in (8.75) is calculated by applying the out-of-balance force $Q_i^{(s)}$ to the ith nodal degree of freedom, with all other nodal displacements kept fixed. The process is therefore identical to the moment-distribution procedure, which has been used extensively in hand calculation analysis of frames. However, faster convergence is achieved if the acceleration factor β is used.

Although direct solution methods are used almost exclusively in finite element analysis programs, it is important to recognize some advantages of the Gauss-Seidel method. The solution scheme might be used effectively in the problem areas of reanalysis and optimization, although a form of Gauss elimination may still be more efficient.[45-50] Namely, if in reanalysis a structure is changed only slightly, the previous solution is a good starting vector for the Gauss-Seidel iterative solution of the new structure.

Another advantage of the Gauss-Seidel method can be that the assembled stiffness matrix need not be formed and thus out-of-core solution is avoided, because all matrix multiplications can be carried out on the element level. For example, instead of calculating $\mathbf{K}_L\mathbf{U}^{(s+1)}$, we can evaluate $\sum_m \mathbf{K}_L^{(m)}\,\mathbf{U}^{(s+1)}$, where the summation goes over all finite elements and $\mathbf{K}_L^{(m)}$ is the contribution of the mth element to $\mathbf{K}_L$. The same procedure is generally used in the implementation of the central difference scheme for transient analysis (see Section 9.2.1).

Finally, we should also recognize that there is the possibility of combining the direct and iterative solution schemes discussed above. As an example, the substructuring described in Section 8.2.4 might be used to assemble a coefficient matrix of a system of equations that are solved using Gauss-Seidel iteration. Such combinations can lead to a variety of procedures that may display considerable advantage in specific applications.

8.5 SOLUTION ERRORS

In the preceding sections we presented various algorithms for the solution of $\mathbf{KU} = \mathbf{R}$. We applied the solution procedures for instructive purposes to some small problems, but in practical analysis the methods are employed on large systems of equations using the electronic digital computer. However, when using the computer it is important to note that the elements of the matrices can only be represented to a fixed number of digits, and this means that errors will be introduced in the solution.[1-10, 51-59] The aim in this section is to discuss the solution errors that can occur in Gauss elimination, i.e., when any one of the algorithms described in Section 8.2 is employed, and to give guidelines for avoiding the introduction of large errors.

In order to identify the source of the errors, let us assume that we use a computer in which a number is represented using t digits in single precision. Then, to increase accuracy, double precision arithmetic may be specified, in which case each number is approximately represented using $2t$ digits. Table 8.1 gives the number t for some typical computers.

Considering the finite digit arithmetic computers, it is found that the t digits are used quite differently in different machines. However, most computers, in effect, perform the arithmetic operations and afterward "chop off" all digits beyond the number of digits carried. Therefore, for demonstration

TABLE 8.1 *Number of digits used to represent floating-point number.*

Computer	*Single Precision Arithmetic*	*Double Precision Arithmetic*	t
UNIVAC	8	16	8
IBM	6	16	6
CDC	14	28	14

purposes we assume in this section that the computer at hand first adds, subtracts, multiplies, and divides two numbers exactly, and then to obtain the finite precision results, chops off all digits beyond the t digits used.

In order to demonstrate the finite precision arithmetic, assume that we want to solve the system of equations

$$\begin{bmatrix} 3.42521 & -3.42521 \\ -3.42521 & 101.2431 \end{bmatrix} \begin{bmatrix} U_1 \\ U_2 \end{bmatrix} = \begin{bmatrix} 1.3021 \\ 0.0 \end{bmatrix} \tag{8.76}$$

where $\mathbf{K}$ and $\mathbf{R}$ are given "exactly." The exact solution is (to 10 digits)

$$U_1 = 0.3934633449; \qquad U_2 = 0.0133114709 \tag{8.77}$$

Assume now that for a (hypothetical) computer at hand, $t = 3$; i.e., each number is represented to only 3 digits. In this case the solution to the equations in (8.76) would be obtained by first representing $\mathbf{K}$ and $\mathbf{R}$ using only the first three digits in each number, and then calculating the solution by always using only three-digit representations with chopping off of the additional digits. Employing the basic Gauss elimination algorithm (see Section 8.2.1), the solution would be as follows:

$$\begin{bmatrix} 3.42 & -3.42 \\ -3.42 & 101 \end{bmatrix} \begin{bmatrix} \hat{U}_1 \\ \hat{U}_2 \end{bmatrix} = \begin{bmatrix} 1.30 \\ 0 \end{bmatrix} \tag{8.78}$$

where the hats over U_1 and U_2 indicate that the solution of (8.78) is different from the solution of (8.76). Using that with chopping to three digits,

$$101 - \left(\frac{-3.42}{3.42}\right)(-3.42) = 101 - (-1)(-3.42) = 97.5$$

we obtain

$$\begin{bmatrix} 3.42 & -3.42 \\ 0.0 & 97.5 \end{bmatrix} \begin{bmatrix} \bar{U}_1 \\ \bar{U}_2 \end{bmatrix} = \begin{bmatrix} 1.30 \\ 1.30 \end{bmatrix} \tag{8.79}$$

where the bars over U_1 and U_2 indicate that we solve (8.78) approximately. Continuing to use three-digit chopping arithmetic, we have

$$\left.\begin{aligned} \bar{U}_2 &= \frac{1.30}{97.5} = 0.0133 \\ \text{and} \qquad \bar{U}_1 &= \frac{1}{3.42}[1.30 - (-3.42)(0.0133)] = \frac{1}{3.42}(1.34) \\ &= 0.391 \end{aligned}\right\} \tag{8.80}$$

Referring to the above example, we can identify two kinds of errors: a *truncation error* and *a round-off error*. The truncation error is the error arising because the exact matrix $\mathbf{K}$ and load vector $\mathbf{R}$ in (8.76) are only represented to three-digit precision, as given in (8.78). The round-off error is the error that arises in the solution of (8.78) because only three-digit arithmetic is used. Considering the situations in which each type of error would be large, we note that the truncation error can be large if the absolute magnitude of the elements in the matrix $\mathbf{K}$ including the diagonal elements varies by a large amount. The round-off error can be large if a small diagonal element d_{ii} is used that creates a large multiplier l_{ij}. The reason for the truncation and round-off errors to be large under the above conditions is that the basic operation in the factorization is a subtraction of a multiple of the pivot row from the rows below it. If in this operation numbers of widely different magnitudes are subtracted—that have, however, only been represented to a fixed number of digits—the errors in this basic operation can be relatively large. For general systems, matrix scaling may be used in some cases, in which the complete rows and columns of the matrix are scaled down or up to obtain matrix elements that are more equal in absolute magnitude.

As was pointed out above, the total error in the displacement solution is the sum of the truncation and the round-off errors. To identify the round-off errors and the truncation errors individually in the above example, we need to solve (8.78) exactly, in which case we obtain

$$\begin{bmatrix} 3.42 & -3.42 \\ 0 & 97.58 \end{bmatrix} \begin{bmatrix} \hat{U}_1 \\ \hat{U}_2 \end{bmatrix} = \begin{bmatrix} 1.30 \\ 1.30 \end{bmatrix} \tag{8.81}$$

and

$$\begin{aligned} \hat{U}_1 &= 0.3934393613 \\ \hat{U}_2 &= 0.0133224020 \end{aligned} \tag{8.82}$$

The error in the solution arising from initial truncation is therefore

$$\hat{\mathbf{r}} = \begin{bmatrix} U_1 \\ U_2 \end{bmatrix} - \begin{bmatrix} \hat{U}_1 \\ \hat{U}_2 \end{bmatrix} = \begin{bmatrix} 0.0000239836 \\ -0.0000109311 \end{bmatrix} \tag{8.83}$$

and the round-off error is

$$\bar{\mathbf{r}} = \begin{bmatrix} \hat{U}_1 \\ \hat{U}_2 \end{bmatrix} - \begin{bmatrix} \bar{U}_1 \\ \bar{U}_2 \end{bmatrix} = \begin{bmatrix} 0.0024393613 \\ 0.0000224020 \end{bmatrix} \tag{8.84}$$

The total error $\mathbf{r}$ is the sum of $\bar{\mathbf{r}}$ and $\hat{\mathbf{r}}$, or

$$\mathbf{r} = \begin{bmatrix} U_1 \\ U_2 \end{bmatrix} - \begin{bmatrix} \bar{U}_1 \\ \bar{U}_2 \end{bmatrix} = \begin{bmatrix} 0.0024633449 \\ 0.0000114709 \end{bmatrix} \tag{8.85}$$

In the above evaluation of the solution errors, we used the exact solutions to (8.76) and (8.81). In practical analyses, these exact solutions cannot be obtained; instead, double precision arithmetic could be employed to calculate close approximations to them.

Consider now that in a specific analysis the solution obtained to the equations $\mathbf{KU} = \mathbf{R}$ is $\bar{\mathbf{U}}$; i.e., because of truncation and round-off errors, $\bar{\mathbf{U}}$ is calculated instead of $\mathbf{U}$. It appears that the error in the solution can be obtained by evaluating a residual $\Delta\mathbf{R}$, where

$$\Delta\mathbf{R} = \mathbf{R} - \mathbf{K}\bar{\mathbf{U}} \tag{8.86}$$

In practice, $\Delta\mathbf{R}$ would be calculated using double precision arithmetic. Substituting $\mathbf{KU}$ for $\mathbf{R}$ into (8.86), we have for the solution error $\mathbf{r} = \mathbf{U} - \bar{\mathbf{U}}$,

$$\mathbf{r} = \mathbf{K}^{-1}\Delta\mathbf{R} \tag{8.87}$$

meaning that *although* $\Delta\mathbf{R}$ *may be small, the error in the solution may still be large.* On the other hand, for an accurate solution $\Delta\mathbf{R}$ must be small. Therefore, a small residual $\Delta\mathbf{R}$ is a necessary but not a sufficient condition for an accurate solution.

EXAMPLE 8.18: Calculate $\Delta\mathbf{R}$ and $\mathbf{r}$ for the introductory example considered above.

Using the values for $\mathbf{R}$, $\mathbf{K}$, and $\bar{\mathbf{U}}$ given in (8.76) and (8.80), respectively, we obtain, using (8.86),

$$\Delta\mathbf{R} = \begin{bmatrix} 1.3021 \\ 0 \end{bmatrix} - \begin{bmatrix} 3.42521 & -3.42521 \\ -3.42521 & 101.2431 \end{bmatrix} \begin{bmatrix} 0.391 \\ 0.0133 \end{bmatrix}$$

or

$$\Delta\mathbf{R} = \begin{bmatrix} 0.00839818 \\ -0.00042520 \end{bmatrix}$$

Hence, using (8.87), we have

$$\mathbf{r} = \begin{bmatrix} 0.00253338 \\ 0.00008151 \end{bmatrix}$$

In this case, $\Delta\mathbf{R}$ and $\mathbf{r}$ are both relatively small, because $\mathbf{K}$ is well-conditioned.

EXAMPLE 8.19: Consider the system of equations

$$\begin{bmatrix} 4.855 & -4 & 1 & 0 \\ -4 & 5.855 & -4 & 1 \\ 1 & -4 & 5.855 & -4 \\ 0 & 1 & -4 & 4.855 \end{bmatrix} \begin{bmatrix} U_1 \\ U_2 \\ U_3 \\ U_4 \end{bmatrix} = \begin{bmatrix} -1.59 \\ 1 \\ 1 \\ -1.64 \end{bmatrix}$$

Use six-digit arithmetic with chopping to calculate the solution.

Following the basic Gauss elimination process, we have

$$\begin{bmatrix} 4.85500 & -4 & 1 & 0 \\ 0 & 2.55944 & -3.17610 & 1 \\ 0 & -3.17610 & 5.64902 & -4 \\ 0 & 1 & -4 & 4.85500 \end{bmatrix} \begin{bmatrix} \bar{U}_1 \\ \bar{U}_2 \\ \bar{U}_3 \\ \bar{U}_4 \end{bmatrix} = \begin{bmatrix} -1.59000 \\ -0.309980 \\ 1.32719 \\ -1.64000 \end{bmatrix}$$

$$\begin{bmatrix} 4.85500 & -4 & 1 & 0 \\ 0 & 2.55944 & -3.17610 & 1 \\ 0 & 0 & 1.70771 & -2.75907 \\ 0 & 0 & -2.75907 & 4.46429 \end{bmatrix} \begin{bmatrix} \bar{U}_1 \\ \bar{U}_2 \\ \bar{U}_3 \\ \bar{U}_4 \end{bmatrix} = \begin{bmatrix} -1.59000 \\ -0.309980 \\ 0.942827 \\ -1.51888 \end{bmatrix}$$

$$\begin{bmatrix} 4.85500 & -4 & 1 & 0 \\ & 2.55944 & -3.17610 & 1 \\ & & 1.70771 & -2.75907 \\ & & & 0.006600 \end{bmatrix} \begin{bmatrix} \bar{U}_1 \\ \bar{U}_2 \\ \bar{U}_3 \\ \bar{U}_4 \end{bmatrix} = \begin{bmatrix} -1.59000 \\ -0.309980 \\ 0.942827 \\ 0.004390 \end{bmatrix}$$

The back-substitution yields

$$\bar{\mathbf{U}} = \begin{bmatrix} 0.686706 \\ 1.63768 \\ 1.62674 \\ 0.665151 \end{bmatrix}$$

The exact answer (to seven digits) is

$$\mathbf{U} = \begin{bmatrix} 0.7037247 \\ 1.6652256 \\ 1.6542831 \\ 0.6821567 \end{bmatrix}$$

Evaluating $\mathbf{\Delta R}$ as given in (8.86), we obtain

$$\mathbf{\Delta R} = \begin{bmatrix} -1.59 \\ 1 \\ 1 \\ -1.64 \end{bmatrix} - \begin{bmatrix} -1.59002237 \\ 0.99998340 \\ 0.99994470 \\ -1.639971895 \end{bmatrix} = \begin{bmatrix} 0.00002237 \\ 0.00001660 \\ 0.00005530 \\ -0.000028105 \end{bmatrix}$$

Also evaluating $\mathbf{r}$, we have $\mathbf{r} = \begin{bmatrix} 0.01702 \\ 0.02756 \\ 0.02754 \\ 0.01701 \end{bmatrix}$

We therefore see that $\mathbf{\Delta R}$ is relatively much smaller than $\mathbf{r}$. Indeed, the displacement errors are of the order 1 to 2% although the load errors seem to indicate an accurate solution of the equations.

Considering (8.87), we may expect that solution accuracy is difficult to obtain when the smallest eigenvalue of $\mathbf{K}$ is very small or nearly zero; i.e., the system can almost undergo rigid body motion. Namely, in that case the elements in $\mathbf{K}^{-1}$ are large and the solution errors may be large although $\mathbf{\Delta R}$ is small. Also, to substantiate this conclusion, we may realize that if λ_1 of $\mathbf{K}$ is small, the solution $\mathbf{KU} = \mathbf{R}$ may be thought of as one step of inverse iteration with a shift close to λ_1. But the analysis in Section 11.2.1 shows that in such a case the solution tends to have components of the corresponding eigenvector. These components now appear as solution errors.

To obtain more information on the solution errors, an analysis can be performed which shows that it is not only a small (near zero) eigenvalue λ_1 but a large ratio of the largest to the smallest eigenvalues of $\mathbf{K}$ that determines the solution errors. Namely, in the solution of $\mathbf{KU} = \mathbf{R}$, owing to truncation and round-off errors, we may assume that we in fact solve

$$(\mathbf{K} + \delta\mathbf{K})(\mathbf{U} + \delta\mathbf{U}) = \mathbf{R} \tag{8.88}$$

Assuming that $\delta\mathbf{K}\,\delta\mathbf{U}$ is small in relation to the other terms, we have approximately

$$\delta\mathbf{U} = -\mathbf{K}^{-1}\,\delta\mathbf{K}\,\mathbf{U} \tag{8.89}$$

or, taking norms,

$$\frac{\|\delta\mathbf{U}\|}{\|\mathbf{U}\|} \leq \text{cond}\,(\mathbf{K})\frac{\|\delta\mathbf{K}\|}{\|\mathbf{K}\|} \tag{8.90}$$

where cond $(\mathbf{K})$ is the *condition number* of $\mathbf{K}$,

$$\text{cond}\,(\mathbf{K}) = \frac{\lambda_n}{\lambda_1} \tag{8.91}$$

Therefore, *a large condition number means that solution errors are more likely.* Experience and tests have shown that in finite element analysis, round-off errors are not as serious as the initial truncation errors, and indeed it appears that for an error estimate only initial truncation errors must be considered.[59] To evaluate an estimate of the initial truncation errors, we can assume that for a t-digit precision computer,

$$\frac{\|\delta\mathbf{K}\|}{\|\mathbf{K}\|} = 10^{-t} \tag{8.92}$$

Also, assuming s-digit precision in the solution, we have

$$\frac{\|\delta \mathbf{U}\|}{\|\mathbf{U}\|} = 10^{-s} \tag{8.93}$$

Substituting (8.92) and (8.93) into (8.90), we obtain as an estimate of the number of accurate digits obtained in the solution,

$$s \geq t - \log_{10} (\text{cond } (\mathbf{K})) \tag{8.94}$$

EXAMPLE 8.20: Calculate the condition number of the matrix **K** used in Example 8.19. Then estimate the accuracy that can be expected in the equation solution.

In this case we have

$$\lambda_1 = 0.000898$$
$$\lambda_4 = 12.9452$$

Hence

$$\text{cond } (\mathbf{K}) = 14415.6$$

and

$$\log_{10} (\text{cond } (\mathbf{K})) = 4.15883$$

Thus the number of accurate digits using six-digit arithmetic predicted using (8.94) are

$$s \geq 6 - 4.16$$

or one- to two-digit accuracy can be expected. Comparing this result with the results obtained in Example 8.19, we observe that, indeed, only one- to two-digit accuracy was obtained.

The condition number of **K** can in practice be evaluated approximately by calculating an upper bound for λ_n, say λ_n^u,

$$\lambda_n^u = \|\mathbf{K}\| \tag{8.95}$$

where any matrix norm may be used (see Example 8.21), and evaluating a lower bound for λ_1, say λ_1^l, using inverse iteration (see Section 11.2.1). We thus have

$$\text{cond } (\mathbf{K}) \doteq \frac{\lambda_n^u}{\lambda_1^l} \tag{8.96}$$

EXAMPLE 8.21: Calculate an estimate on the condition number of the matrix **K** used in Example 8.19.

Here we have, using the ∞ norm (see Section 2.9),

$$\|\mathbf{K}\|_\infty = 14.855$$

and by inverse iteration we obtain $\lambda_1 = 0.0009$. Hence

$$\log_{10} (\text{cond } (\mathbf{K})) = 4.2176$$

and the conclusions reached in Example 8.20 are still valid.

Although much additional research is required in the area of finite element solution accuracy, the above considerations on round-off and truncation errors yield the following two important results:

1. Both types of errors can be expected to be large if structures with widely varying stiffness are analyzed. Large stiffness differences may be due to different material moduli, or they may be the result of the finite element modeling used, in which case a more effective model can fre-

quently be chosen. This may be achieved by the use of finite elements that are nearly equal in size and have almost the same lengths in each dimension, the use of master-slave degrees of freedom, i.e., constraint equations (see Section 4.2.2 and Example 8.23), and relative degrees of freedom (see Example 8.24).

2. Since truncation errors are most significant, to improve the solution accuracy it is necessary both to evaluate the stiffness matrix **K** and also the solution of $\mathbf{KU} = \mathbf{R}$ in double precision. It is not sufficient (a) to evaluate **K** in single precision and then solve the equations in double precision (see Example 8.22), or (b) to evaluate **K** in single precision, solve the equations in single precision using a Gauss elimination procedure, and then iterate for an improvement in the solution employing, for example, the Gauss-Seidel method.

We demonstrate the two conclusions above by means of some simple examples.

EXAMPLE 8.22: Consider the simple spring system in Fig. 8.11. Calculate the displacements when $k = 1$, $K = 10{,}000$ using four-digit arithmetic. The equilibrium equations of the system are

$$\begin{bmatrix} K & -K & 0 \\ -K & 2K & -K \\ 0 & -K & K+k \end{bmatrix} \begin{bmatrix} U_1 \\ U_2 \\ U_3 \end{bmatrix} = \begin{bmatrix} 1 \\ 0 \\ 1 \end{bmatrix}$$

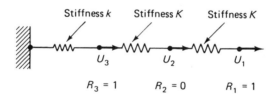

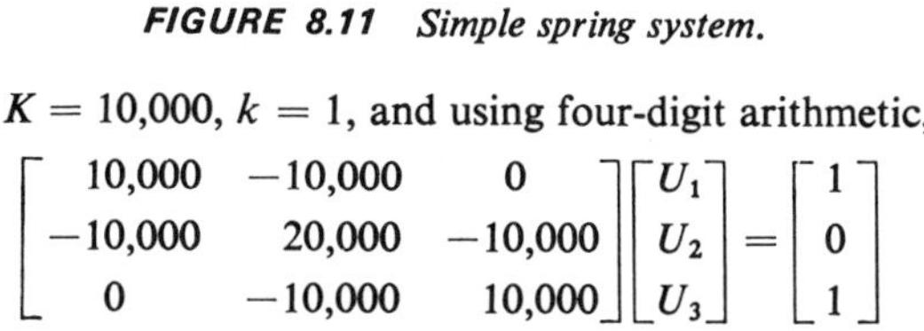

FIGURE 8.11 *Simple spring system.*

Substituting $K = 10{,}000$, $k = 1$, and using four-digit arithmetic, we have

$$\begin{bmatrix} 10{,}000 & -10{,}000 & 0 \\ -10{,}000 & 20{,}000 & -10{,}000 \\ 0 & -10{,}000 & 10{,}000 \end{bmatrix} \begin{bmatrix} U_1 \\ U_2 \\ U_3 \end{bmatrix} = \begin{bmatrix} 1 \\ 0 \\ 1 \end{bmatrix}$$

The triangularization of the coefficient matrix gives

$$\begin{bmatrix} 10{,}000 & -10{,}000 & 0 \\ 0 & 10{,}000 & -10{,}000 \\ 0 & 0 & 0 \end{bmatrix} \begin{bmatrix} U_1 \\ U_2 \\ U_3 \end{bmatrix} = \begin{bmatrix} 1.0 \\ 1.0 \\ 2.0 \end{bmatrix}$$

Hence a solution is not possible, because $d_{nn} = 0.0$.

To obtain a solution we may employ higher-digit arithmetic. In practice, this would mean that double precision arithmetic would be used, i.e., in this case, eight- instead of four-digit arithmetic.

Using eight-digit arithmetic (indeed five-digits would be sufficient), we obtain the exact solution as follows:

$$\begin{bmatrix} 10{,}000 & -10{,}000 & 0 \\ -10{,}000 & 20{,}000 & -10{,}000 \\ 0 & -10{,}000 & 10{,}001 \end{bmatrix} \begin{bmatrix} U_1 \\ U_2 \\ U_3 \end{bmatrix} = \begin{bmatrix} 1 \\ 0 \\ 1 \end{bmatrix}$$

$$\begin{bmatrix} 10{,}000 & -10{,}000 & 0 \\ 0 & 10{,}000 & -10{,}000 \\ 0 & 0 & 1 \end{bmatrix} \begin{bmatrix} U_1 \\ U_2 \\ U_3 \end{bmatrix} = \begin{bmatrix} 1.0 \\ 1.0 \\ 2.0 \end{bmatrix}$$

Hence
$$\mathbf{U} = \begin{bmatrix} 2.0002 \\ 2.0001 \\ 2.0 \end{bmatrix}$$

This example shows that a sufficient number of digits carried in the arithmetic may be vital for the solution not to break down.

EXAMPLE 8.23: Use the master-slave solution procedure to analyze the system considered in Fig. 8.11.

The basic assumption in the master-slave analysis is the use of the constraint equations

$$U_1 = U_2 = U_3$$

The equilibrium equation governing the system is thus

$$kU_1 = 2$$

Substituting for k, we obtain
$$U_1 = 2$$

and the complete solution is
$$\mathbf{U} = \begin{bmatrix} 2.0 \\ 2.0 \\ 2.0 \end{bmatrix}$$

This solution is approximate. However, comparing the solution with the exact result (see Example 8.22), we find that the main response is properly predicted.

EXAMPLE 8.24: Use relative degrees of freedom to analyze the system in Fig. 8.11.

Using relative degrees of freedom, the displacement degrees of freedom defined are U_3, Δ_1, and Δ_2, where

$$U_2 = U_3 + \Delta_2$$
$$U_1 = U_2 + \Delta_1$$

or we have
$$\begin{bmatrix} U_1 \\ U_2 \\ U_3 \end{bmatrix} = \begin{bmatrix} 1 & 1 & 1 \\ 0 & 1 & 1 \\ 0 & 0 & 1 \end{bmatrix} \begin{bmatrix} \Delta_1 \\ \Delta_2 \\ U_3 \end{bmatrix} \qquad \text{(a)}$$

The matrix relating the degrees of freedom Δ_1, Δ_2, and U_3 to the degrees of freedom U_1, U_2, and U_3 is the matrix $\mathbf{T}$. The equilibrium equations for the system using relative degrees of freedom are $(\mathbf{T}^T\mathbf{K}\mathbf{T})\mathbf{U} = \mathbf{T}^T\mathbf{R}$; i.e., the equilibrium equations are now

$$\begin{bmatrix} 10{,}000 & 0 & 0 \\ 0 & 10{,}000 & 0 \\ 0 & 0 & 1.0 \end{bmatrix} \begin{bmatrix} \Delta_1 \\ \Delta_2 \\ U_3 \end{bmatrix} = \begin{bmatrix} 1.0 \\ 1.0 \\ 2.0 \end{bmatrix} \qquad \text{(b)}$$

with the solution $\Delta_1 = 0.0001$

$\Delta_2 = 0.0001$

$U_3 = 2.0$

Hence we obtain $U_1 = 2.0002$

$U_2 = 2.0001$

$U_3 = 2.0000$

Therefore, using four-digit arithmetic, we obtain the exact solution of the system if relative degrees of freedom are used (see Example 8.22). However, it should be noted that the equilibrium equations corresponding to the relative degrees of freedom would have to be formed directly, i.e., without the transformation used in this example.

8.6 SOLUTION OF NONLINEAR EQUATIONS

We discussed in Sections 6.1 and 6.2 that the basic equations to be solved in nonlinear analysis are, at time $t + \Delta t$,

$$ {}^{t+\Delta t}\mathbf{R} - {}^{t+\Delta t}\mathbf{F} = \mathbf{0} \tag{8.97} $$

where the vector ${}^{t+\Delta t}\mathbf{R}$ stores the externally applied nodal loads, and ${}^{t+\Delta t}\mathbf{F}$ is the vector of nodal point forces that are equivalent to the element stresses. Both vectors in (8.97) are evaluated using the principle of virtual displacements. Since the nodal point forces ${}^{t+\Delta t}\mathbf{F}$ depend nonlinearly on the nodal point displacements, it is necessary to iterate in the solution of (8.97). Using a rather physical approach we introduced in Chapter 6 the modified Newton–Raphson iteration, in which we solve for $i = 1, 2, 3, \ldots$

$$ \Delta\mathbf{R}^{(i-1)} = {}^{t+\Delta t}\mathbf{R} - {}^{t+\Delta t}\mathbf{F}^{(i-1)} \tag{8.98} $$

$$ {}^{t}\mathbf{K}\,\Delta\mathbf{U}^{(i)} = \Delta\mathbf{R}^{(i-1)} \tag{8.99} $$

$$ {}^{t+\Delta t}\mathbf{U}^{(i)} = {}^{t+\Delta t}\mathbf{U}^{(i-1)} + \Delta\mathbf{U}^{(i)} \tag{8.100} $$

with

$$ {}^{t+\Delta t}\mathbf{U}^{(0)} = {}^{t}\mathbf{U}; \qquad {}^{t+\Delta t}\mathbf{F}^{(0)} = {}^{t}\mathbf{F} $$

These equations were obtained by linearizing the response of the finite element system about the conditions at time t. In each iteration we calculate in (8.98) an out-of-balance load vector which yields an increment in displacements obtained in (8.99), and we continue the iteration until the out-of-balance load vector $\Delta\mathbf{R}^{(i-1)}$ or the displacement increments $\Delta\mathbf{U}^{(i)}$ are sufficiently small.

The objective in this section is to discuss the above iterative scheme and others for the solution of (8.97) in more detail. Important ingredients of all solution schemes to be presented are the calculation of the vector ${}^{t+\Delta t}\mathbf{F}^{(i)}$ and the tangent stiffness matrix ${}^{t}\mathbf{K}$, and the solution of equations of the form (8.99). The appropriate evaluation of nodal point force vectors and tangent stiffness matrices was discussed in Chapter 6, and the solution of the linearized equations in (8.99) was presented in Sections 8.2 to 8.4; hence the only but very important aspect of concern now is the construction of iterative schemes of the form (8.98) to (8.100) that display good convergence characteristics and can be employed effectively. In fact, the basic problem now under consideration may be understood to be that of accelerating the convergence and overall effectiveness of the method presented in (8.98) to (8.100).

8.6.1 Newton–Raphson Schemes

The most frequently used iteration schemes for the solution of nonlinear finite element equations are some form of Newton–Raphson iteration,[60-66] of which the equations in (8.98) to (8.100) are a special case. The finite element equilibrium requirements amount to finding the solution of the equations

$$\mathbf{f}(\mathbf{U}^*) = \mathbf{0} \tag{8.101}$$

where

$$\mathbf{f}(\mathbf{U}^*) = {}^{t+\Delta t}\mathbf{R}(\mathbf{U}^*) - {}^{t+\Delta t}\mathbf{F}(\mathbf{U}^*) \tag{8.102}$$

Assume that in the iterative solution we have evaluated ${}^{t+\Delta t}\mathbf{U}^{(i-1)}$; then a Taylor series expansion gives

$$\mathbf{f}(\mathbf{U}^*) = \mathbf{f}({}^{t+\Delta t}\mathbf{U}^{(i-1)}) + \left[\frac{\partial \mathbf{f}}{\partial \mathbf{U}}\right]\Bigg|_{{}^{t+\Delta t}\mathbf{U}^{(i-1)}} (\mathbf{U}^* - {}^{t+\Delta t}\mathbf{U}^{(i-1)}) \tag{8.103}$$

where higher-order terms are neglected. Substituting from (8.102) into (8.103) and using (8.101) we obtain

$$\left[\frac{\partial \mathbf{F}}{\partial \mathbf{U}}\right]\Bigg|_{{}^{t+\Delta t}\mathbf{U}^{(i-1)}} (\mathbf{U}^* - {}^{t+\Delta t}\mathbf{U}^{(i-1)}) = {}^{t+\Delta t}\mathbf{R} - {}^{t+\Delta t}\mathbf{F}^{(i-1)} \tag{8.104}$$

where we assumed that the externally applied loads are independent of the displacements. We now define

$$\Delta \mathbf{U}^{(i)} = \mathbf{U}^* - {}^{t+\Delta t}\mathbf{U}^{(i-1)} \tag{8.105}$$

and recognize that

$$\frac{\partial \mathbf{F}}{\partial \mathbf{U}}\Bigg|_{{}^{t+\Delta t}\mathbf{U}^{(i-1)}} = {}^{t+\Delta t}\mathbf{K}^{(i-1)} \tag{8.106}$$

where ${}^{t+\Delta t}\mathbf{K}^{(i-1)}$ is the tangent stiffness matrix in iteration $i - 1$, which we assume to be nonsingular. Thus (8.104) can be written as

$${}^{t+\Delta t}\mathbf{K}^{(i-1)}\,\Delta \mathbf{U}^{(i)} = {}^{t+\Delta t}\mathbf{R} - {}^{t+\Delta t}\mathbf{F}^{(i-1)} \tag{8.107}$$

Since (8.103) represents only a Taylor series approximation, the displacement increment correction $\Delta \mathbf{U}^{(i)}$ is used to obtain the next displacement approximation

$${}^{t+\Delta t}\mathbf{U}^{(i)} = {}^{t+\Delta t}\mathbf{U}^{(i-1)} + \Delta \mathbf{U}^{(i)} \tag{8.108}$$

The relations in (*8.107*) *and* (*8.106*) *constitute the Newton–Raphson solution of* (*8.97*). Since an incremental analysis is performed with time (or load) steps of size Δt (see Chapter 6), the initial conditions in this iteration are ${}^{t+\Delta t}\mathbf{K}^{(0)} = {}^{t}\mathbf{K}$, ${}^{t+\Delta t}\mathbf{F}^{(0)} = {}^{t}\mathbf{F}$ and ${}^{t+\Delta t}\mathbf{U}^{(0)} = {}^{t}\mathbf{U}$. The iteration is continued until appropriate convergence criteria are satisfied as discussed in Section 8.6.3. The Newton–Raphson iteration is demonstrated in the solution of a simple problem below.

EXAMPLE 8.25: For a single degree of freedom system we have

$${}^{t+\Delta t}R = 10; \qquad {}^{t+\Delta t}F = 4 + 2|({}^{t+\Delta t}U)^{1/2}|$$

and ${}^{t}U = 1$. Use the Newton–Raphson iteration to calculate ${}^{t+\Delta t}U$.

In this case we have corresponding to (8.107) as the governing equation

$$\left(\frac{1}{|\sqrt{{}^{t+\Delta t}U^{(i-1)}}|}\right)\Delta U^{(i)} = 6 - 2|\sqrt{{}^{t+\Delta t}U^{(i-1)}}| \tag{a}$$

Using (a) with ${}^{t+\Delta t}U^{(0)} = 1$ we obtain

$$ {}^{t+\Delta t}U^{(1)} = 5.0000; \qquad {}^{t+\Delta t}U^{(2)} = 8.4164 $$
$$ {}^{t+\Delta t}U^{(3)} = 8.9902; \qquad {}^{t+\Delta t}U^{(4)} = 9.0000 $$

and convergence is achieved in four iterations.

Considering the Newton–Raphson iteration it is recognized that in general the major computational cost per iteration lies in the calculation and factorization of the tangent stiffness matrix. Since these calculations can be prohibitively expensive when large-order systems are considered, some modification of the full Newton algorithm is frequently more effective.

One such modification is to use the initial stiffness matrix, ${}^{0}\mathbf{K}$, in (8.107) and thus operate on the equations:

$$ {}^{0}\mathbf{K}\,\Delta\mathbf{U}^{(i)} = {}^{t+\Delta t}\mathbf{R}^{(i-1)} - {}^{t+\Delta t}\mathbf{F}^{(i-1)} \tag{8.109} $$

with the initial conditions ${}^{t+\Delta t}\mathbf{F}^{(0)} = {}^{t}\mathbf{F}$, ${}^{t+\Delta t}\mathbf{U}^{(0)} = {}^{t}\mathbf{U}$. In this case only the matrix ${}^{0}\mathbf{K}$ need be factorized, thus avoiding the expense of recalculating and factorizing many times the coefficient matrix in (8.107). This "*initial stress*" *method* corresponds to a linearization of the response about the initial configuration of the finite element system, and may result into a very slowly or even divergent solution.

In the *modified Newton–Raphson iteration* an approach somewhat in-between the full Newton iteration and the above initial stress method is employed. In this method we use

$$ {}^{\tau}\mathbf{K}\,\Delta\mathbf{U}^{(i)} = {}^{t+\Delta t}\mathbf{R} - {}^{t+\Delta t}\mathbf{F}^{(i-1)} \tag{8.110} $$

with the initial conditions ${}^{t+\Delta t}\mathbf{F}^{(0)} = {}^{t}\mathbf{F}$, ${}^{t+\Delta t}\mathbf{U}^{(0)} = {}^{t}\mathbf{U}$, and τ corresponds to one of the accepted equilibrium configurations at times 0, Δt, $2\,\Delta t, \ldots,$ or t. The modified Newton–Raphson iteration involves fewer stiffness reformations than the full Newton iteration, and bases the stiffness matrix update on an accepted equilibrium configuration. The choice of time steps when the stiffness matrix should be updated depends on the degree of nonlinearity in the system response, i.e. the more nonlinear the response, the more often the updating should be done. Without any *a priori* knowledge of the system behavior, it may be most efficient to update the stiffness matrix at the start of every time step as we did in (8.98) to (8.100), in which case $\tau = t$.

With the very large range of system properties and nonlinearities that may be encountered in engineering analysis, we find that the effectiveness of the above solution approaches depends on the specific problem considered. Provided $\mathbf{f}''$ in (8.102) is continuous and the initial conditions are "close enough" to the solution sought, the full Newton iteration converges quadratically whereas the modified Newton iteration converges in general more slowly depending on the updating used for the coefficient matrix. In practice, considering material nonlinear analysis, the function $\mathbf{f}''$ is frequently not continuous and it is necessary to select from the above iterative schemes an appropriate solution strategy for the specific problem to be solved. Such a solution strategy should generally also include acceleration schemes, e.g., Aitken acceleration, and automatic load step reduction and stiffness reformation methods.[66]

8.6.2 The BFGS Method

As an alternative to forms of Newton iteration, a class of methods known as matrix update methods or quasi-Newton methods has been developed for iteration on nonlinear systems of equations.[67,68] These methods involve updating the coefficient matrix (or rather its inverse) to provide a secant approximation to the matrix from iteration $(i - 1)$ to i. That is, defining a displacement increment

$$\boldsymbol{\delta}^{(i)} = {}^{t+\Delta t}\mathbf{U}^{(i)} - {}^{t+\Delta t}\mathbf{U}^{(i-1)} \tag{8.111}$$

and an increment in the out-of-balance loads, using (8.98),

$$\boldsymbol{\gamma}^{(i)} = \Delta\mathbf{R}^{(i-1)} - \Delta\mathbf{R}^{(i)} \tag{8.112}$$

the updated matrix ${}^{t+\Delta t}\mathbf{K}^{(i)}$ should satisfy the quasi-Newton equation

$${}^{t+\Delta t}\mathbf{K}^{(i)}\,\boldsymbol{\delta}^{(i)} = \boldsymbol{\gamma}^{(i)} \tag{8.113}$$

These quasi-Newton methods provide a compromise between the full reformation of the stiffness matrix performed in the full Newton method and the use of a stiffness matrix from a previous configuration as is done in the modified Newton method. Among the quasi-Newton methods available the BFGS (*B*royden-*F*letcher-*G*oldfarb-*S*hanno) method appears to be most effective, and its use was first suggested for finite element analysis by Matthies and Strang.[69]

In the BFGS method, the following procedure is employed in iteration i to evaluate ${}^{t+\Delta t}\mathbf{U}^{(i)}$ and ${}^{t+\Delta t}\mathbf{K}^{(i)}$, where ${}^{t+\Delta t}\mathbf{K}^{(0)} = {}^{\tau}\mathbf{K}$.[66,69]

Step 1: Evaluate a displacement vector increment:

$$\Delta\bar{\mathbf{U}} = ({}^{t+\Delta t}\mathbf{K}^{-1})^{(i-1)}({}^{t+\Delta t}\mathbf{R} - {}^{t+\Delta t}\mathbf{F}^{(i-1)}) \tag{8.114}$$

This displacement vector defines a "direction" for the actual displacement increment.

Step 2: Perform a line search in the direction $\Delta\bar{\mathbf{U}}$ to satisfy "equilibrium" in this direction. In this line search we evaluate the displacement vector,

$${}^{t+\Delta t}\mathbf{U}^{(i)} = {}^{t+\Delta t}\mathbf{U}^{(i-1)} + \beta\,\Delta\bar{\mathbf{U}} \tag{8.115}$$

where β is a scalar multiplier, and we calculate the out-of-balance loads corresponding to these displacements $({}^{t+\Delta t}\mathbf{R} - {}^{t+\Delta t}\mathbf{F}^{(i)})$. The parameter β is varied until the component of the out-of-balance loads in the direction $\Delta\bar{\mathbf{U}}$, as defined by the inner product $\Delta\bar{\mathbf{U}}^T({}^{t+\Delta t}\mathbf{R} - {}^{t+\Delta t}\mathbf{F}^{(i)})$, is small. This condition is satisfied when, for a convergence tolerance STOL, the following equation is satisfied:

$$\Delta\bar{\mathbf{U}}^T({}^{t+\Delta t}\mathbf{R} - {}^{t+\Delta t}\mathbf{F}^{(i)}) \leq \text{STOL}\ \Delta\bar{\mathbf{U}}^T({}^{t+\Delta t}\mathbf{R} - {}^{t+\Delta t}\mathbf{F}^{(i-1)}) \tag{8.116}$$

The final value of β for which (8.116) is satisfied determines ${}^{t+\Delta t}\mathbf{U}^{(i)}$ in (8.115). We can now calculate $\boldsymbol{\delta}^{(i)}$ and $\boldsymbol{\gamma}^{(i)}$ using (8.111) and (8.112) and proceed with the evaluation of the matrix update that satisfies (8.113).

Step 3: Evaluate the correction to the coefficient matrix. In the BFGS method the updated matrix can be expressed in product form:

$$({}^{t+\Delta t}\mathbf{K}^{-1})^{(i)} = \mathbf{A}^{(i)T}({}^{t+\Delta t}\mathbf{K}^{-1})^{(i-1)}\mathbf{A}^{(i)} \tag{8.117}$$

where the matrix $\mathbf{A}^{(i)}$ is an $(n \times n)$ matrix of the simple form

$$\mathbf{A}^{(i)} = \mathbf{I} + \mathbf{v}^{(i)} \mathbf{w}^{(i)T} \tag{8.118}$$

The vectors $\mathbf{v}^{(i)}$ and $\mathbf{w}^{(i)}$ are calculated from the known nodal point forces and displacements using

$$\mathbf{v}^{(i)} = -\left[\frac{\boldsymbol{\delta}^{(i)T} \boldsymbol{\gamma}^{(i)}}{\boldsymbol{\delta}^{(i)T}\, {}^{t+\Delta t}\mathbf{K}^{(i-1)} \boldsymbol{\delta}^{(i)}}\right]^{1/2} {}^{t+\Delta t}\mathbf{K}^{(i-1)} \boldsymbol{\delta}^{(i)} - \boldsymbol{\gamma}^{(i)} \tag{8.119}$$

and

$$\mathbf{w}^{(i)} = \frac{\boldsymbol{\delta}^{(i)}}{\boldsymbol{\delta}^{(i)T} \boldsymbol{\gamma}^{(i)}} \tag{8.120}$$

The vector ${}^{t+\Delta t}\mathbf{K}^{(i-1)} \boldsymbol{\delta}^{(i)}$ in (8.119) is equal to $\beta({}^{t+\Delta t}\mathbf{R} - {}^{t+\Delta t}\mathbf{F}^{(i-1)})$ and was already computed.

Since the product defined in (8.117) is positive definite and symmetric, to avoid numerically dangerous updates, the condition number $c^{(i)}$ of the updating matrix $\mathbf{A}^{(i)}$ is calculated:

$$c^{(i)} = \left[\frac{\boldsymbol{\delta}^{(i)T} \boldsymbol{\gamma}^{(i)}}{\boldsymbol{\delta}^{(i)T}\, {}^{t+\Delta t}\mathbf{K}^{(i-1)} \boldsymbol{\delta}^{(i)}}\right]^{1/2} \tag{8.121}$$

This condition number is then compared with some preset tolerance, say 10^5, and the updating is not performed if the condition number exceeds this tolerance.

Considering the actual computations involved, it should be recognized that using the matrix updates defined above, the calculation of the search direction in (8.114) can be rewritten as

$$\Delta\bar{\mathbf{U}} = (\mathbf{I} + \mathbf{w}^{(i-1)} \mathbf{v}^{(i-1)T}) \ldots (\mathbf{I} + \mathbf{w}^{(1)} \mathbf{v}^{(1)T})\, {}^{\tau}\mathbf{K}^{-1} (\mathbf{I} + \mathbf{v}^{(1)} \mathbf{w}^{(1)T}) \ldots (\mathbf{I} + \mathbf{v}^{(i-1)} \mathbf{w}^{(i-1)T})[{}^{t+\Delta t}\mathbf{R} - {}^{t+\Delta t}\mathbf{F}^{(i-1)}] \tag{8.122}$$

Hence the search direction can be computed without explicitly calculating the updated matrices, or performing any additional costly matrix factorizations as required in the full Newton–Raphson method.

The above BFGS method has already been used effectively in the solution of a variety of nonlinear problems,[69,70] but the technique constitutes a relatively recent development and significant further experiences in the use of quasi-Newton methods and improvements in their effectiveness can be expected. These methods together with the Newton and other iteration schemes should finally lead to self-adaptive techniques that adjust tolerances and choose solution approaches automatically to obtain an optimum solution of a set of nonlinear finite element equations.

We demonstrate the BFGS iteration in the following simple example.

EXAMPLE 8.26: Use the BFGS iteration method to solve for ${}^{t+\Delta t}U$ of the system considered in Example 8.25. Omit the line searches in the solution.

We should realize that because this is a single degree of freedom system the solution for ${}^{t+\Delta t}U$ could be evaluated using only line searches, i.e. by applying (8.116), provided STOL is a tight enough convergence tolerance. However, to demonstrate in this example the basic steps of the BFGS method more clearly (the use of relations (8.111) to (8.113)), we do not include line searches in the iterative solution.

In this analysis (8.114) reduces to

$$\Delta \bar{U} = ({}^{t+\Delta t}K^{-1})^{(i-1)}\{6 - 2|\sqrt{{}^{t+\Delta t}U^{(i-1)}}|\}$$

with $({}^{t+\Delta t}K^{-1})^{(0)} = 1$, ${}^{t+\Delta t}U^{(0)} = 1$, and using $\beta = 1.0$ we obtain the following values:

i	${}^{t+\Delta t}U^{(i-1)}$	$\Delta \bar{U} = \delta^{(i)}$	${}^{t+\Delta t}U^{(i)}$	$\gamma^{(i)}$	$({}^{t+\Delta t}K^{-1})^{(i)}$
1	1.000	4.000	5.000	2.472	1.618
2	5.000	2.472	7.472	0.995	2.485
3	7.472	1.324	8.796	0.465	2.850
4	8.796	0.194	8.991	0.065	2.982
5	8.991	0.009	9.000	0.003	2.999

and convergence is achieved after five iterations.

8.6.3 Convergence Criteria

If an incremental solution strategy based on iterative methods is to be effective, realistic criteria should be used for the termination of the iteration. At the end of each iteration, the solution obtained should be checked to see whether it has converged within preset tolerances or whether the iteration is diverging. If the convergence tolerances are too loose, inaccurate results are obtained, and if the tolerances are too tight, much computational effort is spent to obtain needless accuracy. Similarly, an ineffective divergence check can terminate the iteration when the solution is not actually diverging or force the iteration to search for an unattainable solution. The objective in this section is to discuss briefly some convergence criteria.

Since we are seeking the displacement configuration corresponding to time $t + \Delta t$, it is natural to require that the displacements at the end of each iteration be within a certain tolerance of the true displacement solution. Hence, a realistic convergence criterion is

$$\frac{\|\Delta \mathbf{U}^{(i)}\|_2}{\|{}^{t+\Delta t}\mathbf{U}\|_2} \leq \epsilon_D \tag{8.123}$$

where ϵ_D is a displacement convergence tolerance. The vector ${}^{t+\Delta t}\mathbf{U}$ is not known and must be approximated. In the solution of some problems it is appropriate to use in (8.123) the last calculated value ${}^{t+\Delta t}\mathbf{U}^{(i)}$ as an approximation to ${}^{t+\Delta t}\mathbf{U}$. However, in other analyses the actual solution may still be far from the value obtained when convergence is measured using (8.123) with ${}^{t+\Delta t}\mathbf{U}^{(i)}$. This is the case when the calculated displacements change only little in each iteration, but continue to change for many iterations, as for example in elastic-plastic analysis under loading conditions.

A second convergence criterion is obtained by measuring the out-of-balance load vector. For example, we may require that the norm of the out-of-balance load vector be within a preset tolerance, ϵ_F, of the original load increment

$$\|{}^{t+\Delta t}\mathbf{R} - {}^{t+\Delta t}\mathbf{F}^{(i)}\|_2 \leq \epsilon_F \|{}^{t+\Delta t}\mathbf{R} - {}^{t}\mathbf{F}\|_2 \tag{8.124}$$

Considering a force check special attention may need to be given to the inconsistencies in units that can appear in the force vector, e.g. forces and moments in beam elements, and that the displacement solution does not enter the termination criterion. As an illustration of the latter difficulty, consider an elastic-plastic truss with a very small strain-hardening modulus entering the plastic region. In this case, the out-of-balance loads may be very small while the displacements may still be grossly in error.

In order to provide some indication of when both the displacements and the forces are near their equilibrium values, a third convergence criterion may be useful, in which the increment in internal energy during each iteration (i.e. the amount of work done by the out-of-balance loads on the displacement increments) is compared to the initial internal energy increment. Convergence is assumed to be reached when, with ϵ_E a preset energy tolerance,

$$\Delta \mathbf{U}^{(i)T}({}^{t+\Delta t}\mathbf{R} - {}^{t+\Delta t}\mathbf{F}^{(i-1)}) \leq \epsilon_E (\Delta \mathbf{U}^{(1)T}({}^{t+\Delta t}\mathbf{R} - {}^{t}\mathbf{F})) \qquad (8.125)$$

Some experiences with these tolerances are given in ref. [66].

In the above discussion we did not specifically address the solution difficulties that arise in passing over limit (or bifurcation) points. The coefficient matrix in (8.107) then becomes indefinite but (as in eigenvalue calculations, see Section 12.2) its $\mathbf{LDL}^T$ factorization can be used provided there is no significant multiplier growth; hence a factorization close to a limit point must be avoided. In addition, it is then necessary to iterate with both, the displacement and the load levels, and the development of such schemes is currently under active research.[71]

REFERENCES

1. V. N. Faddeeva, *Computational Methods of Linear Algebra*, Dover Publications, Inc., New York, N.Y., 1959.
2. G. E. Forsythe and C. B. Moler, *Computer Solution of Linear Algebraic Systems*, Prentice-Hall, Inc., Englewood Cliffs, N.J., 1967.
3. V. V. Klyuyev and N. I. Kokovkin-Scherbak, "On the Minimization of the Number of Arithmetic Operations for the Solution of Linear Algebraic Systems of Equations," Technical Report CS 24, Computer Science Department, Stanford University, 1965.
4. B. Noble, *Applied Linear Algebra*, Prentice-Hall, Inc., Englewood Cliffs, N.J., 1969.
5. C. E. Fröberg, *Introduction to Numerical Analysis*, Addison-Wesley Publishing Company, Inc., Reading, Mass., 1969.
6. S. H. Crandall, *Engineering Analysis*, McGraw-Hill Book Company, New York, N.Y., 1956.
7. J. H. Wilkinson, *The Algebraic Eigenvalue Problem*, Oxford University Press, Inc., London, 1965.
8. R. S. Martin, G. Peters, and J. H. Wilkinson, "Symmetric Decomposition of a Positive Definite Matrix," *Numerische Mathematik*, Vol. 7, 1965, pp. 362–383.
9. C. Lanczos, *Applied Analysis*, Prentice-Hall, Inc., Englewood Cliffs, N.J., 1956.
10. R. Zurmühl, *Matrizen*, Springer-Verlag, Berlin, 1964.
11. P. D. Crout, "A Short Method for Evaluating Determinants and Solving Systems of Linear Equations with Real or Complex Coefficients," *A.I.E.E. Transactions*, Vol. 60, 1941, pp. 1235–1240.
12. E. L. Wilson, K. J. Bathe, and W. P. Doherty, "Direct Solution of Large Systems of Linear Equations," *Computers and Structures*, Vol. 4, pp. 363–372.
13. R. W. Clough and J. Penzien, *Dynamics of Structures*, McGraw-Hill Book Company, New York, N.Y., 1975.
14. E. L. Wilson, "The Static Condensation Algorithm," *International Journal for Numerical Methods in Engineering*, Vol. 8, 1974, pp. 199–203.

15. E. L. Wilson, "Structural Analysis of Axisymmetric Solids," *A.I.A.A. Journal*, Vol. 3, 1965, pp. 2269–2274.

16. G. Kron, "Solving Highly Complex Elastic Structures in Easy States," *Journal of Applied Mechanics*, Vol. 22, 1955, pp. 235–244.

17. J. S. Przemieniecki, "Matrix Structural Analysis of Substructures," *A.I.A.A. Journal*, Vol. 1, 1963, pp. 138–147.

18. J. S. Przemieniecki, *Theory of Matrix Structural Analysis*, McGraw-Hill Book Company, New York, N.Y., 1968.

19. M. F. Rubinstein, "Combined Analysis by Substructures and Recursion," *A.S.C.E., Journal of the Structural Division*, Vol. 93, No. ST2, Apr. 1967, pp. 231–235.

20. T. Furnike, "Computerized Multiple Level Substructuring Analysis," *Computers and Structures*, Vol. 2, 1972, pp. 1063–1073.

21. E. L. Wilson and H. H. Dovey, "Static and Earthquake Analysis of Three-Dimensional Frame and Shear Wall Buildings," Report EERC 72-1, College of Engineering, University of California, Berkeley, 1972.

22. H. A. Kamel, D. Liu, M. W. McCabe, and V. Philippopoulos, "Some Developments in the Analysis of Complex Ship Structures," in *Advances in Computational Methods in Structural Mechanics and Design*, (J. T. Oden, R. W. Clough, Y. Yamamoto, eds.), University of Alabama Press, University of Alabama, Huntsville, Ala., 1972.

23. D. P. Mondkar and G. H. Powell, "Large Capacity Equation Solver for Structural Analysis," *Computers and Structures*, Vol. 4, 1974, pp. 699–728.

24. G. Cantin, "An Equation Solver of Very Large Capacity," *International Journal for Numerical Methods in Engineering*, Vol. 3, 1971, pp. 379–388.

25. K. J. Bathe, E. L. Wilson, and F. E. Peterson, "SAP IV—A Structural Analysis Program for Static and Dynamic Response of Linear Systems," Report EERC 73–11, College of Engineering University of California, Berkeley, June 1973, revised Apr. 1974.

26. K. J. Bathe, "ADINA—A Finite Element Program for Automatic Dynamic Incremental Nonlinear Analysis," Report 82448–1, Acoustics and Vibration Laboratory, Department of Mechanical Engineering, Massachusetts Institute of Technology, Cambridge, Mass., 1975, rev. 1978.

27. EAC/EASE2—User Information Manual/Theoretical, Control Data Corporation Publication 84002700, Minneapolis, Minn., 1973.

28. W. Weaver, Jr., *Computer Programs for Structural Analysis*, Van Nostrand Reinhold Company, New York, N.Y., 1967.

29. P. V. Marcal (ed.), "General Purpose Finite Element Computer Programs," Proceedings of Seminar A.S.M.E. Winter Annual Meeting, New York, Nov., 1970.

30. S. J. Fenves, N. Perrone, J. Robinson, and W. C. Schnobrich, *Numerical and Computer Methods in Structural Mechanics*, Academic Press, Inc., New York, N.Y., 1973.

31. W. Pilkey, K. Saczalski, and H. Schaeffer (eds.), *Structural Mechanics Computer Programs*, University Press of Virginia, Charlottesville, Va., 1974.

32. B. M. Irons, "A Frontal Solution Program for Finite Element Analysis," *International Journal for Numerical Methods in Engineering*, Vol. 2, 1970, pp. 5–32.

33. R. J. Melosh and R. M. Bamford, "Efficient Solution of Load-Deflection Equations," *A.S.C.E., Journal of the Structural Division*, Vol. 95, No. ST4, Apr. 1969, pp. 661–676.

34. T. K. Hellen, "A Frontal Solution for Finite Element Techniques," Central Electricity Generating Board, Report RD/B/N1459, Berkeley Nuclear Laboratories, Berkeley, Gloucestershire, 1969.

35. R. W. Clough and E. L. Wilson, "Stress Analysis of a Gravity Dam by the Finite Element Method," Proceedings, Symposium on the Use of Computers in Civil Engineering, Lisbon, Portugal, Oct. 1962.

36. V. B. Venkayya, "Iterative Method for the Analysis of Large Structural Systems," Air Force Flight Dynamics Laboratory, Wright-Patterson A.F.B., Ohio, Report AFFDL-TR-67-194, Apr. 1968.

37. R. V. Southwell, *Relaxation Methods in Theoretical Physics*, Oxford University Press, New York, 1946.

38. E. Stiefel, "On Some Relaxation Methods," *Zeitschrift für angewandte Mathematik und Physik*, Vol. 3, 1952, pp. 1–33.

39. M. F. Rubinstein and D. E. Wikholm, "Analysis by Group Iteration Using Substructures," *A.S.C.E., Journal of the Structural Division*, Vol. 94, No. ST2, Feb. 1968, pp. 363–375.

40. R. S. Varga, *Matrix Iterative Analysis*, Prentice-Hall, Inc., Englewood Cliffs, N.J., 1962.

41. R. Van Norton, "The Solution of Linear Equations by the Gauss–Seidel Method," in *Mathematical Methods for Digital Computers*, Vol. 1 (A. Ralston and H. S. Wilf, eds.), John Wiley & Sons, Inc., New York, N.Y., 1960.

42. F. Beckman, "The Solution of Linear Equations by the Conjugate Gradient Method," in *Mathematical Methods for Digital Computers*, Vol. 1 (A. Ralston and H. S. Wilf, eds.), John Wiley & Sons, Inc., New York, N.Y., 1960.

43. J. K. Reid, "On the Method of Conjugate Gradients for the Solution of Large Sparse Systems of Linear Equations," Conference on Large Sparse Sets of Linear Equations, St. Catherine's College, Oxford, Apr. 1970, pp. 231–254.

44. I. Fried, "A Gradient Computational Procedure for the Solution of Large Problems Arising from the Finite Element Discretization Method," *International Journal for Numerical Methods in Engineering*, Vol. 2, 1970, pp. 477–494.

45. J. H. Argyris, "The Matrix Analysis of Structures with Cutouts and Modifications," Proceedings, 9th International Congress of Applied Mechanics, Brussels, Vol. 67, 1956, pp. 131–142.

46. J. M. Bennett, "Triangular Factors of Modified Matrices," *Numerische Mathematik*, Vol. 7, 1965, pp. 217–221.

47. J. Sobieszczanski, "Structural Modification by Perturbation Method," *A.S.C.E., Journal of the Structural Division*, Vol. 94, No. ST12, Dec. 1968, pp. 2799–2816.

48. D. Kavlie and G. H. Powell, "Efficient Reanalysis of Modified Structures," *A.S.C.E., Journal of the Structural Division*, Vol. 97, No. ST1, Jan. 1971, pp. 377–392 ("Discussion" by N. E. Wiberg, Oct. 1971).

49. J. H. Argyris and J. R. Roy, "General Treatment of Structural Modifications," *A.S.C.E., Journal of the Structural Division*, Vol. 98, No. ST2, Feb. 1972, pp. 465–492.

50. B. Mohraz and R. N. Wright, "Solving Topologically Modified Structures," paper presented at the National Symposium on Computerized Structural Analysis and Design, George Washington University, Washington, D.C., Mar. 1972.

51. J. H. Wilkinson, *Rounding Errors in Algebraic Processes*, Prentice-Hall, Inc., Englewood Cliffs, N.J., 1962.

52. J. von Neumann and H. H. Goldstine, "Numerical Inverting of Matrices of High Order," *Bulletin of the American Mathematical Society*, Vol. 53, 1947, pp. 1021–1099; and *Proceedings of the American Mathematical Society*, Vol. 2, 1951, pp. 188–202.

53. R. S. Martin, G. Peters, and J. H. Wilkinson, "Iterative Refinement of the Solu-

tion of a Positive Definite System of Equations," *Numerische Mathematik*, Vol. 8, 1966, pp. 203–216.

54. J. H. Wilkinson, "The Solution of Ill-Conditioned Linear Equations," in *Mathematical Methods for Digital Computers*, Vol. 2 (A. Ralston and H. S. Wilf, eds.), John Wiley & Sons, Inc., New York, N.Y., 1967.

55. F. L. Bauer, "Optimally Scaled Matrices," *Numerische Mathematik*, Vol. 5, No. 1, 1963, pp. 73–87.

56. B. M. Irons, "Roundoff Criteria in Direct Stiffness Solutions," *A.I.A.A. Journal*, Vol. 6, No. 7, July 1968, pp. 1308–1312.

57. R. A. Rosanoff, J. F. Gloudeman, and S. Levy, "Numerical Conditioning of Stiffness Matrix Formulations for Frame Structures," *Proceedings, 2nd Conference on Matrix Methods in Structural Mechanics*, AFFDL-TR-68-150, Wright-Patterson A.F.B., Ohio, 1968, pp. 1029–1060.

58. R. J. Melosh, "Manipulation Errors in Finite Element Analysis," in *Recent Advances in Matrix Methods of Structural Analysis and Design* (R. H. Gallagher, Y. Yamada, and J. T. Oden, eds.), University of Alabama Press, Huntsville, Ala., 1971.

59. J. R. Roy, "Numerical Error in Structural Solutions," *A.S.C.E., Journal of the Structural Division*, Vol. 97, No. ST4, Apr. 1971, pp. 1039–1054.

60. J. A. Stricklin, W. E. Haisler, and W. A. von Riesemann, "Evaluation of Solution Procedures for Material and/or Geometrically Nonlinear Structural Analysis," *A.I.A.A. Journal*, Vol. 11, 1973, pp. 292–299.

61. J. T. Oden, *Finite Elements of Nonlinear Continua*, McGraw-Hill Book Company, New York, N.Y., 1972.

62. C. A. Felippa, "Procedures for Computer Analysis of Large Nonlinear Structural Systems," *Proceedings, Int. Symp. on Large Engineering Systems*, Manitoba, Winnipeg, Canada, Aug. 1976.

63. K. J. Bathe, "An Assessment of Current Solution Capabilities for Nonlinear Problems in Solid Mechanics," in *Numerical Methods for Partial Differential Equations*-III, (B. Hubbard, ed.) Academic Press, New York, N.Y., 1976.

64. P. G. Bergan, G. Horrigmoe, B. Kråkeland, and T. H. Søreide, "Solution Techniques for Nonlinear Finite Element Problems," *International Journal for Numerical Methods in Engineering*, Vol. 12, 1978, pp. 1677–1696.

65. J. H. Argyris, L. E. Vaz, and K. J. Willam, "Improved Solution Methods for Inelastic Rate Problems," *Journal Computer Methods in Applied Mechanics and Engineering*, Vol. 16, 1978, pp. 231–277.

66. K. J. Bathe and A. P. Cimento, "Some Practical Procedures for the Solution of Nonlinear Finite Element Equations," *Journal Computer Methods in Applied Mechanics and Engineering*, Vol. 22, 1980, pp. 59–85.

67. W. Murray (ed.), *Numerical Methods for Unconstrained Optimization*, Academic Press, New York, N.Y., 1972.

68. J. E. Dennis, Jr., "A Brief Survey of Convergence Results for Quasi-Newton Methods," *SIAM-AMS Proceedings*, Vol. 9, 1976, pp. 185–200.

69. H. Matthies and G. Strang, "The Solution of Nonlinear Finite Element Equations," *International Journal for Numerical Methods in Engineering*, Vol. 14, 1979, pp. 1613–1626.

70. K. J. Bathe, (ed), *Nonlinear Finite Element Analysis and ADINA*, *J. Computers and Structures*, Vol. 13, No. 5/6, June 1981.

71. W. Wunderlich, E. Stein, and K. J. Bathe, (eds.), *Nonlinear Finite Element Analysis in Structural Mechanics*, Springer-Verlag, Berlin, 1981.

9

SOLUTION OF EQUILIBRIUM EQUATIONS IN DYNAMIC ANALYSIS

9.1 INTRODUCTION

In Section 4.2.1 we derived the equations of equilibrium governing the linear dynamic response of a system of finite elements

$$\mathbf{M}\ddot{\mathbf{U}} + \mathbf{C}\dot{\mathbf{U}} + \mathbf{K}\mathbf{U} = \mathbf{R} \tag{9.1}$$

where $\mathbf{M}$, $\mathbf{C}$, and $\mathbf{K}$ are the mass, damping, and stiffness matrices; $\mathbf{R}$ is the external load vector; and $\mathbf{U}$, $\dot{\mathbf{U}}$, and $\ddot{\mathbf{U}}$ are the displacement, velocity, and acceleration vectors of the finite element assemblage. It should be recalled that (9.1) was derived from considerations of statics at time t; i.e., (9.1) may be written

$$\mathbf{F}_I(t) + \mathbf{F}_D(t) + \mathbf{F}_E(t) = \mathbf{R}(t) \tag{9.2}$$

where $\mathbf{F}_I(t)$ are the inertia forces, $\mathbf{F}_I(t) = \mathbf{M}\ddot{\mathbf{U}}$, $\mathbf{F}_D(t)$ are the damping forces, $\mathbf{F}_D(t) = \mathbf{C}\dot{\mathbf{U}}$, and $\mathbf{F}_E(t)$ are the elastic forces, $\mathbf{F}_E(t) = \mathbf{K}\mathbf{U}$, all of them being time-dependent. Therefore, in dynamic analysis, in principle, static equilibrium at time t, which includes the effect of acceleration-dependent inertia forces and velocity-dependent damping forces, is considered. Vice versa, in static analysis the equations of motion in (9.1) are considered, with inertia and damping effects neglected.

The choice for a static or dynamic analysis (i.e., for including or neglecting velocity- and acceleration-dependent forces in the analysis) is usually decided by engineering judgment, the objective thereby being to reduce the analysis effort required. However, it should be realized that the assumptions of a static analysis should be justified, since otherwise the analysis results are meaningless. Indeed, in nonlinear analysis the assumption of neglecting inertia and damping forces may be so severe that a solution may be difficult or not possible to obtain.

Mathematically, (9.1) represents a system of linear differential equations of second order and, in principle, the solution to the equations can be obtained by standard procedures for the solution of differential equations with constant coefficients.[1-2] However, the procedures proposed for the solution of general systems of differential equations can become very expensive if the order of the matrices is large—unless specific advantage is taken of the special characteristics of the coefficient matrices **K**, **C**, and **M**. In practical finite element analysis, we are therefore mainly interested in a few effective methods and we will concentrate in the next sections on the presentation of those techniques. The procedures that we will consider are divided into two methods of solution: direct integration and mode superposition. Although the two techniques may at first sight appear to be quite different, in fact, they are closely related, and the choice for one method or the other is determined only by their numerical effectiveness.

In the following we consider first (see Sections 9.2 to 9.4) the solution of the linear equilibrium equations (9.1), and then we discuss the solution of the nonlinear equations of finite element systems idealizing structures and solids (see Section 9.5). Finally, we briefly point out in Section 9.6 that the basic concepts discussed are also directly applicable to the analysis of fluid flow, heat transfer, and other problems.

9.2 DIRECT INTEGRATION METHODS

In direct integration the equations in (9.1) are integrated using a numerical step-by-step procedure, the term "direct" meaning that prior to the numerical integration, no transformation of the equations into a different form is carried out. In essence, direct numerical integration is based on two ideas. First, instead of trying to satisfy (9.1) at any time t, it is aimed to satisfy (9.1) only at discrete time intervals Δt apart. This means that, basically, (static) equilibrium, which includes the effect of inertia and damping forces, is sought at discrete time points within the interval of solution. Therefore, it appears that all solution techniques employed in static analysis can probably also be used effectively in direct integration. The second idea on which a direct integration method is based is that a variation of displacements, velocities, and accelerations within each time interval Δt is assumed. As will be discussed in detail, it is the form of the assumption on the variation of displacements, velocities, and accelerations within each time interval that determines the accuracy, stability, and cost of the solution procedure.

In the following, assume that the displacement, velocity, and acceleration vectors at time 0, denoted by ${}^0\mathbf{U}$, ${}^0\dot{\mathbf{U}}$, and ${}^0\ddot{\mathbf{U}}$, respectively, are known, and let the solution to (9.1) be required from time 0 to time T. In the solution the time span under consideration, T, is subdivided into n equal time intervals Δt (i.e., $\Delta t = T/n$), and the integration scheme employed establishes an approximate solution at times 0, Δt, $2\,\Delta t$, $3\,\Delta t$, . . . , t, $t + \Delta t$, . . . , T. Since an algorithm calculates the solution at the next required time from the solutions at the previous times considered, we derive the algorithms by assuming that the solutions at times 0, Δt, $2\Delta t$, . . . , t are known and that the solution at time $t + \Delta t$ is required next. The calculations performed to obtain the solution at time $t + \Delta t$ are typical for calculating the solution at time Δt later than considered so far, and thus establish the general algorithm which can be used to calculate the solution at all discrete time points.

In the following sections a few commonly used effective direct integration methods are presented. Considerations of accuracy, selection of time step size, and a discussion of the advantages of one method over the other are postponed to Section 9.4.

9.2.1 The Central Difference Method

If the equilibrium relation in (9.1) is regarded as a system of ordinary differential equations with constant coefficients, it follows that any convenient finite difference expressions to approximate the accelerations and velocities in terms of displacements can be used. Therefore, theoretically a large number of different finite difference expressions could be employed. However, the solution scheme should be effective, and it follows that only a few schemes need to be considered. One procedure that can be very effective in the solution of some problems is the central difference method,[1] in which it is assumed that

$$ {}^{t}\ddot{\mathbf{U}} = \frac{1}{\Delta t^2}\{{}^{t-\Delta t}\mathbf{U} - 2\,{}^{t}\mathbf{U} + {}^{t+\Delta t}\mathbf{U}\} \tag{9.3} $$

The error in the expansion (9.3) is of order $(\Delta t)^2$, and to have the same order of error in the velocity expansion, we can use

$$ {}^{t}\dot{\mathbf{U}} = \frac{1}{2\Delta t}(-{}^{t-\Delta t}\mathbf{U} + {}^{t+\Delta t}\mathbf{U}) \tag{9.4} $$

The displacement solution for time $t + \Delta t$ is obtained by considering (9.1) at time t, i.e.,

$$ \mathbf{M}\,{}^{t}\ddot{\mathbf{U}} + \mathbf{C}\,{}^{t}\dot{\mathbf{U}} + \mathbf{K}\,{}^{t}\mathbf{U} = {}^{t}\mathbf{R} \tag{9.5} $$

Substituting the relations for ${}^{t}\ddot{\mathbf{U}}$ and ${}^{t}\dot{\mathbf{U}}$ in (9.3) and (9.4), respectively, into (9.5), we obtain

$$ \left(\frac{1}{\Delta t^2}\mathbf{M} + \frac{1}{2\Delta t}\mathbf{C}\right){}^{t+\Delta t}\mathbf{U} = {}^{t}\mathbf{R} - \left(\mathbf{K} - \frac{2}{\Delta t^2}\mathbf{M}\right){}^{t}\mathbf{U} - \left(\frac{1}{\Delta t^2}\mathbf{M} - \frac{1}{2\Delta t}\mathbf{C}\right){}^{t-\Delta t}\mathbf{U} \tag{9.6} $$

from which we can solve for ${}^{t+\Delta t}\mathbf{U}$. It should be noted that the solution of ${}^{t+\Delta t}\mathbf{U}$ is thus based on using the equilibrium conditions at time t; i.e., ${}^{t+\Delta t}\mathbf{U}$ is calculated by using (9.5). For this reason the integration procedure is called an *explicit integration method*, and it is noted that such integration schemes do not require a factorization of the (effective) stiffness matrix in the step-by-step solution. On the other hand, the Houbolt, Wilson, and Newmark methods, considered in the next sections, use the equilibrium conditions at time $t + \Delta t$ and are called *implicit integration methods*.

A second observation is that using the central difference method, the calculation of ${}^{t+\Delta t}\mathbf{U}$ involves ${}^{t}\mathbf{U}$ and ${}^{t-\Delta t}\mathbf{U}$. Therefore, to calculate the solution at time Δt, a special starting procedure must be used. Since ${}^{0}\mathbf{U}$, ${}^{0}\dot{\mathbf{U}}$, and ${}^{0}\ddot{\mathbf{U}}$ are known [note that with ${}^{0}\mathbf{U}$ and ${}^{0}\dot{\mathbf{U}}$ known, ${}^{0}\ddot{\mathbf{U}}$ can be calculated using (9.1) at time 0; see Example 9.1], the relations in (9.3) and (9.4) can be used to obtain ${}^{-\Delta t}\mathbf{U}$; i.e., we have

$$ {}^{-\Delta t}U_i = {}^{0}U_i - \Delta t\,{}^{0}\dot{U}_i + \frac{\Delta t^2}{2}\,{}^{0}\ddot{U}_i \tag{9.7} $$

where the superscript (i) indicates the ith element of the vector considered.

Table 9.1 summarizes the time integration scheme as it might be implemented in the computer.

TABLE 9.1 *Step-by-step solution using central difference method (general mass and damping matrices).*

A. *Initial Calculations:*

1. Form stiffness matrix $\mathbf{K}$, mass matrix $\mathbf{M}$, and damping matrix $\mathbf{C}$.
2. Initialize ${}^0\mathbf{U}$, ${}^0\dot{\mathbf{U}}$, and ${}^0\ddot{\mathbf{U}}$.
3. Select time step Δt, $\Delta t < \Delta t_{cr}$, and calculate integration constants:

$$a_0 = \frac{1}{\Delta t^2}; \qquad a_1 = \frac{1}{2\Delta t}; \qquad a_2 = 2a_0; \qquad a_3 = \frac{1}{a_2}$$

4. Calculate ${}^{-\Delta t}\mathbf{U} = {}^0\mathbf{U} - \Delta t\, {}^0\dot{\mathbf{U}} + a_3\, {}^0\ddot{\mathbf{U}}$.
5. Form effective mass matrix $\hat{\mathbf{M}} = a_0\mathbf{M} + a_1\mathbf{C}$.
6. Triangularize $\hat{\mathbf{M}}$: $\hat{\mathbf{M}} = \mathbf{L}\mathbf{D}\mathbf{L}^T$.

B. *For Each Time Step:*

1. Calculate effective loads at time t:

$${}^t\hat{\mathbf{R}} = {}^t\mathbf{R} - (\mathbf{K} - a_2\mathbf{M})\, {}^t\mathbf{U} - (a_0\mathbf{M} - a_1\mathbf{C})\, {}^{t-\Delta t}\mathbf{U}$$

2. Solve for displacements at time $t + \Delta t$:

$$\mathbf{L}\mathbf{D}\mathbf{L}^T\, {}^{t+\Delta t}\mathbf{U} = {}^t\hat{\mathbf{R}}$$

3. If required, evaluate accelerations and velocities at time t:

$${}^t\ddot{\mathbf{U}} = a_0({}^{t-\Delta t}\mathbf{U} - 2\, {}^t\mathbf{U} + {}^{t+\Delta t}\mathbf{U})$$
$${}^t\dot{\mathbf{U}} = a_1(-{}^{t-\Delta t}\mathbf{U} + {}^{t+\Delta t}\mathbf{U})$$

Assume that the system has no physical damping; i.e., $\mathbf{C}$ is zero. In this case (9.6) reduces to

$$\left(\frac{1}{\Delta t^2}\mathbf{M}\right) {}^{t+\Delta t}\mathbf{U} = {}^t\hat{\mathbf{R}} \tag{9.8}$$

where

$${}^t\hat{\mathbf{R}} = {}^t\mathbf{R} - \left(\mathbf{K} - \frac{2}{\Delta t^2}\mathbf{M}\right) {}^t\mathbf{U} - \left(\frac{1}{\Delta t^2}\mathbf{M}\right) {}^{t-\Delta t}\mathbf{U} \tag{9.9}$$

Therefore, if the mass matrix is diagonal, the system of equations in (9.1) can be solved without factorizing a matrix; i.e., only matrix multiplications are required to obtain the right-hand side effective load vector ${}^t\hat{\mathbf{R}}$ after which the displacement components are obtained, using

$${}^{t+\Delta t}U_i = {}^t\hat{R}_i\left(\frac{\Delta t^2}{m_{ii}}\right) \tag{9.10}$$

where ${}^{t+\Delta t}U_i$ and ${}^t\hat{R}_i$ denote the ith components of the vectors ${}^{t+\Delta t}\mathbf{U}$ and ${}^t\hat{\mathbf{R}}$, respectively, and m_{ii} is the ith diagonal element of the mass matrix, and it is assumed that $m_{ii} > 0$.

If neither the stiffness nor the mass matrix of the element assemblage is to be triangularized, it is also not necessary to assemble $\mathbf{K}$ and $\mathbf{M}$. We have shown in Section 4.2.1 that

$$\mathbf{K} = \sum_i \mathbf{K}^{(i)}; \qquad \mathbf{M} = \sum_i \mathbf{M}^{(i)} \tag{9.11}$$

which means that $\mathbf{K}\, {}^t\mathbf{U}$, $(2\mathbf{M}/\Delta t^2)\, {}^t\mathbf{U}$, and $(\mathbf{M}/\Delta t^2)\, {}^{t-\Delta t}\mathbf{U}$, as required in (9.9), can be calculated on the element level by summing the contributions from each element to the effective load vector. Hence ${}^t\hat{\mathbf{R}}$ is evaluated using

$$
{}^t\hat{\mathbf{R}} = {}^t\mathbf{R} - \sum_i (\mathbf{K}^{(i)}\,{}^t\mathbf{U}) - \sum_i \frac{1}{\Delta t^2}\mathbf{M}^{(i)}({}^{t-\Delta t}\mathbf{U} - 2\,{}^t\mathbf{U}) \tag{9.12}
$$

where it is important to note that the products $\mathbf{K}^{(i)}\,{}^t\mathbf{U}$, and $\mathbf{M}^{(i)}\,({}^{t-\Delta t}\mathbf{U} - 2\,{}^t\mathbf{U})$ are evaluated using $\mathbf{K}^{(i)}$ and $\mathbf{M}^{(i)}$ in compacted form; i.e., $\mathbf{K}_i^e$ and $\mathbf{M}_i^e$ are employed (see Section 4.2 and the Appendix). Furthermore, we also recognize that $\mathbf{K}\,{}^t\mathbf{U}$ is equal to the nodal point forces corresponding to the element stresses at time t, ${}^t\mathbf{F}$, and could also be evaluated as described in Section 6.3, see (6.57), and ${}^t\mathbf{F}$ is used as such in (9.103).

The advantage of using the central difference method in the form given in (9.10) and (9.12) now becomes apparent. Since no stiffness and mass matrices of the complete element assemblage need to be calculated, the solution can essentially be carried out on the element level and relatively little high-speed storage is required. The method becomes even more effective if element stiffness and mass matrices of subsequent elements are the same, because in that case it is only necessary to calculate or read from back-up storage the matrices corresponding to the first element in the series. Using the central difference scheme as given in (9.10) and (9.12), systems of very large order have been solved effectively.

Considering the shortcomings of the central difference method, it must be recognized that the effectiveness of the procedure depends on the use of a diagonal mass matrix and the neglect of general velocity-dependent damping forces. If only a diagonal damping matrix is included, the benefits of performing the solution on the element level are preserved. For practical purposes the disadvantage that only a diagonal mass matrix can be employed is usually not very serious, because good accuracy of solution can be obtained by using a fine-enough finite element discretization.

A second very important consideration in the use of the central difference scheme is that the integration method requires that the time step Δt is smaller than a critical value, Δt_{cr}, which can be calculated from the mass and stiffness properties of the complete element assemblage. More specifically, we will show in Section 9.4.2 that to obtain a valid solution,

$$
\Delta t \leq \Delta t_{cr} = \frac{T_n}{\pi} \tag{9.13}
$$

where T_n is the smallest period of the finite element assemblage with n degrees of freedom. The period T_n could be calculated using one of the techniques discussed in Chapters 11 and 12, or a lower bound on T_n may be evaluated using norms (see Section 2.9). In practice we frequently estimate an appropriate time step Δt using the considerations given in Section 9.4.4.

In the solution using (9.10), it was assumed that $m_{ii} > 0$ for all i. The relation in (9.13) states this requirement once more, because a zero diagonal element in a diagonal mass matrix means that the element assemblage has a zero period (see Section 10.2.4). In general, all diagonal elements of the mass matrix can be assumed to be larger than zero, in which case (9.13) gives a limit on the magnitude of the time step Δt that can be used in the integration. In the analysis of some problems (9.13) may not require an unduly small time step; but, in general, the time step that should be small enough for accuracy of integration may be many times larger than Δt_{cr} obtained from (9.13). Since the total cost of analysis is approximately inversely proportional to the magnitude of time step used, it follows that if the time step can be m times as large, the cost would be reduced by a factor of m.

We discuss the selection of the time step Δt for direct integration in Section 9.4. However, the reason for Δt to be artificially small in some analyses can already be explained using a simple argument. Assume that we consider the direct integration of the equilibrium equations in (9.1) when the order n of the matrices is relatively large, say n is 100, at least. The time step for the integration would be selected using (9.13). Assume that we now change the smallest diagonal element of the mass matrix to become very small and, in fact, nearly zero. As enumerated above, a diagonal element in the mass matrix cannot be exactly zero, because T_n would then be zero and the integration would not be possible. However, as the diagonal element in the mass matrix approaches zero, the smallest period of the system, and hence Δt_{cr}, approaches zero. Therefore, the reduction of one mass element necessitates a severe reduction in the time step size that can be used in the integration. On the other hand, since the order of the system is large, we would hardly expect that the dynamic response of the element assemblage changes very much when the smallest mass element is reduced, even to become zero. Hence the cost of analysis would be unduly large, only because of one very small mass element. The same condition is also reached when the stiffness of only one element is changed to become large.

Integration schemes that require the use of a time step Δt smaller than a critical time step Δt_{cr}, such as the central difference method, are said to be conditionally stable. If a time step is used larger than Δt_{cr}, the integration is unstable, meaning that any errors resulting from the numerical integration or round-off in the computer grow and make the response calculations worthless in most cases. The concept of stability of integration is very important, and we will discuss it further in Section 9.4. However, at this stage it is useful to consider the following example.

EXAMPLE 9.1: Consider a simple system for which the governing equilibrium equations are

$$\begin{bmatrix} 2 & 0 \\ 0 & 1 \end{bmatrix}\begin{bmatrix} \ddot{U}_1 \\ \ddot{U}_2 \end{bmatrix} + \begin{bmatrix} 6 & -2 \\ -2 & 4 \end{bmatrix}\begin{bmatrix} U_1 \\ U_2 \end{bmatrix} = \begin{bmatrix} 0 \\ 10 \end{bmatrix} \tag{a}$$

The free vibration periods of the system are given in Example 9.6, where we find that $T_1 = 4.45$, $T_2 = 2.8$. Use the central difference method in direct integration with time steps (1) $\Delta t = T_2/10$ and (2) $\Delta t = 10T_2$ to calculate the response of the system for 12 steps. Assume that ${}^0\mathbf{U} = \mathbf{0}$ and ${}^0\dot{\mathbf{U}} = \mathbf{0}$.

The first step is to calculate ${}^0\ddot{\mathbf{U}}$ using the equations in (a) at time 0; i.e., we use

$$\begin{bmatrix} 2 & 0 \\ 0 & 1 \end{bmatrix}{}^0\ddot{\mathbf{U}} + \begin{bmatrix} 6 & -2 \\ -2 & 4 \end{bmatrix}\begin{bmatrix} 0 \\ 0 \end{bmatrix} = \begin{bmatrix} 0 \\ 10 \end{bmatrix}$$

Hence

$${}^0\ddot{\mathbf{U}} = \begin{bmatrix} 0 \\ 10 \end{bmatrix}$$

Now we follow the calculations in Table 9.1.

Consider case (1), in which $\Delta t = 0.28$. We then have

$$a_0 = \frac{1}{(0.28)^2} = 12.8; \qquad a_1 = \frac{1}{(2)(0.28)} = 1.79$$

$$a_2 = 2a_0 = 25.5; \qquad a_3 = \frac{1}{a_2} = 0.0392$$

Hence

$$^{-\Delta t}\mathbf{U} = \begin{bmatrix} 0 \\ 0 \end{bmatrix} - 0.28\begin{bmatrix} 0 \\ 0 \end{bmatrix} + 0.0392\begin{bmatrix} 0 \\ 10 \end{bmatrix} = \begin{bmatrix} 0 \\ 0.392 \end{bmatrix}$$

$$\hat{\mathbf{M}} = 12.8\begin{bmatrix} 2 & 0 \\ 0 & 1 \end{bmatrix} + 1.79\begin{bmatrix} 0 & 0 \\ 0 & 0 \end{bmatrix} = \begin{bmatrix} 25.5 & 0 \\ 0 & 12.8 \end{bmatrix}$$

The effective loads at time t are

$$^t\hat{\mathbf{R}} = \begin{bmatrix} 0 \\ 10 \end{bmatrix} + \begin{bmatrix} 45.0 & 2 \\ 2 & 21.5 \end{bmatrix} {}^t\mathbf{U} - \begin{bmatrix} 25.5 & 0 \\ 0 & 12.8 \end{bmatrix} {}^{t-\Delta t}\mathbf{U}$$

Hence we need to solve the following equations for each time step,

$$\begin{bmatrix} 25.5 & 0 \\ 0 & 12.8 \end{bmatrix} {}^{t+\Delta t}\mathbf{U} = {}^t\hat{\mathbf{R}} \qquad \text{(b)}$$

It should be noted that the solution of the equations in (b) is trivial because the coefficient matrix is diagonal.

Calculating the solution to (b) for each time step, we obtain

Time	Δt	$2\Delta t$	$3\Delta t$	$4\Delta t$	$5\Delta t$	$6\Delta t$	$7\Delta t$	$8\Delta t$	$9\Delta t$	$10\Delta t$	$11\Delta t$	$12\Delta t$
$^t\mathbf{U}$	0	0.0307	0.168	0.487	1.02	1.70	2.40	2.91	3.07	2.77	2.04	1.02
	0.392	1.45	2.83	4.14	5.02	5.26	4.90	4.17	3.37	2.78	2.54	2.60

The solution obtained is compared with the exact results in Example 9.7.

Consider now case (2), in which $\Delta t = 28$. Following through the same calculations, we find that

$$^{\Delta t}\mathbf{U} = \begin{bmatrix} 0 \\ 3.83 \times 10^3 \end{bmatrix}; \qquad {}^{2\Delta t}\mathbf{U} = \begin{bmatrix} 3.03 \times 10^6 \\ -1.21 \times 10^7 \end{bmatrix}$$

and the calculated displacements continue to increase. Since the time step Δt is about 6 times larger than T_1 and 10 times larger than T_2, we can certainly not expect accuracy in the numerical integration. But of particular interest is whether the calculated values decrease or increase. The increase in the values as observed in this example is a consequence of the time integration scheme not being stable. As pointed out above, the time step Δt must not be larger than Δt_{cr} for stability in the integration using the central difference method, where $\Delta t_{cr} = (1/\pi)T_2$. In this case the time step Δt is much larger, and the calculated response increases without bound. This is the typical phenomenon of instability. We shall see in Examples 9.2, 9.3, and 9.4 that the response predicted using $\Delta t = 28$ with the unconditionally stable Houbolt, Wilson θ, and Newmark methods is also very inaccurate, but does not increase.

We discussed above the main disadvantage of the central difference method: the scheme is only conditionally stable. Various other integration methods are also conditionally stable.[1] Since the effective use of conditionally stable methods is limited to certain problems, we consider in the following sections commonly employed integration schemes which are unconditionally stable. The effectiveness of unconditionally stable integration schemes derives from the fact that to obtain accuracy in the integration, the time step Δt can be selected without a requirement such as (9.13), and in many cases Δt can be orders of magnitudes larger than (9.13) would allow. However, the integration methods discussed in the following are implicit; i.e., a triangularization of the stiffness matrix $\mathbf{K}$, or rather of an effective stiffness matrix, is required for solution.

9.2.2 The Houbolt Method

The Houbolt integration scheme is somewhat related to the previously discussed central difference method in that standard finite difference expressions are used to approximate the acceleration and velocity components in terms of the displacement components. The following finite difference expansions are employed in the Houbolt integration method:[3]

$$ {}^{t+\Delta t}\ddot{\mathbf{U}} = \frac{1}{\Delta t^2}\{2\,{}^{t+\Delta t}\mathbf{U} - 5\,{}^{t}\mathbf{U} + 4\,{}^{t-\Delta t}\mathbf{U} - {}^{t-2\Delta t}\mathbf{U}\} \tag{9.14} $$

and

$$ {}^{t+\Delta t}\dot{\mathbf{U}} = \frac{1}{6\Delta t}\{11\,{}^{t+\Delta t}\mathbf{U} - 18\,{}^{t}\mathbf{U} + 9\,{}^{t-\Delta t}\mathbf{U} - 2\,{}^{t-2\Delta t}\mathbf{U}\} \tag{9.15} $$

which are two backward-difference formulas with errors of order $(\Delta t)^2$.

In order to obtain the solution at time $t + \Delta t$, we now consider (9.1) at time $t + \Delta t$ (and not at time t as for the central difference method), which gives

$$ \mathbf{M}\,{}^{t+\Delta t}\ddot{\mathbf{U}} + \mathbf{C}\,{}^{t+\Delta t}\dot{\mathbf{U}} + \mathbf{K}\,{}^{t+\Delta t}\mathbf{U} = {}^{t+\Delta t}\mathbf{R} \tag{9.16} $$

Substituting (9.14) and (9.15) into (9.16) and arranging all known vectors on the right-hand side, we obtain for the solution of ${}^{t+\Delta t}\mathbf{U}$,

$$ \begin{aligned} \left(\frac{2}{\Delta t^2}\mathbf{M} + \frac{11}{6\Delta t}\mathbf{C} + \mathbf{K}\right){}^{t+\Delta t}\mathbf{U} &= {}^{t+\Delta t}\mathbf{R} + \left(\frac{5}{\Delta t^2}\mathbf{M} + \frac{3}{\Delta t}\mathbf{C}\right){}^{t}\mathbf{U} \\ &\quad - \left(\frac{4}{\Delta t^2}\mathbf{M} + \frac{3}{2\Delta t}\mathbf{C}\right){}^{t-\Delta t}\mathbf{U} + \left(\frac{1}{\Delta t^2}\mathbf{M} + \frac{1}{3\Delta t}\mathbf{C}\right){}^{t-2\Delta t}\mathbf{U} \end{aligned} \tag{9.17} $$

As shown in (9.17), the solution of ${}^{t+\Delta t}\mathbf{U}$ requires knowledge of ${}^{t}\mathbf{U}$, ${}^{t-\Delta t}\mathbf{U}$, and ${}^{t-2\Delta t}\mathbf{U}$. Although the knowledge of ${}^{0}\mathbf{U}$, ${}^{0}\dot{\mathbf{U}}$, and ${}^{0}\ddot{\mathbf{U}}$ is useful to start the Houbolt integration scheme, it is more accurate to calculate ${}^{\Delta t}\mathbf{U}$ and ${}^{2\Delta t}\mathbf{U}$ by some other means; i.e., we employ special starting procedures. One way of proceeding is to integrate (9.1) for the solution of ${}^{\Delta t}\mathbf{U}$ and ${}^{2\Delta t}\mathbf{U}$ using a different integration scheme, possibly a conditionally stable method such as the central difference scheme with a fraction of Δt as the time step (see Example 9.2). Table 9.2 summarizes the Houbolt integration procedure for use in a computer program.

A basic difference between the Houbolt method in Table 9.2 and the central difference scheme in Table 9.1 is the appearance of the stiffness matrix $\mathbf{K}$ as a factor to the required displacements ${}^{t+\Delta t}\mathbf{U}$. The term $\mathbf{K}\,{}^{t+\Delta t}\mathbf{U}$ appears because in (9.16) equilibrium is considered at time $t + \Delta t$ and not at time t as in the central difference method. The Houbolt method is, for this reason, an implicit integration scheme, whereas the central difference method was an explicit proce-

TABLE 9.2 *Step-by-step solution using Houbolt integration method.*

A. *Initial Calculations:*

1. Form stiffness matrix $\mathbf{K}$, mass matrix $\mathbf{M}$, and damping matrix $\mathbf{C}$.
2. Initialize ${}^0\mathbf{U}$, ${}^0\dot{\mathbf{U}}$, and ${}^0\ddot{\mathbf{U}}$.
3. Select time step Δt and calculate integration constants:

$$a_0 = \frac{2}{\Delta t^2}; \quad a_1 = \frac{11}{6\Delta t}; \quad a_2 = \frac{5}{\Delta t^2}; \quad a_3 = \frac{3}{\Delta t}; \quad a_4 = -2a_0;$$

$$a_5 = \frac{-a_3}{2}; \quad a_6 = \frac{a_0}{2}; \quad a_7 = \frac{a_3}{9}$$

4. Use special starting procedure to calculate ${}^{\Delta t}\mathbf{U}$ and ${}^{2\Delta t}\mathbf{U}$.
5. Calculate effective stiffness matrix $\hat{\mathbf{K}}$: $\hat{\mathbf{K}} = \mathbf{K} + a_0\mathbf{M} + a_1\mathbf{C}$.
6. Triangularize $\hat{\mathbf{K}}$: $\hat{\mathbf{K}} = \mathbf{LDL}^T$.

B. *For Each Time Step:*

1. Calculate effective loads at time $t + \Delta t$:

$${}^{t+\Delta t}\hat{\mathbf{R}} = {}^{t+\Delta t}\mathbf{R} + \mathbf{M}(a_2\,{}^t\mathbf{U} + a_4\,{}^{t-\Delta t}\mathbf{U} + a_6\,{}^{t-2\Delta t}\mathbf{U}) + \mathbf{C}(a_3\,{}^t\mathbf{U} + a_5\,{}^{t-\Delta t}\mathbf{U} + a_7\,{}^{t-2\Delta t}\mathbf{U})$$

2. Solve for displacements at time $t + \Delta t$:

$$\mathbf{LDL}^T\,{}^{t+\Delta t}\mathbf{U} = {}^{t+\Delta t}\hat{\mathbf{R}}$$

3. If required, evaluate accelerations and velocities at time $t + \Delta t$:

$${}^{t+\Delta t}\ddot{\mathbf{U}} = a_0\,{}^{t+\Delta t}\mathbf{U} - a_2\,{}^t\mathbf{U} - a_4\,{}^{t-\Delta t}\mathbf{U} - a_6\,{}^{t-2\Delta t}\mathbf{U}$$
$${}^{t+\Delta t}\dot{\mathbf{U}} = a_1\,{}^{t+\Delta t}\mathbf{U} - a_3\,{}^t\mathbf{U} - a_5\,{}^{t-\Delta t}\mathbf{U} - a_7\,{}^{t-2\Delta t}\mathbf{U}$$

dure. With regard to the time step Δt that can be used in the integration, there is no critical time step limit, and Δt can in general be selected much larger than given in (9.13) for the central difference method.

A noteworthy point is that the step-by-step solution scheme based on the Houbolt method reduces directly to a static analysis, if mass and damping effects are neglected, whereas the central difference method solution in Table 9.1 could not be used. In other words, if $\mathbf{C} = \mathbf{0}$ and $\mathbf{M} = \mathbf{0}$, the solution method in Table 9.2 yields the static solution for time-dependent loads.

EXAMPLE 9.2: Use the Houbolt direct integration scheme to calculate the response of the system considered in Example 9.1.

First we consider the case $\Delta t = 0.28$. We then have, following Table 9.2:

$$a_0 = 25.5; \quad a_1 = 6.55; \quad a_2 = 63.8; \quad a_3 = 10.7;$$
$$a_4 = -51.0; \quad a_5 = -5.36; \quad a_6 = 12.8; \quad a_7 = 1.19$$

To start the integration we need ${}^{\Delta t}\mathbf{U}$ and ${}^{2\Delta t}\mathbf{U}$. Let us here use simply the values calculated with the central difference method in Example 9.1, i.e.,

$${}^{\Delta t}\mathbf{U} = \begin{bmatrix} 0.0 \\ 0.392 \end{bmatrix}; \qquad {}^{2\Delta t}\mathbf{U} = \begin{bmatrix} 0.0307 \\ 1.45 \end{bmatrix}$$

Next we calculate $\hat{\mathbf{K}}$ and obtain

$$\hat{\mathbf{K}} = \begin{bmatrix} 6 & -2 \\ -2 & 4 \end{bmatrix} + 25.5\begin{bmatrix} 2 & 0 \\ 0 & 1 \end{bmatrix} = \begin{bmatrix} 57 & -2 \\ -2 & 29.5 \end{bmatrix}$$

For each time step we need ${}^{t+\Delta t}\hat{\mathbf{R}}$, which is in this case

$$ {}^{t+\Delta t}\hat{\mathbf{R}} = \begin{bmatrix} 0 \\ 10 \end{bmatrix} + \begin{bmatrix} 2 & 0 \\ 0 & 1 \end{bmatrix} \{63.8\,{}^{t}\mathbf{U} - 51.0\,{}^{t-\Delta t}\mathbf{U} + 12.8\,{}^{t-2\Delta t}\mathbf{U}\} $$

Solving $\hat{\mathbf{K}}\,{}^{t+\Delta t}\mathbf{U} = {}^{t+\Delta t}\hat{\mathbf{R}}$ for 12 time steps, we obtain

Time	Δt	$2\Delta t$	$3\Delta t$	$4\Delta t$	$5\Delta t$	$6\Delta t$	$7\Delta t$	$8\Delta t$	$9\Delta t$	$10\Delta t$	$11\Delta t$	$12\Delta t$
${}^{t}\mathbf{U}$	0 0.392	0.0307 1.45	0.167 2.80	0.461 4.08	0.923 5.02	1.50 5.43	2.11 5.31	2.60 4.77	2.86 4.01	2.80 3.24	2.40 2.63	1.72 2.28

The solution obtained is compared with the exact results in Example 9.7.

Next we consider the case $\Delta t = 28$ in order to observe the unconditional stability of the Houbolt operator. To start the integration we use the exact response at times Δt and $2\Delta t$ (see Example 9.7),

$$ {}^{\Delta t}\mathbf{U} = \begin{bmatrix} 2.19 \\ 2.24 \end{bmatrix}; \qquad {}^{2\Delta t}\mathbf{U} = \begin{bmatrix} 2.92 \\ 3.12 \end{bmatrix} $$

It is interesting to compare $\hat{\mathbf{K}}$ with $\mathbf{K}$,

$$ \hat{\mathbf{K}} = \begin{bmatrix} 6 & -2 \\ -2 & 4 \end{bmatrix} + 0.00255 \begin{bmatrix} 2 & 0 \\ 0 & 1 \end{bmatrix} = \begin{bmatrix} 6.0051 & -2.0000 \\ -2.0000 & 4.00255 \end{bmatrix} $$

where it is noted that $\hat{\mathbf{K}}$ is almost equal to $\mathbf{K}$. The displacement response over 12 time steps is given in the following table:

Time	Δt	$2\Delta t$	$3\Delta t$	$4\Delta t$	$5\Delta t$	$6\Delta t$	$7\Delta t$	$8\Delta t$	$9\Delta t$	$10\Delta t$	$11\Delta t$	$12\Delta t$
${}^{t}\mathbf{U}$	2.19 2.24	2.92 3.12	1.00 3.00	1.00 3.00	1.00 3.00	1.00 3.00	1.00 3.00	1.00 3.00	1.00 3.00	1.00 3.00	1.00 3.00	1.00 3.00

It should be noted that the static solution is

$$ {}^{t}\mathbf{U} = \begin{bmatrix} 1.0 \\ 3.0 \end{bmatrix} $$

Therefore, the displacement response very rapidly approaches the static solution.

9.2.3 The Wilson θ Method

The Wilson θ method is essentially an extension of the linear acceleration method, in which a linear variation of acceleration from time t to time $t + \Delta t$ is assumed. Referring to Fig. 9.1, in the Wilson θ method the acceleration is assumed to be linear from time t to time $t + \theta\,\Delta t$, where $\theta \geq 1.0$.[4] When $\theta = 1.0$, the method reduces to the linear acceleration scheme, but we will show in Section 9.4 that for unconditional stability we need to use $\theta \geq 1.37$, and usually we employ $\theta = 1.40$.

Let τ denote the increase in time, where $0 \leq \tau \leq \theta\,\Delta t$; then for the time interval t to $t + \theta\,\Delta t$, it is assumed that

$$ {}^{t+\tau}\ddot{\mathbf{U}} = {}^{t}\ddot{\mathbf{U}} + \frac{\tau}{\theta\Delta t}({}^{t+\theta\Delta t}\ddot{\mathbf{U}} - {}^{t}\ddot{\mathbf{U}}) \tag{9.18} $$

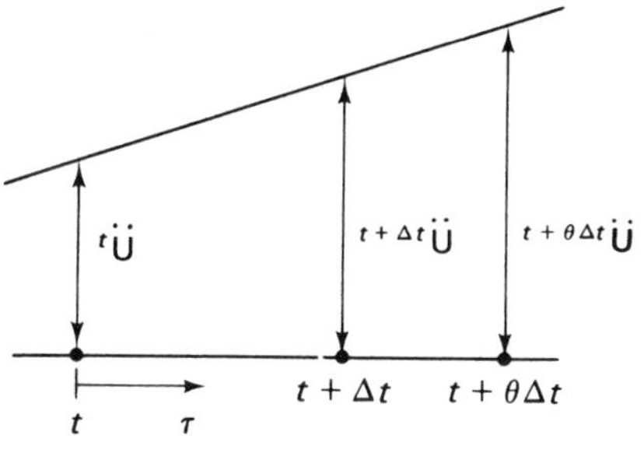

FIGURE 9.1 *Linear acceleration assumption of Wilson θ method.*

Integrating (9.18), we obtain

$$ {}^{t+\tau}\dot{\mathbf{U}} = {}^t\dot{\mathbf{U}} + {}^t\ddot{\mathbf{U}}\tau + \frac{\tau^2}{2\theta\Delta t}({}^{t+\theta\Delta t}\ddot{\mathbf{U}} - {}^t\ddot{\mathbf{U}}) \tag{9.19} $$

and

$$ {}^{t+\tau}\mathbf{U} = {}^t\mathbf{U} + {}^t\dot{\mathbf{U}}\tau + \frac{1}{2}{}^t\ddot{\mathbf{U}}\tau^2 + \frac{1}{6\theta\Delta t}\tau^3({}^{t+\theta\Delta t}\ddot{\mathbf{U}} - {}^t\ddot{\mathbf{U}}) \tag{9.20} $$

Using (9.19) and (9.20), we have at time $t + \theta\Delta t$,

$$ {}^{t+\theta\Delta t}\dot{\mathbf{U}} = {}^t\dot{\mathbf{U}} + \frac{\theta\,\Delta t}{2}({}^{t+\theta\Delta t}\ddot{\mathbf{U}} + {}^t\ddot{\mathbf{U}}) \tag{9.21} $$

$$ {}^{t+\theta\Delta t}\mathbf{U} = {}^t\mathbf{U} + \theta\,\Delta t\,{}^t\dot{\mathbf{U}} + \frac{\theta^2\,\Delta t^2}{6}({}^{t+\theta\Delta t}\ddot{\mathbf{U}} + 2\,{}^t\ddot{\mathbf{U}}) \tag{9.22} $$

from which we can solve for ${}^{t+\theta\Delta t}\ddot{\mathbf{U}}$ and ${}^{t+\theta\Delta t}\dot{\mathbf{U}}$ in terms of ${}^{t+\theta\Delta t}\mathbf{U}$:

$$ {}^{t+\theta\Delta t}\ddot{\mathbf{U}} = \frac{6}{\theta^2\,\Delta t^2}({}^{t+\theta\Delta t}\mathbf{U} - {}^t\mathbf{U}) - \frac{6}{\theta\,\Delta t}{}^t\dot{\mathbf{U}} - 2\,{}^t\ddot{\mathbf{U}} \tag{9.23} $$

and

$$ {}^{t+\theta\Delta t}\dot{\mathbf{U}} = \frac{3}{\theta\,\Delta t}({}^{t+\theta\Delta t}\mathbf{U} - {}^t\mathbf{U}) - 2\,{}^t\dot{\mathbf{U}} - \frac{\theta\,\Delta t}{2}{}^t\ddot{\mathbf{U}} \tag{9.24} $$

To obtain the solution for the displacements, velocities, and accelerations at time $t + \Delta t$, the equilibrium equations (9.1) are considered at time $t + \theta\,\Delta t$. However, because the accelerations are assumed to vary linearly, a linearly projected load vector is used; i.e., the equation employed is

$$ \mathbf{M}\,{}^{t+\theta\Delta t}\ddot{\mathbf{U}} + \mathbf{C}\,{}^{t+\theta\Delta t}\dot{\mathbf{U}} + \mathbf{K}\,{}^{t+\theta\Delta t}\mathbf{U} = {}^{t+\theta\Delta t}\bar{\mathbf{R}} \tag{9.25} $$

where

$$ {}^{t+\theta\Delta t}\bar{\mathbf{R}} = {}^t\mathbf{R} + \theta({}^{t+\Delta t}\mathbf{R} - {}^t\mathbf{R}) \tag{9.26} $$

Substituting (9.23) and (9.24) into (9.25), an equation is obtained from which ${}^{t+\theta\Delta t}\mathbf{U}$ can be solved. Then substituting ${}^{t+\theta\Delta t}\mathbf{U}$ into (9.23) we obtain ${}^{t+\theta\Delta t}\ddot{\mathbf{U}}$, which is used in (9.18), (9.19), and (9.20), all evaluated at $\tau = \Delta t$ to calculate ${}^{t+\Delta t}\ddot{\mathbf{U}}$, ${}^{t+\Delta t}\dot{\mathbf{U}}$, and ${}^{t+\Delta t}\mathbf{U}$. The complete algorithm used in the integration is given in Table 9.3.

As pointed out earlier, the Wilson θ method is also an implicit integration method, because the stiffness matrix $\mathbf{K}$ is a coefficient matrix to the unknown displacement vector. It may also be noted that no special starting procedures are needed, since the displacements, velocities, and accelerations at time $t + \Delta t$ are expressed in terms of the same quantities at time t only.

TABLE 9.3 *Step-by-step solution using Wilson θ integration method.*

A. *Initial Calculations:*

1. Form stiffness matrix $\mathbf{K}$, mass matrix $\mathbf{M}$, and damping matrix $\mathbf{C}$.
2. Initialize ${}^0\mathbf{U}$, ${}^0\dot{\mathbf{U}}$, and ${}^0\ddot{\mathbf{U}}$.
3. Select time step Δt and calculate integration constants, $\theta = 1.4$ (usually):

$$a_0 = \frac{6}{(\theta\,\Delta t)^2}; \quad a_1 = \frac{3}{\theta\,\Delta t}; \quad a_2 = 2a_1; \quad a_3 = \frac{\theta\,\Delta t}{2}; \quad a_4 = \frac{a_0}{\theta};$$

$$a_5 = \frac{-a_2}{\theta}; \quad a_6 = 1 - \frac{3}{\theta}; \quad a_7 = \frac{\Delta t}{2}; \quad a_8 = \frac{\Delta t^2}{6}$$

4. Form effective stiffness matrix $\hat{\mathbf{K}}$: $\hat{\mathbf{K}} = \mathbf{K} + a_0\mathbf{M} + a_1\mathbf{C}$.
5. Triangularize $\hat{\mathbf{K}}$: $\hat{\mathbf{K}} = \mathbf{LDL}^T$.

B. *For Each Time Step:*

1. Calculate effective loads at time $t + \theta\,\Delta t$:

$${}^{t+\theta\Delta t}\hat{\mathbf{R}} = {}^t\mathbf{R} + \theta({}^{t+\Delta t}\mathbf{R} - {}^t\mathbf{R}) + \mathbf{M}(a_0\,{}^t\mathbf{U} + a_2\,{}^t\dot{\mathbf{U}} + 2\,{}^t\ddot{\mathbf{U}}) + \mathbf{C}(a_1\,{}^t\mathbf{U} + 2\,{}^t\dot{\mathbf{U}} + a_3\,{}^t\ddot{\mathbf{U}})$$

2. Solve for displacements at time $t + \theta\,\Delta t$:

$$\mathbf{LDL}^T\,{}^{t+\theta\Delta t}\mathbf{U} = {}^{t+\theta\Delta t}\hat{\mathbf{R}}$$

3. Calculate displacements, velocities, and accelerations at time $t + \Delta t$:

$$\begin{aligned}
{}^{t+\Delta t}\ddot{\mathbf{U}} &= a_4({}^{t+\theta\Delta t}\mathbf{U} - {}^t\mathbf{U}) + a_5\,{}^t\dot{\mathbf{U}} + a_6\,{}^t\ddot{\mathbf{U}} \\
{}^{t+\Delta t}\dot{\mathbf{U}} &= {}^t\dot{\mathbf{U}} + a_7({}^{t+\Delta t}\ddot{\mathbf{U}} + {}^t\ddot{\mathbf{U}}) \\
{}^{t+\Delta t}\mathbf{U} &= {}^t\mathbf{U} + \Delta t\,{}^t\dot{\mathbf{U}} + a_8({}^{t+\Delta t}\ddot{\mathbf{U}} + 2\,{}^t\ddot{\mathbf{U}})
\end{aligned}$$

EXAMPLE 9.3: Calculate the displacement response of the system considered in Examples 9.1 and 9.2 using the Wilson θ method. Use $\theta = 1.4$.

First we consider the case $\Delta t = 0.28$. Following the steps of calculations in Table 9.3, we have

$${}^0\mathbf{U} = \begin{bmatrix} 0 \\ 0 \end{bmatrix}; \quad {}^0\dot{\mathbf{U}} = \begin{bmatrix} 0 \\ 0 \end{bmatrix}; \quad {}^0\ddot{\mathbf{U}} = \begin{bmatrix} 0 \\ 10 \end{bmatrix}$$

where ${}^0\ddot{\mathbf{U}}$ was evaluated in Example 9.1. Then

$$a_0 = 39.0; \quad a_1 = 7.65; \quad a_2 = 15.3; \quad a_3 = 0.196; \quad a_4 = 27.9;$$
$$a_5 = -10.9; \quad a_6 = -1.14; \quad a_7 = 0.14; \quad a_8 = 0.0131$$

and

$$\hat{\mathbf{K}} = \begin{bmatrix} 6 & -2 \\ -2 & 4 \end{bmatrix} + 39.0\begin{bmatrix} 2 & 0 \\ 0 & 1 \end{bmatrix} = \begin{bmatrix} 84.1 & -2 \\ -2 & 43.0 \end{bmatrix}$$

For each time step we need to evaluate

$${}^{t+\theta\Delta t}\hat{\mathbf{R}} = \begin{bmatrix} 0 \\ 10 \end{bmatrix} + \begin{bmatrix} 0 \\ 0 \end{bmatrix} + \begin{bmatrix} 2 & 0 \\ 0 & 1 \end{bmatrix}\{39.0\,{}^t\mathbf{U} + 15.3\,{}^t\dot{\mathbf{U}} + 2\,{}^t\ddot{\mathbf{U}}\}$$
$$\hat{\mathbf{K}}\,{}^{t+\theta\Delta t}\mathbf{U} = {}^{t+\theta\Delta t}\hat{\mathbf{R}}$$

and then calculate

$$\begin{aligned}
{}^{t+\Delta t}\ddot{\mathbf{U}} &= 27.9({}^{t+\theta\Delta t}\mathbf{U} - {}^t\mathbf{U}) - 10.9\,{}^t\dot{\mathbf{U}} - 1.14\,{}^t\ddot{\mathbf{U}} \\
{}^{t+\Delta t}\dot{\mathbf{U}} &= {}^t\dot{\mathbf{U}} + 0.14({}^{t+\Delta t}\ddot{\mathbf{U}} + {}^t\ddot{\mathbf{U}}) \\
{}^{t+\Delta t}\mathbf{U} &= {}^t\mathbf{U} + 0.28\,{}^t\dot{\mathbf{U}} + 0.0131({}^{t+\Delta t}\ddot{\mathbf{U}} + 2\,{}^t\ddot{\mathbf{U}})
\end{aligned}$$

Time	Δt	$2\Delta t$	$3\Delta t$	$4\Delta t$	$5\Delta t$	$6\Delta t$	$7\Delta t$	$8\Delta t$	$9\Delta t$	$10\Delta t$	$11\Delta t$	$12\Delta t$
${}^t\mathbf{U}$	0.00605	0.0525	0.196	0.490	0.952	1.54	2.16	2.67	2.92	2.82	2.33	1.54
	0.366	1.34	2.64	3.92	4.88	5.31	5.18	4.61	3.82	3.06	2.52	2.29

The solution obtained is compared with the exact results in Example 9.7.

Consider now the direct integration with a time step $\Delta t = 28$. In this case we have

$$\hat{\mathbf{K}} = \begin{bmatrix} 6 & -2 \\ -2 & 4 \end{bmatrix} + 0.00392 \begin{bmatrix} 2 & 0 \\ 0 & 1 \end{bmatrix} = \begin{bmatrix} 6.00784 & -2 \\ -2 & 4.00392 \end{bmatrix}$$

where we note that $\hat{\mathbf{K}}$ is nearly equal to $\mathbf{K}$, as in the integration using the Houbolt method.

The displacement response obtained over 12 time steps is

Time	Δt	$2\Delta t$	$3\Delta t$	$4\Delta t$	$5\Delta t$	$6\Delta t$	$7\Delta t$	$8\Delta t$	$9\Delta t$	$10\Delta t$	$11\Delta t$	$12\Delta t$
${}^t\mathbf{U}$	1.09	2.82	−2.61	5.86	−4.47	6.59	−4.38	5.97	−3.46	4.92	−2.39	3.89
	1123.	−834.	674.	−519.	406.	−308.	242.	−181.	144.	−105.	86.1	−60.9

Here it should be noted that the initial conditions on the accelerations, i.e., ${}^0\ddot{\mathbf{U}} = \begin{bmatrix} 0 \\ 10 \end{bmatrix}$, cause the large initial displacement response. This response is damped out with increasing time. If we use ${}^0\ddot{\mathbf{U}} = \begin{bmatrix} 0 \\ 0 \end{bmatrix}$, the calculated displacement response is

Time	Δt	$2\Delta t$	$3\Delta t$	$4\Delta t$	$5\Delta t$	$6\Delta t$	$7\Delta t$	$8\Delta t$	$9\Delta t$	$10\Delta t$	$11\Delta t$	$12\Delta t$
${}^t\mathbf{U}$	0.363	1.44	0.632	1.29	0.782	1.17	0.875	1.09	0.929	1.05	0.960	1.03
	1.09	4.33	1.89	3.87	2.32	3.52	2.60	3.31	2.77	3.18	2.86	3.11

where it is observed that the static solution is approached (see Example 9.2).

9.2.4 The Newmark Method

The Newmark integration scheme can also be understood to be an extension of the linear acceleration method. The following assumptions are used:[5]

$$ {}^{t+\Delta t}\dot{\mathbf{U}} = {}^t\dot{\mathbf{U}} + [(1-\delta)\,{}^t\ddot{\mathbf{U}} + \delta\,{}^{t+\Delta t}\ddot{\mathbf{U}}]\,\Delta t \tag{9.27}$$

$$ {}^{t+\Delta t}\mathbf{U} = {}^t\mathbf{U} + {}^t\dot{\mathbf{U}}\,\Delta t + [(\tfrac{1}{2}-\alpha)\,{}^t\ddot{\mathbf{U}} + \alpha\,{}^{t+\Delta t}\ddot{\mathbf{U}}]\,\Delta t^2 \tag{9.28}$$

where α and δ are parameters that can be determined to obtain integration accuracy and stability. When $\delta = \frac{1}{2}$ and $\alpha = \frac{1}{6}$, relations (9.27) and (9.28) correspond to the linear acceleration method (which is also obtained using $\theta = 1$ in the Wilson θ method). Newmark originally proposed as an unconditionally stable scheme the constant-average-acceleration method (also called trapezoidal rule), in which case $\delta = \frac{1}{2}$ and $\alpha = \frac{1}{4}$ (see Fig. 9.2).

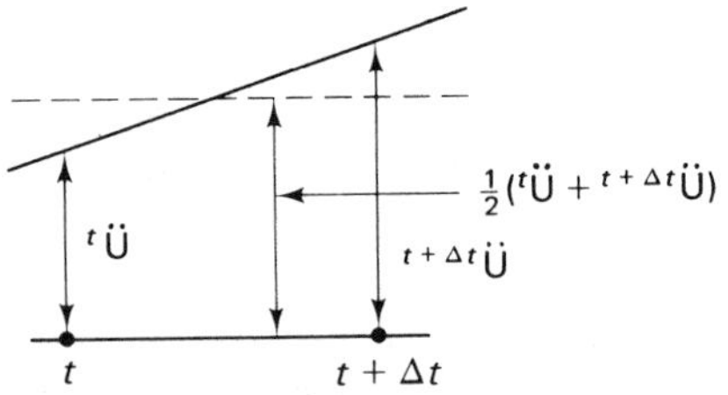

FIGURE 9.2 *Newmark's constant-average-acceleration scheme.*

In addition to (9.27) and (9.28), for solution of the displacements, velocities, and accelerations at time $t + \Delta t$, the equilibrium equations (9.1) at time $t + \Delta t$ are also considered:

$$\mathbf{M}\,{}^{t+\Delta t}\ddot{\mathbf{U}} + \mathbf{C}\,{}^{t+\Delta t}\dot{\mathbf{U}} + \mathbf{K}\,{}^{t+\Delta t}\mathbf{U} = {}^{t+\Delta t}\mathbf{R} \tag{9.29}$$

Solving from (9.28) for ${}^{t+\Delta t}\ddot{\mathbf{U}}$ in terms of ${}^{t+\Delta t}\mathbf{U}$, and then substituting for ${}^{t+\Delta t}\ddot{\mathbf{U}}$ into (9.27), we obtain equations for ${}^{t+\Delta t}\ddot{\mathbf{U}}$ and ${}^{t+\Delta t}\dot{\mathbf{U}}$, each in terms of the unknown displacements ${}^{t+\Delta t}\mathbf{U}$ only. These two relations for ${}^{t+\Delta t}\dot{\mathbf{U}}$ and ${}^{t+\Delta t}\ddot{\mathbf{U}}$ are substituted into (9.29) to solve for ${}^{t+\Delta t}\mathbf{U}$, after which, using (9.27) and (9.28), ${}^{t+\Delta t}\ddot{\mathbf{U}}$ and ${}^{t+\Delta t}\dot{\mathbf{U}}$ can also be calculated.

The complete algorithm using the Newmark integration scheme is given in Table 9.4. The close relationship between the computer implementation of the Newmark and the Wilson method should be noted, which makes it possible to conveniently use both integration schemes in one single computer program.[6]

TABLE 9.4 *Step-by-step solution using Newmark integration method.*

A. *Initial Calculations:*

1. Form stiffness matrix $\mathbf{K}$, mass matrix $\mathbf{M}$, and damping matrix $\mathbf{C}$.
2. Initialize ${}^{0}\mathbf{U}$, ${}^{0}\dot{\mathbf{U}}$, and ${}^{0}\ddot{\mathbf{U}}$.
3. Select time step size Δt, parameters α and δ, and calculate integration constants:

$$\delta \geq 0.50; \qquad \alpha \geq 0.25(0.5 + \delta)^2$$

$$a_0 = \frac{1}{\alpha\,\Delta t^2}; \qquad a_1 = \frac{\delta}{\alpha\,\Delta t}; \qquad a_2 = \frac{1}{\alpha\,\Delta t}; \qquad a_3 = \frac{1}{2\alpha} - 1;$$

$$a_4 = \frac{\delta}{\alpha} - 1; \qquad a_5 = \frac{\Delta t}{2}\left(\frac{\delta}{\alpha} - 2\right); \qquad a_6 = \Delta t(1 - \delta); \qquad a_7 = \delta\,\Delta t$$

4. Form effective stiffness matrix $\hat{\mathbf{K}}$: $\hat{\mathbf{K}} = \mathbf{K} + a_0\mathbf{M} + a_1\mathbf{C}$.
5. Triangularize $\hat{\mathbf{K}}$: $\hat{\mathbf{K}} = \mathbf{L}\mathbf{D}\mathbf{L}^T$.

B. *For Each Time Step:*

1. Calculate effective loads at time $t + \Delta t$:

$${}^{t+\Delta t}\hat{\mathbf{R}} = {}^{t+\Delta t}\mathbf{R} + \mathbf{M}(a_0\,{}^{t}\mathbf{U} + a_2\,{}^{t}\dot{\mathbf{U}} + a_3\,{}^{t}\ddot{\mathbf{U}}) + \mathbf{C}(a_1\,{}^{t}\mathbf{U} + a_4\,{}^{t}\dot{\mathbf{U}} + a_5\,{}^{t}\ddot{\mathbf{U}})$$

2. Solve for displacements at time $t + \Delta t$:

$$\mathbf{L}\mathbf{D}\mathbf{L}^T\,{}^{t+\Delta t}\mathbf{U} = {}^{t+\Delta t}\hat{\mathbf{R}}$$

3. Calculate accelerations and velocities at time $t + \Delta t$:

$${}^{t+\Delta t}\ddot{\mathbf{U}} = a_0({}^{t+\Delta t}\mathbf{U} - {}^{t}\mathbf{U}) - a_2\,{}^{t}\dot{\mathbf{U}} - a_3\,{}^{t}\ddot{\mathbf{U}}$$
$${}^{t+\Delta t}\dot{\mathbf{U}} = {}^{t}\dot{\mathbf{U}} + a_6\,{}^{t}\ddot{\mathbf{U}} + a_7\,{}^{t+\Delta t}\ddot{\mathbf{U}}$$

EXAMPLE 9.4: Calculate the displacement response of the system considered in Examples 9.1 to 9.3 using the Newmark method. Use $\alpha = 0.25$, $\delta = 0.5$.

Consider first the case $\Delta t = 0.28$. Following the steps of calculations given in Table 9.4, we have

$$ {}^0\mathbf{U} = \begin{bmatrix} 0 \\ 0 \end{bmatrix}; \qquad {}^0\dot{\mathbf{U}} = \begin{bmatrix} 0 \\ 0 \end{bmatrix}; \qquad {}^0\ddot{\mathbf{U}} = \begin{bmatrix} 0 \\ 10 \end{bmatrix} $$

The integration constants are

$$ a_0 = 51.0; \quad a_1 = 7.14; \quad a_2 = 14.3; \quad a_3 = 1.00; $$
$$ a_4 = 1.00; \quad a_5 = 0.00; \quad a_6 = 0.14; \quad a_7 = 0.14 $$

Thus the effective stiffness matrix is

$$ \hat{\mathbf{K}} = \begin{bmatrix} 6 & -2 \\ -2 & 4 \end{bmatrix} + 51.0 \begin{bmatrix} 2 & 0 \\ 0 & 1 \end{bmatrix} = \begin{bmatrix} 108 & -2 \\ -2 & 55 \end{bmatrix} $$

For each time step we need to evaluate

$$ {}^{t+\Delta t}\hat{\mathbf{R}} = \begin{bmatrix} 0 \\ 10 \end{bmatrix} + \begin{bmatrix} 2 & 0 \\ 0 & 1 \end{bmatrix} (51\,{}^t\mathbf{U} + 14.3\,{}^t\dot{\mathbf{U}} + 1.0\,{}^t\ddot{\mathbf{U}}) $$

Then

$$ \hat{\mathbf{K}}\,{}^{t+\Delta t}\mathbf{U} = {}^{t+\Delta t}\hat{\mathbf{R}} $$

and

$$ {}^{t+\Delta t}\ddot{\mathbf{U}} = 51.0({}^{t+\Delta t}\mathbf{U} - {}^t\mathbf{U}) - 14.3\,{}^t\dot{\mathbf{U}} - 1.0\,{}^t\ddot{\mathbf{U}} $$
$$ {}^{t+\Delta t}\dot{\mathbf{U}} = {}^t\dot{\mathbf{U}} + 0.14\,{}^t\ddot{\mathbf{U}} + 0.14\,{}^{t+\Delta t}\ddot{\mathbf{U}} $$

Performing the above calculations, we obtain

Time	Δt	$2\Delta t$	$3\Delta t$	$4\Delta t$	$5\Delta t$	$6\Delta t$	$7\Delta t$	$8\Delta t$	$9\Delta t$	$10\Delta t$	$11\Delta t$	$12\Delta t$
${}^t\mathbf{U}$	0.00673	0.0505	0.189	0.485	0.961	1.58	2.23	2.76	3.00	2.85	2.28	1.40
	0.364	1.35	2.68	4.00	4.95	5.34	5.13	4.48	3.64	2.90	2.44	2.31

The solution obtained is compared with the exact results in Example 9.7.

Next we employ the time step $\Delta t = 28.0$. In this case we have

$$ \hat{\mathbf{K}} = \begin{bmatrix} 6 & -2 \\ -2 & 4 \end{bmatrix} + 0.0051 \begin{bmatrix} 2 & 0 \\ 0 & 1 \end{bmatrix} = \begin{bmatrix} 6.0102 & -2.0000 \\ -2.0000 & 4.0051 \end{bmatrix} $$

Therefore, $\hat{\mathbf{K}}$ is, as in the integration using the Houbolt or Wilson θ method, nearly equal to $\mathbf{K}$.

Using the initial conditions ${}^0\ddot{\mathbf{U}} = \begin{bmatrix} 0 \\ 10 \end{bmatrix}$, we obtain as the displacement response

Time	Δt	$2\Delta t$	$3\Delta t$	$4\Delta t$	$5\Delta t$	$6\Delta t$	$7\Delta t$	$8\Delta t$	$9\Delta t$	$10\Delta t$	$11\Delta t$	$12\Delta t$
${}^t\mathbf{U}$	1.99	0.028	1.94	0.112	1.83	0.248	1.67	0.429	1.47	0.648	1.23	0.894
	5.99	0.045	5.90	0.177	5.72	0.393	5.47	0.685	5.14	1.04	4.76	1.45

However, using as initial conditions ${}^0\ddot{\mathbf{U}} = \begin{bmatrix} 0 \\ 0 \end{bmatrix}$, we obtain

Time	Δt	$2\Delta t$	$3\Delta t$	$4\Delta t$	$5\Delta t$	$6\Delta t$	$7\Delta t$	$8\Delta t$	$9\Delta t$	$10\Delta t$	$11\Delta t$	$12\Delta t$
${}^t\mathbf{U}$	0.996 2.99	1.01 3.02	0.982 2.97	1.02 3.04	0.969 2.95	1.04 3.06	0.957 2.93	1.05 3.08	0.947 2.91	1.06 3.09	0.940 2.90	1.06 3.11

and the solution oscillates about the static response.

9.2.5 The Coupling of Different Integration Operators

So far we have assumed that all dynamic equilibrium equations are solved using the same time integration scheme. As discussed in Section 9.4, the choice of which operator to use for an effective solution depends on the problem to be analyzed. However, for certain kinds of problems it may be advantageous to use different operators to integrate the response in different regions of the total element assemblage (see, e.g., ref. [7-10]). This is particularly the case when the stiffness and mass characteristics (i.e., the characteristic time constants) of the total element assemblage are quite different in different parts of the element assemblage. An example is the analysis of fluid-structure systems, in which the fluid is very flexible when measured on the stiffness of the structure. Here the explicit time integration of the fluid response using the conditionally stable central difference method and an implicit unconditionally stable time integration of the structural response (using, for example, the Newmark method) may be a natural choice (see Section 9.4). The reasons are that, firstly, the physical phenomenon to be analyzed may be a wave propagation in the fluid and a structural vibration of the structure, and secondly, the critical time step size for an explicit time integration of the fluid response is usually much larger than the time step required for explicit time integration of the structural response. The result may be that by proper choice of the finite element idealizations of the fluid and the structure, the explicit time integration of the fluid response and implicit time integration of the structural response can be performed with a time step that is relatively large but small enough to yield a stable and accurate solution.*

The use of a combination of operators for the integration of dynamic response raises the questions of which operators to choose and how to couple them. There are a large number of possibilities, but in general the selection of the operators depends on their stability and accuracy characteristics, including the effects due to the operator coupling, and the overall effectiveness of the resulting time integration. We demonstrate the use of explicit–implicit time integration in the analysis of the simple problem considered already in Examples 9.1 to 9.4.

EXAMPLE 9.5: Solve for U_1 and U_2 of the simple system considered in Example 9.1 using the explicit central difference method for U_1 and the implicit trapezoidal rule (Newmark's method with $\alpha = \frac{1}{4}$, $\delta = \frac{1}{2}$) for U_2.

In the explicit integration we consider the equilibrium at time t to calculate the

* It may also be efficient to use different time step sizes for the explicit and implicit integrations, with the one time step size being a multiple of the other.

displacement for time $t + \Delta t$. Considering degree of freedom 1 we have

$$2\,{}^t\ddot{U}_1 + 6\,{}^tU_1 - 2\,{}^tU_2 = 0 \tag{a}$$

In the implicit integration we consider the equilibrium at time $t + \Delta t$ to calculate the displacement for time $t + \Delta t$. Thus, we have for degree of freedom 2,

$${}^{t+\Delta t}\ddot{U}_2 - 2\,{}^{t+\Delta t}U_1 + 4\,{}^{t+\Delta t}U_2 = 10 \tag{b}$$

For (a) we now use the central difference method

$${}^t\ddot{U}_1 = \frac{{}^{t+\Delta t}U_1 - 2\,{}^tU_1 + {}^{t-\Delta t}U_1}{(\Delta t)^2} \tag{c}$$

and for (b) we use the trapezoidal rule

$$\begin{aligned} {}^{t+\Delta t}\dot{U}_2 &= {}^t\dot{U}_2 + \frac{\Delta t}{2}({}^t\ddot{U}_2 + {}^{t+\Delta t}\ddot{U}_2) \\ {}^{t+\Delta t}U_2 &= {}^tU_2 + {}^t\dot{U}_2\,\Delta t + \frac{(\Delta t)^2}{4}[{}^t\ddot{U}_2 + {}^{t+\Delta t}\ddot{U}_2] \end{aligned} \tag{d}$$

The initial conditions are

$${}^0U_1 = {}^0\dot{U}_1 = {}^0\ddot{U}_1 = {}^0U_2 = {}^0\dot{U}_2 = 0, \qquad {}^0\ddot{U}_2 = 10$$

Hence, using (9.7) to obtain the starting value ${}^{-\Delta t}U_1$ we obtain ${}^{-\Delta t}U_1 = 0$.

We can now use, for each time step, (a) and (c) to solve for ${}^{t+\Delta t}U_1$ and then (b) and (d) to solve for ${}^{t+\Delta t}U_2$. We should note that in this solution we evaluate ${}^{t+\Delta t}U_1$ by projecting ahead from the equilibrium configuration at time t of the degree of freedom 1, and we then accept this value of ${}^{t+\Delta t}U_1$ to evaluate ${}^{t+\Delta t}U_2$ implicitly. Using this procedure we obtain with $\Delta t = 0.28$ the following data:

Time	Δt	$2\Delta t$	$3\Delta t$	$4\Delta t$	$5\Delta t$	$6\Delta t$	$7\Delta t$	$8\Delta t$	$9\Delta t$	$10\Delta t$	$11\Delta t$	$12\Delta t$
${}^t\mathbf{U}$	0.	.0285	.156	.457	.962	1.63	2.33	2.88	3.11	2.90	2.24	1.25
	.364	1.35	2.68	3.98	4.93	5.32	5.12	4.50	3.70	2.99	2.54	2.39

This solution compares with the response calculated in Example 9.1. If, however, we now try to obtain a solution with $\Delta t = 28$ we find that the solution is unstable, i.e. the predicted displacements very rapidly grow out of bound.

As demonstrated in the above example, the use of explicit–implicit time integration requires that the time step be small enough for stability in the explicitly integrated domain, while special considerations need to be given to the interface conditions. In an alternative but closely related solution approach, a single implicit time integration scheme is used for the complete domain but the stiffness of the very flexible part is not added to the coefficient matrix and dynamic equilibrium is satisfied by iteration[11] (see Section 9.5.2).

9.3 MODE SUPERPOSITION

Tables 9.1 to 9.4, summarizing the direct integration schemes, show that if a diagonal mass matrix and no damping is assumed, the number of operations for one time step are—as a very rough estimate—somewhat larger than $2nm_K$, where n and m_K are the order and half-bandwidth of the stiffness matrix considered, respectively (assuming constant column heights, see Section 8.2.3). In the central difference method $2nm_K$ operations are required for the product of the stiffness matrix times the displacement vector (assuming in addition that the elements of the matrix under the skyline are largely nonzero), and in the Houbolt, Wilson, and Newmark methods about $2nm_K$ operations are required for the solution of the system equations in each time step.* The initial triangular factorization of the effective stiffness matrix requires additional operations. Furthermore, if a consistent mass matrix is used, or a damping matrix is included in the analysis, an additional number of operations proportional to nm_K is required per time step in either case. Therefore, neglecting the operations for the initial calculations, a total number of about $\alpha nm_K s$ operations are required in the complete integration, where α depends on the characteristics of the matrices used, $\alpha \geq 2$, and s is the number of time steps.

The above considerations show that the number of operations required in the direct integration are directly proportional to the number of time steps used in the analysis. Therefore, in general, the use of direct integration can be expected to be effective when the response for a relatively short duration (i.e., for a few time steps) is required. However, if the integration must be carried out for many time steps, it may be more effective to first transform the equilibrium equations in (9.1) into a form in which the step-by-step solution is less costly. In particular, since the number of operations required is directly proportional to the half-bandwidth m_K of the stiffness matrix, a reduction in m_K would decrease proportionally the cost of the step-by-step solution.

It is important at this stage to fully recognize what we have proposed to pursue. We recall that (9.1) are the equilibrium equations obtained when the finite element interpolation functions are used in the evaluation of the virtual work equation (4.5) (see Section 4.2.1). The resulting matrices **K**, **M**, and **C** have a bandwidth that is determined by the numbering of the finite element nodal points (see the Appendix). Therefore, the topology of the finite element mesh determines the order and bandwidth of the system matrices. In order to reduce the bandwidth of the system matrices, we may rearrange the nodal point numbering; however, there is a limit on the minimum bandwidth that can be obtained in this way, and we therefore set out to follow a different procedure.

9.3.1 Change of Basis to Modal Generalized Displacements

We propose to transform the equilibrium equations into a more effective form for direct integration by using the following transformation on the finite element nodal point displacements **U**,

$$\mathbf{U}(t) = \mathbf{P}\mathbf{X}(t) \tag{9.30}$$

where **P** is a square matrix and $\mathbf{X}(t)$ is a time-dependent vector of order n. The transformation matrix **P** is still unknown and will have to be determined. The components of **X** are referred to as *generalized displacements*. Substituting (9.30)

* Note that in practice an "effective m_K" must be used in the formula and this value can be much smaller for the central difference method than for the implicit methods because there may be many zero elements within the actual band.

into (9.1) and premultiplying by $\mathbf{P}^T$, we obtain

$$\tilde{\mathbf{M}}\ddot{\mathbf{X}}(t) + \tilde{\mathbf{C}}\dot{\mathbf{X}}(t) + \tilde{\mathbf{K}}\mathbf{X}(t) = \tilde{\mathbf{R}}(t) \tag{9.31}$$

where $\tilde{\mathbf{M}} = \mathbf{P}^T\mathbf{M}\mathbf{P}; \quad \tilde{\mathbf{C}} = \mathbf{P}^T\mathbf{C}\mathbf{P}; \quad \tilde{\mathbf{K}} = \mathbf{P}^T\mathbf{K}\mathbf{P}; \quad \tilde{\mathbf{R}} = \mathbf{P}^T\mathbf{R}$ (9.32)

It should be noted that this transformation is obtained by substituting (9.30) into (4.8) to express the element displacements in terms of the generalized displacements,

$$\mathbf{u}^{(m)}(x, y, z, t) = \mathbf{H}^{(m)}\mathbf{P}\mathbf{X}(t) \tag{9.33}$$

and then using (9.33) in the virtual work equation (4.12). Therefore, in essence, to obtain (9.31) from (9.1), a change of basis from the finite element displacement basis to a generalized displacement basis has been performed (see Section 2.5).

The objective of the transformation is to obtain new system stiffness, mass, and damping matrices, $\tilde{\mathbf{K}}$, $\tilde{\mathbf{M}}$, and $\tilde{\mathbf{C}}$, which have a smaller bandwidth than the original system matrices, and the transformation matrix $\mathbf{P}$ should be selected accordingly. In addition, it should be noted that $\mathbf{P}$ must be nonsingular (i.e., the rank of $\mathbf{P}$ must be n) in order to have a unique relation between any vectors $\mathbf{U}$ and $\mathbf{X}$ as expressed in (9.30).

In theory, there can be many different transformation matrices $\mathbf{P}$, which would reduce the bandwidth of the system matrices. However, in practice, an effective transformation matrix is established using the displacement solutions of the free vibration equilibrium equations with damping neglected,

$$\mathbf{M}\ddot{\mathbf{U}} + \mathbf{K}\mathbf{U} = \mathbf{0} \tag{9.34}$$

The solution to (9.34) can be postulated to be of the form

$$\mathbf{U} = \boldsymbol{\phi} \sin \omega(t - t_0) \tag{9.35}$$

where $\boldsymbol{\phi}$ is a vector of order n, t the time variable, t_0 a time constant, and ω a constant identified to represent the frequency of vibration (rad/sec) of the vector $\boldsymbol{\phi}$.

Substituting (9.35) into (9.34), we obtain the generalized eigenproblem, from which $\boldsymbol{\phi}$ and ω must be determined,

$$\mathbf{K}\boldsymbol{\phi} = \omega^2\mathbf{M}\boldsymbol{\phi} \tag{9.36}$$

The eigenproblem in (9.36) yields the n eigensolutions $(\omega_1^2, \boldsymbol{\phi}_1)$, $(\omega_2^2, \boldsymbol{\phi}_2)$, . . . , $(\omega_n^2, \boldsymbol{\phi}_n)$, where the eigenvectors are $\mathbf{M}$-orthonormalized (see Section 10.2.1); i.e.,

$$\boldsymbol{\phi}_i^T\mathbf{M}\boldsymbol{\phi}_j \begin{cases} = 1; & i = j \\ = 0; & i \neq j \end{cases} \tag{9.37}$$

and

$$0 \leq \omega_1^2 \leq \omega_2^2 \leq \omega_3^2 \cdots \leq \omega_n^2 \tag{9.38}$$

The vector $\boldsymbol{\phi}_i$ is called the ith-mode shape vector, and ω_i is the corresponding frequency of vibration (rad/sec). It should be emphasized that (9.34) is satisfied using any of the n displacement solutions $\boldsymbol{\phi}_i \sin \omega_i(t - t_0)$, $i = 1, 2, \ldots, n$. For a physical interpretation of ω_i and $\boldsymbol{\phi}_i$ see Example 9.6.

Defining a matrix $\mathbf{\Phi}$ whose columns are the eigenvectors $\boldsymbol{\phi}_i$ and a diagonal matrix $\mathbf{\Omega}^2$ which stores the eigenvalues ω_i^2 on its diagonal; i.e.,

$$\mathbf{\Phi} = [\boldsymbol{\phi}_1, \boldsymbol{\phi}_2, \cdots, \boldsymbol{\phi}_n]; \qquad \mathbf{\Omega}^2 = \begin{bmatrix} \omega_1^2 & & & & \\ & \omega_2^2 & & & \\ & & \cdot & & \\ & & & \cdot & \\ & & & & \cdot \\ & & & & & \omega_n^2 \end{bmatrix} \tag{9.39}$$

we can write the n solutions to (9.36) as

$$\mathbf{K\Phi} = \mathbf{M\Phi\Omega}^2 \tag{9.40}$$

Since the eigenvectors are $\mathbf{M}$-orthonormal, we have

$$\mathbf{\Phi}^T\mathbf{K\Phi} = \mathbf{\Omega}^2; \qquad \mathbf{\Phi}^T\mathbf{M\Phi} = \mathbf{I} \tag{9.41}$$

It is now apparent that the matrix $\mathbf{\Phi}$ would be a suitable transformation matrix $\mathbf{P}$ in (9.30). Using

$$\mathbf{U}(t) = \mathbf{\Phi X}(t) \tag{9.42}$$

we obtain equilibrium equations that correspond to the modal generalized displacements

$$\ddot{\mathbf{X}}(t) + \mathbf{\Phi}^T\mathbf{C\Phi}\dot{\mathbf{X}}(t) + \mathbf{\Omega}^2\mathbf{X}(t) = \mathbf{\Phi}^T\mathbf{R}(t) \tag{9.43}$$

The initial conditions on $\mathbf{X}(t)$ are obtained using (9.42) and the $\mathbf{M}$-orthonormality of $\mathbf{\Phi}$; i.e., at time 0 we have

$${}^0\mathbf{X} = \mathbf{\Phi}^T\mathbf{M}\,{}^0\mathbf{U}; \qquad {}^0\dot{\mathbf{X}} = \mathbf{\Phi}^T\mathbf{M}\,{}^0\dot{\mathbf{U}} \tag{9.44}$$

The equations in (9.43) show that if a damping matrix is not included in the analysis, the finite element equilibrium equations are decoupled when using in the transformation matrix $\mathbf{P}$ the free vibration mode shapes of the finite element system. Since the derivation of the damping matrix can in many cases not be carried out explicitly[12,13] but the damping effects can only be included approximately, it is reasonable to use a damping matrix that includes all required effects but at the same time allows an effective solution of the equilibrium equations. In many analyses damping effects are neglected altogether, and it is this case that we shall discuss first.

EXAMPLE 9.6: Calculate the transformation matrix $\mathbf{\Phi}$ for the problem considered in Examples 9.1 to 9.4, and thus establish the decoupled equations of equilibrium in the basis of mode shape vectors.

For the system under consideration we have

$$\mathbf{K} = \begin{bmatrix} 6 & -2 \\ -2 & 4 \end{bmatrix}; \qquad \mathbf{M} = \begin{bmatrix} 2 & 0 \\ 0 & 1 \end{bmatrix}; \qquad \mathbf{R} = \begin{bmatrix} 0 \\ 10 \end{bmatrix}$$

The generalized eigenproblem to be solved is therefore

$$\begin{bmatrix} 6 & -2 \\ -2 & 4 \end{bmatrix}\boldsymbol{\phi} = \omega^2\begin{bmatrix} 2 & 0 \\ 0 & 1 \end{bmatrix}\boldsymbol{\phi}$$

The solution is obtained by one of the methods given in Chapters 10 to 12. Here we simply give the two solutions without derivations:

$$\omega_1^2 = 2, \qquad \boldsymbol{\phi}_1 = \begin{bmatrix} \dfrac{1}{\sqrt{3}} \\ \dfrac{1}{\sqrt{3}} \end{bmatrix}$$

$$\omega_2^2 = 5, \qquad \boldsymbol{\phi}_2 = \begin{bmatrix} \dfrac{1}{2}\sqrt{\dfrac{2}{3}} \\ -\sqrt{\dfrac{2}{3}} \end{bmatrix}$$

Therefore, considering the free vibration equilibrium equations of the system

$$\begin{bmatrix} 2 & 0 \\ 0 & 1 \end{bmatrix} \ddot{\mathbf{U}}(t) + \begin{bmatrix} 6 & -2 \\ -2 & 4 \end{bmatrix} \mathbf{U}(t) = \mathbf{0} \tag{a}$$

the following two solutions are possible:

$$\mathbf{U}_1(t) = \begin{bmatrix} \dfrac{1}{\sqrt{3}} \\ \dfrac{1}{\sqrt{3}} \end{bmatrix} \sin \sqrt{2}\,(t - t_0^1) \quad \text{and} \quad \mathbf{U}_2(t) = \begin{bmatrix} \dfrac{1}{2}\sqrt{\dfrac{2}{3}} \\ -\sqrt{\dfrac{2}{3}} \end{bmatrix} \sin \sqrt{5}\,(t - t_0^2)$$

That the vectors $\mathbf{U}_1(t)$ and $\mathbf{U}_2(t)$ indeed satisfy the relation in (a) can be verified simply by substituting $\mathbf{U}_1$ or $\mathbf{U}_2$ into the equilibrium equations. The actual solution to the equations in (a) is of the form

$$\mathbf{U}(t) = \alpha \begin{bmatrix} \dfrac{1}{\sqrt{3}} \\ \dfrac{1}{\sqrt{3}} \end{bmatrix} \sin \sqrt{2}\,(t - t_0^1) + \beta \begin{bmatrix} \dfrac{1}{2}\sqrt{\dfrac{2}{3}} \\ -\sqrt{\dfrac{2}{3}} \end{bmatrix} \sin \sqrt{5}\,(t - t_0^2)$$

where α, β, t_0^1, and t_0^2 are determined by the initial conditions on $\mathbf{U}$ and $\dot{\mathbf{U}}$. In particular, if we impose initial conditions corresponding to α (or β) only, we have that the system vibrates in the corresponding eigenvector with frequency $\sqrt{2}$ rad/sec (or $\sqrt{5}$ rad/sec). The general procedure of solution for α, β, t_1^0, and t_2^0 is discussed in Section 9.3.2.

Having evaluated $(\omega_1^2, \boldsymbol{\phi}_1)$ and $(\omega_2^2, \boldsymbol{\phi}_2)$ for the problem in Examples 9.1 to 9.4, we arrive at the following equilibrium equations in the basis of eigenvectors:

$$\ddot{\mathbf{X}}(t) + \begin{bmatrix} 2 & 0 \\ 0 & 5 \end{bmatrix} \mathbf{X}(t) = \begin{bmatrix} \dfrac{1}{\sqrt{3}} & \dfrac{1}{\sqrt{3}} \\ \dfrac{1}{2}\sqrt{\dfrac{2}{3}} & -\sqrt{\dfrac{2}{3}} \end{bmatrix} \begin{bmatrix} 0 \\ 10 \end{bmatrix}$$

or

$$\ddot{\mathbf{X}}(t) + \begin{bmatrix} 2 & 0 \\ 0 & 5 \end{bmatrix} \mathbf{X}(t) = \begin{bmatrix} \dfrac{10}{\sqrt{3}} \\ -10\sqrt{\dfrac{2}{3}} \end{bmatrix}$$

9.3.2 Analysis with Damping Neglected

If velocity-dependent damping effects are not included in the analysis, (9.43) reduces to

$$\ddot{\mathbf{X}}(t) + \mathbf{\Omega}^2 \mathbf{X}(t) = \mathbf{\Phi}^T \mathbf{R}(t) \tag{9.45}$$

i.e., n individual equations of the form

where

$$\left.\begin{aligned} \ddot{x}_i(t) + \omega_i^2 x_i(t) &= r_i(t) \\ r_i(t) &= \boldsymbol{\phi}_i^T \mathbf{R}(t) \end{aligned}\right\} i = 1,2, \ldots, n \tag{9.46}$$

We note that the ith typical equation in (9.46) is the equilibrium equation of a single degree of freedom system with unit mass and stiffness ω_i^2. The initial conditions on the motion of this system are obtained from (9.44):

$$\begin{aligned} x_i|_{t=0} &= \boldsymbol{\phi}_i^T \mathbf{M}\, {}^0\mathbf{U} \\ \dot{x}_i|_{t=0} &= \boldsymbol{\phi}_i^T \mathbf{M}\, {}^0\dot{\mathbf{U}} \end{aligned} \tag{9.47}$$

The solution to each equation in (9.46) can be obtained using the integration algorithms in Tables 9.1 to 9.4 or can be calculated using the Duhamel integral:

$$x_i(t) = \frac{1}{\omega_i} \int_0^t r_i(\tau) \sin \omega_i(t - \tau)\, d\tau + \alpha_i \sin \omega_i t + \beta_i \cos \omega_i t \tag{9.48}$$

where α_i and β_i are determined from the initial conditions in (9.47). The Duhamel integral in (9.48) may have to be evaluated numerically. In addition, it should be noted that various other integration methods could also be used in the solution of (9.46).

For the complete response, the solution to all n equations in (9.46), $i = 1, 2, \ldots, n$, must be calculated and then the finite element nodal point displacements are obtained by superposition of the response in each mode; i.e., using (9.42), we obtain

$$\mathbf{U}(t) = \sum_{i=1}^{n} \boldsymbol{\phi}_i x_i(t) \tag{9.49}$$

Therefore, in summary, the response analysis by mode superposition requires, first, the solution of the eigenvalues and eigenvectors of the problem in (9.36), then the solution of the decoupled equilibrium equations in (9.46), and, finally, the superposition of the response in each eigenvector as expressed in (9.49). In the analysis, the eigenvectors are the free vibration mode shapes of the finite element assemblage. As mentioned earlier, the choice between mode superposition analysis and direct integration described in Section 9.2 is merely one of numerical effectiveness. The solutions obtained using either procedure are identical within the numerical errors of the time integration schemes used [if the same time integration methods are used in direct integration and the solution of (9.46), the same numerical errors are present] and the round-off errors in the computer.

EXAMPLE 9.7: Use mode superposition to calculate the displacement response of the system considered in Examples 9.1 to 9.4 and 9.6. (1) Calculate the exact response by integrating each of the two decoupled equilibrium equations exactly. (2) Use the Newmark method with time step $\Delta t = 0.28$ for the time integration.

We established the decoupled equilibrium equations of the system under consideration in Example 9.6; i.e., the two equilibrium equations to be solved are

$$\begin{aligned} \ddot{x}_1 + 2x_1 &= \frac{10}{\sqrt{3}} \\ \ddot{x}_2 + 5x_2 &= -10\sqrt{\frac{2}{3}} \end{aligned} \tag{a}$$

The initial conditions on the system are $\mathbf{U}|_{t=0} = \mathbf{0}$, $\dot{\mathbf{U}}|_{t=0} = \mathbf{0}$, and hence, using (9.47), we have

$$x_1|_{t=0} = 0 \qquad \dot{x}_1|_{t=0} = 0$$
$$x_2|_{t=0} = 0 \qquad \dot{x}_2|_{t=0} = 0 \tag{b}$$

Also, to obtain $\mathbf{U}$ we need to use the relation in (9.42), which, using the eigenvectors calculated in Example 9.6, gives

$$\mathbf{U}(t) = \begin{bmatrix} \dfrac{1}{\sqrt{3}} & \dfrac{1}{2}\sqrt{\dfrac{2}{3}} \\ \dfrac{1}{\sqrt{3}} & -\sqrt{\dfrac{2}{3}} \end{bmatrix} \mathbf{X}(t) \tag{c}$$

The exact solutions to the equations in (a) and (b) are

$$x_1 = \frac{5}{\sqrt{3}}(1 - \cos\sqrt{2}\,t)$$
$$x_2 = 2\sqrt{\frac{2}{3}}(-1 + \cos\sqrt{5}\,t) \tag{d}$$

Hence using (c) we have

$$\mathbf{U}(t) = \begin{bmatrix} \dfrac{1}{\sqrt{3}} & \dfrac{1}{2}\sqrt{\dfrac{2}{3}} \\ \dfrac{1}{\sqrt{3}} & -\sqrt{\dfrac{2}{3}} \end{bmatrix} \begin{bmatrix} \dfrac{5}{\sqrt{3}}(1 - \cos\sqrt{2}\,t) \\ 2\sqrt{\dfrac{2}{3}}(-1 + \cos\sqrt{5}\,t) \end{bmatrix} \tag{e}$$

Evaluating the displacements from (e) for times $\Delta t, 2\Delta t, \ldots, 12\Delta t$, where $\Delta t = 0.28$, we obtain

Time	Δt	$2\Delta t$	$3\Delta t$	$4\Delta t$	$5\Delta t$	$6\Delta t$	$7\Delta t$	$8\Delta t$	$9\Delta t$	$10\Delta t$	$11\Delta t$	$12\Delta t$
${}^t\mathbf{U}$	0.003	0.038	0.176	0.486	0.996	1.66	2.338	2.861	3.052	2.806	2.131	1.157
	0.382	1.41	2.78	4.09	5.00	5.29	4.986	4.277	3.457	2.806	2.484	2.489

The results obtained are compared in Fig. 9.3 with the response predicted using the central difference, Houbolt, Wilson θ, and Newmark methods in Examples 9.1 to 9.4, respectively. The discussion in Section 9.4 will show that the time step Δt selected for the direct integrations is relatively large, and with this in mind it may be noted that the direct integration schemes predict a fair approximation to the exact response of the system.

Instead of evaluating the exact response as given in (d) we could use a numerical integration scheme to solve the equations in (a). Here we employ the Newmark method and obtain:

Time	Δt	$2\Delta t$	$3\Delta t$	$4\Delta t$	$5\Delta t$	$6\Delta t$
$x_1(t)$	0.2258	0.8199	1.807	2.379	4.123	5.064
$x_2(t)$	−0.3046	−0.7920	−2.1239	−2.939	−3.258	−2.632

	$7\Delta t$	$8\Delta t$	$9\Delta t$	$10\Delta t$	$11\Delta t$	$12\Delta t$
$x_1(t)$	5.579	5.774	5.521	4.855	3.866	2.773
$x_2(t)$	−2.161	−1.156	−0.3307	−0.004083	−0.2482	−1.088

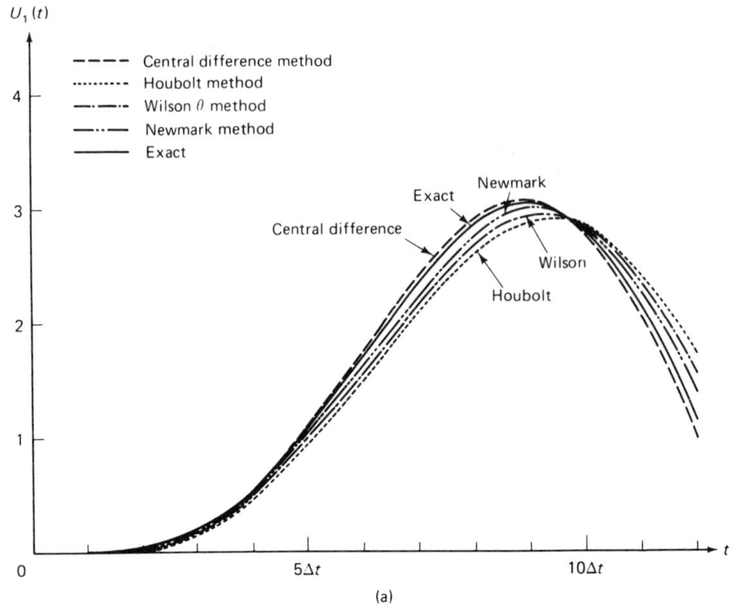

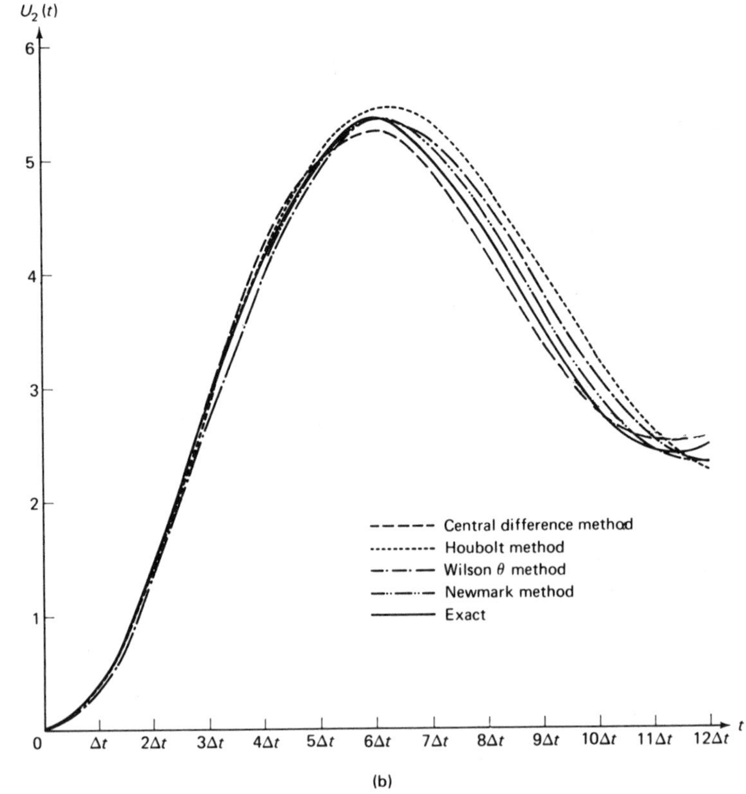

FIGURE 9.3 *Displacement response of system considered in Examples 9.1, 9.2, 9.3, 9.4, and 9.7.*

The solution for $U_1(t)$ and $U_2(t)$ is now evaluated by substituting for $\mathbf{X}(t)$ in the relation given in (c). As expected, it is found that the displacement response thus predicted is the same as the response obtained when using the Newmark method in direct integration.

As discussed so far, the only difference between a mode superposition and a direct integration analysis is that prior to the time integration, a change of basis is carried out, namely from the finite element coordinate basis to the basis of eigenvectors of the generalized eigenproblem $\mathbf{K}\boldsymbol{\phi} = \omega^2\mathbf{M}\boldsymbol{\phi}$. Since mathematically the same space is spanned by the n eigenvectors as by the n nodal point finite element displacements, the same solution must be obtained in both analyses. The choice of whether to use direct integration or mode superposition will therefore be decided by considerations of effectiveness only. However, this choice can only be made once an important additional aspect of mode superposition, as used in practical analysis, has been presented. In fact, it is this aspect that makes mode superposition analysis of some structures feasible with regard to cost, whereas direct integration would be prohibitively expensive.

Consider the decoupled equilibrium equations in (9.46). We note that if $r_i(t)$ is zero and the initial conditions on x_i and $\dot{x}_i$ are zero, then x_i is zero at all time t. These specific conditions would be realized, for example, if all finite element nodal point displacements and velocities are zero at time 0 and the applied loading is of the form $\mathbf{R}(t) = \mathbf{M}\boldsymbol{\phi}_j f(t)$, where $f(t)$ is an arbitrary function of t. In such a case, since $\boldsymbol{\phi}_i^T\mathbf{M}\boldsymbol{\phi}_j = \delta_{ij}$ (δ_{ij} = Kronecker delta), we would have that only $x_j(t)$ is nonzero. These are rather stringent conditions, and in general analysis can hardly be expected to apply exactly to many of the n equations in (9.46), because the loading is in general arbitrary. However, in addition to the fact that the loading may be nearly orthogonal to $\boldsymbol{\phi}_i$, it is also the frequency content of the loading that determines whether the ith equation in (9.46) will contribute significantly to the response. Namely, the response $x_i(t)$ is relatively large if the excitation frequency contained in r_i lies near ω_i.

To demonstrate this basic consideration we introduce the following example.

EXAMPLE 9.8: Consider a one degree of freedom system with the equilibrium equation

$$\ddot{x}(t) + \omega^2 x(t) = R \sin pt$$

and initial conditions $\quad x|_{t=0} = 0, \qquad \dot{x}|_{t=0} = 1 \qquad$ (a)

Use the Duhamel integral to calculate the displacement response.

We note that the system is subjected to a periodic force input and a nonzero initial velocity. Using the relation in (9.48), we obtain

$$x(t) = \frac{R}{\omega}\int_0^t \sin p\tau \sin \omega(t-\tau)\, d\tau + \alpha \sin \omega t + \beta \cos \omega t$$

Evaluating the integral, we obtain

$$x(t) = \frac{R/\omega^2}{1 - p^2/\omega^2} \sin pt + \alpha \sin \omega t + \beta \cos \omega t \qquad \text{(b)}$$

We now need to use the initial conditions to evaluate α and β. The solution at time $t = 0$ is

$$x\big|_{t=0} = \beta$$

$$\dot{x}\big|_{t=0} = \frac{Rp/\omega^2}{1 - p^2/\omega^2} + \alpha\omega$$

Using the conditions in (a), we obtain

$$\beta = 0; \qquad \alpha = \frac{1}{\omega} - \frac{Rp/\omega^3}{1 - p^2/\omega^2}$$

Substituting for α and β into (b), we thus have

$$x(t) = \frac{R/\omega^2}{1 - p^2/\omega^2} \sin pt + \left(\frac{1}{\omega} - \frac{Rp/\omega^3}{1 - p^2/\omega^2}\right) \sin \omega t$$

which may also be written as

$$x(t) = Dx_{\text{stat}} + x_{\text{trans}} \tag{d}$$

where x_{stat} is the static response of the system,

$$x_{\text{stat}} = \frac{R}{\omega^2} \sin pt$$

x_{trans} is the transient response

$$x_{\text{trans}} = \left(\frac{1}{\omega} - \frac{Rp/\omega^3}{1 - p^2/\omega^2}\right) \sin \omega t$$

and D *is the dynamic load factor,*

$$D = \frac{1}{1 - p^2/\omega^2}$$

The analysis of the response of the single degree of freedom system considered in the above example showed that the complete response is the sum of two contributions:

1. a dynamic response obtained by multiplying the static response with a dynamic load factor (this is the particular solution of the governing differential equation), and
2. an additional dynamic response which we called the transient response.

These observations pertain also to an actual practical analysis of a multiple degree of freedom system, because, firstly the complete response is obtained as a superposition of the response measured in each modal degree of freedom, and secondly, the actual loading can be represented in a Fourier decomposition as a superposition of harmonic sine and cosine contributions. Therefore, the above two observations apply to each modal response corresponding to each Fourier component of the loading.

An important difference between an actual practical response analysis and the solution in Example 9.8 is, however, that in practice the effect of damping must be included as discussed in Section 9.3.3. The presence of damping reduces the dynamic load factor (which then cannot be infinite) and damps out the transient response.

Figure 9.4 shows the dynamic load factor as a function of p/ω (and the damping ratio ξ discussed in Section 9.3.3).[12] The information in Fig. 9.4 is

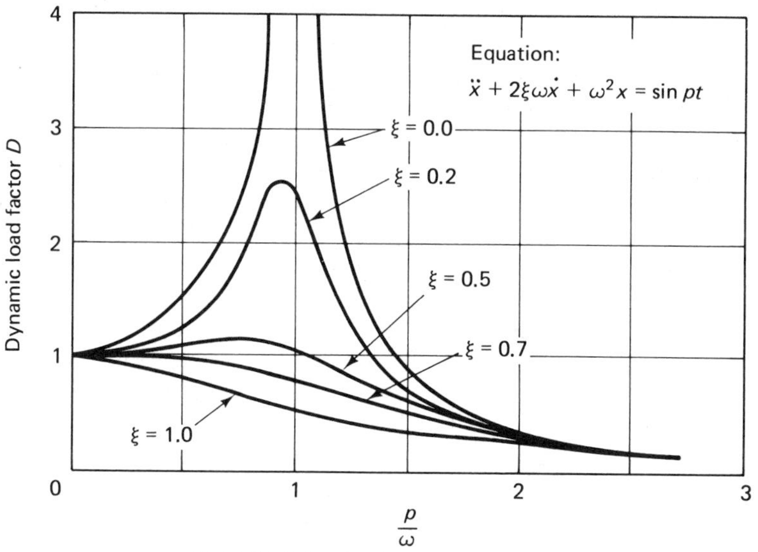

FIGURE 9.4 *The dynamic load factor.*

obtained by solving (9.54) as in Example 9.8. If we apply the information given in this figure to the analysis of an actual practical system we recognize that the response in the modes with p/ω large is negligible (the loads vary so rapidly that the system cannot respond), and that the static response is measured when p/ω is close to zero (the loads vary so slowly that the system simply follows the loads statically). Therefore, in the analysis of a multiple degree of freedom system, the response in the high frequencies of the system (that are much larger than the highest frequencies contained in the loads) is simply a static response.

The essence of a mode superposition solution of a dynamic response is that frequently only a small fraction of the total number of decoupled equations need be considered, in order to obtain a good approximate solution to the actual response of the system. Most frequently, only the first p equilibrium equations need be used; i.e., we need to include in the analysis the equations (9.46) for $i = 1, 2, \ldots, p$, where $p \ll n$, in order to obtain a good approximate solution. This means that we also need to solve only for the lowest p eigenvalues and corresponding eigenvectors of the problem in (9.36) and we only sum in (9.49) the response in the first p modes.

The reason that only the lowest modes are considered in a practical finite element analysis lies in the complete modeling process for dynamic analysis. Namely, so far, we have only been concerned with the exact solution of the finite element system equilibrium equations in (9.1). However, what we really aim to obtain is a good approximation to the actual exact response of the structure under consideration. We showed in Section 4.2.5 that under certain conditions the finite element analysis can be understood to be a Ritz analysis. Referring to the discussion in Sections 4.2.5 and 10.3.2, in that case, in finite element analysis, upper bounds to the "exact" frequencies of the actual structure are obtained. Moreover, in general, even when the monotonic convergence conditions are not satisfied, the finite element analysis approximates the lowest exact frequencies best, and little or no accuracy can be expected in approximating the higher frequencies and mode shapes. Therefore, there is usually little justification for including the dynamic response in the mode shapes with the high frequencies in the analysis. In fact, the finite element mesh should be chosen such that all important exact frequencies and vibration mode shapes are well approximated, and then the solution need only be calculated including the response in these modes. However, this can be achieved precisely using mode superposition analysis by considering only the important modes of the finite element system.

It is primarily for the fact that in a mode superposition analysis only a few modes may need to be considered that the mode superposition procedure can be much more effective than direct integration. However, it also follows that the effectiveness of mode superposition depends on the number of modes that must be included in the analysis. *In general, the structure considered and the spatial distribution and frequency content of the loading determine the number of modes to be used.* For earthquake loading, in some cases only the 10 lowest modes need be considered, although the order of the system n may be larger than 1000. On the other hand, for blast or shock loading, many more modes need generally be included, and p may be as large as $2n/3$. Finally, in vibration excitation analysis, only a few intermediate frequencies may be excited, such as all frequencies between the lower and upper frequency limits ω_l and ω_u, respectively.

Considering the problem of selecting the number of modes to be included in the mode superposition analysis, it should always be kept in mind that an approximate solution to the dynamic equilibrium equations in (9.1) is sought. Therefore, if not enough modes are considered, the equations in (9.1) are not solved accurately enough. But this means, in effect, that equilibrium, including inertia forces, is not satisfied for the approximate response calculated. Denoting by $\mathbf{U}^p$ the response predicted by mode superposition when p modes are considered, an indication of the accuracy of analysis at any time t is obtained by calculating an error measure ϵ^p, such as

$$\epsilon^p(t) = \frac{\|\mathbf{R}(t) - [\mathbf{M}\ddot{\mathbf{U}}^p(t) + \mathbf{K}\mathbf{U}^p(t)]\|_2}{\|\mathbf{R}(t)\|_2} \tag{9.50}$$

where we assume that $\|\mathbf{R}(t)\|_2 \neq 0$. If a good approximate solution of the system equilibrium equations in (9.1) has been obtained, $\epsilon^p(t)$ will be small at any time t. But it must be noted that $\mathbf{U}^p(t)$ must have been obtained by an accurate calculation of the response in each of the p modes considered, because this way the only error is due to not including enough modes in the analysis.

It should be noted that the error measure ϵ^p calculated in (9.50) determines how well equilibrium including inertia forces is satisfied, and is a measure of the nodal point loads not balanced by inertia and elastic nodal point forces (see (9.2)). Alternatively, we may say that ϵ^p is a measure of that part of the external load vector that has not been included in the mode superposition analysis,

$$\Delta\mathbf{R} = \mathbf{R} - \sum_{i=1}^{p} r_i(\mathbf{M}\boldsymbol{\phi}_i) \tag{9.51}$$

Since for a properly modeled problem the response to $\Delta\mathbf{R}$ should be at most a static response, a good correction $\Delta\mathbf{U}$ to the mode superposition solution $\mathbf{U}^p$ can be obtained from

$$\mathbf{K}\,\Delta\mathbf{U}(t) = \Delta\mathbf{R}(t) \tag{9.52}$$

where the solution of (9.52) may only be required for certain times at which the maximum response is measured.

Considering the error measure ϵ_p we should note that in a direct integration analysis $\epsilon_p(t)$ is zero for all times.

In summary, therefore, assuming that the decoupled equations in (9.46) have been solved accurately, the errors in a mode superposition analysis using $p < n$ are due to the fact that not enough modes have been used, whereas the errors in a direct integration analysis arise because too large a time step is employed.

From the above discussion it may appear that the mode superposition procedure has an inherent advantage over direct integration in that the response corresponding to the higher, probably inaccurate frequencies of the finite element system is not included in the analysis. However, assuming that in the finite element analysis all important frequencies are predicted accurately, and that little response is calculated in the higher modes of the system (which are presumably not important), the inclusion of the finite element system high-frequency response will not seriously affect the accuracy of the solution. In addition, we will discuss in Section 9.4 that in direct integration, advantage can be taken also of integrating accurately only the first p equations in (9.46) and neglecting the high-frequency response of the finite element system. This is achieved by using an unconditionally stable direct integration method and selecting an appropriate integration time step Δt, which, in general, is much larger than the integration step used with a conditionally stable integration scheme.

9.3.3 Analysis with Damping Included

The general form of the equilibrium equations of the finite element system in the basis of the eigenvectors $\boldsymbol{\phi}_i$, $i = 1, \ldots, n$ was given in (9.43), which shows that providing damping effects are neglected, the equilibrium equations decouple and the time integration can be carried out individually for each equation. Considering the analysis of systems in which damping effects cannot be neglected, we still would like to deal with decoupled equilibrium equations in (9.43), merely to be able to use essentially the same computational procedure whether damping effects are included or neglected. In general, the damping matrix $\mathbf{C}$ cannot be constructed from element damping matrices, such as the mass and stiffness matrices of the element assemblage, and its purpose is to approximate the overall energy dissipation during the system response. The mode superposition analysis is particularly effective if it can be assumed that damping is proportional, in which case

$$\boldsymbol{\phi}_i^T\mathbf{C}\boldsymbol{\phi}_j = 2\omega_i\xi_i\,\delta_{ij} \tag{9.53}$$

where ξ_i is a modal damping parameter and δ_{ij} is the Kronecker delta ($\delta_{ij} = 1$ for $i = j$, $\delta_{ij} = 0$ for $i \neq j$). Therefore, using (9.53) it is assumed that the eigenvectors $\boldsymbol{\phi}_i$, $i = 1, 2, \ldots, n$, are also $\mathbf{C}$-orthogonal and the equations in (9.43) reduce to n equations of the form

$$\ddot{x}_i(t) + 2\omega_i\xi_i\dot{x}_i(t) + \omega_i^2x_i(t) = r_i(t) \tag{9.54}$$

where $r_i(t)$ and the initial conditions on $x_i(t)$ have already been defined in (9.46) and (9.47). We note that (9.54) is the equilibrium equation governing motion of the single degree of freedom system considered in (9.46) when ξ_i is the damping ratio.

If the relation in (9.53) is used to account for damping effects, the procedure of solution of the finite element equilibrium equations in (9.43) is the same as in the case when damping is neglected (see Section 9.3.2) except that the response in each mode is obtained by solving (9.54). This response can be calculated using an integration scheme such as those given in Tables 9.1 to 9.4 or by evaluating the Duhamel integral to obtain

$$x_i(t) = \frac{1}{\bar{\omega}_i} \int_0^t r_i(\tau) e^{-\xi_i \omega_i (t-\tau)} \sin \bar{\omega}_i (t - \tau)\, d\tau + e^{-\xi_i \omega_i t} \{\alpha_i \sin \bar{\omega}_i t + \beta_i \cos \bar{\omega}_i t\} \tag{9.55}$$

where
$$\bar{\omega}_i = \omega_i \sqrt{1 - \xi_i^2}$$

and α_i and β_i are calculated using the initial conditions in (9.47).

In considering the implications of using (9.53) to take account of damping effects, the following observations are made. Firstly, the assumption in (9.53) means that the total damping in the structure is the sum of individual damping in each mode. The damping in one mode could be observed, for example, by imposing initial conditions corresponding to that mode only (i.e., ${}^0\mathbf{U} = \boldsymbol{\phi}_i$ for mode i) and measuring the amplitude decay during the free damped vibration. In fact, the ability to measure values for the damping ratios ξ_i, and thus approximate in many cases in a realistic manner the damping behavior of the complete structural system, is an important consideration. A second observation relating to the mode superposition analysis is that for the numerical solution of the finite element equilibrium equations in (9.1) using the decoupled equations in (9.54), we do not calculate the damping matrix $\mathbf{C}$, but only the stiffness and mass matrices $\mathbf{K}$ and $\mathbf{M}$.

As discussed, damping effects can readily be taken into account in mode superposition analysis provided that (9.53) is satisfied. However, assume that it would be numerically more effective to use direct step-by-step integration and that realistic damping ratios ξ_i, $i = 1, \ldots, p$ are known. In that case, it is necessary to evaluate the matrix $\mathbf{C}$ explicitly, which when substituted into (9.53) yields the established damping ratios ξ_i. If $p = 2$, Rayleigh damping can be assumed, which is of the form

$$\mathbf{C} = \alpha \mathbf{M} + \beta \mathbf{K} \tag{9.56}$$

where α and β are constants to be determined from two given damping ratios that correspond to two unequal frequencies of vibration.

EXAMPLE 9.9: Assume that for a multiple degree of freedom system $\omega_1 = 2$ and $\omega_2 = 3$, and that in those two modes we require 2% and 10% critical damping, respectively; i.e., we require $\xi_1 = 0.02$ and $\xi_2 = 0.10$. Establish the constants α and β for Rayleigh damping in order that a direct step-by-step integration can be carried out.

In Rayleigh damping we have

$$\mathbf{C} = \alpha \mathbf{M} + \beta \mathbf{K} \tag{a}$$

But using the relation in (9.53) we obtain using (a)

$$\boldsymbol{\phi}_i^T(\alpha\mathbf{M} + \beta\mathbf{K})\boldsymbol{\phi}_i = 2\omega_i\xi_i$$

or

$$\alpha + \beta\omega_i^2 = 2\omega_i\xi_i \tag{b}$$

Using this relation for ω_1, ξ_1 and ω_2, ξ_2, we obtain two equations for α and β:

$$\begin{aligned} \alpha + 4\beta &= 0.08 \\ \alpha + 9\beta &= 0.60 \end{aligned} \tag{c}$$

The solution of (c) is $\alpha = -0.336$ and $\beta = 0.104$. Thus the damping matrix to be used is

$$\mathbf{C} = -0.336\mathbf{M} + 0.104\mathbf{K} \tag{d}$$

With the damping matrix given, we can now establish the damping ratio that is specified at any value of ω_i, when the Rayleigh damping matrix in (d) is used. Namely, the relation in (b) gives

$$\xi_i = \frac{-0.336 + 0.104\omega_i^2}{2\omega_i}$$

for all values of ω_i.

In actual analysis it may well be that the damping ratios are known for many more than two frequencies. In that case two average values, say $\bar{\xi}_1$ and $\bar{\xi}_2$, are used to evaluate α and β. Consider the following example.

EXAMPLE 9.10: Assume that the approximate damping to be specified for a multiple degree of freedom system is as follows:

$$\begin{array}{llll} \xi_1 = 0.002; & \omega_1 = 2; & \xi_2 = 0.03; & \omega_2 = 3 \\ \xi_3 = 0.04; & \omega_3 = 7; & \xi_4 = 0.10; & \omega_4 = 15 \\ \xi_5 = 0.14; & \omega_5 = 19 & & \end{array}$$

Choose appropriate Rayleigh damping parameters α and β.

As in Example 9.9, we determine α and β from the relation

$$\alpha + \beta\omega_i^2 = 2\omega_i\xi_i \tag{a}$$

However, only two pairs of values $(\bar{\xi}_1, \bar{\omega}_1)$ and $(\bar{\xi}_2, \bar{\omega}_2)$ determine α and β. Considering the spacing of the frequencies, we use

$$\begin{array}{ll} \bar{\xi}_1 = 0.03; & \bar{\omega}_1 = 4 \\ \bar{\xi}_2 = 0.12; & \bar{\omega}_2 = 17 \end{array} \tag{b}$$

For the values in (b) we obtain, using (a),

$$\begin{aligned} \alpha + 16\beta &= 0.24 \\ \alpha + 289\beta &= 4.08 \end{aligned}$$

Hence $\alpha = 0.01498$, $\beta = 0.01405$, and we obtain

$$\mathbf{C} = 0.01498\mathbf{M} + 0.01405\mathbf{K} \tag{c}$$

We can now calculate which actual damping ratios are employed when the damping matrix $\mathbf{C}$ in (c) is used. Namely, from (a) we obtain

$$\xi_i = \frac{0.01498 + 0.01405\omega_i^2}{2\omega_i}$$

Figure 9.5 shows the relation of ξ_i as a function of ω_i.

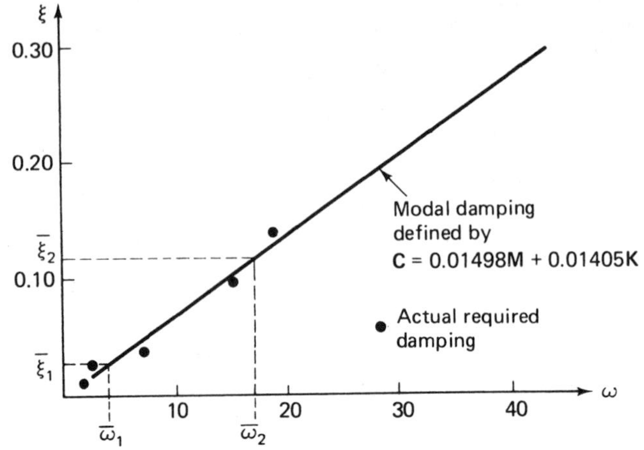

FIGURE 9.5 *Damping as a function of frequency.*

The procedure of calculating α and β in Examples 9.9 and 9.10 may suggest the use of a more complicated damping matrix if more than only two damping ratios are used to establish **C**. Assume that the p damping ratios ξ_i, $i = 1, 2, \ldots, p$ are given to define **C**. Then a damping matrix that satisfies the relation in (9.53) is obtained using the Caughey series,

$$\mathbf{C} = \mathbf{M} \sum_{k=0}^{p-1} a_k [\mathbf{M}^{-1} \quad \mathbf{K}]^k \tag{9.57}$$

where the coefficients a_k, $k = 1, \ldots, p$, are calculated from the p simultaneous equations

$$\xi_i = \frac{1}{2}\left(\frac{a_0}{\omega_i} + a_1\omega_i + a_2\omega_i^3 + \ldots + a_{p-1}\omega_i^{2p-3}\right)$$

We should note that with $p = 2$, (9.57) reduces to Rayleigh damping, as presented in (9.56). An important observation is that if $p > 2$, the damping matrix **C** in (9.57) is, in general, a full matrix. Since the cost of analysis is increased by a very significant amount if the damping matrix is not banded, in most practical analyses using direct integration, Rayleigh damping is assumed. A disadvantage of Rayleigh damping is that the higher modes are considerably more damped than the lower modes, for which the Rayleigh constants have been selected (see Example 9.10).

In practice, reasonable Rayleigh coefficients in the analysis of a specific structure may often be selected using available information on the damping characteristics of a typical similar structure; i.e., approximately the same α and β values are used in the analysis of similar structures. The magnitude of the Rayleigh coefficients is to a large extent determined by the energy dissipation characteristics of the structural materials.

In the above discussion we assumed that the damping characteristics of the structure can be represented appropriately using proportional damping, either in a mode superposition analysis or in a direct integration procedure. In many analyses, the assumption of proportional damping [i.e., that (9.53) is satisfied] is adequate. However, in the analysis of structures with widely varying material properties, nonproportional damping may need to be used. For example, in the analysis of foundation-structure interaction problems, significantly more damping may be observed in the foundation than in the surface structure. In this case it may be reasonable to assign in the construction of the damping matrix different Rayleigh coefficients α and β to different parts of the structure, which results into a damping matrix that does not satisfy the relation in (9.53). Another case of nonproportional damping is encountered when concentrated dampers corresponding to specific degrees of freedom (e.g., at the support points of a structure) are specified.

The solution of the finite element system equilibrium equations with nonproportional damping can be obtained using the direct integration algorithms in Tables 9.1 to 9.4 without modifications, because the property of the damping matrix did not enter the derivation of the solution procedures. On the other hand considering mode superposition analysis using the free vibration mode shapes with damping neglected as base vectors, we find that $\mathbf{\Phi}^T\mathbf{C}\mathbf{\Phi}$ in (9.43) is in the case of nonproportional damping a full matrix. In other words, the equilibrium equations in the basis of mode shape vectors are no longer decoupled. But, if it can be assumed that the primary response of the system is still contained in the subspace spanned by $\boldsymbol{\phi}_1, \ldots, \boldsymbol{\phi}_p$, it is only necessary to consider the first p equations in (9.43). Assuming that the coupling in the damping matrix $\mathbf{\Phi}^T\mathbf{C}\mathbf{\Phi}$ between x_i, $i = 1, \ldots, p$, and x_i, $i = p + 1, \ldots, n$, can be neglected, the first p equations in (9.43) decouple from the equations $(p + 1)$ to n and can be solved by direct integration using the algorithms in Tables 9.1 to 9.4 (see Example 9.11). In an alternative analysis procedure, the decoupling of the finite element equilibrium equations is achieved by solving a quadratic eigenproblem, in which case complex frequencies and vibration mode shapes are calculated.

EXAMPLE 9.11: Consider the solution of the equilibrium equations

$$\begin{bmatrix} \frac{1}{2} & & \\ & 1 & \\ & & \frac{1}{2} \end{bmatrix} \ddot{\mathbf{U}} + \begin{bmatrix} 0.1 & & \\ & 0 & \\ & & 0.5 \end{bmatrix} \dot{\mathbf{U}} + \begin{bmatrix} 2 & -1 & 0 \\ -1 & 4 & -1 \\ 0 & -1 & 2 \end{bmatrix} \mathbf{U} = \mathbf{R}(t) \qquad \text{(a)}$$

The free vibration mode shapes with damping neglected and corresponding frequencies of vibration are calculated in Example 10.4 and are

$$\mathbf{\Phi} = \begin{bmatrix} \sqrt{\frac{1}{2}} & 1 & -\sqrt{\frac{1}{2}} \\ \sqrt{\frac{1}{2}} & 0 & \sqrt{\frac{1}{2}} \\ \sqrt{\frac{1}{2}} & -1 & -\sqrt{\frac{1}{2}} \end{bmatrix}; \qquad \mathbf{\Omega}^2 = \begin{bmatrix} 2 & & \\ & 4 & \\ & & 6 \end{bmatrix}$$

Transform the equilibrium equations in (a) to equilibrium relations in the mode shape basis.

Using $\mathbf{U} = \mathbf{\Phi}\mathbf{X}$, we obtain corresponding to (9.43) the equilibrium relations

$$\ddot{\mathbf{X}}(t) + \begin{bmatrix} 0.3 & -0.2\sqrt{2} & -0.3 \\ -0.2\sqrt{2} & 0.6 & 0.2\sqrt{2} \\ -0.3 & 0.2\sqrt{2} & 0.3 \end{bmatrix} \dot{\mathbf{X}}(t) + \begin{bmatrix} 2 & & \\ & 4 & \\ & & 6 \end{bmatrix} \mathbf{X}(t)$$

$$
= \begin{bmatrix} \sqrt{\frac{1}{2}} & \sqrt{\frac{1}{2}} & \sqrt{\frac{1}{2}} \\ 1 & 0 & -1 \\ -\sqrt{\frac{1}{2}} & \sqrt{\frac{1}{2}} & -\sqrt{\frac{1}{2}} \end{bmatrix} \mathbf{R}(t) \tag{b}
$$

If it were now known that because of the specific loading applied, the primary response lies in the first mode only, we may obtain an approximate response by solving only

$$
\ddot{x}_1(t) + 0.3\dot{x}(t) + 2x_1(t) = \begin{bmatrix} \frac{1}{\sqrt{2}} & \frac{1}{\sqrt{2}} & \frac{1}{\sqrt{2}} \end{bmatrix} \mathbf{R}(t) \tag{c}
$$

and then calculating

$$
\mathbf{U}(t) = \begin{bmatrix} \frac{1}{\sqrt{2}} \\ \frac{1}{\sqrt{2}} \\ \frac{1}{\sqrt{2}} \end{bmatrix} x_1(t)
$$

However, it should be noted that because $\mathbf{\Phi}^T\mathbf{C}\mathbf{\Phi}$ in (b) is full, the solution of $x_1(t)$ from (c) does not give the actual response in the first mode, because the damping coupling has been neglected.

9.4 ANALYSIS OF DIRECT INTEGRATION METHODS

In the preceding sections we presented the two principal procedures used for the solution of the dynamic equilibrium equations

$$
\mathbf{M}\ddot{\mathbf{U}}(t) + \mathbf{C}\dot{\mathbf{U}}(t) + \mathbf{K}\mathbf{U}(t) = \mathbf{R}(t) \tag{9.58}
$$

where the matrices and vectors have been defined in Section 9.1. The two procedures were mode superposition and direct integration. The integration schemes considered were the central difference method, the Houbolt method, the Wilson θ method, and the Newmark integration procedure (see Tables 9.1 to 9.4). We stated that using the central difference scheme, a time step Δt smaller than a critical time step Δt_{cr} has to be used; but when employing the other three integration schemes, a similar time step limitation is not applicable.

An important observation was that the cost of a direct integration analysis (i.e., the number of operations required) is directly proportional to the number of time steps required for solution. It follows that the selection of an appropriate time step in direct integration is of much importance. On one hand, the time step must be small enough to obtain accuracy in the solution; but, on the other hand, the time step must not be smaller than necessary, because this would mean that the solution is more costly than actually required. The aim in this section is to discuss in detail the problem of selecting an appropriate time step Δt for direct integration. The two fundamental concepts to be considered are those of stability and accuracy of the integration schemes. The analysis of the stability and accuracy characteristics of the integration methods results into guidelines for the selection of an appropriate time step.

A first fundamental observation for the analysis of a direct integration method is the relation between mode superposition and direct integration. We pointed out in Section 9.3 that, in essence, using either procedure the solution is obtained by numerical integration. However, in the mode superposition analysis, a change of basis from the finite element coordinate basis to the basis of eigenvectors of the generalized eigenproblem,

$$\mathbf{K}\boldsymbol{\phi} = \omega^2\mathbf{M}\boldsymbol{\phi} \tag{9.59}$$

is performed prior to the time integration. Writing

$$\mathbf{U}(t) = \boldsymbol{\Phi}\mathbf{X}(t) \tag{9.60}$$

where the columns in $\boldsymbol{\Phi}$ are the $\mathbf{M}$-orthonormalized eigenvectors (free vibration modes) $\boldsymbol{\phi}_1, \ldots, \boldsymbol{\phi}_n$, and substituting for $\mathbf{U}(t)$ into (9.58) we obtain

$$\ddot{\mathbf{X}}(t) + \boldsymbol{\Delta}\dot{\mathbf{X}}(t) + \boldsymbol{\Omega}^2\mathbf{X}(t) = \boldsymbol{\Phi}^T\mathbf{R}(t) \tag{9.61}$$

where $\boldsymbol{\Omega}^2$ is a diagonal matrix listing the eigenvalues of (9.59) (free vibration frequencies squared) $\omega_1^2, \ldots, \omega_n^2$. Assuming that the damping is proportional, $\boldsymbol{\Delta}$ is a diagonal matrix, $\boldsymbol{\Delta} = \text{diag}\,(2\omega_i\xi_i)$, where ξ_i is the damping ratio in the ith mode.

The equation in (9.61) consists of n uncoupled equations, which can be solved, for example, using the Duhamel integral. Alternatively, one of the numerical integration schemes discussed as direct integration procedures may be used. Because the periods of vibration T_i, $i = 1, \ldots, n$, are known, where $T_i = 2\pi/\omega_i$, we can choose in the numerical integration of each equation in (9.61) an appropriate time step that assures a required level of accuracy. On the other hand, if all n equations in (9.61) are integrated using the same time step Δt, then the mode superposition analysis is completely equivalent to a direct integration analysis, in which the same integration scheme and the same time step Δt are used. In other words, the solution of the finite element system equilibrium equations would be identical using either procedure. Therefore, *to study the accuracy of direct integration, we may focus attention on the integration of the equations in (9.61) with a common time step Δt instead of considering (9.58). In this way, the variables to be considered in the stability and accuracy analysis of the direct integration method are only Δt, ω_i, and ξ_i, $i = 1, \ldots, n$, and not all elements of the stiffness, mass, and damping matrices. Furthermore, because all n equations in (9.61) are similar, we only need to study the integration of one typical row in (9.61) which may be written*

$$\ddot{x} + 2\xi\omega\dot{x} + \omega^2 x = r \tag{9.62}$$

and is the equilibrium equation governing motion of a single degree of freedom system with free vibration period T, damping ratio ξ, and applied load r.

It may be mentioned here that the above procedure of changing basis, i.e., using the transformation in (9.60) is also used in the convergence analysis of eigenvalue and eigenvector solution methods (see Chapter 11). The reason for carrying out the transformation is the same as above; namely, many fewer variables need to be considered in the analysis.

Considering the solution characteristics of a direct integration method, the problem is, therefore, to estimate the integration errors in the solution of (9.62) as a function of $\Delta t/T$, ξ, and r. Various procedures could be followed.[1-2, 14-23] However, in the following discussion we employ a relatively simple and constructive procedure in which the first step is to evaluate an approximation and load operator which relates explicitly the unknown required variables at time $t + \Delta t$ to previously calculated quantities.[18]

9.4.1 Direct Integration Approximation and Load Operators

In analogy to the derivation of the direct integration methods (see Section 9.2), assume that we have obtained the required solution for the discrete times $0, \Delta t, 2\Delta t, 3\Delta t, \ldots, t - \Delta t, t$, and that the solution for time $t + \Delta t$ is required next. Then for the specific integration method considered, we aim to establish the following recursive relationship:

$$^{t+\Delta t}\hat{\mathbf{X}} = \mathbf{A}\, {}^t\hat{\mathbf{X}} + \mathbf{L}\,({}^{t+\nu}r) \tag{9.63}$$

where $^{t+\Delta t}\hat{\mathbf{X}}$ and $^t\hat{\mathbf{X}}$ are vectors storing the solution quantities (e.g., displacements, velocities, etc.) and $^{t+\nu}r$ is the load at time $t + \nu$. We will see that ν may be 0, Δt, or $\theta\, \Delta t$ for the integration methods considered. The matrix $\mathbf{A}$ and vector $\mathbf{L}$ are the *integration approximation and load operators*, respectively. Each quantity in (9.63) depends on the specific integration scheme employed. However, before deriving the matrices and vectors corresponding to the different integration procedures, we note that (9.63) can be used to calculate the solution at any time $t + n\Delta t$; namely, applying (9.63) recursively, we obtain

$$\begin{aligned} ^{t+n\Delta t}\hat{\mathbf{X}} = {} & \mathbf{A}^n\, {}^t\hat{\mathbf{X}} + \mathbf{A}^{n-1}\, \mathbf{L}\,({}^{t+\nu}r) + \mathbf{A}^{n-2}\, \mathbf{L}\,({}^{t+\Delta t+\nu}r) + \ldots \\ & + \mathbf{A}\, \mathbf{L}\,({}^{t+(n-2)\,\Delta t+\nu}r) + \mathbf{L}\,({}^{t+(n-1)\,\Delta t+\nu}r) \end{aligned} \tag{9.64}$$

It is this relation that we will use for the study of the stability and accuracy of the integration methods. In the following sections we derive the operators $\mathbf{A}$ and $\mathbf{L}$ for the different integration methods considered, where we refer to the presentations in Sections 9.2.1 to 9.2.4.

The central difference method: In the central difference integration scheme we use (9.3) and (9.4) to approximate the acceleration and velocity at time t, respectively. The equilibrium equation (9.62) is considered at time t; i.e., we use

$$^t\ddot{x} + 2\xi\omega\, {}^t\dot{x} + \omega^2\, {}^t x = {}^t r \tag{9.65}$$

$$^t\ddot{x} = \frac{1}{\Delta t^2}\{{}^{t-\Delta t}x - 2\, {}^t x + {}^{t+\Delta t}x\} \tag{9.66}$$

$$^t\dot{x} = \frac{1}{2\Delta t}\{-{}^{t-\Delta t}x + {}^{t+\Delta t}x\} \tag{9.67}$$

Substituting (9.66) and (9.67) into (9.65) and solving for $^{t+\Delta t}x$, we obtain

$$^{t+\Delta t}x = \frac{2 - \omega^2\, \Delta t^2}{1 + \xi\omega\, \Delta t}\, {}^t x - \frac{1 - \xi\omega\, \Delta t}{1 + \xi\omega\, \Delta t}\, {}^{t-\Delta t}x + \frac{\Delta t^2}{1 + \xi\omega\, \Delta t}\, {}^t r \tag{9.68}$$

The solution (9.68) can now be written in the form (9.63), i.e., we have

$$\begin{bmatrix} ^{t+\Delta t}x \\ ^t x \end{bmatrix} = \mathbf{A} \begin{bmatrix} ^t x \\ ^{t-\Delta t}x \end{bmatrix} + \mathbf{L}\, {}^t r \tag{9.69}$$

where
$$\mathbf{A} = \begin{bmatrix} \dfrac{2 - \omega^2 \Delta t^2}{1 + \xi\omega \Delta t} & -\dfrac{1 - \xi\omega \Delta t}{1 + \xi\omega \Delta t} \\ 1 & 0 \end{bmatrix} \tag{9.70}$$

and
$$\mathbf{L} = \begin{bmatrix} \dfrac{\Delta t^2}{1 + \xi\omega \Delta t} \\ 0 \end{bmatrix} \tag{9.71}$$

The Houbolt method: In the Houbolt integration scheme the equilibrium equation (9.62) is considered at time $t + \Delta t$ and two backward difference formulas are used for the acceleration and velocity at time $t + \Delta t$; i.e., we use

$$ {}^{t+\Delta t}\ddot{x} + 2\xi\omega\, {}^{t+\Delta t}\dot{x} + \omega^2\, {}^{t+\Delta t}x = {}^{t+\Delta t}r \tag{9.72}$$

$$ {}^{t+\Delta t}\ddot{x} = \frac{1}{\Delta t^2}(2\, {}^{t+\Delta t}x - 5\, {}^{t}x + 4\, {}^{t-\Delta t}x - {}^{t-2\Delta t}x) \tag{9.73}$$

$$ {}^{t+\Delta t}\dot{x} = \frac{1}{6\Delta t}(11\, {}^{t+\Delta t}x - 18\, {}^{t}x + 9\, {}^{t-\Delta t}x - 2\, {}^{t-2\Delta t}x) \tag{9.74}$$

Substituting (9.73) and (9.74) into (9.72), we can establish the relation

$$\begin{bmatrix} {}^{t+\Delta t}x \\ {}^{t}x \\ {}^{t-\Delta t}x \end{bmatrix} = \mathbf{A} \begin{bmatrix} {}^{t}x \\ {}^{t-\Delta t}x \\ {}^{t-2\Delta t}x \end{bmatrix} + \mathbf{L}\, {}^{t+\Delta t}r \tag{9.75}$$

where
$$\mathbf{A} = \begin{bmatrix} \dfrac{5\beta}{\omega^2 \Delta t^2} + 6\kappa & -\left(\dfrac{4\beta}{\omega^2 \Delta t^2} + 3\kappa\right) & \dfrac{\beta}{\omega^2 \Delta t^2} + \dfrac{2\kappa}{3} \\ 1 & 0 & 0 \\ 0 & 1 & 0 \end{bmatrix} \tag{9.76}$$

$$\beta = \left(\frac{2}{\omega^2 \Delta t^2} + \frac{11\xi}{3\omega \Delta t} + 1\right)^{-1}; \qquad \kappa = \frac{\xi\beta}{\omega \Delta t} \tag{9.77}$$

and
$$\mathbf{L} = \begin{bmatrix} \dfrac{\beta}{\omega^2} \\ 0 \\ 0 \end{bmatrix} \tag{9.78}$$

The Wilson θ method: The basic assumption in the Wilson θ method is that the acceleration varies linearly over the time interval from t to $t + \theta\,\Delta t$, where $\theta \geq 1$ and is determined to obtain optimum stability and accuracy characteristics. Let τ denote the increase in time from time t, where $0 \leq \tau \leq \theta\,\Delta t$; then for the time interval t to $t + \theta\,\Delta t$, we have

$$ {}^{t+\tau}\ddot{x} = {}^{t}\ddot{x} + ({}^{t+\Delta t}\ddot{x} - {}^{t}\ddot{x})\frac{\tau}{\Delta t} \tag{9.79}$$

$$ {}^{t+\tau}\dot{x} = {}^{t}\dot{x} + {}^{t}\ddot{x}\,\tau + ({}^{t+\Delta t}\ddot{x} - {}^{t}\ddot{x})\frac{\tau^2}{2\Delta t} \tag{9.80}$$

$$ {}^{t+\tau}x = {}^{t}x + {}^{t}\dot{x}\,\tau + \tfrac{1}{2}\,{}^{t}\ddot{x}\,\tau^2 + ({}^{t+\Delta t}\ddot{x} - {}^{t}\ddot{x})\frac{\tau^3}{6\Delta t} \tag{9.81}$$

At time $t + \Delta t$ we have

$$ {}^{t+\Delta t}\dot{x} = {}^{t}\dot{x} + ({}^{t+\Delta t}\ddot{x} + {}^{t}\ddot{x})\frac{\Delta t}{2} \tag{9.82}$$

$$ {}^{t+\Delta t}x = {}^{t}x + {}^{t}\dot{x}\,\Delta t + (2\,{}^{t}\ddot{x} + {}^{t+\Delta t}\ddot{x})\frac{\Delta t^2}{6} \tag{9.83}$$

Furthermore, in the Wilson θ method the equilibrium equation (9.62) is considered at time $t + \theta\,\Delta t$ [with a projected load, see (9.25)], which gives

$$^{t+\theta\Delta t}\ddot{x} + 2\xi\omega\, ^{t+\theta\Delta t}\dot{x} + \omega^2\, ^{t+\theta\Delta t}x = \, ^{t+\theta\Delta t}r \tag{9.84}$$

Using (9.79), (9.80), and (9.81) at time $\tau = \theta\,\Delta t$ to substitute into (9.84), an equation is obtained with $^{t+\Delta t}\ddot{x}$ as the only unknown. Solving for $^{t+\Delta t}\ddot{x}$ and substituting into (9.82) and (9.83), the following relationship of the form (9.63) is established:

$$\begin{bmatrix} ^{t+\Delta t}\ddot{x} \\ ^{t+\Delta t}\dot{x} \\ ^{t+\Delta t}x \end{bmatrix} = \mathbf{A} \begin{bmatrix} ^{t}\ddot{x} \\ ^{t}\dot{x} \\ ^{t}x \end{bmatrix} + \mathbf{L}\, ^{t+\theta\Delta t}r \tag{9.85}$$

where

$$\mathbf{A} = \begin{bmatrix} 1 - \dfrac{\beta\theta^2}{3} - \dfrac{1}{\theta} - \kappa\theta & \dfrac{1}{\Delta t}(-\beta\theta - 2\kappa) & \dfrac{1}{\Delta t^2}(-\beta) \\ \Delta t\left(1 - \dfrac{1}{2\theta} - \dfrac{\beta\theta^2}{6} - \dfrac{\kappa\theta}{2}\right) & 1 - \dfrac{\beta\theta}{2} - \kappa & \dfrac{1}{\Delta t}\left(-\dfrac{\beta}{2}\right) \\ \Delta t^2\left(\dfrac{1}{2} - \dfrac{1}{6\theta} - \dfrac{\beta\theta^2}{18} - \dfrac{\kappa\theta}{6}\right) & \Delta t\left(1 - \dfrac{\beta\theta}{6} - \dfrac{\kappa}{3}\right) & 1 - \dfrac{\beta}{6} \end{bmatrix} \tag{9.86}$$

$$\beta = \left(\frac{\theta}{\omega^2\,\Delta t^2} + \frac{\xi\theta^2}{\omega\,\Delta t} + \frac{\theta^3}{6}\right)^{-1}; \qquad \kappa = \frac{\xi\beta}{\omega\,\Delta t} \tag{9.87}$$

and

$$\mathbf{L} = \begin{bmatrix} \dfrac{\beta}{\omega^2\,\Delta t^2} \\ \dfrac{\beta}{2\omega^2\,\Delta t} \\ \dfrac{\beta}{6\omega^2} \end{bmatrix} \tag{9.88}$$

The Newmark method: In the Newmark integration scheme the equilibrium equation (9.62) is considered at time $t + \Delta t$; i.e., we use

$$^{t+\Delta t}\ddot{x} + 2\xi\omega\, ^{t+\Delta t}\dot{x} + \omega^2\, ^{t+\Delta t}x = \, ^{t+\Delta t}r \tag{9.89}$$

and the following expansions are employed for the velocity and displacement at time $t + \Delta t$:

$$^{t+\Delta t}\dot{x} = \, ^{t}\dot{x} + [(1 - \delta)\, ^{t}\ddot{x} + \delta\, ^{t+\Delta t}\ddot{x}]\,\Delta t \tag{9.90}$$

$$^{t+\Delta t}x = \, ^{t}x + \, ^{t}\dot{x}\,\Delta t + [(\tfrac{1}{2} - \alpha)\, ^{t}\ddot{x} + \alpha\, ^{t+\Delta t}\ddot{x}]\,\Delta t^2 \tag{9.91}$$

where δ and α are parameters to be chosen to obtain optimum stability and accuracy. Newmark proposed as an unconditionally stable scheme the constant-average-acceleration method, in which case $\delta = \frac{1}{2}$ and $\alpha = \frac{1}{4}$.

Substituting for $^{t+\Delta t}\dot{x}$ and $^{t+\Delta t}x$ into (9.89), we can solve for $^{t+\Delta t}\ddot{x}$ and then use (9.90) and (9.91) to calculate $^{t+\Delta t}\dot{x}$ and $^{t+\Delta t}x$. We thus can establish the relation

$$\begin{bmatrix} ^{t+\Delta t}\ddot{x} \\ ^{t+\Delta t}\dot{x} \\ ^{t+\Delta t}x \end{bmatrix} = \mathbf{A} \begin{bmatrix} ^{t}\ddot{x} \\ ^{t}\dot{x} \\ ^{t}x \end{bmatrix} + \mathbf{L}\, ^{t+\Delta t}r \tag{9.92}$$

where

$$
\mathbf{A} = \begin{bmatrix} -(\frac{1}{2}-\alpha)\beta - 2(1-\delta)\kappa & \frac{1}{\Delta t}(-\beta - 2\kappa) & \frac{1}{\Delta t^2}(-\beta) \\ \Delta t[1-\delta-(\frac{1}{2}-\alpha)\delta\beta - 2(1-\delta)\delta\kappa] & 1-\beta\delta-2\delta\kappa & \frac{1}{\Delta t}(-\beta\delta) \\ \Delta t^2[\frac{1}{2}-\alpha-(\frac{1}{2}-\alpha)\alpha\beta - 2(1-\delta)\alpha\kappa] & \Delta t(1-\alpha\beta-2\alpha\kappa) & (1-\alpha\beta) \end{bmatrix} \tag{9.93}
$$

$$
\beta = \left(\frac{1}{\omega^2\,\Delta t^2} + \frac{2\xi\delta}{\omega\,\Delta t} + \alpha\right)^{-1}; \qquad \kappa = \frac{\xi\beta}{\omega\,\Delta t} \tag{9.94}
$$

and

$$
\mathbf{L} = \begin{bmatrix} \dfrac{\beta}{\omega^2\,\Delta t^2} \\ \dfrac{\beta\delta}{\omega^2\,\Delta t} \\ \dfrac{\alpha\beta}{\omega^2} \end{bmatrix} \tag{9.95}
$$

We should note the close relationship between the Newmark operators in (9.93) and (9.95) and the Wilson θ method operators in (9.86) and (9.88). That is, using $\delta = \frac{1}{2}$, $\alpha = \frac{1}{6}$, and $\theta = 1.0$, the same approximation and load operators are obtained in both methods. This should be expected because for the above parameters both methods assume a linear variation of acceleration over the time interval t to $t + \Delta t$.

9.4.2 Stability Analysis

The aim in the numerical integration of the finite element system equilibrium equations is to evaluate a good approximation to the actual dynamic response of the structure under consideration. In order to predict the dynamic response of the structure accurately, it would seem that all system equilibrium equations in (9.58) must be integrated to high precision, and this means that all n equations of the form (9.62) need be integrated accurately. Since in direct integration the same time step is used for each equation of the form (9.62), Δt would have to be selected corresponding to the smallest period in the system, which may mean that the time step is very small indeed. As an estimate of Δt thus required, it appears that if the smallest period is T_n, Δt would have to be about $T_n/10$. However, we discussed in Section 9.3.2 that in many analyses the primary response lies in only some modes of vibration and that for this reason, only some mode shapes are considered in mode superposition analysis. In addition, it was pointed out that in many analyses there is little justification to include the response predicted in the higher modes, because the frequencies and mode shapes of the finite element mesh can only be crude approximations to the "exact" quantities. Therefore, the finite element idealization has to be chosen in such a way that the lowest p frequencies and mode shapes of the structure are predicted accurately, where p is determined by the distribution and frequency content of the loading.

Considering direct integration, it can therefore be concluded that in many analyses we are only interested in integrating the first p equations of the n equations in (9.61) accurately. This means that we would be able to revise Δt to be about $T_p/10$, i.e., T_p/T_n times larger than our first estimate. In practical analysis the ratio T_p/T_n can be very large, say of the order 1000, meaning that the analysis would be much more effective using $\Delta t = T_p/10$. However, assuming that we select a time step Δt of magnitude $T_p/10$, we realize that in the direct integration also the response in the higher modes is automatically integrated with the same time step. Since we cannot possibly integrate accurately the response in those modes for which Δt is larger than half the natural period T, an important question is: What "response" is predicted in the numerical integration of (9.62) when $\Delta t/T$ is large? This is, in essence, the question of stability of an integration scheme. Stability of an integration method means that the initial conditions for the equations with a large value $\Delta t/T$ must not be amplified artificially and thus make the integration of the lower modes worthless. Stability also means that any errors in the displacements, velocities, and accelerations at time t, which may be due to round-off in the computer, do not grow in the integration. Stability is assured if the time step is small enough to integrate accurately the response in the highest frequency component. But this may require a very small time step, and, as was pointed out above, the accurate integration of the high-frequency response predicted by the finite element assemblage is in many cases not justified and therefore not necessary.

The stability of an integration method is determined by examining the behavior of the numerical solution for arbitrary initial conditions. Therefore, we consider the integration of (9.62) when no load is specified; i.e., $r = 0$. The solution for prescribed initial conditions only as obtained from (9.64) is, hence,

$$^{t+n\Delta t}\hat{\mathbf{X}} = \mathbf{A}^n \, {}^t\hat{\mathbf{X}} \tag{9.96}$$

Considering the stability of integration methods, we have procedures that are unconditionally stable and that are only conditionally stable. *An integration method is unconditionally stable if the solution for any initial conditions does not grow without bound for any time step* Δt, *in particular when* $\Delta t/T$ *is large. The method is only conditionally stable if the above only holds provided that* $\Delta t/T$ *is smaller than a certain value, usually called the stability limit.*

To investigate the stability of an integration method we may conveniently use the relation in (9.96), where the approximation operators corresponding to some popular integration schemes have been presented in Section 9.4.1.

For the analysis we use the spectral decomposition of $\mathbf{A}$ and have

$$\mathbf{A}^n = \mathbf{P}\mathbf{J}^n\mathbf{P}^{-1} \tag{9.97}$$

where $\mathbf{P}$ is the matrix of eigenvectors of $\mathbf{A}$ and $\mathbf{J}$ is the Jordan form of $\mathbf{A}$ with eigenvalues λ_i of $\mathbf{A}$ on its diagonal. We considered in Section 2.7 the case of $\mathbf{A}$ being symmetric (see Example 2.23), however, the approximation operator is in general nonsymmetric and we therefore must use the more general decomposition $\mathbf{A} = \mathbf{P}\mathbf{J}\mathbf{P}^{-1}$. Let $\rho(\mathbf{A})$ be the spectral radius of $\mathbf{A}$ defined as

$$\rho(\mathbf{A}) = \max |\lambda_i|; \qquad i = 1, 2, \ldots \tag{9.98}$$

then $\mathbf{J}^n$ *is bounded for* $n \longrightarrow \infty$ *if and only if* $\rho(\mathbf{A}) \leq 1$. *This is our stability criterion.* Furthermore, $\mathbf{J}^n \longrightarrow \mathbf{0}$ if $\rho(\mathbf{A}) < 1$, and the smaller $\rho(\mathbf{A})$, the more rapid is the convergence.

Since the stability of an integration method depends only on the eigenvalues of the approximation operator, it may be convenient to apply a similarity transformation on $\mathbf{A}$ before evaluating the eigenvalues. In the case of the Newmark method and Wilson θ method we apply the similarity transformation $\mathbf{D}^{-1}\mathbf{A}\mathbf{D}$, where $\mathbf{D}$ is a diagonal matrix with $d_{ii} = (\Delta t)^i$. As might be expected, we then find that the spectral radii and therefore the stability of the integration methods depend only on the time ratio $\Delta t/T$, the damping ratio ξ, and the integration parameters used. Therefore, for given $\Delta t/T$ and ξ, it is possible in the Wilson θ method and in the Newmark method to vary the parameters θ and α, δ, repectively, to obtain optimum stability and accuracy characteristics.

Consider as a simple example the stability analysis of the central difference method.

EXAMPLE 9.12: Analyze the central difference method for its integration stability. Consider the case $\xi = 0.0$.

We need to calculate the spectral radius of the approximation operator given in (9.70) when $\xi = 0$. The eigenvalue problem $\mathbf{A}\mathbf{u} = \lambda\mathbf{u}$ to be solved is

$$\begin{bmatrix} 2 - \omega^2\,\Delta t^2 & -1 \\ 1 & 0 \end{bmatrix}\mathbf{u} = \lambda\mathbf{u} \qquad \text{(a)}$$

The eigenvalues are the roots of the characteristic polynomial $p(\lambda)$ (see Section 2.7) defined as

$$p(\lambda) = (2 - \omega^2\,\Delta t^2 - \lambda)(-\lambda) + 1$$

Hence

$$\lambda_1 = \frac{2 - \omega^2\,\Delta t^2}{2} + \sqrt{\frac{(2 - \omega^2\,\Delta t^2)^2}{4} - 1}$$

$$\lambda_2 = \frac{2 - \omega^2\,\Delta t^2}{2} - \sqrt{\frac{(2 - \omega^2\,\Delta t^2)^2}{4} - 1}$$

For stability we need that the absolute value of λ_1 and λ_2 be smaller or equal to 1; i.e., the spectral radius $\rho(\mathbf{A})$ of the matrix $\mathbf{A}$ in (a) must satisfy the condition $\rho(\mathbf{A}) \leq 1$. Figure 9.6 shows $\rho(\mathbf{A})$ as a function of $\Delta t/T$, where $\omega = 2\pi/T$. The spectral radius is smaller than 1 for $\Delta t/T < 1/\pi$, and hence the central difference method is stable provided that $\Delta t \leq \Delta t_{cr}$, where $\Delta t_{cr} = T_n/\pi$.

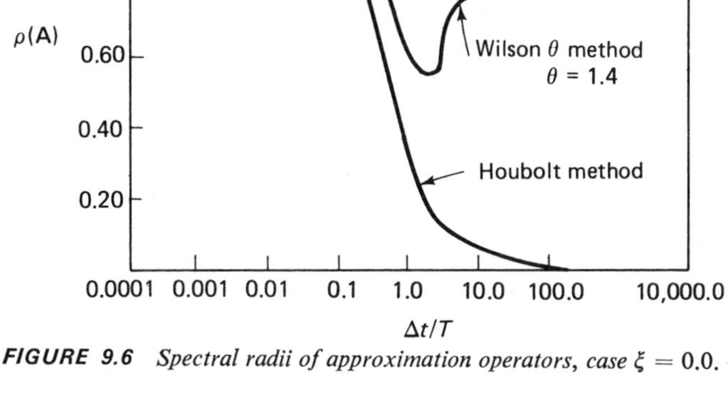

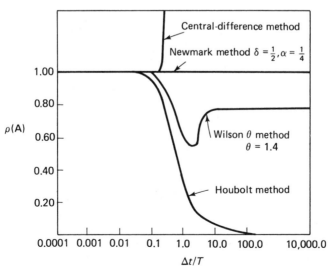

FIGURE 9.6 *Spectral radii of approximation operators, case* $\xi = 0.0$.

By the same procedure as employed in Example 9.12 the Wilson θ method, Newmark, and Houbolt methods can be analyzed for stability using the corresponding approximation operators. Assuming no physical damping (i.e., $\xi = 0.0$) Fig. 9.6 shows the stability characteristics of the methods considered in Section 9.2. It is noted that the central difference method is only conditionally stable as evaluated in Example 9.12 and that the Newmark, Wilson, and Houbolt methods are unconditionally stable. For the case $\xi > 0$, the stability analysis would include ξ as an additional variable. However, in general, small values of ξ do not change the overall stability characteristics of an integration scheme.

In order to evaluate an optimum value of θ for the Wilson θ method, the variation of the spectral radius of the approximation operator as a function of θ must be calculated as shown in Fig. 9.7. It is seen that unconditional stability is obtained when $\theta \geq 1.37$. This information must be supplemented by the accuracy of the method in order to arrive at the optimum value of θ (see Section 9.4.3).

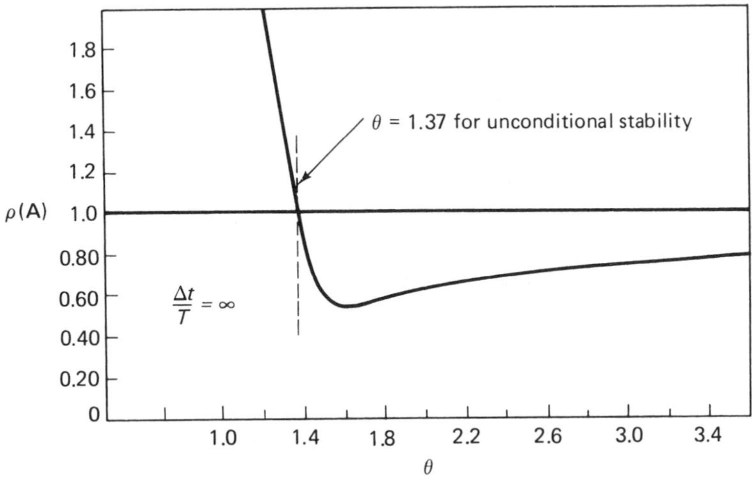

FIGURE 9.7 *Spectral radius $\rho(A)$ as a function of θ in the Wilson θ method.*

Considering the Newmark method, the two parameters α and δ can be varied to obtain optimum stability and accuracy. The integration scheme is unconditionally stable provided that $\delta \geq 0.5$ and $\alpha \geq 0.25(\delta + 0.5)^2$. We will see in the next section that the method corresponding to $\delta = 0.5$ and $\alpha = 0.25$ has the most desirable accuracy characteristics.

Although only the most widely used integration schemes have been discussed, the above considerations already show that a large number of integration methods are available.[17-24] Hence, the analyst has to make a choice as to which method to use. This choice is much influenced by the accuracy characteristics of the method, i.e., the accuracy that can be obtained in the integration for a given time step Δt.

9.4.3 Accuracy Analysis

The decision as to which integration operator to use in a practical analysis is governed by the cost of solution, which in turn is determined by the number of time steps required in the integration. If a conditionally stable algorithm such as the central difference method is employed, the time step size, and hence the number of time steps for a given time range considered, is determined by the critical time step Δt_{cr}, and not much choice is available. However, using an unconditionally stable operator, the time step has to be chosen to yield an accurate and effective solution. Because the direct integration of the equilibrium equations in (9.58) is equivalent to integrating simultaneously all n decoupled equations of the form (9.62) we can study the integration accuracy obtained in the solution of (9.58) by assessing the accuracy obtained in the integration of (9.62) as a function of $\Delta t/T$, ξ, and r. The solution to (9.62) was given in (9.64) and it is this equation that we use to assess the integration errors.

Various studies of the integration accuracies of the Newmark, Wilson, Houbolt, and central difference methods have been reported,[17-23] and because the accuracy of a specific procedure depends on a number of factors, in the following we only summarize some important solution characteristics. For this purpose we consider the solution of the initial value problem defined by

$$\left.\begin{aligned} \ddot{x} + \omega^2 x &= 0 \\ \text{and} \qquad {}^0x = 1.0; \quad {}^0\dot{x} &= 0.0; \quad {}^0\ddot{x} = -\omega^2 \end{aligned}\right\} \tag{9.99}$$

for which the exact solution is $x = \cos \omega t$. For a complete analysis we would have to consider also (1) the initial value problem corresponding to ${}^0x = 0.0$, ${}^0\dot{x} = \omega$, ${}^0\ddot{x} = 0$ with the exact solution $x = \sin \omega t$, and (2) the solution for a general loading condition. In addition, the influence of the damping parameter ξ would need to be investigated. However, the significant solution characteristics can already be demonstrated by considering only the numerical solution of the problem in (9.99).

The Newmark and Wilson methods can be used directly with the initial conditions given in (9.99). However, in the Houbolt method, the initial conditions are only defined by initial displacements, and in the following study the exact displacement values for ${}^{\Delta t}x$ and ${}^{2\Delta t}x$ obtained using the solution $x = \cos \omega t$ have been employed.

The numerical solution of (9.99) using the different integration methods shows that the errors in the integrations can be measured in terms of period elongation and amplitude decay. Figure 9.8 shows the percentage period elongations and amplitude decays in the implicit integration schemes discussed as a function of $\Delta t/T$. The relationships have been obtained by evaluating (9.64) using the computer and comparing the exact solution $x = \cos \omega t$ with the numerical solutions. It should be noted that (9.64) only yields the solution at the discrete time points Δt apart. To obtain maximum displacements, as required in Fig. 9.8, the relations in (9.80), (9.81) and (9.90), (9.91), respectively, have been employed in the Wilson θ and Newmark methods, and an interpolating polynomial of order 3, which fits the displacements at the discrete time points Δt apart, has been used in the Houbolt method.

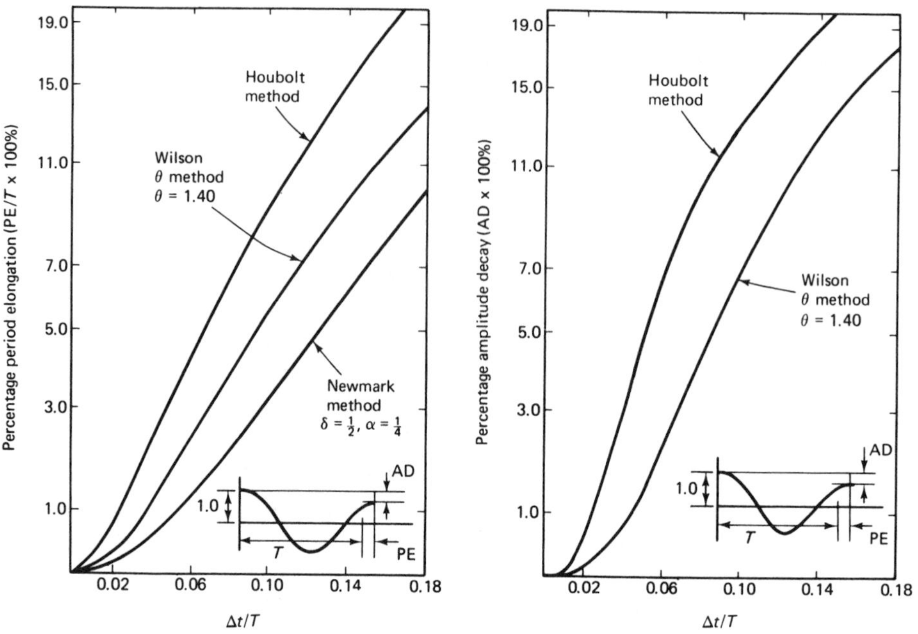

FIGURE 9.8 *Percentage period elongations and amplitude decays.*

The curves in Fig. 9.8 show that, in general, the numerical integrations using any of the methods are accurate when $\Delta t/T$ is smaller than about 0.01. However, when the time step-to-period ratio is larger, the various integration methods exhibit quite different characteristics. Notably, for a given ratio $\Delta t/T$, the Wilson θ method with $\theta = 1.4$ introduces less amplitude decay and period elongation than the Houbolt method, and the Newmark constant-average-acceleration method introduces only period elongation and no amplitude decay.

The characteristics of the integration errors exhibited in Fig. 9.8 are used in the discussion of the simultaneous integration of all n equations of form (9.62) as required in the solution of (9.61) and thus in the solution of (9.58). We observe that the equations for which the time step-to-period ratio is small are integrated accurately, but that the response in the equations for which $\Delta t/T$ is large is not obtained with any precision.

An important problem is the choice of an appropriate time step Δt. Using the central difference method, the time step has to be chosen such that Δt is smaller than Δt_{cr} evaluated in Example 9.12 which means that Δt will, in general, be small enough to obtain accuracy in the integration of practically all n equations in (9.61). However, using one of the unconditionally stable schemes, we realize that the time step Δt can be much larger and should only be small enough that the response in all modes which significantly contribute to the total structural response is calculated accurately. The other modal response components are not evaluated accurately, but the errors are unimportant because the response measured in those components is negligible. However, it need be noted again that this applies only if the integration scheme is unconditionally stable, i.e., if it is assumed that the amplitudes in the components which are negligible do not grow.

Using one of the unconditionally stable algorithms the time step Δt can be chosen by referring to the integration errors displayed in Fig. 9.8. Since one important phenomenon is the amplitude decay observed in the numerical integration of the modes for which $\Delta t/T$ is large, consider the following demonstrative case. Assume that using the Wilson method with $\theta = 1.4$, a time step is selected which gives $\Delta t/T_1 = 0.01$, where T_1 is the fundamental period of a six-degree-of-freedom system and $T_{i+1} = T_i/10$, $i = 1, \ldots, 5$. Let the initial conditions in each mode be those given in (9.99) and let the integration be performed for 100 time steps. Figure 9.9 shows the response calculated in the fundamental and higher modes. It is observed that the amplitude decay caused by the numerical integration errors effectively "filters" the high mode response out of the solution. The same effect is obtained using the Houbolt method, whereas when the Newmark constant-average-acceleration scheme is employed, which does not introduce amplitude decay, the high frequency response is retained in the solution. In order to obtain amplitude decay using the Newmark method, it is necessary to employ $\delta > 0.5$ and correspondingly $\alpha = \frac{1}{4}(\delta + 0.5)^2$.

It should be noted that the above considerations may be considered to be rather theoretical, because in a practical analysis the initial conditions on the second to sixth mode should be negligible for the above integration time step to be realistic. However, integration accuracy cannot possibly be obtained in the response of the modes for which the time step-to-period ratio is large, and it

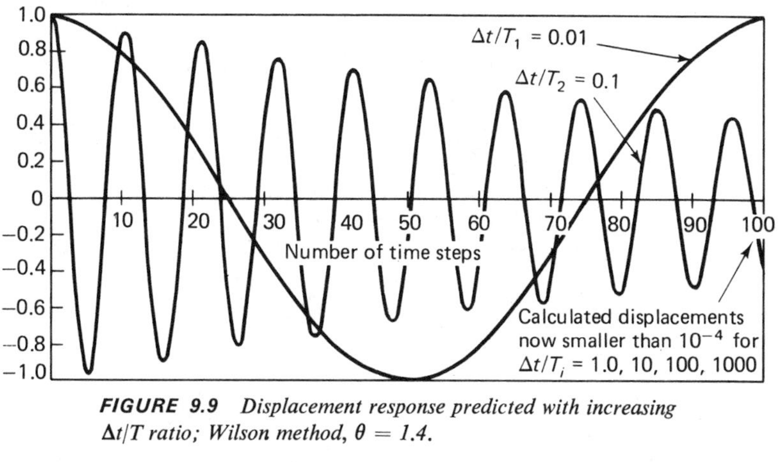

FIGURE 9.9 *Displacement response predicted with increasing $\Delta t/T$ ratio; Wilson method, $\theta = 1.4$.*

might be best to simply suppress any response calculations in these modes. On the other hand, since the response in those modes for which the time step-to-period ratio is large is supposed to be negligible, the filtering should not actually be required, and because of the smallest numerical errors, the Newmark method appears to be most effective.

In most practical analyses, little difference in the computational effort using either the Newmark or Wilson θ method is observed, because the errors resulting in the period elongations require about the same time step using either procedure. On the other hand, the Houbolt method has the disadvantage that special starting procedures must be used.

It should be realized that the stability and accuracy analyses given in this section assume that a linear finite element analysis is carried out. Some important additional considerations are required in the analysis of nonlinear response (see Section 9.5).

9.4.4 Some Practical Considerations

In order to obtain an effective solution of a dynamic response, it is important to choose an appropriate time integration scheme. This choice depends on the finite element idealization upon which the integration scheme is operating. However, the finite element idealization to be chosen depends in turn upon the actual physical problem to be analyzed. It follows therefore that the selection of an appropriate finite element idealization of a problem and the choice of an effective integration scheme for the response solution are closely related and must be considered together. The finite element model and time integration scheme are chosen differently depending on whether a structural dynamics or a wave propagation problem is solved.

The basic consideration in the selection of an appropriate finite element model of a structural dynamics problem is that only the lowest modes (or only a few intermediate modes) of a physical system are being excited by the load vector. We discussed this consideration already in Section 9.3.2, where we addressed the problem of how many modes need be included in a mode superposition analysis. Referring to Section 9.3.2 we can conclude that if a Fourier analysis of the dynamic load input shows that only frequencies below ω_u are contained in the loading, then the finite element mesh should at most represent accurately the frequencies to about $\omega_{co} = 4\omega_u$ of the actual system. There is no need to represent the higher frequencies of the actual system accurately in the finite element system, because the dynamic response contribution in those frequencies is negligible; i.e., for values of p/ω in Fig. 9.4 smaller than 0.25, practically a static response is measured and this response is directly included in the direct integration step-by-step dynamic response calculations. Also, the value of ω_{co} should be reduced appropriately if the loading is practically orthogonal to the mode shapes.*

The complete procedure for the modeling of a structural vibration problem is therefore:

1. Identify the frequencies contained in the loading, using a Fourier analysis if necessary.
2. Choose a finite element mesh that can accurately represent the static response and accurately represents all frequencies up to about ω_{co}.
3. Perform the direct integration analysis. The time step Δt for this solution should equal about $\frac{1}{20}T_{co}$, where $T_{co} = 2\pi/\omega_{co}$, or be smaller for stability reasons.†

* This reduction of ω_{co} may be considerable, for example, in earthquake response analysis of certain structures.

† Note that if a mode superposition solution is used for this step (as described in Section 9.3), then ω_{co} would be the highest frequency to be included in the solution.

When analyzing a structural dynamics problem, in most cases, an implicit time integration is most effective. In this integration the time step Δt need generally only be $\frac{1}{20}T_{co}$ (and not smaller, unless convergence difficulties are encountered in the iteration in a nonlinear response calculation; see Section 9.5.2). If an implicit time integration is employed, it is frequently effective to use higher-order finite elements, for example, the eight- and twenty-node elements of Figs. 5.5 and 5.6 in two- and three-dimensional analysis, respectively, and a consistent mass idealization. The higher-order elements are effective in the representation of bending behavior, but need generally be employed with a consistent load vector, so that the midside and the corner nodes are subjected to their appropriate load contributions in the analysis.

The observation that the use of higher-order elements can be effective with implicit time integration in the analysis of structural dynamics problems is consistent with the fact that higher-order elements have generally been found to be efficient in static analysis, and structural dynamics problems can be thought of as "static problems including inertia effects." If, on the other hand, the finite element idealization is so large that a multiple block out-of-core solution is necessary with a large bandwidth it can be more efficient to use explicit time integration with a lumped mass matrix, in which case no effective stiffness matrices are assembled and triangularized but a smaller time step Δt must generally be employed in the solution.

Considering next the analysis of wave propagation problems, the major difference between a structural dynamics problem and a wave propagation problem can be understood to be that in a wave propagation problem a large number of frequencies are excited in the system. It follows that one way to analyze a wave propagation problem is to use a sufficiently high cut-off frequency ω_{co} to obtain enough solution accuracy. The difficulties are in identifying the cut-off frequency to be used and in establishing a corresponding finite element model.

Instead of using the above considerations to obtain an appropriate finite element mesh for the analysis of a wave propagation problem, it is generally more effective to employ the concepts used in finite difference analysis and the method of characteristics[2] in order to establish an appropriate finite element mesh and time step Δt for the analysis.

If we assume that the critical wave length to be represented is L_w, the total time for this wave to travel past a point is

$$t_w = \frac{L_w}{c} \tag{9.100}$$

where c is the wave speed. Assuming that n time steps are necessary to represent the travel of the wave

$$\Delta t = \frac{t_w}{n} \tag{9.101}$$

and the "effective length" of a finite element should be

$$L_e = c\,\Delta t \tag{9.102}$$

This effective length and corresponding time step must be able to represent the complete wave travel accurately and is chosen differently depending on the kind of element idealization and time integration scheme used.

Although a special case, the effectiveness of using (9.102) becomes apparent when a one-dimensional analysis of a bar is performed with a lumped mass idealization and the central difference method. If a uniform bar free at both ends and subjected to a sudden constant step load is idealized as an assemblage of two-node truss elements each of length $c\,\Delta t$, the exact wave propagation response is obtained in the solution of the model. It is also interesting to note that the time step Δt given in (9.102) corresponds to the stability limit T_n/π derived in Example 9.12, i.e. $\omega_n = 2c/L_e$, and the nonzero (highest) frequency of a single unconstraint element is ω_n. Hence, the most accurate solution is obtained by integrating with a time step equal to the stability limit and the solution is less accurate when a *smaller* time step is employed! This deterioration in the accuracy of the predicted solution when Δt is smaller than Δt_{cr} is most pronounced when a relatively coarse spatial discretization is used.

In more complex two- and three-dimensional analyses, the exact solution is generally not obtained, and L_e must be chosen depending on whether the central difference method or an implicit method is employed for solution. If the explicit central difference method is used, a lumped mass matrix should be employed and in this case low-order finite elements in uniform meshes are probably most effective, i.e. the four- and eight-node elements of Figs. 5.5 and 5.6 are frequently employed in two- and three-dimensional analysis, respectively. Using these elements we can chose L_e to be equal to the smallest distance between any two of the nodes of the mesh employed, and this length determines Δt as given by (9.102). However, for higher-order (parabolic or cubic) continuum elements the time step has to be further reduced, because the interior nodes "are stiffer" than the corner nodes. Also, if structural (beam, plate or shell) elements are included in the mesh, the time step size Δt may be governed by the flexural modes in these elements so that the distances between nodes do not alone determine Δt.[24] Since the condition is always that $\Delta t \leq T_n/\pi$, where T_n is the smallest period of the mesh, we need to establish a lower bound on T_n. This bound is given by the smallest period $T_n^{(m)}$ of any element, considered individually, measured over all elements m in the mesh.* For some elements, this period can be established exactly in closed form, whereas for more complex (distorted and curved) elements, a lower bound on $T_n^{(m)}$ may have to be employed.

The choice of the effective length L_e and hence time step Δt is considerably simpler if an implicit unconditionally stable time integration method is used. In this case, L_e can be equal to the smallest distance between any two of the nodes that lie in the direction of the wave travel.

The above considerations have been put forward for linear dynamic analysis, but are largely also applicable in nonlinear analysis. An important point in nonlinear analysis is that the periods and wave velocities modeled in the finite ele-

* This follows because using (4.16), (4.23) and (10.59) we have

$$\omega_n^2 = \frac{\boldsymbol{\phi}_n^T \sum_m \mathbf{K}^{(m)} \boldsymbol{\phi}_n}{\boldsymbol{\phi}_n^T \sum_m \mathbf{M}^{(m)} \boldsymbol{\phi}_n} = \frac{\sum_m \mathcal{U}^{(m)}}{\sum_m \mathcal{g}^{(m)}}$$

and

$$\mathcal{U}^{(m)} \leq \omega_n^{2\,(m)} \mathcal{g}^{(m)},$$

where $\omega_n^{(m)}$ is the largest frequency (rad/time) of element m.

ment system change during its response. Therefore, the selection of the time step size must take into account that in a structural dynamics problem T_u changes, and that in a wave propagation problem the value of c in (9.102) is not constant. To obtain a conservative estimate for Δt in a wave propagation problem, the wave velocity c is best chosen as the maximum value that can be reached. Also, to determine the effective length L_e, the distance between integration points should be considered instead of distances between nodes. Since the physical properties of the system change during its response, it may also be effective to change the time step size during the response calculations.

We discuss additional considerations pertaining to nonlinear analysis in the following section.

9.5 SOLUTION OF NONLINEAR EQUATIONS IN DYNAMIC ANALYSIS

The solution of the nonlinear dynamic response of a finite element system is, in essence, obtained using the procedures already discussed: the incremental formulations presented in Chapter 6, the iterative solution procedures discussed in Section 8.6 and the time integration algorithms presented in this chapter.[23-29] Hence, the major basic procedures used in a nonlinear dynamic response solution have already been presented, and we only need to briefly summarize in the following how these procedures are employed together in a nonlinear dynamic analysis.

9.5.1 Explicit Integration

The most common explicit time integration operator used in nonlinear dynamic analysis is probably the central difference operator. As in linear analysis (see Section 9.2), the equilibrium of the finite element system is considered at time t in order to calculate the displacements at time $t + \Delta t$. Neglecting the effect of a damping matrix we operate for each discrete time step solution on the equations, see (6.52),

$$\mathbf{M}\,{}^t\ddot{\mathbf{U}} = {}^t\mathbf{R} - {}^t\mathbf{F} \tag{9.103}$$

where the nodal point force vector ${}^t\mathbf{F}$ is evaluated as discussed in Section 6.3. The solution for the nodal point displacements at time $t + \Delta t$ is obtained using the central difference approximation for the accelerations (given in (9.3)) to substitute for ${}^t\ddot{\mathbf{U}}$ in (9.103). Thus, as in linear analysis, if we know ${}^{t-\Delta t}\mathbf{U}$ and ${}^t\mathbf{U}$, the relations in (9.3) and (9.103) are employed to calculate ${}^{t+\Delta t}\mathbf{U}$. The solution therefore simply corresponds to a forward marching in time; the main advantage of the method is that with $\mathbf{M}$ a diagonal matrix the solution of ${}^{t+\Delta t}\mathbf{U}$ does not involve a triangular factorization of a coefficient matrix.

The shortcoming in the use of the central difference method lies in the severe time step restriction: for stability, the time step size Δt must be smaller than a critical time step, Δt_{cr}, which is equal to T_n/π, where T_n is the smallest period in the finite element system. This time step restriction was derived considering a linear system (see Section 9.4.2), but the result is also applicable to nonlinear analysis, since for each time step the nonlinear response calculation may be thought of—in an approximate way—as a linear analysis. However, whereas in a linear analysis the stiffness properties remain constant, in a nonlinear analysis these properties change during the response calculations. These changes in the material and/or geometric conditions enter in the evaluation of the force vector ${}^t\mathbf{F}$ as discussed in Chapter 6. Since therefore the value of T_n is not constant during the response calculation, the time step Δt needs to be decreased if the system stiffens, and this time step adjustment must be performed in a conservative manner, so that with certainty the condition $\Delta t \leq T_n/\pi$ is satisfied at all times.

To emphasize the above point, consider an analysis in which the time step is always smaller than the critical time step except for a few successive solution steps, and for these solution steps the time step Δt is just slightly larger than the critical time step. In such a case, the analysis results may not show an "obvious" solution instability, but instead a significant solution error is accumulated over the solution steps for which the time step size was larger than the critical value for stability. The situation is quite different from what is observed in linear analysis, where the solution quickly "blows up" if the time step is larger than the critical time step size for stability. This phenomenon is somewhat demonstrated in the response predicted for the simple one degree of freedom spring-mass system shown in Fig. 9.10. In the solution the time step Δt is slightly larger than the critical time step for stability in the stiff region of the spring. Since the time step corresponds to a stable time step for small spring displacements, the response calculations are partly stable and partly unstable. The calculated response is shown in Fig. 9.10 and it is observed that although the predicted displacements are grossly in error, the solution does not blow up. Hence, if this single degree of freedom system would correspond to a higher frequency in a large finite element model, a significant error accumulation could take place without an obvious blow-up of the solution.

The proper choice of the time step Δt is therefore a most important ingredient of the explicit time integration. A few guidelines for the choice of Δt have been given in Section 9.4.4.

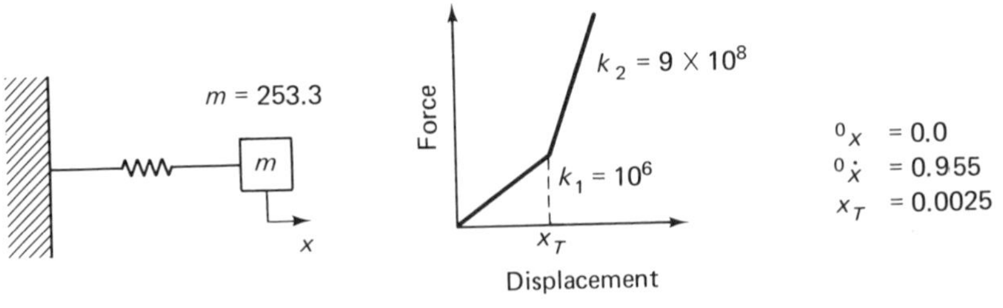

FIGURE 9.10 Response of bilinear elastic system as predicted using the central difference method; $\Delta t_{cr} = 0.001061027$; the accurate response, dark line with displacement $\ll 0.1$, was calculated with $\Delta t = 0.000106103$; the "unstable" response was calculated with $\Delta t = 0.00106103$.

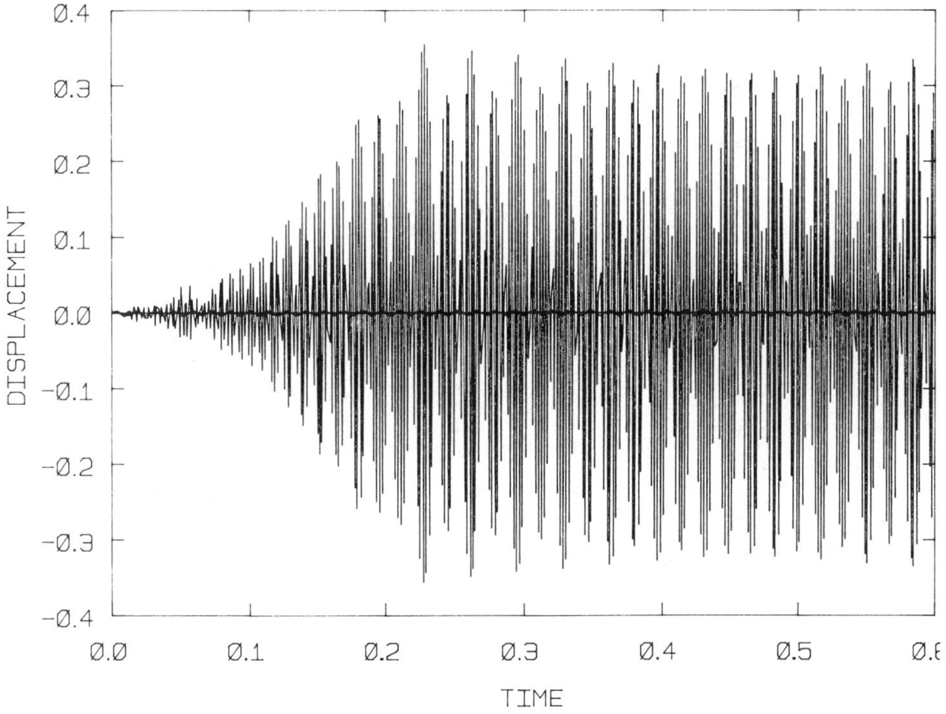

FIGURE 9.10 (*cont.*)

9.5.2 Implicit Integration

All the implicit time integration schemes discussed previously for linear dynamic analysis can also be employed in nonlinear dynamic response calculations. A very common technique used is the trapezoidal rule, which is Newmark's method with $\delta = \frac{1}{2}$ and $\alpha = \frac{1}{4}$, and we use this method to demonstrate the basic additional considerations involved in a nonlinear analysis.

As in linear analysis, using implicit time integration we consider the equilibrium of the system at time $t + \Delta t$. This requires in nonlinear analysis that an iteration be performed. Using the modified Newton–Raphson iteration the governing equilibrium equations have been derived in Chapter 6 and are (neglecting the effects of a damping matrix):

$$\mathbf{M}\,{}^{t+\Delta t}\ddot{\mathbf{U}}^{(k)} + {}^{t}\mathbf{K}\,\Delta\mathbf{U}^{(k)} = {}^{t+\Delta t}\mathbf{R} - {}^{t+\Delta t}\mathbf{F}^{(k-1)} \tag{9.104}$$

$${}^{t+\Delta t}\mathbf{U}^{(k)} = {}^{t+\Delta t}\mathbf{U}^{(k-1)} + \Delta\mathbf{U}^{(k)} \tag{9.105}$$

Using the trapezoidal rule of time integration, the following assumptions are employed:

$${}^{t+\Delta t}\mathbf{U} = {}^{t}\mathbf{U} + \frac{\Delta t}{2}({}^{t}\dot{\mathbf{U}} + {}^{t+\Delta t}\dot{\mathbf{U}}) \tag{9.106}$$

$${}^{t+\Delta t}\dot{\mathbf{U}} = {}^{t}\dot{\mathbf{U}} + \frac{\Delta t}{2}({}^{t}\ddot{\mathbf{U}} + {}^{t+\Delta t}\ddot{\mathbf{U}}) \tag{9.107}$$

Using the relations in (9.105) to (9.107) we thus obtain

$$^{t+\Delta t}\ddot{\mathbf{U}}^{(k)} = \frac{4}{\Delta t^2}({}^{t+\Delta t}\mathbf{U}^{(k-1)} - {}^{t}\mathbf{U} + \Delta\mathbf{U}^{(k)}) - \frac{4}{\Delta t}{}^{t}\dot{\mathbf{U}} - {}^{t}\ddot{\mathbf{U}} \tag{9.108}$$

and substituting into (9.104) we have

$$^{t}\hat{\mathbf{K}}\,\Delta\mathbf{U}^{(k)} = {}^{t+\Delta t}\mathbf{R} - {}^{t+\Delta t}\mathbf{F}^{(k-1)} - \mathbf{M}\left(\frac{4}{\Delta t^2}({}^{t+\Delta t}\mathbf{U}^{(k-1)} - {}^{t}\mathbf{U}) - \frac{4}{\Delta t}{}^{t}\dot{\mathbf{U}} - {}^{t}\ddot{\mathbf{U}}\right) \tag{9.109}$$

where

$$^{t}\hat{\mathbf{K}} = {}^{t}\mathbf{K} + \frac{4}{\Delta t^2}\mathbf{M} \tag{9.110}$$

We now notice that the iterative equations in dynamic nonlinear analysis using implicit time integration are of the same form as the equations that we considered in static nonlinear analysis, except that both the coefficient matrix and the nodal point force vector contain contributions from the inertia of the system. We can therefore directly conclude that all iterative solution strategies discussed in Section 8.6 for static analysis are also directly applicable to the solution of (9.109). However, since the inertia of the system renders its dynamic response, in general, "more smooth" than its static response, convergence of the iteration can, in general, be expected to be more rapid than in static analysis. In addition, convergence is always achieved in dynamic analysis provided Δt is small enough, whereas we observed earlier that in nonlinear static analysis the modified Newton–Raphson iteration does not converge in certain loading conditions (see Section 8.6). The numerical reason for the better convergence characteristics in a dynamic analysis lies in the contribution of the mass matrix to the coefficient matrix. Indeed, this contribution becomes dominant when the time step is small. This fact may be used in dynamic analysis of systems with flexible and stiff domains (e.g. fluid-structure interaction problems) by neglecting the stiffness contributions of the flexible domains in the coefficient matrix. This solution approach is effective if a diagonal mass matrix is employed, and the bandwidth of the coefficient matrix in (9.109) is significantly decreased so that the decrease in the number of numerical operations required per iteration outweighs any increase in the number of iterations needed for convergence.[11]

It is interesting to note that in the first solutions of nonlinear dynamic finite element response, equilibrium iterations were not performed in the step-by-step incremental analysis;[4,25,26] i.e., the relation in (9.109) was simply solved for $i = 1$ and the incremental displacement $\Delta\mathbf{U}^{(1)}$ was accepted as an accurate approximation to the actual displacement increment from time t to time $t + \Delta t$. However, it was then recognized that *the iteration can actually be of utmost importance,*[27] *since any error admitted in the incremental solution at a particular time directly affects in a path-dependent manner the solution at any subsequent time.* Indeed, because any nonlinear dynamic response is highly path-dependent, the analysis of a nonlinear dynamic problem requires iteration more stringently at each time step than does a static analysis.

A simple demonstration of the above observations is given in Fig. 9.11. This figure shows the results obtained in the analysis of a simple pendulum which was idealized as a truss element with a concentrated mass at its free end. The pendulum was released from a horizontal position and the response was calculated for about one period of oscillation. In the analysis the convergence toler-

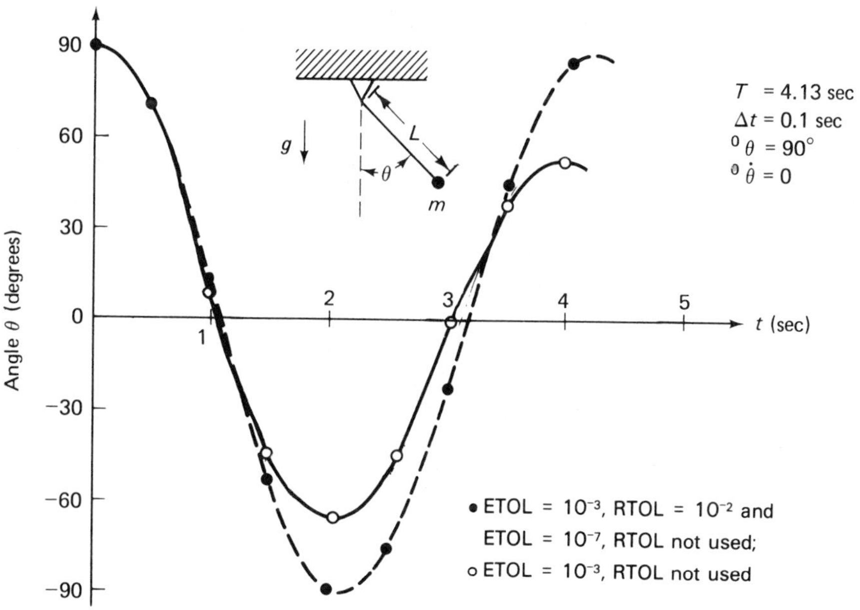

FIGURE 9.11 *Analysis of simple pendulum using trapezoidal rule, RNORM = mg.*

ances already discussed in Section 8.6 were used, but including the effect of inertia, i.e., convergence is reached when the following conditions are satisfied:

$$\frac{\| {}^{t+\Delta t}\mathbf{R} - {}^{t+\Delta t}\mathbf{F}^{(i-1)} - \mathbf{M}\,{}^{t+\Delta t}\ddot{\mathbf{U}}^{(i-1)} \|_2}{\text{RNORM}} \leq \text{RTOL} \tag{9.111}$$

and

$$\frac{\Delta\mathbf{U}^{(i)T}({}^{t+\Delta t}\mathbf{R} - {}^{t+\Delta t}\mathbf{F}^{(i-1)} - \mathbf{M}\,{}^{t+\Delta t}\ddot{\mathbf{U}}^{(i-1)})}{\Delta\mathbf{U}^{(1)T}({}^{t+\Delta t}\mathbf{R} - {}^{t}\mathbf{F} - \mathbf{M}\,{}^{t}\ddot{\mathbf{U}})} \leq \text{ETOL} \tag{9.112}$$

in which RTOL is a force tolerance and ETOL is an energy tolerance. Figure 9.11 demonstrates the importance of iterating and doing so with a sufficiently tight convergence tolerance. In this analysis energy is lost if the convergence tolerance is not tight enough, but depending on the problem being considered the predicted response may also blow up if iteration is not used. In practice, it is frequently the case that only a few iterations per time step are required to obtain a stable solution.[28]

Because of the importance of equilibrium iterations in nonlinear dynamic analysis, the dynamic analysis capabilities in the finite element computer programs NONSAP[27] and ADINA[6,29] were constructed specifically to include the possibility of equilibrium iterations during the response calculations. Various analyses of the characteristics of time integration schemes in nonlinear dynamic response calculations, that have been made more recently, have substantiated that iteration for dynamic equilibrium is an important part of an implicit time integration.[30-34] In view of the theoretical analyses conducted and the experience obtained, it may finally be recommended that, in summary, for a nonlinear dynamic analysis using implicit time integration the analyst should employ an operator that is unconditionally stable in linear analysis (a good choice is the trapezoidal rule), use equilibrium iterations with tight enough convergence tolerances, and select the time step size based on the guidelines given in Section 9.4 and on the fact that convergence in the equilibrium iterations must be achieved.

9.5.3 Solution Using Mode Superposition

In considering linear analysis we discussed in Section 9.3 that the essence of mode superposition is a transformation from the element nodal point degrees of freedom to the generalized degrees of freedom of the vibration mode shapes. Since the dynamic equilibrium equations in the basis of the mode shape vectors decouple (assuming proportional damping), mode superposition analysis can be very effective in linear analysis if only some vibration modes are excited by the loading. The same basic principles are also applicable to nonlinear analysis; however, in this case the vibration mode shapes and frequencies change, and to transform the coefficient matrix in (9.109) into diagonal form the free vibration mode shapes of the system at time t need to be used in the transformation. The calculation of the vibration mode shapes and frequencies at time t, when these quantities have been calculated at a previous time, could be achieved economically using the subspace iteration method (see Section 12.3). However, the complete mode superposition analysis of nonlinear dynamic response is generally only effective when the solution can be obtained without updating the stiffness matrix too frequently. In this case, the governing finite element equilibrium equations for the solution of the response at time $t + \Delta t$ are

$$\mathbf{M}\,{}^{t+\Delta t}\ddot{\mathbf{U}}^{(k)} + {}^{\tau}\mathbf{K}\,\Delta\mathbf{U}^{(k)} = {}^{t+\Delta t}\mathbf{R} - {}^{t+\Delta t}\mathbf{F}^{(k-1)} \qquad k = 1, 2, \ldots \tag{9.113}$$

where ${}^{\tau}\mathbf{K}$ is the stiffness matrix corresponding to the configuration at some previous time τ. In the mode superposition analysis we now use

$${}^{t+\Delta t}\mathbf{U} = \sum_{i=r}^{s} \boldsymbol{\phi}_i\,{}^{t+\Delta t}x_i \tag{9.114}$$

where ${}^{t+\Delta t}x_i$ is the ith generalized modal displacement at time $t + \Delta t$, and

$${}^{\tau}\mathbf{K}\boldsymbol{\phi}_i = \omega_i^2\mathbf{M}\boldsymbol{\phi}_i; \qquad i = r, \ldots, s \tag{9.115}$$

that is, $(\omega_i, \boldsymbol{\phi}_i)$ are free vibration frequencies (rad/sec) and mode shape vectors of the system at time τ. Using (9.114) in the usual way, the equations in (9.113) are transformed to

$${}^{t+\Delta t}\ddot{\mathbf{X}}^{(k)} + \mathbf{\Omega}^2\,\Delta\mathbf{X}^{(k)} = \mathbf{\Phi}^T({}^{t+\Delta t}\mathbf{R} - {}^{t+\Delta t}\mathbf{F}^{(k-1)}) \qquad k = 1, 2, \ldots \tag{9.116}$$

where

$$\mathbf{\Omega}^2 = \begin{bmatrix} \omega_r^2 & & & \\ & \cdot & & \\ & & \cdot & \\ & & & \cdot \\ & & & & \omega_s^2 \end{bmatrix}; \qquad \mathbf{\Phi} = [\boldsymbol{\phi}_r, \ldots, \boldsymbol{\phi}_s]; \qquad {}^{t+\Delta t}\mathbf{X} = \begin{bmatrix} {}^{t+\Delta t}x_r \\ \cdot \\ \cdot \\ \cdot \\ {}^{t+\Delta t}x_s \end{bmatrix} \tag{9.117}$$

The relations in (9.116) are the equilibrium equations at time $t + \Delta t$ in the generalized modal displacements of time τ; the corresponding mass matrix is an identity matrix, the stiffness matrix is $\mathbf{\Omega}^2$, the external load vector is $\mathbf{\Phi}^T\,{}^{t+\Delta t}\mathbf{R}$ and the generalized force vector corresponding to the element stresses in iteration $(k - 1)$ is $\mathbf{\Phi}^T\,{}^{t+\Delta t}\mathbf{F}^{(k-1)}$. The solution of (9.116) can be obtained using, for example, the trapezoidal rule (see Section 9.5.2).

In general, the use of mode superposition in nonlinear dynamic analysis can be effective if only a relatively few mode shapes need to be considered in the analysis, and if the system is only locally nonlinear. Such conditions may be encountered, for example, in the analysis of earthquake response and vibration excitation, and it is in these areas that the technique has been employed.[35,36]

9.6 SOLUTION OF NONSTRUCTURAL PROBLEMS

Although we considered in the previous sections the solution of the dynamic response of structures and solids, it should be recognized that many of the basic concepts discussed are also directly applicable to the analysis of other types of problems. Namely, in the solution of a nonstructural problem, the choice lies again between the use of explicit or implicit time integration, mode superposition analysis needs to be considered, and it may also be advantageous to use different time integration schemes for different domains of the complete element assemblage (see Section 9.2.5). The stability and accuracy properties of the time integration schemes used are analyzed in basically the same way as in structural analysis, and the important basic observations made concerning nonlinear structural analysis are also applicable to the analysis of nonstructural problems.

The nonstructural problems that we have in mind are heat transfer, field problems and fluid flow (see Chapter 7). The major difference in the time integration of the governing equations of these problems, when compared to structural analysis, is that we now only deal with first derivatives in time. Therefore time integration operators different from those discussed in the previous sections are employed.

Based on the discussion in Section 9.4 we can present a time integration scheme used in the analysis of heat transfer and fluid flow by considering a typical one degree of freedom equilibrium equation,

$$\dot{x} + \lambda x = r; \qquad x|_{t=0} = {}^0x \tag{9.118}$$

where, for example, considering a heat transfer problem, x is the unknown temperature, λ the conductivity, and r is the heat input to the system. A family of integration operators that can be employed effectively for the solution of (9.118) is given by the following assumptions:

$$\begin{aligned} {}^{t+\alpha\Delta t}\dot{x} &= ({}^{t+\Delta t}x - {}^{t}x)/\Delta t \\ {}^{t+\alpha\Delta t}x &= (1-\alpha)\,{}^{t}x + \alpha\,{}^{t+\Delta t}x \end{aligned} \tag{9.119}$$

where α is a constant which is chosen to yield optimum stability and accuracy properties. To solve for ${}^{t+\Delta t}x$ we proceed as described in Section 9.2. Namely, if ${}^{t}x$ is known, we can use (9.118) at time $t + \alpha\Delta t$ and (9.119) to solve for ${}^{t+\Delta t}x$, and so on (see Example 9.13). This α-method was already introduced in Section 6.4.2 for the solution of the creep response of finite element systems.

The properties of the integration procedure depend on the value of α that is employed. The following procedures are in frequent use:[1,37,38]

$\alpha = 0$, explicit Euler forward method, stable provided $\Delta t \leq 2/\lambda$, first-order accurate in Δt;

$\alpha = \frac{1}{2}$, implicit trapezoidal rule, unconditionally stable, second-order accurate in Δt;

$\alpha = 1$, implicit Euler backward method, unconditionally stable, first-order accurate in Δt.

We demonstrate the evaluation of the stability and accuracy properties and the use of the integration procedure defined in (9.119) in the following examples.

EXAMPLE 9.13: Analyze the α-method for its stability properties.

The stability of the α-integration scheme is investigated by using (9.118) at $t + \alpha\Delta t$, with $r = 0$, substituting (9.119) and solving for the variable x at time $t + \Delta t$,

$$^{t+\Delta t}x = \frac{1 - (1 - \alpha)\,\Delta t\,(^{t+\alpha\Delta t}\lambda)}{1 + \alpha\,\Delta t\,(^{t+\alpha\Delta t}\lambda)}\;{}^{t}x$$

For stability we need
$$\left|\frac{1 - (1 - \alpha)\,\Delta t\,(^{t+\alpha\Delta t}\lambda)}{1 + \alpha\,\Delta t\,(^{t+\alpha\Delta t}\lambda)}\right| \leq 1 \tag{a}$$

But this relation is satisfied for any Δt provided $\alpha \geq \frac{1}{2}$; hence the scheme is unconditionally stable when $\alpha \geq \frac{1}{2}$. If $\alpha < \frac{1}{2}$ the relation in (a) is satisfied provided

$$\Delta t \leq \frac{2}{(1 - 2\alpha)(^{t+\alpha\Delta t}\lambda)}$$

meaning that the method is conditionally stable for $\alpha < \frac{1}{2}$.

EXAMPLE 9.14: Develop the equations to be solved in a nonlinear transient heat transfer analysis using the Euler backward method and the modified Newton-Raphson method.

The governing equations in a nonlinear heat transfer analysis using an implicit time integration procedure are (see Section 7.2.3):

$$^{t+\Delta t}\mathbf{C}^{(i)}\;{}^{t+\Delta t}\dot{\boldsymbol{\theta}}^{(i)} + {}^{t}\mathbf{K}\,\Delta\boldsymbol{\theta}^{(i)} = {}^{t+\Delta t}\tilde{\mathbf{Q}}^{(i-1)} \tag{a}$$

where $^{t+\Delta t}\tilde{\mathbf{Q}}^{(i-1)}$ is a vector of nodal point heat flows corresponding to time $t + \Delta t$ and iteration $(i - 1)$. Using the Euler backward method we have

$$^{t+\Delta t}\dot{\boldsymbol{\theta}}^{(i)} = \frac{^{t+\Delta t}\boldsymbol{\theta}^{(i-1)} + \Delta\boldsymbol{\theta}^{(i)} - {}^{t}\boldsymbol{\theta}}{\Delta t}$$

and hence (a) reduces to

$$\left({}^{t}\mathbf{K} + \frac{1}{\Delta t}\,{}^{t+\Delta t}\mathbf{C}^{(i)}\right)\Delta\boldsymbol{\theta}^{(i)} = {}^{t+\Delta t}\tilde{\mathbf{Q}}^{(i-1)} - {}^{t+\Delta t}\mathbf{C}^{(i)}\;{}^{t+\Delta t}\dot{\boldsymbol{\theta}}^{(i-1)} \tag{b}$$

where
$$^{t+\Delta t}\dot{\boldsymbol{\theta}}^{(i-1)} = \frac{^{t+\Delta t}\boldsymbol{\theta}^{(i-1)} - {}^{t}\boldsymbol{\theta}}{\Delta t}$$

For solution the relation (b) is further linearized (corresponding to a modified Newton–Raphson iteration) using

$$\left({}^{t}\mathbf{K} + \frac{1}{\Delta t}\,{}^{t}\mathbf{C}\right)\Delta\boldsymbol{\theta}^{(i)} = {}^{t+\Delta t}\tilde{\mathbf{Q}}^{(i-1)} - {}^{t+\Delta t}\mathbf{C}^{(i-1)}\;{}^{t+\Delta t}\dot{\boldsymbol{\theta}}^{(i-1)}$$

REFERENCES

1. L. Collatz, *The Numerical Treatment of Differential Equations*, Springer-Verlag, New York, N.Y., 1966.
2. S. H. Crandall, *Engineering Analysis*, McGraw-Hill Book Company, New York, N.Y., 1956.

3. J. C. Houbolt, "A Recurrence Matrix Solution for the Dynamic Response of Elastic Aircraft," *Journal of Aeronautical Science*, Vol. 17, 1950, pp. 540–550.
4. E. L. Wilson, I. Farhoomand, and K. J. Bathe, "Nonlinear Dynamic Analysis of Complex Structures," *International Journal of Earthquake Engineering and Structural Dynamics*, Vol. 1, 1973, pp. 241–252.
5. N. M. Newmark, "A Method of Computation for Structural Dynamics," A.S.C.E., *Journal of Engineering Mechanics Division*, Vol. 85, 1959, pp. 67–94.
6. K. J. Bathe, "ADINA—A Finite Element Program for Automatic Dynamic Incremental Nonlinear Analysis," Report 82448–1, Acoustics and Vibration Laboratory, Department of Mechanical Engineering, Massachusetts Institute of Technology, Cambridge, Mass., 1975, rev. Dec. 1978.
7. F. H. Harlow and A. A. Amsden, "A Numerical Fluid Dynamics Calculation Method for All Flow Speeds," *J. Comput. Phys.* Vol. 8, 1971, pp. 197–213.
8. T. Belytschko and R. Mullen, "Mesh Partitions of Explicit-Implicit Time Integrations," in *Formulations and Computational Algorithms in Finite Element Analysis* (K. J. Bathe, J. T. Oden, and W. Wunderlich, eds.), M.I.T. Press, 1977.
9. T. Belytschko, H. J. Yen, and R. Mullen, "Mixed Methods for Time Integration," *J. Computer Methods in Applied Mechanics and Engineering*, Vol. 17/18, pp. 259–275 (1979).
10. T. J. R. Hughes and W. K. Liu, "Implicit-Explicit Finite Elements in Transient Analysis: Stability Theory," *Journal of Applied Mechanics*, Vol. 45 (June 1978).
11. K. J. Bathe and V. Sonnad, "On Effective Implicit Time Integration of Fluid-Structure Problems," *International Journal for Numerical Methods in Engineering*, Vol. 15, pp. 943–948, 1980.
12. R. W. Clough and J. Penzien, *Dynamics of Structures*, McGraw-Hill Book Company, New York, N.Y., 1974.
13. J. M. Biggs, *Introduction to Structural Dynamics*, McGraw-Hill Book Company, New York, N.Y., 1964.
14. P. E. Lax and R. D. Richtmyer, "Survey of the Stability of Finite Difference Equations," *Communications in Pure Applied Mathematics*, Vol. 9, 1956, pp. 267–293.
15. P. Henrici, *Error Propagation for Difference Methods*, John Wiley & Sons, Inc., New York, N.Y., 1963.
16. G. P. Destefano, "Causes of Instabilities in Numerical Integration Techniques," *International Journal of Computational Mathematics*, Vol. 2, 1968, pp. 123–142.
17. R. E. Nickell, "On the Stability of Approximation Operators in Problems of Structural Dynamics," *International Journal of Solids and Structures*, Vol. 7, 1971, pp. 301–319.
18. K. J. Bathe and E. L. Wilson, "Stability and Accuracy Analysis of Direct Integration Methods," *International Journal of Earthquake Engineering and Structural Dynamics*, Vol. 1, 1973, pp. 283–291.
19. G. L. Goudreau and R. L. Taylor, "Evaluation of Numerical Integration Methods in Elastodynamics," *J. Computer Methods in Applied Mechanics and Engineering*, Vol. 2, No. 1, 1973, pp. 69–97.
20. R. D. Krieg, "Unconditional Stability in Numerical Time Integration Methods," Trans. A.S.M.E., *Journal of Applied Mechanics*, June 1973, pp. 417–421.
21. R. E. Nickell, "Direct Integration in Structural Dynamics," A.S.C.E., *Journal of Engineering Mechanics Division*, Vol. 99, 1973, pp. 303–317.
22. H. M. Hilber, T. J. R. Hughes, and R. L. Taylor, "Improved Numerical Dissipation for Time Integration Algorithms in Structural Mechanics," *International*

Journal of Earthquake Engineering and Structural Dynamics, Vol. 5, 1977, pp. 283–292.

23. G. E. Weeks, "Temporal Operators for Nonlinear Structural Dynamics Problems," A.S.C.E., *Journal of Engineering Mechanics Division*, Vol. 98, No. EM5, Proc. Paper 9260, Oct. 1972, pp. 1087–1104.
24. T. Belytschko, "Fluid-Structure Interaction," *J. Computers and Structures*, Vol. 12, 1980, pp. 459–469.
25. J. A. Stricklin, J. E. Martinez, J. R. Tillerson, J. H. Hong, and W. E. Haisler, "Nonlinear Dynamic Analysis of Shells of Revolution by the Matrix Displacement Method," *A.I.A.A. Journal*, Vol. 9, No. 4, 1971.
26. J. F. McNamara and P. V. Marcal, "Incremental Stiffness Method for Finite Element Analysis of the Nonlinear Dynamic Problem," in *Numerical and Computer Methods in Structural Mechanics* (S. J. Fenves, N. Perrone, J. Robinson, and W. C. Schnobrich, eds.), Academic Press, Inc., New York, 1973.
27. K. J. Bathe and E. L. Wilson, "NONSAP—A General Finite Element Program for Nonlinear Dynamic Analysis of Complex Structures," Paper M3-1, *Proceedings, 2nd Int. Conf. on Structural Mechanics in Reactor Tech.*, Berlin, Germany, Sept. 1973.
28. K. J. Bathe and W. Hahn, "On Transient Analysis of Fluid-Structure Systems," *J. Computers and Structures*, Vol. 10, 1979, pp. 383–391.
29. K. J. Bathe, "Static and Dynamic Geometric and Material Nonlinear Analysis using ADINA," Report 82448–2, Acoustics and Vibration Lab., Dept of Mech. Eng., M.I.T., May 1976, (rev. May 1977).
30. T. Belytschko and D. F. Schoeberle, "On the Unconditional Stability of an Implicit Algorithm for Nonlinear Structural Dynamics," *Journal of Applied Mechanics*, Vol. 42, Dec. 1975, pp. 865–969.
31. T. J. R. Hughes, "A Note on the Stability of Newmark's Algorithm in Nonlinear Structural Dynamics," *International Journal for Numerical Methods in Engineering*, Vol. 11, No. 2, 1977, pp. 383–386.
32. K. C. Park, "An Improved Stiffly Stable Method for Direct Integration of Nonlinear Structural Dynamics Equations," *Journal of Applied Mechanics*, Vol. 42, June 1975, pp. 464–470.
33. E. Haug, O. S. Nguyen, and A. L. DeRouvray, "An Improved Energy Conserving Implicit Time Integration Algorithm for Nonlinear Dynamic Structural Analysis," *Proceedings, 4th Int. Conf. on Struct. Mech. in Reactor Technology*, San Francisco, Aug. 1977.
34. T. J. R. Hughes, T. K. Caughey, and W. K. Liu, "Finite Element Methods for Nonlinear Elastodynamics which Conserve Energy," *Journal of Applied Mechanics*, Vol. 45, June 1978, pp. 366–370.
35. R. E. Nickell, "Nonlinear Dynamics by Mode Superposition," *J. Computer Methods in Applied Mechanics and Engineering*, Vol. 7, pp. 107–129, 1976.
36. K. J. Bathe and S. Gracewski, "On Nonlinear Dynamic Analysis Using Substructuring and Mode Superposition," *J. Computers and Structures*, Vol. 13, 1981, pp. 699–707.
37. T. J. R. Hughes, "Unconditionally Stable Algorithms for Nonlinear Heat Conduction," *J. Computer Methods in Applied Mechanics and Engineering*, Vol. 10, 1977, pp. 135–141.
38. K. J. Bathe and M. R. Khoshgoftaar, "Finite Element Formulation and Solution of Nonlinear Heat Transfer," *J. Nuclear Engineering and Design*, Vol. 51, 1979, pp. 389–401.

10

PRELIMINARIES TO THE SOLUTION OF EIGENPROBLEMS

10.1 INTRODUCTION

In various sections of the preceding chapters we encountered eigenproblems and the statement of their solutions. We did not at that time discuss how to obtain the required eigenvalues and eigenvectors. It is the purpose of this and the next chapters to describe the actual solution procedures used to solve the eigenproblems of interest. Before presenting the algorithms, we discuss in this chapter some important basic considerations for the solution of eigenproblems.[1-7]

First, let us briefly summarize the eigenproblems that we want to solve. The simplest problem encountered is the standard eigenproblem,

$$\mathbf{K}\boldsymbol{\phi} = \lambda\boldsymbol{\phi} \tag{10.1}$$

where $\mathbf{K}$ is the stiffness matrix of a single finite element or of an element assemblage. We recall that $\mathbf{K}$ has order n and half-bandwidth $m_{\mathbf{K}}$ (i.e., the total bandwidth is $2m_{\mathbf{K}} + 1$) and that $\mathbf{K}$ is positive semidefinite or positive definite. There are n eigenvalues and corresponding eigenvectors satisfying (10.1). The ith eigenpair is denoted as $(\lambda_i, \boldsymbol{\phi}_i)$, where the eigenvalues are ordered according to their magnitudes:

$$0 \leq \lambda_1 \leq \lambda_2 \leq \ldots \leq \lambda_{n-1} \leq \lambda_n \tag{10.2}$$

The solution for p eigenpairs can be written

$$\mathbf{K}\boldsymbol{\Phi} = \boldsymbol{\Phi}\boldsymbol{\Lambda} \tag{10.3}$$

where $\boldsymbol{\Phi}$ is an $n \times p$ matrix with its columns equal to the p eigenvectors and $\boldsymbol{\Lambda}$ is a $p \times p$ diagonal matrix listing the corresponding eigenvalues. As an example, (10.3) may represent the solution to the lowest p eigenvalues and corresponding eigenvectors of $\mathbf{K}$, in which case $\boldsymbol{\Phi} = [\boldsymbol{\phi}_1, \ldots, \boldsymbol{\phi}_p]$ and $\boldsymbol{\Lambda} = \text{diag}(\lambda_i)$, $i = 1, \ldots, p$. We recall that if $\mathbf{K}$ is positive definite, $\lambda_i > 0$, $i = 1, \ldots, n$, and if $\mathbf{K}$ is positive semidefinite, $\lambda_i \geq 0$, $i = 1, \ldots, n$, where the number of zero eigenvalues is equal to the number of rigid body modes in the system.

The solution of the eigenvalue problem in (10.1) is, for example, sought in the evaluation of an element stiffness matrix or in the calculation of the condition number of a structure stiffness matrix. We discussed in Section 4.2.5 that the representation of the element stiffness matrix in its canonical form (i.e., in the eigenvector basis) is used to evaluate the effectiveness of the element. In this case all eigenvalues and vectors of **K** must be calculated. On the other hand, to evaluate the condition number of a stiffness matrix, only the smallest and largest eigenvalues are required (see Section 8.5).

Before proceeding to the generalized eigenproblems, we should mention that other standard eigenproblems may also need to be solved. For example, we may require the eigenvalues of the mass matrix **M**, in which case **M** replaces **K** in (10.1). Similarly, we may want to solve for the eigenvalues of a conductivity or heat capacity matrix in heat flow analysis (see Section 7.2).

A very frequently considered eigenproblem is the one to be solved in vibration mode superposition analysis (see Section 9.3). In this case we consider the generalized eigenproblem

$$\mathbf{K}\boldsymbol{\phi} = \lambda \mathbf{M}\boldsymbol{\phi} \tag{10.4}$$

where **K** and **M** are, respectively, the stiffness matrix and mass matrix of the finite element assemblage. The eigenvalues λ_i and eigenvectors $\boldsymbol{\phi}_i$ are the free vibration frequencies (rad/sec) squared, ω_i^2, and corresponding mode shape vectors, respectively. The properties of **K** are as discussed above. The mass matrix may be banded, in which case its half-bandwidth m_{M} is equal to m_{K}, or **M** may be diagonal with $m_{ii} \geq 0$; i.e., some diagonal elements may possibly be zero. A banded mass matrix, obtained in a consistent mass analysis, is always positive definite, whereas a lumped mass matrix is positive definite only if all diagonal elements are larger than zero. In general, a diagonal mass matrix is positive semidefinite.

In analogy to (10.3), the solution for p eigenvalues and corresponding eigenvectors of (10.4) can be written

$$\mathbf{K}\boldsymbol{\Phi} = \mathbf{M}\boldsymbol{\Phi}\boldsymbol{\Lambda} \tag{10.5}$$

where the columns in $\boldsymbol{\Phi}$ are the eigenvectors and $\boldsymbol{\Lambda}$ is a diagonal matrix listing the corresponding eigenvalues.

We should note that the generalized eigenproblem in (10.4) reduces to the standard eigenproblem in (10.1) if **M** is an identity matrix. In other words, the eigenvalues and eigenvectors in (10.3) can also be thought of as frequencies squared and vibration mode shapes of the system when unit mass is specified at each degree of freedom. Corresponding to the possible eigenvalues in the solution of (10.1), the generalized eigenproblem in (10.4) has eigenvalues $\lambda_i \geq 0$, $i = 1, \ldots, n$, where the number of zero eigenvalues is again equal to the number of rigid body modes in the system.

Two other generalized eigenproblems should be mentioned briefly. The first problem is solved in linearized buckling analysis, in which case we consider (see Section 6.5.2)

$$(-\mathbf{K}_G)\boldsymbol{\phi} = \lambda \mathbf{K}\boldsymbol{\phi} \tag{10.6}$$

where $\mathbf{K}_G$ is the nonlinear strain (geometric or initial stress) stiffness matrix and **K** is the linear strain stiffness matrix. The nonlinear strain stiffness matrix is banded with the same bandwidth as **K** and is, in general, indefinite. The eigenvalues of (10.6) can be positive and negative and they give the buckling loads with the largest eigenvalue defining the smallest critical load.

A third generalized eigenproblem is encountered in heat transfer analysis, where we consider the equation

$$\mathbf{K}\boldsymbol{\phi} = \lambda \mathbf{C}\boldsymbol{\phi} \tag{10.7}$$

where $\mathbf{K}$ is the heat conductivity matrix and $\mathbf{C}$ is the heat capacity matrix. The eigenvalues and eigenvectors are the thermal eigenvalues and mode shapes, respectively. The solution of (10.7) is required in heat transfer analysis using mode superposition. The matrices $\mathbf{K}$ and $\mathbf{C}$ in (10.7) are positive definite or positive semidefinite, so that the eigenvalues of (10.7) are $\lambda_i \geq 0$, $i = 1, \ldots, n$.

In this and the next chapter we discuss the solution of the eigenproblems $\mathbf{K}\boldsymbol{\phi} = \lambda\boldsymbol{\phi}$ and $\mathbf{K}\boldsymbol{\phi} = \lambda\mathbf{M}\boldsymbol{\phi}$ in (10.1) and (10.4). These eigenproblems are encountered frequently in practice. However, it should be realized that all algorithms to be presented are also applicable to the solution of other eigenproblems, provided they are of the same form and the matrices satisfy the appropriate conditions of positive definiteness, semidefiniteness, and so on. For example, to solve the problem in (10.7), the mass matrix $\mathbf{M}$ need simply be replaced by the heat capacity matrix $\mathbf{C}$, and the matrix $\mathbf{K}$ is the heat conductivity matrix.

Considering the actual computer solution of the required eigenproblems, we recall that in the introduction to equation solution procedures in static analysis (see Section 8.1), we observed the importance of using effective calculation procedures. This is even more so in eigensystem calculations, because the solution of eigenvalues and corresponding eigenvectors requires, in general, much more computer effort than the solution of static equilibrium equations. A particularly important consideration is that the solution algorithms must be stable, which is more difficult to achieve than in static analysis.

A variety of eigensystem solution methods have been developed and are reported in the literature.[1-7] Most of the techniques have been devised for rather general matrices. However, in finite element analysis we are concerned with the solution of the specific eigenproblems summarized above, in which each of the matrices has specific properties, such as being banded, positive definite, and so on. The eigensystem solution algorithms should take advantage of these properties in order to make a more economical solution possible.

The objective in this chapter is to lay the foundation for a thorough understanding of effective eigensolution methods. This is accomplished by first discussing the properties of the matrices, eigenvalues, and eigenvectors of the problems of interest, and then presenting some approximate solution techniques. The actual solution methods recommended for use are presented in Chapters 11 and 12.

10.2 FUNDAMENTAL FACTS USED IN THE SOLUTION OF EIGENSYSTEMS

Before the working of any eigensystem solution procedure can be properly studied, it is necessary first to thoroughly understand the different properties of the matrices and eigenvalues and eigenvectors considered.[1-7] In particular, we will find that all solution methods are, in essence, based on these fundamental properties. We therefore want to summarize in this section the important properties of the matrices and their eigensystems, although some of the material has already been presented in other sections of the book. As pointed out in Section 10.1, we consider the eigenproblem $\mathbf{K}\boldsymbol{\phi} = \lambda\mathbf{M}\boldsymbol{\phi}$, which reduces to $\mathbf{K}\boldsymbol{\phi} = \lambda\boldsymbol{\phi}$ when $\mathbf{M} = \mathbf{I}$, but the observations made are equally applicable to other eigenproblems of interest.

10.2.1 Properties of the Eigenvectors

It was stated that the solution of the generalized eigenproblem $\mathbf{K}\boldsymbol{\phi} = \lambda\mathbf{M}\boldsymbol{\phi}$ yields n eigenvalues $\lambda_1, \ldots, \lambda_n$, ordered as shown in (10.2), and corresponding eigenvectors $\boldsymbol{\phi}_1, \ldots, \boldsymbol{\phi}_n$. Each eigenpair $(\lambda_i, \boldsymbol{\phi}_i)$ satisfies (10.4); i.e., we have

$$\mathbf{K}\boldsymbol{\phi}_i = \lambda_i\mathbf{M}\boldsymbol{\phi}_i; \qquad i = 1, \ldots, n \tag{10.8}$$

The significance of (10.8) should be well understood. The equation says that if we establish a vector $\lambda_i\mathbf{M}\boldsymbol{\phi}_i$ and use it as a load vector $\mathbf{R}$ in the equation $\mathbf{KU} = \mathbf{R}$, then $\mathbf{U} = \boldsymbol{\phi}_i$. This thought may immediately suggest the use of static solution algorithms for the calculation of an eigenvector. We will see later that the $\mathbf{LDL}^T$ decomposition algorithm is indeed an important part of eigensolution procedures.

The equation in (10.8) also shows that an eigenvector is only defined within a multiple of itself; i.e., we also have

$$\mathbf{K}(\alpha\boldsymbol{\phi}_i) = \lambda_i\mathbf{M}(\alpha\boldsymbol{\phi}_i) \tag{10.9}$$

where α is a nonzero constant. Therefore, with $\boldsymbol{\phi}_i$ being an eigenvector, $\alpha\boldsymbol{\phi}_i$ is also an eigenvector, and we say that an eigenvector is only defined by its direction in the n-dimensional space considered. However, in our discussion we refer to *the* eigenvectors $\boldsymbol{\phi}_i$ as satisfying (10.8) and also the relation $\boldsymbol{\phi}_i^T\mathbf{M}\boldsymbol{\phi}_i = 1$, which fixes the lengths of the eigenvectors, i.e., the absolute magnitude of the elements in each eigenvector. However, we may note that the eigenvectors are still only defined within a multiplier of -1.

An important relation which the eigenvectors satisfy is that of $\mathbf{M}$-orthonormality; i.e., we have

$$\boldsymbol{\phi}_i^T\mathbf{M}\boldsymbol{\phi}_j = \delta_{ij} \tag{10.10}$$

where δ_{ij} is the Kronecker delta. This relation follows from the orthonormality of the eigenvectors of standard eigenproblems (see Section 2.7) and is discussed further in Section 10.2.5. Premultiplying (10.8) by $\boldsymbol{\phi}_j$ transposed and using the condition in (10.10), we obtain

$$\boldsymbol{\phi}_i^T\mathbf{K}\boldsymbol{\phi}_j = \lambda_i\delta_{ij} \tag{10.11}$$

meaning that the eigenvectors are also $\mathbf{K}$-orthogonal. When using the relations in (10.10) and (10.11), it should be kept in mind that the $\mathbf{M}$- and $\mathbf{K}$-orthogonality follow from (10.8), and that (10.8) is the basic equation to be satisfied. In other words, if we believe that we have an eigenvector and eigenvalue, then as a check we should substitute into (10.8) (see Example 10.3).

So far we have made no mention of multiple eigenvalues and corresponding eigenvectors. It is important to realize that in this case the eigenvectors are not unique but that we can always choose a set of $\mathbf{M}$-orthonormal eigenvectors, which span the subspace that corresponds to a multiple eigenvalue (see Section

2.7). In other words, assume that λ_i has multiplicity m (i.e., $\lambda_i = \lambda_{i+1} = \ldots = \lambda_{i+m-1}$), then we can choose m eigenvectors $\boldsymbol{\phi}_i, \ldots, \boldsymbol{\phi}_{i+m-1}$, which span the m-dimensional subspace corresponding to the eigenvalues of magnitude λ_i and which satisfy the orthogonality relations in (10.10) and (10.11). However, the eigenvectors are not unique; instead, the eigenspace corresponding to λ_i is unique. We demonstrate the results by means of some examples.

EXAMPLE 10.1: The stiffness matrix and mass matrix of a two degree of freedom system are

$$\mathbf{K} = \begin{bmatrix} 5 & -2 \\ -2 & 2 \end{bmatrix}; \qquad \mathbf{M} = \begin{bmatrix} \frac{5}{4} & 0 \\ 0 & \frac{1}{5} \end{bmatrix}$$

It is believed that the two eigenpairs of the problem $\mathbf{K}\boldsymbol{\phi} = \lambda\boldsymbol{\phi}$ are

$$(d_1, \mathbf{v}_1) = \left(1, \begin{bmatrix} \dfrac{1}{\sqrt{5}} \\ \dfrac{2}{\sqrt{5}} \end{bmatrix}\right); \qquad (d_2, \mathbf{v}_2) = \left(6, \begin{bmatrix} \dfrac{2}{\sqrt{5}} \\ -\dfrac{1}{\sqrt{5}} \end{bmatrix}\right) \tag{a}$$

and the two eigenpairs of the problem $\mathbf{K}\boldsymbol{\phi} = \lambda\mathbf{M}\boldsymbol{\phi}$ are

$$(g_1, \mathbf{w}_1) = \left(2, \begin{bmatrix} \frac{4}{5} \\ 1 \end{bmatrix}\right); \qquad (g_2, \mathbf{w}_2) = \left(12, \begin{bmatrix} \frac{2}{5} \\ -2 \end{bmatrix}\right) \tag{b}$$

Verify that we indeed have in (a) and (b) the eigensolutions of the problems $\mathbf{K}\boldsymbol{\phi} = \lambda\boldsymbol{\phi}$ and $\mathbf{K}\boldsymbol{\phi} = \lambda\mathbf{M}\boldsymbol{\phi}$, respectively.

Consider first the problem $\mathbf{K}\boldsymbol{\phi} = \lambda\boldsymbol{\phi}$. The values given in (a) are indeed the eigensolution if they satisfy the relation in (10.8) with $\mathbf{M} = \mathbf{I}$ and, to fix the lengths of the vectors, the orthonormality relations in (10.10) with $\mathbf{M} = \mathbf{I}$. Substituting into (10.3), which expresses the relation in (10.8) for all eigenpairs, we have

$$\begin{bmatrix} 5 & -2 \\ -2 & 2 \end{bmatrix} \begin{bmatrix} \dfrac{1}{\sqrt{5}} & \dfrac{2}{\sqrt{5}} \\ \dfrac{2}{\sqrt{5}} & -\dfrac{1}{\sqrt{5}} \end{bmatrix} = \begin{bmatrix} \dfrac{1}{\sqrt{5}} & \dfrac{2}{\sqrt{5}} \\ \dfrac{2}{\sqrt{5}} & -\dfrac{1}{\sqrt{5}} \end{bmatrix} \begin{bmatrix} 1 & 0 \\ 0 & 6 \end{bmatrix}$$

or

$$\begin{bmatrix} \dfrac{1}{\sqrt{5}} & \dfrac{12}{\sqrt{5}} \\ \dfrac{2}{\sqrt{5}} & -\dfrac{6}{\sqrt{5}} \end{bmatrix} = \begin{bmatrix} \dfrac{1}{\sqrt{5}} & \dfrac{12}{\sqrt{5}} \\ \dfrac{2}{\sqrt{5}} & -\dfrac{6}{\sqrt{5}} \end{bmatrix}$$

Evaluating (10.10) we obtain

$$\mathbf{v}_1^T\mathbf{v}_1 = \begin{bmatrix} \dfrac{1}{\sqrt{5}} & \dfrac{2}{\sqrt{5}} \end{bmatrix} \begin{bmatrix} \dfrac{1}{\sqrt{5}} \\ \dfrac{2}{\sqrt{5}} \end{bmatrix} = 1$$

$$\mathbf{v}_2^T\mathbf{v}_1 = \begin{bmatrix} \dfrac{2}{\sqrt{5}} & -\dfrac{1}{\sqrt{5}} \end{bmatrix} \begin{bmatrix} \dfrac{1}{\sqrt{5}} \\ \dfrac{2}{\sqrt{5}} \end{bmatrix} = 0$$

$$\mathbf{v}_2^T\mathbf{v}_2 = \begin{bmatrix} \dfrac{2}{\sqrt{5}} & -\dfrac{1}{\sqrt{5}} \end{bmatrix} \begin{bmatrix} \dfrac{2}{\sqrt{5}} \\ -\dfrac{1}{\sqrt{5}} \end{bmatrix} = 1$$

$$\mathbf{v}_1^T\mathbf{v}_2 = \mathbf{v}_2^T\mathbf{v}_1 = 0$$

Therefore, the relations in (10.3) and (10.10) are satisfied and we have $\lambda_1 = d_1$, $\lambda_2 = d_2$, $\boldsymbol{\phi}_1 = \mathbf{v}_1$, and $\boldsymbol{\phi}_2 = \mathbf{v}_2$.

To check whether we have in (b) the eigensolution to $\mathbf{K}\boldsymbol{\phi} = \lambda\mathbf{M}\boldsymbol{\phi}$, we proceed in an analogous way. Substituting into (10.5) we have

$$\begin{bmatrix} 5 & -2 \\ -2 & 2 \end{bmatrix}\begin{bmatrix} \frac{4}{5} & \frac{2}{5} \\ 1 & -2 \end{bmatrix} = \begin{bmatrix} \frac{5}{4} & 0 \\ 0 & \frac{1}{5} \end{bmatrix}\begin{bmatrix} \frac{4}{5} & \frac{2}{5} \\ 1 & -2 \end{bmatrix}\begin{bmatrix} 2 & 0 \\ 0 & 12 \end{bmatrix}$$

or

$$\begin{bmatrix} 2 & 6 \\ \dfrac{2}{5} & -\dfrac{24}{5} \end{bmatrix} = \begin{bmatrix} 2 & 6 \\ \dfrac{2}{5} & -\dfrac{24}{5} \end{bmatrix}$$

and evaluating (10.10) we obtain

$$\mathbf{w}_1^T\mathbf{M}\mathbf{w}_1 = [\tfrac{4}{5} \quad 1]\begin{bmatrix} \frac{5}{4} & 0 \\ 0 & \frac{1}{5} \end{bmatrix}\begin{bmatrix} \frac{4}{5} \\ 1 \end{bmatrix} = 1$$

$$\mathbf{w}_2^T\mathbf{M}\mathbf{w}_1 = [\tfrac{2}{5} \quad -2]\begin{bmatrix} \frac{5}{4} & 0 \\ 0 & \frac{1}{5} \end{bmatrix}\begin{bmatrix} \frac{4}{5} \\ 1 \end{bmatrix} = 0$$

$$\mathbf{w}_2^T\mathbf{M}\mathbf{w}_2 = [\tfrac{2}{5} \quad -2]\begin{bmatrix} \frac{5}{4} & 0 \\ 0 & \frac{1}{5} \end{bmatrix}\begin{bmatrix} \frac{2}{5} \\ -2 \end{bmatrix} = 1$$

$$\mathbf{w}_1^T\mathbf{M}\mathbf{w}_2 = \mathbf{w}_2^T\mathbf{M}\mathbf{w}_1 = 0$$

Hence the relations in (10.5) and (10.10) are satisfied and we have

$$\lambda_1 = g_1,\ \lambda_2 = g_2,\ \boldsymbol{\phi}_1 = \mathbf{w}_1, \text{ and } \boldsymbol{\phi}_2 = \mathbf{w}_2.$$

EXAMPLE 10.2: Consider the eigenproblem

$$\mathbf{K}\boldsymbol{\phi} = \lambda\boldsymbol{\phi} \qquad \text{with } \mathbf{K} = \begin{bmatrix} 2 & & \\ & 2 & \\ & & 3 \end{bmatrix}$$

and show that the eigenvectors corresponding to the multiple eigenvalue are not unique.

The eigenvalues of $\mathbf{K}$ are $\lambda_1 = 2$, $\lambda_2 = 2$, and $\lambda_3 = 3$, and a set of eigenvectors is

$$\boldsymbol{\phi}_1 = \begin{bmatrix} 1 \\ 0 \\ 0 \end{bmatrix}; \qquad \boldsymbol{\phi}_2 = \begin{bmatrix} 0 \\ 1 \\ 0 \end{bmatrix}; \qquad \boldsymbol{\phi}_3 = \begin{bmatrix} 0 \\ 0 \\ 1 \end{bmatrix} \tag{a}$$

where $\boldsymbol{\phi}_3$ is unique. These values could be checked as in Example 10.1. However, any linear combinations of $\boldsymbol{\phi}_1$ and $\boldsymbol{\phi}_2$ given in (a) that satisfy the orthonormality conditions in (10.10) with $\mathbf{M} = \mathbf{I}$ would also be eigenvectors. For example, we could use

$$\boldsymbol{\phi}_1 = \begin{bmatrix} \dfrac{1}{\sqrt{2}} \\ \dfrac{1}{\sqrt{2}} \\ 0 \end{bmatrix} \quad \text{and} \quad \boldsymbol{\phi}_2 = \begin{bmatrix} \dfrac{1}{\sqrt{2}} \\ -\dfrac{1}{\sqrt{2}} \\ 0 \end{bmatrix}$$

That $\boldsymbol{\phi}_1$ and $\boldsymbol{\phi}_2$ are indeed eigenvectors corresponding to $\lambda_1 = \lambda_2 = 2$ can again be checked as in Example 10.1. It should be noted that any eigenvectors $\boldsymbol{\phi}_1$ and $\boldsymbol{\phi}_2$ provide a basis for the unique two-dimensional eigenspace that corresponds to λ_1 and λ_2.

The solution to (10.4) for all p required eigenvalues and corrseponding eigenvectors was established in (10.5). Using the relations in (10.10) and (10.11), we may now write

$$\mathbf{\Phi}^T\mathbf{K}\mathbf{\Phi} = \mathbf{\Lambda} \tag{10.12}$$

and

$$\mathbf{\Phi}^T\mathbf{M}\mathbf{\Phi} = \mathbf{I} \tag{10.13}$$

where the p columns of $\mathbf{\Phi}$ are the eigenvectors. *It is very important to note that (10.12) and (10.13) are conditions which the eigenvectors must satisfy, but that if the* **M**-*orthonormality and* **K**-*orthogonality are satisfied, the p vectors need not necessarily be eigenvectors, unless* $p = n$. In other words, assume that $\mathbf{X}$ stores p vectors, $p < n$, and that $\mathbf{X}^T\mathbf{K}\mathbf{X} = \mathbf{D}$ and $\mathbf{X}^T\mathbf{M}\mathbf{X} = \mathbf{I}$; then the vectors in $\mathbf{X}$ and the diagonal elements in $\mathbf{D}$ may or may not be eigenvectors and eigenvalues of (10.4). However, if $p = n$, then $\mathbf{X} = \mathbf{\Phi}$ and $\mathbf{D} = \mathbf{\Lambda}$, because only the eigenvectors span the complete n-dimensional space and diagonalize the matrices $\mathbf{K}$ and $\mathbf{M}$. The above aspect is very important and is frequently not observed in practice (see Section 10.3.2). To underline the concept we present the following example.

EXAMPLE 10.3: Consider the eigenproblem $\mathbf{K}\boldsymbol{\phi} = \lambda\mathbf{M}\boldsymbol{\phi}$, where

$$\mathbf{K} = \begin{bmatrix} 2 & -1 & 0 \\ -1 & 4 & -1 \\ 0 & -1 & 2 \end{bmatrix}; \qquad \mathbf{M} = \begin{bmatrix} \frac{1}{2} & 0 & 0 \\ 0 & 1 & 0 \\ 0 & 0 & \frac{1}{2} \end{bmatrix}$$

and the two vectors

$$\mathbf{v}_1 = \begin{bmatrix} 1 \\ \frac{1}{\sqrt{2}} \\ 0 \end{bmatrix}; \qquad \mathbf{v}_2 = \begin{bmatrix} 1 \\ -\frac{1}{\sqrt{2}} \\ 0 \end{bmatrix}$$

Show that the vectors $\mathbf{v}_1$ and $\mathbf{v}_2$ satisfy the orthogonality relations in (10.12) and (10.13) [i.e., the relations in (10.11) and (10.10)] but that they are not eigenvectors.

For the check we let $\mathbf{v}_1$ and $\mathbf{v}_2$ be the columns in $\mathbf{\Phi}$, and we evaluate (10.12) and (10.13). Thus we obtain

$$\begin{bmatrix} 1 & \frac{1}{\sqrt{2}} & 0 \\ 1 & -\frac{1}{\sqrt{2}} & 0 \end{bmatrix} \begin{bmatrix} 2 & -1 & 0 \\ -1 & 4 & -1 \\ 0 & -1 & 2 \end{bmatrix} \begin{bmatrix} 1 & 1 \\ \frac{1}{\sqrt{2}} & -\frac{1}{\sqrt{2}} \\ 0 & 0 \end{bmatrix} = \begin{bmatrix} (4 - \sqrt{2}) & 0 \\ 0 & (4 + \sqrt{2}) \end{bmatrix} \tag{a}$$

and

$$\begin{bmatrix} 1 & \frac{1}{\sqrt{2}} & 0 \\ 1 & -\frac{1}{\sqrt{2}} & 0 \end{bmatrix} \begin{bmatrix} \frac{1}{2} & & \\ & 1 & \\ & & \frac{1}{2} \end{bmatrix} \begin{bmatrix} 1 & 1 \\ \frac{1}{\sqrt{2}} & -\frac{1}{\sqrt{2}} \\ 0 & 0 \end{bmatrix} = \begin{bmatrix} 1 & 0 \\ 0 & 1 \end{bmatrix}$$

Hence the orthogonality relations are satisfied. To show that $\mathbf{v}_1$ and $\mathbf{v}_2$ are not eigenvectors, we employ (10.8). For example,

$$\mathbf{K}\mathbf{v}_1 = \begin{bmatrix} 2 - \frac{1}{\sqrt{2}} \\ -1 + \frac{4}{\sqrt{2}} \\ -\frac{1}{\sqrt{2}} \end{bmatrix}; \qquad \mathbf{M}\mathbf{v}_1 = \begin{bmatrix} \frac{1}{2} \\ \frac{1}{\sqrt{2}} \\ 0 \end{bmatrix}$$

However, the vector $\mathbf{Kv}_1$ cannot be equal to the vector $\alpha\mathbf{Mv}_1$, where α is a scalar; i.e., $\mathbf{Kv}_1$ is not parallel to $\mathbf{Mv}_1$, and therefore $\mathbf{v}_1$ is not an eigenvector. Similarly, $\mathbf{v}_2$ is not an eigenvector and the values $(4 - \sqrt{2})$ and $(4 + \sqrt{2})$ calculated in (a) are not eigenvalues. The actual eigenvalues and corresponding eigenvectors are given in Example 10.4.

In the preceding presentation we considered the properties of the eigenvectors of the problem $\mathbf{K}\boldsymbol{\phi} = \lambda\mathbf{M}\boldsymbol{\phi}$, and we should now briefly comment on the properties of the eigenvectors calculated in the solution of the other eigenvalue problems of interest. The comment is simple: the orthogonality relations discussed here hold equally for the eigenvectors of the problems encountered in buckling analysis and heat transfer analysis. That is, we also have in buckling analysis, using the notation in (10.6),

$$\left.\begin{aligned} \boldsymbol{\phi}_i^T\mathbf{K}\boldsymbol{\phi}_j &= \delta_{ij}; & \boldsymbol{\phi}_i^T\mathbf{K}_G\boldsymbol{\phi}_j &= \lambda_i\delta_{ij} \\ \boldsymbol{\Phi}^T\mathbf{K}\boldsymbol{\Phi} &= \mathbf{I}; & \boldsymbol{\Phi}^T\mathbf{K}_G\boldsymbol{\Phi} &= \boldsymbol{\Lambda} \end{aligned}\right\} \tag{10.14}$$

and in heat transfer analysis, using the notation in (10.7), we have

$$\left.\begin{aligned} \boldsymbol{\phi}_i^T\mathbf{C}\boldsymbol{\phi}_j &= \delta_{ij}; & \boldsymbol{\phi}_i^T\mathbf{K}\boldsymbol{\phi}_j &= \lambda_i\delta_{ij} \\ \boldsymbol{\Phi}^T\mathbf{C}\boldsymbol{\Phi} &= \mathbf{I}; & \boldsymbol{\Phi}^T\mathbf{K}\boldsymbol{\Phi} &= \boldsymbol{\Lambda} \end{aligned}\right\} \tag{10.15}$$

As for the eigenproblem $\mathbf{K}\boldsymbol{\phi} = \lambda\mathbf{M}\boldsymbol{\phi}$, the proof of the relations in (10.14) and (10.15) depends on the fact that the generalized eigenproblems can be transformed to a standard form. We discuss this matter further in Section 10.2.5.

10.2.2 The Characteristic Polynomials of the Eigenproblem $K\phi = \lambda M\phi$ and of Its Associated Constraint Problems

An important property of the eigenvalues of the problem $\mathbf{K}\boldsymbol{\phi} = \lambda\mathbf{M}\boldsymbol{\phi}$ is that they are the roots of the characteristic polynomial,

$$p(\lambda) = \det(\mathbf{K} - \lambda\mathbf{M}) \tag{10.16}$$

We can show that this property derives from the basic relation in (10.8). Rewriting (10.8) in the form

$$(\mathbf{K} - \lambda_i\mathbf{M})\boldsymbol{\phi}_i = \mathbf{0} \tag{10.17}$$

we observe that (10.8) can only be satisfied for nontrivial $\boldsymbol{\phi}_i$ (i.e., $\boldsymbol{\phi}_i$ not being equal to a null vector) provided that the matrix $\mathbf{K} - \lambda_i\mathbf{M}$ is singular. This means that if we factorize $\mathbf{K} - \lambda_i\mathbf{M}$ into a unit lower triangular matrix $\mathbf{L}$ and an upper triangular matrix $\mathbf{S}$ using Gauss elimination, we have that $s_{nn} = 0.0$. However, since

$$p(\lambda_i) = \det \mathbf{LS} = \prod_{i=1}^{n} s_{ii} \tag{10.18}$$

it follows that $p(\lambda_i) = 0.0$. Furthermore, if λ_i has multiplicity m we also have $s_{n-1,\,n-1} = \ldots = s_{n-m+1,\,n-m+1} = 0.0$. We should note that in the factorization of $\mathbf{K} - \lambda_i\mathbf{M}$, interchanges may be needed, in which case the factorization of $\mathbf{K} - \lambda_i\mathbf{M}$ with its rows and possibly its columns interchanged is obtained [each row and each column interchange then introduces a sign change in the determinant which must be taken into account (see Section 1.8)]. If no interchanges are carried out, or row and corresponding column interchanges are performed, which in practice is nearly always possible (but see Example 10.4 for a case where it is not possible), the coefficient matrix remains symmetric. In this case we can write for (10.18),

$$p(\lambda_i) = \det \mathbf{LDL}^T = \prod_{i=1}^{n} d_{ii} \tag{10.19}$$

where $\mathbf{LDL}^T$ is the factorization of $\mathbf{K} - \lambda_i \mathbf{M}$ or of the matrix derived from it by interchanging rows and corresponding columns, i.e., using a different ordering for the system degrees of freedom (see Section 8.2.5). The condition $s_{nn} = 0.0$ is now $d_{nn} = 0.0$, and in case λ_i has multiplicity m, the last m elements of $\mathbf{D}$ are zero.

In Section 8.2.5 we discussed the Sturm sequence property of the characteristic polynomials of the constraint problems associated with the problem $\mathbf{K}\boldsymbol{\phi} = \lambda\boldsymbol{\phi}$. The same properties that we observed in that discussion are applicable also to the characteristic polynomials of the constraint problems associated with the problem $\mathbf{K}\boldsymbol{\phi} = \lambda\mathbf{M}\boldsymbol{\phi}$. The proof follows from the fact that the generalized eigenproblem $\mathbf{K}\boldsymbol{\phi} = \lambda\mathbf{M}\boldsymbol{\phi}$ can be transformed to a standard eigenproblem, for which the Sturm sequence property of the characteristic polynomials holds. Referring the proof to Section 10.2.5, Example 10.11, let us summarize the important result.

The eigenproblem of the rth associated constraint problem corresponding to $\mathbf{K}\boldsymbol{\phi} = \lambda\mathbf{M}\boldsymbol{\phi}$ is given by

$$\mathbf{K}^{(r)}\boldsymbol{\phi}^{(r)} = \lambda^{(r)}\mathbf{M}^{(r)}\boldsymbol{\phi}^{(r)} \tag{10.20}$$

where all matrices are of order $n - r$, and $\mathbf{K}^{(r)}$ and $\mathbf{M}^{(r)}$ are obtained by deleting from $\mathbf{K}$ and $\mathbf{M}$ the last r rows and columns. The characteristic polynomial of the rth associated constraint problem is

$$p^{(r)}(\lambda^{(r)}) = \det(\mathbf{K}^{(r)} - \lambda^{(r)}\mathbf{M}^{(r)}) \tag{10.21}$$

and as for the special case $\mathbf{M} = \mathbf{I}$, we have that the eigenvalues of the $(r+1)$st constraint problem separate those of the rth constraint problem; i.e., as stated in (8.38), we again have

$$\lambda_1^{(r)} \le \lambda_1^{(r+1)} \le \lambda_2^{(r)} \le \lambda_2^{(r+1)} \ldots \le \lambda_{n-r-1}^{(r)} \le \lambda_{n-r-1}^{(r+1)} \le \lambda_{n-r}^{(r)} \tag{10.22}$$

EXAMPLE 10.4: Consider the eigenproblem $\mathbf{K}\boldsymbol{\phi} = \lambda\mathbf{M}\boldsymbol{\phi}$, where

$$\mathbf{K} = \begin{bmatrix} 2 & -1 & 0 \\ -1 & 4 & -1 \\ 0 & -1 & 2 \end{bmatrix}; \qquad \mathbf{M} = \begin{bmatrix} \frac{1}{2} & & \\ & 1 & \\ & & \frac{1}{2} \end{bmatrix}$$

(a) Calculate the eigenvalues using the characteristic polynomial as defined in (10.16).

(b) Solve for the eigenvectors $\boldsymbol{\phi}_i$, $i = 1, 2, 3$, by using the relation in (10.17) and the $\mathbf{M}$-orthonormality condition of the eigenvectors.

(c) Calculate the eigenvalues of the associated constraint problems and show that the eigenvalue separation property given in (10.22) holds.

Using (10.16) we obtain the characteristic polynomial

$$p(\lambda) = (2 - \tfrac{1}{2}\lambda)(4 - \lambda)(2 - \tfrac{1}{2}\lambda) - (-1)(-1)(2 - \tfrac{1}{2}\lambda) - (-1)(-1)(2 - \tfrac{1}{2}\lambda)$$

Hence
$$p(\lambda) = -\tfrac{1}{4}\lambda^3 + 3\lambda^2 - 11\lambda + 12$$

and we have
$$\lambda_1 = 2; \qquad \lambda_2 = 4; \qquad \lambda_3 = 6$$

To obtain the corresponding eigenvectors, we use the relation in (10.17). For λ_1 we have

$$\begin{bmatrix} 1 & -1 & 0 \\ -1 & 2 & -1 \\ 0 & -1 & 1 \end{bmatrix} \boldsymbol{\phi}_1 = \mathbf{0} \tag{a}$$

The coefficient matrix $\mathbf{K} - \lambda_1\mathbf{M}$ in (a) can be factorized into $\mathbf{LDL}^T$ without interchanges. Using the procedure described in Section 8.2.2, we obtain

$$\begin{bmatrix} 1 & & \\ -1 & 1 & \\ 0 & -1 & 1 \end{bmatrix} \begin{bmatrix} 1 & & \\ & 1 & \\ & & 0 \end{bmatrix} \begin{bmatrix} 1 & -1 & 0 \\ & 1 & -1 \\ & & 1 \end{bmatrix} \boldsymbol{\phi}_1 = \mathbf{0} \tag{b}$$

We note that $d_{33} = 0.0$. To evaluate $\boldsymbol{\phi}_1$, we obtain from (b),

$$\begin{bmatrix} 1 & -1 & 0 \\ & 1 & -1 \\ & & 0 \end{bmatrix} \boldsymbol{\phi}_1 = \mathbf{0}$$

Using also that $\boldsymbol{\phi}_1^T\mathbf{M}\boldsymbol{\phi}_1 = 1$ we have

$$\boldsymbol{\phi}_1^T = \begin{bmatrix} \frac{1}{\sqrt{2}} & \frac{1}{\sqrt{2}} & \frac{1}{\sqrt{2}} \end{bmatrix}$$

To obtain $\boldsymbol{\phi}_2$ and $\boldsymbol{\phi}_3$, we proceed in an analogous way. Evaluating $\mathbf{K} - \lambda_2\mathbf{M}$, we obtain from (10.17),

$$\begin{bmatrix} 0 & -1 & 0 \\ -1 & 0 & -1 \\ 0 & -1 & 0 \end{bmatrix} \boldsymbol{\phi}_2 = \mathbf{0}$$

In this case we cannot factorize the coefficient matrix preserving symmetry; i.e., we need to interchange the first and second rows only (and not the corresponding columns). This row interchange results into the relation

$$\begin{bmatrix} -1 & 0 & -1 \\ 0 & -1 & 0 \\ 0 & -1 & 0 \end{bmatrix} \boldsymbol{\phi}_2 = \mathbf{0}$$

Factorizing the coefficient matrix into a lower unit triangular matrix $\mathbf{L}$ and an upper triangular matrix $\mathbf{S}$, we obtain

$$\begin{bmatrix} 1 & & \\ 0 & 1 & \\ 0 & 1 & 1 \end{bmatrix} \begin{bmatrix} -1 & 0 & -1 \\ & -1 & 0 \\ & & 0 \end{bmatrix} \boldsymbol{\phi}_2 = \mathbf{0}$$

and hence $s_{33} = 0.0$. To solve for $\boldsymbol{\phi}_2$, we use

$$\begin{bmatrix} -1 & 0 & -1 \\ & -1 & 0 \\ & & 0 \end{bmatrix} \boldsymbol{\phi}_2 = \mathbf{0}$$

and that $\boldsymbol{\phi}_2^T\mathbf{M}\boldsymbol{\phi}_2 = 1$. Thus we obtain

$$\boldsymbol{\phi}_2^T = [-1 \quad 0 \quad 1]$$

To calculate $\boldsymbol{\phi}_3$, we evaluate $\mathbf{K} - \lambda_3\mathbf{M}$ and have

$$\begin{bmatrix} -1 & -1 & 0 \\ -1 & -2 & -1 \\ 0 & -1 & -1 \end{bmatrix} \boldsymbol{\phi}_3 = \mathbf{0}$$

The coefficient matrix can be factorized into $\mathbf{LDL}^T$ without interchanges; i.e., we have

$$\begin{bmatrix} 1 & & \\ 1 & 1 & \\ 0 & 1 & 1 \end{bmatrix} \begin{bmatrix} -1 & & \\ & -1 & \\ & & 0 \end{bmatrix} \begin{bmatrix} 1 & 1 & 0 \\ & 1 & 1 \\ & & 1 \end{bmatrix} \boldsymbol{\phi}_3 = \mathbf{0}$$

We note that $d_{33} = 0.0$. To calculate $\boldsymbol{\phi}_3$ we use

$$\begin{bmatrix} -1 & -1 & 0 \\ 0 & -1 & -1 \\ 0 & 0 & 0 \end{bmatrix} \boldsymbol{\phi}_3 = \mathbf{0}$$

and $\boldsymbol{\phi}_3^T\mathbf{M}\boldsymbol{\phi}_3 = 1$. Hence

$$\boldsymbol{\phi}_3^T = \begin{bmatrix} \dfrac{1}{\sqrt{2}} & -\dfrac{1}{\sqrt{2}} & \dfrac{1}{\sqrt{2}} \end{bmatrix}$$

The eigenvalues of the first associated constraint problem are obtained from the solution of

$$\begin{bmatrix} 2 & -1 \\ -1 & 4 \end{bmatrix} \boldsymbol{\phi}^{(1)} = \lambda^{(1)} \begin{bmatrix} \frac{1}{2} & 0 \\ 0 & 1 \end{bmatrix} \boldsymbol{\phi}^{(1)}$$

Hence

$$p^{(1)}(\lambda^{(1)}) = \tfrac{1}{2}\lambda^{(1)^2} - 4\lambda^{(1)} + 7$$

and

$$\lambda_1^{(1)} = 4 - \sqrt{2}; \qquad \lambda_2^{(1)} = 4 + \sqrt{2}$$

Also $\lambda_1^{(2)} = 4$, and hence the eigenvalue separation property given in (10.22) is in this case:

1. For the eigenvalues of the first and second associated constraint problems,

$$4 - \sqrt{2} < 4 < 4 + \sqrt{2}$$

2. For the eigenvalues of $\mathbf{K}\boldsymbol{\phi} = \lambda\mathbf{M}\boldsymbol{\phi}$ and the first associated constraint problem,

$$2 < (4 - \sqrt{2}) < 4 < (4 + \sqrt{2}) < 6$$

An important fact that follows from the property of the separation of eigenvalues as expressed in (10.22) is the following. Assume that we can factorize the matrix $\mathbf{K} - \mu\mathbf{M}$ into $\mathbf{LDL}^T$; i.e., none of the associated constraint problems has a zero eigenvalue. For simplicity of discussion let us first assume that all eigenvalues are distinct; i.e., there are no multiple eigenvalues. *The important fact is that in the decomposition of* $\mathbf{K} - \mu\mathbf{M}$, *the number of negative elements in* $\mathbf{D}$ *is equal to the number of eigenvalues smaller than* μ. Conversely, if $\lambda_i < \mu < \lambda_{i+1}$, there are exactly i negative diagonal elements in $\mathbf{D}$. The proof is obtained using the separation property in (10.22), and is relatively easily outlined by the following considerations. Referring to Fig. 10.1, assume that in

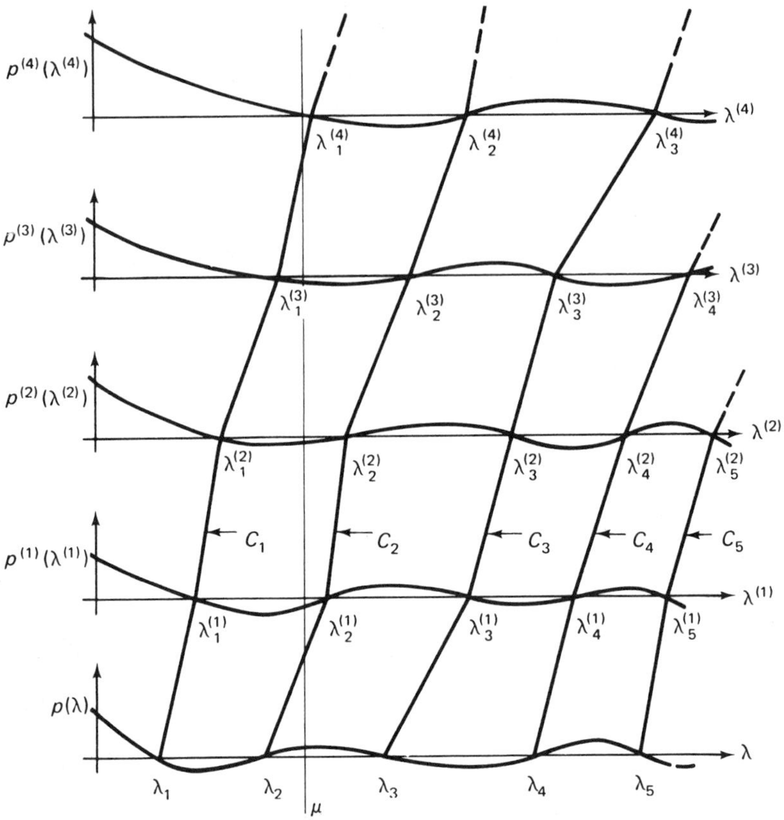

FIGURE 10.1 *Construction of curves C_i for the characteristic polynomials of the problem $\mathbf{K}\boldsymbol{\phi} = \lambda\mathbf{M}\boldsymbol{\phi}$ and of the associated constraint problems.*

the sketch of the characteristic polynomials we connect by straight lines all eigenvalues $\lambda_1^{(r)}$, $r = 0, 1, \ldots,$ with $\lambda_1^{(0)} = \lambda_1$, and call the resulting curve C_1. Similarly, we establish curves $C_2, C_3, \ldots$ as indicated in Fig. 10.1. Consider now that $\lambda_i < \mu < \lambda_{i+1}$ and draw a vertical line corresponding to μ in the figure of the characteristic polynomials; i.e., this line establishes where μ lies in relation to the eigenvalues of the associated constraint problems. We note that the line corresponding to μ must cross the curves $C_1, \ldots, C_i$ and because of the eigenvalue separation property cannot cross the curves $C_{i+1}, \ldots, C_n$. However, since

$$p^{(r)}(\mu) = \prod_{i=1}^{n-r} d_{ii} \tag{10.23}$$

and since each crossing of μ with an envelope C_k corresponds to a negative element appearing in $\mathbf{D}$, we have exactly i negative elements in $\mathbf{D}$.

The above considerations also hold in the case of multiple eigenvalues; i.e., in Fig. 10.1 we would merely have that some eigenvalues are equal, but the argument given above would not change.

The property that the number of negative elements in $\mathbf{D}$ is equal to the eigenvalues smaller than μ can be used directly in the solution of eigenvalues (see Section 11.5). Namely, by assuming a shift μ and checking whether μ is smaller or larger than the required eigenvalue, we can successively reduce the interval in which the eigenvalue must lie. We demonstrate the solution procedure in the following example.

EXAMPLE 10.5: Use the fact that the number of negative elements in $\mathbf{D}$, where $\mathbf{LDL}^T = \mathbf{K} - \mu\mathbf{M}$, is equal to the number of eigenvalues smaller than μ in order to calculate λ_2 of $\mathbf{K}\boldsymbol{\phi} = \lambda\mathbf{M}\boldsymbol{\phi}$, where

$$\mathbf{K} = \begin{bmatrix} 2 & -1 & 0 \\ -1 & 4 & -1 \\ 0 & -1 & 2 \end{bmatrix}; \qquad \mathbf{M} = \begin{bmatrix} \frac{1}{2} & & \\ & 1 & \\ & & \frac{1}{2} \end{bmatrix}$$

The three eigenvalues of the problem have already been calculated in Example 10.4. We now proceed in the following systematic steps:

1. Let us assume $\mu = 1$ and evaluate $\mathbf{LDL}^T$ of $\mathbf{K} - \mu\mathbf{M}$,

$$\mathbf{K} - \mu\mathbf{M} = \begin{bmatrix} \frac{3}{2} & -1 & 0 \\ -1 & 3 & -1 \\ 0 & -1 & \frac{3}{2} \end{bmatrix}$$

Hence $$\mathbf{LDL}^T = \begin{bmatrix} 1 & & \\ -\frac{2}{3} & 1 & \\ 0 & -\frac{3}{7} & 1 \end{bmatrix} \begin{bmatrix} \frac{3}{2} & & \\ & \frac{7}{3} & \\ & & \frac{15}{14} \end{bmatrix} \begin{bmatrix} 1 & -\frac{2}{3} & 0 \\ & 1 & -\frac{3}{7} \\ & & 1 \end{bmatrix}$$

Since all elements in $\mathbf{D}$ are larger than zero, we have $\lambda_1 > 1$.

2. We now try $\mu = 8$, where

$$\mathbf{K} - \mu\mathbf{M} = \begin{bmatrix} -2 & -1 & 0 \\ -1 & -4 & -1 \\ 0 & -1 & -2 \end{bmatrix}$$

and $$\mathbf{LDL}^T = \begin{bmatrix} 1 & & \\ \frac{1}{2} & 1 & \\ 0 & \frac{2}{7} & 1 \end{bmatrix} \begin{bmatrix} -2 & & \\ & -\frac{7}{2} & \\ & & -\frac{12}{7} \end{bmatrix} \begin{bmatrix} 1 & \frac{1}{2} & 0 \\ & 1 & \frac{2}{7} \\ & & 1 \end{bmatrix}$$

Since all three diagonal elements are smaller than zero, it follows that $\lambda_3 < 8$.

3. The next estimate μ should logically lie between 1 and 8; we choose $\mu = 5$, for which

$$\mathbf{K} - \mu\mathbf{M} = \begin{bmatrix} -\frac{1}{2} & -1 & 0 \\ -1 & -1 & -1 \\ 0 & -1 & -\frac{1}{2} \end{bmatrix}$$

and $$\mathbf{LDL}^T = \begin{bmatrix} 1 & & \\ 2 & 1 & \\ 0 & -1 & 1 \end{bmatrix} \begin{bmatrix} -\frac{1}{2} & & \\ & +1 & \\ & & -\frac{3}{2} \end{bmatrix} \begin{bmatrix} 1 & 2 & 0 \\ & 1 & -1 \\ & & 1 \end{bmatrix}$$

Since two negative elements are in $\mathbf{D}$, we have that $\lambda_2 < 5$.

4. The next estimate must lie between 1 and 5. Let us use $\mu = 3$, in which case

$$\mathbf{K} - \mu\mathbf{M} = \begin{bmatrix} \frac{1}{2} & -1 & 0 \\ -1 & 1 & -1 \\ 0 & -1 & \frac{1}{2} \end{bmatrix}$$

and

$$\mathbf{LDL}^T = \begin{bmatrix} 1 & & \\ -2 & 1 & \\ 0 & 1 & 1 \end{bmatrix} \begin{bmatrix} \frac{1}{2} & & \\ & -1 & \\ & & \frac{3}{2} \end{bmatrix} \begin{bmatrix} 1 & -2 & 0 \\ & 1 & 1 \\ & & 1 \end{bmatrix}$$

Hence $\lambda_2 > 3$, because there is only one negative element in $\mathbf{D}$.

The pattern of the solution procedure has now been established. So far we know that $3 < \lambda_2 < 5$. In order to obtain a closer estimate on λ_2 we would continue choosing a shift μ in the interval 3 to 5 and investigate whether the new shift is smaller or larger than λ_2. By always choosing an appropriate new shift the required eigenvalue can be determined very accurately (see Section 11.5). It should be noted that we did not need to use interchanges in the factorizations of $\mathbf{K} - \mu\mathbf{M}$ carried out above.

10.2.3 Shifting

An important procedure that is used extensively in the solution of eigenvalues and eigenvectors is shifting. The purpose of shifting is to accelerate the calculations of the required eigensystem. In the solution of $\mathbf{K}\boldsymbol{\phi} = \lambda\mathbf{M}\boldsymbol{\phi}$, we perform a shift ρ on $\mathbf{K}$ by calculating

$$\hat{\mathbf{K}} = \mathbf{K} - \rho\mathbf{M} \tag{10.24}$$

and we then consider the eigenproblem

$$\hat{\mathbf{K}}\boldsymbol{\psi} = \mu\mathbf{M}\boldsymbol{\psi} \tag{10.25}$$

To identify how the eigenvalues and eigenvectors of $\mathbf{K}\boldsymbol{\phi} = \lambda\mathbf{M}\boldsymbol{\phi}$ are related to those of the problem $\hat{\mathbf{K}}\boldsymbol{\psi} = \mu\mathbf{M}\boldsymbol{\psi}$ we rewrite (10.25) in the form

$$\mathbf{K}\boldsymbol{\psi} = \gamma\mathbf{M}\boldsymbol{\psi} \tag{10.26}$$

where $\gamma = \rho + \mu$. However, (10.26) is, in fact, the eigenproblem $\mathbf{K}\boldsymbol{\phi} = \lambda\mathbf{M}\boldsymbol{\phi}$, and since the solution of this problem is unique, we have

$$\lambda_i = \rho + \mu_i; \qquad \boldsymbol{\phi}_i = \boldsymbol{\psi}_i \tag{10.27}$$

In other words, the eigenvectors of $\hat{\mathbf{K}}\boldsymbol{\psi} = \mu\mathbf{M}\boldsymbol{\psi}$ *are the same as the eigenvectors of* $\mathbf{K}\boldsymbol{\phi} = \lambda\mathbf{M}\boldsymbol{\phi}$, *but the eigenvalues have been decreased by* ρ. A frequent application of shifting occurs in the calculation of rigid body modes when an algorithm is to be used that is not designed explicitly to calculate zero eigenvalues. We illustrate such an application in the following example.

EXAMPLE 10.6: Consider the eigenproblem

$$\begin{bmatrix} 3 & -3 \\ -3 & 3 \end{bmatrix}\boldsymbol{\phi} = \lambda\begin{bmatrix} 2 & 1 \\ 1 & 2 \end{bmatrix}\boldsymbol{\phi} \tag{a}$$

Calculate the eigenvalues and eigenvectors. Then impose a shift $\rho = -2$ and solve again for the eigenvalues and corresponding eigenvectors.

To calculate the eigenvalues we use the characteristic polynomial

$$p(\lambda) = \det(\mathbf{K} - \lambda\mathbf{M}) = 3\lambda^2 - 18\lambda$$

and thus obtain $\lambda_1 = 0$, $\lambda_2 = 6$. To calculate $\boldsymbol{\phi}_1$ and $\boldsymbol{\phi}_2$ we use the relation in (10.17) and the mass orthonormality condition $\boldsymbol{\phi}_i^T\mathbf{M}\boldsymbol{\phi}_i = 1$. We have

$$\begin{bmatrix} 3 & -3 \\ -3 & 3 \end{bmatrix}\boldsymbol{\phi}_1 = \mathbf{0}; \qquad \text{hence } \boldsymbol{\phi}_1 = \begin{bmatrix} \dfrac{1}{\sqrt{6}} \\ \dfrac{1}{\sqrt{6}} \end{bmatrix} \tag{b}$$

and

$$\begin{bmatrix} -9 & -9 \\ -9 & -9 \end{bmatrix}\boldsymbol{\phi}_2 = \mathbf{0}; \qquad \text{hence } \boldsymbol{\phi}_2 = \begin{bmatrix} \dfrac{1}{\sqrt{2}} \\ -\dfrac{1}{\sqrt{2}} \end{bmatrix} \tag{c}$$

Imposing a shift of $\rho = -2$, we obtain the problem

$$\begin{bmatrix} 7 & -1 \\ -1 & 7 \end{bmatrix}\boldsymbol{\phi} = \lambda\begin{bmatrix} 2 & 1 \\ 1 & 2 \end{bmatrix}\boldsymbol{\phi} \tag{d}$$

Proceeding as before, we have

$$p(\lambda) = \lambda^2 - 10\lambda + 16$$

and obtain as the roots $\lambda_1 = 2$, $\lambda_2 = 8$. Hence the eigenvalues have increased by 2; i.e., they have decreased by ρ.

The eigenvectors would be calculated using (10.17). However, we note that this relation again yields the equations in (b) and (c), and therefore the eigenvectors of the problem in (d) are those of the problem in (a).

An important observation resulting from the above discussion is that, in principle, we only need solution algorithms to calculate the eigenvalues and corresponding eigenvectors of the problem $\mathbf{K}\boldsymbol{\phi} = \lambda\mathbf{M}\boldsymbol{\phi}$ when all eigenvalues are larger than zero. This follows, because if rigid body modes are present, we may always operate on a shifted stiffness matrix that renders all eigenvalues positive.

Extensive applications of shifting are given in Chapters 11 and 12, where the various eigensystem solution algorithms are discussed.

10.2.4 Effect of Zero Mass

Considering a lumped mass analysis, we pointed out that the mass matrix $\mathbf{M}$ is diagonal with positive and possibly some zero diagonal elements. If all elements m_{ii} are larger than zero, the eigenvalues λ_i cannot be obtained without the use of an eigenvalue solution algorithm as described in Chapter 11. However, if $\mathbf{M}$ has some zero diagonal elements, say r diagonal elements in $\mathbf{M}$ are zero, we can immediately say that the problem $\mathbf{K}\boldsymbol{\phi} = \lambda\mathbf{M}\boldsymbol{\phi}$ has the eigenvalues $\lambda_n = \lambda_{n-1} = \ldots = \lambda_{n-r+1} = \infty$, and can also construct the corresponding eigenvectors by inspection.

To obtain the above result, let us recall the fundamental objective in an eigensolution. It is important to remember that all we require is a vector $\boldsymbol{\phi}$

and scalar λ that satisfy the equation

$$\mathbf{K}\boldsymbol{\phi} = \lambda\mathbf{M}\boldsymbol{\phi} \tag{10.4}$$

where $\boldsymbol{\phi}$ is nontrivial; i.e., $\boldsymbol{\phi}$ is a vector with at least one element in it nonzero. In other words, if we have a vector $\boldsymbol{\phi}$ and scalar λ that satisfy (10.4), then λ and $\boldsymbol{\phi}$ are an eigenvalue λ_i and eigenvector $\boldsymbol{\phi}_i$, respectively, where it should be noted that it does not matter how $\boldsymbol{\phi}$ and λ have been obtained. If, for example, we can guess $\boldsymbol{\phi}$ and λ, we should certainly take advantage of it. This may be the case when rigid body modes are present in the structural element assemblage. Thus, if we know that the element assemblage can undergo a rigid body mode, we have $\lambda_1 = 0$ and need to seek $\boldsymbol{\phi}_1$ to satisfy the equation $\mathbf{K}\boldsymbol{\phi}_1 = \mathbf{0}$. In general, the solution of $\boldsymbol{\phi}_1$ must be obtained using an equation solver, but in a simple finite element assemblage, we may be able to identify $\boldsymbol{\phi}_1$ by inspection.

In the case of r zero diagonal elements in a diagonal mass matrix $\mathbf{M}$, we can always immediately establish r eigenvalues and corresponding eigenvectors. Namely, rewriting the eigenproblem in (10.4) in the form

$$\mathbf{M}\boldsymbol{\phi} = \mu\mathbf{K}\boldsymbol{\phi} \tag{10.28}$$

where $\mu = \lambda^{-1}$, we find that if $m_{kk} = 0.0$, we have an eigenpair $(\mu_i, \boldsymbol{\phi}_i) = (0.0, \mathbf{e}_k)$; i.e.,

$$\boldsymbol{\phi}_i^T = [0 \quad 0 \quad \dots \quad 0 \quad \underset{\uparrow k\text{th element}}{1} \quad 0 \quad \dots \quad 0]; \qquad \mu_i = 0.0 \tag{10.29}$$

That $\boldsymbol{\phi}_i$ and μ_i in (10.29) are indeed an eigenvector and eigenvalue of (10.28) is verified by simply substituting into (10.28) and noting that $(\mu_i, \boldsymbol{\phi}_i)$ is a nontrivial solution. Since $\mu = \lambda^{-1}$, we therefore found that an eigenpair of $\mathbf{K}\boldsymbol{\phi} = \lambda\mathbf{M}\boldsymbol{\phi}$ is given by $(\lambda_i, \boldsymbol{\phi}_i) = (\infty, \mathbf{e}_k)$. Considering the case of r zero diagonal elements in $\mathbf{M}$, it follows that there are r infinite eigenvalues, and the corresponding eigenvectors can be taken as unit vectors with the 1 in a location corresponding to a zero mass element in $\mathbf{M}$. We should note that since λ_n is then an eigenvalue of multiplicity r, the corresponding eigenvectors are not unique (see Section 10.2.1). In addition we note that the length of an eigenvector cannot be fixed using the condition of $\mathbf{M}$-orthonormality. We demonstrate how we establish the eigenvalues and eigenvectors by means of a small example.

EXAMPLE 10.7: Consider the eigenproblem

$$\begin{bmatrix} 2 & -1 & & \\ -1 & 2 & -1 & \\ & -1 & 2 & -1 \\ & & -1 & 1 \end{bmatrix} \boldsymbol{\phi} = \lambda \begin{bmatrix} 0 & & & \\ & 2 & & \\ & & 0 & \\ & & & 1 \end{bmatrix} \boldsymbol{\phi}$$

Establish λ_3, λ_4 and $\boldsymbol{\phi}_3$, $\boldsymbol{\phi}_4$.

There are two zero diagonal elements in $\mathbf{M}$; hence $\lambda_3 = \infty$, $\lambda_4 = \infty$. As corresponding eigenvectors we can use

$$\boldsymbol{\phi}_3 = \begin{bmatrix} 1 \\ 0 \\ 0 \\ 0 \end{bmatrix}; \qquad \boldsymbol{\phi}_4 = \begin{bmatrix} 0 \\ 0 \\ 1 \\ 0 \end{bmatrix} \tag{a}$$

Alternatively, any linear combination of $\boldsymbol{\phi}_3$ and $\boldsymbol{\phi}_4$ given in (a) would represent an eigenvector. We should note that $\boldsymbol{\phi}_i^T\mathbf{M}\boldsymbol{\phi}_i = 0$ for $i = 3, 4$, and therefore the magnitude of the elements in $\boldsymbol{\phi}_i$ cannot be fixed using the **M**-orthonormality condition.

10.2.5 Transformation of the Generalized Eigenproblem $K\phi = \lambda M\phi$ to a Standard Form

The most common eigenproblems that are encountered in general scientific analysis are standard eigenproblems, and most other eigenproblems can be reduced to a standard form. For this reason, the solution of standard eigenproblems has attracted much attention in numerical analysis, and many solution algorithms are available. The purpose of this section is to show how the eigenproblem $\mathbf{K}\boldsymbol{\phi} = \lambda\mathbf{M}\boldsymbol{\phi}$ can be reduced to a standard form. The implications of the transformation are twofold. First, because the transformation is possible, use can be made of the various solution algorithms available for standard eigenproblems. We will see that the effectiveness of the eigensolution procedure employed depends to a large degree on the decision whether or not to carry out a transformation to a standard form. Second, *if a generalized eigenproblem can be written in standard form, the properties of the eigenvalues, eigenvectors, and characteristic polynomials of the generalized eigenproblem can be deduced from the properties of the corresponding quantities of the standard eigenproblem.* Realizing that the properties of standard eigenproblems are more easily assessed, it is to a large extent for the second reason that the transformation to a standard eigenproblem is important to be studied. Indeed, after having presented the transformation procedures, we will show how the properties of the eigenvectors (see Section 10.2.1) and the properties of the characteristic polynomials (see Section 10.2.2) of the problem $\mathbf{K}\boldsymbol{\phi} = \lambda\mathbf{M}\boldsymbol{\phi}$ are derived from the corresponding properties of the standard eigenproblem.

In the following we assume that **M** is positive definite. This is the case when **M** is diagonal with $m_{ii} > 0$, $i = 1, \ldots, n$, or **M** is banded, as in a consistent mass analysis. If **M** is diagonal with some zero diagonal elements, we first need to perform static condensation on the massless degrees of freedom as described in Section 10.3.1. Assuming that **M** is positive definite, we can transform the generalized eigenproblem $\mathbf{K}\boldsymbol{\phi} = \lambda\mathbf{M}\boldsymbol{\phi}$ given in (10.4) by using a decomposition of **M** of the form

$$\mathbf{M} = \mathbf{S}\mathbf{S}^T \tag{10.30}$$

where **S** is any nonsingular matrix. Namely, substituting for **M** into (10.4), we have

$$\mathbf{K}\boldsymbol{\phi} = \lambda\mathbf{S}\mathbf{S}^T\boldsymbol{\phi} \tag{10.31}$$

Premultiplying both sides of (10.31) by $\mathbf{S}^{-1}$ and defining a vector

$$\tilde{\boldsymbol{\phi}} = \mathbf{S}^T\boldsymbol{\phi} \tag{10.32}$$

we obtain the standard eigenproblem

$$\tilde{\mathbf{K}}\tilde{\boldsymbol{\phi}} = \lambda\tilde{\boldsymbol{\phi}} \tag{10.33}$$

where

$$\tilde{\mathbf{K}} = \mathbf{S}^{-1}\mathbf{K}\mathbf{S}^{-T} \tag{10.34}$$

One of two decompositions of $\mathbf{M}$ is used in general: the Cholesky factorization or the spectral decomposition of $\mathbf{M}$. The Cholesky factorization of $\mathbf{M}$ is obtained as described in Section 8.2.4 and yields $\mathbf{M} = \tilde{\mathbf{L}}_M \tilde{\mathbf{L}}_M^T$. In (10.30) to (10.34) we therefore have

$$\mathbf{S} = \tilde{\mathbf{L}}_M \tag{10.35}$$

The spectral decomposition of $\mathbf{M}$ requires solution of the complete eigensystem of $\mathbf{M}$. Denoting the matrix of orthonormal eigenvectors by $\mathbf{R}$ and the diagonal matrix of eigenvalues by $\mathbf{D}^2$, we have

$$\mathbf{M} = \mathbf{R}\mathbf{D}^2\mathbf{R}^T \tag{10.36}$$

and we use in (10.30) to (10.34),

$$\mathbf{S} = \mathbf{R}\mathbf{D} \tag{10.37}$$

It should be noted that when $\mathbf{M}$ is diagonal, the matrices $\mathbf{S}$ in (10.35) and (10.37) are the same, but when $\mathbf{M}$ is banded, they are different.

Considering the effectiveness of the solution of the required eigenvalues and eigenvectors of (10.33), it is most important that $\tilde{\mathbf{K}}$ has the same bandwidth as $\mathbf{K}$ when $\mathbf{M}$ is diagonal. However, in case $\mathbf{M}$ is banded, $\tilde{\mathbf{K}}$ in (10.33) is in general a full matrix, which makes the transformation ineffective in almost all large-order finite element analyses. This will become more apparent in Chapters 11 and 12 when various eigensystem solution algorithms are discussed.

Comparing the Cholesky factorization and the spectral decomposition of $\mathbf{M}$, it may be noted that the use of the Cholesky factors is in general computationally more efficient than the use of the spectral decomposition because fewer operations are involved in calculating $\tilde{\mathbf{L}}_M$ than $\mathbf{R}$ and $\mathbf{D}$. However, the spectral decomposition of $\mathbf{M}$ may yield a more accurate solution of $\mathbf{K}\boldsymbol{\phi} = \lambda\mathbf{M}\boldsymbol{\phi}$. Assume that $\mathbf{M}$ is ill-conditioned with respect to inversion; then the transformation process to the standard eigenproblem is also ill-conditioned. In that case it is important to employ the more stable transformation procedure. Using the Cholesky factorization of $\mathbf{M}$ without pivoting, we find that $\tilde{\mathbf{L}}_M^{-1}$ has large elements in many locations because of the coupling in $\mathbf{M}$ and $\tilde{\mathbf{L}}_M^{-1}$. Consequently, $\tilde{\mathbf{K}}$ is calculated with little precision, and the lowest eigenvalues and corresponding eigenvectors are determined inaccurately.

On the other hand, using the spectral decomposition of $\mathbf{M}$, good accuracy may be obtained in the elements of $\mathbf{R}$ and $\mathbf{D}^2$, although some elements in $\mathbf{D}^2$ are small in relation to the other elements. The ill-conditioning of $\mathbf{M}$ is now only concentrated in the small elements of $\mathbf{D}^2$, and considering $\tilde{\mathbf{K}}$, only those rows and columns which correspond to the small elements in $\mathbf{D}$ have large elements, and the eigenvalues of normal size are more likely to be preserved accurately.

Consider the following examples of transforming the generalized eigenvalue problem $\mathbf{K}\boldsymbol{\phi} = \lambda\mathbf{M}\boldsymbol{\phi}$ to a standard form.

EXAMPLE 10.8: Consider the problem $\mathbf{K}\boldsymbol{\phi} = \lambda\mathbf{M}\boldsymbol{\phi}$, where

$$\mathbf{K} = \begin{bmatrix} 3 & -1 & 0 \\ -1 & 2 & -1 \\ 0 & -1 & 1 \end{bmatrix}; \qquad \mathbf{M} = \begin{bmatrix} 2 & 1 & 0 \\ 1 & 3 & 1 \\ 0 & 1 & 2 \end{bmatrix}$$

Use the Cholesky factorization of $\mathbf{M}$ to calculate the matrix $\tilde{\mathbf{K}}$ of a corresponding standard eigenproblem.

We first calculate the $\mathbf{LDL}^T$ decomposition of $\mathbf{M}$,

$$\mathbf{M} = \begin{bmatrix} 1 & & \\ \frac{1}{2} & 1 & \\ 0 & \frac{2}{5} & 1 \end{bmatrix} \begin{bmatrix} 2 & & \\ & \frac{5}{2} & \\ & & \frac{8}{5} \end{bmatrix} \begin{bmatrix} 1 & \frac{1}{2} & 0 \\ & 1 & \frac{2}{5} \\ & & 1 \end{bmatrix}$$

Hence the Cholesky factor of $\mathbf{M}$ is (see Section 8.2.4)

$$\tilde{\mathbf{L}}_{\mathbf{M}} = \begin{bmatrix} \sqrt{2} & & \\ \frac{1}{\sqrt{2}} & \sqrt{\frac{5}{2}} & \\ 0 & \sqrt{\frac{2}{5}} & \sqrt{\frac{8}{5}} \end{bmatrix}$$

and

$$\tilde{\mathbf{L}}_{\mathbf{M}}^{-1} = \begin{bmatrix} \frac{1}{\sqrt{2}} & & \\ -\frac{1}{\sqrt{10}} & \sqrt{\frac{2}{5}} & \\ \frac{1}{\sqrt{40}} & -\frac{1}{\sqrt{10}} & \sqrt{\frac{5}{8}} \end{bmatrix}$$

The matrix of the standard eigenproblem $\tilde{\mathbf{K}} = \tilde{\mathbf{L}}_{\mathbf{M}}^{-1}\mathbf{K}\tilde{\mathbf{L}}_{\mathbf{M}}^{-T}$ is in this case,

$$\tilde{\mathbf{K}} = \begin{bmatrix} \frac{3}{2} & -\frac{\sqrt{5}}{2} & \sqrt{\frac{5}{16}} \\ -\frac{\sqrt{5}}{2} & \frac{3}{2} & -\frac{5}{4} \\ \sqrt{\frac{5}{16}} & -\frac{5}{4} & \frac{3}{2} \end{bmatrix}$$

EXAMPLE 10.9: Consider the generalized eigenproblem of Example 10.8. Use the spectral decomposition of $\mathbf{M}$ to calculate the matrix $\tilde{\mathbf{K}}$ of a corresponding standard eigenproblem.

The eigenvalues and corresponding eigenvectors of the problem $\mathbf{M}\boldsymbol{\phi} = \lambda\boldsymbol{\phi}$ can be calculated as shown in Example 10.4. We obtain $\lambda_1 = 1$, $\lambda_2 = 2$, $\lambda_3 = 4$, and

$$\boldsymbol{\phi}_1 = \begin{bmatrix} \frac{1}{\sqrt{3}} \\ -\frac{1}{\sqrt{3}} \\ \frac{1}{\sqrt{3}} \end{bmatrix}; \quad \boldsymbol{\phi}_2 = \begin{bmatrix} \frac{1}{\sqrt{2}} \\ 0 \\ -\frac{1}{\sqrt{2}} \end{bmatrix}; \quad \boldsymbol{\phi}_3 = \begin{bmatrix} \frac{1}{\sqrt{6}} \\ \frac{2}{\sqrt{6}} \\ \frac{1}{\sqrt{6}} \end{bmatrix}$$

Hence the decomposition $\mathbf{M} = \mathbf{RD}^2\mathbf{R}^T$ is

$$\mathbf{M} = \begin{bmatrix} \frac{1}{\sqrt{3}} & \frac{1}{\sqrt{2}} & \frac{1}{\sqrt{6}} \\ -\frac{1}{\sqrt{3}} & 0 & \frac{2}{\sqrt{6}} \\ \frac{1}{\sqrt{3}} & -\frac{1}{\sqrt{2}} & \frac{1}{\sqrt{6}} \end{bmatrix} \begin{bmatrix} 1 & & \\ & 2 & \\ & & 4 \end{bmatrix} \begin{bmatrix} \frac{1}{\sqrt{3}} & -\frac{1}{\sqrt{3}} & \frac{1}{\sqrt{3}} \\ \frac{1}{\sqrt{2}} & 0 & -\frac{1}{\sqrt{2}} \\ \frac{1}{\sqrt{6}} & \frac{2}{\sqrt{6}} & \frac{1}{\sqrt{6}} \end{bmatrix}$$

Noting that $\mathbf{S} = \mathbf{RD}$ and $\mathbf{S}^{-1} = \mathbf{D}^{-1}\mathbf{R}^T$ because $\mathbf{RR}^T = \mathbf{I}$, we obtain

$$\mathbf{S}^{-1} = \begin{bmatrix} \frac{1}{\sqrt{3}} & -\frac{1}{\sqrt{3}} & \frac{1}{\sqrt{3}} \\ \frac{1}{2} & 0 & -\frac{1}{2} \\ \frac{1}{2\sqrt{6}} & \frac{1}{\sqrt{6}} & \frac{1}{2\sqrt{6}} \end{bmatrix}$$

The matrix of the standard eigenproblem is $\tilde{\mathbf{K}} = \mathbf{S}^{-1}\mathbf{K}\mathbf{S}^{-T}$; i.e.,

$$\tilde{\mathbf{K}} = \begin{bmatrix} \frac{10}{3} & \frac{1}{\sqrt{3}} & -\frac{1}{3\sqrt{2}} \\ \frac{1}{\sqrt{3}} & 1 & \frac{1}{2\sqrt{6}} \\ -\frac{1}{3\sqrt{2}} & \frac{1}{2\sqrt{6}} & \frac{1}{6} \end{bmatrix}$$

We should note that the matrix $\tilde{\mathbf{K}}$ obtained here is different from the matrix $\tilde{\mathbf{K}}$ derived in Example 10.8.

In the above discussion we considered only the factorization of $\mathbf{M}$ into $\mathbf{M} = \mathbf{SS}^T$ and then the transformation of $\mathbf{K}\boldsymbol{\phi} = \lambda\mathbf{M}\boldsymbol{\phi}$ into the form given in (10.33). We pointed out that this transformation can yield inaccurate results if $\mathbf{M}$ is ill-conditioned. In such a case it seems natural to avoid the decomposition of $\mathbf{M}$ and instead use a factorization of $\mathbf{K}$. Rewriting $\mathbf{K}\boldsymbol{\phi} = \lambda\mathbf{M}\boldsymbol{\phi}$ in the form $\mathbf{M}\boldsymbol{\phi} = (1/\lambda)\mathbf{K}\boldsymbol{\phi}$, we can use an analogous procedure to obtain the eigenproblem

$$\tilde{\mathbf{M}}\tilde{\boldsymbol{\phi}} = \frac{1}{\lambda}\tilde{\boldsymbol{\phi}} \tag{10.38}$$

where

$$\tilde{\mathbf{M}} = \mathbf{S}^{-1}\mathbf{M}\mathbf{S}^{-T} \tag{10.39}$$

$$\mathbf{K} = \mathbf{SS}^T \tag{10.40}$$

$$\tilde{\boldsymbol{\phi}} = \mathbf{S}^T\boldsymbol{\phi} \tag{10.41}$$

and $\mathbf{S}$ is obtained using the Cholesky factor or spectral decomposition of $\mathbf{K}$. If $\mathbf{K}$ is well-conditioned, the transformation is also well-conditioned. However, since $\mathbf{K}$ is always banded, we have that $\tilde{\mathbf{M}}$ is always a full matrix, and the transformation is usually inefficient for the solution of $\mathbf{K}\boldsymbol{\phi} = \lambda\mathbf{M}\boldsymbol{\phi}$.

As we pointed out earlier, the possibility of actually solving a generalized eigenproblem by first transforming it into a standard form is only one reason why we considered the above transformations. The second reason is that the properties of the eigensolution of the generalized eigenproblem can be deduced from the properties of the solution of the corresponding standard eigenproblem. Specifically, we can derive the orthogonality properties of the eigenvectors as given in (10.10) and (10.11), and the Sturm sequence property of the characteristic polynomials of the eigenproblem $\mathbf{K}\boldsymbol{\phi} = \lambda\mathbf{M}\boldsymbol{\phi}$ and of its associated constraint problems as given in (10.22). In both cases no fundamentally new concepts need be proved; instead; the corresponding properties of the standard eigenproblem, which is obtained from the generalized eigenproblem, are used. We give the proofs in the following examples as an application of the transformation of a generalized eigenproblem to a standard form.

EXAMPLE 10.10: Show that the eigenvectors of the problem $\mathbf{K}\boldsymbol{\phi} = \lambda\mathbf{M}\boldsymbol{\phi}$ are **M**- and **K**-orthogonal, and discuss the orthogonality of the eigenvectors of the problems $\mathbf{K}_G\boldsymbol{\phi} = \lambda\mathbf{K}\boldsymbol{\phi}$ given in (10.6) and $\mathbf{K}\boldsymbol{\phi} = \lambda\mathbf{C}\boldsymbol{\phi}$ given in (10.7).

The eigenvector orthogonality is proved by transforming the generalized eigenproblem to a standard form and using the fact that the eigenvectors of a standard eigenproblem with a symmetric matrix are orthogonal. Consider first the problem $\mathbf{K}\boldsymbol{\phi} = \lambda\mathbf{M}\boldsymbol{\phi}$ and assume that **M** is positive definite. Then we can use the transformation in (10.30) to (10.34) to obtain, as an equivalent eigenproblem,

$$\tilde{\mathbf{K}}\tilde{\boldsymbol{\phi}} = \lambda\tilde{\boldsymbol{\phi}}$$

where $$\mathbf{M} = \mathbf{S}\mathbf{S}^T; \qquad \tilde{\mathbf{K}} = \mathbf{S}^{-1}\mathbf{K}\mathbf{S}^{-T}; \qquad \tilde{\boldsymbol{\phi}} = \mathbf{S}^T\boldsymbol{\phi}$$

But since the eigenvectors $\tilde{\boldsymbol{\phi}}_i$ of the problem $\tilde{\mathbf{K}}\tilde{\boldsymbol{\phi}} = \lambda\tilde{\boldsymbol{\phi}}$ have the properties (see Section 2.7)

$$\tilde{\boldsymbol{\phi}}_i^T\tilde{\boldsymbol{\phi}}_j = \delta_{ij}; \qquad \tilde{\boldsymbol{\phi}}_i^T\tilde{\mathbf{K}}\tilde{\boldsymbol{\phi}}_j = \lambda_i\delta_{ij}$$

we have, substituting $\tilde{\boldsymbol{\phi}}_i = \mathbf{S}^T\boldsymbol{\phi}_i$, $\tilde{\boldsymbol{\phi}}_j = \mathbf{S}^T\boldsymbol{\phi}_j$,

$$\boldsymbol{\phi}_i^T\mathbf{M}\boldsymbol{\phi}_j = \delta_{ij}; \qquad \boldsymbol{\phi}_i^T\mathbf{K}\boldsymbol{\phi}_j = \lambda_i\delta_{ij} \tag{a}$$

In case **M** is not positive definite, we consider the eigenproblem $\mathbf{M}\boldsymbol{\phi} = (1/\lambda)\mathbf{K}\boldsymbol{\phi}$ (with **K** positive definite or a shift must be imposed; see Section 10.2.3). We now use the transformation

$$\tilde{\mathbf{M}}\tilde{\boldsymbol{\phi}} = \left(\frac{1}{\lambda}\right)\tilde{\boldsymbol{\phi}}$$

where $$\mathbf{K} = \mathbf{S}\mathbf{S}^T; \qquad \tilde{\mathbf{M}} = \mathbf{S}^{-1}\mathbf{M}\mathbf{S}^{-T}; \qquad \tilde{\boldsymbol{\phi}} = \mathbf{S}^T\boldsymbol{\phi}$$

and the properties $$\tilde{\boldsymbol{\phi}}_i^T\tilde{\boldsymbol{\phi}}_j = \delta_{ij}; \qquad \tilde{\boldsymbol{\phi}}_i^T\tilde{\mathbf{M}}\tilde{\boldsymbol{\phi}}_j = \left(\frac{1}{\lambda_i}\right)\delta_{ij}$$

Substituting for $\tilde{\boldsymbol{\phi}}_i$ and $\tilde{\boldsymbol{\phi}}_j$, we obtain

$$\boldsymbol{\phi}_i^T\mathbf{K}\boldsymbol{\phi}_j = \delta_{ij}; \qquad \boldsymbol{\phi}_i^T\mathbf{M}\boldsymbol{\phi}_j = \left(\frac{1}{\lambda_i}\right)\delta_{ij} \tag{b}$$

with the eigenvectors now being **K**- orthonormalized, because the problem $\mathbf{M}\boldsymbol{\phi} = (1/\lambda)\mathbf{K}\boldsymbol{\phi}$ was considered. To obtain the same vectors as in the problem $\mathbf{K}\boldsymbol{\phi} = \lambda\mathbf{M}\boldsymbol{\phi}$, we need to multiply the eigenvectors $\boldsymbol{\phi}_i$ of the problem $\mathbf{M}\boldsymbol{\phi} = (1/\lambda)\mathbf{K}\boldsymbol{\phi}$ by the factors $\sqrt{\lambda_i}$, $i = 1, \ldots, n$.

Considering these proofs, we note that we arranged the eigenproblems in such a way that the matrix associated with the eigenvalue, i.e., the matrix on the right-hand side of the eigenvalue problem, is positive definite. This is necessary to be able to carry out the transformation of the generalized eigenproblem to a standard form and thus derive the eigenvector orthogonality properties in (a) and (b). However, considering the problems $\mathbf{K}_G\boldsymbol{\phi} = \lambda\mathbf{K}\boldsymbol{\phi}$ and $\mathbf{K}\boldsymbol{\phi} = \lambda\mathbf{C}\boldsymbol{\phi}$ given in (10.6) and (10.7), we can proceed in a similar manner. The results would be the eigenvector orthogonality properties given in (10.14) and (10.15).

EXAMPLE 10.11: Prove the Sturm sequence property of the characteristic polynomials of the problem $\mathbf{K}\boldsymbol{\phi} = \lambda\mathbf{M}\boldsymbol{\phi}$ and the associated constraint problems. Demonstrate the proof for the following matrices:

$$\mathbf{K} = \begin{bmatrix} 3 & -1 & \\ -1 & 2 & -1 \\ & -1 & 1 \end{bmatrix}; \qquad \mathbf{M} = \begin{bmatrix} 4 & 4 & \\ 4 & 8 & 4 \\ & 4 & 8 \end{bmatrix} \tag{a}$$

The proof that we consider here is based on the transformation of the eigenproblems $\mathbf{K}\boldsymbol{\phi} = \lambda\mathbf{M}\boldsymbol{\phi}$ and $\mathbf{K}^{(r)}\boldsymbol{\phi}^{(r)} = \lambda^{(r)}\mathbf{M}^{(r)}\boldsymbol{\phi}^{(r)}$ to standard eigenproblems, for which the characteristic polynomials are known to form a Sturm sequence (see Sections 2.8 and 8.2.5).

As in Example 10.10, we assume first that $\mathbf{M}$ is positive definite. In this case we can transform the problem $\mathbf{K}\boldsymbol{\phi} = \lambda\mathbf{M}\boldsymbol{\phi}$ into the form

$$\tilde{\mathbf{K}}\tilde{\boldsymbol{\phi}} = \lambda\tilde{\boldsymbol{\phi}}$$

where $$\tilde{\mathbf{K}} = \tilde{\mathbf{L}}_{\mathbf{M}}^{-1}\mathbf{K}\tilde{\mathbf{L}}_{\mathbf{M}}^{-T}; \qquad \mathbf{M} = \tilde{\mathbf{L}}_{\mathbf{M}}\tilde{\mathbf{L}}_{\mathbf{M}}^{T}; \qquad \tilde{\boldsymbol{\phi}} = \tilde{\mathbf{L}}_{\mathbf{M}}^{T}\boldsymbol{\phi}$$

and $\tilde{\mathbf{L}}_{\mathbf{M}}$ is the Cholesky factor of $\mathbf{M}$.

Considering the eigenproblems $\tilde{\mathbf{K}}\tilde{\boldsymbol{\phi}} = \lambda\tilde{\boldsymbol{\phi}}$ and $\tilde{\mathbf{K}}^{(r)}\tilde{\boldsymbol{\phi}}^{(r)} = \lambda^{(r)}\tilde{\boldsymbol{\phi}}^{(r)}$, $r = 1, \ldots, n-1$, (see (8.37)), we know that the characteristic polynomials form a Sturm sequence. On the other hand, if we consider the eigenproblem $\mathbf{K}\boldsymbol{\phi} = \lambda\mathbf{M}\boldsymbol{\phi}$ and the eigenproblems of its associated constraint problems, i.e., $\mathbf{K}^{(r)}\boldsymbol{\phi}^{(r)} = \lambda^{(r)}\mathbf{M}^{(r)}\boldsymbol{\phi}^{(r)}$, (see 10.20), we note that *the problems* $\tilde{\mathbf{K}}^{(r)}\tilde{\boldsymbol{\phi}}^{(r)} = \lambda^{(r)}\tilde{\boldsymbol{\phi}}^{(r)}$ *and* $\mathbf{K}^{(r)}\boldsymbol{\phi}^{(r)} = \lambda^{(r)}\mathbf{M}^{(r)}\boldsymbol{\phi}^{(r)}$ *have the same eigenvalues*. Namely, $\tilde{\mathbf{K}}^{(r)}\tilde{\boldsymbol{\phi}}^{(r)} = \lambda^{(r)}\tilde{\boldsymbol{\phi}}^{(r)}$ is a standard eigenproblem corresponding to $\mathbf{K}^{(r)}\boldsymbol{\phi}^{(r)} = \lambda^{(r)}\mathbf{M}^{(r)}\boldsymbol{\phi}^{(r)}$; i.e., instead of eliminating the r rows and columns from $\tilde{\mathbf{K}}$ (to obtain $\tilde{\mathbf{K}}^{(r)}$) we can also calculate $\tilde{\mathbf{K}}^{(r)}$ as follows:

$$\tilde{\mathbf{K}}^{(r)} = \tilde{\mathbf{L}}_{\mathbf{M}}^{(r)-1}\mathbf{K}^{(r)}\tilde{\mathbf{L}}_{\mathbf{M}}^{(r)-T}; \qquad \mathbf{M}^{(r)} = \tilde{\mathbf{L}}_{\mathbf{M}}^{(r)}\tilde{\mathbf{L}}_{\mathbf{M}}^{(r)T}; \qquad \tilde{\boldsymbol{\phi}}^{(r)} = \tilde{\mathbf{L}}_{\mathbf{M}}^{(r)T}\boldsymbol{\phi}^{(r)} \tag{b}$$

Note that $\tilde{\mathbf{L}}_{\mathbf{M}}^{(r)}$ and $\tilde{\mathbf{L}}_{\mathbf{M}}^{(r)-1}$ can simply be obtained by deleting the last r rows and columns of $\tilde{\mathbf{L}}_{\mathbf{M}}$ and $\tilde{\mathbf{L}}_{\mathbf{M}}^{-1}$, respectively.

Hence the Sturm sequence property also holds for the characteristic polynomials of $\mathbf{K}\boldsymbol{\phi} = \lambda\mathbf{M}\boldsymbol{\phi}$ and the associated constraint problems.

For the example to be considered we have

$$\tilde{\mathbf{L}}_{\mathbf{M}} = \begin{bmatrix} 2 & 0 & 0 \\ 2 & 2 & 0 \\ 0 & 2 & 2 \end{bmatrix} \tag{c}$$

Hence

$$\tilde{\mathbf{K}} = \begin{bmatrix} \frac{1}{2} & 0 & 0 \\ -\frac{1}{2} & \frac{1}{2} & 0 \\ \frac{1}{2} & -\frac{1}{2} & \frac{1}{2} \end{bmatrix} \begin{bmatrix} 3 & -1 & 0 \\ -1 & 2 & -1 \\ 0 & -1 & 1 \end{bmatrix} \begin{bmatrix} \frac{1}{2} & -\frac{1}{2} & \frac{1}{2} \\ 0 & \frac{1}{2} & -\frac{1}{2} \\ 0 & 0 & \frac{1}{2} \end{bmatrix} = \begin{bmatrix} \frac{3}{4} & -1 & 1 \\ -1 & \frac{7}{4} & -2 \\ 1 & -2 & \frac{5}{2} \end{bmatrix} \tag{d}$$

Using $\tilde{\mathbf{K}}$ in (d) to obtain $\tilde{\mathbf{K}}^{(1)}$ and $\tilde{\mathbf{K}}^{(2)}$, we have

$$\tilde{\mathbf{K}}^{(1)} = \begin{bmatrix} \frac{3}{4} & -1 \\ -1 & \frac{7}{4} \end{bmatrix}; \qquad \tilde{\mathbf{K}}^{(2)} = [\tfrac{3}{4}]$$

On the other hand, we can obtain the same matrices $\tilde{\mathbf{K}}^{(1)}$ and $\tilde{\mathbf{K}}^{(2)}$ using the relations in (b),

$$\tilde{\mathbf{K}}^{(1)} = \tilde{\mathbf{L}}_{\mathbf{M}}^{(1)-1}\mathbf{K}^{(1)}\tilde{\mathbf{L}}_{\mathbf{M}}^{(1)-T}; \qquad \tilde{\mathbf{K}}^{(2)} = \tilde{\mathbf{L}}_{\mathbf{M}}^{(2)-1}\mathbf{K}^{(2)}\tilde{\mathbf{L}}_{\mathbf{M}}^{(2)-T}$$

where $\mathbf{K}^{(r)}$ and $\mathbf{M}^{(r)}$ (to calculate $\tilde{\mathbf{L}}_{\mathbf{M}}^{(r)}$) are obtained from $\mathbf{K}$ and $\mathbf{M}$ in (a).

In the preceding discussion we assumed that $\mathbf{M}$ is positive definite. If $\mathbf{M}$ is positive semidefinite, we can consider the problem $\mathbf{M}\boldsymbol{\phi} = (1/\lambda)\mathbf{K}\boldsymbol{\phi}$ instead, in which $\mathbf{K}$ is positive definite (this may mean that a shift has to be imposed; see Section 10.2.3) and thus show that the Sturm sequence property still holds.

It may be noted that it follows from the above discussion that the characteristic polynomials of the eigenproblems $\mathbf{K}_G\boldsymbol{\phi} = \lambda\mathbf{K}\boldsymbol{\phi}$ and $\mathbf{K}\boldsymbol{\phi} = \lambda\mathbf{C}\boldsymbol{\phi}$ given in (10.6) and (10.7) and of their associated constraint problems also form a Sturm sequence.

10.3 APPROXIMATE SOLUTION TECHNIQUES

It is apparent from the nature of a dynamic problem that a dynamic response calculation must be substantially more costly than a static analysis. Namely, whereas in a static analysis the solution is obtained in one step, in dynamics the solution is required at a number of discrete time points over the time interval considered. Indeed, we found that in a direct step-by-step integration solution, an equation of statics, which includes the effects of inertia and damping forces, is considered at the end of each discrete time step (see Section 9.2). Considering a mode superposition analysis, the main computational effort is spent in the calculation of the required frequencies and mode shapes, which also requires considerably more effort than a static analysis. It is therefore natural that much attention has been directed toward effective algorithms for the calculation of the required eigensystem in the problem $\mathbf{K}\boldsymbol{\phi} = \lambda\mathbf{M}\boldsymbol{\phi}$. In fact, because the "exact" solution of the required eigenvalues and corresponding eigenvectors can be prohibitively expensive when the order of the system is large and a "conventional" technique is used, approximate techniques of solution have been developed. The purpose of this section is to present the major approximate methods that have been designed and are currently still in use.[8-29]

The approximate solution techniques have primarily been developed to calculate the lowest eigenvalues and corresponding eigenvectors in the problem $\mathbf{K}\boldsymbol{\phi} = \lambda\mathbf{M}\boldsymbol{\phi}$, when the order of the system is large. Most programs use exact solution techniques in the analysis of small-order systems. However, the problem of calculating the few lowest eigenpairs of relatively large-order systems is very important and is encountered in all branches of structural engineering and in particular in earthquake response analysis. In the following sections we present four major techniques. *The aim in the presentation is not to advocate the implementation of any one of these methods, but rather to describe their practical use, the limitations, and the assumptions employed. Moreover, the relationships between the approximate techniques are described, and in Chapter 12 we will, in fact, find that the approximate techniques considered here can be understood to be a first iteration (and may be used as such) in the subspace iteration algorithm (see Section 12.3).*

10.3.1 Static Condensation

We encountered the procedure of static condensation already in the solution of static equilibrium equations, where we showed that static condensation is, in fact, an application of Gauss elimination (see Section 8.2.4). In static condensation we eliminated those degrees of freedom that are not required to appear in the global finite element assemblage. For example, the displacement degrees of freedom at the internal nodes of a finite element can be condensed out, because they do not take part in imposing interelement continuity. We mentioned already in Section 8.2.4 that the name "static condensation" was actually coined in dynamic analysis.

The basic assumption of static condensation in the calculation of frequencies and mode shapes is that the mass of the structure can be lumped at only some specific degrees of freedom without much effect on the accuracy of the frequencies and mode shapes of interest. In case of a lumped mass matrix with some zero diagonal elements, some of the mass lumping has already been carried out. However, additional mass lumping is in general required. Typically, we may have that the ratio of mass degrees of freedom to the total number of degrees of freedom is somewhere between $\frac{1}{2}$ and $\frac{1}{10}$. The more mass lumping is performed, the less computer effort is required in the solution, however, the more probable it is also that the required frequencies and mode shapes are not predicted accurately. We shall have more to say about this later.

Assume that the mass lumping has been carried out. By partitioning **K** and **M**, we can then write the eigenproblem $\mathbf{K}\boldsymbol{\phi} = \lambda\mathbf{M}\boldsymbol{\phi}$ in the form

$$\begin{bmatrix} \mathbf{K}_{aa} & \mathbf{K}_{ac} \\ \mathbf{K}_{ca} & \mathbf{K}_{cc} \end{bmatrix} \begin{bmatrix} \boldsymbol{\phi}_a \\ \boldsymbol{\phi}_c \end{bmatrix} = \lambda \begin{bmatrix} \mathbf{M}_a & \mathbf{0} \\ \mathbf{0} & \mathbf{0} \end{bmatrix} \begin{bmatrix} \boldsymbol{\phi}_a \\ \boldsymbol{\phi}_c \end{bmatrix} \tag{10.42}$$

where $\boldsymbol{\phi}_a$ and $\boldsymbol{\phi}_c$ are the displacements at the mass and the massless degrees of freedom, respectively, and $\mathbf{M}_a$ is a diagonal mass matrix. The relation in (10.42) gives the condition

$$\mathbf{K}_{ca}\boldsymbol{\phi}_a + \mathbf{K}_{cc}\boldsymbol{\phi}_c = \mathbf{0} \tag{10.43}$$

which can be used to eliminate $\boldsymbol{\phi}_c$. From (10.43) we obtain

$$\boldsymbol{\phi}_c = -\mathbf{K}_{cc}^{-1}\mathbf{K}_{ca}\boldsymbol{\phi}_a \tag{10.44}$$

and substituting into (10.42), we obtain the reduced eigenproblem

$$\mathbf{K}_a\boldsymbol{\phi}_a = \lambda\mathbf{M}_a\boldsymbol{\phi}_a \tag{10.45}$$

where

$$\mathbf{K}_a = \mathbf{K}_{aa} - \mathbf{K}_{ac}\mathbf{K}_{cc}^{-1}\mathbf{K}_{ca} \tag{10.46}$$

The solution of the generalized eigenproblem in (10.45) is in most cases obtained by transforming the problem first into a standard form as described in Section 10.2.5. Since $\mathbf{M}_a$ is a diagonal mass matrix with all its diagonal elements positive and probably not small, the transformation is in general well-conditioned.

The analogy to the use of static condensation in static analysis should be noted. Namely, realizing that the right-hand side of (10.42) may be understood to be a load vector **R**, where

$$\mathbf{R} = \begin{bmatrix} \lambda\mathbf{M}_a\boldsymbol{\phi}_a \\ \mathbf{0} \end{bmatrix} \tag{10.47}$$

we can use Gauss elimination on the massless degrees of freedom in the same way as we do on the degrees of freedom associated with the interior nodes of an element or a substructure (see Section 8.2.4).

One important aspect should be observed when comparing the static condensation procedure on the massless degrees of freedom in (10.42) to (10.46) on the one side, with Gauss elimination or static condensation in static analysis on the other side. Considering (10.47), we find that the loads at the $\boldsymbol{\phi}_a$ degrees of freedom depend on the eigenvalue (free vibration frequency squared) and eigenvector (mode shape displacements). This means that in (10.45) a further reduction of the number of degrees of freedom to be considered is not possible. This is a basic difference to static condensation as applied in static analysis, where the loads are given explicitly and their effect can be carried over to the remaining degrees of freedom.

EXAMPLE 10.12: Use static condensation to calculate the eigenvalues and eigenvectors of the problem $\mathbf{K}\boldsymbol{\phi} = \lambda\mathbf{M}\boldsymbol{\phi}$, where

$$\mathbf{K} = \begin{bmatrix} 2 & -1 & 0 & 0 \\ -1 & 2 & -1 & 0 \\ 0 & -1 & 2 & -1 \\ 0 & 0 & -1 & 1 \end{bmatrix}; \qquad \mathbf{M} = \begin{bmatrix} 0 & & & \\ & 2 & & \\ & & 0 & \\ & & & 1 \end{bmatrix}$$

First we rearrange columns and rows to obtain the form given in (10.42), which is

$$\begin{bmatrix} 2 & 0 & -1 & -1 \\ 0 & 1 & 0 & -1 \\ -1 & 0 & 2 & 0 \\ -1 & -1 & 0 & 2 \end{bmatrix} \begin{bmatrix} \boldsymbol{\phi}_a \\ \boldsymbol{\phi}_c \end{bmatrix} = \lambda \begin{bmatrix} 2 & & & \\ & 1 & & \\ & & 0 & \\ & & & 0 \end{bmatrix} \begin{bmatrix} \boldsymbol{\phi}_a \\ \boldsymbol{\phi}_c \end{bmatrix}$$

Hence $\mathbf{K}_a$ given in (10.46) is in this case,

$$\mathbf{K}_a = \begin{bmatrix} 2 & 0 \\ 0 & 1 \end{bmatrix} - \begin{bmatrix} -1 & -1 \\ 0 & -1 \end{bmatrix} \begin{bmatrix} \frac{1}{2} & 0 \\ 0 & \frac{1}{2} \end{bmatrix} \begin{bmatrix} -1 & 0 \\ -1 & -1 \end{bmatrix} = \begin{bmatrix} 1 & -\frac{1}{2} \\ -\frac{1}{2} & \frac{1}{2} \end{bmatrix}$$

The eigenproblem $\mathbf{K}_a\boldsymbol{\phi}_a = \lambda\mathbf{M}_a\boldsymbol{\phi}_a$ is, therefore,

$$\begin{bmatrix} 1 & -\frac{1}{2} \\ -\frac{1}{2} & \frac{1}{2} \end{bmatrix} \boldsymbol{\phi}_a = \lambda \begin{bmatrix} 2 & \\ & 1 \end{bmatrix} \boldsymbol{\phi}_a$$

and we have $\qquad \det(\mathbf{K}_a - \lambda\mathbf{M}_a) = 2\lambda^2 - 2\lambda + \frac{1}{4}$

Hence $\qquad \lambda_1 = \frac{1}{2} - \frac{\sqrt{2}}{4}; \qquad \lambda_2 = \frac{1}{2} + \frac{\sqrt{2}}{4}$

The corresponding eigenvectors are calculated using

$$(\mathbf{K}_a - \lambda_i\mathbf{M}_a)\boldsymbol{\phi}_{a_i} = \mathbf{0}; \qquad \boldsymbol{\phi}_{a_i}^T\mathbf{M}_a\boldsymbol{\phi}_{a_i} = 1$$

Hence $\qquad \boldsymbol{\phi}_{a_1} = \begin{bmatrix} \frac{1}{2} \\ \frac{\sqrt{2}}{2} \end{bmatrix}; \qquad \boldsymbol{\phi}_{a_2} = \begin{bmatrix} -\frac{1}{2} \\ \frac{\sqrt{2}}{2} \end{bmatrix}$

Using (10.44), we obtain

$$\boldsymbol{\phi}_{c_1} = -\begin{bmatrix} \frac{1}{2} & 0 \\ 0 & \frac{1}{2} \end{bmatrix} \begin{bmatrix} -1 & 0 \\ -1 & -1 \end{bmatrix} \begin{bmatrix} \frac{1}{2} \\ \frac{\sqrt{2}}{2} \end{bmatrix} = \begin{bmatrix} \frac{1}{4} \\ \frac{1+\sqrt{2}}{4} \end{bmatrix}$$

$$\boldsymbol{\phi}_{c_2} = -\begin{bmatrix} \frac{1}{2} & 0 \\ 0 & \frac{1}{2} \end{bmatrix} \begin{bmatrix} -1 & 0 \\ -1 & -1 \end{bmatrix} \begin{bmatrix} -\frac{1}{2} \\ \frac{\sqrt{2}}{2} \end{bmatrix} = \begin{bmatrix} -\frac{1}{4} \\ \frac{-1+\sqrt{2}}{4} \end{bmatrix}$$

Therefore, the solution to the eigenproblem $\mathbf{K}\boldsymbol{\phi} = \lambda\mathbf{M}\boldsymbol{\phi}$ is

$$\lambda_1 = \frac{1}{2} - \frac{\sqrt{2}}{4}; \qquad \boldsymbol{\phi}_1 = \begin{bmatrix} \frac{1}{4} \\ \frac{1}{2} \\ \frac{1+\sqrt{2}}{4} \\ \frac{\sqrt{2}}{2} \end{bmatrix}$$

$$\lambda_2 = \frac{1}{2} + \frac{\sqrt{2}}{4}; \qquad \boldsymbol{\phi}_2 = \begin{bmatrix} -\frac{1}{4} \\ -\frac{1}{2} \\ \frac{-1+\sqrt{2}}{4} \\ \frac{\sqrt{2}}{2} \end{bmatrix}$$

$$\lambda_3 = \infty; \qquad \boldsymbol{\phi}_3 = \begin{bmatrix} 1 \\ 0 \\ 0 \\ 0 \end{bmatrix}$$

$$\lambda_4 = \infty; \qquad \boldsymbol{\phi}_4 = \begin{bmatrix} 0 \\ 0 \\ 1 \\ 0 \end{bmatrix}$$

In the above discussion we gave the formal matrix equations for carrying out static condensation. The main computational effort is in calculating $\mathbf{K}_a$ given in (10.46), where it should be noted that in practice a formal inversion of $\mathbf{K}_{cc}$ is not performed. Instead, $\mathbf{K}_a$ can be obtained conveniently using the Cholesky factor $\tilde{\mathbf{L}}_c$ of $\mathbf{K}_{cc}$. If we factorize $\mathbf{K}_{cc}$,

$$\mathbf{K}_{cc} = \tilde{\mathbf{L}}_c\tilde{\mathbf{L}}_c^T \tag{10.48}$$

we can calculate $\mathbf{K}_a$ in the following way:

$$\mathbf{K}_a = \mathbf{K}_{aa} - \mathbf{Y}^T\mathbf{Y} \tag{10.49}$$

where $\mathbf{Y}$ is solved from

$$\tilde{\mathbf{L}}_c\mathbf{Y} = \mathbf{K}_{ca} \tag{10.50}$$

As was pointed out earlier, the above procedure is, in fact, Gauss elimination of the massless degrees of freedom, i.e., elimination of those degrees of freedom at which no external forces (mass effects) are acting. Therefore, an alternative procedure to the one given in (10.42) to (10.50) is to directly use Gauss elimination on the $\boldsymbol{\phi}_c$ degrees of freedom without partitioning $\mathbf{K}$ into the submatrices $\mathbf{K}_{aa}$, $\mathbf{K}_{cc}$, $\mathbf{K}_{ac}$, and $\mathbf{K}_{ca}$, because Gauss elimination can be performed in any order (see Example 8.1, Section 8.2.1). However, the bandwidth of the stiffness matrix will then, in general, increase during the reduction process, and problems of storage need be considered.

For the solution of the eigenproblem $\mathbf{K}_a\boldsymbol{\phi}_a = \lambda\mathbf{M}_a\boldsymbol{\phi}_a$, it is important to note that $\mathbf{K}_a$ is, in general, a full matrix, and the solution is relatively expensive unless the order of the matrices is small.

Instead of calculating the matrix $\mathbf{K}_a$, it may be preferable to evaluate the flexibility matrix $\mathbf{F}_a = \mathbf{K}_a^{-1}$, which is obtained using

$$\begin{bmatrix} \mathbf{K}_{aa} & \mathbf{K}_{ac} \\ \mathbf{K}_{ca} & \mathbf{K}_{cc} \end{bmatrix} \begin{bmatrix} \mathbf{F}_a \\ \mathbf{F}_c \end{bmatrix} = \begin{bmatrix} \mathbf{I} \\ \mathbf{0} \end{bmatrix} \tag{10.51}$$

where $\mathbf{I}$ is a unit matrix of the same order as $\mathbf{K}_{aa}$. Therefore, in (10.51) we solve for the displacements of the structure when unit loads are applied in turn at the mass degrees of freedom. Although the degrees of freedom have been partitioned in (10.51), there is no need for it in this analysis (see Example 10.13). Having solved for $\mathbf{F}_a$, we now consider instead of (10.45) the eigenproblem

$$\left(\frac{1}{\lambda}\right)\boldsymbol{\phi}_a = \mathbf{F}_a\mathbf{M}_a\boldsymbol{\phi}_a \tag{10.52}$$

Although this eigenproblem is of a slightly different form than the generalized problem $\mathbf{K}\boldsymbol{\phi} = \lambda\mathbf{M}\boldsymbol{\phi}$, the transformation to a standard problem proceeds in much the same way (see Section 10.2.5). For the transformation we define

$$\tilde{\boldsymbol{\phi}}_a = \mathbf{M}_a^{1/2}\boldsymbol{\phi}_a \tag{10.53}$$

where $\mathbf{M}_a^{1/2}$ is a diagonal matrix with its ith diagonal element equal to the root of the ith diagonal element of $\mathbf{M}_a$. Premultiplying both sides of (10.52) by $\mathbf{M}_a^{1/2}$ and substituting the relation in (10.53), we obtain

$$\tilde{\mathbf{F}}_a\tilde{\boldsymbol{\phi}}_a = \left(\frac{1}{\lambda}\right)\tilde{\boldsymbol{\phi}}_a \tag{10.54}$$

$$\tilde{\mathbf{F}}_a = \mathbf{M}_a^{1/2}\,\mathbf{F}_a\,\mathbf{M}_a^{1/2} \tag{10.55}$$

Once the displacements $\boldsymbol{\phi}_a$ have been calculated, we obtain the complete displacement vector using

$$\begin{bmatrix} \boldsymbol{\phi}_a \\ \boldsymbol{\phi}_c \end{bmatrix} = \begin{bmatrix} \mathbf{I} \\ \mathbf{F}_c\mathbf{K}_a \end{bmatrix}\boldsymbol{\phi}_a \tag{10.56}$$

where $\mathbf{F}_c$ was calculated in (10.51). The relation in (10.56) is arrived at by realizing that the forces applied at the mass degrees of freedom to impose $\boldsymbol{\phi}_a$ are $\mathbf{K}_a\boldsymbol{\phi}_a$. Using (10.51), the corresponding displacements at all degrees of freedom are given in (10.56).

EXAMPLE 10.13: Use the procedure given in (10.51) to (10.56) to calculate the eigenvalues and eigenvectors of the problem $\mathbf{K}\boldsymbol{\phi} = \lambda\mathbf{M}\boldsymbol{\phi}$ considered in Example 10.12.

The first step is to solve the equations

$$\begin{bmatrix} 2 & -1 & 0 & 0 \\ -1 & 2 & -1 & 0 \\ 0 & -1 & 2 & -1 \\ 0 & 0 & -1 & 1 \end{bmatrix} [\mathbf{v}_1 \quad \mathbf{v}_2] = \begin{bmatrix} 0 & 0 \\ 1 & 0 \\ 0 & 0 \\ 0 & 1 \end{bmatrix} \tag{a}$$

where we did not interchange rows and columns in $\mathbf{K}$ in order to obtain the form in (10.51).

For the solution of the equations in (a), we use the $\mathbf{LDL}^T$ decomposition of $\mathbf{K}$, where

$$\mathbf{L} = \begin{bmatrix} 1 & & & \\ -\frac{1}{2} & 1 & & \\ 0 & -\frac{2}{3} & 1 & \\ 0 & 0 & -\frac{3}{4} & 1 \end{bmatrix}; \qquad \mathbf{D} = \begin{bmatrix} 2 & & & \\ & \frac{3}{2} & & \\ & & \frac{4}{3} & \\ & & & \frac{1}{4} \end{bmatrix}$$

Hence we obtain $\quad \mathbf{v}_1^T = [1 \quad 2 \quad 2 \quad 2]; \qquad \mathbf{v}_2^T = [1 \quad 2 \quad 3 \quad 4]$

and hence $\quad \mathbf{F}_a = \begin{bmatrix} 2 & 2 \\ 2 & 4 \end{bmatrix}; \qquad \mathbf{F}_c = \begin{bmatrix} 1 & 1 \\ 2 & 3 \end{bmatrix}$

$$\tilde{\mathbf{F}}_a = \begin{bmatrix} \sqrt{2} & 0 \\ 0 & 1 \end{bmatrix} \begin{bmatrix} 2 & 2 \\ 2 & 4 \end{bmatrix} \begin{bmatrix} \sqrt{2} & 0 \\ 0 & 1 \end{bmatrix} = \begin{bmatrix} 4 & 2\sqrt{2} \\ 2\sqrt{2} & 4 \end{bmatrix}$$

The solution of the eigenproblem

$$\tilde{\mathbf{F}}_a \tilde{\boldsymbol{\phi}}_a = \mu \tilde{\boldsymbol{\phi}}_a$$

gives

$$\mu_1 = 4 - 2\sqrt{2}; \qquad \tilde{\boldsymbol{\phi}}_{a_1} = \begin{bmatrix} -\dfrac{1}{\sqrt{2}} \\ \dfrac{1}{\sqrt{2}} \end{bmatrix}$$

$$\mu_2 = 4 + 2\sqrt{2}; \qquad \tilde{\boldsymbol{\phi}}_{a_2} = \begin{bmatrix} \dfrac{1}{\sqrt{2}} \\ \dfrac{1}{\sqrt{2}} \end{bmatrix} \tag{b}$$

Since $\boldsymbol{\phi}_a = \mathbf{M}_a^{-1/2} \tilde{\boldsymbol{\phi}}_a$, we have

$$\boldsymbol{\phi}_{a_1} = \begin{bmatrix} \dfrac{1}{\sqrt{2}} & 0 \\ 0 & 1 \end{bmatrix} \begin{bmatrix} -\dfrac{1}{\sqrt{2}} \\ \dfrac{1}{\sqrt{2}} \end{bmatrix} = \begin{bmatrix} -\dfrac{1}{2} \\ \dfrac{1}{\sqrt{2}} \end{bmatrix}; \qquad \boldsymbol{\phi}_{a_2} = \begin{bmatrix} \dfrac{1}{2} \\ \dfrac{1}{\sqrt{2}} \end{bmatrix} \tag{c}$$

The vectors $\boldsymbol{\phi}_{c_1}$, and $\boldsymbol{\phi}_{c_2}$ are calculated using (10.56); hence

$$\boldsymbol{\phi}_{c_1} = \begin{bmatrix} -\dfrac{1}{4} \\ \dfrac{-1 + \sqrt{2}}{4} \end{bmatrix}; \qquad \boldsymbol{\phi}_{c_2} = \begin{bmatrix} \dfrac{1}{4} \\ \dfrac{1 + \sqrt{2}}{4} \end{bmatrix} \tag{d}$$

Since $\mu = 1/\lambda$ we realize that in (b), (c), and (d) we have the same solution as obtained in Example 10.12.

Considering the different procedures of eliminating the massless degrees of freedom, the results of the eigensystem analysis are the same irrespective of the procedure followed, i.e., whether $\mathbf{K}_a$ or $\mathbf{F}_a$ is established and whether the eigenproblem in (10.45) or in (10.52) is solved. The basic assumption in the analysis is that resulting from mass lumping. As we discussed in Section 10.2.4, each zero mass corresponds to an infinite frequency in the system. Therefore, in approximating the equation $\mathbf{K}\boldsymbol{\phi} = \lambda\mathbf{M}\boldsymbol{\phi}$ by the equation in (10.42), we replace, in fact, some of the frequencies of $\mathbf{K}\boldsymbol{\phi} = \lambda\mathbf{M}\boldsymbol{\phi}$ by infinite frequencies and assume that the lowest frequencies solved from either equation are not much different. The accuracy with which the lowest frequencies of $\mathbf{K}\boldsymbol{\phi} = \lambda\mathbf{M}\boldsymbol{\phi}$ are approximated by solving $\mathbf{K}_a\boldsymbol{\phi}_a = \lambda\mathbf{M}_a\boldsymbol{\phi}_a$ depends on the specific mass lumping chosen and may be adequate or crude indeed. In general, more accuracy can be expected if more mass degrees of freedom are included. However, realizing that the static condensation results in $\mathbf{K}_a$ to have a larger bandwidth than $\mathbf{K}$ (and $\mathbf{F}_a$ is certainly full), the computational effort required in the solution of the reduced eigenproblem increases rapidly as the order of $\mathbf{K}_a$ becomes large (see Section 11.3). On the other hand, if sufficient mass degrees of freedom for accuracy of solution are selected, we may no longer want to calculate the complete eigensystem of $\mathbf{K}_a\boldsymbol{\phi}_a = \lambda\mathbf{M}_a\boldsymbol{\phi}_a$ but only the smallest eigenvalues and corresponding vectors. However, in this case we may just as well consider the problem $\mathbf{K}\boldsymbol{\phi} = \lambda\mathbf{M}\boldsymbol{\phi}$ without mass lumping and solve directly only for the eigenvalues and vectors of interest using one of the algorithms described in Chapter 12.

In summary, the main shortcoming of the mass lumping procedure followed by static condensation is that the accuracy of solution depends to a large degree on the experience of the analyst to distribute the mass appropriately, and that the solution accuracy is actually not assessed. We consider the following example to show the approximation that can typically result.

EXAMPLE 10.14: In Example 10.4 we calculated the eigensystem of the problem $\mathbf{K}\boldsymbol{\phi} = \lambda\mathbf{M}\boldsymbol{\phi}$, where $\mathbf{K}$ and $\mathbf{M}$ are given in the example. To evaluate an approximation to the smallest eigenvalue and corresponding eigenvector, consider instead the following eigenproblem, in which the mass is lumped:

$$\begin{bmatrix} 2 & -1 & 0 \\ -1 & 4 & -1 \\ 0 & -1 & 2 \end{bmatrix} \boldsymbol{\phi} = \lambda \begin{bmatrix} 0 & & \\ & 2 & \\ & & 0 \end{bmatrix} \boldsymbol{\phi} \tag{a}$$

Using the procedure given in (10.51) to (10.56), we obtain

$$\mathbf{F}_a = \tfrac{1}{3}; \qquad \mathbf{F}_c = \begin{bmatrix} \frac{1}{6} \\ \frac{1}{6} \end{bmatrix}$$

Hence $\lambda_1 = \frac{3}{2}$, $\boldsymbol{\phi}_{a_1} = 1/\sqrt{2}$, and

$$\boldsymbol{\phi}_{c_1} = \begin{bmatrix} \dfrac{1}{2\sqrt{2}} \\ \dfrac{1}{2\sqrt{2}} \end{bmatrix}$$

The solution of the eigenproblem in (a) for the smallest eigenvalue and corresponding eigenvector is hence

$$\lambda_1 = \tfrac{3}{2}; \qquad \boldsymbol{\phi}_1 = \begin{bmatrix} \dfrac{1}{2\sqrt{2}} \\ \dfrac{1}{\sqrt{2}} \\ \dfrac{1}{2\sqrt{2}} \end{bmatrix}$$

whereas the solution of the original problem (see Example 10.4) is

$$\lambda_1 = 2; \qquad \boldsymbol{\phi}_1 = \begin{bmatrix} \frac{1}{\sqrt{2}} \\ \frac{1}{\sqrt{2}} \\ \frac{1}{\sqrt{2}} \end{bmatrix}$$

It should be noted that using the mass lumping procedure, the eigenvalues can be smaller—as in this example—or larger than the eigenvalues of the original system.

10.3.2 Rayleigh–Ritz Analysis

A most general technique for finding approximations to the lowest eigenvalues and corresponding eigenvectors of the problem $\mathbf{K}\boldsymbol{\phi} = \lambda\mathbf{M}\boldsymbol{\phi}$ is the Rayleigh-Ritz analysis. *The static condensation procedure in Section 10.3.1, the component mode synthesis described in the next section, and various other methods can be understood to be Ritz analyses.* As we will see, the techniques differ only in the choice of the Ritz basis vectors assumed in the analysis. In the following we first present the Rayleigh-Ritz analysis procedure in general and then show how other techniques relate to it.

The eigenproblem that we consider is

$$\mathbf{K}\boldsymbol{\phi} = \lambda\mathbf{M}\boldsymbol{\phi} \tag{10.4}$$

where we now first assume for clarity of presentation that $\mathbf{K}$ and $\mathbf{M}$ are both positive definite, which assures that the eigenvalues are all positive; i.e., $\lambda_1 > 0$. As we pointed out in Section 10.2.3, $\mathbf{K}$ can be assumed positive definite, because a shift can always be introduced to obtain a shifted stiffness matrix which satisfies this condition. As for the mass matrix, we now assume that $\mathbf{M}$ is a consistent mass matrix or a lumped mass matrix with no zero diagonal elements, which is a condition that we shall later relax.

Consider first the Rayleigh minimum principle, which states that

$$\lambda_1 = \min \rho(\boldsymbol{\phi}) \tag{10.57}$$

where the minimum is taken over all possible vectors $\boldsymbol{\phi}$, and $\rho(\boldsymbol{\phi})$ is the Rayleigh quotient

$$\rho(\boldsymbol{\phi}) = \frac{\boldsymbol{\phi}^T\mathbf{K}\boldsymbol{\phi}}{\boldsymbol{\phi}^T\mathbf{M}\boldsymbol{\phi}} \tag{10.58}$$

This Rayleigh quotient is obtained from the Rayleigh quotient of the standard eigenvalue problem $\tilde{\mathbf{K}}\tilde{\boldsymbol{\phi}} = \lambda\tilde{\boldsymbol{\phi}}$ (see Sections 2.8 and 10.2.5). Since both $\mathbf{K}$ and $\mathbf{M}$ are positive definite, $\rho(\boldsymbol{\phi})$ has finite values for all $\boldsymbol{\phi}$. Referring to Section 2.8 the bounds on the Rayleigh quotient are

$$0 < \lambda_1 \leq \rho(\boldsymbol{\phi}) \leq \lambda_n < \infty \tag{10.59}$$

In the Ritz analysis we consider a set of vectors $\bar{\boldsymbol{\phi}}$, which are linear combinations of the Ritz basis vectors $\boldsymbol{\psi}_i$, $i = 1, \ldots, q$; i.e., a typical vector is given by

$$\bar{\boldsymbol{\phi}} = \sum_{i=1}^{q} x_i \boldsymbol{\psi}_i \tag{10.60}$$

where the x_i are the Ritz coordinates. Since $\bar{\boldsymbol{\phi}}$ is a linear combination of the Ritz basis vectors, $\bar{\boldsymbol{\phi}}$ cannot be any arbitrary vector, but instead lies in the subspace spanned by the Ritz basis vectors, which we call V_q (see Sections 2.2 and 12.3). It should be noted that the vectors $\boldsymbol{\psi}_i$, $i = 1, \ldots, q$, must be linearly independent; therefore, the subspace V_q has dimension q. Also, denoting the n-dimensional space in which the matrices $\mathbf{K}$ and $\mathbf{M}$ are defined by V_n, we have that V_q is contained in V_n.

In the Rayleigh-Ritz analysis we aim to determine the specific vectors $\bar{\boldsymbol{\phi}}_i$, $i = 1, \ldots, q$, which, with the constraint of lying in the subspace spanned by the Ritz basis vectors, "best" approximate the required eigenvectors. For this purpose we invoke the Rayleigh minimum principle. The use of this principle determines in what sense the solution does "best" approximate the eigenvectors sought, an aspect that we shall point out during the presentation of the solution procedure.

To invoke the Rayleigh minimum principle on $\bar{\boldsymbol{\phi}}$, we first evaluate the Rayleigh quotient,

$$\rho(\bar{\boldsymbol{\phi}}) = \frac{\sum_{j=1}^{q} \sum_{i=1}^{q} x_i x_j \tilde{k}_{ij}}{\sum_{j=1}^{q} \sum_{i=1}^{q} x_i x_j \tilde{m}_{ij}} = \frac{\tilde{k}}{\tilde{m}} \tag{10.61}$$

where

$$\tilde{k}_{ij} = \boldsymbol{\psi}_i^T \mathbf{K} \boldsymbol{\psi}_j \tag{10.62}$$

$$\tilde{m}_{ij} = \boldsymbol{\psi}_i^T \mathbf{M} \boldsymbol{\psi}_j \tag{10.63}$$

The necessary condition for a minimum of $\rho(\bar{\boldsymbol{\phi}})$ given in (10.61) is $\partial \rho(\bar{\boldsymbol{\phi}})/\partial x_i = 0$, $i = 1, \ldots, q$, because the x_i are the only variables. However,

$$\frac{\partial \rho(\bar{\boldsymbol{\phi}})}{\partial x_i} = \frac{2\tilde{m} \sum_{j=1}^{q} x_j \tilde{k}_{ij} - 2\tilde{k} \sum_{j=1}^{q} x_j \tilde{m}_{ij}}{\tilde{m}^2} \tag{10.64}$$

and using $\rho = \tilde{k}/\tilde{m}$, the condition for a minimum of $\rho(\bar{\boldsymbol{\phi}})$ is

$$\sum_{j=1}^{q} (\tilde{k}_{ij} - \rho \tilde{m}_{ij}) x_j = 0 \qquad \text{for } i = 1, \ldots, q \tag{10.65}$$

In actual analysis we write the q equations in (10.65) in matrix form, thus obtaining the eigenproblem

$$\tilde{\mathbf{K}} \mathbf{x} = \rho \tilde{\mathbf{M}} \mathbf{x} \tag{10.66}$$

where $\tilde{\mathbf{K}}$ and $\tilde{\mathbf{M}}$ are $q \times q$ matrices with typical elements defined in (10.62) and (10.63), respectively, and $\mathbf{x}$ is a vector of the Ritz coordinates sought:

$$\mathbf{x}^T = [x_1 \quad x_2 \quad \ldots \quad x_q] \tag{10.67}$$

The solution to (10.66) yields q eigenvalues $\rho_1, \ldots, \rho_q$, which are approximations to $\lambda_1, \ldots, \lambda_q$, and q eigenvectors,

$$\begin{aligned} \mathbf{x}_1^T &= [x_1^1 \quad x_2^1 \quad \ldots \quad x_q^1] \\ \mathbf{x}_2^T &= [x_1^2 \quad x_2^2 \quad \ldots \quad \mathbf{x}_q^2] \\ &\;\;\vdots \\ \mathbf{x}_q^T &= [x_1^q \quad x_2^q \quad \cdots \quad x_q^q] \end{aligned} \tag{10.68}$$

The eigenvectors $\mathbf{x}_i$ are used to evaluate the vectors $\bar{\boldsymbol{\phi}}_1, \ldots, \bar{\boldsymbol{\phi}}_q$, which are approximations to the eigenvectors $\boldsymbol{\phi}_1, \ldots, \boldsymbol{\phi}_q$. Using (10.68) and (10.60), we have

$$\bar{\boldsymbol{\phi}}_i = \sum_{j=1}^{q} x_j^i \ \boldsymbol{\psi}_j; \qquad i = 1, \ldots, q \tag{10.69}$$

An important feature of the eigenvalue approximations calculated in the analysis is that they are upper bound approximations to the eigenvalues of interest; i.e.,

$$\lambda_1 \leq \rho_1; \qquad \lambda_2 \leq \rho_2; \qquad \lambda_3 \leq \rho_3; \ldots; \qquad \lambda_q \leq \rho_q \leq \lambda_n \tag{10.70}$$

meaning that since $\mathbf{K}$ and $\mathbf{M}$ are assumed to be positive definite, $\tilde{\mathbf{K}}$ and $\tilde{\mathbf{M}}$ are also positive definite matrices.

The proof of the inequality in (10.70) shows the actual mechanism that is used to obtain the eigenvalue approximations ρ_i. To calculate ρ_1 we search for the minimum of $\rho(\bar{\boldsymbol{\phi}})$ that can be reached by linearly combining all available Ritz basis vectors. The inequality $\lambda_1 \leq \rho_1$ follows from the Rayleigh minimum principle in (10.57) and because V_q is contained in the n-dimensional space V_n, in which $\mathbf{K}$ and $\mathbf{M}$ are defined.

The condition that is employed to obtain ρ_2 is typical of the mechanism used to calculate the approximations to the higher eigenvalues. First we observe that for the eigenvalue problem $\mathbf{K}\boldsymbol{\phi} = \lambda\mathbf{M}\boldsymbol{\phi}$, we have

$$\lambda_2 = \min \rho(\boldsymbol{\phi}) \tag{10.71}$$

where the minimum is now taken over all possible vectors $\boldsymbol{\phi}$ in V_n that satisfy the orthogonality condition (see Section 2.8)

$$\boldsymbol{\phi}^T\mathbf{M}\boldsymbol{\phi}_1 = 0 \tag{10.72}$$

Considering the approximate eigenvectors $\bar{\boldsymbol{\phi}}_i$ obtained in the Rayleigh–Ritz analysis, we observe that

$$\bar{\boldsymbol{\phi}}_i^T\mathbf{M}\bar{\boldsymbol{\phi}}_j = \delta_{ij} \tag{10.73}$$

where δ_{ij} = Kronecker delta, and that, therefore, in the above Rayleigh–Ritz analysis we obtained ρ_2 by evaluating

$$\rho_2 = \min \rho(\bar{\boldsymbol{\phi}}) \tag{10.74}$$

where the minimum was taken over all possible vectors $\bar{\boldsymbol{\phi}}$ in V_q that satisfy the orthogonality condition

$$\bar{\boldsymbol{\phi}}^T\mathbf{M}\bar{\boldsymbol{\phi}}_1 = 0 \tag{10.75}$$

To show that $\lambda_2 \leq \rho_2$, we consider an auxiliary problem; i.e., assume that we evaluate

$$\tilde{\rho}_2 = \min \rho(\bar{\boldsymbol{\phi}}) \tag{10.76}$$

where the minimum is taken over all vectors $\bar{\boldsymbol{\phi}}$ that satisfy the condition

$$\bar{\boldsymbol{\phi}}^T\mathbf{M}\boldsymbol{\phi}_1 = 0 \tag{10.77}$$

The problem defined in (10.76) and (10.77) is the same as the problem in (10.71) and (10.72), except that in the latter case the minimum is taken over all $\boldsymbol{\phi}$, whereas in the problem of (10.76) and (10.77) we consider all vectors $\bar{\boldsymbol{\phi}}$ in V_q. Then since V_q is contained in V_n, we have $\lambda_2 \leq \tilde{\rho}_2$. On the other hand, $\tilde{\rho}_2 \leq \rho_2$, because the most severe constraint on $\bar{\boldsymbol{\phi}}$ in (10.77) is $\bar{\boldsymbol{\phi}}_1$. Therefore,

we have

$$\lambda_2 \leq \tilde{\rho}_2 \leq \rho_2 \tag{10.78}$$

The basis for the calculation of $\bar{\boldsymbol{\phi}}_2$ and hence ρ_2 is that the minimum of $\rho(\bar{\boldsymbol{\phi}})$ is sought with the orthogonality condition in (10.75) on $\bar{\boldsymbol{\phi}}_1$, Similarly, to obtain ρ_i and $\bar{\boldsymbol{\phi}}_i$, we in fact minimize $\rho(\bar{\boldsymbol{\phi}})$ with the orthogonality conditions $\bar{\boldsymbol{\phi}}^T\mathbf{M}\bar{\boldsymbol{\phi}}_j = 0$ for $j = 1, \ldots, i-1$. Accordingly, the inequality on ρ_i in (10.70) can be proved in an analogous manner to the procedure used above for ρ_2, but all $(i-1)$ constraint equations need to be satisfied.

The observation that $(i-1)$ constraint equations need to be fulfilled in the evaluation of ρ_i also indicates that we can expect less accuracy in the approximation of the higher eigenvalues than in the approximation of the lower eigenvalues, for which less constraints are imposed. This is generally also observed in actual analysis.

Considering the procedure in practical dynamic analysis, the Ritz basis functions may be calculated from a static solution in which q load patterns are specified in $\mathbf{R}$; i.e., we consider

$$\mathbf{K}\boldsymbol{\Psi} = \mathbf{R} \tag{10.79}$$

where $\boldsymbol{\Psi}$ is an $n \times q$ matrix storing the Ritz basis vectors; i.e., $\boldsymbol{\Psi} = [\boldsymbol{\psi}_1, \ldots, \boldsymbol{\psi}_q]$. The analysis is continued by evaluating the projections of $\mathbf{K}$ and $\mathbf{M}$ onto the subspace V_q spanned by the vectors $\boldsymbol{\psi}_i$, $i = 1, \ldots, q$; i.e., we calculate

$$\tilde{\mathbf{K}} = \boldsymbol{\Psi}^T\mathbf{K}\boldsymbol{\Psi} \tag{10.80}$$

and

$$\tilde{\mathbf{M}} = \boldsymbol{\Psi}^T\mathbf{M}\boldsymbol{\Psi} \tag{10.81}$$

where because of (10.79) we have

$$\tilde{\mathbf{K}} = \boldsymbol{\Psi}^T\mathbf{R} \tag{10.82}$$

Next we solve the eigenproblem $\tilde{\mathbf{K}}\mathbf{x} = \rho\tilde{\mathbf{M}}\mathbf{x}$, the solution of which can be written

$$\tilde{\mathbf{K}}\mathbf{X} = \tilde{\mathbf{M}}\mathbf{X}\boldsymbol{\rho} \tag{10.83}$$

where $\boldsymbol{\rho}$ is a diagonal matrix listing the eigenvalue approximations ρ_i, $\boldsymbol{\rho} = \text{diag}\,(\rho_i)$, and $\mathbf{X}$ is a matrix storing the $\tilde{\mathbf{M}}$-orthonormal eigenvectors $\mathbf{x}_1, \ldots, \mathbf{x}_q$. The approximations to the eigenvectors of the problem $\mathbf{K}\boldsymbol{\phi} = \lambda\mathbf{M}\boldsymbol{\phi}$ are then

$$\bar{\boldsymbol{\Phi}} = \boldsymbol{\Psi}\mathbf{X} \tag{10.84}$$

So far we assumed that the mass matrix of the finite element system is positive definite; i.e., $\mathbf{M}$ is not a diagonal mass matrix with some zero diagonal elements. The reason for this assumption was to avoid the case $\bar{\boldsymbol{\phi}}^T\mathbf{M}\bar{\boldsymbol{\phi}}$ equal to zero in the calculation of the Rayleigh quotient, in which case $\rho(\bar{\boldsymbol{\phi}})$ gives an infinite eigenvalue. However, the Rayleigh–Ritz analysis can be carried out as described above when $\mathbf{M}$ is a diagonal matrix with some zero diagonal elements, provided the Ritz basis vectors are selected to lie in the subspace that corresponds to the finite eigenvalues. In addition, the Ritz basis vectors must be linearly independent when considering only the mass degrees of freedom, in order to obtain a positive definite matrix $\tilde{\mathbf{M}}$. One way of achieving this in practice is to excite in each of the load vectors in $\mathbf{R}$ in (10.79) different mass degrees of freedom (see Section 12.3.3 and Example 10.16).

Of particular interest are the errors that we may expect in the solution. *Although we have shown that an eigenvalue calculated from the Ritz analysis is an upper bound on the corresponding exact eigenvalue of the system, we did not establish anything about the actual error in the eigenvalue.* This error depends on the Ritz basis vectors used, because the vectors $\bar{\boldsymbol{\phi}}$ are linear combinations of the Ritz basis vectors $\boldsymbol{\psi}_i$, $i = 1, \ldots, q$. We can only obtain good results if the vectors $\boldsymbol{\psi}_i$ span a subspace V_q that is close to the least dominant subspace of $\mathbf{K}$ and $\mathbf{M}$ spanned by $\boldsymbol{\phi}_1, \ldots, \boldsymbol{\phi}_q$. It should be noted that this does not mean that the Ritz basis vectors should each be close to an eigenvector sought, but rather that linear combinations of the Ritz basis vectors can establish good approximations of the required eigenvectors of $\mathbf{K}\boldsymbol{\phi} = \lambda\mathbf{M}\boldsymbol{\phi}$. We further discuss the selection of good Ritz basis vectors and the approximations involved in the analysis in Section 12.3, when we present the subspace iteration method, because this method uses the Ritz analysis technique.

To demonstrate the Rayleigh-Ritz analysis procedure, consider the following examples.

EXAMPLE 10.15: Obtain approximate solutions to the eigenproblem $\mathbf{K}\boldsymbol{\phi} = \lambda\mathbf{M}\boldsymbol{\phi}$ considered in Example 10.4, where

$$\mathbf{K} = \begin{bmatrix} 2 & -1 & 0 \\ -1 & 4 & -1 \\ 0 & -1 & 2 \end{bmatrix}; \qquad \mathbf{M} = \begin{bmatrix} \frac{1}{2} & & \\ & 1 & \\ & & \frac{1}{2} \end{bmatrix}$$

The exact eigenvalues are $\lambda_1 = 2$, $\lambda_2 = 4$, $\lambda_3 = 6$.

1. Use the following load vectors to generate the Ritz basis vectors

$$\mathbf{R} = \begin{bmatrix} 1 & 0 \\ 0 & 0 \\ 0 & 1 \end{bmatrix}$$

2. Then use a different set of load vectors to generate the Ritz basis vectors

$$\mathbf{R} = \begin{bmatrix} 1 & 0 \\ 1 & 1 \\ 1 & 0 \end{bmatrix}$$

In the Ritz analysis we employ the relations in (10.79) to (10.84), and obtain, in case (1),

$$\begin{bmatrix} 2 & -1 & 0 \\ -1 & 4 & -1 \\ 0 & -1 & 2 \end{bmatrix} \boldsymbol{\Psi} = \begin{bmatrix} 1 & 0 \\ 0 & 0 \\ 0 & 1 \end{bmatrix}$$

Hence

$$\boldsymbol{\Psi} = \begin{bmatrix} \frac{7}{12} & \frac{1}{12} \\ \frac{1}{6} & \frac{1}{6} \\ \frac{1}{12} & \frac{7}{12} \end{bmatrix}$$

and

$$\tilde{\mathbf{K}} = \frac{1}{12}\begin{bmatrix} 7 & 1 \\ 1 & 7 \end{bmatrix}; \qquad \tilde{\mathbf{M}} = \frac{1}{144}\begin{bmatrix} 29 & 11 \\ 11 & 29 \end{bmatrix}$$

The solution of the eigenproblem $\tilde{\mathbf{K}}\mathbf{x} = \rho\tilde{\mathbf{M}}\mathbf{x}$ is

$$(\rho_1, \mathbf{x}_1) = \left(2.4004, \begin{bmatrix} 1.3418 \\ 1.3418 \end{bmatrix}\right); \qquad (\rho_2, \mathbf{x}_2) = \left(4.0032, \begin{bmatrix} 2.0008 \\ -2.0008 \end{bmatrix}\right)$$

Hence we have as eigenvalue approximations $\rho_1 = 2.40$, $\rho_2 = 4.00$, and evaluating

$$\bar{\mathbf{\Phi}} = \begin{bmatrix} \frac{7}{12} & \frac{1}{12} \\ \frac{1}{6} & \frac{1}{6} \\ \frac{1}{12} & \frac{7}{12} \end{bmatrix} \begin{bmatrix} 1.342 & 2.000 \\ 1.342 & -2.000 \end{bmatrix} = \begin{bmatrix} 0.895 & 1.00 \\ 0.447 & 0 \\ 0.895 & -1.00 \end{bmatrix}$$

we have $$\bar{\boldsymbol{\phi}}_1 = \begin{bmatrix} 0.895 \\ 0.447 \\ 0.895 \end{bmatrix}; \qquad \bar{\boldsymbol{\phi}}_2 = \begin{bmatrix} 1.00 \\ 0.00 \\ -1.00 \end{bmatrix}$$

Next we assume the load vectors in (2) and solve

$$\begin{bmatrix} 2 & -1 & 0 \\ -1 & 4 & -1 \\ 0 & -1 & 2 \end{bmatrix} \mathbf{\Psi} = \begin{bmatrix} 1 & 0 \\ 1 & 1 \\ 1 & 0 \end{bmatrix}$$

Hence $$\mathbf{\Psi} = \begin{bmatrix} \frac{5}{6} & \frac{1}{6} \\ \frac{2}{3} & \frac{1}{3} \\ \frac{5}{6} & \frac{1}{6} \end{bmatrix}$$

and $$\tilde{\mathbf{K}} = \begin{bmatrix} \frac{7}{3} & \frac{2}{3} \\ \frac{2}{3} & \frac{1}{3} \end{bmatrix}; \qquad \tilde{\mathbf{M}} = \frac{1}{36}\begin{bmatrix} 41 & 13 \\ 13 & 5 \end{bmatrix}$$

The solution of the eigenproblem $\tilde{\mathbf{K}}\mathbf{x} = \rho\tilde{\mathbf{M}}\mathbf{x}$ gives

$$(\rho_1, \mathbf{x}_1) = \left(2.0000, \begin{bmatrix} 0.70711 \\ 0.70711 \end{bmatrix}\right); \qquad (\rho_2, \mathbf{x}_2) = \left(6.0000, \begin{bmatrix} -2.1213 \\ 6.3640 \end{bmatrix}\right)$$

Hence we have as eigenvalue approximations $\rho_1 = 2.00$, $\rho_2 = 6.00$, and evaluating

$$\bar{\mathbf{\Phi}} = \begin{bmatrix} \frac{5}{6} & \frac{1}{6} \\ \frac{2}{3} & \frac{1}{3} \\ \frac{5}{6} & \frac{1}{6} \end{bmatrix} \begin{bmatrix} 0.70711 & -2.1213 \\ 0.70711 & 6.3640 \end{bmatrix} = \begin{bmatrix} 0.70711 & -0.70708 \\ 0.70711 & 0.70713 \\ 0.70711 & -0.70708 \end{bmatrix}$$

we have $$\bar{\boldsymbol{\phi}}_1 = \begin{bmatrix} 0.70711 \\ 0.70711 \\ 0.70711 \end{bmatrix}; \qquad \bar{\boldsymbol{\phi}}_2 = \begin{bmatrix} -0.70708 \\ 0.70713 \\ -0.70708 \end{bmatrix}$$

Comparing the results with the exact solution, it is interesting to note that in case (1), $\rho_1 > \lambda_1$ and $\rho_2 = \lambda_2$, whereas in case (2), $\rho_1 = \lambda_1$ and $\rho_2 = \lambda_3$. In both cases we did not obtain good approximations to the lowest two eigenvalues, and it is clearly demonstrated that the results depend completely on the initial Ritz basis vectors chosen.

EXAMPLE 10.16: Use the Rayleigh–Ritz analysis to calculate an approximation to λ_1 and $\boldsymbol{\phi}_1$ of the eigenproblem considered in Example 10.12.

We note that in this case **M** is positive semidefinite. Therefore, to carry out the Ritz analysis we need to choose a load vector in **R** that excites at least one mass. Assume that we use

$$\mathbf{R}^T = [0 \quad 1 \quad 0 \quad 0]$$

Then the solution of (10.79) yields (see Example 10.13)

$$\mathbf{\Psi}^T = [1 \quad 2 \quad 2 \quad 2]$$

Hence $$\tilde{\mathbf{K}} = [2]; \qquad \tilde{\mathbf{M}} = [12]$$

$$\rho_1 = \frac{1}{6}; \qquad \mathbf{x}_1 = \left[\frac{1}{2\sqrt{3}}\right]$$

and $$\bar{\boldsymbol{\phi}}_1^T = \begin{bmatrix} \frac{1}{2\sqrt{3}} & \frac{1}{\sqrt{3}} & \frac{1}{\sqrt{3}} & \frac{1}{\sqrt{3}} \end{bmatrix}$$

Hence we have as expected that $\rho_1 > \lambda_1$.

The Ritz analysis procedure presented above is a very general tool, and, as pointed out earlier, various analysis methods known under different names can actually be shown to be Ritz analyses. In Section 10.3.3 we present the component mode synthesis as a Ritz analysis. In the following we briefly want to show that the technique of static condensation as described in Section 10.3.1 is, in fact, also a Ritz analysis.

In the static condensation analysis we assumed that all mass can be lumped at q degrees of freedom. Therefore, as an approximation to the eigenproblem $\mathbf{K}\boldsymbol{\phi} = \lambda\mathbf{M}\boldsymbol{\phi}$, we obtained the following problem:

$$\begin{bmatrix} \mathbf{K}_{aa} & \mathbf{K}_{ac} \\ \mathbf{K}_{ca} & \mathbf{K}_{cc} \end{bmatrix} \begin{bmatrix} \boldsymbol{\phi}_a \\ \boldsymbol{\phi}_c \end{bmatrix} = \lambda \begin{bmatrix} \mathbf{M}_a & \mathbf{0} \\ \mathbf{0} & \mathbf{0} \end{bmatrix} \begin{bmatrix} \boldsymbol{\phi}_a \\ \boldsymbol{\phi}_c \end{bmatrix} \tag{10.42}$$

with q finite and $(n - q)$ infinite eigenvalues, which correspond to the massless degrees of freedom (see Section 10.2.4). To calculate the finite eigenvalues, we used static condensation on the massless degrees of freedom and arrived at the eigenproblem

$$\mathbf{K}_a\boldsymbol{\phi}_a = \lambda\mathbf{M}_a\boldsymbol{\phi}_a \tag{10.45}$$

where $\mathbf{K}_a$ is defined in (10.46). However, this solution is actually a Ritz analysis of the lumped mass model considered in (10.42). The Ritz basis vectors are the displacement patterns associated with the $\boldsymbol{\phi}_a$ degrees of freedom when the $\boldsymbol{\phi}_c$ degrees of freedom are released. Solving the equations

$$\begin{bmatrix} \mathbf{K}_{aa} & \mathbf{K}_{ac} \\ \mathbf{K}_{ca} & \mathbf{K}_{cc} \end{bmatrix} \begin{bmatrix} \mathbf{F}_a \\ \mathbf{F}_c \end{bmatrix} = \begin{bmatrix} \mathbf{I} \\ \mathbf{0} \end{bmatrix} \tag{10.51}$$

in which $\mathbf{F}_a = \mathbf{K}_a^{-1}$, we have that the Ritz basis vectors to be used in (10.80), (10.81), and (10.84) are

$$\boldsymbol{\Psi} = \begin{bmatrix} \mathbf{I} \\ \mathbf{F}_c\mathbf{K}_a \end{bmatrix} \tag{10.85}$$

To verify that a Ritz analysis with the base vectors in (10.85) yields in fact (10.45), we evaluate (10.80) and (10.81). Substituting for $\boldsymbol{\Psi}$ and $\mathbf{K}$ in (10.80), we obtain

$$\tilde{\mathbf{K}} = [\mathbf{I} \quad (\mathbf{F}_c\mathbf{K}_a)^T] \begin{bmatrix} \mathbf{K}_{aa} & \mathbf{K}_{ac} \\ \mathbf{K}_{ca} & \mathbf{K}_{cc} \end{bmatrix} \begin{bmatrix} \mathbf{I} \\ \mathbf{F}_c\mathbf{K}_a \end{bmatrix} \tag{10.86}$$

which, using (10.51), reduces to

$$\tilde{\mathbf{K}} = \mathbf{K}_a \tag{10.87}$$

Similarly, substituting for $\boldsymbol{\Psi}$ and $\mathbf{M}$ in (10.81), we have

$$\tilde{\mathbf{M}} = [\mathbf{I} \quad (\mathbf{F}_c\mathbf{K}_a)^T] \begin{bmatrix} \mathbf{M}_a & \mathbf{0} \\ \mathbf{0} & \mathbf{0} \end{bmatrix} \begin{bmatrix} \mathbf{I} \\ \mathbf{F}_c\mathbf{K}_a \end{bmatrix} \tag{10.88}$$

or

$$\tilde{\mathbf{M}} = \mathbf{M}_a \tag{10.89}$$

Hence, in the static condensation we actually perform a Ritz analysis of the lumped mass model. It should be noted that in the analysis we calculate the q

finite eigenvalues exactly (i.e., $\rho_i = \lambda_i$ for $i = 1, \ldots, q$), because the Ritz basis vectors span the q-dimensional subspace corresponding to the finite eigenvalues. In practice, the evaluation of the vectors $\mathbf{\Psi}$ in (10.85) is not necessary (and would be costly), and instead the Ritz analysis is better carried out using

$$\mathbf{\Psi} = \begin{bmatrix} \mathbf{F}_a \\ \mathbf{F}_c \end{bmatrix} \tag{10.90}$$

Since the vectors in (10.90) span the same subspace as the vectors in (10.85), the same eigenvalues and eigenvectors are calculated employing either set of base vectors. Specifically, using (10.90) we obtain in the Ritz analysis the reduced eigenproblem

$$\mathbf{F}_a\mathbf{x} = \lambda \mathbf{F}_a\mathbf{M}_a\mathbf{F}_a\mathbf{x} \tag{10.91}$$

To show that this eigenproblem is indeed equivalent to the problem in (10.45), we premultiply both sides in (10.91) by $\mathbf{K}_a$ and use the transformation $\mathbf{x} = \mathbf{K}_a\tilde{\mathbf{x}}$, giving $\mathbf{K}_a\tilde{\mathbf{x}} = \lambda\mathbf{M}_a\tilde{\mathbf{x}}$, i.e., the problem in (10.45).

EXAMPLE 10.17: Use the Ritz analysis procedure to perform static condensation of the massless degrees of freedom in the problem $\mathbf{K}\boldsymbol{\phi} = \lambda\mathbf{M}\boldsymbol{\phi}$ considered in Example 10.12.

We first need to evaluate the Ritz basis vectors given in (10.90). This has been done in Example 10.13, where we found that

$$\mathbf{F}_a = \begin{bmatrix} 2 & 2 \\ 2 & 4 \end{bmatrix}; \qquad \mathbf{F}_c = \begin{bmatrix} 1 & 1 \\ 2 & 3 \end{bmatrix}$$

The Ritz reduction given in (10.91) thus yields the eigenproblem

$$\begin{bmatrix} 2 & 2 \\ 2 & 4 \end{bmatrix}\mathbf{x} = \lambda \begin{bmatrix} 12 & 16 \\ 16 & 24 \end{bmatrix}\mathbf{x}$$

Finally, we should note that the use of the Ritz basis vectors in (10.85) (or in (10.90)) is also known as the *Guyan reduction*.[12,28] In the Guyan scheme the Ritz vectors are used to operate on a lumped mass matrix with zero elements on the diagonal as in (10.88) or on general full lumped or consistent mass matrices. In this reduction the $\boldsymbol{\phi}_a$ degrees of freedom are frequently referred to as dynamic degrees of freedom.

10.3.3 Component Mode Synthesis

As for the static condensation procedure, the component mode synthesis is, in fact, a Ritz analysis, and the method might have been presented in the previous section as a specific application. However, as was repeatedly pointed out, the most important aspect in Ritz analysis is the selection of appropriate Ritz basis vectors, because the results can only be as good as the Ritz basis vectors allow them to be. The specific scheme used in the component mode synthesis is of particular interest, which is the reason we want to devote a separate section to the discussion of the method.

The component mode synthesis has been developed to a large extent as a natural consequence of the analysis procedure followed in practice when large and complex structures are analyzed. The general practical procedure is that different groups perform the analyses of different components of the structure under consideration. For example, in a reactor analysis, one group may analyze a main pipe and another group a piping system attached to it. In a first preliminary analysis, both groups work separately and model the effects of the other components on the specific component that they consider in an approximate manner. For example, in the analysis of the two piping systems referred to above, the group analyzing the side branch may assume full fixity at the point of intersection with the main pipe, and the group analyzing the main pipe may introduce a concentrated spring and mass to allow for the side branch. The advantage of considering separately the components of the structure is primarily one of time scheduling; i.e., the separate groups can work on the analyses and designs of the components at the same time. It is primarily for this reason that the component mode synthesis is very appealing in the analysis and design of large structural systems.

Assume that the preliminary analyses of the components have been carried out and that the complete structure shall now be analyzed. It is at this stage that the component mode synthesis is a natural procedure to use. Namely, with the mode shape characteristics of each component known, it appears natural to use this information in estimating the frequencies and mode shapes of the complete structure. The specific procedure may vary according to the specific information available, but, in essence, the mode shapes of the components are used in a Rayleigh–Ritz analysis to calculate approximate mode shapes and frequencies of the complete structure.

Consider for illustration that each component structure was obtained by fixing all its boundary degrees of freedom, and denote the stiffness matrices of the component structures by $\mathbf{K}_{\mathrm{I}}, \mathbf{K}_{\mathrm{II}}, \ldots, \mathbf{K}_{\mathrm{M}}$ (see Example 10.18). Assume that only component structures $L-1$ and L connect, $L = 2, \ldots, M$; then we can write for the stiffness matrix of the complete structure,

$$\mathbf{K} = \begin{bmatrix} \mathbf{K}_{\mathrm{I}} & \cdot & & & \\ \cdot & \cdot\mathbf{K}_{\mathrm{II}}\cdot & \cdot & & \\ & \cdot & \cdot & & \\ & & & \ddots & \\ & & & & \mathbf{K}_{\mathrm{M}} \end{bmatrix} \tag{10.92}$$

Using an analogous notation for the mass matrices, we also have

$$\mathbf{M} = \begin{bmatrix} \mathbf{M}_{\mathrm{I}} & \cdot & & & \\ \cdot & \cdot\;\mathbf{M}_{\mathrm{II}}\;\cdot & \cdot & & \\ & \cdot & \cdot & & \\ & & & \ddots & \\ & & & & \mathbf{M}_{\mathrm{M}} \end{bmatrix} \tag{10.93}$$

Assume that the lowest eigenvalues and corresponding eigenvectors of each component structure have been calculated; i.e., we have for each component structure,

$$
\left.\begin{aligned}
\mathbf{K}_{\mathrm{I}}\mathbf{\Phi}_{\mathrm{I}} &= \mathbf{M}_{\mathrm{I}}\mathbf{\Phi}_{\mathrm{I}}\mathbf{\Lambda}_{\mathrm{I}} \\
\mathbf{K}_{\mathrm{II}}\mathbf{\Phi}_{\mathrm{II}} &= \mathbf{M}_{\mathrm{II}}\mathbf{\Phi}_{\mathrm{II}}\mathbf{\Lambda}_{\mathrm{II}} \\
&\ \ \vdots \\
\mathbf{K}_{\mathrm{M}}\mathbf{\Phi}_{\mathrm{M}} &= \mathbf{M}_{\mathrm{M}}\mathbf{\Phi}_{\mathrm{M}}\mathbf{\Lambda}_{\mathrm{M}}
\end{aligned}\right\} \tag{10.94}
$$

where $\mathbf{\Phi}_L$ and $\mathbf{\Lambda}_L$ are the matrices of calculated eigenvectors and eigenvalues of the Lth component structure.

In a component mode synthesis, approximate mode shapes and frequencies can be obtained by performing a Rayleigh–Ritz analysis with the following assumed loads on the right-hand side of (10.79),

$$
\mathbf{R} = \begin{bmatrix}
\mathbf{\Phi}_{\mathrm{I}} & \mathbf{0} & \mathbf{0} & \cdots \\
\mathbf{0} & \mathbf{I}_{\mathrm{I,II}} & \mathbf{0} & \\
\mathbf{\Phi}_{\mathrm{II}} & \mathbf{0} & \mathbf{0} & \cdots \\
\mathbf{0} & \mathbf{0} & \mathbf{I}_{\mathrm{II,III}} & \\
\vdots & \vdots & \vdots & \cdots \\
\mathbf{\Phi}_{\mathrm{M}} & \mathbf{0} & &
\end{bmatrix} \tag{10.95}
$$

where $\mathbf{I}_{L-1,L}$ is a unit matrix of order equal to the connection degrees of freedom between component structures $L - 1$ and L. The unit matrices correspond to loads that are applied to the connection degrees of freedom of the component structures. Since in the derivation of the mode shape matrices used in (10.95), the component structures were fixed at their boundaries, the unit loads have the effect of releasing these connection degrees of freedom. If, on the other hand, the connection degrees of freedom have been included in the analysis of the component structures, we may dispense with the unit matrices in **R**.

An important consideration is the accuracy that can be expected in the above component mode synthesis. Since a Ritz analysis is performed, all accuracy considerations discussed in Section 10.3.2 are directly applicable; i.e., the analysis yields upper bounds to the exact eigenvalues of the problem $\mathbf{K}\boldsymbol{\phi} = \lambda\mathbf{M}\boldsymbol{\phi}$. However, the actual accuracy achieved in the solution is not known, although it can be evaluated, for example, as described in Section 10.4. The fact that the solution accuracy is highly dependent on the vectors used in **R** (i.e., the Ritz basis vectors) is, as in all Ritz analyses, the main defect of the method. However, in practice, reasonable accuracy can often be obtained because the eigenvectors corresponding to the smallest eigenvalues of the component structures are used in **R**. We demonstrate the analysis procedure in the following example.

EXAMPLE 10.18: Consider the eigenproblem $\mathbf{K}\boldsymbol{\phi} = \lambda\mathbf{M}\boldsymbol{\phi}$, where

$$
\mathbf{K} = \begin{bmatrix}
2 & -1 & & & \\
-1 & 2 & -1 & & \\
 & -1 & 2 & -1 & \\
 & & -1 & 2 & -1 \\
 & & & -1 & 1
\end{bmatrix}; \qquad
\mathbf{M} = \begin{bmatrix}
1 & & & & \\
 & 1 & & & \\
 & & 1 & & \\
 & & & 1 & \\
 & & & & \tfrac{1}{2}
\end{bmatrix}
$$

Use the substructure eigenproblems indicated by the dashed lines in **K** and **M** to establish the load matrix given in (10.95) for a component mode synthesis analysis. Then calculate eigenvalue and eigenvector approximations.

Here we have for substructure I,

$$\mathbf{K}_{\mathrm{I}} = \begin{bmatrix} 2 & -1 \\ -1 & 2 \end{bmatrix}; \qquad \mathbf{M}_{\mathrm{I}} = \begin{bmatrix} 1 & 0 \\ 0 & 1 \end{bmatrix}$$

with the eigensolution

$$\lambda_1 = 1, \quad \lambda_2 = 3; \qquad \boldsymbol{\phi}_1 = \begin{bmatrix} \frac{\sqrt{2}}{2} \\ \frac{\sqrt{2}}{2} \end{bmatrix}, \qquad \boldsymbol{\phi}_2 = \begin{bmatrix} -\frac{\sqrt{2}}{2} \\ \frac{\sqrt{2}}{2} \end{bmatrix}$$

and for substructure II,

$$\mathbf{K}_{\mathrm{II}} = \begin{bmatrix} 2 & -1 \\ -1 & 1 \end{bmatrix}; \qquad \mathbf{M}_{\mathrm{II}} = \begin{bmatrix} 1 & 0 \\ 0 & \frac{1}{2} \end{bmatrix}$$

with the eigensolution

$$\lambda_1 = 2 - \sqrt{2}, \quad \lambda_2 = 2 + \sqrt{2}; \qquad \boldsymbol{\phi}_1 = \begin{bmatrix} \frac{\sqrt{2}}{2} \\ 1 \end{bmatrix}, \qquad \boldsymbol{\phi}_2 = \begin{bmatrix} -\frac{\sqrt{2}}{2} \\ 1 \end{bmatrix}$$

Thus we have for the matrix **R** in (10.95),

$$\mathbf{R} = \begin{bmatrix} \frac{\sqrt{2}}{2} & -\frac{\sqrt{2}}{2} & 0 \\ \frac{\sqrt{2}}{2} & \frac{\sqrt{2}}{2} & 0 \\ 0 & 0 & 1 \\ \frac{\sqrt{2}}{2} & -\frac{\sqrt{2}}{2} & 0 \\ 1 & 1 & 0 \end{bmatrix}$$

Now performing the Ritz analysis as given in (10.79) to (10.84), we obtain

$$\tilde{\mathbf{K}} = \begin{bmatrix} 22.40 & 5.328 & 7.243 \\ 5.328 & 2.257 & 1.586 \\ 7.243 & 1.586 & 3 \end{bmatrix}$$

$$\tilde{\mathbf{M}} = \begin{bmatrix} 222.4 & 50.69 & 77.69 \\ 50.69 & 11.94 & 17.59 \\ 77.69 & 17.59 & 27.5 \end{bmatrix}$$

and hence

$$\boldsymbol{\rho} = \begin{bmatrix} 0.098 & & \\ & 2.83 & \\ & & 1.82 \end{bmatrix}$$

$$\bar{\boldsymbol{\Phi}} = \begin{bmatrix} 0.207 & -0.773 & 0.00690 \\ 0.181 & 0.0984 & -0.0655 \\ 0.509 & 1.47 & 0.443 \\ 0.594 & -0.385 & -0.166 \\ 0.655 & 0.574 & -0.978 \end{bmatrix}$$

The exact eigenvalues are

$$\lambda_1 = 0.098; \qquad \lambda_2 = 0.824; \qquad \lambda_3 = 2.00; \qquad \lambda_4 = 3.18; \qquad \lambda_5 = 3.90$$

and hence we note that we obtained in ρ_1 a good approximation to λ_1, but ρ_2 and ρ_3 do not represent approximations to eigenvalues.

10.3.4 Lanczos Method

The Lanczos method was originally proposed for the tridiagonalization of matrices.[29] Once the coefficient matrices of the generalized eigenproblem have been tridiagonalized, the eigenvalues and vectors can be calculated effectively using the techniques described in Chapter 11.

Let $\mathbf{x}$ be an arbitrary starting vector and let us normalize this vector with respect to the matrix $\mathbf{M}$ to obtain $\mathbf{x}_1$;

$$\mathbf{x}_1 = \frac{\mathbf{x}}{\gamma}; \qquad \gamma = (\mathbf{x}^T\mathbf{M}\mathbf{x})^{1/2}$$

Also, let $\beta_1 = 0$. In the Lanczos algorithm we then calculate the vectors $\mathbf{x}_2, \ldots, \mathbf{x}_q$ using the following equations for $i = 2, \ldots, q$:

$$\mathbf{K}\bar{\mathbf{x}}_i = \mathbf{M}\mathbf{x}_{i-1} \tag{10.96}$$

$$\alpha_{i-1} = \bar{\mathbf{x}}_i^T\mathbf{M}\mathbf{x}_{i-1} \tag{10.97}$$

$$\tilde{\mathbf{x}}_i = \bar{\mathbf{x}}_i - \alpha_{i-1}\mathbf{x}_{i-1} - \beta_{i-1}\mathbf{x}_{i-2} \tag{10.98}$$

$$\beta_i = (\tilde{\mathbf{x}}_i^T\mathbf{M}\tilde{\mathbf{x}}_i)^{1/2} \tag{10.99}$$

$$\mathbf{x}_i = \tilde{\mathbf{x}}_i/\beta_i \tag{10.100}$$

Theoretically, the sequence of vectors $\mathbf{x}_i$, $i = 1, \ldots, q$, generated using the above relations are $\mathbf{M}$-orthonormal and the matrix $\mathbf{X} = [\mathbf{x}_1, \ldots, \mathbf{x}_q]$ satisfies the following relation:

$$\mathbf{X}^T(\mathbf{M}\mathbf{K}^{-1}\mathbf{M})\mathbf{X} = \mathbf{T}_q \tag{10.101}$$

where $\mathbf{T}_q$ is tridiagonal and of order q

$$\mathbf{T}_q = \begin{bmatrix} \alpha_1 & \beta_2 & & & & \\ \beta_2 & \alpha_2 & \beta_3 & & & \\ & & \cdot & & & \\ & & & \cdot & & \\ & & & & \cdot & \\ & & & & \alpha_{q-1} & \beta_q \\ & & & & \beta_q & \alpha_q \end{bmatrix} \tag{10.102}$$

Using (10.101) we can now relate the eigenvalues and vectors of $\mathbf{T}_q$ when $q = n$ to those of the problem $\mathbf{K}\boldsymbol{\phi} = \lambda\mathbf{M}\boldsymbol{\phi}$, which may also be written as

$$\frac{1}{\lambda}\mathbf{M}\boldsymbol{\phi} = \mathbf{M}\mathbf{K}^{-1}\mathbf{M}\boldsymbol{\phi} \tag{10.103}$$

Namely, using the transformation

$$\boldsymbol{\phi} = \mathbf{X}\tilde{\boldsymbol{\phi}} \tag{10.104}$$

and (10.101) we obtain from (10.103)

$$\mathbf{T}_n\tilde{\boldsymbol{\phi}} = \frac{1}{\lambda}\tilde{\boldsymbol{\phi}} \tag{10.105}$$

Hence, the eigenvalues of $\mathbf{T}_n$ are the reciprocals of the eigenvalues of $\mathbf{K}\boldsymbol{\phi} = \lambda\mathbf{M}\boldsymbol{\phi}$ and the eigenvectors of the two problems are related as given in (10.104).

In practice, we find that in the solution of large systems the vectors $\mathbf{x}_i$ may not be $\mathbf{M}$-orthogonal due to round-off in the calculations. For this and other reasons, a considerable amount of research has been spent on the basic Lanczos method[30-32].

The practical use of the Lanczos algorithm lies in that if q is smaller than n, the eigenvalues of $\mathbf{T}_q$ can give good approximations to the smallest eigenvalues of the problem $\mathbf{K}\boldsymbol{\phi} = \lambda\mathbf{M}\boldsymbol{\phi}$. However, the actual accuracy with which the smallest eigenvalues of $\mathbf{K}\boldsymbol{\phi} = \lambda\mathbf{M}\boldsymbol{\phi}$ are approximated is not known and neither do we know whether a required eigenvalue has been missed completely. In order to assure an accurate eigensolution, the Lanczos procedure must be combined with an iterative scheme, and eigenvalue error bounds with Sturm sequence checks must be calculated (see Section 12.3.3). To demonstrate the Lanczos procedure we consider the following example.

EXAMPLE 10.19: Use the Lanczos algorithm to calculate the two smallest eigenvalues of the eigenproblem $\mathbf{K}\boldsymbol{\phi} = \lambda\mathbf{M}\boldsymbol{\phi}$ considered already in Example 10.18.

Using the algorithm in (10.96) to (10.100) with $\mathbf{x}$ a full unit vector we obtain

$$\gamma_1 = 2.121; \qquad \mathbf{x}_1 = 0.4714 \begin{bmatrix} 1 \\ 1 \\ 1 \\ 1 \\ 1 \end{bmatrix}$$

$$\text{for } i = 2: \quad \bar{\mathbf{x}}_2 = \begin{bmatrix} 2.121 \\ 3.771 \\ 4.950 \\ 5.657 \\ 5.893 \end{bmatrix}; \qquad \alpha_1 = 9.167; \qquad \tilde{\mathbf{x}}_2 = \begin{bmatrix} -2.200 \\ -0.5500 \\ 0.6285 \\ 1.336 \\ 1.571 \end{bmatrix};$$

$$\beta_2 = 2.925; \qquad \mathbf{x}_2 = \begin{bmatrix} -.7521 \\ -.1880 \\ .2149 \\ .4566 \\ .5372 \end{bmatrix}$$

$$\text{for } i = 3: \quad \bar{\mathbf{x}}_3 = \begin{bmatrix} 0.000 \\ 0.7521 \\ 1.692 \\ 2.417 \\ 2.686 \end{bmatrix}; \qquad \alpha_2 = 2.048$$

Hence we have

$$\mathbf{T}_2 = \begin{bmatrix} 9.167 & 2.925 \\ 2.925 & 2.048 \end{bmatrix}$$

Approximations to the eigenvalues of $\mathbf{K}\boldsymbol{\phi} = \lambda\mathbf{M}\boldsymbol{\phi}$ are obtained by solving

$$\mathbf{T}_2\bar{\boldsymbol{\phi}} = \frac{1}{\rho}\bar{\boldsymbol{\phi}}$$

which gives $\qquad \rho_1 = 0.0979, \quad \rho_2 = 1.0001$

Comparing these values with the exact eigenvalues of $\mathbf{K}\boldsymbol{\phi} = \lambda\mathbf{M}\boldsymbol{\phi}$ (see Example 10.18) we find that ρ_1 is a good approximation to λ_1 but ρ_2 is not very close to λ_2.

The smallest eigenvalue is well predicted in this solution because the starting vector $\mathbf{x}$ is relatively close to $\boldsymbol{\phi}_1$.

10.4 SOLUTION ERRORS

An important part of an eigenvalue and vector solution is to estimate the accuracy with which the required eigensystem has been calculated. Since an eigensystem solution is necessarily iterative, the solution is terminated once convergence within the prescribed tolerances has been obtained. Although a good approximation to the required eigensystem should have been obtained at convergence, no definite conclusion can in most cases be drawn about the actual accuracy of the calculated values. Furthermore, in case one of the approximate solution techniques outlined in Section 10.3 is used, an estimate on the actual solution accuracy obtained is even more important.

In order to identify the accuracy that has been obtained in an eigensolution, we recall that the equation to be solved is

$$\mathbf{K}\boldsymbol{\phi} = \lambda\mathbf{M}\boldsymbol{\phi} \tag{10.106}$$

Assume that using any one solution procedure we obtained an approximation $\bar{\lambda}$ and $\bar{\boldsymbol{\phi}}$ to an eigenpair. It is important to note that at this stage it does not matter how the values $\bar{\lambda}$ and $\bar{\boldsymbol{\phi}}$ have been evaluated and indeed they may have been "guessed." Then, similar to the error analysis employed in the solution of the static equilibrium equations (see Section 8.5), we can evaluate a residual vector, $\mathbf{r}$, which gives important information about the accuracy with which $\bar{\lambda}$ and $\bar{\boldsymbol{\phi}}$ approximate the eigenpair, where

$$\mathbf{r} = \mathbf{K}\bar{\boldsymbol{\phi}} - \bar{\lambda}\mathbf{M}\bar{\boldsymbol{\phi}} \tag{10.107}$$

Let us assume first that $\mathbf{M} = \mathbf{I}$. In that case we can write, using the relations in (10.12) and (10.13),

$$\mathbf{r} = \boldsymbol{\Phi}(\boldsymbol{\Lambda} - \bar{\lambda}\mathbf{I})\boldsymbol{\Phi}^T\bar{\boldsymbol{\phi}} \tag{10.108}$$

or because $\bar{\lambda}$ is not equal but only close to an eigenvalue, we have

$$\bar{\boldsymbol{\phi}} = \boldsymbol{\Phi}(\boldsymbol{\Lambda} - \bar{\lambda}\mathbf{I})^{-1}\boldsymbol{\Phi}^T\mathbf{r} \tag{10.109}$$

Hence, because $\|\bar{\boldsymbol{\phi}}\|_2 = 1$, taking norms we obtain

$$1 \le \|(\boldsymbol{\Lambda} - \bar{\lambda}\mathbf{I})^{-1}\|_2 \, \|\mathbf{r}\|_2 \tag{10.110}$$

But since
$$\|(\boldsymbol{\Lambda} - \bar{\lambda}\mathbf{I})^{-1}\|_2 = \max_i \frac{1}{|\lambda_i - \bar{\lambda}|}$$

we have
$$\min_i |\lambda_i - \bar{\lambda}| \le \|\mathbf{r}\|_2 \tag{10.111}$$

Therefore, a conclusive statement can be made about the accuracy with which $\bar{\lambda}$ approximates an eigenvalue λ_i by evaluating $\|\mathbf{r}\|_2$ as expressed in (10.111). This is quite different from the information that could be obtained from the evaluation of the residual vector $\mathbf{r}$ *in the solution of the static equilibrium equations.*

Although the relation in (10.111) establishes that $\bar{\lambda}$ is close to an eigenvalue provided that $\|\mathbf{r}\|_2$ is small, it should be recognized that the relation does not tell which eigenvalue is approximated. In fact, to identify which specific eigenvalue has been approximated, it is necessary to use the Sturm sequence property (see Section 10.2.2 and the following example).

EXAMPLE 10.20: Consider the eigenproblem $\mathbf{K}\boldsymbol{\phi} = \lambda\boldsymbol{\phi}$, where

$$\mathbf{K} = \begin{bmatrix} 3 & -1 & 0 \\ -1 & 2 & -1 \\ 0 & -1 & 3 \end{bmatrix}$$

The eigensolution is $\lambda_1 = 1$, $\lambda_2 = 3$, $\lambda_3 = 4$, and

$$\boldsymbol{\phi}_1 = \frac{1}{\sqrt{6}}\begin{bmatrix} 1 \\ 2 \\ 1 \end{bmatrix}; \qquad \boldsymbol{\phi}_2 = \frac{1}{\sqrt{2}}\begin{bmatrix} 1 \\ 0 \\ -1 \end{bmatrix}; \qquad \boldsymbol{\phi}_3 = \frac{1}{\sqrt{3}}\begin{bmatrix} 1 \\ -1 \\ 1 \end{bmatrix}$$

Assume that we calculated

$$\bar{\lambda} = 3.1 \quad \text{and} \quad \bar{\boldsymbol{\phi}} = \begin{bmatrix} 0.7 \\ 0.1414 \\ -0.7 \end{bmatrix}$$

as approximations to λ_2 and $\boldsymbol{\phi}_2$. Apply the error bound relation in (10.111).

We have in this case

$$\mathbf{r} = \begin{bmatrix} 3 & -1 & 0 \\ -1 & 2 & -1 \\ 0 & -1 & 3 \end{bmatrix}\begin{bmatrix} 0.7 \\ 0.1414 \\ -0.7 \end{bmatrix} - 3.1\begin{bmatrix} 1 & & \\ & 1 & \\ & & 1 \end{bmatrix}\begin{bmatrix} 0.7 \\ 0.1414 \\ -0.7 \end{bmatrix}$$

Hence
$$\mathbf{r} = \begin{bmatrix} -0.2114 \\ -0.1555 \\ -0.2114 \end{bmatrix}; \qquad \|\mathbf{r}\|_2 = 0.3370$$

The relation in (10.111) now gives

$$|\lambda_2 - \bar{\lambda}| \leq 0.3370$$

which is indeed true because $\bar{\lambda} - \lambda_2 = 0.1$.

Assume now that we only have calculated $\bar{\lambda}$ and $\bar{\boldsymbol{\phi}}$ and do not know which eigenvalue and eigenvector they approximate. In this case we can use the relation in (10.111) to establish bounds on the unknown exact eigenvalue in order to apply Sturm sequence checks (see Section 10.2.2).

For the example considered here we have

$$2.7630 \leq \lambda_i \leq 3.4370$$

Let us use as a lower bound 2.7 and as an upper bound 3.5. The $\mathbf{LDL}^T$ triangular factorization of $\mathbf{K} - \mu\mathbf{I}$ gives, at $\mu = 2.7$,

$$\begin{bmatrix} 0.3 & -1 & 0 \\ -1 & -0.7 & -1 \\ 0 & -1 & 0.3 \end{bmatrix}$$

$$= \begin{bmatrix} 1 & & \\ -3.333 & 1 & \\ 0 & 0.2479 & 1 \end{bmatrix}\begin{bmatrix} 0.3 & & \\ & -4.033 & \\ & & 0.3248 \end{bmatrix}\begin{bmatrix} 1 & -3.333 & 0 \\ & 1 & 0.2479 \\ & & 1 \end{bmatrix} \qquad \text{(a)}$$

and at $\mu = 3.5$,

$$\begin{bmatrix} -0.5 & -1 & 0 \\ -1 & -1.5 & -1 \\ 0 & -1 & -0.5 \end{bmatrix} = \begin{bmatrix} 1 & & \\ 2 & 1 & \\ 0 & -2 & 1 \end{bmatrix}\begin{bmatrix} -0.5 & & \\ & 0.5 & \\ & & -2.5 \end{bmatrix}\begin{bmatrix} 1 & 2 & 0 \\ & 1 & -2 \\ & & 1 \end{bmatrix} \qquad \text{(b)}$$

But there is one negative element in **D** of (a) and there are two negative elements in **D** of (b); hence we can conclude that $2.7 < \lambda_2 < 3.5$. Furthermore, it follows that $\bar{\lambda}$ and $\bar{\boldsymbol{\phi}}$ are approximations to λ_2 and $\boldsymbol{\phi}_2$.

Considering now the accuracy with which $\bar{\boldsymbol{\phi}}$ approximates an eigenvector, an analysis equivalent to the one above not only requires the evaluation of $\|\mathbf{r}\|_2$, but also the spacing between the individual eigenvalues is needed. In actual analysis this spacing is only known approximately, because the eigenvalues have only been evaluated to a specific accuracy.

Assume that $\bar{\lambda}$ and $\bar{\boldsymbol{\phi}}$ have been calculated, where $\|\bar{\boldsymbol{\phi}}\|_2 = 1$, and that $\bar{\lambda}$ approximates the eigenvalues λ_i, $i = p, \ldots, q$. For the error analysis we also assume that the eigenvalues λ_i for all i but $i \neq p, \ldots, q$ are known (although we would need to use here the calculated eigenvalues). The final result of the accuracy analysis is that if $|\lambda_i - \bar{\lambda}| \leq \|\mathbf{r}\|_2$ for $i = p, \ldots, q$ and $|\lambda_i - \bar{\lambda}| \geq s$ for all $i, i \neq p, \ldots, q$, then there is a vector $\tilde{\boldsymbol{\phi}} = \alpha_p \boldsymbol{\phi}_p + \ldots + \alpha_q \boldsymbol{\phi}_q$, for which $\|\bar{\boldsymbol{\phi}} - \tilde{\boldsymbol{\phi}}\|_2 \leq \|\mathbf{r}\|_2 / s$ (see Example 10.21). Therefore, if $\bar{\lambda}$ is an approximation to a single eigenvalue λ_i, the corresponding vector $\bar{\boldsymbol{\phi}}$ is an approximation to $\boldsymbol{\phi}_i$, where

$$\|\bar{\boldsymbol{\phi}} - \alpha_i \boldsymbol{\phi}_i\|_2 \leq \frac{\|\mathbf{r}\|_2}{s}; \qquad s = \min_{\substack{\text{all } j \\ j \neq i}} |\lambda_j - \bar{\lambda}| \tag{10.112}$$

However, if $\bar{\lambda}$ is close to a number of eigenvalues $\lambda_p, \ldots, \lambda_q$, then the analysis only shows that the corresponding vector $\bar{\boldsymbol{\phi}}$ is close to a vector that lies in the subspace corresponding to $\boldsymbol{\phi}_p, \ldots, \boldsymbol{\phi}_q$. In practical analysis (i.e., mode superposition in dynamic response calculations) this is most likely all that is required, because the close eigenvalues may almost be dealt with as equal eigenvalues, in which case the calculated eigenvectors would also not be unique but lie in the subspace corresponding to the equal eigenvalues. In the following we first give the proof for the accuracy with which $\bar{\boldsymbol{\phi}}$ approximates an eigenvector and then demonstrate the results by means of examples.

EXAMPLE 10.21: Assume that we have calculated $\bar{\lambda}$, $\bar{\boldsymbol{\phi}}$, with $\|\bar{\boldsymbol{\phi}}\|_2 = 1$, as eigenvalue and eigenvector approximations and that $\mathbf{K}\bar{\boldsymbol{\phi}} - \bar{\lambda}\bar{\boldsymbol{\phi}} = \mathbf{r}$. Consider the case in which $|\lambda_i - \bar{\lambda}| \leq \|\mathbf{r}\|_2$ for $i = 1, \ldots, q$ and $|\lambda_i - \bar{\lambda}| \geq s$ for $i = q + 1, \ldots, n$. Show that $\|\bar{\boldsymbol{\phi}} - \tilde{\boldsymbol{\phi}}\|_2 \leq \|\mathbf{r}\|_2 / s$, where $\tilde{\boldsymbol{\phi}}$ is a vector in the subspace that corresponds to $\boldsymbol{\phi}_1, \ldots, \boldsymbol{\phi}_q$.

The calculated eigenvector approximation $\bar{\boldsymbol{\phi}}$ can be written as

$$\bar{\boldsymbol{\phi}} = \sum_{i=1}^{n} \alpha_i \boldsymbol{\phi}_i$$

Using $\tilde{\boldsymbol{\phi}} = \sum_{i=1}^{q} \alpha_i \boldsymbol{\phi}_i$ we have

$$\|\bar{\boldsymbol{\phi}} - \tilde{\boldsymbol{\phi}}\|_2 = \left\| \sum_{i=q+1}^{n} \alpha_i \boldsymbol{\phi}_i \right\|_2$$

or, because $\boldsymbol{\phi}_i^T \boldsymbol{\phi}_j = \delta_{ij}$,

$$\|\bar{\boldsymbol{\phi}} - \tilde{\boldsymbol{\phi}}\|_2 = \left\{ \sum_{i=q+1}^{n} \alpha_i^2 \right\}^{1/2} \tag{a}$$

But

$$\|\mathbf{r}\|_2 = \|\mathbf{K}\bar{\boldsymbol{\phi}} - \bar{\lambda}\bar{\boldsymbol{\phi}}\|_2$$

$$= \left\| \sum_{i=1}^{n} \alpha_i (\lambda_i - \bar{\lambda}) \boldsymbol{\phi}_i \right\|_2$$

or $$\|\mathbf{r}\|_2 = \left\{\sum_{i=1}^{n} \alpha_i^2(\lambda_i - \bar{\lambda})^2\right\}^{1/2}$$

which gives $$\|\mathbf{r}\|_2 \geq s\left\{\sum_{i=q+1}^{n} \alpha_i^2\right\}^{1/2} \tag{b}$$

Hence combining (a) and (b), we obtain

$$\|\bar{\boldsymbol{\phi}} - \tilde{\boldsymbol{\phi}}\|_2 \leq \frac{\|\mathbf{r}\|_2}{s}$$

EXAMPLE 10.22: Consider the eigenproblem of Example 10.20. Assume that λ_1 and λ_3 are known (i.e., $\lambda_1 = 1$, $\lambda_3 = 4$) and that $\bar{\lambda}$ and $\bar{\boldsymbol{\phi}}$ given in Example 10.20 have been evaluated. (In actual analysis we would only have approximations to λ_1 and λ_3, and all error bound calculations would be approximate.) Estimate the accuracy with which $\bar{\boldsymbol{\phi}}$ approximates $\boldsymbol{\phi}_2$.

For the estimate we use the relation in (10.112). In this case we have

$$\|\bar{\boldsymbol{\phi}} - \alpha_2\boldsymbol{\phi}_2\|_2 \leq \frac{0.3370}{s}$$

with $$s = \min_{i=1,3} |\lambda_i - \bar{\lambda}|$$

Hence, since $\bar{\lambda} = 3.1$, we have $s = 0.9$ and

$$\|\bar{\boldsymbol{\phi}} - \alpha_2\boldsymbol{\phi}_2\|_2 \leq 0.3744$$

Evaluating $\|\bar{\boldsymbol{\phi}} - \boldsymbol{\phi}_2\|_2$ exactly, we have

$$\|\bar{\boldsymbol{\phi}} - \boldsymbol{\phi}_2\|_2 = \left\{\left(0.7 - \frac{1}{\sqrt{2}}\right)^2 + (0.1414 - 0)^2 + \left(-0.7 + \frac{1}{\sqrt{2}}\right)^2\right\}^{1/2} = 0.1418$$

EXAMPLE 10.23: Consider the eigenproblem $\mathbf{K}\boldsymbol{\phi} = \lambda\boldsymbol{\phi}$, where

$$\mathbf{K} = \begin{bmatrix} 100 & -1 \\ -1 & 100 \end{bmatrix}$$

The eigenvalues and eigenvectors of the problem are

$$\lambda_1 = 99, \quad \boldsymbol{\phi}_1 = \frac{1}{\sqrt{2}}\begin{bmatrix} 1 \\ 1 \end{bmatrix}; \qquad \lambda_2 = 101, \quad \boldsymbol{\phi}_2 = \frac{1}{\sqrt{2}}\begin{bmatrix} 1 \\ -1 \end{bmatrix}$$

Assume that we have calculated eigenvalue and eigenvector approximations $\bar{\lambda} = 100$, $\bar{\boldsymbol{\phi}} = \begin{bmatrix} 1 \\ 0 \end{bmatrix}$. Evaluate $\mathbf{r}$ and thus establish the relations given in (10.111) and (10.112).

First we calculate $\mathbf{r}$ as given in (10.107)

$$\mathbf{r} = \begin{bmatrix} 100 & -1 \\ -1 & 100 \end{bmatrix}\begin{bmatrix} 1 \\ 0 \end{bmatrix} - 100\begin{bmatrix} 1 \\ 0 \end{bmatrix} = \begin{bmatrix} 0 \\ -1 \end{bmatrix}$$

Hence $\|\mathbf{r}\|_2 = 1$ and (10.111) yields

$$\min_i |\lambda_i - \bar{\lambda}| \leq 1 \tag{a}$$

Therefore, we can conclude that an eigenvalue has been approximated with about 1% or less error. Since we know λ_1 and λ_2, we can compare $\bar{\lambda}$ with λ_1 or λ_2 and find that (a) does indeed hold.

Considering now the eigenvector approximation $\bar{\boldsymbol{\phi}}$, we note that $\bar{\boldsymbol{\phi}}$ does not approximate either $\boldsymbol{\phi}_1$ or $\boldsymbol{\phi}_2$. This is also reflected by evaluating the relation (10.112). Assuming that $\bar{\boldsymbol{\phi}}$ is an approximation to $\boldsymbol{\phi}_1$, which gives $s = 1$, we have

$$\|\bar{\boldsymbol{\phi}} - \alpha_1 \boldsymbol{\phi}_1\|_2 \leq 1$$

Similarly, assuming that $\bar{\boldsymbol{\phi}}$ is an approximation to $\boldsymbol{\phi}_2$, we obtain

$$\|\bar{\boldsymbol{\phi}} - \alpha_2 \boldsymbol{\phi}_2\|_2 \leq 1$$

and in both cases the bound obtained is very large (note that $\|\boldsymbol{\phi}_1\|_2 = 1$ and $\|\boldsymbol{\phi}_2\|_2 = 1$), indicating that $\bar{\boldsymbol{\phi}}$ does not approximate an eigenvector.

So far we dealt with the accuracy of an eigensolution obtained for the problem $\mathbf{K}\boldsymbol{\phi} = \lambda\boldsymbol{\phi}$. In order to estimate the accuracy obtained in the solution of a generalized eigenproblem $\mathbf{K}\boldsymbol{\phi} = \lambda\mathbf{M}\boldsymbol{\phi}$ we can use the results obtained by transforming the generalized eigenproblem to the standard form. Using the information given in Section 10.2.5, we can write the generalized problem $\mathbf{K}\boldsymbol{\phi} = \lambda\mathbf{M}\boldsymbol{\phi}$ in the form $\tilde{\mathbf{K}}\tilde{\boldsymbol{\phi}} = \lambda\tilde{\boldsymbol{\phi}}$, where $\tilde{\boldsymbol{\phi}} = \mathbf{S}^T\boldsymbol{\phi}$. Assume that we calculated as an approximation to λ_i and $\boldsymbol{\phi}_i$ the values $\bar{\lambda}$ and $\bar{\boldsymbol{\phi}}$. Then, in analogy to the calculations performed above, we can evaluate an error vector, $\mathbf{r}_M$, where

$$\mathbf{r}_M = \mathbf{K}\bar{\boldsymbol{\phi}} - \bar{\lambda}\mathbf{M}\bar{\boldsymbol{\phi}} \tag{10.113}$$

In order to relate the error vector in (10.113) to the error vector that corresponds to the standard eigenproblem, we use that $\mathbf{M} = \mathbf{S}\mathbf{S}^T$, and then

$$\mathbf{r} = \tilde{\mathbf{K}}\tilde{\bar{\boldsymbol{\phi}}} - \bar{\lambda}\tilde{\bar{\boldsymbol{\phi}}} \tag{10.114}$$

where $\mathbf{r} = \mathbf{S}^{-1}\mathbf{r}_M$, $\tilde{\bar{\boldsymbol{\phi}}} = \mathbf{S}^T\bar{\boldsymbol{\phi}}$, and $\tilde{\mathbf{K}} = \mathbf{S}^{-1}\mathbf{K}\mathbf{S}^{-T}$. It is the vector $\mathbf{S}^{-1}\mathbf{r}_M$ that we would need to use, therefore, to calculate an actual error bound, as given in (10.111). It should be noted that these error bound calculations would require the factorization of $\mathbf{M}$ into $\mathbf{S}\mathbf{S}^T$, where it is assumed that $\mathbf{M}$ is positive definite.*

Although exact error bound calculations can be carried out as discussed above, in practical finite element analysis it is expedient to evaluate for the calculated eigenvalue and vector approximations $\bar{\lambda}$ and $\bar{\boldsymbol{\phi}}$ the following error measure (see Tables 12.1 and 12.2):

$$\epsilon = \frac{\|\mathbf{K}\bar{\boldsymbol{\phi}} - \bar{\lambda}\mathbf{M}\bar{\boldsymbol{\phi}}\|_2}{\|\mathbf{K}\bar{\boldsymbol{\phi}}\|_2} \tag{10.115}$$

Since, physically, $\mathbf{K}\bar{\boldsymbol{\phi}}$ represents the elastic nodal point forces and $\bar{\lambda}\mathbf{M}\bar{\boldsymbol{\phi}}$ represents the inertia nodal point forces when the finite element assemblage is vibrating in the mode $\bar{\boldsymbol{\phi}}$, we evaluate in (10.115) the norm of out-of-balance nodal point forces divided by the norm of elastic nodal point forces. This quantity should be small if $\bar{\lambda}$ and $\bar{\boldsymbol{\phi}}$ are an accurate solution of an eigenpair.

If $\mathbf{M} = \mathbf{I}$, it should be noted that we can write

$$\bar{\lambda}\epsilon = \|\mathbf{r}\|_2 \tag{10.116}$$

and hence

$$\epsilon \geq \min_i \frac{|\lambda_i - \bar{\lambda}|}{\bar{\lambda}} \tag{10.117}$$

EXAMPLE 10.24: Consider the eigenproblem $\mathbf{K}\boldsymbol{\phi} = \lambda\mathbf{M}\boldsymbol{\phi}$, where

$$\mathbf{K} = \begin{bmatrix} 10 & -10 \\ -10 & 100 \end{bmatrix}; \qquad \mathbf{M} = \begin{bmatrix} 2 & 1 \\ 1 & 4 \end{bmatrix}$$

* To avoid the factorization of $\mathbf{M}$ we may instead consider the problem $\mathbf{M}\boldsymbol{\phi} = \lambda^{-1}\mathbf{K}\boldsymbol{\phi}$ if the factorization of $\mathbf{K}$ is already available, and then establish bounds on λ^{-1}.

The exact eigenvalues and eigenvectors to 12-digit precision are

$$\lambda_1 = 3.863385512876; \qquad \boldsymbol{\phi}_1 = \begin{bmatrix} 0.640776011246 \\ 0.105070337503 \end{bmatrix}$$

$$\lambda_2 = 33.279471629982; \qquad \boldsymbol{\phi}_2 = \begin{bmatrix} -0.401041986380 \\ 0.524093989558 \end{bmatrix}$$

Assume that $\bar{\boldsymbol{\phi}} = (\boldsymbol{\phi}_1 + \delta\boldsymbol{\phi}_2)c$, where c is such that $\bar{\boldsymbol{\phi}}^T\mathbf{M}\bar{\boldsymbol{\phi}} = 1$ and $\delta = 10^{-1}$, 10^{-3}, and 10^{-6}. For each value of δ evaluate $\bar{\lambda}$ as the Rayleigh quotient of $\bar{\boldsymbol{\phi}}$, and calculate the exact error bound on $|\lambda_1 - \bar{\lambda}|$ and the error value ϵ given in (10.115).

The following table summarizes the results obtained. The equations used to evaluate the error measures are given in (10.113) to (10.115). The results in the table show that for each value of δ, the relation in (10.111) is satisfied and that ϵ is also small for an accurate solution.

δ	10^{-1}	10^{-3}	10^{-6}
$\bar{\boldsymbol{\phi}}$	0.597690792656 0.156698194481	0.640374649073 0.105594378695	0.640775610204 0.105070861597
$\bar{\boldsymbol{\phi}}^T\mathbf{K}\bar{\boldsymbol{\phi}}$	4.154633890275	3.863414928932	3.863385512905
$\bar{\lambda}$	4.154633890275	3.863414928932	3.863385512905
$\mathbf{r}_M$	−1.207470493734 4.605630581124	−0.008218153965 0.049838803226	−0.000008177422 0.000049870085
$\mathbf{r}$	1.634419466242 1.411679295681	0.021106743617 0.015042545327	0.000021152364 0.000015049775
$\lvert\lambda_1 - \bar{\lambda}\rvert$	0.291248377399	0.000029416056	0.000000000029
$\lVert\mathbf{r}\rVert_2$	2.159667897036	0.025918580132	0.000025959936
ϵ	0.447113235813	0.007458208660	0.000007491764

REFERENCES

1. B. Noble, *Applied Linear Algebra*, Prentice-Hall, Inc., Englewood Cliffs, N.J., 1969.
2. J. H. Wilkinson, *The Algebraic Eigenvalue Problem*, Clarendon Press, Oxford, 1965.
3. R. Zurmühl, *Matrizen*, Springer-Verlag, Berlin, 1964.
4. H. R. Schwarz, H. Rutishauser, and E. Stiefel, *Matrizen-Numerik*, B. G. Teubner, Stuttgart, 1972.
5. C. Lanczos, *Applied Analysis*, Prentice-Hall, Inc., Englewood Cliffs, N.J. 1956.
6. L. Collatz, *The Numerical Treatment of Differential Equations*, Springer-Verlag, New York, N.Y., 1966.
7. S. H. Crandall, *Engineering Analysis*, McGraw-Hill Book Company, New York, N.Y., 1956.

8. K. J. Bathe, "Solution Methods for Large Generalized Eigenvalue Problems in Structural Engineering," Report UC SESM 71–20, Civil Engineering Department, University of California, Berkeley, 1971.
9. E. L. Wilson, "Earthquake Analysis of Reactor Structures," A.S.M.E. First National Congress on Pressure Vessels and Piping Components, San Francisco, May 1971.
10. B. M. Irons, "Eigenvalue Economisers in Vibration Problems," *Journal of the Royal Aeronautical Society*, Vol. 67, 1963, p. 526.
11. B. M. Irons, "Structural Eigenvalue Problems: Elimination of Unwanted Variables," *A.I.A.A., Journal*, Vol. 3, No. 5, July 1965.
12. R. J. Guyan, "Reduction of Stiffness and Mass Matrices," *A.I.A.A. Journal*, Vol. 3, No. 2, 1965.
13. R. Uhrig, "Reduction of the Number of Unknowns in the Displacement Method Applied to Kinetic Problems," *Journal of Sound and Vibration*, Vol. 4, No. 2, 1966, pp. 149–155.
14. R. G. Anderson, B. M. Irons, and O. C. Zienkiewicz, "Vibration and Stability of Plates Using Finite Elements," *International Journal of Solids and Structures*, Vol. 4, 1968, pp. 1031–1055.
15. J. N. Ramsden and J. R. Stoker, "Mass Condensation; a Semi-automatic Method for Reducing the Size of Vibration Problems," *International Journal for Numerical Methods in Engineering*, Vol. 1, 1969, pp. 333–349.
16. R. W. Clough, "Analysis of Structural Vibrations and Dynamic Response," in *Recent Advances in Matrix Methods of Structural Analysis and Design* (R. H. Gallagher, Y. Yamada, and J. T. Oden, eds.), University of Alabama Press, University of Alabama, Huntsville, Ala., 1971.
17. G. C. Wright and G. A. Miles, "An Economical Method for Determining the Smallest Eigenvalues of Large Linear Systems," *International Journal for Numerical Methods in Engineering*, Vol. 3, 1971, pp. 25–33.
18. M. Geradin, "Error Bounds for Eigenvalue Analysis by Elimination of Variables," *Journal of Sound and Vibration*, Vol. 19, 1971, pp. 111–132.
19. W. L. Craver, Jr., and D. M. Egle, "A Method for Selection of Significant Terms in the Assumed Solution in a Rayleigh–Ritz Analysis," *Journal of Sound and Vibration*, Vol. 22, 1972, pp. 133–142.
20. W. C. Hurty, "Dynamic Analysis of Structural Systems by Component Modal Synthesis," Report 32-530, Jet Propulsion Laboratory Report, Pasadena, Calif., 1964.
21. W. C. Hurty, "Dynamic Analysis of Structural Systems Using Component Modes," *A.I.A.A. Journal*, Vol. 3, 1965, pp. 678–685.
22. R. L. Goldman, "Vibration Analysis by Dynamic Partitioning, *A.I.A.A. Journal*, Vol. 7, No. 6, 1969.
23. S. Hon, "Review of Modal Synthesis Techniques and a New Approach," *Shock and Vibration Bulletin*, Vol. 40, No. 4, 1969.
24. R. L. Bajan, C. C. Feng, and T. J. Jaszlics, "Vibration Analysis of Complex Structural Systems by Modal Substitution," *Shock and Vibration Bulletin*, Vol. 39, No. 3, 1969.
25. R. R. Craig, Jr., and M. C. C. Bampton, "Coupling of Substructures for Dynamic Analysis," *A.I.A.A. Journal*, Vol. 6, No. 7, 1968.
26. R. L. Smith, "Substructuring Techniques to Analyze Complex Structures," Graduate Student Report, Structural Engineering and Structural Mechanics, University of California, Department of Civil Engineering, Berkeley, California, 1974.

27. R. R. Craig, Jr., and C.-J. Chang, "A Review of Substructure Coupling Methods for Dynamic Analysis," Advances in Engineering Science, NASA CP-2001, Nov. 1976, pp. 393–408.
28. C. P. Johnson, R. R. Craig, Jr., A. Yargicoglu, R. Rajatabhothi, "Quadratic Reduction for the Eigenproblem," *International Journal for Numerical Methods in Engineering*, Vol. 15, pp. 911–923, 1980.
29. C. Lanczos, "An Iteration Method for the Solution of the Eigenvalue Problem of Linear Differential and Integral Operators," *J. Research Natl. Bur. Standards*, Vol. 45, 1950, pp. 255–282.
30. C. C. Paige, "Computational Variants of the Lanczos Method for the Eigenproblem," *J. of the Institute of Mathematics and Its Applications*, Vol. 10, 1972, pp. 373–381.
31. T. Ericsson and A. Ruhe, "The Spectral Transformation Lanczos Method for the Numerical Solution of Large Sparse Generalized Symmetric Eigenvalue Problems," Institute of Information Processing, Univ. of Umea, Report UMINF-76.79, 1979.
32. B. N. Parlett and S. N. Scott "The Lanczos Algorithm with Selective Orthogonalization," *Math. Comp.*, Vol. 33, 1979, pp. 217–238.

11

SOLUTION METHODS FOR EIGENPROBLEMS

11.1 INTRODUCTION

In Chapter 10 we discussed the basic facts that are used in eigenvalue and vector solutions and some techniques for the calculation of approximations to the required eigensystem. The purpose of this and the next chapter is to present effective eigensolution techniques. The methods considered here are based on the fundamental aspects discussed in Chapter 10. Therefore, for a thorough understanding of the solution techniques to be presented in this chapter, it is necessary to be very familiar with the material discussed in Chapter 10. In addition, we also employ the notation that was defined in Chapter 10.

As before, we concentrate on the solution of the eigenproblem

$$\mathbf{K}\boldsymbol{\phi} = \lambda\mathbf{M}\boldsymbol{\phi} \tag{11.1}$$

and, in particular, on the calculation of the smallest eigenvalues $\lambda_1, \ldots, \lambda_q$ and corresponding eigenvectors $\boldsymbol{\phi}_1, \ldots, \boldsymbol{\phi}_q$. *The solution methods under consideration can be subdivided into four groups, corresponding to which basic property is used as the basis of the solution algorithm.*[1-11]

The vector iteration methods make up the first group, in which the basic property used is that

$$\mathbf{K}\boldsymbol{\phi}_i = \lambda_i\mathbf{M}\boldsymbol{\phi}_i \tag{11.2}$$

The transformation methods constitute the second group, using that

$$\boldsymbol{\Phi}^T\mathbf{K}\boldsymbol{\Phi} = \boldsymbol{\Lambda} \tag{11.3}$$

and

$$\boldsymbol{\Phi}^T\mathbf{M}\boldsymbol{\Phi} = \mathbf{I} \tag{11.4}$$

where $\boldsymbol{\Phi} = [\boldsymbol{\phi}_1, \ldots, \boldsymbol{\phi}_n]$ and $\boldsymbol{\Lambda} = \text{diag}(\lambda_1)$, $i = 1, \ldots, n$. The solution methods of the third group are polynomial iteration techniques which operate on the fact that

$$p(\lambda_i) = 0 \tag{11.5}$$

where

$$p(\lambda) = \det(\mathbf{K} - \lambda\mathbf{M}) \tag{11.6}$$

The solution methods of the fourth group employ the Sturm sequence property of the characteristic polynomials

$$p(\lambda) = \det(\mathbf{K} - \lambda\mathbf{M}) \tag{11.7}$$

and

$$p^{(r)}(\lambda^{(r)}) = \det(\mathbf{K}^{(r)} - \lambda^{(r)}\mathbf{M}^{(r)}); \qquad r = 1, \ldots, n-1 \tag{11.8}$$

where $p^{(r)}(\lambda^{(r)})$ is the characteristic polynomial of the rth associated constraint problem corresponding to $\mathbf{K}\boldsymbol{\phi} = \lambda\mathbf{M}\boldsymbol{\phi}$.

In addition to using an algorithm that is based on the basic properties summarized above, techniques are under investigation which are based on minimizing the Rayleigh quotient[12-14] (see Section 2.7).

A number of solution algorithms have been developed within each of the above four groups of solution methods.[1-10] However, for an effective calculation of the required eigensystem of $\mathbf{K}\boldsymbol{\phi} = \lambda\mathbf{M}\boldsymbol{\phi}$, only a few techniques need to be considered. Corresponding to the four groups above, we present in Sections 11.2 to 11.5 those techniques that are most important in finite element analysis. In addition to those techniques that can be classified to fall into one of the four groups, we discuss in Chapter 12 the determinant search technique and the subspace iteration method, both of which use a combination of the fundamental properties given in (11.2) to (11.8).

Before presenting the solution techniques of interest, a few basic additional points should be noted. It is important to realize that *all solution methods must be iterative in nature, because, basically, solving the eigenvalue problem* $\mathbf{K}\boldsymbol{\phi} = \lambda\mathbf{M}\boldsymbol{\phi}$ *is equivalent to calculating the roots of the polynomial* $p(\lambda)$, *which has order equal to the order of* **K** *and* **M**. Since there are for the general case no explicit formulas available for the calculation of the roots of $p(\lambda)$ when the order of p is larger than 4, an iterative solution method has to be used. However, before iteration is started, we may choose to transform the matrices **K** and **M** into a form that allows a more economical solution of the required eigensystem (see Section 11.3.3).

Although iteration is needed in the solution of an eigenpair $(\lambda_i, \boldsymbol{\phi}_i)$, it should be noted that once one member of the eigenpair has been calculated, we can obtain the other member without further iteration. Assume that λ_i has been evaluated by iteration; then we can obtain $\boldsymbol{\phi}_i$ using (11.2); i.e., $\boldsymbol{\phi}_i$ is calculated by solving

$$(\mathbf{K} - \lambda_i\mathbf{M})\boldsymbol{\phi}_i = \mathbf{0} \tag{11.9}$$

On the other hand, if we have evaluated $\boldsymbol{\phi}_i$ by iteration, we can obtain the required eigenvalue from the Rayleigh quotient; i.e., using (11.3) and (11.4), we have

$$\lambda_i = \boldsymbol{\phi}_i^T\mathbf{K}\boldsymbol{\phi}_i; \qquad \boldsymbol{\phi}_i^T\mathbf{M}\boldsymbol{\phi}_i = 1 \tag{11.10}$$

Therefore, when considering the design of an effective solution method, a basic question is whether we should first solve for the eigenvalue λ_i and then calculate the eigenvector $\boldsymbol{\phi}_i$, or vice versa, or whether it is most economical to solve for both λ_i and $\boldsymbol{\phi}_i$ simultaneously. The answer to this question depends on the solution requirements and the properties of the matrices **K** and **M**, i.e., such factors as the number of required eigenpairs, the order of **K** and **M**, the bandwidth of **K**, and whether **M** is banded.

The above discussion already leads us to realize that there is no single algorithm that always provides an efficient solution. Indeed, an algorithm that is effective in the solution of a specific problem may be totally inadequate for the solution of another problem. Therefore, an important aspect in the following sections is to give guidelines for the selection of the appropriate solution method corresponding to a given problem.

The effectiveness of a solution method depends largely on two factors: firstly, the possibility of a reliable use of the procedure and, secondly, the cost of solution. The solution cost is essentially determined by the number of high-speed storage operations and an efficient use of back-up storage devices. However, it is most important that a solution method can be employed in a reliable manner. This means that for well-defined stiffness and mass matrices the solution is always obtained to the required precision without solution breakdown. In practice, a solution is then only interrupted when the problem is ill-defined, for example, due to a data input error the stiffness and mass matrices are not properly defined. This solution interruption then occurs best as early as possible during the calculations, i.e., prior to any large computational expense. We should study the algorithms presented in the following with these considerations.

11.2 VECTOR ITERATION METHODS

As has been pointed out already, in the solution of an eigenvector or an eigenvalue we do need to use iteration. In Section 11.1 we classed the solution methods according to the basic relation upon which they operate. In the vector iteration methods the basic relation considered is[1-9]

$$\mathbf{K}\boldsymbol{\phi} = \lambda \mathbf{M}\boldsymbol{\phi} \tag{11.1}$$

The aim is to satisfy the equation in (11.1) by directly operating upon it. Consider that we assume a vector for $\boldsymbol{\phi}$, say $\mathbf{x}_1$, and assume a value for λ, say $\lambda = 1$. We can then evaluate the right-hand side of (11.1); i.e., we may calculate

$$\mathbf{R}_1 = (1)\mathbf{M}\mathbf{x}_1 \tag{11.11}$$

Since $\mathbf{x}_1$ is an arbitrarily assumed vector, we do not have, in general, that $\mathbf{K}\mathbf{x}_1 = \mathbf{R}_1$. If we had that $\mathbf{K}\mathbf{x}_1 = \mathbf{R}_1$, then $\mathbf{x}_1$ would be an eigenvector and, except for trivial cases, our assumptions would have been extremely lucky. Instead, we have an equilibrium equation as encountered in static analysis (see Section 8.2), which we may write

$$\mathbf{K}\mathbf{x}_2 = \mathbf{R}_1, \qquad \mathbf{x}_2 \neq \mathbf{x}_1 \tag{11.12}$$

where $\mathbf{x}_2$ is the displacement solution corresponding to the applied forces $\mathbf{R}_1$. Since we know that we have to use iteration to solve for an eigenvector, we may now intuitively feel that $\mathbf{x}_2$ may be a better approximation to an eigenvector than was $\mathbf{x}_1$. This is indeed the case, and by repeating the cycle we obtain an increasingly better approximation to an eigenvector.

The procedure described above is the basis of inverse iteration. We will see that other vector iteration techniques work in a similar way. Specifically, in forward iteration, the iterative cycle is reversed; i.e., in the first step we evaluate $\mathbf{R}_1 = \mathbf{K}\mathbf{x}_1$ and then obtain the improved approximation, $\mathbf{x}_2$, to an eigenvector by solving $\mathbf{M}\mathbf{x}_2 = \mathbf{R}_1$.

The basic steps of the vector iteration schemes and of the other solution methods that we consider later can be introduced using intuition. Namely, we need to satisfy one of the basic relations summarized in Section 11.1 and try to do so by some iterative cycle. However, the real justification for using any one of the methods derives from the fact that they do work and that they can be used economically.

11.2.1 Inverse Iteration

The technique of inverse iteration is very effectively used to calculate an eigenvector, and at the same time the corresponding eigenvalue can also be evaluated. Inverse iteration is employed in various important iteration procedures, including the determinant search and subspace iteration methods described in Sections 12.2 and 12.3. It is therefore important that we discuss the method in detail.

In this section we assume that **K** is positive definite, whereas **M** may be a digonal mass matrix with or without zero diagonal elements or may be a banded mass matrix. If **K** is only positive semidefinite, a shift should be used prior to the iteration (see Section 11.2.3).

In the following we first consider the basic equations used in inverse iteration and then present a more effective form of the technique. In the solution we assume a starting iteration vector $\mathbf{x}_1$ and then evaluate in each iteration step $k = 1, 2, \ldots$:

$$\mathbf{K}\bar{\mathbf{x}}_{k+1} = \mathbf{M}\mathbf{x}_k \tag{11.13}$$

and

$$\mathbf{x}_{k+1} = \frac{\bar{\mathbf{x}}_{k+1}}{(\bar{\mathbf{x}}_{k+1}^T\mathbf{M}\bar{\mathbf{x}}_{k+1})^{1/2}} \tag{11.14}$$

where provided that $\mathbf{x}_1$ is not **M**-orthogonal to $\boldsymbol{\phi}_1$, meaning that $\mathbf{x}_1^T\mathbf{M}\boldsymbol{\phi}_1 \neq 0$, we have that

$$\mathbf{x}_{k+1} \longrightarrow \boldsymbol{\phi}_1 \qquad \text{as } k \longrightarrow \infty$$

The basic step in the iteration is the solution of the equations in (11.13) in which we evaluate a vector $\bar{\mathbf{x}}_{k+1}$ with a direction closer to an eigenvector than the previous iteration vector $\mathbf{x}_k$. The calculation in (11.14) merely assures that the **M**-weighted length of the new iteration vector $\mathbf{x}_{k+1}$ is unity; i.e., we want that $\mathbf{x}_{k+1}$ satisfies the mass orthonormality relation

$$\mathbf{x}_{k+1}^T\mathbf{M}\mathbf{x}_{k+1} = 1 \tag{11.15}$$

Substituting for $\mathbf{x}_{k+1}$ from (11.14) into (11.15), we find that (11.15) is indeed satisfied. If the scaling in (11.14) is not included in the iteration, the elements of the iteration vectors grow (or decrease) in each step and the iteration vectors do not converge to $\boldsymbol{\phi}_1$ but to a multiple of it. We illustrate the procedure by means of the following example.

EXAMPLE 11.1: Consider the eigenproblem $\mathbf{K}\boldsymbol{\phi} = \lambda\mathbf{M}\boldsymbol{\phi}$, where

$$\mathbf{K} = \begin{bmatrix} 2 & -1 & 0 & 0 \\ -1 & 2 & -1 & 0 \\ 0 & -1 & 2 & -1 \\ 0 & 0 & -1 & 1 \end{bmatrix}; \qquad \mathbf{M} = \begin{bmatrix} 0 & & & \\ & 2 & & \\ & & 0 & \\ & & & 1 \end{bmatrix}$$

The eigenvalues and corresponding vectors of the problem have been evaluated in Examples 10.12 and 10.13. Use two steps of inverse iteration to evaluate an approximation to $\boldsymbol{\phi}_1$.

The first step is to decompose $\mathbf{K}$ into $\mathbf{LDL}^T$ in order to be able to solve the equations in (11.13). We obtained the triangular factors of $\mathbf{K}$ in Example 10.13.

As starting iteration vector we need a vector that is not orthogonal to $\boldsymbol{\phi}_1$. Since we do not know $\boldsymbol{\phi}_1$, we cannot make sure that $\boldsymbol{\phi}_1^T\mathbf{M}\mathbf{x}_1 \neq 0$, but we want to pick a vector that is not likely to be orthogonal to $\boldsymbol{\phi}_1$. Experience has shown that a good starting vector is in many cases a unit full vector (but see Example 11.6 for a case in which a unit full vector is a bad choice). In this example we use

$$\mathbf{x}_1^T = [1 \quad 1 \quad 1 \quad 1]$$

and then obtain, for $k = 1$,

$$\begin{bmatrix} 2 & -1 & 0 & 0 \\ -1 & 2 & -1 & 0 \\ 0 & -1 & 2 & -1 \\ 0 & 0 & -1 & 1 \end{bmatrix} \bar{\mathbf{x}}_2 = \begin{bmatrix} 0 & & & \\ & 2 & & \\ & & 0 & \\ & & & 1 \end{bmatrix} \begin{bmatrix} 1 \\ 1 \\ 1 \\ 1 \end{bmatrix}$$

Hence

$$\bar{\mathbf{x}}_2 = \begin{bmatrix} 3 \\ 6 \\ 7 \\ 8 \end{bmatrix}; \qquad \bar{\mathbf{x}}_2^T\mathbf{M}\bar{\mathbf{x}}_2 = 136$$

and

$$\mathbf{x}_2 = \frac{1}{\sqrt{136}} \begin{bmatrix} 3 \\ 6 \\ 7 \\ 8 \end{bmatrix}$$

Note that the zero diagonal elements in $\mathbf{M}$ do not introduce solution difficulties. Proceeding to the next iteration, $k = 2$, we use

$$\begin{bmatrix} 2 & -1 & 0 & 0 \\ -1 & 2 & -1 & 0 \\ 0 & -1 & 2 & -1 \\ 0 & 0 & -1 & 1 \end{bmatrix} \bar{\mathbf{x}}_3 = \begin{bmatrix} 0 & & & \\ & 2 & & \\ & & 0 & \\ & & & 1 \end{bmatrix} \begin{bmatrix} \dfrac{3}{\sqrt{136}} \\ \dfrac{6}{\sqrt{136}} \\ \dfrac{7}{\sqrt{136}} \\ \dfrac{8}{\sqrt{136}} \end{bmatrix}$$

Hence

$$\bar{\mathbf{x}}_3 = \frac{1}{\sqrt{136}} \begin{bmatrix} 20 \\ 40 \\ 48 \\ 56 \end{bmatrix}; \qquad \bar{\mathbf{x}}_3^T\mathbf{M}\bar{\mathbf{x}}_3 = \frac{6336}{136}$$

and

$$\mathbf{x}_3 = \frac{1}{\sqrt{6336}} \begin{bmatrix} 20 \\ 40 \\ 48 \\ 56 \end{bmatrix}$$

Comparing $\mathbf{x}_3$ with the exact solution (see Example 10.12) we have

$$\mathbf{x}_3 = \begin{bmatrix} 0.251 \\ 0.503 \\ 0.603 \\ 0.704 \end{bmatrix} \quad \text{and} \quad \boldsymbol{\phi}_1 = \begin{bmatrix} 0.250 \\ 0.500 \\ 0.602 \\ 0.707 \end{bmatrix}$$

Hence with only two iterations we have already obtained a fair approximation to $\boldsymbol{\phi}_1$.

The relations in (11.13) and (11.14) state the basic inverse iteration algorithm. However, in actual computer implementation it is more effective to iterate as follows. Assuming that $\mathbf{y}_1 = \mathbf{M}\mathbf{x}_1$, we evaluate for $k = 1, 2, \ldots$,

$$\mathbf{K}\bar{\mathbf{x}}_{k+1} = \mathbf{y}_k \tag{11.16}$$

$$\bar{\mathbf{y}}_{k+1} = \mathbf{M}\bar{\mathbf{x}}_{k+1} \tag{11.17}$$

$$\rho(\bar{\mathbf{x}}_{k+1}) = \frac{\bar{\mathbf{x}}_{k+1}^T \mathbf{y}_k}{\bar{\mathbf{x}}_{k+1}^T \bar{\mathbf{y}}_{k+1}} \tag{11.18}$$

$$\mathbf{y}_{k+1} = \frac{\bar{\mathbf{y}}_{k+1}}{(\bar{\mathbf{x}}_{k+1}^T \bar{\mathbf{y}}_{k+1})^{1/2}} \tag{11.19}$$

where, provided that $\mathbf{y}_1^T \boldsymbol{\phi}_1 \neq 0$,

$$\mathbf{y}_{k+1} \longrightarrow \mathbf{M}\boldsymbol{\phi}_1 \quad \text{and} \quad \rho(\bar{\mathbf{x}}_{k+1}) \longrightarrow \lambda_1 \qquad \text{as } k \longrightarrow \infty$$

It should be noted that we essentially dispense in (11.16) to (11.19) with the calculation of the matrix product $\mathbf{M}\mathbf{x}_k$ in (11.13) by iterating on $\mathbf{y}_k$ rather than on $\mathbf{x}_k$. But the value of $\bar{\mathbf{y}}_{k+1}$ is evaluated in either procedure; i.e., $\bar{\mathbf{y}}_{k+1}$ must be calculated in (11.14) and is evaluated in (11.17). Using the second iteration procedure, we obtain in (11.18) an approximation to the eigenvalue λ_1 given by the Rayleigh quotient $\rho(\bar{\mathbf{x}}_{k+1})$. It is this approximation to λ_1 which is conveniently used to determine convergence in the iteration. Denoting the current approximation to λ_1 by $\lambda_1^{(k+1)}$ [i.e., $\lambda_1^{(k+1)} = \rho(\bar{\mathbf{x}}_{k+1})$], we have convergence when

$$\frac{|\lambda_1^{(k+1)} - \lambda_1^{(k)}|}{\lambda_1^{(k+1)}} \leq \text{tol} \tag{11.20}$$

where tol should be 10^{-2s} or smaller when the eigenvalue λ_1 is required to $2s$-digit accuracy. The eigenvector will then be accurate to about s or more digits (see (11.33)). Let l be the last iteration; then we have

$$\lambda_1 \doteq \rho(\bar{\mathbf{x}}_{l+1}) \tag{11.21}$$

and

$$\boldsymbol{\phi}_1 \doteq \frac{\bar{\mathbf{x}}_{l+1}}{(\bar{\mathbf{x}}_{l+1}^T \bar{\mathbf{y}}_{l+1})^{1/2}} \tag{11.22}$$

EXAMPLE 11.2: Use the inverse iteration procedure given in (11.16) to (11.19) to evaluate an approximation to λ_1 and $\boldsymbol{\phi}_1$ of the eigenproblem $\mathbf{K}\boldsymbol{\phi} = \lambda\mathbf{M}\boldsymbol{\phi}$ considered in Example 11.1. Use tol $= 10^{-6}$ (i.e., $s = 3$) in (11.20), in order to measure convergence.

As in Example 11.1 we start the iteration with

$$\mathbf{x}_1 = \begin{bmatrix} 1 \\ 1 \\ 1 \\ 1 \end{bmatrix}$$

Proceeding as given in (11.16) to (11.19), we obtain for $k = 1$,

$$\mathbf{y}_1 = \begin{bmatrix} 0 \\ 2 \\ 0 \\ 1 \end{bmatrix}; \qquad \bar{\mathbf{x}}_2 = \begin{bmatrix} 3 \\ 6 \\ 7 \\ 8 \end{bmatrix}; \qquad \bar{\mathbf{y}}_2 = \begin{bmatrix} 0 \\ 12 \\ 0 \\ 8 \end{bmatrix}$$

$$\rho(\bar{\mathbf{x}}_2) = \frac{\bar{\mathbf{x}}_2^T \mathbf{y}_1}{\bar{\mathbf{x}}_2^T \bar{\mathbf{y}}_2} = 0.1470588$$

and

$$\mathbf{y}_2 = \begin{bmatrix} 0.0 \\ 1.02899 \\ 0.0 \\ 0.68599 \end{bmatrix}$$

TABLE 11.1

k	$\bar{\mathbf{x}}_{k+1}$	$\bar{\mathbf{y}}_{k+1}$	$\rho(\bar{\mathbf{x}}_{k+1})$	$\frac{\lvert \lambda_1^{(k+1)} - \lambda_1^{(k)} \rvert}{\lambda_1^{(k+1)}}$	$\mathbf{y}_k$
1	3	0	0.1470588	—	0
	6	12			1.02899
	7	0			0
	8	8			0.68599
2	1.71499	0	0.1464646	0.004056795132	0
	3.42997	6.85994			1.00504
	4.11597	0			0
	4.80196	4.80196			0.70353
3	1.70856	0	0.1464471	0.000119538581	0
	3.41713	6.83426			1.00087
	4.12066	0			0
	4.82418	4.82418			0.70649
4	1.70736	0	0.1464466	0.000003518989	0
	3.41472	6.82944			1.00015
	4.12121	0			0
	4.82771	4.82771			0.70700
5	1.70715	0	0.1464466	0.000000103589	0
	3.41430	6.82860			1.00003
	4.12130	0			0
	4.82830	4.82830			0.70709

The next iterations are carried out in the same way. The results are summarized in Table 11.1. It is seen that after five interations, convergence has been achieved. It should be noted that the Rayleigh quotient $\rho(\bar{\mathbf{x}}_{k+1})$ converges much faster than the vector $\bar{\mathbf{x}}_{k+1}$ (see Example 11.3) and converges from above to λ_1. Using (11.21) and (11.22), we have

$$\lambda_1 \doteq 0.146447; \qquad \boldsymbol{\phi}_1 \doteq \begin{bmatrix} 0.25001 \\ 0.50001 \\ 0\ 60355 \\ 0.70709 \end{bmatrix}$$

In the above discussion we have merely stated the iteration scheme and its convergence. We then applied the method in two examples, but did not formally prove convergence. In the following we derive the convergence properties, because we believe that the proof is very instructive.

The first step in the proof of convergence and of the convergence rate given here is similar to the procedure used in the analysis of direct integration methods (see Section 9.4). The fundamental equation used in inverse iteration is the relation in (11.13). Neglecting the scaling of the elements in the iteration vector, we basically use for $k = 1, 2, \ldots,$

$$\mathbf{K}\mathbf{x}_{k+1} = \mathbf{M}\mathbf{x}_k \tag{11.23}$$

where we stated that $\mathbf{x}_{k+1}$ will now converge to a multiple of $\boldsymbol{\phi}_1$. To show convergence it is convenient (as in the analysis of direct integration procedures) to change basis from the finite element coordinate basis to the basis of eigenvectors; namely, we can write for any iteration vector $\mathbf{x}_k$,

$$\mathbf{x}_k = \boldsymbol{\Phi}\mathbf{z}_k \tag{11.24}$$

where $\boldsymbol{\Phi}$ is the matrix of eigenvectors $\boldsymbol{\Phi} = [\boldsymbol{\phi}_1, \ldots, \boldsymbol{\phi}_n]$. It should be realized that because $\boldsymbol{\Phi}$ is nonsingular, there is a unique vector $\mathbf{z}_k$ for any vector $\mathbf{x}_k$. Substituting for $\mathbf{x}_k$ and $\mathbf{x}_{k+1}$ from (11.24) into (11.23), premultiplying by $\boldsymbol{\Phi}^T$ and using the orthogonality relations $\boldsymbol{\Phi}^T\mathbf{K}\boldsymbol{\Phi} = \boldsymbol{\Lambda}$ and $\boldsymbol{\Phi}^T\mathbf{M}\boldsymbol{\Phi} = \mathbf{I}$, we obtain

$$\boldsymbol{\Lambda}\mathbf{z}_{k+1} = \mathbf{z}_k \tag{11.25}$$

where $\boldsymbol{\Lambda} = \text{diag}(\lambda_i)$. Comparing (11.25) with (11.23) we find that the iterations are of the same form with $\mathbf{K} = \boldsymbol{\Lambda}$ and $\mathbf{M} = \mathbf{I}$. We may wonder why the transformation in (11.24) is used, since $\boldsymbol{\Phi}$ is unknown. However, we should realize that the transformation is only employed to investigate the convergence behavior of inverse iteration. Namely, because *in theory* (11.25) is equivalent to (11.23), the convergence properties of (11.25) are also those of (11.23). But the convergence characteristics of (11.25) are relatively easy to investigate, since the eigenvalues are the diagonal elements of $\boldsymbol{\Lambda}$ and the eigenvectors are the unit vectors $\mathbf{e}_i$, where

$$\mathbf{e}_i^T = [0 \ldots 0 \ \underset{\uparrow\, i\text{th location}}{1} \ 0 \ldots 0] \tag{11.26}$$

In the presentation of the inverse iteration algorithms given in (11.13) to (11.14) and (11.16) to (11.22), we stated that the starting iteration vector $\mathbf{x}_1$ must not be $\mathbf{M}$-orthogonal to $\boldsymbol{\phi}_1$. Equivalently, in (11.25) the iteration vector $\mathbf{z}_1$ must not be orthogonal to $\mathbf{e}_1$. Assume that we use

$$\mathbf{z}_1^T = [1 \quad 1 \quad 1 \ldots 1] \tag{11.27}$$

We discuss the effect of this assumption in Section 11.2.6. Then using (11.25) for $k = 1, \ldots, l$, we obtain

$$\mathbf{z}_{l+1}^T = \left[\left(\frac{1}{\lambda_1}\right)^l \left(\frac{1}{\lambda_2}\right)^l \cdots \left(\frac{1}{\lambda_n}\right)^l\right] \tag{11.28}$$

Let us first assume that $\lambda_1 < \lambda_2$. To show that $\mathbf{z}_{l+1}$ converges to a multiple of $\mathbf{e}_1$ as $l \longrightarrow \infty$, we multiply $\mathbf{z}_{l+1}$ in (11.28) by $(\lambda_1)^l$ to obtain

$$\bar{\mathbf{z}}_{l+1} = \begin{bmatrix} 1 \\ (\lambda_1/\lambda_2)^l \\ \cdot \\ \cdot \\ \cdot \\ (\lambda_1/\lambda_n)^l \end{bmatrix} \tag{11.29}$$

and observe that $\bar{\mathbf{z}}_{l+1}$ converges to $\mathbf{e}_1$ as $l \longrightarrow \infty$. Hence $\mathbf{z}_{l+1}$ converges to a multiple of $\mathbf{e}_1$ as $l \longrightarrow \infty$.

To evaluate the order and rate of convergence, we use the convergence definition given in Section 2.9. For the iteration here under consideration we obtain

$$\lim_{l \to \infty} \frac{\| \bar{\mathbf{z}}_{l+1} - \mathbf{e}_1 \|_2}{\| \bar{\mathbf{z}}_l - \mathbf{e}_1 \|_2} = \frac{\lambda_1}{\lambda_2} \tag{11.30}$$

Hence convergence is linear and the rate of convergence is λ_1/λ_2. This convergence rate is also shown in the iteration vector $\bar{\mathbf{z}}_{l+1}$ in (11.29); i.e., those elements in the iteration vector that should tend to zero do so with at least the ratio λ_1/λ_2 in each additional iteration. Thus if $\lambda_2 > \lambda_1$, it is the relative magnitude of λ_1 to λ_2 that determines how fast the iteration vector converges to the eigenvector $\boldsymbol{\phi}_1$.

In the above discussion we assumed that $\lambda_1 < \lambda_2$. Let us now consider the case of a multiple eigenvalue, namely $\lambda_1 = \lambda_2 = \ldots = \lambda_m$. Then we have in (11.29),

$$\bar{\mathbf{z}}_{l+1}^T = \begin{bmatrix} 1 & 1 & \ldots & 1 & \left(\frac{\lambda_1}{\lambda_{m+1}}\right)^l & \ldots & \left(\frac{\lambda_1}{\lambda_n}\right)^l \end{bmatrix} \tag{11.31}$$

and the convergence rate of the iteration vector is λ_1/λ_{m+1}. Therefore, in general, the rate of convergence of the iteration vector in inverse iteration is given by the ratio of λ_1 to the next distinct eigenvalue.

In the iteration given in (11.16) to (11.22), we obtain an approximation to the eigenvalue λ_1 by evaluating the Rayleigh quotient. Corresponding to (11.18) the Rayleigh quotient calculated in the iteration of (11.25) would be

$$\rho(\mathbf{z}_{k+1}) = \frac{\mathbf{z}_{k+1}^T \mathbf{z}_k}{\mathbf{z}_{k+1}^T \mathbf{z}_{k+1}} \tag{11.32}$$

Assume that we consider the last iteration in which $k = l$. Then substituting for $\mathbf{z}_l$ and $\mathbf{z}_{l+1}$ from (11.28) into (11.32), we obtain

$$\rho(\mathbf{z}_{l+1}) = \frac{\lambda_1 \sum_{i=1}^{n} (\lambda_1/\lambda_i)^{2l-1}}{\sum_{i=1}^{n} (\lambda_1/\lambda_i)^{2l}} \tag{11.33}$$

Hence we have for λ_1 being a simple or multiple eigenvalue,

$$\rho(\mathbf{z}_{l+1}) \longrightarrow \lambda_1 \qquad \text{as } l \longrightarrow \infty$$

Also, convergence is linear with the rate equal to $(\lambda_1/\lambda_{m+1})^2$, where λ_{m+1} is defined as in (11.31). This convergence rate substantiates the observation that if an eigenvector is known with an error ϵ, then the Rayleigh quotient yields an approximation to the corresponding eigenvalue with error ϵ^2 (see Section 2.8).

Before demonstrating the results by means of a small example, it should be recalled that we assumed in the above analysis a full unit starting iteration vector as given in (11.27). The convergence properties derived hold for any starting iteration vector that is not orthogonal to the eigenvector of interest, but the convergence rates can in many practical analyses only be observed as the number of iterations becomes large. The same observation also holds for any of the other convergence analyses that are presented in the following sections. We discuss this observation with other important practical aspects in Section 11.2.6.

EXAMPLE 11.3: For the problem considered in Example 11.2, calculate the ultimate convergence rates of the iteration vector and the Rayleigh quotient. Compare the ultimate convergence rates with those actually observed in the inverse iteration carried out in Example 11.2.

For the evaluation of the theoretical convergence rates, we need λ_1 and λ_2. We calculated the eigenvalues in Example 10.12 and found

$$\lambda_1 = \frac{1}{2} - \frac{\sqrt{2}}{4}$$
$$\lambda_2 = \frac{1}{2} + \frac{\sqrt{2}}{4}$$

Hence the ultimate convergence rate of the iteration vector is

$$\frac{\lambda_1}{\lambda_2} = 0.17$$

and the ultimate convergence rate of the Rayleigh quotient is

$$\left(\frac{\lambda_1}{\lambda_2}\right)^2 = 0.029$$

The actual vector convergence obtained is observed by evaluating the ratio $r_{k+1}, k = 1, 2, \ldots,$ where

$$r_{k+1} = \frac{\| \mathbf{x}_{k+1} - \boldsymbol{\phi}_1 \|_2}{\| \mathbf{x}_k - \boldsymbol{\phi}_1 \|_2}$$

and we assume that $\boldsymbol{\phi}_1$ is obtained in the last iteration [see (11.22)].

For the iteration in Example 11.2, we thus obtain

$$r_2 = 0.026083; \qquad r_3 = 0.170559; \qquad r_4 = 0.167134; \qquad r_5 = 0.144251$$

Ignoring r_2 because the iteration just started, we see that the theoretical and actual convergence rates compare quite well.

Similarly, the actual convergence of the Rayleigh quotient calculated in Example 11.2 is observed by evaluating

$$\epsilon_{k+1} = \frac{|\rho(\bar{\mathbf{x}}_{k+1}) - \lambda_1|}{|\rho(\bar{\mathbf{x}}_k) - \lambda_1|}$$

where we use the converged value of the Rayleigh quotient for λ_1. In the iteration of Example 11.2 we have

$$\epsilon_3 = 0.028768; \qquad \epsilon_4 = 0.027778; \qquad \epsilon_5 = 0$$

Hence, we see that the theoretical and observed convergence rates agree again quite well in this solution.

11.2.2 Forward Iteration

The method of forward iteration is complementary to the inverse iteration technique, in that the method yields the eigenvector corresponding to the largest eigenvalue. Whereas we assumed in inverse iteration that $\mathbf{K}$ be positive definite, we assume in this section that $\mathbf{M}$ is positive definite; otherwise, a shift must be used (see Section 11.2.3). Having chosen a starting iteration vector $\mathbf{x}_1$, in forward iteration we evaluate, for $k = 1, 2, \ldots,$

$$\mathbf{M}\bar{\mathbf{x}}_{k+1} = \mathbf{K}\mathbf{x}_k \tag{11.34}$$

and

$$\mathbf{x}_{k+1} = \frac{\bar{\mathbf{x}}_{k+1}}{(\bar{\mathbf{x}}_{k+1}^T \mathbf{M}\bar{\mathbf{x}}_{k+1})^{1/2}} \tag{11.35}$$

where provided that $\mathbf{x}_1$ is not $\mathbf{M}$-orthogonal to $\boldsymbol{\phi}_n$, we have that

$$\mathbf{x}_{k+1} \longrightarrow \boldsymbol{\phi}_n \qquad \text{as } k \longrightarrow \infty$$

The analogy to inverse iteration should be noted; the only difference is that we solve (11.34) rather than (11.13) to obtain an improved eigenvector. This means, in practice, that in the inverse iteration we need to triangularize the matrix $\mathbf{K}$ and in the forward iteration we decompose $\mathbf{M}$.

A more effective forward iteration procedure than in (11.34) and (11.35) would be obtained by using equations that are analogous to those in (11.16) to (11.22). Assuming that $\mathbf{y}_1 = \mathbf{K}\mathbf{x}_1$, we evaluate for $k = 1, 2, \ldots,$

$$\mathbf{M}\bar{\mathbf{x}}_{k+1} = \mathbf{y}_k \tag{11.36}$$

$$\bar{\mathbf{y}}_{k+1} = \mathbf{K}\bar{\mathbf{x}}_{k+1} \tag{11.37}$$

$$\rho(\bar{\mathbf{x}}_{k+1}) = \frac{\bar{\mathbf{x}}_{k+1}^T \bar{\mathbf{y}}_{k+1}}{\bar{\mathbf{x}}_{k+1}^T \mathbf{y}_k} \tag{11.38}$$

$$\mathbf{y}_{k+1} = \frac{\bar{\mathbf{y}}_{k+1}}{(\bar{\mathbf{x}}_{k+1}^T \mathbf{y}_k)^{1/2}} \tag{11.39}$$

where provided that $\boldsymbol{\phi}_n^T \mathbf{y}_1 \neq 0$,

$$\mathbf{y}_{k+1} \longrightarrow \mathbf{K}\boldsymbol{\phi}_n \quad \text{and} \quad \rho(\bar{\mathbf{x}}_{k+1}) \longrightarrow \lambda_n \qquad \text{as } k \longrightarrow \infty$$

Convergence in the iteration could again be measured as given in (11.20), and denoting by l the last iteration, we have

$$\lambda_n \doteq \rho(\bar{\mathbf{x}}_{l+1}) \tag{11.40}$$

and

$$\boldsymbol{\phi}_n \doteq \frac{\bar{\mathbf{x}}_{l+1}}{(\bar{\mathbf{x}}_{l+1}^T \mathbf{y}_l)^{1/2}} \tag{11.41}$$

Considering the analysis of convergence of the iteration vector to $\boldsymbol{\phi}_n$, it can be carried out following the same procedure that was used in the evaluation of the convergence characteristics of inverse iteration. Alternatively, we may use the results that we obtained in the analysis of inverse iteration. Namely, assume that we write the eigenproblem $\mathbf{K}\boldsymbol{\phi} = \lambda\mathbf{M}\boldsymbol{\phi}$ in the form $\mathbf{M}\boldsymbol{\phi} = \lambda^{-1}\mathbf{K}\boldsymbol{\phi}$; then using inverse iteration to solve for an eigenvector and corresponding eigenvalue is equivalent to performing forward iteration on the problem $\mathbf{K}\boldsymbol{\phi} = \lambda\mathbf{M}\boldsymbol{\phi}$. But since we converge in the inverse iteration of (11.16) to (11.22) to the smallest eigenvalue and corresponding eigenvector, and since for the problem $\mathbf{M}\boldsymbol{\phi} = \lambda^{-1}\mathbf{K}\boldsymbol{\phi}$, this eigenvalue is λ_n^{-1}, where λ_n is the largest eigenvalue of $\mathbf{K}\boldsymbol{\phi} = \lambda\mathbf{M}\boldsymbol{\phi}$, we converge in the forward iteration of (11.36) to (11.41) to λ_n and $\boldsymbol{\phi}_n$ and the convergence rate of the iteration vector is λ_{n-1}/λ_n. We should note that the Rayleigh quotient evaluated in (11.38) is $(\bar{\mathbf{x}}_{k+1}^T\mathbf{K}\bar{\mathbf{x}}_{k+1})/(\bar{\mathbf{x}}_{k+1}^T\mathbf{M}\bar{\mathbf{x}}_{k+1})$, i.e., just the inverse of the Rayleigh quotient to calculate an approximation to λ_n^{-1} in the problem $\mathbf{M}\boldsymbol{\phi} = \lambda^{-1}\mathbf{K}\boldsymbol{\phi}$.

We demonstrate the iteration and convergence in the following example.

EXAMPLE 11.4: Use forward iteration as given in (11.36) to (11.41) with tol 10^{-6} in (11.20) to evaluate λ_4 and $\boldsymbol{\phi}_4$ of the eigenproblem $\mathbf{K}\boldsymbol{\phi} = \lambda\mathbf{M}\boldsymbol{\phi}$, where

$$\mathbf{K} = \begin{bmatrix} 5 & -4 & 1 & 0 \\ -4 & 6 & -4 & 1 \\ 1 & -4 & 6 & -4 \\ 0 & 1 & -4 & 5 \end{bmatrix}; \qquad \mathbf{M} = \begin{bmatrix} 2 & & & \\ & 2 & & \\ & & 1 & \\ & & & 1 \end{bmatrix}$$

The physical problem considered in this example is the free vibration response of the simply supported beam shown in Fig. 8.1 with the above mass matrix.

Starting the iteration with

$$\mathbf{x}_1 = \begin{bmatrix} 1 \\ 1 \\ 1 \\ 1 \end{bmatrix}$$

we calculate in the inverse iteration the values summarized in Table 11.2.

TABLE 11.2

k	$\bar{\mathbf{x}}_{k+1}$	$\bar{\mathbf{y}}_{k+1}$	$\rho(\bar{\mathbf{x}}_{k+1})$	$\mathbf{y}_{k+1}$	$\frac{\lvert\lambda_1^{(k+1)} - \lambda_1^{(k)}\rvert}{\lambda_1^{(k+1)}}$
1	1	6	5.93333	2.1909	—
	−0.5	−1		−0.3652	
	−1	−11		−4.0166	
	2	13.5		4.9295	
2	1.0955	2.1909	8.57886	0.3345	0.3084
	−0.1826	15.5188		2.3694	
	−4.0166	−41.9921		−6.4112	
	4.9295	40.5315		6.1882	
3	0.1672	−10.3137	10.15966	−1.1372	0.1556
	1.1847	38.2720		4.2198	
	−6.4112	−67.7914		−7.4748	
	6.1882	57.7704		6.3696	
4	−0.5686	−18.7569	10.55204	−1.8219	0.03719
	2.1099	51.2010		4.9733	
	−7.4745	−79.3332		−7.7059	
	6.3696	63.8557		6.2025	
5	−0.9110	−22.2074	10.62367	−2.0995	0.006743
	2.4867	55.5901		5.2556	
	−7.7059	−89.9033		−7.7433	
	6.2025	64.3230		6.0812	
6	−1.0498	−23.5033	10.63595	−2.2115	0.001154
	2.6278	57.0203		5.3651	
	−7.7433	−82.3457		−7.7480	
	6.0812	64.0072		6.0225	
7	−1.1057	−24.0068	10.63802	−2.2570	0.0001954
	2.6826	57.5326		5.4089	
	−7.7479	−82.4138		−7.7481	
	6.0225	63.7869		5.9969	
8	−1.1285	−24.2083	10.63838	−2.2756	0.00003304
	2.7044	57.7298		5.4267	
	−7.7481	−82.4222		−7.7478	
	5.9969	68.6811		5.9861	
9	−1.1378	−24.2902	10.63844	−2.2833	0.000005584
	2.7133	57.8086		5.4340	
	−7.7478	−82.4224		−7.7476	
	5.9861	63.6351		5.9816	
10	−1.1416	−24.3237	10.63845	−2.2864	0.0000009437
	2.7170	57.8405		5.4369	
	−7.7476	−82.4219		−7.7476	
	5.9816	63.6157		5.9898	

Hence, we need 10 iterations for a convergence tolerance of 10^{-6} in (11.20), and we then use, as given in (11.40) and (11.41),

$$\lambda_4 \doteq 10.63845; \qquad \boldsymbol{\phi}_4 \doteq \begin{bmatrix} -0.10731 \\ 0.25539 \\ -0.72827 \\ 0.56227 \end{bmatrix}$$

11.2.3 Shifting in Vector Iteration

The convergence analysis of inverse iteration in Section 11.2.1 showed that assuming $\lambda_1 < \lambda_2$, the iteration vector converges with a rate λ_1/λ_2 to the eigenvector $\boldsymbol{\phi}_1$. Therefore, depending on the magnitude of λ_1 and λ_2, the convergence rate can be arbitrarily low, say $\lambda_1/\lambda_2 = 0.99999$, or can be very high, say $\lambda_1/\lambda_2 = 0.01$. Similarly, in forward iteration the convergence rate can be low or high. Therefore, a natural question must be how to improve the convergence rate in the vector iterations. We show in this section that the convergence rate can be much improved by shifting. In addition, a shift can be used to obtain convergence to an eigenpair other than $(\lambda_1, \boldsymbol{\phi}_1)$ and $(\lambda_n, \boldsymbol{\phi}_n)$ in inverse and forward iterations, respectively, and a shift is used effectively in inverse iteration when $\mathbf{K}$ is positive semidefinite and in forward iteration when $\mathbf{M}$ is diagonal with some zero diagonal elements (see Example 11.6).

Assume that a shift μ is applied as described in Section 10.2.3; then we consider the eigenproblem

$$(\mathbf{K} - \mu\mathbf{M})\boldsymbol{\phi} = \eta\mathbf{M}\boldsymbol{\phi} \tag{11.42}$$

where the eigenvalues of the original problem $\mathbf{K}\boldsymbol{\phi} = \lambda\mathbf{M}\boldsymbol{\phi}$ and of the problem in (11.42) are related by $\eta_i = \lambda_i - \mu$, $i = 1, \ldots, n$. To analyze the convergence properties of inverse and forward iteration when applied to the problem in (11.42), we follow in all respects the procedure used in Section 11.2.1. The first step is to consider the problem in the basis of eigenvectors $\boldsymbol{\Phi}$. Using the transformation

$$\boldsymbol{\phi} = \boldsymbol{\Phi}\boldsymbol{\psi} \tag{11.43}$$

we obtain for the convergence analysis the equivalent eigenproblem

$$(\boldsymbol{\Lambda} - \mu\mathbf{I})\boldsymbol{\psi} = \eta\boldsymbol{\psi} \tag{11.44}$$

Consider first inverse iteration and assume that all eigenvalues are distinct. In that case we obtain, using the notation in Section 11.2.1,

$$\mathbf{z}_{l+1}^T = \begin{bmatrix} \dfrac{1}{(\lambda_1 - \mu)^l} & \dfrac{1}{(\lambda_2 - \mu)^l} & \cdots & \dfrac{1}{(\lambda_n - \mu)^l} \end{bmatrix} \tag{11.45}$$

where it is assumed that all $(\lambda_i - \mu)$ are nonzero, but they may be positive or negative. Assume that $(\lambda_i - \mu)$ is smallest in absolute magnitude when $i = j$; then multiplying $\mathbf{z}_{l+1}$ by $(\lambda_j - \mu)^l$, we obtain

$$\bar{\mathbf{z}}_{l+1} = \begin{bmatrix} \left(\dfrac{\lambda_j - \mu}{\lambda_1 - \mu}\right)^l \\ \cdot \\ \cdot \\ \cdot \\ \left(\dfrac{\lambda_j - \mu}{\lambda_{j-1} - \mu}\right)^l \\ 1 \\ \left(\dfrac{\lambda_j - \mu}{\lambda_{j+1} - \mu}\right)^l \\ \cdot \\ \cdot \\ \cdot \\ \left(\dfrac{\lambda_j - \mu}{\lambda_n - \mu}\right)^l \end{bmatrix} \tag{11.46}$$

where $|(\lambda_j - \mu)/(\lambda_p - \mu)| < 1$ for all $p \neq j$. Hence, in the iteration we have $\bar{\mathbf{z}}_{l+1} \longrightarrow \mathbf{e}_j$, meaning that in inverse iteration to solve (11.42), the iteration vector converges to $\boldsymbol{\phi}_j$. Furthermore, we obtain $\lambda_j = \eta_j + \mu$. The convergence rate in the iteration is determined by the element $(\lambda_j - \mu)/(\lambda_p - \mu)$, which is largest in absolute magnitude, $p \neq j$; i.e., the convergence rate, r, is

$$r = \max_{p \neq j} \left| \frac{\lambda_j - \mu}{\lambda_p - \mu} \right| \tag{11.47}$$

Since λ_j is nearest μ, the convergence rate of the iteration vector in (11.42) to the eigenvector $\boldsymbol{\phi}_j$ is either

$$\left| \frac{\lambda_j - \mu}{\lambda_{j-1} - \mu} \right| \quad \text{or} \quad \left| \frac{\lambda_j - \mu}{\lambda_{j+1} - \mu} \right|$$

whichever is larger. The convergence rate for a typical case is shown in Fig. 11.1.

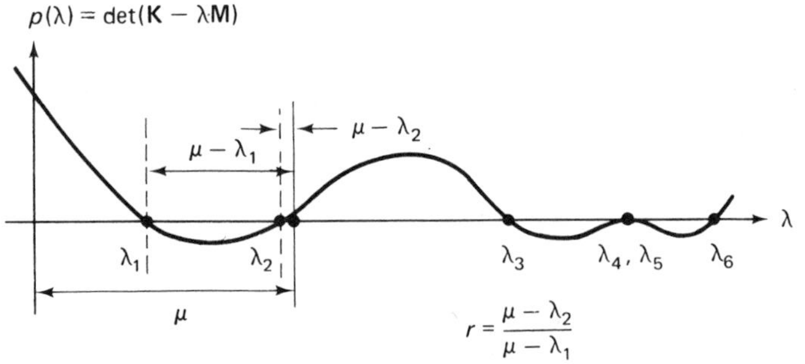

FIGURE 11.1 *Example of vector convergence rate r in inverse iteration.*

Using the results of the above convergence analysis and of the analysis of inverse iteration without shifting (see Section 11.2.1), two additional conclusions are reached. Firstly, we observe that the convergence rate of the Rayleigh quotient, which for μ nearest to λ_j converges to $(\lambda_j - \mu)$, is

$$\left| \frac{\lambda_j - \mu}{\lambda_{j-1} - \mu} \right|^2 \quad \text{or} \quad \left| \frac{\lambda_j - \mu}{\lambda_{j+1} - \mu} \right|^2$$

whichever is larger.

The second observation concerns the case λ_j being a multiple eigenvalue. The analysis in Section 11.2.1 and the conclusions above show that if $\lambda_j = \lambda_{j+1} = \ldots = \lambda_{j+m-1}$, the rate of convergence of the iteration vector is

$$\max_{p \neq j, j+1, \ldots, j+m-1} \left| \frac{\lambda_j - \mu}{\lambda_p - \mu} \right|$$

and convergence occurs to a vector in the subspace corresponding to λ_j.

The important point in inverse iteration with shifting is that by choosing a shift near enough the specific eigenvalue of interest, we can, in theory, have a convergence rate that is as high as required; i.e., we would only need to make $|\lambda_j - \mu|$ small enough in relation to $|\lambda_p - \mu|$ defined above. However, in an actual solution scheme the difficulty is to find an appropriate μ, for which we consider various methods in the next sections.

EXAMPLE 11.5: Use inverse iteration as given in (11.16) to (11.22) in order to calculate $(\lambda_1, \boldsymbol{\phi}_1)$ of the problem $\mathbf{K}\boldsymbol{\phi} = \lambda \mathbf{M}\boldsymbol{\phi}$, where $\mathbf{K}$ and $\mathbf{M}$ are given in Example 11.4. Then impose the shift $\mu = 10$ and show that in the inverse iteration convergence occurs toward λ_4 and $\boldsymbol{\phi}_4$.

Using inverse iteration on $\mathbf{K}\boldsymbol{\phi} = \lambda \mathbf{M}\boldsymbol{\phi}$ as in Example 11.2 gives convergence after three iterations with a tolerance of 10^{-6},

$$\lambda_1 \doteq 0.09654; \qquad \boldsymbol{\phi}_1 \doteq \begin{bmatrix} 0.3126 \\ 0.4955 \\ 0.4791 \\ 0.2898 \end{bmatrix}$$

Now imposing a shift of $\mu = 10$, we obtain

$$\mathbf{K} - \mu \mathbf{M} = \begin{bmatrix} -15 & -4 & 1 & 0 \\ -4 & -14 & -4 & 1 \\ 1 & -4 & -4 & -4 \\ 0 & 1 & -4 & -5 \end{bmatrix}$$

Using inverse iteration on the problem $(\mathbf{K} - \mu \mathbf{M})\boldsymbol{\phi} = \eta \mathbf{M}\boldsymbol{\phi}$, we obtain convergence after six iterations with

$$\rho(\bar{\mathbf{x}}_7) = 0.6385; \qquad \mathbf{x}_7 = \begin{bmatrix} -0.1076 \\ 0.2556 \\ -0.7283 \\ 0.5620 \end{bmatrix}$$

Since we imposed a shift, we do know that $\mu + \rho(\bar{\mathbf{x}}_7)$ is an approximation to an eigenvalue and $\mathbf{x}_7$ is an approximation to the corresponding eigenvector. But we do not know which eigenpair has been approximated. By comparing $\mathbf{x}_7$ with the results obtained in Example 11.4 we find that we have

$$\lambda_4 \doteq \mu + \rho(\mathbf{x}_7) \doteq 10.6385; \qquad \boldsymbol{\phi}_4 \doteq \mathbf{x}_7$$

EXAMPLE 11.6: Consider the unsupported beam element depicted in Fig. 8.10. Show that the usual inverse iteration algorithm of calculating λ_1 and $\boldsymbol{\phi}_1$ does not work, but that after imposing a shift the standard algorithm can again be applied.

The first step in the inverse iteration defined in (11.16) is, in this case with $\mathbf{M} = \mathbf{I}$ and $\mathbf{x}_1$ a full unit vector,

$$\begin{bmatrix} 12 & -6 & -12 & -6 \\ -6 & 4 & 6 & 2 \\ -12 & 6 & 12 & 6 \\ -6 & 2 & 6 & 4 \end{bmatrix} \bar{\mathbf{x}}_2 = \begin{bmatrix} 1 \\ 1 \\ 1 \\ 1 \end{bmatrix} \tag{a}$$

Using Gauss elimination to solve the equations, we arrive at

$$\begin{bmatrix} 12 & -6 & -12 & -6 \\ & 1 & 0 & -1 \\ & & 0 & 0 \\ & & & 0 \end{bmatrix} \bar{\mathbf{x}}_2 = \begin{bmatrix} 1 \\ \frac{3}{2} \\ 2 \\ \frac{7}{2} \end{bmatrix}$$

and hence the equations in (a) have no solution. They only have a solution if the right-hand side [i.e., $\mathbf{x}_1$ in (11.16)] is a null vector. There would be no difficulty in modifying the solution procedure when a singular coefficient matrix is encountered, and the advantage would be that the eigenvector would be calculated in one iteration. On the other hand, if we impose a shift, we can use the standard iteration procedure, and stability problems are avoided in the calculation of other eigenvalues and eigenvectors. Assume that we use $\mu = -6$ so that all λ_i are positive. Then we have

$$\mathbf{K} - \mu\mathbf{I} = \begin{bmatrix} 18 & -6 & -12 & -6 \\ -6 & 10 & 6 & 2 \\ -12 & 6 & 18 & 6 \\ -6 & 2 & 6 & 10 \end{bmatrix}$$

The inverse iteration can now be performed in the standard manner using a full unit starting iteration vector. Convergence is achieved after five iterations to a tolerance of 10^{-6}, and we have

$$\rho(\bar{\mathbf{x}}_6) = 6.000000; \qquad \mathbf{x}_6 = \begin{bmatrix} 0.73784 \\ 0.42165 \\ 0.31625 \\ 0.42165 \end{bmatrix}$$

Hence taking account of the shift, we have

$$\lambda_1 \doteq 0.0; \qquad \boldsymbol{\phi}_1 \doteq \mathbf{x}_6$$

We showed above that the rate of convergence in inverse iteration can be greatly increased by shifting. We may now wonder whether the convergence rate in forward iteration can be increased in a similar way. In analogy to the convergence proof of inverse iteration with a shift, we can generalize the convergence analysis of forward iteration when a shift μ is used. The final result is that the iteration vector converges to the eigenvector $\boldsymbol{\phi}_j$ which corresponds to the largest eigenvalue $|\lambda_j - \mu|$ of the problem in (11.42), where

$$|\lambda_j - \mu| = \max_{\text{all } i} |\lambda_i - \mu| \tag{11.48}$$

The convergence rate of the iteration vector is given by

$$r = \max_{p \neq j} \left| \frac{\lambda_p - \mu}{\lambda_j - \mu} \right| \tag{11.49}$$

which, in fact, is the ratio of the second largest eigenvalue to the largest eigenvalue (both measured in absolute values) of the problem $(\mathbf{K} - \mu\mathbf{M})\boldsymbol{\phi} = \eta\mathbf{M}\boldsymbol{\phi}$.

In case λ_j is a multiple eigenvalue, say $\lambda_j = \lambda_{j+1} = \ldots = \lambda_{j+m-1}$, the iteration vector converges to a vector in the subspace corresponding to λ_j, and the rate of convergence is

$$\max_{p \neq j, j+1, \ldots, j+m-1} \left| \frac{\lambda_p - \mu}{\lambda_j - \mu} \right|$$

The main difference between the convergence rate in (11.47) and (11.49) is that, in (11.47), λ_p is in the denominator, whereas in (11.49), λ_p is in the numerator. This limits the convergence rate in forward iteration and by means of shifting convergence can be obtained only to the eigenpair $(\lambda_n, \boldsymbol{\phi}_n)$ or to the eigenpair $(\lambda_1, \boldsymbol{\phi}_1)$. To achieve the highest convergence rates to $\boldsymbol{\phi}_n$ and $\boldsymbol{\phi}_1$, we need to choose $\mu = (\lambda_1 + \lambda_{n-1})/2$ and $\mu = (\lambda_2 + \lambda_n)/2$, respectively, and have the corresponding convergence rates

$$\left| \frac{\lambda_{n-1} - \dfrac{\lambda_1 + \lambda_{n-1}}{2}}{\lambda_n - \dfrac{\lambda_1 + \lambda_{n-1}}{2}} \right| \quad \text{and} \quad \left| \frac{\lambda_2 - \dfrac{\lambda_2 + \lambda_n}{2}}{\lambda_1 - \dfrac{\lambda_2 + \lambda_n}{2}} \right|$$

(see Fig. 11.2). Therefore, a much higher convergence rate can be obtained with shifting in inverse iteration than using forward iteration. For this reason and because a shift can be chosen to converge to any eigenpair, inverse iteration is much more important in practical analysis, and in the algorithms presented later, we always use inverse iteration whenever vector iteration is required.

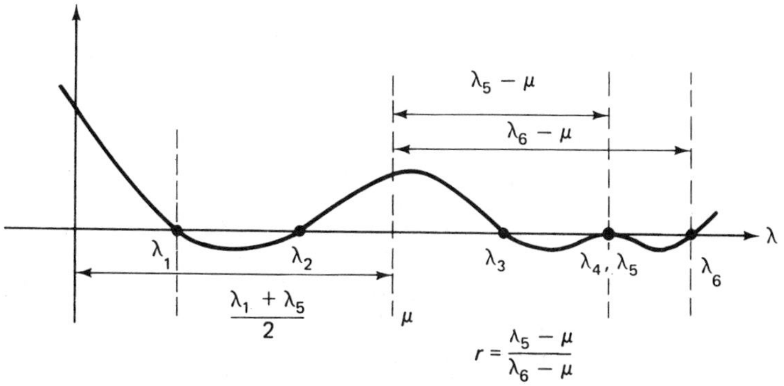

FIGURE 11.2 *Shifting to obtain best convergence rate r in forward iteration for λ_6 (λ_6 = largest eigenvalue).*

11.2.4 Rayleigh Quotient Iteration

We discussed in Section 11.2.3 that the convergence rate in inverse iteration can be much improved by shifting. In practice, the difficulty lies in choosing an appropriate shift. One possibility is to use as shift value the Rayleigh quotient calculated in (11.18), which is an approximation to the eigenvalue sought. If a new shift using (11.18) is evaluated in each iteration, we have the Rayleigh quotient iteration.[15-18] In this procedure we assume a starting iteration vector $\mathbf{x}_1$, hence $\mathbf{y}_1 = \mathbf{M}\mathbf{x}_1$, a starting shift, $\rho(\bar{\mathbf{x}}_1)$, which is usually zero, and then evaluate for $k = 1, 2, \ldots$:

$$(\mathbf{K} - \rho(\bar{\mathbf{x}}_k)\mathbf{M})\bar{\mathbf{x}}_{k+1} = \mathbf{y}_k \tag{11.50}$$

$$\bar{\mathbf{y}}_{k+1} = \mathbf{M}\bar{\mathbf{x}}_{k+1} \tag{11.51}$$

$$\rho(\bar{\mathbf{x}}_{k+1}) = \frac{\bar{\mathbf{x}}_{k+1}^T \mathbf{y}_k}{\bar{\mathbf{x}}_{k+1}^T \bar{\mathbf{y}}_{k+1}} + \rho(\bar{\mathbf{x}}_k) \tag{11.52}$$

$$\mathbf{y}_{k+1} = \frac{\bar{\mathbf{y}}_{k+1}}{(\bar{\mathbf{x}}_{k+1}^T \bar{\mathbf{y}}_{k+1})^{1/2}} \tag{11.53}$$

where now $\mathbf{y}_{k+1} \longrightarrow \mathbf{M}\boldsymbol{\phi}_i$ and $\rho(\bar{\mathbf{x}}_{k+1}) \longrightarrow \lambda_i$ as $k \longrightarrow \infty$

The eigenvalue λ_i and corresponding eigenvector $\boldsymbol{\phi}_i$ to which the iteration converges depend on the starting iteration vector $\mathbf{x}_1$ and the initial shift $\rho(\bar{\mathbf{x}}_1)$. If $\mathbf{x}_1$ has strong components of an eigenvector, say $\boldsymbol{\phi}_k$, and $\rho(\bar{\mathbf{x}}_2)$ provides a sufficiently close shift to the corresponding eigenvalue λ_k, then the iteration converges to the eigenpair $(\lambda_k, \boldsymbol{\phi}_k)$, and the ultimate order of convergence for both λ_k and $\boldsymbol{\phi}_k$ is cubic. Otherwise, convergence will occur to another eigenpair (disregarding a special case that has a very small probability), but *convergence will always be cubic.*[17] This excellent convergence behavior is a most important observation. We may intuitively explain it by the fact that in inverse iteration the vector converges linearly, and with an error of order ϵ in the vector the Rayleigh quotient predicts the eigenvalue with an error of order ϵ^2. Since the eigenvalue approximation used as a shift has a direct effect on the approximation to be obtained for the eigenvector, and vice versa, it seems probable that in Rayleigh quotient iteration, the order of convergence is cubic for both the eigenvalue and eigenvector.

To analyze the convergence characteristics of Rayleigh quotient iteration we may proceed in the same way as in the analysis of inverse iteration; i.e., we consider the iteration in the basis of the eigenvectors. In this case we use the transformation in (11.24) and write the two basic equations of Rayleigh quotient iteration [i.e., (11.50) and (11.52), respectively] in the following form:

$$(\boldsymbol{\Lambda} - \rho(\mathbf{z}_k)\mathbf{I})\mathbf{z}_{k+1} = \mathbf{z}_k \tag{11.54}$$

and

$$\rho(\mathbf{z}_{k+1}) = \frac{\mathbf{z}_{k+1}^T \mathbf{z}_k}{\mathbf{z}_{k+1}^T \mathbf{z}_{k+1}} + \rho(\mathbf{z}_k) \tag{11.55}$$

where the length normalization of the iteration vector has been omitted.

To consider the convergence characteristics of the iteration vector, let us perform an approximate convergence analysis that gives insight into the working of the algorithm. Assume that the current iteration vector $\mathbf{z}_l$ is already close to the eigenvector $\mathbf{e}_1$; i.e., we have

$$\mathbf{z}_l^T = [1 \quad o(\epsilon) \quad o(\epsilon) \ldots o(\epsilon)] \tag{11.56}$$

where $o(\epsilon)$ denotes "of order ϵ" and $\epsilon \ll 1$. We then obtain

$$\rho(\mathbf{z}_l) = \lambda_1 + o(\epsilon^2) \tag{11.57}$$

Solving from (11.54) for $\mathbf{z}_{l+1}$, we thus have

$$\mathbf{z}_{l+1}^T = \left[\frac{1}{o(\epsilon^2)} \; \frac{o(\epsilon)}{\lambda_2 - \lambda_1} \cdots \frac{o(\epsilon)}{\lambda_n - \lambda_1}\right] \tag{11.58}$$

In order to assess the convergence of the iteration vector we normalize to 1 the first component of $\mathbf{z}_{l+1}$, to obtain

$$\bar{\mathbf{z}}_{l+1}^T = [1 \quad o(\epsilon^3) \quad o(\epsilon^3) \quad \ldots \quad o(\epsilon^3)] \tag{11.59}$$

Hence the elements that in $\mathbf{z}_l$ have been of order ϵ are now of order ϵ^3, indicating cubic convergence.

Consider the following example to demonstrate the characteristics of Rayleigh quotient iteration.

EXAMPLE 11.7: Perform the Rayleigh quotient iteration on the problem $\mathbf{\Lambda}\boldsymbol{\phi} = \lambda\boldsymbol{\phi}$, where

$$\mathbf{\Lambda} = \begin{bmatrix} 2 & 0 \\ 0 & 6 \end{bmatrix}$$

Use as starting iteration vectors $\mathbf{x}_1$ the vectors

$$(1) \quad \mathbf{x}_1 = \begin{bmatrix} 1 \\ 1 \end{bmatrix}; \qquad (2) \quad \mathbf{x}_1 = \begin{bmatrix} 1 \\ 0.1 \end{bmatrix}$$

Using the relations given in (11.50) to (11.53) [with $\rho(\bar{\mathbf{x}}_1) = 0.0$], we obtain, in case (1),

$$\bar{\mathbf{x}}_2 = \begin{bmatrix} 0.500 \\ 0.166667 \end{bmatrix}; \qquad \rho(\bar{\mathbf{x}}_2) = 2.40$$

$$\mathbf{y}_2 = \begin{bmatrix} 0.94868 \\ 0.31623 \end{bmatrix}$$

$$\bar{\mathbf{x}}_3 = \begin{bmatrix} -2.37171 \\ 0.08784 \end{bmatrix}; \qquad \rho(\bar{\mathbf{x}}_3) = 2.00548$$

$$\mathbf{y}_3 = \begin{bmatrix} -0.99931 \\ 0.03701 \end{bmatrix}$$

$$\bar{\mathbf{x}}_4 = \begin{bmatrix} 182.37496 \\ 0.00927 \end{bmatrix}; \qquad \rho(\bar{\mathbf{x}}_4) = 2.000000$$

and

$$\mathbf{y}_4 = \begin{bmatrix} 1.0000 \\ 0.00005 \end{bmatrix}$$

Hence we see that in three steps of iteration we obtained a good approximation to the required eigenvalue and eigenvector.

In case (2) we have

$$\bar{\mathbf{x}}_2 = \begin{bmatrix} 0.50000 \\ 0.016667 \end{bmatrix}; \qquad \rho(\bar{\mathbf{x}}_2) = 2.00444$$

$$\mathbf{y}_2 = \begin{bmatrix} 0.99944 \\ 0.033315 \end{bmatrix}$$

and then

$$\bar{\mathbf{x}}_3 = \begin{bmatrix} -225.125 \\ 0.00834 \end{bmatrix}; \qquad \rho(\bar{\mathbf{x}}_3) = 2.000001$$

$$\mathbf{y}_3 = \begin{bmatrix} -1.00000 \\ 0.000037 \end{bmatrix}$$

We observe that in this case two iterations are sufficient to obtain a good approximation to the required eigenvalue and eigenvector, because the starting iteration vector was already closer to the required eigenvector.

As was pointed out in the preceding discussion, the Rayleigh quotient iteration can, in principle, converge to any eigenpair. Therefore, if we are interested in the smallest p eigenvalues and corresponding eigenvectors, we need to supplement the Rayleigh quotient iterations by another technique to assure convergence to one of the eigenpairs sought. For example, to calculate the smallest eigenvalue and corresponding eigenvector, we may first use the inverse iteration in (11.16) to (11.19) without shifting to obtain an iteration vector that is a good approximation of $\boldsymbol{\phi}_1$, and only then start with Rayleigh quotient iteration.[22] However, the difficulty lies in assessing how many inverse iterations must be performed before Rayleigh quotient shifting can be started and yet convergence to $\boldsymbol{\phi}_1$ and λ_1 is achieved. Unfortunately, this question can in general not be resolved, and it is necessary to use the Sturm sequence property to make sure that the required eigenvalue and corresponding eigenvector have indeed been calculated (see Section 11.5).

11.2.5 Matrix Deflation and Gram–Schmidt Orthogonalization

In Sections 11.2.1 to 11.2.4 we discussed how an eigenvalue and corresponding eigenvector can be calculated using vector iteration. The basic inverse iteration technique converges to λ_1 and $\boldsymbol{\phi}_1$ (see Section 11.2.1), the basic forward iteration can be used to calculate λ_n and $\boldsymbol{\phi}_n$ (see Section 11.2.2), but the methods can also be employed with shifting to calculate other eigenvalues and corresponding eigenvectors (see Section 11.2.3). Assume now that we have calculated a specific eigenpair, say $(\lambda_k, \boldsymbol{\phi}_k)$, using either method and that we require the solution of another eigenpair. To assure that we do not converge again to λ_k and $\boldsymbol{\phi}_k$, we need to deflate either the matrices or the iteration vectors.[1-10,19,20]

Matrix deflation has been applied extensively in the solution of standard eigenproblems. The problem may be $\mathbf{K}\boldsymbol{\phi} = \lambda\boldsymbol{\phi}$, i.e., when $\mathbf{M}$ is the identity matrix in $\mathbf{K}\boldsymbol{\phi} = \lambda\mathbf{M}\boldsymbol{\phi}$, or may be $\tilde{\mathbf{K}}\tilde{\boldsymbol{\phi}} = \lambda\tilde{\boldsymbol{\phi}}$, which is obtained by transforming the generalized eigenproblem into a standard form (see Section 10.2.5). We recall that this transformation is effective when $\mathbf{M}$ is diagonal and all diagonal elements are larger than zero, because in such a case $\tilde{\mathbf{K}}$ has the same bandwidth as $\mathbf{K}$.

Consider the deflation of $\mathbf{K}\boldsymbol{\phi} = \lambda\boldsymbol{\phi}$, because the deflation of $\tilde{\mathbf{K}}\tilde{\boldsymbol{\phi}} = \lambda\tilde{\boldsymbol{\phi}}$ would be obtained in the same way. A stable matrix deflation can be carried out by finding an orthogonal matrix $\mathbf{P}$ whose first column is the calculated eigenvector $\boldsymbol{\phi}_k$.

Writing $\mathbf{P}$ as

$$\mathbf{P} = [\boldsymbol{\phi}_k, \mathbf{p}_2, \ldots, \mathbf{p}_n] \tag{11.60}$$

we need to have $\boldsymbol{\phi}_k^T\mathbf{p}_i = 0$ for $i = 2, \ldots, n$. It then follows that

$$\mathbf{P}^T\mathbf{K}\mathbf{P} = \begin{bmatrix} \lambda_k & \mathbf{0} \\ \mathbf{0} & \mathbf{K}_1 \end{bmatrix} \tag{11.61}$$

because $\boldsymbol{\phi}_k^T\boldsymbol{\phi}_k = 1$. The important point is that $\mathbf{P}^T\mathbf{K}\mathbf{P}$ has the same eigenvalues as $\mathbf{K}$, and therefore $\mathbf{K}_1$ must have all eigenvalues of $\mathbf{K}$ except λ_k. In addition, denoting the eigenvectors of $\mathbf{P}^T\mathbf{K}\mathbf{P}$ by $\bar{\boldsymbol{\phi}}_i$, we have that

$$\boldsymbol{\phi}_i = \mathbf{P}\bar{\boldsymbol{\phi}}_i \tag{11.62}$$

It is important to note that the matrix $\mathbf{P}$ is not unique and that various techniques can be used to construct an appropriate transformation matrix.

Since $\mathbf{K}$ is banded we would like to have that the transformation does not destroy the bandform, and a very effective procedure can be used.[21,22]

From this discussion it follows that once a second required eigenpair using $\mathbf{K}_1$ has been evaluated, the process of deflation could be repeated by working with $\mathbf{K}_1$ rather than with $\mathbf{K}$. Therefore, we may continue deflating until all required eigenvalues and eigenvectors have been calculated. The disadvantage of matrix deflation is that the eigenvectors have to be calculated to very high precision to avoid the accumulation of errors introduced in the deflation process.

Instead of matrix deflation, we may deflate the iteration vector in order to converge to an eigenpair other than $(\lambda_k, \boldsymbol{\phi}_k)$. The basis of vector deflation is that in order for an iteration vector to converge in forward or inverse iteration to a required eigenvector, the iteration vector must not be orthogonal to it. Hence, conversely, if the iteration vector is orthogonalized to the eigenvectors already calculated, we eliminate the possibility that the iteration converges to any one of them, and, as we will see, convergence occurs instead to another eigenvector.

A particular vector orthogonalization procedure that is employed extensively is the Gram-Schmidt method. The procedure can be used in the solution of the generalized eigenproblem $\mathbf{K}\boldsymbol{\phi} = \lambda\mathbf{M}\boldsymbol{\phi}$, where $\mathbf{M}$ can take the different forms that we encounter in finite element analysis.

In order to consider a general case, assume that we have calculated in inverse iteration the eigenvectors $\boldsymbol{\phi}_1, \boldsymbol{\phi}_2, \ldots, \boldsymbol{\phi}_m$, and that we want to $\mathbf{M}$-orthogonalize $\mathbf{x}_1$ to these eigenvectors. In Gram-Schmidt orthogonalization a vector $\tilde{\mathbf{x}}_1$, which is $\mathbf{M}$-orthogonal to the eigenvectors $\boldsymbol{\phi}_i$, $i = 1, \ldots, m$, is calculated using

$$\tilde{\mathbf{x}}_1 = \mathbf{x}_1 - \sum_{i=1}^{m} \alpha_i \boldsymbol{\phi}_i \tag{11.63}$$

where the coefficients α_i are obtained using the conditions that $\boldsymbol{\phi}_i^T\mathbf{M}\tilde{\mathbf{x}}_1 = 0$, $i = 1, \ldots, m$, and $\boldsymbol{\phi}_i^T\mathbf{M}\boldsymbol{\phi}_j = \delta_{ij}$. Premultiplying both sides of (11.63) by $\boldsymbol{\phi}_i^T\mathbf{M}$, we therefore obtain

$$\alpha_i = \boldsymbol{\phi}_i^T\mathbf{M}\mathbf{x}_1; \qquad i = 1, \ldots, m \tag{11.64}$$

In the inverse iteration we would now use $\tilde{\mathbf{x}}_1$ as the starting iteration vector instead of $\mathbf{x}_1$ and provided that $\mathbf{x}_1^T\mathbf{M}\boldsymbol{\phi}_{m+1} \neq 0$, convergence occurs (at least in theory, see Section 11.2.6) to $\boldsymbol{\phi}_{m+1}$ and λ_{m+1}.

To prove the convergence given above, we consider as before the iteration process in the basis of eigenvectors; i.e., we analyze the iteration given in (11.25) when the Gram-Schmidt orthogonalization is included. In this case the eigenvectors corresponding to the smallest eigenvalues are $\mathbf{e}_i$, $i = 1, \ldots, m$. Carrying out the deflation of the starting iteration vector $\mathbf{z}_1$ in (11.27), we obtain

$$\tilde{\mathbf{z}}_1 = \mathbf{z}_1 - \sum_{i=1}^{m} \alpha_i \mathbf{e}_i \tag{11.65}$$

with

$$\alpha_i = \mathbf{e}_i^T\mathbf{z}_1 = 1; \qquad i = 1, \ldots, m \tag{11.66}$$

Hence

$$\tilde{\mathbf{z}}_1^T = [0 \;\ldots\; 0 \;\; \underset{\text{element } m+1}{1} \;\ldots\; 1] \tag{11.67}$$

Using now $\tilde{\mathbf{z}}_1$ as the starting iteration vector and performing the convergence analysis as discussed in Section 11.2.1, we find that if $\lambda_{m+2} > \lambda_{m+1}$, we have that $\bar{\mathbf{z}}_{l+1} \longrightarrow \mathbf{e}_{m+1}$, as was required to prove. Furthermore,we find that the rate of convergence of the eigenvector is $\lambda_{m+1}/\lambda_{m+2}$, and in case λ_{m+1} is a multiple eigenvalue, the rate of convergence is given by the ratio of λ_{m+1} to the next distinct eigenvalue.

Although so far we have discussed Gram–Schmidt orthogonalization in connection with vector inverse iteration, it should be realized that the orthogonalization procedure can be used equally well in the other vector iteration methods. All convergence considerations discussed in the presentation of inverse iteration, forward iteration, and Rayleigh quotient iteration are also applicable when Gram–Schmidt orthogonalization is included if it is taken into account that convergence to the eigenvectors already calculated is not possible.

EXAMPLE 11.8: Calculate, using Gram–Schmidt orthogonalization, an appropriate starting iteration vector for the solution of the problem $\mathbf{K}\boldsymbol{\phi} = \lambda\mathbf{M}\boldsymbol{\phi}$, where **K** and **M** are given in Example 11.4. Assume that the eigenpairs $(\lambda_1, \boldsymbol{\phi}_1)$ and $(\lambda_4, \boldsymbol{\phi}_4)$ are known as obtained in Example 11.5 and that convergence to another eigenpair is sought.

To determine an appropriate starting iteration vector, we want to deflate the unit full vector of the iteration vectors $\boldsymbol{\phi}_1$ and $\boldsymbol{\phi}_4$; i.e., (11.63) reads

$$\tilde{\mathbf{x}}_1 = \begin{bmatrix} 1 \\ 1 \\ 1 \\ 1 \end{bmatrix} - \alpha_1\boldsymbol{\phi}_1 - \alpha_4\boldsymbol{\phi}_4$$

where α_1 and α_4 are obtained using (11.64):

$$\alpha_1 = \boldsymbol{\phi}_1^T\mathbf{M}\mathbf{x}_1; \qquad \alpha_4 = \boldsymbol{\phi}_4^T\mathbf{M}\mathbf{x}_1$$

Substituting for **M**, $\boldsymbol{\phi}_1$, and $\boldsymbol{\phi}_4$, we obtain

$$\alpha_1 = 2.385; \qquad \alpha_4 = 0.1299$$

Then, to a few-digit accuracy,

$$\tilde{\mathbf{x}}_1 = \begin{bmatrix} 0.2683 \\ -0.2149 \\ -0.04812 \\ 0.2358 \end{bmatrix}$$

11.2.6 Some Practical Considerations Concerning Vector Iterations

So far we have discussed the theory used in vector iteration techniques. However, for a proper computer implementation of the methods, it is important to interpret the results obtained concerning the iteration procedures. Of particular importance are practical convergence and stability considerations when any one of the techniques is used.

A first important point is that the convergence rates of the iterations may turn out when measured in practice to be rather theoretical, Namely, we assumed that the starting iteration vector $\mathbf{z}_1$ in (11.27) is a unit full vector, which corresponds to a vector $\mathbf{x}_1 = \sum_{i=1}^{n} \boldsymbol{\phi}_i$. This means that the starting iteration vector is equally strong in each of the eigenvectors $\boldsymbol{\phi}_i$. We chose this starting iteration vector to identify easily the theoretical convergence rate with which the iteration vector approaches the required eigenvector. However, in practice it is hardly possible to pick $\mathbf{x}_1 = \sum_{i=1}^{n} \boldsymbol{\phi}_i$ as the starting iteration vector, and instead we have

$$\mathbf{x}_1 = \sum_{i=1}^{n} \alpha_i \boldsymbol{\phi}_i \tag{11.68}$$

where the α_i are arbitrary constants. This vector $\mathbf{x}_1$ corresponds to the following vector in the basis of eigenvectors:

$$\mathbf{z}_1 = \begin{bmatrix} \alpha_1 \\ \cdot \\ \cdot \\ \cdot \\ \alpha_n \end{bmatrix} \tag{11.69}$$

To identify the effect of the constants α_i, consider as an example the convergence analysis of inverse iteration without shifting when the starting vector in (11.68) is used and $\lambda_2 > \lambda_1$. The conclusions reached will be equally applicable to other vector iteration methods. As before, we consider the iteration in the basis of eigenvectors $\boldsymbol{\Phi}$ and require $\alpha_1 \neq 0$ in order to have $\mathbf{x}_1^T \mathbf{M} \boldsymbol{\phi}_1 \neq 0$. After l inverse iterations we now have instead of (11.29)

$$\bar{\mathbf{z}}_{l+1} = \begin{bmatrix} 1 \\ \beta_2 \left(\dfrac{\lambda_1}{\lambda_2}\right)^l \\ \cdot \\ \cdot \\ \beta_n \left(\dfrac{\lambda_1}{\lambda_n}\right)^l \end{bmatrix} \tag{11.70}$$

$$\beta_i = \frac{\alpha_i}{\alpha_1}; \qquad i = 2, \ldots, n \tag{11.71}$$

Therefore, the iteration vector obtained now has the multipliers β_i in its last $(n - 1)$ components. In the iteration the ith component is still decreasing with each iteration, as in (11.29) by the factor λ_1/λ_i, $i = 2, \ldots, n$, and the rate of convergence is λ_1/λ_2, as was already derived in Section 11.2.1. However, in practical analysis the unknown coefficients β_i may produce the result that the theoretical convergence rate is not observed for many iterations. In practice, therefore, not only the order and rate of convergence, but equally important the "quality" of the starting iteration vector determines the number of iterations required for convergence. Furthermore, it is important to use a high-enough convergence tolerance to prevent premature acceptance of the iteration vector as an approximation to the required eigenvector.

Together with the vector iterations, we may use a matrix deflation procedure or Gram-Schmidt vector orthogonalization to obtain convergence to an eigenpair not already calculated (see Section 11.2.5). We mentioned already that for matrix deflation, the eigenvectors have to be evaluated to relatively high precision to preserve stability. Considering the Gram-Schmidt orthogonalization, the method must also be used with care. Namely, if the technique is employed in inverse or forward iteration without shifting, it is necessary to calculate the eigenvectors to high precision in order that Gram-Schmidt orthogonalization will work. In addition, the iteration vector should be orthogonalized in each iteration to the eigenvectors already calculated.

Let us now draw an important conclusion. We pointed out earlier in the presentation of the vector iteration techniques that it is difficult (and indeed theory shows impossible) to *assure* convergence to a specific (but arbitrarily selected) eigenvalue and corresponding eigenvector. The discussion concerning practical aspects in this section substantiates those observations, and it is concluded that the vector iteration procedures and the Gram-Schmidt orthogonalization process must be employed with care if a specific eigenvalue and corresponding eigenvector are required. We will see in Chapter 12 that, in fact, both techniques are best employed and are used very effectively in conjunction with other solution strategies.

11.3 TRANSFORMATION METHODS

We pointed out in Section 11.1 that the transformation methods comprise a group of eigensystem solution procedures that employ the basic properties of the eigenvectors in the matrix $\mathbf{\Phi}$,

$$\mathbf{\Phi}^T\mathbf{K}\mathbf{\Phi} = \mathbf{\Lambda} \tag{11.3}$$

and

$$\mathbf{\Phi}^T\mathbf{M}\mathbf{\Phi} = \mathbf{I} \tag{11.4}$$

Since the matrix $\mathbf{\Phi}$, of order $n \times n$, which diagonalizes $\mathbf{K}$ and $\mathbf{M}$ in the way given in (11.3) and (11.4) is unique, we can try to construct it by iteration.[1-9] The basic scheme is to reduce $\mathbf{K}$ and $\mathbf{M}$ into diagonal form using successive pre- and postmultiplication by matrices $\mathbf{P}_k^T$ and $\mathbf{P}_k$, respectively, where $k = 1, 2, \ldots$. Specifically, if we define $\mathbf{K}_1 = \mathbf{K}$ and $\mathbf{M}_1 = \mathbf{M}$, we form

$$\left.\begin{aligned} \mathbf{K}_2 &= \mathbf{P}_1^T\mathbf{K}_1\mathbf{P}_1 \\ \mathbf{K}_3 &= \mathbf{P}_2^T\mathbf{K}_2\mathbf{P}_2 \\ &\vdots \\ \mathbf{K}_{k+1} &= \mathbf{P}_k^T\mathbf{K}_k\mathbf{P}_k \\ &\vdots \end{aligned}\right\} \tag{11.72}$$

Similarly,

$$\left.\begin{aligned} \mathbf{M}_2 &= \mathbf{P}_1^T\mathbf{M}_1\mathbf{P}_1 \\ \mathbf{M}_3 &= \mathbf{P}_2^T\mathbf{M}_2\mathbf{P}_2 \\ &\vdots \\ \mathbf{M}_{k+1} &= \mathbf{P}_k^T\mathbf{M}_k\mathbf{P}_k \\ &\vdots \end{aligned}\right\} \tag{11.73}$$

where the matrices $\mathbf{P}_k$ are selected to bring $\mathbf{K}_k$ and $\mathbf{M}_k$ closer to diagonal form. Then for a proper procedure we apparently need to have

$$\mathbf{K}_{k+1} \longrightarrow \mathbf{\Lambda} \quad \text{and} \quad \mathbf{M}_{k+1} \longrightarrow \mathbf{I} \qquad \text{as } k \longrightarrow \infty$$

in which case, with l being the last iteration,

$$\mathbf{\Phi} = \mathbf{P}_1\mathbf{P}_2 \ldots \mathbf{P}_l \tag{11.74}$$

In practice, it is not necessary that $\mathbf{M}_{k+1}$ converges to $\mathbf{I}$ and $\mathbf{K}_{k+1}$ to $\mathbf{\Lambda}$, but they only need to converge to diagonal form. Namely, if

$$\mathbf{K}_{k+1} \longrightarrow \operatorname{diag}(K_r) \quad \text{and} \quad \mathbf{M}_{k+1} \longrightarrow \operatorname{diag}(M_r) \qquad \text{as } k \longrightarrow \infty$$

then with l indicating the last iteration and disregarding that the eigenvalues and eigenvectors may not be in the usual order,

$$\mathbf{\Lambda} = \operatorname{diag}\left(\frac{K_r^{(l+1)}}{M_r^{(l+1)}}\right) \tag{11.75}$$

and

$$\mathbf{\Phi} = \mathbf{P}_1\mathbf{P}_2 \ldots \mathbf{P}_l \operatorname{diag}\left(\frac{1}{\sqrt{M_r^{(l+1)}}}\right) \tag{11.76}$$

Using the basic idea described above, a number of different iteration methods have been proposed. We shall discuss in the next sections only the Jacobi and the Householder-QR methods, which are believed to be most effective in finite element analysis. However, before presenting the techniques in detail we should point out one important aspect. In the above introduction it was implied that iteration is started with pre- and postmultiplication by $\mathbf{P}_1^T$ and $\mathbf{P}_1$, respectively, which is indeed the case in the Jacobi solution methods. However, it should be noted that, alternatively, we may first aim to transform the eigenvalue problem $\mathbf{K}\boldsymbol{\phi} = \lambda\mathbf{M}\boldsymbol{\phi}$ into a form that is more economical to use in the iteration. In particular, when $\mathbf{M} = \mathbf{I}$, the first m transformations in (11.72) may be used to reduce $\mathbf{K}$ into tridiagonal form without iteration, after which the matrices $\mathbf{P}_i$, $i = m + 1, \ldots, l$, are applied in an iterative manner to bring $\mathbf{K}_{m+1}$ into diagonal form. In such a case the first matrices $\mathbf{P}_1, \ldots, \mathbf{P}_m$ may be of different form than the later applied matrices $\mathbf{P}_{m+1}, \ldots, \mathbf{P}_l$. An application of this procedure is the Householder-QR method, in which Householder matrices are used to first transform $\mathbf{K}$ into tridiagonal form and then rotation matrices are employed in the QR transformations. The same solution strategy can also be used to solve the generalized eigenproblem $\mathbf{K}\boldsymbol{\phi} = \lambda\mathbf{M}\boldsymbol{\phi}$, $\mathbf{M} \neq \mathbf{I}$, provided that the problem is first transformed into the standard form.

11.3.1 The Jacobi Method

The basic Jacobi solution method has been developed for the solution of standard eigenproblems (i.e., $\mathbf{M}$ being the identity matrix), and we consider it in this section. The method was proposed over a century ago and has been used extensively.[1-9,23-27] A major advantage of the procedure is its simplicity and stability. Since the eigenvector properties in (11.3) and (11.4) (with $\mathbf{M} = \mathbf{I}$) are applicable to all symmetric matrices $\mathbf{K}$ with no restriction on the eigenvalues, the Jacobi method can be used to calculate negative, zero, or positive eigenvalues.

Considering the standard eigenproblem $\mathbf{K}\boldsymbol{\phi} = \lambda\boldsymbol{\phi}$, the kth iteration step defined in (11.72) reduces to

$$\mathbf{K}_{k+1} = \mathbf{P}_k^T\mathbf{K}_k\mathbf{P}_k \tag{11.77}$$

where $\mathbf{P}_k$ is an orthogonal matrix; i.e., (11.73) gives

$$\mathbf{P}_k^T\mathbf{P}_k = \mathbf{I} \tag{11.78}$$

In the Jacobi solution the matrix $\mathbf{P}_k$ is a rotation matrix which is selected in such way that an off-diagonal element in $\mathbf{K}_k$ is zeroed. If element (i, j) is to be reduced to zero, the corresponding orthogonal matrix $\mathbf{P}_k$ is

$$\mathbf{P}_k = \left[\begin{array}{ccccccccccc}
1 &&&&&&&&&& \\
& \ddots &&&&&&&&& \\
&& 1 &&&&&&&& \\
&&& \overset{i\text{th}}{\cos\theta} &&&& \overset{j\text{th column}}{-\sin\theta} &&& \\
&&&& 1 &&&&&& \\
&&&&& \ddots &&&&& \\
&&&&&& 1 &&&& \\
&&& \sin\theta &&&& \cos\theta &&& \\
&&&&&&&& 1 && \\
&&&&&&&&& \ddots & \\
&&&&&&&&&& 1
\end{array}\right]
\begin{array}{l}
\\ \\ \\ \text{—}i\text{th} \\ \\ \\ \\ \text{—}j\text{th row} \\ \\ \\ \\
\end{array} \tag{11.79}$$

where θ is selected from the condition that element (i, j) in $\mathbf{K}_{k+1}$ be zero. Denoting element (i, j) in $\mathbf{K}_k$ by $k_{ij}^{(k)}$, we use

$$\tan 2\theta = \frac{2k_{ij}^{(k)}}{k_{ii}^{(k)} - k_{jj}^{(k)}} \qquad \text{for } k_{ii}^{(k)} \neq k_{jj}^{(k)} \tag{11.80}$$

and

$$\theta = \frac{\pi}{4} \qquad \text{for } k_{ii}^{(k)} = k_{jj}^{(k)} \tag{11.81}$$

It should be noted that the numerical evaluation of $\mathbf{K}_{k+1}$ in (11.77) requires only the linear combination of two rows and two columns. In addition, advantage should also be taken of the fact that $\mathbf{K}_k$ is symmetric for all k; i.e., we should only work on the upper (or lower) triangular part of the matrix, including its diagonal elements.

An important point to emphasize is that although the transformation in (11.77) reduces an off-diagonal element in $\mathbf{K}_k$ to zero, this element will again become nonzero during the transformations that follow. Therefore, for the design of an actual algorithm, we have to decide which element to reduce to zero. One choice is to always zero the largest off-diagonal element in $\mathbf{K}_k$. However, the search for the largest element is time consuming, and it may be preferable to simply carry out the Jacobi transformations systematically, row-by-row or column-by-column, which is known as the *cyclic Jacobi procedure*. Running once over all off-diagonal elements is one *sweep*. The disadvantage of this procedure is that regardless of its size, an off-diagonal element is always zeroed; i.e., the element may already be nearly zero and a rotation is still applied.

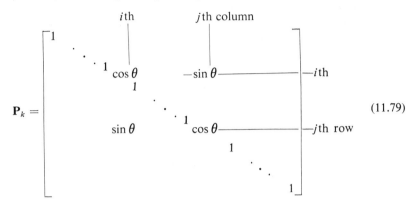

A procedure that has been used very effectively is the *threshold Jacobi method*, in which the off-diagonal elements are tested sequentially, namely row-by-row (or column-by-column), and a rotation is only applied if the element is larger than the threshold for that sweep. To define an appropriate threshold we note that, physically, in the diagonalization of $\mathbf{K}$ we want to reduce the coupling between the degrees of freedom i and j. A measure of this coupling is given by $(k_{ij}^2/k_{ii}k_{jj})^{1/2}$, and it is this factor that can be used effectively in deciding whether to apply a rotation. In addition to having a realistic threshold tolerance, it is also necessary to measure convergence. As described above, $\mathbf{K}_{k+1} \longrightarrow \mathbf{\Lambda}$ as $k \longrightarrow \infty$, but in the numerical computations we seek only a close-enough approximation to the eigenvalues and corresponding eigenvectors. Let l be the last iteration; i.e., we have, to the precision required,

$$\mathbf{K}_{l+1} \doteq \mathbf{\Lambda} \tag{11.82}$$

Then we say that convergence to a tolerance s has been achieved provided that

$$\frac{|k_{ii}^{(l+1)} - k_{ii}^{(l)}|}{k_{ii}^{(l+1)}} \leq 10^{-s}; \qquad i = 1, \ldots, n \tag{11.83}$$

and

$$\left[\frac{(k_{ij}^{(l+1)})^2}{k_{ii}^{(l+1)}k_{jj}^{(l+1)}}\right]^{1/2} \leq 10^{-s}; \qquad \text{all } i, j;\ i < j \tag{11.84}$$

The relation in (11.83) has to be satisfied because the element $k_{ii}^{(l+1)}$ is the current approximation to an eigenvalue, and the relation states that the current and last approximations to the eigenvalues did not change in the first s digits. This convergence measure is essentially the same as the one used in vector iteration in (11.20). The relation in (11.84) assures that the off-diagonal elements are indeed small.

Having discussed the main aspects of the iteration, we may now summarize the actual solution procedure. The following steps have been used in a threshold Jacobi iteration:

1. Initialize the threshold for the sweep. Typically the threshold used for sweep m may be 10^{-2m}.
2. For all (i, j) with $i < j$ calculate the coupling factor $((k_{ij}^{(k)})^2/k_{ii}^{(k)}k_{jj}^{(k)})^{1/2}$ and apply a transformation if the factor is larger than the current threshold.
3. Use (11.83) to check for convergence. If the relation in (11.83) is not satisfied, continue with the next sweep; i.e., go to (1). If (11.83) is satisfied, check if (11.84) is also satisfied; if "yes," the iteration converged; if "no," continue with the next sweep.

So far we have stated the algorithm but we have not shown that convergence will indeed always occur. The proof of convergence has been given elsewhere[23-27] and will not be repeated here, because little additional insight into the working of the solution procedure would be gained. However, one important point should be noted—that convergence is quadratic once the off-diagonal elements are small. Since rapid convergence is obtained once the off-diagonal elements are small, little extra cost is involved in solving for the eigensystem to high accuracy when an approximate solution has been obtained. In practical solutions we use $m = 2$ and $s = 12$, and about six sweeps are required for solution of the eigensystem to high accuracy. A program used is given in the next section, when we discuss the solution of the generalized eigenproblem $\mathbf{K}\boldsymbol{\phi} = \lambda\mathbf{M}\boldsymbol{\phi}$.

EXAMPLE 11.9: Calculate the eigensystem of the matrix **K**, where

$$\mathbf{K} = \begin{bmatrix} 5 & -4 & 1 & 0 \\ -4 & 6 & -4 & 1 \\ 1 & -4 & 6 & -4 \\ 0 & 1 & -4 & 5 \end{bmatrix}$$

Use the threshold Jacobi iteration described above.

To demonstrate the solution algorithm we give one sweep in detail and then the results obtained in the next sweeps.

For sweep 1 we have as a threshold 10^{-2}. We therefore obtain the following results. For $i = 1, j = 2$:

$$\cos\theta = 0.7497; \qquad \sin\theta = 0.6618$$

and thus

$$\mathbf{P}_1 = \begin{bmatrix} 0.7497 & -0.6618 & 0 & 0 \\ 0.6618 & 0.7497 & 0 & 0 \\ 0 & 0 & 1 & 0 \\ 0 & 0 & 0 & 1 \end{bmatrix}$$

$$\mathbf{P}_1^T\mathbf{K}\mathbf{P}_1 = \begin{bmatrix} 1.469 & 0 & -1.898 & 0.6618 \\ 0 & 9.531 & -3.661 & 0.7497 \\ -1.898 & -3.661 & 6 & -4 \\ 0.6618 & 0.7497 & -4 & 5 \end{bmatrix}$$

For $i = 1, j = 3$:

$$\cos\theta = 0.9398; \qquad \sin\theta = 0.3416$$

$$\mathbf{P}_2 = \begin{bmatrix} 0.9398 & 0 & -0.3416 & 0 \\ 0 & 1 & 0 & 0 \\ 0.3416 & 0 & 0.9398 & 0 \\ 0 & 0 & 0 & 1 \end{bmatrix}$$

$$\mathbf{P}_2^T\mathbf{P}_1^T\mathbf{K}\mathbf{P}_1\mathbf{P}_2 = \begin{bmatrix} 0.7792 & -1.250 & 0 & -0.7444 \\ -1.250 & 9.531 & -3.440 & 0.7497 \\ 0 & -3.440 & 6.690 & -3.986 \\ -0.7444 & 0.7497 & -3.986 & 5 \end{bmatrix}$$

$$\mathbf{P}_1\mathbf{P}_2 = \begin{bmatrix} 0.7046 & -0.6618 & -0.2561 & 0 \\ 0.6220 & 0.7497 & -0.2261 & 0 \\ 0.3416 & 0 & 0.9398 & 0 \\ 0 & 0 & 0 & 1 \end{bmatrix}$$

For $i = 1, j = 4$:

$$\cos\theta = 0.9857; \qquad \sin\theta = 0.1687$$

$$\mathbf{P}_3 = \begin{bmatrix} 0.9857 & 0 & 0 & -0.1687 \\ 0 & 1 & 0 & 0 \\ 0 & 0 & 1 & 0 \\ 0.1687 & 0 & 0 & 0.9857 \end{bmatrix}$$

$$\mathbf{P}_3^T\mathbf{P}_2^T\mathbf{P}_1^T\mathbf{K}\mathbf{P}_1\mathbf{P}_2\mathbf{P}_3 = \begin{bmatrix} 0.6518 & -1.106 & -0.6725 & 0 \\ -1.106 & 9.531 & -3.440 & 0.9499 \\ -0.6725 & -3.440 & 6.690 & -3.928 \\ 0 & 0.9499 & -3.928 & 5.127 \end{bmatrix}$$

$$\mathbf{P}_1\mathbf{P}_2\mathbf{P}_3 = \begin{bmatrix} 0.6945 & -0.6618 & -0.2561 & -0.1189 \\ 0.6131 & 0.7497 & -0.2261 & -0.1050 \\ 0.3367 & 0 & 0.9398 & -0.0576 \\ 0.1687 & 0 & 0 & 0.9857 \end{bmatrix}$$

For $i = 2, j = 3$:

$$\cos\theta = 0.8312; \qquad \sin\theta = -0.5560$$

$$\mathbf{P}_4 = \begin{bmatrix} 1 & 0 & 0 & 0 \\ 0 & 0.8312 & 0.5560 & 0 \\ 0 & -0.5560 & 0.8312 & 0 \\ 0 & 0 & 0 & 1 \end{bmatrix}$$

$$\mathbf{P}_4^T\mathbf{P}_3^T\mathbf{P}_2^T\mathbf{P}_1^T\mathbf{K}\mathbf{P}_1\mathbf{P}_2\mathbf{P}_3\mathbf{P}_4 = \begin{bmatrix} 0.6518 & 0.5453 & -1.174 & 0 \\ -0.5453 & 11.83 & 0 & 2.974 \\ -1.174 & 0 & 4.388 & -2.737 \\ 0 & 2.974 & -2.737 & 5.127 \end{bmatrix}$$

$$\mathbf{P}_1\mathbf{P}_2\mathbf{P}_3\mathbf{P}_4 = \begin{bmatrix} 0.6945 & -0.4077 & -0.5808 & -0.1189 \\ 0.6131 & 0.7488 & 0.2289 & -0.1050 \\ 0.3367 & -0.5226 & 0.7812 & -0.0576 \\ 0.1682 & 0 & 0 & 0.9857 \end{bmatrix}$$

For $i = 2, j = 4$:

$$\cos\theta = 0.9349; \qquad \sin\theta = 0.3549$$

$$\mathbf{P}_5 = \begin{bmatrix} 1 & 0 & 0 & 0 \\ 0 & 0.9349 & 0 & -0.3549 \\ 0 & 0 & 1 & 0 \\ 0 & 0.3549 & 0 & 0.9349 \end{bmatrix}$$

$$\mathbf{P}_5^T\mathbf{P}_4^T\mathbf{P}_3^T\mathbf{P}_2^T\mathbf{P}_1^T\mathbf{K}\mathbf{P}_1\mathbf{P}_2\mathbf{P}_3\mathbf{P}_4\mathbf{P}_5 = \begin{bmatrix} 0.6518 & 0.5098 & -1.174 & 0.1935 \\ -0.5098 & 12.96 & 0.9713 & 0 \\ -1.174 & -0.9713 & 4.388 & -2.559 \\ 0.1935 & 0 & -2.559 & 3.999 \end{bmatrix}$$

$$\mathbf{P}_1\mathbf{P}_2\mathbf{P}_3\mathbf{P}_4\mathbf{P}_5 = \begin{bmatrix} 0.6945 & -0.4233 & -0.5808 & 0.0335 \\ 0.6131 & 0.6628 & 0.2289 & -0.3639 \\ 0.3367 & 0.5090 & 0.7812 & 0.1316 \\ 0.1687 & 0.3498 & 0 & 0.9213 \end{bmatrix}$$

To complete the sweep, we zero element (3, 4), using

$$\cos\theta = 0.7335; \qquad \sin\theta = -0.6797$$

$$\mathbf{P}_6 = \begin{bmatrix} 1 & 0 & 0 & 0 \\ 0 & 1 & 0 & 0 \\ 0 & 0 & 0.7335 & 0.6797 \\ 0 & 0 & -0.6797 & 0.7335 \end{bmatrix}$$

and hence the approximations obtained for $\mathbf{\Lambda}$ and $\mathbf{\Phi}$ are

$$\mathbf{\Lambda} \doteq \mathbf{P}_6^T \ldots \mathbf{P}_1^T\mathbf{K}\mathbf{P}_1 \ldots \mathbf{P}_6$$

i.e.,

$$\mathbf{\Lambda} \doteq \begin{bmatrix} 0.6518 & -0.5098 & -0.9926 & -0.6560 \\ -0.5098 & 12.96 & -0.7124 & -0.6602 \\ -0.9926 & -0.7124 & 6.7596 & 0 \\ -0.6560 & -0.6602 & 0 & 1.6272 \end{bmatrix}$$

and

$$\mathbf{\Phi} \doteq \mathbf{P}_1 \ldots \mathbf{P}_6$$

i.e.,

$$\mathbf{\Phi} \doteq \begin{bmatrix} 0.6945 & -0.4233 & -0.4488 & -0.3702 \\ 0.6131 & 0.6628 & 0.4152 & -0.1113 \\ 0.3367 & -0.5090 & 0.4835 & 0.6275 \\ 0.1687 & 0.3498 & -0.6264 & 0.6759 \end{bmatrix}$$

After the second sweep we obtain

$$\mathbf{\Lambda} \doteq \begin{bmatrix} 0.1563 & -0.3635 & 0.0063 & -0.0176 \\ -0.3635 & 13.08 & -0.0020 & 0 \\ 0.0063 & -0.0020 & 6.845 & 0 \\ -0.0176 & 0 & 0 & 1.910 \end{bmatrix}$$

$$\mathbf{\Phi} \doteq \begin{bmatrix} 0.3875 & -0.3612 & -0.6017 & -0.5978 \\ 0.5884 & 0.6184 & 0.3710 & -0.3657 \\ 0.6148 & -0.5843 & 0.3714 & 0.3777 \\ 0.3546 & 0.3816 & -0.6020 & 0.6052 \end{bmatrix}$$

And after the third sweep we have

$$\mathbf{\Lambda} \doteq \begin{bmatrix} 0.1459 & & & \\ & 13.09 & & \\ & & 6.854 & \\ & & & 1.910 \end{bmatrix}$$

$$\mathbf{\Phi} \doteq \begin{bmatrix} 0.3717 & -0.3717 & -0.6015 & -0.6015 \\ 0.6015 & 0.6015 & 0.3717 & -0.3717 \\ 0.6015 & -0.6015 & 0.3717 & 0.3717 \\ 0.3717 & 0.3717 & -0.6015 & 0.6015 \end{bmatrix}$$

The approximation for $\mathbf{\Lambda}$ is diagonal to the precision given and we can use

$$\lambda_1 \doteq 0.1459; \qquad \boldsymbol{\phi}_1 \doteq \begin{bmatrix} 0.3717 \\ 0.6015 \\ 0.6015 \\ 0.3717 \end{bmatrix}$$

$$\lambda_2 \doteq 1.910; \qquad \boldsymbol{\phi}_2 \doteq \begin{bmatrix} -0.6015 \\ -0.3717 \\ 0.3717 \\ 0.6015 \end{bmatrix}$$

$$\lambda_3 \doteq 6.854; \qquad \boldsymbol{\phi}_3 \doteq \begin{bmatrix} -0.6015 \\ 0.3717 \\ 0.3717 \\ -0.6015 \end{bmatrix}$$

$$\lambda_4 \doteq 13.09; \qquad \boldsymbol{\phi}_4 \doteq \begin{bmatrix} -0.3717 \\ 0.6015 \\ -0.6015 \\ 0.3717 \end{bmatrix}$$

It should be noted that the eigenvalues and eigenvectors did not appear in the usual order in the approximation for $\mathbf{\Lambda}$ and $\mathbf{\Phi}$.

In the following example we demonstrate the quadratic convergence when the off-diagonal elements are already small.

EXAMPLE 11.10: Consider the Jacobi solution of the eigenproblem $\mathbf{K}\boldsymbol{\phi} = \lambda\boldsymbol{\phi}$, where

$$\mathbf{K} = \begin{bmatrix} k_{11} & o(\epsilon) & o(\epsilon) \\ o(\epsilon) & k_{22} & o(\epsilon) \\ o(\epsilon) & o(\epsilon) & k_{33} \end{bmatrix}$$

The symbol $o(\epsilon)$ signifies "of order ϵ," where $\epsilon \ll k_{ii}$, $i = 1, 2, 3$. Show that after one complete sweep, all off-diagonal elements are of order ϵ^2, meaning that convergence is quadratic.

Since the rotations to be applied are small, we make the assumption that $\sin\theta = \theta$ and $\cos\theta = 1$. Hence the relation in (11.80) gives

$$\theta = \frac{k_{ij}^{(k)}}{k_{ii}^{(k)} - k_{jj}^{(k)}}$$

In one sweep we need to set to zero, in succession, once all off-diagonal elements. Using $\mathbf{K}_1 = \mathbf{K}$ we obtain $\mathbf{K}_2$ by zeroing element (1, 2) in $\mathbf{K}_1$,

$$\mathbf{K}_2 = \mathbf{P}_1^T\mathbf{K}_1\mathbf{P}_1$$

where

$$\mathbf{P}_1 = \begin{bmatrix} 1 & \dfrac{-o(\epsilon)}{k_{11} - k_{22}} & 0 \\ \dfrac{o(\epsilon)}{k_{11} - k_{22}} & 1 & 0 \\ 0 & 0 & 1 \end{bmatrix}$$

Hence

$$\mathbf{K}_2 = \begin{bmatrix} k_{11} + o(\epsilon^2) & 0 & o(\epsilon) \\ 0 & k_{22} + o(\epsilon^2) & o(\epsilon) \\ o(\epsilon) & o(\epsilon) & k_{33} \end{bmatrix}$$

Similarly, we zero element (1, 3) in $\mathbf{K}_2$ to obtain $\mathbf{K}_3$,

$$\mathbf{K}_3 = \begin{bmatrix} k_{11} + o(\epsilon^2) & o(\epsilon^2) & 0 \\ o(\epsilon^2) & k_{22} + o(\epsilon^2) & o(\epsilon) \\ 0 & o(\epsilon) & k_{33} + o(\epsilon^2) \end{bmatrix}$$

Finally, we zero element (2, 3) in $\mathbf{K}_3$ and have

$$\mathbf{K}_4 = \begin{bmatrix} k_{11} + o(\epsilon^2) & o(\epsilon^2) & o(\epsilon^2) \\ o(\epsilon^2) & k_{22} + o(\epsilon^2) & 0 \\ o(\epsilon^2) & 0 & k_{33} + o(\epsilon^2) \end{bmatrix}$$

with all off-diagonal elements at least $o(\epsilon^2)$.

11.3.2 The Generalized Jacobi Method

In the previous section we discussed the solution of the standard eigenproblem $\mathbf{K}\boldsymbol{\phi} = \lambda\boldsymbol{\phi}$ using the conventional Jacobi rotation matrices in order to reduce $\mathbf{K}$ into diagonal form. To solve the generalized problem $\mathbf{K}\boldsymbol{\phi} = \lambda\mathbf{M}\boldsymbol{\phi}$, $\mathbf{M} \neq \mathbf{I}$, using the standard Jacobi method it would be necessary to first transform the problem into a standard form. However, this transformation can be dispensed with by using a generalized Jacobi solution method which operates directly on $\mathbf{K}$ and $\mathbf{M}$.[11,28] The algorithm proceeds as summarized in (11.72) to (11.76) and is a natural extension of the standard Jacobi solution scheme; i.e., the generalized method reduces to the scheme presented for the problem $\mathbf{K}\boldsymbol{\phi} = \lambda\boldsymbol{\phi}$ when $\mathbf{M}$ is an identity matrix.

Referring to the discussion in the previous section, in the generalized Jacobi iteration we use the following matrix $\mathbf{P}_k$:

$$
\mathbf{P}_k = \begin{array}{c} \begin{matrix} & & i\text{th} & & j\text{th column} & \end{matrix} \\ \begin{bmatrix} 1 & & & & & & \\ & \ddots & & & & & \\ & & 1 & & \alpha & & \\ & & & & & & \\ & & \gamma & & 1 & & \\ & & & & & \ddots & \\ & & & & & & 1 \end{bmatrix} \end{array} \begin{matrix} \\ \\ \text{—}i\text{th} \\ \\ \text{—}j\text{th row} \\ \\ \\ \end{matrix} \tag{11.85}
$$

where the constants α and γ are selected in such a way as to reduce to zero simultaneously elements (i, j) in $\mathbf{K}_k$ and $\mathbf{M}_k$. Therefore, the values of α and γ are a function of the elements $k_{ij}^{(k)}$, $k_{ii}^{(k)}$, $k_{jj}^{(k)}$, $m_{ij}^{(k)}$, $m_{ii}^{(k)}$, and $m_{jj}^{(k)}$, where the superscript (k) indicates that the kth iteration is considered. Performing the multiplications $\mathbf{P}_k^T\mathbf{K}_k\mathbf{P}_k$ and $\mathbf{P}_k^T\mathbf{M}_k\mathbf{P}_k$ and using the condition that $k_{ij}^{(k+1)}$ and $m_{ij}^{(k+1)}$ shall be zero, we obtain the following two equations for α and γ:

$$
\alpha k_{ii}^{(k)} + (1 + \alpha\gamma)k_{ij}^{(k)} + \gamma k_{jj}^{(k)} = 0 \tag{11.86}
$$

and

$$
\alpha m_{ii}^{(k)} + (1 + \alpha\gamma)m_{ij}^{(k)} + \gamma m_{jj}^{(k)} = 0 \tag{11.87}
$$

If

$$
\frac{k_{ii}^{(k)}}{m_{ii}^{(k)}} = \frac{k_{jj}^{(k)}}{m_{jj}^{(k)}} = \frac{k_{ij}^{(k)}}{m_{ij}^{(k)}}
$$

(i.e., the submatrices considered are scalar multiples, which may be regarded to be a trivial case), we use $\alpha = 0$ and $\gamma = -k_{ij}^{(k)}/k_{jj}^{(k)}$. In general, to solve for α and γ from (11.86) and (11.87) we define

$$
\left.\begin{aligned} \bar{k}_{ii}^{(k)} &= k_{ii}^{(k)}m_{ij}^{(k)} - m_{ii}^{(k)}k_{ij}^{(k)} \\ \bar{k}_{jj}^{(k)} &= k_{jj}^{(k)}m_{ij}^{(k)} - m_{jj}^{(k)}k_{ij}^{(k)} \\ \bar{k}^{(k)} &= k_{ii}^{(k)}m_{jj}^{(k)} - k_{jj}^{(k)}m_{ii}^{(k)} \end{aligned}\right\} \tag{11.88}
$$

and

$$
\gamma = -\frac{\bar{k}_{ii}^{(k)}}{x}; \qquad \alpha = \frac{\bar{k}_{jj}^{(k)}}{x} \tag{11.89}
$$

The value of x to obtain α and γ is then to be determined using

$$
x = \frac{\bar{k}^{(k)}}{2} + \text{sign}\,(\bar{k}^{(k)})\sqrt{\left(\frac{\bar{k}^{(k)}}{2}\right)^2 + \bar{k}_{ii}^{(k)}\bar{k}_{jj}^{(k)}} \tag{11.90}
$$

The above relations for α and γ are used and have primarily been developed for the case when $\mathbf{M}$ is a positive definite full or banded mass matrix. In that case (and, in fact, also under less restrictive conditions) we have that

$$
\left(\frac{\bar{k}^{(k)}}{2}\right)^2 + \bar{k}_{ii}^{(k)}\bar{k}_{jj}^{(k)} > 0,
$$

and hence x is always nonzero. In addition, also, det $\mathbf{P}_k \neq 0$, which indeed is the necessary condition for the algorithm to work.

The generalized Jacobi solution procedure has been used a great deal in the subspace iteration method (see Section 12.3) and when a consistent mass idealization is employed. However, other situations may arise as well. Assume that $\mathbf{M}$ is a diagonal mass matrix, $\mathbf{M} \neq \mathbf{I}$, and $m_{ii} > 0$, in which case we employ in (11.88)

$$\bar{k}_{ii}^{(k)} = -m_{ii}^{(k)} k_{ij}^{(k)}; \qquad \bar{k}_{jj}^{(k)} = -m_{jj}^{(k)} k_{ij}^{(k)} \tag{11.91}$$

and otherwise (11.85) to (11.90) are used as before. However, if $\mathbf{M} = \mathbf{I}$, the relation in (11.87) yields that $\alpha = -\gamma$, and we recognize that $\mathbf{P}_k$ in (11.85) is a multiple of the rotation matrix defined in (11.79) (see Example 11.11). In addition, it should be mentioned that the solution procedure can be adapted to solve the problem $\mathbf{K}\boldsymbol{\phi} = \lambda\mathbf{M}\boldsymbol{\phi}$, when $\mathbf{M}$ is a diagonal matrix with some zero diagonal elements.

The complete solution process is analogous to the Jacobi iteration in the solution of the problem $\mathbf{K}\boldsymbol{\phi} = \lambda\boldsymbol{\phi}$, which was presented in the preceding section. The differences lie in that now a mass coupling factor $((m_{ij}^{(k)})^2/m_{ii}^{(k)}m_{jj}^{(k)})^{1/2}$ must also be calculated, unless $\mathbf{M}$ is diagonal, and the transformation is applied to $\mathbf{K}_k$ and $\mathbf{M}_k$. Convergence is measured by comparing successive eigenvalue approximations and by testing if all off-diagonal elements are small enough; i.e., with l being the last iteration, convergence has been achieved if

$$\frac{|\lambda_i^{(l+1)} - \lambda_i^{(l)}|}{\lambda_i^{(l+1)}} \le 10^{-s}; \qquad i = 1, \ldots, n \tag{11.92}$$

where

$$\lambda_i^{(l)} = \frac{k_{ii}^{(l)}}{m_{ii}^{(l)}}; \qquad \lambda_i^{(l+1)} = \frac{k_{ii}^{(l+1)}}{m_{ii}^{(l+1)}} \tag{11.93}$$

and

$$\left[\frac{(k_{ij}^{(l+1)})^2}{k_{ii}^{(l+1)}k_{jj}^{(l+1)}}\right]^{1/2} \le 10^{-s}; \qquad \left[\frac{(m_{ij}^{(l+1)})^2}{m_{ii}^{(l+1)}m_{jj}^{(l+1)}}\right]^{1/2} \le 10^{-s}; \qquad \text{all } i, j; \qquad i < j \tag{11.94}$$

where 10^{-s} is the convergence tolerance.

Table 11.3 summarizes the solution procedure for the case $\mathbf{M}$ full (or banded) and positive definite. The relations given in Table 11.3 are employed directly in the Subroutine JACOBI, which is presented at the end of this section. We should note that Table 11.3 also gives an operation count of the solution process and the storage requirements. The total number of operations in one sweep as given in the table are an upper bound because it is assumed that both matrices are full and that all off-diagonal elements are zeroed; i.e., the threshold tolerance is never passed. Considering the number of sweeps required for solution, the same experience as with the solution of standard eigenproblems holds; i.e., with $m = 2$ and $s = 12$ in the iteration (see Section 11.3.1) about six sweeps are required for solution of the eigensystem to high accuracy.

The following examples demonstrate some of the characteristics of the generalized Jacobi solution algorithm.

TABLE 11.3 *Summary of generalized Jacobi solution.*

Operation	*Calculation*	*Number of Operations*	*Required Storage*
Calculation of coupling factors	$\dfrac{k_{ij}^{(k)2}}{k_{ii}^{(k)}k_{jj}^{(k)}}; \quad \dfrac{m_{ij}^{(k)2}}{m_{ii}^{(k)}m_{jj}^{(k)}}$	6	
Transformation to zero elements (i, j)	$\bar{k}_{ii}^{(k)} = k_{ii}^{(k)}m_{ij}^{(k)} - m_{ii}^{(k)}k_{ij}^{(k)}$ $\bar{k}_{jj}^{(k)} = k_{jj}^{(k)}m_{ij}^{(k)} - m_{jj}^{(k)}k_{ij}^{(k)}$ $\bar{k}^{(k)} = k_{ii}^{(k)}m_{jj}^{(k)} - k_{jj}^{(k)}m_{ii}^{(k)}$ $x = \dfrac{\bar{k}^{(k)}}{2} + (\text{sign } \bar{k}^{(k)})\sqrt{\left(\dfrac{\bar{k}^{(k)}}{2}\right)^2 + \bar{k}_{ii}^{(k)}\bar{k}_{jj}^{(k)}}$ $\gamma = -\dfrac{\bar{k}_{ii}^{(k)}}{x}, \quad \alpha = \dfrac{\bar{k}_{jj}^{(k)}}{x}$ $\mathbf{K}_{k+1} = \mathbf{P}_k^T\mathbf{K}_k\mathbf{P}_k, \quad \mathbf{M}_{k+1} = \mathbf{P}_k^T\mathbf{M}_k\mathbf{P}_k$	$4n + 12$	Using symmetry of matrices $n(n + 2)$
Calculation of eigenvectors	$(\mathbf{P}_1 \ldots \mathbf{P}_{k-1})\mathbf{P}_k$	$2n$	n^2
	Total for one sweep	$3n^3 + 6n^2$	$2n^2 + 2n$

EXAMPLE 11.11: Prove that the generalized Jacobi method reduces to the standard technique when $\mathbf{M} = \mathbf{I}$.

For the proof we need only consider the calculation of the transformation matrices that would be used to zero a typical off-diagonal element. We want to show that the transformation matrices obtained in the standard and generalized Jacobi methods are multiples of each other, namely in that case we could, by proper scaling, obtain the standard iteration scheme. Since each step of iteration consists of applying a rotation in the (i, j)th plane, we can without loss of generality consider the solution of the problem

$$\begin{bmatrix} k_{11} & k_{12} \\ k_{12} & k_{22} \end{bmatrix} \boldsymbol{\phi} = \lambda \boldsymbol{\phi}$$

Using (11.88) to (11.90) we thus obtain

$$\alpha = -\gamma; \qquad \mathbf{P}_1 = \begin{bmatrix} 1 & -\gamma \\ \gamma & 1 \end{bmatrix} \tag{a}$$

and

$$\gamma = \frac{-k_{11} + k_{22} \pm \sqrt{(k_{11} - k_{22})^2 + 4k_{12}^2}}{2k_{12}}$$

On the other hand, in the standard Jacobi solution we use

$$\mathbf{P}_1 = \begin{bmatrix} \cos\theta & -\sin\theta \\ \sin\theta & \cos\theta \end{bmatrix}$$

which may be written as

$$\mathbf{P}_1 = \cos\theta \begin{bmatrix} 1 & -\tan\theta \\ \tan\theta & 1 \end{bmatrix} \tag{b}$$

Thus, $\mathbf{P}_1$ in (b) would be a multiple of $\mathbf{P}_1$ in (a) if $\tan\theta = \gamma$. In the standard Jacobi method we obtain $\tan 2\theta$ using (11.80). In this case, we have

$$\tan 2\theta = \frac{2k_{12}}{k_{11} - k_{22}} \tag{c}$$

We also have, by simple trigonometry,

$$\tan 2\theta = \frac{2\tan\theta}{1 - \tan^2\theta} \tag{d}$$

Using (c) and (d) we can solve for $\tan\theta$ to be used in (b) and obtain

$$\tan\theta = \frac{-k_{11} + k_{22} \pm \sqrt{(k_{11} - k_{22})^2 + 4k_{12}^2}}{2k_{12}}$$

Hence $\gamma = \tan\theta$ and the generalized Jacobi iteration is equivalent to the standard method when $\mathbf{M} = \mathbf{I}$.

EXAMPLE 11.12: Use the generalized Jacobi method to calculate the eigensystem of the problem $\mathbf{K}\boldsymbol{\phi} = \lambda\mathbf{M}\boldsymbol{\phi}$.

(1) In the first case let

$$\mathbf{K} = \begin{bmatrix} 1 & -1 \\ -1 & 1 \end{bmatrix}; \qquad \mathbf{M} = \begin{bmatrix} 2 & 1 \\ 1 & 2 \end{bmatrix}$$

We note that $\mathbf{K}$ is singular and hence we expect a zero eigenvalue.

(2) Then let

$$\mathbf{K} = \begin{bmatrix} 2 & 1 \\ 1 & 2 \end{bmatrix}; \qquad \mathbf{M} = \begin{bmatrix} 2 & 0 \\ 0 & 0 \end{bmatrix}$$

in which case we have an infinite eigenvalue.

For solution, we use the relations in (11.85) to (11.90). Considering the problem in case (1), we obtain

$$\bar{k}_{11}^{(1)} = 3; \qquad \bar{k}_{22}^{(1)} = 3; \qquad \bar{k}^{(1)} = 0$$
$$x = 3; \qquad \gamma = -1; \qquad \alpha = 1$$
$$\mathbf{P}_1 = \begin{bmatrix} 1 & 1 \\ -1 & 1 \end{bmatrix}$$

Hence $$\mathbf{P}_1^T\mathbf{K}\mathbf{P}_1 \begin{bmatrix} 4 & 0 \\ 0 & 0 \end{bmatrix}; \qquad \mathbf{P}_1^T\mathbf{M}\mathbf{P}_1 = \begin{bmatrix} 2 & 0 \\ 0 & 6 \end{bmatrix}$$

To obtain $\mathbf{\Lambda}$ and $\mathbf{\Phi}$ we use (11.75) and (11.76) and arrange the columns in the matrices in the appropriate order. Hence

$$\mathbf{\Lambda} = \begin{bmatrix} 0 & \\ & 2 \end{bmatrix}; \qquad \mathbf{\Phi} = \begin{bmatrix} \dfrac{1}{\sqrt{6}} & \dfrac{1}{\sqrt{2}} \\ \dfrac{1}{\sqrt{6}} & -\dfrac{1}{\sqrt{2}} \end{bmatrix}$$

Now consider the problem in case (2). Here we have

$$\bar{k}_{11}^{(1)} = -2; \qquad \bar{k}_{22}^{(1)} = 0; \qquad \bar{k}^{(1)} = -4$$
$$x = -4; \qquad \alpha = 0; \qquad \gamma = -\tfrac{1}{2}$$
$$\mathbf{P}_1 = \begin{bmatrix} 1 & 0 \\ -\frac{1}{2} & 1 \end{bmatrix}$$

Hence $$\mathbf{P}_1^T\mathbf{K}\mathbf{P}_1 = \begin{bmatrix} \frac{3}{2} & 0 \\ 0 & 2 \end{bmatrix}; \qquad \mathbf{P}_1^T\mathbf{M}\mathbf{P}_1 = \begin{bmatrix} 2 & 0 \\ 0 & 0 \end{bmatrix}$$

and $$\mathbf{\Lambda} = \begin{bmatrix} \frac{3}{4} & \\ & \infty \end{bmatrix}; \qquad \boldsymbol{\phi}_1 = \begin{bmatrix} \dfrac{1}{\sqrt{2}} \\ -\dfrac{1}{2\sqrt{2}} \end{bmatrix}$$

The above discussion of the generalized Jacobi solution method has already indicated in some way the advantages of the solution technique. Firstly, the transformation of the generalized eigenproblem to the standard form is avoided. This is particularly advantageous when (1) the matrices are ill-conditioned, and (2) the off-diagonal elements in **K** and **M** are already small or, equivalently, when there are only a few nonzero off-diagonal elements. In the first case the direct solution of $\mathbf{K}\boldsymbol{\phi} = \lambda\mathbf{M}\boldsymbol{\phi}$ avoids the solution of a standard eigenproblem of a matrix with very large and small elements (see Section 10.2.5). In the second case the eigenproblem is already nearly solved, because the zeroing of small or only a few off-diagonal elements in **K** and **M** will not result into a large change of the diagonal elements of the matrices, the ratios of which are the eigenvalues. In addition fast convergence can be expected when the off-diagonal elements are small (see Section 11.3.1). We will see that this case arises in the subspace iteration method described in Section 12.3, which is one reason why the generalized Jacobi method is used effectively in that technique.

It should be noted that the Jacobi solution methods solve simultaneously for all eigenvalues and corresponding eigenvectors. However, in finite element analysis we require in most cases only some eigenpairs, and the use of a Jacobi solution procedure can be very inefficient, in particular when the order of **K** and **M** is large. In such cases much more effective solution methods are available, which only solve for the specific eigenvalues and eigenvectors that are actually required. However, the generalized Jacobi solution method presented in this section can be used very effectively as part of those solution strategies (see Section 12.3). When the order of the matrices **K** and **M** is relatively small, the solution of the eigenproblem is not very expensive, and the Jacobi iteration may also be attractive because of its simplicity and elegance of solution.

Subroutine JACOBI: Program JACOBI is used to calculate all eigenvalues and corresponding eigenvectors of the generalized eigenproblem $\mathbf{K}\boldsymbol{\phi} = \lambda\mathbf{M}\boldsymbol{\phi}$. The argument variables and use of the subroutine are defined using comment cards in the program.

```
      SUBROUTINE JACOBI (A,B,X,EIGV,D,N,RTOL,NSMAX,IFPR,IOUT)              JCBI0001
C .................................................................       JCBI0002
C .                                                               .       JCBI0003
C .    P R O G R A M                                              .       JCBI0004
C .        TO SOLVE THE GENERALIZED EIGENPROBLEM USING THE        .       JCBI0005
C .        GENERALIZED JACOBI ITERATION                           .       JCBI0006
C .                                                               .       JCBI0007
C . - - INPUT VARIABLES - -                                       .       JCBI0008
C .      A (N,N)    = STIFFNESS MATRIX (ASSUMED POSITIVE DEFINITE) .       JCBI0009
C .      B (N,N)    = MASS MATRIX (ASSUMED POSITIVE DEFINITE)      .       JCBI0010
C .      X (N,N)    = MATRIX STORING EIGENVECTORS ON SOLUTION EXIT .       JCBI0011
C .      EIGV (N)   = VECTOR STORING EIGENVALUES ON SOLUTION EXIT  .       JCBI0012
C .      D (N)      = WORKING VECTOR                               .       JCBI0013
C .      N          = ORDER OF MATRICES A AND B                    .       JCBI0014
C .      RTOL       = CONVERGENCE TOLERANCE (USUALLY SET TO 10.**-12).     JCBI0015
C .      NSMAX      = MAXIMUM NUMBER OF SWEEPS ALLOWED             .       JCBI0016
C .                              (USUALLY SET TO 15)               .       JCBI0017
C .      IFPR       = FLAG FOR PRINTING DURING ITERATION           .       JCBI0018
C .               EQ.0    NO PRINTING                              .       JCBI0019
C .               EQ.1    INTERMEDIATE RESULTS ARE PRINTED         .       JCBI0020
C .      IOUT       = OUTPUT DEVICE NUMBER                         .       JCBI0021
C .                                                               .       JCBI0022
C . - - OUTPUT - -                                                .       JCBI0023
C .      A (N,N)    = DIAGONALIZED STIFFNESS MATRIX               .       JCBI0024
C .      B (N,N)    = DIGONALIZED MASS MATRIX                     .       JCBI0025
C .      X (N,N)    = EIGENVECTORS STORED COLUMNWISE              .       JCBI0026
C .      EIGV (N)   = EIGENVALUES                                 .       JCBI0027
C .                                                               .       JCBI0028
C .................................................................       JCBI0029
      IMPLICIT REAL*8 (A-H,O-Z)                                           JCBI0030
      ABS (X) =DABS (X)                                                   JCBI0031
      SQRT (X) =DSQRT (X)                                                 JCBI0032
C .................................................................       JCBI0033
C .        THIS PROGRAM IS USED IN SINGLE PRECISION ARITHMETIC ON  .       JCBI0034
C .        CDC EQUIPMENT AND DOUBLE PRECISION ARITHMETIC ON IBM   .       JCBI0035
C .        OR UNIVAC MACHINES .ACTIVATE,DEACTIVATE OR ADJUST ABOVE .       JCBI0036
C .        CARDS FOR SINGLE OR DOUBLE PRECISION ARITHMETIC        .       JCBI0037
C .................................................................       JCBI0038
      DIMENSION A (N,N) ,B (N,N) ,X (N,N) ,EIGV (N) ,D (N)               JCBI0039
C                                                                         JCBI0040
C     INITIALIZE EIGENVALUE AND EIGENVECTOR MATRICES                      JCBI0041
C                                                                         JCBI0042
      DO 10 I=1,N                                                         JCBI0043
      IF (A (I,I).GT.0. .AND. B (I,I).GT.0.) GO TO 4                      JCBI0044
      WRITE (IOUT,2020)                                                   JCBI0045
      STOP                                                                JCBI0046
    4 D (I) =A (I,I) /B (I,I)                                             JCBI0047
   10 EIGV (I) =D (I)                                                     JCBI0048
      DO 30 I=1,N                                                         JCBI0049
      DO 20 J=1,N                                                         JCBI0050
   20 X (I,J) =0.                                                         JCBI0051
   30 X (I,I) =1.                                                         JCBI0052
      IF (N.EQ.1) RETURN                                                  JCBI0053
C                                                                         JCBI0054
C     INITIALIZE SWEEP COUNTER AND BEGIN ITERATION                        JCBI0055
C                                                                         JCBI0056
      NSWEEP=0                                                            JCBI0057
      NR=N-1                                                              JCBI0058
   40 NSWEEP=NSWEEP+1                                                     JCBI0059
      IF (IFPR.EQ.1) WRITE (IOUT,2000) NSWEEP                             JCBI0060
C                                                                         JCBI0061
C     CHECK IF PRESENT OFF-DIAGONAL ELEMENT IS LARGE ENOUGH TO REQUIRE ZEROING  JCBI0062
C                                                                         JCBI0063
      EPS= (.01**NSWEEP) **2                                             JCBI0064
      DO 210 J=1,NR                                                       JCBI0065
      JJ=J+1                                                              JCBI0066
      DO 210 K=JJ,N                                                       JCBI0067
      EPTOLA= (A (J,K) *A (J,K) ) / (A (J,J) *A (K,K) )                   JCBI0068
      EPTOLB= (B (J,K) *B (J,K) ) / (B (J,J) *B (K,K) )                   JCBI0069
      IF ((EPTOLA.LT.EPS) .AND. (EPTOLB.LT.EPS) ) GO TO 210              JCBI0070
C                                                                         JCBI0071
```

```
C      IF ZEROING IS REQUIRED, CALCULATE THE ROTATION MATRIX ELEMENTS CA AND CG      JCBI0072
C                                                                                     JCBI0073
       AKK=A(K,K)*B(J,K)-B(K,K)*A(J,K)                                                JCBI0074
       AJJ=A(J,J)*B(J,K)-B(J,J)*A(J,K)                                                JCBI0075
       AB=A(J,J)*B(K,K)-A(K,K)*B(J,J)                                                 JCBI0076
       CHECK=(AB*AB+4.*AKK*AJJ)/4.                                                    JCBI0077
       IF(CHECK)50,60,60                                                              JCBI0078
    50 WRITE(IOUT,2020)                                                               JCBI0079
       STOP                                                                           JCBI0080
    60 SQCH=SQRT(CHECK)                                                               JCBI0081
       D1=AB/2.+SQCH                                                                  JCBI0082
       D2=AB/2.-SQCH                                                                  JCBI0083
       DEN=D1                                                                         JCBI0084
       IF(ABS(D2).GT.ABS(D1))DEN=D2                                                   JCBI0085
       IF(DEN)80,70,80                                                                JCBI0086
    70 CA=0.                                                                          JCBI0087
       CG=-A(J,K)/A(K,K)                                                              JCBI0088
       GO TO 90                                                                       JCBI0089
    80 CA=AKK/DEN                                                                     JCBI0090
       CG=-AJJ/DEN                                                                    JCBI0091
C                                                                                     JCBI0092
C      PERFORM THE GENERALIZED ROTATION TO ZERO THE PRESENT OFF-DIAGONAL ELEMENT      JCBI0093
C                                                                                     JCBI0094
    90 IF(N-2)100,190,100                                                             JCBI0095
   100 JP1=J+1                                                                        JCBI0096
       JM1=J-1                                                                        JCBI0097
       KP1=K+1                                                                        JCBI0098
       KM1=K-1                                                                        JCBI0099
       IF(JM1-1)130,110,110                                                           JCBI0100
   110 DO 120 I=1,JM1                                                                 JCBI0101
       AJ=A(I,J)                                                                      JCBI0102
       BJ=B(I,J)                                                                      JCBI0103
       AK=A(I,K)                                                                      JCBI0104
       BK=B(I,K)                                                                      JCBI0105
       A(I,J)=AJ+CG*AK                                                                JCBI0106
       B(I,J)=BJ+CG*BK                                                                JCBI0107
       A(I,K)=AK+CA*AJ                                                                JCBI0108
   120 B(I,K)=BK+CA*BJ                                                                JCBI0109
   130 IF(KP1-N)140,140,160                                                           JCBI0110
   140 DO 150 I=KP1,N                                                                 JCBI0111
       AJ=A(J,I)                                                                      JCBI0112
       BJ=B(J,I)                                                                      JCBI0113
       AK=A(K,I)                                                                      JCBI0114
       BK=B(K,I)                                                                      JCBI0115
       A(J,I)=AJ+CG*AK                                                                JCBI0116
       B(J,I)=BJ+CG*BK                                                                JCBI0117
       A(K,I)=AK+CA*AJ                                                                JCBI0118
   150 B(K,I)=BK+CA*BJ                                                                JCBI0119
   160 IF(JP1-KM1)170,170,190                                                         JCBI0120
   170 DO 180 I=JP1,KM1                                                               JCBI0121
       AJ=A(J,I)                                                                      JCBI0122
       BJ=B(J,I)                                                                      JCBI0123
       AK=A(I,K)                                                                      JCBI0124
       BK=B(I,K)                                                                      JCBI0125
       A(J,I)=AJ+CG*AK                                                                JCBI0126
       B(J,I)=BJ+CG*BK                                                                JCBI0127
       A(I,K)=AK+CA*AJ                                                                JCBI0128
   180 B(I,K)=BK+CA*BJ                                                                JCBI0129
   190 AK=A(K,K)                                                                      JCBI0130
       BK=B(K,K)                                                                      JCBI0131
       A(K,K)=AK+2.*CA*A(J,K)+CA*CA*A(J,J)                                            JCBI0132
       B(K,K)=BK+2.*CA*B(J,K)+CA*CA*B(J,J)                                            JCBI0133
       A(J,J)=A(J,J)+2.*CG*A(J,K)+CG*CG*AK                                            JCBI0134
       B(J,J)=B(J,J)+2.*CG*B(J,K)+CG*CG*BK                                            JCBI0135
       A(J,K)=0.                                                                      JCBI0136
       B(J,K)=0.                                                                      JCBI0137
C                                                                                     JCBI0138
C      UPDATE THE EIGENVECTOR MATRIX AFTER EACH ROTATION                              JCBI0139
C                                                                                     JCBI0140
       DO 200 I=1,N                                                                   JCBI0141
       XJ=X(I,J)                                                                      JCBI0142
       XK=X(I,K)                                                                      JCBI0143
       X(I,J)=XJ+CG*XK                                                                JCBI0144
   200 X(I,K)=XK+CA*XJ                                                                JCBI0145
   210 CONTINUE                                                                       JCBI0146
C                                                                                     JCBI0147
C      UPDATE THE EIGENVALUES AFTER EACH SWEEP                                        JCBI0148
C                                                                                     JCBI0149
       DO 220 I=1,N                                                                   JCBI0150
       IF(A(I,I).GT.C. .AND. B(I,I).GT.0.)GO TO 220                                   JCBI0151
       WRITE(IOUT,2020)                                                               JCBI0152
       STOP                                                                           JCBI0153
   220 EIGV(I)=A(I,I)/B(I,I)                                                          JCBI0154
       IF(IFPR.EQ.0)GO TO 230                                                         JCBI0155
       WRITE(IOUT,2030)                                                               JCBI0156
       WRITE(IOUT,2010) (EIGV(I),I=1,N)                                               JCBI0157
C                                                                                     JCBI0158
C      CHECK FOR CONVERGENCE                                                          JCBI0159
C                                                                                     JCBI0160
   230 DO 240 I=1,N                                                                   JCBI0161
       TOL=RTOL*D(I)                                                                  JCBI0162
       DIF=ABS(EIGV(I)-D(I))                                                          JCBI0163
       IF(DIF.GT.TOL)GO TO 280                                                        JCBI0164
   240 CONTINUE                                                                       JCBI0165
C                                                                                     JCBI0166
```

```
C     CHECK ALL CFF-DIAGONAL ELEMENTS TO SEE IF ANOTHER SWEEP IS REQUIRED        JCBI0167
C                                                                                JCBI0168
      EPS=RTOL**2                                                                JCBI0169
      DO 250 J=1,NR                                                              JCBI0170
      JJ=J+1                                                                     JCBI0171
      DO 250 K=JJ,N                                                              JCBI0172
      EPSA=(A(J,K)*A(J,K))/(A(J,J)*A(K,K))                                       JCBI0173
      EPSB=(B(J,K)*B(J,K))/(B(J,J)*B(K,K))                                       JCBI0174
      IF((EPSA.LT.EPS).AND.(EPSB.LT.EPS))GO TO 25)                               JCBI0175
      GO TO 280                                                                  JCBI0176
  250 CONTINUE                                                                   JCBI0177
C                                                                                JCBI0178
C     FILL OUT BCTTOM TRIANGLE CF RESULTANT MATRICES AND SCALE EIGENVECTORS      JCBI0179
C                                                                                JCBI0180
  255 DO 260 I=1,N                                                               JCBI0181
      DO 260 J=1,N                                                               JCBI0182
      A(J,I)=A(I,J)                                                              JCBI0183
  260 B(J,I)=B(I,J)                                                              JCBI0184
      DO 270 J=1,N                                                               JCBI0185
      BB=SQRT(B(J,J))                                                            JCBI0186
      DO 270 K=1,N                                                               JCBI0187
  270 X(K,J)=X(K,J)/BB                                                           JCBI0188
      RETURN                                                                     JCBI0189
C                                                                                JCBI0190
C     UPDATE D MATRIX AND START NEW SWEEP, IF ALLOWED                            JCBI0191
C                                                                                JCBI0192
  280 DO 290 I=1,N                                                               JCBI0193
  290 D(I)=EIGV(I)                                                               JCBI0194
      IF(NSWEEP.LT.NSMAX)GO TO 40                                                JCBI0195
      GO TO 255                                                                  JCBI0196
 2000 FORMAT(27H0SWEEP NUMBER IN *JACCBI* = ,I4)                                 JCBI0197
 2010 FORMAT(1H0,6B20.12)                                                        JCBI0198
 2020 FORMAT (25H0*** ERROR  SOLUTICN STOP  /                                    JCBI0199
     1        30H MATRICES NOT POSITVE DEFINITE)                                 JCBI0200
 2030 FORMAT(36H0CURRENT EIGENVALUES IN *JACOBI* ARE,/)                          JCBI0201
      END                                                                        JCBI0202
```

11.3.3 The Householder-QR-Inverse Iteration Solution

Another most important transformation solution technique is the House-holder-QR-inverse iteration (HQRI) method, although this procedure is restricted to the solution of the standard eigenproblem.[1,17,18,29-36] Therefore, if the generalized eigenproblem $K\phi = \lambda M\phi$ is considered, it must first be transformed into the standard form before the HQRI solution technique can be used. As pointed out in Section 10.2.5, this transformation is only effective in some cases.

In the following discussion, we consider the problem $K\phi = \lambda\phi$, in which K may have zero (and could also have negative) eigenvalues. Therefore, it is not necessary to impose a shift prior to applying the HQRI algorithm in order to solve for only positive eigenvalues (see Section 10.2.3). The name "HQRI solution method" stands for the following three solution steps:

1. Householder transformations are employed to reduce the matrix K into tridiagonal form.
2. QR iteration yields all eigenvalues.
3. Using inverse iteration, the required eigenvectors of the tridiagonal matrix are calculated and transformed to obtain the eigenvectors of K.

A basic difference to the Jacobi solution method is therefore that the matrix is first transformed without iteration into a tridiagonal form. This matrix can then be used effectively in the QR iterative solution, in which all eigenvalues are calculated. Finally, only those eigenvectors that are actually requested are evaluated. We will note that unless many eigenvectors must be calculated, the transformation of K into tridiagonal form requires most of the numerical operations. In the following we consider in detail the three distinct steps carried out in the HQRI solution.

The Householder reduction: The Householder reduction to tridiagonal form involves $(n - 2)$ transformations of the form (11.72); i.e., using $\mathbf{K}_1 = \mathbf{K}$, we calculate

$$\mathbf{K}_{k+1} = \mathbf{P}_k^T \mathbf{K}_k \mathbf{P}_k; \qquad k = 1, \ldots, n-2 \tag{11.95}$$

where the $\mathbf{P}_k$ are Householder transformation matrices (see Sections 2.5 and 8.3.2):

$$\mathbf{P}_k = \mathbf{I} - \theta \mathbf{w}_k \mathbf{w}_k^T \tag{11.96}$$

$$\theta = \frac{2}{\mathbf{w}_k^T \mathbf{w}_k} \tag{11.97}$$

To show how the vector $\mathbf{w}_k$ which defines the matrix $\mathbf{P}_k$ is calculated, we consider the case $k = 1$, which is typical. First, we partition $\mathbf{K}_1$, $\mathbf{P}_1$, and $\mathbf{w}_1$ into submatrices as follows:

$$\left.\begin{aligned} \mathbf{P}_1 &= \left[\begin{array}{c:c} 1 & \mathbf{0} \\ \hdashline \mathbf{0} & \bar{\mathbf{P}}_1 \end{array}\right]; \qquad \mathbf{w}_1 = \left[\begin{array}{c} 0 \\ \hdashline \bar{\mathbf{w}}_1 \end{array}\right] \\ \mathbf{K}_1 &= \left[\begin{array}{c:c} k_{11} & \mathbf{k}_1^T \\ \hdashline \mathbf{k}_1 & \mathbf{K}_{11} \end{array}\right] \end{aligned}\right\} \tag{11.98}$$

where $\mathbf{K}_{11}$, $\bar{\mathbf{P}}_1$, and $\bar{\mathbf{w}}_1$ are of order $(n - 1)$. In the general case of step k, we have corresponding matrices of order $(n - k)$. Performing the multiplication in (11.95), we obtain, using the notation of (11.98),

$$\mathbf{K}_2 = \left[\begin{array}{c:c} k_{11} & \mathbf{k}_1^T \bar{\mathbf{P}}_1 \\ \hdashline \bar{\mathbf{P}}_1^T \mathbf{k}_1 & \bar{\mathbf{P}}_1^T \mathbf{K}_{11} \bar{\mathbf{P}}_1 \end{array}\right] \tag{11.99}$$

The condition is now that the first column and row of $\mathbf{K}_2$ be in tridiagonal form; i.e., we want that $\mathbf{K}_2$ is of the form

$$\mathbf{K}_2 = \left[\begin{array}{c:cccc} k_{11} & \times & 0 & \ldots & 0 \\ \hdashline \times & & & & \\ 0 & & & & \\ \vdots & & & \bar{\mathbf{K}}_2 & \\ 0 & & & & \end{array}\right] \tag{11.100}$$

where $\times$ indicates a nonzero value, and

$$\bar{\mathbf{K}}_2 = \bar{\mathbf{P}}_1^T \mathbf{K}_{11} \bar{\mathbf{P}}_1 \tag{11.101}$$

The form of $\mathbf{K}_2$ in (11.100) is achieved by realizing that $\bar{\mathbf{P}}_1$ is a reflection matrix (see Section 2.5). Therefore, we can use $\bar{\mathbf{P}}_1$ to reflect the vector $\mathbf{k}_1$ of $\mathbf{K}_1$ in (11.98) into a vector that has only its first component nonzero. Since the length of the new vector must be the length of $\mathbf{k}_1$, we determine $\bar{\mathbf{w}}_1$ from the condition

$$(\mathbf{I} - \theta \bar{\mathbf{w}}_1 \bar{\mathbf{w}}_1^T)\mathbf{k}_1 = \pm \|\mathbf{k}_1\|_2 \mathbf{e}_1 \tag{11.102}$$

where $\mathbf{e}_1$ is a unit vector of dimension $(n - 1)$; i.e., $\mathbf{e}_1^T = [1 \quad 0 \quad 0 \ldots 0]$, and the $+$ or $-$ sign can be selected to obtain best numerical stability. Noting that we only need to solve for a multiple of $\bar{\mathbf{w}}_1$ [i.e., only the direction of the vector is important (see Section 2.5)], we obtain from (11.102) as a suitable value for $\bar{\mathbf{w}}_1$,

$$\bar{\mathbf{w}}_1 = \mathbf{k}_1 + \text{sign}\,(k_{21})\,\|\mathbf{k}_1\|_2\,\mathbf{e}_1 \tag{11.103}$$

where k_{21} is element (2, 1) of $\mathbf{K}_1$.

With $\bar{\mathbf{w}}_1$ defined in (11.103), the first Householder transformation, $k = 1$ in (11.95), can be carried out. In the next step, $k = 2$, we can consider the matrix $\bar{\mathbf{K}}_2$ in (11.100) in the same way as we considered $\mathbf{K}_1$ in (11.98) to (11.103), because the reduction of the first column and row of $\bar{\mathbf{K}}_2$ does not affect the first column and row in $\mathbf{K}_2$. Thus the general algorithm for the transformation of $\mathbf{K}$ into tridiagonal form is established. We demonstrate the procedure in the following example.

EXAMPLE 11.13: Use Householder transformation matrices to reduce $\mathbf{K}$ to tridiagonal form, where

$$\mathbf{K} = \begin{bmatrix} 5 & -4 & 1 & 0 \\ -4 & 6 & -4 & 1 \\ 1 & -4 & 6 & -4 \\ 0 & 1 & -4 & 5 \end{bmatrix}$$

Here we have using (11.95) to (11.103) to reduce column 1,

$$\bar{\mathbf{w}}_1 = \begin{bmatrix} -4 \\ 1 \\ 0 \end{bmatrix} - 4.123 \begin{bmatrix} 1 \\ 0 \\ 0 \end{bmatrix} = \begin{bmatrix} -8.1231 \\ 1 \\ 0 \end{bmatrix}$$

Hence

$$\mathbf{w}_1 = \begin{bmatrix} 0 \\ -8.1231 \\ 1 \\ 0 \end{bmatrix}; \qquad \theta_1 = 0.0298575$$

$$\mathbf{P}_1 = \begin{bmatrix} 1 & 0 & 0 & 0 \\ 0 & -0.9701 & 0.2425 & 0 \\ 0 & 0.2425 & 0.9701 & 0 \\ 0 & 0 & 0 & 1 \end{bmatrix}$$

and

$$\mathbf{K}_2 = \begin{bmatrix} 5 & 4.1231 & 0 & 0 \\ 4.1231 & 7.8823 & 3.5294 & -1.9403 \\ 0 & 3.5294 & 4.1177 & -3.6380 \\ 0 & -1.9403 & -3.6380 & 5 \end{bmatrix}$$

Next we reduce column 2,

$$\bar{\mathbf{w}}_2 = \begin{bmatrix} 3.5294 \\ -1.9403 \end{bmatrix} + 4.0276 \begin{bmatrix} 1 \\ 0 \end{bmatrix} = \begin{bmatrix} 7.5570 \\ -1.9403 \end{bmatrix}$$

$$\mathbf{w}_2 = \begin{bmatrix} 0 \\ 0 \\ 7.5570 \\ -1.9403 \end{bmatrix}; \qquad \theta_2 = 0.0328553$$

$$\mathbf{P}_2 = \begin{bmatrix} 1 & 0 & 0 & 0 \\ 0 & 1 & 0 & 0 \\ 0 & 0 & -0.8763 & 0.4817 \\ 0 & 0 & 0.4817 & 0.8763 \end{bmatrix}$$

Hence $$\mathbf{K}_3 = \begin{bmatrix} 5 & 4.1231 & 0 & 0 \\ 4.1231 & 7.8823 & -4.0276 & 0 \\ 0 & -4.0276 & 7.3941 & 2.3219 \\ 0 & 0 & 2.3219 & 1.7236 \end{bmatrix}$$

Some important numerical aspects should be noted. First, the reduced matrices $\mathbf{K}_2, \mathbf{K}_3, \ldots, \mathbf{K}_{n-1}$ are symmetric. This means that in the reduction we need only store the lower symmetric part of $\mathbf{K}$. Furthermore, to store the $\bar{\mathbf{w}}_k$, $k = 1, 2, \ldots, n-2$, we can use the storage locations below the subdiagonal elements in the currently being reduced matrix.

A disadvantage of the Householder transformations is that the bandwidth is increased in the unreduced part of $\mathbf{K}_{k+1}$. Hence, in the reduction, essentially no advantage can be taken of the bandedness of $\mathbf{K}$.

An important aspect in the transformation is the evaluation of the matrix product $\bar{\mathbf{P}}_1^T\mathbf{K}_{11}\bar{\mathbf{P}}_1$ and the similar products required in the next steps. In the most general case a triple matrix product with matrices of order n requires $2n^3$ operations, and if this many operations were required, the Householder reduction would be quite uneconomical. However, by taking advantage of the special nature of the matrix $\bar{\mathbf{P}}_1$, we can evaluate the product $\bar{\mathbf{P}}_1^T\mathbf{K}_{11}\bar{\mathbf{P}}_1$ by calculating

$$\left.\begin{aligned} \mathbf{v}_1 &= \mathbf{K}_{11}\bar{\mathbf{w}}_1 \\ \mathbf{p}_1^T &= \theta_1\mathbf{v}_1^T \\ \beta_1 &= \mathbf{p}_1^T\bar{\mathbf{w}}_1 \\ \mathbf{q}_1 &= \mathbf{p}_1 - \theta_1\beta_1\bar{\mathbf{w}}_1 \end{aligned}\right\} \tag{11.104}$$

and then
$$\bar{\mathbf{P}}_1^T\mathbf{K}_{11}\bar{\mathbf{P}}_1 = \mathbf{K}_{11} - \bar{\mathbf{w}}_1\mathbf{p}_1^T - \mathbf{q}_1\bar{\mathbf{w}}_1^T \tag{11.105}$$

which requires only about $3m^2 + 3m$ operations, where m is the order of $\bar{\mathbf{P}}_1$ and $\mathbf{K}_{11}$ (i.e., $m = n - 1$ in this case). Hence the multiplication $\bar{\mathbf{P}}_1^T\mathbf{K}_{11}\bar{\mathbf{P}}_1$ requires a number of operations of the order m-squared rather than m-cubed, which is a very significant reduction. We demonstrate the procedure given in (11.104) and (11.105) by reworking Example 11.13.

EXAMPLE 11.14: Use the relations given in (11.104) and (11.105) to reduce the matrix $\mathbf{K}$ of Example 11.13 into tridiagonal form.

Here we obtain for the reduction of column 1 using $\bar{\mathbf{w}}_1$ and θ_1 calculated in Example 11.13,

$$\mathbf{v}_1 = \begin{bmatrix} 6 & -4 & 1 \\ -4 & 6 & -4 \\ 1 & -4 & 5 \end{bmatrix} \bar{\mathbf{w}}_1 = \begin{bmatrix} -52.738 \\ 38.4924 \\ -12.1231 \end{bmatrix}$$

$$\mathbf{p}_1^T = [-1.5746 \quad 1.1493 \quad -0.36197]; \qquad \beta_1 = 13.9403; \qquad \mathbf{q}_1 = \begin{bmatrix} 1.8064 \\ 0.7331 \\ -0.3620 \end{bmatrix}$$

$$\bar{\mathbf{P}}_1^T\mathbf{K}_{11}\bar{\mathbf{P}}_1 = \begin{bmatrix} 6 & -4 & 1 \\ -4 & 6 & -4 \\ 1 & -4 & 5 \end{bmatrix} - \begin{bmatrix} 12.7910 & -9.335 & 2.9403 \\ -1.5746 & 1.1493 & -0.3620 \\ 0 & 0 & 0 \end{bmatrix}$$
$$- \begin{bmatrix} -14.6734 & 1.8064 & 0 \\ -5.9548 & 0.73307 & 0 \\ 2.9403 & -0.3620 & 0 \end{bmatrix}$$

$$\text{or} \qquad \bar{\mathbf{P}}_1^T\mathbf{K}_{11}\bar{\mathbf{P}}_1 = \begin{bmatrix} 7.8823 & 3.5294 & -1.9403 \\ 3.5294 & 4.1177 & -3.6380 \\ -1.9403 & -3.6380 & 5 \end{bmatrix}$$

$$\text{and hence} \qquad \mathbf{K}_2 = \begin{bmatrix} 5 & 4.1231 & 0 & 0 \\ 4.1231 & 7.8823 & 3.5294 & -1.9403 \\ 0 & 3.5294 & 4.1177 & -3.6380 \\ 0 & -1.9403 & -3.6380 & 5 \end{bmatrix}$$

Next we reduce the second column

$$\mathbf{v}_2 = \begin{bmatrix} 4.1177 & -3.6380 \\ -3.6380 & 5 \end{bmatrix}; \quad \bar{\mathbf{w}}_2 = \begin{bmatrix} 38.1759 \\ -37.1941 \end{bmatrix}$$

$$\mathbf{p}_2^T = [1.2543 \quad -1.2220]; \qquad \beta_2 = 11.8497; \qquad \mathbf{q}_2 = \begin{bmatrix} -1.6878 \\ -0.4666 \end{bmatrix}$$

$$\bar{\mathbf{P}}_2^T\mathbf{K}_{22}\bar{\mathbf{P}}_2 = \begin{bmatrix} 4.1177 & -3.6380 \\ -3.6380 & 5 \end{bmatrix} - \begin{bmatrix} 9.4786 & -9.2348 \\ -2.4337 & 2.3711 \end{bmatrix} - \begin{bmatrix} -12.7550 & 3.2749 \\ -3.5263 & 0.90538 \end{bmatrix}$$

$$= \begin{bmatrix} 7.3941 & 2.3219 \\ 2.3219 & 1.7236 \end{bmatrix}$$

$$\text{and hence} \qquad \mathbf{K}_3 = \begin{bmatrix} 5 & 4.1231 & 0 & 0 \\ 4.1231 & 7.9823 & -4.0276 & 0 \\ 0 & -4.0276 & 7.3941 & 2.3219 \\ 0 & 0 & 2.3219 & 1.7236 \end{bmatrix}$$

The QR iteration: In the HQRI solution procedure, the QR iteration is applied to the tridiagonal matrix obtained by the Householder transformation of **K**. However, it should be realized that the QR iteration could be applied to the original matrix **K** as well, and that the transformation of **K** into tridiagonal form prior to the iteration is merely carried out to improve the efficiency of solution. In the following we therefore consider first how the iteration is applied to a general symmetric matrix **K**.

The name "QR iteration" derives from the notation used in the algorithm. Namely, the basic step in the iteration is to decompose **K** into the form

$$\mathbf{K} = \mathbf{QR} \tag{11.106}$$

where **Q** is an orthogonal and **R** is an upper triangular matrix. We then form

$$\mathbf{RQ} = \mathbf{Q}^T\mathbf{KQ} \tag{11.107}$$

Therefore, by calculating **RQ** we in fact carry out a transformation of the form (11.72).

The factorization in (11.106) could be obtained by applying the Gram-Schmidt process to the columns of **K**. In practice, it is more effective to reduce **K** into upper triangular form using Jacobi rotation matrices; i.e., we evaluate (see Section 8.3.1)

$$\mathbf{P}_{n,n-1}^T \ldots \mathbf{P}_{3,1}^T\mathbf{P}_{2,1}^T\mathbf{K} = \mathbf{R} \tag{11.108}$$

where the matrix $\mathbf{P}_{j,i}^T$ is given in (8.53) and is selected to zero element (j, i).

Using (11.108), we have, corresponding to (11.106),

$$\mathbf{Q} = \mathbf{P}_{2,1}\mathbf{P}_{3,1} \dots \mathbf{P}_{n,n-1} \tag{11.109}$$

The QR iteration algorithm is obtained by repeating the process given in (11.106) and (11.107). Using the notation $\mathbf{K}_1 = \mathbf{K}$, we form

$$\mathbf{K}_k = \mathbf{Q}_k\mathbf{R}_k \tag{11.110}$$

and then

$$\mathbf{K}_{k+1} = \mathbf{R}_k\mathbf{Q}_k \tag{11.111}$$

where, then, disregarding that eigenvalues and eigenvectors may not be in the usual order,

$$\mathbf{K}_{k+1} \longrightarrow \mathbf{\Lambda} \quad \text{and} \quad \mathbf{Q}_1 \dots \mathbf{Q}_{k-1}\mathbf{Q}_k \longrightarrow \mathbf{\Phi} \qquad \text{as } k \longrightarrow \infty$$

We demonstrate the iteration process in the following example.

EXAMPLE 11.15: Use the QR iteration with $\mathbf{Q}$ obtained as a product of Jacobi rotation matrices to calculate the eigensystem of $\mathbf{K}$, where

$$\mathbf{K} = \begin{bmatrix} 5 & -4 & 1 & 0 \\ -4 & 6 & -4 & 1 \\ 1 & -4 & 6 & -4 \\ 0 & 1 & -4 & 5 \end{bmatrix}$$

In the solution we want to employ the Jacobi rotation matrices to establish the matrices $\mathbf{Q}_k$. Using the matrix $\mathbf{P}_{j,i}^T$ given in (8.53) to reduce to zero element (j, i) in the current matrix, we have

$$\sin\theta = \frac{\bar{k}_{ji}}{\sqrt{\bar{k}_{ii}^2 + \bar{k}_{ji}^2}}$$

$$\cos\theta = \frac{\bar{k}_{ii}}{\sqrt{\bar{k}_{ii}^2 + \bar{k}_{ji}^2}}$$

where the bar indicates that the elements of the current matrix are used.

Proceeding as given in (11.108), we obtain for element (2, 1),

$$\sin\theta = -0.6247; \qquad \cos\theta = 0.7809$$

$$\mathbf{P}_{2,1} = \begin{bmatrix} 0.7809 & 0.6247 & 0 & 0 \\ -0.6247 & 0.7809 & 0 & 0 \\ 0 & 0 & 1 & 0 \\ 0 & 0 & 0 & 1 \end{bmatrix}$$

and

$$\mathbf{P}_{2,1}^T\mathbf{K} = \begin{bmatrix} 6.403 & -6.872 & 3.280 & -0.6247 \\ 0 & 2.186 & -2.499 & 0.7809 \\ 1 & -4 & 6 & -4 \\ 0 & 1 & -4 & 5 \end{bmatrix}$$

Next we zero element (3, 1),

$$\mathbf{P}_{3,1}^T\mathbf{P}_{2,1}^T\mathbf{K} = \begin{bmatrix} 6.481 & -7.407 & 4.166 & -1.234 \\ 0 & 2.186 & -2.499 & 0.7809 \\ 0 & -2.892 & 5.422 & -3.856 \\ 0 & 1 & -4 & 5 \end{bmatrix}$$

Noting that element (4, 1) is zero already, we continue with the factorization by zeroing element (3, 2),

$$\mathbf{P}_{3,2}^T\mathbf{P}_{3,1}^T\mathbf{P}_{2,1}^T\mathbf{K} = \begin{bmatrix} 6.481 & -7.407 & 4.166 & -1.234 \\ 0 & 3.625 & -5.832 & 3.546 \\ 0 & 0 & 1.277 & -1.703 \\ 0 & 1 & -4 & 5 \end{bmatrix}$$

Proceeding in the same manner, we obtain

$$\mathbf{P}_{4,2}^T\mathbf{P}_{3,2}^T\mathbf{P}_{3,1}^T\mathbf{P}_{2,1}^T\mathbf{K} = \begin{bmatrix} 6.481 & -7.407 & 4.166 & -1.234 \\ 0 & 3.761 & -6.686 & 4.748 \\ 0 & 0 & 1.277 & -1.703 \\ 0 & 0 & -2.305 & 3.877 \end{bmatrix}$$

and finally $\mathbf{R}_1 = \mathbf{P}_{4,3}^T\mathbf{P}_{4,2}^T\mathbf{P}_{3,2}^T\mathbf{P}_{3,1}^T\mathbf{P}_{2,1}^T\mathbf{K}$

$$\mathbf{R}_1 = \begin{bmatrix} 6.481 & -7.407 & 4.166 & -1.234 \\ 0 & 3.761 & -6.686 & 4.748 \\ 0 & 0 & 2.635 & -4.216 \\ 0 & 0 & 0 & 0.3892 \end{bmatrix}$$

Also, we have $\mathbf{Q}_1 = \mathbf{P}_{2,1}\mathbf{P}_{3,1}\mathbf{P}_{3,2}\mathbf{P}_{4,2}\mathbf{P}_{4,3}$

$$\mathbf{Q}_1 = \begin{bmatrix} 0.7715 & 0.4558 & 0.3162 & 0.3114 \\ -0.6172 & 0.3799 & 0.4216 & 0.5449 \\ 0.1543 & -0.7597 & 0.1054 & 0.6228 \\ 0 & 0.2659 & -0.8433 & 0.4671 \end{bmatrix}$$

The first iteration step of QR is completed by calculating

$$\mathbf{K}_2 = \mathbf{R}_1\mathbf{Q}_1$$

to obtain

$$\mathbf{K}_2 = \begin{bmatrix} 10.21 & -3.353 & 0.4066 & 0 \\ -3.353 & 7.771 & -3.123 & 0.1035 \\ 0.4066 & -3.123 & 3.833 & -0.3282 \\ 0 & 0.1035 & -0.3282 & 0.1818 \end{bmatrix}$$

The following results are obtained in the next steps of QR. For $k = 2$:

$$\mathbf{R}_2 = \begin{bmatrix} 10.76 & -5.723 & 1.504 & -0.0446 \\ 0 & 6.974 & -4.163 & 0.2284 \\ 0 & 0 & 2.265 & -0.2752 \\ 0 & 0 & 0 & 0.1471 \end{bmatrix}$$

$$\mathbf{Q}_1\mathbf{Q}_2 = \begin{bmatrix} 0.6024 & 0.4943 & 0.5084 & 0.3665 \\ -0.6885 & -0.0257 & 0.4099 & 0.5978 \\ 0.3873 & -0.6409 & -0.2715 & 0.6046 \\ -0.1147 & 0.5867 & 0.7070 & 0.3779 \end{bmatrix}$$

$$\mathbf{K}_3 = \mathbf{R}_2\mathbf{Q}_2 = \begin{bmatrix} 12.05 & -2.331 & 0.0856 & 0 \\ -2.331 & 7.726 & -0.9483 & 0.0022 \\ 0.0856 & -0.9483 & 2.0740 & -0.0173 \\ 0 & 0.0022 & -0.0173 & 0.1461 \end{bmatrix}$$

For $k = 3$:

$$\mathbf{R}_3 = \begin{bmatrix} 12.28 & -3.761 & 0.2785 & -0.0005 \\ 0 & 7.202 & -1.173 & 0.0044 \\ 0 & 0 & 1.938 & -0.0182 \\ 0 & 0 & 0 & 0.1459 \end{bmatrix}$$

$$\mathbf{Q}_1\mathbf{Q}_2\mathbf{Q}_3 = \begin{bmatrix} 0.5011 & 0.5302 & 0.5743 & 0.3713 \\ -0.6682 & -0.2076 & 0.3860 & 0.6012 \\ 0.5000 & -0.5157 & -0.3492 & 0.6018 \\ -0.2290 & 0.6401 & -0.6319 & 0.3722 \end{bmatrix}$$

$$\mathbf{K}_4 = \mathbf{R}_3\mathbf{Q}_3 = \begin{bmatrix} 12.77 & -1.375 & 0.0135 & 0 \\ -1.375 & 7.162 & -0.2481 & 0 \\ 0.0135 & -0.2481 & 1.922 & -0.0013 \\ 0 & 0 & -0.0013 & 0.1459 \end{bmatrix}$$

And after nine iterations, we have:

$$\mathbf{R}_9 = \begin{bmatrix} 13.09 & -0.0869 & 0 & 0 \\ 0 & 6.854 & -0.0005 & 0 \\ 0 & 0 & 1.910 & 0 \\ 0 & 0 & 0 & 0.1459 \end{bmatrix}$$

$$\mathbf{Q}_1\mathbf{Q}_2\mathbf{Q}_3\mathbf{Q}_4\mathbf{Q}_5\mathbf{Q}_6\mathbf{Q}_7\mathbf{Q}_8\mathbf{Q}_9 = \begin{bmatrix} 0.3746 & 0.5997 & 0.6015 & 0.3717 \\ -0.6033 & -0.3689 & 0.3718 & 0.6015 \\ 0.5997 & -0.3746 & -0.3717 & 0.6015 \\ -0.3689 & 0.6033 & -0.6015 & 0.3718 \end{bmatrix}$$

$$\mathbf{K}_{10} = \begin{bmatrix} 13.09 & -0.0298 & 0 & 0 \\ -0.0298 & 6.8542 & -0.0001 & 0 \\ 0 & -0.0001 & 1.910 & 0 \\ 0 & 0 & 0 & 0.1459 \end{bmatrix}$$

Thus we have, after nine steps of QR iteration,

$$\lambda_1 \doteq 0.1459; \qquad \boldsymbol{\phi}_1 \doteq \begin{bmatrix} 0.3717 \\ 0.6015 \\ 0.6015 \\ 0.3718 \end{bmatrix}$$

$$\lambda_2 \doteq 1.910; \qquad \boldsymbol{\phi}_2 \doteq \begin{bmatrix} 0.6015 \\ 0.3718 \\ -0.3717 \\ -0.6015 \end{bmatrix}$$

$$\lambda_3 \doteq 6.854; \qquad \boldsymbol{\phi}_3 \doteq \begin{bmatrix} 0.5997 \\ -0.3689 \\ -0.3746 \\ 0.6033 \end{bmatrix}$$

$$\lambda_4 \doteq 13.09; \qquad \boldsymbol{\phi}_4 \doteq \begin{bmatrix} 0.3746 \\ -0.6033 \\ 0.5997 \\ -0.3689 \end{bmatrix}$$

These results can be compared with the results obtained using the Jacobi method in Example 11.9. It is interesting to note that in the above solution, λ_1 and $\boldsymbol{\phi}_1$ converged first and indeed were well predicted after only three QR iterations. This is a consequence of the fact that QR iteration is closely related to inverse iteration (see Example 11.16), in which the lowest eigenvalues and corresponding vectors converge first.

Although the QR iteration may look similar to the Jacobi solution procedure, the method is, in fact, completely different. This may be observed by studying the convergence characteristics of the QR solution procedure, because it is then found that *the QR method is intimately related to inverse iteration.* In Example 11.16 we compare QR and inverse iteration, where it is assumed that the matrix **K** is nonsingular. As we may recall, this assumption is necessary for inverse iteration and can always be satisfied by using a shift (see Section 10.2.3).

EXAMPLE 11.16: Show the theoretical relationship between QR and inverse iteration.

In the QR method, we obtain after l steps of iteration,

$$\mathbf{K}_{l+1} = \mathbf{Q}_l^T\mathbf{Q}_{l-1}^T \ldots \mathbf{Q}_1^T\mathbf{K}_1\mathbf{Q}_1 \ldots \mathbf{Q}_{l-1}\mathbf{Q}_l$$

or

$$\mathbf{K}_{l+1} = \mathbf{P}_l^T\mathbf{K}_1\mathbf{P}_l; \qquad \mathbf{P}_l = \mathbf{Q}_1 \ldots \mathbf{Q}_l$$

Let us define

$$\mathbf{S}_l = \mathbf{R}_l \ldots \mathbf{R}_1$$

Then we have

$$\begin{aligned}\mathbf{P}_l\mathbf{S}_l &= \mathbf{P}_{l-1}\mathbf{Q}_l\mathbf{R}_l\mathbf{S}_{l-1}\\ &= \mathbf{P}_{l-1}\mathbf{K}_l\mathbf{S}_{l-1}\end{aligned}$$

If we note that

$$\mathbf{K}_1\mathbf{P}_{l-1} = \mathbf{P}_{l-1}\mathbf{K}_l$$

we get

$$\mathbf{P}_l\mathbf{S}_l = \mathbf{K}_1\mathbf{P}_{l-1}\mathbf{S}_{l-1}$$

In an analogous manner we obtain $\mathbf{P}_{l-1}\mathbf{S}_{l-1} = \mathbf{K}_1\mathbf{P}_{l-2}\mathbf{S}_{l-2}$, and so on, and thus conclude that

$$\mathbf{P}_l\mathbf{S}_l = \mathbf{K}^l \tag{a}$$

Assuming that **K** is nonsingular, we have from (a)

$$\mathbf{P}_l = \mathbf{K}^{-l}\mathbf{S}_l^T$$

or equating columns on both sides,

$$\mathbf{P}_l\mathbf{E} = \mathbf{K}^{-l}\mathbf{S}_l^T\mathbf{E} \tag{b}$$

where **E** consists of the last p columns of **I**.

Consider now inverse iteration on p vectors. This iteration process can be written

$$\mathbf{K}\mathbf{X}_k = \mathbf{X}_{k-1}\mathbf{L}_k; \qquad k = 1, 2, \ldots$$

where $\mathbf{L}_k$ is a lower triangular matrix chosen so that $\mathbf{X}_k^T\mathbf{X}_k = \mathbf{I}$. The matrix $\mathbf{L}_k$ can be determined using the Gram–Schmidt process on the iteration vectors. Hence after l steps we have

$$\mathbf{X}_l = \mathbf{K}^{-l}\mathbf{X}_0\bar{\mathbf{L}}_l; \qquad \bar{\mathbf{L}}_l = \mathbf{L}_1 \ldots \mathbf{L}_l \tag{c}$$

On the other hand, the relation in (b) can be written

$$\mathbf{P}_l\mathbf{E} = \mathbf{K}^{-l}\mathbf{E}\bar{\mathbf{S}}_l \tag{d}$$

where $\bar{\mathbf{S}}_l$ consists of the last p columns and rows of $\mathbf{S}_l^T$. Using that $\mathbf{X}_l^T\mathbf{X}_l = \mathbf{I}$ and

$(\mathbf{P}_l\mathbf{E})^T(\mathbf{P}_l\mathbf{E}) = \mathbf{I}$, we obtain, from (c) and (d), respectively,

$$\bar{\mathbf{L}}_l^{-T}\bar{\mathbf{L}}_l^{-1} = \mathbf{X}_0^T\mathbf{K}^{-2l}\mathbf{X}_0 \tag{e}$$

and

$$\bar{\mathbf{S}}_l^{-T}\bar{\mathbf{S}}_l^{-1} = \mathbf{E}^T\mathbf{K}^{-2l}\mathbf{E} \tag{f}$$

The equations in (e) and (f) can now be used to show the relationship between inverse iteration and the QR solution procedure. Namely, if we choose $\mathbf{X}_0 = \mathbf{E}$, we find from (e) and (f) that $\bar{\mathbf{L}}_l = \bar{\mathbf{S}}_l$, because these matrices are the Cholesky factors of the same positive definite matrix. However, referring then to (c) and (d), we can conclude that the inverse iteration yields vectors $\mathbf{X}_l$ which are the last p columns in $\mathbf{P}_l$ of the QR solution.

The relationship between the QR solution method and simple inverse iteration suggests that an acceleration of convergence in the QR iteration described in (11.110) and (11.111) should be possible. This is indeed the case and, in practice, QR iteration is used with shifting; i.e., instead of (11.110) and (11.111), the following decompositions are employed:

$$\mathbf{K}_k - \mu_k\mathbf{I} = \mathbf{Q}_k\mathbf{R}_k \tag{11.112}$$

$$\mathbf{K}_{k+1} = \mathbf{R}_k\mathbf{Q}_k + \mu_k\mathbf{I} \tag{11.113}$$

where then, as before,

$$\mathbf{K}_{k+1} \longrightarrow \mathbf{\Lambda} \quad \text{and} \quad \mathbf{Q}_1 \ldots \mathbf{Q}_{k-1}\mathbf{Q}_k \longrightarrow \mathbf{\Phi} \qquad \text{as } k \longrightarrow \infty$$

However, *if* μ_k *is element* (n, n) *of* $\mathbf{K}_k$, *the* QR *iteration corresponds to Rayleigh quotient iteration, ultimately giving cubic convergence.*

As pointed out earlier, in practice the QR iteration should be applied after reduction of $\mathbf{K}$ into tridiagonal form using Householder transformation matrices; i.e., the QR solution should be applied to the matrix $\mathbf{K}_{n-1}$ in (11.95), which we now call $\mathbf{T}_1$. When the matrix is tridiagonal, the QR process is very effective; i.e., by experience about $9n^2$ operations are required for solution of all eigenvalues. It is also not necessary to formally go through the procedure discussed above and demonstrated in Example 11.15; instead, we may use explicit formulas that relate the elements in $\mathbf{T}_{k+1}$ to the elements of $\mathbf{T}_k$, $k = 1, 2, \ldots$.[1,36]

Calculation of eigenvectors: The eigenvalues are generally calculated to full machine precision, because convergence is very rapid in the QR iteration with shifting. Once the eigenvalues have been evaluated very accurately, we calculate only the required eigenvectors of the tridiagonal matrix $\mathbf{T}_1$ by simple inverse iteration with shifts equal to the corresponding eigenvalues. Two steps of inverse iteration starting with a full unit vector are usually sufficient. The eigenvectors of $\mathbf{T}_1$ then need to be transformed with the Householder transformations used to obtain the eigenvectors of $\mathbf{K}$; i.e., denoting the ith eigenvector of $\mathbf{T}_1$ by $\boldsymbol{\psi}_i$, we have, using the transformation matrices $\mathbf{P}_k$ in (11.95),

$$\boldsymbol{\phi}_i = \mathbf{P}_1\mathbf{P}_2 \ldots \mathbf{P}_{n-2}\boldsymbol{\psi}_i \tag{11.114}$$

With the three basic steps of the HQRI solution method described above, Table 11.4 summarizes the complete procedure and presents the high-speed storage needed and the number of operations required. It is noted that the greater part of the total number of operations is used for the Householder transformations in (11.95) and, in case many eigenvectors need be calculated,

TABLE 11.4 *Summary of Householder-QR-inverse iteration solution.*

Operation	*Calculation*	*Number of Operations*	*Required Storage*
Householder transformations	$\mathbf{K}_{k+1} = \mathbf{P}_k^T \mathbf{K}_k \mathbf{P}_k;\ k = 1, 2, \ldots, n-2$ $\mathbf{K}_1 = \mathbf{K}$	$\frac{2}{3}n^3 + \frac{3}{2}n^2$	
QR iterations	$\mathbf{T}_{k+1} = \mathbf{Q}_k^T \mathbf{T}_k \mathbf{Q}_k;\quad k = 1, 2, \ldots$ $\mathbf{T}_1 = \mathbf{K}_{n-1}$	$9n^2$	Using symmetry of matrix
Calculation of p eigenvectors	$(\mathbf{K}_{n-1} - \lambda_i \mathbf{I})\mathbf{x}_i^{(k+1)} = \mathbf{x}_i^{(k)};\quad k = 1, 2,\quad i = 1, 2, \ldots, p$	$10pn$	$\frac{n}{2}(n+1) + 6n$
Transformation of eigenvectors	$\boldsymbol{\phi}_i = \mathbf{P}_1 \ldots \mathbf{P}_{n-2}\mathbf{x}_i^{(3)};\ i = 1, 2, \ldots, p$	$pn(n-1)$	
Total for all eigenvalues and p eigenvectors		$\frac{2}{3}n^3 + \frac{21}{2}n^2 + pn^2 + 9pn$	

for the eigenvector transformations in (11.114). Therefore, it is seen that the calculation of the eigenvalues of $\mathbf{T}_1$ is not very expensive, but the preparation of $\mathbf{K}$ into a form that can be used effectively for the iteration process requires most of the numerical effort.

It should be noted that Table 11.4 does not include the operations required for transforming a generalized eigenproblem into a standard form. If this transformation is carried out, the eigenvectors calculated in Table 11.4 must also be transformed to the eigenvectors of the generalized eigenproblem as discussed in Section 10.2.5.

11.4 POLYNOMIAL ITERATION TECHNIQUES

The close relationship between the calculation of the zeros of a polynomial and the evaluation of eigenvalues has been discussed in Section 10.2.2. Namely, defining the characteristic polynomial $p(\lambda)$, where

$$p(\lambda) = \det (\mathbf{K} - \lambda \mathbf{M}) \tag{11.6}$$

the zeros of $p(\lambda)$ are the eigenvalues of the eigenproblem $\mathbf{K}\boldsymbol{\phi} = \lambda \mathbf{M}\boldsymbol{\phi}$. To calculate the eigenvalues, we therefore may operate on $p(\lambda)$ to extract the zeros of the polynomial. A number of solution procedures have been proposed.[1-11] However, basically there are two strategies: explicit and implicit evaluation procedures, both of which may use the same basic iteration schemes.

In the discussion of the polynomial iteration schemes, we assume that the solution is carried out directly using $\mathbf{K}$ and $\mathbf{M}$ of the finite element assemblage, i.e., without transforming the problem into a different form. For example, if $\mathbf{M}$ is the identity matrix, we could transform $\mathbf{K}$ first into tridiagonal form as is done in the HQRI solution (see Section 11.3.3). In case $\mathbf{M} \neq \mathbf{I}$, we would need to transform the generalized eigenproblem into a standard form (see Section 10.2.5), before the Householder reduction into a tridiagonal matrix could be performed. Whichever problem we finally consider in the iterative solution of the required eigenvalues, the solution startegy would not be changed. However, if only a few eigenvalues are to be calculated, the direct solution using $\mathbf{K}$ and $\mathbf{M}$ is nearly always most effective (see Section 12.2).

It should be noted that using a polynomial iteration method, only the eigenvalues are calculated. The corresponding eigenvectors can then be obtained effectively by using inverse iteration with shifting; i.e., each required eigenvector is obtained by inverse iteration at a shift equal to the corresponding eigenvalue.

11.4.1 Explicit Polynomial Iteration

In the explicit polynomial iteration methods, the first step is to write $p(\lambda)$ in the form

$$p(\lambda) = a_0 + a_1\lambda + a_2\lambda^2 + \ldots + a_n\lambda^n \tag{11.115}$$

and evaluate the polynomial coefficients $a_0, a_1, \ldots, a_n$. The second step is to calculate the roots of the polynomial. We demonstrate the procedure by means of an example.

EXAMPLE 11.17: Establish the coefficients of the characteristic polynomial of the problem $\mathbf{K}\boldsymbol{\phi} = \lambda\mathbf{M}\boldsymbol{\phi}$, where **K** and **M** are the matrices used in Example 11.4.

The problem is here to evaluate the expression

$$p(\lambda) = \det\begin{bmatrix} 5-2\lambda & -4 & 1 & 0 \\ -4 & 6-2\lambda & -4 & 1 \\ 1 & -4 & 6-\lambda & -4 \\ 0 & 1 & -4 & 5-\lambda \end{bmatrix}$$

Following the rules given in Section 1.8 for the evaluation of a determinant, we obtain

$$\begin{aligned} p(\lambda) = (5-2\lambda)\det&\begin{bmatrix} 6-2\lambda & -4 & 1 \\ -4 & 6-\lambda & -4 \\ 1 & -4 & 5-\lambda \end{bmatrix} \\ + (4)\det&\begin{bmatrix} -4 & -4 & 1 \\ 1 & 6-\lambda & -4 \\ 0 & -4 & 5-\lambda \end{bmatrix} \\ + (1)\det&\begin{bmatrix} -4 & 6-2\lambda & 1 \\ 1 & -4 & -4 \\ 0 & 1 & 5-\lambda \end{bmatrix} \end{aligned}$$

Hence

$$\begin{aligned} p(\lambda) = {} & (5-2\lambda)\{(6-2\lambda)[(6-\lambda)(5-\lambda)-16] \\ & + 4[-4(5-\lambda)+4] + 16 - (6-\lambda)\} \\ & + 4\{-4[(6-\lambda)(5-\lambda)-16] + 4(5-\lambda) - 4\} \\ & + \{-4[(-4)(5-\lambda)+4] - (6-2\lambda)(5-\lambda) + 1\} \end{aligned}$$

and the expression finally reduces to

$$p(\lambda) = 4\lambda^4 - 66\lambda^3 + 276\lambda^2 - 285\lambda + 25$$

In the general case when n is large, we cannot evaluate the polynomial coefficients as easily as in the example above. The expansion of the determinant would require about n-factorial operations, which are far too many operations to make the method practical. However, other procedures have been developed; for example, the Newton identities may be used. Once the coefficients have been evaluated, it is necessary to employ a standard polynomial root finder, using, for example, a Newton iteration or secant iteration to evaluate the required eigenvalues.

Although the procedure seems most natural to use, one difficulty has caused the method to be almost completely abandoned for the solution of eigenvalue problems. A basic defect of the method is that small errors in the coefficients $a_0, \ldots, a_n$ cause large errors in the roots of the polynomial. But small errors are almost unavoidable, owing to round-off in the computer. Therefore, an explicit evaluation of the coefficients $a_0, \ldots, a_n$ from **K** and **M** with subsequent solution of the required eigenvalues is not effective in general analysis.

11.4.2 Implicit Polynomial Iteration

In an implicit polynomial iteration solution we evaluate the value of $p(\lambda)$ directly without calculating first the coefficients $a_0, \ldots, a_n$ in (11.115). The value of $p(\lambda)$ can be obtained effectively by decomposing $\mathbf{K} - \lambda\mathbf{M}$ into a lower unit triangular matrix **L** and an upper triangular matrix **S**; i.e., we have

$$\mathbf{K} - \lambda\mathbf{M} = \mathbf{L}\mathbf{S} \tag{11.116}$$

where then

$$\det(\mathbf{K} - \lambda\mathbf{M}) = \prod_{i=1}^{n} s_{ii} \tag{11.117}$$

The decomposition of $\mathbf{K} - \lambda\mathbf{M}$ is obtained as discussed in Section 8.2, but, as pointed out in Section 8.2.5, may require interchanges when $\lambda > \lambda_1$. In case row and corresponding column interchanges are carried out, the coefficient matrix remains symmetric and, in effect, the degree of freedom numbering has then merely been rearranged. In other words, the stiffness and mass matrices of the finite element system actually used correspond to a different degree of freedom numbering than originally specified. On the other hand, if row interchanges only are employed, a nonsymmetric coefficient matrix is obtained. However, in either case it is important to note that the required row and column interchanges could have been carried out prior to the Gauss elimination process to obtain an "effective" coefficient matrix $\mathbf{K} - \lambda\mathbf{M}$ that is considered in (11.116), and no more interchanges would then be needed. Each row or column interchange introduced merely effects a change in sign of the determinant. In practice, we do not know the actual row and column interchanges that will be required, but the realization that all interchanges could have been carried out prior to the factorization shows that we can always use the Gauss elimination procedure given in (8.10) to (8.14), provided that we admit that the "effective" initial coefficient matrix may be nonsymmetric. Consider the following example.

EXAMPLE 11.18: Use Gauss elimination to evaluate $p(\lambda) = \det(\mathbf{K} - \lambda\mathbf{M})$, where

$$\mathbf{K} = \begin{bmatrix} 2 & -1 & 0 \\ -1 & 4 & -1 \\ 0 & -1 & 2 \end{bmatrix}; \qquad \mathbf{M} = \begin{bmatrix} 1 & & \\ & 1 & \\ & & \frac{1}{2} \end{bmatrix}; \qquad \lambda = 2$$

In this case we have

$$\mathbf{K} - \lambda\mathbf{M} = \begin{bmatrix} 0 & -1 & 0 \\ -1 & 2 & -1 \\ 0 & -1 & 1 \end{bmatrix}$$

Since the first diagonal element is zero, we need to use interchanges. Assume that we interchange the first and second rows (and not the corresponding columns); then we effectively factorize $\bar{\mathbf{K}} - \lambda\bar{\mathbf{M}}$, where

$$\bar{\mathbf{K}} = \begin{bmatrix} -1 & 4 & -1 \\ 2 & -1 & 0 \\ 0 & -1 & 2 \end{bmatrix}; \qquad \bar{\mathbf{M}} = \begin{bmatrix} 0 & 1 & 0 \\ 1 & 0 & 0 \\ 0 & 0 & \frac{1}{2} \end{bmatrix}$$

The factorization of $\bar{\mathbf{K}} - \lambda\bar{\mathbf{M}}$ is now obtained in the usual way (see Examples 10.4 and 10.5),

$$\bar{\mathbf{K}} - \lambda\bar{\mathbf{M}} = \begin{bmatrix} 1 & & \\ 0 & 1 & \\ 0 & 1 & 1 \end{bmatrix} \begin{bmatrix} -1 & 2 & -1 \\ & -1 & 0 \\ & & 1 \end{bmatrix}$$

Hence

$$\det(\bar{\mathbf{K}} - \lambda\bar{\mathbf{M}}) = (-1)(-1)(1) = 1$$

and taking account of the fact that the row interchange changed the sign of the determinant (see Section 1.8), we have

$$\det(\mathbf{K} - \lambda\mathbf{M}) = -1$$

As pointed out above, if in the Gauss elimination no interchanges have been carried out or each row interchange was accompanied by a corresponding column interchange, the coefficient matrix $\mathbf{K} - \lambda\mathbf{M}$ in (11.116) is symmetric. In this case we have as in Section (8.2.2) that $\mathbf{S} = \mathbf{D}\mathbf{L}^T$ and hence

$$\det(\mathbf{K} - \lambda\mathbf{M}) = \prod_{i=1}^{n} d_{ii} \tag{11.118}$$

In the determinant search solution method (see Section 12.2) we only use a factorization if it can be completed without interchanges and therefore always employ (11.118). In this case, one polynomial evaluation requires about $\frac{1}{2}nm_{\mathbf{K}}^2$ operations, where n is the order of $\mathbf{K}$ and $\mathbf{M}$ and $m_{\mathbf{K}}$ is the half-bandwidth of $\mathbf{K}$.

With a scheme available for an accurate evaluation of $p(\lambda)$, we can now use a number of iteration schemes to calculate a root of the polynomial. One commonly used and simple technique is the secant method, in which a linear interpolation is employed; i.e., let $\mu_{k-1} < \mu_k$, then we iterate using

$$\mu_{k+1} = \mu_k - \frac{p(\mu_k)}{p(\mu_k) - p(\mu_{k-1})}(\mu_k - \mu_{k-1}) \tag{11.119}$$

where μ_k is the kth iterate (see Fig. 11.3). We may note that the secant method

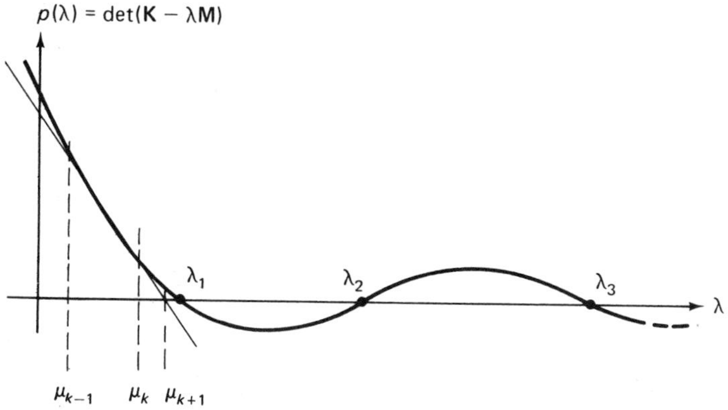

FIGURE 11.3 *Secant iteration for calculation of λ_1.*

in (11.119) is an approximation to the Newton iteration,

$$\mu_{k+1} = \mu_k - \frac{p(\mu_k)}{p'(\mu_k)} \tag{11.120}$$

where $p'(\mu_k)$ is approximated using

$$p'(\mu_k) \doteq \frac{p(\mu_k) - p(\mu_{k-1})}{\mu_k - \mu_{k-1}} \tag{11.121}$$

An actual scheme to accurately evaluate $p'(\mu_k)$ has been tested but was not found to be effective.[11]

Another technique commonly used in the solution of complex eigenvalue problems is Muller's method, in which case a quadratic interpolation is employed. A disadvantage of Muller's method in the solution of $\mathbf{K}\boldsymbol{\phi} = \lambda\mathbf{M}\boldsymbol{\phi}$ is that although the starting values μ_k, μ_{k-1}, and μ_{k-2} are real, the calculated value μ_{k+1} may be complex.

It should be noted that so far we did not discuss to which eigenvalue any one of the iteration strategies converges. This depends on the starting iteration values. Using μ_{k-1} and μ_k both smaller than λ_1, the Newton and secant iterations converge montonically to λ_1, as shown in Fig. 11.3, and the ultimate order of convergence is quadratic and linear, respectively. In effect, convergence is achieved in the iterations because $p''(\lambda) > 0$ for $\lambda < \lambda_1$, which is always the case for any order and bandwidth of **K** and **M**. However, convergence cannot be guaranteed for arbitrary starting values μ_{k-1} and μ_k. Therefore, a solution algorithm to calculate a set of required eigenvalues using the secant iteration must be designed with care to assure convergence to the actually required roots.

Consider the following example of a secant iteration.

EXAMPLE 11.19: Use the secant iteration to calculate λ_1 of the eigenproblem $\mathbf{K}\boldsymbol{\phi} = \lambda\mathbf{M}\boldsymbol{\phi}$, where

$$\mathbf{K} = \begin{bmatrix} 2 & -1 & 0 \\ -1 & 4 & -1 \\ 0 & -1 & 2 \end{bmatrix}; \qquad \mathbf{M} = \begin{bmatrix} \frac{1}{2} & & \\ & 1 & \\ & & \frac{1}{2} \end{bmatrix}$$

For the secant iteration we need two starting values for μ_1 and μ_2 that are lower bounds to λ_1. Let $\mu_1 = -1$ and $\mu_2 = 0$ Then we have

$$p(-1) = \det \begin{bmatrix} \frac{5}{2} & -1 & 0 \\ -1 & 5 & -1 \\ 0 & -1 & \frac{5}{2} \end{bmatrix}$$

$$= \det \begin{bmatrix} 1 & & \\ -\frac{2}{5} & 1 & \\ 0 & -\frac{5}{23} & 1 \end{bmatrix} \begin{bmatrix} \frac{5}{2} & & \\ & \frac{23}{5} & \\ & & \frac{105}{46} \end{bmatrix} \begin{bmatrix} 1 & -\frac{2}{5} & 0 \\ & 1 & -\frac{5}{23} \\ & & 1 \end{bmatrix}$$

Hence $$p(-1) = (\tfrac{5}{2})(\tfrac{23}{5})(\tfrac{105}{46}) = 26.25$$

Similarly, $$p(0) = \det \begin{bmatrix} 2 & -1 & 0 \\ -1 & 4 & -1 \\ 0 & -1 & 2 \end{bmatrix} = 12$$

Now using (11.119) we obtain as the next shift,

$$\mu_3 = 0 - \frac{12}{12 - 26.25}[0 - (-1)]$$

Hence $$\mu_3 = 0.8421$$

Continuing in the same way, we obtain

$$\begin{aligned} p(0.8421) &= 4.7150 \\ \mu_4 &= 1.3871 \\ p(1.3871) &= 1.8467 \\ \mu_5 &= 1.7380 \\ p(1.7380) &= 0.63136 \\ \mu_6 &= 1.9203 \\ p(1.9203) &= 0.16899 \\ \mu_7 &= 1.9870 \\ p(1.9870) &= 0.026347 \\ \mu_8 &= 1.9993 \end{aligned}$$

Hence after six iterations we have as an approximation to λ_1, $\lambda_1 \doteq 1.9993$.

11.5 METHODS BASED ON THE STURM SEQUENCE PROPERTY

In Section 10.2.2 we discussed the Sturm sequence property of the characteristic polynomials of the problem $\mathbf{K}\boldsymbol{\phi} = \lambda\mathbf{M}\boldsymbol{\phi}$ and of its associated constraint problems. The main result was the following. Assume that for a shift μ_k, the Gauss factorization of $\mathbf{K} - \mu_k\mathbf{M}$ into $\mathbf{LDL}^T$ can be obtained. Then the number of negative elements in $\mathbf{D}$ is equal to the number of eigenvalues smaller than μ_k. This result can directly be used to construct an algorithm for the calculation of eigenvalues and corresponding eigenvectors.[1,11,37-40] As in the discussion of the polynomial iteration methods, we assume in the following that the solution is performed directly using $\mathbf{K}$ and $\mathbf{M}$, although the same strategies could be used after the generalized eigenproblem has been transformed into a different form. Also, as in Section 11.4, the solution method to be presented only solves for the eigenvalues, and the corresponding eigenvectors would be calculated using inverse iteration with shifting (see Section 11.2.3).

Consider that we want to solve for all eigenvalues between λ_l and λ_u, where λ_l and λ_u are the lower and upper limits, respectively. For example, we may have a case such as depicted in Fig. 11.4 or λ_l may be zero, in which case

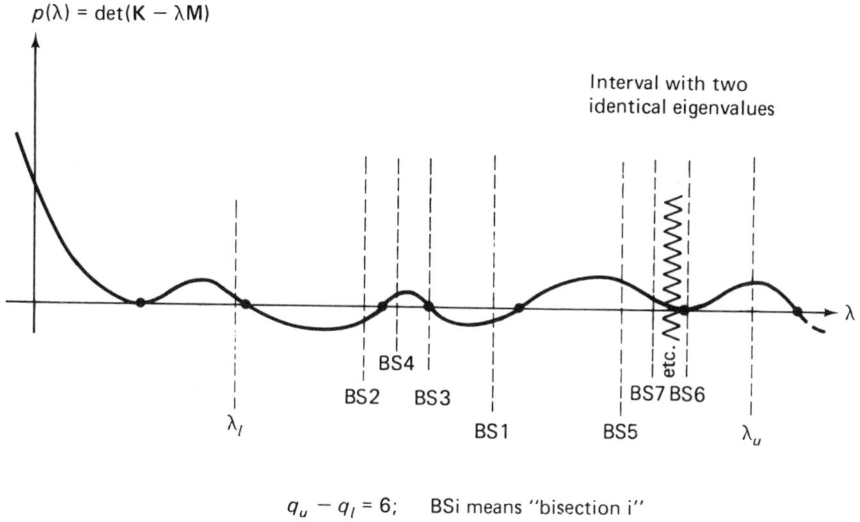

FIGURE 11.4 *Use of Sturm sequence property to isolate eigenvalues.*

we would need to solve for all eigenvalues up to the value λ_u. The basis of the solution procedure is the triangular factorization of $\mathbf{K} - \mu_k\mathbf{M}$, where μ_k is selected in such way as to obtain from the positive or negative signs of the diagonal elements in the factorization meaningful information about the unknown and required eigenvalues. The following solution procedure, known as the *bisection method*, may be used (refer to Fig. 11.4 for a typical example):

1. Factorize $\mathbf{K} - \lambda_l\mathbf{M}$ and hence find how many eigenvalues, say q_l, are smaller than λ_l.
2. Apply the Sturm sequence check at $\mathbf{K} - \lambda_u\mathbf{M}$ and hence find how

many eigenvalues, say q_u, are smaller than λ_u, There are thus $q_u - q_l$ eigenvalues between λ_u and λ_l.

3. Use a simple scheme of bisection to identify intervals in which the individual eigenvalues lie. In this process, those intervals in which more than one eigenvalue are known to lie are successively bisected and the Sturm sequence check is carried out until all eigenvalues are isolated.
4. Calculate the eigenvalues to the accuracy required and then obtain the corresponding eigenvectors by inverse iteration.

To obtain the eigenvalues accurately in step 4, the method of bisection is abandoned and a more efficient procedure is generally used. For example, the secant iteration presented in Section 11.4.2 can be employed once the eigenvalues have been isolated (see Example 11.20).

The above technique to calculate the required eigenvalues is straightforward. However, the method can be quite inefficient, because each iteration in step 3 requires a triangular factorization and because many iterations may be needed. The number of iterations can be particularly high when multiple eigenvalues (see Fig. 11.4) or eigenvalue clusters must be evaluated, in which case additional strategies to accelerate the process must be included. In general, the method is effective if only a few factorizations are needed to identify intervals of individual eigenvalues, and the smallest required eigenvalue is much larger than λ_1.

EXAMPLE 11.20: Use the bisection method followed by secant iteration to calculate λ_2 of the problem $\mathbf{K}\boldsymbol{\phi} = \lambda\mathbf{M}\boldsymbol{\phi}$, where

$$\mathbf{K} = \begin{bmatrix} 2 & -1 & 0 \\ -1 & 4 & -1 \\ 0 & -1 & 2 \end{bmatrix}; \qquad \mathbf{M} = \begin{bmatrix} \frac{1}{2} & & \\ & 1 & \\ & & \frac{1}{2} \end{bmatrix}$$

We already considered this problem in Example 10.5, where we isolated the eigenvalues using the bisection technique. Specifically, we found that

$$\lambda_1 < 3 < \lambda_2 < 5 < \lambda_3$$

Hence we can start the secant iteration with $\mu_1 = 3$, $\mu_2 = 5$, and using the results of Example 10.5:

$$p(\mu_1) = -\tfrac{3}{4}; \qquad p(\mu_2) = \tfrac{3}{4}$$

Using (11.119), we obtain

$$\mu_3 = 5 - \frac{\frac{3}{4}}{(\frac{3}{4}) - (-\frac{3}{4})}(5 - 3)$$

or

$$\mu_3 = 4$$

Next we need to evaluate $\quad p(\mu_3) = \det(\mathbf{K} - \mu_3\mathbf{M})$

and we find that $p(\mu_3) = 0.0$. Hence $\lambda_2 = \mu_3 = 4$. Therefore, in this case one step of secant iteration was sufficient to calculate λ_2. But it may be noted that to evaluate $p(\mu_3)$ by Gauss factorization of $\mathbf{K} - \mu_3\mathbf{M}$, a row interchange is needed (see Example 10.4).

It is important to note that we assumed in the above presentation that the factorization of $\mathbf{K} - \mu_k\mathbf{M}$ into $\mathbf{LDL}^T$ can be obtained. However, we discussed in Sections 8.2.5 and 11.4.2 that if $\mu_k > \lambda_1$, interchanges may be required. If each row interchange is accompanied by a corresponding column interchange, the coefficient matrix remains symmetric and the interchanges correspond merely to a rearranging of the degree of freedom numbering. Hence the $\mathbf{LDL}^T$ factorization of a coefficient matrix that corresponds to a different degree of freedom numbering is obtained. However, if row interchanges only are used, a nonsymmetric coefficient matrix is obtained. In this case the basic algorithm described above can still be used, but proper account has to be taken of the effect of each row interchange on the number of negative pivot elements that will be encountered.

The important point is that in the bisection algorithm, the accuracy with which the eigenvalues are calculated depends on the possibility of obtaining an accurate factorization of $\mathbf{K} - \mu_k\mathbf{M}$. Therefore, it is necessary in practical calculations to allow interchanges in the Gauss elimination, and it has been found that it is most effective to work with row interchanges only. There is no need to immediately start row interchanges, except when multiplier growth is encountered (see Section 8.5). However, the possible need for row interchanges means that more high-speed storage must be allocated, because the half-bandwidth may increase from $m_{\mathbf{K}}$ to $2m_{\mathbf{K}}$; and if row interchanges are carried out, the number of operations increases also.

The bisection method has two major disadvantages: (1) it is necessary to complete the factorization of $\mathbf{K} - \mu_k\mathbf{M}$ with as much accuracy as possible, although the coefficient matrix may be ill-conditioned (indeed, this may still be the case even when row interchanges are included), and (2) convergence can be very slow when a cluster of eigenvalues has to be solved. However, the Sturm sequence property can be employed in conjunction with other solution strategies, and it is in such cases that the property can be extremely useful. In particular, the Sturm sequence property is employed in the determinant search algorithm (see Section 12.2), in which the factorization of $\mathbf{K} - \mu_k\mathbf{M}$ is carried out without interchanges but is completed only provided that no instability occurs. If the factorization proves to be unstable, a different μ_k is selected. This is possible because the final accuracy with which the eigenvalues and eigenvectors are calculated does not depend on the specific shift μ_k used in the solution.

REFERENCES

1. J. H. Wilkinson, *The Algebraic Eigenvalue Problem*, Oxford University Press, Inc., London, 1965.
2. V. N. Faddeeva, *Computational Methods of Linear Algebra*, Dover Publications, Inc., New York, N.Y., 1959.
3. B. Noble, *Applied Linear Algebra*, Prentice-Hall, Inc., Englewood Cliffs, N.J., 1969.
4. C. E. Fröberg, *Introduction to Numerical Analysis*, Addison-Wesley Publishing Company, Inc., Reading, Mass., 1969.
5. C. Lanczos, *Applied Analysis*, Prentice-Hall, Inc., Englewood Cliffs, N.J., 1956.
6. S. H. Crandall, *Engineering Analysis*, McGraw-Hill Book Company, New York, N.Y., 1956.
7. R. Zurmühl, *Matrizen*, Springer-Verlag, Berlin, 1964.

8. H. R. Schwarz, H. Rutishauser, and E. Stiefel, *Matrizen-Numerik*, B. G. Teubner, Stuttgart, 1972.
9. A. S. Householder, *Principles of Numerical Analysis*, McGraw-Hill Book Company, New York, N.Y., 1963.
10. G. Peters and J. H. Wilkinson, "Ax = λBx and the Generalized Eigenproblem," *S.I.A.M., Journal of Numerical Analysis*, Vol. 7, 1970, pp. 479–492.
11. K. J. Bathe, "Solution Methods for Large Generalized Eigenvalue Problems in Structural Engineering," Report UC SESM 71-20, Civil Engineering Department, University of California, Berkeley, California, 1971.
12. W. W. Bradbary and R. Fletcher, "New Iterative Methods for Solution of the Eigenproblem Arising in Structural Dynamics," *Proceedings, 2nd Conference on Matrix Methods in Structural Mechanics*, Wright-Patterson A.F.B., Ohio, 1968.
13. R. Fox and M. Kapoor, "A Minimization Method for the Solution of the Eigenproblem Arising in Structural Dynamics," *Proceedings, 2nd Conference on Matrix Methods in Structural Mechanics*, Wright-Patterson A.F.B., Ohio, 1968.
14. I. Fried, "Optimal Gradient Minimization Scheme for Finite Element Eigenproblems," *Journal of Sound and Vibration*, Vol. 20, No. 3, 1972, pp. 333–342.
15. S. H. Crandall, "Iterative Procedures Related to Relaxation Methods for Eigenvalue Problems," *Proceedings, Royal Society*, London, Vol. A207, 1951, p. 416.
16. A. M. Ostrowski, "On the Convergence of the Rayleigh Quotient Iteration for the Computation of the Characteristic Roots and Vectors, Parts I–VI," *Archives for Rational Mechanics and Analysis*, Vols. 1–4, 1958–1959.
17. B. N. Parlett and W. Kahan, "On the Convergence of a Practical QR Algorithm," *Proceedings, I.F.I.P. Congress*, 1968.
18. W. G. Poole, Jr., "A Geometric Convergence Theory for the Eigenvalues of a Symmetric Tridiagonal Matrix," Technical Report CS 43, Computer Science Department, Stanford University, California, 1966.
19. H. Wielandt, "Bestimmung höherer Eigenwerte durch gebrochene Iteration," Bericht B44/J/37 der Aerodynamischen Versuchsanstalt, Göttingen, 1944.
20. E. E. Osborne, "On Acceleration and Matrix Deflation Processes Used with the Power Method," *Journal of the Society for Industrial and Applied Mathematics*, Vol. 6, 1958, pp. 279–287.
21. H. Rutishauser, "Deflation bei Bandmatrizen," *Z.A.M.P.*, Vol. 10, 1959, pp. 314–319.
22. C. A. Felippa, "BANEIG—Eigenvalue Routine for Symmetric Band Matrices," Computer Programming Series, SESM, Department of Civil Engineering, University of California, Berkeley, California, 1966.
23. C. G. J. Jacobi, "Über ein leichtes Verfahren die in der Theorie der Säculärstörungen vorkommenden Gleichungen numerisch aufzulösen," *Crelle's Journal*, Vol. 30, 1846, pp. 51–94.
24. H. H. Goldstine, F. J. Murray, and J. von Neumann, "The Jacobi Method for Real Symmetric Matrices," *Journal of the Association for Computing Machinery*, Vol. 6, 1959, pp. 59–96.
25. A. Schönhage, "Zur Konvergenz des Jacobi-Verfahrens," *Numerische Mathematik*, Vol. 3, 1961, pp. 374–380.
26. G. E. Forsythe and P. Henrici, "The Cyclic Jacobi Method for Computing the Principal Values of a Complex Matrix," *Transactions of the American Mathematical Society*, Vol. 94, 1960, pp. 1–23.
27. J. H. Wilkinson, "Note on the Quadratic Convergence of the Cyclic Jacobi Process," *Numerische Mathematik*, Vol. 4, 1962, pp. 296–300.

28. S. Falk and P. Langemeyer, "Das Jacobische Rotationsverfahren für reellsymmetrische Matrizenpaare," *Elektronische Datenverarbeitung*, 1960, pp. 30–34.
29. R. S. Martin, C. Reinsch, and J. H. Wilkinson, "Householder's Tridiagonalization of a Symmetric Matrix," *Numerische Mathematik*, Vol. 11, 1968, pp. 181–195.
30. J. G. F. Francis, "The QR Transformation, Parts I and II," *Computer Journal*, Vol. 4, 1961–1962, pp. 265, 332.
31. V. N. Kublanovskaya, "On Some Algorithms for the Solution of the Complete Problem of Proper Values," *Journal of Computer Mathematics and Mathematical Physics*, Vol. 1, 1961.
32. B. N. Parlett, "Convergence of the QR Algorithm," *Numerische Mathematik*, Vol. 7, 1965, p. 187; Vol. 10, 1967, p. 163.
33. B. N. Parlett, "Global Convergence of the Basic QR Algorithm on Hessenberg Matrices," *Mathematical Computation*, Vol. 22, 1968, p. 803.
34. J. H. Wilkinson, "The QR Algorithm for Real Symmetric Matrices with Multiple Eigenvalues," *Computer Journal*, Vol. 8, 1965, p. 85.
35. W. Kahan and J. Varah, "Two Working Algorithms for the Eigenvalues of a Symmetric Tridiagonal Matrix," Technical Report CS 43, Computer Science Department, Stanford University, California, 1966.
36. J. M. Ortega and H. F. Kaiser, "The LL^T and QR Methods for Symmetric Tridiagonal Matrices," *Computer Journal*, Vol. 6, 1963, pp. 99–101.
37. J. W. Givens, "A Method of Computing Eigenvalues and Eigenvectors Suggested by Classical Results on Symmetric Matrices," *National Bureau of Standards Applied Mathematics Series 29*, Government Printing Office, Washington, D.C., L953, pp. 117–122.
38. W. Barth, R. S. Martin, and J. H. Wilkinson, "Calculation of the Eigenvalues of a Symmetric Tridiagonal Matrix by the Method of Bisection," *Numerische Mathematik*, Vol. 9, 1967, pp. 386–393.
39. K. K. Gupta, "Vibration of Frames and Other Structures with Banded Stiffness Matrix," *International Journal for Numerical Methods in Engineering*, Vol. 2, 1970, pp. 221–228.
40. K. K. Gupta, "Solution of Eigenvalue Problems by the Sturm Sequence Method," *International Journal for Numerical Methods in Engineering*, Vòl. 4, 1972, pp. 379–404.

12

SOLUTION OF LARGE EIGENPROBLEMS

12.1 INTRODUCTION

In Chapter 11 we described individually four basic groups of solution techniques to calculate either an eigenvalue or an eigenvector, or all eigenvalues and eigenvectors simultaneously. The solution methods described are based on some fundamental facts as summarized in Section 11.1. Considering the effectiveness of the solution procedures, none of the methods is always most efficient, but the solution technique to be used should be selected according to the specific problem to be solved.

As was discussed in Section 9.3, the efficient solution of the required eigenvalues and eigenvectors is of much importance in mode superposition analysis of large finite element systems. Indeed, because procedures for the exact solution of large eigenproblems have been expensive for a long time, approximate solution techniques have been and are still used. Some commonly employed approximate solution methods have been discussed in Section 10.3. *The final conclusion was that the approximate techniques must be used with great care and cannot be recommended for general use.*

Since it is very important to employ an optimum algorithm for the calculation of the required eigenvalues and eigenvectors in the problem $\mathbf{K}\boldsymbol{\phi} = \lambda\mathbf{M}\boldsymbol{\phi}$ when the order of $\mathbf{K}$ and $\mathbf{M}$ is large, it appears natural to try and combine the basic solution methods presented in Sections 11.2 to 11.5 in an effort to design more effective techniques. The solution methods to be designed should take specific advantage of the solution requirements and exploit the advantages of each basic solution procedure when taking account of the specific characteristics of $\mathbf{K}$ and $\mathbf{M}$. In the following, we describe in detail two solution methods, the determinant search technique and the subspace iteration method, which are believed to be very effective in finite element analysis.[1-4] We do not want to imply that these methods yield, at present and in the future, in each analysis, the most effective solution because there is no single algorithm that is always most effective. Indeed, given a specific eigenproblem, which method is used most appropriately depends on the characteristics of $\mathbf{K}$ and $\mathbf{M}$ and the number of eigenpairs required. However, the determinant search and subspace iteration methods have been used extensively in a number of general-purpose finite element analysis programs and have proven reliable and cost-effective. We assume in the description of the algorithms that the material covered in the preceding sections is known, and we make extensive reference to it.

Before presenting the solution techniques, we should define what we mean by a large eigenproblem. As we discussed in Section 9.3, in finite element analysis we are mainly interested in the smallest eigenvalues. Correspondingly, we define an eigenproblem as large if it is much cheaper to solve for only the required eigenvalues and eigenvectors instead of simply calculating all. In general, the system will be large if the high-speed core storage of a reasonably sized computer is too small for using an in-core Householder-QR-inverse iteration method or the generalized Jacobi solution technique. In that case it is most probably more effective to employ the determinant search or subspace iteration methods described in the following sections, because out-of-core solutions using transformation methods that calculate all eigenvalues require much use of back-up storage and are expensive. However, even for in-core solutions, the determinant search and subspace iteration techniques are in most cases more effective than the HQRI or Jacobi solution methods, because they calculate directly (i.e., without a transformation to the standard eigenproblem) only those eigenvalues and corresponding eigenvectors that are actually required. In addition, during the solution process optimum advantage is taken of the special characteristics of the matrices $\mathbf{K}$ and $\mathbf{M}$.

The determinant search technique and the subspace iteration method have both been developed for the solution of the p smallest eigenvalues and corresponding eigenvectors as mostly required in dynamic analysis. However, the subspace iteration method has also been extended to calculate very effectively all eigenvalues in a specified interval (see Section 12.3.6). Furthermore, the determinant search and subspace iteration methods can also be employed in buckling analysis for the problem $(-\mathbf{K}_G)\boldsymbol{\phi} = \lambda\mathbf{K}\boldsymbol{\phi}$, where $\mathbf{K}$ and $\mathbf{K}_G$ are the linear and nonlinear strain stiffness matrices and we require the largest eigenvalue and corresponding eigenvector. In this case, prior to the solution, a shift is imposed that is larger than the largest eigenvalue, and the algorithms are modified to evaluate the smallest negative eigenvalue. Namely, the smallest negative eigenvalue of the shifted system added to the shift gives the required largest eigenvalue of $(-\mathbf{K}_G)\boldsymbol{\phi} = \lambda\mathbf{K}\boldsymbol{\phi}$ (see Section 10.2.3).

12.2 THE DETERMINANT SEARCH METHOD

The fundamental techniques used in the determinant search solution method are: an iteration with the characteristic polynomial $p(\lambda)$, the use of the Sturm sequence property of the characteristic polynomials $p(\lambda)$ and $p^{(r)}(\lambda^{(r)})$, and vector inverse iteration. These techniques have been described in Sections 11.4.2, 11.5, and 11.2.3, respectively, where also the advantages and disadvantages of each procedure have been discussed. The specific aim in the development of the determinant search method was to combine and extend these techniques in an optimum manner for the solution of the smallest eigenvalues and corresponding eigenvectors in the problem $\mathbf{K}\boldsymbol{\phi} = \lambda\mathbf{M}\boldsymbol{\phi}$. It should be noted that each of the fundamental techniques can be used directly for the solution of the eigenproblem $\mathbf{K}\boldsymbol{\phi} = \lambda\mathbf{M}\boldsymbol{\phi}$; i.e., no transformation to the standard eigenproblem is required. This is a very important point, because difficulties that may arise from ill-conditioning or loss of bandwidth due to the transformation are avoided.

12.2.1 Preliminary Considerations

We recall that the iteration on the polynomial $p(\lambda)$ or the use of the Sturm sequence property are basic techniques to calculate an eigenvalue, whereas inverse iteration is a basic technique for evaluating an eigenvector. Also, once an eigenvalue or an eigenvector has been calculated, the other member of the eigenpair can be evaluated without further iteration, i.e.:

1. Given λ_i, the eigenvector $\boldsymbol{\phi}_i$ is obtained by solving $(\mathbf{K} - \lambda_i\mathbf{M})\boldsymbol{\phi}_i = \mathbf{0}$.
2. Given $\boldsymbol{\phi}_i$, the eigenvalue λ_i is obtained from the Rayleigh quotient $\lambda_i = \boldsymbol{\phi}_i^T\mathbf{K}\boldsymbol{\phi}_i/\boldsymbol{\phi}_i^T\mathbf{M}\boldsymbol{\phi}_i$.

If, however, only approximations to $\boldsymbol{\phi}_i$ or λ_i are known, we need further iteration to evaluate one member of the eigenpair $(\lambda_i, \boldsymbol{\phi}_i)$ accurately, after which the other member can be obtained as in 1 or 2 above.

Assume that we have a shift μ, for example, obtained by polynomial iteration, where $|\lambda_i - \mu|/|\lambda_j - \mu|$ is small for all j, $j \neq i$. In this case we can use inverse iteration with shift μ to calculate $\boldsymbol{\phi}_i$, as discussed in Section 11.2.3. In the iteration we may use Rayleigh quotient shifting to accelerate convergence, in which case we would obtain λ_i and $\boldsymbol{\phi}_i$ simultaneously. On the other hand, assume that we have a vector $\boldsymbol{\phi}$, which is a close approximation to the eigenvector $\boldsymbol{\phi}_i$. In such case we may use the Rayleigh quotient to obtain a shift near λ_i and then proceed again with inverse iteration and possibly Rayleigh quotient shifting. But it is important to note that a Rayleigh quotient shift can only be applied once the iteration vector is a close enough approximation to the eigenvector sought, because only in such a case will convergence occur to the required eigenpair (see Section 11.2.4).

Basically, therefore, the problem reduces to finding an approximation to either the next required eigenvalue or the corresponding eigenvector. Once an acceptable approximation has been obtained, the eigenpair is evaluated accurately by vector inverse iteration or Rayleigh quotient iteration. However, the choice of whether to find an approximation to the unknown eigenvalue or the corresponding eigenvector or to both simultaneously depends on the relative cost of the evaluations. Considering vector iteration as the basic technique to obtain an eigenvector, we discussed in Section 11.2 the difficulties of obtaining by simple inverse iteration (without shifting) an approximation to an eigenvector $\boldsymbol{\phi}_i$, where $i > 1$. In the vector iteration it is necessary to either deflate the matrix or the iteration vectors by the eigenpairs already calculated, which requires that the eigenpairs have been evaluated very accurately. Another disadvantage is that convergence can be very slow, namely, when eigenvalues are of nearly equal magnitude. *These difficulties suggest that it is more effective to calculate first an approximation to the unknown eigenvalue using a polynomial iteration scheme, and switch to inverse iteration only when a shift close to the required eigenvalue has been obtained. This is exactly the basic strategy followed in the determinant search solution method.*

12.2.2 The Solution Algorithm

Consider the iteration for the eigenpair $(\lambda_1, \boldsymbol{\phi}_1)$, which, as is pointed out later, is typical. Let μ_{k-1} and μ_k be two approximations to λ_1, where $\mu_{k-1} < \mu_k \leq \lambda_1$, as shown in Fig. 12.1. The first objective in the algorithm is to obtain as economically as possible a shift near λ_1. This is accomplished by using an accelerated secant iteration in which the next shift μ_{k+1} is calculated using

$$\mu_{k+1} = \mu_k - \eta \frac{p(\mu_k)}{p(\mu_k) - p(\mu_{k-1})}(\mu_k - \mu_{k-1}) \tag{12.1}$$

where η is a constant. When $\eta = 1$ this is the standard secant iteration described in Section 11.4.2, in which $\mu_{k+1} \longrightarrow \lambda_1$ as $k \longrightarrow \infty$, and convergence can be slow. However, since the objective is to merely obtain a shift near λ_1 the algorithm uses $\eta \geq 2$. For $\eta = 2$ it is known that $\mu_{k+1} \leq \mu_a$, where μ_a is the smallest stationary point of p (see Fig. 12.1). Thus the iteration with $\eta = 2$ can only jump over one root, which would be detected by a sign change in p. But it must be observed that in case λ_1 is a multiple root, the iteration can still be slow.

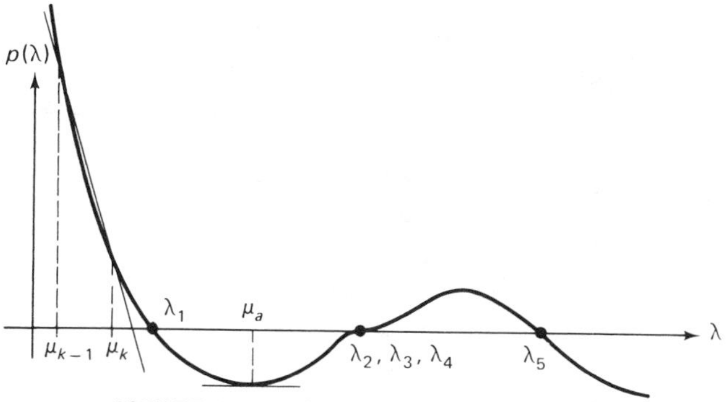

FIGURE 12.1 *Characteristic polynomial $p(\lambda)$.*

If the only information available were to be the value of p, it would be vitally important that the iteration not be able to jump over more than one root, since it would be impossible to detect when a number of roots had been passed. However, the algorithm also uses the Sturm sequence property, namely that the number of negative diagonal elements in the triangular factorization at μ_{k+1} is equal to the number of eigenvalues smaller than μ_{k+1} (see Section 11.5). Therefore, there is no restriction to $\eta = 2$, and the iteration can be accelerated further. In the polynomial iteration η is doubled after each step in which the iterates did not change in their two most significant digits, i.e., when the correction to the current shift is already relatively small. The iteration may thus jump over a single root, a multiple root, or into a cluster of roots, but this is always detected by counting the number of negative diagonal elements in the triangular factorization. Also, with the strategy adopted, the algorithm cannot jump far beyond the unknown roots.

Once it has been established that the secant iteration jumped over one or more unknown eigenvalues, vector inverse iteration with Gram-Schmidt orthogonalization is used to calculate the eigenvalues and eigenvectors accurately, as described in Section 11.2.5. Except in rare cases, no Rayleigh quotient shift is carried out, because once the iteration vector is an approximation to the eigenvector sought, convergence is usually rapid and another factorization would be inefficient. It should be noted that in this iteration, eigenvectors corresponding to eigenvalues larger than μ_{k+1} may be calculated if μ_{k+1} lies in an eigenvalue cluster. This is hardly a disadvantage, because these eigenvalues and eigenvectors are probably required anyway.

As the aim is merely to obtain a shift near the next unknown eigenvalue, it may happen that before any jump occurs, two successive iterates in (12.1) do not differ any more in their six most significant digits. The last iterate is then accepted as the place at which to switch to inverse iteration with Gram-Schmidt orthogonalization.

Of particular importance in the algorithm is the effective evaluation of $p(\lambda)$ (see Section 11.4.2). It should be noted that in the determinant search algorithm, the triangular factorization of $\mathbf{K} - \lambda\mathbf{M}$ into $\mathbf{LDL}^T$ is calculated using Gauss elimination without interchanges. Therefore, the stability of the decomposition is theoretically not guaranteed when $\lambda > \lambda_1$; i.e., multiplier growth could occur, which, however, would always be detected. It is now important to note that the algorithm only aims to find in the polynomial iteration a modest approximation to the required eigenvalue; therefore, if unacceptable multiplier growth does occur, λ is slightly changed in the algorithm, and another factorization is carried out. This is justified because in practice this condition is rarely reached. In summary, therefore, *the triangular factorization can be carried out without interchanges, because the evaluation of $p(\lambda)$ is only completed if no instability occurs, and the specific value used for the current shift does not affect the final accuracy of the eigenvalues and corresponding eigenvectors.*

The iteration described above for λ_1 and $\boldsymbol{\phi}_1$ is typical because the advantage of the one-sided approach to the eigenvalue is also obtained for any other eigenpair, say $(\lambda_{j+1}, \boldsymbol{\phi}_{j+1})$, by using instead of $p(\lambda)$ in (12.1) the deflated polynomial $p_j(\lambda)$ (Fig. 12.2), where

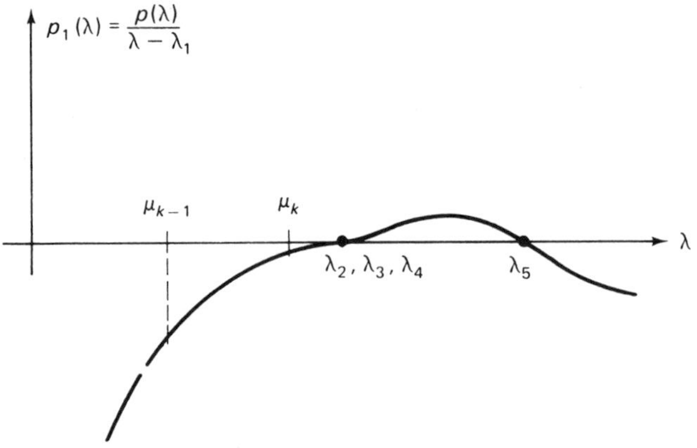

FIGURE 12.2 *Deflated characteristic polynomial with λ_1 suppressed.*

$$p_j(\lambda) = \frac{p(\lambda)}{\prod_{i=1}^{j} (\lambda - \lambda_i^{(l+1)})} \tag{12.2}$$

where $\lambda_i^{(l+1)}$ is the accepted approximation to the eigenvalue λ_i. Therefore, to iterate towards the next unknown eigenvalue, (12.2) is used to suppress all calculated eigenvalues from two previously obtained polynomial values which are the starting values in (12.1). It should be noted that in order not to divide in (12.2) by values close to zero, two λ values need be selected which are far enough from the calculated roots.

12.2.3 Some Remarks on the Implementation of the Determinant Search Method

A few practical details are worth mentioning:

1. A principal advantage of this solution procedure is that each eigenpair is essentially obtained independently from all those previously calculated. Eigenvalues and eigenvectors, therefore, need not be calculated to very high precision, as is the case when an explicit matrix deflation procedure is used (see Section 11.2.5). If not very high accuracy is required, the errors in earlier computed eigenpairs will not prevent attaining the required accuracy in the remaining eigenvalues and eigenvectors.
2. At a shift near the next required eigenvalue, Gram-Schmidt orthogonalization to all previously calculated eigenvectors is not necessary. Allowing for multiple eigenvalues, in the program developed, the iteration vector is only orthogonalized to the last six calculated eigenvectors.
3. It should be noted that in practical examples $p(\lambda)$ can be much larger than the overflow of the computer; therefore, scale factors which keep the implementation machine-independent must be introduced in the evaluation of $p(\lambda)$.[7]
4. To start the secant iteration, two lower bounds on λ_1 are needed. The first bound is $\mu_1 = 0.0$; the other bound is obtained from an approximation to $\boldsymbol{\phi}_1$ calculated by vector inverse iteration at λ equal to zero. Assume that $\mathbf{x}_k$ is the iteration vector after $(k - 1)$ inverse iterations; then we can use

$$\mu_2 = (1 - 0.01) \frac{\mathbf{x}_k^T \mathbf{K} \mathbf{x}_k}{\mathbf{x}_k^T \mathbf{M} \mathbf{x}_k} \tag{12.3}$$

   It may happen that μ_2 is larger than λ_1. However, this is detected by the Sturm sequence count in the factorization at μ_2. Let γ be the number of negative pivots in the triangular factorization. Then we divide μ_2 by $(\gamma + 1)$ until γ equals zero.
5. The algorithm is most efficiently used in an in-core solution routine. Because a few triangular factorizations are required in the calculation of each eigenpair, relatively much peripheral processor (tape and disc

reading and writing) time must be expected in out-of-core solutions. Since the algorithm has been designed for the analysis of small banded systems, on reasonably sized computers, relatively large-order systems can be solved in high-speed storage.

In order to assess the effectiveness of the algorithm in the solution of a specific problem, Table 12.1 summarizes the steps in the solution procedure and gives the total number of operations used and the storage requirements. The number of iterations required for solution depends on the system considered, but, as assumed in Table 12.1, experience has shown that an average of about six secant steps and six inverse iterations are needed for the solution of one eigenpair. However, when solving for a cluster of eigenvalues, more iterations are probably required for the calculation of the first eigenpair, but then the next eigenpairs are obtained with only a few more iterations.

Assume that the determinant search algorithm has calculated all required eigenpairs. The solution is completed by calculating error estimates on the precision with which the eigenvalues and eigenvectors have been evaluated. The specific error quantities calculated have been presented in Section 10.4 and are, for each eigenpair,

$$\epsilon_i = \frac{\|\mathbf{K}\boldsymbol{\phi}_i^{(l+1)} - \lambda_i^{(l+1)}\mathbf{M}\boldsymbol{\phi}_i^{(l+1)}\|_2}{\|\mathbf{K}\boldsymbol{\phi}_i^{(l+1)}\|_2} \tag{12.4}$$

where $\boldsymbol{\phi}_i^{(l+1)}$ and $\lambda_i^{(l+1)}$ are the ith eigenvector and eigenvalue predicted in the last iteration.

As mentioned already, the determinant search method is used in a number of finite element computer programs, and actual subroutines have been published (see References 1, 5, and 7). However, the method is most effective as an in-core solver and considerable limitations may exist on the system size that can be considered. A more generally applicable solution procedure for large finite element systems is the subspace iteration method, which is presented next.

12.3 THE SUBSPACE ITERATION METHOD

The important idea in the determinant search method presented in the previous section was to combine polynomial iteration and vector inverse iteration in an optimum manner to obtain an effective algorithm for the solution of small banded systems. However, when the bandwidth of the matrices **K** and **M** increases, polynomial iterations are much more costly than vector inverse iterations. Figure 12.3 shows the number of inverse iterations that can be carried out for one polynomial iteration, when measured in terms of high-speed storage operations. The relationship is shown as a function of the half-bandwidth m_K of the stiffness matrix, and it is assumed that in the vector inverse iterations, Gram-Schmidt orthogonalization is performed to the last six eigenvectors calculated. The figure shows that for a large bandwidth, many inverse iterations are equivalent to one polynomial evaluation. The cost of a polynomial evaluation when measured on vector inverse iterations increases still further when the operations have to be performed out-of-core, because a great deal more back-up storage reading and writing is needed in a polynomial

TABLE 12.1 *Summary of determinant search solution.*

Operation	Calculation	Number of Operations: $m = m_{\mathbf{K}} = m_{\mathbf{M}}$	Number of Operations: $m = m_{\mathbf{K}},\ m_{\mathbf{M}} = 0$	Required Storage $m = m_{\mathbf{K}} = m_{\mathbf{M}};\ m = m_{\mathbf{K}},\ m_{\mathbf{M}} = 0$
Secant iteration	$\bar{\mathbf{K}} = \mathbf{K} - \mu_k \mathbf{M}$	$n(m+1)$	n	
	$\bar{\mathbf{K}} = \mathbf{LDL}^T$	$\frac{1}{2}nm^2 + \frac{3}{2}nm$	$\frac{1}{2}nm^2 + \frac{3}{2}nm$	
	$p(\mu_k) = \prod_{i=1}^{n} d_{ii}$	n	n	
Inverse iteration	$\bar{\mathbf{K}}\bar{\mathbf{x}}_{k+1} = \mathbf{y}_k$	$n(2m+1)$	$n(2m+1)$	
	$\bar{\mathbf{y}}_{k+1} = \mathbf{M}\bar{\mathbf{x}}_{k+1}$	$n(2m+1)$	n	Using symmetry of matrices
	$\rho(\bar{\mathbf{x}}_{k+1}) = \dfrac{\bar{\mathbf{x}}_{k+1}^T \mathbf{y}_k}{\bar{\mathbf{x}}_{k+1}^T \bar{\mathbf{y}}_{k+1}}$	$2n$	$2n$	$2n(m+1) + 9n$ $n(m+1) + 10n$
	$\mathbf{y}_{k+1} = \dfrac{\bar{\mathbf{y}}_{k+1} - \sum_{j=1}^{6} \alpha_{i-j}\bar{\boldsymbol{\phi}}_{i-j}}{(\bar{\mathbf{x}}_{k+1}^T \bar{\mathbf{y}}_{k+1})^{1/2}}$ where $\bar{\boldsymbol{\phi}}_j = \mathbf{M}\boldsymbol{\phi}_j,\ \alpha_j = \bar{\mathbf{x}}_{k+1}^T \bar{\boldsymbol{\phi}}_j$	$13n$	$13n$	(Algorithm has also been implemented as out-of-core solver)[7]
Error estimates	$\dfrac{\lVert \mathbf{K}\boldsymbol{\phi}_i^{(l+1)} - \lambda_i^{(l+1)} \mathbf{M}\boldsymbol{\phi}_i^{(l+1)} \rVert_2}{\lVert \mathbf{K}\boldsymbol{\phi}_i^{(l+1)} \rVert_2}$	$4nm + 5n$	$2nm + 5n$	
Total for p lowest eigenvalues and associated eigenvectors assuming six secant and six inverse iterations per eigenpair		$(3nm^2 + 43nm + 119n)p$	$(3nm^2 + 23nm + 119n)p$	

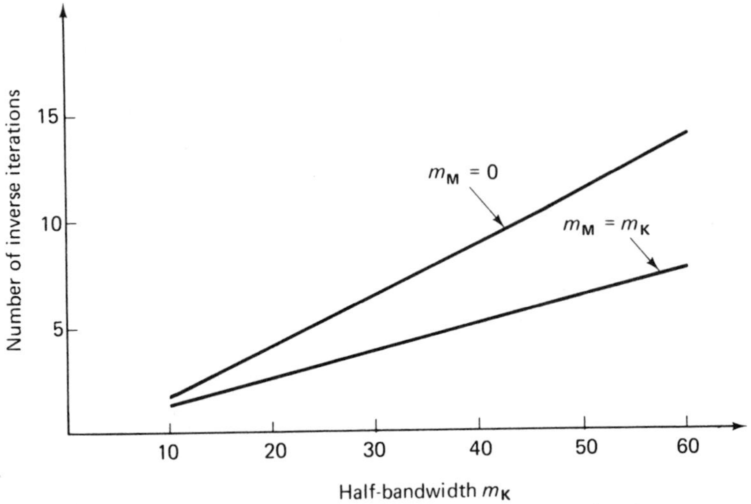

FIGURE 12.3 *Number of inverse iterations per polynomial evaluation (Gram–Schmidt orthogonalization to last six eigenvectors).*

evaluation than in vector inverse iteration. These considerations suggest that for the solution of out-of-core systems and systems with large bandwidths, vector inverse iterations should be used primarily. In addition, if inverse iteration is performed simultaneously for a number of the required eigenvectors, the solution of out-of-core systems may be further enhanced, because the back-up storage reading and writing of the triangularized matrix and the iteration vectors is reduced. It is specifically with those ideas in mind that the subspace iteration method has been designed.

In essence, the subspace iteration method developed by Bathe[1] consists of the following three steps:

1. Establish q starting iteration vectors, $q > p$, where p is the number of eigenvalues and vectors to be calculated.
2. Use simultaneous inverse iteration on the q vectors and Ritz analysis to extract the "best" eigenvalue and eigenvector approximations from the q iteration vectors.
3. After iteration convergence, use the Sturm sequence check to verify that the required eigenvalues and corresponding eigenvectors have been calculated.

The solution procedure was named (in ref. [1]) subspace iteration method, because the iteration is equivalent to iterating with a q-dimensional subspace and should not be regarded as a simultaneous iteration with q individual iteration vectors. Specifically, as we discuss in the following sections, it must be noted that the selection of the starting iteration vectors in step (1) and the Sturm sequence check in step (3) are considered very important parts of the iteration procedure. Altogether, the subspace iteration method is largely based on various techniques that have been used earlier,[9–16] namely simultaneous vector iteration, Sturm sequence information (see Section 10.2.2), and Rayleigh–Ritz analysis (see Section 10.3.2), but most enlightening has been the work of Rutishauser.[17]

12.3.1 Preliminary Considerations

The basic objective in the subspace iteration method is to solve for the lowest p eigenvalues and corresponding eigenvectors satisfying

$$\mathbf{K\Phi} = \mathbf{M\Phi\Lambda} \tag{12.5}$$

where $\mathbf{\Lambda} = \text{diag}(\lambda_i)$ and $\mathbf{\Phi} = [\boldsymbol{\phi}_1, \ldots, \boldsymbol{\phi}_p]$.

In addition to the relation in (12.5), the eigenvectors also satisfy the orthogonality conditions

$$\mathbf{\Phi}^T\mathbf{K\Phi} = \mathbf{\Lambda}; \qquad \mathbf{\Phi}^T\mathbf{M\Phi} = \mathbf{I} \tag{12.6}$$

where $\mathbf{I}$ is a unit matrix of order p because $\mathbf{\Phi}$ stores only p eigenvectors. It is important to note that the relation in (12.5) is a necessary and sufficient condition for the vectors in $\mathbf{\Phi}$ to be eigenvectors but that the eigenvector orthogonality conditions in (12.6) are necessary but not sufficient. In other words, if we have p vectors that satisfy (12.6), $p < n$, then these vectors are not necessarily eigenvectors. However, if the p vectors satisfy (12.5), they are definitely eigenvectors, although we still need to make sure that they are indeed the p specific eigenvectors sought (see Section 10.2.1).

The essential idea of the subspace iteration method uses the fact that the eigenvectors in (12.5) form an $\mathbf{M}$-orthonormal basis of the p-dimensional least-dominant subspace of the operators $\mathbf{K}$ and $\mathbf{M}$, which we will now call E_∞ (see Section 2.2). In the solution the iteration with p linearly independent vectors can therefore be regarded as an iteration with a subspace. The starting iteration vectors span E_1 and iteration continues until, to sufficient accuracy, E_∞ is spanned. The fact that iteration is performed with a subspace has some important consequences. The total number of required iterations depends on how "close" E_1 is to E_∞, and not on how close each iteration vector is to an eigenvector. Hence the effectiveness of the algorithm lies in that it is much easier to establish a p-dimensional starting subspace which is close to E_∞ than to find p vectors that each are close to a required eigenvector. The specific algorithm used to establish the starting iteration vectors is described later. Also, *because iteration is performed with a subspace, convergence of the subspace is all that is required and not of individual iteration vectors to eigenvectors. In other words, if the iteration vectors are linear combinations of the required eigenvectors, the solution algorithm converges in one step.*

To demonstrate the essential idea, we first consider simultaneous vector inverse iteration on p vectors. Let $\mathbf{X}_1$ store the p starting iteration vectors, which span the starting subspace E_1. Simultaneous inverse iteration on the p vectors can be written

$$\mathbf{KX}_{k+1} = \mathbf{MX}_k; \qquad k = 1, 2, \ldots \tag{12.7}$$

where we now observe that the p iteration vectors in $\mathbf{X}_{k+1}$ span a p-dimensional subspace E_{k+1}, and the sequence of subspaces generated converges to E_∞, provided the starting vectors are not orthogonal to E_∞. This seems to contradict the fact that in this iteration each column in $\mathbf{X}_{k+1}$ is known to converge to the least dominant eigenvector unless the column is deficient in $\boldsymbol{\phi}_1$ (see Section 11.2.1). Actually, there is no contradiction. Although in exact arithmetic the vectors in $\mathbf{X}_{k+1}$ span $\mathbf{E}_{k+1}$, they do become more and more parallel and therefore a poorer and poorer basis for E_{k+1}. One way to preserve numerical stability is to generate orthogonal bases in the subspaces E_{k+1} using the Gram-Schmidt process (see Section 11.2.5). In this case we iterate for $k = 1, 2, \ldots$ as follows:

$$\mathbf{K}\bar{\mathbf{X}}_{k+1} = \mathbf{M}\mathbf{X}_k \tag{12.8}$$

$$\mathbf{X}_{k+1} = \bar{\mathbf{X}}_{k+1}\mathbf{R}_{k+1} \tag{12.9}$$

where $\mathbf{R}_{k+1}$ is an upper-triangular matrix chosen in such way that $\mathbf{X}_{k+1}^T\mathbf{M}\mathbf{X}_{k+1} = \mathbf{I}$. Then provided that the starting vectors in $\mathbf{X}_1$ are not deficient in the eigenvectors $\boldsymbol{\phi}_1, \boldsymbol{\phi}_2, \ldots, \boldsymbol{\phi}_p$, we have

$$\mathbf{X}_{k+1} \longrightarrow \boldsymbol{\Phi}; \qquad \mathbf{R}_{k+1} \longrightarrow \boldsymbol{\Lambda}$$

It is important to note that the iteration in (12.7) generates the same sequence of subspaces as the iteration in (12.8) and (12.9). However, the ith column in $\mathbf{X}_{k+1}$ of (12.8) and (12.9) converges linearly to $\boldsymbol{\phi}_i$ with the convergence rate equal to max $\{\lambda_{i-1}/\lambda_i, \lambda_i/\lambda_{i+1}\}$. To demonstrate the solution procedure, consider the following example.

EXAMPLE 12.1: Consider the eigenproblem $\mathbf{K}\boldsymbol{\phi} = \lambda\mathbf{M}\boldsymbol{\phi}$, where

$$\mathbf{K} = \begin{bmatrix} 2 & -1 & 0 \\ -1 & 4 & -1 \\ 0 & -1 & 2 \end{bmatrix}; \qquad \mathbf{M} = \begin{bmatrix} \frac{1}{2} & & \\ & 1 & \\ & & \frac{1}{2} \end{bmatrix}$$

The two lowest eigenvalues and corresponding eigenvectors are (see Example 10.4)

$$\lambda_1 = 2, \quad \boldsymbol{\phi}_1 = \begin{bmatrix} \frac{1}{\sqrt{2}} \\ \frac{1}{\sqrt{2}} \\ \frac{1}{\sqrt{2}} \end{bmatrix}; \qquad \lambda_2 = 4, \quad \boldsymbol{\phi}_2 = \begin{bmatrix} -1 \\ 0 \\ 1 \end{bmatrix}$$

Use the simultaneous vector iteration with Gram–Schmidt orthogonalization given in (12.8) and (12.9) with starting vectors

$$\mathbf{X}_1 = \begin{bmatrix} 0 & 2 \\ 1 & 1 \\ 2 & 0 \end{bmatrix}$$

to calculate approximations to λ_1, $\boldsymbol{\phi}_1$ and λ_2, $\boldsymbol{\phi}_2$.

Solving for $\bar{\mathbf{X}}_2$ the relation $\mathbf{K}\bar{\mathbf{X}}_2 = \mathbf{M}\mathbf{X}_1$ gives

$$\bar{\mathbf{X}}_2 = \begin{bmatrix} 0.25 & 0.75 \\ 0.50 & 0.50 \\ 0.75 & 0.25 \end{bmatrix}$$

$\mathbf{M}$-orthonormalizing $\bar{\mathbf{X}}_2$ gives

$$\mathbf{X}_2 = \begin{bmatrix} 0.3333 & 1.179 \\ 0.6667 & 0.2357 \\ 1.000 & -0.7071 \end{bmatrix}; \qquad \mathbf{R}_2 = \begin{bmatrix} 1.333 & -1.650 \\ 0 & 2.121 \end{bmatrix}$$

Proceeding similarly, we obtain the following results:

$$\mathbf{X}_3 = \begin{bmatrix} 0.5222 & 1.108 \\ 0.6963 & 0.1231 \\ 0.8704 & -0.8614 \end{bmatrix}; \qquad \mathbf{R}_3 = \begin{bmatrix} 2.089 & -0.9847 \\ 0 & 3.830 \end{bmatrix}$$

$$\mathbf{X}_4 = \begin{bmatrix} 0.6163 & 1.058 \\ 0.7044 & 0.0622 \\ 0.7924 & -0.9339 \end{bmatrix}; \qquad \mathbf{R}_4 = \begin{bmatrix} 2.023 & -0.5202 \\ 0 & 3.954 \end{bmatrix}$$

$$\mathbf{X}_5 = \begin{bmatrix} 0.6623 & 1.030 \\ 0.7064 & 0.0312 \\ 0.7506 & -0.9678 \end{bmatrix}; \qquad \mathbf{R}_5 = \begin{bmatrix} 2.006 & -0.2639 \\ 0 & 3.988 \end{bmatrix}$$

$$\mathbf{X}_6 = \begin{bmatrix} 0.6848 & 1.015 \\ 0.7069 & 0.0156 \\ 0.7290 & -0.9841 \end{bmatrix}; \qquad \mathbf{R}_6 = \begin{bmatrix} 2.001 & -0.1324 \\ 0 & 3.997 \end{bmatrix}$$

$$\mathbf{X}_7 = \begin{bmatrix} 0.6960 & 1.008 \\ 0.7071 & 0.0078 \\ 0.7181 & -0.9921 \end{bmatrix}; \qquad \mathbf{R}_7 = \begin{bmatrix} 2.000 & -0.0663 \\ 0 & 3.999 \end{bmatrix}$$

$$\mathbf{X}_8 = \begin{bmatrix} 0.7016 & 1.004 \\ 0.7071 & 0.0039 \\ 0.7126 & -0.9961 \end{bmatrix}; \qquad \mathbf{R}_8 = \begin{bmatrix} 2.000 & -0.0331 \\ 0 & 4.000 \end{bmatrix}$$

$$\mathbf{X}_9 = \begin{bmatrix} 0.7043 & 1.002 \\ 0.7071 & 0.0020 \\ 0.7099 & -0.9980 \end{bmatrix}; \qquad \mathbf{R}_9 = \begin{bmatrix} 2.000 & -0.0166 \\ 0 & 4.000 \end{bmatrix}$$

$$\mathbf{X}_{10} = \begin{bmatrix} 0.7057 & 1.001 \\ 0.7071 & 0.0010 \\ 0.7085 & -0.9990 \end{bmatrix}; \qquad \mathbf{R}_{10} = \begin{bmatrix} 2.000 & -0.0083 \\ 0 & 4.000 \end{bmatrix}$$

and after nine iterations we have

$$\boldsymbol{\phi}_1 \doteq \begin{bmatrix} 0.7057 \\ 0.7071 \\ 0.7085 \end{bmatrix}; \qquad \lambda_1 \doteq 2.000$$

$$\boldsymbol{\phi}_2 \doteq \begin{bmatrix} 1.001 \\ 0.0010 \\ -0.9990 \end{bmatrix}; \qquad \lambda_2 \doteq 4.000$$

It should be noted that although the vectors in $\mathbf{X}_1$ already span the space of $\boldsymbol{\phi}_1$ and $\boldsymbol{\phi}_2$, we need a relatively large number of iterations for convergence.

The solution in the preceding example demonstrates the iteration procedure in (12.8) and (12.9) and also shows the main deficiency of the method. Namely, each iteration vector is forced to converge to a different eigenvector by orthogonalizing the ith iteration vector to the $(i - 1)$ iteration vectors that have been orthogonalized already without allowing a more effective linear combination of the vectors to take place. In the preceding example, the iteration was started with two iteration vectors that were linear combinations of the required eigenvectors and this did not yield any advantage. In general, if the iteration vectors in $\mathbf{X}_{k+1}$ span E_∞ but are not eigenvectors (i.e., the vectors in $\mathbf{X}_{k+1}$ are linear combinations of the eigenvectors $\boldsymbol{\phi}_1, \ldots, \boldsymbol{\phi}_p$), then, although the subspace E_{k+1} already converged, many more iterations may be needed in order to turn the orthogonal basis of iteration vectors into the basis of eigenvectors.

12.3.2 Subspace Iteration

The following algorithm, which we call *subspace iteration*, finds an orthogonal basis of vectors in E_{k+1}, thus preserving numerical stability in the iteration of (12.7), and also calculates in one step the required eigenvectors when E_{k+1} converged to E_∞. This algorithm is the iteration used in the subspace iteration method, i.e. step 2 of the complete solution phase (see Section 12.3).

For $k = 1, 2, \ldots$, iterate from E_k to E_{k+1}:

$$\mathbf{K}\bar{\mathbf{X}}_{k+1} = \mathbf{M}\mathbf{X}_k \tag{12.10}$$

Find the projections of the operators **K** and **M** onto E_{k+1}:

$$\mathbf{K}_{k+1} = \bar{\mathbf{X}}_{k+1}^T \mathbf{K}\bar{\mathbf{X}}_{k+1} \tag{12.11}$$

$$\mathbf{M}_{k+1} = \bar{\mathbf{X}}_{k+1}^T \mathbf{M}\bar{\mathbf{X}}_{k+1} \tag{12.12}$$

Solve for the eigensystem of the projected operators:

$$\mathbf{K}_{k+1}\mathbf{Q}_{k+1} = \mathbf{M}_{k+1}\mathbf{Q}_{k+1}\mathbf{\Lambda}_{k+1} \tag{12.13}$$

Find an improved approximation to the eigenvectors:

$$\mathbf{X}_{k+1} = \bar{\mathbf{X}}_{k+1}\mathbf{Q}_{k+1} \tag{12.14}$$

Then, provided that the vectors in $\mathbf{X}_1$ are not orthogonal to one of the required eigenvectors, we have

$$\mathbf{\Lambda}_{k+1} \longrightarrow \mathbf{\Lambda} \quad \text{and} \quad \mathbf{X}_{k+1} \longrightarrow \mathbf{\Phi} \qquad \text{as } k \longrightarrow \infty$$

In the subspace iteration, it is implied that the iteration vectors are ordered in an appropriate way; i.e., the iteration vectors converging to $\boldsymbol{\phi}_1, \boldsymbol{\phi}_2, \ldots$ are stored as the first, second, ... columns of $\mathbf{X}_{k+1}$. We demonstrate the iteration procedure by calculating the solution to the problem considered in Example 12.1.

EXAMPLE 12.2: Use the subspace iteration algorithm to solve the problem considered in Example 12.1.

Using the relations in (12.10) to (12.14) with **K**, **M**, and $\mathbf{X}_1$ given in Example 12.1, we obtain

$$\bar{\mathbf{X}}_2 = \frac{1}{4}\begin{bmatrix} 1 & 3 \\ 2 & 2 \\ 3 & 1 \end{bmatrix}$$

$$\mathbf{K}_2 = \frac{1}{4}\begin{bmatrix} 5 & 3 \\ 3 & 5 \end{bmatrix}; \qquad \mathbf{M}_2 = \frac{1}{16}\begin{bmatrix} 9 & 7 \\ 7 & 9 \end{bmatrix}$$

Hence

$$\mathbf{\Lambda}_2 = \begin{bmatrix} 2 & 0 \\ 0 & 4 \end{bmatrix}; \qquad \mathbf{Q}_2 = \begin{bmatrix} \dfrac{1}{\sqrt{2}} & 2 \\ \dfrac{1}{\sqrt{2}} & -2 \end{bmatrix}$$

and

$$\mathbf{X}_2 = \begin{bmatrix} \dfrac{1}{\sqrt{2}} & -1 \\ \dfrac{1}{\sqrt{2}} & 0 \\ \dfrac{1}{\sqrt{2}} & 1 \end{bmatrix}$$

Comparing the results with the solution obtained in Example 12.1, we observe that we have calculated the exact eigenvalues and eigenvectors in the first subspace iteration. This must be the case because the starting iteration vectors in $\mathbf{X}_1$ span the subspace defined by $\boldsymbol{\phi}_1$ and $\boldsymbol{\phi}_2$.

Considering the subspace iteration, a first observation is that $\mathbf{K}_{k+1}$ and $\mathbf{M}_{k+1}$ in (12.11) and (12.12), respectively, tend toward diagonal form as the number of iterations increases; i.e., $\mathbf{K}_{k+1}$ and $\mathbf{M}_{k+1}$ are diagonal matrices when the columns in $\mathbf{X}_{k+1}$ store multiples of eigenvectors. Hence it follows from the discussion in Section 11.3.2 that the generalized Jacobi method can be used very effectively for the solution of the eigenproblem in (12.13).

An important aspect is the convergence of the method. Assuming that in the iteration the vectors in $\mathbf{X}_{k+1}$ are ordered in such way that the ith diagonal element in $\mathbf{\Lambda}_{k+1}$ is larger than the $(i-1)$st element, $i = 2, \ldots, p$, then *the ith column in $\mathbf{X}_{k+1}$ converges linearly to $\boldsymbol{\phi}_i$ and the convergence rate is λ_i/λ_{p+1}*[17,18]. Although this is an asymptotic convergence rate, it indicates that the smallest eigenvalues converge fastest. In addition, *a higher convergence rate can be obtained by using q iteration vectors, with $q > p$. However, using more iteration vectors will also increase the computer effort for one iteration. In practice, we have found that $q = \min\{2p, p + 8\}$ is, in general, effective and we use this value.* Considering the convergence rate, it should be noted that multiple eigenvalues do not decrease the rate of convergence, provided that $\lambda_{q+1} > \lambda_p$.

As for the iteration schemes presented earlier, the theoretical convergence behavior can only be observed in practice when the iteration vectors are already relatively close to eigenvectors. However, in practice, we are very interested in knowing what happens already in the first few iterations when E_{k+1} is not yet "close" to E_∞. Indeed, the effectiveness of the algorithm lies to a large extent in that a few iterations can already give good approximations to the required eigenpairs, the reason being that one subspace iteration given in (12.10) to (12.14) is, in fact, a Ritz analysis, as described in Section 10.3.2. Therefore, *all characteristics of the Ritz analysis pertain also to the subspace iteration; i.e., the smallest eigenvalues are approximated best in the iteration and all eigenvalue approximations are upper bounds on the actual eigenvalues sought.* It follows that we may think of the subspace iteration as a repeated application of the Ritz analysis method of Section 10.3.2, in which the eigenvector approximations calculated in the previous iteration are used to form the right-hand side load vectors in the current iteration.

It is important to realize that using either one of the iteration procedures given in (12.7), (12.8) and (12.9), or (12.10) to (12.14), the same subspace E_{k+1} is spanned by the iteration vectors. Therefore, there is no need to iterate always as given in (12.10) to (12.14), but we may first use the simple inverse iteration in (12.7) or inverse iteration with Gram–Schmidt orthogonalization as given in (12.8) and (12.9), and finally use the subspace iteration scheme given in (12.10) to (12.14). The calculated results would be the same in theory as those obtained when subspace iterations only are performed. However, the difficulty then lies in deciding at what stage to orthogonalize the iteration vectors by using (12.11) to (12.14), because the iteration in (12.10) yields vectors that are more and more parallel. Also, the Gram–Schmidt orthogonalization is numerically not very stable. If the iteration vectors have become "too close to each other" because either the initial assumptions gave iteration vectors that were almost parallel or by iterating without orthogonalization, it may be impossible to orthogonalize them because of finite precision arithmetic in the computer. Unfortunately, considering large finite element systems, the starting iteration vectors can in some cases be nearly parallel, although they span a subspace that is close to E_∞, and it is best to immediately orthogonalize the iteration vectors using the projections of $\mathbf{K}$ and $\mathbf{M}$ onto E_2. In addition, using the subspace iteration, we obtain in each iteration "best" approximations to the required eigenvalues and eigenvectors and can measure convergence in each iteration. It may also be noted that because of back-up storage reading and writing required, Gram–Schmidt orthogonalization can be ineffective in out-of-core solution.

12.3.3 Starting Iteration Vectors

The first step in the subspace iteration method is the selection of the starting iteration vectors in $\mathbf{X}_1$ (see (12.10)). As pointed out earlier, if starting vectors are used that span the least dominant subspace, the iteration converges in one step. This is the case, for example, when there are only p nonzero masses in a diagonal mass matrix and the starting vectors are unit vectors $\mathbf{e}_i$ with the entries $+1$ corresponding to the mass degrees of freedom. In this case one subspace iteration is in effect a static condensation analysis or a Guyan reduction. This follows from the discussion in Section 10.3.2 and because a subspace iteration embodies a Ritz analysis. Consider the following example.

EXAMPLE 12.3: Use the subspace iteration to calculate the eigenpairs $(\lambda_1, \boldsymbol{\phi}_1)$ and $(\lambda_2, \boldsymbol{\phi}_2)$ of the problem $\mathbf{K}\boldsymbol{\phi} = \lambda\mathbf{M}\boldsymbol{\phi}$, where

$$\mathbf{K} = \begin{bmatrix} 2 & -1 & 0 & 0 \\ -1 & 2 & -1 & 0 \\ 0 & -1 & 2 & -1 \\ 0 & 0 & -1 & 1 \end{bmatrix}; \qquad \mathbf{M} = \begin{bmatrix} 0 & & & \\ & 2 & & \\ & & 0 & \\ & & & 1 \end{bmatrix}$$

As suggested above, we use as starting vectors the unit vectors $\mathbf{e}_2$ and $\mathbf{e}_4$. Following (12.10) to (12.14), we then obtain

$$\begin{bmatrix} 2 & -1 & 0 & 0 \\ -1 & 2 & -1 & 0 \\ 0 & -1 & 2 & -1 \\ 0 & 0 & -1 & 1 \end{bmatrix} \bar{\mathbf{X}}_2 = \begin{bmatrix} 0 & 0 \\ 2 & 0 \\ 0 & 0 \\ 0 & 1 \end{bmatrix}$$

$$\bar{\mathbf{X}}_2 = \begin{bmatrix} 2 & 1 \\ 4 & 2 \\ 4 & 3 \\ 4 & 4 \end{bmatrix}$$

and $$\mathbf{K}_2 = 4\begin{bmatrix} 2 & 1 \\ 1 & 1 \end{bmatrix}; \qquad \mathbf{M}_2 = 8\begin{bmatrix} 6 & 4 \\ 4 & 3 \end{bmatrix}$$

Hence

$$\mathbf{\Lambda}_2 = \begin{bmatrix} \left(\frac{1}{2} - \frac{\sqrt{2}}{4}\right) & 0 \\ 0 & \left(\frac{1}{2} + \frac{\sqrt{2}}{4}\right) \end{bmatrix}; \qquad \mathbf{Q}_2 = \begin{bmatrix} \frac{1}{8 + 4\sqrt{2}} & \frac{1}{4\sqrt{2} - 8} \\ \frac{1}{4 + 4\sqrt{2}} & \frac{1}{4\sqrt{2} - 4} \end{bmatrix}$$

and

$$\mathbf{X}_2 = \begin{bmatrix} \frac{1}{4} & -\frac{1}{4} \\ \frac{1}{2} & -\frac{1}{2} \\ \frac{1 + \sqrt{2}}{4} & \frac{-1 + \sqrt{2}}{4} \\ \frac{\sqrt{2}}{2} & \frac{\sqrt{2}}{2} \end{bmatrix}$$

Comparing these results with the solution calculated in Example 10.12, we observe that we obtained the exact results with one subspace iteration.

A second case for which the subspace iteration can converge in one step arises when **K** and **M** are both diagonal. This is a rather trivial case, but it is considered for the development of an effective procedure for the selection of starting iteration vectors when general matrices are involved in the analysis. When **K** and **M** are diagonal, the iteration vectors should be unit vectors with the entries +1 corresponding to those degrees of freedom that have the smallest ratios k_{ii}/m_{ii}. These vectors are, in effect, the eigenvectors corresponding to the smallest eigenvalues, and this is why convergence is achieved in one step. We demonstrate the procedure in the following example.

EXAMPLE 12.4: Construct starting iteration vectors for the solution of the lowest two eigenvalues by subspace iteration when considering the problem $\mathbf{K}\boldsymbol{\phi} = \lambda\mathbf{M}\boldsymbol{\phi}$, where

$$\mathbf{K} = \begin{bmatrix} 3 & & & \\ & 2 & & \\ & & 4 & \\ & & & 8 \end{bmatrix}; \qquad \mathbf{M} = \begin{bmatrix} 2 & & & \\ & 0 & & \\ & & 4 & \\ & & & 1 \end{bmatrix}$$

The ratios k_{ii}/m_{ii} are here $\frac{3}{2}$, ∞, 1, and 8 for $i = 1, \ldots, 4$, and indeed these are the eigenvalues of the problem. The starting vectors to be used are therefore

$$\mathbf{X}_1 = \left[\begin{array}{c|c} 0 & 1 \\ 0 & 0 \\ 1 & 0 \\ 0 & 0 \end{array}\right]$$

The vectors $\mathbf{X}_1$ are multiples of the required eigenvectors, and hence convergence occurs in the first step of iteration.

The two cases that we dealt with above involved rather special matrices; i.e., in the first case static condensation could be performed and in the second case the matrices **K** and **M** were diagonal. In both cases unit vectors $\mathbf{e}_i$ were employed, where $i = r_1, r_2, \ldots, r_p$, and the $r_j, j = 1, 2, \ldots, p$, corresponded to the p smallest values of k_{ii}/m_{ii} over all i. Using this notation we have $r_1 = 2$ and $r_2 = 4$ for Example 12.3 and $r_1 = 3, r_2 = 1$ for Example 12.4.

Although such specific matrices will hardly be encountered in general practical analysis, the results concerning the construction of the starting iteration vectors indicate how in general analysis effective starting iteration vectors may be selected. A general observation is that starting vectors that span the least dominant subspace E_∞ could be selected in the above cases, because the mass and stiffness properties have been lumped to a sufficient degree. In a general case, such lumping is not possible or would result in inaccurate stiffness and mass representations of the actual structure. However, although the matrices $\mathbf{K}$ and $\mathbf{M}$ do not have the exact same form used above, the discussion shows that the starting iteration vectors should be constructed to excite those degrees of freedom with which large mass and small stiffness are associated. Based on this observation, in essence, the following algorithm has been used effectively for the selection of the starting iteration vectors. The first column in $\mathbf{MX}_1$ is simply the diagonal of $\mathbf{M}$. This assures that all mass degrees of freedom are excited. The other columns in $\mathbf{MX}_1$ are unit vectors $\mathbf{e}_i$ with entries $+1$ at the coordinates with the smallest ratios k_{ii}/m_{ii}. We note that except for the first iteration vector, this scheme would construct the same starting subspace as used in Examples 12.3 and 12.4.* (In the analysis of large systems, an appropriate spacing between the unit entries in the starting vectors is also important and taken into account; see program SSPACE.)

The starting subspace described above is, in general, only an approximation to the actual subspace required, which we denoted as E_∞; however, the closer $\mathbf{K}$ and $\mathbf{M}$ are to the matrix forms used in Examples 12.3 and 12.4, the "better" is the starting subspace; i.e., the fewer iterations can be expected until convergence. In practice, the number of iterations required for convergence depends on the matrices $\mathbf{K}$ and $\mathbf{M}$ and on the accuracy required in the eigenvalues and eigenvectors. Experience has shown that with the starting subspace described above and $q = \min\ (2p, p + 8)$, of the order ten iterations are needed to calculate the largest eigenvalue λ_p to about six digit precision, with the smaller eigenvalues being predicted more accurately.

As an alternative to using the above starting subspace, it can also be effective to employ the Lanczos procedure to generate the starting iteration vectors. In this case, we use the equations given in Section 10.3.4 with $\mathbf{x}$ being a full unit vector. Numerical experience has shown that with these starting vectors it may be effective to use q considerably larger than p, say $q = 2p$.

The starting subspaces above have proven by experience to be effective. However, "better" starting vectors may still be available in specific applications. For instance, in dynamic optimization, as the structure is modified in small steps, the eigensystem of the previous structure may be a good approximation to the eigensystem of the new structure. Similarly, if some eigenvectors have already been evaluated and we now want to solve for more eigenvectors, the already calculated eigenvectors would effectively be used in $\mathbf{X}_1$. If the eigenvectors of substructures have already been obtained, these may also be used in establishing the starting iteration vectors in $\mathbf{X}_1$ (see Section 10.3.3).

* It is also reasonable to use as the last column in $\mathbf{MX}_1$ a random vector.

12.3.4 Convergence

In the subspace iteration it is necessary to measure convergence. Assume that in the iterations $(k - 1)$ and k, eigenvalue approximations $\lambda_i^{(k)}$ and $\lambda_i^{(k+1)}$, $i = 1, \ldots, p$, respectively, have been calculated. Convergence is measured as in inverse iteration (see Section 11.2.1); i.e., we require that

$$\frac{|\lambda_i^{(k+1)} - \lambda_i^{(k)}|}{\lambda_i^{(k+1)}} \leq \text{tol}; \qquad i = 1, \ldots, p \tag{12.15}$$

where tol should be 10^{-2s} when the eigenvalues shall be accurate to about $2s$ digits. For example, if we iterate until all p ratios in (12.15) are smaller than 10^{-6}, we very likely have that λ_p has been approximated to about six-digit accuracy, and the smaller eigenvalues have been evaluated more accurately. Since the eigenvalue approximations are calculated using a Rayleigh quotient, the eigenvector approximations are accurate to only about s (or more) digits. It should be noted that the iteration is performed with q vectors, $q > p$, but convergence is measured only on the approximations obtained for the p smallest eigenvalues.

Another important aspect when using the subspace iteration technique is that of verifying that in fact the required eigenvalues and vectors have been calculated since the relations in (12.5) and (12.6) are satisfied by any eigenpairs. This verification is the third important phase of the subspace iteration method (see Section 12.3). As pointed out, the iteration in (12.10) to (12.14) converges in the limit to the eigenvectors $\boldsymbol{\phi}_1, \ldots, \boldsymbol{\phi}_p$, provided the starting iteration vectors in $\mathbf{X}_1$ are not orthogonal to any one of the required eigenvectors. The starting subspaces described above have proven, by experience, to be very satisfactory in this regard, although there is no formal mathematical proof available that convergence will indeed always occur. However, once the convergence limit in (12.15) is satisfied, with s being at least equal to 3, we can make sure that the smallest eigenvalues and corresponding eigenvectors have indeed been calculated. For the check we use the Sturm sequence property of the characteristic polynomials of the problems $\mathbf{K}\boldsymbol{\phi} = \lambda\mathbf{M}\boldsymbol{\phi}$ and $\mathbf{K}^{(r)}\boldsymbol{\phi}^{(r)} = \lambda^{(r)}\mathbf{M}^{(r)}\boldsymbol{\phi}^{(r)}$ at a shift μ, where μ is just to the right of the calculated value for λ_p (see Fig. 12.4). The Sturm sequence property yields that in the Gauss factorization of $\mathbf{K} - \mu\mathbf{M}$ into $\mathbf{LDL}^T$, the number of negative elements in $\mathbf{D}$ is equal to the number of eigenvalues smaller than μ. Hence, in the case considered, we should

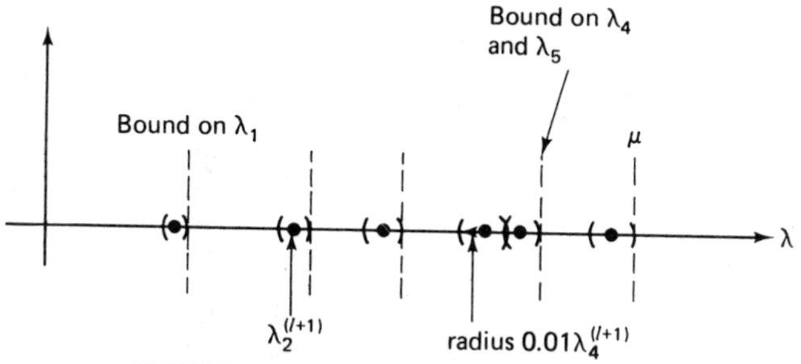

FIGURE 12.4 *Bounds on eigenvalues to apply Sturm sequence check, $p = 6$.*

TABLE 12.2 *Summary of subspace iteration solution.*

Operation	Calculation	Number of Operations: $m = m_K = m_M$	Number of Operations: $m = m_K$, $m_M = 0$	Required Storage
Factorization of **K**	$\mathbf{K} = \mathbf{LDL}^T$	$\frac{1}{2}nm^2 + \frac{3}{2}nm$	$\frac{1}{2}nm^2 + \frac{3}{2}nm$	
Subspace iteration	$\mathbf{K}\bar{\mathbf{X}}_{k+1} = \mathbf{Y}_k$	$nq(2m + 1)$	$nq(2m + 1)$	
	$\mathbf{K}_{k+1} = \bar{\mathbf{X}}_{k+1}^T \mathbf{Y}_k$	$\frac{1}{2}nq(q + 1)$	$\frac{1}{2}nq(q + 1)$	
	$\bar{\mathbf{Y}}_{k+1} = \mathbf{M}\bar{\mathbf{X}}_{k+1}$	$nq(2m + 1)$	nq	
	$\mathbf{M}_{k+1} = \bar{\mathbf{X}}_{k+1}^T \bar{\mathbf{Y}}_{k+1}$	$\frac{1}{2}nq(q + 1)$	$\frac{1}{2}nq(q + 1)$	Algorithm is effectively implemented as out-of-core solver
	$\mathbf{K}_{k+1}\mathbf{Q}_{k+1} = \mathbf{M}_{k+1}\mathbf{Q}_{k+1}\mathbf{\Lambda}_{k+1}$	$O(q^3)$ neglected		
	$\mathbf{Y}_{k+1} = \bar{\mathbf{Y}}_{k+1}\mathbf{Q}_{k+1}$	nq^2	nq^2	
Sturm sequence check	$\bar{\mathbf{K}} = \mathbf{K} - \mu\mathbf{M}$	$n(m + 1)$	n	
	$\bar{\mathbf{K}} = \mathbf{LDL}^T$	$\frac{1}{2}nm^2 + \frac{3}{2}nm$	$\frac{1}{2}nm^2 + \frac{3}{2}nm$	
Error estimates	$\dfrac{\lVert \mathbf{K}\boldsymbol{\phi}_i^{(l+1)} - \lambda_i^{(l+1)}\mathbf{M}\boldsymbol{\phi}_i^{(l+1)} \rVert_2}{\lVert \mathbf{K}\boldsymbol{\phi}_i^{(l+1)} \rVert_2}$	$4nm + 5n$	$2nm + 5n$	
Total for solution of p lowest eigenvalues and associated eigenvectors, assuming that ten iterations are required and $q = \min\{2p, p + 8\}$		$nm^2 + nm(4 + 4p) + 5np + 20nq(2m + q + \frac{3}{2})$	$nm^2 + nm(3 + 2p) + 5np + 20nq(m + q + \frac{3}{2})$	

have p negative elements in $\mathbf{D}$. However, in order to apply the Sturm sequence check, a meaningful μ must be used that takes account of the fact that we have obtained only approximations to the exact eigenvalues of the problem $\mathbf{K}\boldsymbol{\phi} = \lambda\mathbf{M}\boldsymbol{\phi}$. Let l be the last iteration, so that the calculated eigenvalues are $\lambda_1^{(l+1)}$, $\lambda_2^{(l+1)}, \ldots, \lambda_p^{(l+1)}$. A conservative estimate for a region in which the exact eigenvalues of the problem $\mathbf{K}\boldsymbol{\phi} = \lambda\mathbf{M}\boldsymbol{\phi}$ lie is given by

$$0.99\lambda_i^{(l+1)} < \lambda_i < 1.01\lambda_i^{(l+1)} \tag{12.16}$$

where only the smallest eigenvalues that converged to a tolerance of 10^{-2s} should be included. The relation in (12.16) can be used to establish bounds on all exact eigenvalues, and hence a realistic Sturm sequence check can be applied.

12.3.5 Implementation of the Subspace Iteration Method

The equations of subspace iteration have been presented in (12.10) to (12.14). However, in actual implementation, the solution can be performed more effectively as summarized in Table 12.2, which also gives the corresponding number of operations used (including those in error estimates, see Sections 12.2.3 and 10.4).

The solution method is presented in a compact manner in the computer program SSPACE. This program can directly be employed, but when a larger number of eigenpairs are sought, the acceleration procedures mentioned in Section 12.3.6 are very useful.

Subroutine SSPACE: Program SSPACE is an implementation of the subspace iteration method presented above for the solution of the smallest eigenvalues and corresponding eigenvectors of the generalized eigenproblem $\mathbf{K}\boldsymbol{\phi} = \lambda\mathbf{M}\boldsymbol{\phi}$. The argument variables and use of the subroutine are defined using comment cards in the program.

```
      SUBROUTINE SSPACE (A,B,MAXA,R,EIGV,TT,W,AR,BR,VEC,D,RTOLV,BUP,BLO,
     1 BUPC,NN,NNM,NWK,NWM,NROOT,RTOL,NC,NNC,NITEM,IFSS,IFPR,NSTIF,IOUT)
C . . . . . . . . . . . . . . . . . . . . . . . . . . . . . . . . . . .
C .                                                                   .
C .   P R O G R A M                                                   .
C .        TO SOLVE FOR THE SMALLEST EIGENVALUES AND CORRESPONDING    .
C .        EIGENVECTORS IN THE GENERALIZED EIGENPROBLEM USING THE     .
C .        SUBSPACE ITERATION METHOD                                  .
C .                                                                   .
C .  - - INPUT VARIABLES - -                                          .
C .        A(NWK)     = STIFFNESS MATRIX IN COMPACTED FORM (ASSUMED   .
C .                     POSITIVE DEFINITE)                            .
C .        B(NWM)     = MASS MATRIX IN COMPACTED FORM                 .
C .        MAXA(NNM)  = VECTOR CONTAINING ADDRESSES OF DIAGONAL       .
C .                     ELEMENTS OF STIFFNESS MATRIX A                .
C .        R(NN,NC)   = EIGENVECTORS ON SOLUTION EXIT                 .
C .        EIGV(NC)   = EIGENVALUES ON SOLUTION EXIT                  .
C .        TT(NN)     = WORKING VECTOR                                .
C .        W(NN)      = WORKING VECTOR                                .
C .        AR(NNC)    = WORKING MATRIX STORING PROJECTION OF K        .
C .        BR(NNC)    = WORKING MATRIX STORING PROJECTION OF M        .
C .        VEC(NC,NC)= WORKING MATRIX                                 .
C .        D(NC)      = WORKING VECTOR                                .
C .        RTOLV(NC)  = WORKING VECTOR                                .
C .        BUP(NC)    = WORKING VECTOR                                .
C .        BLO(NC)    = WORKING VECTOR                                .
C .        BUPC(NC)   = WORKING VECTOR                                .
C .        NN         = ORDER OF STIFFNESS AND MASS MATRICES          .
C .        NNM        = NN + 1                                        .
C .        NWK        = NUMBER OF ELEMENTS BELOW SKYLINE OF           .
C .                     STIFFNESS MATRIX                              .
C .        NWM        = NUMBER OF ELEMENTS BELOW SKYLINE OF           .
C .                     MASS MATRIX                                   .
C .                       I. E. NWM=NWK FOR CONSISTENT MASS MATRIX    .
C .                             NWM=NN  FOR LUMPED MASS MATRIX        .
```

```
C .       NROOT      = NUMBER OF REQUIRED EIGENVALUES AND EIGENVECTORS.
C .       RTOL       = CONVERGENCE TOLERANCE ON EIGENVALUES           .
C .                    ( 1.E-06 OR SMALLER )                          .
C .       NC         = NUMBER OF ITERATION VECTORS USED               .
C .                    (USUALLY SET TO MIN(2*NROOT, NROOT+8), BUT NC  .
C .                    CANNOT BE LARGER THAN THE NUMBER OF MASS       .
C .                    DEGREES OF FREEDOM)                            .
C .       NNC        = NC*(NC+1)/2 DIMENSION OF STORAGE VECTORS AR,BR .
C .       NITEM      = MAXIMUM NUMBER OF SUBSPACE ITERATIONS PERMITTED.
C .                    (USUALLY SET TO 16)                            .
C .                    THE PARAMETERS NC AND/OR NITEM MUST BE         .
C .                    INCREASED IF A SOLUTION HAS NOT CONVERGED      .
C .       IFSS       = FLAG FOR STURM SEQUENCE CHECK                  .
C .                      EQ.0  NO CHECK                               .
C .                      EQ.1  CHECK                                  .
C .       IFPR       = FLAG FOR PRINTING DURING ITERATION             .
C .                      EQ.0  NO PRINTING                            .
C .                      EQ.1  PRINT                                  .
C .       NSTIF      = SCRATCH FILE TO STORE STIFFNESS MATRIX         .
C .       IOUT       = OUTPUT PRINTING FILE                           .
C .                                                                   .
C .  - - OUTPUT - -                                                   .
C .       EIGV(NROOT) = EIGENVALUES                                   .
C .       R(NN,NROOT) = EIGENVECTORS                                  .
C .                                                                   .
C . . . . . . . . . . . . . . . . . . . . . . . . . . . . . . . . . . .
      IMPLICIT REAL*8 (A-H,O-Z)
      INTEGER MAXA(NNM)
      DIMENSION A(NWK),B(NWM),R(NN,NC),TT(NN),W(NN),EIGV(NC),
     1          D(NC),VEC(NC,NC),AR(NNC),BR(NNC),RTOLV(NC),BUP(NC),
     2          BLO(NC),BUPC(NC)
      SQRT(X)=DSQRT(X)
      ABS(X)=DABS(X)
C . . . . . . . . . . . . . . . . . . . . . . . . . . . . . . . . . . .
C .       THIS PROGRAM IS USED IN SINGLE PRECISION ARITHMETIC ON      .
C .       CDC EQUIPMENT AND DOUBLE PRECISION ARITHMETIC ON IBM        .
C .       OR UNIVAC MACHINES .ACTIVATE,DEACTIVATE OR ADJUST ABOVE     .
C .       CARDS FOR SINGLE OR DOUBLE PRECISION ARITHMETIC             .
C . . . . . . . . . . . . . . . . . . . . . . . . . . . . . . . . . . .
C
C     SET TOLERANCE FOR JACOBI ITERATION
      TOLJ=0.000000000001
C
C     INITIALIZATION
C
      ICONV=0
      NSCH=0
      NSMAX=12
      N1=NC + 1
      NC1=NC - 1
      REWIND NSTIF
      WRITE (NSTIF) A
      DO 60 I=1,NC
   60 D(I)=0.
C
C     ESTABLISH STARTING ITERATION VECTORS
C
      ND=NN/NC
      IF (NWM.GT.NN) GO TO 4
      J=0
      DO 2 I=1,NN
      II=MAXA(I)
      R(I,1)=B(I)
      IF (B(I).GT.0) J=J+1
    2 W(I)=B(I)/A(II)
      IF (NC.LE.J) GO TO 16
      WRITE (IOUT,1007)
      STOP
    4 DO 10 I=1,NN
      II=MAXA(I)
      R(I,1)=B(II)
   10 W(I)=B(II)/A(II)
   16 DO 20 J=2,NC
      DO 20 I=1,NN
   20 R(I,J)=0.
C
      L=NN-ND
      DO 30 J=2,NC
      RT=0.
      DO 40 I=1,L
      IF (W(I).LT.RT) GO TO 40
      RT=W(I)
      IJ=I
   40 CONTINUE
      DO 50 I=L,NN
      IF (W(I).LE.RT) GO TO 50
      RT=W(I)
      IJ=I
```

```
      50 CONTINUE                                                          SPCE0124
         TT(J)=FLOAT(IJ)                                                   SPCE0125
         W(IJ)=0.                                                          SPCE0126
         L=L-ND                                                            SPCE0127
      30 R(IJ,J)=1.                                                        SPCE0128
   C                                                                       SPCE0129
         WRITE (IOUT,1008)                                                 SPCE0130
         WRITE (IOUT,1002) (TT(J),J=2,NC)                                  SPCE0131
   C                                                                       SPCE0132
   C     FACTORIZE MATRIX A INTO (L)*(D)*(L(T))                            SPCE0133
   C                                                                       SPCE0134
         ISH=0                                                             SPCE0135
         CALL DECOMP (A,MAXA,NN,ISH,IOUT)                                  SPCE0136
   C                                                                       SPCE0137
   C - - -S T A R T   O F   I T E R A T I O N   L O O P                    SPCE0138
   C                                                                       SPCE0139
         NITE=0                                                            SPCE0140
     100 NITE=NITE + 1                                                     SPCE0141
         IF (IFPR.EQ.0) GO TO 90                                          SPCE0142
         WRITE (IOUT,1010) NITE                                            SPCE0143
   C                                                                       SPCE0144
   C     CALCULATE THE PROJECTIONS OF A AND B                             SPCE0145
   C                                                                       SPCE0146
      90 IJ=0                                                              SPCE0147
         DO 110 J=1,NC                                                     SPCE0148
         DO 120 K=1,NN                                                     SPCE0149
     120 TT(K)=R(K,J)                                                      SPCE0150
         CALL REDBAK (A,TT,MAXA,NN)                                        SPCE0151
         DO 130 I=J,NC                                                     SPCE0152
         ART=0.                                                            SPCE0153
         DO 140 K=1,NN                                                     SPCE0154
     140 ART=ART + R(K,I)*TT(K)                                            SPCE0155
         IJ=IJ + 1                                                         SPCE0156
     130 AR(IJ)=ART                                                        SPCE0157
         DO 150 K=1,NN                                                     SPCE0158
     150 R(K,J)=TT(K)                                                      SPCE0159
     110 CONTINUE                                                          SPCE0160
         IJ=0                                                              SPCE0161
         DO 160 J=1,NC                                                     SPCE0162
         CALL MULT (TT,B,R(1,J),MAXA,NN,NWM)                               SPCE0163
         DO 180 I=J,NC                                                     SPCE0164
         BRT=0.                                                            SPCE0165
         DO 190 K=1,NN                                                     SPCE0166
     190 BRT=BRT + R(K,I)*TT(K)                                            SPCE0167
         IJ=IJ + 1                                                         SPCE0168
     180 BR(IJ)=BRT                                                        SPCE0169
         IF (ICONV.GT.0) GO TO 160                                        SPCE0170
         DO 200 K=1,NN                                                     SPCE0171
     200 R(K,J)=TT(K)                                                      SPCE0172
     160 CONTINUE                                                          SPCE0173
   C                                                                       SPCE0174
   C     SOLVE FOR EIGENSYSTEM OF SUBSPACE OPERATORS                      SPCE0175
   C                                                                       SPCE0176
         IF (IFPR.EQ.0) GO TO 320                                        SPCE0177
         IND=1                                                             SPCE0178
     210 WRITE (IOUT,1020)                                                 SPCE0179
         II=1                                                              SPCE0180
         DO 300 I=1,NC                                                     SPCE0181
         ITEMP=II+NC-I                                                     SPCE0182
         WRITE(IOUT,1005) (AR(J),J=II,ITEMP)                              SPCE0183
     300 II=II + N1 - I                                                   SPCE0184
         WRITE (IOUT,1030)                                                 SPCE0185
         II=1                                                             SPCE0186
         DO 310 I=1,NC                                                     SPCE0187
         ITEMP=II+NC-I                                                     SPCE0188
         WRITE(IOUT,1005) (BR(J),J=II,ITEMP)                              SPCE0189
     310 II=II + N1 - I                                                   SPCE0190
         IF(IND.EQ.2) GO TO 350                                          SPCE0191
   C                                                                       SPCE0192
     320 CALL JACOBI (AR,BR,VEC,EIGV,W,NC,NNC,TOLJ,NSMAX,IFPR,IOUT)       SPCE0193
   C                                                                       SPCE0194
         IF (IFPR.EQ.0) GO TO 350                                        SPCE0195
         WRITE (IOUT,1040)                                                 SPCE0196
         IND=2                                                             SPCE0197
         GO TO 210                                                         SPCE0198
   C                                                                       SPCE0199
   C     ARRANGE EIGENVALUES IN ASCENDING ORDER                          SPCE0200
   C                                                                       SPCE0201
     350 IS=0                                                              SPCE0202
         II=1                                                             SPCE0203
         DO 360 I=1,NC1                                                    SPCE0204
         ITEMP=II + N1 - I                                                SPCE0205
         IF (EIGV(I+1).GE.EIGV(I)) GO TO 360                              SPCE0206
         IS=IS+1                                                           SPCE0207
         EIGVT=EIGV(I+1)                                                   SPCE0208
         EIGV(I+1)=EIGV(I)                                                 SPCE0209
         EIGV(I)=EIGVT                                                     SPCE0210
         BT=BR(ITEMP)                                                      SPCE0211
         BR(ITEMP)=BR(II)                                                  SPCE0212
```

```
      BR(II)=BT
      DO 370 K=1,NC
      RT=VEC(K,I+1)
      VEC(K,I+1)=VEC(K,I)
  370 VEC(K,I)=RT
  360 II=ITEMP
      IF (IS.GT.0) GO TO 350
      IF (IFPR.EQ.0) GO TO 375
      WRITE (IOUT,1035)
      WRITE (IOUT,1006) (EIGV(I),I=1,NC)
C
C     CALCULATE B TIMES APPROXIMATE EIGENVECTORS (ICONV.EQ.0)
C        OR      FINAL EIGENVECTOR APPROXIMATIONS (ICONV.GT.0)
C
  375 DO 420 I=1,NN
      DO 422 J=1,NC
  422 TT(J)=R(I,J)
      DO 424 K=1,NC
      RT=0.
      DO 430 L=1,NC
  430 RT=RT + TT(L)*VEC(L,K)
  424 R(I,K)=RT
  420 CONTINUE
      IF (ICONV.GT.0) GO TO 500
C
C     CHECK FOR CONVERGENCE OF EIGENVALUES
C
      DO 380 I=1,NC
      DIF=ABS(EIGV(I)-D(I))
  380 RTOLV(I)=DIF/EIGV(I)
      IF (IFPR.EQ.0) GO TO 385
      WRITE (IOUT,1050)
      WRITE (IOUT,1005) (RTOLV(I),I=1,NC)
C
  385 DO 390 I=1,NROOT
      IF (RTOLV(I).GT.RTOL) GO TO 400
  390 CONTINUE
      WRITE (IOUT,1060) RTOL
      ICONV=1
      GO TO 100
  400 IF (NITE.LT.NITEM) GO TO 410
      WRITE (IOUT,1070)
      ICONV=2
      IFSS=0
      GO TO 100
C
  410 DO 440 I=1,NC
  440 D(I)=EIGV(I)
      GO TO 100
C
C - - - E N D   O F   I T E R A T I O N   L O O P
C
  500 WRITE (IOUT,1100)
      WRITE (IOUT,1006) (EIGV(I),I=1,NROOT)
      WRITE (IOUT,1110)
      DO 530 J=1,NROOT
  530 WRITE (IOUT,1005) (R(K,J),K=1,NN)
C
C     CALCULATE AND PRINT ERROR NORMS
C
      REWIND NSTIF
      READ(NSTIF) A
C
      DO 580 L=1,NROOT
      RT=EIGV(L)
      CALL MULT(TT,A,R(1,L),MAXA,NN,NWK)
      VNORM=0.
      DO 590 I=1,NN
  590 VNORM=VNORM+TT(I)*TT(I)
      CALL MULT(W,B,R(1,L),MAXA,NN,NWM)
      WNORM=0.
      DO 600 I=1,NN
      TT(I)=TT(I) - RT*W(I)
  600 WNORM=WNORM + TT(I)*TT(I)
      VNORM=SQRT(VNORM)
      WNORM=SQRT(WNORM)
      D(L)=WNORM/VNORM
  580 CONTINUE
      WRITE(IOUT,1115)
      WRITE(IOUT,1006) (D(I),I=1,NROOT)
C
C     APPLY STURM SEQUENCE CHECK
C
      IF (IFSS.EQ.0) GO TO 700
      CALL SCHECK (EIGV,RTOLV,BUP,BLO,BUPC,D,NC,NEI,RTOL,SHIFT)
C
      WRITE (IOUT,1120) SHIFT
C
C     SHIFT MATRIX A
C
```

```
      REWIND NSTIF
      READ (NSTIF) A
      IF (NWM.GT.NN) GO TO 645
      DO 640 I=1,NN
      II=MAXA(I)
  640 A(II)=A(II) - B(I)*SHIFT
      GO TO 660
  645 DO 650 I=1,NWK
  650 A(I)=A(I) - B(I)*SHIFT
C
C     FACTORIZE SHIFTED MATRIX
C
  660 ISH=1
      CALL DECOMP (A,MAXA,NN,ISH,IOUT)
C
C     COUNT NUMBER OF NEGATIVE DIAGONAL ELEMENTS
C
      NSCH=0
      DO 664 I=1,NN
      II=MAXA(I)
      IF (A(II).LT.O.) NSCH=NSCH + 1
  664 CONTINUE
      IF (NSCH.EQ.NEI) GO TO 670
      NMIS=NSCH - NEI
      WRITE (IOUT,1130) NMIS
      GO TO 700
  670 WRITE (IOUT,1140) NSCH
  700 RETURN
C
 1002 FORMAT (1HO,10F10.0)
 1005 FORMAT (1H ,12E11.4)
 1006 FORMAT (1HO,6E22.14)
 1007 FORMAT (///63H STOP, NC IS LARGER THAN THE NUMBER OF MASS DEGREES
     1OF FREEDOM  )
 1008 FORMAT ( ///,62H DEGREES OF FREEDOM EXCITED BY UNIT STARTING ITERA
     1TION VECTORS)
 1010 FORMAT (1H1,32HI T E R A T I O N   N U M B E R ,I4)
 1020 FORMAT (28HOPROJECTION OF A (MATRIX AR) )
 1030 FORMAT (28HOPROJECTION OF B (MATRIX BR) )
 1035 FORMAT (30HOEIGENVALUES OF AR-LAMBDA*BR  )
 1040 FORMAT (40HOAR AND BR AFTER JACOBI DIAGONALIZATION  )
 1050 FORMAT (43HORELATIVE TOLERANCE REACHED ON EIGENVALUES  )
 1060 FORMAT (///,30H CONVERGENCE REACHED FOR RTOL  ,E10.4)
 1070 FORMAT (1H1,51H*** NO CONVERGENCE IN MAXIMUM NUMBER OF ITERATIONS
     1               ,9HPERMITTED/35H WE ACCEPT CURRENT ITERATION VALUES/
     2               42H THE STURM SEQUENCE CHECK IS NOT PERFORMED)
 1100 FORMAT (///,31H THE CALCULATED EIGENVALUES ARE  )
 1115 FORMAT (//1X,36HPRINT ERROR NORMS ON THE EIGENVALUES  )
 1110 FORMAT (//, 32H THE CALCULATED EIGENVECTORS ARE  //)
 1120 FORMAT  (///,23H CHECK APPLIED AT SHIFT   ,E22.14)
 1130 FORMAT (// 10H THERE ARE  ,I4,21H EIGENVALUES MISSING  )
 1140 FORMAT (// 20H WE FOUND THE LOWEST  ,I4,12H EIGENVALUES  )
C
      END
      SUBROUTINE DECOMP (A,MAXA,NN,ISH,IOUT)
C . . . . . . . . . . . . . . . . . . . . . . . . . . . . . . . . . . .
C .                                                                   .
C .   P R O G R A M                                                   .
C .        TO CALCULATE (L)*(D)*(L)(T) FACTORIZATION OF               .
C .        STIFFNESS MATRIX                                           .
C .                                                                   .
C . . . . . . . . . . . . . . . . . . . . . . . . . . . . . . . . . . .
C
      IMPLICIT REAL*8 (A-H,O-Z)
      DIMENSION A(1),MAXA(1)
      ABS(X)=DABS(X)
      IF (NN.EQ.1) RETURN
C
      DO 200 N=1,NN
      KN=MAXA(N)
      KL=KN + 1
      KU=MAXA(N+1) - 1
      KH=KU - KL
      IF (KH) 304,240,210
  210 K=N - KH
      IC=0
      KLT=KU
      DO 260 J=1,KH
      IC=IC + 1
      KLT=KLT - 1
      KI=MAXA(K)
      ND=MAXA(K+1) - KI - 1
      IF (ND) 260,260,270
  270 KK=MINO(IC,ND)
      C=0.
      DO 280 L=1,KK
  280 C=C + A(KI+L)*A(KLT+L)
      A(KLT)=A(KLT) - C
  260 K=K + 1
  240 K=N
```

```
              B=O.                                                        DCMP0037
              DO 300 KK=KL,KU                                             DCMP0038
              K=K - 1                                                     DCMP0039
              KI=MAXA(K)                                                  DCMP0040
              C=A(KK)/A(KI)                                               DCMP0041
              IF (ABS(C).LT.1.E07) GO TO 290                              DCMP0042
              WRITE (IOUT,2010) N,C                                       DCMP0043
              STOP                                                        DCMP0044
          290 B=B + C*A(KK)                                               DCMP0045
          300 A(KK)=C                                                     DCMP0046
              A(KN)=A(KN) - B                                             DCMP0047
          304 IF (A(KN)) 310,310,200                                      DCMP0048
          310 IF (ISH.EQ.0) GO TO 320                                     DCMP0049
              IF (A(KN).EQ.0.) A(KN)=-1.E-16                              DCMP0050
              GO TO 200                                                   DCMP0051
          320 WRITE(IOUT,2000) N,A(KN)                                    DCMP0052
              STOP                                                        DCMP0053
          200 CONTINUE                                                    DCMP0054
      C                                                                   DCMP0055
              RETURN                                                      DCMP0056
         2000 FORMAT(//48H STOP - STIFFNESS MATRIX NOT POSITIVE DEFINITE  ,//  DCMP0057
             1           32H NONPOSITIVE PIVOT FOR EQUATION  ,I4,//       DCMP0058
             2           10H PIVOT =  ,E20.12 )                           DCMP0059
         2010 FORMAT (//47H STOP - STURM SEQUENCE CHECK FAILED BECAUSE OF ,  DCMP0060
             135HMULTIPLIER GROWTH FOR COLUMN NUMBER,I4,//12H MULTIPLIER=,E20.8)  DCMP0061
              END                                                         DCMP0062
              SUBROUTINE REDBAK (A,V,MAXA,NN)                             RDBK0001
      C . . . . . . . . . . . . . . . . . . . . . . . . . . . . . . . .  RDBK0002
      C .                                                             .   RDBK0003
      C .   P R O G R A M                                             .   RDBK0004
      C .        TO REDUCE AND BACK-SUBSTITUTE ITERATION VECTORS      .   RDBK0005
      C .                                                             .   RDBK0006
      C . . . . . . . . . . . . . . . . . . . . . . . . . . . . . . . .  RDBK0007
      C                                                                   RDBK0008
              IMPLICIT REAL*8 (A-H,O-Z)                                   RDBK0009
              DIMENSION A(1),V(1),MAXA(1)                                 RDBK0010
      C                                                                   RDBK0011
              DO 400 N=1,NN                                               RDBK0012
              KL=MAXA(N) + 1                                              RDBK0013
              KU=MAXA(N+1) - 1                                            RDBK0014
              IF (KU-KL) 400,410,410                                      RDBK0015
          410 K=N                                                         RDBK0016
              C=0.                                                        RDBK0017
              DO 420 KK=KL,KU                                             RDBK0018
              K=K - 1                                                     RDBK0019
          420 C=C + A(KK)*V(K)                                            RDBK0020
              V(N)=V(N) - C                                               RDBK0021
          400 CONTINUE                                                    RDBK0022
      C                                                                   RDBK0023
              DO 480 N=1,NN                                               RDBK0024
              K=MAXA(N)                                                   RDBK0025
          480 V(N)=V(N)/A(K)                                              RDBK0026
              IF (NN.EQ.1) RETURN                                         RDBK0027
              N=NN                                                        RDBK0028
              DO 500 L=2,NN                                               RDBK0029
              KL=MAXA(N) + 1                                              RDBK0030
              KU=MAXA(N+1) - 1                                            RDBK0031
              IF (KU-KL) 500,510,510                                      RDBK0032
          510 K=N                                                         RDBK0033
              DO 520 KK=KL,KU                                             RDBK0034
              K=K - 1                                                     RDBK0035
          520 V(K)=V(K) - A(KK)*V(N)                                      RDBK0036
          500 N=N - 1                                                     RDBK0037
      C                                                                   RDBK0038
              RETURN                                                      RDBK0039
              END                                                         RDBK0040
              SUBROUTINE MULT (TT,B,RR,MAXA,NN,NWM)                       MULT0001
      C . . . . . . . . . . . . . . . . . . . . . . . . . . . . . . . .  MULT0002
      C .                                                             .   MULT0003
      C .   P R O G R A M                                             .   MULT0004
      C .        TO EVALUATE PRODUCT OF B TIMES RR AND STORE RESULT IN TT  .  MULT0005
      C .                                                             .   MULT0006
      C . . . . . . . . . . . . . . . . . . . . . . . . . . . . . . . .  MULT0007
      C                                                                   MULT0008
              IMPLICIT REAL*8 (A-H,O-Z)                                   MULT0009
              DIMENSION TT(1),B(1),RR(1),MAXA(1)                          MULT0010
      C                                                                   MULT0011
              IF (NWM.GT.NN) GO TO 20                                     MULT0012
              DO 10 I=1,NN                                                MULT0013
           10 TT(I)=B(I)*RR(I)                                           MULT0014
              RETURN                                                      MULT0015
      C                                                                   MULT0016
           20 DO 40 I=1,NN                                               MULT0017
           40 TT(I)=0.                                                   MULT0018
              DO 100 I=1,NN                                               MULT0019
              KL=MAXA(I)                                                  MULT0020
              KU=MAXA(I+1) - 1                                            MULT0021
              II=I + 1                                                    MULT0022
```

```
        CC=RR(I)                                                        MULT0023
        DO 100 KK=KL,KU                                                 MULT0024
        II=II - 1                                                       MULT0025
    100 TT(II)=TT(II) + B(KK)*CC                                        MULT0026
        IF (NN.EQ.1) RETURN                                            MULT0027
        DO 200 I=2,NN                                                   MULT0028
        KL=MAXA(I) + 1                                                  MULT0029
        KU=MAXA(I+1) - 1                                                MULT0030
        IF (KU-KL) 200,210,210                                          MULT0031
    210 II=I                                                            MULT0032
        AA=0.                                                           MULT0033
        DO 220 KK=KL,KU                                                 MULT0034
        II=II - 1                                                       MULT0035
    220 AA=AA + B(KK)*RR(II)                                           MULT0036
        TT(I)=TT(I) + AA                                                MULT0037
    200 CONTINUE                                                        MULT0038
C                                                                       MULT0039
        RETURN                                                          MULT0040
        END                                                             MULT0041
        SUBROUTINE SCHECK (EIGV,RTOLV,BUP,BLO,BUPC,NEIV,NC,NEI,RTOL,SHIFT)  SCHK0001
C . . . . . . . . . . . . . . . . . . . . . . . . . . . . . . . . .     SCHK0002
C .                                                                 .   SCHK0003
C .     P R O G R A M                                               .   SCHK0004
C .          TO EVALUATE SHIFT FOR STURM SEQUENCE CHECK             .   SCHK0005
C .                                                                 .   SCHK0006
C . . . . . . . . . . . . . . . . . . . . . . . . . . . . . . . . .     SCHK0007
C                                                                       SCHK0008
        IMPLICIT REAL*8 (A-H,O-Z)                                       SCHK0009
        DIMENSION EIGV(NC),RTOLV(NC),BUP(NC),BLO(NC),BUPC(NC),NEIV(NC)  SCHK0010
C                                                                       SCHK0011
        FTOL=0.01                                                       SCHK0012
C                                                                       SCHK0013
        DO 100 I=1,NC                                                   SCHK0014
        BUP(I)=EIGV(I)*(1. + FTOL)                                      SCHK0015
    100 BLO(I)=EIGV(I)*(1. - FTOL)                                      SCHK0016
        NROOT=0                                                         SCHK0017
        DO 120 I=1,NC                                                   SCHK0018
    120 IF (RTOLV(I).LT.RTOL) NROOT=NROOT+1                            SCHK0019
        IF (NROOT.GE.1) GO TO 200                                       SCHK0020
        WRITE (6,1010)                                                  SCHK0021
        STOP                                                            SCHK0022
C                                                                       SCHK0023
C       FIND UPPER BOUNDS ON EIGENVALUE CLUSTERS                        SCHK0024
C                                                                       SCHK0025
    200 DO 240 I=1,NROOT                                                SCHK0026
    240 NEIV(I)=1                                                       SCHK0027
        IF (NROOT.NE.1) GO TO 260                                       SCHK0028
        BUPC(1)=BUP(1)                                                  SCHK0029
        LM=1                                                            SCHK0030
        L=1                                                             SCHK0031
        I=2                                                             SCHK0032
        GO TO 295                                                       SCHK0033
    260 L=1                                                             SCHK0034
        I=2                                                             SCHK0035
    270 IF (BUP(I-1).LE.BLO(I)) GO TO 280                              SCHK0036
        NEIV(L)=NEIV(L)+1                                               SCHK0037
        I=I+1                                                           SCHK0038
        IF (I.LE.NROOT) GO TO 270                                       SCHK0039
    280 BUPC(L)=BUP(I-1)                                                SCHK0040
        IF (I.GT.NROOT) GO TO 290                                       SCHK0041
        L=L+1                                                           SCHK0042
        I=I+1                                                           SCHK0043
        IF (I.LE.NROOT) GO TO 270                                       SCHK0044
        BUPC(L)=BUP(I-1)                                                SCHK0045
    290 LM=L                                                            SCHK0046
        IF (NROOT.EQ.NC) GO TO 300                                      SCHK0047
    295 IF (BUP(I-1).LE.BLO(I)) GO TO 300                              SCHK0048
        IF (RTOLV(I).GT.RTOL) GO TO 300                                SCHK0049
        BUPC(L)=BUP(I)                                                  SCHK0050
        NEIV(L)=NEIV(L)+1                                               SCHK0051
        NROOT=NROOT+1                                                   SCHK0052
        IF (NROOT.EQ.NC) GO TO 300                                      SCHK0053
        I=I+1                                                           SCHK0054
        GO TO 295                                                       SCHK0055
C                                                                       SCHK0056
C       FIND SHIFT                                                      SCHK0057
C                                                                       SCHK0058
    300 WRITE (6,1020)                                                  SCHK0059
        WRITE (6,1005) (BUPC(I),I=1,LM)                                SCHK0060
        WRITE (6,1030)                                                  SCHK0061
        WRITE (6,1006) (NEIV(I),I=1,LM)                                SCHK0062
        LL=LM-1                                                         SCHK0063
        IF (LM.EQ.1) GO TO 310                                          SCHK0064
    330 DO 320 I=1,LL                                                   SCHK0065
    320 NEIV(L)=NEIV(L)+NEIV(I)                                        SCHK0066
        L=L-1                                                          SCHK0067
        LL=LL-1                                                         SCHK0068
        IF (L.NE.1) GO TO 330                                           SCHK0069
    310 WRITE (6,1040)                                                  SCHK0070
```

```
      WRITE (6,1006) (NEIV(I),I=1,LM)                                    SCHK0071
      L=0                                                                SCHK0072
      DO 340 I=1,LM                                                      SCHK0073
      L=L+1                                                              SCHK0074
      IF (NEIV(I).GE.NROOT) GO TO 350                                    SCHK0075
  340 CONTINUE                                                           SCHK0076
  350 SHIFT=BUPC(L)                                                      SCHK0077
      NEI=NEIV(L)                                                        SCHK0078
C                                                                        SCHK0079
      RETURN                                                             SCHK0080
C                                                                        SCHK0081
 1005 FORMAT (1H0,6E22.14)                                              SCHK0082
 1006 FORMAT (1H0,6I22)                                                 SCHK0083
 1010 FORMAT (37H0***ERROR   SOLUTION STOP IN *SCHECK*, / 12X,           SCHK0084
     1       21HNO EIGENVALUES FOUND., / 1X)                            SCHK0085
 1020 FORMAT (//,37H UPPER BOUNDS ON EIGENVALUE CLUSTERS   )            SCHK0086
 1030 FORMAT (34H0NO OF EIGENVALUES IN EACH CLUSTER   )                 SCHK0087
 1040 FORMAT (42H0NO OF EIGENVALUES LESS THAN UPPER BOUNDS   )          SCHK0088
      END                                                               SCHK0089
      SUBROUTINE JACOBI (A,B,X,EIGV,D,N,NWA,RTOL,NSMAX,IFPR,IOUT)        JCBI0001
C ......................................................................  JCBI0002
C .                                                                   .  JCBI0003
C .                                                                   .  JCBI0004
C .   P R O G R A M                                                   .  JCBI0005
C .      TO SOLVE THE GENERALIZED EIGENPROBLEM USING THE             .  JCBI0006
C .      GENERALIZED JACOBI ITERATION                                .  JCBI0007
C ......................................................................  JCBI0008
      IMPLICIT REAL*8(A-H,O-Z)                                          JCBI0009
      DIMENSION A(NWA),B(NWA),X(N,N),EIGV(N),D(N)                       JCBI0010
      ABS(X)=DABS(X)                                                     JCBI0011
      SQRT(X)=DSQRT(X)                                                   JCBI0012
C                                                                        JCBI0013
C     INITIALIZE EIGENVALUE AND EIGENVECTOR MATRICES                    JCBI0014
C                                                                        JCBI0015
      N1=N + 1                                                           JCBI0016
      II=1                                                               JCBI0017
      DO 10 I=1,N                                                        JCBI0018
      IF(A(II).GT.O. .AND. B(II).GT.O.) GO TO 4                          JCBI0019
      WRITE(IOUT,2020) II,A(II),B(II)                                    JCBI0020
      STOP                                                               JCBI0021
    4 D(I)=A(II)/B(II)                                                   JCBI0022
      EIGV(I)=D(I)                                                       JCBI0023
   10 II=II + N1 - I                                                     JCBI0024
      DO 30 I=1,N                                                        JCBI0025
      DO 20 J=1,N                                                        JCBI0026
   20 X(I,J)=0.                                                          JCBI0027
   30 X(I,I)=1.                                                          JCBI0028
      IF(N.EQ.1) RETURN                                                  JCBI0029
C                                                                        JCBI0030
C     INITIALIZE SWEEP COUNTER AND BEGIN ITERATION                      JCBI0031
C                                                                        JCBI0032
      NSWEEP=0                                                           JCBI0033
      NR=N-1                                                             JCBI0034
   40 NSWEEP=NSWEEP+1                                                    JCBI0035
      IF(IFPR.EQ.1)WRITE(IOUT,2000)NSWEEP                                JCBI0036
C                                                                        JCBI0037
C     CHECK IF PRESENT OFF-DIAGONAL ELEMENT IS LARGE ENOUGH TO REQUIRE  JCBI0038
C     ZEROING                                                            JCBI0039
C                                                                        JCBI0040
      EPS=(.01**NSWEEP)**2                                               JCBI0041
      DO 210 J=1,NR                                                      JCBI0042
      JP1=J+1                                                            JCBI0043
      JM1=J-1                                                            JCBI0044
      LJK=JM1*N - JM1*J/2                                                JCBI0045
      JJ=LJK + J                                                         JCBI0046
      DO 210 K=JP1,N                                                     JCBI0047
      KP1=K+1                                                            JCBI0048
      KM1=K-1                                                            JCBI0049
      JK=LJK + K                                                         JCBI0050
      KK=KM1*N - KM1*K/2 + K                                             JCBI0051
      EPTOLA=(A(JK)*A(JK))/(A(JJ)*A(KK))                                 JCBI0052
      EPTOLB=(B(JK)*B(JK))/(B(JJ)*B(KK))                                 JCBI0053
      IF((EPTOLA.LT.EPS).AND.(EPTOLB.LT.EPS))GO TO 210                   JCBI0054
C                                                                        JCBI0055
C     IF ZEROING IS REQUIRED, CALCULATE THE ROTATION MATRIX ELEMENTS CA JCBI0056
C     AND CG                                                             JCBI0057
C                                                                        JCBI0058
      AKK=A(KK)*B(JK)-B(KK)*A(JK)                                        JCBI0059
      AJJ=A(JJ)*B(JK)-B(JJ)*A(JK)                                        JCBI0060
      AB=A(JJ)*B(KK)-A(KK)*B(JJ)                                         JCBI0061
      CHECK=(AB*AB+4.*AKK*AJJ)/4.                                        JCBI0062
      IF(CHECK)50,60,60                                                  JCBI0063
   50 WRITE(IOUT,2020)                                                   JCBI0064
      STOP                                                               JCBI0065
   60 SQCH=SQRT(CHECK)                                                   JCBI0066
      D1=AB/2.+SQCH                                                      JCBI0067
      D2=AB/2.-SQCH                                                      JCBI0068
      DEN=D1                                                             JCBI0069
      IF(ABS(D2).GT.ABS(D1))DEN=D2                                       
```

```
          IF(DEN)80,70,80                                          JCBIO070
    70 CA=0.                                                        JCBIO071
       CG=-A(JK)/A(KK)                                              JCBIO072
       GO TO 90                                                     JCBIO073
    80 CA=AKK/DEN                                                   JCBIO074
       CG=-AJJ/DEN                                                  JCBIO075
C                                                                   JCBIO076
C      PERFORM THE GENERALIZED ROTATION TO ZERO THE PRESENT OFF-DIAGONAL  JCBIO077
C      ELEMENT                                                      JCBIO078
C                                                                   JCBIO079
    90 IF(N-2)100,190,100                                           JCBIO080
   100 IF(JM1-1)130,110,110                                         JCBIO081
   110 DO 120 I=1,JM1                                               JCBIO082
       IM1=I - 1                                                    JCBIO083
       IJ=IM1*N - IM1*I/2 + J                                       JCBIO084
       IK=IM1*N - IM1*I/2 + K                                       JCBIO085
       AJ=A(IJ)                                                     JCBIO086
       BJ=B(IJ)                                                     JCBIO087
       AK=A(IK)                                                     JCBIO088
       BK=B(IK)                                                     JCBIO089
       A(IJ)=AJ+CG*AK                                               JCBIO090
       B(IJ)=BJ+CG*BK                                               JCBIO091
       A(IK)=AK+CA*AJ                                               JCBIO092
   120 B(IK)=BK+CA*BJ                                               JCBIO093
   130 IF(KP1-N)140,140,160                                         JCBIO094
   140 LJI=JM1*N - JM1*J/2                                          JCBIO095
       LKI=KM1*N - KM1*K/2                                          JCBIO096
       DO 150 I=KP1,N                                               JCBIO097
       JI=LJI + I                                                   JCBIO098
       KI=LKI + I                                                   JCBIO099
       AJ=A(JI)                                                     JCBIO100
       BJ=B(JI)                                                     JCBIO101
       AK=A(KI)                                                     JCBIO102
       BK=B(KI)                                                     JCBIO103
       A(JI)=AJ+CG*AK                                               JCBIO104
       B(JI)=BJ+CG*BK                                               JCBIO105
       A(KI)=AK+CA*AJ                                               JCBIO106
   150 B(KI)=BK+CA*BJ                                               JCBIO107
   160 IF(JP1-KM1)170,170,190                                       JCBIO108
   170 LJI=JM1*N - JM1*J/2                                          JCBIO109
       DO 180 I=JP1,KM1                                             JCBIO110
       JI=LJI + I                                                   JCBIO111
       IM1=I - 1                                                    JCBIO112
       IK=IM1*N - IM1*I/2 + K                                       JCBIO113
       AJ=A(JI)                                                     JCBIO114
       BJ=B(JI)                                                     JCBIO115
       AK=A(IK)                                                     JCBIO116
       BK=B(IK)                                                     JCBIO117
       A(JI)=AJ+CG*AK                                               JCBIO118
       B(JI)=BJ+CG*BK                                               JCBIO119
       A(IK)=AK+CA*AJ                                               JCBIO120
   180 B(IK)=BK+CA*BJ                                               JCBIO121
   190 AK=A(KK)                                                     JCBIO122
       BK=B(KK)                                                     JCBIO123
       A(KK)=AK+2.*CA*A(JK)+CA*CA*A(JJ)                             JCBIO124
       B(KK)=BK+2.*CA*B(JK)+CA*CA*B(JJ)                             JCBIO125
       A(JJ)=A(JJ)+2.*CG*A(JK)+CG*CG*AK                             JCBIO126
       B(JJ)=B(JJ)+2.*CG*B(JK)+CG*CG*BK                             JCBIO127
       A(JK)=0.                                                     JCBIO128
       B(JK)=0.                                                     JCBIO129
C                                                                   JCBIO130
C      UPDATE THE EIGENVECTOR MATRIX AFTER EACH ROTATION           JCBIO131
C                                                                   JCBIO132
       DO 200 I=1,N                                                 JCBIO133
       XJ=X(I,J)                                                    JCBIO134
       XK=X(I,K)                                                    JCBIO135
       X(I,J)=XJ+CG*XK                                              JCBIO136
   200 X(I,K)=XK+CA*XJ                                              JCBIO137
   210 CONTINUE                                                     JCBIO138
C                                                                   JCBIO139
C      UPDATE THE EIGENVALUES AFTER EACH SWEEP                      JCBIO140
C                                                                   JCBIO141
       II=1                                                         JCBIO142
       DO 220 I=1,N                                                 JCBIO143
       IF (A(II).GT.O. .AND. B(II).GT.O.) GO TO 215                 JCBIO144
       WRITE(IOUT,2020) II,A(II),B(II)                             JCBIO145
       STOP                                                         JCBIO146
   215 EIGV(I)=A(II)/B(II)                                          JCBIO147
   220 II=II + N1 - I                                               JCBIO148
       IF(IFPR.EQ.O)GO TO 230                                       JCBIO149
       WRITE(IOUT,2030)                                             JCBIO150
       WRITE(IOUT,2010) (EIGV(I),I=1,N)                             JCBIO151
C                                                                   JCBIO152
C      CHECK FOR CONVERGENCE                                        JCBIO153
C                                                                   JCBIO154
   230 DO 240 I=1,N                                                 JCBIO155
       TOL=RTOL*D(I)                                                JCBIO156
       DIF=ABS(EIGV(I)-D(I))                                        JCBIO157
       IF(DIF.GT.TOL)GO TO 280                                      JCBIO158
```

```
  240 CONTINUE
C
C     CHECK ALL OFF-DIAGONAL ELEMENTS TO SEE IF ANOTHER SWEEP IS
C     REQUIRED
C
      EPS=RTOL**2
      DO 250 J=1,NR
      JM1=J-1
      JP1=J+1
      LJK=JM1*N - JM1*J/2
      JJ=LJK + J
      DO 250 K=JP1,N
      KM1=K-1
      JK=LJK + K
      KK=KM1*N - KM1*K/2 + K
      EPSA=(A(JK)*A(JK))/(A(JJ)*A(KK))
      EPSB=(B(JK)*B(JK))/(B(JJ)*B(KK))
      IF((EPSA.LT.EPS).AND.(EPSB.LT.EPS))GO TO 250
      GO TO 280
  250 CONTINUE
C
C     FILL OUT BOTTOM TRIANGLE OF RESULTANT MATRICES AND SCALE
C     EIGENVECTORS
C
  255 II=1
      DO 275 I=1,N
      BB=SQRT(B(II))
      DO 270 K=1,N
  270 X(K,I)=X(K,I)/BB
  275 II=II + N1 - I
      RETURN
C
C     UPDATE  D  MATRIX AND START NEW SWEEP, IF ALLOWED
C
  280 DO 290 I=1,N
  290 D(I)=EIGV(I)
      IF(NSWEEP.LT.NSMAX)GO TO 40
      GO TO 255
 2000 FORMAT(27H0SWEEP NUMBER IN *JACOBI* = ,I4)
 2010 FORMAT(1H0,6E20.12)
 2020 FORMAT (25H0*** ERROR  SOLUTION STOP  /
     1        31H MATRICES NOT POSITIVE DEFINITE /
     2 4H II=,I4,6HA(II)=,E20.12,6HB(II)=,E20.12)
 2030 FORMAT(36H0CURRENT EIGENVALUES IN *JACOBI* ARE,/)
      END
```

12.3.6 Some Remarks on the Subspace Iteration Method

The subspace iteration method as presented above was developed for the solution of the smallest eigenvalues and corresponding eigenvectors, where p was assumed to be small (say $p < 20$). Considering the solution of problems for a larger number of eigenpairs, say $p > 40$, experience shows that the cost of solution using the subspace iteration method rises rapidly as the number of eigenpairs considered is increased. This rapid increase in cost is due to a number of factors that can be neglected when the solution of only a few eigenpairs is required. An important point is that a relatively large number of subspace iterations may be required if $q = p + 8$. Namely, in this case, when p is large, the convergence rate to $\boldsymbol{\phi}_p$, equal to λ_p/λ_{q+1}, can be close to one. On the other hand, if q is increased the numerical operations per subspace iteration are increased significantly. Another shortcoming of the solution with q large is that a relatively large number of iteration vectors is used throughout *all* subspace iterations, although convergence to the smallest eigenvalues of those required is generally already achieved in the first few iterations. A further important consideration pertains to the high-speed core and low-speed back-up storage requirements. As the number of iteration vectors q increases, the number of matrix blocks that need be used in an out-of-core solution can also increase significantly and the peripheral processing expenditures can be large. Finally, we should note that the number of numerical operations required in the solution of the reduced eigenproblem in (12.13) becomes significant when q is large, and these numerical operations cannot be neglected in the operation count as was done in Table 12.2.

The above considerations show that procedures to accelerate the basic subspace iteration solution, presented in Sections 12.3.2 to 12.3.4, are very desirable, in particular, when a larger number of eigenpairs are to be calculated. Various acceleration procedures for the basic subspace iteration method have indeed been discussed,[18-20] and a comprehensive solution algorithm which can be very much more effective than the basic method is presented in ref. [21]. In this program the solution steps of the basic subspace iteration method are accelerated using vector overrelaxation, shifting techniques and the Lanczos method to generate the starting subspace. Using the method it can be effective to employ less iteration vectors than the number of eigenpairs to be calculated, i.e., q can be smaller than p, and the solution procedure can also be employed to calculate eigenvalues and corresponding eigenvectors in a specified interval.

This accelerated subspace iteration method represents the final result of a research effort in which various individual solution strategies are combined to increase the overall effectiveness of subspace iterations. The method represents at present a very effective and versatile solution procedure, but—because the development of improved eigensolution techniques represents such an exciting challenge—further improvements in the subspace iteration solution technique and other methods must still be expected. These improvements will likely be a result of a further merging and development of various solution approaches and the optimum implementation of the resulting techniques.

REFERENCES

1. K. J. Bathe, "Solution Methods of Large Generalized Eigenvalue Problems in Structural Engineering," Report UC SESM 71–20, Civil Engineering Department, University of California, Berkeley, 1971.
2. K. J. Bathe and E. L. Wilson, "Large Eigenvalue Problems in Dynamic Analysis," *A.S.C.E., Journal of Engineering Mechanics Division*, Vol. 98, 1972, pp. 1471–1485.
3. K. J. Bathe and E. L. Wilson, "Eigensolution of Large Structural Systems with Small Bandwidth," *A.S.C.E., Journal of Engineering Mechanics Division*, Vol. 99, 1973, pp. 467–479.
4. K. J. Bathe and E. L. Wilson, "Solution Methods for Eigenvalue Problems in Structural Mechanics," *International Journal for Numerical Methods in Engineering*, Vol. 6, 1973, pp. 213–226.
5. K. J. Bathe, E. L. Wilson, and F. E. Peterson, "SAP IV—A Structural Analysis Program for Static and Dynamic Response of Linear Systems," Report EERC 73–11, College of Engineering, University of California, Berkeley, June 1973, revised Apr. 1974.
6. EAC/EASE2 Dynamics—User Information Manual/Theoretical, Control Data Corporation, Minneapolis, Minn. (in press).
7. K. J. Bathe, "ADINA—A Finite Element Program for Automatic Dynamic Incremental Nonlinear Analysis," Report 82448–1, Acoustics and Vibration

Laboratory, Department of Mechanical Engineering, Massachusetts Institute of Technology, 1975 (rev. 1978).

8. K. J. Bathe, "An Assessment of Current Finite Element Analysis of Nonlinear Problems in Solid Mechanics," in *Numerical Solution of Partial Differential Equations—III*, (B. Hubbard, ed.), Academic Press, Inc., New York, N.Y., 1976.

9. J. H. Wilkinson, *The Algebraic Eigenvalue Problem*, Clarendon Press, Oxford, 1965.

10. H. R. Schwarz, H. Rutishauser, and E. Stiefel, *Matrizen-Numerik*, B. G. Teubner, Stuttgart, 1972.

11. F. L. Bauer, "Das Verfahren der Treppen-Iteration und Verwandte Verfahren zur Lösung Algebraischer Eigenwertprobleme," *Z.A.M.P.*, Vol. 8, 1957, pp. 214–235.

12. O. E. Brönlund, "Eigenvalues of Large Matrices," Symposium on Finite Element Techniques at the Institut für Statik und Dynamik der Luft und Raumfahrtskonstruktionen, University of Stuttgart, Germany, 1969.

13. A. Jennings, "A Direct Iteration Method of Obtaining Latent Roots and Vectors of a Symmetric Matrix," *Proceedings of the Cambridge Philosophical Society*, Vol. 63, 1967, pp. 755–765.

14. A. Jennings and D. R. L. Orr, "Application of the Simultaneous Iteration Method to Undamped Vibration Problems," *International Journal for Numerical Methods in Engineering*, Vol. 3, 1971, pp. 13–24.

15. S. B. Dong, J. A. Wolf, and F. E. Peterson, "On a Direct-Iterative Eigensolution Technique," *International Journal for Numerical Methods in Engineering*, Vol. 4, 1972, pp. 155–162.

16. P. S. Jensen, "The Solution of Large Symmetric Eigenproblems by Sectioning," *S.I.A.M. Journal of Numerical Analysis*, Vol. 9, 1972, pp. 534–545.

17. H. Rutishauser, "Computational Aspects of F. L. Bauer's Simultaneous Iteration Method," *Numerische Mathematik*, Vol. 13, 1969, pp. 4–13.

18. K. J. Bathe, "Convergence of Subspace Iteration," in *Formulations and Numerical Algorithms in Finite Element Analysis*, (K. J. Bathe, J. T. Oden, and W. Wunderlich, eds.), M.I.T. Press, Cambridge, Mass., 1977.

19. Y. Yamamoto, and H. Ohtsubo, "Subspace Iteration Accelerated by Using Chebyshev Polynomials for Eigenvalue Problems with Symmetric Matrices," *International Journal for Numerical Methods in Engineering*, Vol. 10, 1976, pp. 935–944.

20. A. Jennings, and T.J.A. Agar, "Hybrid Sturm Sequence and Simultaneous Iteration Methods," *Proceedings, Symposium on Applications of Computer Methods in Engineering*, Univ. of Southern California, Los Angeles, Aug. 1977.

21. K. J. Bathe, and S. Ramaswamy, "An Accelerated Subspace Iteration Method," *J. Computer Methods in Applied Mechanics and Engineering*, Vol. 23, 1980, pp. 313–331.

APPENDIX

IMPLEMENTATION OF THE FINITE ELEMENT METHOD

A.1 INTRODUCTION

In this book we have presented formulations, general theories and numerical methods of finite element analysis. The objective in this appendix is to discuss some important computational aspects pertaining to the implementation of finite element procedures. Although the implementation of the displacement-based finite element analysis is discussed, it should be noted that most of the concepts presented in this appendix can also be employed in finite element analysis using hybrid and mixed formulations.

The main advantage that the finite element method has over other analysis techniques is its large generality. Normally, as was pointed out, it seems possible, by using many elements, to virtually approximate any continuum with complex boundary and loading conditions to such a degree that an accurate analysis can be carried out. In practice, however, obvious engineering limitations arise, a most important one being the cost of the analysis. As the number of elements used is increased, the manpower required to prepare the relevant data and interpret the results increases, and also a larger amount of computer time is needed for the analysis. Furthermore, the limitations of the program and the computer may prevent the use of a larger number of finite elements to idealize the continuum. The limitations may be due to the high-speed and back-up (low-speed) storage available, or the round-off and truncation errors occurring in the analysis because of finite precision arithmetic. Also, the malfunctioning of a hardware component, if the analysis is carried out using many computer hours, is possible. It is therefore desirable to use efficient finite element programs.

The effectiveness of a program depends essentially on the following factors. Firstly, the use of efficient finite elements is important. Secondly, efficient programming methods and sophisticated use of the available computer hardware and software are important. Although this aspect of program development is computer-dependent, using standard FORTRAN IV and high- and low-speed storage in a system-independent manner, very effective computer programs can be developed. If such a program is to be permanently installed on a specific computer, its efficiency may normally be increased with relatively little effort by making use of the specific hardware and software options available. In the following, we therefore discuss the design of finite element programs in which largely computer-independent procedures are used. Also, numerical algorithms are presented using standard FORTRAN IV.

The third very important aspect in the development of a finite element program is the use of appropriate numerical techniques. As an example, if inappropriate techniques for the solution of the frequencies of a system in a dynamic analysis are employed, the cost may be many times larger than with effective techniques, and, even worse, a solution may not be possible at all if an unstable algorithm is employed. In order to implement the finite element method in practice, we need to use the digital computer. However, even with a relatively large capacity computer available, the feasibility of a problem solution and the effectiveness of an analysis depend directly on the numerical procedures employed.

Assume that an actual structure has been idealized as an assemblage of finite elements. The stress analysis process can be understood to consist of essentially three phases:

1. Calculation of structure matrices **K**, **M**, **C**, and **R**, whichever are applicable.
2. Solution of equilibrium equations.
3. Evaluation of element stresses.

In case of the analysis of a heat transfer, field or fluid mechanics problem, the steps are identical, but the pertinent matrices and solution quantities need be considered.

The objective in this appendix is to describe a program implementation of the first and third phases and to present a small computer program that has all the important features of a general code. Although the total solution may be subdivided into the above three phases, it should be realized that the specific implementation of one phase can have a pronounced effect on the efficiency of the other phases, and, indeed, in some programs the first two phases are carried out simultaneously (e.g., when using the frontal solution method, see Section 8.2.4)

As might be imagined, there is no unique optimum program organization for the evaluation of the structure matrices, and various different and effective strategies are currently followed.[1-4] In addition, new ideas may be developed, particularly as new computer equipment becomes available. However, although a program design might at first sight be quite different, in effect, the same basic steps are followed. For this reason, it is deemed to be most instructive to discuss in detail all the important features of one specific implementation, and we choose to discuss an implementation that uses many procedures employed in the program SAP[5-7], but in total is very close to the implementation of the program ADINA[8]. In the first part of this appendix, the program algorithms used are discussed, and in the second part a small program is presented that uses all the important features of the above general-purpose programs.

A.2 COMPUTER PROGRAM ORGANIZATION FOR CALCULATION OF STRUCTURE MATRICES

The final results of this phase are the required structure matrices for the solution of the system equilibrium equations. In a static analysis the computer program needs to calculate the structure stiffness matrix and the load vectors. In a dynamic analysis, the program must also establish the system mass and damping matrices. In the implementation to be described here, the calculation of the structure matrices is performed as follows:

1. The nodal point and element information are read and/or generated.
2. The element stiffness matrices, mass and damping matrices, and equivalent nodal loads are calculated.
3. The structure matrices **K**, **M**, **C**, and **R**, whichever are applicable, are assembled.

A.2.1 Nodal Point and Element Information Read-In

Consider first the data that correspond to the nodal points. Assume that the program is set up to allow a maximum of six degrees of freedom at each node, three translational and three rotational degrees of freedom, as shown in Fig. A.1. Corresponding to each nodal point, it must then be identified which of these degrees of freedom shall actually be used in the analysis, i.e., which of the six possible nodal degrees of freedom correspond to degrees of freedom of the finite element assemblage. This is achieved by defining an identification array, the array ID, of dimension 6 times NUMNP, where NUMNP is equal to the number of nodal points in the system. Element (i, j) in the ID matrix corresponds to the ith degree of freedom at the nodal point j. If ID(I, J) = 0, the corresponding degree of freedom is defined in the element system, and if ID(I, J) = 1, the degree of freedom is not defined. It should be noted that using the same procedure, an ID array for more (or less) than 6 degrees of freedom per nodal point could be established, and, indeed, the number of degrees of freedom per nodal point could be a variable. Consider the following simple example.

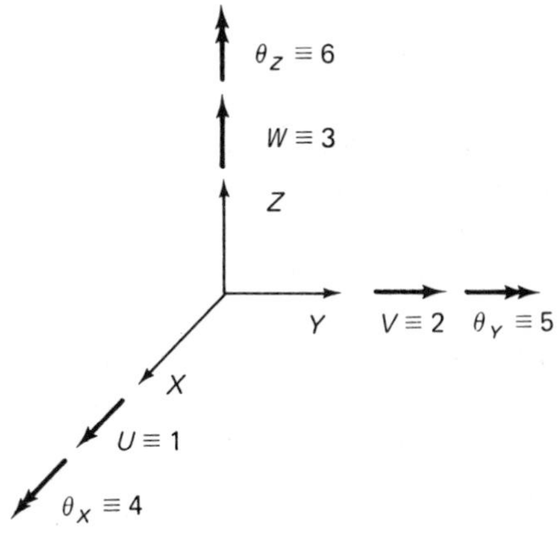

FIGURE A.1 *Possible degrees of freedom at a nodal point.*

EXAMPLE A.1: Establish the ID array for the plane stress element idealization of the cantilever in Fig. A.2 in order to define the active and nonactive degrees of freedom.

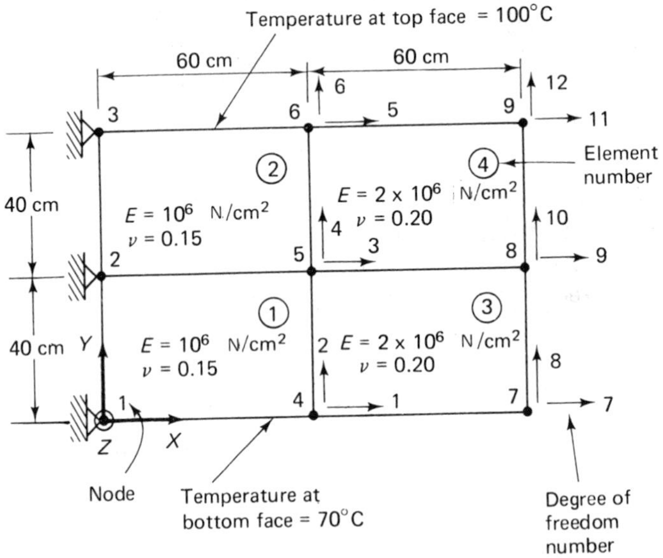

FIGURE A.2 *Finite element cantilever idealization.*

The active degrees of freedom are defined by ID(I, J) = 0, and the nonactive degrees of freedom are defined by ID(I, J) = 1. Since the cantilever is in the X–Y plane, and plane stress elements are used in the idealization, only X and Y translational degrees of freedom are active. By inspection, the ID array is given by

$$\text{ID} = \begin{bmatrix} 1 & 1 & 1 & 0 & 0 & 0 & 0 & 0 & 0 \\ 1 & 1 & 1 & 0 & 0 & 0 & 0 & 0 & 0 \\ 1 & 1 & 1 & 1 & 1 & 1 & 1 & 1 & 1 \\ 1 & 1 & 1 & 1 & 1 & 1 & 1 & 1 & 1 \\ 1 & 1 & 1 & 1 & 1 & 1 & 1 & 1 & 1 \\ 1 & 1 & 1 & 1 & 1 & 1 & 1 & 1 & 1 \end{bmatrix}$$

Once all active degrees of freedom have been defined by zeros in the ID array, the equation numbers corresponding to these degrees of freedom are assigned. The procedure is to simply scan column after column through the ID array and replace each zero by an equation number, which increases successively from 1 to the total number of equations. At the same time the entries corresponding to the non-active degrees of freedom are set to zero.

EXAMPLE A.2: Modify the ID array obtained in Example A.1 for the analysis of the cantilever plate in Fig. A.2 to obtain the ID array that defines the equation numbers corresponding to the active degrees of freedom.

As explained above, we simply replace the zeros, column by column, in succession by equation numbers to obtain

$$
\text{ID} = \begin{bmatrix} 0 & 0 & 0 & 1 & 3 & 5 & 7 & 9 & 11 \\ 0 & 0 & 0 & 2 & 4 & 6 & 8 & 10 & 12 \\ 0 & 0 & 0 & 0 & 0 & 0 & 0 & 0 & 0 \\ 0 & 0 & 0 & 0 & 0 & 0 & 0 & 0 & 0 \\ 0 & 0 & 0 & 0 & 0 & 0 & 0 & 0 & 0 \\ 0 & 0 & 0 & 0 & 0 & 0 & 0 & 0 & 0 \end{bmatrix}
$$

Apart from the definition of all active degrees of freedom, we also need to read the X, Y, Z global coordinates and, if required, the temperature corresponding to each nodal point. For the cantilever beam of Fig. A.2, the X, Y, Z coordinate arrays and nodal point temperature array T would be as follows:

$$
\left.\begin{aligned}
\mathbf{X}^T &= [0.0 \quad 0.0 \quad 0.0 \quad 60.0 \quad 60.0 \quad 60.0 \quad 120.0 \quad 120.0 \quad 120.0] \\
\mathbf{Y}^T &= [0.0 \quad 40.0 \quad 80.0 \quad 0.0 \quad 40.0 \quad 80.0 \quad 0.0 \quad 40.0 \quad 80.0] \\
\mathbf{Z}^T &= [0.0 \quad 0.0 \quad 0.0 \quad 0.0 \quad 0.0 \quad 0.0 \quad 0.0 \quad 0.0 \quad 0.0] \\
\mathbf{T}^T &= [70.0 \quad 85.0 \quad 100.0 \quad 70.0 \quad 85.0 \quad 100.0 \quad 70.0 \quad 85.0 \quad 100.0]
\end{aligned}\right\} \qquad \text{(A.1)}
$$

At this stage, with all the nodal point data known, the program may read and generate the element information. It is expedient to consider each element type in turn. For example, in the analysis of a container structure, all beam elements, all plane stress elements, and all shell elements are read and generated together. This is efficient, because specific information need be provided for each element of a certain type, which, because of its repetitive nature, can be generated to some degree if all elements of the same type are specified together. Furthermore, the element routine for an element type that reads the element data and calculates the element matrices need only be called once.

The required data corresponding to an element depend on the specific element type. In general, the information required for each element is the element node numbers that correspond to the nodal point numbers of the complete element assemblage, the element material properties, and the surface and body forces applied to the element. Since the element material properties and the element loading are the same for many elements, it is efficient to define material property sets and load sets pertaining to an element type. These sets are specified at the beginning of each group of element data. Therefore, any one of the material property sets and element load sets can be assigned to an element at the same time the element node numbers are read.

EXAMPLE A.3: Consider the analysis of the cantilever plate shown in Fig. A.2 and the local element node numbering defined in Fig. 5.5. For each element give the node numbers that correspond to the nodal point numbers of the complete element assemblage. Also indicate the use of material property sets.

In this analysis we define two material property sets: material property set 1 for $E = 10^7 \text{N/cm}^2$ and $\nu = 0.15$, and material property set 2 for $E = 2 \times 10^7 \text{N/cm}^2$ and $\nu = 0.20$. We then have the following node numbers and material property sets for each element:

Element 1: node numbers: 5, 2, 1, 4; material property set: 1
Element 2: node numbers: 6, 3, 2, 5; material property set: 1
Element 3: node numbers: 8, 5, 4, 7; material property set: 2
Element 4: node numbers: 9, 6, 5, 8; material property set: 2

A.2.2 Calculation of Element Stiffness, Mass, and Equivalent Nodal Loads

The general procedure to calculate element matrices was discussed in Chapters 4 and 5 and a computer implementation was presented in Section 5.9. The program organization during this phase consists of calling the appropriate element subroutines for each element. During the element matrix calculations, the element coordinates, properties, and load sets, which have been read and stored in the preceding phase (Section A.2.1), are needed. After calculation, an element matrix may either be stored on back-up storage, because the assemblage to the structure matrices is carried out later,[7] or the element matrix may be added immediately to the appropriate structure matrix.[8]

A.2.3 Assemblage of Structure Matrices

The assemblage process to obtain the structure stiffness matrix $\mathbf{K}$ is symbolically written

$$\mathbf{K} = \sum_i \mathbf{K}^{(i)} \tag{A.2}$$

where the matrix $\mathbf{K}^{(i)}$ is the stiffness matrix of the ith element and the summation goes over all elements in the assemblage. In an analogous manner the structure mass matrix and load vectors are assembled from the element mass matrices and element load vectors, respectively. In addition to the element stiffness, mass, and load matrices, concentrated stiffnesses, masses, and loads corresponding to specific degrees of freedom can also be added.

It should be noted that the element stiffness matrices $\mathbf{K}^{(i)}$ in (A.2) are of the same order as the structure stiffness matrix $\mathbf{K}$. However, considering the internal structure of the matrices $\mathbf{K}^{(i)}$, nonzero elements are only in those rows and columns that correspond to element degrees of freedom (see Section 4.2). Therefore, in practice, we only need to store the compacted element stiffness matrix, which is of order equal to the number of element degrees of freedom, together with an array that relates to each element degree of freedom the corresponding assemblage degree of freedom. This vector is conveniently a connectivity array LM, in which entry i gives the equation number that corresponds to the element degree of freedom i.

EXAMPLE A.4: Using the convention for the element degrees of freedom shown in Fig. 5.5, establish the connectivity arrays defining the assemblage degrees of freedom of the elements in the assemblage of Fig. A.2.

Consider element 1 in Fig. A.2. For this element, nodes 5, 2, 1, and 4 of the element assemblage correspond to the element nodes 1, 2, 3, and 4, respectively (Fig. 5.5). Using the ID array, the equation numbers corresponding to the nodes 5, 2, 1, and 4 of the element assemblage are obtained, and hence the relation between the column (and row) numbers of the compacted, or local, element stiffness matrix and the global stiffness matrix is as follows:

Corresponding column and row numbers

For compacted matrix	1	2	3	4	5	6	7	8
For $\mathbf{K}^{(1)}$	3	4	0	0	0	0	1	2

The array LM, storing the assemblage degrees of freedom of this element, is therefore

$$\mathrm{LM}^T = [3 \quad 4 \quad 0 \quad 0 \quad 0 \quad 0 \quad 1 \quad 2]$$

where a zero means that the corresponding column and row of the compacted element stiffness are ignored and do not enter the global structure stiffness matrix.

Similarly, we can obtain the LM arrays that correspond to the elements 2, 3, and 4. We have for element 2,

$$\mathrm{LM}^T = [5 \quad 6 \quad 0 \quad 0 \quad 0 \quad 0 \quad 3 \quad 4]$$

for element 3,
$$\mathrm{LM}^T = [9 \quad 10 \quad 3 \quad 4 \quad 1 \quad 2 \quad 7 \quad 8]$$

and for element 4,
$$\mathrm{LM}^T = [11 \quad 12 \quad 5 \quad 6 \quad 3 \quad 4 \quad 9 \quad 10]$$

As shown in the example above, the connectivity array of an element is determined from the nodal points to which the element is connected and the equation numbers that have been assigned to these nodal points. Once the array LM has been defined, the corresponding element stiffness matrix can be added in compact form to the structure stiffness matrix **K**, but the process must take due account of the specific storage scheme used for **K**. As already pointed out in Section 1.3, an effective storage scheme for the structure stiffness matrix is to store only the elements below the skyline of the matrix **K** (i.e., the active columns of **K**) in a one-dimensional array A. However, together with the active column storage scheme, we also need a specific procedure of addressing the elements of **K** in A when they are stored as indicated in Section 1.3. Thus, before we are able to proceed with the assemblage of the element stiffness matrices, it is necessary to establish the addresses of the stiffness matrix elements in the one-dimensional array A.

Figure A.3 shows the element pattern of a typical stiffness matrix. Let us derive the storage scheme and addressing procedure that we propose to use. Since the matrix is symmetric, we choose to only store and work on the part above and including the diagonal. However, in addition, we observe that the elements (i, j) of **K** (i.e., k_{ij}) are zero for $j > i + m_K$. The value m_K is known as the *half-bandwidth* of the matrix. Defining by m_i the row number of the first nonzero element in column i (Fig. A.3), the variables m_i, $i = 1, \ldots, n$, define the *skyline* of the matrix, and the variables $(i - m_i)$ are the *column heights.* Furthermore, the half-bandwidth of the stiffness matrix, m_K, equals max $\{i - m_i\}$, $i = 1, \ldots, n$; i.e., m_K *is equal to the maximum difference in global degrees of freedom pertaining to any one of the finite elements in the mesh.* In many finite element analyses, the column heights vary with i, and it is important that all zero elements outside the skyline not be included in the equation solution (see Section 8.2.3). On the other hand, zero elements within the skyline of the matrix are stored and operated upon, since they will become, in many cases, nonzero elements during the matrix reduction.

The column heights are determined from the connectivity arrays, LM, of the elements; i.e., by evaluating m_i, we also obtain the column height $(i - m_i)$. Consider, as an example, that m_{10} of the stiffness matrix that corresponds to the element assemblage in Fig. A.2 is required. The LM arrays of the four elements have been calculated above. We note that only elements 3 and 4

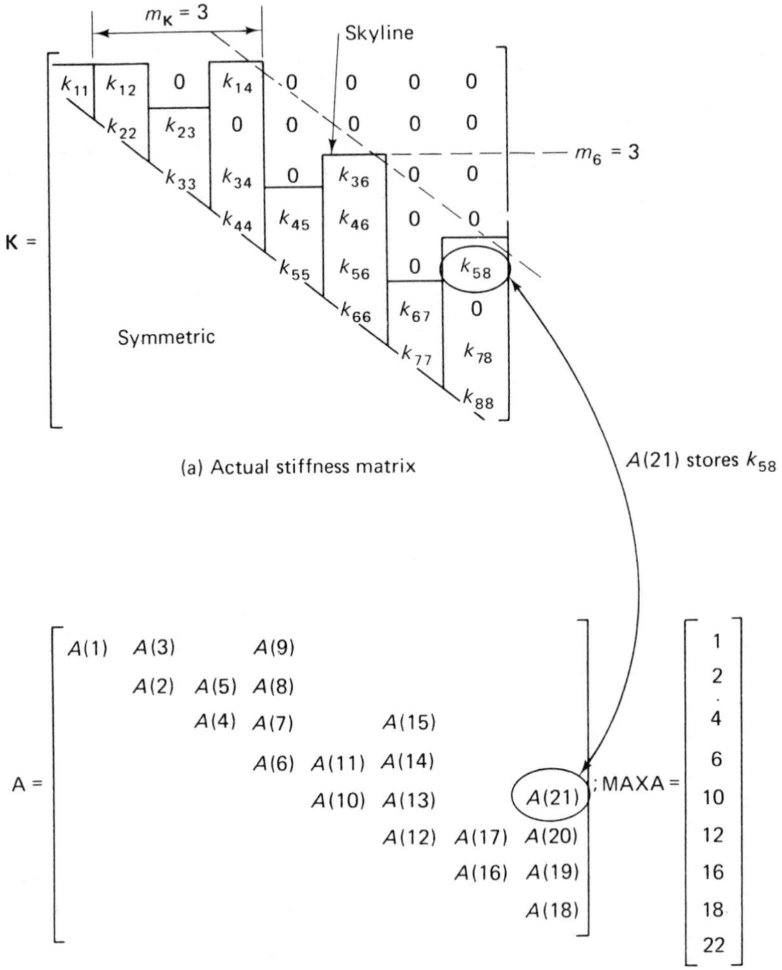

(b) Array A storing elements of K

FIGURE A.3 *Storage scheme used for a typical stiffness matrix.*

couple into degree of freedom 10, and that the smallest number of degree of freedom in the LM array of these elements is 1; hence $m_{10} = 1$ and the column height of column 10 is 9.

With the column heights of a stiffness matrix defined, we can now store all elements below the skyline of **K** as a one-dimensional array in A; i.e., the active columns of **K** including the diagonal elements are stored consecutively in A. Figure A.3 shows which storage locations the elements of the matrix **K** given in the figure would take in A. In addition to A, we also define an array MAXA, which stores the addresses of the diagonal elements of **K** in A; i.e., the address of the ith diagonal element of **K**, k_{ii}, in A is MAXA(I). Referring to Fig. A.3, it is noted that MAXA(I) is equal to the sum of the column heights up to the $(i-1)$st column plus I. Hence the number of nonzero elements in the ith column of **K** is equal to MAXA(I+1) − MAXA(I), and the element addresses are MAXA(I), MAXA(I) +1, MAXA(I)+2, . . . , MAXA(I+1)−1. It follows that using this storage scheme of **K** in A together with the address array MAXA each element of **K** in A can be addressed easily.

The storage scheme described above is used in the computer program STAP presented in Section A.4 and in the equation and eigenvalue solution subroutines of Sections 8.2.3 and 12.3.5. It may also be noted that this active column storage scheme is employed in the ADINA program.[8] The effectiveness of the scheme lies essentially in that no elements outside the skyline are stored and processed in the calculations.

In the discussion of algorithms for the solution of the equations $\mathbf{KU} = \mathbf{R}$, where **K**, **U**, and **R** are the stiffness matrix, the displacement vector, and the load vector of the element assemblage, respectively, we pointed out that the solution procedures most commonly used require about $\frac{1}{2}nm_K^2$ operations, where n is the order of the stiffness matrix, m_K is its half-bandwidth, and we assume constant column heights, i.e., $(i - m_i) = m_K$ for nearly all i. Therefore, it can be important to minimize m_K both from considerations of storage requirements and number of operations. In case the column heights vary, a mean or "effective" value for m_K must be used (see Section 8.2.3). In practice we can frequently determine a reasonable nodal point numbering by inspection. However, this nodal point numbering may not be particularly easy to generate, and various automatic schemes are currently used for bandwidth reduction.[9,10] Figure A.4 shows a typical good and bad nodal point numbering.

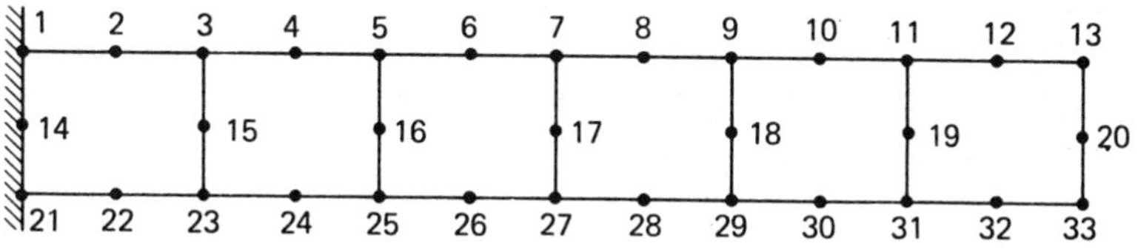

(a) Bad nodal point numbering, $m_K + 1 = 46$

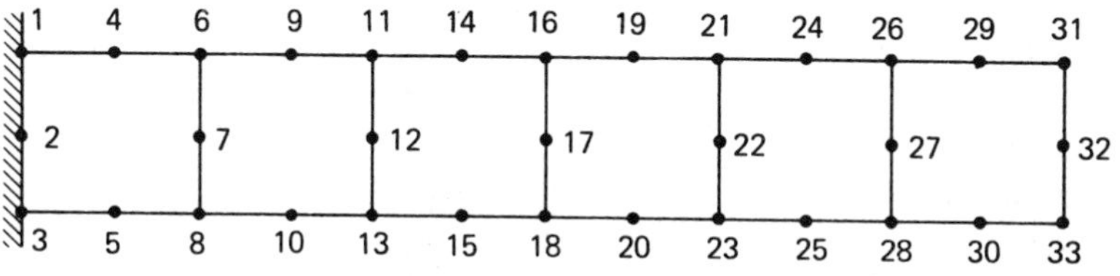

(b) Good nodal point numbering, $m_K + 1 = 16$

FIGURE A.4 *Bad and good nodal point numbering for finite element assemblage.*

It should be pointed out that in the discussion of the above storage scheme, we implicitly assumed that the entire array A (i.e., the sum of all active columns of the matrix **K**) does fit into the available high speed storage of the computer. For instructional purposes it is deemed to be most appropriate to concentrate on in-core solution, although in practice large systems are solved by storing the matrices in blocks.[5-8,11] Considering out-of-core solution, in principle, the same storage scheme as for in-core solutions is effective.[8] The main additional problem is one of program logistics; namely, the individual blocks of the matrices must be stored on back-up storage and called into high-speed storage in an effective manner. Various procedures are in use, and in all cases, specific attention has been given to minimize the amount of tape or disc reading required in the assemblage. However, once an effective in-core finite element solution has been studied, little difficulty should be encountered in understanding specific out-of-core implementations.

A.3 CALCULATION OF ELEMENT STRESSES

In the previous section we described the process of assembling the individual finite element matrices into the total structure matrices. The next step is the calculation of nodal point displacements using the procedures discussed in Chapters 8 and 9. Once the nodal point displacements have been obtained, element stresses are calculated in the final phase of the analysis.

The equations employed in the element stress calculations are (4.10) and (4.11). However, as in the assemblage of the structure matrices, it is again effective to manipulate compacted finite element matrices, i.e., only deal with the nonzero columns of $\mathbf{B}^{(m)}$ in (4.10). Using the implementation described in the previous section, we calculate the element compacted strain-displacement transformation matrix and then extract the element nodal point displacements from the total displacement vector using the LM array of the element. The procedure is implemented in program STAP described next. It should be noted that finite element stresses can be calculated at any desired location by simply establishing the strain-displacement transformation matrix for the point under consideration in the element. In isoparametric finite element analysis we use the procedures given in Chapter 5.

A.4 EXAMPLE PROGRAM STAP

Probably the best way of getting familiar with the implementation of finite element analysis is to study an actual computer program that, although simplified in various areas, shows all the important features of more general codes. The following program, STAP (STatic Analysis Program), is a simple computer program for static linear elastic finite element analysis.

The main objective in the presentation of the program is to show the overall flow of a typical finite element analysis program, and for this reason only a truss element has been made available in STAP. However, the code can generally be used for one-, two- and three-dimensional analysis, and additional elements can be added with relative ease.*

Figure A.5 shows a flow chart of the program, and Figure A.6 gives the storage allocations used during the various program phases. Before presenting the listing of the program, we give the input manual. It may be noted that the coding and the input manual of program STAP are similar to the procedures used in ADINA.[8]

* The program has been developed on a CDC Cyber 175 and an IBM 370 machine, and to implement it on other equipment some changes may be necessary.

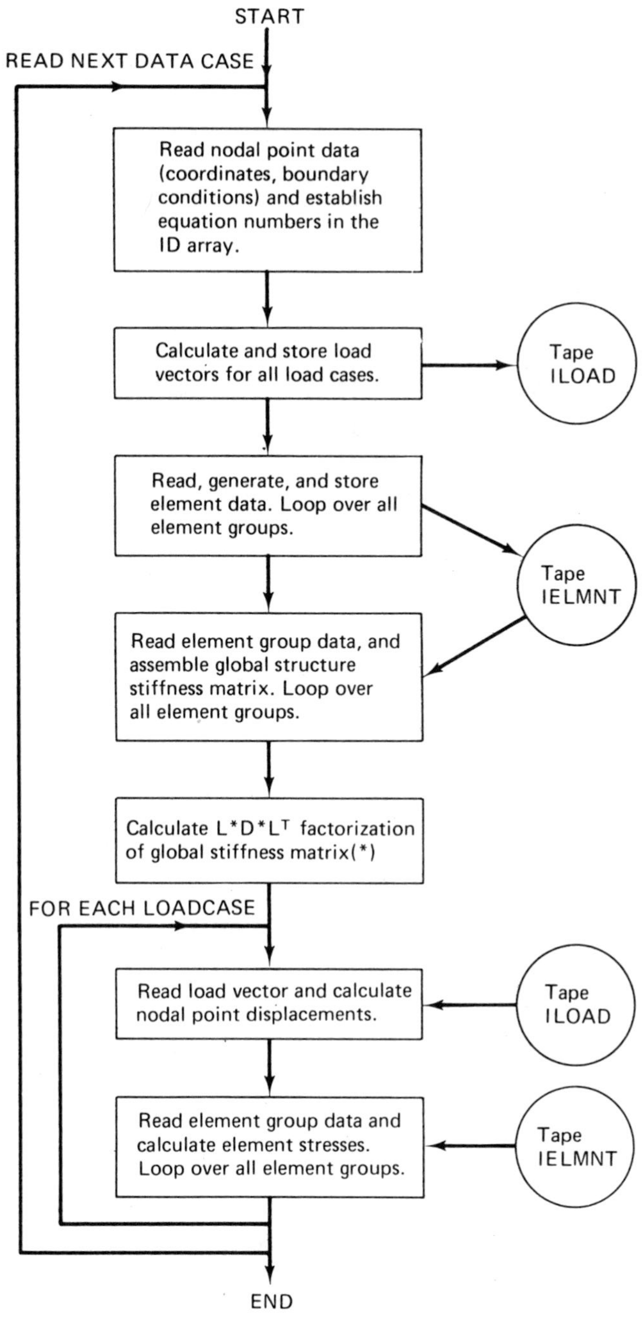
START
READ NEXT DATA CASE
Read nodal point data (coordinates, boundary conditions) and establish equation numbers in the ID array.
Calculate and store load vectors for all load cases.
Tape ILOAD
Read, generate, and store element data. Loop over all element groups.
Tape IELMNT
Read element group data, and assemble global structure stiffness matrix. Loop over all element groups.
Calculate L*D*L^T factorization of global stiffness matrix(*)
FOR EACH LOADCASE
Read load vector and calculate nodal point displacements.
Tape ILOAD
Read element group data and calculate element stresses. Loop over all element groups.
Tape IELMNT
END

FIGURE A.5 *Flow chart of program STAP.*

* The equation solver used is COLSOL described in Section 8.2.2.

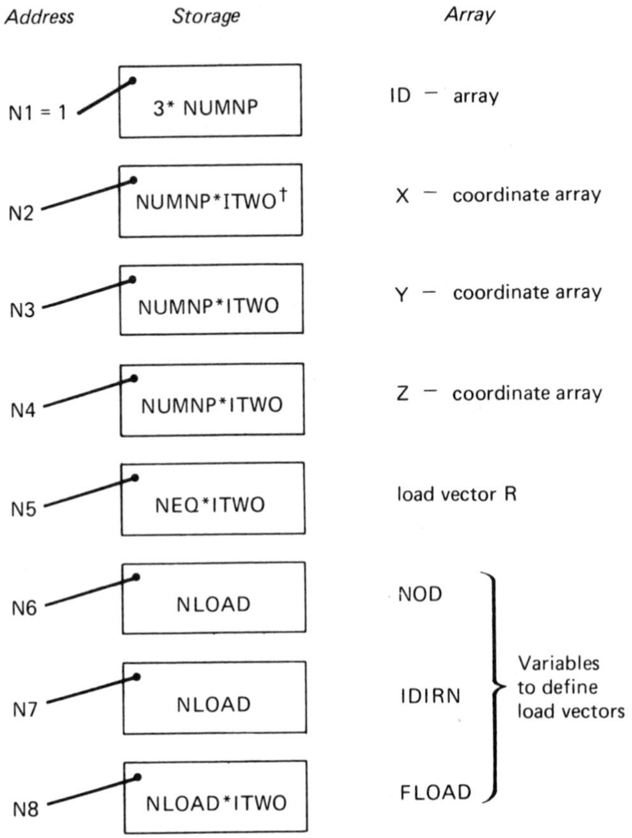

a. Input of ID array, nodal point coordinates, and load vectors.

FIGURE A.6 *High-speed storage allocation in program STAP. (See user's manual for definition of variables.) ITWO = 1 in single precision arithmetic. ITWO = 2 in double precision arithmetic.*

A.4.1 Data Input to Computer Program STAP

I. HEADING CARD (20A4)

Note	*Columns*	*Variable*	*Entry*
(1)	1–80	HED (20)	Enter the master heading information for use in labeling the output

NOTES/

1. Begin each new data case with a new heading card. Two blank cards must be input after the last data case.

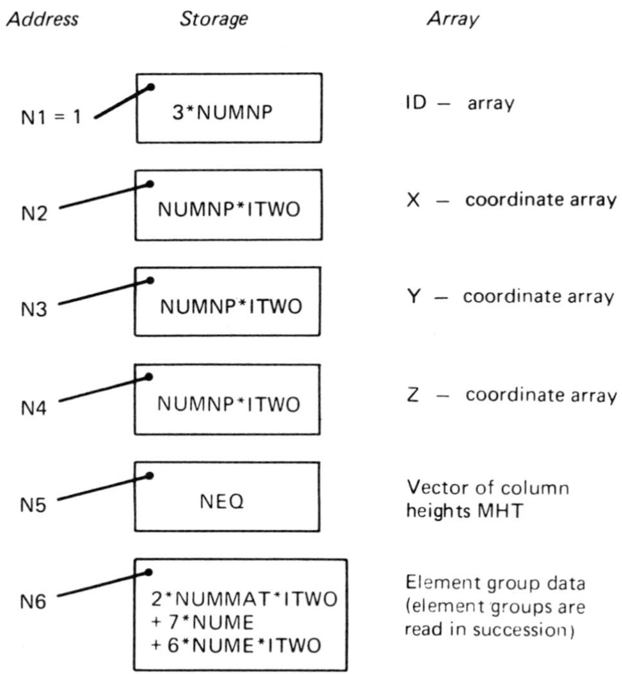

b. Element data input.

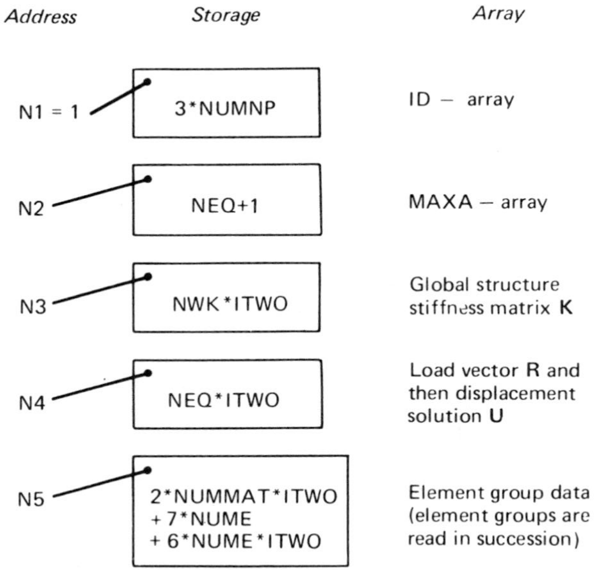

c. Assemblage of global structure stiffness; displacement and stress solution phase.

FIGURE A.6 *Continued.*

II. CONTROL CARD (4I5)

Note	*Columns*	*Variable*	*Entry*
(1)	1–5	NUMNP	Total number of nodal points; EQ.0, program stop
(2)	6–10	NUMEG	Total number of element groups, GT.0
(3)	11–15	NLCASE	Number of load cases, GT.0
(4)	16–20	MODEX	Flag indicating solution mode; EQ.0, data check only EQ.1, execution

NOTES/

1. The total number of nodes (NUMNP) controls the amount of data to be read in Section III. If NUMNP.EQ.0 the program stops.
2. The total number of elements are dealt with in element groups. An element group consists of a convenient collection of elements. Each element group is input as given in Section V. There must be at least one element per element group, and there must be at least one element group.
3. The number of load cases (NLCASE) gives the number of load vectors for which the displacement and stress solution is sought.
4. The MODEX parameter determines whether the program is to check the data without executing the analysis (i.e., MODEX.EQ.0) or if the program is to solve the problem (i.e., MODEX.EQ.1). In the data check only mode, the program only reads and prints all data.

III. NODAL POINT DATA CARDS (4I5, 3F10.0, I5)

Note	*Columns*	*Variable*	*Entry*
(1)	1–5	N	Node (joint) number; GE.1 and LE.NUMNP
(2)	6–10	ID (1, N)	X—translation boundary code
	11–15	ID (2, N)	Y—translation boundary code
	16–20	ID (3, N)	Z—translation boundary code
(3)	21–30	X(N)	X—coordinate
	31–40	Y(N)	Y—coordinate
	41–50	Z(N)	Z—coordinate
(4)	51–55	KN	Node number increment for node data generation; EQ.0, no generation

NOTES/

1. Nodal data must be defined for all (NUMNP) nodes. Node data may be input directly (i.e., each node on its own individual card) or the generation option may be used if applicable (see note 4 below). Admissible node numbers range from 1 to the total number of nodes (NUMNP). The last node that is input must be NUMNP.

2. Boundary condition codes can only be assigned the following values (M = 1, 2, 3)

   ID(M,N) = 0; unspecified (free) displacement.

   ID(M,N) = 1; deleted (fixed) displacement.

   An unspecified [ID(M,N) = 0] degree of freedom is free to translate as the solution dictates. Concentrated forces may be applied in this degree of freedom.

   One system equilibrium equation is established for each unspecified degree of freedom in the model. The total number of equilibrium equations is defined as NEQ and is always less than three times the total number of nodes in the system.

   Deleted [ID(M,N) = 1] degrees of freedom are removed from the final set of equilibrium equations. Deleted degrees of freedom are used to define fixities (points of external reaction), and any loads applied in these degrees of freedom are ignored by the program.
3. The geometric location of each node is specified by its X, Y, and Z coordinates.
4. Node cards need not be input in node order sequence; eventually, however, all nodes in the set [1,NUMNP] must be defined. Node data for a series of nodes

   $$[N_1, N_1 + 1*KN_1, N_1 + 2*KN_1, \ldots, N_2]$$

   may be generated from information given on two cards in sequence—

   CARD 1 – N_1, ID(1, N_1), . . . , X(N_1), . . . , KN_1

   CARD 2 – N_2, ID(1, N_2), . . . , X(N_2), . . . , KN_2

   KN_1 is the node generation parameter given on the first card in the sequence. The first generated node is N_1+1*KN_1; the second generated node is N_1+2*KN_1, etc. Generation continues until node number N_2-KN_1 is established. Note that the node difference N_2-N_1 must be evenly divisible by KN_1.

   In the generation the boundary condition codes [ID(L,J) values] of the generated nodes are set equal to those of node N_1. The coordinate values of the generated nodes are interpolated linearly.

IV. LOAD DATA CARDS

Each load case requires the following set of cards. The total number of load cases were defined on the CONTROL CARD (Section II).

CARD 1 (2I5)

Note	*Columns*	*Variable*	*Entry*
(1)	1–5	LL	Enter the load case number
(2)	6–10	NLOAD	Enter the number of concentrated loads applied in this load case

NOTES/

1. Load cases must be input in ascending sequence beginning with 1.
2. The variable NLOAD defines the number of cards to be read next for this load case.

NEXT CARDS (2I5, F10.0)

Note	Columns	Variable	Entry
(1)	1–5	NOD	Node number to which this load is applied; GE.1 and LE.NUMNP
(2)	6–10	IDIRN	Degree of freedom number for this load component; EQ.1, X—direction EQ.2, Y—direction EQ.3, Z—direction
	11–20	FLOAD	Magnitude of load

NOTES/

1. For each concentrated load applied in this load case, one card must be supplied.
2. All loads must be acting into the global X, Y, or Z direction.

V. TRUSS ELEMENTS

TRUSS elements are two-node members allowed arbitrary orientation in the X, Y, Z system. The truss transmits axial force only, and in general is a six degree of freedom element (i.e., three global translation components at each end of the member). The following sequence of cards is input for each element group. The total number of element groups (NUMEG) was defined on the CONTROL CARD (Section II).

V.1 Element Group Control Card (3I5)

Note	Columns	Variable	Entry
	1–5	NPAR(1)	Enter the number 1
(1)	6–10	NPAR(2)	Number of TRUSS elements in this group; NPAR(2) = NUME GE.1
(2)	11–15	NPAR(3)	Number of different sets of section/material properties; NPAR(3) = NUMMAT GE.1 EQ.0, default set to 1

NOTES/

1. TRUSS element numbers begin with one (1) and end with the total number of elements in this group, NPAR(2). Element data are input in Section V.3.

2. The variable NPAR(3) defines the number of sets of material/section properties to be read in Section V.2 below.

V.2 Material/Section Property Cards (I5, 2F10.0)

NUMMAT cards are read in this section.

Note	*Columns*	*Variable*	*Entry*
(1)	1–5	N	Number of property set
	6–15	E(N)	Young's modulus
	16–25	AREA(N)	Section area

NOTES/

1. Property sets are input in ascending sequence beginning with 1 and ending with NUMMAT. The Young's modulus and section area of each TRUSS element defined below is identified by one of the property sets input here.

V.3 Element Data Cards (5I5)

NUME elements must be input and/or generated in this section in ascending sequence beginning with one (1).

Note	*Columns*	*Variable*	*Entry*
	1–5	M	TRUSS element number; GE.1 and LE.NUME
	6–10	II	Node number at one end
	11–15	JJ	Node number at other end; GE.1 and LE.NUMNP
(1)	16–20	MTYP	Material property set; GE.1 and LE. NUMMAT
(2)	21–25	KG	Node generation increment used to compute node numbers for missing elements; EQ.0, default set to 1

NOTES/

1. The material/section property sets have been defined in Section V.2 above.
2. Elements must be input in increasing element number order. If cards for elements [M+1, M+2, . . . , M+J] are omitted, these "J" missing elements are generated using MTYP of element "M" and by incrementing the node numbers of successive elements with the value KG; KG is taken from the first card of the element generation sequence (i.e., from the Mth element card). The last element (NUME) must always be input.

A.4.2 Listing of Program STAP

```
C . . . . . . . . . . . . . . . . . . . . . . . . . . . . . . . . . . . .
C .                                                                  .
C .                           S T A P                                .
C .                                                                  .
C .         AN IN-CORE SOLUTION STATIC ANALYSIS PROGRAM              .
C .                                                                  .
C . . . . . . . . . . . . . . . . . . . . . . . . . . . . . . . . . . . .
      COMMON /SOL/ NUMNP,NEQ,NWK,NUMEST,MIDEST,MAXEST,MK
      COMMON /DIM/ N1,N2,N3,N4,N5,N6,N7,N8,N9,N10,N11,N12,N13,N14,N15
      COMMON /EL/ IND,NPAR(10),NUMEG,MTOT,NFIRST,NLAST,ITWO
      COMMON /VAR/ NG,MODEX
      COMMON /TAPES/ IELMNT,ILOAD,IIN,IOUT
C
      DIMENSION TIM(5), HED(20)
      DIMENSION IA(1)
      EQUIVALENCE (A(1),IA(1))
C . . . . . . . . . . . . . . . . . . . . . . . . . . . . . . . . . . . .
C     THE FOLLOWING TWO CARDS ARE USED TO DETERMINE THE MAXIMUM HIGH
C     SPEED STORAGE THAT CAN BE USED FOR SOLUTION. TO CHANGE THE HIGH
C     SPEED STORAGE AVAILABLE FOR EXECUTION CHANGE THE VALUE OF MTOT
C     AND CORRESPONDINGLY COMMON A(MTOT)
C . . . . . . . . . . . . . . . . . . . . . . . . . . . . . . . . . . . .
      COMMON A(10000)
      MTOT=10000
C
C . . . . . . . . . . . . . . . . . . . . . . . . . . . . . . . . . . . .
C     DOUBLE PRECISION CARD
C       ITWO = 1 SINGLE PRECISION ARITHMETIC
C       ITWO = 2 DOUBLE PRECISION ARITHMETIC
C . . . . . . . . . . . . . . . . . . . . . . . . . . . . . . . . . . . .
C
      ITWO=2
C
C     THE FOLLOWING SCRATCH FILES ARE USED
C        IELMNT = TAPE STORING ELEMENT DATA
C        ILOAD  = TAPE STORING LOAD VECTORS
C        IIN    = INPUT TAPE
C        IOUT   = OUTPUT TAPE
C
      IELMNT = 1
      ILOAD = 2
      IIN = 5
      IOUT = 6
C
  200 NUMEST=0
      MAXEST=0
C     * * * * * * * * * * * * * * * * * * * * * * * * *
C
C     * * *   I N P U T   P H A S E   * * *
C
C     * * * * * * * * * * * * * * * * * * * * * * * * *
      CALL SECOND (TIM(1))
C
C
C
C     R E A D   C O N T R O L   I N F O R M A T I O N
C
C
      READ (IIN,1000)   HED,NUMNP,NUMEG,NLCASE,MODEX
      IF (NUMNP.EQ.0) STOP
      WRITE (IOUT,2000) HED,NUMNP,NUMEG,NLCASE,MODEX
C
C
C     R E A D   N O D A L   P O I N T   D A T A
C
C
      N1= 1
      N2=N1 + 3 * NUMNP
      N3=N2 + NUMNP *ITWO
      N4=N3 + NUMNP*ITWO
      N5=N4 + NUMNP*ITWO
      IF (N5.GT.MTOT) CALL ERROR (N5 - MTOT,1)
C
      CALL INPUT (A(N1),A(N2),A(N3),A(N4),NUMNP,NEQ)
C
      NEQ1=NEQ + 1
C
C     C A L C U L A T E   A N D   S T O R E   L O A D   V E C T O R S
C
C
      N6=N5 + NEQ*ITWO
      WRITE(IOUT,2005)
C
      REWIND ILOAD
C
      DO 300 L=1,NLCASE
C
      READ (IIN,1010)   LL,NLOAD
C
```

```
      WRITE (IOUT,2010) LL,NLCAD
      IF (LL.EQ.L) GO TO 310
      WRITE (IOUT,2020)
      STOP
  310 CONTINUE
C
      N7=N6 + NLOAD
      N8=N7 + NLOAD
      N9=N8 + NLOAD*ITWO
C
      IF (N9.GT.MTOT) CALL ERROR (N9 - MTOT,2)
C
      CALL LOADS (A(N5),A(N6),A(N7),A(N8),A(N1),NLOAD,NEQ)
C
  300 CONTINUE
C
C
C     R E A D , G E N E R A T E   A N D   S T O R E
C     E L E M E N T   D A T A
C
C     CLEAR STORAGE
C
      N6=N5 + NEQ
      DC 10 I=N5,N6
   10 IA(I)=0
      IND=1
C
      CALL ELCAL
C
      CALL SECOND (TIM(2))
C     * * * * * * * * * * * * * * * * * * * * * *
C
C     * * *   S O L U T I O N   P H A S E   * * *
C
C     * * * * * * * * * * * * * * * * * * * * * *
C
C     A S S E M B L E   S T I F F N E S S   M A T R I X
C
C
      CALL ADDRES (A(N2),A(N5))
C
      MM=NWK/NEQ
      N3=N2 + NEQ + 1
      N4=N3 + NWK*ITWO
      N5=N4 + NEQ*ITWO
      N6=N5 + MAXEST
      IF (N6.GT.MTOT) CALL ERROR (N6 - MTOT,4)
C
C     WRITE TOTAL SYSTEM DATA
C
      WRITE(ICUT,2025) NEQ,NWK,MK,MM
C
C     IN DATA CHECK ONLY MODE WE SKIP ALL FURTHER CALCULATIONS
C
      IF (MODEX.GT.0) GO TO 100
      CALL SECOND (TIM(3))
      CALL SECOND (TIM(4))
      CALL SECOND (TIM(5))
      GO TO 120


C
C     CLEAR STORAGE
C
  100 NNL=NWK + NEQ
      CALL CLEAR(A(N3),NNL)
C
C
      IND=2
C
      CALL ASSEM (A(N5))
C
      CALL SECOND (TIM(3))
C
C     T R I A N G U L A R I Z E   S T I F F N E S S   M A T R I X
C
      KTR=1
      CALL COLSCL (A(N3),A(N4),A(N2),NEQ,NWK,NEQ1,KTR)
C
   35 CALL SECCND (TIM(4))
C
      KTR=2
      IND=3
C
      REWIND ILCAD
      DO 400 L=1,NLCASE
C
      CALL LOADV (A(N4),NEQ)
C
C
C     C A L C U L A T I O N   C F   D I S P L A C E M E N T S
C
C
```

```
      CALL COLSCL (A(N3),A(N4),A(N2),NEQ,NWK,NEQ1,KTR)          STAPO181
C                                                               STAPO182
      WRITE(IOUT,2015) L                                        STAPO183
      CALL WRITE (A(N4),A(N1),NEQ,NUMNP)                        STAPO184
C                                                               STAPO185
C                                                               STAPO186
C     C A L C U L A T I O N   O F   S T R E S S E S             STAPO187
C                                                               STAPO188
C                                                               STAPO189
      CALL STRESS (A(N5))                                       STAPO190
C                                                               STAPO191
  400 CONTINUE                                                  STAPO192
C                                                               STAPO193
      CALL SECCND (TIM(5))                                      STAPO194
C                                                               STAPO195
C     PRINT SOLUTION TIMES                                      STAPO196
C                                                               STAPO197
  120 TT=0.                                                     STAPO198
      DO 500 I=1,4                                              STAPO199
      TIM(I)=TIM(I+1) - TIM(I)                                  STAPO200
  500 TT=TT + TIM(I)                                            STAPO201
      WRITE (IOUT,2030) HED,(TIM(I),I=1,4),TT                   STAPO202
C                                                               STAPO203
C     READ NEXT ANALYSIS CASE                                   STAPO204
C                                                               STAPO205
      GO TO 200                                                 STAPO206
C                                                               STAPO207
 1000 FORMAT (20A4/4I5)                                         STAPO208
 1010 FORMAT (2I5)                                              STAPO209
C                                                               STAPO210
C                                                               STAPO211
 2000 FORMAT (1H1,20A4 ///                                      STAPO212
     155H C O N T R O L   I N F O R M A T I O N       //5X,     STAPO213
     255HNUMBER OF NODAL POINTS . . . . . . . . . (NUMNP)  =,I5//5X,  STAPO214
     355HNUMBER OF ELEMENT GROUPS . . . . . . . . (NUMEG)  =,I5//5X,  STAPO215
     455HNUMBER OF LOAD CASES . . . . . . . . . . (NLCASE)  =,I5//5X, STAPO216
     555HSOLUTION MODE . . . . . . . . . . . . . (MODEX)  =,I5 /5X,   STAPO217
     655H      EQ.0, DATA CHECK                       /5X,    STAPO218
     755H      EQ.1, EXECUTION                           )    STAPO219
 2005 FORMAT (1H1,26H L O A D   C A S E   D A T A )            STAPO220
 2010 FORMAT (////4X,33H LOAD CASE NUMBER . . . . . . . =,I5//5X,  STAPO221
     132HNUMBER OF CONCENTRATED LOADS . = ,I5)                 STAPO222
 2015 FORMAT(1H1,9HLOAD CASE,I3)                               STAPO223
 2020 FORMAT (1X,40H*** ERROR  LOAD CASES ARE NOT IN ORDER  )  STAPO224
 2025 FORMAT(1H1,                                              STAPO225
     155HTOTAL SYSTEM DATA                        ///5X,       STAPO226
     255HNUMBER OF EQUATIONS . . . . . . . . . . . (NEQ)  =,I5//5X,  STAPO227
     355HNUMBER OF MATRIX ELEMENTS . . . . . . . . (NWK)  =,I5//5X,  STAPO228
     455HMAXIMUM HALF BANDWIDTH . . . . . . . . . (MK )  =,I5//5X,   STAPO229
     555HMEAN HALF BANDWIDTH . . . . . . . . . . . (MM )  =,I5)      STAPO230
 2030 FORMAT (1H1,48H S O L U T I O N   T I M E   L O G   I N   S E C // STAPO231
     112X,11HFOR PROBLEM//1X,20A4 ////5X,                      STAPO232
     251HTIME FOR INPUT PHASE . . . . . . . . . . . = ,F12.2//5X,  STAPO233
     351HTIME FCR CALCULATION OF STRUCTURE STIFFNESS MATRIX= ,F12.2//5X, STAPO234
     451HTRIANGULARIZATION OF STIFFNESS MATRIX . . . . . =,F12.2//5X,  STAPO235
     551HTIME FOR LOAD CASE SOLUTICNS . . . . . . . . . =,F12.2///5X,  STAPO236
     651H T O T A L   S O L U T I C N   T I M E . . . . . = ,F12.2)   STAPO237
C                                                               STAPO238
      END                                                      STAPO239


      SUBROUTINE ERROR(N,I)                                     EROR0001
C . . . . . . . . . . . . . . . . . . . . . . . . . . . . . . . EROR0002
C .                                                           : EROR0003
C .   P R O G R A M                                           . EROR0004
C .        TO PRINT MESSAGES WHEN HIGH-SPEED STORAGE IS EXCEEDED . EROR0005
C :                                                           , EROR0006
C . . . . . . . . . . . . . . . . . . . . . . . . . . . . . . . EROR0007
      CCMMON /TAPES/ IELMNT,ILOAD,IIN,IOUT                      EROR0008
C                                                               EROR0009
      GO TO (1,2,3,4),I                                         EROR0010
C                                                               EROR0011
    1 WRITE(IOUT,2000)                                          EROR0012
      GO TO 6                                                   EROR0013
    2 WRITE(IOUT,2010)                                          EROR0014
      GO TO 6                                                   EROR0015
    3 WRITE(IOUT,2020)                                          EROR0016
      GO TO 6                                                   EROR0017
    4 WRITE(IOUT,2030)                                          EROR0018
C                                                               EROR0019
    6 WRITE (IOUT,2050) N                                       EROR0020
      STOP                                                      EROR0021
 2000 FORMAT (//48H NOT ENOUGH STORAGE FOR READ-IN OF ID ARRAY AND EROR0022
     123HNODAL POINT COORDINATES)                              EROR0023
 2010 FORMAT (//50H NOT ENOUGH STORAGE FOR DEFINITION OF LOAD VECTORS) EROR0024
 2020 FORMAT (//42H NOT ENOUGH STORAGE FOR ELEMENT DATA INPUT ) EROR0025
 2030 FORMAT (//53H NOT ENOUGH STORAGE FOR ASSEMBLAGE OF GLOBAL STRUCTUR EROR0026
     155HE STIFFNESS, AND DISPLACEMENT AND STRESS SOLUTION PHASE ) EROR0027
 2050 FORMAT (// 32H *** ERROR   STORAGE EXCEEDED BY, I9)       EROR0028
      END                                                      EROR0029

                                                               INPT0001
      SUBROUTINE INPUT (ID,X,Y,Z,NUMNP,NEQ)
```

```
C
C .  .  .  .  .  .  .  .  .  .  .  .  .  .  .  .  .  .  .  .  .  .  .  .  .  .  .  .  .  .  .  .  .  .  .
C .                                                                                              .
C .      P R O G R A M                                                                           .
C .         .TO READ, GENERATE, AND PRINT NODAL POINT INPUT DATA                                 .
C .          . TO CALCULATE EQUATION NUMBERS AND STORE THEM IN ID ARRRAY                         .
C .                                                                                              .
C .            N=ELEMENT NUMBER                                                                  .
C .            ID=BOUNDARY CONDITICN CODES (0=FREE,1=DELETED)                                    .
C .            X,Y,Z= COORDINATES                                                                .
C .            KN= GENERATION CODE                                                               .
C .                      I.E. INCREMENT ON NODAL POINT NUMBER                                    .
C .                                                                                              .
C .  .  .  .  .  .  .  .  .  .  .  .  .  .  .  .  .  .  .  .  .  .  .  .  .  .  .  .  .  .  .  .  .  .  .
C
      IMPLICIT REAL*8(A-H,O-Z)
C .  .  .  .  .  .  .  .  .  .  .  .  .  .  .  .  .  .  .  .  .  .  .  .  .  .  .  .  .  .  .  .  .
C .           THIS PROGRAM IS USED IN SINGLE PRECISION ARITHMETIC ON              .
C .           CDC EQUIPMENT AND DCUBLE PRECISION ARITHMETIC ON IBM                .
C .           OR UNIVAC MACHINES .ACTIVATE,DEACTIVATE OR ADJUST ABOVE             .
C .           CARD FOR SINGLE OR DOUBLE PRECISION ARITHMETIC                      .
C .  .  .  .  .  .  .  .  .  .  .  .  .  .  .  .  .  .  .  .  .  .  .  .  .  .  .  .  .  .  .  .  .
      COMMON /TAPES/ IELMNT,ILOAD,IIN,IOUT
      DIMENSION X(1),Y(1),Z(1),ID(3,NUMNP)
C
C
C       READ AND GENERATE NODAL POINT DATA
C
      WRITE (IOUT,2000)
      WRITE (IOUT,2010)
      WRITE(IOUT,2020)
      KNOLD=0
      NOLD=0
C
   10 READ   (IIN,1000)  N,(ID(I,N),I=1,3),X(N),Y(N),Z(N),KN
      WRITE (IOUT,2030) N,(ID(I,N),I=1,3),X(N),Y(N),Z(N),KN
      IF (KNOLD.EQ.0) GO TO 50
      NUM=(N-NOLD) / KNOLD
      NUMN=NUM-1
      IF(NUMN.LT.1) GO TO 50
      XNUM=NUM
      DX=(X(N)-X(NOLD))/XNUM
      DY=(Y(N)-Y(NOLD))/XNUM
      DZ=(Z(N)-Z(NOLD))/XNUM
      K=NOLD
      DO 30 J=1,NUMN
      KK=K
      K=K + KNOLD
      X(K)=X(KK)+DX
      Y(K)=Y(KK)+DY
      Z(K)=Z(KK)+DZ
      DO 30 I=1,3
      ID(I,K)=IC(I,KK)
   30 CONTINUE
C
   50 NOLD=N
      KNOLD=KN
      IF(N.NE.NUMNP) GO TO 10
C
C       WRITE COMPLETE NODAL DATA
C
      WRITE (IOUT,2015)
      WRITE (IOUT,2020)
      DC 200 N=1,NUMNP
  200 WRITE (IOUT,2030) (N,(IC(I,N),I=1,3),X(N),Y(N),Z(N),KN)
C
C       NUMBER UNKNOWNS
C
      NEQ=0
      DO 100 N=1,NUMNP
      DO 100 I=1,3
      IF (ID(I,N)) 110,120,110
  120 NEC=NEQ + 1
      IC(I,N)=NEQ
      GO TO 100
  110 ID(I,N)=0
  100 CONTINUE
C
C       WRITE EQUATION NUMBERS
C
      WRITE (IOUT,2040) (N,(IC(I,N),I=1,3),N=1,NUMNP)
      RETURN
C
 1000 FORMAT (4I5,3F10.0,I5)
 2000 FORMAT(1H1,33H N O D A L   P O I N T   D A T A    //)
 2010 FORMAT(18H INPUT NODAL CATA //)
 2015 FORMAT(///22H GENERATED NODAL DATA //)
 2020 FORMAT(7H  NODE  ,9X,8HBCUNDARY,25X,11HNODAL POINT,17X,
     1 4HMESH/7H NUMBER,5X,16HCONDITION  CODES,21X,11HCOORDINATES,14X,
     2 1CHGENERATING/77X,4HCODE//15X,1HX,4X,1HY,4X,1HZ,15X,1HX,12X,1HY,
     3 12X,1HZ,10X,2HKN)
```

Right margin line numbers:
INPT0002
INPT0003
INPT0004
INPT0005
INPT0006
INPT0007
INPT0008
INPT0009
INPT0010
INPT0011
INPT0012
INPT0013
INPT0014
INPT0015
INPT0016
INPT0017
INPT0018
INPT0019
INPT0020
INPT0021
INPT0022
INPT0023
INPT0024
INPT0025
INPT0026
INPT0027
INPT0028
INPT0029
INPT0030
INPT0031
INPT0032
INPT0033
INPT0034
INPT0035
INPT0036
INPT0037
INPT0038
INPT0039
INPT0040
INPT0041
INPT0042
INPT0043
INPT0044
INPT0045
INPT0046
INPT0047
INPT0048
INPT0049
INPT0050
INPT0051
INPT0052
INPT0053
INPT0054
INPT0055
INPT0056
INPT0057
INPT0058
INPT0059
INPT0060
INPT0061
INPT0062
INPT0063
INPT0064
INPT0065
INPT0066
INPT0067
INPT0068
INPT0069
INPT0070
INPT0071
INPT0072
INPT0073
INPT0074
INPT0075
INPT0076
INPT0077
INPT0078
INPT0079
INPT0080
INPT0081
INPT0082
INPT0083
INPT0084
INPT0085
INPT0086
INPT0087
INPT0088
INPT0089
INPT0090
INPT0091
INPT0092
INPT0093

```
 2030 FORMAT (I5,6X,3I5,6X,3F13.3,3X,I6)
 2040 FCRMAT(//17H EQUATION NUMBERS//,4X,4HNODE,9X,
     1 17HDEGREE OF FREEDOM/3X,6HNUMBER//,
     2 5X,1HN,13X,1HX,4X,1HY,4X,1HZ/(1X,I5,9X,3I5))
C
      END

      SUBROUTINE LOADS (R,NOD,IDIRN,FLOAD,ID,NLOAD,NEQ)
C
C
C . . . . . . . . . . . . . . . . . . . . . . . . . . . . . . . . . . . .
C .                                                                      .
C .   P R O G R A M                                                      .
C .      . TO READ NODAL LOAD DATA                                       .
C .      . TO CALCULATE THE LOAD VECTOR R FOR EACH LOAD CASE AND         .
C .        WRITE ONTO TAPE ILOAD                                         .
C .                                                                      .
C . . . . . . . . . . . . . . . . . . . . . . . . . . . . . . . . . . . .
C
C
      IMPLICIT REAL*8(A-H,O-Z)
C . . . . . . . . . . . . . . . . . . . . . . . . . . . . . . . . . . . .
C .        THIS PROGRAM IS USED IN SINGLE PRECISION ARITHMETIC ON        .
C .        CDC EQUIPMENT AND DOUBLE PRECISION ARITHMETIC ON IBM          .
C .        OR UNIVAC MACHINES .ACTIVATE,DEACTIVATE OR ADJUST ABOVE       .
C .        CARD FOR SINGLE OR DOUBLE PRECISION ARITHMETIC                .
C . . . . . . . . . . . . . . . . . . . . . . . . . . . . . . . . . . . .
      CCMMCN /VAR/ NG,MODEX
      COMMON /TAPES/ IELMNT,ILOAD,IIN,IOUT
      DIMENSION R(NEQ),NOD(1),IDIRN(1),FLOAD(1)
      DIMENSICN ID(3,1)
C
      WRITE (ICUT,2000)
      READ (IIN,1000) (NOD(I),IDIRN(I),FLOAD(I),I=1,NLOAD)
      WRITE (IOUT,2010) (NOD(I),IDIRN(I),FLOAD(I),I=1,NLOAD)
      IF (MODEX.EQ.0) RETURN
C
      DO 210 I=1,NEQ
  210 R(I)=0.
C
      DC 220 L=1,NLOAD
      LN=NOD(L)
      LI=IDIRN(L)
      II=ID(LI,LN)
      IF (II) 220,220,240
  240 R(II)=R(II) + FLOAD(L)
C
  220 CONTINUE
C
      WRITE (ILCAD) R
C
  200 CCNTINUE
C
 1000 FORMAT (2I5,F10.0)
 2000 FCRMAT (////4X,30HNODE        DIRECTICN      LOAD/
     1 3X,6HNUMBER,19X,9HMAGNITUDE )
 2010 FCRMAT (1H0,I6,9X,I4,7X,E12.5)
      RETURN
      END
      SUBROUTINE ELCAL
C
C . . . . . . . . . . . . . . . . . . . . . . . . . . . . . . . . . . . .
C .                                                                      .
C .   P R O G R A M                                                      .
C .        TO LOOP OVER ALL ELEMENT GROUPS FOR READING,                  .
C .        GENERATING AND STORING THE ELEMENT DATA                       .
C .                                                                      .
C . . . . . . . . . . . . . . . . . . . . . . . . . . . . . . . . . . . .
C
      CCMMON /SOL/ NUMNP,NEQ,NWK,NUMEST,MIDEST,MAXEST,MK
      CCMMON /EL/ IND,NPAR(10),NUMEG,MTOT,NFIRST,NLAST,ITWO
      COMMON /TAPES/ IELMNT,ILCAD,IIN,IOUT
      CCMMON A(1)
C
C
      REWIND IELMNT
      WRITE (IOUT,2000)
C
C
C     LOOP OVER ALL ELEMENT GROUPS
C
C
      DO 100 N=1,NUMEG
      IF (N.NE.1) WRITE (IOUT,2010)
C
      READ (IIN,1000) NPAR
C
C
      CALL ELEMNT
C
      IF (MIDEST.GT.MAXEST) MAXEST=MIDEST
C
      WRITE (IELMNT) MIDEST,NPAR,(A(I),I=NFIRST,NLAST)
C
```

```
C                                                                ELCL0036
  100 CONTINUE                                                    ELCL0037
C                                                                ELCL0038
      RETURN                                                     ELCL0039
C                                                                ELCL0040
 1000 FORMAT (10I5)                                              ELCL0041
 2000 FORMAT (1H1,36HE L E M E N T   G R O U P   D A T A   ///)  ELCL0042
 2010 FORMAT (1H1)                                               ELCL0043
C                                                                ELCL0044
      END                                                        ELCL0045


      SUBROUTINE ELEMNT                                          ELMN0001
C                                                                ELMN0002
C . . . . . . . . . . . . . . . . . . . . . . . . . . . . . .   ELMN0003
C .                                                          .  ELMN0004
C .   P R O G R A M                                          .  ELMN0005
C .       TO CALL THE APPROPRIATE ELEMENT SUBROUTINE         .  ELMN0006
C .                                                          .  ELMN0007
C . . . . . . . . . . . . . . . . . . . . . . . . . . . . . .   ELMN0008
C                                                                ELMN0009
      COMMON /EL/ IND,NPAR(10),NUMEG,MTOT,NFIRST,NLAST,ITWO      ELMN0010
C                                                                ELMN0011
      NPAR1=NPAR(1)                                              ELMN0012
C                                                                ELMN0013
      GO TO (1,2,3),NPAR1                                        ELMN0014
C                                                                ELMN0015
    1 CALL TRUSS                                                 ELMN0016
      RETURN                                                     ELMN0017
C                                                                ELMN0018
C     OTHER ELEMENT TYPES WOULD BE CALLED HERE, IDENTIFYING EACH ELMN0019
C     ELEMENT TYPE BY A DIFFERENT NPAR(1) PARAMETER              ELMN0020
C                                                                ELMN0021
    2 RETURN                                                     ELMN0022
C                                                                ELMN0023
    3 RETURN                                                     ELMN0024
C                                                                ELMN0025
      END                                                        ELMN0026


      SUBROUTINE COLHT (MHT,NC,LM)                               CLHT0001
C . . . . . . . . . . . . . . . . . . . . . . . . . . . . . .   CLHT0002
C .                                                          .  CLHT0003
C .   P R O G R A M                                          .  CLHT0004
C .       TO CALCULATE COLUMN HEIGHTS                        .  CLHT0005
C .                                                          .  CLHT0006
C . . . . . . . . . . . . . . . . . . . . . . . . . . . . . .   CLHT0007
C                                                                CLHT0008
      COMMON /SOL/ NUMNP,NEQ,NWK,NUMEST,MIDEST,MAXEST,MK         CLHT0009
      DIMENSION LM(1),MHT(1)                                     CLHT0010
C                                                                CLHT0011
      LS=100000                                                  CLHT0012
      DO 100 I=1,ND                                              CLHT0013
      IF (LM(I)) 110,100,110                                     CLHT0014
  110 IF (LM(I)-LS) 120,100,100                                  CLHT0015
  120 LS=LM(I)                                                   CLHT0016
  100 CONTINUE                                                   CLHT0017
C                                                                CLHT0018
      DO 200 I=1,ND                                              CLHT0019
C                                                                CLHT0020
      II=LM(I)                                                   CLHT0021
      IF (II.EQ.0) GO TO 200                                     CLHT0022
      ME=II - LS                                                 CLHT0023
      IF (ME.GT.MHT(II)) MHT(II)=ME                              CLHT0024
  200 CONTINUE                                                   CLHT0025
C                                                                CLHT0026
      RETURN                                                     CLHT0027
      END


      SUBROUTINE ADDRES(MAXA,MHT)                                ADRS0001
C                                                                ADRS0002
C . . . . . . . . . . . . . . . . . . . . . . . . . . . . . .   ADRS0003
C :                                                             ADRS0004
C .   P R O G R A M                                          .  ADRS0005
C .       TO CALCULATE ADDRESSES OF DIAGONAL ELEMENTS IN BANDED ADRS0006
C .       MATRIX WHOSE COLUMN HEIGHTS ARE KNOWN              .  ADRS0007
C .                                                          .  ADRS0008
C .       MHT  = ACTIVE COLUMN HEIGHTS                       .  ADRS0009
C .       MAXA = ADDRESSES OF DIAGONAL ELEMENTS              .  ADRS0010
C .                                                          .  ADRS0011
C . . . . . . . . . . . . . . . . . . . . . . . . . . . . . .   ADRS0012
C                                                                ADRS0013
      COMMON /SOL/ NUMNP,NEQ,NWK,NUMEST,MIDEST,MAXEST,MK         ADRS0014
      DIMENSION MAXA(1),MHT(1)                                   ADRS0015
C                                                                ADRS0016
C     CLEAR ARRAY MAXA                                           ADRS0017
C                                                                ADRS0018
      NN=NEQ + 1                                                 ADRS0019
      DO 20 I=1,NN                                               ADRS0020
   20 MAXA(I)=0.0                                                ADRS0021
C                                                                ADRS0022
      MAXA(1)=1                                                  ADRS0023
      MAXA(2)=2                                                  ADRS0024
      MK=0                                                       ADRS0025
      IF (NEQ.EQ.1) GO TO 100                                    ADRS0026
      DO 10 I=2,NEQ                                              ADRS0027
```

```
      IF (MHT(I).GT.MK) MK=MHT(I)                                    ADRS0028
   10 MAXA(I+1)=MAXA(I) + MHT(I) + 1                                 ADRS0029
  100 MK=MK + 1                                                      ADRS0030
      NWK=MAXA(NEQ+1) - MAXA(1)                                      ADRS0031
C                                                                    ADRS0032
      RETURN                                                         ADRS0033
      END                                                            ADRS0034


      SUBROUTINE CLEAR(A,N)                                          CLER0001
C . . . . . . . . . . . . . . . . . . . . . . . . . . . . . . . .   CLER0002
C .                                                             .   CLER0003
C .                                                             .   CLER0004
C .    P R O G R A M                                            .   CLER0005
C .        TO CLEAR ARRAY A                                     .   CLER0006
C .                                                             .   CLER0007
C . . . . . . . . . . . . . . . . . . . . . . . . . . . . . . . .   CLER0008
      IMPLICIT REAL*8(A-H,O-Z)                                      CLER0009
C .                                                             .   CLER0010
C .        THIS PROGRAM IS USED IN SINGLE PRECISION ARITHMETIC ON .   CLER0011
C .        CDC EQUIPMENT AND DOUBLE PRECISION ARITHMETIC ON IBM .   CLER0012
C .        OR UNIVAC MACHINES .ACTIVATE,DEACTIVATE OR ADJUST ABOVE .   CLER0013
C .        CARD FOR SINGLE OR DOUBLE PRECISION ARITHMETIC       .   CLER0014
C . . . . . . . . . . . . . . . . . . . . . . . . . . . . . . . .   CLER0015
      DIMENSION A(1)                                                CLER0016
      DO 10 I=1,N                                                   CLER0017
   10 A(I)=0.                                                       CLER0018
      RETURN                                                        CLER0019
      END


      SUBROUTINE ASSEM (AA)                                         ASEM0001
C                                                                   ASEM0002
C . . . . . . . . . . . . . . . . . . . . . . . . . . . . . . . .   ASEM0003
C .                                                             .   ASEM0004
C .    P R O G R A M                                            .   ASEM0005
C .        TO CALL ELEMENT SUBROUTINES FOR ASSEMBAGE OF THE     .   ASEM0006
C .        STRUCTURE STIFFNESS MATRIX                           .   ASEM0007
C .                                                             .   ASEM0008
C . . . . . . . . . . . . . . . . . . . . . . . . . . . . . . . .   ASEM0009
C                                                                   ASEM0010
      COMMON /EL/ IND,NPAR(10),NUMEG,MTOT,NFIRST,NLAST,ITWO          ASEM0011
      COMMON /TAPES/ IELMNT,ILOAD,IIN,IOUT                           ASEM0012
      DIMENSION AA(1)                                               ASEM0013
C                                                                   ASEM0014
      REWIND IELMNT                                                 ASEM0015
C                                                                   ASEM0016
      DO 200 N=1,NUMEG                                              ASEM0017
      READ (IELMNT)NUMEST,NPAR,(AA(I),I=1,NUMEST)                   ASEM0018
C                                                                   ASEM0019
      CALL ELEMNT                                                   ASEM0020
C                                                                   ASEM0021
  200 CONTINUE                                                      ASEM0022
      RETURN                                                        ASEM0023
C                                                                   ASEM0024
C                                                                   ASEM0025
      END                                                           ASEM0026


      SUBROUTINE ADDBAN (A,MAXA,S,LM,ND)                            ADBN0001
C                                                                   ADBN0002
C . . . . . . . . . . . . . . . . . . . . . . . . . . . . . . . .   ADBN0003
C .                                                             .   ADBN0004
C .    P R O G R A M                                            .   ADBN0005
C .        TO ASSEMBLE UPPER TRIANGULAR ELEMENT STIFFNESS INTO  .   ADBN0006
C .        COMPACTED GLOBAL STIFFNESS                           .   ADBN0007
C .                                                             .   ADBN0008
C .        A = GLOBAL STIFFNESS                                 .   ADBN0009
C .        S = ELEMENT STIFFNESS                                .   ADBN0010
C .        ND = DEGREES OF FREEDOM IN ELEMENT STIFFNESS         .   ADBN0011
C .                                                             .   ADBN0012
C .            S(1)        S(2)       S(3)      . . .           .   ADBN0013
C .        S =             S(ND+1)    S(ND+2)   . . .           .   ADBN0014
C .                                   S(2*ND)   . . .           .   ADBN0015
C .                                             . . .           .   ADBN0016
C .                                                             .   ADBN0017
C .                                                             .   ADBN0018
C .            A(1)        A(3)       A(6)      . . .           .   ADBN0019
C .        A =             A(2)       A(5)      . . .           .   ADBN0020
C .                                   A(4)      . . .           .   ADBN0021
C .                                             . . .           .   ADBN0022
C .                                                             .   ADBN0023
C .                                                             .   ADBN0024
C .                                                             .   ADBN0025
C . . . . . . . . . . . . . . . . . . . . . . . . . . . . . . . .   ADBN0026
      IMPLICIT REAL*8(A-H,O-Z)                                      ADBN0027
C .                                                             .   ADBN0028
C .        THIS PROGRAM IS USED IN SINGLE PRECISION ARITHMETIC ON .   ADBN0029
C .        CDC EQUIPMENT AND DOUBLE PRECISION ARITHMETIC ON IBM .   ADBN0030
C .        OR UNIVAC MACHINES .ACTIVATE,DEACTIVATE OR ADJUST ABOVE .   ADBN0031
C .        CARD FOR SINGLE OR DOUBLE PRECISION ARITHMETIC       .   ADBN0032
C . . . . . . . . . . . . . . . . . . . . . . . . . . . . . . . .   ADBN0033
C                                                                   ADBN0034
      DIMENSION A(1),MAXA(1),S(1),LM(1)                             ADBN0035
C                                                                   ADBN0036
```

```
      NDI=0                                                              ADBN0037
      DO 200 I=1,ND                                                      ADBN0038
      II=LM(I)                                                           ADBN0039
      IF (II) 200,200,100                                                ADBN0040
  100 MI=MAXA(II)                                                        ADBN0040
      KS=I                                                               ADBN0041
      DC 220 J=1,ND                                                      ADBN0042
      JJ=LM(J)                                                           ADBN0043
      IF (JJ) 220,220,110                                                ADBN0044
  110 IJ=II - JJ                                                         ADBN0045
      IF (IJ) 220,210,210                                                ADBN0046
  210 KK=MI + IJ                                                         ADBN0047
      KSS=KS                                                             ADBN0048
      IF (J.GE.I) KSS=J + NDI                                            ADBN0049
      A(KK)=A(KK) + S(KSS)                                               ADBN0050
  220 KS=KS + ND - J                                                     ADBN0051
  200 NDI=NDI + ND - I                                                   ADBN0052
C                                                                        ADBN0053
      RETURN                                                             ADBN0054
      END                                                               ADBN0055
                                                                         ADBN0056

      SUBROUTINE COLSOL(A,V,MAXA,NN,NWK,NNM,KKK)                         CLSL0001
C . . . . . . . . . . . . . . . . . . . . . . . . . . . . . . . . .     CLSL0002
C .                                                              .      CLSL0003
C .      P R O G R A M                                           .      CLSL0004
C .          TO SOLVE FINITE ELEMENT STATIC EQUILIBRIUM EQUATIONS IN .  CLSL0004
C .          CORE, USING COMPACTED STORAGE AND COLUMN REDUCTION SCHEME . CLSL0005
C .                                                              .      CLSL0006
C .      - - INPUT VARIABLES - -                                 .      CLSL0007
C .          A(NWK)    = STIFFNESS MATRIX STORED IN COMPACTED FORM .    CLSL0008
C .          V(NN)     = RIGHT-HAND-SIDE LOAD VECTOR             .      CLSL0009
C .          MAXA(NNM) = VECTOR CONTAINING ADDRESSES OF DIAGONAL  .     CLSL0010
C .                      ELEMENTS OF STIFFNESS MATRIX IN A       .      CLSL0011
C .          NN        = NUMBER OF EQUATIONS                      .     CLSL0012
C .          NWK       = NUMBER OF ELEMENTS BELOW SKYLINE OF MATRIX .   CLSL0013
C .          NNM       = NN + 1                                  .      CLSL0014
C .          KKK       = INPUT FLAG                              .      CLSL0015
C .              EQ. 1 TRIANGULARIZATION OF STIFFNESS MATRIX      .     CLSL0016
C .              EQ. 2 REDUCTION AND BACK-SUBSTITUTION OF LOAD VECTOR . CLSL0017
C .          IOUT      = NUMBER OF OUTPUT DEVICE                 .      CLSL0018
C .                                                              .      CLSL0019
C .      - - OUTPUT - -                                          .      CLSL0020
C .          A(NWK)    = D AND L - FACTORS OF STIFFNESS MATRIX    .     CLSL0021
C .          V(NN)     = DISPLACEMENT VECTOR                     .      CLSL0022
C .                                                              .      CLSL0023
C . . . . . . . . . . . . . . . . . . . . . . . . . . . . . . . . .     CLSL0024
      IMPLICIT REAL*8(A-H,C-Z)                                          CLSL0025
C . . . . . . . . . . . . . . . . . . . . . . . . . . . . . . . . .     CLSL0026
C .          THIS PROGRAM IS USED IN SINGLE PRECISION ARITHMETIC ON ,  CLSL0027
C .          CDC EQUIPMENT AND DOUBLE PRECISION ARITHMETIC ON IBM .    CLSL0028
C .          OR UNIVAC MACHINES .ACTIVATE,DEACTIVATE OR ADJUST ABOVE . CLSL0029
C .          CARD FOR SINGLE OR DOUBLE PRECISION ARITHMETIC      .      CLSL0030
C . . . . . . . . . . . . . . . . . . . . . . . . . . . . . . . . .     CLSL0031
      COMMON /TAPES/ IELMNT,ILOAD,IIN,IOUT                              CLSL0032
      DIMENSION A(NWK),V(1),MAXA(1)                                     CLSL0033
C                                                                        CLSL0034
                                                                         CLSL0035
C                                                                        CLSL0036
C     PERFORM L*D*L(T) FACTORIZATION OF STIFFNESS MATRIX
      IF(KKK-2)40,150,150                                               CLSL0037
   40 DC 140 N=1,NN                                                     CLSL0038
      KN=MAXA(N)                                                        CLSL0039
      KL=KN+1                                                           CLSL0040
      KU=MAXA(N+1) - 1                                                  CLSL0041
      KH=KU - KL                                                        CLSL0042
      IF(KH)110,90,50                                                   CLSL0043
   50 K=N-KH                                                            CLSL0044
      IC=0                                                              CLSL0045
      KLT=KU                                                            CLSL0046
      DO 80 J=1,KH                                                      CLSL0047
      IC=IC + 1                                                         CLSL0048
      KLT=KLT - 1                                                       CLSL0049
      KI=MAXA(K)                                                        CLSL0050
      ND=MAXA(K+1) - KI - 1                                             CLSL0051
      IF(ND)80,80,60                                                    CLSL0052
   60 KK=MIN0(IC,ND)                                                    CLSL0053
      C=0.                                                              CLSL0054
      DO 70 L=1,KK                                                      CLSL0055
   70 C=C+A(KI+L)*A(KLT+L)                                              CLSL0056
      A(KLT)=A(KLT) - C                                                 CLSL0057
   80 K=K+1                                                             CLSL0058
   90 K=N                                                               CLSL0059
      B=0.                                                              CLSL0060
      DO 100 KK=KL,KU                                                   CLSL0061
      K=K - 1                                                           CLSL0062
      KI=MAXA(K)                                                        CLSL0063
      C=A(KK)/A(KI)                                                     CLSL0064
      B=B + C*A(KK)                                                     CLSL0065
  100 A(KK)=C                                                           CLSL0066
      A(KN)=A(KN) - B                                                   CLSL0067
  110 IF (A(KN))120,120,140                                            CLSL0068
  120 WRITE(IOUT,2000) N,A(KN)                                          CLSL0069
      STOP                                                              CLSL0070
  140 CONTINUE                                                          CLSL0071
      RETURN                                                            CLSL0072
                                                                         CLSL0073
```

```
C                                                                    CLSL0074
C          REDUCE RIGHT-HAND-SIDE LOAD VECTOR                        CLSL0075
C                                                                    CLSL0076
      150 DO 180 N=1,NN                                              CLSL0077
          KL=MAXA(N) + 1                                            CLSL0078
          KU=MAXA(N+1) - 1                                          CLSL0079
          IF(KU-KL)180,160,160                                      CLSL0080
      160 K=N                                                        CLSL0081
          C=0.                                                       CLSL0082
          DO 170 KK=KL,KU                                           CLSL0083
          K=K - 1                                                   CLSL0084
      170 C=C+A(KK)*V(K)                                            CLSL0085
          V(N)=V(N) - C                                            CLSL0086
      180 CONTINUE                                                  CLSL0087
C                                                                    CLSL0088
C          BACK-SUBSTITUTE                                          CLSL0089
C                                                                    CLSL0090
          DO 200 N=1,NN                                             CLSL0091
          K=MAXA(N)                                                CLSL0092
      200 V(N)=V(N)/A(K)                                           CLSL0093
          IF (NN.EQ.1) RETURN                                      CLSL0094
          N=NN                                                     CLSL0095
          DO 230 L=2,NN                                            CLSL0096
          KL=MAXA(N) + 1                                          CLSL0097
          KU=MAXA(N+1) - 1                                        CLSL0098
          IF(KU-KL)230,210,210                                    CLSL0099
      210 K=N                                                      CLSL0100
          DO 220 KK=KL,KU                                         CLSL0101
          K=K - 1                                                 CLSL0102
      220 V(K)=V(K)-A(KK)*V(N)                                    CLSL0103
      230 N=N-1                                                    CLSL0104
          RETURN                                                  CLSL0105
     2000 FORMAT(//48H STOP - STIFFNESS MATRIX NOT POSITIVE DEFINITE   ,// CLSL0106
     1             32H NONPOSITIVE PIVOT FOR EQUATION  ,I4,//          CLSL0107
     2             10H PIVOT =   ,E20.12 )                             CLSL0108
          END                                                        CLSL0109


      SUBROUTINE LOADV (R,NEQ)                                       LODV0001
C                                                                    LODV0002
C                                                                    LODV0003
C . . . . . . . . . . . . . . . . . . . . . . . . . . . . . . . . .  LODV0004
C .                                                              .  LODV0005
C .   P R O G R A M                                              .  LODV0006
C .        TO OBTAIN THE LOAD VECTOR                             .  LODV0007
C .                                                              .  LODV0008
C . . . . . . . . . . . . . . . . . . . . . . . . . . . . . . . . .  LODV0009
      IMPLICIT REAL*8(A-H,C-Z)                                       LODV0010
C . . . . . . . . . . . . . . . . . . . . . . . . . . . . . . . . .  LODV0011
C .       THIS PROGRAM IS USED IN SINGLE PRECISION ARITHMETIC ON .  LODV0012
C .       CDC EQUIPMENT AND DOUBLE PRECISION ARITHMETIC ON IBM   .  LODV0013
C .       OR UNIVAC MACHINES .ACTIVATE,DEACTIVATE OR ADJUST ABOVE .  LODV0014
C .       CARD FOR SINGLE OR DOUBLE PRECISION ARITHMETIC         .  LODV0015
C . . . . . . . . . . . . . . . . . . . . . . . . . . . . . . . . .  LODV0016
C                                                                    LODV0017
      COMMON /TAPES/ IELMNT,ILCAD,IIN,IOUT                           LODV0018
      DIMENSION R(NEQ)                                               LODV0019
C                                                                    LODV0020
      READ (ILOAD) R                                                 LODV0021
C                                                                    LODV0022
      RETURN                                                         LODV0023
      END


      SUBROUTINE WRITE (DISP,ID,NEQ,NUMNP)                           WRTE0001
C                                                                    WRTE0002
C . . . . . . . . . . . . . . . . . . . . . . . . . . . . . . . . .  WRTE0003
C .                                                              .  WRTE0004
C .   P R O G R A M                                              .  WRTE0005
C .        TO PRINT DISPLACEMENTS                                .  WRTE0006
C . . . . . . . . . . . . . . . . . . . . . . . . . . . . . . . . .  WRTE0007
      IMPLICIT REAL*8(A-H,C-Z)                                       WRTE0008
C . . . . . . . . . . . . . . . . . . . . . . . . . . . . . . . . .  WRTE0009
C .       THIS PROGRAM IS USED IN SINGLE PRECISION ARITHMETIC ON .  WRTE0010
C .       CDC EQUIPMENT AND DOUBLE PRECISION ARITHMETIC ON IBM   .  WRTE0011
C .       OR UNIVAC MACHINES .ACTIVATE,DEACTIVATE OR ADJUST ABOVE .  WRTE0012
C .       CARD FOR SINGLE OR DOUBLE PRECISION ARITHMETIC         .  WRTE0013
C . . . . . . . . . . . . . . . . . . . . . . . . . . . . . . . . .  WRTE0014
C                                                                    WRTE0015
      COMMON /TAPES/ IELMNT,ILOAD,IIN,IOUT                           WRTE0016
      DIMENSION DISP(NEQ),ID(3,NUMNP)                                WRTE0017
      DIMENSION D(3)                                                 WRTE0018
C                                                                    WRTE0019
C          PRINT DISPLACEMENTS                                      WRTE0020
C                                                                    WRTE0021
      WRITE (IOUT,2000)                                              WRTE0022
      IC=4                                                           WRTE0023
C                                                                    WRTE0024
      DO 100 II=1,NUMNP                                              WRTE0025
      IC=IC + 1                                                     WRTE0026
      IF (IC.LT.56) GO TO 105                                        WRTE0027
      WRITE (IOUT,2000)                                              WRTE0028
      IC=4                                                           WRTE0029
  105 DO 110 I=1,3                                                   WRTE0030
  110 D(I)=0.                                                        WRTE0031
```

```
C                                                                        WRTE0032
        CO 120 I=1,3                                                     WRTE0033
        KK=ID(I,II)                                                      WRTE0034
        IL=I                                                             WRTE0035
  120 IF (KK.NE.0) D(IL)=DISP(KK)                                        WRTE0036
C                                                                        WRTE0037
  100 WRITE (IOUT,2010) II,D                                            WRTE0038
C                                                                        WRTE0039
C                                                                        WRTE0040
        RETURN                                                           WRTE0041
C                                                                        WRTE0042
 2000 FORMAT (///, 26H D I S P L A C E M E N T S // 7H NODE   9X         WRTE0043
     114HX-DISPLACEMENT 4X 14HY-DISPLACEMENT 4X 14HZ-DISPLACEMENT)       WRTE0044
 2010 FORMAT (1X,I3, 8X,3E18.6)                                          WRTE0045
C                                                                        WRTE0046
        END                                                              WRTE0047


        SUBROUTINE STRESS (AA)                                           STRS0001
C                                                                        STRS0002
C . . . . . . . . . . . . . . . . . . . . . . . . . . . . . . . . . .    STRS0003
C .                                                                 .    STRS0004
C .   P R O G R A M                                                 .    STRS0005
C .       TO CALL THE ELEMENT SUBROUTINE FOR THE CALCULATION OF     .    STRS0006
C .       STRESSES                                                  .    STRS0007
C .                                                                 .    STRS0008
C . . . . . . . . . . . . . . . . . . . . . . . . . . . . . . . . . .    STRS0009
C                                                                        STRS0010
      COMMON /VAR/ NG,MODEX                                              STRS0011
      COMMON /EL/ IND,NPAR(10),NUMEG,MTOT,NFIRST,NLAST,ITWO              STRS0012
      COMMON /TAPES/ IELMNT,ILOAD,IIN,IOUT                               STRS0013
      DIMENSION AA(1)                                                    STRS0014
C                                                                        STRS0015
C                                                                        STRS0016
C     LOOP OVER ALL ELEMENT GROUPS                                       STRS0017
C                                                                        STRS0018
C                                                                        STRS0019
      REWIND IELMNT                                                      STRS0020
C                                                                        STRS0021
      DO 100 N=1,NUMEG                                                   STRS0022
      NG=N                                                               STRS0023
C                                                                        STRS0024
      READ (IELMNT) NUMEST,NPAR,(AA(I),I=1,NUMEST)                       STRS0025
C                                                                        STRS0026
      CALL ELEMNT                                                        STRS0027
C                                                                        STRS0028
  100 CONTINUE                                                           STRS0029
C                                                                        STRS0030
      RETURN                                                             STRS0031
      END                                                                STRS0032


        SUBROUTINE TRUSS                                                 TRUS0001
C . . . . . . . . . . . . . . . . . . . . . . . . . . . . . . . . . .    TRUS0002
C .                                                                 .    TRUS0003
C .   P R O G R A M                                                 .    TRUS0004
C .       TO SET UP STORAGE AND CALL THE TRUSS ELEMENT SUBROUTINE   .    TRUS0005
C .                                                                 .    TRUS0006
C . . . . . . . . . . . . . . . . . . . . . . . . . . . . . . . . . .    TRUS0007
C                                                                        TRUS0008
      COMMON /SOL/ NUMNP,NEQ,NWK,NUMEST,MIDEST,MAXEST,MK                 TRUS0009
      COMMON /DIM/ N1,N2,N3,N4,N5,N6,N7,N8,N9,N10,N11,N12,N13,N14,N15    TRUS0010
      COMMON /EL/ IND,NPAR(10),NUMEG,MTOT,NFIRST,NLAST,ITWO              TRUS0011
      COMMON /TAPES/ IELMNT,ILOAD,IIN,IOUT                               TRUS0012
      COMMON A(1)                                                        TRUS0013
C                                                                        TRUS0014
      EQUIVALENCE (NPAR(2),NUME),(NPAR(3),NUMMAT)                        TRUS0015
C                                                                        TRUS0016
      NFIRST=N6                                                          TRUS0017
      IF(IND.GT.1) NFIRST=N5                                             TRUS0018
      N101=NFIRST                                                        TRUS0019
      N102=N101 + NUMMAT*ITWO                                            TRUS0020
      N103=N102 + NUMMAT*ITWO                                            TRUS0021
      N104=N103 + 6*NUME                                                 TRUS0022
      N105=N104 + 6*NUME*ITWO                                            TRUS0023
      N106=N105 + NUME                                                   TRUS0024
      NLAST=N106                                                         TRUS0025
C                                                                        TRUS0026
      IF (IND.GT.1) GO TO 100                                            TRUS0027
      IF(NLAST.GT.MTOT) CALL ERROR(NLAST-MTOT,3)                         TRUS0028
      GC TO 200                                                          TRUS0029
  100 IF (NLAST.GT.MTOT) CALL ERROR(NLAST-MTOT,4)                        TRUS0030
C                                                                        TRUS0031
  200 MIDEST=NLAST - NFIRST                                              TRUS0032
C                                                                        TRUS0033
      CALL RUSS (A(N1),A(N2),A(N3),A(N4),A(N4),A(N5),A(N101),A(N10       TRUS0034
     1 2),A(N103),A(N104),A(N105))                                       TRUS0035
C                                                                        TRUS0036
      RETURN                                                             TRUS0037
C                                                                        TRUS0038
      END                                                                TRUS0039


        SUBROUTINE RUSS (ID,X,Y,Z,U,MHT,E,AREA,LM,XYZ,MATP)             RUSS0001
```

```
C . . . . . . . . . . . . . . . . . . . . . . . . . . . . . . . . . . .
C .                                                                   .
C .         TRUSS ELEMENT SUBRCUTINE                                  .
C .                                                                   .
C . . . . . . . . . . . . . . . . . . . . . . . . . . . . . . . . . . .
C
      IMPLICIT REAL*8(A-H,O-Z)
      SQRT(X)=DSQRT(X)
C . . . . . . . . . . . . . . . . . . . . . . . . . . . . . . . . . . .
C .       THIS PROGRAM IS USED IN SINGLE PRECISION ARITHMETIC ON      .
C .       CDC EQUIPMENT AND DOUBLE PRECISION ARITHMETIC ON IBM        .
C .       OR UNIVAC MACHINES .ACTIVATE,DEACTIVATE OR ADJUST ABOVE     .
C .       CARDS FOR SINGLE OR DOUBLE PRECISION ARITHMETIC             .
C . . . . . . . . . . . . . . . . . . . . . . . . . . . . . . . . . . .
      COMMON /SOL/ NUMNP,NEQ,NWK,NUMEST,MIDEST,MAXEST,MK
      COMMON /DIM/ N1,N2,N3,N4,N5,N6,N7,N8,N9,N10,N11,N12,N13,N14,N15
      COMMON /EL/ IND,NPAR(10),NUMEG,MTOT,NFIRST,NLAST,ITWO
      COMMON /VAR/ NG,MODEX
      COMMON /TAPES/ IELMNT,ILOAD,IIN,IOUT
      COMMON A(1)
C
      REAL A
C
      DIMENSION X(1),Y(1),Z(1),ID(3,1),E(1),AREA(1),LM(6,1),
     1          XYZ(6,1),MATP(1),U(1),MHT(1)
      DIMENSICN DR(3),IPS(1)
      DIMENSION S(21),ST(6),D(3)
C
      EQUIVALENCE (NPAR(1),NPAR1),(NPAR(2),NUME),(NPAR(3),NUMMAT)
      ND=6
C
      GO TO (300,610,900),IND
C
C
C     R E A D   A N D   G E N E R A T E   E L E M E N T
C     I N F O R M A T I O N
C
C     READ MATERIAL INFORMATICN
C
  300 WRITE (IOUT,2000) NPAR1,NUME
      IF (NUMMAT.EQ.0) NUMMAT=1
      WRITE (IOUT,2010) NUMMAT
C
      WRITE (IOUT,2020)
      DO 10 I=1,NUMMAT
      READ (IIN,1000) N,E(N),AREA(N)
   10 WRITE (IOUT,2030) N,E(N),AREA(N)
C
C     READ ELEMENT INFORMATION
C

      WRITE (ICUT,2040)
      N=1
  100 READ (IIN,1020) M,II,JJ,MTYP,KG
      IF (KG.EQ.0) KG=1
  120 IF(M.NE.N) GO TO 200
      I=II
      J=JJ
      MTYPE=MTYP
      KKK=KG
C
C     SAVE ELEMENT INFORMATION
C
  200 XYZ(1,N)=X(I)
      XYZ(2,N)=Y(I)
      XYZ(3,N)=Z(I)
C
      XYZ(4,N)=X(J)
      XYZ(5,N)=Y(J)
      XYZ(6,N)=Z(J)
C
C
      MATP(N)=MTYPE
C
      DO 390 L=1,6
  390 LM(L,N)=0.
      DO 400 L=1,3
      LM(L,N)=IC(L,I)
  400 LM(L+3,N)=ID(L,J)
C
C     UPDATE COLUMN HEIGHTS AND BANDWIDTH
C
      CALL COLHT (MHT,ND,LM(1,N))
C
      WRITE (ICUT,2050) N,I,J,MTYPE
      IF (N.EQ.NUME) RETURN
      N=N+1
      I=I+KKK
      J=J+KKK
      IF(N.GT.M) GO TO 100
      GO TO 120
C
C
C
C     A S S E M B L E   S T U C T U R E   S T I F F N E S S   M A T R I X
```

```
C
C
C
  610 DO 500 N=1,NUME
      MTYPE=MATP(N)
      XL2=0.
      DO 505 L=1,3
      D(L)=XYZ(L,N) - XYZ(L+3,N)
  505 XL2=XL2 + D(L)*D(L)
      XL=SQRT(XL2)
      XX=E(MTYPE)*AREA(MTYPE)*XL
      DO 510 L=1,3
      ST(L)=D(L)/XL2
  510 ST(L+3)=-ST(L)
C
      KL=0
      DO 600 L=1,6
      YY=ST(L)*XX
      DO 600 K=L,6
      KL=KL + 1
  600 S(KL)=ST(K)*YY
      CALL ADDBAN (A(N3),A(N2),S,LM(1,N),ND)
  500 CONTINUE
      RETURN
C
C
C     S T R E S S   C A L C U L A T I O N S
C
C
900   IPRINT=0
      DO 830 N=1,NUME
      IPRINT=IPRINT + 1
      IF (IPRINT.GT.50) IPRINT=1
      IF (IPRINT.EQ.1)
     *WRITE (IOUT,2060) NG
      MTYPE=MATP(N)
      XL2=0.
      DO 820 L=1,3
      D(L) = XYZ(L,N) - XYZ(L+3,N)
  820 XL2=XL2+D(L)*D(L)
      DO 814 L=1,3
      ST(L)=(D(L)/XL2)*E(MTYPE)
  814 ST(L+3)=-ST(L)
      STR=0.0
      DO 806 L=1,3
      I=LM(L,N)
      IF (I.LE.0) GO TO 807
      STR=STR+ST(L)*U(I)
  807 J=LM(L+3,N)
      IF (J.LE.0) GO TO 806
      STR=STR+ST(L+3)*U(J)
  806 CONTINUE
      P=STR*AREA(MTYPE)
      WRITE (IOUT,2070) N,P,STR
  830 CONTINUE
C
      RETURN
C
 1000 FORMAT (I5,2F10.0)
 1010 FORMAT (2F10.0)
 1020 FORMAT (5I5)
 2000 FORMAT (36H E L E M E N T   D E F I N I T I O N ///,
     1         14H ELEMENT TYPE ,13(2H .),17H( NPAR(1) ) . . =,I5/,
     2         25H      EQ.1, TRUSS ELEMENTS/,
     3         29H      EQ.2, ELEMENTS CURRENTLY/,
     4         25H      EQ.3, NOT AVAILABLE /,
     5         20H NUMBER OF ELEMENTS.,10(2H .),17H( NPAR(2) ) . . =,
     6         I5//)
 2010 FORMAT (42H M A T E R I A L   D E F I N I T I O N       ///,
     1         37H NUMBER OF DIFFERENT SETS OF MATERIAL
     2            /32H AND CROSS-SECTIONAL  CONSTANTS ,
     3                               4(2H .),17H( NPAR(3) ) . . =,I5//)
 2020 FORMAT (///2X,3HSET,7X,6HYOUNGS,6X,15HCROSS-SECTIONAL/
     1 1X,6HNUMBER,5X,7HMODULUS,10X,4HAREA,/15X,1HE,14X,1HA)
 2030 FORMAT (/I5,4X,E12.5,2X,E14.6)
 2040 FORMAT (1H1,40H E L E M E N T   I N F O R M A T I O N   ///,
     1 8H ELEMENT ,5X,4HNODE,5X,4HNODE,7X,8HMATERIAL/,
     2 9H NUMBER-N,6X,1HI,8X,1HJ,7X,10HSET NUMBER/)
 2050 FORMAT (I5,6X,I5,4X,I5,7X,I5)
 2060 FORMAT (////,46H S T R E S S   C A L C U L A T I O N S   F O R    ,
     1 24HE L E M E N T   G R O U P,I4//,2X,
     27HELEMENT,12X,5HFORCE,12X,6HSTRESS /,2X,6HNUMBER/)
 2070 FORMAT (1X,I5,11X,E13.6,4X,E13.6)
C
      END

      SUBROUTINE SECOND(TIM)
C
C     SUBROUTINE TO OBTAIN TIME
C     THIS SUBROUTINE HAS BEEN USED ON THE IBM 370 AT M.I.T.
C
      CALL TIMING(II)
      TIM=FLOAT(II)/100.
C
      RETURN
      END
```

REFERENCES

1. P. V. Marcal (ed.), *General Purpose Finite Element Computer Programs*, American Society of Mechanical Engineers, New York, N.Y., 1970.
2. J. T. Oden, R. W. Clough, and Y. Yamamoto (eds.), *Advances in Computational Methods in Structural Mechanics and Design*, University of Alabama Press, Huntsville, Ala., 1972.
3. S. T. Fenves, N. Perrone, J. Robinson, and W. C. Schnobrich, (eds.), *Numerical and Computer Methods in Structural Mechanics*, Academic Press, Inc., New York, N.Y., 1973.
4. W. Pilkey, K. Saczalski, and H. Schaeffer (eds.), *Structural Mechanics Computer Programs*, University Press of Virginia, Charlottesville, Va., 1974.
5. E. L. Wilson, "SAP—A Structural Analysis Program," Report UC SESM 70-20, Department of Civil Engineering, University of California, Berkeley, 1970.
6. E. L. Wilson, "Solid SAP—A Static Analysis Program for Three-Dimensional Solid Structures," Report UC SESM 71-19, Department of Civil Engineering, University of California, Berkeley, 1971.
7. K. J. Bathe, E. L. Wilson, and F. E. Peterson, "SAP IV—A Structural Analysis Program for Static and Dynamic Response of Linear Systems," Report EERC 73-11, College of Engineering, University of California, Berkeley, June 1973, revised Apr. 1974.
8. K. J. Bathe, "ADINA—A Finite Element Program for Automatic Dynamic Incremental Nonlinear Analysis," Report 82448-1, Acoustics and Vibration Laboratory, Department of Mechanical Engineering, Massachusetts Institute of Technology, Cambridge, Mass., 1975 (rev. Dec. 1978).
9. R. Rosen, "Matrix Bandwidth Minimization," *Proceedings, Conference Association for Computing Machinery*, 1968, pp. 585–595.
10. E. H. Cuthill and J. M. McKee, "Reducing the Bandwidth of Sparse Symmetric Matrices," *Proceedings, Conference Association for Computing Machinery*, 1969, pp. 151–172.
11. EAC/EASE2-User Information Manual/Theoretical, Control Data Corporation Publication No. 84002700, Minneapolis, Minn., 1973.

INDEX

INDEX

D

E

F

G

H

I

J

K

L

M

N

O

P

Q

R

S

T

U

V

W

Z